Human Physiology

THE STUDY OF HUMAN PHYSIOLOGY
Second Edition

by
O. C. J. Lippold

The study of human physiology for examination purposes can be greatly simplified by applying the correct learning techniques. A comparison volume to *Human Physiology* entitled *The Study of Human Physiology* is written with various categories of students in mind and gives important information to enable them to become proficient with the minimum expenditure of time and effort.

The method employed consists of making the reader answer questions closely based upon the text of *Human Physiology*, page by page, as he reads through the text book. According to his answers, he proceeds at a pace set by his own progress in the subject. Answers are given in such a way that the maximum of information is obtained from working through the questions; topics, known to be difficult for students to understand are given especial attention.

An examination has three main functions. The first is obviously to act as an academic hurdle, or measure of a student's attainment in a subject. The second function is to be part of the ongoing learning process in a subject both as a stimulation for work and in terms of feedback from the actual material set in the paper itself. The third is to delineate the boundaries of the subject matter required. In physiology it is unusual for a detailed syllabus to be provided; past examination papers usually serve this purpose. An important part of the book, therefore, is the multiple choice questionnaire which can be used for all three of the functions enumerated above.

The previous edition of this book appeared under the title of *Questions and Answers in Physiology*.

BioMed
0491003-x

```
WINTON, FRANK ROBERT
HUMAN PHYSIOLOGY
              000491003
```

QP   34.L76.7

THE UNIVERSITY OF LIVERPOOL

HAROLD COHEN LIBRARY

Please return or renew, on or before the last date below. A fine is payable on late returned items. Books may be recalled after one week for the use of another reader. Books may be renewed by telephone: 0151 794 5412.

For conditions of borrowing, see Library Regulation

Human Physiology

Revised and Edited by

O. C. J. LIPPOLD
M.D.
Reader in Physiology, University of London

and

F. R. WINTON
M.D., D.Sc.
Sometime Professor of Pharmacology, University of London

Seventh Edition

CHURCHILL LIVINGSTONE
EDINBURGH LONDON AND NEW YORK 1979

CHURCHILL LIVINGSTONE
Medical Division of Longman Group Limited

Distributed in the United States of America by
Longman Inc., 19 West 44th Street, New York,
N.Y. 10036 and by associated companies,
branches and representatives throughout
the world.

© Longman Group Limited 1979

All rights reserved. No part of this publication
may be reproduced, stored in a retrieval system,
or transmitted in any form or by any means,
electronic, mechanical, photocopying, recording
or otherwise, without the prior permission of the
publishers (Churchill Livingstone, 23 Ravelston
Terrace, Edinburgh, EH4 3TL)

First Edition 1930
   Reprinted 1932
Second Edition 1935
   Reprinted 1936
Third Edition 1948
   Reprinted 1949
Fourth Edition 1955
Fifth Edition 1962
Sixth Edition 1968
   Reprinted 1972
Seventh Edition 1979

ISBN 0 443 01751 4

British Library Cataloguing in Publication Data

Winton, Frank Robert
   Human physiology. 7th ed.
   1. Physiology
   I. Title II. Bayliss, Leonard Ernest
   III. Lippold, Olof Conrad John
   612   QP34.5   78–40109

Typeset in Great Britain at The Pitman Press, Bath
and printed in Singapore by Huntsmen Offset Printing (Pte) Ltd.

Preface to Seventh Edition

This textbook of *Human Physiology* is for the use of students of medicine, dentistry and the biological sciences. It has been written in a clear and simple manner with the objective of providing an understandable and easily assimilated account of the physiological knowledge that is required as a basis for the practice of medicine and dentistry. For students, it is necessary to limit the subject matter; in this book the authors have aimed at a readable and balanced exposition of the general principles of physiology and the experimental research that has led to the establishment of these principles. Where appropriate, proof is given in support of statements made and clinical conditions that illustrate the physiology are described.

Throughout, the intention has been to inculcate the habit of critical throught. It is vital to the process of becoming fully conversant with a subject such as physiology, that the general principles are first clearly understood; it is this understanding that requires active critical participation by the reader.

London, 1978

O. C. J. L.
F. R. W.

Historical Note

In 1928, Professor (late Sir Charles) Lovatt Evans proposed to two younger members of his staff in the Department of Physiology at University College London that they should write a comparatively short textbook of Physiology, to run as a companion to the longer textbook by Professor E. H. Starling who died in 1927. Lovatt Evans had begun his first revision of "Starling", to appear as the fifth edition in 1930, as did the first edition of "Winton and Bayliss". Both were tactfully and very effectively helped through teething troubles by the late Mr. J. Rivers of J. & A. Churchill Ltd. The two authors wrote all of the shorter book except for the chapter on special senses, written by their fellow lecturer, the late Dr. R. J. Lythgoe. Dr. Grace Eggleton, in the same department, read the proofs and prepared the index.

Physiology developed quickly in every direction, and in subsequent editions certain chapters or parts of chapters were contributed by colleagues with corresponding experience. Warm appreciation is due to them for the ready way in which they allowed the authors to edit and particularly to curb the length of their contributions.

Then, alas, Leonard Bayliss died in 1965. Son of Sir William Bayliss (*General Physiology*) and nephew of Starling, Leonard was one of the most thoughtful and original physiologists and had put an immense amount of work into preparation of successive editions of *Human Physiology*. His co-author has retired from University work and was fortunate in inducing Dr. O. C. J. Lippold to join him in preparing the sixth and seventh editions and indeed, to take the lead in doing so.

1978   F.R.W.

Acknowledgements

We would like to record our thanks to the many individuals who have helped in the production of this edition, particularly our students and their tutors who have made many valuable suggestions regarding both the general layout and specific items in the text.

Dr. Lippold wishes to thank Professor Jack Sinclair of the University of Auckland, New Zealand for great kindness in providing facilities in his Department of Physiology where the writing of this book was done. It would be very difficult to bring out a book as large as this one, working only in spare time, so the provision of a sabbatical year's leave from University College London is gratefully acknowledged.

We thank many of our colleagues for much help of various kinds including the critical review of newly written material. Help with the planning of the book was given by Professors Jack Sinclair and Royce Farelly of the University of Auckland; specific chapters were read by Drs. Lynn Bindman, Alex Milne, Bruce Lynn and Barbara Cogdell. Their suggestions have been incorporated in the final version. The index was prepared by Dr. Barbara Cogdell. The typing was excellently done by Ruth Milne and Mollie Kirk.

This edition contains many new illustrations; some of these have been specially prepared, while others have been taken unchanged from original publications. For all these whether modified or unchanged, we would like to thank the authors concerned and to record our gratitude for the co-operation of the editors of the following journals: *Acta physiologica Scandinavica; American Journal of Physiology; Brain; British Journal of Anaesthesia; British Medical Bulletin; Bulletin of the Johns Hopkins Hospital; Clinical Science; Nature; Journal of the Acoustical Society of America; Journal of the American Medical Association; Journal of Biophysical and Biochemical Cytology; Journal of General Physiology; Journal of Neurophysiology; Journal of Physiology; Proceedings of the Royal Society* and the *Scientific American*.

The following publishers have been kind enough to grant permission for the use of certain illustrations from their publications and we thank them for this facility: The D. Appleton-Century Co.; Edward Arnold Ltd.; Bailliere, Tindall & Cox; G. Bell & Sons; Cambridge University Press; W. Heffer & Sons Ltd.; Hinrichson Editions; Lloyd-Luke (Medical Books) Ltd.; W. H. Freeman; Macmillan & Co. Ltd.; Oxford University Press; The W. B. Saunders Co.; Shaw & Sons; Charles C. Thomas; and the Williams & Wilkins Co.

Finally, we would like to express our gratitude to Anne Ashford, Lynn Baxter and Tim Hailstone of Churchill Livingstone for the great deal of work they have done in the publishing of this book.

Contributors

Professor M. de Burgh Daly M.A., M.D., Sc.D., M.R.C.P.,
Department of Physiology,
The Medical College of St Bartholomew's Hospital,
Charterhouse Square,
London

Professor S. E. Dicker Ph.D., D.Sc., MD.,
Chemistry Department,
University College London,
London

Professor R. A. Gregory C.B.E., F.R.S.,
Physiological Laboratory,
Brownlow Hill,
P.O. Box 147,
Liverpool

Professor B. R. Jewell Ph.D., M.B., B.S.,
Department of Physiology,
The University of Leeds,
Leeds

Dr. O. C. J. Lippold M.D. (Lond.), M.B., B.S.,
Department of Physiology,
University College London,
London

Dr. P. A. Merton M.B.,
The Physiological Laboratory,
Downing Street,
Cambridge

Professor D. H. Smyth M.D., D.Sc., F.R.S.,
Unit for Metabolic Studies in Psychiatry,
Middlewood Hospital,
P.O. Box 134,
Sheffield

Biographical Note

Michael de Burgh Daly, M.A., M.D., Sc.D., M.R.C.P., has been Professor of Physiology at St Bartholomew's Hospital Medical College, Charterhouse Square, London, since 1958. He was educated at Gonville and Caius College, Cambridge and St Bartholomew's Hospital, London, and is formerly Rockefeller Foundation Travelling Fellow in Medicine and Locke Research Fellow of The Royal Society.

Sebastian E. Dicker, Ph.D., D.Sc., M.D., is Emeritus Professor of Physiology at the University of London and Honorary Research Fellow at University College London. He was formerly Head of the Department of Physiology, Chelsea College, University of London.

Roderick A. Gregory, C.B.E., F.R.S., was educated at University College London where he took a B.Sc. Honours degree in physiology in 1934 and subsequently became first Bayliss-Starling Scholar and then Sharpey Scholar and also became medically qualified from University College Hospital. He studied with A. C. Ivy at Northwestern University, Chicago and F. C. Mann at the Mayo Clinic, Rochester, Minnesota. He has been Professor of Physiology at the University of Liverpool since 1948. He was elected F.R.S. in 1965. He has worked in the field of gastro-intestinal physiology since 1939.

Brian R. Jewell, Ph.D., M.B., B.Sc., is Professor of Physiology, the University of Leeds. He was formerly Reader in Experimental Physiology in the University of London at University College. He is a member of the Editorial Board of the *Journal of Physiology* (1972–79; Distributing Editor, 1976–78; Chairman, 1978–79). His main research interest is in the physiology of cardiac muscle.

Olof Lippold is Reader in Physiology, University of London and works in the Physiology Department at University College. He qualified in Medicine at University College Hospital and has pursued a career in medical research. A neurophysiologist, his present interests are in the neurological mechanisms underlying normal muscular tremor and the tremor of Parkinson's disease.

Patrick A. Merton is Reader in Human Physiology, University of Cambridge. He was educated at Beaumont College; Trinity College, Cambridge; St Thomas's Hospital. M.B. Cambridge 1946. He was on the staff of the Medical Research Council, Neurological Research Unit, National Hospital for Nervous Diseases, Queen Square, London from 1946–1957. He visited the laboratory of Professor Ragnar Granit in Stockholm, 1952–1954. He was Lecturer in the Physiological Laboratory, Cambridge, from 1957–1977, and Fellow and Lecturer in Medical Sciences, Trinity College, Cambridge, in 1962.

David H. Smyth, M.D., D.Sc., F.R.S., was Professor of Physiology in the University of Sheffield from 1946 to 1973 and is now Emeritus Professor. His main scientific interest is in the mechanisms of absorption of nutrients from the intestine.

Contents

Preface to Seventh Edition ... v
Prologue ... xiii

Section 1: Specialised Cells and Tissues

1. The structure and function of the cell ... 1
2. Membrane transport: passage of water and solutes across membranes ... 7
3. Specialised cells and tissues ... 10
4. Nervous tissue ... 21
5. Neuromuscular and synaptic transmission ... 39
6. Muscle: introduction ... 54
7. The structure and function of skeletal muscle ... 56
8. The structure and function of cardiac muscle ... 70
9. The structure and function of smooth muscle ... 76
10. The composition of blood ... 82
11. Blood: oxygen transport ... 89
12. Oxygen transport in man and its disorders ... 95
13. The carriage of carbon dioxide ... 99

Section 2: The Cellular Environment and its Control

14. Control systems: theory ... 104
15. Control systems: operation ... 109
16. Properties of body fluids ... 114
17. The distribution of body fluids ... 118
18. The formation of body fluids ... 122
19. The control of water distribution in the body ... 127
20. The kidney and control of body fluids ... 130
21. The formation of urine ... 134
22. The physiology of tubules ... 139
23. The control of kidney function ... 146
24. Micturition ... 149
25. Acid-base regulation ... 152
26. The physiological control of reaction ... 156
27. Body temperature and its control ... 161
28. Vascular mechanisms in body temperature control ... 165
29. Performance characteristics of the thermoregulatory system ... 168
30. The autonomic nervous system ... 172
31. Chemical transmission in the autonomic system ... 179
32A The liver ... 185
32B Bloodstream communication ... 189
32C The adrenal gland ... 191
32D The thyroid gland ... 198
32E The pituitary gland ... 202
32F Endocrine control of plasma constituents ... 206

Section 3: Systems for Transport in the Body

33. The general function of the circulation ... 210
34. The design of the circulation ... 212
35. Some general circulatory responses ... 222
36. The control of the vascular system ... 225
37. Hormonal regulation of the circulation ... 232
38. Characteristics of the circulation in various organs ... 235
39. The heart: general characteristics as a pump ... 248
40. The control of cardiac output ... 255
41. Cardiac electrophysiology ... 261
42. Clinical electrocardiography ... 266
43. Mechanics of breathing ... 272
44. Lung volumes, ventilation and diffusion ... 277
45. The regulation of breathing ... 285
46. Examples of respiratory adjustments ... 299
47. Artificial respiration and cardiac resuscitation ... 308
48. Digestion in the mouth and stomach ... 312
49. Intestinal digestion ... 319
50. General metabolism ... 325
51. The metabolic history of the food substances ... 332
52. Nutrition ... 341

Section 4: Information: Its Reception and Processing

53. General properties of receptors ... 349
54. Cutaneous receptors ... 357
55. Muscle receptors ... 361
56. The auditory system ... 364
57. The labyrinth ... 373
58. Vision ... 377
59. The central nervous system: I ... 392
60. The central nervous system: II ... 406
61. The central nervous system: III ... 423

Section 5: Growth and Reproductive Physiology

62. Human growth — 429
63. The physiology of the female reproductive system — 433
64. The oestrous and menstrual cycles — 437
65. The hormonal control of sexual cycles — 440
66. Fertility control — 446
67. Pregnancy, parturition and lactation — 448
68. The male reproductive system — 455
69. Coitus (Copulation) — 459
70. Sex determination and differentiation — 461

Section 6: Compensatory Ajustments and Reactions to Injury

71. Defence reactions (inflammation, immunity, healing) — 465
72. The coagulation of blood — 470
73. The blood groups — 474
74. The pathophysiology of shock — 477
75. High blood pressure — 481
76. Hepatic failure — 483
77. Transplantation — 485
78. Physiological tremor — 486
79. Changes in the fetal circulation at birth — 490
80. Pain — 491

Epilogue — 496
Appendix I: Units and normal values — 498
Appendix II: Formulae — 500
Bibliography — 501
Index — 502

Prologue

The dead and the living body differ. How they differ and how the living body sees, moves, digests, keeps warm, and so on, is the province of physiology. We see a man lift an arm or walk without staggering and as physiologists we wonder what is going on in his muscles and nervous system. To discover things of this kind, we have first to take the machine to pieces, much as we should have to take a motor car to pieces to explain its varied performance on the road. You cannot, of course, take a man to pieces, so the corresponding tissues of animals, say the limb muscles, intestine or heart, have to be examined one by one after removal from animals immediately after death. If the property of an organ is found to be similar in a number of animals, such as the frog, rabbit, cat, dog and monkey, we presume that it may well be much the same in man. This basic knowledge about individual organs can be extended to a study of their functions and interactions in anaesthetised animals. Only then, as a rule, can painless methods be devised for similar and further studies on unanaesthetised animals and man. Thus physiology extends from physical and chemical processes in cells and tissues to elaborate performances of whole animals, but it does not extend very far into studies of behaviour or mental affairs which are the province of psychology.

Physiology has another boundary; it is primarily concerned with normal animals. Disease processes belong to the subject matter of pathology and medicine. Practising physicians, however, need a vivid awareness of physiological processes in the body for both diagnosis and treatment. Normally they lack the leisure to submit disorders presented by individual patients to extensive physiological analysis; moreover, the treatment of the sick cannot be delayed till the outcome of such analysis is known. Treatment depends on speedy diagnosis which is based on the symptoms reported by the patient, on physical signs observed by the physician and perhaps on a few simple tests such as X-ray, histological or bio-chemical examinations. Such slender elements of evidence must be fitted into the whole relevant knowledge of physiological processes in the body before they have much value beyond rule-of-thumb indicators, yielding "slot-machine diagnosis". So it happens that important advances in physiology are made in the research departments of medicine, particularly in fields somewhat neglected by physiologists who may be preoccupied with studies without such applications to medicine. Certain diseases such as diabetes mellitus have, however, been extensively studied by physiologists sometimes with dramatic improvements in treatment as a consequence.

The most noticeable and measurable sign of deterioration in ill-health is often the reduction of the amount of muscular exercise that can be endured without excessive breathlessness and thumping of the heart. If a sedentary clerk and an athlete run side by side to catch a train, the first may arrive panting, sweating and exhausted and take quite some time to recover, while the second may suffer little discomfort and that soon over. Neither would be regarded as ill unless his performance deteriorated to the point of interfering with his normal occupation. A third person may find that, for example, climbing slowly up a staircase may engender such breathlessness as to demand a rest on reaching each floor, yet the physiological mechanisms controlling breathing are much the same in all three. Thus physiology is essential to the understanding of both normal and disease processes.

The line dividing the living and the dead is not easy to draw. If you prod a live animal it is apt to move, but neither anaesthetised animals nor most plants move actively when disturbed. Growth and reproduction are regarded as characteristic of the living state, but many adults past growth and beyond child-bearing are far from dead and still appear irritable to prodding.

Men and large animals are certainly dead when the heart has stopped beating for so long that there is no hope of recovery. In them, the blood pumped by the heart is essential for carrying foodstuffs and their products in the intestine to organs like muscles, to which, also, oxygen is carried by blood from the lungs. Energy derived from chemical reactions resembling burning can then be converted into mechanical or other forms of useful energy, while the waste products of the combustion are carried away, again by the blood, the carbon dioxide to the lungs and most soluble products to the kidneys, where they are eliminated from the body. Perhaps the continued use of oxygen as part source of chemical energy is one of the more widespread features of life, but anaerobic organisms provide exceptions; moreover, power for industry and traffic is still derived from fuels oxidised in apparatus which is indubitably dead.

In animals of microscopic size, in contrast to larger animals, blood circulation is not needed. Every part is so near the surface of the cell or organism that it can derive oxygen and foodstuffs by direct diffusion from the surface. A blood circulation is, however, essential in larger animals and this may be illustrated by some features of muscular exercise in man.

Muscle occupies much of the total body weight and, when active, each gram of muscle uses oxygen and produces waste products more rapidly than any other tissue. The large output of carbon dioxide itself is one of the factors producing the well-known increase in breathing during strenuous exercise, and this also secures the increase in oxygen intake. A large increase in blood flow is needed to carry the extra carbon dioxide and oxygen to and from the lungs and is produced by the heart which accordingly beats at a higher rate and puts out more blood per beat. Moreover, muscles resemble manmade engines in being unable to convert more than, at best, about one-quarter of the chemical energy into mechanical work. Three-quarters or more, therefore, of the chemical energy is converted to heat. Again, the blood is essential for transporting the extra heat from muscles to the skin from which the heat can be lost by radiation and conduction. Indeed, in healthy people the severity of the exercise they can take is generally limited, not by their muscles, but by the rate at which the heart can pump blood round the body.

Cardiac output can be measured in man, but not easily enough to be a useful test in clinical practice. Heart rate, easily measured as the pulse of the radial artery at the wrist, is a most useful quick guide to change in cardiac output, but when the frequency reaches about 180 per minute the time available for the heart to refill between contractions becomes too short and the output per beat falls. Below this limit, pulse-rate can be used as a guide to the degree of physical fitness in two ways. First, during moderate exertion the pulse-rate will rise more in sick people than would be expected in normally fit people. Secondly, during rest following exertion, the pulse-rate will return to normal more slowly.

Undue fatigue is well-known to accompany ill-health. Fatigue is a state which may reduce or end longer-lasting muscular exercise or other forms of activity. It is said to supervene when an activity, which has been sustained for some time, diminishes although the incentive to maintain it remains unchanged and effective, the standard of performance being completely restored after a period of rest. Formerly, fatigue was attributed to the accumulation in the body of metabolic products of activity known as "fatigue products". Though this may happen in a few kinds of activity, it is now considered incapable of explaining most forms of fatigue. Different kinds of activity may be impaired by different processes which may concern primarily almost any of the physiological systems in the body.

An isolated frog muscle contracts less and less in response to repeated electrical stimulation of constant strength; if given a rest, the contractions regain their original size. By analogy, it is said to have become "fatigued". This is attributed to inadequate supply of oxygen diffusing from the surface of the muscle to the fibres and leading to an accumulation of lactic acid. Accordingly, it is difficult to fatigue a single muscle in an anaesthetised mammal by repeated electrical stimulation because it is well-provided with oxygen from its own blood circulation. In very severe exercise in man involving many muscles, however, the oxygen supply to muscles is restricted by the cardiac output, as mentioned earlier, and lactic acid correspondingly accumulates in the body and reduces muscular activity. In less severe exercise the lactic acid is oxidised to carbon dioxide and water; but if such exercise continues till the animal, say a dog running on a treadmill, is exhausted, the fatigue is due to lack of fuel. If glucose is administered periodically, the dog can continue to run for much longer without exhaustion.

Another factor which produces fatigue after even short spells of moderate exercise, is the undue rise in body temperature which occurs in a hot moist atmosphere, or if evaporation is reduced by wearing a rubber or oilskin coat. Reduction in blood volume which reduces blood circulation in the brain often makes people feel tired and "go slow" accordingly. When men stand strictly to attention for half an hour or so, they may even faint. The loss in blood volume is here due to high blood pressure in the veins and capillaries of the immobile legs with outward filtration of water into the tissue fluid. There are many ways other than blood or oxygen lack in which the brain can become fatigued, mental work among them as some students know from experience.

The activity, and even survival, of the cells and tissues of an animal or man depend on the properties of the fluids surrounding them. For example, chemical processes in cells are catalysed by enzymes, at varying rates according to temperature and acidity; but some vary more than others. Particularly in the more complex species, therefore, body temperature, the acidity and osmolar concentration of body fluids and many other factors must remain nearly constant if physiological processes are to continue in normal balance. Such ideas were crystallised by Claude Bernard (1830); the higher animals were said to have an *internal environment* which is maintained remarkably constant, even during great changes of the external environment, such as climate, the amount and kinds of food and water taken in, and the extent of muscular activity. The pattern of regulatory processes in the body which operate to control the internal environment within so narrow a range of variation was called *physiological homeostasis* by W. B. Cannon (1929) and is one of the essential factors in maintenance of health and, indeed, in survival.

The following chapters include many descriptions of such regulatory processes appertaining to the physiological systems concerned. The interactions between these processes, to achieve homeostasis, will be considered further in the epilogue following the chapters devoted to the individual physiological systems.

Section 1:

Specialised Cells and Tissues

1. The Structure and Function of the Cell

Cells

Nearly all tissues and organs in the body are composed of *cells*, held together by various intercellular supporting substances. The cells of the body have certain basic characteristics common to all kinds of cell but apart from these, cells display an enormous variety in form, each type being specially adapted to carry out its particular function. Red blood cells, for example, are specialised in structure to enable them to carry oxygen from the lungs to the tissues. Muscle fibres are cells which have the ability to contract. Although the various cells of the body have such diverse functions and structure, they all consist basically of a fluid system, protoplasm, and are surrounded by a membrane which acts as a barrier between the fluid within the cell and the fluid outside it.

During the course of normal cellular activity molecules and ions pass across this membrane and there is a dynamic interchange between substances in solution in the extracellular and intracellular spaces.

Cell structure

Within the boundaries of the cell membrane, the microscope shows that the cell consists of cytoplasm, a nucleus and certain other structures such as the Golgi apparatus, the centrosome and various types of granule, such as mitochondria.

The use of the electron microscope, with its high powers of magnification and the discovery of chemical methods for analysing the materials within cells, have led to great advances in our knowledge of cell biology over the past twenty years or more. Cells are characterised by the ability to transform energy, as in the output of work by muscle cells or electrical energy of the nerve impulse, both of which involve the utilisation of chemical energy contained in molecules brought to the cell. They are also characterised by their synthesis of macromolecules. It would be safe to say that macromolecules are only found in conjunction with cells either living or dead. Figure 1.1 shows an elementary typical animal cell diagrammatically.

Mitochondria

These structures within most cells, are found in the cytoplasm and are concerned with the energy turnover during functioning of the cell. Skeletal muscle for example, with its high energy transformation rate, has many mitochondria.

They are 0·5 to 12 μm in length and about 0·5 to 1 μm in diameter. They are bounded by a double membrane, the inner layer of which has many folded projections into the interior of the mitochondria. Mitochondria provide the basic energy transformations involved in cellular respiration. By oxidative phosphorylation, the energy contained in materials brought to the cell is turned into adenosine triphosphate, or ATP, which is the primary energy source in living cells. Within mitochondria therefore, we find the enzymes of the tricarboxylic acid cycle, those for catalysing fatty acids and those for coupling electron transport to the cycle.

The endoplasmic reticulum

Within the cytoplasm of most cells is found a complicated network of internal membranes making up vesicles and tubules.

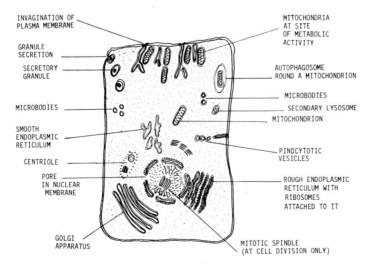

Fig. 1.1 Diagram of a typical animal cell. The major structures found in cells are shown although it is most unlikely that any one cell would contain all these organelles simultaneously. Notice that the endoplasmic reticulum opens onto the cell's surface and the interior of this system is continuous with the external environment of the cell.

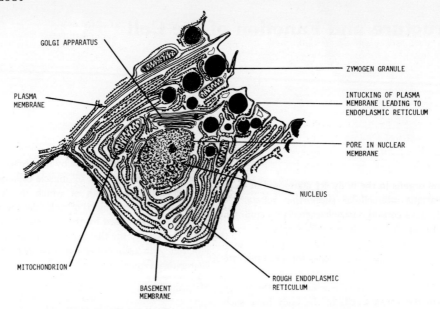

Fig. 1.2 Drawing from electron micrograph of a secretory pancreatic cell. The rough endoplasmic reticulum is a net of connecting tubules, 30 to 80 nm diameter, bearing particles on their *outer* surfaces called ribosomes. These are round, electron-dense structures about 15 nm diameter made of RNA and ribonuclear protein. At ribosomes, amino acids are assembled into proteins which pass into the reticular lumen; in secretory cells such as this one, there is a dense reticulum. The Golgi apparatus is a plate-like reticulum of smooth membranes, 7 nm thick and in layers about 15 to 20 nm apart. The enzymes found here are for linking sugars and protein to form glycoproteins.

Through this system, called the endoplasmic reticulum, various substances are transported from the outer surface of the cell, onto which some of the tubules open, to the nucleus or to other parts of the cell such as mitochondria (see Figs. 1.2 and 1.3).

As we can see from Figs. 1.2, 1.3, the membrane of the endoplasmic reticulum is continuous with that bounding the whole cell, and the reticulum can be thought of as a complex invagination with the contents of the tubules being in continuity with the exterior environment of the cell. The nucleus, which is a specialised part of the cytoplasm containing genetic material (DNA and RNA), is surrounded by the so-called nuclear membrane, a system of sacs of the endoplasmic reticulum, as can be seen in the figure.

The endoplasmic reticulum is either *smooth* or *rough*, the latter term referring to the electron-microscopic appearance of granules along the cytoplasmic border of the tubules. The granules are *ribosomes*, composed of ribonucleic acid, RNA, having the function of synthesising protein. Hence one finds that cells which produce large quantities of protein, e.g. liver cells, contain very many ribosomes.

The Golgi bodies are found near the nucleus and are part of the endoplasmic reticulum in the form of smooth vesicles, possibly concerned with the formation of membrane material although we do not as yet know for certain what their function is.

Lysosomes

These are roughly spherical organelles, 250 to 750 nm in diameter, and contain protein and various hydrolytic enzymes such as ribonuclease, deoxyribonuclease, phosphatases and proteolytic enzymes. One would predict such a mixture of lytic enzymes to have a remarkably destructive effect if let loose within any cell, but fortunately they are contained within a membrane. When the animal dies, however, the integrity of the membrane is destroyed and rapid post-mortem autolysis occurs.

Normal function of lysosomes is concerned with the digestion of substances gaining entry to the cell. Such substances are enclosed within the cell in a vacuole to which a lysosome fuses and digestion occurs. The digested products diffuse out via the lysosome membrane into the cytoplasm of the cell.

White cells, whose function is to remove foreign materials from the plasma and interstitial fluid, contain many lysosomes. When first formed, white cells have lysosomes with many granules within them; during their life span these granules, which are the enzyme molecules, are used up and the granular content progressively decreases.

Cell damage, as a result of oxygen lack or the action of cellular poisons, ruptures the lysosomal membrane thus releasing the hydrolytic enzymes. This destroys the cell, and the escaping enzymes diffuse to other regions and may also cause destruction there.

The nucleus

This is an approximately spherical body, up to 10 μm in diameter, about three-quarters of its dry weight being protein. The remainder is DNA and RNA. The nuclei in cells are stained selectively by basic dyes (e.g. toluidine blue) because such dyes bind with the nucleic acids, DNA and RNA. A network of densely stained regions within the nucleus can be seen and is termed *chromatin*; it consists of the chromosomes which appear at the time of cell division. The most dense region within the nucleus is called the *nucleolus* and is made up of spiral fibres of RNA. It appears to be concerned with protein synthesis (see Fig. 1.3).

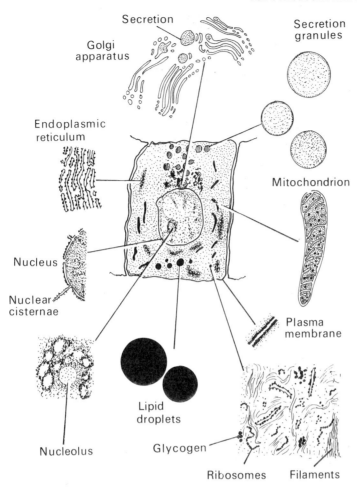

Fig. 1.3 Section of mouse pancreatic cell fixed with osmium. Part of two mitochondria can be seen, as well as elements of the endoplasmic reticulum. The cristae mitochondriales can be seen running into the substance of each mitochondrion.

Centrosomes

These are paired structures which appear only when cell division is about to occur. They are at the opposite ends of the nucleus, attached to the chromatin, and are thought to be the motile apparatus which pulls the chromatin apart. It is interesting to note that bodies of similar structure (cylindrical; containing eleven fibres) are found immediately below the cilia in ciliated cells.

SPECIALISED STRUCTURES IN CELLS

The cells which make up the whole body are very different in structure according to the functions they must carry out. Some of the general features of cells described in the previous paragraphs are exaggerated; others may appear, and yet others be absent entirely. These specialised features confer upon tissues their ability to undertake particular functions such as the transmission of information by nerve fibres, the production of external work by contracting muscle cells, or the secretion of materials by glands.

The cell boundary

We have already seen that the boundary of all cells consists of a highly specialised membrane. It will be appreciated that a thin, pliable structure like this can have little mechanical strength; most cells therefore are modified over all or part of their membrane surfaces. Usually, the membrane has a mucopolysaccharide coat (carbohydrate–protein complexes which form the basis of the ground-substance of connective tissue).

Cells in tissues do not usually touch each other; there is a space about 10–20 nm wide between them, filled with interstitial fluid. It is via this route that water, ions etc. can penetrate to the surface of the cell. Often, the adhesion between cells is increased by *reticulin* fibres in the surrounding coat of mucopolysaccharide. Reticulin is a fibrous protein and where it is thickly applied to a cell, is said to be a *basement membrane*. The latter can be stained with silver salts to make it visible under the light microscope.

Different parts of cells have different thicknesses of such extra-cellular material. For example, an epithelial cell may have a thick basement membrane, a polysaccharide-covered side wall and a simple plasma membrane on its surface.

Adhesion between cells

To maintain the structural integrity of the body and the tissues of which it is composed, it is necessary for cells to adhere to each other. Cells are attached to each other in complex ways to form masses of tissue, such as in the liver or pancreas, tubes to enclose fluids, such as the ureters or blood vessels, and in specialised organs such as muscles which move and slide with respect to other structures.

This ability of cells to stick to each other is one of the important features of the process of differentiation; indeed, when cells become malignant this ability is usually lost, resulting in the formation of secondary cancerous deposits scattered throughout the body.

There are specialised structures which are responsible for firm fixation of one cell with another. These are shown diagrammatically in Fig. 1.4, and are found in three main configurations, the *desmosome* (macula adherens), the *intermediate junction* (zonula adherens) and the *tight junction* (zonula occludens).

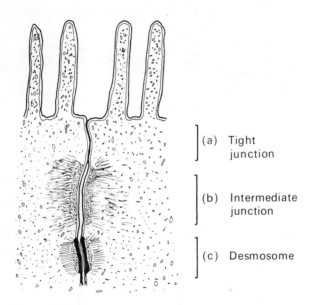

Fig. 1.4 Diagram showing the structure of (a) a tight junction, (b) an intermediate junction and (c) a desmosome as seen under the electron microscope. The section is of the junctional complex found between two columnar epithelial cells (see text).

THE DESMOSOME
This is a double structure consisting of thickening of the opposed membrane surfaces of two cells adjacent to each other. Subjacent to each thickened layer, is a fine feltwork of filaments embedded in condensed cytoplasm. In plan view these structures are discoid.

No filaments or other transverse structures have been seen to connect the two halves of a desmosome; we do not yet know the way in which cells are held together tightly at these points.

THE INTERMEDIATE JUNCTION
Intermediate junctions are similar to desmosomes but are more extensive. For example in epithelial cells they often form a girdle completely round the cell.

THE TIGHT JUNCTION
Here the intercellular space is obliterated, the outer layers of the plasma membranes of the two cells concerned being fused. Such fusion presents a diffusion barrier and can act as a seal between one body cavity and another. It is found in epithelial cell layers and offers a watertight boundary, for example in organs such as the bladder.

In some cells an even more extensive fusion of the plasma membrane is found, so that the membrane disappears and the cytoplasm is contiguous between cells. Such groups of cells are termed *syncytial*; examples are to be found in heart muscle, large osteoclasts etc.

Pigment granules

Certain tissues, for example the retina and the skin, contain cells having large numbers of dark brown or black granules within them. These cells are called *melanocytes*, and the pigment *melanin* is contained in *melanosomes*. The various shades of colouring of the hair, skin and even of the iris are due to interference patterns formed by it. Skin and mucous membranes, of course, receive part of their distinctive colouring from the haemoglobin within the visible small blood vessels.

Melanin is synthesised from tyrosine which is first converted to dihydroxyphenylalanine (DOPA).

The function of the pigment is to reflect ultraviolet radiation and hence protect the skin from sunburn; in the iris to cut down light transmission into the eye, and in the retina to absorb scattered light, thus enhancing contrast in the retinal transmitted image.

Pinocytosis

As we have described, passive diffusion and active transport mechanisms account for the passage of water, gases, ions and small molecules across the cell membrane. Larger molecules and particulate matter are enabled to pass through the membrane by the process of pinocytosis. This involves the molecule or particle first becoming attached to the outer surface of the cell membrane. Then the region of membrane involved invaginates to form a minute vesicle. This is closed off, enters the cytoplasm and eventually the vesicular membrane is disrupted, leaving the ingested material free within the cell (Fig. 1.5).

Formation of the pinocytotic vesicle, and the subsequent movement of it, depends on energy production by the cell; the evidence for this is that ATPase can be found in the vesicle wall. It is also interesting to note that selectivity of the process could possibly be achieved in terms of the nature of which particles adhere to the membrane in the first place. All cells are capable of pinocytosis.

Phagocytosis

Macrophages, granulocytes and *histiocytes* are capable of ingesting relatively large particles, such as bacteria, by utilising the process of phagocytosis. This process is fundamentally the same as pinocytosis but on a larger scale. During phagocytosis, the specific granules in the cell's cytoplasm break down and disappear, in so doing liberating hydrolytic enzymes which break down the ingested bacterium. This mechanism is important in the defence of the body against infection and can be investigated *in vitro* when leucocytes and bacteria are brought together. *In vivo*, phagocytosis is more effective in tissue spaces than it is in circulating blood.

Secretion

Secretion is the formation and release by cells of a fluid containing substances such as mucin, enzymes, hormones etc. The process requires the expenditure of energy. Secretory cells (as found in glands) usually contain specific granules or droplets which are intracytoplasmic accumulations of the precursors of their secretion, and which first appear in the cell close to the Golgi apparatus.

In the process of secretion, the secretion droplets are formed within membrane-limited vacuoles. These then leave the region of the Golgi apparatus and move towards the apex of the cell. The membrane enclosing the droplet then contacts the plasma membrane, coalesces with it and becomes continuous with the cell surface. The secretory product flows out and is discharged without any rupture of the membrane being caused. It is the reverse of the pinocytosis shown in Fig. 1.5.

brush-border. Often, the microvillae have specific enzymes adsorbed upon them and their function is thought to be more complex than a mere multiplication of surface area.

Cilia and flagellae are actively motile projections of the cell and have the function of moving the fluid in contact with the cell surface (Fig. 1.6).

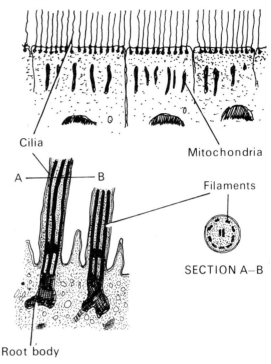

Fig. 1.6 Ultrastructure of cilia. The internal structure of each cilium consists of nine paired filaments and two more larger central filaments. These filaments arise from basal bodies, and often cross-striated rootlets are visible attached to the basal bodies. A cilium is usually about 4μm in length.

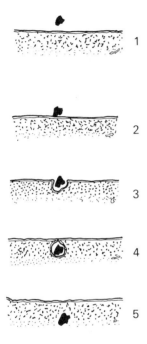

Fig. 1.5 Stages in the transport of a large particle (shown as an amorphous black dot) through the cell membrane by pinocytosis.

The reverse process is used for the discharge of secretion of macromolecules into the extracellular space or lumen of a gland. Examples are secretion of enzymes by enzyme-secreting cells or the release of transmitter at nerve terminals.

A double process of this kind enables particulate matter to be moved directly from one cell to an adjacent one. For example, melanocytes producing melanin transfer it direct to the epithelial cells where it is then stored. Pinocytosis also enables transfer of materials directly through a cell without them being in contact with the contents of the cell. Capillary permeability and certain parts of the absorption process in the gut utilise this mechanism.

Current views on secretion indicate that the ribosomes on the endoplasmic reticulum synthesise the secretion; the material is then transported to the Golgi apparatus where it is concentrated and given its membrane coating. Secretion is discussed in more detail later.

Microvillae, cilia and flagellae

These are projections of the cell membrane, finger-like in structure which have the function of increasing the surface area of the membrane (microvillae) or have motile properties.

Microvillae are characteristically small finger-like projections on the external membrane surface, perhaps up to 1 μm in length; when closely packed they give the appearance of a so-called

The fine structure, as seen under the electron microscope, consists of nine paired filaments in a ring with two central ones, as can be seen in Fig. 1.6, attached to a root-body.

Ciliary motion resembles the lashing of a whip, being forceful in the one direction only. All cilia on a mucous surface (for example) are unidirectional and beat in coordinated waves as shown in Fig. 1.7.

Flagellae are longer than cilia, being up to 50 μm long, as seen in the spermatozoan. The internal structure is the same.

The cell membrane

Cell function depends on the integrity of the cell membrane and its property of allowing certain molecules and ions to cross it

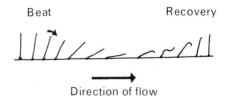

Fig. 1.7 Mechanism of action of a cilium. A stiff, forceful stroke from left to right moves fluid along in direction of arrow. Slack recovery shown on the right. All cilia on a given sheet of mucosa beat in the same direction and are synchronised in waves, much like the movements of ears of wheat in a wind-blown field.

The way in which each cilium is made to move is not known.

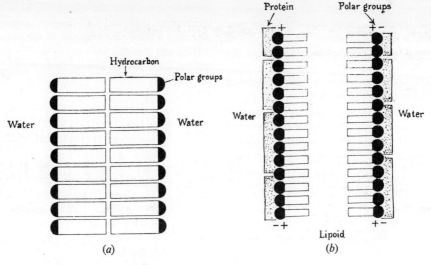

Fig. 1.8 Hypothetical structure of the cell membrane: (a) red cell membrane; (b) general pattern. (Davson and Danielli, 1952, *Permeability of Natural Membranes*. Cambridge University Press.)

while preventing the passage of others. The membrane, on analysis, is found to consist largely of proteins and lipids. The protein is a fibrous substance called *stromatin* and the lipids are *cholesterol, lecithin* and *cephalin*. The lipoid substances form a bimolecular layer round the cell as shown in Fig. 1.8. Since the membrane is largely lipid in composition, fat-soluble substances penetrate into the cell easily, whereas non-fat-soluble ones enter only slowly or not at all. Lipids are good electrical insulators, a fact that we will discuss later in connection with the high electrical impedance of membranes, characteristic of all cells.

One can visualise the membrane as having a sieve-like or porous structure, the openings (or pores) being of molecular size, so that small molecules will travel through them. The pores have not been seen under the electron microscope—indeed, they probably do not exist in a simple form—but rather they represent preferential pathways within the complex molecular structure of the membrane along which diffusion can take place.

The electron microscope shows that the membrane is typically about 7·5 nm* thick, and it appears to have a triple-layered structure, an inner and outer dense layer being separated by a less dense region between them. The outer layer is polysaccharide and the inner one is protein, both being one molecule thick. The central region is a bimolecular arrangement of phospholipid. In Fig. 1.8, the black dots are the polar groups of the molecule which are *hydrophilic* (attracted to water) and consist of the nitrogen–carbon–phosphorus groupings of the phospholipid. The rest of each molecule, a fatty acid (long-chain hydrocarbon—shown as white bars), is *hydrophobic*.

These long molecules are arranged at right angles to the membrane surface, in pairs, as shown in Fig. 1.8.

---

* 1 nm (nanometre) is 10^{-9} m. The metre is 1 650 763·73 wavelengths of the orange–red spectral line of the atom of krypton-86 in a vacuum ($\lambda\ ^{86}Kr$).

Functions of the cell membrane

In addition to delimiting the boundaries of the cell, the membrane is a highly specialised structure for controlling the passage of substances into and out of the cell. In some specialised cells such as nerve and muscle cells, the membrane is also *excitable* and can respond to stimulation with an electrical response (an *action potential*, which is a sudden change in potential across the membrane). This response may be propagated along the cell membrane as in nerve fibres, to transmit information, or it may trigger the contractile mechanism in muscle.

Membrane permeability

Many substances enter and leave the cell while it is active by passing through the membrane. A membrane which allows materials to pass through it is termed *permeable* (to the materials). A membrane which will let only water through is called *semipermeable*; it must be noted that in fact such membranes do not exist in biological systems. Most cell membranes are *selectively permeable*, a term denoting that in addition to water, certain substances can traverse the membrane whilst others cannot. The reader will soon discover that common usage tends to confuse the distinction between the latter two terms. The word "permeability" is also applied to describe the properties of blood vessel walls, intestinal cells and other regions which separate two fluid compartments.

Permeability of the cell membrane either is *active*, i.e. metabolic processes are involved, carrier mechanisms are used and energy is consumed in the process, or it is *passive*, i.e. no energy is required to move particles across the membrane, only a concentration gradient being involved.

2. Membrane Transport: Passage of Water and Solutes across Membranes

It is a fundamental law of physical chemistry, based on the universal properties of matter and energy, that the components of a solution tend to move from regions where they are in high concentration, to regions where they are in low concentration. They will, in fact, move in this direction unless they are prevented from doing so by some obstruction, or are driven in the opposite direction by some other force. This force may be electrical; it may be mechanical, such as a hydrostatic pressure; or it may be chemical, such as would be produced by interactions with other substances present in the solution or in contact with it. This general law applies to the water as well as to the substances dissolved in it. The concentration of the water becomes smaller as the concentration of the dissolved substances becomes larger, since, in a given volume, water molecules are replaced by solute molecules. But when we refer to the "concentration" of a solution, we ordinarily mean the *solute* concentration: water will thus move from a less concentrated solution (in this sense) to a more concentrated solution. The concentrations, however, must be expressed in terms of "osmolarity", that is, of the sum of the (molar) concentrations of all the solutes present, each ion of an electrolyte being considered as a separate substance. Unless the solution is very dilute, we must then multiply by the appropriate value of the "osmotic coefficient", which may be found in tables of physical and chemical constants. For mammalian body fluids, the value is about 0.9. The osmolar concentration of any solution may be measured directly in terms of its freezing point, or vapour pressure.

The fact, as indicated in Fig. 2.1, that the fluids in the

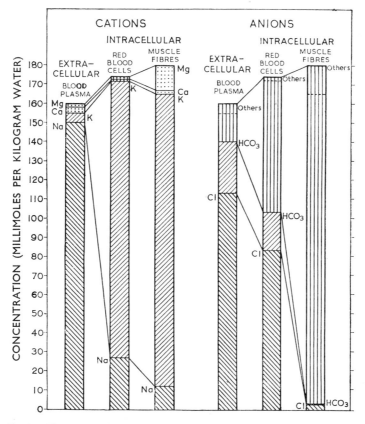

Fig. 2.1 The concentration of electrolytes in the extracellular and intracellular fluids of man.

    The values plotted may be taken as representative, but there are appreciable variations between one individual and another, and in any one individual according to circumstances.

    The concentrations are given in millimoles per kilogram of water; those of the "other anions" are deduced from the necessity for the total anion concentration to be electrically equivalent to the total cation concentration. The intracellular fluid appears to be more concentrated than the extracellular fluid; but actually the two fluids are osmotically equivalent, as some of the intracellular constituents are not osmotically active.

different compartments have different compositions, shows that there must be obstructions, or barriers of some kind at their junctions. Such barriers exist between the intracellular and interstitial fluids (cell membranes) and between interstitial fluid and plasma (walls of capillary blood vessels). The properties of these barriers determine the nature of the interchanges which take place between these different compartments.

The Donnan Membrane Equilibrium

Suppose we have two solutions of sodium chloride, and add to one of them the sodium salt of a protein or of any of the organic anions found within living cells. The two solutions are now put one on each side of a boundary, or membrane, through which sodium and chloride ions can penetrate but the organic anions cannot. The concentration of sodium ions is greater on one side of the membrane than on the other, but their natural tendency to diffuse in the direction of the concentration gradient is immediately opposed by an electrical force set up by the indiffusible anions which are left behind; this pulls them in again. Chloride ions, on the other hand, are driven out, so as to create a concentration gradient by the same electrical force; a force which pulls in positively charged ions will push out negatively charged ions. The freely diffusible ions thus become distributed unevenly between the two solutions, and an electrical potential difference is developed across the membrane. In these conditions, all freely diffusible ions which may be present (sodium, potassium, chloride and bicarbonate, for example) will move from one solution to the other until, for each kind of ion, the concentration in one solution is related to that in the other by a general equation defining what is known as the "Donnan Membrane Equilibrium". Using plasma (pl) and interstitial fluid (int) separated by the capillary wall as concrete examples, the equation is:

$$\frac{[Na^+]pl}{[Na^+]int} = \frac{[K^+]pl}{[K^+]int} = r$$

and

$$\frac{[Cl^-]pl}{[Cl^-]int} = \frac{[HCO_3^-]pl}{[HCO_3^-]int} = \frac{1}{r}$$

the square brackets indicating concentrations. The electrical potential difference between the two fluids is given by the Nernst equation:

$$E = \frac{RT}{ZF} \ln r$$
$$= 61 \cdot 5 \log r \text{ at } 37°C \text{ (millivolts)}^*$$
$$= 57 \cdot 2 \log r \text{ at } 15°C \text{ (millivolts)}$$

---

* Precise details of this equation are unimportant at the present juncture, but for information, R is the universal gas constant ($8 \cdot 31$ J mol^{-1}), T is absolute temperature, F is the Faraday (electric charge per gram equivalent of univalentions: 96 500 C mol^{-1}) and Z is the charge on the ion. When these values are substituted at 18°C for example, the relation is:

$$E(mV) = \frac{60}{Z} \log r$$

Hence the potential in millivolts across a membrane is determined by the ratio of the concentrations of the ions across it (see also p. 29).

The value of the "distribution ratio", denoted by r, increases if the concentration of the indiffusible (e.g. protein) ions is increased, or if the concentration of the diffusible (e.g. chloride) ions is decreased. In the plasma, the indiffusible ions are anions, and r is greater than 1, the concentrations of sodium and potassium are greater in the plasma than in the interstitial fluid (by about 5 per cent) and the concentrations of chloride and bicarbonate are smaller; the plasma is electrically negative to the interstitial fluid by about 1 millivolt (mV).

The equation, as just given, is an approximation only, since instead of concentrations we should, strictly, use "activities". By direct analysis of plasma and of interstitial fluid, or of a simple solution in equilibrium with plasma at a membrane impermeable to proteins (an "ultrafiltrate" or "dialysate" of plasma), it is found that the value of r for sodium ions is 1·06, and that for chloride ions is 1·04. The apparent discrepancy is due to the fact that the activity coefficients of sodium and chloride ions are slightly smaller in the plasma than in a protein-free fluid.

Membrane transport

We are still far from understanding the nature and properties of the cell membranes. Certain facts, however, which are of importance in the present connection are well established. Water can penetrate relatively freely into and out of the cells; certain substances in solution, notably oxygen and carbon dioxide, can penetrate nearly as freely; other substances such as urea and certain organic compounds (known as "non-polar" or "lipoid-soluble" compounds) whose molecular weight is not too large, and small univalent ions, can penetrate, but much less freely.

The non-polar substances are those which are soluble in ether, benzene and other liquid hydrocarbons, in which the "lipoid" materials, such as long-chain fatty acids, cholesterol and lecithin, are also soluble. It is partly because of the relative ease with which non-polar substances penetrate, that the cell membrane is thought to be composed chiefly of lipoid material. For this reason also, non-polar substances are used for determining the total volume of water in the body.

From the compositions of the intracellular and extracellular fluid we see that the concentration gradients of potassium ions and of chloride ions are in the direction which is to be expected if there is a Donnan equilibrium that is set up by the presence of indiffusible anions within the cells. There is no reason to suppose that these ions (and bicarbonate ions) cannot penetrate the cell membranes. Indeed, the use of radioactive isotopes and the existence of the "chloride shift" in the red blood cells (as discussed in Chapter 13) show that they do penetrate. There is also an electrical potential difference across the cell membranes of the sign to be expected (intracellular fluid negative to extracellular fluid) and of about the right size (10 to 100 mV). But the concentration gradient of sodium ions is in the opposite direction from that of potassium ions and it might seem, at first sight, that the cell membranes must be totally impermeable to sodium ions. But the use of isotopes has shown that this is not so, and it is impossible to avoid the conclusion that sodium ions are being expelled from the interior of the cell by some active process, the "sodium pump", which is only kept going by a continuous supply of energy from metabolic reactions. In effect, the cell membranes behave as if they were impermeable to sodium ions, since any that enter are immediately expelled.

The sodium pump

A "sodium pump" which expels positively charged ions will generate an electrical potential difference between the intracellular fluid and the extracellular fluid, and this will draw out chloride ions as well. The concentration ratios of all the ions which can diffuse freely will still be defined by the membrane equilibrium equation, and the value of the distribution ratio r will be related to the electrical potential difference across the membrane whether this is due to the presence of indiffusible ions or to the action of the "sodium pump". But the outward movement of sodium ions through the membranes of most kinds of cell is at least partly "coupled" to the inward movement of potassium ions, one potassium ion going in for each sodium ion "pumped" out. (Such a "pump" will not generate an electrical potential difference.) Owing to this restraint on the movement of potassium ions, the concentration ratio (intracellular fluid)/(extracellular fluid) is not precisely the inverse of that of chloride ions, nor precisely that to be expected from the electrical potential difference across the cell membrane. The distribution of potassium ions, as well as that of sodium ions, between the intracellular fluid and the extracellular fluid is not that to be expected if there were a Donnan equilibrium between the fluids, and the left-hand part of the membrane equilibrium equation does not apply. In red blood cells, the discrepancy is large, for both sodium ions and potassium ions, as may be seen in Fig. 2.1; but in muscle and nerve cells the discrepancy is small for potassium ions, though large for sodium ions.

The cell membranes thus allow many substances in solution to pass through them. But if there is a change in the total osmolar concentration of either the intracellular fluid or of the extracellular fluid, water is found to pass from one to the other much more rapidly than any of the substances in solution. The cells swell or shrink until the two fluids are once more in equilibrium with one another. Since the cells are not rigid, and can swell or shrink quite freely, changes in the concentration of the extracellular fluid are accompanied by equal changes in the concentration of the intracellular fluid, and vice versa; the consequent shifts of water from one to the other may, even in the normal living animal, be of quite considerable magnitude. For example, the activity of almost any organ is accompanied by the production of metabolites which are mainly of smaller molecular weight than the parent substances from which they are derived; the osmolar concentration of the active cells rises, therefore, and the cells swell. Muscular exercise, in particular, causes a rise in the volume of the intracellular fluid which can be observed, in man, by the methods of measurement already described.

Owing to the maintenance of osmotic equilibrium between the intracellular and extracellular fluids by a shift of water from one to the other, it is important that the substances used for measuring the volume of the extracellular fluid should be such that they can be analysed accurately even in low concentration. If the substance added to the extracellular fluid increases the osmolar concentration significantly, enough water will be drawn out from the intracellular fluid to produce a significant error in the estimated volume of the extracellular fluid. Indeed, by suitable calculation, it is possible to deduce the total volume of water in the body from the concentration of an added substance which does not enter the cells at all.

3. Specialised Cells and Tissues

I. CONNECTIVE TISSUE

Structure of connective tissue

Connective tissue consists of cells and extracellular fibres embedded in an amorphous ground substance (mucopolysaccharides). It normally contains tissue fluid.

Development is from the mesenchyme, an interlacing weave of *reticular fibres* being deposited between the mesenchyme cells. The fibres aggregate and form coarser bundles with the staining reactions of *collagen*. The cells gradually elongate and stretch out along the bundles of fibres and are called *fibroblasts*. The whole structure consists of fibres, cells and many "potential" spaces between them which may, on occasion, be filled with fluid. Other cells such as macrophages, formed elements from plasma, and elastic fibres are also found.

Collagen

Fibres in connective tissue are mainly constructed of collagen; its abundance varies in different situations, dependent upon the required functions of the connective tissue there. Collagen fibres appear under light microscopy as colourless strands, 1–20 μm thick and of indeterminate length. Usually running in all directions within the tissue, they are wavy unless the connective tissue is held under tension.

Collagen is flexible; it has considerable resistance to extension, and its breaking point is 300 or 400 kg cm^{-2}. Denaturing collagen, by boiling gives gelatin. Chemical treatment can give collagen *fibrils* each about 300 nm long and 1·5 nm thick, consisting of molecules of *tropocollagen* (three polypeptide chains of helical form). Hydroxyproline, one of the amino acids concerned, does not occur in any other animal protein and is therefore used as a measure of the collagen content of a tissue. Figure 3.1 is a diagram showing the general structure of loose connective tissue.

Functions of connective tissue

Connective tissue is the structure which surrounds and supports the other tissues and organs. In addition, it necessarily is concerned in the exchange of materials between blood and tissues. It is also important in protection against infection, repair after injury, and as a storage organ of fat—in the form of adipose cells deposited within it.

The mechanical properties of connective tissue are largely dependent on the fibres it has within it. Collagen gives tensile strength without appreciable distortion under load; fibres of *elastin* (a fibrous protein much like collagen, but differing in its amino acid content) confer the property of elasticity. Collagenous tissue is therefore found at sites where tensile strength is important, such as in tendons, capsules of organs or joints and in various septa. Elastic tissue occurs for example in

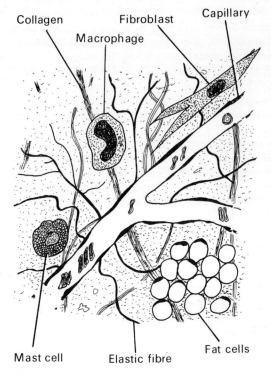

Fig. 3.1 Diagram to show the general structure of loose fibrous connective tissue. Collagenous and elastic fibres can be seen together with a fibroblast, macrophage, mast cell and a capillary.

Mast cells used to be thought of as phagocytic (German *masten,* to feed). It is now known that they manufacture carbohydrates and the mucopolysaccharide *heparin,* an anticoagulant. Histamine and 5-HT are also produced.

hollow organs which are distensible; the bladder and the walls of the aorta thus contain much elastic connective tissue. The fibres, whether collagenous or elastic, are aligned in such a way as to resist the mechanical stress imposed on the tissue. Loose connective tissue, sometimes termed areolar tissue, is commonly found between muscles, around the intestines, and in other places where considerable relative movement between adjacent organs occurs. It has an abundant, hydrated ground substance with relatively sparsely distributed collagen or elastic fibres, usually arranged in random fashion.

Connective tissue is important in the nutrition of the tissues that it surrounds, because the passage of substances to and from the bloodstream must often be through it. Metabolites presumably diffuse via the aqueous phase of the ground substance and along the fluid coating the fibres.

The protein polysaccharide complexes in the ground substance are polyelectrolytes and it has been suggested that connective tissue has a role in the maintenance of water and electrolyte balance. In addition to the storage of energy in the form of fat (to be discussed in detail: Chapter 51), about half of the circulating protein in the form of albumen and globulin, is at any one time in the interstitial spaces of the connective tissue. Again, some controlling function is suggested by the observation that the composition of proteins in connective tissue differs from that in the plasma.

Inflammation and repair

Damage to tissue elicits the inflammatory response. Destruction of the foreign agents, e.g. bacteria etc., and repair are brought about by mobilisation of the white cells in the blood and similar cells in the tissues. These cells actively phagocytose the infective agents, other types of white cell release antibodies which kill them off, and mast cells liberate histamine which causes local vasodilation and *oedema* (accumulation of tissue fluid) in order to dilute the irritant and help the action of the antibodies.

Repair, after injury, is carried out by the activity of the fibroblasts which produce a connective tissue scar consisting of masses of collagen fibres, disposed so as to make good any deficit. The detail of these processes will be discussed in Chapter 71.

Abnormalities of connective tissue

Abnormalities in connective tissue can arise in terms of the cells present, the fibrous and elastic matrix, and the ground substance. Ascorbic acid is necessary, in the human, for the normal formation of collagen. *In vitro* experiments have shown that the essential conversion of proline to hydroxyproline does not occur in fibroblasts of animals deprived of ascorbic acid. Scorbutic animals therefore display a syndrome characterised by the weakness of all connective tissue and an inability to form new functional tissue. Thus widespread haemorrhages are found, due to weakened intercellular cement in capillary endothelium, bones and teeth are malformed and weakened, and wound healing is interfered with because collagen fibres cannot be laid down by the fibroblasts.

II. ADIPOSE TISSUE

The fat depots of the body are not the inert masses of tissue that once they were thought to be. Although it was appreciated that they represented an energy store, an insulating layer, and possibly gave structural support in some locations, it was not until recently that adipose tissue was recognised as an organ having great importance in metabolism.

Energy store

Animals need a reservoir of energy-producing material because feeding is intermittent and does not usually coincide with the times during which energy is required. Fat is a suitable substance for this role, since it needs a smaller amount of it to store a given quantity of energy, than would be true for either carbohydrate or protein. Ten per cent of the total body weight is fat in an average man; this would be six weeks' energy reserve. Obese persons may store up to ten times this amount (Fig. 3.2).

Adipose tissue cells can actively synthesise fat from carbohydrate; they are controlled by a nerve supply and by hormones.

The fat cells are widely distributed throughout the body, especially in subcutaneous tissue. In young children it is uniformly thick over the body, but in adults it has a characteristic pattern. In females, subcutaneous fat is thickest over the breasts, buttocks and anterior parts of the thighs. In males, fat is usually present over the deltoid, triceps, lumbosacral area and the buttocks.

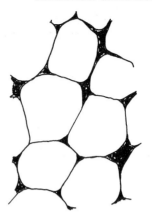

Fig. 3.2 Ordinary adipose tissue as seen under light microscopy. In each cell the fat droplet which occupies most of the cytoplasm has been extracted. Remaining are the thin, darkly stained bars of cytoplasm and occasional nuclei.

Fat balance

The fat depots can be considered as a metabolic energy bank, a dynamic equilibrium existing at any one time between deposits and removals. Fat is laid down from (1) *chylomicrons* containing fats and fatty acids absorbed from the intestine, (2) triglycerides synthesised in the adipose cells themselves from carbohydrate and (3) fatty acids produced from glucose in the liver.

Fat deposits are utilised by the hydrolysis of triglycerides, which release fatty acids into the circulation. This is called *lipolysis*, and so long as the blood-sugar level is normal, does not occur on a major scale. Lipolysis is increased many times during fasting. It is controlled by hormones and by the nervous system.

The control of lipid metabolism is described in detail in Chapter 51 so only a brief summary will be given here.

Hormonal control

The main factor controlling the uptake of glucose and its conversion to fat by adipose tissue, is the hormone *insulin* secreted by the pancreas (Chapter 49). It acts by accelerating the transport of glucose into the cell and by enhancing the rate of conversion of glucose to glycogen.

Autonomic nervous system

Adipose tissue has an abundant supply from the autonomic nervous system, a fact demonstrable experimentally by denervating the subscapular region on one side. The fat cells on the denervated side soon become larger than those on the control side. If the experimental animal is starved, only the adipose tissue on the control, normally innervated side is depleted of fat.

Noradrenaline is the chemical neurotransmitter involved; the injection of small amounts of it will rapidly increase the amount of free fatty acid circulating in the bloodstream—an action due to the transmitter stimulating the action of lipase which hydrolyses the adipose tissue triglycerides.

III. SKIN

Surprisingly, the skin is one of the largest organs in the body, for it accounts for no less than 15 per cent of the total body weight in the adult human subject. Its main function is to encase and protect the rest of the body structures. It also has other important functions including acting as a receptor for external stimuli (such as tactile, thermal etc.), taking part in temperature regulation, water balance, the secretion of various substances and, lastly, functioning as a storage organ for fat.

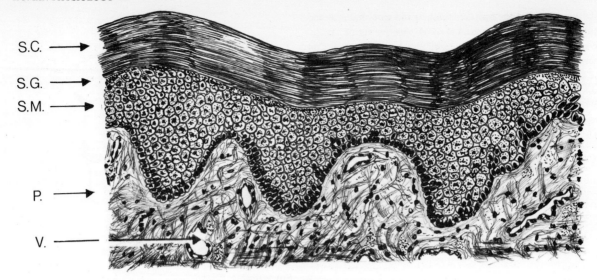

Fig. 3.3 Diagram of the basic structure of the skin.
S.C. Stratum corneum
S.G. Stratum germinativum
S.M. Stratum Malpighii
P. Papilla
V. Blood vessel

Structure of skin

The skin is composed of two layers, the *epidermis* and the *corium* (Fig. 3.3).

The epidermis consists of many layers of epithelial cells, moulded over the corium, the superficial layers being the hard, horny *cuticle* of keratinised cells, the deeper layers forming the soft mucous layer called the *Malpighian layer*. The horny layer is composed of flattened cornified cells which have lost their nuclei. The cells just superficial to the deepest part are attached to each other with fine protoplasmic bridges (*prickle cells*). The corium is made of connective tissue and merges into the subcutaneous connective tissue. The elasticity of skin is due to elastic fibres; in old age, or prolonged exposure to the sun and in long-standing smokers these diminish in number and wrinkles appear.

Blood vessels in the skin are in the form of a capillary network in the corium and loops enter the papillae. There are no vessels in the epidermis.

Mechanical properties of the skin

The mechanical properties which determine the structure of skin in relation to its functions, as mentioned above, are (1) *tension*, (2) *compliance* and (3) *tensile strength*.

The tension exhibited by a sample of skin, can be measured by determining the maximum deforming force it exerts, and is expressed in newtons per centimetre ($N\,cm^{-1}$). It is dependent upon the content of elastic tissue; it is greater in the young than in the old. For this reason, an excised piece of skin from a young person shrinks more than does a similar piece from an adult.

The compliance of skin is a measure of its ability to return to its original condition after deformation. It can be measured by determining the depression in the skin resulting from the application of weights. This property is also dependent upon the amount of elastic tissue present and to some extent upon the state of hydration of the skin and subcutaneous tissue.

Tensile strength is the resistance of the skin to rupture by external force and is a function of the amount of collagen present. It is measured in excised strips of skin; an average adult value would be 1 to 2 $kg\,mm^{-2}$.

Adaptation to environmental factors

As can be seen from the foregoing considerations, the mechanical properties of the skin are mainly determined by factors in the corium. The epidermis, by its thickness, hardness and flexibility, protects the body against injury. After repeated mechanical stimulation, such as friction, the epidermis is thickened. An example is the callous to be found on the middle finger of textbook authors.

Repeated rubbing of the skin (10 min per day for 30 days) gives rise to up to 70 per cent thickening of the epidermis in the region concerned. The Malpighian layer does not become thickened in this process, so it follows that the thickening is due to decreased shedding of the keratinised epidermis rather than to increased formation of it.

Keratinisation

Keratin is the fibrous protein which comprises most of the epidermis and related structures such as hairs and nails. It is composed of 18 amino acids, there being a disulphide bond between the cystine molecules of adjacent polypeptide chains (which accounts for the strength of the material).

The first stage in the keratinisation process is the appearance within a cell of *tonofibrils* initially about 5 nm in diameter but increasing to 10 nm and then becoming coated with an electron-dense sheath. Eventually the cell, when fully keratinised, is filled with compact bundles of fibrils, closely cemented together. At this stage the cell, as such, is dead. The sheath mentioned above is a thin layer of lipid, and disorders of lipid metabolism lead to imperfect keratinisation. For instance, drugs employed to lower blood cholesterol interfere with keratinisation.

Colour of the skin

The skin colouration comes from four sources:
1. Melanin, as melanocytes in the deep layers of the stratum germinativum.
2. Oxyhaemoglobin and reduced haemoglobin in the skin capillaries.
3. Carotene in the skin adipose tissue.
4. Melanoid—a degradation product of melanin.

Thus red and blue skins depend on factors such as skin vasodilation (which gives a more saturated colour) and the degree of oxygen saturation (which determines the hue). Melanin, being in the deep layers of the epidermis, reflects light and this passes through the keratinised layers twice, during which the shorter wavelengths tend to be scattered. The deeper the melanin therefore, the more blue the skin colour will be.

Racial differences in skin colour depend on the total amount of pigment present; in an adult Negro this may be of the order of 1 g. The degree of pigmentation is increased by exposure to ultraviolet irradiation; it is an adaptation to sunlight in the sense that pigmentation increases the reflectivity of the skin to these wavelengths, and hence minimises sunburn.

The control of pigmentation appears to be mainly hormonal. In many animal species, rapid changes in colouration are possible because the pigment is contained in chromatophores in the form of granules which can be concentrated or dispersed throughout the cytoplasm. In the human, the control of pigmentation is a much slower process and is achieved by alterations in the numbers of melanocytes in the skin.

Melanocyte stimulating hormone (MSH) is secreted by the pituitary gland and this stimulates melanocyte formation. In Addison's disease (a primary adrenal failure) the diffuse melanosis characteristic of the condition is due to the release of adrenal inhibition of pituitary MSH output.

It is worth noting that secondary adrenal cortical insufficiency arising from hypopituitarism is usually characterised by a decrease in pigmentation. Also, the giving of ACTH systemically often results in melanosis. A direct effect of the application of adrenocortical hormones has been reported; if these are locally put on hair (in the rat), a loss of pigmentation ensues.

The epidermal appendages

The structures arising from the skin are hairs, nails, sweat glands and sebaceous glands.

These structures arise during fetal life as downgrowths from the underneath of the epidermis.

Hair

The *hair follicles* are tubular invaginations, out of which grow the hairs themselves; each hair is a dead, keratinised shaft arising from the hair-bulb, as shown in Fig. 3.4. Attached to the follicle walls is involuntary muscle, the *arrector pili* muscle. This muscle, on contraction makes hair stand upright (goose-pimples) and is supplied by adrenergic nerve fibres. In exposure to cold, erection of hair traps a layer of air next to the skin which acts as a thermal barrier.

The growth of hair occurs in cycles. After a period of rapid growth, the deepest part of the follicle degenerates and the hair loosens and may fall off. The follicle then begins a resting phase, being shorter and thicker in shape. Growth is later resumed and a new hair takes the place of the old one. In the human, the growing phase may be as long as three years and the resting interval is of about three months' duration. Individual follicles behave randomly in their cycles. In other mammals the activity is synchronised in all follicles and periodic moulting occurs when the old hair is shed. Usually this is seasonal and allows for adjustment in insulating properties of the pelt with climatic conditions.

Sebaceous glands

In man, sebaceous glands are found everywhere on the skin apart from the palms and soles. They are *holocrine* glands, i.e.

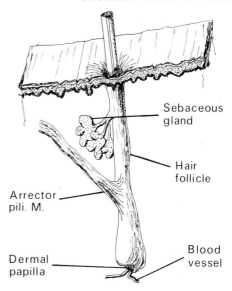

Fig. 3.4 Diagram of growing hair follicle and its associated structures.

their secretion is composed of disintegrated cells and their contents, discharged directly into the sebaceous duct.

The function of *sebum* is protective and in the lubrication of skin and hair. It consists of a mixture of fatty acids, triglycerides, cholesterol esters and esters of the higher aliphatic alcohols. Ovarian and adrenal hormones control sebaceous secretion.

Sweat glands

There are two types of sweat gland, *eccrine* and *apocrine*.

Eccrine glands (ordinary sweat glands) are found over the whole of the skin in the human except for the lips, nail-beds and glans penis. They are simple coiled tubular glands, in the dermis, about 0.3 to 0.4 mm diameter, and appear in histological sections as a close-knit mass of sections of tubes. Parallel to the lumen of the gland are myoepithelial cells which, on contraction, expel sweat onto the skin surface.

Secretion of sweat

Moisture upon the skin cools the body as it evaporates; sweat secretion is an important mechanism in the control of body temperature under warm environmental conditions. See Chapter 29 for the details of this process.

Thermal sweating involves the skin over most of the body surface; emotional sweating tends to occur only on palms, soles, and in the axillae.

Insensible perspiration

In addition to sweat secretion, a continuous low-level water loss takes place from the skin due to the passage of water through it by diffusion.

Control of sweating

Cholinergic fibres from the sympathetic nervous system supply the sweat glands. Hence cholinergic drugs induce sweat secretion while sympathomimetics have no action. The central control mechanism is found in the thermoregulatory centres in the hypothalamus (Chapter 27).

Apocrine glands

These glands are similar in structure to sweat glands but many times larger in size. They are innervated by sympathetic adrenergic fibres.

Apocrine secretion is a yellowish-tinted thick, milky fluid, having a characteristic pungent odour, a large part of which is often contributed by bacterial decomposition. The activity of these glands is stimulated by adrenergic factors such as fear, sexual excitement etc. and can be artificially induced by the injection of adrenaline.

The precise function of this secretion is still speculative but we may consider it to be a *pheromone*, or olfactory chemical messenger, concerned with sexual attraction. The glands are inactive until puberty.

Underarm deodorants largely derive their efficacy from their content of aluminium hydroxide which causes the keratinised layers surrounding the mouths of the glands to swell, thus blocking the flow of secretion. Some also contain bactericidal substances which inhibit the malodorous decomposition of the secretion.

It would seem an illogical procedure to remove the naturally occurring pheromones and then to replace them with artificially synthesised scents.

IV. BONE

Cartilage and bone make up the skeleton of the body, giving it the mechanical properties of rigidity, strength in tension, compression and shear and serving as anchoring structures for muscle systems. Bone also has functions in calcium, phosphate and magnesium metabolism and bone marrow is the site of formation of many of the cells found in blood. During life, bone is continually being broken down and renewed.

Structure of bone

Bone consists of two main materials: an organic part, largely collagen; and a mineral part which consists of a crystalline matrix of calcium, phosphate and other elements. The tensile strength of bone is due to its collagen content; the minerals account for its strength in shear and compression.

The organic, or *osteoid*, component comprises about one-quarter of the weight of bone (rather more by volume) and 95 per cent of this is collagen, 5 per cent being interfibrillar ground substance (mucopolysaccharides and protein). Bone collagen is able to form deposits of minerals in conjunction with its fibres during the formation of bone.

The mineral part of bone is of rigid crystalline form called *hydroxyapatite* [$Ca_{10}(PO_4)_6(OH)_2$; plus minute additions of Mg^{2+}, Na^+, CO_3^{2-}, Citrate^{4-} and F^-].

Two forms of bone are distinguishable with the naked eye, *spongy bone* and *compact bone*. Spony bone consists of a three-dimensional lattice of branching bony spicules (*trabeculae*) which enclose a system of communicating spaces which contain bone marrow. Compact bone is a solid continuous mass having microscopic spaces within it. There is not a sharp gradation between the two types of bone.

A typical long bone (femur, humerus) is a thick-walled, hollow, cylindrical shaft (the *diaphysis*) of compact bone; the central marrow cavity is termed the *medulla*. The ends of the shaft, mainly spongy bone covered by a thin cortex of compact bone, are termed the *epiphyses*. In growth, the epiphysis and diaphysis are separated by columns of spongy bone (the *metaphysis*) where the actual growth is taking place. The articular surfaces of bones are covered by a layer of hyaline cartilage, the *articular cartilage*.

Most bones are covered by the *periosteum*, a specialised connective tissue layer which can, when necessary, form new bone. In thoracic surgery, access to the lung is obtained by "subperiosteal resection" of one or more ribs. The periosteum remaining allows new bone later to form and repair the defect.

Microscopic structure of bone

A section of the shaft of a long bone can be ground down and examined under the microscope. We find that compact bone consists of largely the calcified interstitial bone matrix condensed into layers or *lamellae*, 3 to 7 μm thick.

Uniformly spaced within these are the *lacunae*, spaces each filled with a bone cell or *osteocyte*. Radiating in all directions from the lacunae, are very fine, branching *canaliculi* penetrating the substance of the lamellae and anastomosing with the nearby canaliculi derived from other lacunae. The function of this interconnecting system of canaliculi is thought to be the nutrition of bone cells; diffusion of metabolites etc. between perivascular spaces and the cells occurs along these channels. Figure 3.5 is a diagrammatic summary of the microscopic structure of a typical bone.

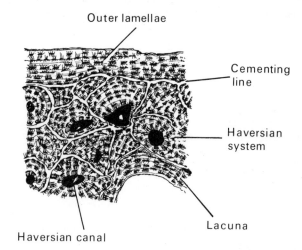

Fig. 3.5(a) Diagram of cross section of bone showing lamellar patterns and Haversian systems.

Fig. 3.5(b) Diagram of a typical Haversian system showing the lacunae and canaliculi. Higher magnification than in Fig. 3.5(a). The lacunae and canaliculi, here shown filled with a dye, in life contain the osteocytes and their processes.

The cells in bone

Actively growing bone contains the three types of bone cells: *osteoblasts, osteocytes, osteoclasts*. Since there is good evidence that transformation occurs between these types of cell, they may be regarded as different functional states of one cell type. In addition, we find that other cells in bone have the potentiality to develop into such osteoid cells, namely hemopoietic cells, cells in the periosteum and even cells in the walls of blood vessels in bone.

Osteoblasts are associated with the formation of new bone; they are found where bone surfaces are advancing. They form a layer of cuboidal or low columnar cells in regions of new bone matrix and are strongly basophilic owing to their content of RNA; they also have a high content of alkaline phosphatase in their cytoplasm.

Osteocytes are the main cells in fully formed bone; they live in the lacunae between the bone matrix and send thin protoplasmic processes into the canaliculi. The osteocyte is in essence an osteoblast which has become trapped by the bone matrix that it has formed. It is no longer forming new matrix rapidly but presumably does continue to function in the maintenance of it. It is thought that osteocytes are continually modifying the matrix, bone salt being removed or added; they are therefore playing an active role in the regulation of calcium in body fluids (and to a lesser extent a number of other ions—see Chapter 16).

Osteoclasts are found in close contact with areas of bone resorption. They are giant cells having up to twenty nuclei, and are present in surface depressions of bone or where bony spicules are being absorbed. This clear relationship has led to the usually-held view that osteoclasts are playing the major role in the removal of bone. However, an equally plausible interpretation would be that osteoclasts merely represent the coalesced osteocytes liberated as the bone resorption process opens up the lacunae.

Electron microscopy of the membrane of an osteoclast reveals an elaborate, specialised structure where it abuts the bone. There is a deep infolding of the membrane and often, in these so-called *lobopodia*, can be observed small crystals of bone mineral, presumably liberated from the bone matrix. There is, however, little other evidence that osteoclasts are actively erosive of bone, or that they are phagocytic.

The architecture of bone

The lamellae of compact bone occur in three different patterns: (1) Concentrically arranged around longitudinal vascular channels within the bone. These are really longitudinal cylindrical units and are termed *Haversian systems* (see Fig. 3.5). (2) Between adjacent Haversian systems are varying shaped and sized (often triangular) systems, filling in. These are the *interstitial systems*. (3) At the external bone surface are found sheets or lamellae, arranged as a continuous cover around the bone (*circumferential lamellae*).

The central Haversian canal contains small blood vessels ensheathed in connective tissue. In diameter the canals are about 100 μm across.

The medullary parts of the bones have compact bone arranged regularly in the form of *trabeculae* which form in the direction of the forces most often to be resisted by the bone concerned. This fact that the internal architecture of bone develops in a manner best fitted to withstand mechanical stress has been known for a long time.

It can, for example, be shown that if the stresses applied to a bone are altered, the trabeculae then get rearranged to counter the new forces. This can be seen where joints are artificially fused because quite different patterns of trabeculae are generated in the bones thus immobilised. After fractures, the callus is altered with time to restore the original architecture of the bone.

One attractive theory to account for this adaptation to function in bone is that the crystals of hydroxyapatite are piezoelectric crystals. When subjected to stress, a potential is generated at the crystal faces and this is the stimulus to which the osteoblasts and osteoclasts respond when they either build up or resorb bone as it is being remodelled to the new stresses. The experimental evidence for this is indirect, but it is suggestive that a polarising current applied between an electrode thrust into the bone and an indifferent electrode at a distance, results in either the formation of a mass of new bone, or in the resorption of bone. Which of these two processes occurs at and around the bone electrode, depends on its polarity.

It has been speculated that the effect is due to the current flow either helping or impeding the transport of metabolites in tissue fluid according to its direction.

Calcium metabolism and bone

The details of calcium deposition in and mobilisation from bone are little understood. Local factors and humoral factors are involved.

It has been suggested that the initial "seeding" of crystals of apatite needs a particular stereochemical configuration to be present in the collagen. Collagen with the banding at 64 nm intervals, as found in bone, when put into solutions of calcium and phosphate ions will induce the formation of crystals of hydroxyapatite. On the other hand, tropocollagen unstructured fibrils and other forms of collagen are found to be ineffective in this regard. This theory does not account for the lack of calcification in places like tendons, where the collagen also has the common 64 nm period, so one reluctantly must postulate a further specific inhibiting substance to be present in the ground substance there.

Calcium is constantly being interchanged between bone and blood; calcium ion concentration in the plasma is maintained relatively constant as a result of these operations. There seems to be a dual mechanism involved. Simple diffusion between bone and blood gives a constant blood level of about $1 \cdot 8$ mmol l^{-1} of calcium, which comes mainly from relatively newly laid-down bone (*metabolic bone* as opposed to the less actively exchanging older bone, or *structural bone*). The remaining $0 \cdot 7$ mmol l^{-1} of calcium in the blood is controlled by parathyroid hormone and involves this hormone activating osteoclasts. For details of the action of parathyroid hormone and its opposing hormone, calcitonin, see Chapter 32.

Hormones and bone

The parathyroid hormone functions to keep blood calcium at a constant level; it does this by means of a feedback mechanism, the blood level controlling its release from the glands. In opposition, the hormone calcitonin secreted by the thyroid gland, prevents bone resorption and hence tends to lower blood calcium.

In the clinical state of hyperparathyroidism (von Recklinghausen's disease) the oversecretion of parathormone results in bone being removed at a great rate and being replaced by fibrous tissue containing many osteoclasts.

Gonadal hormones affect bones. In general, testicular and ovarian hormones promote bone deposition. Their absence may lead to *osteoporosis* (rarefied bone). The main effect of these hormones appears to be on the rate of skeletal maturation. In precocious sexual development, the maturation of bone is earlier.

Growth hormone (from the anterior lobe of the pituitary

gland) controls skeletal growth. In growing animals all the bones are increased in size by the hormone; in adults only those without fused epiphyses can increase in size (see Chapter 32).

Nutrition and bone

Vitamin A is necessary for normal bone development, and growth. Deficiency gives rise to the appearance of large numbers of osteoblasts. Ossification is depressed, failure of remodelling occurs, yet periosteal bone is deposited normally. Bones are therefore of abnormal shape. For example, skull bones are thicker and coarser and foramina may be constricted so that nerves and blood vessels passing through them are obstructed. Excess of vitamin A gives rise to large numbers of osteoclasts and hence bone resorption. Bones, such as the skull, become thinner.

Vitamin C is essential for the development of collagen. It has no effect upon calcification, but of course normal collagen tissue is essential before ossification and mineral deposition can occur. Deficiency of vitamin C leads to an absence of the cartilage matrix in which early bone is laid down. Osteoblasts lose their characteristic form and since the calcifiable matrix is lacking, bone deposition does not occur. Compact bone already formed gets thinner. It may be replaced by spongy bone. Bony trabeculae are thin. Healing after fracture does not take place if there is vitamin C deficiency.

Vitamin D has its action on the regulation of calcium and phosphorus metabolism. Of especial importance in infancy, lack of this vitamin causes *rickets*, a condition of inadequate calcification of both cartilage and bone. There are characteristic changes to be observed at the epiphyseal disc (lack of calcification, failure of the cells to mature and an accumulation of cartilage cells). It becomes swollen, elongated and invaded by blood vessels. Bones are, as a result, deformed in a characteristic way.

In the adult, lack of vitamin D gives rise to *osteomalacia* (decalcification, replacement by osteoid tissue).

Large excess of vitamin D gives rise to a demineralisation of the whole skeleton.

Response of bones to injury

The healing of a fracture is dependent mainly upon the response of the osteogenic cells in the nearby periosteum to turn into osteoblasts. The line of the fracture usually bleeds, resulting in a clot between the bone ends, and in addition the lack of blood supply to the site of the fracture leads to death of the osteocytes near the break.

A collar of callus tissue is formed by the osteogenic cells of the fractured ends near to the break. The initial bone formation occurs in the deepest part of this collar, some distance from the break. An internal callus is generated likewise by the endosteum. Bone trabeculae then appear in the callus and any loose fragments and dead bone is resorbed. The nature of the stimulus giving rise to this process is at present unknown.

Friction between two bones gives rise to an area of dense and polished bone at areas of contact. This is called *ebumated* bone and can be found where articular cartilage has disappeared or when an ununited fracture has led to friction between bones.

Inadequate blood supply to bone leads to sclerosis and atrophy. Hyperaemia gives rarefaction.

Denervation of a limb leads to maldevelopment of bone. This is due not to a direct effect, for there are no trophic effects of nerves upon bone but is due to the loss of muscle pull.

Radiation (X-rays, radioactive elements etc.) results in local damage to bone. This is of importance since certain elements such as lead, plutonium, strontium and radium are concentrated in bone. In areas where these elements (if radioactive) are concentrated, the osteocytes die, leaving empty lacunae.

In acute illness, a transitory cessation of bone browth occurs. These show as "lines of arrested growth" appearing in X-ray photographs as horizontal striae in the long bones.

V. TEETH

There are thirty-two permanent teeth in the human adult (2 incisors, 1 canine, 2 premolars and 3 molars in each half dental-arcade).

The structure of a typical adult molar tooth is shown in Fig. 3.6.

The teeth mince up food into particles small enough to form a bolus capable of passing down the oesophagus. The jaw muscles can exert forces of up to 10^3 N between the molar teeth. Upper and lower teeth are serrated and fit into each other enabling small particles of food to be ground up. This fitting is termed *occlusion*.

Dentine

The main part of the tooth is made of dentine, composed of calcium phosphate salts deposited in a collagen meshwork very much like compact bone. However, dentine does not contain osteoblasts or osteoclasts or lacunae. There are no spaces for blood vessels and nerves. Dentine is deposited and maintained by a layer of cells termed *odontoblasts* lining the inner surface, where it adjoins the pulp. The strength in compression and shear of teeth is due to its content of calcium salts; the strength in tension and resistance to fracture is due to the collagen content.

Enamel

The outer surface of a tooth is sheathed in enamel, the hardest substance in the body, which is formed prior to eruption by epithelial cells called *ameloblasts*. Once formed, no new enamel can be added. Enamel consists of hydroxyapatite crystals embedded in a fine mesh of fibres of keratin. Since the crystals are small and keratin is a very insoluble and resistant protein, enamel is a hard and resistant material, ideally suited to the function it has to perform.

Cementum

The *periodontal* membrane which lines the tooth alveolus (socket), secretes a bony substance, the cementum. Collagen fibres pass from the bone of the jaw, through the periodontal membrane and into the cementum. These fibres hold the tooth in place and are anchored at either end by the alveolar bone and the cementum. Physiological adaptation to forces applied to the tooth occurs; if these increase, the thickness of the cementum increases and the number of collagen fibres becomes greater.

A rich supply of stretch-sensitive endings exists in this region (in the periodontal membrane) subserving reflexes involved in chewing, the contraction of jaw muscles etc.

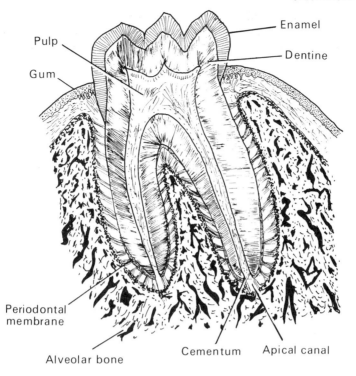

Fig. 3.6 Diagram of sagittal section of a human molar tooth.

The pulp cavity

The pulp, filling the tooth, is composed of connective tissue, nerves (with abundant pain endings as the majority of persons are aware), blood vessels and lymphatics. As mentioned already, odontoblasts line the cavity and during early stages in the life of the tooth continue to lay down dentine and thus make the pulp cavity smaller. In adult life this activity ceases but damage to the dentine either by infection or by the dental surgeon's drill can be repaired by their further activity. The odontoblasts have fine cytoplasmic projections which penetrate right through the dentine within small *dentinal tubules*, and it is thought that this is a system for providing the dentine with nutrition.

Toothache

Nerves in the pulp cavity do not pass into the dentine. Even so, a tooth having had its enamel layer damaged is extremely sensitive to pain. This fact has been adduced as evidence that the odontoblast projections transmit the pain in some way but this idea seems unlikely to be true. Usually, the pain in a defective tooth is initiated by osmotic stimuli (such as eating confectionery or drinking sweetened beverages), in which case it is likely that diffusion of the offending sugar can take place into the pulp itself and fire off the pain receptors. Also thermal stimuli can produce non-permanent tooth pain, again presumably because damaged enamel decreases the thermal insulation between the inside of the tooth and its external environment within the mouth.

Permanent toothache results from a pulpitis, or infection of the pulp cavity, usually as a result of micro-organisms entering it via the damaged enamel. If the infection spreads from the pulp cavity via the apical canal into the tooth alveolus, a dental abscess forms and by local pressure within the bony structure of the jaw causes great pain.

Eruption of teeth

In early life, teeth start to move upward from the bony structure of the jaw, passing through the oral epithelium, to reach their adult configuration. The mechanisms responsible for this eruption are not known, although a number of theories have been put forward. One theory is that the ascending tooth is 'jet-propelled", i.e. the excess growth of tissue in the pulp cavity causes this tissue to be forced out of the apical canal and fill up the alveolar space, thus pushing up the tooth. A more plausible hypothesis is that the bone surrounding the roots of the tooth progressively enlarges, forcing the tooth out. In either case the remarkably precise controlling systems for eruption are as entirely mysterious as are the maladjustments which result in some teeth being "impacted", or erupting in the wrong direction.

Mineral exchange and the teeth

Since the construction of teeth is similar to that of bone, we might predict that the mineral turnover would be similar (i.e. for ordinary bone about 0·5 to 1·0 g Ca per day is exchanged between blood and bone). However, since dentine has no osteoblasts or osteoclasts within it, we find the exchange to be much slower than in ordinary bone. Cementum, though, behaves like ordinary bone, as it does contain the same cellular elements as bone. Enamel, having no cellular and indeed practically no protein content, is very stable and one ends ones' life with the same molecules of calcium in it as were laid down when the tooth was formed in the first place.

Applied physiology of teeth

There are two common and important physiopathological problems involving the teeth; *caries* (dental decay) and *malocclusions*, or the failure of opposing teeth to meet when biting.

DENTAL CARIES

There are two currently held theories to account for tooth decay:
1. Acids in the mouth erode the protein (keratin) matrix of the

enamel. The acids are formed by bacterial action in the crevices in the teeth and gums. Since a common source of lactic acid could be carbohydrate splitting by bacteria, it is usually emphasised that a diet including much refined carbohydrate, such as sugar, will lead to considerable dental decay.
2. Proteolytic enzymes secreted by bacteria in tooth crevices, attack the keratin in the enamel. Then the hydroxyapatite is gradually attacked by the saliva, since its protective keratin has disappeared.

It is possible that a developmental abnormality may lead to dental caries, for it has been noted that minute fissures or canals are sometimes seen in the enamel and it is postulated that bacteria may gain ingress to the dentine by multiplying along the length of these minute channels.

The teeth of some individuals are more resistant to decay than others, and it has been found that children brought up in a region with a high content of fluorine in the drinking water have a lower-than-average incidence of dental caries. Water may have up to 10 p.p.m. of fluoride in it. Epidemiological studies show that where the drinking water contains less than 1 p.p.m. of fluoride, dental caries is much more common than elsewhere. Research shows that the artificial addition of fluoride in these areas, reduces the incidence considerably.

With fluoride contents of over about 3 p.p.m., brown mottling of the teeth occurs, but this, although unsightly, is not associated with any disease. When the content of fluoride is 10 p.p.m. or over, the condition of fluorosis is likely to develop, in which the density of bone in all parts of the body is increased, ligaments become calcified, and the spine becomes stiff and immobile. Campaigns for and against the artificial addition of fluoride to drinking water (to bring it up to a concentration of 1 p.p.m.) have been fought with great political vehemence and as a rule with little regard to scientific truth.

Fluorine, in the small doses mentioned, does not give rise to harder enamel than usual, and the mechanism by which decay is kept at bay is a mystery. It is possibly due to the fact that fluoride inactivates proteolytic enzymes before they are able to digest the keratin of the enamel.

The most promising development in research upon the prevention of dental caries is the finding that vaccines can be made against some common mouth micro-organisms, and that the local injection of these into the gums prevents decay entirely thereafter.

MALOCCLUSION

A congenital defect (or imposed unfair stress on growing teeth such as might be due to persistent thumb-sucking) leads to the teeth in one or both jaws growing in the wrong direction. The misalignment in turn causes the cutting and grinding actions to be inadequate. It may even cause displacement of the jaw.

Orthodontic treatment, directed towards applying continuous gentle pressure on the offending teeth, can correct considerable abnormalities by inducing the teeth and structure of the jaw to move into a new direction.

The physiological interest of these procedures lies in the clear evidence of the influence of imposed mechanical stresses (of a long-term nature) upon growth and development.

VI. GLANDS

All cells in the body are able to pass various substances, ions, water, etc., through their membranes. Certain cells elaborate specific secretions which are passed from the interior of the cell into either a collecting duct, a body cavity or into the bloodstream. Such cells are termed *secretory*, or glandular, cells.

They may be part of an organ, such as the kidney, or they may be a mass of cells forming a gland.

Secretory cells

These cells follow the normal pattern of construction for the cell (see Chapter 1) with the addition of specific granules to be found in the cytoplasm which represent the secretory precursors. If one observes a gland cell during different stages of secretory activity, a sequence of cytological events can be made out. When the cell is stimulated, the number of specific granules decreases and water-filled vacuoles appear. After the granules have been depleted, the endoplasmic reticulum becomes more prominent and the Golgi apparatus is hypertrophied. Mitochondria become larger and more numerous. When the secretory granules reappear, they do so near to the Golgi apparatus, a fact which has led to the current view that the Golgi apparatus is the site at which secretion products are concentrated within the cell prior to its being released through the cell membrane.

Electron-microscopic study of the secretory process, mainly using the acinar cells in the pancreas, has shown that the nature of the secretion is determined by the activity of the chromosomal DNA in the cell's nucleus. The "blueprint" is encoded in the nuclear DNA; the synthesis of the specific proteins to be secreted occurs in the cytoplasm. The information is transferred from the one to the other by *messenger RNA, transfer RNA* and *ribosomal RNA*, possibly via "pores" in the nuclear membrane.

The messenger RNA is formed upon the nuclear DNA; it then moves out into the cytoplasm and can determine the correct sequence for the assembly of the amino acids required for secretion. Further RNA from the nucleus forms ribosomes in the cytoplasm. Several ribosomes attach themselves to each messenger RNA molecule forming the so-called *polysomes*, which are attached to the endoplasmic reticulum. Each type of transfer RNA molecule becomes attached to its specific amino acid molecule and then transports it to a ribosome. At the ribosome, the transfer RNA sticks its amino acid into the appropriate site in the developing molecule. Eventually complete protein molecules are synthesised by the replication of this process, and then are passed across the lining membrane of the endoplasmic reticulum to accumulate there.

This secretion product then travels within the reticulum to the Golgi complex. Here small vesicles of reticulum are "budded-off" later to coalesce into larger membrane-surrounded vesicles. These are the "prosecretory granules" but droplets would more accurately describe them since they contain fluid.

During secretion these droplets then move to the cell membrane, their own membrane becomes continuous with that of the cell and eventually the contents are extruded (see Figs. 1.3 and 1.5).

Experiments with tritium-labelled amino acids show that it takes about 20 min for the formation to occur and the secretion product to reach the Golgi region; the mean time that the labelled molecules spend within the cell during secretion was found in one series of experiments to be 53 min.

Types of glandular cell

There are several ways in which a gland may secrete material:
1. The *merocrine* cell discharges its content through the cell membrane as described above, the membrane and cell remaining essentially intact.
2. *Apocrine* secretion involves the loss of part of the apical cytoplasm of the cell, i.e. this is the secretory product.
3. *Holocrine* cells are completely discharged by the gland, after they have become filled with the active material (e.g. sebum).

New cells are continually being produced to form this secretion. On the whole, these three types are indistinct and the viscosity of the secretion product seems to be the major differentiating factor in determining the way in which the material is discharged.

Type of gland

Glands are classified as *exocrine* when they discharge their contents into a duct or the lumen of a hollow structure. They are *endocrine* when the secretion is made into the bloodstream.

EXOCRINE GLANDS

Unicellular exocrine glands are found in mammals among the columnar cells of many epithelial membranes, as *mucous* or *goblet* cells. They secrete mucus, a protein-polysaccharide (water and mucin), which has lubricating functions. Such a cell will secrete more or less continuously during the whole of its lifespan, which would be four or five days in the alimentary canal.

Where greater volumes of secretion are required, secretory cells are grouped together to increase the surface area of the secretory membrane available.

Multicellular glands are found in several anatomical forms, as illustrated in Fig. 3.7.

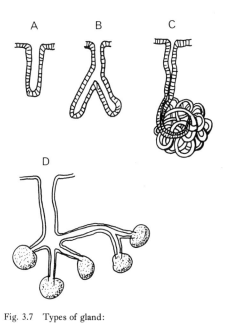

Fig. 3.7 Types of gland:
A: Simple; intestinal crypt
B: Branched; gastric gland } tubular glands
C: Coiled; sweat gland
D: Compound acinar gland.

Secretory processes

Energy is required in order to maintain secretion by a cell. It is an active process. Not only are substances moved against concentration gradients, or moved about within the cytoplasm, they may be synthesised by processes involving energy consumption.

The work done in secretion can be roughly estimated (for simple substances) if the free energy changes in the chemical reactions involved are known. For example, the transport of H^+ ions from blood to gastric juice has been estimated to require approximately 36 MJ mol^{-1} HCl.

The formula used was,

$$W = RT \ln \frac{C_1}{C_2}$$

where W = work per mole secreted, R = gas constant, T = absolute temperature, C_1 and C_2 the concentrations of H^+ ion (i.e. pH).

Blood flow to glands

Most glands elaborate their secretory products from the blood flowing through them. Thus in glands secreting large quantities of watery fluids, e.g. sweat glands, we might expect considerable increases in blood flow through the gland when it is active. Experiment has shown this to be true—in fact, it may increase up to ten times the resting values.

The mechanism of this raised blood flow has been investigated, and it appears (in salivary and sweat glands) that during glandular activity a certain proteolytic enzyme is released by the gland and this reacts with tissue fluid protein to form *bradykinin*. The latter is a powerful vasodilator and therefore has a local effect in increasing the blood flow through the gland.

Control of gland secretion

Some glands function continuously; others only secrete in response to a particular stimulus. In either case, however, the rate of secretion is controlled in a number of ways. In general, the nerve supply to a gland activates it with a rapid time-course; the control over long periods of time is usually carried out by the activity of hormones.

Nervous control of secretion

The precise way in which stimulation of the nerve supply to a gland causes its cells to secrete is unknown. Various theories have been proposed however.

Activation of the nerve fibre leading to a secretory cell produces a specific response in the membrane at the basal regions of the cell. The permeability to chloride is increased and an active flux of chloride ions takes place into the cell. This results in the interior of the cell becoming more electronegative with respect to the extracellular fluid (i.e. it is hyperpolarised). By the Donnan equilibrium requirements (see p. 8), positive ions will now enter the cell to balance the excess.

There are now more ions within the activated cell so that an osmotic force is generated which results in a net flow of water into the cell. The hydrostatic pressure rises as the cell swells and this is postulated to result in the formation of small discontinuities in the free cell membrane of the glandular cell through which water, electrolytes and the secreted materials then pass into the lumen of the gland.

Such an armchair theory may sound somewhat complex and overimaginative but there is suggestive evidence for part of it. A microelectrode tip within a secretory cell shows the normal resting potential of the cell to be around 45 mV, with the interior negative to the exterior. Nerve stimulation causes a hyperpolarisation of up to 25 mV, and usually if the gland has a dual nerve supply the parasympathetic is more effective in producing the hyperpolarisation than is stimulation of the sympathetic division.

There is a delay of about one to two seconds before stimulation causes hyperpolarisation, and the increased membrane polarisation (secretomotor potential) lasts several seconds after stimulation has ceased.

Hormonal control of secretion

Hormones have a slowly acting effect on the rate of secretion of a glandular cell. It is thus not feasible to see whether or not a membrane hyperpolarisation is again involved in this mechanism, for d.c. amplifiers do not have the necessary long-term stability for the question to be settled beyond doubt. In some cases (e.g. secretion of gastric juice) both rapid onset and prolonged secretion are needed; here the gland activity is initiated by nerve impulses and its secretion is then maintained by a hormone.

Local factors
Secretory cells may also be activated by local mechanical or chemical stimuli. For instance, the secretion of mucus in the gastro-intestinal tract is thought to be in part due to simple mechanical stimulation of the mucosa. A similar mechanism may explain the mucous response to foreign bodies in the respiratory tract, although it is possible that a chemical mediator such as bradykinin or histamine may be acting as an intermediary.

4. Nervous Tissue

P. A. Merton

Messages can be sent in the body by one of two systems, rapidly along the nerves, or slowly by hormones, "chemical messengers", in the blood. All messages from sensory receptors on the surface of the body, or deep inside, reach the brain along nerve fibres, and the detailed orders which the brain issues to the muscles and to most other effector organs (glands etc.) are also sent along nerves. Hormones in the blood act more slowly and are used for more generalised and diffuse types of control (e.g. thyroid hormone) or in the control of specific and localised mechanisms when the pathway of control does not need to include the central nervous system (e.g. the stimulation of acid secretion in the stomach by gastrin).

In the study of peripheral nerve the ultimate object is to give a complete account in terms of physics and chemistry of how signals are transmitted along individual nerve fibres during normal activity in the living body. Direct investigation of single fibres in intact animals is, however, almost always impracticable. The majority of experiments have of necessity been done on isolated nerve trunks containing many nerve fibres. Fortunately there is good evidence that the mode of functioning of nerve fibres is not seriously upset when they are carefully dissected out of the body. Also, the disadvantages of using a multifibre preparation are offset by the well-founded belief that individual nerve fibres function very independently; the sciatic nerve, for instance, contains motor fibres to very many widely separated muscles, and sensory fibres from a large area of skin. The fact that we can move a toe without any other muscles in the leg contracting, and that sensations from the toe are never confused with those from elsewhere in the leg, shows that the insulation between fibres must be excellent. These and other even stronger but more elaborate pieces of evidence give us every reason to suppose that when all the fibres in a large nerve trunk are excited together by an electric shock their individual behaviour will not be very different from usual, while the summing of all their effects makes observations much easier.

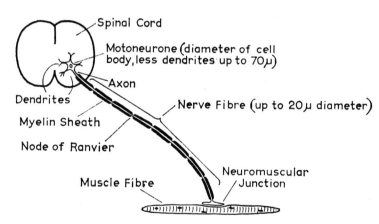

Fig. 4.1 Diagram of a nerve fibre.

A nerve fibre is a tubular process, often very long, arising from a nerve cell. The nerve cell protoplasm inside the tube is called axoplasm. In the diagram diameters are grossly exaggerated, with the true sizes roughly indicated ($1\mu = 0.001$ mm).

The cell depicted is a motor nerve cell, or motoneurone, whose nerve fibre, or axon, ends on a skeletal muscle fibre. Only a single muscle fibre is shown but, in fact, a single motor nerve fibre, by extensive branching inside the muscle, may supply 50 or more muscle fibres. A motor nerve fibre with all the muscle fibres belonging to it is called a "motor unit".

Sensory nerve fibres, running to sense-endings in the skin and elsewhere, look precisely similar to motor fibres. Their cell bodies, however, are in the dorsal root ganglia (see Fig. 59.1), not in the spinal cord. A peripheral nerve trunk, such as the sciatic nerve, is a bundle of large numbers of motor and sensory fibres (together with sympathetic fibres; see Chapter 30) enclosed in connective tissue.

In vertebrates all nerve fibres (except the very smallest) are covered for most of their length by an electrically insulating fatty layer, the myelin sheath, the cell membrane being exposed only at the nodes of Ranvier, spaced about 1 mm apart. To start with, however, the complications introduced by the myelin sheath can safely be ignored and the nerve fibre considered simply as a tube of cell membrane containing axoplasm.

The nerve fibres of invertebrates, often used for experiment because of their large size, are non-myelinated but are otherwise essentially similar in structure and mechanism to vertebrate myelinated fibres.

The nerve–muscle preparation

Many of the most important properties of excitable tissues were first discovered, and are still most conveniently demonstrated, upon a nerve–muscle preparation from a frog, usually the gastrocnemius muscle with the sciatic nerve attached. Ever since the end of the eighteenth century when Galvani discovered the electric current with his celebrated experiments on the excitation of the frog's leg by contact with metals, the frog's nerve–muscle preparation, stimulated electrically, has been the conventional object of neurophysiological enquiry. Many agencies other than electricity can be used to stimulate the nerve: local heating, pinching or tapping, the application of very hypertonic solutions and of various chemical substances, and in fact almost anything which is likely eventually to injure it. But the electrical stimulus has always been preferred to these, not only because of the convenience with which its timing and its intensity and duration can be controlled, but because unless quite unnecessarily strong it appears to cause no injury to the nerve. We now know that this is no coincidence because, as we shall see later, nervous conduction is itself electrical so that the electric current is the nerve's natural stimulus.

The simplest method of exciting a nerve electrically is to pass low-voltage direct current into it, using a circuit such as that shown in Fig. 4.2 to make and break the current and to alter its strength. With this equipment the main facts of nerve excitation are readily demonstrated.

be investigated by connecting a condenser across the electrodes, as shown dotted in Fig. 4.2. A larger current is then necessary to obtain the same size of contraction as before, or, with a sufficiently large condenser, no contraction can be obtained at all. The nerve is said to "accommodate" to slowly rising currents. With peripheral nerve "slowly" means times of the order of 1/50 second; when the switch is closed without the condenser in circuit the rate of rise of current is some hundreds of times faster than that.

Threshold

Another characteristic of nerve excitation by all kinds of stimuli is that nothing happens at all in the muscle unless the stimulus reaches a certain strength, no matter how abruptly it is applied. This is expressed by saying that the nerve has a "threshold" for the stimulus and unless the threshold is exceeded the nerve does not conduct. Such behaviour is quite different from that of many more familiar kinds of conduction; for example, if the output from the potentiometer of Fig. 4.2 is connected to an ordinary galvanometer, some current will flow in the galvanometer however small the voltage applied. It may be difficult to detect but it will be there all right. If, however, the nerve of a gastrocnemius–sciatic preparation is laid on the electrodes as before and the effect of making and breaking the current tried while the voltage applied to the nerve is gradually increased, nothing whatever happens in the muscle until the potentiometer

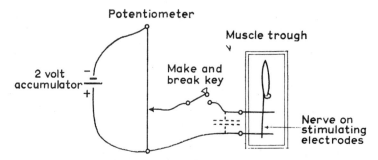

Fig. 4.2 Apparatus for stimulation of nerve by interrupted direct current.
The muscle and the nerve in the trough are covered with Ringer's fluid. Contractions of the muscle are detected by watching it. The electrodes are of silver wire rendered nonpolarisable by coating electrolytically with silver chloride. If ordinary wire is used with direct current the back electromotive force (e.m.f.) due to electrolytic polarisation of the wires causes the threshold to wander unpredictably.
To slow the rate of rise of current a condenser (shown dotted) may be connected across the electrodes. With an ordinary low resistance potentiometer a condenser of the order of 10 000 microfarads (μF) would be necessary, but smaller and more convenient values could be used if a resistance were inserted in one of the leads between the potentiometer and the condenser.

Accommodation

In the first place it is observed that contractions can only be obtained at the moment when the stimulating current is made or broken and not when it flows continuously. This illustrates a very important general property which nerve shares with many irritable tissues: that a sudden change of stimulus is more effective than a slow or maintained alteration. We are familiar enough with this in the case of our sensations; it is always difficult to detect a gradual change in the brightness of a light or in the loudness of a sound. With the nerve the effect of slowing down the rate of change of current when the circuit is closed can

is set to give perhaps half a volt, after which the muscle contraction increases rapidly in size as the voltage is further raised. The absence of response below half a volt is genuine and not merely due to a difficulty in detecting very small contractions. No matter how sensitive the equipment no response at all can be detected below the threshold.

The threshold of a nerve is not fixed but depends among other things, on the rate of rise of the current, and also on its duration if this is less than about 10 milliseconds (1/100 second). As the duration of the current pulse is reduced below 10 milliseconds, the strength of current necessary to excite rises progressively. The experimentally determined relationship between pulse duration and threshold current is known as a strength–duration curve. With pulses shorter than about 1 millisecond the quantity of electricity (i.e. the product of current strength and pulse duration) necessary to excite becomes a constant. This is because the nerve membrane behaves like a leaky electrical capacity. It is known that the condition for excitation by short pulses is that a certain length of nerve membrane should be depolarised by a certain number of millivolts. Hence, for durations of pulse so short that leakage current does not significantly discharge the capacity, a fixed quantity of electricity must be supplied in order to reach threshold.

Site of excitation

A third fundamental property is that only current leaving the nerve excites it. In the ordinary convention current flows from anode to cathode. Current will thus enter the nerve in the neighbourhood of the anode, flow along the nerve in between the electrodes and leave under the cathode, which in Fig. 4.2 is the electrode nearer the muscle. It is found that the nerve may be cooled, anaesthetised or crushed, either at the anode or between the electrodes, without affecting the threshold when the current is switched on. But any of these operations at the cathode causes a very large increase in threshold, showing that excitation occurs under the cathode.

Break excitation. As mentioned above, the muscle also contracts when the current is broken. Similar evidence shows that in this instance excitation occurs under the anode. Anode break excitation is not a general phenomenon. Many tissues, such as freshly dissected frog muscle, do not show it; even the frog's sciatic nerve does not do so if its circulation is intact. It seems to occur in deteriorating preparations in which the fibres are depolarised—i.e. have a low resting potential. The explanation appears to be that the anodal current repolarises the fibres, but when the current stops they depolarise again rapidly and this excites them (see p. 27).

Conduction velocity

So far we have only described how the nerve is excited and said nothing about transmission of the excitation to the muscle. Modern understanding of the nerve impulse began with Helmholtz's demonstration in 1850 that the influence, whatever it was, that passed down the nerve from the point of stimulation to the muscle and made it contract, travelled at a definite velocity, which he measured. In the frog it is about 60 miles per hour (25 metres per second). For one form of this experiment Helmholtz invented a myograph similar to that still widely used in the classroom (Fig. 4.3). The muscle pulled upon a lever to which was attached a long pointer writing on a revolving smoked drum. Thus a graph of contraction against time during a twitch was automatically drawn in the lamp-black. The nerve was stimulated by an induction coil, it being arranged that the stimulus occurred at the same point on the tracing on each occasion. The stimulus was first applied to the nerve at a point near the muscle and a tracing of a twitch obtained. It was then moved to the end of the nerve distant from the muscle and a twitch again recorded. The second twitch was the same shape and size as the first, but the whole tracing was shifted along the drum away from the point corresponding to the time of opening of the contact breaker. Clearly, the second twitch arises after a longer latency than the first and this is due to the time taken for the

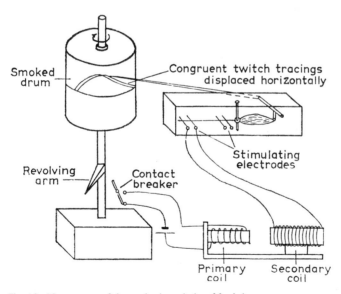

Fig. 4.3 Measurement of the conduction velocity of frog's leg.
  A sciatic-gastrocnemius preparation lies in the trough covered in Ringer's fluid. The tendon of the muscle pulls on a pivoted lever to which is attached a writing arm; the inner end of the muscle is fixed by a pin through the knee.
  Stimuli are given by an induction coil, a device similar to a motor-car ignition coil, having a primary coil of few turns and a secondary coil of many turns, which delivers a brief high voltage pulse when the current in the primary is suddenly broken by a contact breaker. The strength of shock is adjustable by altering the separation of the primary and secondary coils.
  Stimuli can be applied to the nerve either through a pair of electrodes on the end of the nerve, or through another pair close to the muscle. The contact breaker of the induction coil is knocked open by an arm attached to the shaft of the myograph drum. Thus the stimulus is always delivered at the same point of rotation of the drum.

nervous influence set up by the shock to travel from the end of the nerve to the point near the muscle where the first stimulus was applied.

Helmholtz's experiment was of importance because it showed that the process of conduction in the nerve was amenable to physical measurement and was not some mysterious and intangible influence passing with "the speed of thought". Helmholtz soon extended his measurements to both motor and sensory nerves in healthy human beings. For the motor fibres he stimulated the ulnar nerve at the shoulder and at the elbow and measured the difference in the latency of contraction of the muscles in the ball of the thumb. For sensory fibres he measured the difference in the time a subject takes to react to electric shocks applied, say, to the foot and the loin. This method involved the unproven assumption that no factors other than nerve conduction time cause the difference in reaction time from the two sites, but the answer it gave has been confirmed by modern methods. The conduction velocities of the fastest motor and sensory fibres are roughly similar and are about 50 metres per second in human limbs. Helmholtz also discovered, on the human subject, that nerve conduction is much slowed in the cold. A 10°C fall in temperature slows conduction by a factor of 1·7. Human limbs are often surprisingly cold in winter and even in summer seldom reach "body temperature" (37°C, or 98·4°F).

The resting potential

At about the same time that Helmholtz was measuring conduction velocities, du Bois Reymond and others were investigating the electrical phenomena of nerve and muscle which led eventually to the electrical theory of nerve conduction. It was found that when a nerve or muscle was cut across, the cut end showed a negative potential difference of a few hundredths of a volt relative to the intact tissue. This is because the interior of nerve and muscle fibres is negative to the exterior with a potential difference across the cell membrane; normally the cell membrane surrounds the whole fibre, so that if two electrodes are placed on the surface at different points no potential difference is found. But cutting the fibre allowed contact to be made with the interior and so we get the "injury potential" (Fig. 4.4). The magnitude of the injury potential recorded in this way is less than the true resting potential because, at the cut end, fluid and dead tissue, inevitably present, provide a current path between the inside and outside of the fibre which short-circuits the resting potential and locally depolarises the fibre. The electrode on the cut end therefore makes contact with the axoplasm at a point where the potential difference is smaller than the true resting potential.

The true resting potential can be measured by thrusting a fine glass capillary electrode along the inside of an axon until the tip is well clear of the depolarised region. This was first achieved in 1939 using giant axons from the squid. The method used is illustrated in Fig. 4.5 and a record of potential made in this way is shown in Fig. 4.6. The resting potential of isolated squid nerve is about 50 millivolts (0·05 V). For fresh vertebrate nerve and muscle fibres the value (obtained by analogous methods) is about 90 millivolts (mV).

The action potential

Du Bois Reymond discovered that when a nerve is stimulated, the resting potential, recorded at the other end of the nerve, diminishes. At the time only slow galvanometers were available, but when rapid recording instruments were developed it was confirmed that the reduction was due to a brief wave of negativity which travelled along the nerve from the stimulating electrodes. As this wave passed under the electrode which was on intact nerve, it caused a temporary reduction in the potential

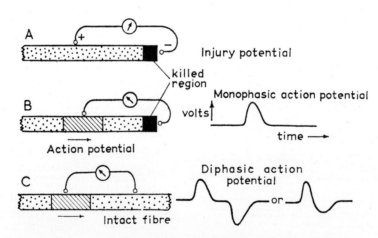

Fig. 4.4 Mode of recording injury and action potentials from a nerve fibre.

A. When the fibre is cut across, the killed end shows a negative potential difference with respect to the intact part.

B. As an action potential (itself a region of negativity) passes under the electrode on intact nerve it causes a brief reduction in the potential difference. The appearance of the action potential recorded in this way on a cathode ray tube oscilloscope is shown to the right. The action potential fades out in the killed region and does not reach the end of the fibre. Hence the action potential is only detected as it passes the electrode on intact nerve, and the record is "monophasic".

C. With electrodes on an intact fibre a potential change, first in one direction and then in the other, is recorded as the action potential passes under each electrode in turn. This gives a "diphasic" record. Often the two phases run into each other giving the appearance shown on the extreme right.

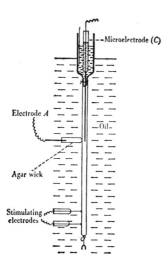

Fig. 4.5 The arrangements for recording resting and action potentials from a giant nerve fibre of a squid.

A cannula is inserted into the cut end of the axon, which hangs freely in oil with a small weight tied to the bottom end. A glass capillary microelectrode (C) is inserted through the cannula as far into the interior of the axon as is desired, usually several centimetres. Contact with the outside of the axon is made by means of a wick soaked in seawater (electrode A).

The photograph shows a microelectrode inside a living squid axon. The axon is seen as a clearer band surrounded by undissected connective tissue and smaller fibres. Each division of the graticule equals 33μm; the axon is thus 500μm ($\frac{1}{2}$ mm) in diameter and the electrode 130μm. (Hodgkin and Huxley Nature, vol. 144, p. 710 (1939).

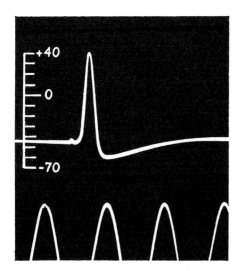

Fog. 4.6 Resting and action potentials from a single nerve fibre of the squid Loligo forbesi recorded with an internal microelectrode by the method illustrated in the previous figure.

The potential difference between the inside and outside electrodes has been amplified by a direct-coupled valve amplifier and displayed on a cathode ray tube oscilloscope. The vertical scale indicates the potential of the internal electrode in millivolts (mV), the seawater outside being taken as zero potential. Time markers at 2 millisecond intervals.

The record reads from left to right. The potential to start with is the resting potential of -45 mV. The first small deflection of 2–3 mV is an artefact caused by the stimulus. It is followed by the action potential which overshoots zero and rises to $+40$ mV. (Hodgkin and Huxley Journal of Physiology, vol. 104, p. 176, 1945.)

difference between that electrode and the one on the cut end (Fig. 4.4). This electrical wave is called the action potential, or, because of its shape, the spike potential. It used to be thought that during the action potential the resting potential was abolished, so that an electrode over the active region and one on the cut end would ideally record zero potential difference. In 1939 it was discovered that, when true potential differences are recorded by an electrode inside the nerve fibre, the action potential is larger than the resting potential (Fig. 4.6). It not only cancels the resting potential but overshoots by some 30–40 mV so that the inside becomes actually positive to the surface of a region at rest. The implications of this revolutionary observation will be considered later.

The class of disturbance to which the nerve impulse belongs. The all-or-none law

A very important question which has attracted much interest from the earliest days is the relationship between the nervous impulse (defined as whatever process it is that passes down the nerve and makes the muscle contract) and the action potential. Is the electrical change an inherent part of the nerve impulse? In spite of much experimentation, no one has ever been able convincingly to separate the two. It is found that they have the same threshold, propagate at the same velocity, are blocked at the same time if the nerve is frozen or compressed, and so on. Evidence of this kind shows that the nerve impulse and the action potential are so closely related that they cannot be dissociated. We now ask a further and a different question. Granted that the action potential is an inherent part of the nerve impulse, is it essential to the actual mechanism of propagation, or is it merely an invariable accompaniment of that mechanism? The evidence so far given permits, and perhaps encourages, this theory, but would be equally consistent with the view that the essential process is some chemical reaction which is always accompanied by electrical changes; the action potential would then be, as it were, merely the noise of the engine. Before describing Hodgkin's experiments which refuted this last possibility and proved that nervous conduction is electrical it will be convenient to bring forward various other facts which throw light on the kind of process that nervous conduction is, and which show how the particular hypothesis of conduction that he tested came to be proposed.

The action potential is, as we have seen, a travelling wave of electric change; in frog's nerve it moves at about 2500 cm/sec and lasts about 1/1000 of a second, so that the active region is confined at any instant to about $2500/1000 = 2\cdot5$ cm length of nerve. Hence in a nerve as long as the sciatic of the frog, stimulated at the end, all activity has ceased at the point of stimulation well before the action potential arrives at the muscle. The action potential is thus quite a localised affair and nervous conduction must be wholly unlike what happens in an ordinary telegraph wire; the common analogy drawn between nerves and telegraph wires is only very superficial.

As this region of electrical disturbance moves down a long nerve in uniform surroundings it neither decreases in size, nor travels more slowly. This suggests that the nerve impulse obtains the energy it needs for propagation locally as it goes along and not, like a rifle bullet, from the stimulus that starts it. This fits in with older observations that the stimulus does not have to be larger when it is applied to the nerve further away from the muscle, and that the impulse cannot be made to go faster by using a larger stimulus than is necessary just to excite.

Not only does the energy for propagation appear to be acquired locally, but local conditions also determine its rate of release. If a section of nerve is cooled the impulse goes more slowly in that length, but speeds up to its original rate when it reaches warm nerve again. Thus the conduction velocity in each stretch is determined only by the conditions in that stretch and not at all by the previous history of the impulse.

Such behaviour would be most simply explained if the impulse was self-propagated by the active region exciting the next section ahead, like a flame passing along a train of gunpowder. The characteristics of the response at any point would depend only on local conditions and not on the response of the next door sections that excited it. This theory is clearly a good one to explain the above observations on conduction velocity but it carries the implication that the size as well as the velocity of the impulse ought to be independent of the size of the stimulus. If release of energy for propagation is a purely local affair, one of the things determined locally should be the size of the impulse. At any point either you should get a full-sized impulse or none at all; it should not be possible to get a half-sized impulse by carefully grading the size of the stimulus. Behaviour of this kind is referred to as "all-or-none", and tissues which display it are said to obey the "all-or-none law". All-or-none behaviour was first observed in the contraction of the heart by Bowditch in 1872.

On the face of it, nerve does not obey the all-or-none law. As we have said earlier, if the nerve of a nerve–muscle preparation is stimulated with stimuli of increasing strength, first threshold is reached, but above threshold the contractions of the muscle for some time increase in size as the stimulus increases; so the preparation as a whole clearly does not behave in an all-or-none manner. An obvious explanation is that each individual motor nerve does obey the all-or-none law, but because the nerve to gastrocnemius contains many fibres with different thresholds they are excited one after another and so the relationship is concealed. This explanation was shown to be correct by Keith Lucas in 1909. He simplified the problem by using a very small muscle, the cutaneous dorsi muscle of the frog, which has only eight or nine motor nerve fibres. When the motor nerve is excited with steadily increasing strengths of shock, the size of the contraction resulting does not increase smoothly but in a series of abrupt steps (Fig. 4.7). The number of steps is never more than the number of motor nerve fibres to the muscle. Each step clearly represents the excitation of one of the motor fibres, and the fact that each fibre has a sharp threshold, and that there is no further increase in the size of contraction until the next step, is interpreted to mean that, in each fibre, the size of the impulse does not

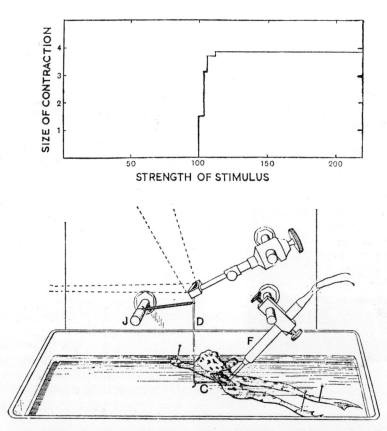

Fig. 4.7 Keith Lucas's apparatus for demonstrating the all-or-none law, using the dorsocutaneous muscle of the frog.

The piece of skin into which the muscle, C, is inserted is cut free and attached to the lever, D, which carries a mirror. The muscle is lightly stretched by a spring between the lever and the support J. F, Stimulating electrodes applied to the motor nerve. Contractions are recorded photographically by means of a beam of light reflected from the mirror.

Above is shown the relationship observed between strength of stimulus (threshold taken as 100) and the size of contraction. The number of steps was never more than the number of motor nerve fibres. (*Journal of Physiology*, vol. 38, p. 113, 1909.)

increase with increasing shock size, once the threshold is passed. This experiment, it should be noted, establishes the all-or-none law for nerve, not muscle. The muscle fibres to which the nerve fibre is connected only come in as indicators of the arrival of a nerve impulse.

Nowadays the all-or-none law is readily observed on the single giant nerve fibres of invertebrates. With the arrangement shown in Fig. 4.5 it is found that, as the stimulus is increased from zero, at first the trace on the cathode-ray tube remains quite flat, except for a small stimulus artefact, until suddenly the picture shown in Fig. 4.6 appears. Further increase in the stimulus causes no increase in the height of the action potential. No amount of adjustment of the stimulus strength ever results in a half-sized action potential. Thus the size of the action potential is independent of the stimulus strength once this has reached threshold.

The all-or-none behaviour of the nerve impulse is important not only because of the light it throws on the nature of nervous transmission, but because it imposes the signalling code for the whole nervous system. Information and instructions can only be sent along a nerve fibre by altering the frequency of nerve impulses and not by altering their size. This system of frequency modulation rather than amplitude modulation, to borrow radio terminology, has very great advantages for long-distance signalling—as radio engineers have discovered. It makes nerve signalling independent of local conditions; all that counts is the number of impulses that get through, and it does not matter if cold or pressure or shortage of oxygen slows them up or reduces their size, as long as conduction does not fail altogether. Likewise it is responsible for the independent functioning of the many fibres bundled together in nerve trunks, for although the electric current generated by an impulse in one fibre will reach its neighbours, it is far too small to excite impulses in them (let alone to extinguish impulses already there) and unless it does this it will not interfere with their signalling at all.

The local circuit theory of transmission

The demonstration of its all-or-none behaviour removed the last difficulties in regarding the nerve impulse as a self-propagating disturbance. We have now to consider by what mechanism the active region excites the region ahead. Many possibilities are ruled out by the observation that nerve fibres (whether sensory or motor) transmit equally well in either direction. If a length of nerve is excited in the middle, the action potential spreads in both directions with equal velocity. Whatever it is that excites locally during propagation must therefore pass with equal ease up or down the fibre. Either diffusion of a chemical substance or the flow of electric current might serve; but diffusion looks like being too slow a process to account for conduction at 60 miles per hour, while there are obvious reasons for preferring an electrical hypothesis: electric currents very easily excite; the action potential produces an electric current; moreover, the polarity is correct, for it is the cathode which excites and likewise the active region is negative.

The hypothesis is, then, that propagation occurs because the active region causes local currents to flow in the inactive region ahead (Fig. 4.8). These currents are in the same sense as the currents that flow when a cathodal stimulus is applied from an external electrode and excite for the same reason. There are no observations which are inconsistent with the local circuit theory,

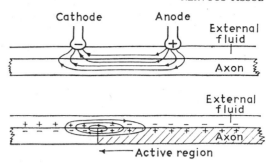

Fig. 4.8 The local-circuit hypothesis of nervous transmission.

The upper diagram shows a single nerve fibre, covered with a layer of fluid, with two stimulating electrodes. Current enters the fibre near the anode and leaves near the cathode, in the manner shown by the lines of current flow.

In the lower diagram an action potential, a region of reversed membrane potential, is travelling from right to left. The surface of the active region is negative with respect to the surface of the resting nerve ahead. Current therefore leaves the resting nerve and flows in the outside fluid down the potential gradient to the active region. The local circuits are completed via the axoplasm, where the same argument applies in reverse.

Thus in front of the active region current leaves the nerve fibre, just as it does under a stimulating cathode. These currents will tend to excite the nerve in front of the active region. The hypothesis is that they do in fact do so, and that this is how the action potential propagates.

(The spatial distribution of local currents along a nerve near a stimulating electrode, or ahead of an action potential, is determined by the cable-like properties of nerve fibres, as is explained in the text. Such a distribution of current spread is termed *electrotonic* to distinguish it from ordinary electrical spread in a homogeneous medium that characterises, for example, the lines of current flow between two electrodes immersed in a beaker of saline. The distinction need not be understood to follow the present argument.)

and since the end of the nineteenth century it has been the most widely accepted hypothesis of propagation, but until 1937 no one knew whether it was true or not. The difficulty is that the action potential recorded from a nerve as it conducts is so very much smaller than the voltage that has to be applied through external electrodes in order to stimulate it. Certainly the local currents flow in the right way to excite, but the unanswered and critical question was: Are they, in fact, large enough to do so?

Hodgkin solved this question in the following way. In a frog's sciatic nerve he blocked conduction by freezing a short length (Fig. 4.9). Immediately after an action potential had arrived at the block, small currents of the kind expected from the local circuit theory could be detected spreading several millimetres beyond the block. Test stimuli showed that at the same time the threshold was reduced by as much as 90 per cent. If the local currents in front of a blocked impulse can lower the threshold by 90 per cent, those from a normal impulse would certainly lower it by 100 per cent, i.e. they would excite. Hence the action potential does constitute a large enough electrical stimulus to excite the nerve ahead of it. It is now known, in fact, that the normal action potential is five to ten times larger than it has to be in order to excite. Electrical propagation has a large safety factor.

Conduction velocity

The velocity of propagation depends on the strength of current flowing in the local circuits ahead of the active region. As described above, because the nerve membrane has capacity, any current must flow for a certain time in order to excite; the larger the current the shorter the time for which it has to flow. Hence, the larger the local currents, the sooner the section of nerve

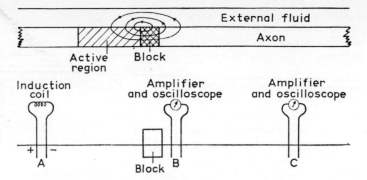

Fig. 4.9 Evidence for electrical transmission in nerve.
A frog's sciatic nerve is blocked by freezing a short length. When the nerve is excited by a shock from an induction coil applied through the electrodes at A, no action potential gets through to the recording electrodes at C, but immediately after an action potential from A arrives at the blocked region, electrodes at B record small local currents spreading through the block. The upper diagram shows, for one fibre, how current spreads through the blocked region, just as it spreads in front of a normal impulse (compare with Fig. 4.8).

The electrodes at B can also be connected to an induction coil and used to measure the threshold of the nerve, the appearance of an action potential at C indicating that the threshold has been reached. Immediately after an action potential arrives at the other side of the block a transient fall in threshold can be detected at B. The fall in threshold occurs at the same time and has the same spatial distribution along the nerve as the local currents set up by the blocked impulse. (After Hodgkin, *Journal of Physiology*, vol. 90, p. 183 and p. 211, 1937.)

ahead of the active region is excited and the higher the conduction velocity. The strength of the local currents depends, among other things, on the electrical resistance of the outgoing path (i.e. the longitudinal resistance of the axon) and the resistance of the return path through the external fluid. The longitudinal resistance of the axon decreases as the cross-sectional area of the axoplasm increases. It is true that an increase in the diameter of the axon also involves an increase in the area and, hence the capacity, of the membrane; but doubling the diameter only increases the capacity by a factor of 2 whereas the resistance (inversely proportional to area) decreases by a factor of 4. Hence the decrease in resistance wins and large nerve fibres conduct faster than small ones. The increase of speed to be obtained by increasing diameter clearly has survival value, for several animals have evolved very large nerve fibres. The squid escapes from its enemies by contracting its mantle, which drives it rapidly backwards through the water. To do this as fast as possible it has developed its giant nerve fibre of up to 1 mm diameter, which conducts at some 25 metres per second.

In the body, nerve fibres are effectively surrounded by a large volume of fluid, so that there is little scope for increasing conduction velocity by lowering the resistance of the return path. In isolated fibres, however, the conduction velocity can be altered by changing the resistance of the return path, an experiment which provides powerful evidence for the electrical theory of propagation. Hodgkin laid a giant axon in air on a grid of platinum strips; electrodes were also provided for stimulating at one end and recording at the other, so that the conduction velocity could be measured. The ends of the platinum strips could be plunged into a trough of mercury, in order to connect them all together electrically. This had the effect of suddenly reducing the longitudinal resistance outside the nerve and, at the same moment, the velocity of the action potential was seen to increase. The fact that conduction velocity can be altered instantaneously by making electrical connections right away from the nerve would be almost impossible to explain unless conduction was itself an electrical process.

THE IONIC BASIS OF NERVE ACTION

The foregoing experiments establish that the action potential propagates by local electric circuits, but although they tell us how it propagates, they reveal nothing about the nature of the potential change itself. By what mechanism does the potential difference across the nerve membrane suddenly reverse and almost equally rapidly return to its resting state again, as it is seen to do in Fig. 4.6? The answer is: by movements of charged ions across the nerve membrane. This essential step of linking nervous activity with the movement of ions was taken by Bernstein in 1902. Bernstein proposed, correctly, that the resting potential, from which the action potential takes off, is a consequence of the unequal distribution of potassium ions across the nerve membrane, and that the action potential is due to movement of sodium and potassium ions down their concentration gradients. He got the details wrong because he thought that the action potential merely cancelled the resting potential whereas, in fact, it reverses it; but the modern Hodgkin–Huxley theory, to be described, rests on Bernstein's principles. Before describing how the potential difference across the membrane is reversed by ionic currents during the action potential, it is first necessary to look into the nature of the resting potential in greater detail.

The resting potential

In sea water and in the blood and extracellular fluid of animals the chief ionic contituents are sodium and chloride ions. There is some potassium, but potassium ions only represent about 2·5 per cent of the total cations. Inside living cells, with few exceptions, the situation is reversed; there is a lot of potassium

Table 4.1 *Approximate concentration of potassium, sodium and chloride in nerve and muscle fibres and in the fluid bathing them (artificial sea water or Ringer's fluid)*

Tissue	Potassium		Sodium		Chloride	
	Inside (mmol l^{-1})	Outside (mmol l^{-1})	Inside (mmol l^{-1})	Outside (mmol l^{-1})	Inside (mmol l^{-1})	Outside (mmol l^{-1})
Squid axon	400	10	50	460	110	540
Frog nerve and muscle	120	2·5	15	120	3	120
Mammalian muscle	140	4	10	150	—	140

1 mmol l^{-1} = 10^{-3} gram-molecules per litre.

(roughly as much as there is sodium outside) but little sodium and chloride (Table 4.1). Although there is this small amount of chloride, by far the greater part of the internal anions consists of organic molecules (in the case of squid nerve mainly amino acids and other small organic acids) carrying net negative charges. Thus, outside cells, we have chiefly sodium ions (Na$^+$) and chloride ions (Cl$^-$) and inside, potassium ions (K$^+$) and organic anions. In solutions of such strength, sodium chloride is completely ionised and does not exist as a compound. Similarly, there is good reason to believe that all, or very nearly all, the internal potassium also exists as free ions. It is misleading and, in a sense, incorrect to speak of cells as containing "potassium proteinate", as is sometimes done.

Leaving to one side the question of how these striking differences in the composition of the cells and the fluid bathing them arise in the first place, let us accept them and consider how they are maintained. One very simple way of keeping the K$^+$ in and the Na$^+$ and Cl$^-$ out would be to have an impermeable cell membrane. But such cells as have been examined prove to be permeable to K$^+$ and Cl$^-$, although they are much less permeable to Na$^+$ and the organic ions. It might be thought that if the cell membrane were permeable to K$^+$ and Cl$^-$ the only way to keep the K$^+$ concentration high and the Cl$^-$ low inside the cells would be to pump the K$^+$ back as it leaks out and pump the Cl$^-$ out as it comes in. But this is not the only way, for the tendency of the K$^+$ to diffuse out can be offset by an electrical potential difference across the membrane. As potassium ions are positively charged, the inside of the cells must be negative relative to the outside in order to attract any potassium ions that try to leave. Since chloride carries an opposite (negative) charge the same potential difference will serve to keep out the chloride ions too. In fact, in all cells on which measurements have been made (mainly nerve and muscle) a resting potential of roughly the expected size has been found. Thus K$^+$ and the Cl$^-$ are (to a first approximation) in equilibrium across the membrane, the tendency of the ions to diffuse down the concentration gradients across the membrane being balanced by the potential gradient.

The value of the "equilibrium potential" which just balances the tendency for a univalent ion of internal concentration C_i to diffuse out into an external fluid containing a concentration C_o is given by the expression

$$E = \frac{RT}{F} \log_e \frac{C_i}{C_o}$$

where T is the absolute temperature, R is the gas constant (as in the gas equation $PV = RT$), and F is Faraday's constant (96 500 coulombs, the amount of electricity needed to deposit 1 gram equivalent of a substance at an electrode). Inserting numerical values of R, T and F and converting to logarithms to base 10 we get:

$$E \text{ (in millivolts)} = 58 \log_{10} \frac{C_i}{C_o} \text{ at } 20°C$$

In fresh squid nerve and in vertebrate nerve and muscle the ratio C_i/C_o for potassium is about 35, giving a potassium equilibrium potential of 58log35 = 90 mV, which is not far from the observed values of resting potential.

An important point that must be grasped in order to understand the ionic mechanism of nerve is that relatively very few ions indeed have to move in order to produce the changes of potential involved. The potential change caused by a certain movement of charge is proportional to the capacity of the structure into which it flows (1 coulomb flowing into a capacity of 1 farad changes the potential by 1 volt). The nerve membrane has a capacity of about 1 microfarad per square centimetre and the resting potential is roughly 100 mV = 1/10 volt. To change the potential difference of 1 cm^2 of membrane by 1/10 volt therefore needs 1/10 microcoulomb = 10^{-7} coulombs of electricity. Now, 1 gram-molecule (mole) of a univalent ion carries a charge of 96 500 coulombs, say 10^5 coulombs, so that 10^{-7} coulombs corresponds to only 10^{-12} mole of a univalent ion. In a 500 μm squid axon the volume of axoplasm covered by 1 cm^2 of membrane is approximately 1/100ml. Squid axoplasm contains 0·4 mole of potassium ions per litre so that the amount of potassium in 1/100 ml is 0·4 × 10^{-5} mole. Thus the axon contains (0·4 × 10^{-5})/10^{-12} = 4 million times more potassium ions than would have to pass across the membrane to alter the potential difference across it by 100 mV.

We are now in a position to understand by means of a hypothetical model just how the resting potential arises (Fig. 4.10). Let us in imagination make up something to represent extracellular fluid by dissolving some sodium chloride and, say, 1/20th of that amount of potassium chloride in water. The inside of the cell will be represented by a solution of the potassium salt of some amino acid (for the sake of definiteness say aspartic acid), together with a small amount of sodium chloride. The total concentrations are chosen so that the two solutions have the same osmotic pressure. We now pour the two solutions into a chamber divided into two parts, which we will call inside and outside, by a membrane permeable to K$^+$ and Cl$^-$, but not to Na$^+$ or aspartate. Because of the concentration differences K$^+$ at once begins to pass from the inside to the outside and Cl$^-$ from the outside to inside. Owing to their large charge, however, few ions have to pass before a potential difference is set up large enough to oppose these movements. (The reader is left to confirm that if, in molar terms, the chosen proportion of KCl outside is the same as the proportion of NaCl inside the same

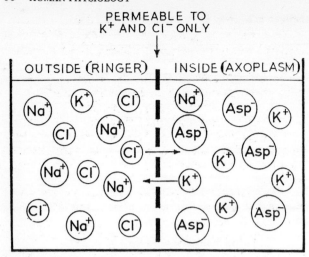

Fig. 4.10 Model illustrating the origin of the resting potential. A hypothetical partition permeable to K^+ and Cl^-, but not to Na^+ nor to organic ions, represents the nerve membrane. The solution in the compartment labelled "outside" is a simplified Ringer's fluid of NaCl with a small proportion of KCl. Axoplasm "inside" is represented by a solution of potassium aspartate with some NaCl, isotonic with Ringer.

potential difference is required to restrain the movement of K^+ out and of Cl^- in.) Thus a membrane potential rapidly builds up, but the movement of ions necessary is so small that if samples of the two fluids were withdrawn and analysed no change in composition could possibly be detected. The movement of ions necessary to give the inside a negative potential relative to the outside would be, as we have seen, of the order of a millionth part of the total.

All cells that have been investigated so far have resting potentials in which the interior of the cells is negative to the surroundings. It is important to remember that this fact could not have been foreseen from a knowledge of the ionic compositions of the cell contents and the external fluid. If instead of being permeable to K^+ and Cl^- and impermeable to Na^+ the reverse had been true, there could have been an equally good equilibrium but the inside would have been positive. This can easily be understood with the above model. When the solutions are poured into the compartments, Na^+ now being the only ion that can move, some sodium ions immediately move inwards. As they carry a positive charge the inside becomes positive and the outside becomes negative. Soon the potential difference becomes sufficient to prevent any further movement of Na^+ down the concentration gradient and equilibrium is established with the inside positive.

This situation is not of merely hypothetical interest. As we shall see later, at the height of the action potential the nerve membrane does become predominantly permeable to Na^+ and the potential difference across it is close to the equilibrium potential for Na^+, with the inside positive. Secondly, even at rest, the membrane is slightly permeable to Na^+. This complicates matters. The simplest way of looking at it is to regard the resting membrane as consisting of a large area of membrane permeable to K^+ and Cl^- and a small area permeable to Na^+. Each area tries to reach its own equilibrium potential but fails. The resultant potential is near the equilibrium potential for K^+ and Cl^-, but shifted some millivolts towards the Na^+ potential, i.e. the inside is less negative. Such a system cannot be in equilibrium. Since the membrane potential is less than the K^+ equilibrium potential, K^+ will diffuse out; and similarly Na^+ enters as the potential is very far from the Na^+ equilibrium potential. Left to itself such a system would slowly run down until the ionic concentration differences levelled out. That it does not do so in living cells is due to an active pump, driven by metabolic energy, which pumps Na^+ out and K^+ in. Thus cells maintain a steady state, but they do not have a simple membrane equilibrium, because energy has to be supplied to maintain their ionic composition.

In general, the potential across the membrane at any time depends on the relative permeabilities to Na^+, K^+ and Cl^-. Other ions to which the membrane is permeable are present in much lower concentrations and make little impression. Normally the membrane of nerves and muscles is moderately permeable to K^+ and Cl^- and almost impermeable to Na^+. So the resting potential is close to the K^+ and Cl^- potential. At the height of the action potential the membrane becomes extremely permeable to Na^+ and the potential difference is the Na^+ potential. Other conditions are also found, as we shall see subsequently; thus the muscle end-plate under the action of acetylcholine becomes very permeable to both Na^+ and K^+ and the membrane potential therefore nearly vanishes.

Bernstein's idea that the resting potential arises because of the K^+ concentration difference for many years rested mainly on the fact that raising the external K^+ concentration reduces the injury potential. It was also known that muscle fibres were permeable to K^+ but not to Na^+; they take up K^+ if the external concentration is raised but not Na^+. Thus if a muscle is placed in hypertonic sodium chloride it shrinks permanently, but if the bathing solution is made hypertonic by dissolving solid potassium chloride there is temporary shrinkage but the fibres soon regain their original volume and are found to have taken up extra K^+ and Cl^-. But quantitative evidence was lacking, partly because the absolute value of the resting potential was uncertain, and partly because in conventional tissues such as frog nerve and muscle it is difficult to get reliable estimates of the Na^+ and Cl^- concentrations inside the fibres; the concentrations inside are rather small, but very large in the extracellular fluid, so that a small error in estimating the amount of the latter produces very large errors in the internal concentrations.

Since 1945, work on single fibres, principally the squid giant axon, has given the necessary data. The permeabilities of the membrane to Na^+ and K^+ have been measured by observing the rate of uptake and of loss of radioactive Na^+ and K^+ by the axon. Furthermore, the axoplasm of squid nerve can be squeezed out from the cut end like toothpaste from a tube, uncontaminated by extracellular fluid, so that accurate values for internal Cl^-, Na^+ and K^+ have been obtained. The value for the resting potential calculated from the permeabilities of the membrane to Na^+ and K^+ and from the concentrations of these ions inside and outside the axon, given in Table 4.1, agrees closely with the actual value, measured with an internal electrode.

The measured internal Cl^- concentration, however, is greater than it should be if the distribution of Cl^- is determined only by the resting potential. It has been shown that the high concentration is because Cl^- is actively pumped into the fibre by a pump similar to, but distinct from the sodium pump described below. Among excitable animal tissues a chloride pump has so far only been demonstrated unequivocally in squid axon. The advantage to the squid of raising the Cl^- concentration in its giant axon is not clear.

Any possible doubt that the resting potential is determined by the concentrations of the ions and the passive permeability properties of the membrane was finally removed when it proved possible to squeeze all the axoplasm out of a squid axon and then to reinflate it and perfuse it with solutions of known composition. With sea water outside and an isotonic solution of a potassium salt inside, the resting potential has normal values (50–70 mV) and the axon conducts action potentials of normal size for many hours. This is so even if the axon is perfused until more than 100 times its volume of isotonic potassium sulphate has passed through. Hence the axon can have no secret soluble ingredients it requires to produce normal resting and action potentials. With isotonic potassium salt outside and isotonic sodium chloride inside the resting potential reverses, the inside of the nerve becoming 50–60 mV positive to the external fluid.

The sodium pump

The mechanism which pumps sodium out of cells and potassium in was first discovered in red cells. In blood that is stored in a refrigerator the red cells gradually lose K^+ and gain Na^+. When they are incubated with some glucose the Na^+ content falls again and the K^+ rises. These movements are against the concentration gradients, and since the movements of two similarly charged ions in opposite directions cannot both be due to the potential difference across the red cell membrane it is clear that an active pump using metabolic energy is involved. A similar mechanism has been shown with great clarity in giant single nerve fibres from the squid. Hodgkin and Keynes pushed a fine glass capillary into a squid axon in the usual way but, instead of recording potential differences, it was used to inject a very small quantity of radioactive Na^+. Within a few seconds radioactivity began to appear in the fluid bathing the fibre. Most of this loss of radioactivity was due to active extrusion of Na^+, for the rate of loss dropped steeply if the pump was poisoned with dinitrophenol (Fig. 4.11) or by cyanide or azide.

Energy for the pump comes from hydrolysing the phosphate bonds of "energy-rich" phosphate compounds such as adenosine triphosphate (ATP). The above poisons stop the production of energy-rich phosphate compounds in the mitochondria. The pump can be restarted again by an injection of ATP down the internal microcapillary. The pump is also poisoned by the cardiac glycosides, the digitalis group of drugs; the one commonly used in experimental work is ouabain. The cardiac glycosides inhibit the adenosine triphosphatase (ATPase) in the membrane. Thus they stop the pump by interfering with the utilisation of ATP.

At one time it was thought that the pump need only extrude Na^+. If we imagine a cell initially containing a lot of Na^+ and Cl^- and a little K^+—similar, that is, in composition to the outside fluid—and proceed to pump out the Na^+, then removing the Na^+ will leave the inside negative, so that K^+ will be drawn in and Cl^- pushed out. Thus by merely pumping Na^+ a situation not unlike that found could certainly be arrived at, with low internal Na^+ and Cl^- and high internal K^+. It appears, however, that the resting potential in nerve and muscle is a little lower than the equilibrium potential for K^+, i.e. there is a little more K^+ inside than could be drawn in by the resting potential. It could only get there against the electrochemical gradient by means of a pump. More direct evidence is provided by the finding that K^+ influx, measured with radioactive K^+, is reduced by poisoning the pump with ouabain.

Uptake of K^+ proves to be coupled to extrusion of Na^+ in a single sodium–potassium exchange pump (Fig. 4.12). The most

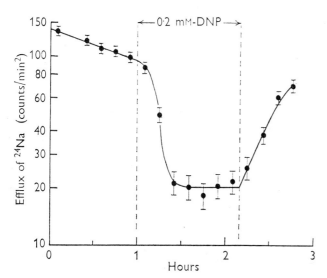

Fig. 4.11 Extrusion of radioactive sodium by a squid giant axon.

An internal microcapillary similar to that shown in Fig. 4.5 was used to inject a small quantity of solution containing ^{24}Na into the axoplasm. The seawater bathing the outside of the fibre was changed at 10 min intervals and the amount of radioactivity in successive samples, due to extruded ^{24}Na, measured in a Geiger counter. Quantity of radioactivity is expressed in counts/min and the units of rate of efflux are, therefore, (counts/min)/min.

The addition of 0·0002 gram-molecules per litre (0·2 mM) dinitrophenol (DNP) to the seawater outside the fibre reduced the efflux by a factor of 5. The pump recovered when the DNP was washed away. (Hodgkin and Keynes, *Journal of Physiology*, vol. 131, p. 592, 1956.)

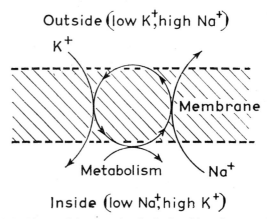

Fig. 4.12 Diagram of the supposed mode of action of the sodium pump. (After Hodgkin and Keynes.)

direct evidence for the coupling of the two processes is that removing K^+ from the external fluid inhibits extrusion of Na^+ from inside the fibre. The only straightforward explanation of this finding is that the pump has to be able to take in K^+ in order to extrude Na^+.

In red cells Glynn and Post and their co-workers have shown that enzymic hydrolysis of ATP not only provides the energy for the pump but is an intimate part of the pumping mechanism itself. The main evidence is: (1) Many cell membranes, including squid axon and red cells, possess a special ATPase which only operates in the presence of K^+ and Na^+. (2) Like the pump, this ATPase is inhibited by ouabain. (3) In red cells the K^+ must be outside the cell and the Na^+ inside, otherwise the ATPase is not activated (i.e. removing K^+ from the external fluid inhibits the hydrolysis of ATP, just as it inhibits the pump). (4) If red cells are prepared with a high K^+ content and no Na^+, and are suspended in saline containing no K^+, the very adverse concentration gradients resulting drive the pump backwards and it synthesises ATP from ADP and inorganic phosphate. This synthesis is ouabain-sensitive. (5) In the last few years biochemists have succeeded in obtaining almost pure preparation of the Na^+- and K^+-activated ATPase. It has a molecular weight of about 250 000, but it has not yet been crystallised or its molecular structure determined. (6) An antiserum prepared to the Na^+- and K^+-activated ATPase from pig kidney inhibits the Na^+ pump of human red cells.

The ionic differences between the cell and its surroundings which the sodium pump maintains, are the basis of the excitable properties of nervous and muscular tissues. Whether in evolution this is the reason for them is not known, but nearly all cells, whether excitable or not, accumulate K^+ and expel Na^+, and a fair proportion of the basal metabolism of the body is devoted to driving the pump by which they do so.

The resting permeability to Na^+ is so low that when the pump is poisoned, large nerve and muscle fibres run down so slowly that the resting and action potentials are not altered significantly for hours. The pump is only necessary for maintaining the composition of the cells over long periods or in recovery after activity. It is important to be clear that the pump is not directly involved in the mechanism of the action potential. The concentration differences of Na^+ and K^+ across the membrane can be regarded as furnishing ionic batteries from which large brief currents can be drawn during the action potential. The pump only recharges the batteries and has nothing to do with turning on the currents during the action potential—which continues quite unaltered if the pump is poisoned.

The action potential

Up to the present we have spoken of the permeabilities of the nerve membrane to Na^+ and K^+ as if they had fixed values. But in fact they depend on the potential difference across the membrane and if this is altered by passing an electric current the permeabilities change. They both change in the same sense, namely, that as the membrane potential is reduced the permeabilities increase, but the time relations differ. When the membrane is suddenly depolarized and held at the new value, the Na^+ permeability rises at once; but the increase is short-lasting and within a few milliseconds the Na^+ permeability has fallen away again. (The process responsible for the fall is called "inactivation".) The K^+ permeability, however, rises only slowly but the increase is then maintained as long as the depolarisation lasts. These facts were discovered by experiments on squid fibres in which the membrane potential was displaced by a known number of millivolts and "clamped" at the new value by an electronic device. The current flowing across the membrane, which is a measure of its permeability to ions, was recorded. If the fibre is in a special solution containing no Na^+, practically all the current is carried by K^+. Hence the record gives the change of K^+ permeability with time. With sea water outside the record gives the sum of the Na^+ and K^+ permeabilities, from which the time course of Na^+ permeability alone can be obtained by subtraction.

When an action potential is set up by an electric stimulus the sequence of events is as follows. Current leaving the fibre under the stimulating cathode causes a voltage drop to develop across the electrical resistance offered by the membrane. This voltage drop is of the opposite polarity to the resting potential and hence it causes a depolarisation of the membrane. The Na^+ permeability at once rises and Na^+ enters, driven both by the concentration gradient and the potential difference. The entry of positively charged sodium ions tends to make the inside of the fibre less negative with respect to the outside, i.e. it still further depolarises the membrane. This depolarisation resulting from Na^+ entry turns on still more Na^+ permeability. Thus a self-regenerative increase in Na^+ current develops, which very rapidly depolarises the membrane and gives the rising phase of the action potential. The inward flow of Na^+ ceases when the membrane potential reaches the equilibrium potential for Na^+ with the inside of the fibre positive to the outside (see p. 30); this marks the peak of the action potential. By this time the Na^+ permeability is already falling, for as we have said the increase caused by depolarisation is only transient. If nothing else happened the potential would stay near the Na^+ potential until Na^+ permeability fell to ordinary levels. The potential difference across the membrane would then be very far from the equilibrium potential for K^+, so that K^+ would leave the fibre. K^+ current would continue to flow until the inside of the fibre was sufficiently negative with respect to the outside to prevent K^+ leaving, i.e. until the resting potential was reached again. In fact the falling phase of the action potential begins earlier and proceeds more rapidly than this, because the delayed rise in K^+ permeability, which begins to make itself felt soon after the peak, allows much larger K^+ currents to flow and greatly accelerates the whole process. The falling phase of the action potential is thus due to an outward movement of K^+. The movements of Na^+ and K^+ during the action potential are shown diagrammatically in Fig. 4.13.

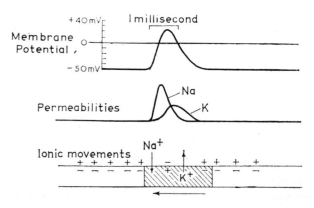

Fig. 4.13 Diagrams showing the movements of Na^+ and K^+ during the action potential.

At the top is an action potential record, as in Fig. 4.6. Below are shown the Na^+ and K^+ permeabilities during the action potential. The rising phase and peak of the action potential are caused by a large but transient rise in Na^+ permeability. The falling phase is due to a delayed rise in K^+ permeability.

In an action potential advancing along a fibre (bottom) Na^+ current entering at the front part of the active region reverses the potential across the membrane; K^+ leaving the fibre in the rear re-establishes the resting potential.

The self-regenerative increase in Na^+ permeability during the rising phase of the action potential is what gives rise to the all-or-none property of nerve. After a stimulus is applied either the membrane goes all the way to the top of the action potential or it sinks quietly back to the resting potential if the stimulus is too small. As the rising phase is self-regenerative it can never go only part of the way. The threshold, or level of depolarisation at which self-regeneration begins, is set by the resting K^+ and Cl^-

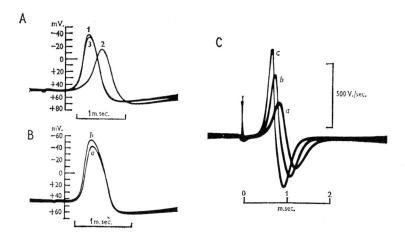

Fig. 4.14 Effect of varying the concentration of Na$^+$ in the external fluid on the height and rate of rise of the action potential in a single giant squid axon.

An internal electrode (Fig. 4.5) was used for recording the potential difference across the nerve membrane.

A. Sodium-deficient solution. Record 1, response in seawater. Record 2, in a solution composed of 50 per cent seawater, 50 per cent isotonic dextrose (Na$^+$ concentration 0·5 times that of seawater). Record 3, after reapplication of seawater.

B. Sodium-rich solution. Record a, response in seawater. Record b, in seawater in which additional sodium chloride has been dissolved to make the Na$^+$ concentration 1·56 times that of seawater.

C. Records showing directly the rate of change of membrane potential, obtained by electronic differentiation of the ordinary action potential record. The height of the first peak in each record is proportional to the maximum rate of rise of the rising phase of the action potential (in volts/second). Record a, fibre in 50 per cent seawater as in A. Record b, in seawater. Record c, in enriched seawater containing 1·56 times the normal Na$^+$ concentration as in B. (Hodgkin and Katz, *Journal of Physiology*, vol. 108, p. 37, 1949.)

permeabilities. If when the stimulus ends the Na$^+$ current inwards tending to depolarise is greater than the sum of the K$^+$ and Cl$^-$ currents tending to repolarise, then the regenerative cycle will occur, otherwise not. If the resting K$^+$ and Cl$^-$ permeabilities are raised, not only is the critical level of depolarisation raised but by Ohm's law a larger stimulating current is needed to produce any given depolarisation, because the resistance of the membrane (which is inversely proportional to the sum of all the ionic permeabilities) is lowered. As we shall see, one form of inhibition in the central nervous system is brought about by increasing the permeability of neurones to K$^+$ and Cl$^-$ thereby making excitatory currents less effective.

The most striking piece of evidence for the sodium hypothesis is the effect of changing the Na$^+$ concentration in the fluid bathing the nerve. It is a direct consequence of the theory that the height and rate of rise of the action potential should vary with the external Na$^+$ concentration. Figure 4.14 shows records of action potentials with low, normal and high Na$^+$ concentrations. The height and rate of rise of the action potential in different Na$^+$ concentrations are in quantitative agreement with the hypothesis. When the Na$^+$ concentrations are the same inside and outside the nerve the action potential should just not over-shoot zero potential difference, for the sodium equilibrium potential would then be zero. This is found to be so. In a sodium-free medium no action potential can be obtained at all. These results have been extended to many other tissues. Frog sciatic nerve will conduct for an hour or more after it is immersed in a sodium-free fluid; but this is because the perineurium (the sheath around the outside of the nerve trunk) is a barrier to diffusion and prevents Na$^+$ getting out. Isolated nerve fibres dissected from the frog's sciatic are blocked within a second in sodium-free media. Block is also rapid in the spinal nerve roots which have no sheath.

These experiments made it highly probable that sodium ions carry the current across the membrane during the rising phase of the action potential, but the matter was not regarded as satisfactory until Na$^+$ had been shown to enter the fibre during activity. This was achieved using radioactive Na$^+$; from the radioactive Na$^+$ entry during a prolonged train of impulses the number of sodium ions entering during a single impulse was calculated and was found to carry a large enough charge to account for the change in potential difference during the action potential.

The final step was to show that the membrane permeability changes for Na$^+$ and K$^+$ as a function of voltage and time, measured by the voltage clamp technique, could by themselves account quantitatively for the shape and conduction velocity of an action potential (and for various other properties: threshold, anode break excitation and refractory period). This involved setting up equations on the basis of the voltage clamp data and laboriously solving them by numerical methods. The results, however, were rewarding, with good agreement between the shape and velocity of the calculated and the experimentally recorded action potential in a squid fibre. Hence the permeability changes to Na$^+$ and K$^+$ offer a full explanation of the action potential.

The individuality of the various elements of permeability change is shown up by specific poisons. Tetrodotoxin (TTX), from the puffer fish of the China seas, in concentrations as low as 10^{-8} M, selectively blocks the increase in Na$^+$ permeability responsible for the rising phase of the action potential. Condylactis toxin, from a sea anemone, allows the Na$^+$

permeability to increase normally with depolarisation, but blocks the inactivation process. The delayed rise in K^+ permeability can be prevented by tetraethylammonium (TEA) ions, which appear to block the K^+ channels.

Calcium entry

In addition to the voltage-dependent Na^+ and K^+ channels responsible for the action potential, many excitable tissues possess a voltage-dependent permeability channel through which Ca^{2+} enters during an action potential. This channel is insensitive to TTX and TEA and is therefore distinct from the Na^+ and K^+ channels. It is blocked by magnesium, manganese or cobalt ions, which have little effect on the Na^+ or K^+ channels, and by specific drugs, e.g. proveratril. First discovered at the neuromuscular junction (p. 39), its properties have been investigated in voltage clamp experiments on TTX-poisoned squid axons which have been injected with the bioluminescent protein aequorin (from a luminescent jellyfish) which emits light in the presence of minute concentrations of Ca^{2+}. Depolarisation of the axon results in a delayed rise in Ca^{2+} entry (detected by an increase in the light output from the aequorin) of a similar time course to the delayed rise in K^+ permeability. This "late" Ca^{2+} channel admits very little Ca^{2+} during an action potential in squid axon, but is highly important in other situations. It appears to carry all the inward current in barnacle muscle, in which the action potential is indifferent to the external Na^+ concentration and insensitive to TTX. In vertebrate cardiac and smooth muscle the inward current is shared between Na^+ and Ca^{2+}, and in some instances the Ca^{2+} component is large enough to give regenerative action potentials on its own after poisoning the Na^+ channel with TTX or removing Na^+ from the external fluid. Ca^{2+} entry in muscle is responsible for excitation–contraction coupling. Even more significantly, Ca^{2+} entry through the late Ca^{2+} channel appears to provide the link between the action potential and the secretion of chemical substances at nerve endings, e.g. at the neuromuscular junction (p. 39), at synapses in the nervous system and in the suprarenal medulla.

Refractory period

A consequence of the regenerative increase in Na^+ current during the action potential is that the nerve cannot be excited again until some time after the regenerative cycle is finished. Obviously a second regenerative process cannot be superimposed on the rising phase. After the peak, the Na^+ permeability mechanism is inactivated and only recovers as the nerve repolarises. The regenerative process can be made to occur again when inactivation has diminished sufficiently for it to be possible by depolarising to obtain a Na^+ current larger than the sum of the K^+ and Cl^- currents tending to repolarise. A larger stimulus than usual will be necessary until the nerve returns to its resting condition for three reasons: firstly, inactivation makes it more difficult to turn on the Na^+ current; secondly, until K^+ permeability returns to normal a larger Na^+ current will be needed to exceed it and cause regenerative action; thirdly, by Ohm's law a larger current than usual will be needed to depolarise by a given amount because the membrane resistance is lowered so long as the K^+ permeability is raised.

The practical effect of these events is that during the spike no stimulus will excite; this is the "absolute refractory period" lasting about 1 millisecond in frog nerve. Then for about another 5 milliseconds (the "relative refractory period") a larger stimulus than usual is necessary. During the relative refractory period the same factors that raise the threshold also decrease the rate of rise of the second spike, reduce its amplitude and cause it to be conducted more slowly.

The refractory period is important because it sets an upper limit to the frequency at which nerve fibres can carry impulses. The limit is about 800 per second in the frog and perhaps 1600 per second in mammals. The frequencies met with in life are usually much lower than this, probably because the above values only apply to the first few impulses. If a prolonged train of impulses is set up cumulative factors enter and later impulses require much larger stimuli. Hence for a sustained discharge, only much lower rates are possible.

Accommodation

If a subthreshold current is applied to a nerve the resulting depolarisation, although insufficient to set off a regenerative change, does result in some inactivation of the Na^+ permeability mechanism and turns on the delayed rise in K^+ permeability. The threshold therefore rises for exactly the same reasons that it is raised during the relative refractory period; these changes are thought to be the cause of accommodation—the process which makes the nerve inexcitable by slowly rising stimuli.

It has been known for a long time that excitable tissues lose their accommodation if placed in solutions without calcium; the threshold falls and small constant currents will cause the discharge of a continuous series of impulses. Eventually impulses arise spontaneously. Calcium ions have been shown to have a profound effect on the permeability mechanisms for sodium and potassium in squid and frog nerve.

Clinically the failure of accommodation in motor fibres is the cause of the spontaneous contractions of muscles, known as "tetany", particularly those of the hand and foot (carpopedal spasms), that occur in states of low blood calcium; for instance when the parathyroid glands are damaged accidentally in the course of an operation on the thyroid. Accommodation also fails in nerves recovering from anoxia. This is the cause of the spotaneous discharge of sensory nerves called "pins and needles" when a tourniquet is removed from a limb, or when the legs are uncrossed after sitting still for some minutes. In the latter case the peroneal (lateral popliteal) nerve is deprived of its blood supply by pressure where it passes around the head of the fibula.

Accommodation is not equally rapid everywhere in an axon. In sensory nerves it is slow near the sensory end organs, so that slowly rising generator potentials arising in the end-organ give rise to trains of nerve impulses. Similarly it must be slow or absent at the origin of the motor fibres in the ventral horn, because a steady depolarisation of the motoneurone cell body causes a maintained discharge of impulses in the motor fibre. The basis of these differences in accommodation at the ends of axons has not yet been investigated in terms of the sodium and potassium permeability mechansisms.

SALTATORY CONDUCTION IN MYELINATED NERVES

We have already seen that increasing the strength of the local currents spreading in front of the action potential by increasing the diameter of the fibre increases the conduction velocity. This is why invertebrates have developed giant fibres. The vertebrates obtain much higher velocities with smaller fibres by another method, making use of the property that local circuits can be made to spread further ahead of the active region by increasing the electrical resistance of the membrane. This can easily be understood by analogy with a submarine telegraph cable; if the insulation is bad the the signal will not reach the other end but will be lost in local circuits through the insulation; the better the insulation, the further the signal will travel. All nerve fibres behave like leaky telegraph cables, with an insulating surface layer surrounding the relatively low resistance core of axoplasm. Because of this the local currents set up by an active region of membrane spread much further along the nerve than they would in a homogeneous system in which the surface layer had the same specific resistance at the axoplasm. Such spread determined by the cable-like properties of the fibres is termed *electrotonic*. (The spread of local currents beyond the frozen region in Fig. 4.9 is electrotonic, and would be abolished if the cable structure were interrupted by squeezing the frozen region with a pair of forceps.) For a nerve axon of given diameter the extent of electrotonic spread clearly depends on how good the insulation is. In the squid giant axon and other invertebrate nerves the insulating layer is always thin. Vertebrates improve the insulation by laying down thick layers of myelin around the fibres, only leaving the membrane exposed at the nodes of Ranvier, spaced roughly every millimetre along the fibre. The action potential is confined to the nodes, the part of the fibre covered in myelin being inexcitable. A high conduction velocity is obtained because the currents produced by the active region, instead of being used to excite the section of nerve immediately ahead, are used to excite the next node some 50 fibre diameters further on. This is possible only because the insulation conferred by the myelin sheath ensures that the local currents get that far and do not leak away through the membrane first. Myelinated fibres transmit about ten times as fast as non-myelinated fibres of the same outside diameter at the same temperature. A typical figure is 90 metres per second (90 m s^{-1}) for a 15 μm myelinated fibre at 37°C.

Evidence that the myelin is an insulator and that activity only occurs at the nodes has been obtained using single fibres dissected from frogs' nerves, a technique that has only been mastered by a handful of physiologists. It has been shown that the threshold is much lower when the stimulating cathode is opposite a node, and that, during the rising phase of the action potential, inward current (i.e. sodium current) only flows

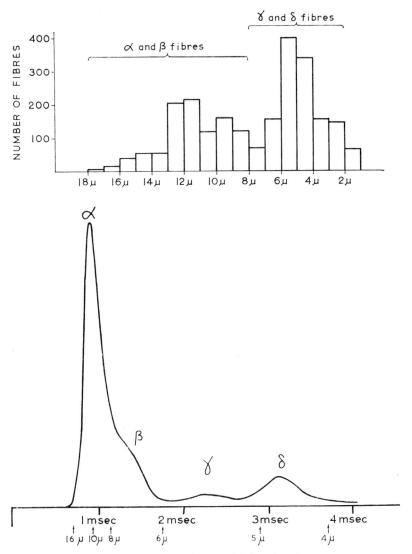

Fig. 4.15 Fibre spectrum and compound action potential of a cat's saphenous nerve.
  The graph at the top of the page shows the numbers of myelinated nerve fibres of different diameters in the saphenous nerve (a cutaneous nerve) of the cat.
  Below is a tracing of the action potential obtained from the same nerve. The conduction distance was 6 cm. The diameters of the fibres responsible for the various parts of the compound action potential are indicated under the scale. (Modified from Gasser and Grundfest, *American Journal of Physiology*, vol. 127, p. 393, 1939.)

Table 4.2 *Size and function of nerve fibres*

Group	Fibre diameter (μm)	Conduction velocity (m s^{-1})	Function
IA IB	12–20 thick myelin	70–120	Somatic (α) motor efferents. Proprioceptive afferents from mechanoreceptors in muscles, tendens and fascia. Golgi tendon organ afferents
II	5–12 moderate myelin	30–70	Kinaesthetic afferents from mechanoreceptors in capsules of joints and from ligaments. Cutaneous afferents from touch and pressure receptors. Some thermal receptors. Pre-ganglionic sympathetic fibres
III	2–5 thin myelin	12–30	Pain afferents. Small motor (γ) efferents to muscle spindles (fusimotor fibres)
IV (or C fibres)	0·1–2 unmyelinated	0·1–12	Pain afferents. Itch Post-ganglionic sympathetic efferents. Tactile, pressure and thermal afferents

through the nodes. By recording from one node after another it was shown that the action potential jumps from one node to the next with a brief delay in between, hence the description "saltatory" from the Latin *saltare*, to leap. The saltatory theory met with opposition from those who believed that there were no nodes in the myelinated nerve fibres of the central nervous system. Ranvier himself said there were none, but they were seen by Cajal and their presence has been amply confirmed by recent investigations.

Another advantage of saltatory conduction, apart from the increased speed, is that the amounts of sodium gained and potassium lost during activity are much decreased, owing to the very much smaller area of active membrane. The amount of sodium entering per impulse is about 300 times smaller in frog's nerve than in an unmyelinated fibre of the same size. Hence the metabolic energy needed for recovery is much less and the fibres are practically indefatigable. There is, however, a debit side to the evolutionary balance sheet. Myelinated fibres, despite their using less oxygen, are, as we shall see shortly, much more easily blocked by lack of oxygen than non-myelinated fibres, and they also appear to fail first in various pathological conditions. We may surmise that even if their nerves were subjected to the insults which human nerves suffer it is unlikely that neuritis would be as common a cause of morbidity in squids as it is in man.

NERVE FIBRE DIAMETER IN RELATION TO FUNCTION

Under the microscope a vertebrate nerve trunk proves to contain myelinated nerve fibres of all sizes from roughly 20 μm down to 1 μm in diameter. There are also numerous small non-myelinated fibres, mostly under 1 μm in diameter, which in many nerves outnumber the myelinated fibres. The relative numbers of myelinated fibres of different diameters varies from nerve to nerve, and between cutaneous and muscular nerves. The fibre spectrum of a cutaneous nerve from a cat is shown in Fig. 4.15.

As we have already seen on p. 28, large nerve fibres conduct faster than small. They also have a lower threshold and give rise to larger action potentials. Both these properties, like their higher conduction velocity, depend on the lower longitudinal resistance of the larger axons. The voltage change across the membrane during the action potential (which is set by the Na$^+$ and K$^+$ concentrations and probably has much the same amplitude in all the myelinated fibres) gives rise to larger currents in the external fluid when the axon resistance is low, and hence to a larger potential recorded with electrodes outside the nerve. Conversely, with the larger axons, a smaller applied voltage is necessary in order to cause a current of threshold strength to flow in them. As a result of these properties, if we take a length of nerve, stimulate it at one end and record the action potential from the other, the action potential just above threshold arrives with the shortest latency and is a simple spike in form; it corresponds to a volley in the largest fibres. A larger shock brings in the next slower fibres which arrive later and produce a hump on the descending phase of the original spike; and so on with larger and larger shocks.

The action potential of the cat's saphenous nerve, with a stimulus large enough to excite all the myelinated fibres, is shown in Fig. 4.15. In the terminology introduced by Erlanger and Gasser, who first investigated the properties of nerve fibres of different sizes in the 1920s, the myelinated fibres of peripheral nerve belong to the A group. The A group action potential is subdivided into α, β, γ and δ elevations, as indicated in the figure. Because of their smaller action potentials the γ and δ fibres, although more numerous than the α and β fibres, make a much smaller showing in the compound action potential. The conduction velocities of the α and β fibres run from 90 m s^{-1} down to, say, 45 m s^{-1}. The γ and δ fibres go from there down to 2 or 3 m s^{-1} for the fibres of about 1 μm; but in the record in Fig. 4.15 the contribution from the smallest fibres is not seen.

The non-meylinated fibres are termed the C group. Their diameters are roughly 0·2 to 1·0 μm, with conduction velocities of about 1 m s^{-1}. If the record in Fig. 4.15 had been taken on a much slower time scale the main C fibre action potential would appear as a very small elevation at about 60 milliseconds.

In cutaneous nerves there are fibres of every size that respond to mechanical stimulation of the skin. The lightest touches (of the kind that evoke sensations of itch and tickle in man) give rise to impulses in δ and C fibres only; firm stimulation brings in A fibres. Painful stimuli appear to be signalled both by slow conducting A fibres and by C fibres. Warming and cooling the skin also give rise to impulses in the slower A fibres and in C fibres.

In muscular nerves the motor fibres that cause the muscle to contract belong to the α group. There are also γ motor fibres that run to supply the small muscle fibres in the muscle spindles—one of the types of sensory ending in muscle. The sensory endings in muscle are connected to both fast- and slow-conducting A fibres.

The myelinated preganglionic fibres in the white sympathetic rami, and elsewhere in the autonomic nervous system, are put in a special group, the B group. They are similar in size and conduction velocity to δ fibres of the A group. The non-myelinated postganglionic sympathetic fibres belong to the C group.

Fast-conducting nerve fibres take up more space and use up more metabolic energy than small ones. This must be one reason why not all nerve fibres are large. It seems clear enough why some touch fibres and motor fibres have to be large, and why fibres conveying ill-localised aching pain can afford to conduct slowly. But much remains to be discovered, particularly about the allocation of different modalities of sensation to the different groups and the reasons for it.

NERVE BLOCK

Conduction in nerve fibres is blocked by many substances with a general anaesthetic action, such as alcohol, morphine, and ether or chloroform vapour. More powerful and convenient than these is cocaine, and its safer derivatives, procaine, etc., dilute solutions of which are widely used in medicine for local anaesthesia. They work by blocking the permeability mechanism for Na^+ entry, like tetrodotoxin. Cocaine blocks small nerve fibres before large; hence the well-known paradox that cocainised surfaces may be insensitive to pain and yet able to feel the brush of a wisp of cotton wool.

Nerves also cease to conduct if deprived of oxygen, but anoxia blocks the myelinated A fibres before the smaller unmyelinated C fibres. Hence in human limbs deprived of blood (ischaemic), touch and motor power fail before sensitivity to pain. When a tourniquet or, better, the pneumatic cuff of a blood-pressure measuring machine (sphygmomanometer), inflated to a pressure above systolic, is applied to the upper arm, the first symptoms of block are numbness of the fingertips coming on after about 14 minutes, followed by loss of touch sensation in the finger and hand and paralysis of the small muscles of the hand. Subsequently, anaesthesia and paralysis slowly ascend the arm. After an hour the whole arm from the cuff down is anaesthetic to touch and paralysed; but pinpricks are still perceived as painful, although the sensation is delayed by a fraction of a second. This persisting, delayed pain is due to unblocked, slowly conducting C fibres. Ischaemia for longer than an hour is not advised (although periods of $1\frac{3}{4}$ hours have been used for experimental purposes, so far with impunity) because of uncertainty as to the time of onset of the disastrous irreversible changes in ischaemic muscles, akin to rigor mortis, occasionally seen in limbs constricted by splinting or plaster. When the circulation is restored after occlusion for an hour or less, recovery of sensation and movement begins within half a minute and proceeds rapidly. The familiar "pins and needles" follows.

It was proved by Thomas Lewis and his co-workers that the cuff blocks conduction because it cuts off the blood supply to the nerve, and not because of the pressure it exerts on the nerve fibres. Thus, provided the pressure is always above systolic blood pressure, block develops in the same time whether the cuff is inflated to 150 or to 300 mm Hg. Again, if a cuff is applied just above the elbow until, say, the hand is anaesthetic to touch, and then a second cuff is inflated on the upper arm above the first and the first cuff be removed, there is no recovery in the hand. The pressure has been removed from the original length of nerve without recovery; hence it cannot be the pressure that is relevant to blocking, but the ischaemia that is maintained by the second cuff.

It was also shown that anaesthesia and paralysis are due to conduction block in the proximal part of the nerve, and not to the effect of ischaemia on sensory endings and on the muscles. Thus if one waits until the hand is anaesthetic and its intrinsic muscles paralysed, and then inflates a cuff round the wrist, subsequent removal of the cuff on the upper arm results in full recovery of sensation and power in the hand, although the hand remains ischaemic. Clearly, the block lay in the stretch of nerve between the upper cuff and the wrist. The reason that the proximal part of a nerve fibre is more susceptible to ischaemia than the distal part is not known; neither is it known why the longest fibres suffer first so that symptoms appear first in the finger tips.

SUMMARY

A nerve trunk is a bundle of nerve fibres. Each fibre can easily be excited by a brief electrical pulse, which must reach a certain strength, the threshold, in order to excite. When a nerve fibre is excited a wave of electrical change, the action potential, passes along the fibre in both directions. In the largest nerve fibres the speed of the action potential is about 20 metres per second in the frog at 20°C, and about 120 metres per second in mammals at 37°C. The action potential lasts about a millisecond.

In each fibre the size of the action potential is independent of the size of the stimulus, once the stimulus reaches threshold. This is called all-or-none behaviour. It implies that the energy for propagation is obtained locally, as it is in the flame that passes along a train of gunpowder.

Propagation proves to be due to local electric currents by which the action potential in one section of the fibre excites the next section ahead. In non-myelinated fibres the process is continuous, but in myelinated fibres the action potential jumps from one node of Ranvier to the next.

Microelectrodes thrust into nerve and muscle fibres show that there is a standing potential difference across the membrane, the resting potential, the inside of the fibre being negative with respect to its surroundings. The resting potential is about 50 mV in squid axons and about 90 mV in vertebrate nerve and muscle.

During the action potential the resting potential is reversed; the inside of the fibre becomes about 40 mV positive with respect to the outside.

The inside of nerve and muscle fibres has a high concentration of potassium ions (K^+) and a low concentration of sodium ions (Na^+). The reverse is true in the extracellular fluid.

The resting potential arises mainly because the fibre in its resting state is permeable to K^+ and relatively impermeable to Na^+. K^+ therefore diffuses out, leaving a negative charge on the inside, until a potential difference large enough to restrain further movement is set up.

The permeability of the fibre membrane to Na^+ and K^+ depends on the potential difference and on time. If the membrane is depolarised (either by an electrical stimulus or by an approaching action potential) there is an immediate but transient rise in Na^+ permeability followed by a delayed rise in K^+ permeability.

With a sufficient depolarisation the increase in Na^+ permeability leads to an inward Na^+ current which itself depolarises. The depolarisation thus becomes self-regenerative; this gives the rising phase of the action potential. Na^+ continues to enter until a potential difference is set up, with the inside positive, sufficient to restrain further movement. This marks the crest of the action potential, by which time, too, Na^+ permeability is spontaneously falling. The delayed K^+ current then restores the resting potential; this is the falling phase of the action potential.

Vertebrate nerves contain myelinated fibres of all sizes from 20 μm down to 1 μm and non-myelinated fibres of 1 μm down to 0.2 μm in diameter. Different functions are served by the different groups of fibre size.

Nerve fibres are blocked by many agencies, but especially easily by local anaesthetics or by cutting off their blood supply.

5. Neuromuscular and Synaptic Transmission

P. A. Merton

The evidence that propagation of the impulse along nerve fibres is electrical is now overwhelming and its mechanism has been described. What happens when an impulse reaches the "end plate" where the nerve terminates on the surface of a muscle fibre by dividing into a spray of fine non-myelinated filaments? After a delay of about a millisecond, an action potential, similar in nature to that in nerve, arises in the neighbourhood of the end plate and propagates by local ciruits in the ordinary way along the muscle fibre membrane in both directions. How does the nerve impulse set off a muscle action potential? The arrival of the active region, which is negative with respect to the rest of the nerve, is equivalent to applying a weak cathodal stimulus to the muscle fibre at the end-plate. This will tend to depolarise and excite the muscle fibre. But does it succeed? The answer is, No. Quantitatively the electrical stimulus is not strong enough. The nerve fibre does not supply nearly sufficient current to depolarise the very much larger area of the muscle fibre membrane. Instead, when the impulse arrives, the nerve ending secretes a very small quantity of the chemical substance acetylcholine. Acetylcholine has practically no effect on nerve or on most of the muscle fibre, but at the end-plate, opposite the nerve ending, the muscle fibre membrane is highly sensitive to acetylcholine, to which it responds by an immediate and very large increase in permeability to both sodium and potassium ions. The effect is equivalent to making a minute hole in the membrane through which ions can pass freely and, as with a mechanical puncture, the resting potential disappears at that point. This depolarisation of the end plate has the same action as applying a strong cathodal electrical stimulus to the muscle membrane around it, which thereupon is rapidly depolarised and excited. Within a millisecond or two the acetylcholine is destroyed by an enzyme, choline-esterase, present in high concentration in the end-plate, which hydrolyses it into inactive acetic acid and choline. Thus acetylcholine and the nerve action potential itself both act in a manner which would tend to depolarise the muscle fibre. The difference is that in practice any depolarisation due to the nerve action potential is too small to be detected, whereas that due to the acetylcholine excites by a comfortable margin. The chemical mechanism can be regarded as a device for amplifying the electrical effect of the nerve impulse by making a temporary chemical puncture in the muscle membrane.

Conclusive evidence for chemical transmission by acetylcholine has only recently been obtained. The story goes back to Claude Bernard's experiments with the South American arrow poison, curare, in 1850. Curare paralyses striated muscles. Bernard showed that it does so by blocking the passage of the nervous impulse from nerve to muscle. After curare, a nerve–muscle preparation no longer contracts when the nerve is stimulated, but it does so again if the stimulating electrodes are moved onto the muscle itself; hence curare does not act by making muscle fibres inexcitable. Nor does it act on nerve fibres, for after painting curare on the nerve alone the muscle contracts in the ordinary way when the nerve is stimulated. Therefore there must be a region with special properties between nerve and muscle, now identified as the end-plate, where curare has its effect. Curare was later found to act by rendering the muscle membrane at the end-plate insensitive to acetylcholine; acetylcholine is still released by the nerve impulse but no depolarisation of the muscle membrane follows. For a hundred years curare remained a physiological curiosity and tool, but recently the purified active constituent, curarine, and other drugs with analogous actions have come into extensive use in surgery for assisting muscular relaxation at operation and thereby permitting a lighter general anaesthesia.

At the beginning of this century Langley showed that the neuro-muscular junction is specially susceptible to chemical excitation as well as to block. In most muscles the end-plates lie near the centre of the fibres, and in parallel-fibred muscles they form a band across the middle of the muscle near the point where the nerve enters. When dilute nictotine solution is applied locally to the end-plate region prolonged twitching of the muscle results, but not when it is put on the nerve-free parts of the muscle or on the nerve itself. The chemically excitable structure revealed by nicotine appears to be the structure that the nerve impulse normally excites, for like the nerve impulse, nicotine action is blocked by curare.

These results would naturally suggest that the nerve fibre itself might excite the muscle by means of a chemical substance liberated at the nerve ending. As a matter of fact, the idea of chemical transmitters had already arisen in connection with the autonomic nervous system. It was known that injection of adrenaline, a substance isolated chemically from the adrenal glands, into the blood, gave rise to many of the effects of sympathetic nerve stimulation (cardiac acceleration, rise of blood pressure etc.). Similarly, injection of muscarine, the poison of a common toadstool, mimicked parasympathetic nerve action (slowing of the heart, salivation etc.).

Definite proof that the nerve impulse does liberate a chemical transmitter was first achieved by Otto Loewi in 1921. He showed that when the isolated frog's heart is slowed by vagal stimulation the perfusion fluid that comes from it will cause slowing of a second heart (Fig. 31.1). The transmitter secreted by vagal endings was identified by Dale and his colleagues as acetylcholine. Refined pharmacological tests had to be developed, for the amounts involved were far too small for chemical methods (Fig. 31.3). Dale and his colleagues afterwards applied similar methods to detect the even more minute amounts of acetylcholine in the blood coming from stimulated skeletal muscle; for this pharmacological *tour de force* the choline-esterase in the muscle, which otherwise would have destroyed the acetylcholine immediately after release, was inhibited by the drug eserine.

These and other experiments made it clear that acetylcholine acts as the transmitter of excitation across the neuromuscular junction. The details of how it works have now been unravelled by modern microelectrode techniques, mainly in the hands of Katz and his co-workers. They employed a method of applying small doses of acetylcholine direct to the motor end-plate by electrophoresis from a micropipette containing the drug. Another, similar, micropipette filled with KCl solution was inserted into the muscle fibre nearby and used as an electrode to record the potential difference between the inside and outside of the end-plate membrane (Fig. 5.1). A squirt of acetylcholine,

a nerve impulse secretes, and the concentration at the membrane rises and falls more slowly. The end plate potential due to a nerve impulse is a much briefer affair and the depolarisation is so rapid that it may be difficult to see where the muscle action potential takes off (Fig. 5.2).

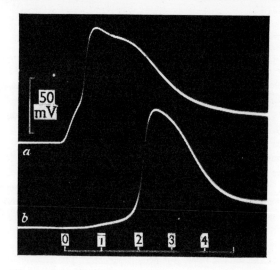

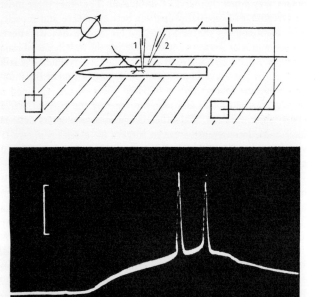

Fig. 5.1 Effect of applying acetylcholine to a motor end-plate in frog muscle.

The diagram shows the experimental arrangements. A glass micropipette (1) containing potassium chloride solution, is inserted into a muscle fibre at the end-plate region and used to record the membrane potential. A second micropipette (2) containing a solution of acetylcholine, is manoeuvred as close as possible to the outside of the end-plate; an outwardly directed pulse of current ejects acetylcholine from the tip of this electrode by electrophoresis. The electronic equipment used to produce brief current pulses is represented in the diagram by a battery in series with a make-and-break switch.

In the record, the step-like deflection in the lower trace signals the current pulse through the acetylcholine pipette. After the pulse, the potential record (upper trace) shows a slow depolarisation of the end-plate membrane (an *end-plate potential*) on which are superimposed two propagated action potentials.

Vertical calibration line, 50 mV. The current pulse lasts 17 milliseconds. (Katz, *Bulletin of the Johns Hopkins Hospital*, vol. 102, p. 275, 1958.)

Fig. 5.2 Neuromuscular transmission in frog muscle recorded with an intracellular microelectrode.

A glass microcapillary is used to record the membrane potential of a single muscle fibre as in Fig. 5.1. The records show what happens after a nerve impulse, set up by stimulating the motor nerve at a distance from the muscle, arrives at the muscle end-plate.

Record *a*, with the microelectrode in the end-plate region, shows a rapidly rising end-plate potential from which an action potential takes off. The point at which the action potential arises is marked by a "step" on the rising phase. (A more conspicuous "step" is to be seen in Fig. 5.3 A.) End-plate action continues during the action potential and causes a "hump" on its falling phase.

In record *b* the microelectrode in the same muscle fibre, 2·5 mm away from the end-plate, picks up the propagated muscle action potential travelling away from the end-plate. The longer latency before the action potential corresponds to the time taken for conduction from the end-plate region. At this distance there is little sign of the end-plate potential.

Such records show, first, that the end-plate potential is a local non-propagated response and, second, that the propagated muscle action potential arises from the end-plate region. Time scale in milliseconds. (Fatt and Katz, *Journal of Physiology*, vol. 115, p. 320, 1951.)

delivered by passing a brief pulse of current through the acetylcholine pipette, gives rise to a depolarisation of the end-plate, the end-plate potential which, if it is large enough, triggers off a muscle action potential (Fig. 5.1). Acetylcholine has no effect if it is applied to the inside of the end-plate membrane after the micropipette has been pushed through the membrane into the muscle fibre. With the pipette outside, the effect falls off rapidly if the pipette is moved along the fibre away from the end-plate. Even with the most careful positioning the pipette cannot be got as close to the muscle membrane as the nerve terminals, which lie in little troughs of membrane on the surface of the fibre, so that a larger quantity of acetylcholine is necessary than

It is not known how acetylcholine causes the greatly increased ionic permeability which depolarises the end plate. The changes in permeability to sodium and potassium ions are certainly quite different from those that occur during an action potential. The basis of the action potential is, first, that the permeabilities to sodium and potassium ions only alter when the potential difference across the membrane alters and, second, that these changes are separated in time, the brief rise in sodium ion permeability preceding the rise in potassium ion permeability. At the end-plate, however, the permeability changes are simultaneous and independent of membrane potential. Thus if a muscle is placed in an isotonic solution of a potassium salt, the resting potential vanishes because the concentration of potassium ions is the same inside and outside. In this condition the muscle and nerve become completely inexcitable, but electrophoretic application of acetylcholine to the end-plate, although it causes no potential change, still results in a large increase in permeability, as evidenced by a fall in the electrical resistance of the membrane. The membrane resistance is measured by passing a known current through the membrane from an internal micropipette, observing the alteration in membrane potential produced and applying Ohm's law.

It is supposed that acetylcholine acts by combining transiently with "receptor" sites on the muscle membrane, each of which when activated opens an ionic permeability "gate". Curare paralyses by combining with the receptors in competition with acetylcholine. Bungarotoxin (from the venom of an Asiatic snake) acts similarly, but binds so tightly to the receptors that it cannot be washed off. By measuring the amount of radioactive bungarotoxin bound to the end-plates of well-washed frog muscle, the number of receptor sites at each end-plate has been estimated at 10^9. Labelling acetylcholine receptors with radioactive bungarotoxin also enables steps to be taken to isolate them from homogenates of muscle, but so far only partial purification has been achieved.

The faint noise of the ion gates opening and closing can be heard by listening in at the end-plate with a microelectrode while a mild depolarisation is produced by bathing the muscle in a dilute solution of acetylcholine. At high gain the record is much rougher than one made without acetylcholine. Statistical analysis reveals that this "acetylcholine noise" is built up of elementary pulses of depolarisation of about 0·0003 mV lasting about a millisecond, each presumably corresponding to the opening of one gate. It follows that only about 10^5 receptor sites out of 10^9 have to be activated to give a normal end-plate potential of about 40 mV.

Quantal release of transmitter

When internal microelectrodes were first used to study electrical events at the end plate it was discovered that, even when no nerve impulses are arriving, the end-plate is not electrically silent: small spontaneous potential changes are recorded from time to time, as shown in Fig. 5.3; they are similar in shape to ordinary end-plate potentials but only about 1 per cent of the size (0·5 mV instead of some 50 mV). The time of occurrence of these miniature end-plate potentials is random (or almost so) as

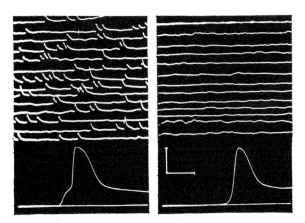

Fig. 5.3 Spontaneous miniature end-plate potentials in frog muscle.

An intracellular microelectrode was used to record the membrane potential of a single muscle fibre as in Figs. 5.1 and 5.2.

In A, the electrode was at the end-plate region. The upper part consists of a number of records taken at slow speed and high amplification, showing the small depolarisations (miniature end-plate potentials) which go on irregularly all the time in the absence of stimulation. In B, the electrode was inside the same muscle fibre 2 mm away from the end-plate. No miniature end-plate potentials are seen.

The lower parts of A and B show the response to nerve stimulation, recorded at a lower gain and on a faster sweep from the same sites. In A there is a conspicuous end-plate potential from which the action potential takes off. No end-plate potential is visible in B. These records, therefore, confirm that the microelectrode was at the end-plate in A (compare Fig. 5.2).

Voltage and time calibrations are given by the L-shaped scales in B. For the upper records the scales represent 3·6 mV and 47 ms.; for the lower records, 50 mV and 2 ms. (Fatt and Katz, *Journal of Physiology*, vol. 117, p. 109, 1952.)

is the time of breakdown of the atoms of a radioactive element. An obvious explanation would be that the miniature potentials are due to accidental leakages of single molecules of acetylcholine from the nerve terminal. But they are about 1000 times larger than the depolarisation from opening a single ion gate. Hence they must represent the simultaneous release of a very large number of acetylcholine molecules in a packet. This is certainly a surprising notion but electron microscopy has revealed that the nerve terminals contain numerous "vesicles", to be seen in Fig. 5.4, which contain the packets of acetylcholine in question. Pharmacological tests have shown that motor nerve fibres contain acetylcholine and also an enzyme *choline acetyltransferase* which synthesises it.

The rate of quantal release depends on the potential difference across the nerve terminal membrane. If the nerve terminal is depolarised by passing a small direct current through it the rate of occurrence of miniature potentials rises steeply; it is estimated that a depolarisation of 15 mV raises the rate by a factor of 10. Whatever the rate, the size of the miniature potentials remains the same; acetylcholine appears always to be released in packets of the same size. When a nerve impulse arrives, it causes a large and rapid depolarisation of the terminal which for a millisecond or so raises the rate of release of acetylcholine packets so high that a hundred or two miniature potentials occur on top of each other. They add up to give a single large end-plate potential.

Depolarisation of the nerve terminal releases transmitter because it allows Ca^{2+} to enter through the "late" Ca channel (p. 34). Lowering the concentration of Ca^{2+} in the extracellular fluid reduces or stops the release of transmitter. Release can also be suppressed by raising the external concentration of magnesium ions or by those other divalent ions, manganese, cobalt etc., which like magnesium are known to block the late Ca channel. On the other hand, the increase in the rate of quantal release during depolarisation by direct current is quantitatively the same after blocking the Na^+ channel with TTX. Sodium entry plays no part.

For a given depolarisation of the terminal the amount of transmitter released increases very steeply with the external Ca^{2+} level. Near the physiological range it is proportional to the fourth power of the external Ca^{2+} concentration. This suggests that, once inside the terminal, calcium ions take part in a process leading to the release of a vesicle of transmitter in which four calcium ions have to co-operate. This hypothetical process is relatively slow, for after Ca^{2+} has entered the terminal there is an irregularly varying latent period, lasting several milliseconds in the frog at low temperatures, before a quantum is released and causes an end-plate potential (Fig. 5.5). Only a little of this latent period can be attributed to events after transmitter release, for when acetylcholine is applied electrophoretically from a similarly placed microelectrode, the first sign of an end-plate response is seen within 0·2 milliseconds at 2·5°C. During release the vesicle comes into contact with the nerve membrane and empties itself through a hole that opens up. It has been caught in the act on electron micrographs. The process is called exocytosis.

Effects of denervation of muscle

When the motor nerve to a muscle is cut and allowed to degenerate, the muscle, over a period of months, wastes away. Denervated mammalian muscle fibres are spontaneously active,

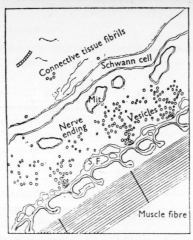

Fig. 5.4A Outline drawing showing the principal structures visible in the electron micrograph, Fig. 5.4B.

The picture shows a fine terminal filament of the motor nerve fibre lying in contact with the surface of the muscle fibre. The area of contact is increased by folding of the muscle fibre membrane at intervals of some $0.4\,\mu$m. A Schwann cell covers the outside of the nerve fibre, and fingers of the Schwann cell (S.F) protrude into the space between nerve ending and muscle fibre. The nerve ending contains vesicles (thought to contain acetylcholine) and mitochondria (Mit). (Birks, Huxley and Katz, *Journal of Physiology*, vol. 150, p. 134, 1960.)

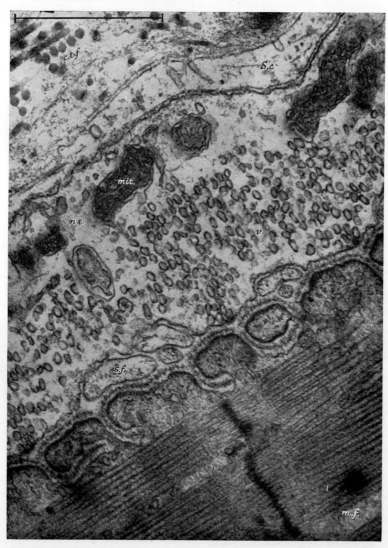

Fig. 5.4B Electron micrograph of the neuromuscular junction of a frog's sartorius muscle. Longitudinal section of the muscle. The scale at the top is $1\,\mu$m long.

individual fibres twitching repetitively and independently. The phenomenon is called fibrillation. It cannot be seen through the skin but if the muscle surface is observed directly it can be seen to be in shimmering movement. The action potentials of fibrillating muscle fibres are easily detected in man by a needle electrode thrust into the muscle and are diagnostic of degenerating muscle.

lasting exotoxin of the bacteria responsible for botulism, a rare but fatal type of food poisoning. Botulinum toxin does not interfere with the response of the end-plate to iontophoretically applied acetylcholine, as does curare, nor does it deplete the stores of acetylcholine in the nerve terminal or interfere with its structure as seen in electron micrographs; but miniature end-plate potentials cease and there is no end-plate potential when a nerve impulse arrives. Thesleff has shown that muscle paralysed with botulinum toxin becomes hypersensitive to acetylcholine

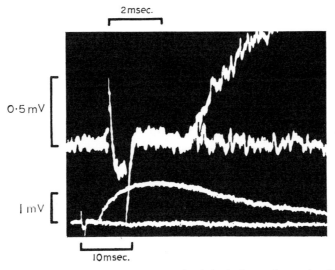

Fig. 5.5 Release of transmitter by an electrical stimulus at the tetrodotoxin poisoned end-plate.

The experimental arrangements were similar to Fig. 5.1, but micropipette 2 contained sodium chloride and was used to depolarise a small area of nerve terminal by passing current through it. The resulting end-plate potentials were again recorded intracellularly by micropipette 1. Temperature 4·5°C.

Recordings were made simultaneously on a fast time-scale at high gain (upper set) and on a slow time-scale at low gain (lower set).

A pulse of current lasting 0·68 milliseconds applied to micropipette 2 caused the "artifact" at the beginning of the records. Two superimposed records were recorded at the same stimulus strength; in one the stimulus caused the release of a single quantum of transmitter after a delay of 3 ms; in the other it had no effect. To facilitate recording, prostigmine, a cholinesterase inhibitor, was used in the bath to increase the size of the end-plate potential.

Repetition of the experiment showed that, although the latency of the end-plate potential varied from stimulus to stimulus, it was never less than 2 ms. (Katz and Miledi, *Proceedings of the Royal Society, Series B*, vol. 167, p. 23, 1967.)

Degenerating muscle is hypersensitive to acetylcholine and gives a larger and more prolonged contraction than usual when a small quantity of acetylcholine is injected into the artery supplying the muscle. (The hypersensitivity to acetylcholine does not appear to be responsible for fibrillation because it is unaffected by curare, which abolishes the action of acetylcholine on muscle.) Electrophoretic application of acetylcholine to hypersensitive muscle has shown that the phenomenon is not due to an increased sensitivity of the end-plate to acetylcholine but to an extension of acetylcholine sensitivity over the whole muscle fibre. In an embryo the muscle fibres are at first sensitive to acetylcholine all over, but later the sensitive region retreats from the ends of the fibres and shrinks onto the end-plate.

Denervation hypersensitivity is due to the production of new receptor sites, the number, as estimated with radio-bungarotoxin, increasing about twentyfold. Hypersensitivity does not develop if protein synthesis is blocked at the mRNA stage with actinomycin D, indicating that denervation acts in some way on the genome of the muscle cell. How it does so is not settled. Release of transmitter from the nerve terminals can be prevented by botulinum toxin, the extremely potent and long-

as fast as muscle that is denervated. Like denervated muscle, it also fibrillates and wastes and will accept innervation from an alien nerve grafted into it. (A supernumerary motor nerve grafted into a normally innervated muscle does not succeed in forming end-plates on it.) Thus all the effects of denervation can be produced without actual degeneration of the nerve or its terminals.

Several groups of workers are now agreed that the onset of hypersensitivity can be delayed by electrical stimulation of denervated muscle fibres; so that mere inactivity is a factor in its development. But other evidence indicates that the lack of substances with a "trophic" action, normally liberated by the nerve ending, plays an important part.

Effect of innervation on speed of contraction of muscle

In adult mammals two kinds of striated muscles are found, "pale" or "fast" muscles with brief twitches and high tetanic fusion frequencies, and "red" or "slow" muscles with long twitches and low fusion frequencies. In man there are slow and fast muscles but the colour differences are inconspicuous. In the newborn kitten all the muscles are equally slow, the adult condition being reached in about four months by a speeding-up of contraction in the fast muscles. In the adult, grafting the motor nerve from a slow muscle (e.g. soleus) into a fast muscle (e.g. flexor digitorum longus) in exchange for its own nerve, effects a fairly complete conversion of the fast muscle into a slow muscle. The opposite cross innervation, e.g. of soleus by the nerve to flexor digitorum longus, results in some speeding up of contraction, but conversion to the fast type is incomplete. It is suspected, but not established, that these highly intriguing phenomena are due to "trophic" substances liberated from the nerve terminals.

Neuromuscular block

When a prolonged train of nerve impulses arrives at a neuromuscular junction the end-plate potential set up by each impulse gradually gets smaller, until it may become inadequate to excite the muscle fibre. The decrease in size of the end-plate potential is primarily due to a decrease in the amount of acetylcholine liberated by each impulse. In healthy subjects there is good evidence that at the frequencies met with during prolonged reflex or voluntary activation of muscle, say 20–40 impulses per second, neuromuscular block is not important, at least until the contractile mechanism is itself almost exhausted. Fatigue in ordinary life is mainly due to failure of the muscle fibres to contract when an action potential passes along them. Electrical excitation of motor nerves at higher frequencies may, however, induce neuromuscular block before the contractile mechanism is exhausted.

Neuromuscular block develops much more quickly in isolated frog muscle than in mammalian muscle with intact blood supply, hence neuromuscular block can readily be demonstrated in fatigued frog muscle in the class room.

SYNAPTIC TRANSMISSION

Neuromuscular transmission is only a special case of the general process of the transmission of electrical activity from one excitable cell to another. In the nervous system, as elsewhere, the living units are cells, here called *neurones*, each a mass of cytoplasm bounded by a cell membrane and possessing a nucleus. When any part of the cell is separated from the nucleus it dies. The cell theory applied to the nervous system is called the *neurone theory*. Nerve fibres are processes of neurones and when separated from the cell body containing the nucleus, e.g. by section of a peripheral nerve, they die and disintegrate. The process is called Wallerian degeneration, after Waller, who first described it in 1850.

For other tissues the cell theory was accepted as soon as the microscope and microscopical staining techniques were sufficiently developed to show clearly the individual cells and the membrane surrounding them. In the nervous system, however, the cell processes may be several feet long, so that it is not possible in general to see the whole of a single cell and be satisfied that it is marked off by a cell membrane from all other cells. As a result, the neurone theory was still in dispute at the beginning of this century. Modern electron microscopy has confirmed that the cell membrane is continuous over the surface of the neurone.

The problem of how activity in one neurone influences activity in another, therefore appears to boil down to the question of how electrical changes at the surface of one neurone can cause electrical changes in another. Such interaction is thought to occur largely, but perhaps not exclusively, at special sites where processes of two neurones come almost into contact with each other. Such a site is called a *synapse*, from the Greek word meaning contact. In the central nervous system synaptic action has chiefly been studied in the neurones of the spinal cord that give rise to the motor nerve fibres to skeletal muscle, the motoneurones. The basic anatomy of the pathways ending synaptically on motoneurones is depicted in Fig. 5.6.

It may seem odd that the discussion of synaptic action, the study of which aims at explaining how the central nervous system works, precedes the chapters describing what the central nervous system does. In fact, although we believe that what we know about synapses will one day take its place in any full explanation of nervous activity, this synthesis is, in most directions, so distant that the observations on synapses stand on their own. It is as if, faced by a large computer, we have probed into its works and discovered the mode of action of its "hardware"; resistances, condensers, transistors etc., and the basic circuits composed of them. This would be important knowledge but, if our object was to understand how the machine extracted a square root, we might not feel much the wiser. On the other hand, a satisfying answer might be immediately obvious to someone who knew nothing whatever about the hardware but who was familiar with the logical structure of the machine—what sort of basic mathematical operations it performed, how numbers were represented in it, and so on—a logical structure conferred on the machine, of course, by the way in which the basic component circuits were connected together. In the central nervous system a beginning has been made on the hardware, but knowledge of the "logical structure" is almost entirely wanting.

Excitatory synaptic action on motoneurones

The synaptic endings of the large spindle afferents on motoneurones (Fig. 5.6) are excitatory and their mode of action is closely similar to that of the neuromuscular junction, according to the evidence obtained by recording the membrane potential of the motoneurone with an internal microelectrode. This technical feat was first achieved by Eccles and his co-workers in 1951. Using anaesthetised cats they first removed the back muscles and the bony roof of the spinal canal to expose the spinal cord; the spinal column of the animal was then held rigidly by clamps to a massive steel frame, on which was also mounted a special micromanipulator carrying a capillary microelectrode filled with potassium chloride solution. The microelectrode was pushed through the spinal cord from the dorsal surface blindly into the ventral horn where the motoneurones lie. The motoneurones are among the largest nerve cells in the mammalian central nervous system, with a diameter of some 70 μm, and they offer a reasonable target. Impalement with a microelectrode of tip diameter about 0·5 μm does not apparently damage them, for they show a steady resting potential of roughly 70 mV. When one of the smaller neurones of the grey matter is entered, however, the resting potential is apt to decline rapidly. A cell is identified as a motoneurone by stimulating the ventral root of that segment. This sends a volley of impulses backwards (antidromically) up the motor nerve fibres into the ventral horn, and if the electrode is in a motoneurone an action potential is recorded (Fig. 5.7G). The action potential of the motoneurone overshoots zero potential and in general appears to be similar in nature to the action potential in the peripheral motor fibre that arises from it.

When a volley of impulses in the large spindle afferents arrives at the synaptic endings, the motoneurone suffers a rapid depolarisation which, provided it is not large enough to excite an action potential in the motoneurone, can be seen to last for some 10 milliseconds. This is called an excitatory post-synaptic potential (EPSP); it looks like, and is similar in mechanism to, an endplate

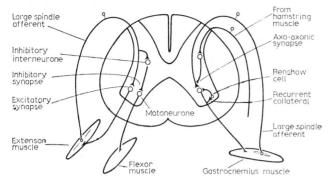

Fig. 5.6 Diagram of the excitatory and inhibitory pathways in the spinal cord studied in this chapter.

On the left is shown a pathway from one of the types of sense organ in muscle, the muscle spindles, to motoneurones. The larger afferent fibres from the muscle spindles are the only fibres entering the cord via the dorsal roots that send branches direct to the motoneurones to make synaptic contact with them. (All other afferents end on interneurones.) This *monosynaptic* pathway is excitatory to the motoneurones of the muscle in which the spindles lie. It is the basis of the stretch reflex, described in Chapter 59 and is most conspicuous in extensor muscles.

The large spindle afferents from extensor muscles also branch to excite interneurones whose axons form inhibitory synapses on the motoneurones of flexor muscles acting at the same joint. (The flexor motoneurones will actually lie in a segment of the spinal cord several segments posterior to the extensor motoneurones in question, although this fact is not conveyed in the diagram.)

On the right two other inhibitory pathways are shown. *Renshaw inhibition* (not confined to gastrocnemius motoneurones) is due to Renshaw interneurones, which are excited by the motoneurone recurrent axon collaterals. *Presynaptic inhibition* is thought to be due to other inhibitory neurones which form axo-axonic synapses on excitatory terminals; the particular connections shown, from hamstring muscles onto large spindle afferents from gastrocnemius are those relevant to Fig. 5.12. This presynaptic inhibitory pathway is shown with one interneurone, but may in fact consist of a chain of two or more interneurones in series.

There are, of course, very many other pathways, some of which involve branches, which are not shown, of the afferent fibres in the diagram. Only what is needed to follow the text has been included.

Each muscle is shown as having only one motoneurone, and each pathway is represented by a single synapse. In fact there are many motoneurones for each muscle, and each excitatory and inhibitory axon divides many times to end in numerous minute synaptic knobs that cover the cell bodies and dendrites of the motoneurones and interneurones. An account of general spinal cord anatomy appears at the beginning of Chapter 59.

potential in muscle. Thus, by passing current through the motoneurone membrane from an internal electrode and measuring the change in potential difference produced, it is found that the membrane resistance falls steeply at the start of the post-synaptic potential. Other evidence suggests that this fall in resistance and the depolarisation that accompanies it are both due to an increase in ionic permeability to sodium, potassium and probably other ions, similar to that which occurs at the endplate.

It is presumed that these permeability changes are due to a similar type of chemical puncture of the membrane under the synaptic endings by a chemical transmitter released from them when a nerve impulse arrives. Electron microscopy reveals numerous "synaptic vesicles" in the synaptic knobs, very similar in appearance to the vesicles in the nerve terminals at the end-plate, which also are suspected to be packets of transmitter. Convincing electrophysiological evidence has also been obtained that release of the excitatory transmitter occurs in "quanta". Spontaneous synaptic potentials, similar in appearance to miniature end-plate potentials, are recorded with a microelectrode in a motoneurone. In the isolated spinal cord of the frog (which survives well in oxygenated Ringer's solution) the frequency of these spontaneous potentials is little altered when all action potentials are abolished by raising the potassium concentration in the bath. Hence, like miniature end-plate potentials, they are believed to represent the spontaneous release of single quanta of transmitter. In the cat the small EPSPs produced by stimulation of one or a few spindle afferents dissected from a muscle nerve have been shown to be compounded of a small number of these unitary synaptic potentials.

The chemical nature of the transmitter from large spindle afferents is not known except that it is not acetylcholine. Acetylcholine is ruled out on strong indirect evidence. It appears that a nerve fibre which liberates acetylcholine at its terminals, a *cholinergic* nerve in Dale's terminology, is cholinergic throughout. Acetylcholine and choline-acetylase, the enzyme that synthesises it, can be detected not only at the terminals but all along the nerve fibre. Thus ventral roots contain acetylcholine and choline acetylase. But dorsal roots contain neither and, since dorsal roots contain large numbers of the fibres that run from muscle receptors to excite motoneurones, it is clear that these fibres are not cholinergic. Iontophoretic application of drugs to the motoneurone from a micropipette has confirmed that it does not respond to acetylcholine or its deriviti es. A large number of other substances have been tried of which derivatives of the acidic amino acids (e.g. cysteic, glutamic and aspartic) were hopeful candidates; but much remains to be done to establish their relation to natural excitatory transmitter substances. A small polypeptide "substance P", detected in 1931 in extracts of brain and intestine by its action on smooth muscle, is currently under consideration; it has a powerful excitatory action on motoneurones and the correct distribution, i.e. it is present in dorsal roots but not in ventral roots.

Although the details of the depolarisation under a single nerve terminal are very similar in both the motoneurone and the end-plate, synaptic transmission in the spinal cord is otherwise arranged very differently from the neuromuscular junction. In mammals each muscle fibre bears only one or two end-plates supplied by a branch or branches of a single motor nerve fibre. Except in extreme fatigue every motor nerve impulse gives rise without fail to a muscle action potential, for each nerve impulse causes an end-plate potential several times larger than is needed to excite the muscle membrane. As the all-or-none law implies, the nervous system can only alter the strength of contraction in a muscle fibre by altering the rate at which nerve impulses reach it. Gradation of response occurs because, although each nerve impulse gives rise to a similar all-or-none action potential in the muscle fibre, the contractile mechanism activated by the muscle action potential results in a partial and long-lasting contraction; a following action potential causes a further contraction which, if it begins with a certain interval, summates with the first. The degree of summation is determined by the frequency of muscle action potentials and thus frequency determines strength of contraction. In the motoneurone, strength of excitation is also related to frequency of arrival of nerve impulses at the synaptic endings; but the graded, non-all-or-none stage where summation takes place is the membrane potential of the motoneurone. In contrast to the end-plate potential the post-synaptic potential, due to an impulse arriving at the synaptic knobs on a motoneurone belonging to a single afferent fibre, is very small, the depolarisation being far below the threshold needed to excite an action potential in the motor axon; but it lasts for some 10

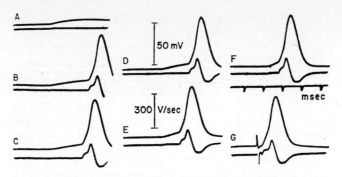

Fig. 5.7 Potentials recorded from a motoneurone in the cat spinal cord with an intracellular microelectrode.

The animal was anaesthetised with pentobarbitone (nembutal). In each of the records A–G the upper trace is the membrane potential. Upward movement of the trace from the resting potential represents a depolarisation of the cell. The 50 mV calibration applies to the upper trace. The lower trace is a differentiated record (cf. Fig. 4.14) showing the rate of change of internal potential. The 300 Vs^{-1} calibration applies to the lower trace.

The cell is identified as a motoneurone in record G, which shows its response to an antidromic impulse set up by stimulation of the ventral root. The shock artifact is followed, after a latency of only a fraction of a millisecond, by a large action potential.

Records A–F show the response to volleys of increasing size in the large spindle afferents, set up by stimulating a muscle nerve. In record A the response is merely a small EPSP (other EPSPs are shown in Fig. 5.12). In records B–F the EPSP rises more and more steeply (due to "spatial summation" of synaptic action) and an action potential takes off from it earlier and earlier, but at a roughly constant level of depolarisation of about 6 mV.

The rising phase of the action potential itself is not smooth but has a knee at about 30 mV depolarisation. The knee separates an initial slower phase, identified on other evidence as the response of the initial segment (the IS spike), from a second more rapid phase (the SD spike). The two phases show up more strikingly in the differentiated records. Both the orthodromic action potentials (B–F) and the antidromic (G) show the IS component, which is thereby clearly distinguished from the EPSP. (Records of Coombs, Curtis and Eccles. In Eccles, *The Physiology of Nerve Cells*. Berlin: Springer Verlag, 1964).

milliseconds so that another impulse arriving within a few milliseconds can add a further post-synaptic potential to it, and so on. Thus the frequency of arrival of nerve impulses at the synapse determines the level of depolarisation of the motoneurone. A similar effect can be obtained at the muscle end-plate by treatment with curare, using a dose which reduces the size of the end-plate potential until with a single nerve impulse it is no longer large enough to excite the muscle fibre. With two or more impulses at intervals of a few milliseconds, the end-plate potentials summate and reach a level sufficient to excite.

Another difference from a muscle fibre is that a motoneurone has synaptic connections with not one, but a large number of afferent fibres. The degree of depolarisation of the motoneurone thus depends both on the number of afferent fibres sending impulses (spatial summation) and on the frequency of impulses in them (temporal summation). In experimental work where the afferent fibres are excited by electrical stimuli applied to a muscular nerve, spatial summation is demonstrated by varying the size of the volley (Fig. 5.7) and temporal summation by varying the interval between two volleys. Whenever in either of these ways the depolarisation of the motoneurone reaches a threshold value an action potential is excited. In more natural circumstances, when, for example, pulling on a muscle sets up a continuous asynchronous discharge of impulse in many afferent fibres, both types of summation occur together and result in a more or less steady depolarisation of the motoneurone; with a sufficient depolarisation a continuous discharge of motor impulses is set up in the axon (Fig. 5.8). The frequency of discharge rises as the level of depolarisation increases. A similar discharge is observed if direct current is passed through a motoneurone in such a direction as to depolarise it. Myelinated nerve fibres do not in general respond to constant current by repetitive firing in this way; they accommodate. Clearly there must be some part of the motoneurone or the axon near it, a "trigger zone", which does not accommodate. The axon near certain sense organs is known also to behave in this way.

There is indirect evidence that the trigger zone is, in fact, the unmyelinated part of the nerve axon as it leaves the motoneurone (referred to by Eccles as the initial segment, or IS) which is thought to have a lower threshold than the cell body proper (the soma) and the dendrites (Eccles's SD). When the neurone is depolarised by synaptic action, or by current injected down the intracellular electrode, the action potential arises first in the initial segment, the threshold depolarisation required in a cat under barbiturate (e.g. pentobarbitone) anaesthesia being about 10 mV. Almost simultaneously the first node of Ranvier on the myelinated part of the axon is excited and an action potential departs down the motor fibre. The action potential then spreads back into the soma and dendrites, the threshold depolarisation for the soma and dendrites being about 30 mV. The purpose of a low threshold trigger zone at the output end of the motoneurone is presumably to allow a consensus of the excitatory and inhibitory actions in the different parts of the soma and dendrites to determine whether the neurone fires. If,

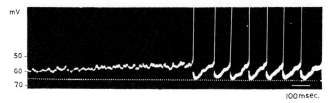

Fig. 5.8 Intracellular records during reflex excitation of a motoneurone, in an un-anaesthetised spinal cat.

The membrane potential (given by the scale at the left) was initially 58 mV (the inside of the cell being negative with respect to its surroundings). Mild reflex activation by moving the hind limb caused a slow depolarisation to 51 mV at which level rhythmical discharge of action potentials began. The top of the action potentials is cut off. Time bar, 100 ms.

At other times during this experiment the resting potential of the cell (i.e. the potential in the absence of action potentials) varied from 65 to 52 mV, the average level for initiating discharge being 51 mV. The dotted line is drawn at 65 mV. After each action potential the membrane potential comes down to about this level. This *after-hyperpolarisation* is due to a temporary increase in potassium permeability, lasting about 100 ms, which renders the cell refractory, i.e. more difficult to depolarise. The frequency of discharge is governed by the time it takes the excitatory synapses to depolarise the cell to the firing level in the face of the after-hyperpolarisation and of any other influences tending to hyperpolarise the cell (e.g. Renshaw inhibition, described below). (Kolmodin and Skoglund, *Acta physiologica scandinavica*, vol. 44, p. 11, 1958.)

for example, the dendrites had the lowest threshold, local excitatory action on one dendrite could fire the whole cell and send an impulse down the axon, regardless of what was going on in the other dendrites. This presumably is undesirable. In computer jargon the motoneurone is required to function as an "and" rather than an "or" device. That is the basis of spatial summation.

This picture of how a motoneurone acts would not be inconsistent with the possibility that action potentials are set up in the finer dendritic branches, provided that they fade out before they reach the soma and leave only a subthreshold contribution to depolarisation of the cell body (like the blocked impulse in Fig. 4.9). Some such elaboration will be necessary to explain how events in distant dendrites, which may be only 1 μm in diameter and more than 1 mm away from the cell body, could cause enough current to flow to exert any influence on the trigger zone.

Inhibitory synaptic action on motoneurones

Synaptic endings which depolarise the motoneurone and tend to excite it are not the only kind. There are also inhibitory synaptic endings on the motoneurone whose action is to cut short or oppose excitation. For instance, impulses in the large spindle afferents from extensor muscles which, as we have just seen, excite the motoneurones of their own muscles, cause inhibition of the motoneurones of flexor muscles acting at the same joint (one of the reflex actions involved in reciprocal innervation; see Chapter 60). When impulses reach the inhibitory synapses the membrane potential of the motoneurone commonly increases; there is a transient hyperpolarisation of the membrane (Fig. 5.9) called an inhibitory post-synaptic potential (IPSP). The IPSP lasts for about the same time as the EPSP: they are, in fact, roughly mirror images of each other. The IPSP is due to a large increase in the permeability of the motoneurone membrane to chloride and potassium ions, but not to sodium ions. As explained on p. 33, the effect of this is to hold the membrane at the resting potential and reduce the efficacy of currents tending to depolarise. Thus inhibitory synaptic action makes the motoneurone more difficult to excite by reducing the effect of excitatory synaptic action.

Inhibitory synaptic action on the motoneurone is blocked by strychnine and by tetanus toxin; this is thought to be why they cause convulsions. Other convulsants do not act in this way.

The intensity of inhibitory action is not gauged, as is excitatory action, by the size of the post-synaptic potential. In cat motoneurones there is normally an IPSP when an inhibitory volley arrives, but if the resting potential happens to be close to the equilibrium potentials for potassium and chloride ions, as it sometimes is, there may be none. Nevertheless, the decrease in membrane resistance due to inhibitory synaptic action still renders excitatory currents less effective, i.e. still inhibits. The decrease in membrane resistance during the IPSP was demonstrated by passing a pulse of current of known intensity through the motoneurone membrane via an internal electrode. The change in membrane potential produced by a pulse during the IPSP was less than that caused by a similar pulse in the resting state; hence, by Ohm's law, the membrane resistance is less during the IPSP.

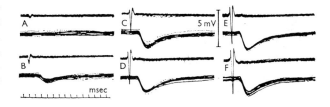

Fig. 5.9 Inhibitory post-synaptic potentials (IPSPs) recorded with an intracellular electrode from a hamstring (knee flexor) motoneurone.

The inhibitory volleys were set up by electrical stimulation of the nerve to quadriceps (the knee extensor). Each record is formed by the superposition of about 40 faint traces. The upper trace in each record shows the size of the afferent volley, as recorded by a surface electrode on the sixth lumbar dorsal root (negativity of this electrode signalled downwards). The lower traces are the intracellular records. Downward deflections signify an increase in membrane potential. The 5 mV calibration applies to the lower traces only. The size of the incoming inhibitory volley was increased progressively from record A to record F. (Records of Coombs, Eccles and Fatt. In Eccles, *The Physiology of Nerve Cells*. Berlin: Springer Verlag, 1964.)

The discovery of the nature of central inhibition and the elucidation of its ionic mechanism was the most spectacular result from Eccles's development of the technique of intracellular recording from motoneurones. The increase in permeability to chloride ions during the IPSP was shown by artificially increasing the Cl⁻ concentration inside a motoneurone by electrophoretic injection of Cl⁻ ions from a KCl-filled microelectrode. If the IPSP is associated with an increase in permeability to Cl⁻ tending to move the membrane potential towards the equilibrium potential for Cl⁻, then raising the internal Cl⁻ concentration (initially low) ought first to arrest this movement and then, when the equilibrium potential for Cl⁻ becomes less than the resting potential, to reverse it. This is precisely what was observed; injection of Cl⁻ ions converted the IPSP to a depolarising potential in effect, an EPSP which, if large enough, excited the neurone.

The injected anion does not have to be Cl⁻. Almost any small anion has the same effect as Cl⁻. Hence the membrane, during the IPSP, must develop a large non-selective permeability to small anions. The evidence for a simultaneous increase in permeability to K⁺ is indirect; it appears to be quantitatively less important than the anion mechanism. The terminals of the large spindle afferents from extensor muscles do not themselves end directly in inhibitory synapses on flexor neurones. Rather, they excite interneurones, which in turn inhibit the flexor neurone, (Fig. 5.6). It is for this reason that, after a volley in the large spindle afferents, the IPSP in a flexor neurone begins about 0.8 milliseconds later (allowing for differences in conduction distance) than the EPSP in an extensor neurone. Interneurones with the appropriate properties have been identified by microelectrode recording. By electrophoresis of the fluorescent dye procion yellow from a microelectrode inside identified interneurones, they can subsequently be seen in histological sections. They lie just dorsal to the motoneurones in the ventral horn. It has been argued that an interneurone is necessary on the inhibitory pathway because the same nerve fibre does not liberate an inhibitory transmitter from some of its terminals and an excitatory transmitter from others (to do so would violate Dale's principle). This line of argument would, however, fall to the ground if it were shown that, in mammals, the same transmitter could both have inhibitory and excitatory effects at different synapses, as acetylcholine has been found to have in molluscs.

The transmitter at inhibitory synapses on the motoneurone is probably glycine. Glycine is present in high relative concentration in the ventral horn; applied electrophoretically to motoneurones it has a powerful inhibitory action and this action is blocked by strychnine which, as we have seen, blocks synaptic inhibitory action on motoneurones. In crustacea and in the mammalian brain γ-aminobutyric acid (GABA) is a proven inhibitory transmitter but, although it has the proper inhibitory action when applied electrophoretically to motoneurones, this action is not antagonised by strychnine.

Renshaw inhibition

Although, as we shall see in the next chapter, stimulation of the central end of a cut ventral root does not cause sensation or reflex movement or any other distant effects, antidromic impulses in motor fibres are by no means without influence in the part of the spinal cord where they arrive, as was first discovered by Renshaw in 1941. The most notable action of antidromic impulses in the motor fibres to a particular muscle is to inhibit the discharge of impulses (reflexly or voluntarily excited) to that muscle or to other muscles with similar actions acting at the same joint (synergists). A convenient muscle in which to demonstrate this effect is soleus. In order to be certain that the effects observed are due to antidromic motor impulses and not to impulses in afferent fibres, which are unavoidably excited when the motor nerve is stimulated, it is necessary to cut all dorsal roots that might contain such fibres. Reflex contractions of soleus can still be set up by stimulation applied to the opposite side of the body. If, during such a contraction, the motor nerve is stimulated, the reflex discharge of motor impulses ceases for perhaps 50 milliseconds (Fig. 5.10A). This will be partly due to a refractory state of the motoneurones consequent on their invasion by antidromic action potentials causing an afterhyperpolarisation (see Fig. 5.8); but that another potent factor is at work is shown by the fact that an antidromic volley in the nerve to the medial gastrocnemius (a synergist of soleus) causes an indistinguishable inhibition of the discharge to soleus (Fig. 5.10B).

As to the mechanism of this inhibition, Renshaw found that an antidromic volley excites a burst of action potentials of extremely high frequency, up to 1500 per second, in interneurones (now known as Renshaw cells) lying in the ventral part of the ventral horn of grey matter (Fig. 5.11). Intracellular recording from the inhibited motoneurones has shown that they display ordinary IPSPs (e.g. reversed by increasing the intracellular chloride and blocked by strychnine), which appear to correspond with Renshaw cell discharges. Hence it is believed that an antidromic volley first excites Renshaw cells which then, in turn, inhibit motoneurones.

The nerve fibres from motoneurones, or at any rate some of them, give off a branch before they leave the grey matter, which turns back into the grey matter. These recurrent collaterals are thought (on not, perhaps, much evidence, but with great plausibility) to be the route by which antidromic impulses excite Renshaw cells (Fig. 5.6). It is worth mentioning that recurrent axon collaterals are very widely distributed in the central nervous system, e.g. the axons both from pyramidal cells in the cerebral cortex and from Purkinje cells in the cerebellum often bear several. Hence mechanisms based on axon collaterals, of which Renshaw inhibition of motoneurones is the only one that has so far proved amenable to analysis, are likely to be of widespread importance.

It is believed that normal, orthodromic, discharge of motoneurones activates the Renshaw cells via the axon collaterals in just the same way that antidromic impulses have been shown to. Hence Renshaw inhibition can be thought of as a local negative feedback loop onto synergic motoneurones, activity in which is a normal accompaniment of motor activity.

The presumed synapse from motoneurone collateral onto Renshaw cell is claimed to be the synapse in the central nervous system for which the chemical transmitter is most securely identified. On Dale's principle it ought to be acetylcholine, and much evidence has been obtained in support of this prediction. Acetylcholine has been detected, by biological assay, in venous blood from the spinal cord after antidromic volleys in motor fibres have been sent in. Renshaw cells are excited by iontophoretically applied acetylcholine, and this action is blocked by dihydro-β-erythroidin (one of the cholinergic-blocking drugs and the most potent of them in this situation). The excitatory action of an antidromic impulse is prolonged by choline-esterase

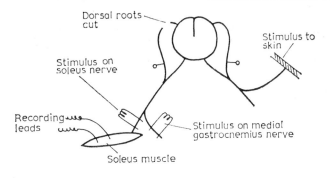
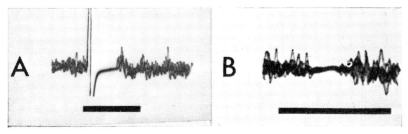

Fig. 5.10 Antidromic block and Renshaw inhibition.
The diagram above gives the experimental arrangements. All dorsal roots supplying one hind limb of a cat are cut and the electrical activity of the soleus muscle is recorded. Contraction of soleus is elicited reflexly by pinching the skin of the opposite flank. An electrical stimulus to the nerve supplying soleus causes a large synchronised action potential in soleus (which is too large to be recorded properly) followed by a period of quiescence (record A) known as antidromic block. A stimulus to the cut nerve to the medial gastrocnemius (record B, from another experiment) results in a similar pause in motor discharge to soleus and cannot therefore be due to a refractory state of soleus motoneurones. This is Renshaw inhibition, which is an important element in antidromic block.
Each record is formed by the superposition of five traces. Time bars 100 ms. (Holmgren and Merton, unpublished records.)

inhibitors and abbreviated by dihydro-β-erythroidin. The inhibitory transmitter released by the Renshaw cells themselves is probably glycine.

The significance of Renshaw inhibition is not known. It is conspicuous in slow acting "tonic" muscles such as soleus and weak or absent in some rapidly acting "phasic" muscles, e.g. the diaphragm. Where it is present, inhibitory feedback is bound to modify and may possibly stabilise in some way the discharge of synergic groups of motoneurones, in a manner analogous to the action of negative feedback in modifying and stabilising the characteristics of an amplifier. But as well as being excited by the motoneurone recurrent collaterals, it is known that Renshaw cells are subject to excitation and inhibition by afferent dorsal root fibres from the skin and muscles of the same and the opposite sides of the body and by fibres from the brain. Hence the Renshaw cells may be involved in the mechanisms that determine which muscles are to contract and how hard they are to contract, as well as in regulating how they do it.

Presynaptic inhibition and dorsal root potentials

In addition to the type of inhibition previously described, effected by inhibitory synapses on the motoneurones themselves, there is another powerful and very much longer-lasting type of inhibition which acts upstream of the motoneurone by reducing the amount of transmitter released from the excitatory synaptic endings and is hence called *presynaptic* inhibition. The evidence for its existence is that stimulation of certain peripheral nerves, which by themselves give rise to no IPSPs, no resistance or threshold change and no other detectable electrical alteration in the motoneurone under observation, causes a prolonged reduction in size of the EPSPs set up in that motoneurone in the ordinary way by stimulation of its large muscle afferents (Fig. 5.12). The technique of minimal afferent stimulation (described in connection with the EPSP), has confirmed that, in the circumstances under consideration, inhibition is truly presynaptic; the number of quanta of transmitter released by a single afferent volley is reduced, while the size of the EPSP due to a single quantum, the "unitary EPSP" is unchanged.

Presynaptic inhibition is prolonged by moderate doses of barbiturate anaesthetics such as pentobarbitone (Nembutal), an important finding to which we return below. The convulsant drug picrotoxin (which is without effect on the IPSP in

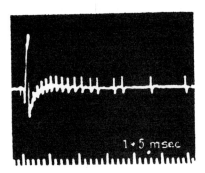

Fig. 5.11 Repetitive discharge in a Renshaw interneurone in the cat spinal cord, following an antidromic volley in motor fibres, recorded with an extracellular electrode in the ventral horn. The large potential at the start is the antidromic action potential in nearby motoneurones picked up by the recording electrode. (Renshaw, *Journal of Neurophysiology*, vol. 9, p. 191, 1946.)

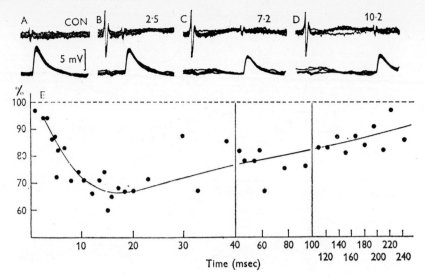

Fig. 5.12 Presynaptic inhibition of excitatory post-synaptic potentials (EPSPs)

In records A to D the lower traces show EPSPs recorded intracellularly from a gastrocnemius (ankle extensor) motoneurone. They were evoked by volleys in the large spindle afferents in the gastrocnemius nerve. Each record is formed by the superposition of five sweeps. In records B, C and D the volley in the gastrocnemius nerve was preceded by a volley in the nerve to a portion of the hamstring muscles (posterior biceps with semitendinosus) at intervals of 2·5, 7·2 and 10·2 ms respectively. These hamstring volleys can be seen to reduce the size of the EPSP without causing any IPSP. A is a control record. The upper traces are from an electrode on the seventh lumbar dorsal root and give an index of the size of the afferent volleys entering the cord (and also of the time scale from the known intervals between volleys in B, C and D).

The graph below plots the time course of presynaptic inhibition for the same experiment from which records A to D were taken. The ordinate gives the size of the EPSP as a percentage of the control size for various intervals between the inhibitory and the excitatory volleys. Inhibition is greatest with an interval of about 15 ms, and persists for at least ¼ second. The vertical lines mark changes in the horizontal scale. (Eccles, Eccles and Magni, *Journal of Physiology*, vol. 159, p. 147, 1961.)

motoneurones) diminishes presynaptic inhibition and also reverses the action of nembutal. (Picrotoxin was formerly used clinically in the treatment of barbiturate poisoning.) Strychnine is without effect on presynaptic inhibition or, rather, it enhances it. Thus the pharmacology of the two kinds of inhibition is quite distinct and so, presumably, are the chemical transmitters involved.

The time course of presynaptic inhibition is similar to that of the large, slow potential changes (dorsal root potentials) that can be recorded from dorsal roots after the arrival of an afferent volley. A causal connection between the two phenomena is strongly suspected but not yet proved. It is, for example, obviously significant that dorsal root potentials, like presynaptic inhibition, are prolonged by barbiturates and reduced in size by picrotoxin.

To record dorsal root potentials a length of root is gently freed and lifted onto two electrodes in air or in a layer of paraffin oil. When an afferent volley set up by an electric shock or by a natural stimulus, such as a tap on the skin, arrives at the cord a monophasic potential change is recorded lasting about a tenth of a second (Fig. 5.13). The electrode closer to the spinal cord goes negative with respect to the distal electrode.

What is the nature of the dorsal root potential? It is not due to action potentials propagating into the root, for it is largest when the proximal electrode is close to the cord and falls off rapidly (roughly by a factor of 2 for each 1½ mm) as the electrodes are moved away from the cord. Propagated action potentials would stay the same size as the electrodes moved. The attenuation of the dorsal root potential along the root suggests that it is due to local currents spreading electrotonically from a site of depolarisation in the cord. This interpretation is confirmed by the observation that the dorsal root potential vanishes if the root is squeezed with forceps between the cord and the recording electrodes (see the explanation on p. 35).

The region of the afferent fibres whose depolarisation is responsible for the dorsal root potential appears to be the fine terminal branches in the grey matter of the cord. Further observations give clues as to the mechanism of this depolarisation. If a row of dorsal roots is cut and a volley sent into the cord by stimulating the end of one of them, dorsal root potentials are observed in roots several segments in front and behind the one stimulated. In the immediately neighbouring roots the potential is about as large as it is in the root stimulated. Hence the slow depolarisation is not confined to the terminals of the fibres along which the volley enters and, indeed, is not specially large in these fibres. Sizeable dorsal root potentials are also produced by stimulating a dorsal root on the opposite side of the cord. Since no fibres entering the cord in a dorsal root cross to the other side, this fact, and other evidence, shows that interneurones are involved in the production of the slow depolarisation. It is thus envisaged that impulses in afferent fibres synaptically excite interneurones, whose axons end on the terminal branches of other afferent fibres and depolarise them (Fig. 5.6). At about the time this proposal was made electron microscopists obligingly revealed suitable axoaxonic synaptic endings in the spinal cord (the ordinary kind are called axodendritic). But a search with

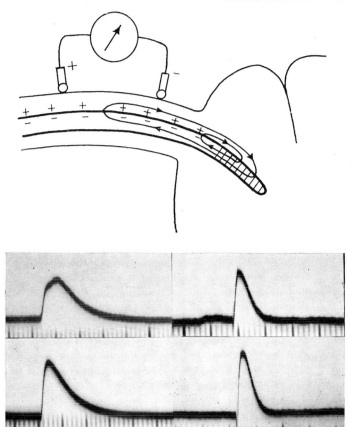

Fig. 5.13 Dorsal root potentials.

Above is shown diagrammatically the position of the recording electrodes on a cut dorsal rootlet. A single nerve fibre is drawn with its terminal portion inside the cord, which is depolarised, shaded. Local currents flow in the sense shown (see also Figs. 4.8 and 4.9) and give rise to a potential difference along the root which is picked up by the electrodes. (In fact, of course, fibres entering the cord divide into many terminal branches as well as giving branches that run up and down the cord, see Chapter 60.

In the records below, the dorsal root potentials illustrated are due to impulses entering by roots adjacent to the cut rootlet to which the recording electrodes are applied. Upward movement of the trace signifies that the electrode near to the cord became negative to the distant electrode. Time markers at 20 and 100 ms intervals.

Top left: Frog, stimulus a light tap on the toes on the same side of the body as the rootlet recorded from. Top right: Cat, the same. Bottom left: Frog, electrical stimulus to a neighbouring rootlet. Bottom right: Cat, the same. (Barron and Matthews. *Journal of Physiology*, vol. 92, p. 276, 1938.)

microelectrodes has not so far revealed interneurones which fill the bill for the dorsal root potential as convincingly as the Renshaw cells do for Renshaw inhibition.

The transmitter released by the responsible interneurones may well be GABA. The main arguments are that GABA is present in high concentration in the dorsal horn and that its proven inhibitory actions elsewhere (e.g. in crustacean muscle) are blocked by picrotoxin which, as we have seen, depresses presynaptic inhibition and the dorsal root potential. Elsewhere, however, GABA is so far only known to cause inhibition associated with increased Cl^- permeability. In order to account for the considerable depolarisation reflected in the dorsal root potential, it is necessary to suppose that GABA has some different action at axoaxonic synapses in the spinal cord, or, for example that presynaptic terminals have a high intracellular Cl^- concentration.

As to how depolarisation of the presynaptic terminal reduces the effectiveness of an afferent impulse arriving there, a plausible hypothesis is that it partially inactivates the Na^+ channel (p. 33) thereby reducing the height of the incoming action potential and so diminishing the amount of transmitter released by it. If the relation between change of membrane potential and rate of transmitter release is as steep as it is at the neuromuscular junction (see p. 43), large inhibitions would only entail a reduction of the action potential by a few millivolts. There is much to be done, however, before this hypothesis can be regarded as established.

Wall claims that impulses in non-myelinated afferent (group C) fibres act in the opposite manner to the larger myelinated fibres we have been considering. They hyperpolarise the terminals of other fibres, appear to increase their synaptic efficacy and, if stimulated alone (a difficult technical feat), give a dorsal root potential of inverted polarity. The significance of these observations is, as yet, uncertain, but they contribute to the belief that the dorsal root potentials are an index of the working of some important nervous mechanism in the cord—a belief reinforced by the discovery that large dorsal root potentials can be evoked by stimulating certain areas of the cerebral cortex and the brain stem.

Synaptic transmission elsewhere in the central nervous system

The types of synaptic action described above cover the main varieties so far found in other parts of the central nervous system. As regards transmitter substances, it is almost certain that some of the endings on cells in the cerebral cortex are cholinergic. There is little adrenaline in the central nervous system, but many neurones containing noradrenaline and dopamine (which is noradrenaline less a hydroxyl group) are present in the brainstem. Because of their catecholamine content these neurones and their processes fluoresce characteristically in ultraviolet light, a property made use of by Fuxe and his colleagues to trace the course of their axons and map the distribution of their synaptic endings. Most of the dopamine in the brain is in a pathway from cells in the substantia nigra to the globus pallidus and putamen (see Chapter 60). Noradrenergic pathways run to the cerebral cortex, to the hypothalamus, to the cerebellar cortex and also down on the spinal cord. The same elegant technique has also revealed 5-hydroxytryptamine in a small midline nucleus in the brainstem. Tryptaminergic pathways run from here upwards to the cortex and down to the cord, but not to the cerebellum. GABA is present in the brain and may well prove to be the main inhibitory transmitter. It has been positively identified as the transmitter at an interneuronal synapse in cerebral cortex, where it produces IPSPs by the Cl^- mechanism. The GABA content of the brain is much diminished in Huntington's chorea. Glutamic acid, widely present, is a candidate as an excitatory transmitter in the brain. Substance P, already mentioned as a possible excitatory transmitter in the spinal cord, is present in many parts of the brain and numerous other new small polypeptides are under active investigation as possible transmitters, their discovery being a quite recent development.

Electrical synapses

It is worth remarking that not all synapses employ chemical transmitters. As we saw, the chemical mechanism at the end-plate is a device to allow a small structure to excite a large one by amplifying the depolarising current. The same is obviously true of the excitatory synaptic knobs on a motoneurone. But if two structures are of comparable size there is no reason why one should not excite the other electrically if a low resistance path for depolarising current is provided where the two membranes come together. This occurs between the cells of heart muscle. In the central nervous system electrical synapses were first demonstrated in crayfish where the giant nerve fibre excites smaller motor nerves; the same fibres have ordinary chemical synapses elsewhere. Synaptic excitation of the giant neurones of goldfish (Mauthner cells) has since been shown to have an electrical component. More remarkably still, an inhibitory electrical synapse has been discovered that operates on the Mauthner cell axon at its origin. Nature may prove to have been equally opportunist in the mammalian central nervous system.

One-way transmission

A characteristic of all synapses in the vertebrate central nervous system yet investigated is that excitation only passes in one direction across them. (Two-way electrical synapses of a highly specialised kind are found at the intersegmental junctions on the giant nerve fibres of the earthworm and in the parasympathetic ciliary ganglion of the chick.) In chemical synapses the transmitter is only liberated by one side of the synapse and only acts on the other side. Electrical back-excitation does not occur because no low resistance path exists across the synapse. In the crayfish electrical synapse, one-way conduction is ensured by an electrical rectifying action of the synaptic membranes; current only flows easily in one direction. It is because of this one-way property of synapses that all nerve fibres yet recorded from normally carry impulses in one direction only. One-way synapses and one-way impulse traffic are only to be expected in the sensory nerves and tracts carrying sensory messages into and within the central nervous system, and in the motor tracts and motor nerves taking orders back to the muscles. These are the parts of the nervous system about which most is known at present. Once again there is no guarantee that elsewhere in the central nervous system all synapses will prove to be one-way or that all nerve fibres will be found to carry one-way traffic. So much remains to be discovered about the nervous system that generalisation beyond what is established by experiment is always unsafe.

SUMMARY

Neuromuscular transmission

Nerve fibres are so much smaller than muscle fibres that when an impulse in a motor nerve arrives at the neuromuscular junction it is not able to cause sufficient current to flow to depolarise the muscle fibre and excite it electrically. Excitation is achieved by a chemical mechanism that has the effect of providing a greatly amplified depolarising current. When the nerve impulse arrives at the end-plate, the nerve terminal secretes a small quantity of acetylcholine. The muscle fibre membrane opposite responds to acetylcholine by a large increase in permeability to both sodium and potassium ions simultaneously. This causes a brisk depolarisation, which excites the surrounding muscle fibre membrane.

The nerve terminal releases acetylcholine in packets, or "quanta", each consiting of many thousand molecules of acetylcholine. A few quanta escape even when no nerve impulses are arriving, causing spontaneous "miniature" end-plate potentials. Depolarisation of the nerve terminal increases the rate of quantal release; the nerve action potential releases a large burst of quanta because it transiently depolarises the terminal and allows calcium to enter.

The muscle fibre membrane is normally only sensitive to acetycholine at the end-plate, but if the muscle is chronically denervated the fibre becomes sensitive to acetylcholine over its whole length.

Synaptic transmission

Excitatory synapses on spinal motoneurones operate similarly to the neuromuscular junction, but on a relatively smaller scale, so that many have to be activated to excite an action potential in a motoneurone. Transmitter release is again in quanta. The chemical nature of the transmitter is not known, but the permeability changes it causes are apparently the same as at the end-plate, and last for about the same time.

Inhibitory synapses on motoneurones operate (by another

transmitter, probably glycine) mainly through a brief selective increase in membrane permeability to chloride and other anions. This has the effect of shifting the membrane potential of the motoneurone towards the chloride equilibrium potential and clamping it there, in antagonism to the depolarising action of excitatory currents.

Apart from reflex and other sources of inhibition, antidromic impulses in motor fibres excite interneurones (Renshaw cells) whose axons terminate in inhibitory synapses on motoneurones (Renshaw inhibition).

Long-lasting inhibition of excitatory action on motoneurones associated with a slow depolarisation of the terminations of dorsal root fibres (recorded as the "dorsal root potential") is the result of a different inhibitory mechanism in which the amount of transmitter released by an excitatory impulse at its synaptic endings is reduced (presynaptic inhibition) without any action on the membrane of the motoneurone itself.

There is evidence that acetylcholine, noradrenaline, dopamine, 5-hydroxytryptamine and γ-aminobutyric acid are chemical transmitters in the brain, but many must await discovery. Small polypeptides, such as "substance P", are recent candidates. Synapses operating electrically are known in fish but not, so far, in mammals.

6. Muscle: Introduction

B. R. Jewell

Muscles are of exceptional importance in the life of higher animals; they are involved either directly or indirectly in virtually every body function. Some forms of muscular activity, such as limb movements and speech, are more obvious than others because they affect the interaction of Man with his environment. Of equal importance are the less overt actions, such as the beating of the heart and the movements of the gut, that affect his internal environment. All these forms of muscular activity are under the partial or complete control of the central nervous system, which influences the muscle directly through its motor nerve supply, and in some cases indirectly through the actions of neurohumoral agents that are released elsewhere and brought to the muscle through its blood supply.

Muscular contraction

The diverse functions performed by muscles in the body all depend on one essential feature of muscle as a tissue—its ability to convert chemical energy into mechanical work. The change in membrane potential (the action potential) brought about by excitation of a muscle cell switches on energy-yielding chemical reactions within the cell by activating the appropriate enzymes. Some of the energy liberated is converted into the mechanical work in an event known as "muscular contraction" and the remainder appears as heat.

Energy sources for muscular contraction

It has been known for a long time that isolated muscles (like intact animals) absorb oxygen and give off carbon dioxide and that these processes are accelerated during activity. The oxygen is used in the oxidation of carbohydrates and other substances (e.g. fatty acids) to carbon dioxide and water. However, oxidative reactions do not provide the *immediate* source of energy for muscular contraction, active transport and many synthetic reactions; the chemical reaction that is coupled directly to these processes is thought to be the hydrolysis of adenosine triphosphate (ATP) to adenosine diphosphate (ADP) and inorganic phosphate (P). This anaerobic reaction is associated with the release of about 40 kilojoules of free energy per mole of ATP hydrolysed. Muscles in fact contain very little ATP (about 3 mmol kg^{-1}), but continuous muscular activity is possible because the ADP produced by the hydrolysis of ATP is rephosphorylated very rapidly by coupled chemical reactions. Anaerobic processes can fulfil this role on a short-term basis, but the eventual restoration of the *status quo* depends on oxidative processes, and these are therefore considered to provide the *ultimate* source of energy used by the muscle.

Experiments on muscle

The behaviour of muscles in the intact animal has been studied extensively and much has been learned, for example, about how limbs are moved, how blood is pumped by the heart and how the uterus expels the fetus during childbirth. The contractions produced by a muscle under these circumstances vary widely in strength and speed, due to the operation of control systems that adjust the activity of the muscle (by neural and sometimes humoral means) to meet the functional requirements of the moment. Contractions that vary in this way are not satisfactory for studies of the basic mechanism of muscular contraction. If one is concerned simply with the muscle as a machine for converting chemical energy into work (as we shall be in the chapters that follow), it is desirable to eliminate these extrinsic control systems; the muscle can then be made to contract in a predictable way by the application of suitable external stimuli (e.g. electric shocks of selected strength, duration and repetition rate). The simplest way of removing the influence of the control systems is to remove the muscle from the animal. Many small muscles will continue to function normally in an artificial environment for several days, provided that their nutritional and excretory needs are satisfied, so this procedure is often a practical proposition. In fact, the bulk of our knowledge of muscle physiology has come from experiments on isolated muscles, especially those of the frog; and in recent years increasing use has been made of single muscle fibres, isolated from a whole muscle by careful microdissection. However, it is important to remember that less drastic methods can sometimes be used; for example, a muscle can be isolated from neural control by severing its motor nerve or by a reversible block of nerve impulse transmission (e.g. by pressure or local anaesthetic). The muscle can then be made to contract by suitable electrical stimuli, which can be applied to the muscle directly or to its motor nerve beyond the point of disconnection. Valuable information about the properties of mammalian muscles *in situ* has been obtained in this way.

CLASSIFICATION OF MUSCLE

Although muscles show wide variations in form, structure and physiological characteristics, it is possible to divide them into a limited number of general types on either anatomical or physiological grounds.

ANATOMICAL CLASSIFICATION OF MUSCLE

Skeletal muscle

Muscles with one or more skeletal attachments account for about 45 per cent of the body weight in man. They are responsible for all body movements, including respiration, and in general they are under voluntary control. The muscle cells (fibres) are of large diameter (typically 50 μm) and they are very elongated; the ratio of length to diameter may be as great as 5000:1. The fibres are multinucleate and they show prominent transverse striations when examined under the microscope.

There are some striated muscles that do not have skeletal attachments (e.g. the muscles of facial expression); they have the same structure and physiological characteristics as skeletal muscle.

Smooth muscle

Smooth muscle accounts for only about 3 per cent of the body weight, but it is found in the walls of almost every hollow tube and organ in the body (e.g. gut, ducts, blood vessels) and at many other sites. The muscle fibres are of small diameter (typically 5 μm) and they are spindle-shaped; the ratio of length to diameter rarely exceeds about 50:1. Each cell contains a single nucleus and there are no transverse striations. This type of muscle is sometimes referred to as *plain* or *non-striated* because of its histological appearance. Smooth muscle is innervated by the autonomic nervous sytem and in general it is not under voluntary control.

Cardiac muscle

This type of muscle is the least abundant in the body (about 0.5 per cent of the body weight), but it is by far the most active. Although the heart has the functional characteristics of a single cell (e.g. activity spreads throughout the muscle) it is not a syncytium from the structural point of view. The myocardial cells are cylindrical in form, with a diameter of 10 to 20 μm and a length/diameter ratio of about 5:1. The cells are arranged in columns that branch and anastomose. Cardiac muscle is like skeletal muscle in that it shows similar transverse striations, but it resembles smooth muscle in that its innervation is derived from the autonomic nervous system.

PHYSIOLOGICAL CLASSIFICATION OF MUSCLE

From the point of view of physiological properties we can divide muscles into single unit and multiunit types. In a *single-unit* muscle (e.g. the heart) activity spreads from cell to cell, and the muscle behaves as a single functional unit. A *multiunit* muscle (e.g. any skeletal muscle) differs in that activity does not spread from cell to cell, and the muscle behaves as a collection of independent functional units. Smooth muscles are mostly of the single unit type, but there are some of the multiunit type and a few with mixed characteristics. The examples given in the diagram (Fig. 6.1) show how this physiological classification cuts

Anatomical classification   *Physiological classification*

Anatomical	Examples	Physiological
SKELETAL		
SMOOTH	pilomotor muscles Intrinsic eye muscles nictitating membrane	MULTIUNIT
	blood vessels vas deferens	MIXED
	Intestinal muscle Uterine muscle Ureteric muscle	
CARDIAC		SINGLE UNIT

FIG.6.1

across the more familiar anatomical divisions of muscle.

In the chapters that follow, skeletal muscle, cardiac muscle and smooth muscle will be considered in turn. Each chapter follows the same general pattern as far as possible. Attention has been concentrated mainly on mammalian muscle because of its particular relevance to human physiology, but where our knowledge of this type of muscle is incomplete, use will be made of information obtained from other sources.

7. The Structure and Function of Skeletal Muscle. Chemical Changes before and after a Tetanic Contraction

B. R. Jewell

The whole muscle

A typical skeletal muscle has a fleshy belly and a tendon at each end. The muscle fibres are organised into bundles known as *fasciculi*. In the human biceps, for example, there are 15 000 of these, each containing 20–60 fibres. In a muscle of the "strap" type the fasciculi lie parallel to its long axis, whereas in a muscle of the "pennate" variety they are arranged obliquely with respect to its long axis.

Almost all skeletal muscles contain specialised structures called muscle spindles, which contain a few poorly striated muscle fibres known as *intrafusal* fibres. Muscle spindles are of great importance in the reflex control of movement (see p. 11), but they make no direct contribution to the mechanical response of the muscle. *The information given in this chapter applies only to the extrafusal fibres that make up the main bulk of the muscle.*

Muscle fibres

Skeletal muscle fibres vary a good deal in size from one muscle to another. In humans the fibre diameter varies from about 10 μm in muscles that produce finely controlled movements to about 100 μm in muscles of coarser action. The fibre length shows even greater variation; for example, in the human sartorius the fibre length approaches 500 mm whereas in the muscles of the middle ear it is only a few millimetres. In all cases the ratio of the fibre length to diameter is very large compared with the corresponding values for smooth and cardiac muscle.

Each muscle fibre is enveloped in a delicate connective tissue sheath, which is intimately related to the plasma membrane of the cell, giving a composite structure known as the *sarcolemma*. The subcellular structure of a skeletal muscle fibre is extremely elaborate; there are internal membrane systems, known as the transverse tubular system (T-system) and the sarcoplasmic reticulum, and each cell contains large numbers of nuclei, mitochondria, and myofibrils. The structure of the myofibrils, the T-system and the sarcoplasmic reticulum will be considered in detail as these are peculiar to the muscle cell.

Myofibrils

Myofibrils are the contractile elements of the muscle. They consist of thread-like structures, about 1 μm in diameter, which probably run the whole length of the fibre. A typical muscle fibre contains 1000 or more of these elements and they account for 70–80 per cent of the fibre volume. Individual myofibrils show transverse striations and the characteristic striated appearance of a skeletal muscle fibre results from the remarkable fact that the striations of the 1000 or so myofibrils are in register across the breadth of the fibre.

Nature of the striated appearance

When the *simple light microscope* is used to look for variations in *light absorption* (i.e. with the condenser aperture wide open), no striations can be seen in living muscle fibres. Striations are visible in fixed and stained material because consecutive regions along the length of the myofibril vary in their affinity for basic dyes. If the microscope is used to look for variations in *refractive index* (i.e. with the condenser stopped down), living muscle fibres have a striated appearance, indicating that the protein concentration varies along the length of the myofibril. Great care is required in the interpretation of the striation pattern seen because this depends critically on the way in which the microscope is set up and focussed. It is often preferable to use a microscope that is designed to look for variations in refractive index, such as the *phase contrast microscope* or the *interference microscope*. If living muscle fibres are examined in the *polarising microscope* the regions of high refractive index are found to be birefringent (anisotropic or A bands) and the regions of low refractive index are non-birefringent (isotropic or I bands).

The main features of the striation pattern are illustrated in Fig. 7.1. The *A bands* are birefringent, they have a high refrac-

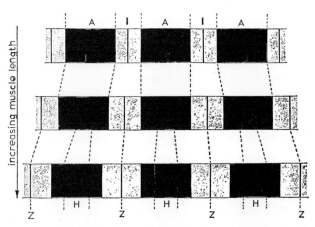

Fig. 7.1 Diagram showing the striation pattern of myofibrils from a skeletal muscle fibre. The three illustrations show how the striation pattern varies with the muscle length. When the muscle is stretched, the A bands remain unchanged in length, but the I bands and H zones increase in length in direct proportion to the amount of stretch. Note that the distance from the Z line to the edge of the H zone remains unchanged.

tive index and they are stained heavily by basic dyes; whereas the *I bands* are non-birefringent, have a low refractive index and do not take up basic dyes. The striation pattern shows additional features, which are illustrated in the Figs. 7.2 and 7.3. The I band is bisected by a narrow region known as the *Z line*, which has a high refractive index and which stains intensely. The A band usually has a central region, known as the *H zone*, in which birefringence, refractive index and affinity for basic dyes are less marked. The Z lines divide the myofibril into units known as *sarcomeres*, which are typically 2 to 3 μm in length.

The striated appearance of the myofibrils is due to the presence of extremely fine thread-like structures known as myofilaments, which are about two orders of magnitude smaller than the myofibrils. These cannot be seen in living muscle fibres

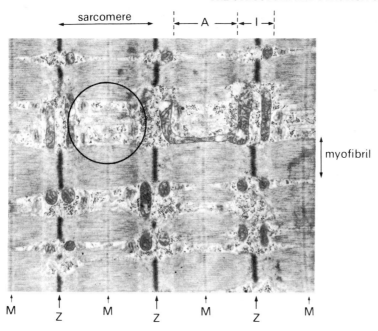

Fig. 7.2 Ultrastructure of mammalian skeletal muscle. Electron micrograph (courtesy Sally G. Page) of a longitudinal section of human laryngeal muscle. The labelling at the top and bottom of the picture give the designations of the transverse striations. Myofilaments are visible within the myofibril at the centre of the picture, and further details of their arrangement are given in Fig. 7.3. Mitochondria are found in the interfibrillar clefts (note especially the U-shaped mitochondrion in the top right-hand quadrant). The main elements of the internal membrane systems are shown in the circled area. Triads are present at 2, 4, 8 and 10 o'clock, and between them in the A-band region elements of the network of tubules making up the sarcotubular system.

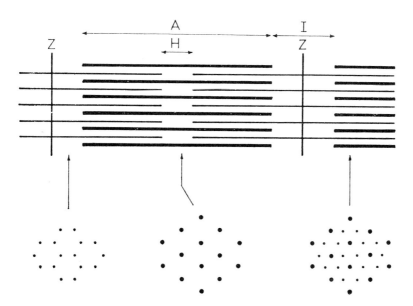

Fig. 7.3 Diagrams illustrating the myofibrillar ultrastructure of skeletal muscle. The upper panel shows the interdigitating of two types of myofilament which are just visible in the electron micrograph shown in Fig. 7.2. There are thick filaments, consisting entirely of myosin, in the A bands and thin filaments which extend from the Z lines into the A bands of the two adjacent sarcomeres; the thin filaments consist of actin and the regulatory proteins, troponin and tropomyosin. The lower panel shows the appearance of the myofilaments in cross sections taken at the points indicated in the upper panel.

because they are much too small to be resolved by any type of light microscope, but they can be demonstrated when thin sections of fixed and stained material are examined in the electron microscope (Fig. 7.2(a)). The A bands are found to contain "thick" filaments that are about $0.01\ \mu m$ (100 Å) in diameter and about $1.5\ \mu m$ in length, and the I bands contain "thin" filaments that are about $0.005\ \mu m$ in diameter and about $1.95\ \mu m$ in length. The thin filaments extend from the Z lines into the adjacent sarcomeres, where they overlap and interdigitate with the thick filaments in the outer parts of the A bands (Fig. 7.3). The amount of overlap depends on the muscle length at the time of fixation. A transverse section through the outer part of the A band shows that the interdigitating filaments form a double hexagonal array (Fig. 7.3). Transverse sections through other parts of the sarcomere show that the I bands contain only thin filaments and that the H zone of the A band contains only thick filaments.

When the resting muscle is stretched and when the active muscle shortens (see Fig. 7.9(a)), the two sets of filaments slide past one another. This is known as the *sliding filament hypothesis*, but there is now so much evidence in support of this hypothesis that it is generally accepted as an established fact.

The filaments consist of aggregations of protein molecules. There is good evidence that the thick filaments are composed entirely of *myosin* (mol. wt ~500 000), and that the thin filaments contain *actin* (mol. wt 57 000), *tropomyosin* (mol. wt 53 000) and *troponin* (mol. wt 50 000). The enzyme that catalyses the breakdown of ATP is a prosthetic group on the myosin molecule. If ATP is added to a solution of myosin (prepared by selective extraction of the protein from muscle), it is rapidly hydrolysed and heat is evolved because the free energy released is all degraded into the heat. Hydrolysis of ATP also occurs with the evolution of heat when ATP is added to a solution containing both actin and myosin (known as an *actomyosin* solution) but in addition the physicochemical properties of the solution change and a precipitate of actomyosin may be formed. These effects result from increased interaction between the actin and myosin molecules, which is brought about by the breakdown of ATP. A dramatic demonstration of the same phenomenon is possible if actomyosin threads are produced by extruding an actomyosin solution through a fine jet into a weak salt solution; when ATP is added to the system it is hydrolysed and the actomyosin thread will either shorten or develop tension depending on the mechanical conditions. The analogy with contraction of the living muscle is obvious.

Internal membrane systems

When longitudinal sections of suitable fixed and stained muscle are examined in the electron microscope, structures known as *triads* are seen between the myofibrils at regular intervals along the length of the muscle fibre (Fig. 7.2). In mammalian skeletal muscle these are found opposite the ends of the A bands (close to the A–I boundaries), but in muscles from other animals they are positioned differently with respect to the striation pattern. The triads are points of conjunction of traversely and longitudinally oriented internal membrane systems. The transverse tubular component (T-system) consists of a network of branching and anastomosing tubules that extends over the entire cross-section of the fibre near each A–I boundary. The tubules are invaginations of the plasma membrane and the space inside them is in direct continuity with the extracellular space through pores arranged around the circumference of the fibre. The longitudinal component is the sarcoplasmic reticulum (sarcotubular system): it consists of an entirely separate system of confluent tubes, which forms an incomplete sheath around each myofibril. Near the A–I boundaries, the tubes are dilated to form *terminal cisternae*, which are in close contact with the transverse tubules, giving the structure known as triads (the transverse tubule, seen in cross-section, is the central element and the terminal cisternae on either side are the lateral elements). These internal membrane systems are concerned with the process of *excitation–contraction coupling*, which is described below.

EXCITATION

Activation of the contractile system in a muscle fibre requires depolarisation of its surface membrane. In experiments on isolated muscle this can be produced in a variety of ways (e.g. by a large increase in the potassium concentration in the bathing solution), but when muscles contract in the body depolarisation of the cell membranes depends on the presence of *action potentials*.

Excitation of a muscle fibre is normally brought about by the arrival of a nerve impulse at its motor nerve terminal. The release of acetylcholine from the nerve terminal results in depolarisation of the muscle fibre membrane in the motor end-plate region (see Chapter 5 on transmission at the neuromuscular junction). The effect of local depolarisation is to generate an action potential in the surrounding membrane and this propagates to the ends of the muscle fibre. Under normal circumstances activity in the motor nerve supply to a muscle originates in the central nervous system, but a volley of nerve impulses (i.e. simultaneous discharge in several nerve fibres) can be produced by the application of an electrical stimulus to the nerve; this is referred to as indirect stimulation of the muscle. As each nerve fibre supplies many muscle fibres (the *motor unit*), the effect of stimulating a single nerve fibre will be to activate all the muscle fibres in that motor unit. It is possible to activate individual muscle fibres by direct stimulation of the muscle and more finely graded contractions can therefore be obtained than would be possible with indirect stimulation; however, this is a non-physiological situation and it must be remembered that the minimum functional unit in muscles contracting under normal circumstances is the motor unit.

Strength–duration relation

For all excitable tissues the stimulus strength required to produce a threshold response depends on the duration of the stimulus, and the relation between these parameters is hyperbolic in form. The curves for motor nerve fibres and skeletal muscle fibres differ as illustrated in Fig. 7.4(a) (curves N and M respectively). The strength–duration relation obtained by *indirect* stimulation is identical to curve N, but the relation obtained by *direct* stimulation is of the form shown in Fig. 7.4(b). The left-hand part of this curve is the same as curve N because strong brief shocks (sometimes called "Faradic" stimuli), which are too short in duration to excite the muscle fibres directly, stimulate intramuscular branches of the motor nerve fibres and thereby excite the muscle fibres indirectly. The right-hand part of the curve is identical to curve M because weak shocks of long duration (sometimes called "galvanic" stimuli) excite the muscle fibres alone. Thus direct stimulation of a muscle only results in direct stimulation of the muscle fibres when the stimuli are of the "galvanic" type (or when the muscle is curarised to block neuromuscular transmission). In experiments on isolated muscles direct stimulation of the muscle fibres is preferred to indirect stimulation because the response evoked is less liable to fatigue, and because the delay between applying a stimulus and obtaining a response is reduced to a minimum (especially if multi-point stimulation is used to activate all parts of the muscle more or less simultaneously).

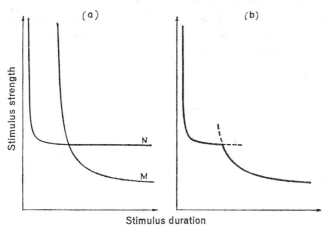

Fig. 7.4 Strength–duration curves. The graphs show how the stimulus strength required to produce a detectable response (threshold strength) varies with the duration of the stimulus: (a) for stimulation of motor nerve fibres (N) and muscle fibres (M), and (b) for stimulation of a whole muscle containing muscle fibres and intramuscular nerve fibres. The curve shown in (b) is a combination of the two curves shown.

Strength–response curve

When a single muscle fibre is stimulated directly, its strength–response curve is of the form shown in Fig. 7.5(a); the threshold and maximal stimulus strengths are the same, showing that the fibre obeys the all-or-nothing law. When a whole muscle is stimulated directly, the strength–response curve is S-shaped and the threshold and maximal stimulus strengths are different (Fig. 7.5(b)). The curve has this characteristic shape because the muscle contains a population of fibres in which the threshold stimulus strength has an approximately Gaussian distribution. The threshold for the whole muscle corresponds with the threshold of the most sensitive muscle fibres, and the maximal stimulus strength for the whole muscle is sufficient to excite the least sensitive fibres in addition to all the rest. In theory the strength–response curve is not a smooth curve as drawn, but a discontinuous line with as many steps on it as there are muscle fibres of different threshold; however, in a muscle containing thousands of fibres these steps are not detectable. When the muscle is stimulated indirectly, the number of theoretical steps is less because the smallest functional unit that can be recruited by a slight increase in stimulus strength is a nerve fibre, and this will bring another *motor unit* into operation. The number of possible steps is therefore equal to the number of motor units and not to the number of muscle fibres. This is, of course, the situation in the body where activation of the muscles is indirect.

Excitation–contraction coupling

The effect of excitation of a muscle fibre is to activate the ATPase sites in the myofibrils by the inward spread of some excitatory influence when the surface membrane of the fibre is depolarised. This inward spread is much too rapid to be explained by the diffusion of a chemical, and in fact what happens is that the action potential is conducted radially along the tubules of the T-system from the surface membrane to the centre of the muscle fibre (Fig. 7.3(a)). It exerts its influence on the myofibrils by triggering the release of calcium from the terminal cisternae of the sacroplasmic reticulum, where it is stored in high concentration.

Under the ionic conditions existing in the intact muscle fibre, the myosin ATPase sites are activated by actin molecules in the regions of overlap between thick and thin filaments (see Fig. 7.2(b)). In the resting muscle, actin is stopped from doing this by the regulatory protein, *troponin*, which exerts its inhibitory influence on actin via *tropomyosin*, the third protein found in the thin filament. The inhibitory influence of troponin on actin ceases when troponin binds calcium, and the crucial link between excitation and contraction is provided by a rise in the sarcoplasmic Ca^{2+} concentration from its very low value of about 10^{-7}M in the resting muscle to about 10^{-5}M in the fully active muscle. The sarcoplasmic Ca^{2+} concentration is low in the resting muscle because most of the intracellular calcium is stored in the terminal cisternae of the sarcoplasmic reticulum, from which it is released by the spread of the action potential through the T-system. The rise in sarcoplasmic Ca^{2+} concentration activates the myosin ATPase as described above and ATP breakdown continues until the Ca^{2+} concentration is reduced to its low resting level again by active uptake of calcium into the sarcoplasmic reticulum.

CONTRACTION

The essential feature of muscular contraction is that the muscle tends to *shorten*, but if shortening is in any way restrained the muscle will also *develop tension*. Muscles often undergo

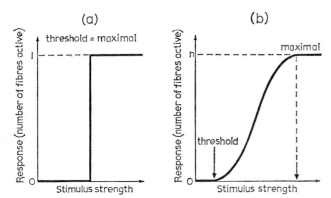

Fig. 7.5 Strength–response curves. (a) This graph shows how the response of a *single muscle fibre* varies with the stimulus strength. The threshold and maximal stimulus strengths are identical, showing that the fibre obeys the "all-or-nothing" law. (b) A similar graph for a *whole muscle* containing *n* fibres.

simultaneous changes in length and tension when they contract naturally in the body, but in experiments on isolated muscle it is customary (in the interests of simplicity) to hold either the length or the tension constant. The conditions are described as *isometric* when changes are measured at constant muscle length, and as *isotonic* when the muscle shortens and the tension in it is constant. When the muscle contracts against a spring it shortens and develops tension at the same time; this is known as an *auxotonic* contraction. As almost all experiments on muscle involve measurements of tension or length changes (sometimes both), the physical principles underlying such measurements will now be outlined.

Measurement of tension

Tension cannot be measured as such; it must first be converted into another physical quantity that can. A device that makes such a conversion is called a *transducer*—in this case a force or tension transducer. The usual procedure for measuring the tension produced by a muscle is to connect it to a strip of spring steel, as illustrated in Fig. 7.6a. Tension in the muscle produces a slight bending of the steel strip, which can be measured either by magnifying the movement with an extension arm that writes on a smoked drum, or by using another transducer (e.g. a strain gauge) to convert the bending of the spring strip into an electrical signal for display on a pen recorder. The transducer must be calibrated by applying known forces to the spring, and the simplest way of doing this is to hang known masses in the place of the muscle; due to the attraction of gravity a mass of M kilograms will produce a force of $M \cdot g$ newtons (N), where g is the acceleration due to gravity ($9 \cdot 8$ ms^{-2}).

Measurement of shortening

The traditional method used for measuring the shortening of the muscle in an isotonic contraction is illustrated in Fig. 7.6b. The free end of the muscle is attached to a simple lever and an extension arm is used to magnify the movement for display on a smoked drum; alternatively, a transducer (e.g. photoelectric device) can be used to convert the movement into an electrical signal for display on a pen recorder. The system must be calibrated by introducing known movements at the point of attachment of the muscle to the lever.

In an isotonic contraction the object is to measure the shortening of the muscle while the tension in it remains constant; such conditions are difficult to achieve in practice because there is no simple way of loading the muscle with a pure force. The method most commonly used is to hang a mass (M kg) on the lever, as shown in Fig. 7.6b. If the mass and the muscle are attached to the pivot at points equidistant from the pivot, then the *load* on the muscle (= *tension* in the muscle = *force* opposing shortening) will by $M \cdot g$ N. Because of the inertia of the mass, the load on the muscle will be constant only when the lever is either stationary or moving at a constant speed. If the speed is increasing (i.e. lever accelerating) or decreasing (i.e. lever decelerating), the force on the muscle will be $M \cdot g \pm M \cdot a$, where a is the acceleration or deceleration of the lever. The inertial component, $M \cdot a$, can be greatly reduced by hanging the mass close to the pivot on a compliant support.

When a muscle shortens by amount x m against a force of $M \cdot g$ N, the *work done* is $M \cdot g \cdot x$ Joules (J). Similarly, when the muscle shortens at a velocity v ms^{-1}, its rate of working, or *power output*, is $M \cdot g \cdot v$ Watts (W) (746 W = 1 horse power (hp)).

Figure 7.6(c) is a schematic diagram of an arrangement for measuring tension and length simultaneously.

The twitch

This is the name given to the mechanical response produced when a single suprathreshold stimulus is applied to a muscle directly or indirectly. Figure 7.7(a) shows the main features of a

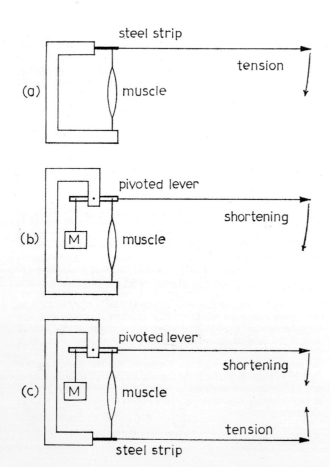

Fig. 7.6 Schematic diagrams of the apparatus used for measuring (a) tension developed in an isometric contraction, (b) shortening of the muscle in an isotonic contraction, and (c) tension and shortening in an isotonic contraction.

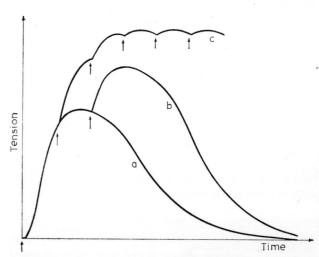

Fig. 7.7 Isometric myograms showing (a) the twitch response produced by a single stimulus, (b) summation produced by applying a second stimulus before the mechanical response produced by the first has disappeared, and (c) the tetanic contraction produced by a series of stimuli. The arrows show the timing of the stimuli in all cases.

typical isometric twitch. There is a *latent period* between the application of the stimulus and the first detectable mechanical response. Its duration depends on the method of stimulation. The minimum value (about 1 per cent of the twitch duration) is achieved by stimulating the muscle directly at several points along its length; this is the time required for the process known as *excitation–contraction coupling*. The latent period is prolonged when stimulation is indirect because of the additional delays introduced by nerve conduction and neuromuscular transmission.

The mechanical response consists of a *phase of contraction*, during which the tension rises to a peak, and a *phase of relaxation*, during which the tension declines to its initial value. The contraction phase occupies about one-third of the total duration of the twitch. Mammalian skeletal muscles vary in their twitch durations; for example, in a cat it is 0·2–0·3 s for the soleus muscle and about half this value for the gastrocnemius. It is important to note that the duration of the action potential (0·002–0·003 s in mammalian muscle) is very short compared with the duration of the mechanical response.

Note that in some animals (e.g. amphibia, birds), the skeletal muscle fibres are of two distinct types, known as "fast" (or "twitch") and "slow" (or "tonic"). These differ not only in speed of contraction, but also in their structure, innervation and electrical properties. The external ocular muscles of mammals contain fibres that closely resemble "slow" amphibian muscle fibres, and the intrafusal fibres found in muscle spindles have some features in common with them. Otherwise the fibres that make up all mammalian skeletal muscles appear to be of the "fast" amphibian type.

Summation

The surface membrane of the muscle fibre is refractory for a short time following the passage of an action potential, but the total duration of the action potential and refractory period is very short compared with the duration of the mechanical response. It is therefore possible to excite the muscle for a second time before the mechanical response produced by the first stimulus has disappeared. If this is done a second mechanical response is added to the first in a process known as *summation* (Fig. 7.7b). The size of the combined response will depend on the timing of the second stimulus.

Tetanus

The summation mechanism can lead to an even greater mechanical response if several stimuli are given instead of just two (Fig. 7.7c). The type of mechanical response produced by repetitive stimulation is called a tetanus. This is said to be a *fused tetanus* when the stimulus frequency is increased to the point where there is no detectable ripple in the mechanical response; otherwise the contraction is called an *unfused tetanus*. The fusion frequency for mammalian muscles at body temperature varies from about 50 s^{-1} for the slowest to about 500 s^{-1} for the fastest muscles.

MECHANICAL PROPERTIES

Mechanical "model" of muscle

From the mechanical point of view we can represent a muscle as a system with three components (Fig. 7.8): a *series elastic component*, a *parallel elastic component* and a *contractile component*. These can be tentatively identified with structures in the actual muscle, as indicated in the caption. This identification is an oversimplification, but the "model" shown is very useful in understanding the mechanical behaviour of a muscle at rest and during activity.

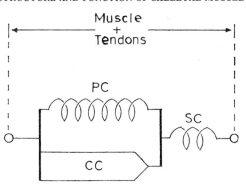

Fig. 7.8 A mechanical "model" of muscle. The mechanical components shown may be tentatively identified as follows: *Series elastic component* (SC) = tendons of muscle; *parallel elastic component* (PC) = connective tissue sheaths, sarcolemma; *contractile component* (CC) = myofibrils.

Resting muscle

The resting muscle has the mechanical properties of an imperfect elastic body. The effect of stretching it is to produce passive (or resting) tension, but the relation between tension produced and the muscle length is non-linear (Fig. 7.11(a), (c)); that is, the resting muscle does not obey Hooke's law. The resistance to stretch is due to the tendons and the connective tissue sheaths, which behave like two springs connected in series (Fig. 7.8).

In the resting muscle the myofibrils offer very little permanent resistance to stretch because the thick and thin filaments can slide freely past one another. The lack of interaction between actin and myosin under these circumstances is due to the "plasticising action" of ATP when it is present but not being hydrolysed because of the low Ca^{2+} concentration. If all the ATP is hydrolysed for example, by stimulating an IAA-poisoned muscle to exhaustion (see p. 66), a powerful interaction between actin and myosin occurs, giving the myofibrils an almost "crystalline" structure. The result is that they become very inextensible and this state is known as *rigor*. (Rigor mortis is a similar condition that appears in the body within a few hours of death as a result of ATP hydrolysis in the muscles.)

Active muscle

When the muscle is stimulated the properties of the elastic components remain unaltered, but the properties of the contractile component change dramatically. It is transformed from a *passive*, freely extensible element into an *active* element that is capable of converting chemical energy into mechanical work. The transition is extremely rapid, requiring only a few milliseconds in a mammalian muscle at body temperature. In each sarcomere of the myofibril the two sets of filaments undergo a relative sliding movement due to the appearance of shearing forces between them when the muscle is excited, and an increase occurs in the amount of overlap between the thick and thin filaments in the outer parts of the A band (Fig. 7.9(a)). The Z lines are drawn together because the filaments behave as extensible rods, and the sarcomere becomes shorter. The same thing happens in sarcomeres along the entire length of the myofibril. The velocity of shortening of the myofibril depends on the force that opposes shortening, as shown in Fig. 7.9(b).

The shearing forces required to produce the relative sliding movement of the two sets of filaments are thought to depend on changes in actin–myosin interaction, but precisely how the free energy released by ATP breakdown is converted into movement remains obscure. The

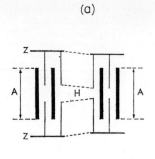
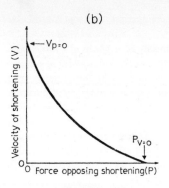

Fig. 7.9 Properties of the contractile component in active muscle.
(a) Diagram showing what happens to the filaments in the myofibril when the muscle is stimulated. A relative sliding movement takes place, increasing the amount of overlap between the two sets of filaments. Note that the length of the A band remains unchanged.
(b) Graph showing how the velocity of shortening of the contractile component varies with the force opposing shortening. The velocity of shortening has its maximum value ($V_p = 0$ or V_{max}) when the force is zero ($p = 0$), and the velocity is zero ($V = 0$) when the force opposing shortening is equal to the tension that can be produced under isometric conditions (p_0 or $P_{V=0}$).

thick filaments have lateral projections that can form cross bridges between the two sets of filaments under some circumstances (e.g. in rigor) and it is thought that similar cross bridges may produce a relative sliding movement of the thin filament with respect to the thick filaments by repeatedly breaking and reattaching at new sites further along the thin filaments.

Isometric contraction

Although the length of the muscle is held constant in an isometric contraction, internal length changes still occur: the muscle fibres shorten because they are in series with extensible tendons (Fig. 7.10), and weak regions of muscle fibres are stretched by the contraction of stronger regions in series with them. The effect of internal length changes is to slow down the rate of rise of tension in an isometric contraction because a muscle fibre that is shortening cannot be developing its maximum tension (see force–velocity relation, Fig. 7.9). The ultimate limit to the rate at which tension rises is set by the time course of activation of the mechanism (e.g. cross bridges) that generates the shearing motion within the sliding filament system.

The tension produced by muscle in an isometric contraction depends on the following factors:

(i) PATTERN OF RESPONSE
The rise of tension in a twitch (Fig. 7.7) is limited by the waning of activity in the contractile system as calcium is pumped back into the sarcoplasmic reticulum (see p. 59), and the tension fails to reach the full tetanic value (P_0). The twitch–tetanus ratio is about 0.5 in mammalian muscle at 37°C, and its value is very dependent on the amount of internal shortening required to develop tension under "isometric" conditions.

(ii) MUSCLE LENGTH
The symbol L_0 will be used in this chapter to denote the *mean body length* of the muscle (i.e. the length midway between the longest and the shortest lengths possible when the muscle is *in situ*). In general, skeletal muscles show no appreciable resting tension at this length, so the tension measured during an isometric contraction at this length (or at shorter lengths) is therefore entirely due to the contractile component. At lengths greater than L_0 the muscle does show resting tension, and the tension measured during a contraction is therefore the sum of the contributions of the contractile component (active tension) and the parallel elastic component (resting tension). Figure 7.11 shows how the total tension and the resting tension vary with the muscle length in muscles with well developed (Fig. 7.11(c)) and poorly developed (Fig. 7.11(a)) connective tissue sheaths. The tension produced by the contractile component can be obtained by subtracting the contribution of the parallel elastic component (resting tension) from the total tension. Figure 7.11(b) and (d) shows that the tension developed by the contractile component varies with the muscle length in exactly the same way in the two types of muscle. The dependence of tension production on the muscle length is related to the amount of overlap between thick and thin filaments in the A band of each sarcomere, as illustrated in Fig. 7.12.

How much the length of a muscle fibre can vary in the body depends on the geometry of the joint (or joints) over which the muscle operates and on the relative lengths of the muscle belly and tendon. However, the behaviour of the muscle at extreme lengths is of purely academic interest because such lengths are unattainable in the body.

(iii) TEMPERATURE
The main effect of temperature is on the time course of the contraction rather than on the tension developed. For example, reducing the temperature of a mammalian muscle from 35° to 25°C has virtually no effect on the tension developed in a twitch or tetanus, but it doubles the duration of the twitch.

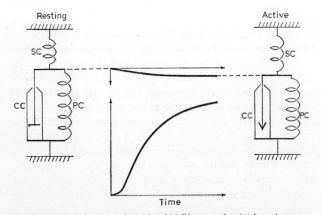

Fig. 7.10 Isometric contraction. (a) and (c) Diagrams showing how the contractile component shortens at the expense of the series elastic component when the muscle contracts under isometric conditions. (b) Graphs showing how the length of the contractile component (upper graph) changes with time as the tension rises (lower graph).

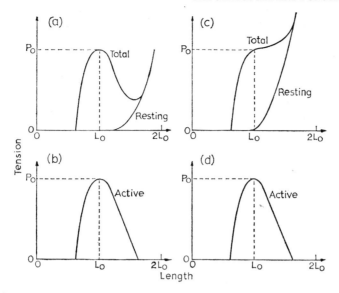

Fig. 7.11 Tension–length curves of resting and active muscles. The upper graphs (a and c) show how the tension in the resting muscle and the total tension in the muscle during an isometric tetanic contraction vary with the muscle length, when the muscle contains little connective tissue (a) or is rich in connective tissue (c). The lower graphs (b and d) show how active tension (i.e. total tension minus resting tension) varies with the muscle length; they are identical for the two types of muscle because the properties of the contractile component are unaffected by the properties of the parallel elastic component. L_0 is the mean body length of the muscles, and p_0 is the tension developed at that length.

(iv) MUSCLE SIZE

The tension developed by a muscle depends on the number of tension-producing elements that are arranged in parallel and not on their length. The tension produced by a muscle will therefore depend on the cross-sectional area of its fibres; mammalian muscles at body temperature produce 0.3–0.35 N mm^{-2} of cross-section. It follows that for muscles of the "strap" type a short fat muscle of a given mass will develop much more tension than a long thin muscle of the same mass. Muscles of the "pennate" variety produce more tension than would be expected from the cross-sectional area of the *muscle*; this is because the total cross-sectional area of the *fibres* (taken at right angles to their long axis) may be much greater than that of the muscle.

Isotonic contraction

The muscle shown in Fig. 7.6(b) and (c) is said to be "preloaded" because the mass that is hung on the lever is supported by the muscle at all times. Preloading has the unfortunate feature that the length of the resting muscle changes when the load is altered. This complication can be avoided by arranging for the load to be supported so that it does not stretch the resting muscle. An adjustable stop is used for this purpose, as shown in Fig. 7.13, and the muscle is said to be "*afterloaded*" because it is not aware of the presence of the load until after the onset of the contraction. The length of the resting muscle can be varied by adjusting the position of the stop, and it simplifies matters if the length is set to L_0 because there is then no resting tension in the muscle.

When the muscle is stimulated (Fig. 7.13) there is the usual latent period (phase A); then the tension in the muscle rises (phase B) with the same time course as in an isometric contraction. However, as soon as the tension in the muscle is equal to the force produced on the lever by the load ($M . g$ N) the muscle begins to shorten (phase C); the tension then remains constant (i.e. *the conditions are isotonic*). If the contraction is a tetanus, as in Fig. 7.13(b), the muscle will continue to shorten until it reaches a limit that depends on the load and on the initial length of the muscle. When stimulation ends the load drops back on to the stop (phase D) and the tension then falls to zero as it does at the end of an isometric contraction (phase E).

The performance of a muscle under isotonic conditions is determined by the following factors:

(i) PATTERN OF RESPONSE

The tension and length changes in an afterloaded isotonic twitch are shown in Fig. 7.13(c). Because the active state begins to disappear very soon after the stimulus, the velocity of shortening (i.e. the slope of the shortening record) is slightly less than it is in a tetanic contraction and the muscle does not have time to shorten as far. In other respects the twitch and tetanic responses are essentially the same.

(ii) LOAD

When the load (i.e. the force opposing shortening) is increased, the isometric phases of the response occupy more and more of the contraction–relaxation cycle; the load is lifted later and dropped earlier and the muscle does not shorten as fast or as far. The relation between the velocity of shortening and the force opposing shortening is the same for the whole muscle (Fig. 7.14(b)) as it is in Fig. 7.9(b). The *work done* (Fig. 7.14(a)) and the *power output* of the muscle are both zero when the force opposing shortening is zero (unloaded shortening) and when the muscle does not shorten (load = P_0 in a tetanus; load = peak tension in a twitch). At intermediate loads the muscle produces power and does work and the maximum performance is obtained when the load is $\frac{1}{3}$ to $\frac{1}{2}$ of P_0 in a tetanus and $\frac{1}{3}$ to $\frac{1}{2}$ of the peak tension in a twitch.

When the load drops during relaxation and the muscle is stretched (as in Fig. 7.13(b) and (c)), all the energy stored during the phase of contraction is dissipated as heat in the muscle. In order to obtain any net work from the muscle over the whole contraction–relaxation cycle, it is necessary to provide some means of supporting the load when it is lifted (e.g. a ratchet mechanism).

(iii) MUSCLE LENGTH

Provided that the muscle is not stretched excessively, the *final length* that it will reach in an isotonic contraction against a given load is in-

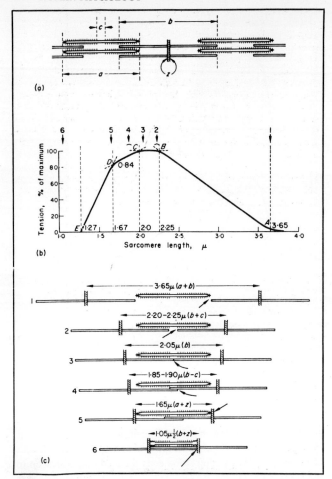

Fig. 7.12 Length–tension relation for single muscle fibres from the frog (courtesy A. M. Gordon, A. F. Huxley and F. J. Julian).

(a) Arrangement of filaments. Length of thick filament, a, = $1.60\,\mu$m; length of thin filament, b, = $2.05\,\mu$m; width of projection-free zone, c, = 0.15–$0.20\,\mu$m; width of Z line, z, = $0.05\,\mu$m. (b) Tension developed at different sarcomere lengths in isometric tetani at 0°C. Numbers along the top indicate critical stages of overlap illustrated in (c).

dependent of its *initial length*. This means that if the length of the resting muscle is increased by altering the position of the afterload stop, the muscle will shorten more when it is stimulated and therefore do more work. This is a fundamental property of muscle and it is of particular importance in cardiac muscle where it provides the basis of Starling's Law of the Heart ("the energy of contraction is a function of the length of the fibres", see Chapters 8 and 39).

(iv) TEMPERATURE

As one might predict from the effects of temperature on the time course of the isometric twitch, changes of temperature also have a marked effect on the velocity of shortening. Lowering the temperature of mammalian muscle from 35° to 25°C reduces the maximum speed of shortening (V_{max}) to about half, but it has no significant effect on the amount of shortening possible.

(v) MUSCLE DIMENSIONS

A skeletal muscle fibre can shorten down to about 60 per cent of its body length in a tetanic contraction against zero load, but the velocity at which it does so (V_{max}) depends on the temperature and the species of animal; for example, V_{max} for human muscle at body temperature is about six muscle lengths per second. Long thin muscles shorten fastest but short fat muscles develop the greatest tensions. The maximum power output seems to be more or less independent of the shape of the muscle; it depends simply on the mass of tissue.

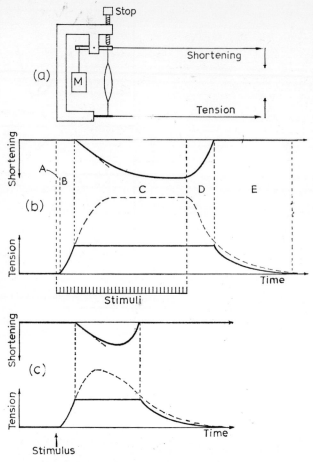

Fig. 7.13 Afterloaded isotonic contractions.

(a) Schematic diagram of the apparatus used. The force opposing shortening is $M \cdot g$ newtons, but the tension in the resting muscle is zero because the load is supported by the afterloaded stop. (b) Tension and length changes in the muscle during an afterloaded isotonic tetanic contraction. There is a latent period (A) and then the tension begins to rise. During this phase (B) the conditions are isometric, but when the tension in the muscle reaches $M \cdot g$ newtons, the muscle shortens (phase C) and the tension then remains constant (i.e. the conditions are isotonic). When stimulation ends, the muscle cannot maintain its shortened state; the load drops (phase D) and the muscle is stretched. When the lever hits the stop, the conditions become isometric once again (phase E) and the tension then falls gradually to zero. The dashed line shows the tension changes that occur when conditions remain isometric throughout the tetanic response. (c) Similar curves showing the tension and length changes in an afterloaded isotonic twitch.

BIOCHEMISTRY

In this section we shall be concerned with the chemical changes associated with muscular activity. The point has already been made (p. 58) that the hydrolysis of ATP is the immediate source of energy for muscular contraction. Early exhaustion of the meagre stores of ATP in the muscle is avoided because the ADP produced by the hydrolysis of ATP is rapidly rephosphorylated by coupled chemical reactions.

Rephosphorylation of ADP

(i) BY THE CREATINE PHOSPHOTRANSFERASE (CPT) REACTION
Although a muscle contains sufficient ATP (2–4 μmol g^{-1}) for only a few contractions, it is impossible to detect any change in the ATP concentration in a normal muscle unless it is stimulated to the point of exhaustion. The reason for this is that the ADP formed is

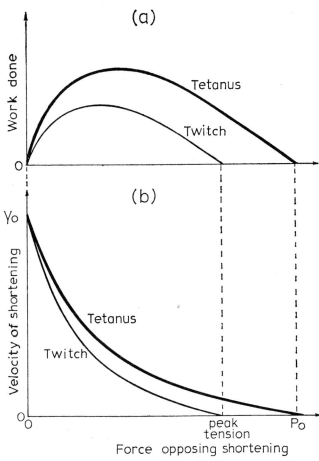

Fig. 7.14 Performance of the muscle under isotonic conditions. Graphs showing how (a) the work done by the muscle and (b) the velocity of shortening of the muscle vary with the force opposing shortening in isotonic twitches and tetani.

rephosphorylated extremely rapidly by the breakdown of phosphorylcreatine (PC) to creatine (C) in a coupled reaction:

$$\begin{array}{ccc} PC & & ATP \\ \downarrow & \text{---(P + free energy)---} & \uparrow \\ C & & ADP \end{array}$$

The substances ATP, ADP, PC and C can be regarded as the reactants in a reversible chemical reaction known as the *CPT reaction*:

$$\frac{ADP}{(0.03)} + \frac{PC}{(22)} = \frac{ATP}{(3)} + \frac{C}{(4)}$$

The numbers in brackets give the concentrations (μmol g^{-1}) of the substances in a resting frog muscle at 0°C. This equilibrium is disturbed when the muscle contracts because the breakdown of ATP causes a reduction in ATP concentration and a rise in ADP concentration. The CPT reaction will proceed from left to right until equilibrium is re-established (Fig. 7.15a) and this happens extremely quickly because the enzyme that catalyses the reaction (creatine phosphotransferase) is very active. The importance of this reaction can be demonstrated by stimulating a muscle that has been poisoned so that PC breakdown provides the only means of resynthesising ATP (see p. 66). The concentration of PC falls as it is used to rephosphorylate the ADP formed during contraction but because of the high equilibrium constant of the CPT reaction ($\sim$20) the ATP concentration changes very little until the store of PC is all but exhausted (Fig. 7.15b). A frog muscle at 0°C can produce about 100 twitches under these circumstances.

(ii) BY OXIDATIVE PROCESSES
When oxygen is available two-carbon (C_2) fragments from any source can be oxidised to carbon dioxide and water via the tricarboxylic acid cycle and the cytochrome chain. The oxidation of each C_2 fragment is coupled to the rephosphorylation of 15 molecules of ADP. In skeletal muscle carbohydrate metabolism is the most important source of C_2 fragments. Each hexose unit is metabolised in two stages: first, it is broken down into two molecules of pyruvic acid via the Embden–Meyerhof pathway (the direct oxidative pathway appears to be unimportant in muscle) with the rephosphorylation of nine molecules of ADP; the two molecules of pyruvic acid are then decarboxylated to give two C_2 fragments, which are oxidised with the rephosphorylation of a further 30 molecules of ADP. Carbohydrate metabolism is accelerated by a rise in ADP concentration in the muscle, but the maximum rate of oxidative rephosphorylation of ADP appears to be limited in practice by the rate at which oxygen can be supplied to the muscle. Because of the very high yield of rephosphorylations per hexose unit (39), the glycogen store in the muscle can provide sufficient energy for many thousands of twitches under aerobic conditions.

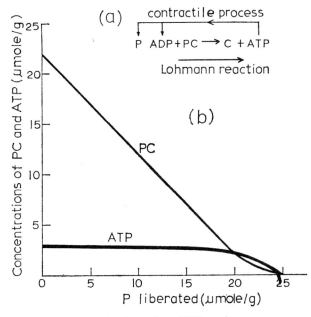

Fig. 7.15 The creatine phosphotransferase (CPT) reaction.
(a) Diagram showing how the ADP produced by the hydrolysis of ATP in the contractile process is rephosphorylated by the CPT reaction. (b) Graphs showing how the PC and ATP concentrations in the iodoacetate-poisoned muscle change as more and more P is formed due to the hydrolysis of ATP. Note how the ATP concentration is maintained at the expense of the PC concentration (Courtesy of D. R. Wilkie.)

(iii) GLYCOLYSIS (ANAEROBIC BREAKDOWN OF GLYCOGEN)
The splitting of glycogen into hexose units and the formation of pyruvic acid are anaerobic processes but the coenzyme NAD is reduced to NADH by one reaction in the Embden–Meyerhof pathway. When oxygen is available the NADH is reoxidised via the cytochrome system with the rephosphorylation of three molecules of ADP per molecule of NADH (i.e. six molecules of ADP rephosphorylated per hexose unit). In the absence of oxygen the NADH is reoxidised by the reduction of pyruvic acid to lactic acid, but this reaction is not accompanied by any rephosphorylation of ADP. Thus when the conditions are anaerobic, only three molecules of ADP are rephosphorylated per hexose unit. This is only one-third of the yield under aerobic conditions, so the effective energy store represented by the glycogen content is considerably diminished if the muscle is deprived of oxygen; in an isolated frog muscle at 0°C it is sufficient to permit about 1600 twitches provided that the lactic acid produced can escape from the muscle into an adequate volume of bathing solution. If the lactate cannot escape (e.g. if the muscle is stimulated in an atmosphere of nitrogen), then the muscle ceases to respond to stimulation much sooner (after about 300 twitches) because the accumulation of lactic acid in the muscle reduces the pH to the point where contraction is no longer possible.

Use of poisons

If the muscle is kept in *oxygen-free nitrogen* the glycogen breakdown under anaerobic conditions can be determined from the change in either glycogen or lactic acid concentrations. If, in addition, the muscle is treated with *iodoacetate* (IAA), glycogen breakdown is impossible because the IAA blocks the enzyme triophosphate dehydrogenase in the Embden–Meyerhof pathway and because the direct oxidative pathway cannot be used in the absence of oxygen. In these circumstances the CPT reaction provides the only means of resynthesising ATP and the only net chemical change that takes place during contraction is PC breakdown. If the CPT reaction is prevented by fluorodinitrobenzene (FDNB), which poisons the coupling enzyme (creatine phosphotransferase), then ATP breakdown is the only net chemical change that takes place during a brief contraction.

Chemical changes during and after a single twitch

ATP is hydrolysed during the contraction and the ADP formed is rephosphorylated very rapidly by the CPT reaction, which proceeds from left to right (Fig. 7.16(a)); these events are often

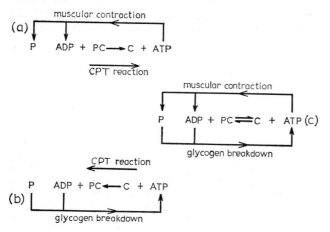

Fig. 7.16 ATP breakdown and resynthesis during contraction and recovery. (a) Chemical changes taking place during a twitch ("initial processes"). (b) Chemical changes taking place during recovery ("recovery processes"). (c) Steady state established during prolonged activity, in which the rate of ATP breakdown is matched by the rate of ATP resynthesis due to glycogen breakdown.

referred to as the "initial" processes. It is important to realise that although the CPT reaction is extremely effective in maintaining the ATP level at the expense of PC, small changes in the ATP and ADP concentrations must occur. The rise in ADP concentration has an accelerating effect on oxidative phosphorylation and even on the earliest stages of glycogen breakdown, but the reactions involved are sluggish compared to the CPT reaction, and they only account for a small fraction of the ATP resynthesis that takes place during a twitch. At the end of a twitch the C, P and ADP concentrations are above normal and the PC and ATP concentrations are below normal, where "normal" refers to the values observed when the resting muscle is in a steady state (i.e. after a long period of rest). The normal concentrations are subsequently restored by glycogen breakdown, which occurs at an elevated rate in the period following a twitch because of the raised ADP concentration. The glycogen breakdown may be aerobic or anaerobic depending on the availiability of oxygen. The ATP that is produced by the rephosphorylation of ADP is used to rephosphorylate creatine in the CPT reaction, which now proceeds from right to left (Fig. 7.16(b)). The rate of glycogen breakdown gradually falls as the concentrations of the reactants in the CPT reaction (in particular ADP) approach their normal values. Equilibrium is re-established eventually, though this takes at least 30 min following a single twitch of a frog muscle at 0°C. The events illustrated in Fig. 7.16(b) are often referred to as the "recovery processes".

Chemical changes before and after a tetanic contraction

In a twitch there is very little overlap in time between the initial and recovery processes. This is not so in a tetanic contraction where glycogen breakdown makes an important contribution to the ATP resynthesis that takes place during the contraction. In fact if stimulation is continued for long enough, glycogen breakdown will take over this function completely. This is almost certainly what happens during the sustained low grade contractions produced by postural muscles in the body. Initially the CPT reaction is of paramount importance, as it is in a twitch (Fig. 7.16(a)) but as the ADP concentration in the muscle gradually rises, glycogen breakdown is accelerated more and more. Eventually the point is reached where the rate of ATP resynthesis by glycogen breakdown is the same as the rate of ATP breakdown in the contractile process; ATP resynthesis by the CPT reaction then ceases. This state of affairs is illustrated by Fig. 7.16c. The CPT reaction is in equilibrium and any tendency for the ATP concentration to change due to fluctuations in either its rate of breakdown or its rate of resynthesis will be prevented by rapid displacement of the CPT reaction in the appropriate direction. When the contraction ends ATP breakdown by the contractile process ceases, and the ATP and PC concentrations are then restored to normal by recovery processes, as they are in a twitch (Fig. 7.16(b)).

When the rates of ATP breakdown and resynthesis are equal, as in Fig. 7.16(c), the active muscle is in a steady state. The rate of glycogen breakdown that is required to support a given rate of ATP breakdown depends on the availability of oxygen. Under aerobic conditions it is only one-third of the rate under anaerobic conditions. However, the maximum rate of glycogen breakdown under aerobic conditions is limited by the rate at which oxygen can be supplied to the muscle, and if higher rates of ATP resynthesis are required the muscle has to fall back on glycolysis (which is a self-limiting process).

"Red" and "white" muscles

Mammalian muscle fibres have a reddish colour because they contain cytochromes and myoglobin. The fibres can be divided into two main types (red and white) depending on the intensity of their red coloration. All mammalian muscles contain a mixed population of fibres, with red fibres predominating in "red" muscles and white fibres predominating in "white" muscles. Red fibres are capable of much higher rates of oxidative metabolism than white fibres because they contain more cytochrome and myoglobin. The cytochromes are located in the mitochondria and they play a key role in the oxidative rephosphorylation of ADP. The myoglobin is present in the sarcoplasm and its most important property is that it greatly facilitates the diffusion of oxygen from the cell membrane to the mitochondria. (Myoglobin resembles haemoglobin in that it combines with oxygen, but it is not considered to provide a particularly useful oxygen store within the muscle fibre because myoglobin is present in low concentration and it does not give up its oxygen until the partial pressure has fallen to a very low value.)

Red muscles are adapted for continuous activity. As well as a well-developed intracellular oxygen transport system, there is a high capillary density around each fibre; together these provide a good enough supply of oxygen to support aerobic metabolism during continuous muscular activity. White muscles on the other hand do not have well-developed oxygen transport systems and there is a much lower capillary density around each muscle fibre. These muscles can produce high power outputs in phasic contractions, but they are unable to sustain activity for long periods of time.

It used to be thought that red muscles contract and relax more slowly than white muscles, but this is often not the case. Red muscles may be slow contracting (e.g. postural muscles such as the soleus) or fast contracting (e.g. the diaphragm or extraocular muscles); but what they do have in common is that they are active for long periods of time. All mammalian white muscles seem to be fast contracting and the functions they serve in the body require activity for short periods only.

ENERGETICS

All spontaneous chemical reactions release "free energy"; that is, energy that can be converted into work of one sort or another if the chemical reaction that releases it is coupled to a suitable "transformer". Muscles contain transformers of various types: for example, the contractile mechanism can convert free energy into *mechanical work*; active transport mechanisms convert free energy into *osmotic work* or *electrical work*; and coupling enzymes that allow an energy-yielding chemical reaction to drive another chemical reaction in the "backward" direction (i.e. the direction in which it would not proceed spontaneously) convert free energy into *chemical "work"*. Even more exotic transformations are possible in some situations; for example, free energy is converted into light in firefly tails. The coupling between a chemical reaction that releases energy and the mechanism that utilizes it is never perfect, and the *thermodynamic efficiency* of the coupling is defined as the fraction of the free energy released that is converted into work. The free energy that is wasted is degraded into heat immediately, but it is important to remember that the free energy that is successfully converted into work may also be dissipated as heat subsequently (e.g. when an ion that has been transported against a potential gradient diffuses down that gradient again).

Note that although any form of energy may be converted into heat, the reverse energy transformation is impossible in muscle, or in any other system where the temperature is essentially uniform. It can happen in a heat engine only because of the presence of large temperature gradients.

How much free energy is released when a muscle contracts?

When a muscle is poisoned with FDNB so that ATP breakdown is the only chemical reaction that takes place during contraction, all the free energy that is released is either converted into work or degraded into heat. It might seem at first sight that the free energy released could be measured as the sum of the work done and the heat produced by the muscle. Unfortunately it is not as simple as that because the breakdown of ATP is almost certainly associated with an *entropy change*, and this will entail a movement of heat between the muscle and its surroundings that is quite independent of the heat efflux due to the degradation of free energy. Depending on the magnitude and direction of this entropic heat movement, the heat produced by the muscle will be either greater or less than the heat efflux due to the degradation of free energy. As nothing is known about the entropy change that takes place when ATP breakdown actually occurs in the muscle, it is not possible to determine the free energy released during a contraction. However, the sum of the energy appearing as heat and work, which is known as the *enthalpy change*, is worth measuring when ATP breakdown or PC breakdown is the only net chemical change occurring during contraction. The enthalpy change per mole of chemical breakdown ($-\Delta H$) is a constant for a given reaction under given conditions (about $40\,\text{kJ mol}^{-1}$ for ATP breakdown and for PC breakdown *in vivo*), and this figure can be used to calcuate the amount of chemical breakdown (n mol) from the enthalpy change, which must be equal to $n(-\Delta H)\,\text{kJ}$. Furthermore, the enthalpy change per mole of reaction ($-\Delta H$) is probably proportional to the free energy released per mole of reaction ($<\Delta F$) under all circumstances and the proportionality constant ($Y = \Delta F/\Delta H$) may be known one day.

The sum of the energy appearing as heat and work (i.e. the enthalpy change) depends on the following factors:

(i) THE MECHANICAL CONDITIONS
When a muscle is stimulated the rate of release of energy rises rapidly to its maximum value before there is any appreciable mechanical activity. Subsequently the rate of release of energy at each instant depends on the mechanical conditions. The effect of the stimulus is thus not simply to trigger off the release of a fixed amount of energy; it is to switch on energy-yielding chemical reactions, and the amount of energy released depends on the length and tension changes undergone by the muscle during the contraction–relaxation cycle. For example, in twitches of a frog muscle at 0°C the total energy released is greater during an isotonic contraction than it is during an isometric contraction.

(ii) PATTERN OF RESPONSE
When the muscle is kept in the active state by repetitive stimulation, the high rate of energy liberation that is reached soon after a single stimulus may or may not be maintained depending on the type of muscle. As no work is done during the plateau of a tetanic contraction, all the energy that is released during this phase appears as heat (the "maintenance heat").

(iii) THE INITIAL MUSCLE LENGTH
The energy liberation in *isometric contractions* at muscle lengths above L_0 decreases in exactly the same way as the tension developed (see Fig. 7.12). The explanation is thought to be that both the tension developed and the ATP breakdown depend on the number of sites of interaction between actin and myosin, and at lengths above L_0 the amount of overlap between the thick and thin filaments (and therefore the number of sites of interaction per sarcomere) decreases as the muscle is stretched. In contractions at lengths below L_0 the energy liberated does not fall off in proportion to the tension, but it is not clear how the actin-myosin interaction is altered when the H zone disappears and double overlap of thin filaments occurs in the centre of the A band. In *afterloaded isotonic contractions* the amount of shortening and the work done against a given load increase when the initial length of the muscle is increased; so does the energy released during the contraction.

(iv) TEMPERATURE
When the temperature of a frog muscle is raised, the energy liberated in a twitch decreases slightly, but the rate of energy liberation increases because the duration of the twitch is reduced by an increase in temperature. The increased rate of ATP breakdown is associated with an increase in the velocity of shortening of the contractile component when the temperature is raised.

What is the efficiency of the muscle as a machine?

This is a complex question because the free energy is released by several different chemical reactions during contraction and recovery, and it is used for a variety of purposes. The efficiency of the coupling between a chemical reaction that provides free energy and a reaction or process that uses it can be defined in very general terms as the fraction of the free energy released that is put to good use. During contraction and recovery there are four such couplings: (i) the contractile process, (ii) resynthesis of ATP by the CPT reaction, (iii) resynthesis of ATP by glycogen breakdown, and (iv) resynthesis of PC by the CPT reaction. It would be of great interest to know the efficiency of each of these couplings, but in fact the only thing that we can determine is their *overall efficiency* because it is only when the initial and recovery processes are considered together that the free energy released can be determined. Under these circumstances the free energy released is thought to be approximately the same as the total enthalpy change (i.e. work done + heat produced during contraction + heat produced during recovery).

The entropy change that occurs when a reversible chemical reaction takes place in the forward (spontaneous) direction is matched by an equal and opposite entropy change when the reaction is driven in the reverse direction. Any uncertainties about entropy changes related to the breakdown and resynthesis of ATP and PC cease to matter when initial and recovery processes are considered together *because the ATP and PC are completely resynthesised*. The only net reaction that occurs, glycogen breakdown, is not associated with significant changes when it takes place under aerobic conditions. Thus the total enthalpy change during contraction and recovery should be equal to the total free energy released.

The overall efficiency can therefore be calculated as follows:

$$\text{Overall efficiency} = \frac{\text{work}}{\text{free energy released}} \approx \frac{\text{work}}{\text{work} + \text{total heat}}$$

For frog muscle at 0°C, producing its maximum work output, the overall efficiency is 0·2–0·25 (or 20–25 per cent), which compares rather unfavourably with other machines that utilise free energy released by chemical breakdown (e.g. diesel engine 40 per cent, fuel cell 50 per cent, accumulator 80–90 per cent). Unfortunately we do not know whether the main source of inefficiency in the muscle lies in the initial processes or in the recovery processes. It is possible that the efficiency of the contractile process alone compares quite favourably with that of the other machines quoted.

ENERGETICS OF THE RESTING MUSCLE

A continual expenditure of energy is required to keep the resting muscle in a state of readiness for activity. Some of this energy is used by active transport mechanisms to maintain electrical potential gradients and osmotic gradients that are always tending to collapse due to passive diffusion. Energy is also required for various anabolic processes. Although ATP is often involved as an intermediary, the ultimate source of the energy used by the resting muscle is oxidative metabolism. All the free energy released appears as heat sooner or later; some of it is degraded into heat as a result of inefficiency in couplings, but the free energy that is put to good use is also dissipated as heat eventually; for example, when ATP that has been formed by oxidative rephosphorylation of ADP is hydrolysed, and when substances that have been actively transported against a concentration gradient diffuse down that gradient again. This means that the rate of release of free energy in the resting muscle can be measured as the rate of heat production because the only net reaction taking place is oxidative metabolism and this is not associated with any significant entropy changes.

Technically it is difficult to measure the rate of heat production of a resting muscle because the rate is so low. It is easier to measure its oxygen consumption and then to calculate the corresponding rate of heat production from the known calorific value of oxygen. The rate of energy liberation required to maintain the *status quo* in resting muscle depends critically on the temperature (it is increased by a factor of 2·5 for a 10°C rise).

MUSCULAR EXERCISE

Muscular exercise can vary enormously in character; at one extreme we have the spectacular performances of the playing field and gymnasium, and at the other we have the unimposing activities that occupy most of our waking hours. In all cases it is a complicated affair requiring the participation of almost all the systems in the body in one way or another. In this chapter, only the aspects of the subject that concern the muscles directly will be considered.

When a muscle contracts in the body, for example in producing a limb movement, it is never the only muscle active. A coordinated limb movement requires carefully balanced contributions from prive movers, synergists and antagonists. In a given muscle the number of motor units active and the pattern of excitation in each will vary throughout the course of a single movement. It is doubtful whether full activation of a muscle (i.e. tetanic contraction of all the muscle fibres) ever occurs in the body, even in the most violent limb movements. All the muscle fibres that make up a motor unit are of course excited every time that the motor unit fires; but the maximum rate of firing is far below the fusion frequency so the best that they can produce is an unfused tetanic contraction. It appears that the maximum performance that a muscle can be made to produce when isolated from the body (e.g. maximum tension developed, or maximum power output, or maximum work output) never constitutes a limit to its performance *in situ*.

Maximum power output during muscular exercise

All forms of muscular exercise make use of the body as a source of mechanical power. The maximum power that can be produced depends on the time for which it must be maintained, as illustrated by Fig. 7.17(a), in which each circle shows the power produced by a champion cyclist during exercise of the duration indicated. The line shows the maximum mechanical power available from oxidative processes as calculated from the measured oxygen consumption of champion athletes, the known calorific value of oxygen and the estimated overall efficiency of human muscle (20–25 per cent). The observed power output can exceed that available from oxidative processes because a limited amount of additional mechanical work can be obtained from anaerobic sources (phosphorylcreatine breakdown and glycolysis). In a champion athlete this *anerobic reserve* amounts to about 450 W min^{-1} and it can be used at whatever rate is demanded by the exercise. It is of particular importance in exercise of *intermediate duration* (up to 5 min) because oxidative processes are slow to get under way; the anaerobic reserve is used to boost the power output above the level that can be supported by oxidative processes. This is illustrated by Fig. 7.17(b), in which the shaded area shows the estimated contribution of

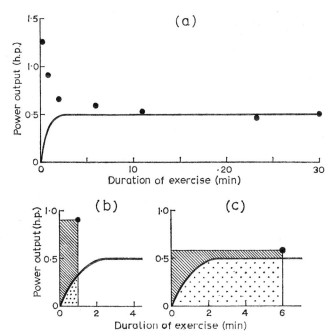

Fig. 7.17 Human power output. The points on these graphs show the power output maintained by champion cyclists during periods of exercise of the duration indicated. (Courtesy of D. R. Wilkie.) Note: 1 horse power (hp) ≈ 746 W. The curve shows the power output that can be obtained by oxidative processes; when the observed power output exceeds this, the extra power is provided by anaerobic hydrolytic processes. In graphs (b) and (c), the rectangles show the work done by the cyclists during exercises lasting 1 min (b) and 6 min (c). The cross-hatched areas show the contribution of hydrolytic processes and the stippled areas show the contribution of oxidative processes.

the anaerobic reserve to the work done by the muscles of a champion cyclist during a sprint lasting 1 min. When the exercise is of slightly longer duration (Fig. 7.17(c)) the relative contribution of the anaerobic reserve to the total work done is less because the oxidative processes have time to get under way. When the anaerobic reserve is utilised it creates what is known as an *oxygen debt*; this must later be repaid because the lactic acid produced as a result of glycolysis must be reconverted to pyruvic acid and then either oxidised (in skeletal muscle, cardiac muscle or liver) or used in the resynthesis of glycogen (liver only). The phosphorylcreatine store also has to be replenished by oxidative processes. The oxygen consumption therefore remains high for some time after exercise ends, and the extra oxygen is used to pay off the oxygen debt incurred during activity.

In exercise of *long duration* (more than 5 min) almost all the energy comes from oxidative processes and a steady state is established, as illustrated in Fig. 7.15(c). The maximum continuous power output is then limited by the rate at which oxygen can be taken up in the lungs and distributed to the active muscles through the circulation; but the anerobic reserve can still be put to good use at any time if a higher power output is required than oxidative processes permit (e.g. in a sprint at the end of a long distance race). Champion athletes can produce a steady power output of about 350 W for long periods and ordinary healthy individuals can manage 150–200 W.

In exercise of *very short duration* (1 s or less) the maximum power output depends on the mass of muscle that can be brought into use, the mechanical properties of the muscles (in particular, their force-velocity curves) and the matching of the load to the muscles through the skeleton. Weight lifting is a good example of exercise of very short duration; it involves a single convulsive movement of the limbs in which a power output of up to about 1500 W may be achieved. This power output is a good deal less than the muscles could produce if optimally loaded because the velocity of movement is very low.

Effect of training

It is uncommon to find athletes who are exceptionally good at exercises of very different durations; thus a weight-lifter rarely makes a good long distance runner and vice versa. The reason is that by training, athletes specialise their bodies for different purposes and weight lifting and long-distance running provide extreme examples of this specialisation. Weight lifting requires afterloaded isotonic contractions of the muscles involved (see Fig. 7.13), and in order to raise a heavy load off the ground very large tensions must be produced. The muscles of a weight-lifter adapt to the requirement for large tensions by increasing their cross-sectional areas enormously. The muscle fibres increase in size (hypertrophy) but not in number. The ratio of myofibrillar protein to sarcoplasmic protein increases because a greater proportion of the cross-section of the fibres is occupied by myofibrils, which increase both in number and size. Similar changes can be induced in selected groups of muscles by means of "isometric" training programmes.

Exercises in which sustained power outputs are required (e.g. long-distance running) lead to a different type of specialisation in the body. The muscles do not increase particularly in bulk but their capacity for oxidative metabolism is greatly enhanced by increases in the number of mitochondria and in the myoglobin content. Concomitant increases occur in the vascularisation of the muscles and in the cardiovascular reserve (i.e. the ability of the heart and circulation to deliver oxygen to the tissues). These adaptive changes allow the muscles to produce greater sustained power outputs than would otherwise be possible and to recover more rapidly after exercise.

8. The Structure and Function of Cardiac Muscle

B. R. Jewell

This chapter is concerned with the properties of cardiac muscle as a tissue; its role as a pump is dealt with in the chapters on the Cardiovascular System. The approach used here is to compare the properties of cardiac muscle with those of skeletal muscle. The former is the prototype of single unit muscles whereas the latter is of multiunit muscles, and some of the differences between them (see Table 8.1 A), reveal the main points of contrast between single unit and multiunit properties. There are other differences that are more specific (Table 8.1 B).

Most of the experimental work on cardiac muscle as a tissue has been done on papillary muscles, which are thin bands of muscle that join the chordae tendinae of the atrioventricular valve cusps to the inner walls of the ventricle. The properties of cardiac muscle described in this chapter are, in the main, those

Table 8.1 *Summary of principal differences between skeletal and cardiac muscle*

	Skeletal	Cardiac
A. *General difference*	*Multiunit muscle*	*Single unit muscle*
Spontaneous activity	None (all activity is neurogenic)	Rhythmic contractions (all activity is myogenic)
Behaviour of muscle as a whole	As collection of independent units, each of which obeys the "all or nothing" law	As single unit obeying the "all or nothing" law. (Muscle is a functional syncytium)
Cell to cell coupling	None	Mechanical and electrical (via intercalated discs)
Function of nerve supply	To initiate contractions (all activity is neurogenic). Excitatory supply only	To modulate myogenic activity (both rate and strength of rhythmic contractions). Excitatory and inhibitory supplies from autonomic nervous system
B. *Particular differences*	*Skeletal*	*Cardiac*
Relation of action potential to mechanical response	Duration of AP is about 1/100 that of MR. Refractory period of fibres is correspondingly short	Duration of AP (and therefore of refractory period) is about the same as that of MR
Role of extracellular Ca in excitation—contraction coupling	None	Important source of Ca for internal store from which Ca is released by action potential
Effect of repetitive stimulation	Summation and tetanus	No appreciable summation (and therefore no tetanus) occurs. Muscle does, however, show interval-strength relation when rate of stimulation varied
Contractility	Not adjustable except through repetitive stimulation	Depends upon extracellular Ca concentration, interval between beats, activity of autonomic nerve supply, effects of drugs
Means available to nervous system for grading contractions	(i) Recruitment of units (ii) Alter rate of firing of each unit (summation and tetanus)	(i) Adjustment of myocardial contractility (ii) Alter heart rate (interval–strength relation)

THE STRUCTURE AND FUNCTION OF CARDIAC MUSCLE 71

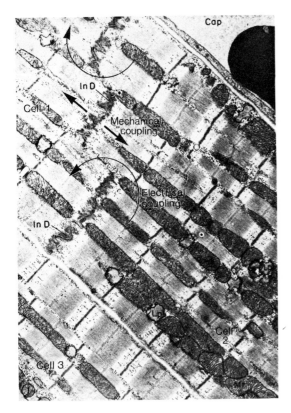

Fig. 8.1 Electron micrograph (Courtesy D. W. Fawcett and N. S. McNutt) of parts of three cardiac muscle fibres and an adjacent capillary (Cap), together with an explanatory diagram. The two upper cells are joined end to end by a typical step-like intercalated disc, which provides mechanical and electrical coupling between the two cells as indicated. Rows of mitochondria, interspersed with lipid droplets (Lp), divide the contractile material into myofibril-like units. The striation pattern and ultrastructure of the contractile material is the same as in skeletal muscle.

of papillary muscle from the cat right ventricle. It should be noted that the properties of atrial muscle are different in some respects, but to consider these would be outside the scope of this book.

STRUCTURE

Cardiac muscle is no longer considered to be a syncytium from the structural point of view. Although light microscope studies indicated that it consists of branching and anastomosing fibres, electron microscope studies have shown that the "fibres" are, in fact, columns of cells, each 10–20 μm in diameter and 50–100 μm in length and that the "intercalated discs" are the closely apposed surface membranes of adjoining cells. The *intercalated disc* is typically a step-like structure (Fig. 8.1) in which there are transverse segments that seem to be specialised to give mechanical continuity from cell to cell, and longitudinal segments containing *nexuses* (see p. 76), which are believed to provide the structural basis for electrical coupling between cells (see below).

The *myofilaments*, their chemical composition and their arrangement with the sarcomere seem to be essentially the same in cardiac and skeletal muscle (see Figs. 8.1 and 7.3), and there are corresponding similarities in their mechanical properties.

There are no true myofibrils, but the contractile material is divided up by *mitochondria*, which are arranged in longitudinal rows (Fig. 8.1). Mitochondria are more abundant in cardiac muscle and there are related differences in the metabolic characteristics of the two types of muscle.

Cardiac muscle fibres have internal membrane systems with transverse and longitudinal components. The transverse tubular system (T-system) consists of wide bore tubes, which are not always transverse in their orientation. These occupy about 5 per cent of the fibre volume (cf. 0.5 per cent in skeletal muscle), and the extracellular calcium in them may be involved in excitation–contraction coupling. The longitudinal component (sarcoplasmic reticulum or sarcotubular system), on the other hand, is much less abundant than in skeletal muscle (about 1 per cent of fibre volume, cf. 5 per cent in the latter). Most of the cisternae are subsarcolemmal in location rather than intimately related to the transverse tubules. The scanty contacts between the transverse and longitudinal components of the sarcoplasmic reticulum appear as two-element structures ("diads") in electron micrographs of cardiac muscle: they are poorly differentiated compared with the triads of skeletal muscle and very haphazardly distributed. Thus there are some important differences in the internal membrane systems of cardiac and skeletal muscle, and these can be correlated to some extent with differences in the excitation–contraction coupling process in the two types of muscle.

EXCITATION

Electrical stimulation

If stimuli of graded strength are applied to an isolated papillary muscle or to the intact heart, the form of the strength–response curve obtained is that of a single muscle fibre (see Fig. 7.5). The muscle behaves as a *functional syncytium* even though it consists of a large number of separate cells. All of the cells are brought into a state of contraction because an action potential generated at any site in the muscle by a threshold stimulus is able to spread from cell to cell as illustrated in Fig. 8.1. The low electrical resistance pathways required for the spread of local currents from one cell to another are thought to be provided by the nexuses in the longitudinal segments of the intercalated discs (Fig. 8.1).

Cardiac action potential

The most important difference between cardiac and skeletal muscle is in the duration of its action potential relative to that of the mechanical response. In skeletal muscle the action potential is over before the contraction even begins, whereas in cardiac muscle (Fig. 8.2) it lasts almost as long as the mechanical response. All the particular differences between the two types of muscle listed in Table 8.1 B depend on this difference in the relative time courses of the electrical and mechanical events.

The characteristic feature of the cardiac action potential is the quasiplateau that separates the rapid depolarisation of the membrane from its repolarisation. The ion movements responsible for this are shown in schematic form in Fig. 8.2. As in the spike type of action potential found in nerve and skeletal muscle, the initial rapid depolarisation of the membrane is due to a fast inward sodium current, which results from a rapid self-regenerative rise in the permeability of the membrane to sodium. In both types of muscle the sodium permeability

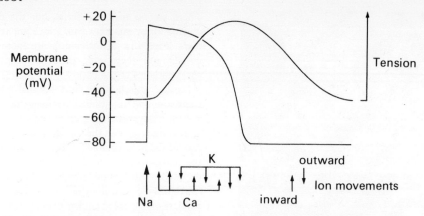

Fig. 8.2 Schematic diagram to show the mechanical response (isometric contraction), action potential, and ion movements across the cell membrane during the action potential in ventricular muscle.

then falls and the membrane begins to repolarise, but in cardiac muscle repolarisation is interrrupted by the quasiplateau. During this period the permeability of the membrane to calcium is increased relative to its resting value and the permeability to potassium is decreased; the balance between the inward calcium current and the outward potassium current is such that the membrane potential remains close to zero during the quasiplateau. As the inward calcium current decays towards zero, the membrane is repolarised increasingly rapidly by the unopposed outward potassium current.

Because the action potential lasts almost as long as the contraction so does the refractory period of the cell membrane. This means that repetitive stimulation of cardiac muscle does not normally result in summation and tetanic contractions (cf. skeletal muscle, Fig. 7.7). The essential requirement for summation is that the cell should be re-excited during the mechanical response produced by the previous stimulus: this cannot happen if the membrane is refractory during that period, as it is in cardiac muscle under normal conditions.

Spontaneous activity

The heart continues to contract rhythmically when it is removed from the body because all activity in the heart is *myogenic* in origin. Action potentials are normally generated by cells in the sinuatrial node in the wall of the right atrium and from there they propagate throughout the entire heart.

Although action potentials spread freely from cell to cell in cardiac muscle, the conduction velocity is rather low (about 1 m s^{-1}). Some muscle fibres in the ventricle are specialised to conduct at higher velocities (about 5 m s^{-1}): these are called Purkinje fibres. They are organised into a conducting system that radiates out from the atrioventricular bundle (bundle of His), branching repeatedly to cover the endocardial surface of the ventricles. The function of this system is to bring all parts of the ventricle into a state of contraction more synchronously than would otherwise be possible.

In contrast to the Purkinje system, there are atrial muscle fibres that are specialised to conduct action potentials at a low velocity (about 0.2 m s^{-1}). Their function is to introduce a delay between excitation of the atria and of the ventricles so that the atria have time to complete their contraction before ventricular systole begins. Further information about the spread of activity in the heart will be found in Chapter 41.

Action potentials recorded from *pacemaker cells* have the form shown in Fig. 8.3(a). The key difference between the membrane of pacemaker cells and that of other myocardial cells is that it undergoes a slow fall in potassium permeability while at rest; this causes a gradual depolarisation of the cell (the pacemaker potential) which results in an action potential when it reaches the threshold value for the cell. The rate of firing of the pacemaker cells is under the control of the autonomic nervous system and the effects of stimulation of the excitatory and inhibitory nerve supplies to the heart are illustrated in Fig. 8.3(b). Although the sinuatrial node is normally the site of the pacemaker in the heart, there are *reserve pacemakers* in the specialised conducting tissues of the heart. As one moves along the conduction path (atrial muscle → atrioventricular node → bundle of His → Purkinje system), the natural rhythm of the reserve pacemaker becomes slower and slower. Under normal conditions, they never become pacemaker sites; but in various disease states—for example, a conduction block between the atria and the ventricles—they can take over the pacemaker role.

Excitation–contraction coupling

The link between electrical events at the cell membrane and the activation of the contractile proteins appears to be provided by calcium in the heart, as it is in skeletal muscle, though a good deal less is known for certain about excitation–contraction coupling in cardiac muscle.

Activation of the contractile system seems to depend upon a rise in the sarcoplasmic Ca^{2+} concentration which is brought about by the action potential; but the source of this calcium is a matter of speculation. The extracellular space is one possibility for, as Sidney Ringer showed in 1883, cardiac muscle soon stops contracting if it is bathed in a calcium-free medium (skeletal muscle does not). During the plateau of the action potential there is a slow inward calcium current across the cell membrane which could be responsible, at least in part, for the rise in sarcoplasmic Ca^{2+} concentration. However, there is evidence suggesting that this calcium enters a store near the surface membrane from which it is released by the upstroke of the next action potential. Relaxation requires a lowering of the sarcoplasmic Ca^{2+} concentration, which is probably brought about by active uptake of calcium by the cisternae of the sarcoplasmic reticulum. It is also important to remember that calcium is extruded from the cell during the interval between beats; when the muscle is in a steady state the amount extruded must equal the amount that entered the cell during the preceding action potential.

The excitation–contraction coupling mechanism seems to be similar in skeletal and cardiac muscle in that the rise in sarcoplasmic Ca^{2+} depends primarily on a release of calcium from an internal store within the cell. The essential difference is in the lability of the internal store; in cardiac muscle, as we shall see later, it seems to be susceptible to all sorts of external influences whereas in skeletal muscle it is not.

THE STRUCTURE AND FUNCTION OF CARDIAC MUSCLE 73

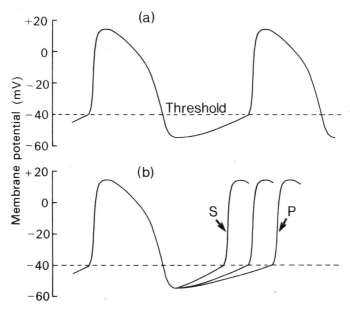

Fig. 8.3 Membrane potential changes in pacemaker cells. (a) Typical record from a cell in the sinu-atrial node. (b) Diagram to illustrate the effects of stimulation of the excitatory (sympathetic) and inhibitory (parasympathetic) nerve supplies to the heart (S and P respectively).

CONTRACTION

Interval–strength relation

Although summation and tetanus are not part of the repertoire of cardiac muscle, its contractile properties are very dependent on the frequency of stimulation. This is illustrated in Fig. 8.4 which shows the interval–strength relation of cat papillary muscle. Each point on the graph shows the tension produced by the muscle in steady-state contractions at a given rate of stimulation; the latter is expressed as the interval between successive stimuli and the scale is logarithmic in order to display to advantage data obtained over a wide range of stimulus frequencies. At very low rates of stimulation, small contractions of a fixed minimal strength are observed; these are called *rested state contractions*. The strength of contraction increases as the stimulus rate is increased until a maximum value is attained at very high rates of stimulation; the maximum strength is 10–20 times the rested state level. The physiological range of heart rates is on the sloping part of the interval–strength relation, so an increase in the heart rate due to increased activity in the sympathetic nerve supply will be accompanied by an increase in the strength of the contractile response because of the interval–strength relation: this will be *in addition* to any direct effect of sympathetic stimulation, on the strength of the heart beat.

Figure 8.4 also shows examples of the *staircase phenomenon*. When there is a sudden change from a slow rate (s) to a fast rate (f) or when stimulation is resumed after a period of rest, the strength of the contraction increases progressively to give an ascending or positive staircase (Fig. 8.4b). A change in rate of stimulation in the opposite direction produces a descending or negative staircase. Figure 8.4c shows the disturbance in contractile response that follows the interpolation of an extra stimulus; this is known as *post-extrasystolic potentiation*.

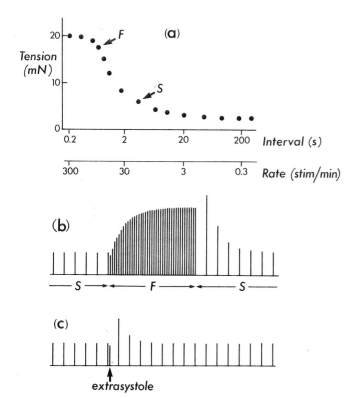

Fig. 8.4 Interval–strength relation. (a) Graph (courtesy J. Koch-Weser and J. R. Blinks) to illustrate how the tension produced in isometric contractions of ventricular muscle varies with the interval between successive stimuli. The latter is plotted on a logarithmic scale and beneath it are shown the equivalent rates of stimulation. (b) Record to show how the tension produced by a muscle changes when the stimulus rate is raised from a slow rate (S) to a faster rate (F) and then changed back again. (c) Record to show the effect of an extrasystole on the subsequent contractile behaviour of the muscle.

The interval–strength relation and its various manifestations are all understandable in terms of what is known about excitation–contraction coupling in cardiac muscle. The calcium content of the cells, and more particularly the internal stores from which calcium is released by the action potential, will depend on the balance between the entry of calcium during the plateau phases of successive action potentials and its extrusion during the intervals between action potentials. A high stimulus rate will cause the calcium content to increase because the proportion of the time that calcium is entering the cell will be increased at the expense of the time available for its extrusion; similarly, a low rate of stimulation will allow the calcium content to fall to a low value because there is ample opportunity in the long intervals between beats for calcium to be extruded from the cells.

Interval–duration relation

The rate of stimulation governs the time course of the contractile response as well as its strength. When the rate of stimulation is very low, both the action potential and the mechanical response are long in duration; but as the rate of stimulation is increased both become shorter. Over the physiological range of heart rate, the interval–duration relation allows systole to occupy an almost constant fraction of the cardiac cycle and this ensures that there is adequate time for ventricular filling during the period of diastole.

Myocardial contractility

The natural response of cardiac muscle is a twitch but, as we have just seen, the strength of the contractile response depends on the rate of stimulation in an isolated preparation and on the spontaneous heart rate *in vivo*. It also depends on a variety of other factors, such as the extracellular Ca^{2+} concentration, the amount of tonic activity in the autonomic nerve supply and the level of circulating catecholamines, none of which has any effect on the contractile behaviour of skeletal muscle. There is thus an additional dimension to the properties of cardiac muscle and the concept of *myocardial contractility* or the *inotropic state* of the heart (inos = strength) has been introduced to denote the ability of the muscle to contract at a given length. Any treatment or manoeuvre that alters this (e.g. change in the rate of stimulation) is called an *inotropic intervention*.

It can be argued that autonomic control of myocardial contractility is essential in cardiac muscle because it is a functional syncytium. The central nervous system can regulate the strength of contraction of a skeletal muscle in the body by varying the number of active motor units; but this is not possible in the heart because all the fibres are activated in every beat. What the autonomic nervous system can do instead is to adjust the contractility of the heart, as well as its spontaneous rate of beating.

Variations in myocardial contractility are thought to reflect variations in the completeness of excitation–contraction coupling. Under normal circumstances (i.e. in the absence of appreciable sympathetic tone or any other inotropic intervention) the rise in sarcoplasmic Ca^{2+} concentration brought about by the action potential produces only partial activation of the contractile system. Inotropic interventions seem to work by allowing the action potential to produce a greater rise in the sarcoplasmic Ca^{2+} concentration and therefore more complete activation of the contractile system. A limit to the positive inotropic effect of any intervention is reached when the rise in Ca^{2+} concentration is sufficient to produce saturation of the calcium binding sites on the contractile proteins.

MECHANICAL PROPERTIES

As one might expect from the similarity in the ultrastructure of the contractile systems of cardiac and skeletal muscles, the mechanical properties of the two types of muscle are also essentially the same. The two fundamental relations that describe the mechanical characteristics of a muscle are its length–tension relation and its force–velocity relation. In this account, the emphasis will be on ways in which these relations differ in the two types of muscle.

Length–tension relation

A typical length–tension relation for an isolated papillary muscle is shown in Fig. 8.5(a), which should be compared with Fig.

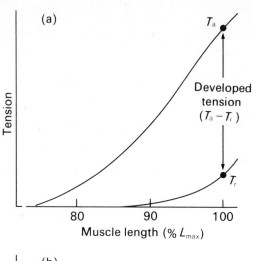

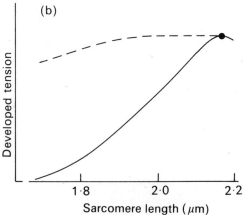

Fig. 8.5 Length–tension relation. (a) Length–tension curves showing how resting tension and the peak tension observed in isometric contractions vary with the muscle length, expressed as a percentage of L_{max}, the length at which the tension *developed* is maximal—i.e. total tension in the active muscle (T_a) minus the tension in the resting muscle (T_r). (b) Length–tension curves showing how tension developed varies with sarcomere length in cardiac muscle (solid line) and skeletal muscle (dashed line).

7.11, which shows comparable curves for skeletal muscle. In cardiac muscle there is resting tension over most of the working range of lengths and the resting length–tension relation rises quite steeply at lengths above L_{max}. In terms of the model shown in Fig. 7.8 (which is equally applicable to cardiac muscle) parallel elastic component of cardiac muscle is relatively stiff and in structural terms this means that the muscle contains a higher proportion of collagen than most skeletal muscles.

In skeletal muscle, the form of the active length–tension relation (Fig. 7.12) is determined mainly by the solid geometry of the sliding filament system. In view of the close similarity of the latter in cardiac and skeletal muscle, one might expect this curve to be equally applicable to cardiac muscle. In fact this is not the case, for the ability of cardiac muscle to develop tension falls off much more steeply at lengths below L_{max} than it does in skeletal muscle (compare solid and dashed curves in Fig. 8.5b). The reason for this is probably that the degree of activation produced by excitation is very length-dependent in cardiac muscle; the geometry of the sliding filament system then becomes a theoretical rather than an actual limitation on tension production at short muscle lengths.

The part of the curve below L_{max} (the "ascending limb") is of special interest to physiologists because it is the most fundamental manifestation of Starling's Law of the Heart ("the energy of contraction, however measured, is a function of the length of the muscle fibres"). The discovery of this basic truth in the heart–lung preparation of the dog and its importance in the intact circulation will be described in Chapter 40.

Force–velocity relation

There is an inverse relation between the velocity of shortening and force exerted by cardiac muscle, as there is in skeletal muscle (Fig. 7.14). However, it is a very difficult relation to study experimentally because it depends on muscle length, the time at which it is determined during the contractile response, and the inotropic state of the preparation (i.e. myocardial contractility). It is not therefore really appropriate to think of the force–velocity relation of cardiac muscle in terms of a single force–velocity curve, but as a family of curves, each of which defines the relation under a particular set of conditions. If we consider the solid line in Fig. 8.6a to be such a curve, then changes in myocardial contractility appear as displacements of the curve either towards or away from the origin of the graph. The important point to note is that changes in contractility lead to changes in both the maximum velocity of shortening (cf. difference between twitch and tetanus in skeletal muscle, Fig. 7.13); because shifts occur in both ends of the curve, changes in myocardial contractility have a large effect on the relation between power output and force exerted by the muscle (Fig. 8.6b) and that is what matters to the heart when it is functioning as a pump.

Energetics

The immediate source of energy for the contractile process in cardiac muscle appears to be the breakdown of ATP and, secondarily, of PC. Rephosphorylation is probably achieved entirely by oxidative processes and cardiac muscle contains an abundance of mitochondria for this purpose (Fig. 8.1). Studies of minced heart muscle have shown that its preferred substrates are lactate, pyruvate and other substances that provide a ready source of two-carbon fragments that can be fed directly into the tricarboxylic acid cycle. During exercise the liberation of lactic acid into the bloodstream by the working muscles provides the heart with a supply of its preferred substrate—a convenient arrangement that helps to minimise the oxygen debt created during exercise.

The energetics of contraction of cardiac muscle can be studied by measuring the oxygen consumption of a preparation beating at a constant rate (isolated muscle or intact heart). If an appropriate value is chosen for the calorific value of oxygen (21 kj litre^{-1} for the oxidation of carbohydrate) the energy expenditure per beat (i.e. the *enthalpy* change) can be calculated from the steady rate of oxygen consumption. It is perhaps surprising to find that when the muscle is loaded so that it is producing maximum external power or work (Fig. 8.6b), its energy expenditure is less than it is under isometric conditions (i.e. when the muscle is doing no external work); the explanation for this is that the energy expenditure of heart muscle is determined almost entirely by the tension it develops. When the muscle is optimally loaded for producing external work, its efficiency is in the region of 10–15 per cent, but it falls off steeply as the load is increased.

The fact that cardiac muscle can deliver less power and do so less efficiently when it is working against a larger than optimal load is one of the reasons why hypertension (a high systemic arterial blood pressure) is a dangerous condition. It places heavy demands on the oxygen supply to the heart which the coronary circulation may not be able to meet, and the problem will be further aggravated by any increase in heart rate (e.g. as a result of exercise, anxiety or excitement).

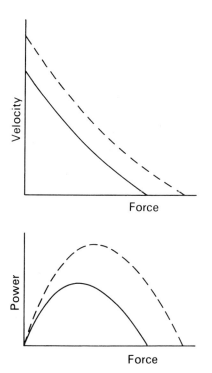

Fig. 8.6 Force–velocity relation. (a) Force–velocity curve obtained under a given set of experimental conditions (solid line) and the shift produced by an inotropic intervention (dashed line). (b) Corresponding curves for the power output of a muscle contracting against different loads (i.e. forces). Similar curves would be obtained if the ordinate was the external work done by the muscle.

9. The Structure and Function of Smooth Muscle

B. R. Jewell

Smooth muscle can be divided into two main types, known as single unit and multiunit smooth muscles. The smooth muscles of the alimentary tract and its associated ducts, the ureters and the uterus are all of the single-unit type, and these make up the main bulk of the smooth muscle in the body. The intrinsic muscles of the eye, the pilomotor muscles of the skin and the nictitating membrane of the cat are examples of the multiunit type. There are also some smooth muscles that seem to occupy an intermediate position such as vascular smooth muscle, which may have mainly single-unit or mainly multi-unit characteristics depending on the species and on its location in the body. The main differences between single-unit and multiunit types of muscle are summarised in Table 8.1 A.

STRUCTURE

Smooth muscle consists of mononucleate, spindle-shaped cells, which are embedded in a three-dimensional network of collagen fibrils. The cells vary a good deal in size; the largest are found in pregnant uterus, where the diameter reaches about 15 μm and their length may be 500 μm, and the smallest are found in the walls of blood vessels (diameter 2 μm and length 20 μm). The cells are organised into bundles, 20–200 μm in diameter, as illustrated in Fig. 9.1(a). When the muscle forms part of the wall of a tube or hollow viscus, the bundles of cells may be arranged as layers (e.g. intestine, bladder) or in a more complex helical form (e.g. blood vessels, uterus of higher vertebrates).

The cytoplasm of smooth muscle cells appears structureless by light microscopy; but when suitable fixed and stained material is examined in the electron microscope *thick and thin myofilaments* can be seen in the ratio of about 1:15 (cf. a ratio of 1:2 in skeletal muscle). In addition, there are intermediate filaments which are not part of the contractile system: these are thought to provide an intracellular "skeleton" for the muscle fibre. The thin filaments (Fig. 9.1(d)) have been shown to contain *actin* and they are virtually indistinguishable from those in skeletal muscle. They are sometimes seen to end in short electron-dense bands, which probably correspond to the Z bands shown in Fig. 7.2. The status of thick filaments is more of a problem. Biochemical tests have demonstrated the presence of *myosin* in smooth muscle and there is evidence from X-ray diffraction studies that myosin is present in organised structures in living muscle, but thick (myosin) filaments cannot always be seen in fixed material. However, it is now generally accepted that smooth muscle cells contain the essential structural components of the contractile system found in skeletal muscle, but arranged differently so that smooth muscle cells do not show transverse striations, and that contraction occurs by a mechanism analogous to the sliding filament system (see p. 61). Smooth muscle has also been shown to contain a troponin-tropomyosin system, which would give calcium sensitivity to its contractile system.

The surface membrane of smooth muscle cells includes many invaginations known as *caveolae* (Fig. 9.1(b)) which are found in association with elements of a well-developed sarcoplasmic reticulum in the superficial regions of the cell. The sarcoplasmic reticulum is thought to be important in excitation-contraction coupling in smooth muscle, as it is in striated muscles, but the role of the caveolae is not clear at present.

Intercellular junctions of two types are found between smooth muscle cells. There are *attachment plaques* (Fig. 9.1(e)), where myofilaments are inserted in a thick layer of electron-dense material on the inner side of the cell membrane; the gap between the adjacent cell membranes is about 10 nm and these junctions are thought to be responsible for mechanical coupling between cells. The other type of junction is the *nexus* (Fig. 9.1(f)) which is characterised by the presence of a gap of only 2 nm between the outer leaflets of the adjoining cell membranes and by the presence of electron-dense particles that "bridge" this narrow gap. This type of close contact is thought to play a part in the electrical coupling between smooth muscle cells.

Innervation

With the exception of amnion muscle, all smooth muscles are innervated to some extent. In the *single-unit* type there are excitatory and inhibitory nerve supplies (as in cardiac muscle), whereas in the *multiunit* type there is only an excitatory supply (as in skeletal muscle). The unmyelinated branches of the autonomic nerve fibres (diameter 0·1–2 μm) are found in intramuscular bundles, each of which is enveloped by a basement lamina and by an extension of the Schwann cell. Each terminal branch makes contact (*en passant*) with several muscle fibres, but the contacts are less specialised and less intimate than the neuromuscular junctions in skeletal muscle. The terminal branches have a beaded appearance because in the regions of contact the axon is dilated and contains numerous synaptic vesicles and mitochondria. There is little or no apparent specialisation of the adjacent muscle cell membrane.

The innervation of single unit and multiunit smooth muscles differs in several respects. In the *single unit* type the axons are found only in bundles and they retain a basement lamina in the regions of contact with muscle fibres. The minimum distance between nerve and muscle cells in these regions is about 80 nm or more, and many muscle cells are probably not innervated at all. In the *multiunit* type of muscle the innervation is more dense and there are numerous single axons (cf. bundles in single unit muscle). There is usually no basement lamina between nerve and muscle cells in the regions of contact and the gap between them may be as small as 20 nm. It is likely that each muscle fibre is innervated by several branches of a given axon and by more than one axon.

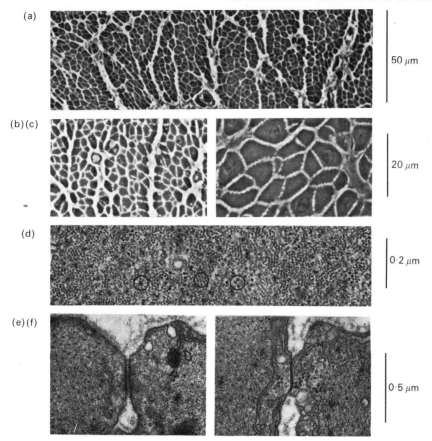

Fig. 9.1 Micrographs of taenia coli of the guinea pig (courtesy of Dr Giorgio Gabella).

(a) Transverse section showing the small size of the muscle cells and the presence of septa of connective tissue separating the cells into bundles.

(b, c) Transverse sections of the same tissue fixed while mildly stretched (b) and after isotonic shortening (c). Note the striking increase in the diameter of the muscle fibres, which is a consequence of the fact that they have shortened to about 20 per cent of their initial resting length.

(d) Electron micrograph showing three types of filament in a transverse section of the muscle. The circle on the left encloses a single thick (myosin) filament, diameter 13–18 nm; the circle in the centre contains about a dozen thin (actin) filaments, diameter 5–7 nm; the circles on the right enclose three intermediate filaments, diameter 10 nm.

(e) Dense patches at the surface of adjacent cells provide attachments for thin filaments. These are thought to transmit force from cell to cell or to neighbouring collagen fibres which provide a framework for the transmission of force through the tissue.

(f) An intimate junction (nexus or gap junction) between two smooth muscle cells, presumed to be the site of electrical coupling between adjacent cells.

EXPERIMENTS ON SMOOTH MUSCLE

Although from the medical point of view smooth muscle is unquestionably of greater importance than skeletal muscle, our knowledge of it has always lagged behind that of skeletal muscle. The main reason for this is that smooth muscle is a rather intractable material for a muscle physiologist to study experimentally. Its awkward features are as follows:

(i) Damaged preparations

Because of the complexity of the tissue architecture it is difficult to make a preparation of smooth muscle in the way that one can of a skeletal muscle, which is a discrete entity with tendons that can be used to couple the muscle to a mechanical recording system. The only way of making a comparable preparation of smooth muscle is to cut out a strip of the tissue and then fix its ends with threads or mechanical clamps. This is rather an unsatisfactory procedure because the damaged cells at the cut edges and clamped ends of the preparation have an unknown influence on its mechanical properties. This particular problem can be avoided if one studies the behaviour of smooth muscle in an intact organ—for example, the bladder, where one would measure intravesical pressure instead of muscle tension and bladder volume instead of muscle length.

(ii) Cell dimensions

The small size of the cells in smooth muscle make it difficult to dissect out single cells for detailed mechanical studies and it also makes it difficult to impale them satisfactorily with microelectrodes without causing extensive damage. In recent years, by using microelectrodes with very small tips (diameter about $0.1\ \mu m$), it has been possible to obtain satisfactory measurements of the membrane potential of smooth muscle cells.

(iii) Spontaneous activity

Spontaneous activity is a characteristic feature of the single unit type of smooth muscle, but it can be a great nuisance when the response of the muscle to controlled external stimulation is being investigated. It is possible to obtain stable electrical and mechanical baselines by cooling the preparation or by treating it with drugs that abolish spontaneous activity (e.g. adrenaline in some situations). The muscle can then be made to contract by electrical stimulation (direct or indirect) and by the application of chemical substances that depolarise the cell membrane.

(iv) Diversity of characteristics

Smooth muscles differ from skeletal muscles in that their properties vary enormously. This means that the experimental observations made on one smooth muscle may not be applicable to another even if this comes from the same animal. The division of smooth muscles into single-unit and multiunit types has rationalised these differences to some extent, but all smooth muscles do not fit neatly into one or other of these categories. The *guinea pig vas deferens* is a good example of a smooth muscle with mixed characteristics; all activity in this tissue is neurogenic in origin (a multiunit characteristic), but activity can spread from one muscle cell to the next (a single unit characteristic). The fact is that if one wants to know something about the physiology of a particular smooth muscle there is no substitute for actual experiments on that material.

EXCITATION PROCESSES IN SINGLE-UNIT SMOOTH MUSCLE

Spontaneous activity

The cells that make up a smooth muscle of the single unit type can contract in the absence of any activity in their autonomic nerve supply. Such contractions are said to be *myogenic* in origin and they occur at a frequency that is the same for all the cells in a given tissue. The frequency varies from one smooth muscle to another and in mammals the range is from about 1 to 10 per minute depending on the species and the location of the smooth muscle.

It is important to make a clear distinction between the behaviour of a smooth muscle (or a piece of smooth muscle) and the behaviour of individual cells within the tissue. A given cell has periods of activity, during which it tends to contract in a rhythmic fashion, and periods of quiescence; but the behaviour of the muscle *as a whole* depends on the extent to which the rhythmic contractions of the active cells are synchronised. (An apt analogy is provided by a large collection of identical clocks: they all tick at the same frequency, but the character of the sound that they generate *en masse* will depend on the degree of synchronisation.) If large groups of cells contract and relax more or less simultaneously the mechanical activity of the muscle consists of *rhythmic contractions* at the same frequency as those produced by an individual cell. On the other hand, if the cells contract asynchronously the mechanical activity of the muscle consists of a sustained contraction, known as *tone* or *tonus*, with little or no superimposed rhythmic component. This picture of spontaneous activity has emerged from microelectrode studies which have shown that the electrical activity of individual cells is essentially the same whatever the nature of the spontaneous activity produced by the whole muscle (i.e. rhythmic contractions or tone).

Spontaneous variations of membrane potential

The variations of membrane potential shown in Fig. 9.2 are typical of those seen in microelectrode studies of individual cells in *guinea pig taenia*. When the microelectrode enters the cell (Fig. 9.2(a)) it reveals the presence of a membrane potential, the inside of the cell being negative with respect to the outside. The alternation between periods of activity and quiescence, each lasting several minutes, seems to depend on very slow swings of membrane potential known as *pendular fluctuations*. Superimposed on these there are membrane potential variations of much shorter duration (0·5–1 s) known as *slow waves*, which are most pronounced during the periods of activity. If the membrane potential falls to the threshold value ("firing level") during a slow wave, then an *action potential* or *spike* is generated (Fig. 9.2(b)). The membrane potential is reversed during the spike, but the peak value reached is rather variable. During the repolarisation phase of the spike, the slow wave is "wiped out" completely; a new slow wave then starts and the cycle may be repeated.

The potential changes shown in Fig. 9.2(b) are characteristic of a cell showing *pacemaker activity*, and they are reminiscent of

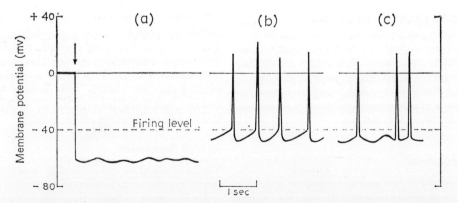

Fig. 9.2 Membrane potential variations in smooth muscle cells. These diagrams are based on microelectrode recordings from cells in guinea pig taenia. (a) Quiescent period: the membrane potential shows slow waves, but the mean potential is well above the firing level. (Arrow shows entry of microelectrode into the cell.) (b) Active period: membrane potential changes in a cell showing pacemaker activity. (c) Active period: activity due to action potentials that have invaded the cell from adjacent active cells.

those seen in the sinuatrial node of the heart. Cells in which the mean membrane potential is too high to allow the slow waves to generate action potentials are quiescent from the mechanical point of view *unless they are invaded by action potentials from a nearby pacemaker cell*. Propagated action potentials (Fig. 9.2(c)) can usually be recognised because they do not necessarily coincide with the peak of a slow wave, and because they do not "wipe out" the slow waves in the way that locally generated spikes do (cf. Fig. 9.2(b) and (c)). The sites of pacemaker activity change continually in the guinea pig taenia. Whether or not a given cell acts as a pacemaker depends on its mean membrane potential. If this is -60 mV or more (i.e. more negative) the magnitude of the slow waves is insufficient to reduce the membrane potential to the firing level, and the cell will be mechanically quiescent unless it is excited by a nearby pacemaker cell. On the other hand, if the mean membrane potential is -50 mV or less (i.e. less negative) the cell becomes a pacemaker because almost every slow wave will reduce the membrane potential to the firing level. Note that the slow waves occur all the time (even when the cell is mechanically quiescent); they reflect the intrinsic ability of the cells to show pacemaker activity.

Other smooth muscles vary in the character of their spontaneous activity, as the following examples show:

URETER

The smooth muscle in the wall of the *rat ureter* bears a particularly close resemblance to cardiac muscle. Action potentials of the "plateau" type are generated by a fixed pacemaker at the renal end of the ureter. These are propagated from cell to cell with the result that well coordinated contractions spread along the length of the ureter, delivering urine into the bladder in a series of spurts. This type of muscular activity is one form of *peristalsis*.

UTERUS

Uterine smooth muscle is particularly interesting because its properties depend on the levels of hormones in the bloodstream. When the uterus is *progesterone-dominated* (e.g. during pregnancy) most of the cells are quiescent and unresponsive to drugs, including oxytocin. There is some tone present as a result of asynchronous firing of scattered pacemaker cells, but the action potentials generated do not propagate far and no rhythmic contractions occur. When the uterus is *oestrogen-dominated* (e.g. during parturition) most of the cells become active, and the contractions that they produce are well coordinated because the number of pacemakers is limited and action potentials propagate for long distances within the tissue. There is a pacemaker site near the insertion of each uterine tube, but it is unusual for both to be active at the same time. The pacemaker cells produce slow waves that last for 10–20 s, and each slow wave generates a long train of spikes. These spread through the muscle and produce rhythmic contractions that are tetanic in character and of great importance in labour.

Effect of stretch on spontaneous activity

A smooth muscle of the single unit type actively resists extension because this causes an increase in the amount of spontaneous activity in the muscle. There is a general reduction of mean membrane potential and an increase in the frequency of slow waves, with the result that quiescent cells become active and the rate of spike discharge in active cells is increased. This property is particularly well-developed in the smooth muscle of certain vascular beds where it is thought to contribute to the autoregulation of blood flow (see Chapter 36).

IONIC BASIS OF MEMBRANE POTENTIAL

In smooth muscle the nearest equivalent to the resting membrane potential seen in nerve and skeletal muscle fibres is the membrane potential observed during periods of mechanical quiescence. This has a maximum value of about -70 mV, which is rather less than the values found in other excitable tissues. It probably has the same ionic basis as the resting membrane potentials found elsewhere (see Chapter 4), but it seems to be influenced to a greater extent by the permeability of the cell membrane to chloride and sodium. The causes of the pendular fluctuations and the slow waves are unknown; variations in the permeability of the membrane to sodium might be the cause of either or both, but this has yet to be proved. The self-regenerative depolarisation of the membrane followed by a reversal of membrane potential during a spike are probably due to the inward movement of sodium and other ions, including calcium. This entry of calcium ions may play a role in excitation–contraction coupling.

Spread of activity from cell to cell

Because activity can spread from one cell to the next in a smooth muscle of the single unit type, bundles of cells (and in some cases the whole muscle) behave as a *functional syncytium*. The extent to which activity spreads in a given muscle is influenced by factors that alter the mean membrane potential (e.g. amount of stretch, local concentrations of humoral agents), but basically it depends on the degree of electrical interaction between adjacent cells. The spread of excitation is thought to depend on local current flow as illustrated in Fig. 9.3. It has been shown that an action potential will propagate from cell to cell only if many of the cells in a transverse plane through a muscle bundle are simultaneously depolarised.

The structural basis of the electrical coupling between cells is generally considered to be the *nexus* (Fig. 9.1(f)) which is a region where the membranes of adjacent cells come into close proximity (2 nm separation). Early studies had shown a good correlation between the incidence of nexuses and the ease with which excitation spread through the tissue, but more recent work has shown a striking lack of correlation in some smooth muscles; thus the nexus may not be the only structural correlate of electrical coupling.

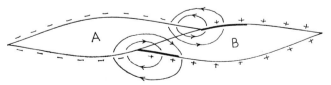

Fig. 9.3 Propagation of an action potenial from cell A to cell B by nexal-syncytial transmission. (See text for explanation.)

Direct electrical stimulation

A strip of smooth muscle responds to a single stimulus by producing a twitch. The strength–response and strength–duration curves are similar to those obtained with skeletal muscle (see Figs. 7.4 and 7.5), but there is no well-defined maximal stimulus because the strength–response curve does not show a plateau at large stimulus strengths. If the muscle is stimulated repetitively summation occurs and a fused tetanus can be produced.

Indirect electrical stimulation

The effect of indirect stimulation has been studied in detail in experiments on *rabbit colon*. The spontaneous activity in this tissue consists of slow waves with bursts of spikes that propagate for short distances. The application of a single shock to the *excitatory nerve supply* (cholinergic) produces a recognisable increment of tension, which is superimposed on the spontaneous activity. Microelectrode studies have shown that the arrival of an excitatory volley results in the appearance of a *junction potential* which is of the same general nature as the endplate potential seen at the neuromuscular junction of skeletal muscle fibres. The effect of this in a particular cell depends on the timing of the junction potential with respect to the slow waves already present; if the cell membrane is sufficiently depolarised an action potential is generated. A single stimulus produces a detectable increment of tension because action potentials are generated more or less synchronously in a certain proportion of the cells. Repetitive stimulation at a low frequency (below 1Hz) gradually increases the proportion of cells that respond synchronously, with the result that the muscle produces well coordinated rhythmic contractions (this process is known as "driving" the tissue). If the frequency is raised above $10\ s^{-1}$ the mechanical response has the appearance of a fused tetanus, but electrical records show summation of junction potentials leading to a maintained depolarisation of the membrane with no superimposed action potentials. The application of a single stimulus to the *inhibitory nerve supply* (adrenergic) has no detectable effect on the spontaneous activity of the muscle, but repetitive stimulation leads to its complete cessation. Microelectrode studies have shown that all action potentials and slow waves are abolished, and the membrane potential is stabilised at a value well above the firing level. A similar effect can be produced by immersing the tissue in a solution containing adrenaline.

Chemical stimulation

Smooth muscles produce a maintained contraction, known as a *contracture*, when the cell membranes are depolarised by immersing the muscle in solutions containing suitable neurohumoral agents or a high concentration of potassium. (Most mammalian skeletal muscles only produce a twitch-like response in these circumstances.)

MECHANICAL PROPERTIES OF SINGLE-UNIT SMOOTH MUSCLE

Spontaneous activity is not suitable for detailed studies of the mechanical properties of smooth muscle. If it is abolished (see p. 78) then direct stimulation can be used to produce twitch and tetanic responses. The mechanical properties of smooth muscle are basically the same as those of skeletal muscle, but there are important quantitative differences, viz:

(i) Speed of contraction

Smooth muscles vary enormously in this respect, but they are all much slower than skeletal muscle. For example the maximum unloaded shortening velocity (V_{max}) of a typical skeletal muscle is about 2000 times that of *cow mesenteric artery* muscle, and about 20 times that of *guinea pig taenia*.

(ii) Tension production

The maximum tension that a smooth muscle can produce ($P_0\ N\ m^{-2}$) is almost twice the corresponding value for skeletal muscle, if the tension produced is related to the cross-sectional area of contractile material.

(iii) Operational range of muscle lengths

In isometric contractions smooth muscle can produce tension over a much wider range of muscle lengths than skeletal muscle; similarly in isotonic contractions it can shorten through much greater distances. Whether this is due to differences in the architecture of the tissue or to fundamental differences in the contractile mechanisms of the two types of muscle is not clear at present. The wide operational range of muscle lengths may well be of importance in the normal functioning of smooth muscle *in situ*, as it is not restricted by skeletal attachments in the way that skeletal muscle is. Figure 9.1 shows micrographs of smooth muscle which was fixed while mildly stretched (Fig. 9.1(b)) and after active shortening (Fig. 9.1(c)).

(iv) Force–velocity curve

The force–velocity curves of smooth muscle and skeletal muscle (see Fig. 7.10(b)) are similar in shape, but the curve for smooth muscle is more convex towards the origin; this suggests that smooth muscle may be more economical in the maintenance of tension (i.e. requires a lower rate of energy expenditure to maintain a given tension), but this prediction has yet to be tested experimentally. The maximum work and power output are produced when the muscle shortens against loads of 0.3 to $0.5\ P_0$, as in skeletal muscle, but the maximum power output is much less because of the lower velocity of shortening of smooth muscle.

ENERGETICS OF SINGLE UNIT SMOOTH MUSCLE

Actin and myosin are present in smooth muscle, and the properties of actomyosin solutions suggest that ATP breakdown provides the free energy used in the contractile process. ATP and PC are present in much smaller amounts than in skeletal muscle and it is possible to detect breakdown of both substances when smooth muscle (*cow mesenteric artery*) contracts under anaerobic conditions. The explanation of this seems to be that the recovery processes are not as effectively stimulated by a rise of ADP concentration as they are in skeletal muscle. It seems likely that excitation–contraction coupling is brought about by the movement of ions that activate the actomyosin ATPase (probably Ca^{2+}) from the extracellular fluid into the cell during the action potential. How the activating ions are removed from the sarcoplasm during relaxation of the muscle is not clear at present.

MULTIUNIT SMOOTH MUSCLE

Multiunit muscles account for a small proportion of the smooth muscle in the body and they have not been widely studied. In fact the only one that has been investigated in any detail from several points of view is the *cat nictitating membrane*. This shows structural features that can be correlated with its physiological properties. Each muscle cell has a complete covering of basement membrane, so regions of close apposition between adjacent cells do not occur. However, the terminal branches of autonomic nerve fibres are found in close contact with the muscle cells (with no intervening basement membrane at most junctions) and multiple innervation of a cell by several axons appear to be common. The gap between adjacent muscle cells usually exceeds $0.1\ \mu m$, whereas the gap between nerve and muscle cells in the junction regions may be as little as 20 nm. These findings are what would be expected from the essential properties of a multiunit muscle: i.e. (i) all activity is neurogenic in origin and (ii) activity does not spread from one muscle cell to the next.

The application of a single shock to the motor nerve supply to the nictitating membrane (cervical sympathetic) produces a twitch response from the muscle, and repetitive stimulation produces a tetanus. Casual observations indicate that the mechanical response of the muscle is slow compared with skeletal muscle, but its mechanical properties have not been examined in any detail.

10. The Composition of Blood

This section of the book deals with systems for transport of materials in the body. Blood, circulating within the vascular system, carries oxygen and nutrient substances to the tissues and carbon dioxide and waste products away from them. We shall here deal with blood as a fluid adapted for this purpose. We will discuss the red cell and the way in which it functions as a carrier and later we will consider in detail oxygen and carbon dioxide carriage by the blood.

In a later chapter other functions of blood will be dealt with, e.g. its part in the defence reactions, the function of white cells and the way in which it clots (Chapter 72).

Fractions of blood: plasma, serum, red-cells

Blood is a thick suspension of cellular elements in a watery electrolyte solution. Centrifugation separates it into plasma and cells (about 50 per cent of each). If the blood is allowed to clot, it spontaneously "retracts" and serum is left. Serum is plasma minus its fibrinogen.

Sedimentation

Since the specific gravity of red cells is 1·10 and that of plasma is 1·03, cells will settle to the bottom of any container at a slow rate, known as the *sedimentation rate*.

Stokes's law deals with the rate at which particles fall, under the influence of gravity, through a viscous fluid. It states that the "terminal velocity" is proportional to the square of the radius of the particles. This leads to the sedimentation rate being a measure of the size of particles in a fluid medium.

Application of this principle to the red cells requires modification, for they are discoid in shape, the cells interact with each other during sedimentation and the orientation of the cells whilst they fall is random. A red cell sediments at about the rate of a sphere of the same volume.

In the pathology laboratory, clinical determinations are made of sedimentation rate using standardised methods (vertical tube, 2·5 mm bore, 10 cm long, blood diluted with citrate). After 1 h a clear band, consisting only of plasma, is seen at the top of the tube; in normal blood up to 3 or 4 mm thick. In various diseases the sedimentation rate may rise to 100 mm per hour and is a useful test for such conditions as rheumatic fever and tuberculosis, where it is often used as a method for following the progress of the disease. It is generally agreed that clumping (rouleaux formation) of the red cells is the causative factor in accelerating the rate. It therefore denotes a change in membrane properties of red cells and in the nature of plasma proteins which are able to affect cell surfaces.

An interesting anomaly is found when the sedimentation tube is not arranged precisely upright. It is found that the sedimentation rate in such a tube is accelerated, a somewhat surprising event as many students, in the old days of physiology practical examinations, found to their cost when asked to explain it.

In the body sedimentation will occur wherever flow of blood is stopped; it occurs also after death.

THE RED BLOOD CORPUSCLES

When observed under the microscope, blood is seen to consist of an enormous number of pale yellow discs, which are the red blood corpuscles (also called *erythrocytes*, or *red cells*), floating in the clear colourless plasma. When seen in bulk, in much thicker layers, the plasma is yellow and the erythrocytes red. The cells are biconcave discs, thicker near the edge than in the middle, their average diameter in human blood is about $8·6\,\mu m$ and their thickness is about $2·6\,\mu m$. Their specific gravity is greater than that of the plasma (about 1·10 against about 1·03), which accounts for the fact that they settle out when blood is allowed to stand, or is spun in a centrifuge. The haemoglobin, which is responsible for the red colour of blood, is contained entirely within the red blood corpuscles. This does not appreciably affect its properties as an efficient carrier of oxygen and carbon dioxide; but if it were in simple (colloidal) solution, it would be lost from the blood through the kidneys. In some circumstances, the viscosity of a suspension of red cells is smaller than that of a solution containing the same quantity of haemoglobin in unit volume, so that less power is needed to drive it round the circulation at the required rate.

Estimation of red cell indices

The important measurements which are carried out in diagnosis are the packed cell volume (PCV), the mean cell haemoglobin (MCH), the mean cell haemoglobin concentration (MCHC) and the mean cell volume (MCV).

The major reason for estimating the PCV is to obtain the MCHC. The PCV is measured by centrifuging blood for 30 min at $1500\,g$. It is the length of the column of red cells expressed as a percentage of the whole length of blood. Normal is 40 to 54 per cent in men; 35 to 47 per cent in women.

The MCHC is the weight in grams of haemoglobin (Hb) in 100 ml of packed red cells, expressed as a percentage. Haemoglobin concentration divided by PCV gives a value correct to within $\pm 2\%$. The lower limit of normal would be about 30 per cent. It is useful in the diagnosis of iron-deficiency anaemia.

An enormous amount of time used to be spent in counting the number of red cells per microlitre of blood, but the method (counting chamber; Neubauer chamber) was very inaccurate. Automatic counters are now used which electronically give rapid values for red cell number, haemoglobin concentration and mean cell volume. From these, the apparatus calculates MCHC, MCH and PCV.

Normal MCV is $75-95\,\mu m^3$. It is lowered in iron deficiency, chronic infections and thalassaemia. Normal MCH is 27–32 pg per cell. It is also low in the above diseases.

The red cell count in healthy men varies from 4·5 to 6 million per μl, the average being about 5·5 million. In women the count is even more variable, but on an average about 10 per cent lower than in men. The uncertainty in the method of obtaining a blood count, as ordinarily performed, is likely to be at least ± 5 per cent.

The reticulocyte count

Reticulocytes are immature red cells, just arrived in the circulation from the bone marrow. They contain RNA left over from haemoglobin synthesis. The reticulocyte count is an estimate of the rate of red cell formation. They are counted by staining blood with a supravital dye such as brilliant cresyl blue. This precipitates the RNA which can be seen when a film is made on a slide and examined under the light microscope. The proportion of reticulocytes to normal red cells can be estimated, which in normals does not exceed 2 per cent.

White cell count

The number of white cells in a blood sample is estimated by diluting blood and counting in a chamber of known dimensions under the light microscope. The diluting fluid consists of a dye to stain the white cells (gentian violet) and dilute acid to cause haemolysis. There are now electronic counting methods available which are quicker and more accurate.

The differential white cell count, however, is still done under the microscope, a film being prepared and stained to enable the proportion of lymphocytes, polymorphs etc. to be estimated.

The normal total white cell count is 10 000 μl^{-1}, polymorphs are 7500 μl^{-1} and 3500 μl^{-1} of lymphocytes.

The platelet count is done in the same way as the white cell count, the diluting fluid being formaldehyde in saline. No dye is used. Normal range for platelets is 150 000 to 400 000 μl^{-1}.

The PCV (packed cell volume or haematocrit)

The red blood corpuscles normally occupy 40 to 54 per cent of the volume of the blood in man (36 to 47 per cent in women). This important ratio can be measured by means of the *haematocrit* (from the Greek for "blood separator"). This is a graduated capillary tube of uniform bore, which is filled with blood and centrifuged at high speed until the column of sedimented red cells shows no further shrinkage; the volume of the tightly packed cells can then be read off and compared with that of the whole blood.

The determination of the haematocrit depends on Stokes's law, just as does sedimentation, but with greater values of g, given by centrifugation. The end-point in this method depends on duration of centrifugation and its speed; thus a standard procedure must be followed in clinical work, e.g. Wintrobe tube bore = 5 mm; 3000 rev/min (=1500 g), for at least 15 min.

Radioactive plasma protein, added to blood, shows that the packed cells still entrap about 5 per cent of plasma. The normal value thus shows close packing of the red cells. In *polycythaemia* (see p. 97) haematocrit values of 60–70 per cent occur. In these patients blood viscosity is unduly high and the circulation is embarrassed.

The microscopic examination of blood

This is conveniently carried out on dried and stained smears on a glass slide. On drying, the red cells shrink so that their diameters are diminished by 8–16 per cent. In such dried preparations of normal human blood the average diameter of the red cells is only 7·4 $\pm$ 0·3 μm. The shrinkage and distortion of the cells so prepared are of no great consequence in the clinical examination of the blood, because the characteristic changes in blood associated with different diseases are commonly described in terms of the changes which appear in the dried smears.

Red blood corpuscles have no nuclei, but are, nevertheless, living cells; they have quite active metabolic processes and require energy for their existence, which they obtain, chiefly by means of the anaerobic breakdown of glucose. But they have a relatively short lifetime and old ones are continually being destroyed and replaced by new ones formed, in post-natal life, in the blood spaces of the bone marrow. In normal individuals, the rate of production is adjusted so as to be equal to the rate of destruction when the number of red cells in the circulating blood is such that the red cell count has its normal value of about 5 million per mm^3; but, following an abnormally large loss of blood, for example after a haemorrhage, the rate of production is greatly increased until the loss is made good. The primary stimulus for this is the reduction in the rate of flow of oxygen to the tissue cells generally; with less haemoglobin in each litre of blood, less oxygen can be carried round from the lungs. The production of red cells is accelerated, also, when the supply of oxygen to the tissues is deficient by virtue of an inadequate supply of oxygen to the lungs, as at high altitudes.

Lifetime of red cells

Many different methods have been used to estimate the average lifetime of a red blood cell and widely different values have been obtained. In principle, a certain proportion of the circulating red cells is "labelled" in some way and the rate of their disappearance measured. (1) Cells may be injected intravenously which contain a different type of agglutinogen from that of the cells normally present, care being taken, of course, that the cells injected are not such as to be affected by agglutinins in the plasma of the recipient. (2) Cells which have been removed from the subject under investigation may be allowed to take up *in vitro* the radioactive isotope of phosphorus, ^{31}P (as phosphate), or that of chromium, ^{55}Cr (as chromate), and then returned to the circulation. (3) The haemoglobin incorporated in the cells formed in the ordinary way in the bone marrow may be "labelled" by injecting intravenously, over a suitable period of time, either (a) salts of the radioactive isotope of iron, ^{59}Fe, or (b) the amino acid glycine in which the ordinary carbon or nitrogen atoms, ^{12}C and ^{11}N, have been replaced by those of the isotopes ^{14}C and ^{15}N, as the case may be; the rate at which these cells appear may be measured as well as the rate at which they disappear. None of the various methods used is without its difficulties and various corrections must be made to the results obtained. The most reliable measurements indicate an average lifetime in man of some 100 to 130 days.

Formation of red cells

In certain abnormal conditions, the rate of production of the red cells may be unduly small, or the rate of destruction unduly large. In either event, in the steady state when the two rates are equal, the red cell count will be less than the normal value, producing *anaemia*.

1. Study of the disease pernicious anaemia (and certain other related kinds of anaemia) has shown that for the normal development of the red cells two specific *vitamins* are necessary: these are *vitamin B_{12}* (which contains the rather uncommon element cobalt, and is also known as cobalamin) and *folic acid* (or pteroyl glutamic acid). Both vitamins are present in adequate quantities in normal protein foods of animal origin, and particularly in liver; folic acid is present also in fresh green vegetables, and vitamin B_{12} is synthesised by many kinds of micro-organism, including some of those in the alimentary tracts of mammals. Anaemia may result from malnutrition but more usually, owing to some diseased condition of the alimentary tract, the vitamins are not properly absorbed and anaemia results even though the diet is fully adequate.

When given as a subcutaneous injection, vitamin B_{12} is highly active in improving the condition of a patient with pernicious anaemia but when given by mouth it is far less active. It would appear that it is not well absorbed from the alimentary canal unless first acted upon by a constituent of the gastric juice, known as the *intrinsic factor*, vitamin B_{12} being the *extrinsic factor*. This intrinsic factor is absent from persons suffering from pernicious anaemia.

2. The cells of the bone marrow, like other actively dividing cells, are very susceptible to the action of *ionising radiations* and particles—X-rays and γ-rays, α-particles, electrons and neutrons emitted during radioactive disintegration of unstable atomic nuclei. After excessive exposure to such radiations, cell division in the bone marrow ceases, or becomes disordered, the rate of production of normal red cells is greatly reduced, and abnormal kinds of cell may appear in the circulation.

3. Iron is needed for the synthesis of the haemoglobin within the red cells; the iron released from destroyed cells is used again in the production of new ones, and ordinary diets contain sufficient to make up for any deficiency. Small amounts of copper are essential for the proper utilisation of the iron. Anaemia, however, may occur in infants during the suckling period, owing to an almost complete lack of iron in the milk; it can be cured by administration of salts of iron.

4. The rate of destruction of the red cells may be abnormally great owing to a congenital defect in their structure or in their metabolic processes, which shortens their lifetime; or to the presence in the plasma, as a result of disease, of a specific "antibody" which destroys them.

Development of red cells

In the earlier stages of their development, red cells possess nuclei. Two stages in the development of nucleated red cells from the mesenchymal parent cell are distinguished; they are known successively as *erythroblasts* and *normoblasts*. The normoblast becomes an adult red cell when it loses its nucleus. For the first few days of its adult life, spots and threads of basophilic material can be demonstrated in the red cell by vital staining with cresyl blue. In this stage it is known as a *reticulocyte*, or reticulated red cell. After the first week of life reticulocytes form only about 1 per cent of the circulating red cells, but they form a larger proportion of the red cells found in bone marrow. The proportion of them in circulating blood rises considerably whenever the new formation of red cells is increased, as after a haemorrhage, or after the successful treatment of pernicious anaemia.

The haemoglobin in the fragments of the old red cells is converted by the reticuloendothelial cells into inorganic iron and the pigment *bilirubin*. These travel in the plasma to the liver, and the bilirubin, after a slight change, is secreted into the bile. In the bowel bilirubin is changed into *stercobilinogen* and *stercobilin*, which gives the faeces their dark colour. Not all the bile pigment is excreted. Some is reabsorbed into the portal bloodstream in the form of a colourless compound called *urobilinogen*, which is picked up by the liver, perhaps to be used in the production of new haemoglobin. When large quantities of bilirubin are formed (as in haemolytic jaundice), and consequently large amounts of urobilinogen absorbed, appreciable amounts of urobilinogen are excreted in the urine. This is to be distinguished from the excretion of bilirubin in the urine which occurs when the bile passages are obstructed, for example, by a gall-stone (obstructive jaundice). Urobilinogen is also found in the urine when the liver function is impaired, and thus unable to deal with the normal quantities carried to it in the portal bloodstream.

The shape of the normal red cell

The physiology of the red cell is a nice easy field for the research worker since an unlimited supply of red cells resides within his own venous system and only a microscope is required in the way of apparatus. Whole textbooks are devoted purely to the red cell.

A typical human red cell is a *biconcave disc* 8.5 ± 0.41 μm in diameter.* The biconcavity is functionally useful in that it allows diffusion of materials, such as oxygen, to take place across the shortest possible distance between the membrane and cell contents. The electron microscope shows no internal structure (or stroma) within the red cell which used to be thought responsible for maintaining its biconcavity. One interesting hypothesis (armchair) to account for the red cell's shape is that the cell membrane attracts the opposing one, thus drawing the cell in like a squashed tennis ball. Such a mechanism might also explain *rouleaux* formation where two or more red cells clump together, like a pile of coins.

ABNORMAL SHAPES

Ovalocytes are elliptical cells as opposed to biconcave ones. *Spherocytes* are more nearly spherical. When cells shrink, the membrane is crumpled or crinkled, e.g. in hypertonic media, a process known as *crenation*.

Sickle cell anaemia is an inherited abnormality in the Negro race, approximately 1 in 600 suffering from the disease. The disease is a chronic hemolytic disorder having periodic crises of bone, joint and abdominal pain. The symptomatology is related to the presence of the abnormal haemoglobin S in the cells; it crystallises under conditions of reduced oxygen tension. When crystalline, the haemoglobin is thought to cause the red cell to become a *meniscocyte* (or *sickle cell*) through the reorientation of molecular forces. On return to an environment having above about 50 mmHg oxygen tension, cells resume their normal biconcave shape.

OSMOTIC PRESSURE, SOLUTE CONCENTRATION AND OSMOTIC EFFECTS ON RED CELLS

When two solutions are separated by a semipermeable membrane, an *osmotic pressure* is set up by a concentration difference between the two, and this pressure is maintained as long as the gradient exists, i.e. until the concentrations of water on either side of the membrane equalise.

The osmotic movement of water may be experimentally halted by applying to the more concentrated solution a pressure equal and opposite to the osmotic pressure. This, indeed, is the principle underlying one form of *osmometer* (for the measurement of osmotic pressure). The solution whose osmotic pressure is to be found is placed in a semipermeable bag having a fine glass capillary sticking up out of it. When placed in water, water enters the bag osmotically. The solution rises up the glass tube until the hydrostatic pressure of this column equals the osmotic pressure. Provided the tube is narrow in bore (so that insufficient water flows into the semipermeable bag to appreciably dilute the solution), the osmotic pressure is simply:

$$\pi = h \cdot \rho \cdot g$$

where π = osmotic pressure; h = height of column; ρ = density of solution and g = gravitational constant.

Osmotic pressure and solute concentration

Van't Hoff's equation gives the quantitative relationship between osmotic pressure and solute concentration:

$$\pi = CRT$$

where π = osmotic pressure, R = the gas constant, T = absolute temperature, and C = solute concentration.

An important feature of this equation is that the solute concentration means more than simply the number of moles of solute per unit volume. This is because the osmotic properties of a solution do not depend upon the nature of the solute particles,

---

* The ± 0.41 μm is the "standard deviation" as found in large numbers of separate measurements of diameter. This means roughly that two-thirds of the cells have a diameter between 8 and 9 μm. Most textbooks give the figure 7.2 μm for mean diameter, but this is too small, being the result of shrinkage in the preparation of the blood smear.

but only upon their number. In other words, in a given volume of solution, equal numbers of particles, whether they are large or small will give equal osmotic pressures.

As an example, a solution containing 0.1 mol l^{-1} sucrose may be compared with one containing 0.1 mol l^{-1} sodium chloride. The chemical concentrations of these in moles per litre are the same. The osmotic pressure of the sodium chloride will, however, be twice that of the sucrose. This is because NaCl is in solution as Na^+ and Cl^- ions, while sucrose molecules are undissociated. For sodium chloride, therefore, twice as many particles are present in a given volume. This factor, relating chemical concentration to osmotic concentration is called G, or the *osmotic coefficient*.

The above explanation of the factor G is simplified, because it only holds in dilute solutions. In practice, it is found that the value for G not only depends upon dissociation as above, but varies with concentration for several reasons.

Osmolarity

From a practical standpoint it is necessary to have a unit of osmotic concentration which is independent of G and its associated difficulties.

A solution which has an osmotic pressure of 22.4 atmospheres (atm) is called *osmolar*.* It has an osmotic concentration of *1 osmole per litre*; this is quite apart from its chemical concentration. The definition of an osmole is therefore the weight of a substance in grams which will produce an osmotic pressure of 22.4 atm if dissolved in 1 litre of solution.

Thus we can express the concentration of a solution in osmoles per litre and this is known as its *osmolarity*.

For those who wish to be rigorous, or who are physical chemists, a preferable measure of osmotic concentration is the *osmolality*, or number of osmoles per kg of solvent (as opposed to "litre of solution"). Physiologically speaking, though, any error due to using osmolarity is so small that it may be neglected; moreover, it is more convenient to use "solution" rather than "water" as the standard, since in most cases we deal with blood, tissue fluid, urine etc. Intracellular fluid, however, contains substances of large molecular weight; its volume is thus appreciably larger than that of intracellular water itself. Osmolality is here the appropriate measure of its osmotic properties. See page 120.

For dilute solutions of non-electrolytes, such as sucrose, a milliosmolar solution is equivalent to a millimolar solution. For electrolytes, however, a milliosmolar solution equals G times millimolar.

PRACTICAL DETERMINATION OF OSMOLARITY (CRYOSCOPY)
The freezing-point of a solution is depressed by an amount which depends on the number of particles in it. An osmolar solution lowers the freezing point by $1.86°C$. Osmometers are now available which rapidly measure freezing point of any solution, e.g. plasma or urine, and the results are given directly in milliosmoles.

Osmosis and red cells

Red cells contain little else apart from haemoglobin (3.4 kg l^{-1}). However, if we consider the actual numbers of molecules it is a different story. Although, for example, the concentration of glucose in red cells is only 750 mg l^{-1} (1/5000 of the concentration of the haemoglobin), the number of molecules is roughly the same. There are 3.4×10^8 molecules of haemoglobin; 2.2×10^8 molecules of glucose.

The total osmolarity of the red cell depends upon its glucose and ions (sodium and chloride) equally with its haemoglobin. However, the red cell membrane is permeable to water, sugars and ions, but not to haemoglobin. The latter, therefore, is the main factor concerned in the osmotic behaviour of the red cell.

Haemolysis

If red cells are put in distilled water (or a hypotonic solution), as one would expect from the previous few paragraphs, water enters them and they swell. The changes are shown in Fig. 10.1.

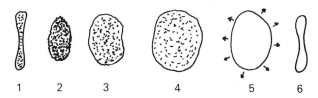

Fig. 10.1 Haemolytic swelling. Stages shown are: (1) Normal red cell ($87 \mu m^3$, $163 \mu m^2$). (2) Cell becomes convex ($87 \mu m^3$, $98 \mu m^2$). (3) When volume has increased by 30 per cent, cell is approximately spherical ($130 \mu m^3$, $113 \mu m^2$). (4) Cell continues to swell as a sphere. (5) When swollen by 100 per cent in volume and with surface area back to its normal value ($196 \mu m^3$, $163 \mu m^2$) haemoglobin leaves the cell. (6) The red cell ghost returns to its normal discoid shape, as in (1).
Values from Ponder, E., *Hemolysis and Related Phenomena*. New York: Grune and Stratton, 1948.

When the volume has increased by about 30 per cent, the cell is almost spherical. It swells more until the membrane suddenly becomes permeable to haemoglobin which then almost completely leaves the cell. This process is called *haemolysis*; the membrane is still intact and can be seen under the microscope as a *ghost* which goes back to the form of a biconcave disc now that its contents are gone.

Osmolarity and tonicity

What is meant by the term *"isotonic"*? Solutions having the same osmotic pressure are said to be *isosmotic*, and those greater or smaller, *hyperosmotic* or *hypo-osmotic*. A solution with 310 mosmol NaCl in it, i.e. a 0.155 M saline solution, is isosmotic, as far as red cells are concerned. Thus red cells will not shrink or swell when put into it, and such a solution is said to be *isotonic*. As before, stronger solutions than this are *hypertonic*; weaker, *hypotonic*.

How does an isotonic solution differ from an isosmotic one? Let us consider the effects of putting red cells in a solution of urea which is able to penetrate the membrane. As explained above, a 310 mosmol urea solution (about 0.31 M) is isosmotic with the contents of the cells. Even so, the cells rapidly haemolyse. This is because urea passes into the cell through the membrane. Cells are more or less impermeable to the osmotically active substances they contain, so the osmotic concentration of the cell thus rises rapidly. This means also that water enters the cell, it swells and ultimately is haemolysed. Of course, any other penetrating solute will have a similar effect, the speed of haemolysis depending upon the rate at which penetration occurs (indeed, haemolysis-time is used as a measure of relative rates of penetration, experimentally).

Thus any solution containing a penetrating solute can never be isotonic, even in a mixture, even in low concentrations. Such

---

* This of course is really strong. The milliosmole (mosmol) is the usual unit (1 osmole = 1000 mosmol).

solutions must always be considered to be hypotonic, no matter what their actual osmolarity is.

Haemolysis in vivo

In normal healthy individuals, continual haemolysis of red cells occurs and terminates their life-span. Various disease processes speed up this haemolysis, such as *haemolytic jaundice*. Jaundice is a yellow colouration of skin, mucosa and particularly the sclera, and is due to the pigmented breakdown products of the haemoglobin released being deposited there. Free haemoglobin is a just small enough molecule to pass the glomerular membrane in the kidney. Excessive haemolysis is therefore detected by finding haemoglobin in the urine.

Osmotic haemolysis does not occur in life since the osmotic pressure of blood and tissue fluids is closely controlled by feedback mechanisms (the osmoreceptors).

A rare condition of excessive haemolsyis is called *paroxysmal nocturnal haemoglobinuria*. In this disease, patients have red cells which haemolyse if the blood becomes slightly more acid than usual, although they are apparently normal at normal pH. At night, the small reduction in alveolar ventilation that always occurs, is enough to cause an accumulation of CO_2 and hence marginally to lower pH. This then leads to the haemolysis. If the patient takes bicarbonate, or is kept awake, the haemoglobinuria does not occur.

Red cell fragility

One may make a test of the "fragility" of red cells by haemolysing them osmotically at various degrees of tonicity. Usually, the cells are separated from the plasma and put into buffered sodium chloride solutions varying between 0·5 to 0·1 tonicity of normal plasma. After 15 min exposure, the cells are centrifuged and the degree of haemolysis in each strength of saline is judged colorimetrically by the depth of haemoglobin pigmentation in the supernatant. Figure 10.2 shows a typical haemolysis curve found in this way.

Normal red cells are all haemolysed in NaCl solutions of less than 0·3 per cent; half at about 0·45 per cent. Fragility may be increased in various diseases. For example, spherocytes haemolyse at lesser increases in cell volume than do normal ones.

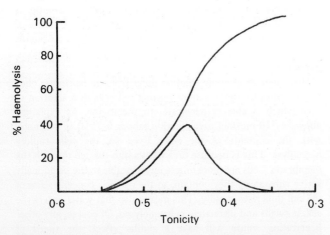

Fig. 10.2 Percentage haemolysis and frequency–distribution curves for normal human red blood corpuscles. About half of the cells will be haemolysed by a NaCl solution of 0·45 tonicity. The cells are put in saline solutions of the tonicities given on the horizontal axis. Percentage haemolysis is determined colorimetrically after centrifugation (top curve); all cells are haemolysed by a tonicity of 0·35 (see frequency–distribution, lower curve).

THE MEMBRANE OF THE RED CELL

Membranes in general

The cell membrane plays a crucial part in the function of all cells. The *cytoplasmic* membrane, which surrounds the cell, transports substances into and out of the cell and, in addition, regulates its internal environment. Inside the cell are other membranes, of much the same structure, which enclose such structures as the nucleus, mitochondria, chloroplasts (in plants) microsomes and other cell organelles. Mitochondrial membrane, for instance, manufactures ATP; chloroplast membrane is the part of the plant cell responsible for photosynthesis.

The basic structure of membranes is now fairly well understood, both the internal and external ones being essentially lipid in composition with scattered protein molecules within them. In mammals, the membranes additionally have carbohydrates in the form of glycoproteins and glycolipids.

The lipids in membranes

About half the average membrane is lipid. Internal membranes are phospholipid; cytoplasmic membranes contain both glycolipid and neutral fat, as well as phospholipid. In the red cell, about one-third of the membrane is cholesterol.

The actual molecules of membrane lipids have a head and two tails (Fig. 10.3), linked at their midpoints by a glycerol grouping. The tails are fatty acid chains having a structure very much like oil molecules and, in the same way that oil and water when mixed tend to separate out, these tails point away from the water within or outside the cell. They are termed *hydrophobic*. The heads of the phospholipids are readily soluble in water and do not show this effect; they are called *hydrophilic*. Such molecules, one end hydrophobic and the other hydrophilic, are said to be *amphipathic*.

Membranes as bilayers
In membranes, lipids form, because of their amphipathic properties, a bilayer with their hydrophilic heads as the inner and outer surfaces and their hydrophobic tails pointed inwards towards each other. This lipid bilayer is around 4–5 nm in thickness, forming the framework of the membrane and the structural support for the other components such as protein. Figure 10.4 is a diagram of a lipid bilayer.

Protein in membranes

The proteins found in cell membranes have various functions:

1. They contribute strength to the membrane.
2. They are enzymes.
3. They have a "pumping" action, e.g. the sodium pump, for transferring materials in or out of the cell.

The precise nature of some of the proteins present has now been determined. In the erythrocyte membrane, polypeptides with molecular weights between 255 000 and 12 500 have been found, the two heaviest being together called *spectrin*.* Spectrin comprises approximately one-third of the red cell membrane protein. A further third is a group of proteins having a molecular

---

* The name "spectrin" was chosen by V. T. Marchesi of Yale, who discovered it in the red cell "ghosts", i.e. the pure membranes left after haemolysis has been induced.

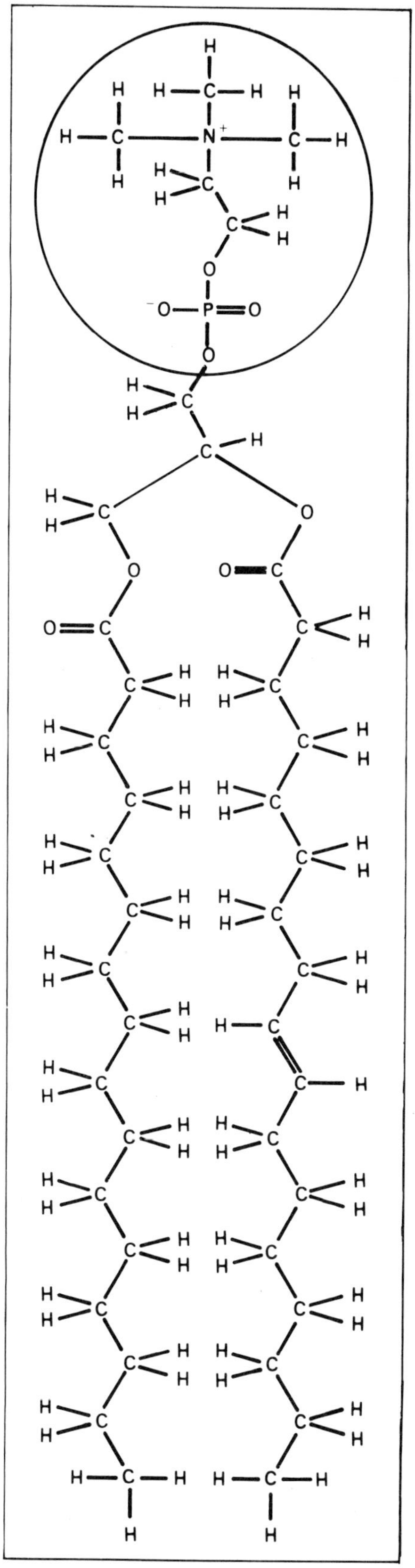

THE COMPOSITION OF BLOOD 87

weight around 90 000; the remainder are proteins of below 70 000 mol. wt.

As can be seen in the diagram (Fig. 10.4), these proteins are either (1) related solely to the inner or outer lipid bilayer: these are termed "extrinsic", or (2) they penetrate the full thickness of the membrane, i.e. they are "intrinsic". Extrinsic proteins can more easily be distributed and even removed from the membrane by chemical means. In electron micrographs, the enzyme ATPase and spectrin can be seen, the ATPase showing up as "headpieces" which stick out of the membrane surface* and the spectrin diffusely lining the interior of the red cell ghost membrane.

AMPHIPATHIC PROTEIN

Intrinsic protein has an unusual environment, as compared with cytoplasmic protein. Only a part of the molecule is in water, the remainder being in oil. Such an arrangement would normally be unstable in terms of molecular forces developed by the proximity of unlike charges and therefore the protein molecule itself would have to be amphipathic in the same way as are the phospholipids. Those parts of the molecule in the water would contain the hydrophilic amino acids (lysine, histidine, arginine, aspartic acid, glutamic acid, serine and threonine). Those within the lipid would have hydrophobic amino acids over their surfaces.

Proteins in the red cell membrane

The glycoprotein found in red cell membranes consists of a single polypeptide chain, one end having the carbohydrate attached consisting mainly of hydrophilic amino acids being on the outer surface and exposed to the extracellular water. The other end also has hydrophilic groupings and this is in the water within the cell. The molecule therefore extends right through the membrane and is known to be made up of about 30 amino acids in a chain. These are hydrophobic and are enclosed by the lipid bilayer itself; this is shown in Fig. 10.5.

The mobility of membrane constituents

One of the most interesting advances made in the study of the cell in recent years is that the structure of cell membranes is not static. The lipids and the proteins move around to a considerable extent.

The lipid mobility depends on how "fluid" it is; this in turn is determined by the degree of saturation of the lipid tails (the extent to which the available C bonds are taken up with hydrogen atoms) and by the temperature. Most of the lipid in mammalian cell membranes is unsaturated and this means that at normal body temperature, it will be fluid (as opposed to solidified) and hence the fatty acid tails can move around.

The movement of fatty acids within the membrane has been studied using electron spin-resonance spectroscopy. In this technique, a "reporter" group (nitroxide) is attached to the fatty acid tail of a "test" molecule which is then substituted into the bilayer. The spectrum of the reporter group varies with the degree of mobility. It is found in this way that the part of the lipid tail closest to the head is the least mobile; the greatest flexibility is at the tip of the tail. Around the protein molecules, the mobility of adjacent lipids is less. It is also found that the lipid molecules move around in the membrane mainly in the plane of the membrane and not up and down in it.

Fig. 10.3 Amphipathic structure of a lipid molecule, with a hydrophilic head and twin hydrophobic tails, is exemplified by this typical phospholipid, specifically a molecule of phosphatidylcholine. Various lipid molecules comprise about half of the mass of mammalian membrane, forming the membrane's structural framework. Their fatty-acid tails may be saturated (left), with a hydrogen atom linked to every carbon bond, or unsaturated (right), with carbons free. (From Capaldi, *Scientific American* March p. 28, 1974).

---

* ATPase is found particularly in the mitochondrial membranes and membranes of chloroplasts in plants (which have a similar function; see p. 1).

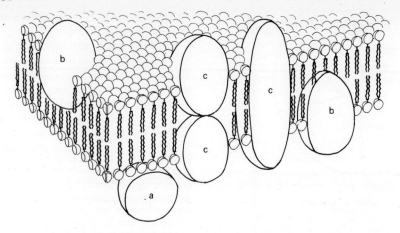

Fig. 10.4 Structural framework typical of cell membranes is made up of a bilayer of lipids with their hydrophilic heads forming outer and inner membrane surfaces and their hydrophobic tails meeting at the centre of the membrane; the bilayer is about 4·5 nm thick. Proteins, the other membrane constituents, are of two kinds. Some (a) lie at or near either membrane surface. The others penetrate the membrane; they may intrude only a short way (b) or may bridge the membrane completely (c), singly or in pairs. (Capaldi, R. A., *Scientific American*, p. 29, March, 1974.)

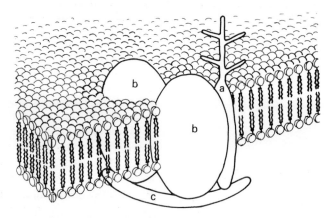

Fig. 10.5 Supramolecular aggregates in red-cell membrane include the two proteins that completely penetrate the membrane. One is a glycoprotein (a); the other is a protein with a molecular weight of 87 000 (b). The third protein in the hypothetical aggregate is spectrin (c). Evidently the three proteins are so linked that if one moves, the others follow. (Capaldi, R. A., *Scientific American*, March, p. 33, 1974.)

The proteins are also found to be free to move around, but being larger than the lipids they move more slowly.

If movement of membrane molecules is random one would expect the ultimate distribution of lipid and protein to be an equalised dispersion throughout the membrane. This is not the case, however; for example in intestinal cells, the glycoproteins are concentrated at one end of the cell and the sodium pump protein at the other. Also in neurones, the enzyme acetylcholinesterase is found only at that part of the cell concerned with synaptic transmission. There must be control systems which determine the precise location of the membrane components but as yet little is known about the nature of them.

11. Blood: Oxygen Transport

The blood, in its circulation, brings to the tissues the foodstuffs and oxygen necessary for their metabolism, and carries away the carbon dioxide and other waste products. The blood, however, is no mere indifferent circulating fluid, but has remarkable "buffering" properties; these enable it to change its composition with respect to the most important substances—oxygen and carbon dioxide—without at the same time bringing about an equally large change in the composition of the fluids surrounding the tissue cells. It can transport much larger quantities of these substances, for example, than could a simple saline solution circulating at the same rate. This property, which is clearly of great importance to the "efficiency" of the blood as a carrier of oxygen and carbon dioxide, is due to the presence of the red "respiratory pigment", *haemoglobin*. Indeed, in the absence of a substance which has the remarkable physical, chemical and physiological properties of haemoglobin, the only animals which could exist would be small and sluggish, living in well-aerated water. One other substance, only, is known which has these properties, and that is *haemocyanin*, the blue "respiratory pigment" of the molluscs and some arthropods.

HAEMOGLOBIN

Chemically, haemoglobin consists of an iron–porphyrin complex, known as reduced haematin, or *haem*, united with a protein, *globin*. Haem is closely related chemically to the prosthetic groups of many enzymes concerned in the oxidation of the foodstuffs, notably the cytochromes, and peroxidase. Haem can unite with a number of nitrogen-containing substances, forming *haemochromogens*; haemoglobin is a special case of a haemochromogen.

Haem and the haemochromogens contain iron in the ferrous (divalent) state. If treated with a suitable oxidising agent, the iron is oxidised to the *ferric* (trivalent) state, with the formation of *haematin* from haem, and *methaemoglobin* from haemoglobin.

Haematin unites with hydrochloric acid to form the hydrochloride *haemin*; this substance crystallises readily from solution, and the character of the crystals is sufficiently distinct to be used as a chemical test for blood pigments. Haemoglobin itself only crystallises easily when derived from certain species, e.g. the horse, rat or guinea-pig. The shape of the haemoglobin crystal varies with the species of blood used. For further information as to the chemical relations of haemoglobin, reference should be made to a textbook of biochemistry. All the compounds concerned have characteristic absorption spectra, which are invaluable for their identification.

Haemoglobin combines with oxygen to form a scarlet compound, *oxyhaemoglobin*; this contains 1 gram-molecule of oxygen for each gram-atom of iron in the haemoglobin. The oxygen can be removed again, just as if it were in simple solution, by shaking the haemoglobin solution repeatedly in a vacuum, or with gas containing no oxygen, such as hydrogen or nitrogen. The colour of the solution changes to purple when the oxygen is removed, and the haemoglobin is then said to be reduced. Haemoglobin, oxygen and oxyhaemoglobin are thus components of a reversible chemical reaction:

$$Hb + O_2 \rightleftharpoons HbO_2$$

(Hb being the symbol generally used for haemoglobin). In this reaction, the oxygen is attached to the haem part of the haemoglobin molecule, but the iron atom remains in the bivalent state; the reaction, therefore, is not one of oxidation, in which the iron would become trivalent, but is referred to as *oxygenation*. Reactions of this kind, in which the oxygen combined with the haem can be removed merely by removing the oxygen dissolved in the solution, occur only when the haem is united to the particular protein, globin; other haemochromogens may be oxidised, but not oxygenated, by oxygen in solution. It is the existence of this easily reversible reaction with oxygen that enables haemoglobin to act as a carrier of oxygen.

In the carriage of carbon dioxide, easily reversible reactions are again involved, but with the globin part of the molecule, rather than the haem part; they are not accompanied by colour changes, and so are less conspicuous. But it is one of the outstanding features in the properties of haemoglobin that the haem and the globin parts interact with one another. The globin not only allows the haem to react reversibly with oxygen, but small changes in its precise chemical composition, or state of ionisation, may affect considerably the affinity of the whole molecule for oxygen, as will be discussed later. Conversely, the state of oxygenation of the haem affects the ionisation of the globin, and its affinity for carbon dioxide.

Oxygen capacity of blood

For an adequate rate of transport of oxygen and carbon dioxide to and from the tissues and the lungs, it is essential that there should be an adequate concentration of haemoglobin in the blood and that the haemoglobin should be able to take up, and lose, oxygen and carbon dioxide. A man may become unduly distressed by going upstairs or running for a bus, not only because his heart or lungs are inadequate, but also because there is not enough haemoglobin in his blood to carry the oxygen and carbon dioxide or to prevent his tissues from becoming too acid. A blood count, of course, will indicate whether he has the normal number of red cells, but these may not contain the normal amount of haemoglobin. The concentration of haemoglobin in the blood may be measured in terms of the maximum quantity of oxygen with which unit volume (usually 100 cm^3) will combine—i.e. the *oxygen capacity* of the blood, as described below or, more simply, in terms of its colour, since haemoglobin has so strong a red colour.

Haemoglobin estimation

The colorimetric (or more strictly absorptiometric) estimation is usually made with the haemoglobin in combination with carbon monoxide (carboxyhaemoglobin); acid haematin and reduced haemoglobin can also be used; all three substances are more stable than oxyhaemoglobin. In the *Haldane haemoglobinometer*, the appropriate volume of the blood under test (20 mm^3) is treated with CO, so as to convert the haemoglobin to carboxyhaemoglobin, and is then diluted in a special tube, of the proper size, until it has the same depth of colour as has an arbitrary standard; this may either be a solution of carboxyhaemoglobin in a sealed tube, or better, a solid rod of suitably coloured glass. The final volume of the diluted unknown blood is then read on special graduations on the tube, which give the haemoglobin concentration as a percentage of that of the standard. This value, sometimes known as the "haemoglobin concentration", does not give the actual haemoglobin concentration of the blood, only the relation to the standard. This latter is conventionally adjusted to contain 15·6 g dl^{-1} blood, which is equivalent to an oxygen capacity of 19·8 ml dl^{-1} blood. These figures, therefore, will also apply to any blood which has "100 per cent haemoglobin".

Alternatively, and more accurately, the haemoglobin concentration may be measured directly in terms of the transmittance of the solution—i.e. the fraction of the incident light which is transmitted through unit depth of solution; the intensities of the incident and transmitted lights being measured by means of a photoelectric cell. It is best to use light of such a colour that it is strongly absorbed by haemoglobin. Solutions of oxyhaemoglobin absorb light in the whole of the blue–violet region of the spectrum (this is why they have a red colour) and in the extreme red end. There are also two well-defined absorption bands, one in the yellow and the other in the green, which are valuable for purposes of identification. Solutions of reduced haemoglobin absorb light more strongly in the extreme red end of the spectrum than do solutions of oxyhaemoglobin, and less strongly in the blue–violet region, and there is only one band in the yellow–green region; this is broad and dim. By the use of suitably coloured filters, transmitting light of certain wavelengths only, and photoelectric cells for measuring the transmittance of the blood or haemoglobin solution at these wavelengths, it is possible to measure, independently, the concentration of oxyhaemoglobin and the concentration of reduced haemoglobin. The complete apparatus for doing this is known as an *oximeter*: it can be made small enough to be attached to the lobe of a man's ear, for example, and will give a continuous record of the degree of oxygenation of his blood.

The colour index

The haemoglobin content of the average red blood cell is expressed in terms of "colour index". The haemoglobin concentration of the blood, as percentage of the standard value, is divided by the red cell count, also as a percentage of the standard value (taken as 5 million per mm^3, so that the percentage red cell count is the actual red cell count multiplied by 20). The colour index is thus ordinarily about unity in healthy persons, but is decreased considerably in *secondary anaemia* (e.g. after repeated haemorrhage), in which the haemoglobin concentration may fall to 50 per cent of the standard value; it is often increased in *pernicious anaemia*, in which the red cell count may fall to one million or less.

COMBINATION OF OXYGEN WITH HAEMOGLOBIN

The amount of oxygen actually combined with the haemoglobin in unit volume of some particular sample of blood—the *oxygen content*—is often expressed as a percentage of the amount that would be combined if it were fully saturated, i.e. as the *percentage saturation*. It is often convenient, also, to refer to the degree of *unsaturation* of a sample of blood; this is a measure of the extent to which the actual oxygen content were found to be, say, 10 ml dl^{-1} blood, and the oxygen *capacity* of the same, or of an exactly similar, sample of blood were found to be, say, 15 ml dl^{-1} blood, the percentage saturation would be 66·7: the sample

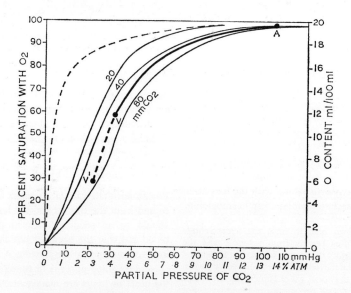

Fig. 11.1 Oxygen equilibrium (dissociation) curves of blood at various partial pressures of carbon dioxide.

The scale of oxygen content on the right of the diagram may be used with reasonable accuracy for normal human blood. If the haemoglobin concentration of the blood considered differs from the normal value, an appropriately different scale must be used.

AVV' is the "physiological oxygen dissociation curve", A being the arterial point, V the venous point at rest, and V' the venous point during exercise.

The broken curve is a rectangular hyperbola; note how little oxygen is given off until the partial pressure of oxygen is reduced to very low values. Such a curve is given by *myoglobin*, which will thus readily withdraw oxygen from haemoglobin in the conditions of venous blood. (After Bock, Field and Adair, and R. Hill.)

would have an unsaturation of 5 ml dl^{-1} and the percentage unsaturation would be 33·3.

If haemoglobin takes part in a reversible chemical reaction with oxygen, to form oxyhaemoglobin, it is to be expected that in any particular sample of blood, the percentage saturation would depend on the concentration of oxygen in solution in the blood (provided, of course, that the haemoglobin is not fully saturated with oxygen). The concentration of oxygen in solution—the amount which is physically dissolved in unit volume—is directly proportional to the partial pressure of oxygen in the gas with which the blood is brought into equilibrium.

The partial pressure of a gas in a mixture of several gases is that fraction of the total pressure which can be considered as being contributed by that gas. This is considered in more detail in the next chapter, where examples are given of the partial pressures of the gases in gas mixtures under various conditions.

Oxygen dissociation curves

It is usual to relate the quantity of oxygen combined with haemoglobin to the partial pressure of oxygen in the gas with which the blood is equilibrated. This relation is shown in Fig. 11.1. Curves such as these are known as *dissociation curves* (or, more properly, *equilibrium curves*). The points through which they are drawn are determined by shaking a sample of blood gently with a suitable mixture of oxygen, nitrogen and, if desired, carbon dioxide, until equilibrium is reached (30 to 45 min), then measuring the oxygen content of the blood and, by analysing the gas mixture, the partial pressure of oxygen. At the partial pressure at which the haemoglobin is fully saturated (100 mmHg; 13·3 kPa), the amount of oxygen dissolved in the blood is about 1/70 of the amount combined with the haemoglobin.

Gas analysis

The type of *gas analysis apparatus* used in establishing the facts given in this chapter was that devised by Haldane. A sample of the gas to be analysed is drawn into a water-jacketed burette, previously completely filled with mercury, and its volume measured. Carbon dioxide is then absorbed by transferring the gas into a vessel from which it displaces a strong solution of caustic potash; it is then brought back into the burette and its volume measured again. Oxygen is absorbed by transferring it in a similar manner into a vessel previously filled with an alkaline solution of pyrogallol, and the residual volume measured once more. Absorption of the carbon dioxide and oxygen is hastened by passing the gas to and from the appropriate absorbing vessels and the burette by raising and lowering the reservoir of mercury connected with the lower end of the burette. Various modifications have been made to the original design in attempts to make its use quicker and simpler. The Scholander apparatus, for example, is specially designed so as to need only 0·5 ml of gas for analysis (the Haldane apparatus needs about 10 ml). It has some advantages over the Haldane apparatus even when plenty of gas is available for analysis; some find it easier to use and keep in proper order.

Modern methods

Modern instrumentation includes the "oxygen meter" which gives a direct reading of O_2 per cent on a scale. The principle of its operation is that oxygen has paramagnetic properties; a little soft-iron dumb-bell, suspended in a magnetic field, is deflected by this field in varying degrees depending on the amount of oxygen present. The infrared CO_2 meter is also a direct reading instrument and works on the principle that infrared waves are absorbed by CO_2. A source and a detector are separated by a tube through which the sample to be analysed is passed and the absorption (proportional to CO_2 concentration) is measured directly on a meter.

Mass spectrometry

Mass spectrometry is the method of choice for gas analysis where accuracy, versatility, rapidity and the simultaneous analysis of several constituents in small samples of gas mixtures is required.

Mass spectrometers analyse mixtures by separating molecules of each gas in accordance with their mass. Gas molecules are first ionised, so that they become electrically charged; they are then accelerated by an electrostatic field and they then pass through an electromagnetic field. The latter deflects the gas molecules according to their charge and mass; the lighter the molecule the greater the deflexion.

The accelerated ions subsequently fall on a collector which gives an output voltage which can be written out on a chart-recorder or oscilloscope face. By varying the accelerating voltage the variously deflected beams of ions will fall in turn upon the collector and a series of peaks will be shown on the recorder, each representing by its size the amount of a particular gas present in the original sample (Fig. 11.2).

Blood–gas analysis

The *amount of oxygen combined with haemoglobin* is estimated by driving off the oxygen from the blood, either: (1) by boiling *in vacuo*; or (2) by adding ferricyanide; or (3) by both together.

The first was the method used by the earliest workers, who evacuated a large vessel by means of a mercury pump, for example, a Töpler pump, and ran a known volume of blood, previous-

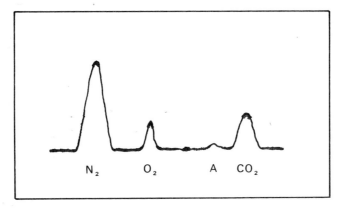

Fig. 11.2 Mass spectrum as seen on oscilloscope screen. Sample of expired air being analysed. Peaks are N_2, O_2, A and CO_2. The "bright-ups" show the position of the peaks selected for tuning pen-recorder to follow variability of gases in sample.

ly warmed to 40°C, into it. The gas liberated was then pumped off and delivered to a gas analysis apparatus, where its total volume and composition were determined.

If oxyhaemoglobin is treated with ferricyanide, the reduced haemoglobin which is always present in small quantities is converted into methaemoglobin, so that the equilibrium between oxyhaemoglobin and reduced haemoglobin is upset; in an attempt to restore equilibrium oxyhaemoglobin gives up its oxygen, and more reduced haemoglobin is formed; this is again removed by the ferricyanide, and more oxyhaemoglobin decomposes, and so on. In this way, the whole of the oxygen in reversible combination can be driven off under suitable conditions, and its volume measured. When this reaction was discovered, it seemed that it would be a much easier way of determining the oxygen combined with haemoglobin than the vacuum pump methods previously used. Haldane, indeed, at once adopted the principle in his blood–gas apparatus. In this, oxyhaemoglobin is mixed with *alkaline* ferricyanide (the alkalinity prevents the escape from the solution of any carbon dioxide gas) and the only gas evolved is oxygen, which is measured directly in a burette over water. Much work has been done by these two methods, but doubt has been cast on their general applicability by the finding that the evolved oxygen is sometimes in part reabsorbed owing to a secondary reaction with the blood in presence of ferricyanide. This can be overcome by taking adequate precautions.

This error is avoided by combination of vacuum extraction and addition of ferricyanide. This principle has been developed by van Slyke to a high pitch of perfection, and his apparatus has been applied to numerous other estimations besides that of oxygen and carbon dioxide in blood. An acid ferricyanide solution and a known volume of the blood under examination are run into an evacuated vessel; the gases given off are extracted by shaking and the residual solution is discharged by means of a two-way tap at the bottom of the extraction vessel. The volume of the extracted gases is determined either by transferring them to a burette, over mercury, and measuring the volume occupied at atmospheric pressure (constant pressure apparatus) or by compressing them to a known volume over mercury and measuring the pressure exerted, with a mercury manometer (constant volume apparatus). This is the more accurate method, since the scale on which the measurements are made is a great deal longer, and no errors can arise from faulty adjustment of the pressure of the gas in the burette. Carbon dioxide is absorbed from the extracted gases by adding a caustic soda solution, and oxygen by adding a solution of pyrogallol or sodium hyposulphite; the contraction in volume (or reduction in pressure) is measured at each stage, and represents the volume of the respective gas present in the extracted mixture.

When plotted in terms of the oxygen content of the blood, the equilibrium curve rises, at large values of the oxygen pressure, to the value of the oxygen capacity. This depends on the haemoglobin concentration of the blood, as measured, for example, by the haemoglobinometer. When plotted in terms of the percentage saturation, the equilibrium curves of all samples of blood rise, of course, to the same maximum value. But such curves may be steeper or flatter, may be compressed or spread out along the axis of oxygen pressure, as the temperature and the acidity of the blood are decreased or increased; and according to the kind of animal, or even particular individual, from which the blood was obtained. Increase of acidity, in particular, decreases the percentage saturation at a given oxygen pressure, i.e.

decreases the affinity of the haemoglobin for oxygen, and thus makes the curve flatter—an effect, as already remarked, which results from the interaction between the haem and the globin. Increased carbon dioxide pressure increases the acidity and hence has the same effect, as is shown by the family of curves at different carbon dioxide pressures given in Fig. 11.1. (It is possible that carbon dioxide also exerts a specific effect due to its combining directly with haemoglobin to a slight extent.) Lastly, the shape of the curve, as indicated by the size of the inflection at low pressures of oxygen, may depend on the electrolyte composition of the blood.

The dissociation curve of whole mammalian blood is always found to be S-shaped, as in Fig. 11.1. This shape is of distinct physiological service, since it enables a large amount of oxygen to dissociate from the haemoglobin without too severe a drop in the oxygen pressure with which it is in equilibrium. This is an important point as regards supply of oxygen to the tissues, for it is the partial pressure of oxygen in the blood, and not the amount of combined oxygen, that determines the rate of diffusion of oxygen from the blood to the tissues. A curve of the rectangular hyperbola type (v broken curve in Fig. 11.1) would, from this point of view, be obviously unserviceable.

Shape of dissociation curve

The shape of the dissociation curve can be explained by applying the Law of Mass Action to the equilibrium between oxygen and haemoglobin; but the reaction is somewhat complicated and is not accurately represented by the simple equation given above. Chemical analysis shows that the weight of haemoglobin which contains 1 atomic weight of iron, i.e. 56 g, is 16 700 g. Direct determinations of the molecular weight of haemoglobin from mammalian blood, by osmotic pressure determinations, or by the ultracentrifuge, give a value of 67 000. Thus each molecule of haemoglobin contains four atoms of iron, and will combine with four molecules of oxygen. These can combine with the Hb_4 molecule one by one, forming the intermediate compounds Hb_4O_2, Hb_4O_4, Hb_4O_6 and, finally, Hb_4O_8. The S shape of the dissociation curve is due to the fact that the affinity of each of the four haem groups for oxygen depends on whether any or all of the other three have already combined with oxygen or not; in Hb_4O_6, in particular, the remaining unoccupied haem must be presumed to have a much greater affinity for oxygen than it had when all four of the haems were unoccupied. The four haem groups must thus interact with each other, presumably through the globin part of the molecule. Since the magnitudes of all these interactions are not yet known, the Mass Action equation must contain four arbitrary constants. But if these are suitably chosen, the rather complicated equation fits the observed curves very accurately.

Factors altering the dissociation curve

The effect of carbon dioxide (and acidity) on the affinity of haemoglobin for oxygen is of considerable importance in enabling the blood to unload oxygen more readily into the tissues. It was first studied by Bohr, Hasselbalch and Krogh in 1904, and is often referred to as the *Bohr effect*. Thus, for example, it can be seen from the curves that at an oxygen partial pressure of 32 mmHg (4.26 kPa) the oxygen content of normal human blood is 13 ml dl^{-1} when the carbon dioxide pressure is 40 mmHg (5.32 kPa), and a little under 12 ml dl^{-1} when the carbon dioxide pressure is increased to 50 mmHg (6.65 kPa). Now, the partial pressure of carbon dioxide is about 40 mmHg in the arterial blood and about 50 mmHg in the venous blood, so that roughly 1 ml of extra oxygen is obtained by the tissues from each decilitre of blood, owing to this displacing action of acid, without lowering the partial pressure of oxygen within them.

Since carbon dioxide is blown off in the lungs, and taken up in the tissues, none of the family of curves shown in Fig. 11.1 accurately represents the "physiological oxygen dissociation curve". This passes from a point, A, corresponding to the conditions met with in the lungs, i.e. 98 per cent saturation with oxygen at a partial pressure of 108 mmHg (14.36 kPa) of oxygen and 40 mmHg of carbon dioxide, to a point, V, corresponding to the conditions met with in the tissues, i.e. 58 per cent saturation with oxygen at a partial pressure of 32 mmHg of oxygen and 50 mmHg of carbon dioxide.

It will be observed that the whole of the oxygen is never removed from the blood. This is due, first, to the fact that a small, but definite, partial pressure of oxygen must exist in the tissue cells and, second, to the fact that oxygen must diffuse from the blood capillaries to the tissue cells, sometimes over quite a long distance. A considerable head of pressure is thus needed to drive the oxygen across at the requisite rate.

Exercise

In exercise, the rate at which oxygen is needed by the muscles is greatly increased, and the amount of oxygen taken from each millilitre of blood is also increased, with the result that the partial pressure of oxygen in the venous blood is decreased (point V' in Fig. 11.1). There is thus a smaller head of pressure available for an increased rate of diffusion. This apparent contradiction is resolved by the observation that the number of capillaries carrying blood, per unit volume of muscle, is enormously increased when the muscle becomes active—possibly becoming 100 times greater. Consequently, not only is the distance decreased over which the oxygen must diffuse, but also the surface is increased over which oxygen can leave the blood.

The fetal blood

The haemoglobin in the blood of the fetus has a greater affinity for oxygen than has that in the maternal blood. Its dissociation curve, even at relatively high partial pressures of carbon dioxide, would lie to the left, in Fig. 11.1, even of that drawn for 20 mmHg (2.66 kPa) of carbon dioxide. When the maternal blood, therefore, is, say, 50 per cent saturated with oxygen, the fetal blood will be, perhaps, 80 per cent saturated. Oxygen will consequently pass readily from the mother to the fetus in the placenta. There is, of course, a corresponding disadvantage, in that the partial pressure of oxygen in the fetal tissues must be small before the fetal haemoglobin will part with its oxygen.

Myoglobin

Most mammalian muscles contain a pigment which is closely related chemically to haemoglobin, and is known as *myohaemoglobin*, or *myoglobin*. It differs from haemoglobin in having an oxygen dissociation curve which is a rectangular hyperbola (broken curve in Fig. 11.1). Myoglobin will become nearly fully saturated with oxygen at partial pressures normally found in the tissues, and at which haemoglobin has parted with most of its oxygen. It is known, also, that the tissue oxidation enzymes will function at oxygen pressures down to 5 mmHg (0.67 kPa) or a little less, and at these pressures myoglobin will part with about 40 per cent of its oxygen. Myoglobin, therefore, can act as an effective reservoir of oxygen in the muscles.

Speed of reactions

Under physiological conditions the combination of oxygen with haemoglobin, and the dissociation of oxyhaemoglobin only take about 1/100 s, and are thus too rapid to limit the rate of exchange of oxygen between the circulating blood in the capillaries (which each corpuscle takes about a second to traverse) and the lungs or tissues. Diffusion through the tissue cells and into the interior of the red blood cells seem to be the main factors which limit the rapidity of oxygen exchange in the animal.

Such very rapid chemical reactions cannot be timed by the ordinary methods—they require the special methods of Hartridge and Roughton for measurement of their rates. To determine the speed of combination, for instance, a solution of reduced haemoglobin and a solution of oxygen in water are driven through separate leads into a small chamber where they mix in less than 1 ms and travel thence into an observation tube. The percentage oxyhaemoglobin in the streaming fluid at various positions along the obervation tube is measured spectroscopically, the fluid being kept in motion whilst the readings are taken. From the rate of flow of the liquid and the distance of the point of observation from the mixing chamber, the time taken by the reaction to reach the oxyhaemoglobin percentage recorded by the spectroscope is simply calculated, and hence the velocity of the reaction can be determined.

THE COMBINATION OF HAEMOGLOBIN WITH CARBON MONOXIDE

Carbon monoxide also combines reversibly with haemoglobin, forming a compound usually known as *carboxyhaemoglobin*, and often written COHb. If a solution of haemoglobin is equilibrated with a mixture of oxygen and carbon monoxide, the partial pressures, P_{O_2} and P_{CO} being so large that no reduced haemoglobin is present, the ratio of carboxyhaemoglobin to oxyhaemoglobin is defined by the equation:

$$\frac{COHb}{O_2Hb} = M \cdot \frac{P_{CO}}{P_{O_2}}$$

where M is about 250.

Thus, if the pressure of carbon monoxide is about 1/250 of that of oxygen, one-half of the haemoglobin will be combined with carbon monoxide, and one-half with oxygen; haemoglobin has about 250 times as great an affinity for carbon monoxide as it has for oxygen. This great affinity is due to the fact that carboxyhaemoglobin dissociates at least 1000 times more slowly than does oxyhaemoglobin. In other respects, the reaction between carbon monoxide and haemoglobin is very like that between oxygen and haemoglobin—the volume of carbon monoxide combined at maximum saturation is the same as that of oxygen; the dissociation curve (in the absence of oxygen) is also S-shaped and similarly affected by acidity and temperature.

Carbon monoxide poisoning

The great affinity between carbon monoxide and haemoglobin accounts for the danger attending the inhalation of small amounts of carbon monoxide or a mixture of gases containing it. Blood can be completely saturated with carbon monoxide at a partial pressure of only 0.5 per cent of an atmosphere (0.6 kPa) at 37°C, whereas it requires at least 15 per cent of oxygen before even approximate saturation is reached. If the whole of the available haemoglobin becomes saturated with carbon monoxide, no oxygen can be carried, and the animal will be asphyxiated.

No marked symptoms are detectable until about 30 per cent of all the haemoglobin in the body is saturated with carbon monoxide. Vision, hearing and intelligence become impaired when the carbon monoxide saturation reaches 50 per cent, and death has been known to occur at 60 per cent saturation; 80 per cent saturation is almost invariably fatal. The average dissociation curve of carboxyhaemoglobin shows that with most people the first symptoms would be observed when the air they were breathing contained about 0·03 per cent (1 part in 3000) of carbon monoxide, while 0·4 per cent (1 part in 250) would be fatal; different people, however, have somewhat different susceptibility.

If a coal-gas leak occurs in an ordinary room (a gas fire is turned on and not lit, for example), it is very unlikely that there will ever be more than 0·25 per cent of gas in the room, owing to leakage up the chimney and through the cracks of the window and diffusion through the walls and the ceiling to adjacent rooms (this would probably not apply to the case of a large gas cooker in a small kitchenette). Ordinary coal gas contains, as a rule, about 10 per cent of carbon monoxide; water gas, however, is now usually added, and the gas as supplied may contain 20 per cent, or more, of carbon monoxide. There might thus be as much as 0·5 per cent of carbon monoxide in the room, so that an ordinary leak in an ordinary room might have fatal consequences, and would almost certainly produce serious symptoms. Coal gas, it must be remembered, is lighter than air, and hence rises to the ceiling; a tall man may thus notice symptoms of anoxaemia before a short man. Natural gas, on the other hand, may contain up to 80 per cent methane, and is much safer.

Tobacco smoke contains appreciable quantities of carbon monoxide, and some of the deleterious effects of oversmoking have been attributed to a chronic anoxaemia produced by the continuous presence of carboxyhaemoglobin in the blood, it is, indeed, quite easy to detect the presence of this compound in the blood of a heavy smoker. The exhaust gas of motor cars also contains considerable amounts of carbon monoxide.

The treatment in all cases of carbon monoxide poisoning should be the administration of oxygen. The high pressure of oxygen not only facilitates the eventual dissociation of the carboxyhaemoglobin, but also enables an appreciable amount of oxygen to be carried in simple solution in the blood.

12. Oxygen Transport in Man and its Disorders

The carriage of oxygen from the lungs to the tissues involves a number of organs, each having its own servocontrol systems. Malfunction of any (or all) of these organs and systems will produce complex interactions and compensations which are worth considering if only from the point of view of the light they throw on the physiological mechanisms involved.

The oxygen supply

The factors concerned in oxygen supply are:

1. Pulmonary gas exchange
2. Pulmonary blood flow
3. Haemoglobin concentration
4. Affinity of haemoglobin for oxygen

In normal life, all these factors are adjusted to the requirements of the tissues, so that the end-capillary oxygen tension is kept at the correct level. Processes involved in tissue oxygen supply are summarised in Fig. 12.1. The whole mechanism has a reserve capacity and the ability to respond quickly to alterations in oxygen requirements.

The lungs

At rest, the lungs take in about 4 ml O_2 min^{-1} kg^{-1}, and can increase this up to twentyfold. The process is regulated by the servosystems outlined in Chapter 45, the receptors for which are the aortic and carotid bodies (sensitive to arterial oxygen tension, P_aO_2) and the brainstem chemoreceptors (responding to arterial carbon dioxide tension, P_aCO_2 and arterial pH). The uptake is not greatly affected by changes in oxygen tension because of the shape of the O_2 dissociation curve, as explained on page 92. Arterial oxygen tension must fall from 100 mmHg (13·3 kPa) to well below 50 mmHg (6·67 kPa) before any obvious increase in pulmonary ventilation is elicited. Contrast this with the small increment in P_aCO_2 which will quickly stimulate respiration.

Cardiovascular system

The level of *cardiac output* and the *distribution* of blood flow in various regions affects oxygen supply. Normal cardiac output is 3–4 litres min^{-1}; maximum levels can be five times as much.

Direct effects of P_aO_2 changes on *cardiac output* do not occur, because other factors are able to effect the required compensation.

Blood flow distribution is variable for functions apart from those concerned in oxygen supply. Separate control mechanisms cater for flow distribution in different organs. For example, skin blood flow is mainly used as a heat-loss regulating system, controlled by hypothalamic centres; yet when the supply of oxygen fails, blood flow is restricted in the skin (and other tissues having a low O_2 extraction) in order to divert it to essential organs such as the brain.

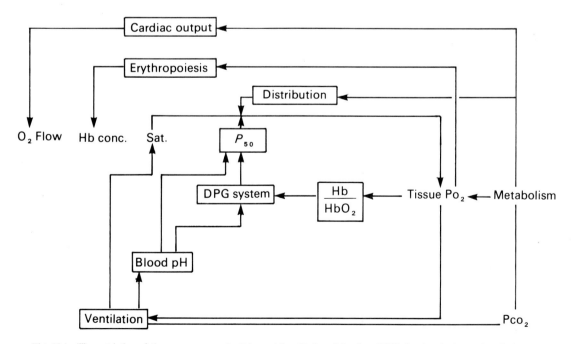

Fig. 12.1 The regulation of tissue oxygen supply. Scheme (after Finch and Lenfant, 1972) showing the interactions between various parameters of the respiratory process and the feedbacks controlling them.

This is a kind of "oxygen reserve" and can compensate for a 50 per cent decrease in oxygen supply (for a limited period).

Oxygen affinity of haemoglobin

This is expressed as the P_{50}, or O_2 tension at which 50 per cent of the haemoglobin is saturated with O_2.* The detailed relationship between PO_2 and percentage saturation is given by the O_2-dissociation curve (see p. 90), the sigmoid shape of which is due to conformational changes in the haemoglobin molecule—the uptake of each molecule of oxygen then leads to an enhanced uptake of the remainder.

Physiologically, this means that the flat top part of the curve maintains a nearly constant arterial oxygen content in the face of a variable alveolar PO_2.

The steep part of the curve allows big changes in oxygen release, but with very small change in the capillary oxygen tension.

CHANGES IN THE O_2-DISSOCIATION CURVE
From the above considerations one would expect any departure from the ideal configuration of the curve to lead to large upsets in oxygen carriage. This will be discussed later. Equally, the portion of the curve upon which a given sample is resting, will greatly influence the amount of oxygen available at the normal tissue PO_2 levels. On the other hand, oxygen uptake is not affected much.

In man, the affinity of pure haemoglobin for oxygen is so marked that oxygen cannot be given off to the tissues, at least around normal tissue oxygen tensions. Various factors promote oxygen release. These are:

1. Hydrogen ions
2. Carbon dioxide
3. 2,3-Diphosphoglycerate (DPG)

If any (or all) of these substances are present, the deoxy configuration of the haemoglobin molecule is stabilised, thus promoting the release of oxygen. Hydrogen ions stabilise the salt bridges; carbon dioxide acts through the formation of carbamino groups and DPG unites the beta-chain subunits.†

The actions of DPG

In each red blood corpuscle there is a mechanism for responding to anoxia. One red cell contains about 15 μmol DPG per g of haemoglobin. The DPG is within the haemoglobin core, binding to the beta-chains, while the haemoglobin is in the *deoxy* form.

If there is increased formation of deoxyhaemoglobin there is increased glycolysis, which raises the concentration of DPG (DPG is formed by the anaerobic glycolytic pathway). This increased DPG level increases the oxygen availability. Other things being equal, if the cell concentration is doubled to, say, 30 μmol g^{-1} there will be an increase in P_{50} from 26·5 to 36·8 mmHg (3·52–4·89 kPa). This is shown in Fig. 12.2. The reader will recognise the similarity between this system and the Bohr effect (p. 92). The unsatisfactory feature of the Bohr effect, as far as being useful is concerned, is that it only operates when acidosis is associated with the hypoxia.

Acidosis, though, inhibits glycolysis in the red cell; a fall of DPG results. This just about counterbalances the pH Bohr effect. Acidosis only promotes oxygen release for a few hours; the DPG mechanism is sustained.

* Note that a high P_{50} means a low affinity of haemoglobin for oxygen, and vice versa.
† The interested reader should consult Perutz, M. F. (1970), Stereochemistry of Cooperative Effects in Haemoglobin, *Nature* **228**, 726.

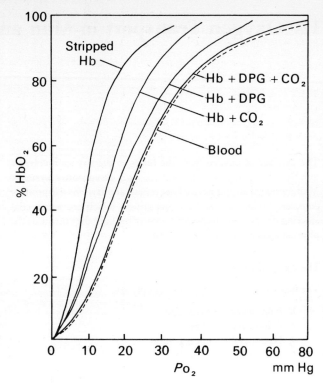

Fig. 12.2 Regulation of oxygen affinity by DPG and CO_2.

Oxygen dissociation curves of blood (extreme right-hand curve) and haemoglobin (left). The effects of adding CO_2, DPG, and both CO_2 and DPG to the stripped haemoglobin are shown in the remaining curves. Neither DPG nor CO_2 alone is enough to lower the oxygen affinity of pure haemoglobin solutions to that of whole blood. The curve for blood (dotted) coincides with that for haemoglobin plus DPG and CO_2.

Details of curves (from left to right): Stripped haemoglobin (2·0 mM in tetramer) at 37°C, 0·1 m KCl; with PCO_2 = 40 mmHg; with DPG (1·2 moles per mole haemoglobin tetramer); with DPG and CO_2; and blood at pCO_2 =40 mmHg. pH of haemoglobin solutions was 7·22 at 50 per cent saturation. Plasma pH of blood at 50 per cent saturation was 7·40, corresponding to a pH of 7·22 inside the red cell. (From Kilmartin and Rossi-Bernardi, *Physiological Reviews*, Vol. 53, p. 884, 1973.)

Haemoglobin concentration

This is regulated by a feedback servomechanism involving the kidney, which in essence is keeping balanced the oxygen supply to, and the oxygen requirement of, renal tissue. Decreases in concentration, or arterial oxygen saturation of, haemoglobin or increased affinity of haemoglobin for oxygen, gives rise to the elaboration of more *erythropoietin* (sometimes called haemopoietin).

There can, of course, be quite large changes in renal blood flow without erythropoietin production occurring, but if these changes result from altered renal oxygen requirements, the ratio between supply and demand remains unaltered. This mechanism, acting on red cell production, can raise the latter from the normal 1 per cent to about 2 per cent of the total number of red cells in the circulation. The normal level 15 g dl^{-1} of haemoglobin in the blood is kept constant in this way. Erythropoiesis is dealt with on page 83.

DISORDERS OF OXYGEN CARRIAGE

The various components of the oxygen carriage mechanism, already described, are all subject to abnormalities of function.

The maximum capacity for oxygen transport is decreased, but usually compensations take place in the rest of the system and basal requirements are usually taken care of.

Disorders of oxygen uptake

If the impairment of oxygen uptake is long-standing, the blood oxygen desaturation leads to the release of erythropoietin and this leads to *polycythaemia* (if the bone marrow is able to respond), and increased haemoglobin concentration (Fig. 12.3).

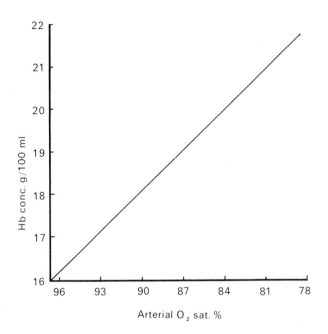

Fig. 12.3 Blood oxygen desaturation and erythropoiesis.
Relation between arterial oxygen saturation (%) and haemoglobin concentration (g 100 ml^{-1} blood) in human subjects at high altitude. The bone marrow responds (via the secretion of erythropoietin) to the decreased arterial oxygen saturation rather than to its partial pressure. Increased DPG levels are also found when arterial saturation is decreased.
(Data of Hurtado *Annals of Internal Medicine*, Vol. 53, p. 247, 1960.)

As would be expected, DPG levels rise when arterial saturation is lowered.

Clinical disorders
Clinical conditions leading to impaired oxygen loading include cardiac and pulmonary shunts, pulmonary disease giving restricted ventilation and in addition, the hypoxia found in living at high altitude.

Adjustments to high altitudes (see also Chapter 46).
Increased pulmonary ventilation occurs, but this leads to excessive washout of carbon dioxide. Thus oxygenation is improved but at the cost of impaired cerebral blood flow, due to the low P_aCO_2. Oxygen release is reduced.

Vasoconstriction takes place peripherally, plasma volume is less, with a shift centrally of blood (which can lead to pulmonary oedema). After a few days, oxygen release is stimulated by a rise in red cell DPG. Also, there are electrolyte changes and a gradual acclimatisation of the respiratory centre which becomes less sensitive.

Cardiovascular failure

The effects of hypoxia include: (1) decreased activity, (2) central nervous system abnormalities, (3) necrosis of the liver, (4) sodium retention by the kidney.

In cardiac failure, it is usually found that the arteriovenous oxygen difference is greater than normal. Compensation is brought about by the redistribution of blood flow (less to skin, splanchnics etc.) and a fall in oxygen affinity. DPG increase gives, as would be predicted, an increase in P_{50}. Normally, the blood haemoglobin does not rise; however, in severe failure when sodium retention and a consequent enlarged plasma volume occur, normal haemoglobin levels are maintained by an increase in the number of circulating red cells. If diuresis should follow (as a result of treatment, for example), the plasma volume will fall leaving a raised haemoglobin.

These adjustments may be upset in the case of febrile illness where both skin blood flow may rise and metabolic requirement for oxygen may be increased by the raised body temperature.

Abnormalities in the amount of haemoglobin

When the concentration of haemoglobin is decreased as in *anaemia*, blood viscosity falls (Chapter 34) and blood volume remains essentially within normal limits. Cardiac output is not altered until the haemoglobin falls to around 10 g dl^{-1}.

The important adjustments are made in terms of a rise in the DPG* content of red cells (thus the haemoglobin oxygen affinity falls), and additionally in terms of a redistribution of blood flow. Since there will be less haemoglobin available for buffering, in exercise there will be a bigger fall in venous pH than usual. This, by the Bohr mechanism, will lead to greater oxygen unloading in tissues (for a given tissue PO_2). These compensatory systems are usually good enough in most patients to ensure that few symptoms accompany anaemia. Only in severe degrees of anaemia, and particularly when it is of a sudden onset (as after a haemorrhage), is dyspnoea and palpitation at all obvious.

When there is an increased number of red cells in the circulation, or *polycythaemia*, oxygen transport is in excess of requirement. Blood viscosity is increased and this may be sufficient to embarrass the circulation. The condition may be compensatory as already mentioned or may arise as an abnormality of erythropoietin production (in kidney tumours).

Abnormal haemoglobin affinity for oxygen

There are various defects in the haemoglobin molecule itself, or in the enzymes which regulate glycolysis, which will give abnormalities in the affinity for oxygen.

Various *abnormalities in the haemoglobin molecule* occur in which one of the amino acids in the molecule is incorrectly substituted by another. This interferes with the allosteric response of the haemoglobin to oxygen loading. Here, it is found that the position and/or shape of the oxygen dissociation curve is altered, in consequence altering the amount of oxygen given up at normal tissue partial pressure, around 40 mmHg O_2 (5·32 kPa)). Since the tissue oxygen requirements are not altered, DPG levels are not affected but there is an alteration in the tissue PO_2 and this in turn affects erythropoiesis. In the long run therefore, either anaemia or polycythaemia results and these abnormalities usually lead to the detection of the underlying haemoglobinopathy. Most are congenital.

---

* It has been calculated that DPG-induced changes in haemoglobin affinity for oxygen may compensate for up to half the oxygen deficit in anaemia.

The *abnormalities in glycolytic enzymes* (also of congenital origin) usually influence oxygen affinity by changing DPG levels.

Enzymatic lesions in the Embden–Meyerhof pathway up to the level of 1,3-DPG, give lower levels of DPG; rate-limiting lesions in the glycolytic pathway following 3-DPG, can either raise or lower DPG levels. In both cases the P_{50} will be affected one way or the other.

Oxygen affinity is altered also in blood with low haemoglobin concentrations. Stored blood loses DPG and therefore has lower oxygen affinity—a fact of clinical importance and a problem to some extent solved by adding inosine to it.

Abnormal pH and electrolyte levels in the blood also affect oxygen affinity.

Tissue abnormalities

Alterations in cellular metabolism will change the oxygen requirement. Demands are met by changes in cardiac output and in blood volume. In athletes, for instance, cardiac reserve is enhanced; blood volume is expanded. Similar changes are seen in thyrotoxicosis. In the latter disease, DPG levels are raised, perhaps due to a specific effect of thyroxin upon glycolysis.

13. The Carriage of Carbon Dioxide

Blood loses oxygen and gains carbon dioxide in its passage through the tissues and undergoes the reverse changes during its passage through the lungs. In the tissues and lungs the changes are roughly equal and opposite, so that it will only be necessary to describe one of them. For convenience, we shall consider the uptake of carbon dioxide by the blood in the tissues. The partial pressure of carbon dioxide in arterial blood is usually about 40 mmHg (5·32 kPa), and the total carbon dioxide content (as estimated by vacuum extraction with acid) is about 50 ml dl^{-1} (at N.T.P.). The partial pressure of carbon dioxide in the tissues is higher than that in the arterial blood, hence carbon dioxide diffuses from the tissues into the blood capillaries. The amount of carbon dioxide taken up by the blood is some twenty times greater than the amount taken up by water under like conditions of carbon dioxide pressure, hence only about 5 per cent of the carbon dioxide uptake by the blood is by simple solution, the remaining 95 per cent being through chemical combination.

The problem then, as in the case of oxygen, is to discover the means whereby each unit volume of blood takes up in the tissues, and gives out in the lungs, so much more carbon dioxide than would water in similar circumstances. We now know that there are two chief means by which this is done; one resulting from the combination of carbon dioxide with water, to form carbonic acid, which is buffered just as is any other acid, and the other resulting from a direct combination of carbon dioxide with haemoglobin.

In forming our ideas as to the transport of carbon dioxide, we must distinguish between the results of experiments in which blood is shaken outside the body with various gas mixtures, and the behaviour of the blood in the circulation. The former usually takes about 15 min or longer; in the latter, the blood is often only in contact with the tissues (where it takes up carbon dioxide) and with the lungs (where it gives up carbon dioxide) for times of the order of 1 s. Attention must therefore be paid to the speed of the various processes, and this has revealed the presence of important factors which would otherwise have been missed.

Carbon dioxide as an acid

Carbon dioxide combines with the water in which it is dissolved, to form carbonic acid; and the buffer substances in the blood (chiefly haemoglobin) prevent the acidity from changing to any serious extent. The reactions involved may be written:

$$CO_2 + H_2O \rightleftharpoons H_2CO_3$$
$$H_2CO_3 \rightleftharpoons H^+ + HCO_3^-$$
$$H^+ + Hb^- \rightleftharpoons HHb$$

These reactions can be combined into the one equation:

$$CO_2 + H_2O + Hb^- \rightleftharpoons HCO_3^- + HHb$$

The net result of the series of reactions therefore, is that almost all the extra carbon dioxide is carried in the blood in the form of HCO_3^- ions, the negative charges being supplied by the Hb^- ions. By this means the amount of carbon dioxide that can be carried in a given volume of blood is increased enormously over that which would be carried at the same partial pressure, in simple solution.

This does not, strictly, express the whole story. There must be some increase in hydrogen ion concentration, otherwise there would be no reason why the ionisation of the haemoglobin should be depressed. This rise in hydrogen ion concentration has the effect of preventing some of the added carbon dioxide from being converted into bicarbonate ions. A little, therefore, remains as carbonic acid (the partial pressure of the carbon dioxide in venous blood is greater than that of arterial blood); the majority forms hydrogen and bicarbonate ions, and most of the hydrogen ions are absorbed by the haemoglobin.

The reaction of carbon dioxide with the plasma proteins may be expressed in a similar way, viz:

$$H_2CO_3 + Pr^- \rightleftharpoons HCO_3^- + HPr$$

and with phosphates by the equation:

$$H_2CO_3 + HPO_4^{2-} \rightleftharpoons HCO_3^- + H_2PO_4^-$$

The plasma proteins contribute only about 7 per cent, and the phosphates less than 3 per cent of the total buffering and carbon dioxide carrying power of the blood. They are, therefore, of minor importance as compared with haemoglobin.

Oxyhaemoglobin and reduced haemoglobin as acids

This superior efficiency of haemoglobin over the plasma proteins is in part due to its higher concentration and its greater ionisation in the physiological pH range; but it is, in the main, due to a much more important factor which has not yet been considered. In the living animal, when blood takes up carbon dioxide it also loses oxygen from combination with haemoglobin, so that the latter is left in the reduced form. Now it has been shown that reduced haemoglobin is a *weaker acid* than oxyhaemoglobin, i.e. at a given pH, reduced haemoglobin is less ionised than oxyhaemoglobin. This means that the reaction

$$H_2CO_3 + Hb^- \rightleftharpoons HCO_3^- + HHb$$

must proceed further to the right at a given partial pressure of carbon dioxide, than does the reaction

$$H_2CO_3 + HbO_2^- \rightleftharpoons HCO_3^- + HHbO_2$$

i.e. the reduced haemoglobin ions do not hold on to their negative charges so firmly as do the oxyhaemoglobin ions, and so transfer them more readily to the bicarbonate ions. Reduced blood, therefore, takes up carbon dioxide in the tissues more readily than does oxygenated blood. Conversely, in the lungs when oxyhaemoglobin is re-formed from reduced haemoglobin the carbon dioxide will be liberated from the blood more readily.

These relationships are well shown by a comparison of the CO_2 dissociation curves of oxygenated and reduced blood. In curves of this kind the total concentration of carbon dioxide (chemically combined and dissolved) is plotted against the partial pressure of carbon dioxide in the gas phase when equilibrium has been reached with the blood. In Fig. 13.1 the point A represents the average condition of the arterial blood, V

The fact that loss of oxygen from the blood, as it passes through the tissues, renders it *ipso facto* more competent to take up carbon dioxide, just where such extra power is needed, and conversely, the fact that the uptake of carbon dioxide *ipso facto* drives off oxygen into the tissues, are beautiful examples of the way in which divers chemical phenomena are coordinated into a harmonious physiological process.

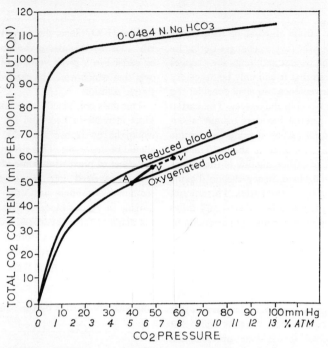

Fig. 13.1 Carbon dioxide equilibrium (dissociation) curves of oxygenated and reduced blood, and of a solution of sodium bicarbonate of the same concentration of total available base as the blood.

AVV is the "physiological carbon dioxide dissociation curve", A being the arterial point, V the venous point at rest, and V' the venous point during exercise. (After Parsons.)

the average condition of the mixed venous blood at rest, V' the condition of the venous blood in exercise. V and V' are both situated between the oxygenated and reduced blood curves, since in the circulation the venous blood is only partially reduced. It will be seen that the line AV is about twice as steep as the carbon dioxide dissociation curve of oxygenated or reduced blood. The change from the fully oxygenated to the partially reduced state enables the blood to take up about twice as much carbon dioxide for a given increase of CO_2 pressure as it would if there were no change in the strength of the haemoglobin as an acid. This phenomenon, first studied by Christiansen, Douglas and Haldane in 1913, is sometimes known as the *Haldane effect*.

The whole difference in CO_2 carrying power of oxygenated and reduced blood must not be attributed to the difference in acid strengths of oxy- and reduced haemoglobin. The formation of carbamino compounds is also partly responsible.

The explanation of how reduced haemoglobin comes to be a weaker acid than oxyhaemoglobin is probably that one of the —NH_2 groups in the molecule, which takes up H^+ ions, is close to the haematin nucleus and that when O_2 combines with the latter, it reduces the ease of binding of the H^+ ion at this neighbouring —NH_2 group. Oxyhaemoglobin, being thus less able to bind H^+ ions, behaves as a stronger acid.

The "chloride shift"*

When carbon dioxide enters the blood from the tissues, some stays in the plasma and the rest passes into the corpuscles. In both media the carbon dioxide combines with water to form carbonic acid, which dissociates into hydrogen ions and bicarbonate ions. Owing to the far greater buffering power of the corpuscle contents, this dissociation will proceed much more extensively in the corpuscle than in the plasma. Consequently, there will be a concentration gradient propelling carbon dioxide (undissociated) from plasma to corpuscle, and another gradient propelling bicarbonate ions from corpuscle to plasma. The membranes of the corpuscles, although permeable to anions such as chloride or bicarbonate, behave as if they were impermeable to cations such as sodium or potassium. Bicarbonate ions will begin to diffuse out under the influence of the concentration gradient, but since they cannot be accompanied by ions of the opposite charge, the corpuscle will immediately develop a net positive charge. The positively charged corpuscle will then draw in negative ions of all kinds from the plasma. Chloride ions, being the most readily

---

* Or also termed "Hamburger interchange", after Dutch physiologist of that name.

available, will predominate, and the process will continue until finally an equilibrium is reached at which,

$$\frac{[HCO_3^-] \text{ in corpuscle}}{[HCO_3^-] \text{ in plasma}} = \frac{[Cl^-] \text{ in corpuscle}}{[Cl^-] \text{ in plasma}}$$

This is a case in which the physicochemical principle known as the Donnan membrane equilibrium applies. The net result is that when the carbon dioxide content of the blood is increased, there is a migration of bicarbonate ions out of the corpuscles into the plasma, and a migration, in exchange, of chloride ions from the plasma into the corpuscles. When the carbon dioxide content is reduced, chloride ions come out of the corpuscles into the plasma, and bicarbonate ions enter instead. The carbon dioxide carrying power of the plasma is thus brought up to the level of, or even beyond that of, the corpuscles. Since the plasma hydrogen ion concentration is determined by the ratio of the carbon dioxide pressure to the concentration of bicarbonate ions, the transfer of bicarbonate ions from the corpuscles will assist in preventing the hydrogen ion concentration of the plasma from rising. The superior buffering power of the corpuscles is thus shared out with the inferior buffering power of the plasma—a process often spoken of as "secondary buffering" of the plasma; it is accompanied by the chloride shift only because the red cell membrane is impermeable to sodium and potassium ions.

If the plasma is replaced by isotonic NaCl solution, the chloride shift again occurs when the blood corpuscle suspension in NaCl is shaken with carbon dioxide, and for the same reasons as above. If, however, isotonic sugar solution is used, there is now no ion in the outside fluid to exchange with HCO_3^- from the corpuscle, so that when the corpuscle suspension is shaken with carbon dioxide the outside sugar solution cannot be secondarily buffered, and, since it contains no intrinsic buffer, it therefore goes very acid.

SEPARATED PLASMA AND TRUE PLASMA

The chloride shift is mainly responsible for the difference between the carbon dioxide dissociation curve of "separated plasma" and that of "true plasma". The curve for *separated plasma* is obtained by centrifuging the blood, removing the supernatant fluid, shaking the latter with various pressures of carbon dioxide and then estimating the total CO_2 content at each CO_2 pressure. Over the physiological range of carbon dioxide pressure the curve is almost as flat as the CO_2 dissociation curve of a solution of sodium bicarbonate (Fig. 13.1): such a curve would be of little physiological service, since the extra amount of carbon dioxide which would be taken up when its pressure is raised from 40 to 50 mmHg (5·32 to 6·65 kPa) would be so small. The dissociation curve of *"true plasma"* is obtained by equilibrating the whole blood with various pressures of carbon dioxide, and then transferring the blood to a centrifuge cup (the escape of carbon dioxide into the air is prevented by a layer of liquid paraffin); the supernatant fluid, separated by centrifuging, is estimated for total CO_2, and the latter plotted against CO_2 pressure. Each sample of plasma thus is in equilibrium, not only with each different pressure of CO_2, but also with the corpuscles—whereas all the samples of separated plasma were in equilibrium with the corpuscles at one particular CO_2 pressure, namely, that which happened to obtain in the blood at the time when the plasma was centrifuged off. The true plasma dissociation curve is clearly quite different from the separated plasma curve; it is, indeed, of the same type as that of whole blood, and like the latter is of a serviceable shape as regards carbon dioxide transport.

The speed of processes in carbon dioxide transport

Most of the individual chemical reactions we have described as occurring in the uptake or output of carbon dioxide are of an ionic type, e.g.

$$H_2CO_3 \rightleftharpoons H^+ + HCO_3^-,$$
$$H_2PO_4^- \rightleftharpoons H^+ + HPO_4^{2-},$$
$$HPr \text{ (protein molecule)} \rightleftharpoons H^+ + Pr^-$$

Simple ionic reactions of this type have been generally supposed to be very rapid—experiments have, indeed, shown that all these reactions reach to within 1 per cent of equilibrium within 1 ms (from whichever side of it they start); they must therefore be too fast to limit the rate of output or uptake of carbon dioxide by the blood. The final chemical reaction which, on the bicarbonate hypothesis, comes just before the evolution of carbon dioxide into the expired air is, however, a non-ionic one, viz. the formation of CO_2 from H_2CO_3.

The value of the velocity constant of this reaction has been measured by several physical chemists; it can thence be calculated that at body pH and temperature, and with the usual concentration of bicarbonate ions in the blood, carbon dioxide could only be evolved in the lungs at about 1/200 the rate at which it actually escapes into the expired air. Either, then, there must be something in the blood which speeds up the reversible reaction between carbon dioxide and water; or else there must be some other reaction, besides the bicarbonate one, which takes part in the transport of carbon dioxide. There are, in fact, both.

Carbonic anhydrase

It can be shown, by quite simple experiments, that the red blood corpuscles contain a substance which catalyses strongly the reaction

$$H_2CO_3 \rightleftharpoons CO_2 + H_2O$$

This substance has been isolated, was found to have all the properties of an enzyme, and was thus given the name "carbonic anhydrase". It is not present in the plasma, but the amount in the corpuscles, if it is as efficient there as in solution, is enough to accelerate the formation of CO_2 from H_2CO_3 about 600 times under body conditions. Presumably during the short time the circulating blood is in the capillaries, the change from CO_2 to bicarbonate and vice versa must occur chiefly in the red corpuscles.

Carbamino haemoglobin

Blood can still react rapidly with a small amount of carbon dioxide even when carbonic anhydrase is absent or incapacitated by addition of some enzyme poison, e.g. KCN; this residual rapid reaction cannot be bicarbonate formation or the reverse (since both these processes, in the absence of carbonic anhydrase, only proceed slowly). There is chemical evidence that it is due to a direct reaction of CO_2 with the $—NH_2$ groups of haemoglobin to form compounds of a carbamino type, e.g.

$$Hb(NH_2) + CO_2 \longrightarrow Hb(NHCOO^-) + H^+$$

This type of reaction is well known in the case of simpler $—NH_2$ containing compounds, such as ammonia and glycine, and is a very rapid one even in the absence of special catalysts.

Carbamino-bound carbon dioxide plays an appreciable role in carbon dioxide transport, even though the absolute amounts of such compounds under physiological conditions form only a small fraction of the total amount of carbon dioxide present in the blood (in the neighbourhood of 5 per cent). Reduced haemoglobin takes up carbon dioxide in the carbamino form more readily than does oxyhaemoglobin. Of the total quantity of carbon dioxide carried from the tissues to the lungs—i.e. of the difference between the carbon dioxide contents of venous and arterial blood—it is estimated that about one-quarter is in the carbamino form.

Other evidence, of a more physicochemical type, suggests that some carbon dioxide combines directly with haemoglobin, not only in the carbamino form, but possibly in some other form.

SUMMARY

The uptake of carbon dioxide by the blood in the tissues is believed to occur as follows:

1. CO_2 diffuses from the tissues into the blood plasma.
2. Some of the CO_2 hydrates slowly in the plasma to form H_2CO_3; the latter then yields its H^+ ions to the plasma proteins and phosphates and forms bicarbonate ions.
3. Most of the CO_2, however, passes into the red corpuscles:
   (a) Some combines directly with haemoglobin to form compounds of a carbamino type—this combination is increased as the haemoglobin loses oxygen in the blood capillary.
   (b) By far the greater part changes over rapidly into H_2CO_3 under the influence of the enzyme carbonic anhydrase; the H_2CO_3 then yields its H^+ ions to the haemoglobin, forming bicarbonate ions. The latter process is much increased as the haemoglobin changes from the oxy- to the reduced form, since reduced haemoglobin, being a weaker acid than oxyhaemoglobin, absorbs H^+ ions more readily and thereby allows H_2CO_3 to become more completely ionised.
4. Bicarbonate ions begin to diffuse out from the corpuscles into the plasma; this sets up an electric field which draws Cl^- ions into the corpuscles from the plasma in place of the HCO_3^- ions, which have diffused out (the chloride shift). The twin process goes on until an equilibrium is reached. The CO_2 carrying power and buffer efficiency of the plasma are thereby brought up to the level of the red corpuscles.

In the lung carbon dioxide is formed and evolved by a reversal of all these processes, as indicated in Fig. 13.2.

THE EFFECTS OF EXERCISE

The changes that take place in the blood as a result of exercise, or of increased activity of any kind, are quantitative rather than qualitative, and their magnitude depends largely upon the magnitude of the simultaneous changes in the respiratory and circulatory systems.

The first result of an increased activity in any group of cells is that more oxygen is used and more carbon dioxide evolved; the partial pressure of oxygen around them falls, and that of carbon dioxide rises. This effect is passed on to the blood in the capillaries, so that the quantities defining the venous points on our dissociation curves are altered, taking up, for example, values somewhat as shown in Figs. 13.1 and 13.2 by the points marked V', i.e. the oxygen pressure falls to 22 mmHg, the carbon dioxide pressure rises to 58 mmHg, the percentage saturation with oxygen falls to 30 per cent, and the total carbon dioxide concentration rises to 60 ml dl^{-1}. The extra carbon dioxide is thus carried away, and the extra oxygen provided, simply by allowing the partial pressure in the blood of the one to rise and of the other to fall. We have seen that this does not necessarily involve similar changes in the tissue cells, owing to the opening up of extra capillaries and the reduction in the distance that has to be traversed by the gases between the blood and the tissue cells. The increased blood flow required is provided for by the vasomotor reflexes and local chemical mechanisms.

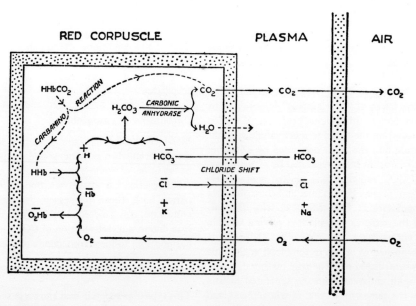

Fig. 13.2 Scheme showing the most important changes involved in the liberation of carbon dioxide from blood into air. (Roughton.)

A limit must come to this, however, and if the activity continues to become greater, the oxygen pressure in the tissue cells begins to fall, and the oxygen supply to be deficient. The cells go into "oxygen debt" and, if this is severe enough, begin to liberate lactic acid. This addition of extra hydrogen ions drives the equilibrium reaction,

$$H_2CO_3 \rightleftharpoons H^+ + HCO_3^-$$

from right to left. The carbonic acid so formed is converted to carbon dioxide and blown off in the lungs (the respiratory centre adjusts the rate of ventilation so as to ensure this). The buffering power of the tissues is reduced, nevertheless, since the buffer salt, bicarbonate, is replaced by the fully ionised lactate, which has no buffer action; the hydrogen ion concentration of the tissues rises. If the exercise continues, there may be so much lactic acid produced that the hydrogen ion concentration of the arterial blood is affected to a significant extent. The carbon dioxide pressure in the lungs may rise also, and these two effects may so alter the oxygen dissociation curve that the per cent saturation of the arterial blood falls. The oxygen and carbon dioxide transport system has now got into a vicious circle, and the exercise must stop. It is, indeed, sometimes terminated by unconsciousness.

It should be remarked that the whole of the body is affected by the changes that result in the blood from one group of muscles going into oxygen debt; the lactic acid, indeed, diffuses into all the wet tissues of the body, and may be oxidised in them—and particularly in the liver and heart—as well as in the muscles in which it was formed. The blood buffers play somewhat the part of a bank which allows one of its clients to overdraw his account at the expense of the rest, but if the overdraft is allowed to increase until it is comparable in size with the whole of the negotiable securities of the bank—and this in spite of the activities of the other clients to redeem it—the bank's failure is inevitable.

Section 2:

The Cellular Environment and its Control

14. Control Systems: Theory

The internal environment

The cells and tissues we have discussed in the previous section are highly specialised examples of primitive cells. No longer are they able to lead an independent existence in the way that an amoeba, or any other single-celled protozoan, might be capable of doing. The advent of specialisation narrows down the range of life-functions that these cells can carry out. To be sure, the simpler cells such as white cells in the blood can perform most of them but a nerve cell, for example, whose main function is to transmit signals, is not able to do much else besides this.

We therefore see other cells and tissues in multicellular organisms adapted to performing the necessary life-functions *en masse*. There are organs for excretion, nutrition, respiration and so on. In the main, they carry out these functions through the intermediary of the tissue fluid surrounding all the cells in the body. The composition of this fluid (*extracellular fluid*) is maintained within fairly narrow limits, so that it contains at all times the necessary supply of nutrient materials, e.g. glucose, oxygen for cell respiration, and is at the same time continuously cleared of metabolites, carbon dioxide etc. produced by active cells. Bathed in such a perfect fluid, all cells, however specialised, can thrive.

The extracellular fluid is constantly circulating within the extracellular spaces throughout the body; it is all the time being mixed by circulating blood and by diffusion between plasma and tissue spaces. All the cells in the body therefore have the same environment, the *internal environment*, or *milieu interieur*, as it was originally termed by Claude Bernard.

Homeostasis

This term, peculiarly beloved by physiologists, indicates the maintenance of constant conditions in the internal environment. The majority of the organs within the body function to keep conditions constant. Examples are the lungs which keep the oxygen supply up to required levels, the kidneys which remove substances, and the intestines which provide nutrient materials.

In the subsequent study of physiology in this book, we shall be dealing with the manner in which organs and systems are organised to perform this function of keeping the internal environment constant. The thread of continuity throughout our study will be the control mechanisms which allow the correct operation of these organs and systems.

Claude Bernard (1878) first clearly and euphoniously enunciated the general principle involved; indeed perhaps it is the only general principle in physiology: "*The fixity of the internal milieu is the necessary condition for free existence*." The great Walter B. Cannon in 1929 went into more detail. It was he who coined the word "homeostasis". He wrote,

> "*The highly developed living being is an open system having many relations to its surroundings—in the respiratory and alimentary tracts and through surface receptors, neuromuscular organs and bony levers. Changes in the surroundings excite reactions in the system, or affect it directly, so that internal disturbances of the system are produced. Such disturbances are normally kept within narrow limits, because automatic adjustments within the system are brought into action, and thereby wide oscillations are prevented and the internal conditions are held fairly constant. . . .*
>
> *The coordinated physiological reactions which maintain most of the steady states of the body are so complex, and are so peculiar to the living organism, that it has been suggested that a specific designation for these states be employed—homeostasis.*"

Physical properties of control systems

In order to understand the way in which control systems in the body operate, we must first consider the general properties of control and, in particular, *feedback* mechanisms. At the outset, we should understand that there are no essential differences between the control systems found in biology and those commonly employed by engineers. In biological systems we are trying to unravel the way in which a life-function is being controlled; the engineer is usually attempting to design a practical and usable mechanism for controlling his machinery or processes.

Until comparatively recent times, the principles of control systems in the body have been investigated purely qualitatively. For example, the regulation of arterial blood pressure has for a long time been known to be under reflex influence; if the pressure rose, a reflex fall was initiated in order to compensate and bring it back to normal once again. Now, it is possible to define quantitatively the parameters of control systems such as this.

Control without feedback

The model T Ford (and other automobiles of that era) had a hand throttle which controlled the engine. On the straight and level, setting this device would cause the car, after a while, to attain a steady velocity. To slow down, the driver could control the engine by closing the throttle. However, even in the days of the famous Henry, roads rarely were straight and level; a constant velocity, even if desirable, could not be maintained without additional complication. The driver, of course, was the controller, and information coming to him via the speedometer and the roadside scenery flashing past was translated into commands for varying the throttle openings. The driver was able, within limits imposed by the power of the engine, the intransigence of pedestrians, or the proximity of other larger vehicles, to keep the speed constant. Of course, it was already known that frequent, rapid and delicate adjustment was vital in this control process, so Ford had thoughtfully also provided the driver with a foot-operated accelerator.

The throttle lever, its cable and the butterfly valve in the engine constituted a control system without feedback. Clearly, it has proved of little practical use and automobiles running on hand throttles are not found today.

Feedback control

One may be forgiven for supposing that the dashing driver with lead in his right foot rarely uses feedback, at least as far as controlling his engine is concerned, but even he must stop occasionally for natural and other reasons. The less brutal among us, particularly if pursued by uniformed gentlemen on motorcycles, will be constantly referring to our speedometers and controlling our velocity to a value as close to that permitted as possible. We can analyse the situation by drawing a flow diagram or model of the controlling system (Fig. 14.1).

From a consideration of the diagram (Fig. 14.2) it is apparent that signals are proceeding around a closed loop as follows:

Accelerator → engine → speedometer → driver's eye → driver's brain → driver's leg muscles → accelerator

This loop is known as a *feedback loop* and has certain well-defined properties which can, if so desired, be analysed mathematically the more accurately to describe them. But

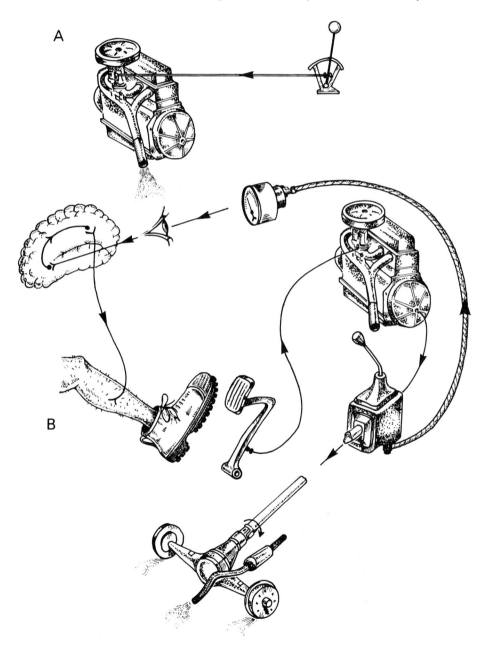

Fig. 14.1 A. Open control loop. Hand throttle controls engine. This will only give constant required speed on straight and level. B. Closed control loop. Driver's eye and brain compute error-signal (misalignment) between required and actual speeds as given by speedo. Brain and leg muscles then adjust accelerator pedal accordingly. To maintain given speed, the accelerator is depressed if the speedo reads too low and released if too high. This is negative feedback.

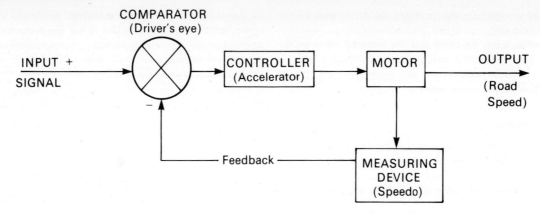

Fig. 14.2 Flow diagram of servoloop shown in Fig. 14.1B. The comparator consists of the driver's eye and brain, including his memory, and also contains the set-point. We might consider the input signal to be a road sign of some kind, i.e. speed limit sign.

mathematics are not necessary to follow the principles involved. The loop is shown in Fig. 14.2.

Negative feedback

In order to keep any system in *status quo*, the feedback we have described has to be negative feedback. We must hope our vehicle driver understands this. He must slow his engine when the speedometer shows him he is proceeding too rapidly; if by mischance he were to apply positive feedback and depress his accelerator when his speedometer increased its reading, the final result would be obvious to predict. And incidentally he will illustrate the important feature of positive feedback all too clearly, namely that it is a once-and-for-all phenomenon.

Bang-bang servo

Engineers, in their down-to-earth manner, describe the simple negative feedback as a *bang-bang servo*, because it just bangs on and bangs off as required. A typical example in everyday use is the thermostat. At a predetermined temperature (the *set-point*) the heater is switched on; when the room is hot enough, off it goes. A simple feedback is no use in piloting a motor vehicle; any driver who acted like one would soon be breathalysed and placed in the cooler.

Proportional control

The learner-driver will find that the accelerator pedal can be depressed through quite a considerable, and possibly alarming, range of movement. With different positions of his right foot, he can apply variable amounts of power from his engine to the road-wheels and can obtain a range of speeds up to the maximum of which the motor is capable. This is termed "proportional control" and represents one stage further in the sophistication and refinement of servomechanisms.

Error-rate (or derivative) control

Feedback loops contain a *comparator*. At its simplest this, for example, will be the knob on the thermostat which is set at the desired room temperature, together with the contacts which open and close when the temperature is right. In the case of our motor-car and driver servosystem, the comparator is within the driver's brain. When pursued by the aforementioned uniformed gentlemen, he compares the actual speedometer reading with the legal requirement displayed at the entry to the restricted zone through which he is currently passing. This comparison between the two signals (present output; actual speed versus desired input; legally required speed) gives the *misalignment*, or *error-signal*.

If our driver has reflexes of lightning and there is absolutely no friction, viscosity, inertia or lost motion in the mechanical arrangements (i.e. the response to the error-signal is truly instantaneous) the motor vehicle could be always kept at its desired speed. However, things are never perfect (Professor Sodde's fifth law) and all practical control systems contain all the defects enumerated above.

Let us consider now what will happen to an inexperienced learner-driver attempting to conduct his vehicle at a steady speed, under the beady eye of the law. He will depress the accelerator, to pick up speed, eyes uncompromisingly glued to his speedometer. Soon the needle will reach the speed limit and then rise above it. At once, realising he is in imminent danger of prosecution, the driver lets his right foot relax. Because of friction, viscosity, inertia and slackness in the accelerator linkage, it is an appreciable time before the speed begins to fall. Eventually, speed is again below the legal limit, and not wishing to be late for his appointment our driver once more depresses the pedal. Again a delay while the speed drops even more, then rises, finally to end well above the limit for a second time.

We can see that this is a highly unsatisfactory mode of progression, and a graph of velocity plotted against time would appear like Fig. 14.3.

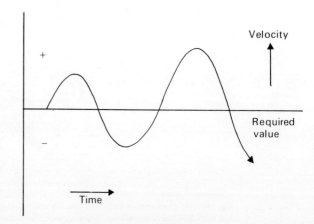

Fig. 14.3 Graph of velocity plotted against time for learner-driver who is ignorant of the principles of derivative feedback.

The cyclical variability of the controlled variable (speed) about its required value (legal speed limit) is a characteristic of proportional and bang-bang systems. The oscillation is termed *hunting*.

Is it possible to eliminate undesirable hunting? The experienced driver, of course, can keep his speed accurately at any desired figure no matter how many traffic cops may be trailing him. How does he do it? It is, in fact, quite simple. He makes his brain perform as an *error-rate* (or *derivative*) controller. Subconsciously, he computes the rate-of-change of the error. He listens to the engine note, observes the flow of the road beneath his wheels, perhaps glances at the speedometer out of the corner of his eye. He applies his negative feedback to the accelerator pedal in amounts proportional to the rate at which the difference between real speed and required speed is changing.

Although worked into all biological servosystems, and nowadays into most engineering ones, error-rate control is only a comparatively recently studied process. In engineering design, it involves a device for obtaining a signal which is the first differential of the error, and then applying it, inverted, to the controller. The process of differentiating can take place in the measuring devices, in the feedback loop or in the comparator—the net result is the same.

Mathematical explanation

The mathematical analysis of control systems is now a large branch of engineering and in recent years biologists have found that mathematical models can be formulated of living control systems and the performance of them predicted.

The classical example used in teaching is a feedback control consisting of a motor driving a pointer on an output shaft which is controlled by the servoloop so as to replicate the movements of a pointer on an input shaft. The shafts have potentiometers on them to give an electrical signal proportional to θ_i, the input angle and to θ_o the output angle. See Fig. 14.4.

The signals produced by the potentiometers are e_i and e_o respectively. Then

$$e_i = K_p \theta_i \quad \text{and} \quad e_o = K_p \theta_o$$

where K_p is a constant dependent upon the potentiometer, various conversion factors etc.

If the two signals e_i and e_o are compared (in the comparator) the difference is the misalignment or error-signal, e,

$$e = e_i - e_o = K_p(\theta_i - \theta_o)$$

Most practical servomechanisms will deal with a signal level at this stage which is too low on its own to have any appreciable effects on the operating conditions. Therefore an amplifier is usually included, as shown in Fig. 14.4, which has an amplification factor of A. Thus the amplified error e_a is

$$e_a = A = AK_p(\theta_i - \theta_o)$$

The output of the system, O, is determined by the output of the amplifier, the amplified error e_a, and modified by a constant due to the motor driving the output shaft K_m. Hence the output is related to the input as follows:

$$O = K_m \times e_a = K_m A K_p(\theta_i - \theta_o) \quad (1)$$

We have so far reached, in this analysis, the situation comparable to that described in the two paragraphs headed "Negative feedback" and "Proportional control".

If friction and inertia are to be considered, as they must be in any practical scheme, two further terms must be included. Let us consider (for simplicity!) that the friction imposed on the motor is proportional to the speed of the motor; it is then

$$F\left(\frac{d\theta_o}{dt}\right)$$

where $d\theta_o/dt$ is the rate of change of the output angle with time t and F is a frictional constant, characteristic of the system we are considering.

The total inertia of the load on the output shaft (this includes the friction) can be represented by

$$I\left(\frac{d^2\theta_o}{dt^2}\right)$$

where

$$\left(\frac{d^2\theta_o}{dt^2}\right) = \frac{d(d\theta/dt)}{dt}$$

in other words, the rate of change of angular velocity, i.e. the angular acceleration.

The output of the motor, minus the frictional load, accelerates this inertial load

$$O - F\frac{d\theta_o}{dt} = I\frac{d^2\theta_o}{dt^2} \quad (2)$$

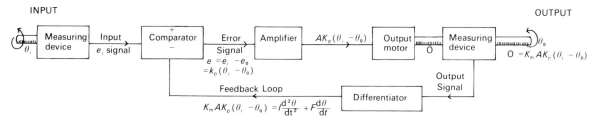

Fig. 14.4 Flow diagram of a typical servocontrol system. It consists of a motor, driving an output shaft on which is a potentiometer recording the rotation of this shaft, θ_0, as the output signal e_0. This is fed back (via the differentiator) via a negative feedback loop to the comparator, which compares the output signal e_0 with the signal from a potentiometer on the input shaft, e_i. The error signal (misalignment) $e = (e_i - e_0)$ is fed to the amplifier, which in turn controls the motor.

Note that the differentiation is shown as occurring in a black box in the feedback loop itself. For simplicity, we have not treated it this way in the text, but it should be obvious that the differential terms may be introduced at any point in the loop. In biological systems, the differentiation usually takes place in the output measuring device (i.e. the output signal is differentiated, just as is shown in this diagram).

If we now combine equations (1) and (2), we obtain a general specification of the servoloop in terms of the relation between its input and its output

$$K_m A K_p (\theta_i - \theta_o) = I \frac{d^2\theta_o}{dt^2} + F \frac{d\theta_o}{dt}$$

If we call the three constants K_m, A, K_p by the one term K

$$(\theta_i - \theta_o) = \frac{I}{K} \cdot \frac{d^2\theta_o}{dt^2} + \frac{F}{K} \cdot \frac{d\theta_o}{dt} \qquad (3)^*$$

---

* Equations such as this one, describing the relation between the input and output of a little black box (or, as here, series of boxes) are called a *transfer function*.

This is a *second-order differential equation* because it contains the term $d^2\theta_o/dt^2$. It is a *linear equation* because it does not have functions and their differentials multiplied together. It is therefore relatively easy to solve.

From cursory perusal of the foregoing analysis we can see that if friction and inertia are present in any feedback control system, it will not work accurately without a proportion of the feedback being a first differential of the input error (to deal with friction; the more the friction, the larger will be this proportion). Also, a proportion of the feedback must be the second differential of the input error (to cancel out the effects of the inertia at the output).

15. Control Systems: Operation

Responses of negative feedback servocontrol

Let us see if we can analyse what happens if a signal is put into one of these control systems. How perfect is the control it exerts?

First of all, we must be clear that the system simply "sees" the misalignment (error-signal), i.e. the difference between input and output, so it is all the same if it arises as a result of a new set-point being fed into it (in our previous analogy, a new speed limit), or as a result of a disturbance to the load at the output (coming to a steep hill).

Engineers customarily investigate the properties (and responses) of servomechanisms by injecting transient or steady signals. Transients might be *step-functions*, *ramps* or *sine-functions*. We will consider, as an illustration, a step-function injected at the input, and the responses we might expect at the output.

Response without derivative feedback

The presence of a *delay* in any closed control loop renders it potentially unstable. If the working round the loop is not instantaneous, a signal put into it will set it into uncontrolled oscillation, much like a plucked violin string or pendulum. Provided that the *gain** in the loop is more than unity, this oscillation will build up. This can be seen in Fig. 15.1, where the response greatly overshoots the required mark.

This situation is sometimes referred to as *underdamping*, or rather *negative damping* if an ever increasing oscillation builds up.

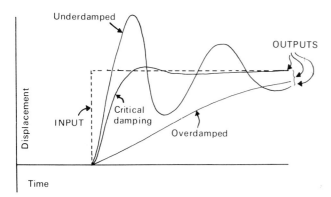

Fig. 15.1 To illustrate underdamping, critical damping and overdamping. The required response (input) is shown dotted and is a step-function displacement. The underdamped response overshoots and oscillates; the overdamped response is very slow.

Negative damping is not shown, but it would lead to the oscillations ever-increasing in amplitude and finally ending in disaster.

---

* The term "gain" refers to the amplification in the loop as a whole. If signals going round and round the loop decrease in size, "oscillation" as such will peter out; the gain is then less than unity.

Response with too much derivative feedback

In most working servomechanisms, the proportion of the negative feedback which consists of either the first or second derivatives of the input-error can be varied at will. For example, in the automatic pilot used on board ship, a variable control labelled "yaw" has this function. In different conditions of wind and sea, it is a matter for trial and error experimentation by the helmsman to determine how big a proportion of derivative control is needed to give the smoothest course, i.e. the result with the least oscillatory error.

With too much derivative feedback, the ship will be very slow to respond, either to changes in course made by the helmsman or, perhaps more important, to compensate for the effects of variations in beam winds or seas tending to shift the vessel off course. The same is true in our model, as can be seen from the second curve in Fig. 15.1. Here, a step-function input gives a slowly climbing response which eventually reaches the required new level after a lengthy delay.

This situation is referred to as *overdamping*.

Critical damping

To the reader who has followed the argument thus far, it will be apparent that some "middle course" is possible between the Scylla of underdamping and the Charybdis of overdamping. Such a condition, in which the "best" response, or the response most nearly coinciding with the waveform of the input, is termed a *critically damped* response. It is represented, somewhat arbitrarily, by the third curve of Fig. 15.1.

Are servosystems critically damped?

It might be thought that all biological servomechanisms would be critically damped, and that would be an end to the matter. On the whole, this is found not to be so and for the following reason.

Most biological systems (and others too) are a compromise. The compromise, as far as damping is concerned, is between speed of response and accuracy. As might be deduced from a brief glance at Fig. 15.1, responses in an overdamped system are slow. In the servocontrol of muscle, to take a common example, speed of response is usually biologically advantageous. Even in the human, quick reflexes help in playing cricket, tennis or squash, in driving cars and so forth, and obviously may also promote survival. Equally, fast responses are inaccurate, containing errors due to overshooting and then oscillation; these are biologically disadvantageous. So we must have a balance between speed of response on the one hand, and accuracy on the other.

In different circumstances a balance tending towards the one or the other might be of greater importance. In the muscle control mentioned already, it would appear that an oscillation of around 2 per cent in amplitude of a maximal voluntary contraction is permitted (as an inaccuracy) in order to gain the maximum speed of response. Being pedantic about it therefore, the muscle servoloop would be considered underdamped to quite a marked extent. Incidentally, this oscillation is familiar to most persons and is called *physiological tremor* (See Chapter 78).

Positive feedback

Positive feedback systems are not utilised in continuous control, for the type of action they produce is explosive or "trigger-like". The error-signal which is fed back to the controller without inversion will give rise to a *vicious spiral*, the bigger the error the larger the output producing it until disaster occurs.

Textbooks often mention the mechanism of circulatory shock as a positive feedback, and although it is undoubtedly true that in shock the arterial pressure falls and, if to low enough levels, then the heart is impaired; pressure falls still lower and so on; one cannot in any sense call this a control mechanism.

A consideration of such vicious spirals will reveal two important factors:

1. It is an all-or-nothing event, i.e. once started, the cycle proceeds to its ultimate end.
2. There is no recovery.

It might be thought, therefore, that there would be no place for positive feedback in biological controls. This is not so, because it is possible to add to the mechanism a *cut-off* which will ensure that it can repeat.

Amplification

The use of positive feedback occurs where amplification is needed. A hormonal example is given in Chapter 65 which will make this point clear.

Another example is to be found in the sodium conductance changes in the generation of the nerve action potential.

Normally, sodium ions do not pass the nerve membrane. However, when an action potential is generated, the membrane depolarises and there is a net flow of potassium ions out of the fibre. At a given level of depolarisation, sodium conductance is increased and sodium ions pass into the cell along their concentration gradient. Any increase in internal sodium concentration will, of course, further depolarise the membrane.

So we have here the vicious spiral, since the greater the depolarisation, the larger the sodium flux (and so on).

Obviously the nerve possessing this positive feedback mechanism would conduct one impulse marvellously—and no more. In fact, a cut-off mechanism exists, the nature of which is as yet not fully understood. There is a time-dependent mechanism to turn off the explosive, *regenerative* sodium flux. It is called the *inactivation of sodium conductance*, and at the end of a period of about half a millisecond, the sodium conductance returns rapidly to very low values.

This inactivation of sodium conductance is accompanied by a state of greatly increased potassium conductance. These are self-repairing events; an accelerated efflux of potassium ions from the axoplasm occurs and leads to a rapid fall of the inside potential to the initial level. The original ion permeabilities are soon reinstated. Thus the increase of potassium conductance which occurs initially is subject to negative feedback; it shuts itself off as it proceeds.

This illustration shows us that positive feedback mechanisms are used when a sudden, explosive kind of change is required. A small initial upset in an equilibrium gives rise to a much larger, regenerative change. The initial change is amplified. In the example given, the sodium conductance increases several thousandfold in the course of a fraction of a millisecond. Positive feedbacks are powerful stuff!

EXAMPLES OF BIOLOGICAL SERVOMECHANISMS

Nearly all the servomechanisms we shall be describing during the remainder of this book follow the general principles already outlined. A few examples, very briefly mentioned, follow.

Control of arterial pressure

The blood supply to tissues is maintained by the flow of blood pumped out by the heart through a series of small vessels or capillaries. The actual amount of blood required by any tissue is highly variable from time to time and also varies as between one part of the body and another. How is the problem of meeting these variable requirements, and controlling the flow rates, solved?

The solution is beautifully simple, so simple in fact that we scarcely recognise that the initial problem ever existed. It is to have a reservoir of blood kept at high pressure and then to tap (bleed) off from it, the variable amounts required by the capillaries. Our problem is then resolved into one of keeping the pressure in this arterial reservoir at a constant and high level.

As the wary reader will by now expect, this control of the arterial pressure is carried out by a number of servomechanisms. In the high pressure system are pressure-sensitive devices, the *baroceptors*. As the pressure in the arterial tree rises, these baroceptors generate impulses which travel to the *vasomotor centre*, in the medulla. When they reach there, their effect is to bring about changes in heart rate (and other parameters of the circulation) tending to lead to a lowered pressure. Thus we have a negative feedback loop which, in fact, exhibits all the characteristics we have described in our model.

The vasomotor centre appears to contain a set-point which determines that the arterial pressure will remain closely controlled at a normal level. Of course, circumstances may require the set-point to alter. Heavy exercise would be an example, during which the requirement of numerous muscles for oxygen (and thus for blood) might increase many times. In order to meet this large increase, the actual pressure level might have to be raised. More about the mechanism can be found in Chapter 36.

Secretion of hormones

The more we discover about the way that hormones work, the more importance do we have to ascribe to feedback controls. As might be expected, hormones are not simply released from their glands of secretion without some sort of control mechanism. Most of these controls are of considerable complexity and involve several glands (and even nervous pathways) in their feedback loops.

We can take the primary sex hormone, oestrogen, as a fairly typical example. Here, the ovary secretes this hormone, in the female, in varying quantities during the sexual cycle. The circulating blood level of the hormone then has a negative feedback action upon certain nuclei in the hypothalamus, a part of the brain, which, as a result, change their secretion of releasing factors. The latter act upon the anterior pituitary gland causing it to secrete either more or less gonadotrophic hormone into the bloodstream. This hormone controls the output of the ovarian hormone, and hence indirectly the blood level of it. There may of course be loops within this loop, but at present we do not have enough detailed information to speculate about this. In any case, it is complex enough without! The matter is dealt with in more detail in Chapter 65.

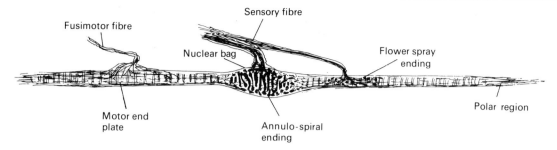

Fig. 15.2 Diagram of light microscopic appearance of a single intrafusal fibre. The annulospiral ending responds to stretch, with adaptation; so does the flower spray but with less adaptation. Motor end-plates, supplied by fusimotor nerves, are found in both polar regions.

The servocontrol of muscle

Our last biological example will concern the control of muscle contraction, a mechanism involving fast nerve pathways very similar to feedbacks encountered in purely engineering applications.

Figure 15.2 shows a muscle spindle as seen under the light microscope. These spindles are attached in parallel with the muscle fibres of the main muscle and they signal the stretch of the muscle by generating impulses in the afferent fibre connected with them. Each spindle consists of two to five small muscle fibres enclosed in a common sheath. These fibres are striated and contract just like ordinary muscle fibres; they have motor end-plates. In the middle of each *intrafusal fibre* is a non-contractile swollen region housing a receptor nerve-ending, which is arranged in a spiral fashion—the *annulospiral* ending.

THE KNEE-JERK

Figure 15.3 illustrates the *reflex arc* in the shape of the familiar "knee-jerk". When the patellar tendon is struck (with the little tendon-hammer used by the neurologist) the quadriceps extensor muscle is stretched. The aforementioned muscle spindles within it are also stretched, as are the annulospiral endings themselves. Impulses, generated by the stretched spindles, pass up to the spinal cord via the afferent limb of the reflex arc (in the posterior ramus) which makes direct contact with motor neurones. The motor neurones, in response, send impulses down their attached motor nerve fibres (in the anterior ramus) to the main muscle which is thereby made to contract. This is a *monosynaptic reflex*; it only lasts 50 ms.

Perusal of Fig. 15.3, or of the previous descriptive paragraph, will lead one to conclude rapidly that a closed loop with negative feedback exists. It is the reflex arc. A lengthening of the muscle (by the patellar tendon tap) leads to a reflex contraction to shorten it, so restoring the *status quo* once again.

When impulses are recorded from a steadily stretched annulo-spiral ending, as long as the muscle is stretched, the ending goes on firing. Thus the muscle, by reflex action, continues to contract; a procedure which is important in maintaining posture (i.e. postural *status quo*).

Derivative control in the muscle servomechanism

As we have mentioned several times already, it is necessary to introduce a term containing a differentiated version of the error-signal in any closed loop that is to be stable, and not go into oscillation at the slightest provocation. How is this differential produced in the muscle control servo?

Figure 15.4 shows the response of an annulospiral ending to three forms of stretch. It can be seen that the frequency of impulses initiated by the ending is in general proportional to the rate of stretching rather than to the actual degree of stretch itself. This is also called *adaptation* of the response to a stimulus.

There are other stretch sensitive endings in the muscle spindle, the so-called *flower spray* endings which show adaptation to a much lesser degree. It is probable that the amount of either component from the two types of ending can be altered to vary the proportion of derivative feedback employed under different circumstances.

The follow-up servo

Sooner or later the student will come upon the term "*positive feedforward*". Is this the same as negative feedback, or if not what is it?

The reflex arc, as is true of most biological systems, is much more complicated than it would appear at first sight. As can be seen from Fig. 15.5, the loop has two further inputs to it than we have mentioned so far. The first is the *alpha motor* pathway by which impulses can travel to the anterior horn cell from the motor cortex and can directly give rise to voluntary contraction of the muscle. The second is a smaller diameter fibre system, originating in the cerebellum and passing via a synapse in the anterolateral columns to the muscle-spindle motor end-plates. This is the *fusimotor* efferent pathway.

At first sight, it might appear a rather useless procedure to activate the fusimotor nerves, for contraction of the intrafusal fibres is known to give rise to no measurable tension in muscle.

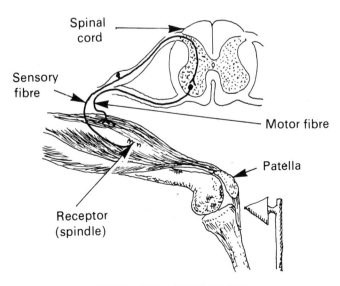

Fig. 15.3 The knee jerk reflex pathway.

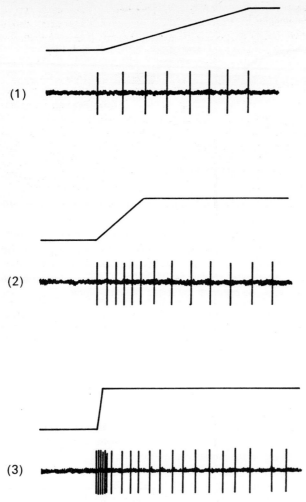

Fig. 15.4 Response of an annulospiral ending to varying rates of stretch: (1) shows (top trace) moderate rate of stretch and (lower trace) action potentials in 1a afferent from spindle; (2) shows more rapid stretch, and (3) very rapid stretch. Note that response frequency varies with rate of stretching.

Here then, we have an indirect way in which contraction can (or theoretically could) be started off. It has been termed the "follow-up servo" and is an example of feedforward.

Another example of feedforward is to be found in the automatic exposure control in modern cameras. A photocell measures incident light (often TTL; through the lens) and electrically opens or closes the lens aperture. Note that feedforward is an *open-loop* condition which does involve some penalties. It is not self-checking, malfunctions are not corrected, levels are not regulated etc.

Advantage of the follow-up servo
Why should muscle control involve a roundabout method of activation such as this? A little thought will show that a movement brought about by the fusimotor route will be positionally accurate (due to the involvement of the stretch reflex servoloop) in spite of variations in load. This is a requirement of a skilled movement. Length of the muscle is controlled irrespective of tension.

In contrast, activation by the alpha route, although having a much shorter latency (larger motor fibres conduct faster; they directly activate the anterior horn cell) cannot include length control. Only tension can be specified. As a first approximation we can say that tension developed will be independent of muscle length.

The view, until recently, has been that any particular movement is brought about by a combination of varying amounts of activation via these two routes. Skilled movements are fusimotor; urgent movements are alpha.

Experimental proof?

In the last year or so, it has become possible to record from peripheral nerves in man during a variety of conditions of muscular contraction. Perhaps one ought to qualify this statement slightly, for it is very much easier to perform the recording in an experimental animal which has nice soft connective tissue round its nerves. In fact, teenage girls were found to be the very best subjects and it may well be that another contributory factor in the success of the experiments was that they were carried out, north of the arctic circle, in winter.

The results showed that no type of muscle contraction was ever preceded by fusimotor fibre discharges: always a motor discharge in the alpha fibres came first, then the muscle contracted and about the same instant came the fusimotor impulses. The only possible interpretation of such findings is that the follow-up servo is just not a real mechanism.

However, when they do contract, the poles pull upon the equator wherein lies the annulospiral ending. The sensitivity to stretch is enhanced; the endings fire off impulses. The fate of these impulses is, of course, to travel around the servoloop as we have already described, activate the main muscle and to produce contraction.

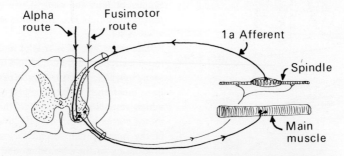

Fig. 15.5 The two routes for excitation of muscle. The α route is an open loop control; impulses from the motor cortex contract the muscle. The spindles would then be incapable of any useful function since they would be slack. Concomitant impulses in the fusimotor route, however, shorten the intrafusal muscle fibres *pari passu* with the main muscle to effect the necessary adjustment. Feedback round the loop is then always possible.

Function of fusimotor system?

Why then do we have such a complex fusimotor system? The answer seems to be quite simple. It is that if the muscle contracts voluntarily, all the sensory proprioceptive elements within it (spindles) will be inoperative since they will be "unloaded". Therefore some provision must be made to shorten, actively, the spindles during a voluntary contraction. A little thought will show that the only practical way to achieve this is to have an ancillary motor system to do it, *pari passu* with the main motor system. It seems that the nerve impulses initiating a voluntary contraction are carefully pre-programmed in the motor cortex, cerebellum etc. to achieve the required skeletal movement and to give exactly the correct amount of shortening in the spindles to keep them working.

These examples of control systems should be enough to make the reader realise their importance and their ubiquitous distribution in biology. An appreciation of the basic principles involved will be a great help in the understanding of the controls to be described in later pages.

16. Properties of Body Fluids

Before detailed consideration of the physiology of body fluids, we shall briefly discuss the properties of various types of saline solution.

Physiological saline and isolated tissues

A great deal may be learnt about the way in which the various organs and tissues of animals perform their functions, by removing them from the whole animal and examining their behaviour in isolation. They will not then be surrounded by their normal interstitial fluid and this must be replaced by some artificial solution; the substances which must be present in such a solution have been discovered largely by trial and error.

Owing to its automatic rhythmic activity, the heart is the most convenient indicator of responses to changes in its perfusion fluid, and the classical experiments of Ringer on this organ form the basis of our knowledge of the subject. If a frog's heart is isolated and placed in a solution, of whatever composition, of which the total osmolar concentration is substantially different from that of the blood, it soon ceases to beat. The requirement that solutions outside cells should be isotonic with the intracellular fluid has already been discussed, and is made obvious by the haemolysis of the red blood cells in hypotonic solutions. That the solution should be isotonic is not, however, sufficient. The heart will not beat in solutions of non-electrolytes, such as glucose. It will do so, however, for a little time in an isotonic solution of sodium chloride (0·65 per cent for the frog); but soon the beats cease, and the heart remains relaxed. If, now, a small amount of calcium chloride is added to the solution, the beats begin again; but after a short while the relaxation after each beat becomes progressively less complete until, at last, the heart remains fully contracted and ceases to beat. If at this stage a suitable small amount of potassium chloride is added to the solution, contractions begin again, and the heart may continue to beat fairly normally for many hours. These observations are illustrated in Fig. 16.1. It is clear, therefore, that at least three salts must be present in an adequate physiological solution—sodium potassium and calcium chlorides—and the heart survives longer still if the solution is made slightly alkaline by adding sodium bicarbonate. Ringer, in his series of experiments, found the concentration of each of these salts which favoured longest survival of the heart-beat of the frog and tortoise. Locke, working with isolated mammalian hearts, found the addition of glucose an advantage and the compositions of these "physiological fluids" are given in Table 16.1. The addition of a small quantity of a magnesium salt (0·01 per cent $MgCl_2$ for example) is beneficial in experiments with many kinds of mammalian tissue (as in Tyrode's and in Krebs' modifications of Ringer's solution), but has little effect on the heart. The concentrations of potassium and calcium may be varied slightly to suit the particular tissue in use.

On the whole, therefore, the necessary composition of an artificial environment for contractile (and other) tissues, as discovered empirically, is very similar to that of the interstitial fluid; this, of course, is hardly surprising. Ringer's and Ringer–Locke's solutions contain smaller concentrations of sodium bicarbonate and, in compensation, greater concentrations of sodium chloride than do the corresponding frog's plasma and mammalian plasma, respectively. Such differences are necessary because the artificial solutions are in equilibrium with room air containing only very small amounts of carbon dioxide, whereas the plasmas contain much larger quantities of free carbon dioxide, mammalian arterial plasma, for example, being in equilibrium with 5 per cent of carbon dioxide. The acidity of the solution is related to the ratio of the concentration of carbon dioxide to that of bicarbonate; reduction in the former, therefore, must be compensated by a reduction in the latter if the solution is not to be too alkaline. In some kinds of experiment it is possible to keep the artificial saline solutions in equilibrium with 5 per cent of carbon dioxide (usually in pure oxygen) without great inconvenience; the bicarbonate concentration is then increased, and the chloride concentration decreased, to about the values found in the plasma (as in one form of

Table 16.1 *Compositions of blood plasma and of physiological saline solutions*

	Ringer's solution (frog's heart) (g)	Frog's blood plasma (g)	Locke's solution (mammalian heart) (g)	Mammalian blood plasma (g)
NaCl	0·65	0·55	0·9	0·7
KCl	0·014	0·023	0·042	0·038
$CaCl_2$*	0·012	0·025	0·024	0·028
$NaHCO_3$	0·02	0·1	0·02	0·23
NaH_2PO_4*	0·001	0·02	—	0·036
Glucose	—	0·04	0·1–0·25	0·07
Water	to 100	(100)	to 100	(100)

* The weights given refer to the anhydrous salts. Appropriately greater weights of the hydrated forms, which are in common use, should be employed in making up Ringer's solutions.

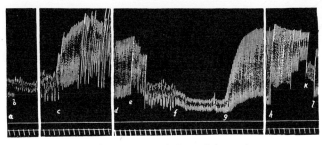

Fig. 16.1 The action of electrolytes on the heart of the tortoise.
 a–b: Beat of excised heart before perfusion. b–c: Perfusion with *sodium chloride* solution, 0·75 g/dl^{-1} with sodium bicarbonate 0·01 g/dl^{-1}. c–d: Addition of *calcium chloride* (3 ml of 0·1 M to 1dl of solution of sodium salts). Contraction improved but relaxation incomplete. d–e: Addition of *potassium chloride* (6 ml of 0·1 M to 1 dl of solution of sodium and calcium salts). Relaxation improved and beats more regular. e–f: Perfusion with solution of *sodium* salts again. f–g: Addition of *potassium chloride* (6 ml of 0·1 M to 1 dl of solution of sodium salts). Beats more regular but contractions small. g–h: Addition of *calcium chloride* (3 ml of 0·1 M to dl of solution of sodium and potassium salts). Contraction greatly improved, with good relaxation. h–k: Addition of excess calcium chloride. 3 ml of 0·1 M solution added to 1 dl of solution of sodium, calcium and potassium salts at each step in the record. Progressive impairment of relaxation. k–l: Potassium chloride added in amount to correspond to the calcium chloride present—i.e. 2 ml 0·1 M KCl for each 1 ml 0·1 M CaCl$_2$. Improved relaxation but poor beat—onwards: Perfusion with normal Ringer's solution.
 The smallness of the beat initially is due to the action of acid metabolites, which are formed during the anoxaemia accompanying dissection; they are rapidly removed by the perfusion fluid. Recording: systole is up-stroke. (From Bayliss, *Principles of General Physiology*, Longmans Green, 1960.)

Krebs–Ringer solution). Alternatively, the bicarbonate and carbon dioxide may be omitted and replaced by an equivalent concentration of a sodium phosphate buffer mixture adjusted to pH 7·4 (as in the other form of Krebs–Ringer solution).

The action of ions on tissues

The subject of the action of individual ions on contractile tissues is inevitably a confused one, owing to the fact that different tissues respond in different fashion to excess or deficiency of any particular ion. There are, however, certain general rules. (1) Sodium salts occupy a unique position. The amount present in Ringer's solution is greatly in excess of the minimal amount necessary. Muscles cease to contract when more than 9/10 is replaced by some non-toxic substance such as glucose or sucrose. (2) The concentrations of calcium and potassium salts are interdependent, and must be in about the right ratio, though the absolute values may vary considerably. (3) Anions are relatively unimportant so long as they are not toxic; chlorides are usually employed, as these salts are all soluble, but the presence of sulphates, bromides, nitrates, bicarbonates etc. has little influence on the survival of activity of an isolated tissue. (4) The hydrogen ion concentration must be not far removed from that of a neutral solution. Most tissues are favoured by a hydrogen ion concentration slightly on the alkaline side of neutrality. Slight acidity produces slowing or arrest of the heartbeat, relaxation of the tone of most unstriated muscles and phenomena analogous to fatigue in striated muscles.

If a muscle or nerve is placed in a solution which contains too high a concentration of potassium ions (say Ringer's solution modified to contain about three times the normal amount of potassium chloride), it becomes inexcitable; the excitability can be restored by washing it with ordinary Ringer's solution. This effect accounts for the practical necessity of washing isolated muscles and nerves with Ringer's solution; this removes the excess of potassium salts which diffuse out of those few fibres which are injured during dissection.

These and many other kinds of experimental study make it clear that for proper functioning of the cells the electrolyte composition of the fluid bathing them must be quite different from that of the fluid within them. The cells contain potassium ions and practically no sodium ions, while the fluid outside them must contain sodium ions and very little potassium ions. Many cells also contain practically no chloride, although this can be replaced by other univalent ions.

Examples of some current perfusion experiments

All organs and systems from the brain downwards have been the subject of experimental investigation whilst isolated and either perfused or superfused.

BRAIN SLICES
The success of many biochemical-type investigations utilising slices of tissue immersed in warmed baths of various types of saline solution has led the neurophysiologist to enquire whether such techniques might not be applied to slices of the nervous system.

In recent years it has, in fact, proved possible to keep slices of brain in a functional state for quite long periods of time in baths of saline. There are two conflicting requirements:

1. The slice must be thin enough to allow adequate diffusion of oxygen to take place so that the neurones do not die (or malfunction).
2. The slice must be thick enough to contain neurones and their processes (axons, dendrites). If these are cut the neurone is abnormal and may well be unresponsive.

A compromise between these two points is usually reached and the freedom from problems of anaesthesia during the experimental period, the lack of movement artefacts due to respiration and the pulse, have made this approach to brain research a most attractive one. Primarily it enables one to approach cellular function, rather than to find out about organisational problems in cortical connectivity.

Organ baths

For many years it has been possible to investigate more or less any contractile tissue or organ by putting in a suitably constructed organ bath and measuring its responses to electrical stimuli, to drugs, ionic concentration changes or to various temperatures. The gut can be experimented upon, the bronchi (here chains are made of several rings taken from bronchial muscle), blood vessels or the uterus may be used.

MUSCLE SPINDLES

A considerable amount of elegant research has been carried out upon the mode of functioning of the muscle spindle, using specimens removed from the body. A muscle, such as the tenuissimus of the kitten, which is small and contains perhaps five spindles, can be used whole, its nerve isolated and recorded from. Furthermore, intricate dissection is possible and a single isolated spindle can be obtained in working condition.

The *advantages* of this approach to experimental physiology are:

1. Lack of the confounding actions of anaesthetics.
2. Good control of variables.
3. Ease of recording, free from movement and other artefacts.
4. There is no need for ancillary apparatus to maintain a whole experimental animal.

The *disadvantages* are:

1. The preparation in isolation may not behave in the typical way it would do *in vivo*.
2. The act of dissecting it out may injure it.
3. There are difficulties associated with supply of oxygen, nutrients etc. The perfusing fluid is usually not strictly comparable with true extracellular fluid.

Isotonic solutions (see also p. 85)

A solution is defined as being *"isotonic"* if, when placed in contact with living cells, no water passes into or out of the cells; *"hypertonic"* if it draws water out of the cells; and *"hypotonic"* if water goes into the cells. If two solutions are separated by a membrane permeable only to water, and impermeable to all the substances in solution, water will move from the solution with the smaller total molecular (osmolar) concentration to the solution with the larger osmolar concentration. But if the membrane is permeable to some of the substances in solution, these substances will move from one solution to the other until their concentrations are the same in both, and they will then have no effect on the movement of water. In calculating whether a solution is isotonic or not, the concentrations of all those substances to which the cell membranes are permeable must be left out of consideration. (This may not be strictly true if the substances are electrolytes, owing to the effect of the Donnan equilibrium, but it is very nearly true.) For example, the red blood corpuscles are impermeable to sodium chloride, but permeable to urea, oxygen and carbon dioxide, so that the addition of, say, urea to an isotonic solution of sodium chloride will not make this solution hypertonic with respect to the red blood corpuscles. Again,

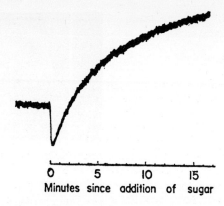

Fig. 16.2 Swelling and shrinking of Human red blood cells.
Changes in the average volume of the cells were recorded directly in terms of the amount of light transmitted through a very dilute suspension in a saline solution. At time 0, 2 ml of a solution of sorbose was added to 10 ml of the suspension, the final concentration of sorbose being 0·3 M. The total osmolarity of the suspending solution was thus approximately doubled. The optical transmittance decreased suddenly, indicating an almost instantaneous shrinkage of the cells: movement of water between the cells and the outside solution is very rapid.

The sorbose then penetrated slowly into the cells, raised the concentration of the intracellular fluid, water was drawn in and the cells swelled. The final optical transmittance is greater than the initial, since (a) the whole suspension has been diluted to 5/6 (the effect of the initial shrinkage is partly masked by this); and (b) the suspending solution has been made hypotonic, since water was added as well as sorbose and the cells have swelled accordingly. (Lefevre and Davies.)

the ordinary collodion membrane, like the glomerular membrane in the kidney, is permeable to all crystalloid substances, so that only the concentrations of the colloidal ones (such as proteins) need be considered. All these membranes, however, are more permeable to water than to any solute, so that transient osmotic effects may be observed even with solutions that are, in the long run, isotonic with respect to these membranes. In physiology, we often use the word "isotonic", without further specification, to mean "isotonic with respect to the red blood corpuscles". For mammalian cells, a solution containing 0·90 to 0·95 g of sodium chloride in 1 dl of water is "isotonic".

The movements of water and dissolved substances into and out of the intracellular fluid are very conveniently studied by using suspensions of red blood corpuscles. Changes in their volume can be measured by the use of the haematocrit in terms of the amount of light transmitted by the suspension. Records from an experiment which illustrates the points just discussed are given in Fig. 16.2. If a solution of the hexose sugar sorbose, of such a concentration as to be apparently hypertonic, is added to a suspension of red blood cells in an isotonic saline solution, there is initially a very rapid withdrawal of water from the cells, which therefore shrink. The sorbose then slowly penetrates into the cells, drawing water with it. Finally, the cells are more swollen than they were initially, since the saline solution has been diluted by the water added with the sorbose, while the sorbose, having the same concentration inside and outside the cells, is osmotically inactive.

An isotonic solution may, alternatively, be defined as a solution which has the same osmotic pressure as the intracellular fluid, a hypertonic solution as one with a greater osmotic pressure, and a hypotonic solution as one with a smaller osmotic pressure. But it must be remembered that the osmotic pressure exerted by a solution depends on the nature of the membrane at which it is developed as well as on the concentration of the solution. The *"tonicity"* of a solution must therefore be defined in terms of the osmotic pressure exerted at the particular membranes considered—e.g. the membranes of certain kinds of cell, the glomerular membranes etc.—and not necessarily in terms of the "ideal" osmotic pressure exerted at a membrane permeable only to water.

Haemolysis

If the cell membranes are destroyed, or sufficiently damaged, the cell contents pass into the surrounding solution. The red blood cells, again, are very useful for studying the conditions in which this occurs; the haemoglobin which they contain has a strong red colour and is thus easily detected in the external solution; and the cells themselves are easily removed by centrifugation.

There are three general ways in which haemolysis may be brought about. In the first, the cells are placed in a solution which has a salt concentration less than that of the plasma; water passes into them and they swell up and finally burst. In the second, the cell membranes are damaged by means of a solution of ether, saponin, bile salts or other surface-active substances. In the third, the blood is repeatedly frozen solid and thawed as rapidly as possible.

When placed in a hypotonic salt solution, the red cells, owing to their biconcave shape (Chapter 10), can increase in volume very considerably before the surface membrane becomes stretched. The amount of further swelling which can take place without rupture of the membrane and haemolysis, varies from one cell to another, even in the same sample of blood from the same individual person; in normal human blood, while the great majority of cells are just haemolysed in about 0.4 per cent NaCl, a small proportion (5 to 10 per cent) will only burst in stronger solutions, and about the same proportion are unaffected unless the concentration is reduced to less than 0.4 per cent. In this way a *resistance* or *fragility curve* can be constructed, showing the proportion of corpuscles haemolysed, against the concentration of the solution in which they are placed.

The proportion of cells with small resistance is greatly increased in the blood of patients suffering from certain diseases such as familial haemolytic jaundice; when these patients have their spleens removed, their red cells attain a resistance which is normal, or even greater than normal. During recovery from anaemia (e.g. pernicious anaemia or after haemorrhage), on the other hand, the proportion with a large resistance is greatly increased. The resistance to haemolysis of a given cell probably depends upon its age, the membrane becoming steadily weaker during its life in the circulation until it finally gives way altogether.

The surface-active substances do not necessarily destroy the cell membrane entirely, but make it permeable to sodium and potassium—perhaps only in certain places. The concentrations of sodium, potassium, chloride and bicarbonate then become more nearly the same inside and outside the cells, and there is an excess osmotic concentration of haemoglobin within the cells: these then take up water, swell and eventually burst, allowing the haemoglobin to escape.

17. The Distribution of Body Fluids

The fluid compartments

The greater part of any animal consists of water solutions; the structural part, which gives the animal its solidity, has less than one half the weight of the fluid part. It is in the body fluids that most of the chemical reactions involved in metabolism take place; and it is by means of diffusion in these fluids, and by their movements as a whole from place to place, that the metabolites reach and enter the active cells and the products of the metabolic reactions leave and are carried away. Changes in the volume and composition of these fluids thus affect the activities of most of the organs and tissues of the body, and it is becoming increasingly apparent that the study of these changes in health and disease is an important branch of physiology.

The composition of this fluid matrix is not uniform throughout, and variations in the concentrations of different kinds of dissolved substances are found in different parts of the body. It is useful to think of the animal body as divided into distinct *fluid compartments* which are more or less separated by various barriers, but at the same time are in equilibrium with each other—or if not in true equilibrium, in a "*steady state*", maintained by appropriate metabolic reactions. The whole body fluid may be conveniently divided into two major compartments: (1) the *intracellular* fluid enclosed by the cell walls and (2) the *extracellular* fluid between and around the cells; this latter can be again divided into (a) the interstitial, or tissue, fluid and (b) the blood plasma. The interstitial fluid, which lies between the cell walls and the walls of the blood vessels, forms the "internal environment" for the cells, and provides the connection between the intracellular fluid, where the metabolic reactions occur, and the plasma.

Table 17.1 *Body fluids (in an average man, weight 70 kg)*

	Weight (kg)	Percentage of body weight
Total body fluids	49	70
1. Intracellular fluid (ICF)	35	50
2. Extracellular fluid (ECF)	14	20
(a) Interstitial fluid	11·2	16
(b) Blood plasma	2·8	4

The plasma, by virtue of its circulation, distributes the local changes in the composition of the interstitial fluid which are produced by metabolic activity, and enables those produced in the muscles, liver etc. to be offset by those produced in the lungs, kidneys and alimentary tract.

Each of the different fluid compartments has its characteristic volume and composition. Their relative sizes are given in Table 17.1 and their compositions are discussed in a later section.

Methods of estimating the fluid volumes

A known quantity of some suitable substance is added to the fluid whose volume is to be estimated and its concentration measured; the quantity added, in grams for example, divided by the concentration, in grams per litre, is the volume of the fluid in which it is dissolved, in litres. The principle of the method is thus simple, but discovering the "suitable substance" is not so simple.

The substance chosen must be harmless and without pharmacological effects which would change the amount or distribution of body fluids. In practice, it must be injected into the blood stream, allowed to become distributed throughout the body, and its concentration estimated in a sample of venous blood, usually taken from an arm vein. Corrections must be made for any production or destruction of the substance in the body, as for example by metabolic processes, and for the quantity eliminated, e.g. in the urine, during the interval between injection and collection of blood for analysis. It is obvious that these corrections should be small if possible. Many of the most accurate methods depend on the use of radioactive isotopes; special facilities for handling and estimating radioactive substances are needed and proper precautions must be taken against radiation hazards.

The volume of the blood

The "reference substance" must be retained within the blood vessels for a period at least long enough to allow it to become uniformly distributed. This is ensured by using a substance which is attached either to the plasma proteins (the volume of the plasma is thus measured) or to the red blood cells (the total volume of these cells, is measured). The volume of the plasma, of the red cells and of the whole blood are related to each other by the *packed cell volume* (volume of red cells as percentage of volume of whole blood; Chapter 10), so that any one may be calculated from any other. The volume of the whole blood, however, is best measured by adding together the volume of the plasma and the volume of the red cells. Three methods have been used most frequently:

(a) *Plasma volume*. A known quantity of the non-toxic dye Evans blue T1824, which becomes bound to the serum albumen, or a known quantity of serum albumen which has been iodinated with radioactive iodine (^{131}I), is injected intravenously. The blood sample is centrifuged and the concentration of dye, or of "labelled" protein, in the plasma is estimated.

(b) *Red cell volume*. A quantity of red blood cells which have been "labelled" by means of radioactive isotopes of iron, phosphorus or chromium, and of known radioactivity, is injected. The radioactivity of a known volume of red cells, packed by centrifuging at high speed, is measured.

(c) *Total blood volume*. A known amount of carbon monoxide mixed with oxygen is rebreathed until all the carbon monoxide is absorbed and combined with haemoglobin. The amount of CO-haemoglobin in unit volume of blood is estimated by means of the reversion spectroscope.

The volume of the extracellular fluid

The substance injected must be able to pass through the walls of the blood capillaries, but unable to pass through the cell membranes. Many different substances have been used, but none satisfies the criteria perfectly and all have some disadvantages. Reliable values seem to have been obtained by the use of inulin, mannitol, sucrose (in man only), sodium sulphate (containing the radioactive isotope of sulphur for ease of analysis), sodium thiosulphate and sodium thiocyanate. The volume which is measured is that of all the water in which the reference substance is dissolved; the volume of the whole extracellular space is larger than this, since there are microscopic and ultramicroscopic structures into which the substance does not penetrate.

The volume of intracellular fluid

This can only be measured as the difference between the total volume of the body fluids and the volume of the extracellular fluid (apart from the volume of the red blood cells), as already described. The measurements are thus subject to considerable uncertainty. The intracellular fluid contains substances of large molecular weight and its total volume is greater than that of the intracellular water.

The total volume of body fluids

Fundamentally, the most perfect method of measuring the volume of water in an animal is to weigh it, kill it, dry the body in an oven, and weigh it again. But if the animal, or man, is not to be destroyed by the method of estimation, a reference substance must be used which penetrates into all the water of the body, i.e. into all three compartments. Very few substances have yet been discovered which are at all suitable. Water containing the isotopes of hydrogen (deuterium or radioactive tritium) is theoretically the best, but the analyses require elaborate apparatus, and tritium is not easily obtained. Urea may be used, although the corrections for metabolism and excretion are rather large and certain other organic compounds (the drug antipyrin, for example) have been found which are better.

THE COMPOSITION OF THE BODY FLUIDS

Water is the largest single component of the body fluids. Its presence in the proper quantity is important for two reasons: first, to maintain the total volume of the body fluids, and the volumes of the various compartments and, second, to maintain the concentrations of the solutes in these fluids. A great many different substances are dissolved in the water and, in general, the different compartments have different compositions. Some substances are present in all the compartments, but few of them have the same concentration in all.

The processes by which the water content of the body is maintained and a "water balance" achieved are discussed in Chapter 19. We shall be concerned here with the nature of the dissolved substances in the various compartments and with the movements of these, and of the water, from one compartment to another.

Extracellular fluid

Apart from proteins, the interstitial fluid has very nearly the same composition as the blood plasma and, since enough plasma for chemical estimation can easily be collected, its composition is usually taken as equivalent to that of the extracellular fluid. The concentrations of the diffusible electrolytes (sodium, potassium, chloride and bicarbonate), however, are not quite the same in the plasma as in the interstitial fluid, as will be discussed later.

The peritoneal, pleural and pericardial fluids, although separable anatomically from the bulk of the interstitial fluid, all have much the same composition and may be regarded as parts of it. Their particular function is largely one of lubrication, enabling the intestines, lungs and heart to move freely in their respective enclosures. Lubrication of the articular surfaces of the bones is performed by the *synovial fluid*; this differs from the other fluids in containing mucin, which helps in the lubricating action.

The *cerebrospinal fluid* and the *intraocular fluids* are usually regarded as part of the extracellular fluid, but differ quantitatively in composition from the general interstitial fluid. They will, therefore, be considered separately, the cerebrospinal fluid in the next chapter, and the intraocular fluids as essential parts of the eye in Chapter 58. Fluid is also present in the inner ear and labyrinth; that which lies between the bony wall and the membranous labyrinth—the *perilymph*—has the composition of the interstitial fluid, but that which lies within the membranous labyrinth—the *endolymph*—has a very different composition.

In spite of the fact that there is no difficulty in obtaining quite large quantities of human blood, a full and exact description of its chemical composition cannot be given. It contains many substances whose identity is known, but in too small a concentration for accurate estimation; and other substances whose presence can be inferred only from their actions, many of them not having been isolated or chemically identified, for example some hormones and enzymes.

Intracellular fluid

The cells which make up the various organs and tissues are intimately surrounded by connective tissue and interstitial fluid; an organ or tissue which has been dissected out, however cleanly, includes both intracellular and extracellular fluids. In order to discover the composition of the intracellular fluid from the gross composition of the whole organ or tissue, it is necessary, therefore, to discover the fraction of the whole organ or tissue which consists of extracellular fluid. This can be done by the use of a reference substance which cannot penetrate into the cells, but since complete accuracy is impossible, we know even less about the exact composition of the intracellular fluid than about that of the extracellular fluid. Moderately complete analyses are available for mammalian striated muscle and much less complete analyses for some other mammalian organs and tissues. The red blood corpuscles do not present this complication, since they can be obtained practically free from plasma by centrifuging the blood, but in some respects their composition is not typical of intracellular fluid in general.

Proteins

The blood plasma contains some 7 to 8 g of protein in solution in each 100 ml. This protein can be split into many different fractions, but for most purposes it is sufficient to consider it as being composed of two parts, serum globulin and serum albumen, accounting for 2 g and 5 g dl^{-1} plasma respectively, in addition to which there is about 0·3 to 0·4 g of fibrinogen. The functional significance of the proteins will be considered in later chapters. The interstitial space, on the other hand, contains only small quantities of protein in solution; but, particularly in the connective tissues, there are substantial quantities of proteins of various kinds, chiefly collagen and elastin, in the form of fine fibres visible under the microscope, which provide rigidity and elasticity to the whole structure. These are surrounded by the "ground substance" which contains mucopolysaccharides (polymerised amino sugars), notably hyaluronic acid, as well as proteins. The "intercellular cement" lies between, and holds together, the individual cells of an organ or tissue, and has the same (or a very similar) composition. The presence of extracellular material which is not in solution is particularly obvious, of course, in such structure as hairs, finger- and toe-nails, tendons, teeth, cartilage and bone.

Proteins in quite considerable concentration also occur within the cells, both in solution and as microscopically visible structures; the distinction, however, is bound to be somewhat vague, since the electron microscope reveals as fibrils, for example, structures invisible in the light microscope. Some kinds of protein, notably the nucleoproteins and those which form many kinds of enzyme, are found in all kinds of cell. The nucleoproteins in many kinds of cell appear ordinarily to be in solution, since they are invisible but during mitosis, when the cell divides and becomes two cells, some of the nucleoproteins become visible as chromatin threads and chromosomes. Other kinds of protein are peculiar to the particular kind of cell considered, and are essential to the function of those cells. The red blood cells, for example, contain haemoglobin in large concentration but apparently in solution; this is essential to their function of transporting oxygen and carbon dioxide. Muscle cells contains special kinds of protein, in relatively small concentration, but in the form of fibrils and filaments; these are essential for the shortening and development of tension characteristic of muscle cells.

The concentration of the proteins in the intracellular fluid and in the blood plasma, in terms of grams per unit volume of fluid, is quite large; the weight (and volume) of the water present is thus correspondingly less than that of the whole fluid. In 1 dl of blood plasma there are 90 to 93 g of water, and in 1 dl of red blood cells or muscle fibre there are 70 to 80 g of water.

Electrolytes

All the body fluids contain salts, and electrolytes, in solution, but their concentrations in the intracellular fluid are very different from those in the extracellular fluid, and it is necessary to consider separately the cations and the anions which together make up these salts. The concentrations of the chief cations and anions in the blood plasma and in the intracellular fluid of red blood cells and muscle fibres are plotted diagrammatically in Fig. 2.1.

When considering the chemical properties and osmotic and electrical equilibria of a solution, the concentrations should be expressed in terms of gram-molecules, rather than in grams, and per kilogram of water, rather than per litre of solution; the substances are dissolved in the water and not in the proteins or other colloidal constituents which may occupy an appreciable fraction of the whole volume of the fluid. A millimolar solution of a substance, say sodium chloride, contains 1/1000 of its molecular weight in grams (e.g. 58·5 mg) in 1 kg (or approximately 1 litre) of water. Concentrations are commonly estimated and expressed, nevertheless, as milligrams per decilitre of solution. They must be expressed like this, in mg/dl, if the precise molecular weight is not known, as in the case of proteins etc (see p. 85).

Cations. The greatest difference between the extracellular fluid and the intracellular fluid is that the former contains chiefly sodium, the molar ratio Na/K being about 30, whereas the latter contains chiefly potassium, the molar ratio Na/K being about 0·18 in the red blood cells and about 0·08 in the muscle fibres. This very large excess of potassium is characteristic of muscle and nerve fibres, but other kinds of cell may be more similar in this respect to the red blood cells.

In some kinds of animal other than man, the red blood cells do not contain an excess of potassium, the Na/K ratio in the cells of the cat and the dog, for example, being about 15.

The only other cations present in significant, though relatively small, concentration are calcium and magnesium. The calcium concentration of the intracellular fluid is somewhat less than that of the extracellular fluid and the magnesium concentration is definitely greater; both are present within the cells largely in combination with organic substances and not in the free state as ions.

Anions. In the extracellular fluid the most important anions are chloride and bicarbonate; the other anions which are present—sulphate, phosphate and lactate—together making up only some 2 per cent of the total concentration. Within the cells, however, the combined concentration of chloride and bicarbonate is far too small to be equivalent to that of the cations. It is necessary to suppose, therefore, that other kinds of anion are present in substantial concentration. In red blood cells, these are provided almost entirely by haemoglobin; in muscle fibres they consist mainly of hexose phosphates, creatine phosphate, adnosine triphosphate and the substance carnosine (β-alanyl histidine), some also being provided by proteins. The concentration of protein ions will depend on the pH of the fluid, since this will affect the strength of the protein as an acid; there will then be inverse changes in the bicarbonate concentration. This effect is of considerable importance in the regulation of the acidity of the blood and the carriage of carbon dioxide.

Other substances

The extracellular fluid contains, in easily analysable concentration, glucose, urea and "lipids" such as lecithin and cholesterol. Urea is the only normal constituent of the body fluids which is present in the same concentration (about 30 mg dl^{-1}, or 5·0 mmol l^{-1}) in both intracellular and extracellular fluids; its concentration depends on the amount of protein taken in the diet, so that there may be quite large differences between different individuals. The glucose concentration in the blood (about 100 mg dl^{-1} or 5·5 mmol l^{-1}) is maintained relatively constant. It is converted in the intracellular fluid into hexose phosphates and glycogen, for example, but glucose may be present in some kinds of cell.

ABNORMALITIES OF FLUID DISTRIBUTION

Water deficiency

There are various causes of lack of water. Ill patients may be unable to drink or to communicate that they feel thirsty (e.g. if they are unconscious). The antidiuretic hormone, ADH (see Chapter 23) may not be secreted in large enough amounts by the pituitary gland thus giving rise to diabetes insipidus. The first result of water lack is the sensation of thirst, the osmolality of the ECF rises and water leaves the ICF to keep the osmotic pressure in the two compartments the same. When about 10 per cent of the total body water is lost, drowsiness and coma result. Plasma sodium levels may rise to 150 or 200 mmol l^{-1}.

Water excess

The increase in total body water again occurs in the ICF as well as the ECF, giving distension of cells. When this swelling occurs in brain tissue surrounded by the inextensible skull, raised intracranial pressure results with the production of such symptoms as nausea, vomiting, headaches, coma and epileptic fits (the so called syndrome of water intoxication, but mostly due to the rise in intracranial pressure).

A simple cause is renal failure, where water is not excreted. A disturbance of the thirst mechanism may result in over drinking of water. Antidiuretic hormone, may be produced in excess of requirements leading to water retention by the kidney.

Sodium Deficiency

There is a close link between sodium levels and the volume of the ECF. Thus sodium depletion leads to a low ECF volume. The clinical picture is initially postural hypotension, i.e. when the patient stands up his blood pressure catastrophically falls and he may become unconscious because of lack of brain oxygen supply. The reason for this effect is that since the volume of circulating blood is reduced, no margin exists for the sudden increase in cardiac output needed when suddenly standing up. Low blood pressure, oliguria, cold sweating extremities and a peripheral cyanosis (due to the oligaemia and poor blood flow) are found.

This condition results typically if fluid is lost from the gastrointestinal tract as in vomiting and diarrhoea, because the fluid lost is approximately isotonic with ECF. Copious sweating in a hot environment leads to sodium loss associated with painful muscle spasms (miner's cramp; page 170).

Sodium excess

Excess of sodium in the body is usually associated with an excess of water, i.e. the body water is kept isotonic and the result is an increase in ECF volume. This gives rise to pitting oedema in subcutaneous tissue (see page 125) and cavities such as the peritoneal space or the pleural space often contain fluid.

Salt and water retention occurs in cardiac failure, cirrhosis of the liver and in the nephrotic syndrome. In cardiac failure the intravascular volume is raised thereby.

18. The Formation of Body Fluids

Capillary interchanges

The walls of the capillary blood vessels are very much more permeable to dissolved substances than are the cell membranes. Of the substances ordinarily present in the extracellular fluid all, except the plasma proteins, can penetrate freely and will diffuse to and fro between the blood and the interstitial fluid. Now, the plasma proteins, like any other substances in solution, will contribute to the total osmolar concentration of the plasma. This, therefore, will be a little greater than that of the interstitial fluid, even though the concentrations of the crystalloid substances are very nearly the same in both. Water will tend to pass into the plasma, accompanied by those substances to which the walls of the capillaries are freely permeable. As already mentioned, however, the tendency of water to move from solutions whose osmolar concentration is small to those whose concentration is large, may be opposed by the application of a hydrostatic pressure. Such a pressure, however, cannot be effective in practice unless there is a rigid membrane separating the two solutions concerned and impermeable to some, at least, of the dissolved substances. If this pressure is just sufficient to prevent any net movement of the water, it is equal to that part of the total osmotic pressure which is contributed by those substances which cannot penetrate the membrane. In the blood vessels there is such a hydrostatic pressure, produced by the action of the heart, and necessary for the circulation of the blood. If the capillary blood pressure exceeds the osmotic pressure due to the proteins (about 25 mm Hg, 3·33 kPa in mammals), fluids will pass out from the plasma into the interstitial fluid. At the arterial ends of the capillaries the pressure is high enough (about 35 mmHg; 4·66 kPa) for this to occur. At the venous ends of the capillaries, the hydrostatic pressure (about 15 mmHg; 2·00 kPa) is less than the osmotic pressure due to the proteins, and fluid will be drawn in again. The net difference between (1) the capillary pressure (less the hydrostatic pressure in the interstitial spaces), and (2) the osmotic pressure difference between the plasma and the tissue fluid, is known as the *effective filtration pressure*. If this is positive, fluid leaves the bloodstream; if it is negative, fluid enters.

This filtration and absorption can be seen in the capillaries of the frog's mesentery. The mesentery is observed under a binocular microscope; a capillary with a rapid circulation is chosen and blocked by pressing a fine blunt glass rod on it by means of a micromanipulator. In some cases the mass of corpuscles remaining in the capillary moves towards the block, indicating that filtration is taking place, and in some cases they move away from the block, indicating absorption. A very fine glass pipette is now inserted into the capillary, connected with a water manometer, and the capillary pressure is measured, the block being still in place. The results of a large number of such experiments are shown in Fig. 18.1, in which the rate of passage of fluid through the capillary wall is plotted against the capillary pressure. It will be seen that, on the average, fluid passes out when the pressure is greater than 11·5 cm of water, (109 Pa) and in when it is less than this; independent

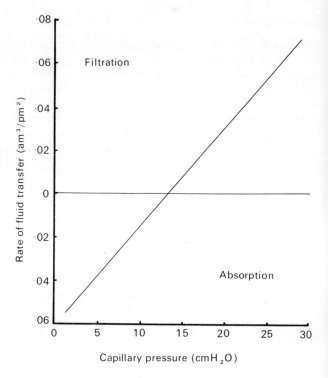

Fig. 18.1 The effect of the pressure within a capillary on the passage of fluid through its walls.

Positive values of the rate of fluid passage indicate filtration of fluid; negative values indicate absorption.

The observations were made on several different frogs, and the straight line, representing the average, passes through the line of zero flow at a hydrostatic pressure of 11·5 cm of water; 109 Pa (about equal to the colloid osmotic pressure of the plasma); fluid passes neither in nor out at this pressure and the rate of flow is directly proportional to the pressure both above and below it.

measurements of the colloid osmotic pressure of frog's plasma indicate that it normally lies between 10 and 12 cm of water.

The same general relation between the rate of filtration or absorption of interstitial fluid, and the excess or deficit of the mean capillary pressure above or below the protein osmotic pressure, can be demonstrated in preparations of the hind-limbs of cats or dogs; but the experimental procedure is more elaborate and the evidence less direct.

Transcapillary exchange

Brief consideration of the foregoing information will indicate that the normal transcapillary interchange of fluid is mainly determined by *capillary pressure*, on the one hand, and by *plasma oncotic pressure*, on the other. Figure 18.2 is a diagrammatic summary of the factors involved.

It is also possible that some capillaries have an outflow of tissue fluid throughout their lengths while others absorb all along their lengths. Net filtration slightly exceeds, in volume, absorption. The fluid left in the extracellular space ultimately

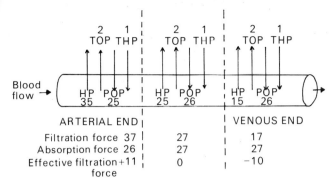

Fig. 18.2 The filtration and absorption of tissue fluid along a capillary. The figures given are mmHg (1 mmHg = 133 Pa). TOP = tissue oncotic pressure. THP = tissue hydrostatic pressure. HP = intravascular hydrostatic pressure. POP = plasma oncotic pressure.

Note that plasma oncotic pressure increases slightly in passage through capillary bed (see text).

returns to the circulation by becoming lymph and travelling to the lymphatic duct in the lymphatics.

FORMATION OF LYMPH

In all the organs of the body, therefore, there is a flow of interstitial fluid out of the capillaries in the parts near the arterioles, and back into the capillaries in the parts towards the venules. The fluid would be expected to collect in the tissue spaces, and build up a hydrostatic pressure, until the inflow and outflow are equal. In fact, the pressure in the tissue spaces is not ordinarily more than 1 or 2 mmHg (0.1–0.25 kPa); the return flow into the plasma is in general rather less rapid than the outflow from the plasma and the excess fluid—which is known as *lymph*—is carried off by the lymphatic system. This consists of very thin-walled vessels provided with valves, which all run towards the thorax, those from the hind-limbs, abdomen, left side of chest and left arm all joining together to form the *thoracic duct*, which empties into the venous system at the junction of the left internal jugular vein with the left subclavian vein. Fluid is propelled along the lymphatics partly by means of the pumping action of muscular movement, as in the veins, and partly by reason of the negative pressure in the thorax.

The walls of the capillaries are not completely impermeable to the plasma proteins. But since the proteins pass through very much less readily than the water and salts, they get left behind during the filtration of the interstitial fluid; the more rapid is the filtration, the more, proportionately, are they left behind. The interstitial fluid, therefore, usually contains some proteins, the concentration varying in different parts of the body, according to the capillary pressure and the permeability of the capillary walls for proteins. Both may vary from part to part and from time to time.

The protein which is filtered off with the interstitial fluid tends to be left behind again when the fluid is reabsorbed into the venous capillaries. Lymph, therefore, has a higher concentration of protein than has the fluid filtered off from the capillaries, the actual value being very variable (1 to 6 per cent); lymph from the intestines and liver has a higher protein concentration (6 per cent) than that from the muscles (2 per cent). Otherwise lymph contains the same substances as does the plasma, and in about the same concentrations. It usually clots if left to stand, but not so rapidly as the blood, owing to the absence of platelets. After a meal there is often a considerable quantity of fat present, giving it a milky appearance; the lymphatic system is the chief route for the absorption of fats from the alimentary canal. It normally also contains considerable numbers of lymphocytes.

Factors affecting the flow of lymph

In the steady state, as already described, any excess of interstitial fluid is carried away in the lymphatics; but any change in the rate of lymph flow is ordinarily associated with an inverse change in the volume of fluid in the interstitial spaces. If, in any organs or tissues, there is a large increase in the rate of production of interstitial fluid, or a large decrease in the rate of its removal, these organs or tissues become swollen and puffy, and a condition of *oedema* is said to have developed. Such an accumulation of interstitial fluid in any part of the body is likely to occur, of course, if there is any obstruction to the lymphatic drainage from that part. It will occur also, more generally, in response to any changes in the difference between the rate of filtration out of the capillaries and the rate of absorption into them.

The factors which influence the flow of lymph may be grouped under three headings: (1) The capillary pressure; (2) The difference between the osmolar concentration of the plasma and that of the tissue fluid; and (3) The permeability of the capillary wall.

THE CAPILLARY PRESSURE
Any rise in capillary pressure occasioned either by dilatation of the arterioles or obstruction of the veins, increases the rate of production of tissue fluid. Such a rise is a common accompaniment of activity in any organ.

Elevation of interstitial fluid pressure over the normal levels increases the passage into lymphatics of interstitial fluid. Lymph flow is increased thereby. Normal lymph flow depends greatly on the *lymphatic pump*. Valves exist in all channels and the massaging effect of muscular contraction, as previously mentioned, also produces lymph flow and increases the pressure differential between inside capillaries and within the tissue spaces.

OSMOLAR CONCENTRATION OF PLASMA AND TISSUE FLUID
Two factors come into the consideration of the osmolar concentration: (a) the concentration of the colloids and (b) the concentration of crystalloids.

The concentration of the plasma proteins, and hence *the colloid osmotic pressure of the plasma*, can be reduced to a very low value by repeatedly bleeding an animal and reinjecting the blood corpuscles suspended in physiological salt solution; the plasma proteins are thus removed, but not the corpuscles, so that the blood can still carry oxygen and carbon dioxide in a more or less normal manner. This process leads to a large increase both in the volume of the interstitial fluid and in the rate of flow of lymph. Such a condition is, of course, highly artificial.

If *the concentration of crystalloids* is suddenly increased by injecting a hypertonic solution intravenously, the absorption of tissue fluid is increased. Although the crystalloids can pass through the walls of the capillaries quite rapidly, water can pass through even more rapidly. Equalisation of the crystalloid concentration in the blood and the interstitial fluid is brought about by a simultaneous passage of water into the blood and passage of crystalloids out of it. The flow of lymph practically ceases for a short time while this is occurring. The total volume of the blood is increased as a result of the inflow of water and thus the concentration of the proteins, and their osmotic pressure, is decreased; the rate of filtration is increased, the lymph flow starts again and continues at a rate larger than normal until the excess fluid has been removed. The net result, therefore, of introducing a hypertonic salt solution is the same as that of removal of the proteins: both act as *lymphagogues* (from "lymph" and the Greek word for "to lead"). The excess fluid is not all returned to the blood by the lymph, however, but mainly stays in the tissues until it is excreted by the kidneys.

The action of the crystalloid concentration is also of importance, during the activity of any organ. The metabolites produced by the active cells will diffuse out into the tissue spaces, raise the osmotic pressure of the interstitial fluid and draw fluid from the capillaries. This, together with the rise in capillary pressure, accounts for the mild degree of oedema that often occurs in muscles after exercise.

PERMEABILITY OF THE CAPILLARY WALL

When a capillary dilates excessively, it allows more fluid to pass through its walls at a given effective filtration pressure and holds back protein less completely. Whether the physiological dilatation of the capillaries that occurs during the activity of the organ supplied by them plays a part in the increased flow of lymph, is disputed, but it is of importance in connection with certain abnormalities. Histamine is one of the most powerful capillary dilators known, and it also injures the capillary endothelium, making it more permeable to large molecules. Histamine, therefore, increases the rate of formation of tissue fluid and of lymph. Its introduction below the surface of the skin through a needle prick results in the formation of a weal, just as if the skin had been burnt. Similar reactions result from contact with the skin of the leaves of the poison ivy, or the stings of stinging nettles or jellyfish; these are known to inject small quantities of histamine (among other substances) into the skin. The release of histamine, or of some substance with similar actions, by damaged tissues will be referred to again in a later chapter, in relation to the defence mechanisms of the body.

The increased permeability of damaged capillaries has been observed by the method described above. When alcohol or mercuric chloride was added to the blood flowing through them, fluid passed out much more rapidly, and there was practically no reabsorption except at very low pressures; they had thus become almost completely permeable to proteins.

There are several pathological conditions in which the rate of elimination of fluid is diminished, and the general water balance of the body disturbed. An excess of fluid accumulates and this is held chiefly in the interstitial spaces, giving rise to oedema. This occurs not only when the excretion of water fails, but also, and more commonly, when the excretion of salts fails; the control systems which preserve the constancy of the osmolar concentration of the body fluids brings about a simultaneous retention of water. An inadequate elimination of fluid will occur, most obviously, when there is a failure of the kidneys themselves, due to disease (e.g. nephritis); oedema, also, not infrequently accompanies disease of the heart. Heart failure is likely to produce a rise in venous pressure, but the consequent direct action on the rate of filtration from the capillaries is only partly responsible for the salt retention and oedema, the origin of which has not yet been explained.

Oedema also occurs in cases of severe malnutrition, when the supply of protein in the diet is grossly inadequate (hunger oedema). The plasma proteins are among the last to be sacrificed for the supply of energy or nitrogen to the rest of the body, so that the oedema is not due, to any considerable extent, to a simple reduction of the colloid osmotic pressure of the plasma. But the proteins of the less essential tissues are broken down to supply energy and to replace proteins unavoidably lost from the essential tissues; the fluid in which they were previously held thus becomes, as it were, "surplus to requirement". For some reasons, which are not obvious, this fluid is not completely eliminated, and accumulates in the interstitial spaces.

Filtration in the lymph nodes

Particulate matter is removed from the lymph as it passes through the lymph nodes (or lymph "glands"). Figure 18.3 illustrates the construction of a typical node.

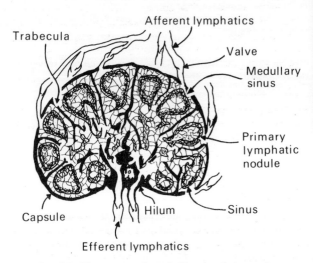

Fig. 18.3 The structure of a typical lymph node (see text).

Lymph reaches the node through the *afferent lymphatics* round its periphery. It leaves via the *efferent lymphatics* at the hilus. The circulation inside the node is through the *medullary sinuses*, lined with *reticulum cells*. The latter are phagocytes which actively take up any particulate matter such as bacteria.

As well as their filtering activity, lymph nodes produce antibodies. Protein of an antigenic nature or bacteria and bacterial products stimulate the formation of appropriate antibodies which are then released to neutralise them. This will be discussed in more detail in Chapter 71.

Thus lymph nodes prevent the spread of infection and foreign materials within the body. When a local infection does develop, it is usual to find that the nodes draining the region involved become enlarged. This is due to hypertrophy of the lymphoid tissue in the node.

Tissue pressure

Most authorities in the past have stated that tissue pressure is about +1 to +4 mmHg (0·1–0·5 kPa). This is because it has been measured by inserting a fine needle into the tissue, putting it under pressure and then determining at what pressure fluid will just commence to leak into the tissue. However, interstitial spaces are rather small compared with the tip diameter of the needle and measurements made with it are suspect. In particular, various factors known to change the volume of fluid in the tissues do not appear to alter the values found for tissue pressure using this technique.

These objections are to some extent overcome by using a small hollow sphere with holes in it, surgically implanted into the tissue. After healing has taken place, the sphere becomes filled with tissue fluid at the same pressure as in the rest of the surrounding tissue and a needle inserted into it will record this tissue pressure. Use of this method shows that small negative pressures, about -5 to -7 mmHg (-0.7 to -0.9 kPa) are nearly always recorded. Thus it seems likely that tissue pressure, contrary to previous opinion, is below atmospheric.

Protein in interstitial fluid

As we have mentioned, protein leaks in small amounts from the capillaries into interstitial spaces. How is it removed? If it is not removed, of course, the colloid osmotic pressure of tissue fluid will rise, eventually to a level causing distressing things to happen to the dynamics of tissue fluid formation. Very little protein from tissue fluid can return to the venous capillary circulation because the concentration gradient is strongly against it. In fact, the tissue protein is removed by the lymph.

As protein tends to collect in the interstitial spaces, the tissue oncotic pressure tends to rise. In turn, this decreases fluid reabsorption at the venous end of the capillaries (as can be seen from Fig. 18.2, if substituted values are made on the right-hand side). Thus the interstitial fluid volume, and hence its pressure, rises. This then results in an increased flow of water and protein into the lymphatics.

There is about 2 per cent of protein in tissue fluid normally, but since the extracellular volume is about four times the plasma volume, the amount of protein in tissue fluid in the whole body is about the same as that in plasma. The composition is similar; tissue fluid will clot but more slowly than blood.

Oedema

When excess fluid accumulates in the tissue spaces, oedema is said to be present. When the increase is of the order of 10 per cent or more, the physical sign of *pitting* can be elicited, i.e. a finger pressed for a short while onto the oedematous skin displaces the excess tissue fluid beneath it, leaving a discernible "pit".

The functional significance of oedema is that the presence of excess interstitial fluid impedes the transfer of oxygen, carbon dioxide, metabolites etc. between blood and cells. Oedema results whenever the formation of tissue fluid exceeds its drainage into the venous capillaries and/or into the lymphatics. This condition can be due to one, or a combination of, various factors:

1. Increased capillary pressure
2. Increased capillary permeability
3. Decreased plasma protein concentration
4. Decreased lymphatic drainage
5. Decreased venous outflow (local obstruction or raised systemic venous pressure).

Clinical syndromes involving permeability increases occur in various disease processes which involve the presence of histamine (anaphylaxis etc.), bacterial toxins; and in various forms of renal disease. Also, clinically, great increases in capillary volume and pressure due to large intravascular infusions and to vasodepressant drugs will have the effect of producing oedema. In the above instances, lymph flow is usually greatly increased, as we might expect.

However, *elephantiasis*, a tropical and subtropical condition, is caused by blockage of lymphatics by mosquito-borne nematode parasites, called *Filaria*. These, in a small proportion of infected cases, give rise to a chronic inflammatory reaction in the lymphatic system (usually in the leg). The oedema, however, is unlike the pitting oedema already described, for the high protein content of the tissue fluid (due to inflammatory processes) stimulates the growth of collagenous tissue. Gradually over a period of years, the enormously enlarged limbs harden and produce chronic elephantiasis.

When the scrotum is involved, this may eventually, without treatment, become so large that the patient has to support it in a wheelbarrow if he wishes to move from place to place.

FORMATION OF CEREBROSPINAL FLUID

There are four membranes covering the central nervous system. They are: (1) the pia mater, which closely invests the nervous substance and carries the blood vessels to it; (2) the arachnoid, separated from the pia by the subarachnoid space, which contains the cerebrospinal fluid; (3) the meningeal layer of the dura mater; and (4) the periosteal layer of the dura mater. The venous sinuses are situated between the two layers of the dura and delicate processes known as the *arachnoid villi* arise from the arachnoid and penetrate into the venous sinuses. The relations between these structures are essentially similar in the skull and in the spinal column; but since the pia ends, with the spinal cord, at the first lumbar vertebra, and the dura and arachnoid extend as far as the second sacral vertebra, the subarachnoid space is of considerable size in this region. Cerebrospinal fluid can, therefore, be readily obtained by inserting a hollow needle (usually between the fourth and fifth lumbar vertebrae) into this space; this procedure is called *lumbar puncture*, and the fluid usually emerges at a rate of about one drop per second.

The cerebrospinal fluid is formed by the plexuses of blood capillaries, known as the *choroid plexuses*, in the ventricles of the brain. This can be shown by the facts that: (a) fluid can be collected from a tube inserted into either of the ventricles; (b) blocking the outflow of a ventricle leads to its distension (hydrocephalus); and (c) if the choroid plexus is first removed, subsequent blocking of the outflow no longer leads to a distension of the ventricle. From the ventricles, the cerebrospinal fluid passes to the *cisterna magna*, an enlargement of the subarachnoid space at the base of the brain, and thence out and up in the subarachnoid spaces around the cerebellum and the cerebral cortex. It returns to the blood by absorption into the venous sinuses through the arachnoid villi. Flow up and down the spaces in and around the spinal cord probably results chiefly from changes in pressure in different parts of the system produced by movements of the head, or sudden changes in blood pressure.

Composition of CSF

The composition of the cerebrospinal fluid is, in most respects, not far from that of an ultrafiltrate of plasma, containing a little protein (0.03 per cent), and is thus very similar to that of the extracellular fluid shown in Fig. 18.1. The actual concentrations of the various constituents are subject to considerable variation, as are those of the blood plasma; but the changes in the two fluids tend to run hand in hand, so that the values of the concentration ratios remain relatively constant. Careful and precise analysis of the fluids drawn from one and the same animal show, however, that the values of these ratios are not quite those to be expected if the fluid were an ultrafiltrate; the choroid plexuses must be regarded as secretory organs, and not merely as membranes which act as simple filters impermeable to proteins. Chloride ions, and to a lesser extent sodium ions, appear to be actively secreted into the cerebrospinal fluid, while potassium ions, urea and glucose are kept out. (The relatively low value of the glucose concentration may be due merely to the fact that it is metabolised by the tissues of the brain, and cannot diffuse out from the plasma very rapidly.)

The blood–brain barrier

That the structures which divide the cerebrospinal fluid from the blood plasma have properties unlike those of the walls of the capillary blood vessels, is shown by the rate at which substances pass across them. Substances with relatively small molecules pass very rapidly through the walls of the capillaries in muscles, for example, and their concentrations in the plasma and the interstitial fluid become identical in a matter of seconds. Passage from the plasma to the cerebrospinal fluid occurs very much more slowly; it takes several minutes for the most rapid of the substances studied, ethyl alcohol, to attain equality of concentration; and after 4 h the concentration of creatine in the cerebrospinal fluid is only about 1/10th of that in the plasma. We are thus led to the idea of the *blood–cerebrospinal fluid barrier*, which controls the passage of substances from one to the other. The rate of penetration of this barrier, moreover, depends markedly on the lipoid solubility of the substances used, so that in its properties it resembles the cell membranes more than the capillary walls. There is also the *blood–brain* barrier between the plasma and the interstitial fluid of the brain, with properties similar to those of the blood–cerebrospinal fluid barrier. Between the interstitial fluid and the cerebrospinal fluid, however, the barrier is less restrictive and substances pass from one to the other relatively freely. It is difficult to get penicillin, for example, into the interstitial fluid of the brain in a useful concentration by injecting it into the blood; it is relatively easy to do so by injecting it into the cerebrospinal fluid.

The total volume of the cerebrospinal fluid in man is about 140 ml, and its rate of formation is about 0.5 ml min^{-1}. The pressure, measured by lumbar puncture, is about 15 cm water when the subject is lying down, and rises to about 28 cm water when he sits up (approximately 0.1 to 0.2 kPa). All these values may vary within wide limits.

Lumbar puncture
Lumbar puncture is a procedure of considerable medical importance, for not only may it relieve increased intracranial pressure by providing an escape for excessive cerebrospinal fluid, but it may yield valuable information enabling a distinction to be made between those diseases which exhibit characteristic changes in the nature and amount of the fluid. Inflammation of the membranes of the brain (meningitis), for example, is accompanied by an abnormally high rate of production of the fluid, by an increase in its protein content (which may be tenfold), by its composition approaching that of an ultrafiltrate of the plasma (except that the glucose concentration falls), and by an enormous increase in the content of leucocytes. Normal fluid contains about 1 to 5 lymphocytes per cubic millimetre; meningitic fluid contains hundreds of cells per cubic millimetre, which are mainly neutrophil leucocytes or lymphocytes.

Local anaesthetics may be injected by lumbar puncture into the subarachnoid space to produce spinal anaesthesia. Air or *lipiodol* (a heavy liquid opaque to X-rays) may be injected into the cisterna magna and the site of a tumour or inflammatory adhesion may then be discovered by the use of X-rays.

Raised intracranial pressure
If the volume of the brain increases, as by the formation of a cerebral tumour, the intracranial pressure rises. This rise is transmitted to all points within the rigid brain-case by the cerebrospinal fluid, with the result that the veins are compressed and the capillary pressure rises. The first structure to suffer is the optic nerve, and the obstruction of the veins running along it causes oedema of the optic disc, which can be observed with an ophthalmoscope; this condition is known as *papilloedema*. The intracranial pressure is also increased in the condition known as *hydrocephalus*, which results from an excessive accumulation of cerebrospinal fluid. This may be caused by obstruction to the fluid pathway from the ventricles to the subarachnoid space, or by blocking or maldevelopment of the arachnoid villi.

Rise in the pressure of the cerebrospinal fluid is also caused by injections of isotonic or hypotonic saline into the blood, showing that its production does depend to some extent upon physicochemical factors. Injections of hypertonic saline, on the other hand, cause a fall in the pressure, probably due chiefly to an osmotic withdrawal of fluid from the brain itself, with consequent diminution in volume. This is a procedure commonly used in brain surgery, where it is often advisable to reduce the volume of the brain before the skull is opened.

19. The Control of Water Distribution in the Body

Introduction

As we have already seen in Chapter 16, constancy in the chemical composition of the extracellular fluid is a condition of continued existence in the face of a changing and hostile environment. We have already discussed many of the control systems and feedbacks concerned in this process of homeostasis. In order to have a stable background upon which to operate these controls, it is also necessary to control, within close limits, the body water in which these chemicals are dissolved.

Apart from considerations of chemical constancy, it is also important to control body water for the maintenance of the circulation. A reduction in extracellular fluid volume leads to reduced plasma volume and this in turn embarrasses the circulation. We need not be surprised to find, therefore, that feedback mechanisms exist to keep extracellular volume nearly constant (by controlling sodium and water excretion).

Water balance

We must first consider the concept of a "water balance" in the body. By this we mean the achieving of a balance between the *intake* of water in the diet and the *output* of water by the kidneys, gastrointestinal tract, skin and lungs. Table 19.1 summarises the actual amounts of water involved in a normal person living in a temperate climate.

Table 19.1 *Water balance (representative values for 24 h in a temperate climate)*

Loss (litres)		Gain (litres)	
Urine	1.5	Food	
Faeces	0.1	Preformed	1.0
Lungs	0.4	Oxidative	1.4
Skin	0.9	Drink	1.5
Total	2.91	Total	2.91

Water can move freely into and out of the cells, so that the concentration of the fluid inside them is also constant. The concentration of the body fluids is determined, of course, by the amount of dissolved substances (chiefly salts) present in them, as well as by the amount of water. The two fluids, inside and outside the cells, contain different kinds of solute, and although salt balance and water balance are interrelated, it is important that the amount of water in the body should not fluctuate very greatly in spite of the unavoidable losses and gains already discussed. In fact, as brought out in Table 19.1, the gains are adjusted so as to equal the losses.

A man weighing 70 kg contains about 14 litres of water outside the cells (in the internal environment) so that about one-fifth of this is ordinarily lost and replaced each day. In exceptional conditions, the daily "turnover" may amount to almost the whole of this extracellular volume. An infant, of 7 kg weight, will have an extracellular fluid volume of about 1.6 litres, but the daily intake and loss of fluid will be about 0.7 litres even in normal circumstances. One-half of the extracellular fluid is thus turned over each day. It is clear that the intake must be accurately controlled if water balance is to be preserved, and that in the infant the margin of reserve is much smaller than in the adult; any variation in the water intake and water loss will cause proportionately greater disturbances in the volume of its body fluids. The relative values of the volume of the extracellular fluid and of the volumes of water gained and lost per day are summarised in Fig. 19.1.

Water balance must ordinarily be maintained by drinking an appropriate amount of water and the need to do this is indicated by the sensation of thirst. There is no special sensation for indicating that too much water has been drunk, but any excess "spills over" rapidly through the kidneys. This "water diuresis" is brought about by an entirely involuntary reflex initiated by "osmoreceptors" sensitive to the concentration of the body fluids and controlling the release of the "antidiuretic hormone" by the neurohypophysis.

Water sources

Most of the daily water intake is in the form of fluids drunk and food eaten. The oxidation of hydrogen in the food gives rise to some water, possibly 100 ml per 24 h. The total water intake would normally be around 2.5 litres per day.

Water output

Water is lost from the body in the urine and as sweat etc. Table 19.2 shows how water is removed from the body under various circumstances.

Table 19.2 *Water output per 24 h (ml)*

	Normal environment	Hot climate	Heavy exercise
Urinary output	1500	1000	500
Skin (insensible)	300	300	300
Skin (sweat)	100	1500	5000
Faeces	200	200	200
Lung (evaporation)	400	200	550
Total	2500 ml	3200 ml	6550 ml

Note the unexpectedly large volume of water lost each day through the skin and by evaporation in the lungs.

CONTROL SYSTEMS AND BODY WATER

Even in 1935, Peters had suggested that some control system existed for changing the excretion of sodium and water in response to the fullness or otherwise of the circulatory vascular bed. We might agree with such a concept because experiments

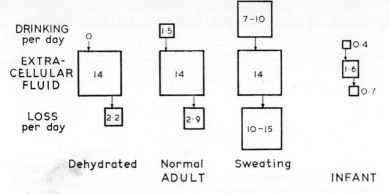

Fig. 19.1 Diagram illustrating water balance.
The volume of water taken in by drinking is the difference between the total volume lost and the volume derived from the food. The dehydrated adult would not be in water balance, the net loss of water being about 0·8 litre/day, even though the urine is as concentrated as possible.

show that there are *volume receptors* to be found within the circulation.

The actions of servosystems on body water will be upon the input volumes and upon the output volumes. In response to deficits (or alternatively a plethora of water) both can be increased or decreased as required. Input control is via the sensations of thirst compelling water intake; output control is mainly via altered kidney function.

Volume receptors

There is now a considerable amount of general evidence for the action of volume receptors. Haemorrhage, even of a very small amount (insufficient to give a measurable fall in glomerular filtration rate), reduces salt and water excretion. Venous pressure cuffs on the thighs will do the same in man; a balloon inflated within the inferior vena cava does it in dogs.

Where are these volume receptors? It does seem likely that there are many such receptors in various parts of the circulation, but our knowledge about them is still rudimentary.

We may regard the circulation as composed of an arterial high-pressure sector, holding about 15 per cent of the whole blood volume, and a venous low-pressure part containing the remainder. Thus 85 per cent of the blood lies within a highly distensible venous reservoir. Veins are several hundreds of times more distensible than arteries. From these facts may be deduced the likelihood that volume receptors will work best in the venous system. Search has revealed this to be true. Localised distension of parts of the low-pressure sector shows that the left atrium when stretched initiates a diuresis. It has, moreover, been shown that the nervous pathways involved go to the medulla via the vagus and that the nerve impulses generated by the stretch eventually reach the supra-optic nuclei.

Right atrial stretching increases sodium excretion due to reduced circulating aldosterone. The injection of *veratridine*, a hypotensive drug, into a vein increases urine flow—an effect abolished by vagal block. Aortic injection is without effect, so it seems likely that the veratridine is acting upon these receptors in the heart.

The fact that cutting both vagi does not abolish these volume effects upon urine flow must mean that there are other endings, elsewhere in the venous system, with the same actions. Indeed, this is a fairly general characteristic of biological control systems—namely that considerable redundancy is incorporated

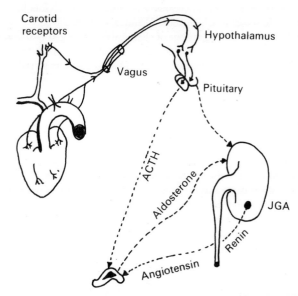

Fig. 19.2 Current views on control system for volume of body fluid. It comprises both nervous (solid lines) and hormonal (dotted lines) components.

into them. Figure 19.2 summarises the current views upon the control systems for body water volume.

It is worth noting that the nervous feedback loops and the hormonal actions can be experimentally blocked simultaneously without seriously upsetting body water regulation. This shows us how little we still know about these control systems.

THE OSMORECEPTORS

The control loop comprising *osmoreceptors, hypothalamus, ADH* and *renal tubules*, when active, leads to increased water retention. The system operates as follows.

ADH and the kidney

ADH in the blood reaching the kidney can reduce urine flow to levels as low as 500 ml per 24 h. In its absence up to 20 l per 24 h of water may be excreted. This is the basic effector mechanism of the control loop and the details of the action of ADH giving this effect is described on page 142.

The control mechanism for this secretion of ADH lies in the supraoptic nuclei of the anterior part of the hypothalamus. Here are found specialised neurones which respond specifically to alterations in the concentration of extracellular fluid. These are termed osmoreceptors. If extracellular osmolarity falls, these osmoreceptors are thought to swell giving rise to an increased rate of impulse generation. The opposite also happens; a rise in osmolarity decreases the firing rate.

These impulses then cause ADH to be released from the posterior pituitary gland (Chapter 32E). This control system works on the usual negative feedback properties of the loop and its sensitivity is such that a change of about one milliosmole in the extracellular fluid will cause measurable retention of, or loss of, water. It is not known for certain if derivative control is involved; it is likely to be.

THIRST

It is generally agreed that water intake is largely governed by thirst; indeed it is difficult to define or measure thirst except in terms of the volume of water required to assuage it. Nevertheless, there is no good agreement about the physiological basis of the sensation of thirst except to the extent that several factors must be involved.

Dehydration of the body tissues becomes manifest most conspicuously as dryness of the mouth and throat and it has been thought that sensations from this region give rise to a drinking reflex. There is now substantial doubt about this simple explanation of thirst, partly because dehydrated animals with oesophageal fistulae in whom no water reaches the stomach drink more than they need to rehydrate the body. It seems that some sensation of fullness in the stomach is concerned in inhibiting drinking. Indeed, dilation of a balloon in the stomach will put a stop to the drinking of such experimental animals. Moreover, a fairly high proportion of normal people claim to feel thirsty when the mouths are not dry, and quite a few fail to feel thirty when their mouths are dried by drugs, such as hyoscine which inhibit salivary secretion.

Drinking centres?

An altogether different approach to the problem of thirst has been the search for a drinking centre in the brain, activated by dehydration of its cells which therefore act as osmoreceptors. These appear to be different from those which are concerned in the onset of water diuresis. Injection into a confined region of the hypothalamus of a goat of a few microlitres of a solution of sodium chloride more concentrated than the body fluids induces excessive drinking; similar injection of a solution of the same concentration as the body fluids, or less concentrated, does not. Moreover, electrical stimulation of the appropriate region induces drinking which begins 20 to 30 s after the stimulus starts and lasts until a few seconds after it stops. Overhydration up to 40 per cent has been produced in goats in this way. Furthermore, destruction of the appropriate region by electrocoagulation has abolished drinking in dehydrated dogs and rats. Clearly, therefore, a region in the hypothalamus is concerned with drinking behaviour and may incorporate the essential osmoreceptors which signal dehydration.

If such a centre were accepted as the primary control of drinking, the usual association of drinking with sensation of dryness in the mouth and throat could well be interpreted as a conditioned reflex, as could the inhibition of drinking by the sensation of fullness in the stomach. On the other hand, these and other receptors which respond to tissue dehydration and the volume of water drunk may give rise to primary reflexes starting and stopping drinking independently of the central receptors associated with the "centre" in the brain.

Profound haemorrhage makes most people thirsty and this has been attributed to reduction of the volume of fluid in the body, exciting some unidentified volume receptors. After haemorrhage, or reduction of the fluid volume by other means, rats have been found to drink more; but no consistent change in drinking pattern has followed substantial haemorrhage in horses or dogs, nor in human donors who have yielded from 5 per cent to 10 per cent of their blood volume to the blood banks. Presumably, therefore, the traditional belief that thirst follows profound haemorrhage from wounds or disease, such as gastric ulcer, is to be explained largely in terms of some other shock-like effect of the injury and not purely as a consequence of the reduction in blood volume.

Disturbance of the water balance

Loss of water from the body may be affected by several kinds of disorder, notably in the activities of certain ductless glands, as discussed more fully in Chapter 32. Deficiency in the secretion of the neurohypophysis (posterior pituitary gland) leads to an excessive excretion of dilute urine (*diabetes insipidus*) and thus to dehydration of the body. Deficiency in the secretion of the adrenal cortex leads to an excessive excretion of sodium chloride and secondarily to a loss of water. Depletion of the body fluids may also result from: (1) loss of fluid from the alimentary tract by persistent vomiting or diarrhoea; and (2) metabolic disturbances such as *diabetes mellitus* (Chapter 32), when there is an unavoidable loss of water and salts accompanying the glucose and ketone bodies excreted by the kidneys. Disorder of the kidneys or of the heart, on the other hand, may so reduce the rate of elimination of water and salts that the volume of the body fluids becomes excessive, as described in Chapter 19.

In abnormal external conditions, the water balance may become disturbed even in normal men. Sustained and profuse sweating, for example, involves large losses of both water and salts, both of which must be replaced. If only water is taken, the body fluids become diluted, leading to water diuresis and loss of much of the water. Correspondingly, many people find that thirst is quenched most effectively after copious sweating if they take enough salt along with the water. If a normal man drinks too much water, and for some reason the excess is not adequately eliminated, "water intoxication" may follow, as in "miner's cramp". On the other hand, if he drinks only sea water, in which salts are more concentrated than they are in the most concentrated urine that can be formed, the body fluids inevitably become excessively concentrated; as they may do in people surviving from shipwreck after the fresh water is exhausted.

Most of the experimental investigations on the control of water and salt balance must be performed on conscious animals or human subjects, because general anaesthetics put the controlling systems out of action. The experimental procedures that can be used are correspondingly limited, and interpretation of the results made more difficult.

20. The Kidney and Control of Body Fluids

S. E. Dicker

Introduction

Hunger and appetite largely control the intake of food; thirst influences the intake of water; taste, to some extent, controls the intake of salt. So might the water content and composition of the body be kept fairly steady, were it not that we like to eat different kinds of food, giving rise to different kinds and quantities of metabolic end-product and affecting differently the water balance of the body, and were it not for extraneous influences such as the social habits of drinking partners and the culinary customs of the home. The ample variation possible in urine formation provides for the spill-over of excess water and salts and indeed of most other soluble constituents of the plasma; so the constituents of the body fluids are kept at the steady levels required for normal physiological working.

Homeostasis and the kidney

Among all the organs controlling homeostasis, the kidneys stand foremost. They are not only responsible for the excretion of non-volatile metabolites, they regulate most efficiently the excretion of most of the components of the body. The power, flexibility and precision of function of the normal kidney is such that there are practically no situations which can overcome its capacity and so disturb the humoral equilibrium of the organism. In the diseased kidney, however, the greater the loss of function, the more easily homeostasis is upset.

Since the main object of this chapter is to show how the kidneys control the maintenance of a constant internal environment, it must be understood that as a study of renal physiology it is deliberately incomplete.

URINE

The daily output of urine varies widely in amount and composition; 1500 ml may be taken as representative for a man under average conditions in this country. Over shorter periods of time, the *rate of production* of urine varies between about 0.3 ml min^{-1} and about 20 ml min^{-1} according to the state of the hydration of the body (1 ml min^{-1} is equivalent to 1450 ml day^{-1}). Normally, urine contains about 4 per cent solids, many of which are included in Table 20.1; but the concentration varies inversely with the rate of water output, and the *total (osmolar) concentration*, as measured for example by the depression of the freezing point, varies from about one-sixth to four times or more that of plasma. The specific gravity, a rough measure of the total concentration, is usually between 1·015 and 1·025, but may fall to 1·002 or rise to 1·030. The *colour* of the urine, due to the presence of urochrome, a pigment of uncertain origin, chemically related to haemoglobin, is also a rough indication of the concentration. Urine is normally somewhat acid, its pH being about 6·0; this may vary between the limits of 4.7 and 8.2 according to the nature of the food and the amount of acid or alkali being excreted.

Quantitatively, the chief constituents of the urine are urea and the chlorides, sulphates and phosphates of sodium and potassium (Table 20.1).

Table 20.1 *Typical Concentrations in Man*

	Plasma g/dl	Urine g/al	Urine/Plasma Concentration ratio
Water	90–93	96	1·05
Proteins	7–9	0	0
Urea	0·03	2	60
Creatinine	0·001	0·15	150
Uric Acid	0·002	0·05	25
Glucose	0·10	0	0
Sodium	0·32	0·35	1
Potassium	0·02	1·15	7
Calcium	0·01	0·015	1·5
Magnesium	0·002	0·01	5
Chloride	0·37	0·6	2
Sulphate	0·003	0·18	60
Phosphate	0·003	0·12	40
Ammonia*	0·0001	0·04	400

* *Ammonia is synthesized in the kidney.*

A concentrated urine often deposits amorphous sodium and potassium urates on cooling. The precipitate is coloured pink by uroerythrin and can be redissolved by warming. Another amorphous deposit may occur in normal urine, namely, phosphates of the alkaline earths; these have a low solubility in alkaline solution, and are precipitated (a) usually as $Ca_3(PO_4)_2$ when the urine is alkaline when voided, or (b) as NH_4MgPO_4—"triple phosphate"—when previously acid or neutral urine becomes alkaline on standing owing to bacterial conversion of the urea into ammonium carbonate. The phosphate is dissolved by the addition of dilute acetic acid. Crystalline deposits are usually associated with abnormal processes. In acid urine, calcium oxalate, cystine, leucine or tyrosine may be found, whereas in alkaline urine, calcium carbonate and phosphates are the commonest.

Urea is the chief nitrogenous end-product of protein metabolism and occurs in large quantities in normal urine. The ability of a kidney to concentrate urea is considered a valuable index of functional activity in disease. It is estimated by administering 15 g of urea in 1 dl of water by mouth and collecting urine at hourly intervals; the first sample may be dilute owing to diuresis, but less than 2 per cent in the second sample would be unusual in normal kidneys.

Other nitrogenous substances in urine occur only in relatively small amounts. Ammonium salts, usually in small concentration, may be increased in acidosis. A large proportion of the ammonia in normal urine is formed in the kidney itself. Uric acid is formed from nucleins and about one-half persists in starvation and is therefore regarded as of endogenous origin; the other half varies in amount with the diet, and is of exogenous origin. Creatinine in the urine is probably formed mainly from the creatine in muscle.

When urine has been allowed to stand for some time, a faint cloud of mucus from the walls of the bladder and urinary passages can often be seen.

Abnormal constituents in urine

Abnormal constituents appear in urine from normal kidneys when soluble and diffusible foreign substances, including many drugs, have been administered. Or again, they appear when certain normal constituents of plasma are present in excessive concentration. For example, the concentration of glucose, normally around 0.10 per cent in plasma, may rise as a result of swallowing 200 g or more within a short time; if the plasma concentration then exceeds a "threshold value" of about 0.18 per cent, some of the glucose is excreted in the urine, a finding known as "alimentary glycosuria". Likewise in some diseases, normal kidneys will eliminate substances not normally secreted, for example, glucose and ketone bodies in diabetes mellitus or bile salts and bile pigments in jaundice. Diseased kidneys may allow substances which are normally retained in the plasma to pass into the urine, and most characteristically so, plasma protein. Proteinuria is characteristic of nephritis, of failure of adequate blood supply to the kidney and of the action of certain poisons on it. A transient appearance of protein occasionally occurs in the urine of healthy people, especially adolescents, after severe exercise and after prolonged standing; the latter may be due to a rise in pressure in the renal vein. Albuminous casts of the tubules may appear in the urine secreted by diseased kidneys, and cells derived from the blood or from the excretory organs are present in the urine in certain pathological circumstances.

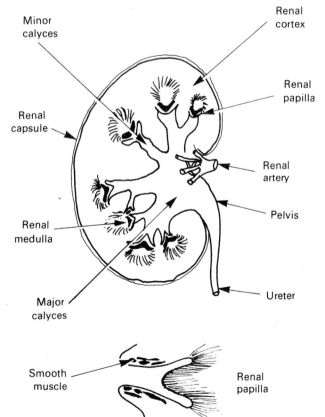

Fig. 20.1 Coronal section through the kidney. The lower diagram shows the renal papilla, slightly enlarged.

THE STRUCTURE OF THE KIDNEY

Urine formation from plasma cannot be understood without reference to the structure of the kidney (Fig. 20.1).

Macroscopic structure of the kidney

A longitudinal section of the kidney gives a macroscopic view of its architecture: (a) the excretory channels occupying the inner and medial part, with the renal pelvis at the hilum; (b) the parenchyma which surrounds the excretory channels and which can be divided into two concentric zones, the medulla and the cortex. The excretory channels form an uninterrupted pathway from the renal parenchyma to the bladder. They can be divided into minor and major calyces, pelvis and ureter. Of these, the calyces and part of the ureter only are in the kidney. The medulla is formed by a series of Malpighian pyramids whose apices project into the lumen of minor calyces. In the human kidney, the number of Malphighian pyramids varies from six to ten.

Renal blood supply

Blood is supplied to the kidneys by the renal arteries which are branches of the aorta; blood leaves the kidneys by the renal veins. The renal arteries enter the kidneys at the hilum; they give rise to the lobar arteries, one for each lobe. The lobar arteries divide into two interlobar arteries which run along the sides of the Malpighian pyramids. When they reach the base of the pyramids they turn and run parallel to their base but go no further than the middle of the base. They are called the arcuate arteries; they then give off at right angles, the interlobular arteries, which lead to the afferent arterioles. In the outer zone of the cortex each afferent arteriole breaks up into a network of capillaries to form the glomerular tuft. These capillaries then unite again to form the efferent arteriole which in turn gives rise to a peritubular net of capillaries, which then run into veins. The veins follow the same course as the arteries, in reverse. In the deeper zone of the cortex, afferent and efferent arteries are in continuity, the glomerulus being supplied by a branch of this arterial axis. The medulla receives its blood supply from straight vessels, the vasa recta, which arise either directly from the arcuate arteries or from the efferent arterioles of the deeper zone of the cortex. The vasa recta pass down through the medulla to the papilla. Straight veins, the ascending vasa recta, run parallel to the straight arteries and joint the arcuate veins. Between the straight arteries and veins there is a network of capillaries whose structure varies with the depth in the kidney. There are nearly twice as many ascending than descending vasa recta. About 90 per cent of the blood that enters the kidney irrigates the cortex, while the remainder goes to the medulla.

Renal blood flow

In adult man, the renal blood flow lies between 1100 and 1300 ml min^{-1} for the two kidneys, which is more than a quarter of the cardiac output. In the kidneys the arteriovenous oxygen difference is only 1–2 ml dl^{-1} as compared 4–5 ml dl^{-1} in the rest of the body. But since the blood flow through the kidneys is so high, the oxygen consumption of the kidneys is about 10 per cent of the total body oxygen consumption.

Innervation of the kidneys

The renal nerves, which are branches of the coeliac ganglia and of the splanchnic nerves, follow the renal arteries as they enter the hilum. Most of the intrarenal nerves are non-myelinated, and innervate the walls of the arteries and arterioles. Occasionally, they reach the glomeruli and spread out on their surface. There are, however, no

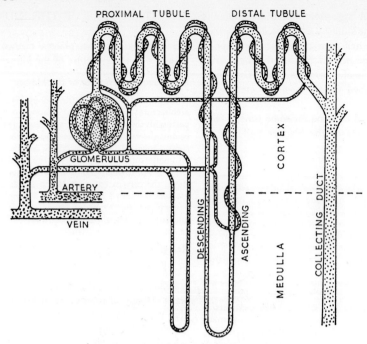

Fig. 20.2 Diagram of a nephron and associated blood vessels.
The proximal and distal tubules have been reduced in length; and the arrangement of these, as well as that of the dual system of capillaries—those within the glomerulus in series with those ramifying over the tubules—has been considerably simplified.

nerves either in the glomeruli or in the epithelial cells of the tubules.

The renal nerves do not seem to affect the elaboration of urine directly, but have a marked vasomotor action. Stimulation of the renal nerves produces renal vasoconstriction and a fall in renal blood flow. Complete denervation, however, does not seem to produce renal vasodilatation. On the other hand, the denervated kidney does not behave like a normal kidney; for instance, hypercapnia, which reduces the blood flow of the intact kidney, does not have this effect in the denervated organ.

Microscopic structure of the renal parenchyma

The structural and functional units of the renal parenchyma are the *nephrons*. The number of nephrons in each human kidney vary, according to the authors, from 1 million (Homer Smith) to 4·5 million (Verney). Each nephron consists of a *glomerulus* and a tube divided into the *proximal convoluted tubule*, the *descending* and *ascending limbs* of the *loop of Henle*, the *distal convoluted tubule* and the *collecting tubule* or duct. These various parts are situated in well-defined places in the kidney: the glomeruli, the proximal and distal convoluted tubules are in the cortex; the loops of Henle, with their limbs, and the collecting ducts are in the medulla.

THE GLOMERULUS

Each nephron begins with a wider blind end, the glomerular capsule (Fig. 20.2) into which protrudes a bunch of blood capillaries known as the glomerular tuft or the glomerulus. The capillary walls are covered by a thin membrane forming the blind end of the tubule and called the glomerular membrane. This is continuous with the outer cup-shaped membrane (Bowman's capsule) which forms the wide end of the tubule, the space between the inner glomerular membrane and the outer funnel-like membrane being called the capsular space which contains glomerular fluid. The idea that the function of the glomerulus is ultrafiltration, that is transport of only the non-colloidal elements of blood into the tubule under influence of blood pressure, was first suggested by the microscopic appearance of Bowman's capsule and has since been proved experimentally.

THE TUBULE

Beyond the neck of the funnel the thin flat cells of the outer capsule change into the columnar or cubical granular cells of the proximal convoluted tubule, resembling in appearance the cells in secretory organs. The glomerulus and proximal convoluted tubule are found in the outer zone, the cortex, of the kidney. The tubule then takes a dip into the medulla before returning to form the distal convoluted tubule near its own glomerulus; the loop formed in this way is known as Henle's loop, and the part it plays in concentrating urine by a countercurrent mechanism has been discovered more recently. Beyond the distal tubule, the tubule plunges once more toward the medulla to enter a collecting duct.

THE BLOOD SUPPLY OF THE NEPHRON

The blood supply to these renal elements is arranged in a peculiar manner. The glomerulus receives blood from branches of the renal artery by a short, wide, "afferent" vessel. From the glomerular capillaries the blood is collected into a longer and narrower "efferent" vessel which, in turn, divides into the capillaries distributed over the tubules. From the peritubular capillaries the blood collects into venous sinuses, where it reaches the radicles of the renal vein (Fig. 20.2). In the mammalian kidney, nearly all the blood which reaches the peritubular capillaries has previously passed through the glomerular capillaries. There are, however, shunts in this circulation which play a disputed part in normal function. Thus the blood normally approaches the glomeruli through the wide channels of the renal arterioles and vasa afferentia, and emerges through the

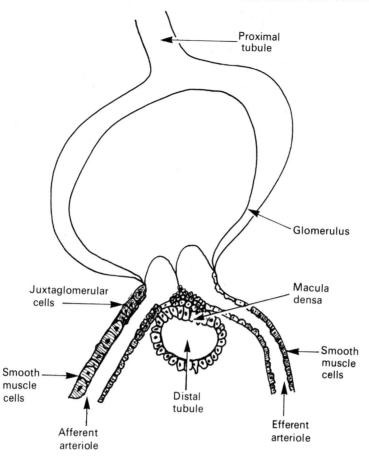

Fig. 20.3 Diagram of juxtaglomerular apparatus.

narrow vasa efferentia to reach the tubular capillaries. Owing to these two anatomical factors—the double capillary system and the difference in diameter of the vasa afferentia and efferentia—the blood pressure in the glomerular capillaries is much higher than in capillaries in other organs. Its pressure in the peritubular capillaries, on the other hand, may be even lower than the normal capillary pressure elsewhere in the body. Thus, on histological grounds alone, it appears that blood at high pressure is separated from the lumen of the renal tubule only by the thin walls of the glomerular capillaries and by the single layer of flattened cells composing the inner membrane of Bowman's capsule. This is most aptly arranged to provide the requisite hydrostatic pressure for the filtration in the glomerulus mentioned earlier. The peritubular capillaries, on the other hand, may be at a lower pressure than capillaries in other organs, and this would facilitate the reabsorption of water and other substances which will be shown later to occur there.

Macula densa

Between the walls of the afferent and efferent arterioles, there is a triangular structure of cells; its apex is the vascular pole of the glomerulus. Its base forms a specialised part of the distal convoluted tubule, the *macula densa*. It is formed of epithelial cells which are characteristically different from the rest of the tubule. It is in close contact with the *juxtaglomerular mass* which consists of a network of foundation substance with small cells between its meshes. The juxtaglomerular mass is the principal source of renin (Fig. 20.3).

21. The Formation of Urine

S. E. Dicker

In 1842, Bowman gave the first description of the main histological features of the kidney. From these he inferred that blood passing through the glomerular capillaries would be likely to filter a watery fluid into the nephron. The tubular cells, mainly in the proximal convoluted tubule, would then add the main constituents of the urine to the fluid passing down from the glomerulus. Two years later, Ludwig suggested that the glomeruli were able to produce a filtrate which contains all the plasma solutes which are found eventually in the urine, but are concentrated by reabsorption of water by the tubules. It is this concept which proved to be nearly correct. All later work by Cushny, by Rehberg, by Starling and Verney and by Winton supported Ludwig's theory of filtration and reabsorption. Clear experimental evidence of the correctness of the theory came in 1924 from Richards, who, using a method of micropuncture (Fig. 21.1), collected glomerular fluid and analysed it. The com-

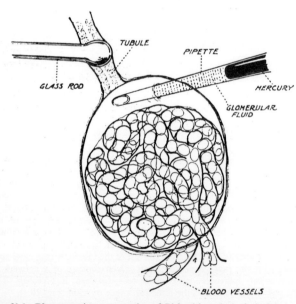

Fig. 21.1 Diagrammatic representation of Richards' method of obtaining a sample of glomerular fluid from Bowman's capsule in the frog.

A very fine pipette (7 to 15 μm dia.), connected with a reservoir of mercury and filled with mercury up to its tip, is thrust through the wall of the capsule and fluid is withdrawn from the capsular space by lowering the mercury reservoir. The tubule is blocked by pressing on it with a fine glass rod, so that no fluid can be sucked back into the capsule.

position of the glomerular filtrate was similar to that of blood plasma. But although Richards showed that urine formation was initiated by a process of filtration, he was unable to measure the actual amount of liquid filtered through the glomeruli. This was essential, since without this knowledge, reabsorption could not be proved. The rate of glomerular filtration was estimated eventually, by Homer W. Smith and his school, using a substance which is freely filtrable but not reabsorbed by any parts of the nephron. Though Ludwig's theory was vindicated, it differs in many respects from what is known now, especially with reference to the role of the tubules and the loops of Henle in the process of elaboration, concentration and dilution of urine. We shall examine first the glomerular stage and then the tubular stage of urine formation.

PHYSIOLOGY OF GLOMERULI

The first results of micropuncture of glomeruli in amphibians were published by Richards in 1924, but it took him and his collaborators another decade before they were able to show by techniques of microanalysis that the osmotic pressure, the electrical conductivity, the pH and the content of sodium, urea, glucose, creatinine and uric acid of the glomerular filtrate are identical with those of the plasma. Although the chloride level within the Bowman's capsule was found to be somewhat higher than that of the plasma, its actual concentration was that to be expected in a process of filtration through a capillary membrane, a situation to which the Donnan–Gibbs equation would apply. Thus it was established that the glomerular fluid was a plasma ultrafiltrate.

"Pores" in the Glomeruli

Confirmation of glomerular filtration is shown by abolishing tubule function by cooling blood reaching the kidneys. An increased flow of isotonic protein-free urine ensues. Poisoning with cyanide also increases urine flow.

The second property of the mammalian kidney indicating a filtration mechanism in the kidney is concerned with the nature of the filter. A filter implies a membrane which will allow the passage of particles below a certain size, but retain larger particles. An ultrafilter, such as the glomerular membrane, should allow the passage of molecules below a certain size, but retain larger molecules. Table 21.1 (after Bayliss, Kerridge and Russell) shows that the kidney differentiates between molecules of different sizes in just such a simple physical way. Figure 21.2 shows that the position of the filter in the kidney is in fact the glomerulus, since the protein with a sufficiently small molecule is shown to pass from the blood into the glomerular space.

The discrimination of the glomerulus between molecules of different size has also been shown in other series of compounds. For example, the dextrans, polysaccharides used therapeutically as blood substitutes, can be hydrolysed in the laboratory to produce molecules of weights from that of glucose up to several millions. Dextrans of different molecular sizes have been injected by Wallenius. In man and the dog, molecules of dextran exceeding about 47 000 molecular weight fail to be excreted in the urine. Molecules with a weight of 5000 to 6000 or lower pass freely into the urine, whereas molecules of intermediate size pass more slowly. Human patients and animals with proteinuria allow the passage of

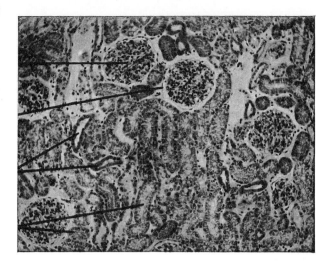

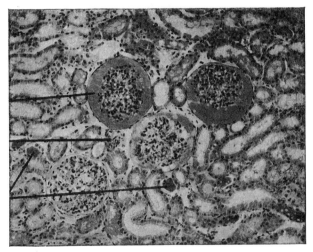

Fig. 21.2 Sections through isolated and perfused kidneys of dogs. (Magnification × 140.)

A. Perfused with normal defibrinated blood for 1¾ quarter hours. B. Perfused with normal defibrinated blood for 1½ h. and with blood containing egg albumin for ½ h. Note the presence of protein in Bowman's capsule, indicating that the glomerular membrane is permeable to proteins of relatively low molecular weight (less than 70 000). (Bayliss, Kerridge and Russell.)

larger dextran molecules (50 000 to 100 000) into the urine indicating an abnormally permeable glomerular membrane. Size as indicated by molecular weight is not the only property of molecules which affects their passage through membranes. The shape of the molecule and presence of electric charges are also concerned.

Hydrostatic pressure across glomerular membrane

The glomerular filtration rate should be fairly linearly related to the hydrostatic pressure minus the colloid osmotic pressure, i.e. the effective filtration pressure. Curiously, this evidence is far from manifest. In normal animals a large rise in arterial pressure, which might be expected to produce a corresponding rise in glomerular capillary pressure, raises the glomerular filtration rate very little. This is attributed to an altogether different mechanism discussed later under the heading "Autoregulation". In the isolated kidney of the dog, the control by autoregulation is less perfect than in the normal kidney and the glomerular filtration rate follows the arterial pressure changes more appropriately. An experiment of this kind is depicted in Fig. 21.3. How to calculate glomerular filtration rate from serum and urine creatinine concentrations and the urine flow will be described later under the heading "Clearances". An important property of the kidney can be seen in Fig. 21.3, namely that an increase in urine flow is accompanied by a change in its composition in the direction of that of the plasma; substances which are more dilute in the urine like chloride, or more concentrated like creatinine, become less so in diuresis.

The pressure fall across the glomerular membrane has been calculated to be about 55 mmHg (7·2 kPa) in a dog kidney, more than half of which is devoted to overcoming the osmotic pressure of the plasma proteins. Under average conditions with mean arterial pressure of, say, 115 mmHg (14 kPa), the glomerular capillary pressure may be about 75 mmHg (10 kPa) and the pressure in Bowman's capsule about 20 mmHg (2·7 kPa). Micropipette measurements indicate pressures of 2 to 3 mmHg (0·3–0·4 kPa) in the distal tubules at resting urine flows, but the pressure rises during diuresis.

Dilution of plasma proteins reduces the colloid osmotic pressure which itself opposes glomerular filtration. The glomerular filtration rate is, therefore, increased if 0·9% NaCl is infused into the renal artery of an experimental animal. This produces "dilution" or "saline diuresis", though this diuresis is augmented by an effect on the tubules. Glomerular capillary pressure may vary although arterial pressure remains unchanged. For example, caffeine increases glomerular pressure by dilatation of the preglomerular arterioles (vasa afferentia).

Filtration in the glomerulus

Filtration (or ultrafiltration) will obtain only if the hydrostatic (or filtration) pressure across the membrane of the capillary is high enough. In the kidney, the effective filtration pressure is equal to the blood pressure in the capillary network of the glomerulus minus the pressure opposing filtration, which is the osmotic pressure of the plasma proteins (the oncotic pressure) and the pressure of the liquid in the Bowman's capsule.

Winton (1931) calculated that under standard conditions with a mean arterial pressure of, say, 115 mmHg (14 kPa) the pressure in the glomerular capillaries was of the order of 75 mmHg (10 kPa), while the pressure in Bowman's capsule was about 20 mmHg (2·7 kPa). These figures were approximate only, but recent direct measurements show that in the dog and the rat glomerular transcapillary hydrostatic pressure is of the order 40–47 mmHg (5·3–6·5 kPa), thus indicating a greater pressure drop along the afferent arterioles than had previously been assumed.

There is no doubt that a sufficient fall of the systemic blood pressure will result in the cessation of glomerular filtration, with resultant anuria. This happens usually when the arterial blood pressure falls below 50–60 mmHg (6–8 kPa). The oncotic pressure exerted by the plasma proteins in the afferent arteriole has been estimated to be of the order of 18 mmHg (2·5 kPa). It is appreciably higher in the efferent arteriole, where it has been estimated to be about 30 mmHg (4 kPa). This is due to water lost during filtration through the capillary bed of the glomerulus. Until recently, capsular pressure had been estimated indirectly only, either by measuring the minimum ureteric pressure needed to diminish urine flow or by equating it with interstitial pressure measured by inserting in the kidney a needle connected to a manometer. The results given by the two techniques differ. Recent direct measurements done by micropuncture show that the pressure in the capsule of Bowman is about 13 mmHg (1·7 kPa).

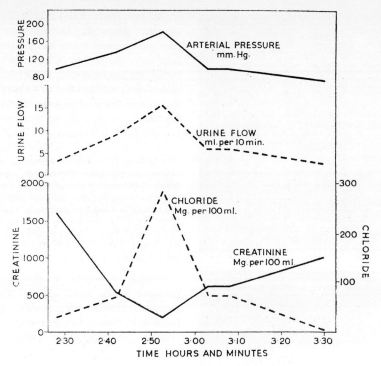

Fig. 21.3 The influence of arterial pressure on the isolated kidney of the dog.
Increase of arterial pressure produces a large increase of urine flow, and a change in its composition (decrease of creatinine concentration and increase of chloride concentration) such that the concentrations of the solutes move in the direction of those in the serum. Serum creatinine 75 mg dl^{-1}. Serum chloride 760 mg dl^{-1} (chloride estimated as NaCl). Temp. 37°C. (Gilson and Winton.)

Effective filtration pressure

These data make it possible to give an estimate of the *effective* filtration pressure. If the glomerular capillary hydrostatic pressure is, say, 45 mmHg (6 kPa), the oncotic pressure 20 mmHg (2·7 kPa) and the capsular pressure 13 mmHg (1·7 kPa), the effective filtration pressure must be of the order of 12 mmHg (1·6 kPa), which is quite low.

We must therefore enquire why it is that a comparatively small drop in the systemic arterial pressure does not result in the arrest of glomerular filtration and hence the cessation of urine flow. There are, in fact, two reasons for this:

(a) Filtration is complicated by the process of *diffusion* which tends to lower the pressures at which filtration will still take place.
(b) There is an extensive *autoregulatory* mechanism peculiar to the kidney which maintains the filtration pressure as high as possible in the face of a falling arterial pressure.

DIFFUSION

While filtration consists essentially of a flow of molecules across pores capable of allowing their passage under the influence of hydrostatic pressure, diffusion which results from molecular motion only, will occur in the absence of a pressure difference across the capillary endothelium. The basement membrane which constitutes the principal barrier of the glomerular capillaries has no pores. It is formed of a gel containing fibrils. This gel imbibes and so allows the passage of water, solutes and particles in suspension at a rate dictated by the coefficient of diffusion of each of these substances as determined by their size, shape and electrical charge of the particular molecule.

AUTOREGULATION OF RENAL BLOOD FLOW (NEGATIVE FEEDBACK)

In most organs, blood flow varies directly with arterial pressure. In the kidney, however, blood flow rises very little, if at all, with increases in arterial pressure so long as the pressure is between the physiological range of 80 to 180 mmHg (10–25 kPa). This feedback control of blood flow is found in denervated and even in isolated perfused kidneys. Equally remarkable is the fact that glomerular filtration remains stable in the face of changes in the systemic blood pressure between 70 to 180 mmHg. Furthermore, results of micropuncture studies in the rat have indicated that pressures in postglomerular efferent vessels remain constant despite variations in arterial pressure, throughout this physiological range. This sheds some light on the site of resistance controlling the blood flow, which in all likelihood resides in the afferent glomerular arterioles. Recently, it has been shown that the autoregulation of glomerular filtration rate is a consequence of the autoregulation of glomerular plasma flow. Whether filtration remains constant or decreases in response to a reduction of the arterial pressure is determined by the changes of effective filtration pressure, membrane permeability and surface area available for filtration relative to the change in plasma flow. Thus constancy of filtration results when net changes in these factors are proportional to the change in plasma flow, and vice versa. Autoregulation exists in the cortex only, not in the medulla, but as only about 10 per cent of the total renal blood flow goes through the medulla, this is relatively

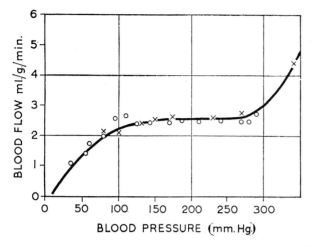

Fig. 21.4 Autoregulation of blood flow between arterial pressures of about 100 and 280 mmHg in a kidney pump—perfused with blood from the carotid artery of the same dog.

Values obtained by continuous increase in pressure (circles) or by sudden increases in pressure (crosses) lie on the same curve, showing that the plain muscle of renal arterioles, unlike many other forms of plain muscle, does not contract after sudden stretch. (Thurau, Kramer and Brechtelsbauer, 1959.)

unimportant (and without effect on the total renal blood flow). Figure 21.4 summarises the process of autoregulation.

Renal plasma clearance

We cannot use direct methods to measure the glomerular filtration rate in the mammalian kidney. Instead, indirect methods have been developed which depend on the concept of *plasma clearance*.

When the functions of the kidney were first investigated, it was discovered that the quantities of a number of different substances excreted in the urine in a given time were fairly constant. This was usually true even if large amounts of water were ingested and there were thus large changes in the rate of urine flow. It was also discovered that for many substances including urea, changing the plasma concentration led to a change in rate of excretion. The rate of excretion was found to be directly proportional to the plasma concentration, so it was concluded that the kidneys were removing the substance from a certain constant volume of plasma in a given time. The volume was called the renal clearance of the material, e.g. urea clearance.

At that time, it was generally supposed that the volume was the amount of glomerular filtrate produced per unit time, i.e. that no urea was either reabsorbed or secreted into the tubules as the filtrate passed along them.

It is of course obvious nowadays that urea clearance is not the glomerular filtration rate; it is not a volume of plasma that is actually cleared of urea. It is however, the volume of plasma that *contains* the quantity of urea excreted in the given time interval. This quantity can easily be found by measuring the concentration of urea in the urine and multiplying it by the volume of urine produced per unit time. It is UV, where U is the urinary concentration of the urea and V is the urine volume per unit time.

This quantity UV must also be the amount removed from the plasma in the same time. This can be found by multiplying the plasma concentration, P by the volume of plasma that contained it, i.e. the urea clearance, C.

Thus, $UV = PC$

and $C = \dfrac{UV}{P}$.

Although plasma clearance does not represent a real volume of plasma that has had all the substance removed from it (in most cases), the concept is of great practical use in examining renal function.

It must be understood that the plasma passing through the glomerulus (and the whole kidney for that matter) is rarely completely cleared of a substance. What usually happens is that a larger volume of plasma is incompletely cleared.

UREA CLEARANCE

An example will make this clearer. Let us find the renal clearance of urea. The plasma concentration (P) of urea is 30 mg dl^{-1}; the amount of urea excreted in the urine (UV) is 20 mg min^{-1}, Thus

$$\begin{aligned} C &= \frac{UV}{P} \\ &= \frac{20 \times 100}{30} \\ &= 67 \text{ ml} \end{aligned}$$

This means therefore that 67 ml of plasma contain the amount of urea which passes into the urine per minute. Seven hundred ml of plasma flow through the kidney per minute; this volume contains 210 mg of urea. The volume filtered is 120 ml (with 36 mg urea in it) and 20 mg of urea get into the urine. The amount of urea in the urine is altering as the glomerular filtrate travels down the tubules.

Inulin clearance (finding glomerular filtration rate)

If we can find a substance which is filtered by the glomeruli, as was urea, but which passes through the tubules untouched (i.e. is neither secreted nor reabsorbed) we should then have a way of estimating the glomerular filtration rate (GFR).

Inulin is such a substance; it is a polymer of fructose and comes from dahlia roots. It is filtered through the glomeruli in the same concentration as it is in the plasma.

As an example, let us inject inulin to achieve a plasma inulin concentration of 1 g l^{-1}. We then find that the inulin excretion (UV) is 125 mg min^{-1}. C now is the same as GFR; thus

$$\begin{aligned} \text{GFR} &= \frac{UV}{P} \\ &= \frac{125 \times 100}{100} \\ &= 125 \text{ ml} \end{aligned}$$

Thus the volume of plasma cleared of inulin per minute is 125 ml. Since inulin is now not further touched in its passage through the remainder of the kidney, the sole mechanism for clearing 125 ml of plasma of inulin is the filtration of 125 ml of plasma through the glomeruli. The glomerular filtration rate can thus be estimated using inulin (or creatinine which behaves somewhat similarly).

PRACTICAL MEASUREMENT OF GFR

1. Concentration of inulin in plasma must be kept constant, so that the value used for P is representative of actual values in the kidney (since a sample from a peripheral vein has to be analysed). This is done by a priming injection followed by a slower infusion to match excretion rates. It is usual to take samples or urine and venous blood every 10 minutes.
2. Subject should be post-prandial and at rest.

3. Urine flow should exceed 3 ml min^{-1} to make analysis as accurate as possible. Urine samples should be obtained by bladder washouts with known volumes of saline.

A method using vitamin B_{12} (cyanocobalamin) is also used; it is labelled with ^{57}Co and can easily be estimated. In humans, however, it is not the method of choice, since it involves giving radioactive material.

EVIDENCE FOR INULIN BEING UNAFFECTED BY TUBULES

1. The renal plasma clearance should be identical for all substances behaving like inulin, i.e. being unaffected by passage through tubules. In the dog, experiments have shown that inulin, creatinine, sodium ferrocyanide and vitamin B_{12} have identical renal clearance volumes. This is unlikely to be a chance finding and must indicate that these different substances are all treated alike by the tubules.

2. *Phloridzin* abolishes the tubular reabsorption of glucose; the latter then (in the phloridzinised dog) behaves just like the substances mentioned in (1) above.

Glomerular filtration rate in man

According to Homer W. Smith, who was the first to measure it, the rate of glomerular filtration in adult man is 130 ± 30 ml min^{-1}. It is slightly smaller in women. It does vary with age: it is smaller in infants and in old people. Taking 130 ml min^{-1} as the average figure of glomerular filtration in man, this means that some 187 litres of plasma are filtered every 24 hours. But since the volume of urine excreted in 24 h is of the order of 1·5–2·0 litres only, it follows that 99 per cent of the glomerular filtrate must have been reabsorbed. It might be worth reflecting on this. The amount of plasma filtered every 24 hours is fifty times that of the whole plasma volume of the body. In man the two kidneys, according to Homer W. Smith, have an average of 2 000 000 glomeruli, with a total filtration area of about 0·76 m^2. This is nearly half the average body surface area of man (1·73 m^2). Since the renal plasma flow is of the order of 650 ml min^{-1}, it can be calculated that some 12 ml of plasma circulate per second over a filter that has a surface of 0·76 m^2, giving a layer of approximately 15 μm deep. With an effective filtration pressure of about 12 mmHg (1·6 kPa), it is not difficult to imagine that 2 ml of water can filter through such a large expanse of endothelium every second. After all, it would amount to no more than one drop of water per second through a membrane of 64 cm^2 in area.

Diodone clearance (finding renal blood flow)

As we have seen, urea and inulin are treated differently by the kidney; but neither substance is completely removed from the blood on a single passage through the kidney, although of course the concentration of both in the renal vein will be reduced. There are substances, however, and *diodone* is one, which are completely removed from the blood on one passage through the kidney. A little thought will show that this gives us a method of determining renal bloodflow, since the volume of plasma completely cleared of a substance going through once, must be the total plasma flow through the kidney.

As an example, diodone injected to give a concentration in the plasma (P) of 1 mg dl^{-1} leads to the excretion of 7 mg in the urine per min (UV). Thus

$$C = \frac{UV}{P}$$
$$= \frac{7 \times 100}{1}$$
$$= 700 \text{ ml}$$

How is this 700 ml clearance made up? The renal plasma flow of 700 ml min^{-1} forms 120 ml of glomerular filtrate. This contains 1·2 mg diodone. Therefore 580 ml min^{-1} of plasma flows into the tubular blood vessels. This will contain 5·8 mg diodone, the whole of which is then actively secreted into the passing urine by the tubular epithelium. In this example the extraction ratio is 100 per cent, i.e. all the diodrast is cleared.

In practice, the extraction ratio is never 100 per cent but of the order of 95 per cent. This is not surprising since only about 95 per cent of the blood entering the kidneys perfuses the glomeruli and the tubules, while the remainder goes to the medulla and the perirenal fat.

22. The Physiology of Tubules

S. E. Dicker

The kidneys greatly modify the composition and volume of the glomerular filtrate as it passes down the tubules. Protein-free plasma, emerging from the glomerulus to enter the tubules at about 125 ml min^{-1}, is transformed into a very much smaller volume of urine leaving the further ends at, say, 1 ml min^{-1}. The main secretory work of the tubules is done in concentrating substances, such as urea, which are present in large amounts and which are relatively highly concentrated in the urine. Important also is the reabsorption of substances valuable to the body, such as glucose, which are present in plasma but normally absent in urine.

The reabsorptive function of the tubules has been unequivocally demonstrated by the experiments of A. N. Richards and his successors, already mentioned, in which a comparison of the composition of the glomerular fluid and bladder urine showed that, in suitable circumstances, certain substances might be present in the former, but absent from the latter. Moreover, the quantity of the glomerular fluid which they collected in a given time, multiplied by the number of glomeruli, was much greater than the volume of urine which appeared in the same time. Water, therefore, is reabsorbed, consequently substances which are not reabsorbed must appear in the urine in a higher concentration than in the plasma (Fig. 22.1).

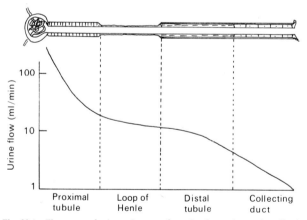

Fig. 22.1 Flow rates of urine as it passes from the glomerulus to the collecting ducts. Glomerular filtration rate is 125 ml min^{-1}; flow falls to 1 ml min^{-1} at the outlet from the collecting ducts, as water is progressively reabsorbed. Pressures within the lumen would be around 15 mmHg in the glomerulus, through 10 mmHg at the end of the loop of Henle to approximately atmospheric pressure at the outlet of the collecting duct.

In man, 125 ml min^{-1}, that is nearly 200 litres in 24 hours, are filtered in about two million glomeruli, producing about 1·5 litres of urine. On the average, therefore, each glomerulus filters 0·1 ml a day, nearly all of which is reabsorbed in passage down its tubule, of average length about 5·5 cm, and with reabsorptive surface greatly increased by microvilli (brush-border) in the proximal segment.

Most products of metabolism, which appear more concentrated in urine than in plasma, are so concentrated because they are less reabsorbed in the tubules than is the water. Nevertheless, some substances are certainly concentrated beyond this level by transfer from peri-tubular capillary blood to the lumen by "secretion" by the tubule cells. Most striking of these are the substances already mentioned whose plasma clearances approach the total plasma flow through the kidney. If nine-tenths of such a substance, say diodone which reaches the kidney in the plasma is excreted in the urine, and if this were derived entirely from the glomerular filtrate, the plasma emerging from the glomerular capillaries would contain only one-tenth of its normal content of water and the blood would be much too viscous to pass through the vasa efferentia. In fact, only about one-fifth of the water in the plasma is removed by glomerular filtration, leaving four-fifths in the plasma in the vasa efferentia. Direct secretion of some substances into the lumen of the tubule is, therefore, certain. Among other such substances are penicillin and the dye, phenol red, which cannot readily pass the glomerular membrane because much of it is bound to plasma protein. Many drugs are similarly protein-bound. Creatinine in some species, such as man, is secreted in small amounts; in others, such as the dog, concentrated only by the reabsorption of water. The plasma clearances of all "secreted" substances are higher than that of inulin but approach this as their concentration in the plasma increases (Fig. 22.2). This is because there is a maximum quantity per minute which the tubule cells can secrete known as the "transport maximum", or Tm; whereas the amount contained in the glomerular filtrate per minute increases directly with concentration in the plasma. The filtered portion, thus, increasingly dwarfs the secreted portion of the substance appearing in the urine.

Tubular reabsorption

Substances whose clearance is smaller than that of inulin pass through the glomeruli and are reabsorbed partially by the tubules. All normal constituents of plasma of small molecular weight, such as sodium, potassium, chloride, phosphate, sulphate, urea, creatinine and water, fall into that category; they are reabsorbed by the tubules to some extent. Thus by using the concept of clearance judiciously, it is possible not only to measure processes of filtration and reabsorption, but to know how the various plasma constituents that are found in the urine are excreted by the kidneys.

This statement requires some qualification. First, the results of clearance study cannot be interpreted correctly without the exact determination of the glomerular filtration rate (inulin clearance) and of the glomerular filtration of the substance under investigation. This means that one has to ascertain that none of the substances are fixed on proteins, since any amount of a substance bound to proteins would not filter through the glomeruli. Second, conclusions drawn from clearance studies must be confirmed by other methods such as the micropuncture technique. For instance, although the clearance of K$^+$ is low, suggesting that it has

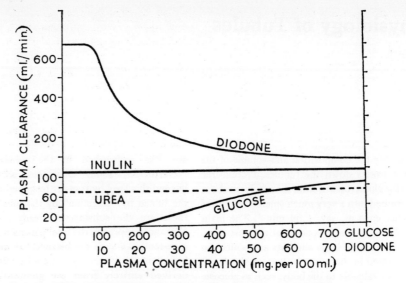

Fig. 22.2 Diagrammatic representation of the effects of plasma concentrations on plasma clearances in the human kidney.

Glucose is "reabsorbed" up to a transport maximum, $T_m = 350$ mg min^{-1}. Diodone is "secreted" up to a $T_m = 57$ mg iodine min^{-1}. Inulin is neither reabsorbed nor secreted in the tubules and the plasma clearance is little affected by the concentration in the plasma.

Some urea diffuses passively out of the tubules in amount about proportional to its concentration in the plasma. (After H. W. Smith.)

been reabsorbed in part only, micropuncture techniques have shown that after K^+ has been filtered by the glomeruli, it is reabsorbed entirely by some parts of the tubules, and that its low clearance represents K^+ excreted into the distal part of the tubules. Third, before concluding that changes observed in the clearance of certain substances are due entirely to altered tubular function, it is essential to ensure that they are not due to changes in glomerular filtration rate. In this respect it is essential to realise that the measuring of inulin clearances is not an absolute measurement of glomerular filtration and that most estimations have a built-in error of the order of 5 to 10 per cent due partly to the difficulty of collecting urine exactly. This error, if ignored, may lead to false conclusions, especially when dealing with substances such as water or sodium whose output in the urine represents only 1 per cent or less of the amount filtered.

Tubular secretion

As we have already seen, there are substances whose clearance is greater than that of inulin. These substances do not usually exist in the body; they can be administered by injection. Such substances, in addition to diodone, are, for instance, phenolsulfonphthalein or phenol red diodone (3,5-dio-O-4-pyridone-N-acetic acid) and derivatives of hippuric acid, penicillin and carinamide. These substances are excreted by the tubules, in addition to being filtered by the glomeruli. Instead of circulating in solution in the plasma water, they circulate bound to proteins. At low concentration, say less than 10 mg dl^{-1} for p-aminohippuric acid (PAH), the amount of PAH excreted by the tubular cells is small but since, in these circumstances, the renal vein plasma contains practically no PAH it follows that all the PAH that enters the kidneys is excreted. When this happens, it is clear that the limit to what can be excreted is determined by the amount of blood that perfuses the kidneys.

The Tubular Transport Maximum, Tm

What happens to substances like diodone, PAH or penicillin when their plasma concentration is increased? Their urinary output increases. But if the amount of any of these substances excreted by the tubules is compared with that filtered by the glomeruli, it will be seen that above a certain level, say 25 mg dl^{-1} of PAH, the tubular excretion reaches a ceiling, which is the *maximal rate of excretion* achieved by the tubular cells. This is called Tm. For PAH, Tm_{PAH} in man is of the order of 77 mg min^{-1}, that of $Tm_{diodone}$ is 42 mg min^{-1} (see Table 22.1). The existence of a maximal tubular excretion, Tm, is a fundamental concept. It is characteristic of renal excretory activity

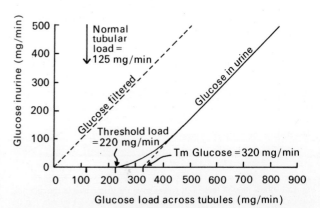

Fig. 22.3 Transport maximum T_m and renal threshold.

The maximum amount of a substance reabsorbed in the tubules is limited (by carrier, specific enzymes available, etc.). The limit, or T_m, for glucose is 320 mg min^{-1} and the excess above this *all* passes into the urine. The dotted line on the graph shows how the amount in the urine of a substance unaffected by passage through the tubules varies with the tubular load of it (i.e. a direct proportionality occurs). The other curve is the relationship between tubular glucose load and rate of glucose appearance in urine. When tubular load is at the normal level of 125 mg min^{-1}, no glucose appears in the urine. The threshold is 220 mg min^{-1}. As the load rises, the amount in the urine stays at 320 mg min^{-1} below the tubular load (320 mg min^{-1} is always being reabsorbed; this is the T_m).

Note that the threshold load of 220 mg min^{-1} is produced by a threshold *concentration* of glucose in the plasma, the *renal threshold*, which for glucose is 180 mg 100 ml^{-1} (for the normal filtration rate of 125 ml min^{-1}).

for substances which are actively secreted. Figure 22.3 is a summary of the concept of Tm as applied to glucose.

Table 22.1 Values of Tm for various substances

Substance	Transport maximum, Tm*
Tubular reabsorption	(mg mm^{-1})
Glucose	320
Plasma protein	30
Ascorbic acid	1.70
Lactate	75
Haemoglobin	1
Phosphate ion	0.10
Sulphate ion	0.06
Tubular secretion	
Diodone	42
PAH	77
Phenol red	56
Creatinine	16

* Tm is the maximum tubular load that can be sustained without overflow of the substance into the urine. Note that sodium ions have no Tm. The term "tubular load" refers to the amount of a substance per min reaching the commencement of the tubules. It is a fraction of the plasma load reaching the kidney (in mg min^{-1}). If 125 ml of glomerular filtrate is formed per minute, containing glucose at 1g l^{-1}, the tubular load of glucose is

$$\frac{125 \times 1000}{1000} = 125 \text{ mg min}^{-1}$$

Sodium load would be about 18 mmol min^{-1} and chloride load 13 mmol min^{-1}.

The proximal convoluted tubule

The clearance of the filterable fraction of all substances normally contained in urine is less than that of inulin or mannitol. Thus for all these substances, glomerular filtration is followed by tubular reabsorption. The percentage reabsorption for a given urinary constituent is obtained by comparing the clearance of the filtered fraction with glomerular filtration. Let us take as an example urea, which is an important constituent of urine. It is completely filterable. Its estimated clearance is 67 ml min^{-1},* while the glomerular filtration rate is 125 ml min^{-1}.* The percentage of urea reabsorbed is therefore about 50 per cent. It is possible to calculate the actual amount of urea reabsorbed (Q). It will be equal to the amount of urea filtered minus the amount excreted in the urine. Thus

$$Q = FP - UV$$

where F is glomerular filtration rate as estimated by means of inulin (or mannitol) clearances, P is the plasma concentration of urea (mg ml^{-1}), U is urinary concentration of urea (mg ml^{-1}) and V the volume of urine excreted (ml min^{-1}). Similar calculations can be used for any constituent of the urine, including water.

Micropuncture of tubules

The actual demonstration that water and solutes are reabsorbed by the proximal tubule came from studies using micropuncture techniques. It can be shown that the volume of glomerular filtrate collected from one glomerulus in amphibians per unit time multiplied by the number of glomeruli in one kidney is about 25 times greater than the volume of definitive urine excreted by that kidney in the same time. In other experiments, in amphibians and rodents, the concentration of inulin or mannitol in the glomerular filtrate was compared with that found in the primitive urine in the proximal tubule. This shows that the concentration of inulin or mannitol in the proximal tubule was much greater than that in the glomerular filtrate. Since neither inulin nor mannitol is excreted by the tubules, the increase in concentration must be the result of water reabsorption.

Micropuncture also shows that glucose is entirely reabsorbed in the proximal tubule, while sodium, potassium, calcium, chloride, bicarbonate, phosphate, creatinine and urea are all reabsorbed to some extent only. The overall effect of the reabsorptive activity of the proximal tubule is to reduce the volume of ultrafiltrate by about 80 per cent and, although some solutes are reabsorbed entirely and others only partially, there is no osmotic gradient along the proximal tubule, the primitive urine remaining isotonic. The mechanism by which 80 per cent of the glomerular filtrate is reabsorbed, irrespective of variations in the volume of primitive urine filtered, is not fully elucidated yet. It is likely, however, that the increased oncotic pressure of the blood in the peritubular vessels may play a role.

The loop of Henle

The loops of Henle, with both descending and ascending limbs, are essentially in the medulla and their function is to produce an osmotic gradient, without which the process of urine concentration cannot operate. The first indication of the role played by loops of Henle in the mechanism of urine concentration came from comparative physiology. In contrast with mammals, vertebrates like frogs or toads have no loops of Henle; their urine is never concentrated; even after severe dehydration, urinary concentration never exceeds that of their plasma. Some birds have no loops of Henle; their urine, collected from the ureters, is not concentrated. Other birds which have developed some loops of Henle can produce moderately concentrated ureteral urine. Furthermore, even among mammals, rodents like *Aplodontia rufa* with ill-developed loops of Henle cannot concentrate their urine well, while others like *Jerboa* (desert rat) in which each nephron has a long loop of Henle can concentrate their urine twentyfold. It would be a mistake, however, to conclude from these observations that the loops of Henle produce concentrated urine. The correct interpretation would be that there is no osmotic gradient in the renal medulla in the absence of the loops of Henle.

The presence of an osmotic gradient in the renal medulla was first demonstrated by taking slices from the kidney and estimating their osmolal concentration. Slices from the cortex containing glomeruli, proximal and convoluted tubules were isotonic with blood plasma, while slices from the medulla containing both limbs of the loops of Henle and collecting ducts were hypertonic; moreover, slices from the inner medulla were more hypertonic than those taken from the subcortical region.

The hairpin countercurrent theory

How can the loops of Henle, with their descending and ascending limbs, produce an osmotic gradient? Two Swiss physicists suggested that it could be explained by the multiplication of a single effect. Let us consider a model of two parallel tubes run-

* These figures of 67 ml, 125 ml etc. are quoted for convenience and conformity with preceding paragraphs. There is, of course, a fairly wide range of values found in normal individuals.

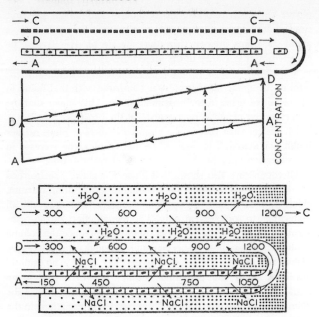

Fig. 22.4 Diagrams illustrating the action of the hairpin countercurrent system.

At the top is an idealized system whose action in producing a great increase in concentration at the tip of the hairpin is described in the text.

At the bottom is shown schematically the system as it exists in the kidney: A, ascending loop of Henle; D, descending loop of Henle; C, collecting duct. A is connected to C by way of the distal tubule, which actively reabsorbs NaCl and allows water to pass from the tubular fluid to the blood so that the two become isotonic. The figures indicate the concentrations (milliosmolar) of the fluid at various points within the loop of Henle and the collecting duct. The stippling indicates the corresponding increase in concentration of the interstitial fluid towards the tip of the loop.

ning close to each other but in opposite directions, A and D, separated by a semipermeable membrane, SPM, i.e. a membrane which is permeable to water only and not to solutes (Fig. 22.4). The liquid which perfuses the two tubes is the same, 0·9 per cent NaCl solution. Let us visualise a small segment cut across the two tubes, A_1 and D_1. If a "force" is now applied on the segment D some water will pass from D_1 to A_1; hence, D_1 will become more concentrated, and A_1 less. This is a *single effect*. As long as the "force" (whether chemical, physical or mechanical) is maintained, the compartment A_1 will be more dilute than D_1. But since the solutions circulate in opposite directions, what happens in the next imaginary segment? At the onset, D_2 will have the same concentration as D_1, and A_2 the same as A_1. If the "force" now acts on D_2, water will enter A_2 and D_2 will become more concentrated than D_1, while A_2 becomes more dilute than A_1. The effect is thus multiplied, and it is clear that the longer the tube D, the more concentrated its contents will become.

We now connect the two tubes A and D with a loop in such a way that the distal end of D, containing a highly concentrated solution, enters the proximal end of A. The liquid entering A will have the same concentration as that leaving D, but as it moves along the length of A it will receive water from D and so become progressively less concentrated, until at its distal end it will be hypotonic. This is how the model works.

In animal biology, there are no semipermeable membranes which allow the passage of water only. So let us replace the semipermeable membrane by a biological membrane, endowed with the ability of transporting ions, such as Na^+, from A to D. No "force" will be required. As Na^+ is transported progressively from A to D, the solution in D will become progressively more concentrated until it reaches the loop, where it enters A and, as it moves along A it becomes progressively less concentrated. This is what has been called a "*hairpin countercurrent system*". The descending and ascending limbs of the loop of Henle are separated by a small amount of interstitial tissue, and it has been shown by micropuncture techniques that Na^+ is actively transported from the ascending loop into the interstitial space and the blood of the vasa recta, from which it enters the descending limb. The net result of the hairpin countercurrent system is the building up of an osmotic gradient, with its maximum concentration in the papilla of the kidney.

Although the present short account provides the bare bones only of how the loops of Henle function, it is essential for the understanding of the mechanism of urine concentration. It is important to realise that the maintenance of a high corticopapillary osmotic gradient is dependent on slow movement of all the fluids in the medulla. Any considerable increase in urinary flow, such as that produced by an osmotic load or by overhydration, tends to abolish the gradient. This is usually accompanied by changes in the circulation in the vasa recta, which suggest that the intramedullary circulation may play an important role in the regulation of renal function, especially with reference to its urinary concentrating mechanism.

The distal convoluted tubule

As it leaves the ascending limb, urine is hypotonic. Micropuncture studies have revealed that the difference in concentration between the isotonic urine that enters the descending limb and that of the urine as it leaves the medulla, averages 116 mosmol kg^{-1} H_2O. As urine proceeds along the distal tubule, it loses its hypotonicity and again becomes isotonic. This is due to a change of permeability of the walls which allows the urine to become isotonic with the rest of the cortex. It is of some interest that urine which enters the distal tubule does not contain any potassium, this ion having been entirely reabsorbed by the cells of the proximal tubule. Potassium is secreted into the urine again in the distal tubule.

The collecting ducts

Urine entering the collecting ducts is isotonic; by the time it leaves the kidney it will be either hypo-, iso- or hypertonic. It will be hypotonic during water diuresis, as a the result of active reabsorption of Na^+ and passive transfer of urea from the lumen into the interstitial tissue, from where these will be removed via the vasa recta. When man is fully hydrated the osmolal concentration of the urine may be as low as 15–20 mosmol kg^{-1} H_2O, and the urine flow may be of the order of 12–15 ml min^{-1}. The highest rate of urine flow cannot exceed that of the primitive urine as it leaves the proximal tubules; and since the glomerular filtration rate is reduced by some 80 per cent in the proximal tubules, the highest rate of urine flow cannot exceed 20 per cent of the filtration rate. The high rates of urine flow observed during full hydration can obtain only when there is no circulating antidiuretic hormone (ADH; q.v.) and the walls of the collecting ducts are virtually impermeable to water. This occurrence is helped by the fact that, at the peak of diuresis, the

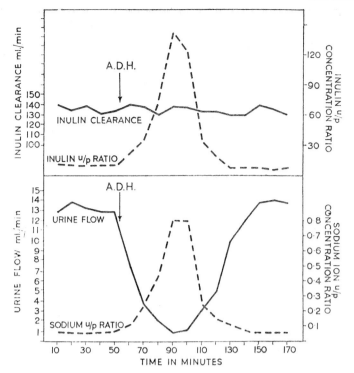

Fig. 22.5 Intravenous injection of 40 milliunits of antidiuretic hormone (ADH) produces in a well-hydrated man (1) antidiuretic action with no change in inulin clearance, i.e. glomerular filtration rate (scale on the left), and (2) increase in urine concentration of sodium and inulin (scales on the right in terms of urine/plasma concentration ratio). Plasma inulin concentration 25 mg dl^{-1} was kept constant by intravenous infusion of 1 g h^{-1} plasma sodium concentration: 325 mg dl^{-1}. (S. E. Dicker.)

osmotic gradient in the medulla is dissipated, very likely as a result of increased medullary blood flow.

The antidiuretic hormone (ADH)

During dehydration, urine flow in man can be as low as 0·2 ml min^{-1} or less, with an osmolal concentration as high as 1200 mosmol kg^{-1} H$_2$O (Fig. 22.5). The osmotic gradient in the medulla is fully restored and the concentration in the loops of Henle and their surrounding tissues will reach 1200 mosmol kg^{-1} H$_2$O in the papilla. As the body loses water, antidiuretic hormone (ADH) is released from the neurohypophysis where it is stored. As circulating ADH reaches the kidneys, it acts on the walls of the collecting ducts, rendering them highly permeable to water, and thus allowing the isotonic urine to equilibrate osmotically with the high concentration of the medulla. The maximum urinary concentration cannot exceed that of the renal papilla. The mechanism of action of ADH on the permeability of the collecting ducts is complicated and not yet fully understood; it must act on the outside of the cells, though its effect is clearly on the lumenal side. The cells most sensitive to its action are in the lower third of the ducts, thus in the region of maximum concentration of the medulla.

Mechanism of action of ADH

According to modern views, ADH acts through a mediator, cyclic AMP or 3′,5′-AMP. It is known that the conversion of ATP to 3′,5′-AMP is catalysed by the enzyme adenylcyclase; and cyclic AMP is inactivated by conversion to 5′-AMP, a reaction catalysed by phosphodiesterase and inhibited by xanthines, such as theophylline. ADH stimulates adenycyclase and so increases the cellular concentration of 3′,5′-AMP.

And since addition of cyclic AMP to the basal (= outside) side of perfused isolated collecting ducts mimics the action of the antidiuretic hormone, present views are that it is the nucleotide which is responsible for the effects of ADH. Whether its action is mediated or not, the changes of permeability of the collecting ducts are related to its concentration. However, once maximum permeability has been achieved, further increase in the concentration of ADH will not have any effect, the final concentration of the urine being determined by the osmolality at the tip of the medulla.

It must be clear by now that the actual mechanism of urine concentration is dictated by changes in the permeability of the walls of the collecting ducts, but that the degree of concentration and of antidiuresis is determined by the concentration achieved by the countercurrent system of the loops of Henle in the medulla.

Mechanisms of secretion and reabsorption of water and ions

The filtration and reabsorption of water and ions are, quantitatively speaking, the most important of kidney functions. In the *proximal tubule,* the reabsorption of both is isosmotic; the fluid reabsorbed and that in the tubule have the same osmolarity. Water and ions are absorbed at the same rate (osmotically) and the proximal tubule is not doing osmotic work. The reabsorption of sodium cannot be accounted for by an electrochemical gradient across the tubular membrane; it is an active process and requires metabolic energy. This is also true for potassium which is all reabsorbed in the proximal tubule. Reabsorption of chloride ions is

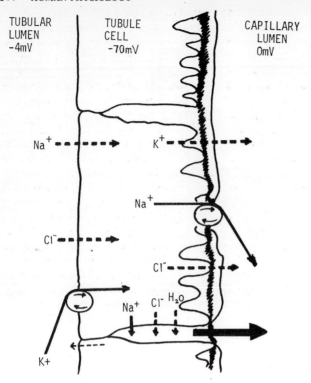

Fig. 22.6 Ion transport in proximal tubule. Large black arrow, filtration of Na$^+$, Cl$^-$ and H$_2$O across epithelial cell basement membrane. Solid lines, active transport; dotted lines passive diffusion down electrochemical gradients. Active transport of Na$^+$ occurs across peritubular membranes and across the intercellular membranes. Cl$^-$ and H$_2$O follow passively.

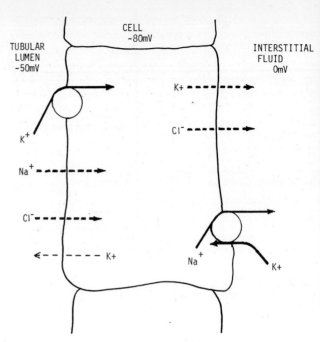

Fig. 22.7. Ion transport across wall of distal tubule. Solid lines are active transport of Na$^+$ and K$^+$ ions. Dotted lines are passive diffusion down an electrochemical gradient.

Intracellular concentration is low and the inside of the cell is electrically negative. This necessitates sodium movement against a gradient i.e. it is active. Also K$^+$ ions move against a steep gradient at the peritubular cell membrane (the concentration gradient is much larger than the electrical gradient of opposite sign).

Passive permeabilities to K$^+$ and Na$^+$ are about equal. (After Pitts R. F. *Physiology of the kidney and body fluids*. Year Book Medical Publishers Chicago 1974.)

passive. Between $\frac{2}{3}$ and $\frac{7}{8}$ of the glomerular filtrate is reabsorbed in the proximal tubule.

The proximal tubular epithelium is highly permeable to water, thus reabsorption of the $\frac{2}{3}$ to $\frac{7}{8}$ of the glomerular filtrate gives rise to an osmotic force which leads to an equivalent fraction of the tubular water being reabsorbed. Since water and ions cross at the same rate, plasma and tubular fluids remain isosmotic.

Reabsorption of salt is in excess of water in the *loop of Henle* and this is the means by which osmotic concentration and dilution of the urine is brought about. The salt reabsorption is proportional to the sodium concentration in tubular fluid leaving the proximal tubule (glomerulotubular balance). The active transport of chloride appears from recent work to give the driving force for sodium reabsorption in the ascending loop of Henle (Fig. 22.6).

The *distal nephron* (distal convoluted tubule plus collecting duct) reabsorbs only a small part of the ions and water from the filtrate ($\frac{1}{5}$ to $\frac{1}{8}$) but in contrast to the proximal tubule, this is very variable and the amounts are dictated by the needs of the body. The distal nephron epithelium establishes considerable ionic concentration gradients between urine and blood e.g. during high salt intake, urinary sodium can be twice the concentration in plasma. It can also produce high osmolar gradients. In water diuresis, the osmolarity of urine may be only one tenth that of plasma; conversely in dehydration it may be four times that of plasma. Antidiuretic hormone controls the permeability and is also said to stimulate the sodium pump. Note that water reabsorption in both the loops of Henle and the collecting ducts is passive.

Distal tubular sodium reabsorbtion is about 10 to 15% of the amount filtered: it is transported against a large concentration

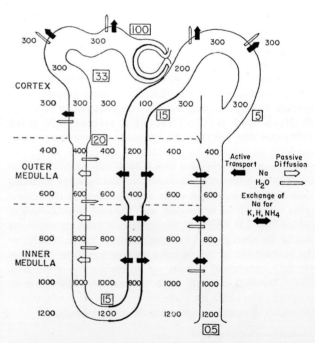

Fig. 22.8 Summary of passive and active exchanges of water and ions in the nephron whilst hypertonic urine is being formed. Concentrations are in mosmol l^{-1}. Large numerals are percentages of glomerular fluid remaining. Chloride is actively transported in ascending loop.

(Pitts R. F. (1974). *Physiology of the kidney and body fluids*. Year Book Medical Publishers. Chicago.)

gradient. At the beginning of the distal tubule, sodium concentration is 60 mmol l^{-1} (i.e. more salt than water has been reabsorbed in the loop of Henle). Along the distal tubule, concentration declines to about 30 mmol l^{-1}, or even lower in diuresis. The gradient across the epithelium can rise to above 140 mmol l^{-1}. Evidently the ion pump is stronger here than in the proximal tubule and/or the water and ion permeability is lower. Distal transtubular sodium gradients depend on the urea concentration of tubular fluid; in the absence of urea the sodium concentration cannot reach the levels mentioned above. Usually chloride is in electrochemical equilibrium across the wall of the distal tubule, but it can in certain circumstances be actively transported. Figure 22.7 illustrates these points.

SUMMARY

In 24h, 150 l of glomerular filtrate is formed. There are 300 mosmol of solute per litre, mainly Na^+, Cl^-, and HCO^-_3 ions. As the filtrate passes down the proximal convoluted tubule Na^+ ions are actively reabsorbed into the cortex, Cl^- ions passively, and H_2O osmotically. The volume of tubular fluid is now only 20 per cent of the amount filtered; it is still 300 mosmol l^{-1}.

In the descending loop of Henle, H_2O passes out into the hypertonic medullary region; Na^+ ions diffuse in. The ascending limbs are impermeable to water, thus osmolar concentration of the tubular fluid here is progressively reduced. There is a gradient across the wall of about 200 mosmol l^{-1} at any given point in the loop. Thus a much bigger gradient is established along the length of the loop (300:1 200 mosmol l^{-1}), the countercurrent multiplier.

Fluid in the distal convoluted tubules is hypertonic to the cortex. During diuresis, ADH is low and the distal nephron epithelium is impermeable to water. Ions are actively reabsorbed, urine volume is large and it is hypotonic. During dehydration, ADH is high causing the epithelium of the distal nephron to be permeable to water. Tubular fluid becomes isotonic with cortical interstitial fluid when the middle of the distal segment is reached. Outward flow of Na^+, Cl^- ions and osmotic diffusion of water, reduce the volume of urine greatly and it gets progressively more concentrated as it flows along the collecting ducts. The water that leaves the distal nephron and the Na^+ ions pumped actively from the ascending limb, are removed by the vasa rectae in the medulla and papillae, i.e. it is a countercurrent exchange reducing the loss of osmotically active solutes. Figure 22.8 summarises these facts.

23. The Control of Kidney Function

S. E. Dicker

Elaboration of urine of variable acidity

Depending on the environmental conditions, the kidneys may excrete alkaline, neutral or acid urine. Since the physiological processes of metabolism lead to the formation of 50 to 80 mmol acid per day, and therefore to a corresponding amount of hydrogen ions, urine is usually acid. To preserve the acid–base balance of the organism and maintain homeostasis, hydrogen ions are excreted in the urine either free or combined. Though the maintenance of the acid–base equilibrium is discussed in detail elsewhere, a few words as to the renal tubular involvement will be given.

Acidification of urine

Acidification of urine occurs as the result of a direct exchange, one H^+ ion being secreted into the tubular lumen for the reabsorption of one C^+ cation. It is linked to the synthesis of carbonic acid molecules from CO_2 and water.

Alkalinisation of urine

Reabsorption of bicarbonate occurs mainly in the proximal tubule. The urine contains all the filtered bicarbonate less an amount which is proportional to the glomerular filtration rate. In man, the amount of bicarbonate reabsorbed is 25 to 27 mmol l^{-1} of filtrate, a figure which is the same as the normal level of bicarbonate in plasma. Thus, in contrast with other substances reabsorbed by the proximal tubules, it would appear that the amount of bicarbonate reabsorbed is determined by its concentration in the reabsorbate. There is evidence, however, suggesting that bicarbonate is not reabsorbed, but that the bicarbonate which disappears from the primitive urine is simply destroyed by the kidney and that the tubular cells then elaborate an amount of bicarbonate identical to that which has been destroyed and which is then passed into the body.

Glomerulotubular balance

So far we have discussed the glomerular and the tubular stages of urine formation, separately. Between the glomerulus and the tubules of the same nephron there is a morphological and functional balance whose importance and mechanism has begun only recently to be appreciated. Nephrons vary in size and for the various sizes of glomeruli there are proportional variations in the tubules. There are at least two different populations of glomeruli, with loops of Henle of different lengths. In any one nephron there seems to be a very precise balance between glomerular and tubular activity. We have seen that the definitive urine represents only a small fraction, about 1 per cent or less, for certain substances such as sodium. If glomerular filtration and tubular reabsorption varied independently of each other, the volume and composition of the final urine would be so disorganised that it would impair homeostasis. Since this does not occur, we have to assume that either glomerular filtration is immutable (which we know it is not) or that changes in glomerular filtration and tubular reabsorption are automatically corrected by each other. So far there is no definitive explanation for glomerulotubular autoregulation. It is possible that autoregulation and oncotic pressure of peritubular plasma and/or the juxtaglomerular apparatus, with the macula densa, play a role in glomerulotubular regulation. Alternatively it is possible that tubular adaptation occurs, not as a direct response to glomerular filtration rate, but to consequent changes in plasma flow. A striking example of this may be the renal adaptation following unilateral nephrectomy.

CIRCULATION AND THE KIDNEY

Under normal conditions, the kidney does not contribute to the variation of peripheral vascular resistance which maintains steady systemic arterial pressure. The sympathetic nerves which supply the organ only come into action when the arterial pressure falls below 60–80 mmHg when profound vasoconstriction and fall in glomerular filtration rate occur. This may be augmented by the effects of circulating adrenaline and noradrenaline under such conditions.

The effects on the secretion of urine of stimulation or blocking of various nervous structures may all be interpreted in terms of the changes induced either on the general arterial pressure or on the calibre of the renal blood vessels, or on both. There is at present no sufficient reason for suspecting a direct nervous influence on the secretory mechanism proper (except, perhaps, for the anatomical fact that nerve fibres do supply tubules). This is illustrated by the following experiment. If one kidney of a dog be completely denervated and the dog allowed to recover from the anaesthetic, the urine can be collected separately from each kidney through exteriorised ureters. The urine coming from the denervated organ is indistinguishable from that coming from its innervated fellow, both as regards rate of flow and composition. The increase of urine flow due to administration of water ("water diuresis") and its inhibition by exercise or stimulation of the skin are equal in both kidneys.

The degree to which circulating adrenaline and noradrenaline normally contribute to maintenance of the tone of renal blood vessels is uncertain. Larger concentrations, due to injection, produce vasoconstriction which leads to a fall in filtration and even anuria.

Aldosterone

Aldosterone and, to a lesser extent, deoxycorticosterone from the adrenal cortex have a profound effect on renal function by promoting reabsorption of sodium and excretion of potassium. Aldosterone inhibitors, e.g. spironolactones, therefore increase sodium output and consequently act as diuretics.

The way in which circulating aldosterone is regulated is interesting. The extracellular volume—and thus sodium concentration in extracellular fluid—in some undefined manner acts on the juxtaglomerular apparatus causing it to release *renin*. The renin then reacts with angiotensinogen in the blood to form angiotensin which controls the secretion of aldosterone from the adrenal cortex (see below).

Any substance which cannot leave the proximal tubules as rapidly as water does will reduce the reabsorption of water and act as an osmotic diuretic; for example, mannitol, as already described, urea, sulphates or an excess of glucose. The diuretics most widely used in medicine are, however, neither water, alcohol, aldosterone inhibitors nor osmotic diuretics but substances, such as chlorothiazide that act by inhibiting chloride or sodium chloride reabsorption. Their diuretic action is sometimes reinforced by combining them with one of the purine diuretics, aminophylline or caffeine, which increases glomerular filtration by dilating the afferent arteriole and may also reduce tubular reabsorption.

Renal regulation of blood pressure (renin)

A decrease in renal blood flow sufficient to give rise to ischaemia of the juxtaglomerular cells, which are in the walls of afferent arterioles next to the glomerular tuft, stimulates these cells to secrete renin.

Renin is an enzyme catalysing the conversion of the non-pressor globulin from the liver, *tensinogen* into *angiotensin I*, a decapeptide. Another enzyme then forms from it the octapeptide *angiotensin II*. Angiotensin II has an action on the arterial pressure through two mechanisms. First, it produces peripheral arteriolar constriction; second, it stimulates the adrenal cortex to secrete aldosterone, which, as described above, causes salt and water retention by the kidney, thus raising the arterial pressure as a result of increasing the extracellular volume. Angiotensin II is destroyed in the plasma by an enzyme, angiotensinase, after about 30 min.

RENAL FEEDBACK CONTROL. THE JUXTAGLOMERULAR APPARATUS

As in most other biological systems, feedback mechanisms are important in the control of function. There are several already known in the kidney. The juxtaglomerular apparatus is found in each nephron. The smooth muscle cells in the afferent arteriole are filled with granules of *renin*, a hormone which raises the blood pressure by mechanisms already discussed.

Osmotic feedback

If glomerular filtrate is formed at too fast a rate, the osmolality of the fluid at the macula densa is below normal. This then gives rise to a constriction of the afferent arteriole, in turn thus decreasing glomerular blood flow. The glomerular filtration rate, (GFR) falls again to normal levels. Here we have a negative feedback loop which essentially maintains GFR at constant levels. Note that it is acting at the single nephron level—each nephron has its own regulating system.

Sodium feedback

When the tubular fluid flow is too high sodium concentration rises in the distal tubule. This also constricts the afferent arteriole, with the result, as outlined above, that the GFR is brought back to normality.

Compensatory renal growth after unilateral nephrectomy

Unilateral nephrectomy halves the total glomerular filtration rate at once, but the tubular activity of the remaining kidney adapts almost immediately to this new situation, so that even in the post-operative hours there is no retention of urinary constituents. The renal tubules allow more of the primitive urine to pass and double the secretion rate of substances that are normally excreted. Though for any given tubule there has been no decrease of glomerular activity, it is the halving of total filtration rate which triggers off this tubular modification. Immediately after the removal of one kidney there is an increase of blood flow in the remaining kidney of about 30 per cent; at the same time there is a commensurate enhancement of glomerular filtration rate. In most mammals where it has been investigated, the functional adaptation of the remaining kidney appears to be complete as early as three hours after unilateral nephrectomy. It is achieved well before any morphological sign of hypertrophy can be detected. Not only are functional adaptation and onset of hypertrophy independent, but compensatory growth will continue for weeks and months, in spite of the fact that functionally the kidney is normal almost immediately after the operation. The rate at which the remaining kidney hypertrophies varies according to species and age. The rate of hypertrophy, when present, is faster in young animals and adolescent men, than in adults; it is faster in rodents than in man. Compensatory renal growth is not produced by an increase in the number of nephrons, which remains unchanged throughout life.

THE KIDNEY AT DIFFERENT AGES

Newborn

The study of renal function in neonates or infants is fraught with difficulties, the most important being the way in which to assess function so as to allow comparison with the adult.

The least controversial method is to adjust measurements of renal function to total body water. But since total body water is not easily estimated, it is customary to adjust the figures to a body surface area which in adult man is about 1.73 m^2.

At birth, glomerular filtration is of the order of 20 per cent of that of adult man. There is an inadequacy of tubular function and an inability to produce concentrated urine. The maximum osmolality of urine during the first days of life seldom exceeds 400 mosmol kg^{-1} H_2O. The kidneys have difficulty in excreting an excess of water or salt.

Old Age

After the age of 50 years in man, kidney function undergoes some degree of involution. The loss of function involves a decrease of glomerular filtration, renal plasma flow, tubular activity and concentrating ability. Kidneys of old men retain their ability to regulate acid-base balance, a property which does not exist in the kidney of neonates.

NON-EXCRETORY FUNCTIONS OF THE KIDNEY

Clinicians who deal with kidney diseases (and others, of course) have long suspected that the healthy kidney has functions other than those involved in the production of urine. Here are a few examples.

Plasma proteins

It has been suggested that in spite of the process of ultrafiltration in the glomeruli, up to 30 g plasma proteins are filtered per 24 hours. But since they do not appear in the urine, they must have been either reabsorbed or denatured and transformed into amino acids.

Fat

The kidney also plays a role in fat metabolism, a view consonant with its high concentration of esterases. Furthermore, it is known that in patients with certain kidney diseases, the fate of plasma lipids is different from that in healthy subjects.

Glucose

Glucose synthesis by the kidneys has been shown both *in vivo* and *in vitro*. For instance, *in vivo*, the blood sugar of dogs whose liver has been removed drops twice as fast if both kidneys are also removed. *In vitro*, it has been shown that renal tissue can synthesise glucose from certain amino acids. The present view is that glucose is produced chiefly in the cortex and metabolised in the medulla.

Erythropoeisis

The polycythaemia of renal tumors, curable by nephrectomy, and the erythroblastopenia of anuria, both suggest that the kidney has a function in erythropoeisis. Plasma of nephrectomised animals loses its ability to stimulate erythropoeisis in animals with depressed bone-marrow. The exact nature of the renal erythropoeitic factor is still unknown.

Prostaglandins

In recent years, research interest has centred upon the antihypertensive function of the normal kidney. When normal kidneys are transplanted into patients suffering from hypertension, there is a return to normal blood pressure. Recently, hypotensive substances have been isolated from the renal medulla and they have been identified as prostaglandins, whose vasodepressor properties have already been well established.

THE INVESTIGATION OF KIDNEY FUNCTION IN MAN

Renal function tests

Very simple tests will tell us a great deal about kidney function, and in particular the presence or absence of disease.

Specific gravity. A rough test of the power of the kidneys to concentrate (i.e. do osmotic work) can be made by measuring urine specific gravity with a hydrometer (urinometer). A morning sample of urine should not fall below a specific gravity of 1·025 (i.e. 1000 mosmol l^{-1}).

Protein. Albumen appears in the urine when the glomerular membrane is damaged and it becomes more permeable. This can be tested using Albustix, paper strips impregnated with bromphenol blue and salicylate buffer which turn from pale yellow to blue if dipped in urine containing protein.

Microscopy. A centrifuged specimen of urine examined under the microscope will show epithelial cell, red cell or protein "casts" indicating tubular damage.

Blood urea. Reduced kidney function allows urea (and other substances) to accumulate in the blood. The normal blood urea is around 30 mg dl^{-1} but a rise above this may not occur in renal failure if the protein intake is low. It should be remembered that extrarenal conditions, that may lower GFR (haemorrhage, shock, vomiting etc.) may cause a raised blood urea in the absence of kidney damage.

Urea clearance tests are not a useful measure of kidney function.

X-ray examination. The secretory parts of the kidney, calyces, pelvis, ureter, can be outlined in X-ray pictures by injecting radio-opaque substances which are concentrated by the kidneys (pyelography). Retrograde pyelography involves catheterisation of the ureteric orifice and injecting the radio-opaque material.

Measurement of Tm. Tubular function is investigated by loading tubular cells with an excess of secreted substance and then determining the actual amount dealt with. If Tm is lower than normal, this is an indication of a loss of tubule cells as a whole or an impairment of tubular carrier enzyme systems.

To determine Tm *diodone,* inulin clearance (i.e. GFR) is first determined. Plasma diodone is analysed following a priming injection and infusion to a level around 40 mg dl^{-1}. Analysis of the urine gives diodone excretion values per minue.

Thus the amount of diodone entering the tubules is known (plasma conc. and GFR); the amount in the urine is known, therefore subtraction gives the Tm for diodone.

Renal blood flow, Tm $_{PAH}$ and GFR may be estimated, but this is not done routinely.

24. Micturition

S. E. Dicker

The urinary excretory channels

Urine is secreted by the kidneys continually, but removed from the body only periodically as and when convenient. From the kidneys, urine travels through excretory channels until it reaches the bladder where it may accumulate for several hours. The excretory channels form an uninterrupted pathway which can be divided into calyces (minor and major), pelvis, ureter and bladder.

The renal calyces

The renal pelvis can contract. If this is so, how is it that urine does not flow back into the collecting ducts which, if it happened, would interfere with normal renal function? It has been postulated that there are some sphincters between the calyces and the pelvis whose contractions are synchronised with those of the renal pelvis.

THE RENAL PELVIS

There is a marked variation in the shape and volume of this structure, though some degree of symmetry is usually present. As a matter of fact, the type described by anatomists as the classical example of normal pelvis is found in a minority of normal individuals only. Some pelves are extrarenal and may expand freely, while others are clearly intrarenal. Not infrequently the renal pelvis is split, the lower part having the characteristic conical shape, while the upper one looks like a dilated ureter. From recordings of the intrapelvic pressure made in man, it would appear that the amplitude of pelvic contractions, when present, is very small and certainly much smaller than that in the ureter. Furthermore, there is no significant variation in the pressure within the pelvis, the resting pressure of the renal pelvis being the same as that in the upper part of the ureter. It is of interest that urine excreted at rates as high as 8 ml min^{-1} can be transported from the pelvis without giving rise to any appreciable pressure fluctuations there. It is possible, therefore, that the low pressure combined with the low amplitude of the contractions of the pelvis explain the absence of urinary reflux into the calyces.

According to pyeloscopy studies, the renal pelvis remains in open communication with the upper part of the ureter, the ureteral cone, both structures being filled as a unit.

The ureters

The upper urinary tract, together with the vesical trigone, is of mesodermal origin. The muscular wall of the ureter is almost solely composed of longitudinal fibres which continue across the vesical trigone where they spread out like a fan.

Peristaltic contractions travel down the ureter at about 2 to 3 cm per second and are repeated one to four times a minute. Consequently, urine enters the bladder in a series of squirts.

From recordings made in man, it is clear that when the intravesical pressure is raised slowly, ureteral activity is unaffected and the pressure in the ureter and the renal pelvis remains low. When the pressure in the bladder is raised quickly, ureteral contractions last longer and are often characterised by pressure spikes. The frequency of the contractions also increases. Although it is generally believed that intravesical pressure (more than intravesical volume) is responsible for changes of ureteral activity, this is not so. During micturition a considerable intravesical pressure develops on account of the contraction of muscles of the bladder; careful studies of ureteral activity during contractions of the bladder in man have shown that, in the absence of abdominal straining, the duration and frequency of ureteral contractions remained unaffected, especially if the urine flow was not interfered with.

The Ureterovesical junction

Urine can pass freely from the ureters into the bladder but not in the reverse direction. This is so because the ureters pass obliquely through the wall of the bladder and any pressure elevation within the bladder will tend to obliterate the lumen of the ureters. This obliqueness is increased as the bladder fills. In some animals, but not man, a mucosal fold also covers the ureteric orifice.

The ureterovesical junction is of considerable clinical importance since its structure prevents reflux of urine into the kidneys and hence limits the spread of urinary tract infection (Fig. 24.1).

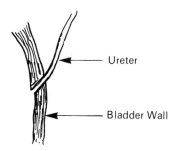

Fig. 24.1 Uretero-vesical junction. The ureter enters the bladder at an oblique angle; when pressure within the bladder rises, this tends to obliterate the lumen of the ureter and this prevents reflux of urine.

The presence of a uretero-vesical sphincter has been claimed, but only longitudinal muscle is found in the intramural part of the ureter.

When the ureters are obstructed by the presence of stones, it has commonly been assumed that hyperperistaltism of the ureter occurs and that this is responsible for the pain of renal colic. The evidence suggests, however, that the response to obstruction is a reduction of phasic activity accompanied by an increase in tone. During the clinical condition of renal colic, stretch of the wall of the urinary tract seems to be the chief stimulus to pain, though the occurrence of pain is independent of ureteral activity.

The bladder

Embryologically, the bladder derives from the cloaca and, but for the trigone, is of endodermic origin. In amphibians, and possibly in the fetus of mammals, the bladder is permeable to ions and water and plays an important role in the maintenance of homeostasis. These functions are lost in man, the bladder acting essentially as a reservoir.

As urine is squirted from the ureters into the bladder, the intravesical pressure in man rises to about 5 cm (0·5 kPa) water at a volume of 200 ml; and remains at that level until the urine volume reaches 400 ml. After this, further filling brings about a steep rise in pressure which reaches 25 cm (2·5 kPa) water for a urine volume of 500 to 600 ml. The maintaining of a nearly constant pressure within the bladder for volumes of urine up to about 400 ml has been attributed to physical properties whereby when a hollow organ is slowly distended (and a length and tension of the muscle fibres in its wall are increased correspondingly), the pressure inside increases very little because the curvature of the wall becomes progressively less (Fig. 24.2).

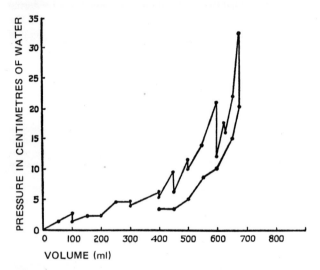

Fig. 24.2 Change in the pressure in the bladder of a man during filling and emptying.

(Upper curve) Water was slowly run into the bladder, and the pressure was observed after the addition of each 50 ml. The inflow was stopped at intervals, so as to allow time for the pressure to approximate to its final value, as shown by the short vertical lines.

(Lower curve) After 700 ml had been run in, the bladder was allowed to empty, 50 ml at a time. The pressures were all lower, indicating that the true equilibrium values had probably not been reached during either filling or emptying.

The pressure is not strictly constant over any range of volumes, but does not vary much between 100 and 400 ml. (Denny-Brown and Robertson.)

Evacuation of the bladder is produced by contraction of the plain muscle of its wall, the *detrusor muscle*. Reflux of urine into the ureters is prevented partly by the contraction of the bladder itself, but mainly by contractions of the muscles of the ureters, which run in the wall of the bladder (see above). The urethra, through which urine from the bladder is expelled, is guarded by two sphincters—the internal being of unstriated, and the external of striated, muscle. As the detrusor muscle contracts, the internal sphincter relaxes, though the act of micturition, which follows the opening of the external sphincter, is entirely voluntary in normal man.

Innervation of the bladder

The bladder has a double autonomic innervation. Stimulation of the parasympathetic pelvic nerves, induces contraction of the detrusor muscle and relaxation of the internal sphincter. This parasympathetic pathway includes both afferent and efferent nerves. Stimulation of the pre-sacral or hypogastric nerves (both sympathetic) causes contraction of the internal sphincter and relaxation of the bladder. Similar effects follow intravenous injections of adrenaline in man (Fig. 24.3).

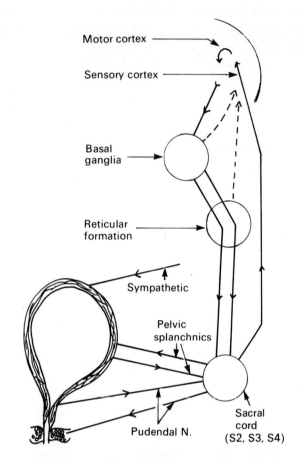

Fig. 24.3 Innervation of the bladder.

Sensations from the bladder

Normally, the desire for micturition is set up when the volume of urine in the bladder reaches 200 to 300 ml. Since the pressure in the bladder has not changed significantly, the desire for micturition must be elicited by the tension in the muscular wall. When the muscular tone of the bladder is increased, either by cold or conditions of emotional stress, the sensation of tension may be reached with a relatively small volume of urine in the bladder. The reverse may happen after drinking alcohol. Up to a point, the sensation from a full bladder can be suppressed from a conscious level in the cerebral cortex, just as it is possible by introspection to become aware of small quantities of urine in the bladder.

Voluntary control of the bladder

Voluntary micturition is brought about by impulses passing from the cerebral cortex, by way of the spinal cord and parasympathetic nerves, to the bladder. The detrusor muscle contracts strongly, raising the pressure in the bladder to 100 cm water (10 kPa), and simultaneously a reciprocal relaxation of the internal sphincter occurs. Voluntary effort to restrain micturition may considerably reduce the pressure within the bladder. Thus we have an example of involuntary muscle under the control of the will. Once micturition has begun, certain reflexes play a part in its completion.

1. Stretching of the bladder wall brings about reflex contraction of the bladder and this reflex can be abolished by cutting both the pelvic nerves, transecting the spinal cord, or cocainising the interior of the bladder. It is unaffected by division of the hypogastric nerves. The reflex arc concerned is along the pelvic nerves to a centre in the hind-brain and back again along the pelvic nerves.

2. The flow of water through the posterior part of the urethra also brings about reflex contraction of the bladder, the reflex arc involving the centre in the hind-brain. Transection of the central nervous system only interferes with reflex micturition if the section is below a plane passing from the inferior colliculi dorsally, to the middle of the pons ventrally.

Diseases of the spinal cord involving the posterior columns may prevent the sensations from the bladder reaching consciousness, although micturition can still be carried out voluntarily. When the pyramidal tracts are interrupted, voluntary micturition is impossible, although the patient may be quite aware of a full bladder. When the voluntary control of micturition is impaired, retention of urine may result in overdistension of the bladder, and reflex passage of small quantities of urine at irregular intervals; this condition is known as "retention with overflow".

25. Acid–Base Regulation

Biological fluids, such as intracellular fluid, tissue fluid and plasma, are kept within fairly narrow limits in terms of their *reaction*, i.e. their degree of acidity or alkalinity. Cells obtain energy from oxidative processes that liberate acidic end-products. These excess acids must be neutralised and subsequently excreted. It is not clear at present why the hydrogen ion content of body fluids must be so closely controlled, for enzymes will work over a far wider range of reaction than is tolerated within the body. One likely possibility is that the lipoprotein structure of the membrane is particularly sensitive, and that undue acidity or alkalinity would alter its properties and hence impair its function.

In this chapter and the next, we will consider the ways in which various control mechanisms act to maintain constancy in the reaction of the body fluids.

The normal reaction of body fluids

Having said, firmly, that the reaction of body fluids must be closely controlled, we now enquire, "How closely?" Normal plasma, in healthy persons, contains between 44 and 36 mmol l^{-1} of H$^+$ ions. This is, of course, the same as saying the pH is between 7.36 and 7.44, and most people regard a figure of 7.4 as easy to remember. The range which the body can withstand, without death supervening, is from pH 7.0 to 7.8. Since the pH notation is a logarithmic scale, this range is rather large in terms of the actual concentrations of H$^+$ ions*, much larger, for instance, than that allowed for either sodium or potassium ions.

Less is known regarding the reaction of the interior of cells. An intracellular glass microelectrode thrust into rat muscle cells recorded pH values of near 6.0, so it is likely that intracellular fluid is more acid than extracellular fluid. If H$^+$ ions were distributed in equilibrium across a cell membrane which had a resting potential of 90 mV, the pH would be 6.0.

Acids

An acid may be defined as a *proton donor*; a base as a *proton acceptor*. Thus

$$AH \rightleftharpoons A^- + H^+$$
Acid → Base + Hydrogen ion (= proton)

Hydrogen ions in water are hydrated (then called hydroxonium ions). The above equation, more accurately, is

$$AH \rightleftharpoons A^- + H_3O^-$$
Acid → Base + Hydroxonium ion

---

* The reader may perhaps be reminded that the pH is a convenient measure of the hydrogen ion concentration, which is usually extremely small. The pH is defined as the negative logarithm of the hydrogen ion concentration; so that a hydrogen ion concentration of, say, 5×10^{-8}, or $10^{-7.3}$ mol l^{-1} corresponds to a pH of 7.3.

The more easily an acid gives up its proton, the stronger it is. Equally, the more strongly a base takes up a proton, the stronger it is. An acid and its equivalent base, AH, is termed an *acid–base pair* (or sometimes a *conjugate pair*). Table 25.1 shows the strengths of some acid–base pairs, and from it we can see that the strengths of the acids and the bases are reciprocally related.

Table 25.1 *Examples of acid–base pairs*

	Approx. pK'	Acid	⇌	Proton	+	Base	
v. strong	−7	HCl	⇌	H$^+$	+	Cl$^-$	v. weak
v. strong	−2	OH$_3^+$	⇌	H$^+$	+	H$_2$O	v. weak
strong	4	H$_2$CO$_3$	⇌	H$^+$	+	HCO$_3^-$	weak
weak	7	H$_2$PO$_4^-$	⇌	H$^+$	+	HPO$_4^{2-}$	strong
weaker	9.5	NH$_4^+$	⇌	H$^+$	+	NH$_3$	stronger
v. weak	16	H$_2$O	⇌	H$^+$	+	OH$^-$	v. strong

pK' is the negative logarithm of the acids' dissociation constant; the stronger the acid, the smaller the pK'. pK' is equal to the pH in a dilute solution in which the weak acid is half neutralised. Note that water can interact in two ways in the table, i.e. it is at one and the same time a very weak base and a very weak acid. The very strong H$_3$O$^+$ and OH$^-$ are only present in very small and equal amounts, so that pure water is, in fact, neutral.
From Robinson, J. R. *Fundamentals of Acid–Base Regulation*. Blackwell. Oxford and Edinburgh. p. 11 (1967).

Bases

Bases found in the body are either anions or uncharged molecules. Cations are not bases in terms of the definition of a base as a proton acceptor. Because they are positively charged, cations must in fact repel protons and not unite with them. Cations (e.g. Na$^+$) will not lose protons either; so the metallic cations are, in fact, not acids nor bases. They do neutralise the charge upon their associated anions, but have no effect on the alkalinity of solutions.

A solution of sodium hydroxide in water contains equal amounts of Na$^+$ ion (as above—not basic nor acidic) and of the strong base OH$^-$. The H$^+$ ions disappear as follows:

$$H^+ + OH^- \rightarrow H_2O$$

Thus the solution is very alkaline because it contains a much lower concentration of H$^+$ ions than does water.

BUFFER ACTION

Most intracellular metabolic processes, for example those involving decarboxylation of keto acids producing CO$_2$, or those involving the oxidation of P and S from dietary fat and protein producing H$^+$ ions, will tend to markedly change tissue fluid H$^+$ ion concentration. Body fluids contain various mechanisms to counter this tendency and to keep the pH constant in the face of added alkali or acid. These are called *buffers*.

Titration curves

If we titrate a weak acid, such as carbonic acid, with a strong base, such as sodium hydroxide, and plot the pH of the solution against the number of moles of base added, we get a curve such as that shown in Fig. 25.1. Clearly, the steeper is the titration

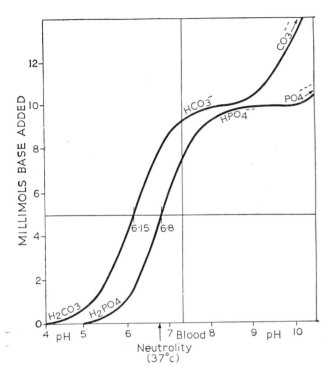

Fig. 25.1 Titration curves of weak acids.
H_2CO_3 and NaH_2PO_4 in 10 mM solution are titrated with NaOH. The value of pK' is given by the pH at which the acid is half titrated. This value varies with the nature and concentration of the other ions in the solution; the curves are drawn approximately for the conditions in plasma. The vertical line is drawn at the pH of blood. (After L. J. Henderson.)

curve, the more base (or, of course, acid) must be added in order to produce a given change in pH, and the more strongly is the solution said to be *buffered*. The salt of the weak acid is very nearly completely ionised, while the free acid is almost completely un-ionised; we can thus regard the variation with pH of the amount of base added, as equivalent to the variation with pH of the degree of ionisation of the acid. The maximum slope, and the strongest buffer action, occurs when the weak acid is exactly half titrated, i.e. when the hydrogen ion concentration is equal to the apparent dissociation constant of the weak acid. In practice, the region of hydrogen ion concentration over which the buffering power is reasonably large is taken to be ten times greater to ten times less than the value of the dissociation constant. Alternatively, we can say that the pH must be within one unit on either side of the value of pK'. The buffering power of a given solution is also, of course, directly proportional to the concentration of those buffer substances whose pK' is within one unit of the pH of the solution. Exactly similar arguments can be used in connection with weak bases, when titrated with strong acids; they also act as buffers, but are not met with in any appreciable concentration in solutions of physiological interest.

The Henderson–Hasselbalch equation

The shape of the titration curves may be deduced from the Law of Mass Action. If we have a weak acid HA which dissociates into H^+ and A^-, the following reaction will be in equilibrium:

$$HA \rightleftharpoons H^+ + A^-$$

whence

$$[H^+] = K_a \frac{[A^-]}{[HA]}$$

or

$$pH = pK_a + \log \frac{[HA]}{[A^-]}$$

This is the famous form of the equation expressed logarithmically by the German physiologist Hasselbalch from the original equation first adapted for physiological use by the American, Henderson, in 1909.

The square brackets denote concentrations. These equations indicate that so long as the H^+ ion concentration is of the same order of magnitude as the dissociation constant K_a, the greater is the H^+ concentration, the less is the acid ionised.

From it, we can determine the pH of any system, provided that the dissociation of the acid–base pair is known, and that the molar ratio of the acid to the salt in the buffer solution can be either specified or determined.

We can also write the equation

$$pH = pK_a + \log \frac{[salt]}{[acid]}$$

or

$$pH = pK_a + \log \frac{[base]}{[acid]}$$

or

$$pH = pK_a + \log \frac{[proton\ acceptor]}{[proton\ donor]}$$

PHYSIOLOGICAL BUFFER SYSTEMS

Buffer solutions, as we have seen, mop up excess protons or provide protons if needed, without allowing any great changes in pH to take place. They consist of weak acids and their conjugate bases, the principal examples in biological systems being as follows.

1. The carbonic acid-bicarbonate system
2. The phosphoric acid system
3. Ammonia.

The bicarbonate system

The concentration of carbonic acid, $[CO_2]$, in water is determined by the partial pressure, PCO_2, of carbon dioxide, as a gas, in equilibrium with it (Henry's Law of Solution):

$$[CO_2] = \alpha PCO_2$$

(where α is the solubility coefficient of CO_2). CO_2 dissolved in water (or plasma) slowly becomes hydrated to form carbonic acid

$$CO_2 + H_2O \rightleftharpoons H_2CO_3$$

In red cells and kidney tubule cells, an enzyme, carbonic anhydrase, catalyses the reaction and its rate is greatly increased. H_2CO_3 is a weak acid; therefore, it is weakly dissociated to H^+ and its conjugate base.

$$H_2CO_3 \rightleftharpoons H^+ + HCO_3^-$$

Thus P_{CO_2} determines the concentration of H_2CO_3.
The equilibrium point is well to the left in this equation.

The bicarbonate buffer system is a solution of a highly dissociated bicarbonate salt in equilibrium with CO_2 as a gas, the P_{CO_2} determining the concentration of H_2CO_3; the bicarbonate concentration depending on the salt. The Henderson–Hasselbalch equation gives us the pH of this buffer:

$$pH = pK + \log \frac{[HCO_3^-]}{[H_2CO_3]}$$

or

$$pH = pK + \log \frac{[HCO_3^-]}{\alpha \times P_{CO_2}}$$

The P_{CO_2} in normal arterial blood is 40 mmHg (5·32 kPa); the solubility coefficient α (plasma at 37°C) is 0·0301 mmol l^{-1}. Thus at equilibrium, the amount of dissolved CO_2 is about 1·3 m mol l^{-1}. A litre of plasma contains approximately 25 mmol HCO_3^-. The ratio of HCO_3^- to H_2CO_3 is, therefore, 20:1.*
The pK of carbonic acid is 6·1 at body temperature.
Substituting

$$pH = 6·1 + \log \frac{20}{1}$$
$$= 7·4$$

This value of pH = 7·4 is the normal pH of blood.

The bicarbonate system is the major buffer in extracellular fluids and its control by the kidney and the lungs is of extreme importance in homeostasis.

In clinical investigation, the detection of acid–base abnormalities may easily be carried out by determining the CO_2 content of arterial blood.

As a rule, arterial CO_2 is measured. Knowing the haematocrit (proportion of cells to plasma) the total CO_2 content of plasma can be found using a nomogram. Since the Henderson–Hasselbalch equation requires knowledge of plasma $[HCO_3^-]$ and $[CO_2]$ separately, there is a problem if we only know total CO_2. It is overcome by substituting in the numerator, "total $CO_2 - [CO_2]$", which is $[HCO_3^-]$. Since $[CO_2] = \alpha \times P_{CO_2}$, the equation becomes

$$pH = pK + \log \frac{\text{total } CO_2 - \alpha P_{CO_2}}{\alpha P_{CO_2}}$$

Arterial blood P_{CO_2} can simply be measured using the CO_2 electrode of Severinghaus.

The phosphoric acid buffer system

In physicochemical terms, the phosphate buffer system should be more effective than the bicarbonate buffer over physiological values of pH. This is because its pK is 6·8 (i.e. nearer the centre of the physiological range in extracellular fluid). In fact there is too little phosphate in blood or tissue fluid for it to play a large part in the regulation of the pH. It is, however, an important mechanism in the kidney.

It operates as follows: the buffer consists of two elements, NaH_2PO_4 and Na_2HPO_4. A strong acid (e.g. hydrochloric) added to an aqueous solution of the two substances gives the reaction

$$HCl + Na_2HPO_4 \rightarrow NaH_2PO_4 + NaCl$$

The operation results in decomposition of the hydrochloric acid, in its place an extra amount of NaH_2PO_4 appearing. Now NaH_2PO_4 is only weakly acidic so a strong acid has been exchanged for a weak one. Thus the pH changes relatively little in the face of added acid.

A strong base provokes the converse reaction when added to the buffer pair:

$$NaOH + NaH_2PO_4 \rightarrow Na_2HPO_4 + H_2O$$

This time NaOH is exchanged for Na_2HPO_4. The latter is a weak base compared with NaOH and again only a small change in pH results.

The phosphate buffer is important in intracellular fluid because its concentration is a good deal higher inside cells than outside. Also the pH inside cells is closer to the pK of the phosphate buffer system than is the pH of extracellular fluid.

The protein buffers

The proteins of the plasma, interstitial fluid and within the cells are the largest buffers in the body. They are buffers because they possess groups which are capable of acting as proton donors or acceptors. These groups, shown in Table 25.2, are amino groups, carboxyl groups, phenolic groups etc.

Table 25.2 *Proton acceptor and donor groupings in proteins*

Conjugate acid	Conjugate base
R—COOH	R—COO$^-$
R—NH$_3^+$	R—NH$_2$
R—OH	R—O$^-$
(guanidinium group)	(guanidino group)
(imidazolium group)	(imidazole group)

Protein molecules are composed of many amino acids and therefore contain very many acidic and basic groups. These will have differing pK's and the whole protein therefore acts as an efficient buffer over quite a wide range of pH.

Ammonia as a buffer

This mechanism for maintaining pH is in a different category from the others. It takes place in the kidney tubules and will be described in detail later. When hydrogen ions appear in tubular fluid, ammonia is formed and then combines as follows in the lumen:

$$NH_3 + H^+ \rightleftharpoons NH_4^+$$
$$\text{Ammonia} \quad \text{Ammonium}$$

Ammonium is an expendable cation; it replaces Na$^+$ which is absorbed.

Relative power of these buffers

Taking into consideration both the pK and concentrations of the buffers in *blood* their buffering power would be as in Table 25.3.

---

* This fraction is termed the "buffer ratio".

Table 25.3 *The relative contribution to buffering of acids*
The number of mmol H^+ required to lower the pH of the amount of each buffer contained in 1 litre of blood from pH = 7·4 to pH = 7·0

Buffer	mmol H^+ ions
Bicarbonate	18
Phosphate	0·3
Plasma protein	1·7 mEq
Haemoglobin	1·7

THE BUFFERS IN BLOOD

The buffer substances present in blood are (a) bicarbonate, (b) haemoglobin, (c) the plasma proteins and (d) phosphates. From Fig. 25.1 we see that the bicarbonate system is not, by itself, a very effective buffer at the normal pH of blood—its pK value is 1·2 units from the pH of blood. It is present, however, in relatively high concentration, so that its effect is quite appreciable. The Na_2HPO_4–NaH_2PO_4 system, on the contrary, is effective as far as buffer power is concerned, but the concentration of phosphates is so low that they play little part in buffering the blood. In the urine, phosphates provide the greater part of the buffering in normal circumstances. The most important buffering substance in the blood is undoubtedly haemoglobin; this, in the corpuscles, exists as a mixture of the potassium salt KHb, which may be regarded as ionised into K^+ and Hb^-, and the free acid HHb, which is un-ionised. Haemoglobin is a complex polybasic acid, and its titration curve consists of a large number of the S-shaped curves of Fig. 25.1 on top of one another, and overlapping. The general conception of buffer action is nevertheless still applicable. The plasma proteins behave in a similar way, but are present in much smaller concentration. If, then, we add a small quantity of an acid to some blood, the hydrogen ion concentration will be increased. This, however, will lead to a reduction in the ionisation of both haemoglobin and carbonic acid, and many of the extra hydrogen ions will be absorbed and tucked away in the undissociated acids. Conversely, if an alkali be added, the extra hydroxyl ions will combine with hydrogen to form water, and the effect will be exactly the opposite of that which occurs when acid is added. Haemoglobin and carbonic acid will dissociate more completely, and provide extra hydrogen ions to replace those which were removed by the alkali. All this is, of course, but a verbal description of the fact illustrated in the titration curves, that in a buffered system the pH changes only slowly with the addition of acid or alkali.

The efficiency of the buffering process in the blood is indicated by the following calculation. If we take blood at a pH 7·4 and add acid until it has a pH 7·2, the increase in free hydrogen ion concentration is

$$10^{-7.2} - 10^{-7.4}, \text{ or } 2\cdot 3 \times 10^{-8} \text{ mol l}^{-1}.$$

During the course of this change in pH, however, the haemoglobin absorbs, by buffer action, no less than 660 000 × 10^{-8} mol l^{-1}, while the carbonic acid absorbs 70 000 × 10^{-8} mol l^{-1}. The total amount of hydrogen ions which must be added to produce this change in pH is thus 730 000 × 10^{-8} per litre of blood, of which only 2·3 × 10^{-8} remain free. If we remove hydrogen ions from normal blood (or add hydroxyl ions) until the pH is 7·6, the decrease in free hydrogen ion concentration is 1·5 × 10^{-8} mol l^{-1}, while haemoglobin releases again 660 000 × 10^{-8} mol l^{-1} and carbonic acid releases 44 000 × 10^{-8} mol l^{-1}. These figures indicate the relative importance of carbonic acid and haemoglobin in buffering the blood. Roughly, 90 per cent of the hydrogen ions added or removed are absorbed or released by haemoglobin, and 10 per cent by the carbonic acid–bicarbonate system.

The above calculations are based on the following equations:
1. The buffering power of haemoglobin is represented by the empirical equation

$$[HbO_2^-] = 0\cdot 22 [HbO_2] \text{ (pH = 6·60)}$$

HbO_2^- is expressed in millimoles per litre and HbO_2 in grams per litre.

2. The buffering power of the carbonic acid–bicarbonate system is represented by the Henderson–Hasselbalch equation

$$\text{pH} = 6\cdot 12 + \log [HCO_3^-] - \log [H_2CO_3]$$

The values taken are:

$$[HbO_2] = 150 \text{ g l}^{-1}$$

$$\text{Total } CO_2 = [H_2CO_3][HCO_3^-]$$

$$= 24\cdot 8 \text{ mmol l}^{-1}$$

$$\text{Total base} = [HCO_3^-] + [HbO_2^-] \text{ at pH 7·4}$$

$$= 50 \text{ mmol l}^{-1}$$

In this example, we have imagined that no carbon dioxide is lost or gained by the blood when the pH is changed. This approximates to the conditions in the capillaries when some acid product of metabolism is formed in the tissues and diffuses into the blood stream. If we consider the body as a whole, however, the conditions are different, since carbon dioxide can be blown off in the lungs and acid or base can be excreted by the kidneys. As we shall see in later chapters, the respiratory centre and the kidneys adjust the ventilation rate and pH of the urine, respectively, in such a way as to keep the pH of the blood constant. The buffering is thus perfect. Confining ourselves for the moment to the actions of the respiratory centre, we see that the buffering results, in the end, entirely from the increased loss, or retention, of carbon dioxide in the lungs. The extra hydrogen ions are entirely absorbed, or released, by the bicarbonate system, and haemoglobin, in the long run, plays no part. There are thus three lines of defence against changes in hydrogen ion concentration: (1) direct buffering in the blood, mainly by haemoglobin; (2) indirect buffering by the action of the respiratory centre in controlling the free carbon dioxide concentration in the blood; and (3) indirect buffering by the kidneys, which more slowly excrete the excess acid or base, as the case may be.

26. The Physiological Control of Reaction

Buffer systems, by themselves, will only minimise shifts in pH in the face of acid metabolites added to extracellular or intracellular fluids. If we require to keep the pH constant, or in other words to control it, some form of negative feedback mechanism must be added to the system.

Bicarbonate is the principal buffer in extracellular fluid, so it should be clear that two possible forms of controlling the reaction of this fluid are available. One is to control the bicarbonate concentration; the other is to control the partial pressure of carbon dioxide. Both these controls do indeed operate in the body.

Bicarbonate control

The kidneys regulate bicarbonate concentration. Tubular epithelium controls the reabsorption of cations and anions from the glomerular filtrate. It normally ensures that there is an excess of cations in the plasma. There is a large source of carbon dioxide in metabolic reactions, which then provides the balance of anions to keep the fluids electrochemically neutral.

$P{CO}_2$ control

The partial pressure of carbon dioxide is kept at a steady level by the respiratory centre. The latter determines the rate at which CO_2 is removed from the body by the lungs relative to the rate at which the body cells are producing it. Normally, $P{CO}_2$ is kept near to 40 mmHg (5·32 kPa).

THE RESPIRATORY CONTROL OF REACTION

As we shall see later, the functions of the respiratory system are to take in oxygen and to eliminate carbon dioxide, the latter being removed from the body in quantities adjusted to keep the partial pressure in blood constant. The ventilation rate is controlled by several factors, an important one being the $P{CO}_2$ in arterial blood perfusing the chemoreceptors and the respiratory centre itself. Raised $P{CO}_2$ stimulates respiration; decreased $P{CO}_2$ leads to a lowered ventilation rate. This is a sensitive feedback control for, as will be explained in the chapter on control of respiration (Chapter 45), an increase in arterial $P{CO}_2$ of only 10 mmHg (1·33 kPa) above the normal of 40 mmHg (5·32 kPa) will give a fourfold increase in alveolar ventilation rate.

With a normal concentration of bicarbonate in the plasma, such a respiratory feedback will keep the pH of the plasma constant (simply by adjusting the $P{CO}_2$ to 40 mmHg all the time). If the bicarbonate level does alter (by the entry of acids or alkalis to the blood), the pH will then vary from 7·4. Now, to keep the pH of blood at 7·4 the respiratory feedback control must have its set-point changed. Indeed, as pointed out in the last chapter, the normal buffer ratio must be restored, by altering the $P{CO}_2$ from 40 mmHg, so that the fraction,

$$\frac{[HCO_3^-]}{[H_2CO_3]} \text{ or } \frac{[HCO_3^-]}{\alpha P{CO}_2} = \frac{20}{1}$$

These adjustments do occur; the respiratory centre is sensitive to pH (as well as $P{CO}_2$) of arterial blood. Pulmonary ventilation is doubled for a fall in pH from 7·4 to 7·3 and halved for a rise from pH 7·4 to 7·5.

This control system can easily be shown in operation by ingesting small amounts of either acids or alkalis. The percentage of CO_2 in alveolar air samples following a dose of sodium bicarbonate, for instance, can easily be shown to increase. If larger amounts of such salts are given, the increased ventilation may be obvious without the use of any measuring devices.

For more details regarding the working of the respiratory control systems, the reader should consult Chapters 45 and 46.

DISTURBANCES OF RESPIRATORY CONTROL

The respiratory centre has a number of functions quite apart from, and indeed often opposed to, that of maintaining blood pH through its effect upon $P{CO}_2$ and the buffer ratio of 20:1. Should these other functions gain the upper hand, disturbances of acid–base regulation may result.

Hypoventilation

Normally, the term hypoventilation would imply a reduced rate of alveolar ventilation, but what level of the latter are we to take as normal? A better definition might be one which included the concept of an impairment of CO_2 elimination (and/or O_2 intake), with a consequent rise in alveolar and arterial $P{CO}_2$. The causes include respiratory obstruction, paralysis and depression of the respiratory centre.

The effects, as one would predict, are a rise in concentration of carbonic acid leading to a fall in pH of the plasma. In extreme cases arterial $P{CO}_2$ may reach 200 mmHg (26·6 kPa) and plasma pH 6·8.

Hyperventilation

Again, we may look upon hyperventilation as a state where alveolar ventilation is continuing at a rate in excess of that required to remove the CO_2 being produced metabolically. Note that the increased rate and volume of breathing due to exercise may sometimes be included in this definition since, although usually $P{CO}_2$ remains constant, it may be diminished in exercise, particularly if this is of short duration and is strenuous.

Voluntary hyperventilation, as in preparing to swim under water for experiments, or in connection with hysterical and emotional upsets, is one cause. High altitudes also induce hyperventilation (see Chapter 46).

Carbon dioxide is removed from the body and hence the

plasma becomes more alkaline. Overbreathing may reduce $P\text{CO}_2$ to as low as 20 to 25 mmHg (2·66 to 3·33 kPa). Blood pH might rise to 7·8 in these circumstances.

THE RENAL CONTROL OF REACTION

The kidney can regulate the loss of hydrogen ions from the body within rather wide limits. The concentration of hydrogen ions in the urine can, if necessary, be raised to 2500 times that in plasma, or it can be decreased to a quarter of the plasma concentration. This represents a change of from pH 4·0 to 8·0. Here, then, we have a feedback mechanism for further control of the reaction of the body.

Production of acid in the body

A mixed diet will give rise to anything up to 100 mmol of H^+ ions per day. As we have seen, these added H^+ ions are buffered in the various body fluids. The conjugate acids formed by the buffer action temporarily remain in the body; carbonic acid, formed in the bicarbonate buffer action, turns into CO_2 which is removed by the lungs. The buffer pairs are re-formed by the excretion in the urine of a quantity of protons equivalent to that produced metabolically.

The kidney regulates body fluid reaction in two ways:
1. It controls the rate of reabsorption by the tubules of bicarbonate which has been filtered out by the glomeruli.
2. It excretes hydrogen ions.

Interaction between buffers in the body

Alterations in the ratio between conjugate acid and conjugate base of any one buffer system in a body fluid, will lead to changes in all the others. Thus extracellular bicarbonate control by the kidney, besides bringing its buffer ratio back to 20:1 (see p. 154), will also have the same effect on all the other buffer systems.

The reabsorption of bicarbonate in the tubules

The reabsorption of bicarbonate (and hence *acidification* of urine) is due to an ion-exchange mechanism in the tubules, whereby H^+ ions produced in tubular cells are exchanged for Na^+ ions in the passing glomerular filtrate.

The details of this mechanism, shown in Fig. 26.1, are that protons (probably from the reaction $CO_2 + H_2O \rightleftharpoons H_2CO_3 \rightleftharpoons HCO_3^- + H^+$, catalysed by carbonic anhydrase) are secreted into tubular fluid. There, they exchange for Na^+ ions which go back into the blood stream, together with the HCO_3^- ions.

The H^+ ions are mopped up by buffers in the tubular fluid, e.g. HCO_3^-, HPO_4^{2-}, creatinine etc. and ammonia (which is manufactured by the kidney).

The end result of this system is the reabsorption of sodium bicarbonate. At low plasma bicarbonate concentrations (10–25 mmol l^{-1}) all the filtered bicarbonate is reabsorbed; the final urine contains more. Above about 30 mmol l^{-1}, bicarbonate appears in the urine in increasing amounts. In other words $T_m HCO_3^- = 28$ mmol l^{-1} glomerular filtrate (see Fig. 26.2).

Formation of an acid urine

As already mentioned, the H^+ ions excreted into the tubular fluid are taken up by HPO_4^{2-} (to form $H_2PO_4^-$), creatinine and small amounts by hydroxybutyrate ions. The most acidic that urine can become is pH 4·0. Obviously the number of H^+ ions which can be dealt with in this way depends on the total amount of base which is in the tubular fluid in the first place, i.e. is filtered.

TITRATABLE ACIDITY
One can easily determine the extent to which this buffering process is going on by measuring the *titratable acidity*. This is the amount of strong base which must be added to the urine to bring its pH back to 7·4, i.e. the plasma pH from which it was filtered. Normal values of titratable acidity would lie between 20 and 40 mmol H^+ per 24 hours.

The renal ammonia system

The other method whereby the kidney transports H^+ ions from blood to urine is by combining them with ammonia. The reader

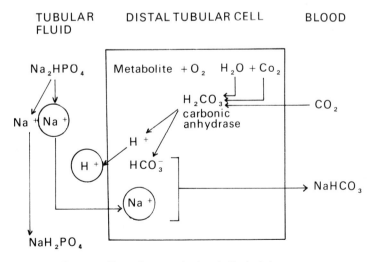

Fig. 26.1 Summary of ion exchange mechanisms in distal tubule.
H^+ secretion system is shown in operation with phosphate buffers in urine. The net effect is the reabsorption of $NaHCO_3$ (as shown at right). In man, the *complete transaction* requires the secretion of around 3500 mEq H^+/24 h.
Note that for each hydrogen ion secreted into tubular fluid, one bicarbonate ion appears in the plasma.

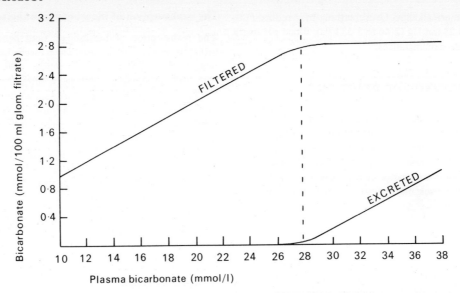

Fig. 26.2 The rate of reabsorption of bicarbonate ions in man and the plasma bicarbonate concentration.

might think that this is a fairly drastic procedure; after all, both acids and ammonia are used by bank robbers and other malefactors to throw in the faces of their pursuers, causing injuries and often disfigurement. The secret, of course, is that both are present together in tubular fluid, although this latter may reach a pH of 4·0 which says a great deal for the acid-resistant properties of the lining of the urinary tract.

The epithelial cells of the proximal tubules, distal tubules and the collecting ducts synthesise ammonia which diffuses out into the lumen. It then reacts with hydrogen ions as shown in Fig. 26.3.

The actual removal of H^+ ions sounds as if it might be quite a problem, if one considers that the major anion in urine is chloride; HCl is a very strong acid and not many H^+ ions could be transported before the pH falls below 4·0 (below which critical value, H^+ ion secretion would stop). How is this problem overcome?

When hydrogen ions combine with ammonia, ammonium ions are formed which then combine with the chloride to give ammonium chloride. The pH does not then fall much because, of course, ammonium chloride is a neutral salt.

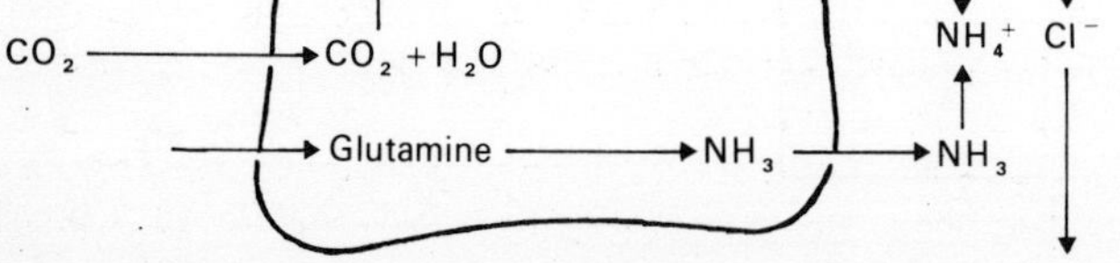

SOURCE OF THE AMMONIA

Roughly 60 per cent of the tubular NH_3 is made from glutamine; the rest from other amino acids. The kidney contains an enzyme, *glutaminase I*, capable of hydrolysing glutamine to glutamic acid and ammonia.

$$\begin{array}{c} NH_2 \\ | \\ COOH\cdot CH \\ | \\ (CH_2)_2 \\ | \\ CONH_2 \end{array} + H_2O \rightarrow \begin{array}{c} NH_2 \\ | \\ COOH\cdot CH \\ | \\ (CH_2)_2 \\ | \\ COOH \end{array} + NH_3$$

Glutamine   Glutamic acid

Proof that the source of kidney NH_3 is indeed the plasma glutamine has been obtained using labelled glutamine containing ^{15}N, infused into a renal artery. Up to 50 per cent of urinary ammonia contained the ^{15}N.

Rate of H^+ ion secretion by the kidney

A rise in arterial PCO_2 (as in respiratory disease) leads to a higher level of plasma bicarbonate occurring. More CO_2 in blood leads to increased bicarbonate reabsorption in the tubules. Less CO_2, likewise, gives a lowered bicarbonate reabsorption. The reason is thought to be that CO_2 diffuses rather rapidly through cell membranes and that the intracellular pH must be altered as a result. Thus the rate of secretion of H^+ ions depends on intracellular pH and indirectly on plasma PCO_2.

ABNORMALITIES OF KIDNEY CONTROL

All other things being equal (which naturally, according to Professor Sodde's Tenth Law, they never are) there are three main factors which lead to abnormality in the function of the control mechanisms just described. These are:

1. Plasma chloride ion concentration
2. Plasma potassium levels
3. Aldosterone secretion.

Fig. 26.3 Secretion of ammonia by tubules, and the reaction between ammonia and hydrogen ions in the lumen.

$$NH_3 + H^+ \rightarrow NH_4^+$$

The ammonia is formed from glutamine, amino acids, etc. The net result of these operations is to increase the sodium bicarbonate concentration in the plasma.

Low plasma chloride levels cause alkalosis

Lowered plasma chloride concentrations are known to cause alkalosis. For once, it is possible correctly to forecast how this will come about, by considering first principles.

The first result of lowered plasma chloride will be a reduction in the number of chloride ions appearing in a given volume of glomerular filtrate. Thus there are fewer of them to be reabsorbed with sodium ions. Hence more hydrogen ions are secreted in exchange for sodium ions.

Now for every hydrogen ion appearing in the tubular fluid, one bicarbonate ion is restored to the plasma (see legend to Fig. 26.1). This shifts the pH of the body fluids towards the alkaline. High chloride levels in plasma, naturally, will give an opposite effect.

High plasma potassium levels cause acidosis

Potassium ions are treated much like hydrogen ions by the tubule cells, and are exchanged similarly for sodium ions. With this in mind, we would expect a high tubular secretion of potassium (which will occur if plasma potassium is raised) to compete with the hydrogen ions in the sodium exchange mechanism. Brief consideration will lead to the conclusion that the hydrogen ion secretion would thereby be decreased. Hence more bicarbonate will get into the urine from plasma and the end result will be an acidosis, or the tendency for an acid-shift of pH in the body fluids.

Excess aldosterone secretion causes alkalosis

Aldosterone is a hormone provoking the reabsorption of sodium by the kidney. This is part of its function in maintaining body electrolyte balance (see Chapter 32). Excess of this hormone gives rise to increased reabsorption of sodium, so that more is present to combine with hydrogen ions. Thus hydrogen ion secretion is increased above normal. As before, this means that bicarbonate return into the plasma is proportionately increased. This means that an alkalosis occurs. Deficiency of aldosterone leads to acidosis.

CLINICAL ABNORMALITY IN THE CONTROL OF REACTION OF BODY FLUIDS

Disturbances in the control of body reaction due to changes in plasma bicarbonate concentration are called *metabolic acidosis* and *metabolic alkalosis*. Those in which changes of carbon dioxide partial pressure are primary, are termed *respiratory acidosis* and *respiratory alkalosis*.

Metabolic acidosis

There are two groups of causes:
1. The addition of acid or the removal of alkali at rates too great for the kidney to keep bicarbonate normal.
2. Kidney impairment, so that normal acid production or base loss cannot be compensated.

In *diabetes mellitus* fat metabolism is abnormal (fat burns in the fires of carbohydrate) with the production of large amounts of β-hydroxybutyric acid and acetoacetic acids. Often dehydration is also present and this impairs the ability of the kidney to remove acid. Ketosis (ketone bodies in extracellular fluid and plasma) also occurs in starvation. Poisoning with substances such as methyl alcohol, salicylates and other acidic radicals also gives metabolic acidosis.

The plasma bicarbonate may also decrease following loss of base from the body. For example, pancreatic and biliary fluids contain high concentrations of bicarbonate.

Impaired ability of the kidney to excrete acid may be a specific defect, or may be part of a general renal failure. The former is termed *renal tubular acidosis*; there is also a syndrome of complex tubular transport defects called the *Fanconi syndrome*.

Carbonic anhydrase inhibitors will obviously seriously impair the hydrogen ion secreting mechanism in renal tubular cells.

Metabolic acidosis will result from implantation of the ureters into the gastrointestinal tract, much to the consternation of the early G.U. surgeons who thought such transplantations might be a good idea in patients with resected bladders, blocked urethras etc. The reason is that the acidity of urine contained in the gut for any length of time disappears and the ammonia is reabsorbed, thereby undoing all the good work done by the tubules.

Metabolic alkalosis

As a primary increase in plasma bicarbonate, metabolic alkalosis is produced by (a) excessive intake of alkali or by (b) abnormal loss of acid.

Persistent alkalosis, severe enough to need treatment, almost invariably occurs when renal function is impaired by dehydration, or when body potassium is depleted. This is because the kidney has an enormous capacity to remove excess bicarbonate (or salts of organic acids, such as acetate, lactate, citrate, which are oxidised to CO_2 in the body). Very large amounts of these must be added before alkalosis ensues.

On the other hand, loss of gastric secretion from the body (persistent vomiting as in *pyloric stenosis*) gives rise to much more significant disturbance of acid–base control. This is because the kidney's power of removing bicarbonate (to compensate for the acid loss—i.e. HCl) is lessened by the dehydration and also because salt intake is usually inadequate in these patients. If the salt intake is not adequate, the bicarbonate excess can not be removed without a drastic reduction in extracellular volume.

As already mentioned, potassium depletion leads to the so-called "hypokalaemic alkalosis". This condition is also due to excessive losses of gastrointestinal secretions.

Respiratory acidosis

In the practice of medicine, it is not always the most obvious and apparently logical mode of treatment that must be employed. A full understanding of the underlying physiology is of the utmost importance in deciding how to treat the patient and his illness.

For a long while it was thought that chronic pulmonary insufficiency should be treated by giving oxygen. After all, it would be logical to suppose that one would treat a condition in which oxygen supply to the body is lacking by administering gas mixtures containing a high proportion of oxygen.

Not before many patients were unnecessarily removed from this world by the action of their medical attendants, was it realised that giving oxygen to those with chronic pulmonary disease is often lethal. In chronic underventilation of the alveoli, the respiratory centre loses its sensitivity to carbon dioxide. Respiration is then mainly driven by the response of the chemoreceptors to anoxia. Thus, when this anoxia is removed (by oxygen tent), the underventilation becomes extreme and severe respiratory acidosis follows. If oxygen administration is continued for any length of time, carbon dioxide narcosis and death ensue.

Respiratory alkalosis

Alkalosis is produced by hyperventilation as described on page 156. The causes are emotional disturbances such as hysteria and somewhat rare disorders of the respiratory centre such as irritative lesions and salicylate poisoning.

Symptomatology and treatment of these disorders of acid–base regulation are not within the scope of this discussion. The interested reader is referred to the bibliography.

27. Body Temperature and its Control

All living matter metabolises; heat is the invariable byproduct of the relatively inefficient life-mechanisms being used. This heat has to be dissipated to the environment somehow, at the same average rate as it is produced, if body temperature is to stay constant. Homeothermous animals, such as mammals and birds, have evolved a feedback control system to achieve this constancy of body temperature. The poikilothermous animals have not, and their body temperature varies with that of their environment. Presumably, a constant body temperature has a survival advantage in that enzymic processes, cell metabolism etc. are not varying all the time with the external weather conditions.

Constancy of the internal environment

The servocontrol system in the warm-blooded animal is designed to maintain a constancy of body temperature in the face of considerable variations in ambient conditions and in the face of large fluctuations of metabolic rate. This regulation is brought about by the active variation of the heat loss characteristics of the system and, to a lesser extent, by an active control of its heat production. One result of the operation of this system is that, as environmental temperatures fall, body metabolism rises—just the opposite to what happens in a cold-blooded animal.

In the human, normal deep body temperature is kept remarkably constant at between 36·5°C (98·4°F) to 37°C. The description of the general properties required by a mammalian temperature controlling system should have by now convinced the reader that a negative feedback servo-mechanism is necessary to do the controlling.

Heat balance in the body

The body is a mass; it has a corresponding heat capacity (i.e. it contains stored thermal energy proportional to its temperature). In temperature equilibrium, heat must be steadily dissipated at a rate equal to the metabolic production of heat. A mismatch of these two rates leads to a rise or fall of body temperature until a new equilibrium is reached. See Fig. 27.1 for a summary of these points.

The processes by which the body temperature is prevented from rising or falling unduly are of two kinds. First, the rate at which heat is lost from the skin is increased or decreased as necessary ("physical" regulation). In man, this is done partly by wearing appropriate clothes but, as in all warm-blooded animals, it is also done automatically. Like any other hot object, the body cools at a rate which depends on the difference between the temperature of its surface and the temperature of the air and other surroundings. It cools more rapidly, also, if it is wet than if it is dry, heat being absorbed by the evaporation of water, and the rate of cooling depends on the difference between the wetness of the body and the wetness (humidity) of the air.

HEAT GAIN	HEAT LOSS
MUSCLE ACTION Exercise Postural Shivering	SWEATING Lungs, panting Insensible perspiration CLOTHING (adjustment) SKIN CIRCULATION Vasodilation. Shunts
SPECIFIC DYNAMIC ACTION OF FOOD	
BASAL HEAT PRODUCTION (from carbohydrate, protein, fat)	BASAL HEAT LOSS
FROM HOT ENVIRONMENT Radiation Conduction Convection	TO COOLER ENVIRONMENT Convection, wind Conduction Radiation

Σ HEAT GAIN $\equiv \Sigma$ HEAT LOSS
(when body temperature is stable, normally at 37° C)

Fig. 27.1 Balance between biological and climatic factors causing heat gain and heat loss.

Physical regulation, accordingly, is exerted in two ways: (a) by raising or lowering the skin temperature through control of the *rate of blood flow through the skin* and thus the rate at which heat is brought out to the surface of the body; and (b) by increasing the wetness of the skin, and thus the rate of cooling, by the *secretion of sweat*. Secondly, if the body temperature falls, in spite of all that can be done to reduce the loss of heat, the metabolic rate is increased, either deliberately by taking muscular exercise, or involuntarily by *shivering* ("chemical" regulation). The temperature of the skin, therefore, and at times that of a large part of the limbs, is subject to considerable variation. What is ordinarily referred to as the "body temperature", which remains nearly constant, is that of the "core", the internal organs of the abdomen, the heart, the brain and the blood within these organs, although they are not all necessarily at exactly the same temperature. The temperature of a thermometer placed in the rectum, or under the tongue, is usually taken as a measure of the body temperature.

THE EFFECTOR MECHANISMS IN TEMPERATURE CONTROL

Before considering the details of the feedback system, we will briefly consider the mechanisms available to the body for (a) altering the *heat production* of the body and for (b) altering the amount of *heat lost* by the body.

Heat production

Heat is produced by body metabolism. In certain organs, e.g. the liver and the heart, this heat production is relatively constant. Skeletal muscle, on the other hand, produces variable amounts of heat; much during exercise, little when at rest.

Basal heat production by the normal human is 0.15–0.17 MJ h^{-1}. This is around 7 MJ day^{-1} in males and 5 MJ in females. Super-added is the heat production due to daily activity, which increases the total to around 10 to 13 MJ day^{-1}. Severe muscular work can raise this figure to 25 MJ day^{-1}.

Some idea of the physiological problems faced by the thermo-regulatory system can be gained by working out what would happen in the absence of any heat loss. The specific heat of the body is approximately 1.0, so a heat output of 4.2×10^3 J kg^{-1} h^{-1} (which is the basal level) will raise the body temperature by 1°C per hour. Normal activity might double or treble this figure.

The heat production by muscle is under control of the thermoregulatory centre and gross movements, or shivering, can occur to increase it substantially.

Heat loss

A hot body loses heat to its cooler surroundings by the physical process of radiation, conduction and convection. The homeothermous animal also loses heat by evaporation of sweat. These mechanisms are controlled by the hypothalamus in ways such as erection of hairs, donning clothes or removing (most of) them and by regulating sweat secretion.

Heat distribution

Heat is generated in muscle and liver. It must be lost from the skin. This entails a heat distribution system which is in fact the circulating blood, whose transfer of heat from the interior to the surface can be largely controlled by opening up or closing down certain vascular channels. Fat persons are at a marked disadvantage here; heat transfer is grossly impeded by a layer of insulating subcutaneous tissue.

The controlled process

The thermoregulatory process involves at least three functionally distinct systems.
1. *Body core*, consisting of all the viscera and CNS (but not skeleton and muscles). The basal heat production largely originates in this core and since it can be hormonally controlled, will to some extent constitute part of a control mechanism.
2. *Voluntary muscles*, surround the core and are of interest because, when the body cools, shivering occurs. Since antagonistic pairs contract simultaneously, no external work is done and all the energy input appears as heat. Note the economy in biological design here; muscle has two functions, one of which utilises its inherent inefficiency in converting energy input to useful work.
3. *Skin* is more than an external bag containing bones and muscles; it is a variable thermal insulator. This property of active control of thermal transmission characteristics, is due to blood flow control. Since blood contains heat generated by the core, transfer of this heat to the environment will depend on how much of the heated blood can reach the skin surface.

The information-flow diagram for human body temperature control is shown in Fig. 27.2.

THE FEEDBACK TRANSDUCERS

A similar control system to that figured in the diagram (Fig. 27.2) is utilised in house-heating. The thermostat is the reference input (set-point), feedback transducer and controller all in one. As we found in our analysis of the action of servo systems (Chapter 14), if a system such as this has a large heat capacity—i.e. we are attempting to install well-controlled heating for a large block of offices (or a Physiology Department building), there will be rather large and inconvenient time-lags in responding to external climatic variability. In such buildings, modern servocontrol systems customarily employ additional "feedforward" of

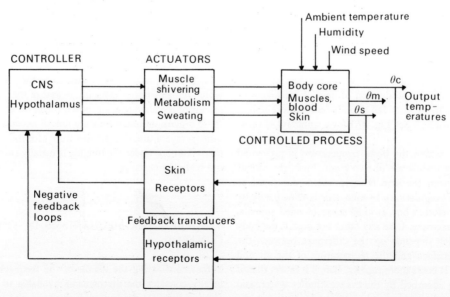

Fig. 27.2 Human thermoregulation. Simplistic information flow diagram. (After Milsum, J. H. (1966) *Biological Control Systems Analysis*. McGraw-Hill, New York.)

information (cf. the muscle servocontrol where feedforward is also used). An additional thermostat is placed outdoors, so that early warning of drastic external climatic changes is given to the controller.

If the outside temperature suddenly falls for example, heat production can immediately be increased in the furnace (to match the increasing heat loss) well in advance of any fall in internal temperature. The room thermostats in the building would have a much longer delay in their response.

Thus such a system reduces the loop delay and reduces oscillation, in fact acting much in the same way as would derivative negative feedback.

Nothing is new; biological servomechanisms were perfected well in advance of those of engineers. We would thus be disappointed if we did not find this kind of feedforward control in the body as well. Of course it does occur and two types of temperature transducer have been clearly found. There are central receptors, obviously analogous with the thermostat, and peripheral receptors in the skin to provide the feedforward.

Cutaneous thermoreceptors

There are two kinds of thermoreceptors. Cold receptors respond mainly to decreasing skin temperatures and it is of considerable interest to note that their maximum discharge rates are found when the temperature is falling rapidly. Their steady-state discharge is trifling compared to their dynamic one. Other receptors respond to a rise above normal body temperature, again being particularly responsive to rate of warming. The skin is more plentifully supplied with cold receptors than with warm ones. Some receptors can be found which combine the two effects, i.e. they will produce action potentials both when cooled and when warmed. For a more detailed account of temperature receptors see Chapter 54.

Hypothalamic thermoreceptors

These neurones live at close quarters to the controlling centre itself. Microelectrode studies reveal that in the hypothalamus are cells which would appear to be histologically just like any other cerebral neurones, but which respond to changes in temperature of arterial blood to the brain or to localised cooling or warming by "thermodes"*. There is some evidence that a localised "slow potential" is generated by temperature change in part of the anterior hypothalamus.

THE CONTROLLER AND COMPARATOR

Latest views on the controlling mechanisms for body temperature indicate that two regions are involved. There is a *heat maintenance* centre in the posterior hypothalamus and a *heat loss* centre in the anterior hypothalamus. The heat loss centre appears to utilise core temperature (mainly) to drive the actuating mechanisms for heat loss. This means that sweating and vasodilatation are actuated by deep body temperature. The heat maintenance centre controls muscle action, shivering, general metabolism and voluntary acts like putting on more warm clothing, on the basis of both skin and core-temperature inputs. There are feedforward pathways also which depend on a link between the voluntary motor cortex and the hypothalamus. For example, increased sweating can occur before any deep or superficial temperature changes can be measured if heavy work is started suddenly.

The Hypothalamic thermoregulatory centres

The results of changing the temperature of the blood flowing to the brain in the carotid arteries of an experimental animal are shown in Fig. 27.3; the changes in the temperature of the tongue give a qualitative indication of those in the temperature of the brain. A rise in the brain temperature is accompanied by a fall in the rectal temperature; and a fall in the brain temperature by a rise in the rectal temperature. These experiments indicate that there are structures in the brain which are sensitive to changes in the temperature of the blood and are capable of so altering the rate of heat loss from the body as a whole that the changes in the blood temperature would be counteracted. Transection of the brain stem at various levels in experimental animals has shown that co-ordinated temperature regulation ceases if the section is made posterior to the hypothalamus, so that the hypothalamus,

---

* A thermode is a probe of small dimensions that can be introduced into the brain substance and its tip can be cooled or warmed.

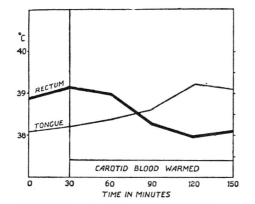
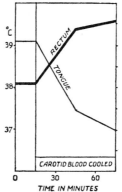

Fig. 27.3 The action of the temperature regulating centres.
The effect on the body temperature of a dog, as measured in the rectum, of warming and cooling the blood flowing in the carotid arteries to the brain. The changes in temperature of the tongue are an indication of those in the brain. Note that the temperature of the rectum moves in the opposite direction to that of the brain, indicating the efforts of the regulating centres to counteract the warming and cooling of the blood supplying it. (From data by Geiger.)

and structures anterior to it, have no connection with the rest of the body; the "thalamic" animal has control of its temperature, but the "decerebrate animal" has little or none. More detailed studies, by localised electrical stimulation, and by warming and cooling small regions of the brain, show that the temperature receptors and the regulating "centres" lie close together in the anterior hypothalamus. It is probable that the organisation of the receptors and regulating centres is essentially similar in man to that in the experimental animals studied.

The regulating centres will initiate responses leading to increased dissipation of heat, or to increased conservation or production of heat, only if they receive appropriate nerve impulses from temperature receptors. The body temperature cannot be regulated unless there is some change in temperature to initiate the regulation. The centres, however, receive nerve impulses from peripheral temperature receptors in the skin and the mucous membranes of the nose and mouth as well as from the central receptors. The temperature of the skin, as we shall see in more detail later, changes quite largely and rapidly in accordance with the temperature of the environment; the peripheral receptors, therefore, will "drive" the regulating centres appropriately even if there is no change in the temperature of the core. A man's body, moreover, has a considerable heat capacity, and appreciable amounts of heat must be gained or lost before there is a sufficient change in the temperature of the core to affect the regulating centres. Even if there were a sudden change in the rate of heat production, or in the temperature of the environment, the temperature of the central receptors would change only slowly, and the measures necessary to restore heat balance would be initiated gradually and after some delay. The peripheral receptors, however, will be affected more rapidly, and being very sensitive to changes in temperature, will immediately "drive" the regulating centres appropriately, thereby increasing the rapidity and precision with which heat balance is restored.

Normally, in a sedentary man, the temperature of the core varies from just below 36·5°C (98°F) to just below 37°C (99°F). Within these limits, there is a regular diurnal variation, the highest value being reached in the late afternoon, and the lowest in the early morning. In women, there may be a rhythmic variation with the menstrual cycle, the temperature being lower during the menstrual period and for some time afterwards. The body temperature rises and falls by about 1°C (2°F) in quite ordinary circumstances according to the temperature of the environment and the amount of clothing worn; on exposure to severe cold it has been known to fall as low as 22·7°C (73°F) with subsequent recovery. It rises during muscular exercise in proportion to the severity of the exercise and may reach 39·5°C (103°F).

The control limits

The amount of control which can be exerted over the rates of heat loss and of heat production, though large, are not unlimited. If the conditions are such that it is impossible to lose heat rapidly enough and the body temperature cannot be prevented from rising, the consequences are likely to be serious. The metabolic rate, like the rates of all chemical reactions, is increased by a rise in temperature. As the body temperature rises, the rate of production of heat will rise with it, and the temperature will rise even more rapidly. A vicious circle is established which will end in heat stroke and death. On the other hand, if in spite of all attempts to retain and produce heat the body temperature continues to fall, the temperature regulating centres, like other nervous structures, will become progressively less active. A time will come when the effort to maintain heat balance is abandoned; the person "basks in the cold" and, unless he is soon removed from the cold environment and warmed, he will die.

28. Vascular Mechanisms in Body Temperature Control

Blood flow through the skin

An ordinary man feels comfortably warm when the temperature of his skin lies roughly between 25° and 35°C, although there are large individual variations. If he were at rest, and had no clothes, the air temperature woud have to lie between about 20°C (68°F) and about 32°C (90°F), higher in dry windy conditions, when heat is carried away more rapidly, than in moist still conditions. In these "neutral" environments (neither hot nor cold) control of the body temperature is carried out entirely by control of the skin circulation. The excitation of the "40°" receptors in the skin falls progressively as the skin temperature falls, and that of the "30°" receptors rises reciprocally, whereas outside the "neutral" conditions, only one or other kind is excited.

If the rate of blood flow through the skin is large, heat is brought out rapidly from the core to the surface of the body, the surface temperature is well above that of the environment and the rate of heat dissipation is large. On the other hand, if no blood were to flow through the skin and superficial layers of the body, heat would reach the surface only by conduction through the layers of tissue, the core, in effect, having retreated away from the surface and become surrounded by an insulating layer. The temperature of the skin would approach that of the environment and the rate of heat loss would be small. The flow of blood through the superficial layers cannot be stopped entirely, however, at least for any considerable period of time, since the tissues continue to use oxygen. But the cold blood returning to the heart from a cold skin flows in veins, the *venae comites*, which form a network round the arteries carrying the blood from the heart; there is an exchange of heat, the venous blood being warmed and the arterial blood cooled (this has been observed experimentally). Heat is thus retained within the core even though the skin receives an adequate supply of blood. If the environment is warm, most of the blood flowing through the skin returned to the heart by way of the superficial veins and this "countercurrent" heat exchange system is largely bypassed.

Magnitude of vasomotor control

The magnitude of the vasomotor control over the vessels of the skin may be judged from the changes in skin temperature in different environments. The temperature of the bare skin is most accurately measured in terms of the rate at which it radiates heat to an appropriately calibrated radiation thermopile held a few millimetres from the skin and protected from draughts. Alternatively, and rather less accurately though often more conveniently, thermocouples may be attached to the skin by adhesive tape or thrust just below the surface of the skin; this method must be used if the skin is covered by clothing. The greatest changes in skin temperature occur in the arms and legs, particularly in the fingers and toes, and the smallest on the head and the trunk. Different parts of the skin thus often have different temperatures, and the thermal insulation provided by the superficial layers varies from part to part. In cold conditions heat is lost more rapidly from the head and trunk than from the arms and legs. When a man moves from warm surroundings, with a maximum flow of blood through the skin, to cold surroundings, with a minimum flow, the average thermal insulation round the whole core increases about five-fold. An average man can thus maintain heat balance, by control of his skin circulation, in spite of a fivefold change in the difference between the body temperature and the air temperature for example, or of a fivefold change in the rate of heat production by metabolism. (There are large variations between different individuals.)

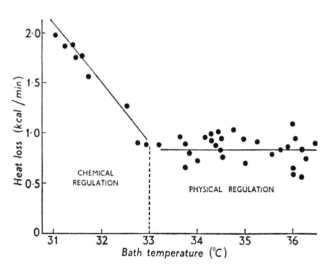

Fig. 28.1 Temperature regulation in a waterbath.

A subject lay in a well-stirred waterbath, maintained at different temperatures on different occasions, and the rate of heat loss to the water measured. Since his temperature was constant during the measurement, this must also have been his rate of heat production. It was constant at temperatures between 36·5° and 33°C, the range of "physical" regulation, by control of the cutaneous blood circulation. It rose steadily as the temperature was reduced from 33° to 31°C, "chemical" regulation, by increase in the metabolic rate, being added (1 kcal = 4·2 × 10³J). (Burton and Bazett, from Burton and Edholm (1955), *Man in a Cold Environment*. Edward Arnold, London.)

The effectiveness of this control is illustrated in Fig. 28.1. A subject was placed in a well-stirred waterbath, maintained at different temperatures on different occasions, and the rate at which heat was transferred from the subject to the bath was measured. At all temperatures between 33°C (91·5°F) and 36·5°C (98°F) this subject was able to keep his rate of heat loss almost unchanged, in spite of the fact that the difference between his body temperature and the temperature of the bath varied by a factor of nearly 10.

Local and reflex vasoconstriction

These changes in the rate of blood flow through the skin are brought about: (a) by direct local action on the calibre of the blood vessels; and (b) by reflex control of the calibre by varying excitation of the vasomotor nerve fibres supplying them.

The immediate effect of exposing some part of the body, say a hand or foot, to a different environment, hotter or colder than before, is a vasodilatation in the part exposed. The rate of blood flow through the hand, for example (almost entirely through the skin circulation) as measured by the rate of heat dissipation in a calorimeter, falls progressively as the temperature of the water is reduced, down to about 10°C. The consequent effect on the temperature of the skin, particularly on the tips of the fingers, when the hand is placed in cold water, is well shown in Fig. 28.2.

was in moderately warm air. On immersing this arm in water at a temperature of 15·5°C, there was an immediate vasoconstriction in the opposite hand, as shown by the decreased rate of heat dissipation to the calorimeter. This must have been due to a nervous reflex, set up by the temperature receptors in the cold arm, since the blood flow through this arm had been arrested before it was cooled, by inflating a sphygmomanometer cuff on the upper arm to 200 mmHg (26·6 kPa).

Reflex vasodilatation

The converse effect, of a reflex vasodilatation when some part of the body is warmed, may also be demonstrated in a similar manner, though not so regularly as the reflex vasoconstriction.

When blood was readmitted to the cold arm, and on returning

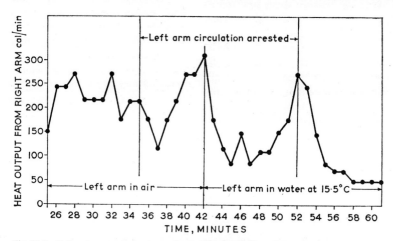

Fig. 28.2 Reflex vasoconstriction in one hand, following cooling of the opposite arm.

The subject's right hand was placed in a water calorimeter at time 0, and the rate of heat elimination calculated from the rate of rise of temperature of the water, after 25 min had been allowed for a steady state to be reached.

Vasoconstriction is indicated by the reduction in the rate of heat output. It occurred: (a) at the 42nd min when the left arm was placed in cold water, even though the circulation through this arm had been arrested; and (b) at the 52nd min when the circulation was restored and cool blood entered the general circulation. (Redrawn from Pickering (1932), Heart, 16, 118.)

Again, the blood flow through the forearm, as measured by venous occlusion plethysmography, depends on the temperature of the water within the plethysmograph. If the temperature lies between about 15° and 20°C, the rate of blood flow is small, and little affected by temperature; but if the temperature is greater than 25°C, the rate of blood flow increases rapidly with rise in temperature, by about fivefold at 35°C and twentyfold at 45°C. These measurements indicate that the blood flow through the muscles of the arm must be affected by temperature, as well as that through the skin.

Effect of sympathetic block

A large part of the local vasomotor response, at least to cold, remains after removal, or blocking, of the sympathetic nerves and is produced either by an axon reflex or by a direct action on the blood vessels. The existence of *reflex* vasomotor control, through the temperature regulating centres, is shown by the response to change in the general body temperature or to change in the skin temperature of some other part of the body. In the experiment illustrated in Fig. 28.2, the rate of heat dissipation from the subject's right hand was measured in a calorimeter at a temperature of about 29°C. To begin with, his bare left forearm

to the general circulation cooled the central receptors, the blood vessels of the hand were again constricted, and remained so for a considerable time. In the absence of circulation through the cold arm, the heat loss from the body was not increased appreciably; there was no "drive" from the central receptors towards increased conservation of heat, the "drive" from the peripheral receptors was overridden and the vasoconstriction was transient, lasting only some seven minutes.

This interaction between signals from peripheral and central receptors is demonstrated, also, by the experiment illustrated in Fig. 28.3. When the subject held his hand in ice-cold water, stimulation of the peripheral temperature receptors led to an immediate constriction of the vessels of the nasal mucosa, as shown by the fall in temperature. The total effect of the reduction in heat loss produced by such vasoconstriction (there may also have been an increase in the metabolic rate) was apparently excessive; the rectal temperature rose slowly, and the vessels in the nasal mucosa dilated again. Later in the experiment the rectal temperature fell, and the mucosal vessels constricted more profoundly than before, the central and peripheral receptors now acting in conjunction.

When some part of the body, say a hand or foot, is placed in a very cold

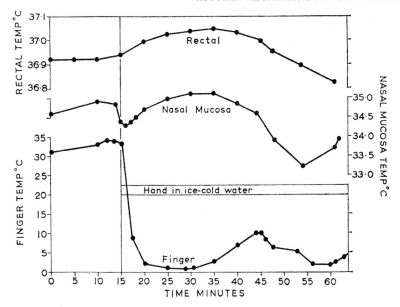

Fig. 28.3 Temperature regulation during extreme cooling of one hand.

The temperatures of the left index finger just beneath the skin, of the nasal mucosa, and within the rectum, are plotted against time. (Note the different scales on which the temperatures are plotted.)

After a control period of 15 min, the whole left hand was immersed in ice-cold water, producing extreme vasoconstriction in the finger, as shown by the great fall in temperature. After 20 min in the cold water, the temperature of the finger began to rise, owing to the onset of "cold vasodilatation".

The fluctuation in the temperature of the nasal mucosa illustrate the vasomotor changes produced by interaction between the peripheral receptors in the cold hand and the central receptors responding to the rectal temperature. (Redrawn from Aschoff, 1944.)

environment (air or water at a temperature below 10°C), the tissues, particularly of the fingers and toes, may become so cold that there is risk of serious damage. This is avoided by the phenomenon known as *cold vasodilatation*, produced by a direct local action on the blood vessels. Although protective locally, this increases the rate of heat loss unless compensated by vasoconstriction elsewhere. As may be seen in Fig. 28.3, after the subject's hand had been in ice-cold water for about twenty minutes, the skin temperature on the finger began to rise, indicating that the blood vessels were opening up again. The rectal temperature fell, and the blood vessels of the nasal mucosa became constricted in compensation. The vasoconstriction extended, also, after some delay, to the blood vessels of the cold hand, and its temperature fell again. Such a "hunting" process, of alternate vasoconstriction and vasodilatation, may continue so long as the hand (or foot) remains in the very cold environment.

The release of a profound vasoconstriction in the skin, either as a result of "cold vasodilatation" or more particularly of a sudden change to a warmer environment, is likely to lead to a rather sudden drop in the general blood temperature, owing to the initial return of cold blood from the skin. It is not uncommon for people to feel much colder, and even to start shivering, on first coming into a warm room after having been out in the cold.

29. Performance Characteristics of the Thermoregulatory System

We now pass on to discuss how the body responds to thermal stress of various kinds.

Reactions to cold: shivering

We have considered so far the effects of exposing only a small part of the body to cold environments. When a large part, or the whole, of the body is so exposed, the rate of heat loss may well be too great even when the blood flow through the skin has been reduced as far as possible everywhere. The first, and normal, reaction of an ordinary civilised man is then to put on more clothes. The effective thermal insulation of the stationary layer of air round the body can be increased, by this means, about sixfold, the limit being set by the fact that if the clothing is too thick one cannot move about; there is no such limit when a man is asleep. Fur-bearing animals have a comparable "pilomotor reflex", brought into action by the temperature regulating centres, by which the hairs are erected and the thickness of the air trapped between them is increased. When the thermal insulation provided by clothes or hair is sufficient, the temperature of the skin itself returns to the "neutral" zone, and further adjustment of the body temperature is made, as before, by control of the skin circulation.

If clothing is inadequate, the excessive heat loss is countered by increasing heat production. As may be seen in Fig. 28.1 this occurred progressively, in the subject studied, when the temperature of the water in which he was immersed fell below 33°C (91°F). The subject's body temperature was constant when the measurements were made, so that the rate of heat loss to the bath must have been equal to the rate of heat production by metabolism. When the temperature of the bath had fallen to 31°C (88°F) the subject had increased his metabolic rate some two and a half times.

The greater part of the increase in metabolic rate results from increased muscular activity; in the absence of obvious muscular exercise, there is greater "tone" and rigidity in the muscles, and individual muscle fibres contract in an uncoordinated and asynchronous manner, producing the irregular tremor of *shivering*. This peculiar kind of muscular contraction is initiated by a "shivering centre" in the posterior hypothalamus; this has direct connections with the temperature regulating centres and is set into action by a sufficient fall in the general body temperature. Shivering may also occur, however, when the skin becomes cold, or when very cold air is breathed, without detectable change in the rectal temperature. Most people, by this means, can increase the metabolic rate to about three times the resting rate. By deliberate muscular exercise, however, the metabolic rate may be increased tenfold, or for a short time, more than a hundredfold.

That part of the increase in metabolic rate which is not due to muscular contractions is thought to be due to increased secretion by the thyroid gland and the adrenal cortex, the temperature regulating centres being connected with the adenohypophysis and increasing the secretion of TSH and of ACTH (Chapter 32); the adrenal medulla also contributes by way of the sympathetic nerve supply. All these glands are known to secrete hormones which increase the metabolic rate of nearly all the organs and tissues of the body. They probably play an important part in the reactions to cold of some kinds of small mammal, but it is doubtful if they are of much importance in man.

Reactions to heat: sweating

The skin is always slightly wet as a result of the "insensible perspiration"; unless the air is saturated with water vapour, or the subject is immersed in water, there is always some loss of heat by the evaporation of water from the skin. The mucous membranes of the respiratory passages, also, are kept wet, and as is pointed out in Chapter 44 the expired air is nearly saturated with water vapour at 37°C. Heat is lost by this route at a rate which increases with increase in the respiratory minute volume, and thus automatically with increase in the metabolic rate; but in man it makes only a small contribution towards preservation of heat balance.

The wetness of the skin, and the loss of heat by evaporation of water, are greatly increased by the onset of active sweating, which occurs when there is a rise in the general body temperature as is shown in Fig. 29.1. How large this rise must be depends on the skin temperature. In a cool environment, when the skin temperature is low, the rise in rectal temperature produced by a moderate amount of muscular exercise, for example, will bring about little or no sweating. In a warm environment, when the skin temperature is high, the same rise in rectal temperature will bring about copious sweating; if it is sufficiently warm, there may be sweating even without muscular exercise.

Sweat is produced by special glands in the skin which are innervated by fibres derived from the sympathetic system (Chapter 30). They are brought into action by means of impulses in these fibres, derived from "centres" in the spinal cord, which may act to some extent independently of the hypothalamic centres; a spinal man may sweat when his skin is heated. The amount of sweat secreted in a given time by, say, an arm or leg may be measured by collecting it in a waterproof bag in which the limb is enclosed. It is not possible, however, to estimate in this way the rate of sweating by the whole body, since evaporation is prevented and heat balance is upset. In studying temperature regulation, the rate at which water is evaporated is of greater interest than the rate at which sweat runs off the surface of the body. The former may be measured by weighing the subject at the beginning and end of the period, and correcting for the volume of fluid drunk, the volume of urine eliminated, and the loss of weight consequent on the loss in the expired air of the carbon of the foodstuffs metabolised—i.e. the difference between the weight of oxygen absorbed and the weight of carbon dioxide eliminated; this may be calculated from the

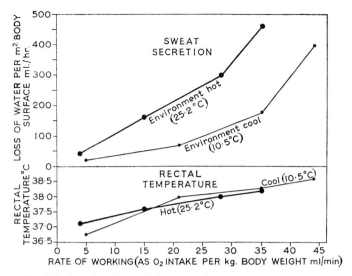

Fig. 29.1 Secretion of sweat during exercise in man.

Exercise was taken on a treadmill running at different speeds and with different gradients on different occasions. The severity of the exercise was measured in terms of the rate of oxygen consumption. Two series of experiments were performed in environments with effective temperatures (p. 171) of 10·5°C (51°F) and 25·2°C (77·5°F) corresponding roughly in England to a mild winter day and a hot summer day respectively.

The rate of sweat secretion (measured as the rate of evaporation of water and including insensible perspiration) increased rapidly with increase in the metabolic rate, and was much greater in the hot environment (skin temperature about 34°C) than in the cool environment (skin temperature about 28·5°C). The skin temperature did not rise when exercise was taken.

The rectal temperature increased progressively with the severity of the exercise; it varied with the environmental temperature when the subject was at rest, but hardly at all when he was taking exercise. (Replotted from data of Robinson et al. in Newburgh (1949), *The Physiology of Heat Regulation*, W. B. Saunders, Philadelphia.)

metabolic rate. In extreme conditions a man may secrete 10–15 litres of sweat a day and, for a short time, evaporate water at a rate of 1 litre an hour.

The sweat glands secrete a vasodilator substance along with the sweat, so that there is a plentiful supply of blood, and thus of heat to evaporate the sweat. In parts of the skin which are particularly well supplied with sweat glands, this active vasodilatation largely replaces the passive vasodilatation, following inhibition of the the vasoconstrictor nerves, which occurs in other parts of the skin.

ECCRINE AND APOCRINE GLANDS

There are two kinds of sweat glands, known as *eccrine* and *apocrine*. The apocrine glands, which occur chiefly in the axillae and pubic regions, are not concerned in temperature regulation, but probably have some secondary sexual function. The eccrine glands may be brought into action, not only by thermal stimuli, but also by emotional and mental stimuli, as in the "cold sweat" of fear and anxiety; the glands in the palms of the hands and sole of the feet, indeed, appear to respond only to such stimuli.

Some kinds of animals (dogs, for example) are not provided with sweat glands. They make use of the evaporation of water for dissipating heat by means of *panting*. When the body is hot, the respiration becomes very fast and shallow, so that a large amount of air passes over the surface of the tongue and outer respiratory passages. The amount of air which enters, and ventilates, the lungs is not increased, since there is no change in the composition of the alveolar air even after several hours of panting.

Effects of humidity

The rate at which heat is lost by evaporation of sweat depends not only on the rate of secretion, but also on the climatic conditions and on the access of air to the skin as affected, for example, by winds and clothing. In hot damp conditions the rate of secretion of sweat may exceed the rate of evaporation, much of the sweat may run off the skin and adequate cooling is difficult. When the atmosphere is dry, on the other hand, the rate of cooling may be large enough to counteract an actual gain of heat from an environment whose temperature is above that of the body. Indeed, a man may keep his temperature constant in an atmosphere with a temperature as high as 120°C (250°F), sufficient to cook meat or to boil an egg, provided that the air is dry.

Composition of sweat

Sweat is not pure water, but is a solution containing, on the average, about 0·3 per cent sodium chloride—about one-half the concentration in plasma; there are very large variations in different circumstances, between different individuals, and between different parts of the same individual. It contains, also, most of the other constituents of the plasma except the proteins; lactate ions and urea, indeed, are usually present in greater concentration than in the plasma. Sweat which runs off the skin without evaporating is not entirely useless, since it washes away the dissolved substances which would otherwise be left on the skin, increasing the concentration of the sweat and reducing the rate of evaporation.

The loss of water by sweating results in the sensation of

thirst, and the loss is made good by drinking. The loss of salt, however, may have important consequences. Failure to take in enought salt when recovering from heavy sweating will result in a dilution of the body fluids; ordinarily the excess water will be rapidly eliminated by the kidneys and the failure to restore the water balance may produce some discomfort but nothing more. If, however, the kidneys are for any reason unable to eliminate the excess water, a more serious condition of *water intoxication* may be produced. Even in the absence of renal insufficiency, this may occur when severe exercise is taken in hot surroundings and pure water is drunk while the exercise continues. Severe exercise largely inhibits water diuresis, partly owing to diversion of the blood from the kidneys to the active muscles and partly by an action on the neurohypophysis leading to a secretion of ADH. The consequent dilution of the body fluids then gives rise to the incapacitating condition known as "miners' (or stokers') cramp". (The condition was first observed in miners and in stokers of coal-fired boilers.) The necessity for taking an adequate quantity of salt along with the water (as occurs naturally in beer, for example) was in fact known empirically among miners and stokers before its rationale was understood (see p. 121).

The loss of substances other than salt in sweat is ordinarily of no consequence, since they would otherwise be excreted by the kidneys. If the kidneys are severely diseased, however, elimination of substances like urea by copious sweating may be of substantial benefit.

The salt concentration of the sweat depends on the state of the salt balance of the whole body. Prolonged existence in hot surroundings, with more or less continuous sweating, is apt to lead to some degree of chronic salt depletion; this is believed to account for the observed decrease in the salt concentration of the sweat as one becomes "acclimatised" to the environment. It does not occur if enough salt is taken to preserve the balance. The rate at which salt is absorbed from the sweat back into the bloodstream, like the rate of sodium reabsorption in the renal tubules, is regulated by means of the hormone *aldosterone*, elaborated by the adrenal cortex; the rate of secretion rises as the sodium content pf the body falls (Chapter 32C).

DISTURBANCES OF THERMOREGULATION

Fever

The clinical thermometer is a vital item in the armamentarium of the compleat physician, not because when in the mouth it blocks unwelcome repartee, but because many disease states are accompanied by fever. This term implies a central temperature raised above normal. It is the result of a disturbance of thermoregulation.

Fever is caused by:
1. Bacterial, viral or protozoal infection
2. Tissue destruction as in infarction, neoplasms etc.

MECHANISM OF FEVER

The mechanism of fever is a "resetting" of the hypothalamic thermostat to a higher value. Heat loss is decreased; heat production is accelerated. As the temperature is rising, the patient usually feels *cold*. This is because the mechanisms preventing heat loss are working all-out (i.e. there is skin vasoconstriction, piloerection and "goosepimples") and skin temperature actually does fall. The patient's friends, and his medical attendant, will observe the resulting skin pallor. In this stage, also, "rigors" occur. A rigor is a bout of severe and uncontrollable shivering, the effect of which is to raise the heat production of the body. The sensations accompanying this shivering are decidedly unpleasant.

All these responses are appropriate to a cold environment although usually it is not cold, and as a result heat-flow balance is disturbed and the patient becomes hotter; the reverse happens when recovery sets in.

Is fever a functional response?

The obvious question is: Does fever serve a useful purpose in the defence mechanisms against infection? There are three possibilities:
1. Fever is a change produced as part of the bodily defence systems.
2. It is a byproduct, and can be tolerated.
3. It is a harmful effect of infection.

At present there is insufficient evidence to support any of these views.

In malaria, heat production may rise to 1 or $1\cdot 5$ MJ h^{-1}; heat loss is unchanged, so the body temperature may well reach 41°C.

Pyrogens

Burden Sanderson used the term "pyrogen" to describe "fever-producing" substances which he found in putrefying meat.

Pyrogens are produced by bacteria (and are usually called endotoxins). They are not destroyed by boiling, and most are lipopolysaccharides, as little as $0\cdot 1$ μg of which will have a pyretic effect.

Antipyretics

If one subscribes to the view that fever of itself is harmful, it will be necessary often to prescribe antipyretic drugs which lower the body temperature. Common examples are acetylsalicylic acid or sodium salicylate.

Interestingly enough, these drugs will not lower the normal body temperature, neither will they affect the hyperthermia of raised environmental temperatures. They must therefore act by neutralising the effects upon the hypothalamic thermostat, of the pyrogen. The microinjection of small amounts of prostaglandin (PGE$_2$) into the hypothalamus raises body temperature, presumably by resetting the thermostat as in fever. It is of interest therefore, to note that aspirin antagonises the actions of prostaglandins.

Artificial hypothermia

An individual's body temperature can be lowered by administration of a sedative to depress the hypothalamic controlling mechanism and immersing him in ice or ice-water. Chlorpromazine is usually used to depress the thermostat.

Deep temperature can be maintained at levels down to 27°C for up to a week if necessary without untoward effects. This technique of artificial cooling is used in heart surgery to enable the heart to be stopped for many minutes at a time since its metabolism is considerably reduced at lower temperatures. Also, the oxygen requirement of the remainder of the body, including the brain, is reduced.

Small laboratory animals, such as mice and rats, may be cooled in a similar manner, down to near freezing point, and held in a state of suspended animation for long periods. Provided rewarming is carried out so that the circulation commences before brain temperature begins to rise, the cooled animal is apparently normal afterwards.

CLIMATE AND ITS MEASUREMENT

It is often useful to be able to measure the properties of an environment which affect the ease with which people living and working in it can regulate their body temperatures. The comfort and efficiency of workers in factories and offices may be affected by the atmosphere in which they have to work; the nature and quantity of clothing, food and drink to be taken on expeditions to polar, mountainous or desert regions are decided by the nature of the climate to be expected.

The *temperature* of the air, other things being equal, determines the rate at which heat is lost from the body by conduction and convection; the *humidity* of the air determines the rate at which heat is lost by the evaporation of water. Both these quantities may easily be measured by means of dry and wet bulb thermometers, sheltered from the sun and the wind.

In given conditions of temperature and humidity, the rate of heat loss is greatly affected by the *wind speed*, which can be measured by instruments known as anemometers. The air in immediate contact with the skin (or the surface of any hot object) forms a "stationary layer", usually a few millimetres thick; beyond this, the air is in motion relative to the skin as a result of thermal currents (hot air is less dense than cold air, and so rises), movements of the person whose temperature is being considered, and above all, the presence of winds. Heat and water vapour must pass through the stationary layer before being carried away by the moving air; the greater the air movement, the thinner is the stationary layer and the more rapid is the passage of heat and water vapour; hair and clothing, conversely, increase the thickness of the stationary layer and decrease the rate of heat loss. In the absence of appreciable evaporation of water, and for a given difference in temperature between the skin and the air, the rate of heat loss in a 15 km h^{-1} wind is about four times as great as it is in still air, and in a 80 km h^{-1} wind about nine times as great. For this reason, the unpleasantness of being in warm moist air may be largely removed by stirring the air with a fan.

Lastly, heat may be lost, or not infrequently gained, by radiation to or from surrounding objects which absorb or emit heat, to interstellar space by night or from the sun by day. The surrounding objects will, in general, be at different temperatures, have different shapes and have surfaces of different radiating efficiencies; measurement of their effect on the heat balance of a man is not easy, and is necessarily somewhat empirical. The "solar heat load", however, can be deduced fairly accurately from the altitude of the sun and the amount of mist or smoke in the atmosphere.

In exceptional circumstances—at least in the ordinary life of civilised man—substantial amounts of heat may be lost to rain or snow which falls on the skin or clothing. If the clothing (or an animal's hair) becomes wet, moreover, the trapped air is replaced by water and the thermal insulation is greatly reduced. Wet clothing is better than nothing, however, since the water within it is trapped and delays the conduction of heat to the rain or snow which falls on the outer surface.

Corrected effective temperature

Various attempts have been made—without marked success—to combine all the relevant properties of the climate into one figure which defines its overall cooling (or heating) power. From such figures, however, it is possible to make useful estimates, for example, of the amount of clothing needed in a given climate by a man taking different amounts of exercise. An empirical figure called the *"effective temperature"* is defined as the temperature of a still atmosphere, saturated with water vapour, which feels, subjectively to an ordinary person, the same as the particular climate under investigation. It may be deduced from measurements of the temperature and humidity of the air and the wind speed (and also the effects of radiation for the *"corrected effective temperature"*), by means of charts constructed from empirical observations on human subjects.

Alternatively, one may use the readings of the katathermometer, suspended in the place where observations are to be made. This is an alcohol thermometer with a large bulb; the cooling power of the air is measured as the reciprocal of the time taken for its temperature to fall from 100° to 95°F. When dry, the katathermometer measures the rate of heat loss by conduction, convection and radiation; when the bulb is moistened it measures also the heat loss by evaporation of water.

The surface of a man is rarely completely wetted (except in swimming) and the cooler and drier is the air, the smaller is the fraction wetted. In ordinary circumstances this fraction lies between 10 per cent and 50 per cent, and an appropriate figure between the two values obtained by the katathermometer should be taken as a measure of the cooling power of the climate.

30. The Autonomic Nervous System

The autonomic (involuntary or vegetative) nervous system is often defined as that part of the peripheral nervous system which is independent of the control of the will. The designation of the system as "autonomic" or "involuntary", stresses this essential feature. The definition, however, like nearly all attempts at definition, is not strictly correct. Reflexes like the knee jerk are involuntary, but do not involve autonomic pathways. On the other hand, by recalling emotions or sensations, the will may exercise more or less control over smooth muscle and glands and in doing so stimulate autonomic fibres. There are, in addition, persons who can control one or another of the autonomic functions. They are able, by effort of will, to slow or to quicken the heart beat, to contract the smooth muscle of the skin or to constrict the pupil.

A modern form of treatment which shows promise in certain cases of hypertension (chronically elevated arterial pressure) is by "feedback" to the patient, via sounds, of his own blood pressure. He is then able, voluntarily, to lower it by control of his autonomic system. Some patients learn to do this very quickly. Unfortunately it has not yet proved possible to make a patient with hypertension maintain the lowered pressure for long enough to obtain a worthwhile therapeutic effect.

For our purpose, the autonomic nervous system is best defined as the efferent pathway to the viscera, including all nerve cells and fibres through which impulses are sent from the central nervous system to glands, smooth muscles and heart, that is, all the efferent fibres in the body except those to the striped (voluntary) muscles. The definition stresses two points: (a) the efferent character of the system; and (b) the fact that we are dealing with a peripheral nervous system. Afferent (sensory) fibres are usually excluded from the autonomic nervous system. When we use the term autonomic nervous system, however, in this restricted sense, we have to realise that most of our so called autonomic nerves, such as the splanchnic, the vagus or the chorda tympani, are mixed nerves containing autonomic (efferent) as well as afferent (sensory) fibres.

Autonomic nerve fibres emerge from the central nervous system, mostly in the motor roots of the spinal cord and in some of the cranial nerves.

The sympathetic and parasympathetic systems

There are two major divisions of the autonomic nervous system, the *sympathetic nervous system* which emerges from the thoracolumbar spinal region, and the *parasympathetic nervous system* which emerges from the cranial and sacral regions. Many organs in the body are influenced by both sympathetic and parasympathetic systems, usually in opposing directions, but the definition of the two systems derives from the anatomical connections with different parts of the central nervous system, not with the way in which physiological function is changed.

There are some organs to which the double antagonistic innervation does not apply. For example, most of the systemic blood vessels are supplied by sympathetic vasoconstrictor fibres only, and centrally induced vasodilatation is brought about by inhibition of vasoconstrictor tone. Some vessels in skeletal muscles are supplied with vasodilator fibres, but these, also, belong anatomically to the sympathetic system. There is no parasympathetic innervation of the smooth muscles of the upper eyelid, of the hair muscles in the skin, of the uterus, of the sweat glands or of the adrenal medulla. Even when a tissue is provided with a dual antagonistic innervation, a too simplified conception of this antagonism is misleading. When light falls into the eye, the pupil constricts as a result of parasympathetic impulses. When the eye is in the shade the pupil dilates, not as a result of stimulation of the sympathetic fibres to the dilator muscle of the pupil, but because of diminished discharge in the parasympathetic system. On the other hand, the dilated pupil of a frightened person is mainly a sympathetically stimulated pupil and dilatation can occur despite the fact that bright light may shine into the eye.

To understand the difference in the working of the different parts of the autonomic nervous system the anatomical arrangements have to be considered. Efferent nerves in general can be subdivided into somatic and autonomic. The somatic nerves pass as medullated nerve fibres from the ventral horn cells in an uninterrupted pathway to the skeletal muscles. In contrast, the autonomic peripheral pathway consists, with one exception, of two neurones. This is illustrated in Fig. 30.1. Impulses emerging from the central nervous system into autonomic fibres thus have to cross a *ganglionic synapse* on their way to the periphery. The first neurone originates in a cell in the central nervous system and the axon (the *pre-ganglionic fibre*) terminates at a synapse near a cell in a ganglion. The ganglion cell with its axon (the *post-ganglionic fibre*), constitutes the second neurone which terminates at the peripheral effector structure. A preganglionic fibre may traverse one or several ganglia without entering into synaptic junction with a ganglion cell; each fibre may form a great number of synapses in a given ganglion or give off collateral branches terminating around the cells of different ganglia. By using *nicotine*, Langley was able to discover the endings of preganglionic fibres and the origin of the postganglionic ones. Nicotine has no action on nerve fibres, but stimulates, and later (or in larger doses) paralyses, the ganglion cells. If it is painted on an autonomic ganglion, therefore, the nerve cells in the ganglion will be excited and give rise to impulses passing along the post-ganglionic fibres, thus indicating the origin of a post-ganglionic fibre to a given peripheral structure. Later on, these cells become paralysed and can no longer be excited when their pre-ganglionic fibres are stimulated. This block will not occur when fibres merely traverse a nicotinised ganglion without entering into synaptic junctions. It is therefore possible by stimulating fibres proximal to a ganglion, painted

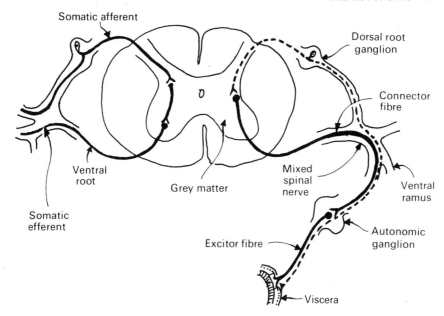

Fig. 30.1 General arrangement of the autonomic nervous system (on right). The somatic afferent and efferent paths are shown on the left.

with nicotine, to find out if they only traverse the ganglion or form synapses with its nerve cells.

Autonomic ganglia

The position of the cell stations or synaptic junctions in the autonomic path varies. Pre-ganglionic fibres may be relatively short, relaying in ganglia near the vertebral column (vertebral or lateral ganglia), to long post-ganglionic fibres. Or the post-ganglionic fibres may be short and the long pre-ganglionic fibres may terminate around ganglion cells situated within the tissue of the innervated organ (terminal or peripheral ganglia). Or the cell stations may have intermediate positions (collateral or pre-vertebral ganglia). There is this general difference. Parasympathetic pre-ganglionic fibres relay in peripheral ganglia or in collateral ganglia (ciliary, sphenopalatine, submaxillary or otic ganglia) situated near the innervated tissue. Terminal ganglia, however, are the exception in the sympathetic path. Its main cell stations are the vertebral ganglia forming the paired sympathetic chains with their adjoining cervical ganglia, in the neck. In addition there are many cell stations in collateral ganglia, like the coeliac and the mesenteric ganglia. There is one exception. The autonomic fibres to the adrenal medulla do not pass a cell station on their way to the periphery. The medullary cells and the sympathetic ganglion cells have a common origin. Both are probably derived from the same primitive masses of neuroblasts, but have followed different paths in their differentiation. The innervation to the adrenal medulla consists of pre-ganglionic fibres of the sympathetic system, the post-ganglionic fibres, so to speak, having been converted into a gland of internal secretion.

Since the pre-ganglionic sympathetic fibres are medullated when they emerge from the lateral horn cells of the cord into the ventral roots and pass as fine filaments to the sympathetic chain, these filaments have a whitish appearance (*white rami communicantes*). The fibres usually lose their myelin sheath near or at the ganglia; thus the connecting filaments containing the non-myelinated post-ganglionic fibres which are sent back to the spinal nerves of the trunk and limbs have a more greyish appearance (*grey rami communicantes*).

The term *intermediate ganglia* is given to sympathetic nerve cells which are not located in the ganglia of the paired sympathetic chains, but in the rami communicantes, often in close proximity to the motor nerves. They are of practical importance in surgery of the sympathetic nervous system. For instance, the operation of thoracolumbar sympathectomy, i.e. removal of the lower ganglia of the sympathetic chains, does not lead to complete sympathetic denervation of the skin in the lower part of the body.

Neuronal relays

The ganglia are distributing centres, in which impulses from a single pre-ganglionic fibre may be relayed to many (twenty or more) post-ganglionic fibres. In the parasympathetic division the relays are situated in or quite near the innervated tissue and this limits the spread of the impulses beyond a restricted area. The position is different in the sympathetic system. Here the anatomical arrangement clearly favours diffuse distribution of the nerve impulse over wide areas. In addition, upon stimulation of the sympathetic, hormones (*adrenaline* and *noradrenaline*) are secreted from the adrenal medulla into the bloodstream, mimicking many of the sympathetic nerve effects, thus emphasising the fact that the organism is not concerned with a limitation of sympathetic effects to restricted areas. The anatomical arrangements of the two divisions reflect the fundamental difference in the function of the two systems. (See Fig. 30.2.)

It is possible that some of the parasympathetic ganglia are weak automatic centres, independent of the central nervous system, from which impulses originate continuously. Such function is attributed to the cells of the nerve plexus in the wall of the digestive tract. To a lesser degree it may be a more general property of parasympathetic ganglia.

The adrenal medullary cells secrete adrenaline and noradrenaline as a result of the release of acetylcholine from the pre-ganglionic nerve endings. We must emphasise again the distinction between transmitter substances and hormones. The former are released from nerve endings and then modify or initiate the activity of cells in close proximity. The

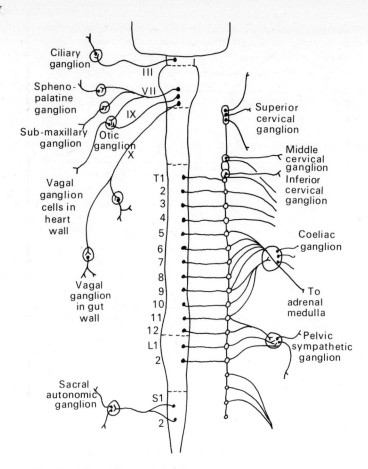

Fig. 30.2 Diagram of connections of the autonomic nervous system: parasympathetic (right); sympathetic (left).

Note fibres to adrenal medulla passing through abdominal ganglion without any synaptic junctions. This ganglion, labelled coeliac, represents all abdominal ganglia, including the inferior mesenteric in which the above phenomenon actually occurs. Note also the fusion between the inferior cervical ganglion and the first thoracic ganglion to form the stellate ganglion.

latter are secreted from gland cells, released into the blood, and act on cells which may be quite remote and in entirely different organs.

THE SYMPATHETIC NERVOUS SYSTEM

The emergency function of the sympathetic

Removal of the paired sympathetic chains with their outlying ganglia, as far as it is technically possible, is compatible with life. Sympathectomised animals show, in fact, no signs of deficiency if kept in sheltered conditions, but when exposed to extreme cold, oxygen lack, carbon dioxide increase, hypoglycaemia, haemorrhage or anaesthesia they may succumb earlier than control animals. The sympathetic innervation fulfils an important function in making the animal fit for states of emergency. There is a widespread discharge of impulses in the sympathetic system in states of physiological stress, during severe muscular work, in situations of danger, in extreme temperatures, asphyxia, haemorrhage, under strong emotions such as fear or rage, or when in pain. The discharge affects also the fibres to the adrenal medulla leading to an output of adrenaline and noradrenaline, as illustrated in Fig. 30.2. This widespread discharge has been likened to a reflex action of the organism with the purpose of strengthening its powers of defence and producing those changes necessary for preparing the organism for the three "Fs" (fright, flight and fight). It is in this connection that Cannon referred to the emergency function of the sympathicoadrenal system. Sympathectomised animals show signs of deficiency in many of these adverse circumstances, but the degree of deficiency varies in different species. The capacity for strenuous muscular exercise is definitely decreased in cats but not in dogs which remain excellent fighters.

Effects of stimulation of the sympathetic

The effects of sympathetic stimulation are easily understood and remembered, when seen in the light of a protective mechanism for emergencies. The dilatation of the pupil (contraction of the radial muscle of the iris), protrusion of the eye or *exophthalmos* (contraction of the smooth muscle at the back of the eye) and opening of the palpebral fissure (contraction of the smooth muscle fibres of the levator palpebrae) increase the perception of light. In animals an alarming appearance is produced by bristling of the hairs of the back and tail. Of this effect, "gooseflesh" alone has survived in man. Bronchodilatation decreases the resistance to the passage of air into the alveoli of the lung. The movements of the digestive tract are inhibited and the sphincters contract. Glucose is mobilised from the liver. The spleen contracts and ejects its store of red blood cells. Fatigue in skeletal muscle may be counteracted. The heart beats more strongly and more frequently; the coronary arteries are dilated. Vasoconstriction occurs in the systemic vessels, mainly in the splanchnic area and in the skin which becomes pale. In the skeletal muscles, however, the blood vessels dilate, thus shifting the flow of blood from regions where it is not urgently needed to the active tissues. The redistribution of blood may occur with little or no rise in arterial blood pressure. No useful purpose would be served if the circulatory effects of sympathetic stimulation, as well as of adrenaline, consisted in an increased activity of the heart in order solely to eject the blood against a greater peripheral resistance. The main effect is redistribution of the blood volume with increased circulation rate. On the other hand, the vasoconstriction following severe haemorrhage tends to keep up or restore an effective arterial blood pressure by adapting the vascular bed to the reduced blood volume.

Adrenaline and noradrenaline

These hormones are apparently not always secreted in a constant proportion. They may be secreted by different cells which can be selectively stimulated from the central nervous system. Insulin hypoglycaemia, for example, increases the proportion of adrenaline secreted, while most other ways of evoking central stimulation of the adrenal medulla produce a preponderance of noradrenaline in the secretion. It is interesting to note in this connection that adrenaline has a much more powerful hyperglycaemic effect than has noradrenaline.

Cannon points out that a general sympathicoadrenal discharge may be harmful unless transformed into action. Heart and circulation may be worked just as hard from an armchair as from a rower's seat. "If no action succeeds the excitement and the emotional stress—even worry and anxiety—persists, then the bodily changes due to the stress are not a preparatory safeguard but may be in themselves profoundly upsetting to the organism as a whole."

Atherosclerosis and sympathetic hyperactivity

Much has been made of the sentiments expressed above by Cannon in recent years as a part explanation for the modern epidemic of coronary heart disease. It is postulated, without direct evidence, that mental stress leads to the deposition of atheromatous plaques in the walls of the coronary vessels. These are fibrinous, fatty deposits which eventually close off the coronary artery, or a main branch of it, and thus deprive heart muscle of its blood supply. Two mechanisms are commonly invoked. One is that a raised arterial pressure, unfairly maintained, leads to damage of the arterial walls; the other postulates that the outpouring of adrenaline mobilises liver fat and raises the circulating level of cholesterol in the blood.

In the view of the present author, however, only two factors have a proven association with coronary heart disease. One is cigarette smoking, the other is that an inverse relation exists between the hardness of drinking water and the incidence of coronary heart disease.

The efferent pathway of reflexes to organs widely distributed in the body

The sympathetic system may act as a unit in conditions of physiological stress, but this is one aspect only of its function. It may also act as the efferent pathway for reflexes in which the blood vessels, sweat glands and hair muscles are the effector organs, widely distributed in the body. Sympathetic fibres are the sole connections between the vasomotor centre in the brain and the blood vessels; regulation of the calibre of the vessels is brought about by increased or decreased sympathetic discharge, the parasympathetic taking no part. It is true that the latter system contains vasodilator fibres to some tissues such as the salivary glands; these fibres, however, are not activated for the purpose of circulatory readjustments but for a specific organ function, salivation. Without a sufficient supply of fluid to the glands, salivation would not continue; accordingly the pattern of the salivary reflex incorporates a localised vasodilatation mediated by parasympathetic nerves. The regulation of heat loss through the temperature regulating centres is almost entirely dependent on the sympathetic system, whether by control of the calibre of the blood vessels in the skin and thus the skin temperature, control of the evaporation of sweat or, in animals, control of the erection of hair or feathers.

Sympathetic discharge to certain tissues is continuous. The heart, the arteries, arterioles, capillaries and probably venules, and the smooth muscles in and around the eye, are kept in a state of continuous although varying tonic contraction as a result of their sympathetic innervation. When this is interrupted, the blood vessels dilate, as was first shown by Claude Bernard when he cut the cervical sympathetic nerves on one side of a rabbit; the vessels of the external ear on the denervated side dilated and the skin temperature rose. When the sympathetic innervation to the eye is interrupted, the pupil contracts, the eye sinks into its socket (enophthalmos) and the upper lid droops (ptosis), giving the eye a sleepy appearance. The smooth muscles of the hairs, the sweat glands, the digestive tract and the medulla of the adrenal gland, on the other hand, receive sympathetic excitation only in special conditions, e.g. those of an "emergency".

It is not surprising, therefore, that different parts of the sympathetic system may act separately from each other and even antagonistically. The following instances will illustrate these points. (1) Emotional blushing is the result of inhibition of sympathetic constrictor tone of the skin vessels of the whole body. In women who blush frequently and vividly, the "blush area" is usually confined to the face and to the V-shaped area in the neck, areas of skin exposed to sunlight by the cut of modern dress. (2) Sweating, limited to the skin around the lips and nose, may be evoked by gustatory stimuli such as chewing spicy foods. (3) When there is a sufficient rise in the environmental temperature, large areas of the skin become flushed and sweat beads appear. Again, the flushing is due to reflex *inhibition* of sympathetic vasoconstrictor tone in the skin but the secretion of sweat is the result of *excitation* of the sympathetic secretory fibres to the sweat glands. When the vasoconstrictor fibres are excited at the same time as the secretory fibres, as in extreme fright, "cold sweat" appears, the sensation of cold being brought about by the restriction of the blood flowing through the skin.

One of the main functions of the sympathetic system is its role in preserving constant internal conditions, the preservation of what Claude Bernard called the *"milieu intérieur"*. The sympathetic system is in part responsible for man's great adaptability to life in different surroundings and for the conservation of his "inner climate" which he carries about with him. The constant changes in the distribution of the circulating blood volume to adapt the organism to changed environmental conditions and to the changing demands created by muscular activity are brought about, as far as nervous mechanisms are involved, through the sympathetic system.

THE PARASYMPATHETIC NERVOUS SYSTEM

Unlike the sympathetic system with its widespread discharge, the parasympathetic system is the main efferent pathway for those reflexes which are more localised and usually influence single organs without affecting others. These reflexes are abolished when the parasympathetic pathway is interrupted; for example, in the eye the pupillary reflexes to light and near vision are no longer obtained. When the parasympathetic pathway to the salivary glands is interrupted, neither the presence of food in the mouth nor its sight or smell will induce salivary secretion. The reflex secretion of gastric and pancreatic juice and of succus entericus are dependent on the integrity of the parasympathetic fibres in the vagus nerve, stimulation of which, in addition, causes increased bile flow and increased activity of the walls of the digestive tract and inhibition of its sphincters. Cutting the parasympathetic fibres to the lacrimal glands abolishes reflex lacrimation. A continuous discharge is exerted through the parasympathetic fibres in the vagus upon the heart's action, as shown by the fact that the heart-rate in man may double when the vagal inhibition is removed, as after atropine. Vagal tone is weak at birth; in a newborn baby, atropine will increase the pulse rate only from about 140 to 160 per minute. The vagal tone to the heart in man is influenced continuously by many reflexes. The significance of this tone depends on the effect of heart rate on mechanical efficiency; the heart uses less oxygen to perform a given amount of work when it is beating slowly than when it is beating quickly.

The parasympathetic fibres from the sacral division are the efferent pathway for the reflex contraction of the urinary bladder and inhibition of its internal sphincter in the micturition reflex. There is no marked sympathetic control of bladder activity, although sympathetic nerves regulate the blood flow in the bladder muscle. The contraction, produced by stimulating the sympathetic nerve of the muscle of the ureteral orifices and of the trigonum, is linked not with the micturition reflex but with the sex function. Section of the hypogastric nerves which contain the sympathetic fibres does not interfere with micturition whereas the bladder becomes paralysed after section of the pelvic nerves.*

The cranial division of the parasympathetic contains vasodilator fibres to the salivary glands and tongue and the sacral division contains similar fibres to the erectile tissue of the external genitalia. The main role of these vasodilators is, as mentioned before, linked with the specific functions of these organs, salivary secretion and erection of the generative organs respectively and not with general circulatory readjustments. Ejaculation is dependent on the integrity of the sympathetic system; its removal causes impotence in the male. Thus both divisions of the autonomic nervous system are involved in the mechanism of coitus (Chapter 69).

Relation between somatic and autonomic systems

A man's conscious activities largely consist in controlling his skeletal ("voluntary") muscles in response to information received through his "special senses" (chiefly vision and hearing) and controlled by his central nervous system. But he cannot do this properly unless his "auxiliary machinery"—cardiovascular system, gastrointestinal tract etc.—is also operating properly, controlled in an appropriate manner by his autonomic nervous system; this, accordingly, is not really autonomous, as the name implies, but co-operates with all the other parts of the nervous system.

Some examples of this have been given in previous chapters. If the spinal cord is severed from "higher" parts of the nervous system, defaecation and micturition are controlled according to the distension of the rectum and bladder respectively. Gunshot wounds may sever completely the spinal cord in men. If this occurs in the cervical or upper thoracic region, disconnecting the sympathetic system from the higher centres, distension of the bladder causes a large reflex rise in blood pressure. If the section through the brain stem is such as to allow the medulla to remain connected with the spinal cord—as by decerebration, or by a gunshot wound in the lower thoracic region of the spinal cord, the blood pressure is well regulated and there is no rise when the bladder is distended. The baroreceptors are connected to the vasoconstrictor fibres of the sympathetic system through the vasomotor centres, and are able to control the cardiovascular system. In the whole normal animal, with intact nervous system, these and other "centres" are subjected to overriding influences from the rest of the nervous system, and their efferent discharges blocked or enhanced.

CONTROL OF THE AUTONOMIC SYSTEM BY THE HYPOTHALAMUS

Just as overriding control of the somatic nervous system is invested in the cerebral cortex and other higher brain centres, the functioning of the autonomic system is directed by centres in the hypothalamus. We may look upon this type of control as partly in the nature of a negative feedback to keep the vegetative functions of the body at a level on which voluntary behaviour may be superimposed with the greatest of economy. Equally, it is partly in the nature of an overall integration of the autonomic nervous system—in servocontrol terms, the hypothalamus is the central nervous region which determines the set-points of the various control systems we have already described.

Little is known about this finer central integration which is necessary for keeping the internal environment constant. The most important structures through which it is exerted, however, lie in the hypothalamus. As mentioned in Chapter 32, the internal secretions of the adenohypophysis (anterior pituitary body), and hence those of many other parts of the endocrine system concerned in stabilising the internal environment, are controlled through the hypothalamus. Here also are the "osmoreceptors" which control drinking and water diuresis through the neurohypophysis (posterior pituitary gland), the centres controlling hunger and appetite and those controlling body temperature and the loss of heat. The role of the hypothalamus in control of *water balance* has been described on page 129, and its function in *temperature regulation* on page 163.

Hypothalamus and sham rage (the limbic system)

Electrical stimulation of the hypothalamus, through implanted electrodes in unanaesthetised animals, results characteristically in excitation of the sympathetic system with rise of blood pressure, dilatation of the pupils, erection of hairs and inhibition of gastrointestinal movements and secretion; but parasym-

---

* Complete removal of the sympathetic supply to the bladder appears at first to result in some weakness of the internal sphincter, but this soon passes off and normal micturition is restored.

Table 30.1 *Summary of the effects of stimulation of the sympathetic and parasympathetic nerves*

Organ	Sympathetic	Receptor type	Parasympathetic
Glands			
Sweat, apocrine glands eccrine glands	Secretion	α	No innervation
Salivary, gastric, intestinal and pancreatic (acini and islets)	Secretion, cholinergic		Secretion
Liver	Glycogenolysis		Increased bile flow
Lacrimal			Secretion
Smooth muscles			
of bronchi	Relaxation	β	Contraction
of oesophagus	Relaxation; usually contraction of cardiac sphincter	β	Contraction; relaxation of cardiac sphincter
of stomach	Usually relaxation	β	Contraction
of intestine	Relaxation	β	Increased tone and motility
of eye iris	Midriasis; contraction of dilator pupillae	α	Miosis; contraction of constrictor pupillae
ciliary	No innervation		Contraction
Internal anal sphincter	Contraction	α	Relaxation
Detrusor of urinary bladder	Relaxation	β	Contraction
Trigone and sphincter of urinary bladder	Contraction	α	Relaxation
Vasa deferentia, seminal vesicles and prostate	Contraction (ejaculation)		No innervation
Uterus	Relaxation; contraction when pregnant	α and β	No innervation
Blood vessels			
of salivary and lacrimal glands	Constriction	α	Dilatation
of abdominal and pelvic viscera	Constriction (dilation)	α (β)	No innervation
of external genitalia	Constriction	α	Dilatation (erection)
of skin and mucosa	Constriction	α	No innervation
of skeletal muscles	Constriction. (Dilation during activity)	α (β)	No innervation
of coronary system	Dilatation		Constriction (?)
Heart			
Frequency of beat	Increased	β	Reduced
Conduction of impulse	Quickened	β	Slowed
Auricular contraction	Strengthened	β	Weakened
Ventricular contraction	Strengthened	β	No innervation in mammals

pathetic effects may also be obtained—contraction of the bladder and increase in gastrointestinal movements—according to the exact position of the stimulus and form of stimulating current. Stimulation of an appropriate area may put the animal into a rage, snarling and biting, with staring eyes, hairs on end and general excitation of the sympathetic system, with perhaps urination and defaecation. A similar condition of "*sham rage*", ill-directed and short-lived, is produced by quite harmless stimuli in animals whose forebrains have been destroyed, leaving the hypothalamus intact.

Some structures in the hypothalamus, therefore, seem to be concerned in producing an abnormally irritable and aggressive type of behaviour, together with excitation and inhibition of many parts of the autonomic nervous system. These structures are normally held in check by higher centres which lie in the rhinencephalon, or "olfactory brain": this consists of structures which form a kind of arch or "limbus" round the rostral brainstem and interhemispheric commissures, and is better called the "*limbic system*" since there is no good evidence that it is concerned with the sense of smell. The most important of these structures, in relation to emotion and temperament, appear to be the *amygdala*. If, instead of the whole forebrain, only the neocortex is removed, leaving the amygdala intact, cats become abnormally placid and show no signs of anger even when ill-treated; if the amygdala are then destroyed bilaterally, the cats become savage and malevolent. The evidence, however, is conflicting, since in other series of experiments, on cats and monkeys, removal of the amygdala has made the animals unusually placid; wild Norway rats also, ordinarily untameable, become gentle. Rather similar conflicting results have followed attempts to improve the condition of assaultive psychotic patients by making lesions in the amygdala. The organisation of the limbic system is obviously very complicated, and much remains to be discovered.

Hypothalamus and cardiovascular regulation

Stimulation of many different regions in the hypothalamus will cause all the known responses of the cardiovascular system to take place. Stimulation of the posterior hypothalamus increases arterial blood pressure (and heart rate). Pre-optic stimulation has an opposite set of effects.

Gastrointestinal function

When certain regions in the hypothalamus are stimulated through microelectrodes, the experimental animal apparently becomes extremely hungry and will go to extreme lengths for food. The *perifornical* area and the *lateral hypothalamus* are associated with appetite and hunger. When these regions are surgically removed, the experimental animal wastes away and dies from malnutrition. There is also a *satiety* centre which, if stimulated, leads to cessation of eating or, if removed, leads to a permanent and voracious appetite. Other regions control gastrointestinal motility.

These observations lead to interesting speculation on the etiology of certain disorders of eating habit in human subjects. A condition known as *anorexia nervosa*, occurring in young women predominantly, leads to gross emaciation, and sometimes to death in the absence of treatment, through lack of appetite for food. Whether this is a disorder of the hypothalamus is yet to be elucidated.

Grossly overweight persons may also be involved in hypothalamic dysfunction, although again there is as yet no proof that this occurs. Some evidence is that these patients often have endocrine abnormalities which can be traced to hypothalamic causes.

The hypothalamic control of endocrine function is discussed in Chapter 32.

31. Chemical Transmission in the Autonomic System

Chemical transmission at nerve endings

The chemical transmission of excitation from motor nerves to skeletal muscles has been discussed in Chapter 5 and its intimate mechanism is most fully understood at these junctions. But historically, the theory of chemical transmission of the nervous impulse arose at the beginning of this century in connection with the autonomic system, to explain the striking similarity between the actions of adrenaline and of sympathetic nerve stimulation, on the one hand, and between the actions of drugs like pilocarpine and muscarine and of parasympathetic nerve stimulation, on the other hand.

Evidence for theory of chemical transmission

Direct experimental evidence in favour of the theory was first produced by Otto Loewi in 1921. Figure 31.1 shows a modifica-

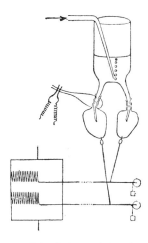

Fig. 31.1 Demonstration of chemical transmission of vagus effect on the frog's heart. Modification of Loewi's original experiment. (After Kahn.)

tion of his original experiments. Two frogs' hearts are supplied with Ringer solution from the same reservoir, the liquid being mixed by the pumping action of the hearts. On stimulating the vagi to the first heart, it is inhibited and may stop beating. When stimulation ceases and the heart starts beating again, a slight but definite inhibition occurs in the second heart which is connected with the first heart only by the Ringer solution. A substance must have been released into the liquid during stimulation of the vagus which on reaching the other heart causes the vagus-like effect. On stimulation of the sympathetic accelerans fibres to the heart an "accelerans substance" is correspondingly released.

ACETYLCHOLINE
Similar experiments with stimulation of the vagus to the mammalian heart perfused with blood failed for a long time to demonstrate the existence of a "vagus substance", for the following reason. We now know that the substance released from the parasympathetic vagus nerve endings is *acetylcholine*. Once released, this is quickly hydrolysed by an enzyme, *cholinesterase*, into choline and acetic acid, both pharmacologically inert substances in comparison with acetylcholine. Choline has, in fact, actions like acetylcholine, but only if given in concentrations several thousand times as great. The amounts of choline set free, therefore, are too small to produce reactions and the hydrolysis may be regarded as an effective mechanism of inactivation. In warm-blooded animals the enzyme will have acted usually before the acetylcholine has had time to enter the capillaries. By the use of a tissue from a cold-blooded animal, and of Ringer solution instead of blood, Loewi had avoided this danger.

ADRENALINE
The effects of sympathetic stimulation have been described as rather, but not quite, like those of adrenaline and the name sympathin was given by Cannon to the "accelerans", adrenaline-like substance, or substances, liberated at sympathetic nerve endings. When noradrenaline (adrenaline without its methyl group) was identified in the adrenal glands, it was found to have actions resembling those of sympathin, particularly in those respects in which the actions of sympathin and adrenaline differed. Sympathin is now generally held to be a mixture of *adrenaline* and *noradrenaline* in different proportions at different sites of action. A third substance, *isoprenaline* (isopropylnoradrenaline), has been found to show those actions of sympathin in which it differs from those of noradrenaline. Isoprenaline has also been found in some mammals.

DETECTION OF ADRENALINE AND NORADRENALINE
Adrenaline is not as rapidly destroyed by blood as is acetylcholine, and as a hormone secreted by the adrenal medulla it is transported by the bloodstream to the tissues on which it acts. It is released from nerve endings and is probably partly destroyed before diffusing into the capillaries. By the action of an enzyme in the tissues, the hydroxyl group in the ortho position on the benzene ring is methylated; the compounds formed are pharmacologically inactive. Another enzyme (monoamine oxidase) may oxidise and deaminate the side-chains. But some of the transmitter certainly reaches the bloodstream in an active form and can, and has been, demonstrated there by its reactions on distant denervated tissues (denervation sensitises the tissues to adrenaline). For instance, when in cats the heart, pupil and nictitating membrane (a third eyelid present in some species) are denervated and the adrenals removed to exclude this source of adrenaline, stimulation of sympathetic fibres to other tissues will cause quickening of the heart, dilatation of the pupil and withdrawal of the nictitating membrane, all of which are typical effects of adrenaline and noradrenaline.

180 HUMAN PHYSIOLOGY

DETECTION OF ACETYLCHOLINE

It would be useless to employ similar methods for the detection of the released acetylcholine. Its enzymatic destruction provides an extremely efficient safeguard against any spread of the effects of the nerve impulse and thus makes acetylcholine particularly suitable as a transmitter for the peripheral effects of the parasympathetic division of the autonomic system with its restricted localised functions. In addition, the quick destruction will also ensure a short duration of the effect not outlasting the nerve impulse for any length of time. How can the acetylcholine be detected if it is so quickly destroyed? This has become possible by the use of eserine (physostigmine), which prevents the hydrolysis by cholinesterase. The released acetylcholine may then escape into the bloodstream and exert effects on distant organs in the same way as sympathin does in the normal course of events. This is illustrated by the experiment shown in Fig. 31.2.

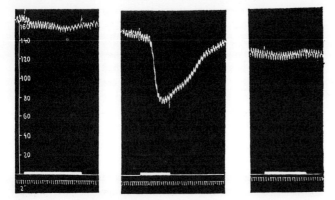

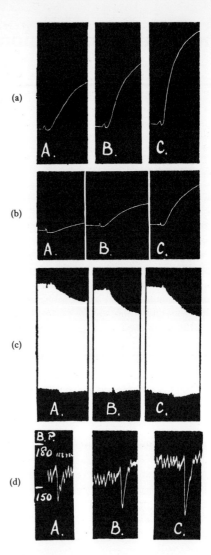

Fig. 31.2 Evidence for the release of acetylcholine.

Effect on the arterial blood pressure of an anaesthetised cat of three stimulations of the chorda-lingual nerve, which contains secretory and vasodilator fibres to the salivary glands and tongue.

In 1, there is no effect since the local vasodilatation is insufficient to affect the general blood pressure. Between 1 and 2, eserine is injected intravenously. In 2, 10 min later, there is general vasodilatation, and a fall of blood pressure, as the released acetylcholine, escaping destruction, diffuses into the blood capillaries. Note the latency of some 10 s, due mainly to the time taken for the blood to travel round the circulation.

Atropine is then given, and this abolishes the vasodilator action of acetylcholine. In 3, therefore, there is no depressor effect.

Before acetylcholine can be acted upon by the cholinesterase, it has to become attached to, or combined with, the enzyme; it is then at once hydrolysed and the enzyme becomes free again. Eserine (and related substances such as prostigmine) also combine with the enzyme and having done so, prevent acetylcholine from combining. Eserine, however, is hydrolysed slowly, if at all, and thus remains fixed to the enzyme. If, therefore, sufficient eserine molecules are available, all the enzyme molecules will, after a time, become blocked and unavailable for acetylcholine. An action such as that of eserine is called *competitive inhibition*. There are certain organic phosphates which also are very potent inhibitors of cholinesterase. Some of them were originally prepared as war gases such as diisopropylfluorophosphonate (DFP), others as insecticides such as tetraethylpyrophosphate (TEPP).

The amounts of acetylcholine released on nerve stimulation and available for analysis are far too small to be detected or identified by our present chemical methods. But acetylcholine has been identified chemically in extracts of the horse spleen, the human placenta and the ox brain, and can be regarded as a substance occurring naturally and being formed in the body. Its

Fig. 31.3 Tests of substance in perfusion fluid emerging from veins of stomach during stimulation of vagus.

From above downwards: (a) eserinised leech muscle; (b) frog's rectus abdominis; (c) frog's heart[1]; (d) cat's blood pressure. In each case, A shows the effect of a suitable dose of acetylcholine, B shows the effect of a dose of the perfusion fluid, adjusted to be proportional to the dose A of acetylcholine. C is the effect of acetylcholine given in twice the concentration of A. In each of the four reactions, the effects of B are intermediate between those in A and those in C.

Concentrations of acetylcholine (A): (a) $1 : 280 \times 10^6$; (b) $1 : 56 \times 10^6$; (c) $1 : 56 \times 10^6$; (d) 1 ml of $1 : 40 \times 10^6$. (After Dale and Feldberg.)

[1]Owing to the slowness of the drum, the individual vertical lines representing heart beats have overlapped. The vertical distances between the upper and lower borders of the white patch nevertheless indicate the relative amplitudes of the heart beat.

identification when released on nerve stimulation is based on pharmacological methods using tissues which respond to minute doses of acetylcholine with characteristic reactions. Some tests in use for this purpose are shown in Fig. 31.3.

Quantitative tests for acetylcholine
(a) Contraction of the muscle of the body wall of the leech, the effect being greatly increased in the presence of eserine. The reaction is very sensitive; it is induced by a concentration of acetylcholine of only 1 to 2 μg l^{-1}.
(b) Contraction of the rectus abdominis muscle of the frog and the sensitising effect of eserine on the action. Sensitive to about 25 μg l^{-1} of acetylcholine.

(c) Inhibition of the beat of the frog's heart. This was the first test used. The action is abolished by atropine.
(d) Depression of cat's blood pressure. The action is sensitised by eserine and abolished by atropine.

In each of these tests, the response to the unknown solution is matched with that of an appropriate dose of a standard solution of pure acetylcholine, and hence the apparent acetylcholine content of the unknown solution is determined. If the apparent acetylcholine content is found to be the same in all four tests and, in addition, the unknown substance is unstable in alkaline solution and destroyed by blood in the absence of eserine, but not in its presence, the identity with acetylcholine is regarded as proved. Other choline esters produce qualitatively similar, but quantitatively different, effects on these tissues, acting relatively more on one than the other compared with acetylcholine.

Criteria for a substance being a transmitter

In order to prove that a nerve impulse acts by the release of acetylcholine, the following three facts must be established.
1. Acetylcholine is released into the tissues when the nerve is stimulated. The eserinised venous blood from a given organ may be collected and tested for acetylcholine, whilst the nerves to the organ are stimulated; or the organ may be perfused with eserinised Ringer's solution, the nerve stimulated and the venous effluent collected and assayed.
2. The action of injected acetylcholine must be identical with, or approximate closely to, that of nerve stimulation, although we have to take into account the fact that the method of injection does not always imitate closely the release of acetylcholine by nerve impulses.
3. Eserine, by delaying the destruction of the released acetylcholine, must potentiate and prolong the effects of nervous stimulation. In some instances, prolonged action of acetylcholine may paralyse a reactive structure; in that case the response to nerve stimulation should be affected similarly after eserine.

The identification of adrenaline is based on similar lines of argument: similarity of the effects of sympathetic nerve stimulation with those of injection of adrenaline and noradrenaline, the actions of various drugs (as mentioned later) in modifying these effects, and pharmacological assay of adrenaline and noradrenaline in venous effluents during stimulation of sympathetic nerves.

Quantitative tests for adrenaline
(a) Rise of blood pressure in the cat or rat after destruction of the spinal cord (to prevent vasomotor reflexes) and treatment with atropine and hexamethonium.
(b) Inhibition of the contractions of the rat's uterus, rat's colon, or hen's colon.

The tests can detect the presence of 1/1000 to 1/100 of a milligram of adrenaline or noradrenaline, but no single test will differentiate between them with certainty. The different preparations, however, have very different sensitivities to adrenaline and noradrenaline, so that by using several tests in parallel the quantity of each in a mixture may be estimated. Adrenaline and noradrenaline may also be separated from one another, and from interfering substances, by the use of paper chromatography.

A systematic analysis of the effects of different nerves soon showed that chemical transmission is not confined to the autonomic nervous system, as is made clear in Chapter 30. Moreover, the peripheral pathway of the autonomic nervous system (except that to the adrenal medulla) consists of two neurones; transmission across the ganglionic synapses must be considered as well as that to the effector organs. All those nerve fibres or neurones from which the nerve impulses are transmitted to the next neurone, or the effector cells, by the action of acetylcholine are described as cholinergic; all those from which transmission occurs by the action of noradrenaline or adrenaline are described as adrenergic.

ADRENERGIC TRANSMISSION

It was found by Oliver and Schafer in 1895 that if an extract of the adrenal gland was injected into an experimental animal its blood pressure would rise considerably. In 1921 Otto Loewi showed that stimulation of the sympathetic nerves to the frog's heart led to the production of an adrenaline-like substance which was the cause of the increase in heart-rate produced. Von Euler in 1946 demonstrated that the transmitter substance in postganglionic sympathetic nerves was *noradrenaline*.

Evidence that noradrenaline is the sympathetic transmitter

(a) The effects of noradrenaline on adrenergic organs are identical with adrenergic nerve stimulation.
(b) Noradrenaline is released at postganglionic sympathetic nerve terminals by impulses reaching them.
(c) Noradrenaline can be detected in adrenergic nerves (biologically or fluorimetrically).
(d) Infused noradrenaline (or released noradrenaline) is inactivated by uptake of the terminals themselves. Substances preventing uptake increase the responses to adrenergic stimulation.

Storage of noradrenaline

Granules, 0.1 to 0.5 μm diameter of dopamine, noradrenaline and adrenaline are found in chromaffin cells in the adrenal medulla. Vesicles (about 50 nm dia.) of noradrenaline are found in adrenergic nerve terminals.

Secretion of adrenal medulla

Acetycholine is released by pre-ganglionic fibres of the splanchnic nerve. Nerve impulses reaching the terminals depolarise the terminal membrane and release acetylcholine. The latter causes calcium ions to enter the post-ganglionic membrane and catecholamines are released from the granules. The release of noradrenaline at adrenergic terminals is thought to be due to a similar mechanism (since anatomically and embryologically the system is the same). Adrenergic fibres contain acetylcholine; this is released by membrane depolarisation at the terminal, calcium entry follows and noradrenaline is released from its vesicles pre-synaptically.

Synthesis of catecholamines in peripheral and central adrenergic neurones and chromaffin cells is shown in Table 31.1.

Post-synaptic actions of noradrenaline

The two types of action must be distinguished. One is adrenaline acting as a hormone when carried in the bloodstream; the other is the synaptic action when noradrenaline is released at the nerve terminal and acts on the post-synaptic membrane. Figure 31.4 gives a rough idea of what happens at the membrane. The system is different from that employed at cholinergic endings where an enzyme destroys the transmitter to limit its action. At adrenergic endings 85 per cent of the noradrenaline is taken up again by the terminal to be reused. This uptake is an active process; how active can be judged from the fact that the concentration gradient is 10 000 to 1!

Table 31.1 *Synthesis of noradrenaline*

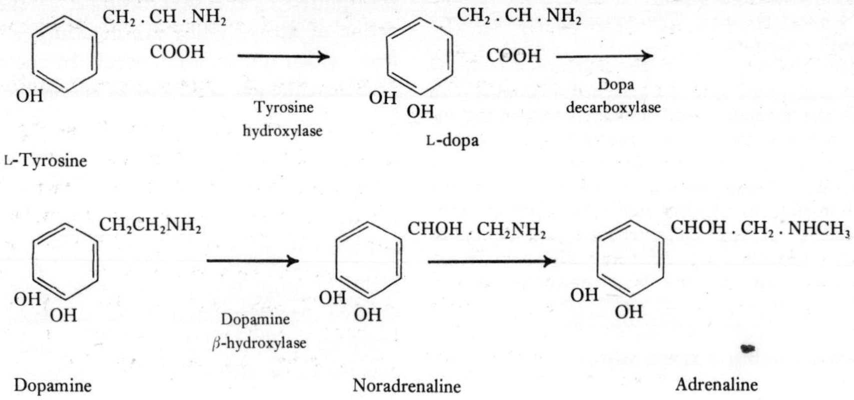

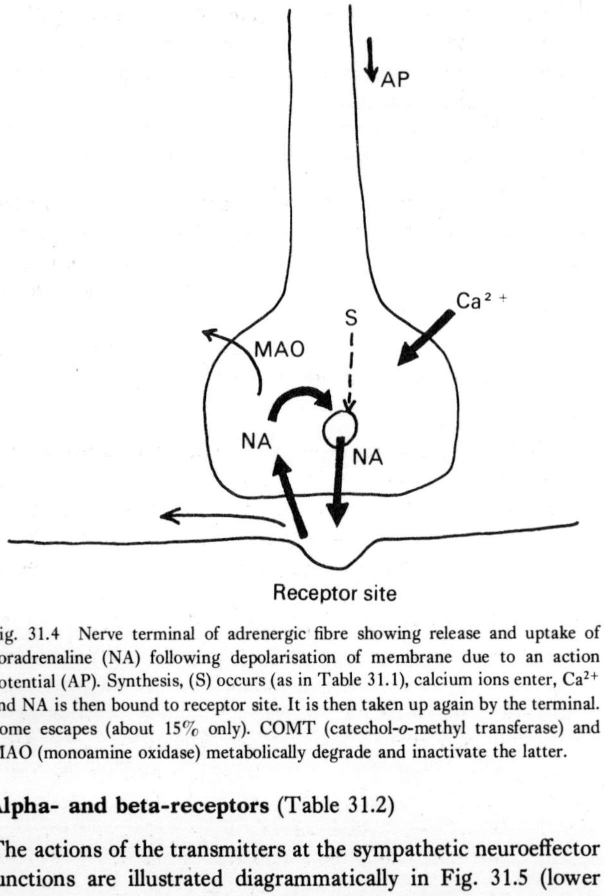

L-Tyrosine →(Tyrosine hydroxylase)→ L-dopa →(Dopa decarboxylase)→

Dopamine →(Dopamine β-hydroxylase)→ Noradrenaline → Adrenaline

Fig. 31.4 Nerve terminal of adrenergic fibre showing release and uptake of noradrenaline (NA) following depolarisation of membrane due to an action potential (AP). Synthesis, (S) occurs (as in Table 31.1), calcium ions enter, Ca^{2+} and NA is then bound to receptor site. It is then taken up again by the terminal. Some escapes (about 15% only). COMT (catechol-*o*-methyl transferase) and MAO (monoamine oxidase) metabolically degrade and inactivate the latter.

Alpha- and beta-receptors (Table 31.2)

The actions of the transmitters at the sympathetic neuroeffector junctions are illustrated diagrammatically in Fig. 31.5 (lower part). Noradrenaline makes smooth muscle contract; for example it produces a vasoconstriction and a rise of arterial pressure. This is termed the alpha (α) effect and the drug is said to act on α receptors in the muscle. (These receptors correspond with the chemically excitable structure, or the "receptive substance", at the neuromuscular junctions and must not be confused with the "sensory" receptors.) Isoprenaline in general produces relaxation of smooth muscle, for example vasodilatation and, more important therapeutically, dilatation of bronchioles. This is termed the beta (β) effect, the drug acting on β receptors. Adrenaline produces a mixed effect since it acts on both kinds of receptor, but in low (i.e. physiological) concentrations chiefly on the β receptors; this action is important in blood vessels of skeletal muscles, but not in those of other organs and tissues. Adrenergic

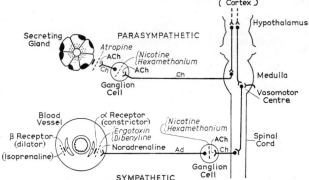

Fig. 31.5 Diagram showing chemical transmission in the autonomic nervous system and the distribution of cholinergic (Ch) and adrenergic (Ad) neurones.

A secreting gland is taken as a typical effector organ innervated from the parasympathetic system; and the smooth muscle in a blood vessel as a typical effector organ innervated from the sympathetic system.

At each junction the transmitter substance liberated at the nerve endings is labelled (ACh = acetylcholine); the receptor is shown by the thick short line; and the site of action of blocking agents, as labelled, shown by the interrupted line. As hormones, secreted by the adrenal glands, noradrenaline acts only on the α receptors and adrenaline, in low concentration, more on the β receptors than the α receptors; in larger concentrations it acts chiefly on the α receptors.

The parasympathetic receptors may be excitatory, as in glands and in smooth muscle cells of the alimentary canal, or inhibitory, as in the heart. The sympathetic β receptors in the blood vessels (when present) are not innervated; but in other kinds of receptor cell, such as the heart, they are thought to be innervated and may be excitatory.

blocking agents, like ergot extracts or dibenzyline, are especially effective in blocking all actions on the α receptors. Adrenaline in larger concentrations acts more on the α receptors than on the β receptors and produces vasoconstriction and rise of blood pressure; but when administered after adrenergic blocking agents, it produces vasodilatation and fall of blood pressure

Table 31.2

α-receptors	β-receptors
Glycogenolysis	Vasodilatation
Dilation of the pupil	Relaxation of bronchi
Cardioacceleration	Cardioacceleration
Vasoconstriction	Increased force of heart-beat
Relaxation of intestines	Relaxation of uterus
Piloerection	Rise in blood sugar

through its action on the β receptors which are still active. The effects of sympathetic stimulation on tissues other than smooth muscle, for example increased frequency of heart-beat or rise in the blood sugar concentration, are regarded as β effects, partly because isoprenaline has more action on them than has noradrenaline, and partly because of their responses to adrenergic blocking agents.

Alpha and beta receptors on a continuum?
Ahlquist's original work separated α and β receptors because of their responses to blocking agents: α receptors were motor and blocked by ergotamine; β were inhibitory and not blocked by ergotamine. Recent work appears to show that α and β receptors are much less distinct. The properties of the two are rarely the same in one tissue as they are in another; there is variation from species to species. One view is that both α receptors and β receptors are in reality parts of an adenyl cyclase system. The two types of receptor represent the extremes of a continuum.

The purinergic nervous system
Survey of the functions of the autonomic nervous system indicates that it has a number of actions that cannot be explained by the action of either acetylcholine or noradrenaline as the transmitter. There are, for example, effects on the calibre of blood vessels. It has been postulated that the transmitter is ATP, hence the term purinergic neurone, but the evidence is not yet complete enough to be certain that this ubiquitous substance is a transmitter in addition to all its other physiological roles.

Specificity of nerve fibres
The adrenergic and cholinergic nature of a nerve is not confined to the endings but is an inherent property of the whole neurone. An adrenergic or cholinergic nerve contains noradrenaline and adrenaline, or acetylcholine, respectively, throughout the whole course of the nerve fibre; when the nerve impulse passes along it, minute amounts of the chemical mediators are released. The difference between the fibre and the ending is only quantitative. At the endings, the process shows a local intensification to ensure transmission to a contiguous cell. No function can yet be postulated for the release along the course of the fibre.

Cholinergic nerves have the ability to synthesise acetylcholine from choline, with the aid of the enzyme *cholineacetylase*, not only at their endings but along the whole course of the fibre. The synthesis is a complex process in which adenosine triphosphate and coenzyme A, the coenzyme for acetylation are involved. The acetylcholine so formed is not free but in loose combination with some cell constituent, probably protein; this complex is pharmacologically inactive and resistant to the action of cholinesterase. At the nerve endings, the nerve impulse releases the acetylcholine from the bound complex, so that it becomes diffusible and pharmacologically active; then it is at once destroyed by the true cholinesterase. About forty-eight hours after a cholinergic nerve is cut, the peripheral end loses its ability to synthesise acetylcholine and the acetylcholine store disappears; at this time the fibre is still able to conduct nerve impulses.

Another observation suggesting that the whole of a neurone is either cholinergic or adrenergic is based on regeneration experiments. When the known facts of regeneration experiments with cross-sutured nerves were reconsidered in the light of the chemical transmission theory, it became evident that cholinergic fibres could replace other cholinergic fibres and enter into functional connections with them and that adrenergic nerves could replace adrenergic ones, but a cholinergic fibre could not enter into functional connection with an adrenergic one or vice versa.

Distribution of adrenergic and cholinergic endings
This is shown diagrammatically in Fig. 31.5. Most of the endings of the sympathetic fibres with their effector cells are adrenergic, which explains the striking similarity between the effects of noradrenaline and adrenaline and of sympathetic stimulation. The endings of the parasympathetic fibres with their effector cells are cholinergic, a fact which explains the equally striking similarity between the effects of parasympathetic stimulation and of drugs like acetylcholine, pilocarpine or muscarine. The secretory nerves to sweat glands, and the vasodilator nerves to blood vessels in skeletal muscles, are peculiar. The endings of the nerve fibres on the effector cells are cholinergic and thus appear to belong to the parasympathetic system; but the nerves arise from ganglion cells which are situated centrally, not peripherally, and anatomically form part of the sympathetic system.

Muscarinic and nicotinic actions of acetylcholine
The synaptic transmission across the ganglia of the autonomic system, both sympathetic and parasympathetic, is cholinergic, and so also is the neuromuscular transmission at the motor end-plates of skeletal muscles (Chapter 5). Two rather different types of cholinergic nerve ending must be distinguished, associated with the two classes into which the pharmacological actions of acetylcholine can be divided.

THE MUSCARINE-LIKE ACTION
Muscarine is a substance of known composition closely related to choline; it is found in extracts of a common toadstool (*Amanita muscaria*). Its effects are the same as those observed on stimulating the post-ganglionic parasympathetic fibres. These actions, whether induced by muscarine or acetylcholine, are abolished by atropine.

The muscarine-like action of acetylcholine is effective in the transmission to the peripheral structures from all post-ganglionic parasympathetic nerves, as well as from the cholinergic post-ganglionic sympathetic nerves; all these nerve effects are blocked by atropine (Fig. 31.5, upper part). After atropine, stimulation of the vagus no longer inhibits the heart, and stimulation of the sympathetic no longer causes sweating in human beings. Nevertheless, the nerve impulses still release their acetylcholine. Atropine, although abolishing the effects of nerve stimulation, has no action on the nerves or nerve endings themselves, but renders the effector structure insensitive to the action of acetylcholine, whether released or artificially applied. Many other drugs can still act after atropine. We do not know why drugs like atropine render the cells insensitive to one kind of drug and not to another.

THE NICOTINE-LIKE ACTION
Nicotine first excites and subsequently paralyses the following structures: (1) the autonomic ganglia; (2) the cells of the adrenal medulla; and (3) the motor end-plates of skeletal muscle. Similar actions can be obtained when acetylcholine is injected. The nicotine-like effects are relatively insensitive to atropine. Acetylcholine has, in addition, stimulating and paralysing effects on cells of the central nervous system.

The nicotine-like action of acetylcholine is effective in the transmission to the cells of the adrenal medulla, the nerve cells of the autonomic ganglia, and the end-plates of skeletal muscle. Transmission at these structures is not blocked by atropine, but is blocked by suitable amounts of nicotine; the motor end-plates, also, are blocked by curarine, the active principle of the poison curare. Nicotine acts on the ganglia, and curare on the end-plates, much as atropine acts on the heart, smooth muscles or gland cells. They prevent the nerve impulses from being transmitted across the junction, but do not prevent the release of acetylcholine at the nerve endings by the incoming impulses. The action is on the ganglion cells, or the end-plates, respectively. The post-ganglionic fibres in the presence of nicotine, and the muscle fibres in the presence of curare, respond as usual to direct electrical stimulation.

Drugs with a curare-like action are widely used in general anaesthesia to produce relaxation of skeletal muscles. Since one cannot breathe without active skeletal muscles, this would be dangerous unless the synthetic curarine substitutes employed had a shorter duration of action than the natural products. Correspondingly, since autonomic ganglia when paralysed have as their most widespread and dramatic consequence a profound vasodilatation and fall in arterial blood pressure, ganglion-blocking drugs are widely used in the treatment of patients with chronic high arterial pressure (hypertension). The actions of the short-lived acetylcholine and the prolonged nicotine would be unsuitable because of their initial stimulating action. Synthetic substances such as hexamethonium bromide have an uncomplicated blocking action which is inconveniently short-lived, but many other synthetic products are now available which have a longer-lived action in blocking the autonomic ganglia in hypertensive patients.

ADRENERGIC BLOCKING DRUGS

An adrenergic receptor blocking substance is one which combines with the α- or β-receptors and hence prevents the action of their transmitters. There are α and β blockers.

Alpha blockers
The ergot alkaloids (e.g. ergotamine, phentolamine and phenoxybenzamine and piperoxan) block α-receptors. They thus produce peripheral vasodilatation. On the whole, they antagonise the pharmacological effects of circulating adrenaline and noradrenaline. They have less effect on the transmitter action of these substances.

Beta blockers
These are usually analogues of isoprenaline, the most commonly used being propranolol which acts chiefly on the heart. Since it prevents the action of transmitter on the β-receptors, the heart is slowed and its force of contraction is diminished. It can be used in the relief of angina and also to abolish cardiac arrhythmias (due to digitalis therapy, for instance).

Adrenergic neurone blockers

Various drugs inhibit the release of noradrenaline at adrenergic nerve terminals. They have no effect on circulating adrenaline or noradrenaline. Guanethidine and bretylium are examples; they are hypotensive drugs because they block all the effects of adrenergic nerves.

Adrenergic transmission is also impaired by the so-called "false transmitters". α-Methyldopa is converted to α-methylnoradrenaline in the nerve terminals and is released when nerve impulses reach the terminal regions. α-Methylnoradrenaline has very much weaker actions than noradrenaline itself, and thus adrenergic activity is considerably reduced.

Reserpine

The Rauwolfia alkaloids deplete tissue stores of 5-hydroxytryptamine and the catecholamines. Reserpine, one of the former alkaloids, has a sedative action probably due to this effect in the hypothalamus or brainstem. It also has a hypotensive action, which may result from removal of catecholamines from the adrenal medulla and from post-ganglionic sympathetic nerve terminals.

The mechanism is thought to be that the drug causes difficulty in storage of transmitter in vesicles, synthesis and uptake being normal. If the transmitter is not enclosed in vesicles, monoamine oxidase quickly destroys it in the mitochondria.

32. Section A: The Liver

R. D. Harkness

Introduction

The word *endocrine*, meaning "inwardly secreting", is used to describe the function of glands that add material to the blood flowing through them; this is in contrast to *exocrine* glands, which make a secretion that is collected by a system of ducts and poured out at a distance from the gland as, for example, saliva into the mouth, and so on. The idea of an endocrine secretion was a late development in the history of physiology. The glands in which this was detected earliest, for example the adrenal medulla, were distinctive small structures producing small quantities of substances named *hormones*, or "chemical messengers", with specific effects, in low concentrations, on other organs. It is customary to consider together as endocrine glands the following: the pituitary gland, the thyroid and parathyroid glands, the islets of Langerhans of the pancreas, the adrenal glands and gonads. However, other organs also produce endocrine secretions, for example the kidney and liver. The liver is also an exocrine gland producing bile, and at the same time in this sense the largest of the internally secreting glands, producing most of the proteins found in plasma other than those concerned with immunity. It also has an important function, in relation to almost all the glands to which the term endocrine is applied by custom, of destroying their secretions. In so far as these glands "control" the functions of other organs and tissues, this control can only be varied if there is some means of removing the active substances. The liver is the principal means of doing this and we will begin by a description of its functions.

THE LIVER

The *exocrine secretion* of the liver goes into the gut, into the duodenum just below the exit from the stomach, and is called *bile*. It is stored and concentrated in man, and most other vertebrates (not mice and horses), in a side branch of the bile duct, the *gall-bladder* (gall is an old-fashioned name for bile). Bile is isotonic with blood but contains more bicarbonate and less chloride ion. Its main constituents are *bile salts*, surface active steroid compounds essentially similar in action to detergents, that disperse, or bring into solution in water, fats and like substances; and *bile pigments* that are the breakdown products of the central, iron-containing, part of the haemoglobin. The products of the liver are plasma proteins, *albumen*, *fibrinogen* and *prothrombin*, and probably all the other factors concerned in the clotting of blood, protein carriers of "hormones" and iron, and a number of metabolic products of importance that effectively it alone produces, notably *glucose* and *urea*. The other major function of the liver is the modification of components of blood plasma, notably "hormones" as noted above. But the liver has a capacity to modify a great variety of substances, including drugs, a process often called "detoxication" though it may, in fact, make them more toxic. This process is analogous to its action on "hormones". Essentially, it consists in the introduction of hydroxyl groups into lipid soluble substances that can penetrate cells and the conjugation (esterification) of these groups with glucuronic acid. The effect is to produce a more water-soluble substance that is excreted into the urine because it no longer penetrates into the cells of the renal tubules, and so diffuses back into the blood after going out through the glomerulus. The liver has an additional function of storing a number of substances, notably *glycogen*, *iron*, *copper* and vitamins A and D.

Structure of the liver

The liver is the largest single individual organ in the body of a mammal (apart from the pregnant uterus), weighing in man 1–2 kg or about 2 per cent of the body weight. There are, of course, other types of tissue that are present in total greater quantity, skeletal muscle making about 40 per cent of body weight, bone about 10 per cent and the dermis of the skin about 10 per cent. In animals of different size the weight of the liver gets proportionately less as the body weight gets larger, according approximately to the following equation:

$$\log \text{liver weight} = a + b \log \text{body weight}$$

where a and b are constants and b is about $\frac{3}{4}$. The resting metabolic rate and oxygen consumption alter in approximately the same way.

The liver contains several sorts of cell but the greater part of its volume, about 95 per cent, is made up of characteristic *hepatocytes* or *parenchymal cells*. These are large cells—in a rat, say, about 20 μm across, with large nuclei, say, 8 μm diameter —and they do most of the work of the liver. The beginnings of the bile ducts start as small blind canaliculi in them, into which they secrete the bile. These canaliculi discharge into the smallest of the bile ducts that eventually form a single duct that discharges into the duodenum, having in most animals a storage vessel, the *gall-bladder*, on a side branch. This, and the ducts, are lined by a single layer of cubical epithelium that can absorb water and salts, and so concentrate the important functional components, bile salts.

Parenchymal cells are arranged around the blood "sinusoids" and these are lined with flat cells many of which (*Kupffer cells*) pick up particulate matter from blood. There are microvilli on the surfaces of these two sorts of cells at the places where they are close together. Although cells other than parenchymal form only a small proportion of the volume of the liver, they form a great proportion numerically (about 40 per cent in the rat).

Blood supply

The proportion of the cardiac output that goes to the liver, about one-fifth to one-quarter, or 1 to $1\frac{1}{2}$ litres per minute in man, gives it even greater importance than one would expect from its weight. The significance of this high rate of flow is that it enables the liver to act upon the blood rapidly, since about a fifth of the total blood volume flows through it every minute. This can be put another way, by saying that if a substance is completely removed from the blood in one passage through the liver (as may be true for aldosterone, an endocrine secretion of the adrenal cortex) its concentration in the general circulation will be halved after every three minutes. Measurements of the total blood flow through the liver in active unanaesthetised animals and people can, in fact, be made by using substances that, if present in low concentrations in blood, are removed almost entirely in one passage, notably *bromsulphthalein*. The difference in concentration in blood going in and out is then effectively equal to the arterial concentration. Rate of blood flow can be calculated from this difference and the total amount of the material removed from the circulation per unit time; the principle is the same as can be used for measurement of renal blood with diodone, and cardiac output (Fick principle).

The blood supply of the liver, in detail, is unusual in two ways. In the first place, there are two supplies: the usual *arterial*, supplying about one-third of the total blood flow, and another from the *portal* vein, which collects the blood from the gut and provides about two-thirds of the liver's blood. The liver is thus in series with the gut and all material absorbed goes there before it reaches any other part of the body. But the portal blood has already supplied oxygen to the gut and so has a lower concentration than arterial blood. The importance of the arterial supply to the liver is that it provides most of the oxygen that the liver needs. (These measurements of blood flow have been made partly on anaesthetised animals in which the vessels can be exposed, and partly on unanaesthetised animals with small flow meters placed at earlier operation in the vessels and connected to the surface.)

The other peculiarity of the liver circulation is that the smallest vessels in which the blood flows past the cells are not like the capillaries in most organs but are longer and more complex. They are called "sinusoids" and form a sort of filter bed. The blood flows out of the liver into the inferior vena cava just below the diaphragm through the hepatic vein. This vein, traced backwards into the liver, forms a tree whose smallest branches run down the centres of small, roughly cylindrical, masses or "lobules" of liver cells, $\frac{1}{2}$–2 mm in diameter and 3–4 mm long in man. Between the lobules are correspondingly small branches of the bile ducts, and of the hepatic artery and portal vein running together.

Less is known about the innervation or flow of blood through the liver and its control in normal life than perhaps one might expect. It seems it is reduced (up to 50 per cent) in exercise and can be increased up to double, for example by glucagon.

Exocrine secretions of the liver, inorganic ions, bile salts and bile pigments

Bile, the exocrine secretion of the liver, contains bile salts (about 1 per cent) and bile pigments (about 0·2 per cent) added to it by hepatocytes.

BILE SALTS AND INORGANIC IONS

Glycocholic and taurocholic acids are conjugated steroid molecules with both lipid hydrophobic regions and hydrophilic groups, used to disperse lipids into an emulsion, especially neutral fat from food in the gut, so that they can be digested and absorbed (see p. 324). They are also required for absorption of fat-soluble vitamins. Their hydrophilic groups enable them to form a surface layer in contact with water, over the lipid which associates with the hydrophobic part of the molecule. When bile is concentrated (up to ten times) in the gall-bladder by absorption of water and salts, the molecules associate in a similar way with one another so that the osmotic pressure rises less than in proportion to the concentration. This makes less work for the concentrating mechanism. Bile pigments are breakdown products of haemoglobin discussed elsewhere (p. 321). These components are carried in a fluid that it seems is provided by the smallest bile ductules. These have a large secretory surface about 100 cm^2 (ignoring the microvilli covering them), and a total length of 2 km or so. This fluid is isotonic with plasma and, like it, contains principally sodium as cation but has relatively more bicarbonate anion. The bile salts are themselves anions and account for the rest.

Bile salts (and, to an extent, pigments) are reabsorbed in the small gut and reused by the liver, which has a large capacity to remove them from blood. Thus a gram of bile salts given intravenously to a man will be virtually completely removed from the circulation in half an hour. With low concentration all is removed in a single passage through the liver, by the same concentrating mechanism that works on substances used to measure liver blood flow, such as bromsulphthalein mentioned above. The re-use of bile salts is known as the *enterohepatic circulation*. How it works quantitatively may be seen from information available for the rat. The whole animal contains a total at any time of about 12 mg of bile salts (body "pool"). The liver output per day in bile is about 50 mg, or about four times greater. Of this, only about 3 mg is newly made, the rest is reused material. Thus more than 90 per cent of the secretion is reabsorbed, mostly (again over 90 per cent) in the small gut, the rest in the large gut. Though normally there is little new bile salt that needs to be made, the capacity of the liver is such that production can be raised by up to twenty times (as, for example, if bile is deliberately drained away by leading the bile duct on to the skin of an experimental animal).

BILE PIGMENTS

The secretion of bile pigments by the liver has the particular interest that it can provide easily detectable evidence of liver function. If bile pigments accumulate in blood and so in extracellular space, they colour the skin and conjunctiva yellow, the condition of *jaundice*.

Bile pigments are made out of the central core of haemoglobin with its ring structure of four pyrrole rings and the iron attached in the middle. The first stage in the breakdown of haemoglobin begins in the reticuloendothelial system in spleen or liver and leads to the formation of *biliverdin*. The ring is opened by oxidation (3 M O$_2$, 6 M NADPH per mole, 1 mole carbon monoxide produced). Biliverdin is then reduced to bilirubin, which is carried on plasma albumen to the liver (2 moles per mole of albumen). Some drugs can displace it, for example salycilates. In the liver, bilirubin is taken up from the albumen on to specific proteins in the hepatocytes called Y and Z. These also bind bromsulphthalein. Deficiency of one (Y) may be associated with jaundice in newborn babies. Bilirubin is then conjugated with glucuronic acid. The enzyme involved, in the endoplasmic reticulum (p. 2), is called UDP glucuronyl transferase. The quantity may be increased by certain drugs, for example, phenobarbitone. An inherited deficiency of it in a strain of rats (Gunn rats) has been useful in investigating its function. The conjugated bile pigment is then excreted into bile and so into faeces, though some is reabsorbed and excreted by the kidney or again by the liver. The capacity of the liver to secrete bile pigments is very great, up to fifty times the normal output in a rat.

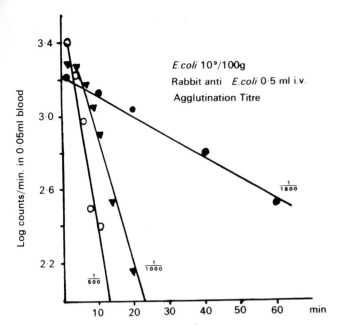

Fig. 32.1 Reticuloendothelial cells in the liver (Kupffer cells) remove particulate matter in the circulation. Bacteria for example are removed, the rate being greatly increased by the presence of antibodies to them (opsonins).

The graph shows the rate at which ^{32}P-labelled *Escherichia coli* are removed from the blood of mice by the reticuloendothelial system. The three curves are obtained when the anti- *E. coli* agglutination titre is 1/500 (fastest), 1/1000 and 1/1800 (slowest). (From B. Benacerraf in (1964) *The Liver*, Ed. C. Ronilla. Academic Press, London.)

Methods of investigating liver function

HEPATECTOMY

A lot has been learned about the functions of the liver by removing it. This requires first that another pathway for the venous blood from the gut be provided by joining the portal vein and vena cava, side by side with an opening between, a so-called *Eck fistula*. The first functional deficiency to become apparent after removal of the liver is the result of a fall in blood glucose level, which is progressive if none is absorbed from the gut.

OTHER METHODS OF INVESTIGATING LIVER FUNCTION

A fundamental requirement for investigation of liver function is knowledge of the composition of the blood going in and out, and of the rate of blood flow. This is difficult in an active animal, particularly to obtain samples of portal blood, but it can be done by having a small tube (catheter) inserted permanently and leading to the skin surface. Because of the difficulties, a lot of work has been done by perfusing the liver with blood after removing it. A large amount of work has been done *in vitro* on every kind of liver preparation such as slices, homogenates etc.

The liver and blood glucose

The liver is the only source of blood glucose in man and many mammals. The importance of glucose in blood and the control of the level is discussed elsewhere. The output of glucose by the liver may be measured from the rate that labelled glucose (^{14}C) put in the bloodstream is diluted by unlabelled glucose. The liver has a store of glucose in the form of glycogen, which may form up to 5 per cent of its weight. Glycogen is not stored alone but with about twice its weight of water; in fact, in a little bit of cytoplasm containing, for example, potassium and other ions. So when the store is used, as in starvation, potassium is liberated into the circulation. It is important to know about this from a practical point of view when one is concerned with the ionic composition of blood of patients. Storage of glucose as glycogen will conversely remove potassium from the circulation.

If the level of glucose in blood going into the liver is raised, the production of new glucose (glyconeogenesis) by the organ is reduced. In a dog about three-quarters of the glucose absorbed from the gut is removed by the liver. Most of the rest goes through and somehow causes the liver to reduce glucose production from other sources (glyconeogenesis) so that the net output is not changed.

Production of urea in mammals

The liver is again virtually the only source of urea formed from the nitrogen removed from amino acids by deamination. It appears that amino acids are destroyed by deamination to form urea if they are not used for protein synthesis by the liver; essential ones that cannot be synthesised are not stored and, if they are missing, then the protein requiring them cannot be synthesised.

Production of proteins: plasma albumen

The human liver produces about 12 g albumen a day for an average man (weight, say, 70 kg). It accounts for about a quarter of the new protein made by the liver daily. The total body pool of albumen is about 180 g, of which about two-thirds is in the blood vascular system, the rest outside. About an eighth of the average dietary requirement of protein (90 g) is used to make albumen, a bit more than this to make enzymes that go into the gut and the rest for cell "turnover". Isotopically labelled albumen is lost by a random process like radioactive decay; that is, the chance of an albumen molecule being destroyed is independent of its age (unlike the state of affairs for haemoglobin).

Little is known about the control of the rate of production of albumen but the rate of breakdown falls with its concentration in plasma (unlike the rate of breakdown of fibrinogen which is unaffected).

Proteins involved in clotting of blood

The liver produces most probably all the proteins involved in clotting of blood (Chapter 72). These, on the whole, turn over faster than plasma albumen. The following is a summary of them, giving the molecular weight and half-life.

The main blood-clotting protein is *fibrinogen*, factor I (mol. wt 340 000; 3–6 days) that is the actual source of the blood clot. The next one, *prothrombin*, factor II (mol. wt 62 000; 3 days) needs the fat-soluble vitamin K, as do factors VII (mol wt 630 000; 2 days), IX (mol. wt 50 000; 1 day) and X (mol. wt 86 000; 2 days), the absorption of which from the gut depends upon the presence of bile salts. If the bile duct is blocked, the accompanying jaundice may be associated with defects of blood clotting resulting from malabsorption of vitamin K. Vitamin K is found in green vegetables, but is also synthesised by micro-organisms in the gut. Upsetting these micro-organisms with antibiotics may also cause a deficiency.

The drug dicoumarol (Warfarin) interferes with blood clotting in the same way as a deficiency of vitamin K, and is destroyed in the liver. The capacity of the liver to do this is affected by certain drugs, notably barbiturates, that increase this capacity by inducing the formation of the enzymes concerned. It is important to realise this, when using dicoumarol to prevent intravascular clotting, if these other drugs are used at the same time.

The liver also produces factors V (mol. wt 29 000; 1½ days) and VIII (mol. wt 180 000; 0·5 day), antihaemophilic globulin. It also probably produces factors XI, XII and XIII. The liver also seems to be concerned in production of factors involved in the reorganisation that follows clotting of blood, fibrinolysis, and in the removal of clotting factors from the blood. Little is known of the factors controlling the production of these materials. Liver disease may be associated with abnormality of clotting and the importance is that one needs to check the absence of such abnormality before doing a liver biopsy for diagnostic purposes.

Other proteins produced by the liver

The liver produces at least some of the transport proteins, for example transferrin (iron) and transcortin (cortisol).

Detoxication by the liver

Many toxic materials, for example those found in plants that may be eaten by mistake, are lipid soluble, and this general property is associated with their capacity to enter cells. They are slowly excreted by the kidney because, having gone out of the blood through the glomeruli, they diffuse back through the cells of the renal tubules. If they can be made more water soluble they become less toxic and more easily excreted. The liver hepatocytes have a non-specific hydroxylase in the endoplasmic reticulum that introduces hydroxyl groups into compounds thus:

$$XH = NADH + H^+ + O_2 \rightarrow XOH + NADP^+ + H_2O$$

It uses a special form of cytochrome (P_{450}), and 95 per cent of the total of this is in the liver (about 2 g in man). Liver oxygen consumption can go up by as much as 40 per cent if a material is being hydroxylated in this way. This enzyme system is peculiar in having a very low turnover number (substrate molecules changed per enzyme mol per min), from 1 to 15 compared to 1000 to millions for most enzymes. The enzyme system may be induced by drugs, with a lag of about 1 day, and a half-life of decay of about half a day when they are removed. It may be lacking, for example, in premature and newborn babies and in starved people, giving undue sensitivity to drugs.

Destruction of hormones

The liver, as mentioned elsewhere, is the main organ removing hormones from the bloodstream. For some, for example steroids, a mechanism like that for detoxication is used; that is, introduction of hydroxyl groups, but with a specific rather than a non-specific enzyme. A measure of the ability of the liver to destroy hormones, now used a good deal, is the *metabolic clearance rate*; that is, the volume of blood completely cleared of material in a certain time. The maximum is about 2200 litres per day; that is, the total hepatic blood flow.

As mentioned elsewhere, many hormones are now known to be carried on specific proteins in plasma. One effect of this is to protect them from destruction by the liver.

Storage of material in the liver

We have already mentioned glycogen as stored in the liver. Other materials stored are iron, copper and vitamins B_{12}, D and A, especially the latter. Over three-quarters of the body store of vitamin A is in the liver. This is of some practical interest as it is a vitamin that can be toxic if taken in excess. It is formed originally in plants and accumulates in increasing quantities in the livers of animals higher in the food chain that eat the lower, for example: plankton → fish → larger fish → seals → polar bears. The livers of polar bears may contain so much vitamin A as to be toxic. But eating too much of the liver of other animals can be too. One of the effects of excessive vitamin A is to stimulate the outgrowth of bony processes from bone (exostoses). These may interfere with movement. Pet cats fed too much liver may get them so that they cannot bend their spines and lick themselves. Vitamin A seems to be stored in Kupffer cells, the other materials in hepatocytes.

The reticuloendothelial system in the liver

We have largely discussed so far, the function of the hepatocytes or parenchymal cells, as most is known about this. The reticuloendothelial system, which removes particulate matter, is also of importance, as evidenced by the fact that (in mice) over 90 per cent of particulate matter put into the bloodstream may be removed by the liver. Although simple visible particles like those of carbon in Indian ink may be used for quantitative measurements, material labelled with radioisotopes is easier to trace; for example, particles of chromium phosphate, denatured albumen labelled with ^{131}I, and bacteria labelled with ^{32}P by growing in a medium containing it. The amount accumulated in the liver can be detected from outside the body by the radiations. The details of what happens to particles have been studied with electron-dense ones like colloidal gold by injecting it into animals and killing them at intervals.

Behaviour of Kupffer cells to particles in the bloodstream

The first thing that happens to particles, at once, is that they stick to the cell surface. In a matter of minutes it folds over them and they are ingested into the cells. They lysosomes with their enzymes break upon them and they are dissolved if they can be attacked. If they do not dissolve, the cell may divide or other things may happen (for example, in the case of particles of asbestos), but it is not possible to go into this very large subject here. The particles that presumably have been of most influence in the evolution of function in the reticuloendothelial system are micro-organisms, and it is of interest that antibody to them in the blood greatly increases the rate at which they are removed (Fig. 32.1).

The removal of particles follows predictable rules. If C_1 and C_2 are particle concentrations at times t_1 and t_2 then

$$\frac{\log C_1/C_2}{t_2 - t_1} = \text{a constant } (K)$$

that has been termed the *phagocyte index*. Particles compete with one another, so one type can prevent another from being taken up, so-called *blockade*.

Abnormality of liver function

The liver, as it lies in series with the gut, receives the full force of any poisonous material that is absorbed. Possibly for this reason it has, like the cells lining the wall of the gut, a remarkable power of regeneration. This is demonstrated if part of the liver is removed surgically, as may be done easily in a rat. The lobes hang in the rat on relatively narrow stalks, and one can tie round these and remove them; it is easy to remove two-thirds of the liver in this way. This induces a massive outburst of cell division in the residual one-third, after a delay of about a day. The residue doubles in weight in a day or two and is well towards the normal size in a week, though in detail all parts do not return as rapidly as this. This growth process is not a true regeneration but a growth (hypertrophy) of the remaining lobes. The nature of the stimulus is still unclear. Little, if any, functional defect is seen after this massive removal of tissue. Similar regenerative processes take place in the human liver but they seem to be much slower, taking up to a year.

The big problem in which liver function is involved is alcoholism. The actual hepatic abnormality that is common is *cirrhosis*, which involves the destruction of the vascular architecture, impeding the flow of portal blood through the liver, and leading to increase in the size of alternative vascular pathways at the places where portal and normal circulation are in contact. For example, at the lower end of the oesophagus enlarged veins (varices) develop which may burst, with fatal loss of blood. Increase in portal venous pressure may lead to accumulation of fluid in the peritoneal cavity (*ascites*). Finally, liver function may fail. This state is characterised by loss of consciousness (coma), the reason being not exactly understood. Various things have been suspected, for example, accumulation of ammonia in blood owing to failure to convert it to urea. Probably a number of factors combine to produce the effect.

Section B: Bloodstream Communication

General features of communication through the bloodstream

BLOOD FLOW THROUGH THE TARGET ORGAN

We may perhaps best begin by considering the simple question of an organ (say A) secreting something into the blood to communicate with another, or "target", organ (say B). What determines how much secretion reaches B? This clearly depends on the *blood flow* through B compared to other organs. If A and B are in a series (as the pancreas and liver, and hypothalamus and pituitary through their portal systems) then all the secretion reaches B. If they are not (and A secretes into the general circulation), B will only receive material at a rate that depends on the proportion of the cardiac output that goes to it. So the rate of blood flow through B will be important.

Endocrine secretions, in some cases, increase the rate of flow of blood through their target organs, so increasing the quantity of secretion received and, presumably, their effect; for example, the secretion going from the pituitary to the thyroid gland, and from the gonads (oestrogens) to the uterus.

DESTRUCTION BY ORGANS OTHER THAN THE "TARGET"

If a secretion were removed only by the target organ, then eventually this would get it all. But this is not so, and the rate at which other tissues receive secretions and what they do to them is important. The organ that seems to be most actively involved is the *liver*, which removes almost all of some hormones that reach it (for example, aldosterone from the adrenal cortex, p. 193). The liver appears, in fact, to destroy almost all hormones and can affect their concentration so as to affect function. For example, when it is damaged severely, effects of excess of hormones may be seen, for example, of oestrogens in men causing the mammary glands to enlarge.

Nature of effects of endocrine secretions

The simplest form of effect of an internal secretion on another organ is to provide something that the organ cannot make itself. This is what the liver does for the brain when it secretes glucose, and may be considered as equivalent to an effect of a "hormone" often referred to as *"permissive"*. For example, the thyroid has "permissive" effects of this sort. This effect may, to some extent, "control" the target but it seems better to keep this term for cases in which there is a continuous quantitative relation between the rate of secretion of hormone and the behaviour of the target. The simplest form of control would be to have a relation between the concentration of the hormone in the blood going through the target and its activity, but it seems that the effect is often not as simple as this. There are delays in effects, so that the concentration may be more rapidly variable than the target organ can follow; and this may react more to change of concentration that concentration *per se*. The reactions of targets may have, in fact, characteristics like sensory organs (Chapter 53).

Endocrine secretions spread through the entire circulation. They are therefore also particularly adapted to produce a widespread effect acting simultaneously (at least within seconds) on a number of different organs. The adrenal medulla exploits this possibility acting on a variety of organs to a common end. Another example is the parathyroid hormone, whose principal target, bone, is a scattered one.

Specificity of hormonal effects

One of the features of hormones is that they reach all organs and tissues but only act on some. And they may act differently on the same type of cell in different sites; for example, adrenaline on smooth muscle cells, contracting one, relaxing another. The specificity seems to depend on specific receptors with specific connections into the organisation of the cell. Thus there are seen to be specific receptors for many hormones in the plasma membrane of the cell; for example, insulin. There are receptors for steroid hormones in the cytoplasm and nucleus.

Amplification by endocrine systems

One of the features of hormonal systems is an arrangement in a series, the most obvious being those that begin in the hypothalamus. A few cells here produce a secretion that goes to the adenohypophysis to produce another secretion that goes to a third gland that produces yet another that, in turn, has an effect. At each stage it seems more material is liberated so that the series can be regarded as an amplifying system.

Feedback of information

In hormonal systems involving "control" there is always a pathway by which information as to the effect of a secretion is fed back to its source. For example, the pituitary anterior lobe that controls the thyroid gland is itself affected by the secretion of the latter. In other cases where the hormone acts upon an "effector" organ, like the kidney, to alter the composition of plasma then a signal denoting composition is fed back through a "detector" to the site of origin of the hormone. Antidiuretic hormone acts on the kidney to change the water content of the plasma, this change being detected in the hypothalamus which controls the output of the hormone. Not enough is known yet for one to be able to describe these systems properly in quantitative detail.

Progress in research on hormones

It is well to discuss briefly the progress of knowledge about endocrine glands. This is an involved subject, in which the nature of the problem may not be clear at first and may only appear as research develops; in contrast to the problem of, say, skeletal muscular contraction which is clear from the beginning.

The first point is to establish that the supposed endocrine gland does secrete into the blood, and the nature of the secretion. Knowledge of the nature of the secretion has usually begun by the purification and identification of materials. This requires a simple method of biological testing sensitive to small quantities of materials; for example, the vaginal smear method for oestrogenic female sex hormones or a method using the ratio of Na to K ions in the urine of adrenalectomised mice for aldosterone (p. 193). Materials have usually been obtained first from extracts of the glands and may not be secreted. For example, there are several active substances in the adrenal cortex that are not secreted. The next stage then is to identify material in greater concentration in venous blood from the gland than in arterial blood going to it. Once the secreted material has been identified, the next stage is to make quantitative measurements of the concentration in the blood under different conditions and of the rate of secretion. Concentrations are usually low and measurement therefore difficult. A method that has been of great value is isotopic dilution, in which a small known amount of isotopically labelled material is added before purification and the amount of dilution measured on a purified sample. Losses in purification do not matter. Another method of relatively recent use for protein hormones is radioimmunoassay using, for example, hormone labelled with ^{131}I. This attaches, for example, to tyrosine without altering the essential properties of the protein. The quantity of material displaced from a complex with antibody is proportional to the amount of unlabelled substance in an added sample. Rate of output of substance is a different problem but can be measured, for example, by arteriovenous differences in concentration if rate of blood flow is known, and by the rate at which labelled material put into the blood vascular system of an animal is diluted by new unlabelled material added by the gland.

The other problems are the mode of biosynthesis of the hormone, the control of its secretion and the mechanism of its action. Biosynthetic pathways form a rather separate and special problem of a primarily biochemical nature. Little is known about mechanisms of release as yet. The mechanism of action, again, is primarily a biochemical problem in a compartment of knowledge that can be, to some extent, separated off without necessarily much loss of understanding of other aspects of the function of the hormone. The difficulty has been to identify the primary effect among the vast number of biochemical changes that many hormones produce. Many have been considered at various times: effects on enzymes, on protein synthesis, liberation of histamine, and so on. The one that seems most universally established is the formation of cyclic AMP (adenosine monophosphate).

At present the structure of most hormones is known. There is some knowledge of their mode of action, and of measurement of variation in blood concentration and output under different physiological conditions. The next stage, when enough of this type of information is available, is to begin to work out quantitatively the function of these glands as control systems.

Section C: The Adrenal Gland

We have considered the liver, which is a large gland secreting into the blood materials, used elsewhere, whose importance is most clearly seen in their absence—for example, glucose. It is of interest to go on to consider a gland that produces, in small quantities, messenger substances active in a positive sense on different organs; and in larger quantities, others that have qualities like some of the substances produced by the liver as well as messenger-like effects.

General structure of the adrenal gland

The adrenal glands get their name from the fact that in mammals they are very near the kidneys, one on either side in the fat behind the abdominal cavity and just to the headward side of the kidney. The glands in man weigh about 10 g, i.e. about 5/1000 of the body weight.

The adrenals consist of two distinct parts, a central medulla (L. *medulla*, marrow) surrounded by the cortex (L. *cortex*, bark) the latter being much larger, making over 90 per cent of the weight of the gland. The adrenal medulla is developed from the nervous system; the cortex, in common with the gonads, from the genital ridge.

The function of the adrenal medulla is relatively simple and will be described first.

THE ADRENAL MEDULLA

NATURE AND EFFECT OF SECRETIONS

In essence the secretory cells of the adrenal medulla correspond to post-ganglionic sympathetic nerve cells. But instead of delivering transmitter substance *noradrenaline* directly and discretely to effector cells (principally smooth muscle of blood vessel walls), they deliver it, or a similar substance (adrenaline) into the bloodstream through which it reaches all the organs and tissues of the body. The secretory cells are controlled by pre-ganglionic sympathetic nerve fibres using the same transmitter substance as elsewhere, namely acetylcholine.

The secretion of the adrenal medulla is, in general, produced discontinuously, suddenly, in a burst of activity under conditions of "stress", and has widespread and immediate results within the time it takes for the blood to travel round the body (10–15 s). The effects are like those of a general stimulation of the sympathetic nervous system, similarly varying according to the type of receptor (α or β, p. 182) present in the tissue or organ affected (commonly called the "target").

Thus in the *cardiovascular system* the secretions cause *arterioles* to constrict in some regions (e.g. skin) and dilate in others (coronary blood vessels) while they have no effect in others (central nervous system). The net effect is usually a rise in arterial blood pressure, and an increase in flow of blood through the central nervous system and heart. The rise of blood pressure is limited by the baroreceptor system (p 227). Noradrenaline has a greater effect in raising blood pressure than adrenaline. On the heart the secretions have the effect of increasing the force exerted at a given length of the muscle so that, for a given state of activity, it works at a shorter length; put another way, the chambers empty more completely when they contract. A related effect is produced on skeletal muscle: the strength of contraction of fatigued muscle is increased. This effect appears to be produced by increase in the ease with which energy is made available to the muscle from the glycogen in its cells. A similar effect is produced in the liver: breakdown of glycogen to glucose, which is put into the circulation in amounts that can raise the concentration in blood by a half or more. This effect is produced, it seems, by liberation in the cell of cyclic adenosine monophosphate (cyclic AMP). This activates an enzyme (phosphorylase) that forms glucose-1-phosphate from glycogen from which, by further steps, glucose is formed.

The secretion of the adrenal medulla has also a number of other effects on smooth muscle. In the *alimentary canal* there is a general relaxation with contraction of the sphincters; the *pupils* of the eye dilate; the smooth muscle in the walls of the small air ducts in the lungs (*bronchioles*) dilate; the small muscles contract that cause hair on the skin to stand up (to produce "goose-flesh", *arrector pili* muscles). This last effect of the "hair standing on end", a time-honoured method of describing fright, gives the clue to the nature of the general effect of the secretion of the adrenal medulla. Its secretion is produced in response to frightening or "stressful" circumstances, and prepares for strong muscular activity, redistributing blood where it is most needed, to the brain, heart and skeletal muscles, and providing a plentiful source of energy in the form of blood glucose, and so on. Organs not immediately needed are inhibited; the alimentary canal, for example; the longitudinal muscle of the stomach relaxes causing this organ to drop down lower in the abdominal cavity—all this perhaps combining to give the sensation described picturesquely as "his heart was in his boots".

The secretions of the adrenal medulla do also, it seems, have some importance in other respects, notably in response to cold; they can increase heat production by the liver by up to nearly half and this effect, it seems, is involved in regulation of body temperature.

Concentration of adrenaline and noradrenaline in blood and in the adrenal medulla

In the human adrenal medulla about 80 per cent of the active material present is adrenaline, 20 per cent noradrenaline, though in fetal life and at birth there is more noradrenaline than adrenaline. The concentration is about 0·05 per cent of the wet weight. It is contained in vesicles 0·1–0·5 μm in diameter in which the concentration is about 5 per cent of wet weight, or 15 per cent of the dry weight. The total amount in the gland is about 6 mg, or 20 000 times the amount normally circulating in the blood plasma; enough to supply 50 to 100 episodes of active secretion. Thus there is a store of material. Normal blood levels are of the order of 0.1–0.3 ng ml^{-1}, or in the range 10^{-9} to 10^{-10} M; about three-quarters of this is inactive as it is carried absorbed on the surface of red blood cells and platelets. The highest blood concentrations, 2000 or more times resting, have been produced by severe lack of oxygen.

Adrenaline disappears rapidly from the circulation with a half-life of minutes. The products of its metabolism, which are inactive, are excreted in the urine. From the amount, 3–7 mg per day, it seems that the store in the gland is enough for one or two days' normal activity.

Abnormal conditions affecting the adrenal medulla

A tumour of adrenal medullary tissue called a *phaechromocytoma* (Gk. *phaeos*, dark; *cytos*, cell) is not uncommon. These tumours secrete noradrenaline into the bloodstream giving rise to periodic "attacks" involving pallor of the skin, awareness of increased heart action ("palpitation"), headache and sweating. Tumours of this nature are hazardous to remove unless under pharmacological control using adrenergic blocking agents, both because of the possibility of liberation of active material into the bloodstream during the process and of drop in blood pressure afterwards. Undetected tumours are also a hazard at operations for unrelated conditions.

THE ADRENAL CORTEX

Structure

The adrenal cortex is not a homogenous collection of cells like the liver but shows a change of structure in successive layers (or zones) as one goes in from the surface capsule to the medulla (Fig. 32.2). The names of the zones are taken from the arrangement of the blood vessels and cells; in the outermost layer they are more or less spherical clumps (zona glomerulosa: L. *glomerulus*, dim. of *glomus*, a ball); on the next, in lines vertical to the surface (z. fasciculata: L. *fasciculus*, dim. *fascis*, a bundle) and in the innermost in a network (z. reticularis: L. *reticula*, dim. of *rete*, a net). The cells are large and many contain conspicuous lipid material from which secretions can be made, as they are reduced by a sudden increase in secretion.

The zonation in structure reflects a zonation in function; the outermost zona glomerulosa is concerned in control of electrolytes, Na$^+$ and K$^+$ principally by the action on the kidney of its secretion aldosterone; the inner zones produce in larger amounts the secretions affecting the metabolic functions with which the gland is concerned and also others, in particular sex hormones (but these are produced mainly by the gonads). The innermost layer is much enlarged in many fetal and newborn mammals and becomes smaller in a matter of days or weeks after birth.

The blood supply to the adrenal cortex resembles to some extent that of a single liver lobule, flowing from the capsular side inwards towards the medulla down long wide vessels (sinusoids) bordered by secretory cells; but again the immediate lining is of reticuloendothelial cells (like the Kupffer cells) that take up particulate matter. The significance of these phagocytic cells is not clear. Possibly their presence explains why tubercle bacilli may lodge specifically in the adrenals. As the blood flows in towards the centre of the gland, there is a connection between the cortex and the medulla through it but none in the reverse direction. An interesting result of this vascular connection, a type of portal system, is that the medullary cells live in concentrations of cortical secretions up to a hundred times higher than cells elsewhere, and it

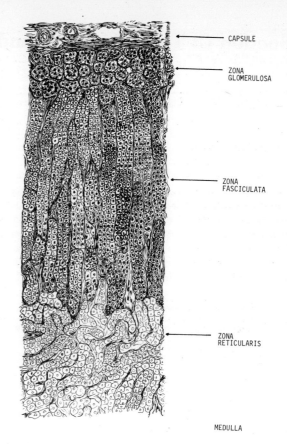

Fig. 32.2 Section of an adrenal gland of a man. Mallory-azan stain. Mag. about 100×. The medulla is not shown. The narrow-zona glomerulosa consists of small columnar cells closely packed in ovoid groups. Outer cell margins are next to capillaries. Nuclei are deeply stained; scanty cytoplasm may contain lipoid droplets. The zona fasciculata is the widest part of the cortex and is made up of polyhedral cells, larger than glomerulosa cells. Cytoplasm is in the form of narrow threads between numerous lipoid droplets. There are numerous mitotic figures. The zona reticularis is composed of cells in the form of anastomosing networks. Cells are 'light' or 'dark' depending on affinity for stain. They have a granular, pale-staining cytoplasm and pale vesicular nuclei. The dark cells are rich in lipoid droplets and have granules of brown pigment in the cytoplasm.

appears that high concentrations are necessary for the production of adrenaline. The rate of flow of blood through the adrenal cortex is about the same as that through the liver, about 1 ml per gram per minute.

A singularity of the adrenal cortex is that it contains a high concentration (up to 0.3 per cent wet weight) of ascorbic acid which is reduced when the secretory activity is increased (as it is also in the ovary after secretory activity). The function of the ascorbic acid is not clear but it seems likely that it is concerned with hydroxylation of steroids in their synthesis.

Unlike the medulla, the cortex has no nerve supply except to blood vessels. Secretory activity is controlled almost entirely by endocrine secretions from the pituitary gland (controlling the production of metabolic hormones) and probably from the kidney (in the case of aldosterone from the zona glomerulosa).

Effects of adrenalectomy

The adrenal cortex is very much more important than the medulla, which can be removed experimentally (in rats) by sucking it out through a fine tube leaving enough cortex to regenerate functionally. The effects of adrenalectomy vary in detail from one species to another. The most general change is an *increased susceptibility to "stress"*, that is to the effects of damaging agents of all sorts, physical injury, poisons, infections and so on. Injuries to which the animal would normally adapt and recover become fatal. This diminished resistance cannot be ascribed to any one single cause. Particular changes are a reduced inability of the kidney to retain sodium and a tendency to retain potassium instead; and a reduced capacity to form glucose from non-carbohydrate precursors (glyconeogenesis) in starvation. But many other changes can be demonstrated in individual tissues. Loss of sodium leads to changes in circulatory and extracellular fluid volumes with eventual failure of the circulatory system to cope with the needs of the body; though this may occur without fluid volume changes. These effects of adrenalectomy can be partially compensated by increased dietary intake of sodium chloride, and animals and people will both respond in this way spontaneously if given access to salt. As we shall see, the effects of adrenalectomy on electrolyte balance (Na, K) and on metabolism (glyconeogenesis etc.) are the results of deprivation of different endosecretions (hormones). We shall discuss the electrolyte effects first, though the active substance concerned was the last to be identified. The other effects are, as yet, still less well understood and seemingly more complex.

The adrenal cortex and sodium and potassium

There is evidence, for example from incubating different parts of the adrenal cortex *in vitro*, that the zona glomerulosa produces the secretory product mainly concerned in the metabolism of sodium and potassium, namely, *aldosterone*.

FUNCTION OF SODIUM

Sodium ions are of great importance as being the principal cations of extracellular fluid and plasma in blood, as contrasted with potassium intracellularly. The total volume of extracellular space depends on the total quantity of sodium ion in the body, and the regulation of this total is, to a great extent, separate from the regulation of the concentration of the ion by the hypothalamus and neurohypophysis, using the antidiuretic hormone (p. 142). This is effected rapidly, in a matter of hours usually, mainly by adjustment of the output of water by the kidneys. Regulation of the total Na^+ is effected more slowly by the kidneys than the intake. The great importance of the total lies in the dependence of extracellular fluid and blood volume upon it. The fatal effects of deprivation of water are to great extent a result of reduction of blood volume and consequent failure of the circulatory system; difficulties arise more commonly from lack of sodium, for example from loss in sweat in a hot environment. Adjustment is on the one hand by reducing output as much as possible, for example the human kidney can reabsorb sodium from glomerular filtrate to give a concentration lower than London tap-water. A balance of loss and intake can be maintained on as little as 1–2 g per day in man. On the other hand, there is clear evidence that the intake is also adjusted according to requirement by a specific appetite for salt, at least in mammals.

ACTIONS OF ALDOSTERONE

The major effect of this substance is on the kidneys, causing sodium to be retained in the body and the output of potassium to be increased. These effects are brought about by action in the cells of the renal tubules, affecting the transport by them of these materials from the tubules back into the blood. The behaviour of other tissues towards these ions is probably affected too but in a way that is, as yet, less well defined. There is a curious delay of about twenty minutes to an hour between the administration of aldosterone and its effect.

Control of aldosterone production

The synthesis and subsequent secretion of aldosterone is controlled, directly or indirectly, via humoral and haemodynamic factors and by the electrolyte content of the blood (see Table 32.1).

Giving ACTH causes a short lasting increase in aldosterone production but hypophysectomy does not usually lead to any decrease in aldosterone production. The most sensitive determinants of aldosterone secretion are the plasma volume and the serum potassium concentration.

In human patients, sodium lack causes about a fivefold increase in aldosterone secretion from the adrenals, i.e. from about 100 to 500 μg per diem. A raised sodium intake lowers the secretion to about 50 μg per diem.

The controlling mechanism is the *renin–angiotensin–aldosterone* system. If plasma volume falls, renal blood flow also falls and with it the renal arterial pressure and pulse pressure. This releases the enzyme renin from the juxtaglomerular apparatus in the kidney (Fig. 32.3). Renin acts on a circulating protein to produce *angiotensin I*, which is then enzymatically converted in circulating blood to *angiotensin II* (see p. 147). Angiotensin II acts directly upon the adrenal cortex, and aldosterone formation and secretion are increased. Aldosterone has its action on the kidney to increase sodium reabsorption by the tubules (distal convoluted tubule and loop of Henle). This then tends to restore the effective plasma volume towards normality.

THE ADRENAL CORTEX AND RESISTANCE TO STRESS

The adrenal cortex has another set of major functions that are, in many respects, less well understood than those we have just been discussing. The heading above is, perhaps, an oversimplified summary because aldosterone may be involved in the changes summarised as "diminished resistance to stress", that is, to strenuous and potentially damaging circumstances. The effect is perhaps best explained by the example of a biological test for activity of adrenal extracts, the so-called Selye–Schenker test, after the originators. Briefly, adrenalectomised rats were put in cold in which normal rats would survive but they would not. The activity of adrenal cortical extract was assayed quantitatively by their ability to preserve life when administered under these circumstances. Part of the defect in the animal that lowers its resistance is a diminished ability to form new glucose from non-carbohydrate precursors (glyconeogenesis). The secretions that remedy this defect are two, *cortisol* and *corticosterone* (called *glucocorticoids* for this reason, but this term is perhaps better not used). They have some activity of the same type as aldosterone but much less. Their most important functions are "enabling" or "permissive"; that is, they are required for a number of processes—for example, growth of the whole animal, development of the mammary gland in pregnancy and lactation. Although there are specific receptors for them in cells, it is difficult to see their function as a "controlling" one, as there seems no clear evidence of a quantitative relation between the

Table 32.1 *Hormones of the adrenal cortex*

Name of hormone	Secretion rate	Plasma concentration	Measurement of urinary output	Substance measured	Stimulus to adrenal	Actions		
Cortisol (from Zona fasciculata)	15–30 mg day^{-1}	a.m. 15–25 μg dl^{-1} p.m. 5–10 μg dl^{-1}	Porter–Silber reaction (colorimetric reaction with phenylhydrazine)	17:21 dihydroxy 20 ketone	ACTH	*Metabolic* Hyperglycemic catabolism of protein. Redistribution of fat. Potassium release from muscle	*Renal* Normal GFR water excretion. Increases Na$^+$ reabsorption at distal tubule. Potassium loss	*Inflammation* Decreases ca[pillary] permeability. resistant to h[yaluroni]dase. Lymph[ocytes] Lysosomes k[ept]
Aldosterone (from Zona glomerulosa)	50–200 μg day^{-1}	0.1–0.15 μg dl^{-1}	Immunoassay or double isotope derivative assay	Aldosterone-18-glucuronide	Change in plasma volume and K$^+$ mediated via renin–angiotensin system. Also other stimuli including ACTH	*Renal* Increases Na$^+$ reabsorption from loop of Henle. Increases Na$^+$ reabsorption by K$^+$ and H$^+$ exchange in distal tubule. Exchange of Na$^+$ for NH$_4^+$ in collecting ducts	*Extrarenal* Decreases Na$^+$/K$^+$ ratio in sweat and saliva	
Adrenal androgens (from Zona reticularis)	DHA 15–25 mg day^{-1}	DHA 0.4–0.5 μg dl^{-1}	Zimmerman reaction (colorimetric dinitrobenzene in alcoholic KOH)	17-Keto group	ACTH. Pituitary trophic hormones. glands.? LH	Growth of body hair. sebaceous Secretion N$_2$ retention. Maturation of skeleton		

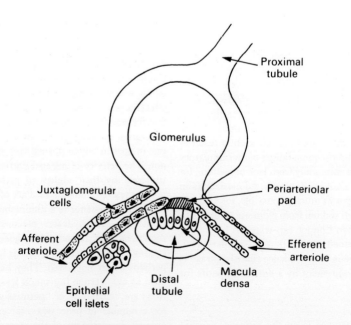

Fig. 32.3 The juxtaglomerular apparatus.
Lowered pressure in the afferent arteriole, or hypoxaemia, results in the secretion of the hormone renin into the venous outflow from the kidney. (From W. F. Cook (1963) in "Hormones and the Kidney", *J. Soc. Endocrinol.* **13,** 251.)

amount secreted and the effect, as in the case of aldosterone. They do have, also, some definite "effects" other than on the production of glucose, though these are curiously negative. Thus they cause the lymph glands to diminish in size, and the thymus and the number of lymphocytes and eosinophil cells in the blood to diminish. Eosinophils, it appears, are made to break up in the bloodstream. These are a part of a more general "anti-inflammatory" effect, for which these substances have been used clinically, though now replaced by synthetic substances of related structure but more active in this respect. This anti-inflammatory effect is so important in treatment in clinical medicine that it needs a separate discussion.

Anti-inflammatory effect of corticosteroids

Inflammation is a general term applied to the changes that follow injuries to tissue, as these have common features whatever the cause; for example, damage by mechanical or chemical means or heat. It is perhaps best summarised by the example of mechanical damage. The prevention of entry of micro-organisms through damaged surfaces and the repair of these and internal injuries have presumably played a predominant part in the evolution of inflammation. The immediate effects of damage are largely on the vascular system (minutes) clotting of blood that has leaked out and construction of small vessels at the actual site of the damage, but followed shortly (tens of minutes) by dilatation of vessels in the tissues round the site of injury (distance of centimetres). This on the skin makes the area red while the increase in blood flow makes it hot, hence the term "inflammation". With the dilation of small vessels there is an increase in permeability with leakage of plasma from blood into the tissues (oedema). The significance of this is probably to pour out antibacterial substances (antibodies among plasma proteins) into the damaged region. The increased blood flow similarly brings an increased number of white blood cells into the region and they go out of the blood vessels (hours to days). These stages are commonly accompanied by discomfort, pain or, on the skin, itching and increased sensitivity ("tenderness") so that pain may be caused by stimuli that would not normally cause it. A later change is often the production of new connective tissue which may join together discontinuities produced by the damage (wound healing) but may also take place when there is no damage of this sort, for example round parasites. This new connective tissue seems to come from cells round small blood vessels and is associated with their growth into the region where it is formed. It is commonly called "granulation tissue", from the fact that in an area of damage to skin it arises in discrete regions round small blood vessels, to give a granular appearance.

The inflammatory response seems to have the function of preventing bacterial invasion, healing the damage and preventing more by drawing the attention of the individual to the injured part. But these biologically "good" effects may be accompanied in some circumstances by "bad". An example, perhaps now the classic one as it involved the first major use of these substances in medicine, is the inflammatory reaction in joints that is found in rheumatoid arthritis. In this there is an inflammatory process in the synovial membrane of the joint and an erosion (destruction from the edge) of the joint cartilage. This inflammatory process is painful and destructive but seemingly useless biologically. Another example of "bad" effects arises from an inflammatory process in the cornea of the eye. Such a process, if continued, involves growth of blood vessels in from the edge of the cornea and new connective tissue is formed. This in the recovery process forms scar tissue. The transparency of the normal cornea appears to be the result of its peculiar structure of unusually thin (20 nm diameter) collagen fibrils of uniform diameter regularly spaced apart. The fibrils of the new formed collagen vary several fold in diameter and include many of larger size than in the normal cornea (Fig. 32.4). They are also irregularly arranged and the combination

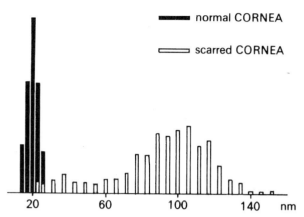

Fig. 32.4 The distribution curves of the diameter of collagen fibrils of a normal and scarred human cornea. Measurements of 1000 fibrils have been made for each curve.

of these causes scattering of light so that the region where they are becomes opaque and interferes with vision. Another example of an effect of inflammation that people want to be without is the discomfort of sunburn. One would suspect a danger, in suppressing inflammation, of weakening resistance of the tissues to bacterial invasion. There is evidence that this does happen with sometimes the effect of producing an overwhelming and fatal infection.

As mentioned already, commonly synthetic steroids are used for their anti-inflammatory effect rather than the actually secreted molecules. One advantage is that many act for longer periods, which may be related to the fact that they are less rapidly destroyed by the enzyme systems that act on the natural substances. The new steroids are also chosen to have less undesirable other effects. One of these side effects is the production of an excessive glucocorticoid effect giving rise to a condition like diabetes mellitus; another that cannot be avoided is an arrest of growth in young people; others are the production of gastric ulcers, thinning of bones (osteoporosis) leading to fractures, particularly to collapse of vertebrae, loss of muscle tissue and weakness. A final complication that cannot be avoided is the suppression of the activity of the person's own adrenal cortex by a mechanism that will be discussed later.

Cortisol and corticosterone

The adrenal cortex contains a large number of steroids some of which are active biologically. Confusion arose at first because it was supposed that these were secreted. It appears now that almost all the materials found are intermediary substances in the synthesis of the secreted substance, whose identity has been established by examination of plasma in the general circulation. In man the secretion of the adrenal cortex that is produced in greatest quantity is *cortisol*, about 10 to 15 mg per day, with about 1 mg per day of *corticosterone*; that is, cortisol, corticosterone and aldosterone are produced in a ratio of about 100:10:1. The concentrations of cortisol and corticosterone in human plasma average normally

100–150 ng ml^{-1} for cortisol (10^{-7}–10^{-6} M) and 10–40 ng ml^{-1} for corticosterone. But there may be considerable diurnal variation (fourfold) with a minimum somewhere near midnight and maximum in early morning (6–9 a.m.). In the fetus a high proportion of this material in blood comes from the mother, and concentrations a few days after birth fall to about a third of adult level, rising thereafter. The "turnover" time of these substances in man is rather longer than for aldosterone, about 2 hours, largely because they are less rapidly destroyed by the liver (for reasons that appear below).

It is interesting to note the small quantity of material produced by the adrenal cortex compared to the weight of the gland: about 2 mg g^{-1} gland per day, or its own weight every year and a bit; this small quantity is perhaps an indication of the difficulty of making the material. There appears to be virtually no store of many of the secretions in the adrenal cortex, or only enough for a few minutes secretion. The concentration of these materials in plasma is increased (up to five times) most notably by stress, for example surgical operations. The effect appears to be the result primarily of the psychological rather than physical stress. For example, the output is increased not only in oarsmen rowing in a race, but also in the cox of the boat.

CARRIAGE OF CORTISOL AND CORTICOSTERONE IN BLOOD AND THEIR DESTRUCTION

Only about 5 per cent of these steroids normally in plasma is free, the rest is bound reversibly mostly to a specific carrier protein, "transcortin" or "corticosteroid-binding glubulin" (CBG), produced, it seems, in the liver. They may also be bound less strongly to plasma albumen. CBG is an β globulin with molecular weight about 50 000 and has one binding site for cortisol per molecule. It turns over more slowly than cortisol (half-life about a week). The quantity does not vary much normally, but in pregnancy there is a rise of twice or more but almost no change in the free steroid. This rise is probably a result, at least partly, of increased circulating oestrogens. These have the same effect in oral contraceptives. There is not much excess of this carrier, the total capacity of it being about half used at normal plasma concentrations. An interesting consequence is that if the plasma concentration rises above the carrying capacity, say about twice the normal concentration, all of the excess is free except a comparatively small amount attached to plasma albumen. As a consequence, a rise of four times in the total concentration can produce a rise of five times more (20×) in the free material. The exact significance of the carrier is not clear but it provides, in effect, a readily available store of material. Material on the carrier is, it seems, not passed through the glomerulus and not immediately available for destruction when it passes through the liver, as this organ only removes 10–15 per cent of the total in one circuit (compared to nearly 100 per cent of aldosterone). The liver destroys these materials in the same way as it does foreign substances, primarily by hydroxylating them (for example the keto group of C-3) and in many cases conjugating the hydroxyl group with glucuronic acid.

CONTROL OF OUTPUT OF CORTISOL AND CORTICOSTERONE

The outputs of cortisol and corticosterone are controlled by an endocrine secretion of the anterior lobe of the pituitary known as *adrenocorticotrophic hormone* (ACTH) or *corticotrophin* (Gk *tropho*, nourish), a peptide of about 40 amino acids. The output of this is, in turn, controlled by a *releasing factor* secreted by the *hypothalamus* into the hypothalamico–hypophyseal portal venous system (*corticotrophin-releasing factor*, or CRF). This system is discussed in more detail on page 203. The primary effect of ACTH is to increase the output of cortisol and corticosterone by the adrenal cortex. This is an immediate effect taking place in minutes. The output can be increased five times or more above normal and removal of the pituitary reduces the output to one-tenth, so that the total range of output that ACTH can control is about fiftyfold. Under continued influence or a raised concentration of ACTH the adrenal cortex gets bigger (up to three times) by increase in size of the individual cells and increase in their number of mitotic division. Conversely, in the absence of ACTH the secretion of the adrenal cortex almost stops and the gland tissue diminishes in size.

It is of interest that this arrangement of several endocrine secretions acting in succession: releasing factor, ACTH and adrenal corticol steroids can act as an amplifier system. Thus approximately 0·1 μg of corticotrophin-releasing factor (CRF—acting on the pituitary anterior lobe) causes the release of about 1 μg ACTH which in turn causes the secretion of about 40 μg of cortisol. Up to this point we can count an amplification of 400. If we include the further stage that 40 μg of cortisol causes the production of about 5600 μg of liver glycogen we get a total amplification of 56 000.

ACTH probably has a short half-life in the blood, a matter of about 10 minutes judging from injected material. The normal concentration in plasma is of the order of 1–200 pg ml^{-1}. As we have already pointed out, the secretion of aldosterone reacts upon its origin through its "effect" acting upon a "detector" system to form a "feedback" loop. A loop of this sort is involved also in the control of the secretion of cortisol and corticosterone but it is made by the concentration of these substances themselves rather than by their "effects". They act back upon the pituitary or hypothalamus and inhibit the production of ACTH. A "negative" feedback system of this sort is also involved in the control of the secretion of the thyroid. Why these two glands? The explanation that makes the most sense to the writer is that the arrangement is an automatic supply system which keeps up the concentration of material in the blood when tissues use more. There is evidence that tissues in general vary the amount that they take out of the blood according to varying circumstances. For example, when animals are exposed to cold more is used, and also in pregnancy.

Although the system may act in this way to provide an automatic supply, the level of this may certainly be varied; for example, the increased secretion in stress is not simply a result of increased demand. The level at which the feedback system operates is raised.

One important consequence of the existence of this negative feedback system is that treatment with artificial corticosteroids suppresses the production of ACTH. And the adrenal cortex then not only stops secreting but loses cellular substance (weeks) and the ability to secrete when stimulated again by ACTH. In fact the ability to respond is greatly reduced in a matter of hours. After a short period of corticosteroid treatment (days) it recovers quite fast (days), but after a more prolonged period it may take months. The consequence of these changes in the ability of the adrenal cortex to secrete is that corticosteroids are in effect drugs of addiction in the sense that they cannot be withdrawn without risk as the natural secretions do not replace the artificial at once. The sort of circumstances in which the withdrawal may occur are likely to be the very ones in which cortical secretions are particularly needed, for example after a car accident or in sudden acute illness.

Other substances secreted by the adrenal cortex

The adrenal cortex secretes small quantities of sex hormones, progesterone, oestrogens and androgens, though the significance of these secretions is not clear. Of particular interest are the androgens as the adrenal is the main source of these in the female and the quantity produced rises at puberty. The metabolism of the androgens produced is complicated by changes and interconversions that occur after they have been secreted, exactly where is not clear. An important metabolic product, quantitatively, is dehydroepiandrosterone, curiously secreted as the sulphate. The oestrogens appear to be the same as produced by the ovary. These are of practical importance after removal of the ovaries to control activity of the sex hormone-dependent mammary gland cancer that has spread to other regions. ACTH appears to increase the production of both androgens and oestrogens by the adrenal cortex.

A curious point of unknown significance is that some adrenal corticosteroids, notably aeticholonone, are pyrogenic—that is, can produce a rise in body temperature.

Abnormality of adrenal function

Diminished adrenal cortical function may result from destruction of the gland by tuberculosis but more commonly for no clear reason. The resulting abnormality is called *Addison's disease*, after the first describer. In addition to the expected changes (resembling those of adrenalectomy in animals), increased pigmentation of the skin is found. The explanation is probably through the breakdown of the negative feedback mechanism leading to increased production of ACTH and coincidental increase in melanophore-stimulating hormone (MSH) with which it has an amino acid sequence in common.

Overproduction of aldosterone is seen in *Conn's syndrome* associated with increased arterial pressure and other changes. Overproduction of cortisol and related substances is found in *Cushing's syndrome* associated with enlargement of the adrenals or a tumour, or a tumour of the anterior lobe of the pituitary, leading to a condition like diabetes mellitus and other changes.

An interesting abnormality is seen in *virilism* in women, that is, development of male secondary sexual characters, such as a beard, because of increased production of androgens. This appears commonly to be the result of an enzymic defect in the adrenals in the synthetic pathway of production of cortisol. This leads to high ACTH production through the feedback pathway in the endeavour of the body to compensate, which leads to the build-up of metabolites before the block. When these are precursors also of androgens there is an increase in their production.

Section D: The Thyroid Gland

The thyroid gland is conveniently described immediately after the adrenal cortex because they have certain features in common, though the nature of the secretions is very different. Like cortisol, the thyroid secretion has a large permissive effect, that is, it enables processes to occur but does not, it seems, directly control them. It appears to be used by tissues and there is a similar "negative feedback" or supply on demand system working through the adenohypophysis.

Structure

The thyroid is the largest of the traditional endocrine glands, weighing in man about 30 g. It lies round the trachea just below the larynx, consisting of two main lobes, one on either side, and a band of tissue joining them across the front of the trachea. Its name is derived from its shape, supposedly resembling a shield (Gk *thureos*). It consists (Fig. 32.5) of small rounded sacs, or *vesicles*, lined with epithelial cells

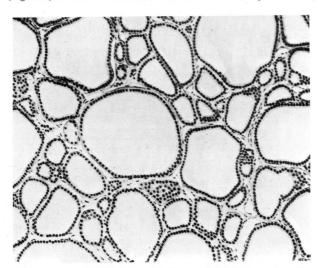

Fig. 32.5 Section of thyroid of adult man. Note the rounded follicles of various sizes, filled with colloid and lined by a single layer of cuboidal epithelial cells, (from Bargmann, 1939, after Herschel.)

and all separate. These vesicles contain stored secretion and the large size of the store (about a month's supply in man) is a unique feature among the endocrine glands. The secretion is stored as part of a protein, *thyroglobulin*. The term "colloid" is used for the whole material stored in the vesicles. The quantity of colloid varies with the degree of activity of the glands, being reduced by active secretion. The vesicles are lined by a single layer of cuboidal epithelial cells. The surface facing the colloid is found in electron microscopic examination to be covered with fine processes (microvilli). Adjacent sides of the cells appear to be fused to form so-called *tight junctions* seen in the electron microscope, a structural feature probably related to the prevention of material diffusing in and out of the vesicles between the cells. One of the features of the gland cells is an ability to concentrate iodide ions to produce a concentration several hundred times higher inside the vesicle than outside. Such a system can clearly only work if there are no leaks.

The thyroid is supported on a connective tissue framework with a plentiful supply of blood vessels. The connective tissue between the vesicles also contains, in some species, endocrine secreting cells (*C cells*) producing *calcitonin* (see p. 208). These cells appear to be evolved from the ultimobranchial bodies of lower vertebrates, while the thyroid vesicle epithelium is evolved from the branchial arch. It develops from the buccal epithelium at the base of the tongue with which it may be connected by a strand of tissue.

Effect of lack of thyroid secretion

The effect of removing the thyroid is not, apparently, very great in adult animals. The principal effect of lack of secretion in man is a fall in basal metabolic rate of 20–40 per cent, accompanied by some changes in metabolism and blood chemistry; and other things less well-defined, such as feeling the cold, and slowed thinking. But effects in the young are severe. Growth and development are stunted, the condition seen in man as *cretinism* (Fig. 32.6) in which thyroid secretion is lacking from birth. The

Fig. 32.6 (a) Cretin 23 months old; (b) the same child 34 months old, after administration of sheep thyroid for 11 months. (Case of W. Osler.)

untreated individual grows slowly, development of bones and teeth are retarded, but most particularly the central nervous system, so that the individual is mentally retarded.

The nature of thyroid secretion

The secretion of the thyroid is derived from two molecules of the amino acid tyrosine with addition of iodine. Two principal compounds are secreted, *thyroxine* with four iodine atoms (and so known as T_4), and *triiodothyronine* with three (T_3, the name *thyronine* being given to the compound without any iodine added). These compounds are amino acids and are made and stored in the peptide chain of the protein *thyroglobulin* in the vesicles of the gland. They appear to be released by proteases digesting the rest of the peptide chain away from them.

The greater part of the secretion, over 90 per cent, is thyroxine (T_4) but there is some evidence that this is converted to triiodothyronine (T_3) before it is "used" in the tissues. In this connection it is interesting that there appear to be specific binding sites in the anterior pituitary for T_3 but not T_4; as we shall see, an important feature of the control of thyroid secretion involves its inhibitory action on the pituitary gland.

Effects of thyroid secretion

The major effects of the thyroid secretions can be described as "permissive"; that is they are needed for growth and development and, in the adult, for a number of processes, for example lactation, but they do not appear to control these processes, though they may in some cases. The classic example is the control of metamorphosis in amphibian tadpoles. This process may be made to take place before the normal time by administration of thyroxine, and the indications are that the production of this substance by the thyroid gland normally determines the onset of metamorphosis. This transition from an enabling to a controlling function may well represent a general process in the evolution of endocrine control. It seems a logical transition to arrange for a material that is necessary for a process to initiate it.

In the adult mammal the principal effect of thyroid secretion is to increase the basal metabolic rate (calorigenic effect). Thyroxine and triiodothyronine produce about the same total increase in heat production, about 4 MJ (1000 Calories) per milligram dose, but the effect of thyroxine is longer lasting and goes up correspondingly to a lower maximum. Both substances act slowly, the effect of thyroxine in a rat taking about three days to reach its peak and lasting about a week. Probably as a result of this slow action the thyroid secretions generally produce no effect on excised tissues *in vitro*. On the other hand, tissues excised from animals previously treated with them continue to show increased oxygen consumption when tested *in vitro*. But not all equally: thus liver and muscle do, but tissues of the central nervous and reticuloendothelial system do not.

MECHANISM OF ACTION
The mechanism of action of thyroid secretions is still obscure, though a great deal of work has been done. Very many changes in enzyme systems can be demonstrated, but the primary effect is still unclear and it seems pointless to describe the confusing mass of information. One general effect that the secretions have is that they seem to uncouple enzyme systems degrading energy producing material for the production of adenosine triphosphate. So it seems muscular activity in people whose thyroids are producing an abnormally large quantity of thyroid secretion (hyperthyroidism) is inefficient, in the sense of using more fuel material than it would normally. This is at least part of the basis for the use of thyroid extracts in slimming to lose weight. Whether they are so used in the lives of animals is another matter about which we do not know. One of the features of the mammalian system of nutrition is that intake of most dietary components up to levels many times normal can be tolerated and balanced by increased output (through the kidneys). Exceptions are those components that produce energy, carbohydrate and fat and the residues of amino acids after the removal of nitrogen. These substances, if not used, are stowed as fat, which is why people get fat from overeating. If a high total food intake were necessary to obtain enough of some limiting factor, a mechanism for increasing the destruction of these might be useful biologically. Special mechanisms for getting rid of excess of such materials are known; for example, the sugary secretions of aphids which take in plant secretions of high carbohydrate and low amino acid content.

Another effect of excessive thyroid secretion in man is to sensitise sympathetic effectors, for example to cardiac muscle, to the transmitter substance, noradrenaline, and to adrenaline. This sensitisation appears to contribute to the dangerous effects of excessive secretion in a so-called "thyroid crisis", that may be fatal.

The formation and metabolism of thyroid secretion

SOURCE OF IODINE
Iodine for the formation of thyroid secretions is obtained from iodide in the diet. The normal intake for man is of the order of 100–200 μg per day (of I^-), the minimum tolerable (with the help of adaptation that will be discussed later) is about 10 μg per day. The total store of iodine in the body of man is about 12 mg, about 8 mg of which (or two-thirds) is in the thyroid gland.

PICK UP OF IODIDE BY THE THYROID GLAND
A great deal is now known about the metabolic pathways of iodine since convenient radioactive isotopes (e.g. ^{131}I) are available to trace it. Inorganic iodide (I^-) is absorbed in the small gut, and conveyed round by the bloodstream to the thyroid, which removes it rapidly, being able to remove from the blood virtually all that comes to it. The thyroid has a *concentrating mechanism* that can produce a concentration difference of as high as 300 times greater inside the vesicles than outside in the extracellular fluid. This mechanism can be blocked specifically by certain ions, thiocyanate (CNS), perchlorate (ClO_4), and permanganate (MnO_4). These complex ions have a size and shape rather similar to iodide and it seems the concentrating mechanism mistakes them for this ion. Blocking the concentrating mechanism effectively stops the thyroid from working.

An equally, or more, important factor in enabling the thyroid to obtain the maximum available iodide is the amount of blood flow through the gland. Iodide is treated by the kidney much like chloride, so the task of the thyroid, under conditions of shortage, is to remove it before it is lost in the urine. It cannot remove more than reaches it in the bloodstream. A minimum value for this may be obtained by measuring the clearance of iodide, that is the imaginary volume of blood "cleared", from which all iodide is removed, in unit time. This can be measured with, for example, ^{131}I, because the radiation penetrates through the tissues and can be collected and measured outside the body. The normal clearance is low, 10–60 ml min^{-1}, but it may go up to more than a litre a minute, or a fifth or more of the cardiac output, so high as to make a noise audible through a stethoscope, and producing an effect in the circulatory system like a direct arteriovenous connection.

Synthesis of thyroid hormones and storage

After the removal of iodide ions into the thyroid the next step is its oxidation to iodine ($I^- \rightarrow I$), and organic combination. This takes place into tyrosine already in the peptide chain of thyroglobin, that is itself, it seems, synthesised newly, at a rate appropriate to the amount of iodide available. The processing of organic combination of iodine can be specifically blocked by a number of substances, for example, thiourea, thiouracil. This block, like that of the concentrating mechanism for iodide, effectively stops the thyroid from working.

Much is known in detail of the biochemical mechanisms involved in the synthesis and metabolism of thyroid hormones, but it is out of place to consider this here.

Secretion and transport of thyroid hormones

As we have already noted, the secretion of thyroid hormones appears to involve the removal of the rest of the thyroglobulin storage molecule by proteolysis, though details are not known.

Thyroglobulin is not found in blood save in exceptional circumstances where there is destruction of thyroid tissue.

The human thyroid secretes about 50–100 μg of active material (T_3 and T_4) a day, or 1/50 to 1/100 of its total store. The secretion turns over slowly, having a half-life in blood of about a week (compare this to minutes for adrenaline, above). It is taken up by tissues generally and the iodine is removed finally returning to the blood (and some to the thyroid) as inorganic iodide again. In some animals (e.g. rats) thyroxine is secreted into the bile by the liver, combined with glucuronic acid or sulphate, and resorbed in the gut, a curious process, the function of which is unclear.

The thyroid secretions are carried in the blood almost entirely in combined inactive form, only about one part in 2000 being free (concentrations of the order 10^{-10}–10^{-11} M). There is then, in effect a large circulating store of material immediately available to tissues. Transport is on specific carrier proteins, *thyroid-binding protein* (TBP), or *thyroid-binding globulin* (TBG), of molecular weight of about 50 000, present in concentrations of 1–2 mg 100 ml^{-1} and carrying normally about one-third its maximum capacity of probably one mole of T_4 or T_3 per mole.

Most of the secretion is carried on this material, a lesser amount on thyroid-binding pre-albumen (TBPA), so-called because it comes in front of albumen in electrophoresis. This is a protein of molecular weight 60 000, about the same as TBP. There seems to be a different binding protein again in the fetus. These carriers occur in mammals, where they have been looked for, but birds appear only to have TBPA, no TBG. The amount of binding protein may vary. For example, in human pregnancy the quantity increases to about double, but the concentration of the free material is not changed. Certain materials including drugs (salicylate) can displace T_4 and T_3 from these carrier proteins.

As well as in the blood, there is a store of thyroxine in the liver reversibly bound to endoplasmic reticulum. This store is in equilibrium with blood and is large, about a third of the total thyroxine of the body, exchangeable with that in blood. It can be increased, by removal of thyroxine from the circulation, by certain drugs that induce formation of an increased quantity of endoplasmic reticulum, for example, phenobarbitone.

Control of thyroid secretion

The rate of secretion of the thyroid is controlled by *thyroid-stimulating hormone* (TSH) produced by the adenohypophysis (anterior lobe of the pituitary gland). TSH is a protein of molecular weight about 10 000. Its production is in turn controlled by a releasing factor (*thyrotropin-releasing factor*, TRF) transmitted from the hypothalamus to the adenohypophysis through the portal system of blood vessels. The hypothalamus can in turn be affected in a variety of ways by other parts of the nervous system, and so by the environment.

The effect of TSH on the gland is first to increase the rate of secretion by the thyroid, in a matter of minutes. Continued stimulation leads to a growth process with cell division and enlargement of the gland up to several times. This enlargement is accompanied by an increase in the rate of flow of blood, concerned presumably with trapping of iodide to provide for the increased rate of secretion.

The control of thyroid secretion by the pituitary gland has the same feature as that of cortisol by the adrenal cortex, namely a negative feedback. Thus a rise in the concentration of thyroid secretion in blood inhibits the production of TSH by the pituitary so again producing an automatic adjustment of supply to demand when utilisation varies, as it does, going up for example in pregnancy and in a cold environment. It seems likely that the range of variation of concentration of secretion in blood between maximal and minimal rates of TSH secretion is not large, and that the setting of the narrow range in which the concentration of secretion is thus regulated, can be altered by the hypothalamus.

Variation in the rate of thyroid secretion in life

We have already mentioned that the requirements of the body for thyroid secretion vary under different circumstances, being increased for example in pregnancy. Another condition in which the rate of secretion is increased is in exposure to cold, though finally when the animal is adapted it drops back to near normal.

Abnormalities of thyroid secretion in man

There is one abnormality of thyroid secretion that is particularly interesting because it illustrates the operation of the negative feedback system under unusual circumstances near to physiological. This is *endemic goitre* (L. *guttur*, throat), which is an enlargement of the thyroid gland found at one time in inland areas where there was a deficiency of iodine in the soil and hence in food grown on it. The lack of iodine prevents the thyroid from producing enough secretion, so that the negative feedback loop is broken. The growth in size of the thyroid gland is a consequence of the increase in TSH production that results. In the first instance, the effect is physiologically valuable because it enables the gland to obtain a larger supply of iodide through the increased blood flow, and so utilise a greater proportion of a restricted dietary supply. This abnormality no longer occurs because it is easily prevented by addition of a small amount of iodide (one part in 10 000–100 000) to table salt. A similar enlargement of the thyroid may occur in animals fed certain foods, for example kale and other plants of the cabbage family that contain antithyroid substances liberated enzymically (from

Fig. 32.7 Exophthalmic goitre.

glycosides) when the plant is chewed up, but not in cooked plants, when the enzyme has been inactivated by heat.

We have already mentioned the effect of deficient thyroid secretion in children, *cretinism*, usually the result of absence or deficiency of thyroid secretion at birth. Deficiency arising in the adult is associated with a condition called *myxoedema*, in which there is accumulation of material under the skin, for example of the face with puffy eyelids. This material is not ordinary oedema with increase of extracellular fluid, but contains hyaluronic acid (Gk *myxa*, mucus). The condition, it appears probable, is an autoimmune disease, that is, it arises from the reaction of the body against its own tissue.

The other common abnormality of thyroid function is an overactivity commonly known by the name of the person who first described it in the country to which one belongs (Graves in England, Parry in Wales, von Basedow in Germany and so on). It is associated with the presence in blood of an immune globulin called *long-acting thyroid stimulator* (LATS), which acts directly on the thyroid gland. The concentration of thyroid-stimulating hormone is actually reduced, through the negative feedback system. This again appears to be an autoimmune disease. Its most obvious feature to an observer is the appearance of protruding eyes. This look is produced partly by a retraction of the upper eyelids, so that the white sclerotic can be seen above the pupil; and partly by a true pushing forward of the eyeballs by accumulation of fat and extracellular material, like that in myxoedema, behind the eyeball. This is described technically as "infiltrative ophthalmopathy" (Fig. 32.7). Afflicted people look nervous and are, but it is beyond the scope of this book to go into the details of this disease.

Section E: The Pituitary Gland

We have described the function of two glands, both "controlled" by the pituitary gland, so it is perhaps a good idea to consider this gland next. The whole gland, like both the previous ones, is a combined one, in which tissues producing different secretions are associated together anatomically. The two main parts of the *pituitary* (L. *pituita*, slime) are the *posterior lobe*, or *neurohypophysis* (Gk *hypo*, under; *phyein*, to grow) so-called because it is derived from a downgrowth from the nervous system (hypothalamus) towards the buccal cavity, and the *anterior lobe*, or *adenohypophysis* (Gk *aden*, acorn), derived from a pouch growing up from the roof of the buccal cavity and wrapping itself round the nervous derivative. The adenohypophysis is itself a gland of multiple functions as it produces at least half a dozen separate secretions, each it seems from a separate cell type.

Structure of the pituitary gland

The human pituitary weighs about half a gram; about three-quarters of this is adenohypophysis, about one-quarter neurohypophysis.

The main features of the structure are shown in Fig. 32.8. The original cavity of the buccal pouch that forms the adenohypophysis remains as a discontinuity separating the larger anterior lobe from the thin *pars intermedia*.

THE ANTERIOR LOBE (ADENOHYPOPHYSIS)

The cells of the anterior lobe have been for a long time subdivided on the basis of the staining reaction of granules in them into *acidophile*, *basophile* and *chromophobe* signifying staining with acid or basic dyes or not staining. It now seems that this only partially represents a subdivision into types producing different hormones. The cells are called after the hormones they make by adding -*troph*—as *thyrotrophs* producing thyrotrophic hormone. The evidence for the identification is based upon the reaction of the pituitary to the hormone of the target gland, in this case the thyroid secretion. Increase or decrease (by removal of the gland) in the amount circulating is associated with changes in the appearance or staining properties of particular identifiable cell types. These cell types are not collected together in one place but generally distributed in the gland, though with a tendency for there to be more in certain areas than others (Fig. 32.9).

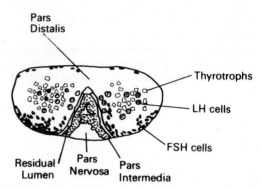

Fig. 32.9 Diagram of horizontal section through the rat hypophysis showing the distribution of the three types of basophil cell.

BLOOD SUPPLY OF THE ANTERIOR LOBE

The small vessels in contact with the cells of the adenohypophysis are more like those in the liver than capillaries, as they are in most tissues and are described usually as sinusoids. Many of the cells lining them, as in the liver, take up particulate matter, though the significance of this habit is quite obscure. The main feature of the blood supply of the pituitary, of great interest and functional importance, is that it is double. A large proportion of the blood comes from the capillary beds in the hypothalamus in a "*portal*" system. This makes a direct vascular connection through which material can be transmitted directly from the hypothalamus to the pituitary. As we shall see it appears to be the main channel through which the secretory activity of the pituitary is controlled (by "*releasing factors*" secreted by hypothalamic cells). There was at one stage some doubt as to the direction of flow in this portal system, but there are techniques now by which it can be directly observed (and samples of blood removed from it). It is also the major source of blood for supplying oxygen to the pituitary. If it is cut off, as it may be by thrombosis after parturition, enough of the pituitary tissue may die to cause a severe loss of function.

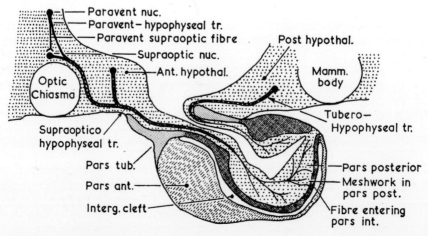

Fig. 32.8 Diagram showing medial sagittal section of pituitary gland of a cat and lower part of the hypothalamus.

THE NEUROHYPOPHYSIS (POSTERIOR LOBE OF THE PITUITARY)
The neurohypophysis consists essentially of the unmyelinated axons of nerve cells of nuclei in the hypothalamus above, ending blindly in tissue consisting otherwise of neuroglial cells (pituicytes). The secretion appears to be made in the nerve cells, and to travel down their axons to be secreted from them into the bloodstream ("*neurosecretion*").

Effect of removal of the pituitary gland

It is difficult to remove only the anterior lobe of the pituitary and in general the whole gland has been removed experimentally ("*hypophysectomy*"). But the effects are largely due to removal of the anterior lobe, as removal of the posterior lobe leaving the nervous tissue above it, has little or no effect.

Removal of the pituitary from an adult animal causes a diminution in size and rate of secretion of the other endocrine glands that it controls, the adrenal cortex, thyroid and gonads. The effect on the reproductive system is to stop effective activity. In the young animal removal of the pituitary additionally reduces growth, the animal growing to only half or less of the normal size.

Secretions of the anterior pituitary lobe

The secretions of the anterior lobe are all polypeptides, or proteins, and they may have up to hundreds of amino acid residues. Many of their amino acid sequences are known. They differ between animal species, and the hormone of one may be less active on another, particularly if this is far away in the animal kingdom; as, for example, a fish is from a mammal. The anterior lobe produces at least six separate hormones: *growth hormone* (GH; somatotropin), with a general action throughout the body; hormones acting on other endocrine glands, *adrenocorticotropic hormone* (ACTH) and *thyroid-stimulating hormone* (TSH; thyrotropin); hormones concerned with reproduction, *follicle-stimulating hormone* (FSH), *luteinising hormone* (LH) or interstitial cell-stimulating hormone, ICSH) and *prolactin*. The functions of the secretions controlling the adrenal and thyroid gland have been considered earlier. The hormones concerned with reproduction, that is, the so-called *gonadatropic hormones* (FSH and LH) and prolactin, will be considered later so that only growth hormone is left to be considered here.

Growth hormone is a protein of molecular weight about 20 000. Administered to young animals repeatedly it will cause them to grow to about double their normal weight. Tissues and organs grow in proportion, following an extension of the normal growth "laws", expressing the way in which different organs grow at different rates, as for example the liver.

In mature or older animals growth hormone has different effects, the one that has attracted most attention perhaps being that it raises the level of plasma glucose. If administered (to dogs) over a long period, it may cause a permanent abnormality of carbohydrate metabolism with raised blood sugar, a condition like diabetes mellitus. Growth hormone can produce structural changes in the cells (β cells) of the islets of Langerhans, which produce the endocrine secretion insulin that lowers blood glucose level. A fall in blood glucose level within the range found in various circumstances of normal life causes a secretion of growth hormone and a rise in the level in the blood. So growth hormone is concerned in the normal control of blood glucose level. The normal level in blood is of the order of 10^{-11}–10^{-12} M in man. "Stress" also causes an increased secretion.

The control of the anterior lobe of the pituitary

Before anything much was understood of the mechanism, it was known that processes that were apparently under the control of the adenohypophysis, for example seasonally variable reproductive function, could be affected by stimuli from diverse sources entering the nervous system. In this case it is the relative duration of light and darkness (which varies with the season). Evidence of a connection of the hypothalamus with the adenohypophysis was that the latter, removed from its normal site and grafted in a remote place in the body, ceased much of its effective function; for example, reproductive cycles (oestrus) stopped. The same effect followed if it was not removed but separated physically from the hypothalamus above by insertion of a layer of non-permeable material like polythene. Activity returned if the barrier was removed and the portal vessels regenerated. Further evidence was that the electrical stimulation of the hypothalamus could affect the adenohypophysis, causing for example ovulation by discharge of gonadotropic hormones.

The most commonly quoted work of this sort is that of the late G. W. Harris done on freely active rabbits. The method, in brief, was to introduce into the nervous system wires insulated except at the tip and coming from a small coil placed, at a preliminary operation, in the head under the skin that healed over it. Stimulating current was induced at will to flow in the secondary coil by placing the animal in a cage in the magnetic field of a large primary coil.

RELEASING FACTORS

The first direct evidence of the presence of substances in the hypothalamus that influenced the adenohypophysis came from two sources. The first was that material could be extracted from this part of the brain, but not others, that caused the release of particular adenohypophyseal hormones when injected into animals. The second and more immediately indicative of a direct link was that pieces of hypothalamus (but not other parts of the nervous system) grown in tissue culture in the same tube as a piece of anterior lobe of the pituitary, caused the latter to produce its endocrine secretion. The first work was done on corticotropin releasing factor (CRF) that caused the anterior lobe of the pituitary to secrete adrenocorticotropic hormone (ACTH). Curiously, the nature of this factor is still not established definitely. But other factors have been isolated and identified, in some cases, actually in the portal blood (in rats). All are small peptides with three to ten amino acid molecules; that is in general a tenth or less the size of the hormones whose secretion they stimulate. These materials, acting as they do in a restricted area directly connected to their site of origin, are needed and present only in small quantities, for example there is about 3 ng only of luteinising hormone-releasing factor (LHRF) in the rat's hypothalamus at puberty; put another way, a million rats would provide 3 mg. Up to half a million hypothalami of larger animals (pigs, cattle) may be needed to get enough material for chemical identification. The following factors have been identified.

Thyrotropin-releasing factor (TRF) causes secretion of thyroid-stimulating hormone (TSH).

Growth hormone-releasing factor (GHRF) causes secretion of growth hormone. It appears that there is also a factor that inhibits the release of growth hormone.

Luteinising hormone-releasing factor (LHRF) causes the release of luteinising hormone (LH). It can also cause the release of follicle-stimulating hormone (FSH). No separate factor releasing this (FSH) has been found, and it seems likely that only this one factor is concerned in the release of both gonadotropic hormones from the pituitary gland. Exactly how it works is not clear, but one can see obvious possibilities of different signals, a small amount over a long period, or a large over a short, as well as of variation of sensitivity of the different cells in the pituitary under the influence of the hormones that the gonads produce; for example oestrogens given a few hours earlier can stop the release of LH by LHRF in dogs.

In addition to these factors, others concerned with release or inhibition of release on melanophore-stimulating hormone (MSH) have been found. This controls the colour of the skin in, for example, frogs, by action of melanophores, but it seems to have no, or only a vestigial, function in mammals. The control of release of prolactin is unclear, but it appears an inhibitor (PIF) is primarily concerned as its secretion is not prevented, indeed increased in glands grafted away from the hypothalamus. The mechanism of action of releasing hormones is no more clearly understood than that of other hormones, but cyclic AMP appears to be involved again.

Work going on at present is concerned with the identification of the cells that produce the releasing hormones, and their production in the hypothalamus; and with the mechanism and control of their release. One of the difficulties is that the small size of the nerve fibres concerned makes it difficult, or impossible, to record from them.

The releasing factors, like other endocrine secretions, are destroyed in the liver and also the kidneys, and in one case at least (CRF) in blood. They are concentrated by the pituitary gland.

Abnormalities of function of the adenohypophysis

The most obvious abnormalities of function of the adenohypophysis concern growth hormone that may be virtually absent as a genetically determined condition in man, and other animals too (mice, for example). The human condition gives rise to the *pituitary dwarf*, about half the normal height when full-grown but normal as regards bodily and mental functions. Pituitary dwarfs often work in circuses.

An excess of growth hormone production, the reason for which is not clear, gives rise to giants up to more than seven feet tall. An excessive quantity of growth hormone may be produced in the adult by a tumour of the cells (acidophil) that make it. This gives rise to growth of the bones and soft tissues of the face and extremities, hands and feet; hence the name of the condition, acromegaly (Gk *akros*, at the end; *megas*, great). The condition may be associated with deficiency of the production of other secretions caused by destruction of tissue by the tumour cells. General deficiency of anterior lobe secretion may also arise, most commonly from thrombosis of the blood vessels leading to death of tissue.

Function of the neurohypophysis

Removal of the posterior lobe of the pituitary (neurohypophysis), as we have already mentioned, produces little if any effect, since the original source of the secretion is not here but in the nerve cells in the hypothalamus above. The secretions of the neurohypophysis are concerned with two functions; control of the water content of the blood plasma and of the activity of smooth muscle in the uterus and mammary glands.

Nature and mechanisms of secretion of the neurohypophysis

The secretions of the neurohypophysis, *antidiuretic hormone* (ADH, vasopressin) and *oxytocin* (Gk *oxus*, quick; *tokos*, birth), are both small peptides like releasing factors (nine amino acids). There are two main varieties of vasopressin differing in a single basic amino acid, arginine or lysine vasopressin, that are found in different animal species.

The secretions of the neurohypophysis appear to be made in nerve cells in the hypothalamus, and passed down their axons as part of a larger protein, to be split off and secreted finally somewhere along the length, into capillaries and the general circulation. The evidence is firstly that a substance of particular staining properties (*Gomori staining substance*) is found in the nerve cells and along their axons and accumulates above an experimental block in them; and secondly that radioactively labelled amino acids (for example, cysteine) are incorporated into material in the cells and move down the axons, as can be shown by autoradiography.

The actual mechanism of secretion is obscure but it appears to be produced by a burst of nerve impulses coming down the axon. One might suppose then that the mechanism is in principle the same as that involved in secretion of neurotransmitters at nerve endings, for example muscle end-plates, with the difference that the secretion is not localised to a particular cell surface but put into the general circulation. One might speculate why the secretion is not put straight into circulation from the nerve cell bodies as with the adrenal medulla. A reasonable explanation is that sending it down the nerve axons is a method of getting it through the "blood–brain barrier" (p. 126), without breaking the latter and exposing the nerve cell bodies to an environment that might upset their function.

The water content of plasma: "antidiuretic hormone"

Development of knowledge came in an unusual way that it is interesting to recount briefly. It was known for a long time that in man tumours causing damage in the pituitary region were associated with a condition called *diabetes insipidus*, in which urine production was at the rate of 20 or more litres a day (Gk *dia*, through; *baino*, go; *insipidus*, insipid, referring to the fact that the urine was watery). The cause was not clear, or even whether it was primarily a renal condition or a psychological one caused by drinking too much. The most obvious effect of extracts of the neurohypophysis appeared unrelated. It raised arterial blood pressure in anaesthetised animals, hence the name *vasopressin* for the active principle. Attempts to reproduce the condition of diabetes insipidus in animals by removing tissue were unsuccessful. The problem was cleared up by S. W. Ranson in the 1930s. He destroyed areas of tissue, without disturbing the surrounding tissues, by pushing in a metal needle coated with electrically insulating material except at the tip and then passing an electric current. The high current density at the tip of the needle destroyed a small area of tissue. The needle was introduced through a hole in the skull in a known direction, for a known distance, to make lesions in a predetermined region, the exact position being determined later after the effects had been observed. Ranson used cats and he found that destruction of the *supraoptic nucleus* of the hypothalamus (so-called because it lies above the optic chiasma) caused a permanent state of diabetes insipidus, after a delay of about a week, which was accounted for

Antidiuretic hormone
(vasopressin, from cattle)

Oxytocin
(from cattle)

Gly = glycine
Arg = arginine
Leu = leucine
Pro = proline
Cys = cysteine

Tyr = tyrosine
Phe = phenylalanine
Ile = isoleucine
Glu = glutamic acid
Asp = aspartic acid

as probably the time required for the nerve axons of the cells in the nucleus to degenerate, with the final loss of all secretory material. At about the same period of time, E. B. Verney re-examined the effects of posterior lobe extracts in un-anaesthetised and undisturbed dogs from whom urine was obtained, directly from the ureters, to provide a continuous record of flow. The effect was to reduce urine flow, hence the name *antidiuretic hormone* (ADH). It is now clear that the blood pressure-raising effect (vasopressin) is produced by doses much higher than the quantity normally secreted.

Verney also found that injection of solutions of sodium chloride of higher concentration than in plasma (hypertonic) into the internal carotid artery would cause a reduction of urine flow that could be mimicked by a suitable dose of antidiuretic hormone. Other hypertonic solutions produced the same effect but not if put into other parts of the circulation. The term "osmoreceptor" was invented for the hypothetical detectors of what was in effect a change in concentration of water in relation to various substances, the most important being sodium chloride. It is to be noted that the "osmotic pressure" that is relevant here is not the so-called "colloid" osmotic pressure of plasma proteins. It is much higher as it is concerned with molecules and ions present in much greater numbers than those of plasma protein. The exact nature of the detectors concerned is still unknown.

The control of activity of the uterus and mammary glands: oxytocin

Vasopressin can contract smooth muscle of blood vessel walls, though its physiological function does not involve the effect. The main function of oxytocin, it seems, is to stimulate the contraction of smooth muscle. The cells particularly concerned in its production lie in the paraventricular nuclei of the hypothalamus. In the mammary glands it causes smooth muscle cells round the secretory alveoli of the lactating gland to contract, and squeeze

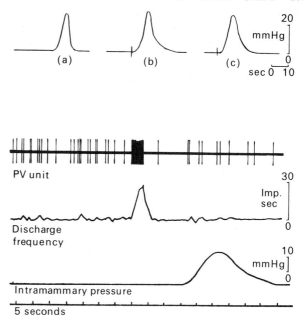

Fig. 32.10 The paraventricular neurosecretory cells in the milk ejection reflex in the rat. An anaesthetised rat is being suckled, and has a stimulating electrode in the neural lobe, a recording microelectrode in the paraventricular nucleus and one mammary gland is cannulated.

Records: Top line shows intramammary pressure in response to (a) suckling, (b) the intravenous injection of 2·5 *mU* oxytocin and (c) electrical stimulation of the neural lobe.

Bottom four recordings show the results of an experiment showing burst discharge from a neurosecretory cell (P.V. unit) about 15 s before the milk ejection reflex occurs indicated by the rise in intramammary pressure. The burst is the result of afferent impulses reaching the paraventricular nucleus from the nipple stimulated by suckling. (Reconstructed from data of Lincoln and Wakereby, 1973.)

milk onwards into bigger ducts, from which it can be sucked out by the young animal (Fig. 32.10). These smooth muscle cells, or *myoepithelial cells*, are of unusual shape, with many processes like the legs of a spider surrounding the alveoli. The secretion of oxytocin by the gland is produced reflexly by stimulation of the nipple by suckling, and it is necessary for the young to get milk. This can be shown simply by putting them to suckle an anaesthetised mother and weighing them to see how much milk they get, then injecting oxytocin into the mother's circulation (an experiment first done on a dog by W. L. Gaines).

The function of oxytocin in relation to the uterus is less well understood. It appears that each main contraction of the uterine muscle at childbirth is a result of secretion of a small quantity of oxytocin. The clearest evidence perhaps has been obtained by the observation by M. Gunther that at each uterine contraction milk was expressed from the breasts of a woman still lactating from a previous pregnancy. Oxytocin appears to be a major factor in the organisation of the process of parturition in man, though clearly other factors must be important, where there is more than one fetus so that only a part of the uterus is active at one time (e.g. in rats). There is some evidence that secretion can be produced reflexly from stimuli arising in the reproductive tract, for example distension of the cervix. What it is that makes the neurophypophysis become active at the start of parturition is unclear.

Oxytocin is destroyed in the liver but there is also an enzyme *oxytocinase*, in blood, increasing in concentration towards the end of human pregnancy.

Section F: Endocrine Control of Plasma Constituents

The endocrine glands we have discussed so far have all been concerned in enabling or controlling the activities of other glands or tissues, with no obvious general feature connecting them. The glands we are now going to consider are concerned, like the neurohypophysis in one of its functions, with the control of the concentration of particular plasma components.

PLASMA GLUCOSE AND ITS REGULATION

The normal level of glucose concentration in human plasma is about 80 mg per 100 ml (4·5 mmol l^{-1}). If it goes up above about double this figure, the capacity of the renal tubules to reabsorb the quantity that comes to them in the glomerular filtrate is exceeded, and the excess is lost in the urine. This is a loss of energy-producing material. But the effects of a fall in blood glucose level are much more disastrous, if this goes below about 40 mg per 100 ml (2·3 mmol l^{-1}). The reason is that glucose is the principal and almost the sole material used by the central nervous system as a source of energy, and it holds no store in the form of glycogen.

It is perhaps not surprising then, that plasma glucose level is subject to elaborate regulatory mechanisms. The principal organ concerned in this regulation is the liver which is the sole source, apart from the diet, of glucose for blood, either from other materials by synthesis or from its store of glycogen. This store is actually quite small, 50–100 g, or enough only for a few hours if it were the sole source of bodily energy. Muscle glycogen cannot be converted to glucose for blood directly, though it can be indirectly if it is converted to lactic acid and conveyed thus through the blood to the liver. Muscle and other tissues, adipose tissue particularly, contribute to regulation of blood glucose level mainly by variation in the rate at which they remove glucose to use it or store it as glycogen or fat.

Endocrine glands in regulation of plasma glucose level

Three endocrine secretions that can, among other actions, affect plasma glucose level, have been mentioned already: adrenaline that can quickly liberate glucose from liver glycogen; cortisol that can raise plasma glucose level by increasing new formation; and growth hormone discussed above. We will discuss here two more hormones that appear to be concerned exclusively with control of plasma glucose level. These are *insulin* and *glucagon*, both produced by the islets of Langerhans in the pancreas. Insulin reduces and glucagon raises the blood glucose level.

THE ISLETS OF LANGERHANS
The major part of the pancreas is exocrine glandular secretory tissue derived from an outgrowth of the epithelium of the alimentary canal. The islets of Langerhans form only about 1 per cent of its bulk, but are of the same origin. They are, however, isolated entirely from their origin into small clumps of cells secreting into the bloodstream. Their total weight amounts to about 1 g in man. Two main sorts of cells, named α and β, can be distinguished by their staining characteristics, producing

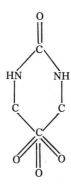

Alloxan

respectively glucagon and insulin. The primary evidence for this is produced by tying the duct of the pancreas, when the exocrine tissue atrophies, that is, in effect, it dies, leaving the endocrine tissue and hormone content unaffected. The opposite type of effect can be produced by administration of, for example, alloxan a pyrimidine like substance that specifically damages the islets of Langerhans. There is other evidence, for example by use of fluorescent antibodies to the secretions, and from fish in which islet tissue is collected into a separate mass.

It is perhaps worth noting here that the pancreatic endocrine secretions share with those of the stomach and of the gut concerned in the control of its activities and those of the pancreas, the peculiarity of being discharged into the portal bloodstream and so have to go through the liver before they reach any other tissue.

Effects of removal of the pancreas: diabetes mellitus

The effects of removal of the pancreas, in for example a dog, apart from those produced by lack of digestive secretions, are now known to be due chiefly to removal of the source of insulin. The first effect is to produce a rise in plasma glucose level, which in a day or so exceeds the renal threshold so that glucose appears in the urine (glycosuria). This was observed a long time ago by von Mering and Minkowski (1890) as a secondary result of the observation that flies collected on the urine of dogs after removal of the pancreas. A further stage in the effect of pancreatectomy is the production of increased quantities of acetone bodies by the liver (acetoacetic acid and β-hydroxybutyric acid). These also get into the urine and being acidic carry with them basic ions, principally sodium. The final effect of loss of base and fluid and accumulation of these materials in blood is fatal.

In man a very similar condition to the above occurs, called diabetes mellitus (L. *mellitus*, sweet; from the sugar in the urine). The cause of this condition is not exactly known because the islets may show no clearly defined abnormality. But it is common, affecting up to half a million people in England, or about 1 per cent of the population. It used to be fatal until insulin was prepared by Banting and Best in 1921. Some idea of the severity of the strain on the metabolic system can be got if one considers that a severe diabetic can lose up to 200 g of sugar a day in the urine, nearly half a pound and the equivalent of about 3·5 MJ of energy, or about a quarter of the ordinary daily requirement; in the acidosis ketone bodies may carry out up to 150 mmol of base a day, or the amount in about a litre of plasma. The severity of the disease varies and the milder cases show only a reduced glucose "tolerance", that is the rise of plasma glucose level after taking a standard quantity of glucose is to a higher level and more prolonged than in a normal person.

Insulin is a peptide of 51 amino acids having the distinction of being the first to have its amino acid sequence established by F. Sanger. Insulin from cattle, which was the source examined, consists of two peptide chains, A and B of respectively 21 and 30 amino acids, joined by a disulphide link between cysteine residues. It now appears that it is actually present in the pancreas as part of a larger inactive molecule, proinsulin. The amino acid sequence varies in individual animals and from animal to animal; for example amino acid no. 30 in the β chain in man is threonine, in the cat serine, in cattle alanine. There are corresponding slight differences in the biological activity of a given insulin in different species. The concentration of insulin in human plasma is usually in the range 10–50 units per ml (unit of cattle (bovine) insulin = 41 μg) or 10^{-9}–10^{-10} M. There is about 8 mg of insulin (200 units) in the human pancreas, a large store representing several hundred to a thousand times the amount circulating in the extracellular space of the body.

The action of insulin

The immediate and most obvious effect of insulin is to reduce the plasma glucose level; when it falls to around 50 mg per 100 ml (3 mmol l^{-1}) there is a reflex secretion of adrenaline, causing the liver to break down glycogen to put glucose into the bloodstream and arrest the fall temporarily. If enough insulin has been given, the fall will continue and at a level of about 40 mg per 100 ml (2·2 mmol l^{-1}) the function of the central nervous system is affected. Insulin is used in normal life in the control of plasma glucose level but appears to have, in addition, a considerable permissive effect. Diabetics on a constant dose can control their plasma glucose levels quite well. This appears to be a function of the liver that it can perform on its own.

The mechanism of action of insulin has been the subject of a great deal of work, but it is still not clearly established, and it would be out of place to discuss in detail here. It appears, among many other things, to affect the permeability of cells to glucose, though this is a difficult thing to investigate directly since glucose, as soon as it gets into the cell, is turned into something else. The effect on plasma glucose level is produced, it seems, by a combination of reduced rate of addition to blood and increased rate of removal and storage, particularly to form fat.

Control of insulin secretion

A good deal is now known about insulin secretion, principally because of the development of methods measuring the concentration in blood using radioactive-labelled antibodies (with ^{131}I or ^{125}I attached to them, radioimmunoassay). In the whole living animal the problem is complicated by the fact that insulin is discharged first into the hepatic portal circulation and the liver removes most of it. Concentrations in the portal vein can be twenty times higher than in the general circulation. The result of this may be that an increase in output of insulin by the pancreas has little detectable effect on the concentration beyond the liver. Factors which affect insulin secretion by the pancreas have been investigated, though for the most part either on perfused pancreas or on isolated islets in tissue culture. It is possible to isolate them if the structure of the pancreas is broken up with collagenase, an enzyme attacking collagen specifically. The most important stimulus to the release of insulin is glucose acting directly on the islets of Langerhans. For it to act, glucose must be metabolised by the islet tissue, it seems. Glucagon also, apparently stimulates release of insulin, though the significance of this is not clear. Vagal stimulation can also cause it to be released. Adrenaline inhibits its release. There is also some evidence that pancreozymin, which is concerned in the control of the secretion of enzymes by the pancreas, can release both insulin and glucagon. How these various effects act together in real life is as yet unclear.

Glucagon is a smaller peptide than insulin containing 29 amino acids. It has received less attention presumably because lack of it produces no disease requiring attention. It seems clear it is produced by the cells in the islets of Langerhans. It acts more rapidly than insulin, in minutes. It may be present as an impurity in samples of insulin. These were known for some time to produce a transient initial rise of plasma glucose level but the cause was not understood. The concentration of glucagon in blood is of the same order of magnitude as that of insulin (10^{-9}–10^{-10} M). The main determining factor in the rate of secretion appears, as in the case of insulin, to be the level of glucose in the blood going through the pancreas.

PLASMA CALCIUM

The concentration of calcium in plasma is approximately 10 mg per 100 ml (2·45 mmol l^{-1}). It seems it is kept constant in an individual to within about 1 per cent. About half this is ionised and half un-ionised, the latter in combination with protein for the most part. The two fractions are in equilibrium so that in a sense the protein could be regarded as a carrier store or buffer of Ca^{2+} ions, which is the form of calcium of physiological importance. The excitability of nervous and muscular tissue depends upon the Ca^{2+} ion concentration. A fall of sufficient magnitude causes spontaneous contraction or spasm of skeletal muscle known as *tetany*. In severe tetany when all muscles contract, the limbs come into position determined by the stronger flexion in the arm and extension in the leg.

There is a huge reservoir of calcium in bone (1–2 kg compared with 1–2 g in the whole extracellular fluid) and the level in blood is controlled to a great extent by taking it in and out of this. As in the case of plasma glucose there is an endocrine secretion that raises plasma calcium and another that lowers it. The first is the secretion of the *parathyroid glands*, the second the secretion of cells evolved from the *ultimobranchial body* (*C cells*), that commonly are found in the connective tissue of the thyroid gland between the vesicles.

The parathyroid glands

There are commonly four parathyroid glands, derivatives of the fourth and fifth branchial arches, that in man lie one at each upper and lower end of the lobes of the thyroid gland. They weigh only about 150 mg or 1/200 as much as the thyroid gland. Two sorts of cells can be distinguished, "chief" cells full of secretory granules and "oxyphil" cells full of mitochondria and nuclear protein.

The function of the parathyroid glands

The parathyroid glands have been removed accidentally with the thyroid when surgical removal of part of the gland was the common treatment of hyperthyroidism. The effect is to lower the plasma calcium level (in a matter of hours) and to produce tetany as described above. Extracts of gland or the active principle (*parathormone*) produce the opposite effect, a rise in plasma calcium level. This effect is largely produced by action on bone and ultimately involves the whole bone tissue including the organic collagenous matrix, not just the calcium salts. So there may be a rise in hydroxyproline excretion in the urine from the destruction of collagen. The effect on bone has been nicely shown by the late N. A. Barnicot, who grafted pieces of parathyroid and other tissue on top of the cranial bones of mice and observed bone to be removed from the neighbourhood of the parathyroid grafts. Bone is removed in the normal way, that is by osteoclasts, though it is not clear how they are stimulated to remove it; by the hormone directly or by its acting on the bone to alter it so that they remove it. There is an interesting relation here between bone as a store of calcium and bone as a skeletal tissue subject to mechanical stimuli that determine, other things being equal, the quantity and arrangement of it. The exact nature of the mechanical stimuli involved is not clear, but immobilisation, of a limb for example, causes bone to be reabsorbed. The calcium liberated in total bodily immobilisation may be in such a quantity as to precipitate in the renal tract as stones (calculi) if nothing is done to prevent this. If there is a conflict between mechanical requirements and a need for calcium, the latter may take priority (as in lactation, when the heaviest requirements for calcium arise in mammals). In this case calcium tends to be reabsorbed from bones subject to least mechanical stress first, like the sternum, rather than from the limb bones. The parathyroid hormone also has an action on the kidney reducing the output of calcium and affects the absorption of calcium in the gut.

CONTROL OF THE PARATHORMONE PRODUCTION

The parathyroids are directly responsive to the level of calcium in plasma. This has been shown in glands transplanted under the skin of the neck (in sheep by D. A. Denton), so that their vascular supply was available for sampling. Raising the level of calcium ion in the blood going into it, or lowering it with a chelating agent like ethylene diaminotetraacetic acid, produce the opposite effects on calcium concentration in the general circulation, after a delay sufficient for changes in secretion to be effective.

Calcitonin (thyrocalcitonin)

Calcitonin is a peptide of known sequence of 33 amino acids. Its effect is to lower the plasma calcium level, that is the opposite of parathormone. It acts more rapidly (minutes). The first evidence for it, obtained by D. H. Copp, was that plasma calcium level in the general circulation was found to fall more rapidly when the level of calcium ion in the blood going to the parathyroids and thyroid was raised, than it did after the parathyroids were removed completely; in other words, there was evidence of some positive factor pushing the plasma calcium level down.

There has been some confusion over the actual source of calcitonin. It now seems that developmentally the source is constant, that is from cells derived from the ultimobranchial body, but the anatomical position of them may vary. Normally they are found in the thyroid gland (C cells) in the interstitial connective tissue.

CONTROL OF CALCITONIN PRODUCTION

The output of calcitonin seems to be controlled like that of parathormone by the level of calcium ion in the blood, and possibly also to some extent by the rate of change of this. The relation of the two hormones is summarised in Fig. 32.11.

HORMONES AND THE ENVIRONMENT: PHEROMONES

As individual organs within an animal can communicate with one another by chemical means transmitted through the bloodstream, so can individual animals communicate through the environmental air or water, the transmitting substances being detected by sense organs. The word *pheromone* was invented (by P. Karlson) to describe the general group of substances used for communication through the environment.

We shall here discuss first the general question of chemical communication through the environment, and second the particular effect that animals can produce on one another by their mere presence; that is the effect of population density, as this is one factor that has a considerable effect on the endocrine system.

Pheromones

Chemical substances conveying information through the environmental medium have received perhaps most attention in insects, for example sex attractants in moths acting through the air over distances of hundreds of metres, and some of the many substances affecting behaviour of ants, trail laying, identification and alarm substances. We will confine ourselves here to pheromones in mammals. Although relatively much less is known about these than about the endocrine system, some effects are clearly established, for example marking of territory by glands in the skin and by urine, and glandular structures associated with the urinary

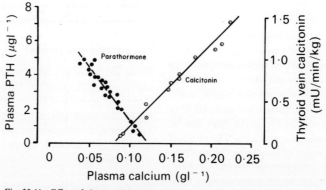

Fig. 32.11 Effect of plasma calcium on plasma levels of parathyroid hormone and calcitonin secretion rate.
(Potts and Buckle, 1968; Care, Cooper, Duncan and Orinio, 1968. Figures replotted.)

tract. There is also clear evidence of sex attractants which may be simple substances, lower fatty acids, present in vaginal secretions, it seems, in the case of rhesus monkeys. More interesting are unexpected effects, for example that in mice recently become pregnant the presence of a new male, other than the original one, in an adjacent cage, may cause the pregnancy to terminate by preventing implantation of the fertilised ova. Other apparently pheromonal effects are the induction of acceleration of the onset of oestrus by a male mouse, the suppression of oestrous cycles and adrenal cortical hypertrophy by other female mice in numbers.

Little is known about pheromones in human communities. As synthetic odoriferous substances are being used increasingly, and as the major problems of the day are those of interactions between people, it seems important that more should be found out. Influence by pheromone, it would seem, is unlikely to be subject to the censorship of rational thought. Indeed, it is possible that a pheromone might not be detectable as a smell at all.

Effects of population density on the endocrine system

If animals are allowed to breed in a limited environment, even with as much food as they require, the density of population does not rise above a certain level which varies with the species. Increase in population density above a certain level produces an increase in fetal and neonatal mortality, and the increase of fetal mortality may be accompanied by an increased incidence of abnormalities of fetal development. A further effect is suppression of gonadal function, so that reproductive cycles and production of fertile ova ceases. More general effects are seen in reduction of resistance to parasites and infections, leading to death at an earlier age. The effect of overcrowding may, however, be much more disastrous and lead to so called "stress death".

APUD cells

The acronym APUD stands for *"Amine content and amine Precursor Uptake and Decarboxylation"* cells. The idea that a set of cells exists in the body having in common the property of elaborating polypeptide hormones has originated in cytochemical and ultrastructural research over the last twenty years. The origin of these cells is supposed to be in the neural crest.

The ultrastructural characteristics are
1. Sparse rough endoplasmic reticulum.
2. Dense smooth endoplasmic reticulum.
3. High content of free ribosomes.
4. Electron dense mitochondria.
5. Membrane bound secretion vesicles.
6. Average cell diameter about 100 to 200 nm.

Table 32.2 gives a list of APUD cells that are known to produce a particular hormone.

APUD cells are thought to have originated some 500 million years ago in early vertebrates, as a separate system of nerve cells. Being motile they have come to be in sites remote from the nervous system, such as entodermal parts of the foregut and have adapted to respond to neurotransmitter amines rather than to any direct nerve supply. They are subject to tumour formation, the so called apudomas.

Table 32.2 *APUD cells and their secretion products*

Origin	Cell	Polypeptide
Anterior pituitary	Corticotrophic	ACTH
	Melanotrophic	MSH
Thyroid	C cells	Calcitonin
Islets of	β	Insulin
Langerhans	α_2	Glucagon
	$\delta\alpha_1$	Gastrin
Stomach	G cells	Gastrin
	A cells	Entero-glucagon
Duodenum and small intestine	S	Secretin
	D_1	Gastrin inhib. polypeptide
	EG (L)	Entero-glucagon

Section 3:

Systems for Transport in the Body

33. The General Function of the Circulation

General requirements

With the increasing complexity of cellular organisation as animals evolved, came the need for a "life-support" system for the individual cells. Cells, differentiated into specialised tissues and organs, were no longer capable of autonomous life without the aid of some system which could distribute oxygen and nutrient materials to them. In this section, we will see how the mammalian circulatory system is designed and operated to function as such a life-support system.

Variability of tissue oxygen requirements

The metabolic processes in cells mostly require oxygen. However, the amount of oxygen needed varies from time to time and from one organ to another. Muscle, for example, might need as little as 0.5 ml g^{-1} wet weight of tissue per hour of oxygen when it is at rest and as much as twenty-five times this value during severe exercise. Bone has a low oxygen requirement (0.03 ml g^{-1} h^{-1}), whilst kidney has a high one (2.5 ml g^{-1} h^{-1}).

The above considerations show that the system for distributing oxygen must be very flexible. Not only does the total amount of oxygen needed by the body change considerably from time to time, but also the regions to which it must be distributed alter.

Other functions of the circulation

Of course, oxygen transport is not the only function that is carried out by the circulation. Metabolic fuels are carried to cells, metabolic end-products are removed, hormones (chemical messengers) are distributed from one part of the body to another and heat is transferred from active regions to the exterior as necessary.

These functions of the circulation are clearly distinct from one another and the requirements for oxygen, for instance, may well conflict with the simultaneous requirements in a tissue for the removal of waste heat. In the kidney, a large blood flow is related to the excretory function; it is greatly in excess of the kidney cells' oxygen requirements. Skin blood flow alters, very largely, in terms of its activities in the transfer of heat across it. The oxygen needs of skin are quite low (0.2 ml g^{-1} h^{-1}) yet on occasion blood flow through it can be rather profuse (see Table 33.1).

The distribution of blood to various organs

As we have seen, the total oxygen requirement of the body is highly variable from time to time, so that a circulatory system capable of varying the total flow of blood to the body is necessary. This alone is not enough however, and in view of the conflicting and variable needs of the different organs, it is also necessary to have mechanisms for distributing blood to those organs needing it. Table 33.2 shows how the blood supply to the body at rest is split up and also how oxygen supply varies.

Table 33.1 *Oxygen consumption of various tissues*

Kidney	2.5 ml g^{-1} h^{-1}	Fat	0.20 ml g^{-1} h^{-1}
Liver	2.0	Bone	0.02
Brain	1.0	Blood	0.01
Heart muscle	0.8*	(Skin)	0.25*
Skeletal muscle	0.5*		

Figures are expressed as ml g^{-1} wet weight of tissue per hour. Those marked * are for conditions of rest. During activity these values may increase by a factor of up to 25 times. The oxygen requirement for skin may change by a factor of 5 depending on its temperature; its blood flow, though, may vary by a factor of 50.

Table 33.2 *The Distribution of Cardiac Output*

Percentage of cardiac output ($=5$ l min^{-1})		Percentage of total oxygen intake ($=250$ ml min^{-1})	
Kidney	25%	Muscle	40%
Liver	15%	Liver	15%
Brain	15%	Kidney	10%
Muscle	15%	Heart	10%
Heart	10%	Brain	10%
Skin	10%	Etc.	10%
Etc.	10%	Skin	5%

Approximate distribution of the output of the heart (at rest) compared with oxygen consumption of various tissues. Notice that the kidney receives more blood than is strictly necessary for supplying its oxygen needs (i.e. it is overperfused), while muscles get less. When they contract, muscles have greatly increased flow rates; in skin in a hot environment, the blood flow also increases.

General organisation of the circulation

We now enquire into the way in which the varying requirements for blood are met by the circulatory system. The mammalian system utilises two pumps (right and left sides of the heart), one pumping blood through the lungs to oxygenate it, the other supplying all the viscera and muscles with the oxygenated blood.

The variations in the total amount of blood required by the body are simply met by alterations in the output of the heart. We will be discussing the control systems which achieve the regulation of the heart for this purpose, at some length.

The variability in *distribution* of the blood, as the requirements of different organs change, is controlled by having a high pressure reservoir of blood (the arterial system) from which adjustable "taps" lead to the organs. These taps are the *arterioles*, small arterial vessels having much smooth muscle in their walls, which is capable of contracting and relaxing to vary the aperture of their lumens. It will be appreciated that the details of the control systems for the arterioles throughout the body will be complicated, because changes in flow to one part will inevitably affect the remainder of the body and this must be compensated.

Other things being equal, the flow through one of these arteriolar taps will be proportional to the pressure difference across it and inversely proportional to the resistance offered by the arteriole.

The reader will naturally be aware that this "self-evident" state of affairs is also found in many inanimate flow systems, including current electricity where it should be familiar as Ohm's law, or

$$I = \frac{E}{R}$$

where I is current, E is potential difference and R is resistance.

We may happily use the same equation in our biological systems, so that

$$F = \frac{BP}{PR}$$

where F is flow rate, BP is arterial pressure and PR is peripheral resistance. Notice that we can apply the relationship both to the flow through a specified tissue as above, or use it in the whole animal as follows

$$CO = \frac{BP}{TPR}$$

where CO is cardiac output, and TPR is total peripheral resistance. We will return to this in greater detail later.

From the above considerations, we can see that there will be two primary ways in which the flow of blood to a tissue can be altered. The first is to alter the calibre of the arterioles supplying the tissue; the second is by changing the level of the general arterial pressure. There will of course be several ways in which each of these two factors themselves can be changed.

Organ priorities

It might be thought that a simple system of local blood flow control by each organ would be sufficient for most purposes but this neglects the fact that certain organs have a priority claim upon the supply of blood. Neuropathologists detect permanent brain damage after the cerebral circulation has been cut off for as little as a minute.* The brain has no oxygen reserve; it must at all times be supplied with an adequate amount of oxygenated blood. The brain's supply of blood must therefore have absolute priority.

This is in contrast to the needs of the skin. Especially if the temperature is low, skin will function normally and survive for many hours of complete *ischaemia* (lack of blood supply). Skeletal muscle occupies an intermediate position, as those medical students, who have inflated cuffs on their arms above arterial pressure to investigate ischaemic pain, will know. The whole arm will quite happily survive periods of up to an hour, and possibly longer, with its arterial supply completely obliterated; certainly if the muscles are not active, there is little problem for many tens of minutes.

---

* This fact is of vital importance for the medical practitioner who may be faced with the resuscitation of drowned or electrocuted persons. If recovery ensues, but only after a period of several minutes' lack of cerebral oxygen supply, it is almost certain that severe, permanent impairment of brain function will follow.

Skeletal muscle, of course, is able to contract in the absence of oxygen, i.e. "anaerobically" until a large *oxygen debt* has accumulated. The debt of course has to be repaid when the circulation is once more re-established. An interesting physiological example of this is to be found in the diving mammals (such as seals, dolphins, whales etc.) who can remain beneath the surface of the sea for quite lengthy periods, yet swimming strongly the while. Apparently the blood supply to their muscles is cut off during the dive and they operate anaerobically. It is known that their muscles are rich in oxygen-containing myoglobin also. This cut-off enables all-important oxygenated blood to be spared for the cerebral circulation.

Control systems for distribution

To provide for these systems of priority and to deal with the variable needs of the different organs, cardiovascular reflexes or negative feedback control mechanisms operate. These are reflex arcs (=feedback loops) involving receptors for arterial pressure, a controlling centre in the medulla and an output link to the heart and to the blood vessels. The continuous and relatively unvarying requirements of the brain are met by a control which keeps the arterial pressure more or less at a constant level (the carotid sinus baroreceptors; see p. 227) and an arteriolar system which maintains a fairly constant resistance to flow. The highly variable flow rates demanded by muscle, low at rest and high in exercise, are achieved by both local and central *vasodilatation*, i.e. an opening-up of vascular channels, as and when more blood is required†.

Summary

Since flow rate is proportional to pressure drop and inversely proportional to resistance, two methods of controlling the circulation, both to individual organs and to the body as a whole are available.

1. Control of arterial pressure. This can be achieved by altering cardiac output, or total peripheral resistance.
2. Control of resistance to flow. This can result from changes in the calibre of arterioles which are small arterial vessels leading from the high-pressure arterial reservoir into the tissue concerned.

Different tissues have differing priorities in their claims for oxygenated blood. Brain which has no reserve of oxygen must at all times be supplied with blood. Skin can go without a blood supply for lengthy periods.

The control of these mechanisms is by means of feedback servoloops, some involving nervous pathways, others being actuated by hormones.

---

† *Teleology*, n. Doctrine of final causes; view that developments are due to the purpose or design that is served by them. (Gk *telos* = end). *Concise Oxford Dictionary*, Eds. Fowler, H. W. and Fowler, F. G. Clarendon Press, Oxford, 5th Ed. 1964.

Although the material in this introduction is presented in a teleological fashion, this way of thinking is in many circumstances useful as a method of understanding the physiology involved. We should not however take it too far, since what may be the best system in the normal state may well not be so in disease or other abnormal conditions.

34. The Design of the Circulation

The circulation of the blood

We owe to Harvey (1628) the conception and proof of the idea that the blood circulates, and as this step marks the beginning of modern physiology, it is of more than usual interest to note his argument. Harvey was able to show that the valves in the heart are so arranged as to allow the passage of blood in only one direction. Further, by watching the motion of the heart in the living animal he concluded that blood is expelled from the ventricles into the pulmonary artery and aorta during systole, and enters the heart from the venae cavae and pulmonary veins during diastole. He calculated that if only a drachm* of blood were expelled at each beat, the heart would in half an hour use up all the blood in the body, and so empty the veins completely and distend the arteries; from this he concluded that the blood must move in a circle, entering the veins from the arteries. Proof that the blood flows continuously in one direction he found in the cutaneous veins of the human arm (Fig. 34.1). It had been shown previously by Fabricius (1603) that if a ligature is tied around the arm the veins swell up distally, and present along their course little swellings which mark the position of valves allowing the passage of blood in only one direction, towards the heart. The significance of this discovery was lost on Fabricius, but was appreciated by Harvey, who made a further observation, which anyone may repeat on his own arm. The middle finger of the left hand is pressed firmly on a prominent cutaneous vein of the right forearm, and blood is massaged out of the vein proximally by rubbing the forefingers firmly up the vein past the next valve. If the forefinger is lifted, the vein fills from above only as far as the valve, but when the middle finger is lifted the whole stretch of collapsed vein rapidly fills from below. This process may be repeated indefinitely, blood always entering from the periphery; we must conclude with Harvey that blood is always entering the veins from the distal side, and the only source of this blood is the arterial system.

The missing link in Harvey's argument, namely, the connection between the arteries and veins, was supplied forty years later by Malpighi, when he discovered the capillaries.

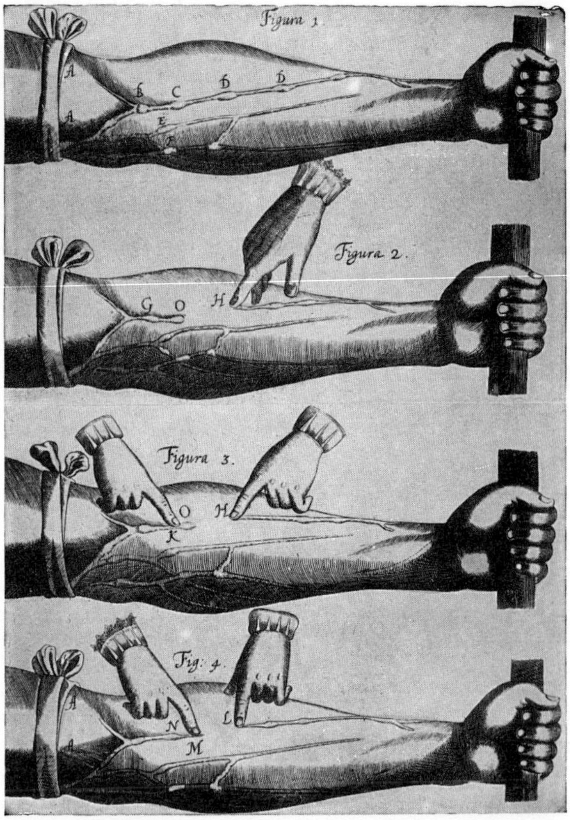

The general arrangement of the circulation

This is shown in Fig. 34.2. In man, as in all mammals, the circulation consists of two circuits connected in series, the greater or *systemic*, and the lesser or *pulmonary* circuit; accordingly the heart consists of two pumps, one for the lungs and the other for the rest of the body. Blood is pumped by the right heart into the pulmonary artery and on through the pulmonary capillaries to the pulmonary veins, and so enters the left heart. The left heart ejects the blood into the aorta, and then into the branching arteries which distribute it to the various organs of the body.

* A drachm is 60 minims; ⅛ ounce; or 3.55×10^{-6} m³.

Fig. 34.1 Harvey's figures illustrating the undirectional flow of blood in the veins. (1) The bandage AA is tied round the arm above the elbow, constricting the veins which become distended. The position of each valve is indicated by a swelling or knot, B, C, D, E, F. (2) The blood is pressed out of a vein from H to O with one finger, while another keeps the vein closed at H. No blood runs back past the valve at O. (3) If the vein is pressed by another finger at K, no blood can be forced backwards past the valve at O. (4) The vein is closed with one finger at L, as before, and emptied by stroking with another finger, M, towards the valve at N; the vein continues empty until the finger at L is removed, when it rapidly fills up from the periphery.

From the capillaries of these organs, the blood is collected into veins, and then returns to the right side of the heart via the inferior and superior venae cavae. During its passage through the pulmonary capillaries, the blood takes up oxygen from the air in the lungs, and loses carbon dioxide; in the capillaries of the other organs, supplied from the left side of the heart, the blood gives up oxygen to the tissues, and takes up carbon dioxide from them.

The blood flow to the various organs and tissues depends on the total output of the heart per minute, and on the proportion of this output which is sent to each of them. The regulation of the cardiac output, and of the eventual distribution of the blood, are thus the central problems of the physiology of the circulation.

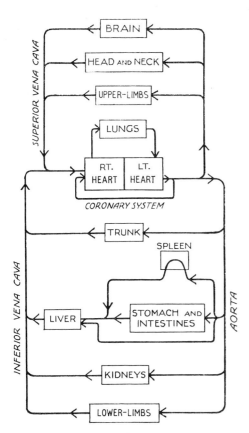

Fig. 34.2 Scheme of the human circulation.

THE BLOOD VESSELS

Routes for flow of blood

There are a multiplicity of routes along which blood can flow from the high-pressure arterial reservoir to the low pressure system, in other words from the aorta to the vena cava. The majority of these routes are *in parallel*; they are therefore only affected indirectly by each other in so far as a flow change through one may alter the general arterial pressure. Some major categories of through passage for blood are given in Fig. 34.2.

There are various possibilities:

1. Simple through passages; most in the body are of this type e.g. limbs, coronary arteries and veins, bone, liver etc.
2. Two capillary beds arranged in series. In the kidney, glomerular capillaries are in series with the tubular capillaries.
3. More than two capillary beds are traversed by the same blood. This happens in the portal system where splenic capillaries are in parallel with mesenteric capillaries both being in series with the liver sinusoids.
4. The pulmonary circuit consists of a single set of capillaries connecting the right ventricle with the left auricle. It has unusual properties for this reason.
5. The bronchial circulation is a direct shunt across the left side of the heart only (it is not shown in Fig. 34.2).

Vessels forming the vascular bed

On each contraction of the heart, blood is expelled from the ventricles at high pressure into the aorta and pulmonary artery. The *aorta* is a wide tube, the thick walls of which are largely composed of elastic tissue; like any other elastic structure, its capacity is determined largely by the pressure of the blood it contains. When blood is expelled during ventricular systole, the aortic pressure rises, the aorta is distended and so accommodates a large part of the blood expelled, the remainder escaping through the arteries. During ventricular diastole, the tension in the aortic walls maintains the flow of blood onwards through the arteries, and the aorta diminishes in size until it is again distended at the next heart beat. In this purely passive way the aorta (and to a less extent its larger branches, which are similar in structure) converts the intermittent flow from the heart into a continuous, though pulsating, flow in the arteries.

Arteries

The arteries have smooth muscle in their walls, which, on suitable stimulation, will contract actively or relax. The proportion of muscular to elastic tissue is greater in small arteries than in large, and the diameter, particularly of the smallest branches, or *arterioles*, can change over a wide range independently of the pressure within the lumen. The state of constriction or dilatation of the arterioles supplying any particular organ largely determines the proportion of the total cardiac output which is sent to it.

Capillaries

The capillaries are about 10 μm in diameter and the blood they contain is separated from the tissues by a single layer of flat endothelial cells which forms the capillary wall; it is accordingly here that the interchange of substances between the blood and the tissues takes place. In spite of their thin walls and the absence of muscle cells, the capillaries are capable of active contraction, and of exerting pressure of 60 mmHg (8 kPa) or more when contracted. In some resting tissues the majority of the capillaries are closed; during activity they open, and thus they also play a part in regulating the distribution of the blood to the organs. Although the capillaries are under nervous control, they are pre-eminently the vessels which react to chemical substances released during the activity of tissues which they supply.

Veins

The blood from the capillaries is collected into *venules*, which join up to form *veins*. These are wide and relatively thin-walled, and offer little resistance to the flow of blood. They are capable of active variation of calibre and are under nervous control. All but the smallest and largest veins contain valves, consisting of a number of semicircular folds of the intima projecting into the lumen. As a rule two such folds are placed opposite one another, and are so formed that when the blood is forced in a direction away from the heart, the folds float out into the bloodstream and block the vein. When the muscles contract the thin-walled veins are squeezed and blood is forced in the only direction it is free to travel, namely towards the heart. When the muscles relax, blood can enter the veins, but only from the arterial side. This "muscle pump" is an important mechanism for facilitating the venous return to the heart, as will be discussed later.

The size of vascular elements in a typical tissue

The artery which supplies an organ, and the vein which drains it, may be some 2 to 5 mm in diameter; there will be 10 000 to

Table 34.1 *Geometry of the dog's mesenteric vascular bed*

Vessel	Diameter (mm)	No.	Total cross-section area (cm²)	Length (cm)	Total volume (cm³)
Aorta	10	1	0.8	40	30
Large arteries	3	40	3.0	20	60
Main arterial branches	1	600	5.0	10	50
Terminal branches	0.6	1 800	5.0	1	25
Arterioles	0.02	40 000 000	125	0.2	25
Capillaries	0.008	1 200 000 000	600	0.1	60
Venules	0.03	80 000 000	570	0.2	110
Terminal veins	1.5	1 800	30	1	30
Main venous branches	2.4	600	27	10	270
Large veins	6.0	40	11	20	220
Vena cava	12.5	1	1.2	40	50

Data taken from F. Mall (1888); quoted by Burton, A. C. (1966). *Physiology and Biophysics of the Circulation.* Chicago. Year Book Publishers.

100 000 arterioles and venules, 0.02 to 0.05 mm in diameter; tens of millions of capillaries, some 0.01 mm in diameter. As the vascular tree branches, the total cross sectional area of the vessels increases: that of all the arterioles is around 10 times that of the artery, and that of all the capillaries some 10 times greater again. Thus the velocity of blood, about 10 cm s^{-1} in the artery, decreases progressively until, in the capillaries, it is less than 1 mm s^{-1}; it then rises again and reaches several centimetres per second in the vein. Even though each capillary is very short (less than 1 mm long) the average particle of blood spends enough time in one (about 1 s, or less if the organ is active) for interchange to take place with the tissues. (All these figures are very approximate, for purposes of illustration only.)

The German histologist Mall, as long ago as 1888 painstakingly quantified the mesenteric vascular bed in dogs and provided data usually quoted nowadays to support the above contentions. It is given, in part, in Table 34.1.

Volumes of various parts of the vascular bed

The distribution of blood volume in a tissue, which may or may not be typical since no one but Mall has had the industry or incentive to find out, is also given in Table 34.1. By far the largest proportion is in the venous system (approximately 80 per cent). The total volume of capillaries is about 10 per cent of the blood volume; this figure is so large because of the astronomical number of capillaries in tissues.

RHEOLOGY OF BLOOD VESSELS

Elastic tissue and blood vessels

Robert Hooke (1635–1703) enunciated the basic law of elastic materials, namely that when elastic substances are stretched, tension develops proportional to the increase in length. *Hooke's law* is shown in Fig. 34.3 and the slope of this line is called *Young's modulus* (Y).

Young's modulus is given by the equation:

$$F = Y \cdot A \frac{L - L_0}{L_0}$$

where F = force; Y = Young's modulus; A = cross sectional area; L = stretched length and L_0 = original length. It is therefore the force that would develop if unit cross-section of the substance is stretched by 100 per

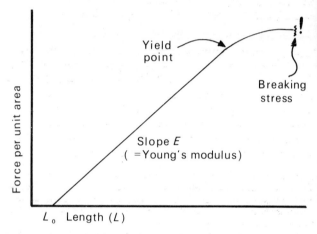

Fig. 34.3 Graph showing the extension of an elastic band (to illustrate Hooke's law). Force per unit cross-section area is plotted against extension $(L - L_0)/(L_0)$ where L is length and L_0 is original length.

The relation is linear and thus the elastic band obeys Hooke's law until the yield point is reached. Here the elastic force decreases for equal increments of length until at! the elastic band breaks.

cent. The actual length of course is immaterial (for conviction, see Fig. 34.3) and in practice the actual stretch of 100 per cent need not be reached in order to calculate Y.

Do arteries and veins obey Hooke's law?

We now come to the all-important question. Do arteries and veins behave like simple elastic structures? We could, of course, take an artery or a bit of an artery and stretch it, measuring the tension developed, to find out experimentally. However, before we do this, let us consider the implications for the function of the circulation, should blood vessels behave as purely elastic structures.

From experience in blowing up balloons and bicycle tyres we know that as the pressure within them rises, they enlarge (until they burst of course). Laplace, in 1821, first put forward his law which relates tension in a curved membrane to its radius of curvature and the pressure contained within it. Laplace's law relates to spherical soap bubbles, but can equally be applied to elastic cylindrical arteries:

$$P = \frac{T}{R}$$

where P is the transmural pressure (N cm^{-2}), T is the circumferential tension in the wall (N cm^{-1}) and R is the radius (cm).

Thus when an artery is distended by the pressure of arterial blood within it, equilibrium is reached when the tension in the wall (T) equals the product of the pressure (P) and the radius (R) of the vessel. If the tension is due entirely to elastic forces in the elastin of the artery wall, then when it is stretched by pressure, Hooke's law will dictate that this tension will increase. Thus, provided there is enough elastin present, the system is stable and will not suffer collapses or blowouts if arterial pressures fall or rise unduly.

On the other hand, arteries without elastin in their walls would be unstable in the face of changes in the pressure within them. If T, for example, did not increase as internal pressure rose, the walls would balloon out and be in danger of a "blowout". And in the opposite sense, as soon as the internal pressure fell, T (in Laplace's equation) would exceed the internal pressure and the vessel would be completely closed off.

LENGTH–TENSION DIAGRAMS OF ARTERIES

Experimental measurement of arterial wall properties does show that they obey Hooke's law; indeed, as can be seen from Fig. 34.4 they do more than do so! The length–tension diagrams show that they curve upwards to the right, i.e. the elastic force upon stretching, progressively exceeds that to be expected on the basis of a linear relation between force and extension. And it should be apparent that this factor will tend to improve vascular stability still more (than simple elastic properties would do).

Probable mechanism for departure from Hooke's law

The arterial wall contains both elastin and collagen, as fibres having simple elastic properties although of very different moduli of elasticity. Either, present in the arterial wall alone (see Fig. 34.4), would lead to the artery following Hooke's law when stretched.

With both present, as stretch increases, more of the far stronger collagen fibres then are pulled out to their unstretched length and add to the total tension. This gives a curve as shown. The initial slope gives a

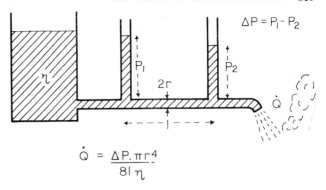

Fig. 34.5 Diagram to illustrate Poiseuille's law. (See text, p. 216)

measure of the elastin present; the final slope indicates how much collagen is present, and the transition level gives the degree of slackness in the collagen fibres.

Age and elasticity in arteries

There is certainly a great deal of truth in the statement that "one is as old as one's arteries", for subclinical or overt arterial disease, especially if it involves the cerebral circulation, leads to the personality changes which we associate with age.

The normal ageing process is characterised by an increasing stiffness of the large arteries. Length tension diagrams show this; there is a change right from birth to old age as follows:

1. The final slope is steeper. This is due to increased fibrous tissue being present.
2. The curves go upward at lower degrees of stretch, i.e. the collagen has less slack.

The results of changes like (1) and (2) above in the circulatory function will be discussed later, but a major effect of the loss of elasticity will be the loss of "smoothing" of the pulse wave with each heart beat. The pulse pressure will be greater, the pulse-wave velocity will be increased and the rise-time of the pulse-wave will be shorter.

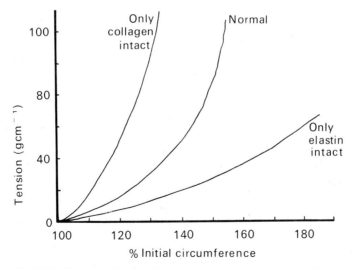

Fig. 34.4 Diagram to show length–tension relations in an artery. Strips of arterial wall are cut circumferentially from the artery and then stretched by hanging weights on them. As can be seen, arteries do not strictly obey Hooke's Law, but resist stretch more strongly the more they are stretched. (Roy, (1905), from Roach, M. R. and Burton, A. C. (1959), *Can. J. Biochem. Physiol.* 37, 557–569).

The left-hand and right-hand curves were obtained respectively by chemical removal of elastin and collagen (see text).

HYDRODYNAMICS: PRESSURE AND FLOW IN FLUIDS

Blood is a fluid, forced by the pumping action of the heart through a series of tubes, the blood vessels. In order to understand the working of the circulation it is necessary to know the physical laws which govern the behaviour of fluids. In applying these laws to conditions as they exist in the living circulation, complexities arise because blood is a fluid having unusual properties while the vessels through which it flows are distensible and often contractile. Moreover, the rates of flow and pressures are by no means uniform since the output of the heart itself is pulsatile.

Poiseuille's experiments

In 1836 and the following years, Poiseuille performed experiments on animals with a view to discovering the relations between blood pressure and flow in the circulation. He found his observations so bewilderingly variable that he turned his attention to the simpler problems concerned with the flow of pure liquids like water through glass tubes. For our purposes, the results of his experiments—which are embodied in Poiseuille's law—may be summarised by the following three statements.

1. *The rate of flow of a fluid through a tube is proportional to the pressure driving it.* The ratio of the pressure to the rate of flow is, of course, the *resistance* of the tube to the flow of the fluid through it.

Mathematically this is

$$Q = \frac{\Delta P}{R}$$

where Q is the flow in ml min^{-1}, ΔP is the pressure gradient (i.e. $P_1 - P_2$) in mmHg, and R is the resistance in mmHg^{-1} ml^{-1} min^{-1}.

A little thought convinces one that this equation

$$Q = \frac{\Delta P}{R}$$

is basically analogous to Ohm's law

$$I = \frac{E}{R}$$

This relationship ceases to be true if the rate of flow is very large and the flow becomes "turbulent". Local turbulence may occur in the blood vascular system, but it is not sufficient to affect the pressure-flow relations.

If we have a number of tubes with different resistances arranged in such a way that all the fluid goes through all of them in succession (in "series"), the pressure drop across each will be proportional to its resistance, and the total resistance of the whole will be the sum of the separate resistances. If the tubes are arranged in such a way that the flow is divided among them (in "parallel"), and all have the same pressure across them, the flow through each will be inversely proportional to its resistance, and the reciprocal of the total resistance of the whole—i.e. the total *conductance*—will be the sum of the separate conductances. Any complicated system of tubes in series and in parallel (such as the blood vascular system) will thus have a resistance to flow, the *total peripheral resistance* (TPR), which will depend on the separate resistances of its component parts; but it may always be measured in terms of the ratio of the pressure applied to the total rate of flow produced. As a rule, the venous pressure is considerably lower than the arterial pressure and in calculating the TPR the former can usually be ignored.

2. *The resistance to flow of any particular fluid is directly proportional to the length of the tube (l), and inversely proportional to the fourth power of its radius (r).* The resistance thus increases very rapidly as the radius becomes smaller. This accounts for the fact that the resistance of the arterioles is much greater than that of the arteries (see p. 229), in spite of the fact that the *total* cross sectional area is greater and the velocity of flow smaller. Flow resistance is therefore a property of the tube (mainly its size), and can be expressed as the hindrance, H.

$$H = \frac{8l}{\pi r^4}$$

For a given pressure gradient, then, fluid will flow faster through a short tube; flow is inversely proportional to the length of the tube.

3. *For any given tube, the flow resistance depends on the nature of the fluid driven through it.* Clearly, the "thicker" the fluid, i.e. the more viscous it is, the greater will be the resistance to flow.

Viscosity

The ratio of the viscosity of a fluid at any temperature, to the viscosity of water at the same temperature is termed the *relative viscosity*, of that fluid. The relative viscosity of most simple fluids such as water or alcohol does not alter, even if the length or diameter of the tube, the rate of flow, or the pressure gradient are varied over wide limits.

Resistance to flow is proportional to the relative viscosity, η.

We can now write the whole of Poiseuille's law as an equation:

$$R = \frac{8\eta l}{\pi r^4}$$

or in a more convenient form, since $Q = \dfrac{\Delta P}{R}$

$$Q = \frac{\Delta P \pi r^4}{8\eta l} \qquad \text{(See Fig. 34.5).}$$

APPARENT VISCOSITY OF BLOOD

Poiseuille's law is not always exactly obeyed by blood, since, unlike that of "perfect" liquids, the value of its viscosity may depend on the rate of flow and on the dimensions of the tube. Strictly, therefore, the idea of viscosity as a "constant" property of a liquid is inapplicable to blood, and the value obtained under any particular set of conditions is called the *apparent viscosity*.

In the blood vessels of a normal animal or man, the velocity is sufficiently great for the apparent viscosity to be independent of the rate of flow, and we are justified in regarding the resistance of the blood vessels as a quantity which varies only with their calibre.

The non-Newtonian nature of blood
(i) If blood is made to flow through a tube at a sufficiently high pressure, its apparent viscosity will be the same whatever actual velocity or pressure is used; it behaves as a "perfect" or "Newtonian" fluid. But if the velocity and pressure are reduced considerably, this is no longer true; the apparent viscosity becomes progressively larger as the velocity and pressure are made smaller, and the flow is now

"anomalous" or "non-Newtonian". It is probable that if the rate of flow is small, the red blood cells stick together and an extra force is required to tear them apart and to allow the blood to flow; the resistance to flow, and the apparent viscosity, are thus increased. If the rate of flow is very large, they move past each other so fast that they cannot stick together, this extra force is no longer needed, and the resistance to flow, and apparent viscosity, become smaller.

Clinical measurements of the viscosity of blood, however, are sometimes made in an "Ostwald viscometer"; the movement of the blood is then so slow that the apparent viscosity is largely affected by the rate at which the blood flows through it—a matter depending on the exact design of the instrument. In a "Hess viscometer" the blood is made to flow at a velocity so great that small variations do not affect the apparent viscosity.

(ii) If the diameter of the tube through which the blood is made to flow is greater than about 0·2 mm, the apparent viscosity (in any conditions of flow) is independent of the actual value of the diameter. But if the diameter is less than about 0·2 mm, the apparent viscosity becomes smaller as the diameter is reduced still further. It is important to remember that the *resistance* to flow becomes greater as the diameter becomes smaller, whether the fluid is "perfect" or has this anomalous property shown by blood; but the increase in resistance, for a given decrease in diameter, is smaller with blood than with a "perfect" fluid. This effect is due to the presence of a narrow layer in contact with the wall of the tube which is deficient in red cells and so has a smaller viscosity than the rest of the blood in the tube. The width of this layer (about the thickness of a red cell) is practically the same whatever the diameter of the tube. In relatively large tubes, its effect is negligible; in small tubes, it provides a "lubricating" layer which becomes progressively more effective in allowing the blood to slip through the tube, as the diameter of the tube becomes progressively smaller.

It is with very small tubes, namely the arterioles, that we are mainly concerned as physiologists; as will be seen in Fig. 34.12 (p. 221), the main pressure fall between the arteries and veins occurs in the small blood vessels. In tubes of such a size (about 0·02 mm diameter) the apparent viscosity has only about one-half the value found in large tubes. Halving the apparent viscosity of the blood would be an important economy in the circulation, for it should halve the work done by the heart in maintaining a given circulation rate; for a given maximum output of work of the heart, it would about double the amount of work a man could perform in violent exercise. Moreover, in the circulation the resistance of the blood vessels is varied by changing their diameters.

Representative values of the apparent viscosity of blood are given in Table 34.2. They include about the smallest and the largest values which have been obtained on a given sample of blood, and show how greatly the viscosity may vary according to the conditions of measurement. In any conditions of flow and size of tube, the viscosity of blood increases rapidly with increase in the haematocrit value, particularly when this exceeds about 50 per cent; the figures given in Table 34.2 show, roughly, the range of physiological interest. This is the most important factor affecting the viscosity of the blood in the circulation.

ABSOLUTE VISCOSITY OF BLOOD

The *absolute viscosity* of blood, and thus the resistance to flow through a given tube, rises as the temperature falls. The effect is not large, and in ordinary conditions of life, the variation never exceeds a few per cent. At the lowest temperature compatible with life (about 23°C), the viscosity is about 40 per cent greater than in normal conditions; and at the highest temperature (about 44°C), it is about 15 per cent smaller. The effect of temperature on the viscosity of blood is nearly the same as that on the viscosity of water, so that the *relative viscosity* of blood is independent of temperature.

Streamline flow, turbulence

Above a certain "critical" velocity, *laminar* or streamline flow becomes turbulent and the relationship found by Poiseuille for flow in tubes between driving force and flow rate is no longer linear. Figure 34.6 illustrates this phenomenon of turbulence.

HAEMODYNAMICS: PRESSURE AND FLOW IN THE CIRCULATION

Pressure–flow relations in blood vessels

We have so far considered the behaviour of Newtonian fluids in a system of rigid tubes; normally a graph of flow versus driving pressure is linear. However, when we come to measure flows and pressures in the vascular system, they are often quite non-linear. A major factor causing this discrepancy is, as one might expect, the distensible nature of the walls of all blood vessels. Blood flow is proportional to the fourth power of the radius of the vessels through which it is flowing; hence anything altering vessel distension will be an important factor in determining the flow.

An example to illustrate the foregoing considerations is given in Fig. 34.7. This shows the non-linear relation found between flow and driving pressure in an experiment upon a perfused rab-

Table 34.2 *Apparent viscosity of blood (relative to water)*

Conditions of measurement	Percentage haematocrit value (%)			Remarks
	30% (anaemia)	45% (normal)	60% (high-altitude acclimatisation)	
In ordinary tubes (dia. > 0·2 mm)	3·0	4·5	6·5	Independent of exact conditions of measurement
In very small tubes, e.g. arterioles	2·0	2·2	3·1	Depends on size of tube
Rate of flow very small (large tubes)	15	40	130	Falls rapidly with increase in rate of flow

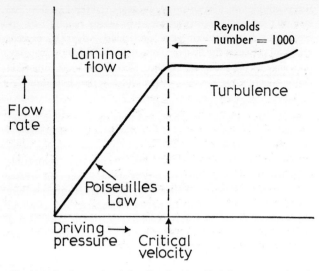

Fig. 34.6 Laminar and turbulent flow in tubes. Turbulence occurs when the Reynold's number, N_e, reaches 1000 (the critical velocity).

$$N_R = \frac{4\dot{Q}\rho}{\eta \pi D}$$

where $\dot{Q}$ is rate of flow, ρ = density, D = dia. of tube, η = viscosity of fluid.

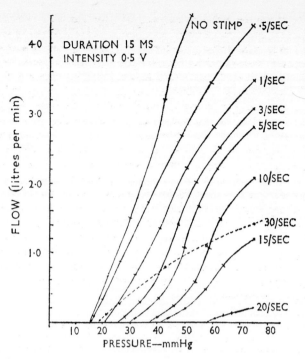

Fig. 34.7 Complex flow–pressure curves found in rabbit ear when cervical sympathetic ganglion is stimulated at various frequencies. For a "perfect" or Newtonian fluid in rigid tubes these would be straight lines going through the origin. Note the "critical closing pressure", i.e. the increasing intercept on the x-axis. (Burton (1952), in *The Visceral Circulation*. Ciba Fdn. Symp. Churchill, London.)

bit's ear. The curves show the effect of increasing the vasomotor tone by electrical stimulation of sympathetic nerves at the frequencies indicated. For a Newtonian fluid in rigid tubes the curves would be straight lines through the origin, but since the vessels are distensible and can also contract when their smooth muscle is stimulated, the complex pressure–flow curves shown are found instead.

Critical closing pressure

The pressure within a blood vessel at which it collapses down completely and is closed to flow is termed the "critical closing pressure". This occurs if and when the forces of muscle contraction together with the elastic forces in the wall of the vessel exceed the pressure within it. This closure occurs when the precapillary sphincters are actively contracting to control capillary flow. It is probably also an important factor during the very low arterial pressures which often occur in circulatory shock.*

If the effects of respiration and of muscular movement on the veins are excluded, the flow of blood through the vessels is produced entirely by the pressure differences established by the heart. Although the blood pressure thus falls continuously from its highest value in the aorta to its lowest in the great veins entering the heart, it is most conveniently measured in three situations, the larger arteries, the capillaries and the great veins.

The arterial blood pressure

The pressure in the arteries varies with each heart-beat, and thus has a *maximum* or *systolic* value and a *minimum* or *diastolic* value; the difference between the two is termed the *pulse pressure*. In the experimental animal the arterial pressure is

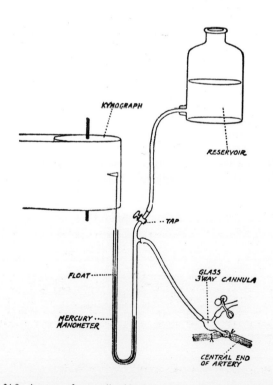

Fig. 34.8 Apparatus for recording blood pressure in the experimental animal.

This figure depicts a typical experimental set-up which could commonly be seen in physiological laboratories at the turn of the century. Now the mercury manometer and kymograph would be superseded by a pressure transducer and electronic chart recorder. The cannula would be a long length of polyvinyl tube introduced into the artery.

---

* As mentioned on p. 215, vessels without elastic tissue in their walls would be unstable in the face of changing internal pressures. Certain arterio-venous anastomoses (glomus bodies), have muscle cells but no elastic fibres. These vessels are either fully open or fully closed; they have no intermediate degree of patency.

commonly measured by inserting into the carotid or femoral artery a glass cannula filled with a solution which prevents the blood from clotting (e.g. half-saturated Na_2SO_4), and connected through pressure tubing filled with the same solution to a mercury manometer (Fig. 34.8). Since the flow is obstructed in the vessel cannulated, the pressure recorded is that at the point of junction with the larger vessel that supplies it; for since there is no flow along the obstructed vessel there is also no fall of pressure. The levels to which the mercury rises in systole and diastole do not accurately represent systolic and diastolic pressures, since they are largely determined by the momentum of the heavy column of mercury. When true systolic and diastolic pressures are required the arterial cannula must be connected through a liquid system to a manometer that will respond quickly enough to record the pressure changes without lag, and in which the actual movement of the recording system is very small, so that a negligible volume of blood is withdrawn from the circulation. A "pick-up" sensitive to changes in electrical resistance ("strain-gauge manometer") or capacitance ("capacitance manometer") is used.

When the arterial cannula is connected to a mercury manometer of very wide bore and hence of very low response frequency, the mercury column does not oscillate and it then records the *mean pressure* of the blood; more usually this value is recorded by using a mercury manometer of narrow bore with very high damping. The mean arterial pressure can be calculated from the systolic and diastolic values if the form of the pulse wave is also known; it approximates more closely to the diastolic than to the systolic pressure, since diastole lasts longer than systole. See Fig. 34.9.

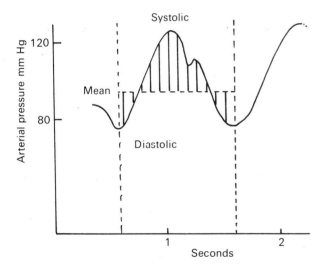

Fig. 34.9 Diagrammatic representation of "mean arterial pressure". Peaks of pulse wave are systolic, troughs of waves are diastolic, and the difference is the pulse pressure. Mean arterial pressure is shown as a dotted line; the area above this value equals the area below it. It is not systolic pressure plus diastolic pressure divided by 2 (!) but is often approximately one-third of the pulse pressure above diastolic.

Measurement of arterial pressure in man

For clinical investigations, the arterial pressure in man may be measured by similar methods. Under local anaesthesia a needle is inserted into the brachial or femoral artery and connected to a capacitance manometer. For routine work an indirect method based on the following principle is used. If an artery is compressed, then the minimum pressure serving completely to stop the flow must be at least as great as the highest pressure (systolic) attained inside the vessel. As the pressure outside the artery is reduced, so blood will flow through the artery for longer and longer periods of the cardiac cycle, until finally when the compressing force is just less than diastolic pressure, blood flow will be unimpeded, and the artery will cease to be deformed at any point of the cardiac cycle. A flat rubber bag contained in a loose but inextensible silk case is wrapped snugly around the upper arm. The interior of the bag is connected to a small hand pump through which air can be introduced or removed, and to a mercury manometer which measures the pressure of air in the cuff. The cuff must be sufficiently wide to transmit the pressure of air it contains to the centre of the limb (12 cm for the upper arm) and sufficiently long to encircle the limb completely. The systolic and diastolic pressures are determined with this sphygmomanometer (Fig. 34.10) as follows.

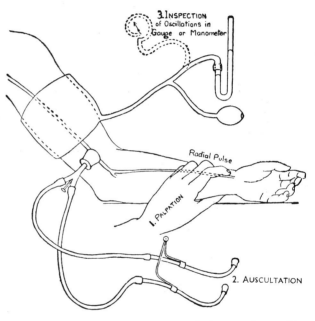

Fig. 34.10 The measurement of the arterial blood pressure in man. (From Harris' *Experimental Physiology*.)

1. *By palpation*. The cuff is inflated until the pulse can no longer be felt at the wrist. Air is allowed to leak out until the pulse returns. The pressure at which the pulse beat can first be felt is taken as systolic pressure.

2. *By auscultation*. The bell of the stethoscope is placed over the brachial artery at the bend of the elbow, and the cuff on the upper arm inflated until all sounds disappear. Air is allowed to leak out until pulse sounds just reappear; this is the systolic pressure. As the pressure is lowered the sounds become louder and louder and then abruptly die away. The point at which the loud sounds begin abruptly to die away is taken as diastolic pressure, for below this point the pressure of the cuff has failed to deform the artery. It is to be noted that while this index is probably a reliable measure of diastolic pressure in most subjects, patients are occasionally encountered in whom the sounds continue to be heard when the pressure in the cuff is reduced to zero; in these cases this index of the diastolic pressure is clearly unreliable.

3. *By the oscillometer.* In some forms of the instrument the bag is connected to a high-frequency diaphragm type of pressure gauge. The oscillations of the diaphragm record the volume changes of the main artery of the limb transmitted to the air in the cuff. When the pressure in the cuff is slowly reduced, the point at which the oscillations of the manometer first increase is taken as systolic, the point of maximum oscillation as diastolic pressure.

The auscultatory method is that most commonly used in the U.K. and the values correspond most closely with direct manometric measurements. The oscillatory method gives values for the systolic and diastolic pressures that are usually 5 to 10 mmHg (0·7 to 1·3 kPa) higher than those obtained by the auscultatory method. Systolic pressures obtained by palpation are, in experienced hands, in fairly close agreement with those obtained by auscultation; owing, however, to the difficulty of feeling the first weak pulse beat, the values particularly in inexperienced hands are frequently some 5 to 10 mmHg lower.

The capillary pressure

Accurate estimation of the capillary pressure in man is extremely difficult. One method is to seal on to the skin a small glass chamber open on the side next to the skin; its interior is connected to a manometer and source of air pressure. The pressure of air necessary to cause a capillary loop (observed microscopically) to disappear is taken as the capillary pressure. This method gives variable results and is less reliable than the following direct method. A finger is immobilised in a bed of plasticine and the cuticle is shaved off from the base of the nail. If a drop of glycerine is placed on the skin, the capillary loops can easily be seen with a binocular microscope and surface illumination. By means of a micromanipulator a fine glass micropipette containing physiological saline and sodium citrate solution is introduced into one of the capillaries, and blood allowed to enter its orifice. The pressure at which the blood neither enters nor leaves the pipette but oscillates with each heart beat is the mean capillary pressure. Owing to difficulties of fixation and of observing sufficiently large capillaries, the base of the nail is as yet the only place where the method is practicable in man.

The venous pressure

The veins offer little frictional resistance, and the blood flows through them with only a small fall of pressure. The venous pressure at some distance from the heart is thus very close to that in the superior vena cava. The most accurate and direct method of determining the venous pressure in man is to introduce into the median basilic vein at the elbow a wide needle connected with a manometer containing a solution of sodium citrate. The solution is allowed to flow into the vein until the meniscus shows small respiratory oscillations about a fairly constant mean value. The height of this meniscus gives the venous pressure in the vein at the point of measurement. To obtain a gauge of general venous pressure such values must be corrected for the difference in level between the vein punctured and the heart, since the pressure in the veins, as in all the vessels, is affected by gravitational forces. It is not easy to say precisely at what level the heart is in man, and so the pressures are usually referred to an easily accessible structure bearing a fairly constant relation to the heart, the junction of the manubrium with the body of the sternum (angle of Louis).

In recumbent healthy subjects when the arm lies at or below the level of the heart, the meniscus in the venous manometer comes to rest at the same level as the angle of Louis, or a centimetre or two below it. Relative to the angle of Louis the venous pressure in health is thus 0 to −2 cm H_2O (0–200 Pa).

Now, the veins, being wide lax vessels, are distended when the pressure of their contained blood is greater than that of the atmosphere. When the venous pressure is a little below that of the atmosphere the veins collapse. If therefore in the recumbent subject we trace a superficial vein, such as the external jugular, from a point well below the level of the manubrium sterni to a point above it, we see that for the lower part of its course the vein is distended, and then at some higher point it collapses and ceases to be visible (Fig. 34.11). At the junction of distended and collapsed vein, pulsations synchronous with the heart beat and respiration will be observed. From what has been said it is clear that the junction of distended and collapsed vein should give the level at which the venous pressure is equal to that of the atmosphere, and in fact it is found that when the venous pressure is measured manometrically, the meniscus lies at the same level as that to which the jugular veins are distended. The point of collapse of the jugular veins is therefore used clinically to measure the venous pressure in man. In so doing it is essential to ensure that the point at which the vein ceases to be visible is not simply the point at which it plunges deeply into the neck; this may be ascertained by noting that the vein fills to a higher level when it is obstructed below by a finger.

In resting man the venous inflow to the heart is small; the healthy heart requires but a small distending pressure to expel this inflow and maintains the venous pressure low. The failing heart, which is working at the limit of its capacity, requires a venous pressure that is several centimetres (up to 1 kPa) higher to accomplish its task; and under such circumstances the venous pressure is raised. It may be readily understood, therefore, that in the resting subject a rise of venous pressure above its normal value is the most usual and most important sign of failure of the heart. For this reason measurement of the venous pressure is of unusual importance clinically.

Normal values for the blood pressure* in resting man at various parts of the vascular circuit are as follows:

Axillary artery	systolic	115 mmHg	(15·3 kPa)
	diastolic	70 mmHg	(9·3 kPa)
	pulse pressure	45 mmHg	(6·0 kPa)
Capillary of nail fold	Arterial limb	32 mmHg	(4·3 kPa)
	summit of loop	20 mmHg	(2·7 kPa)
Superior vena cava		0 to −2 mmHg	(−0·2 kPa)

Of these values the venous pressure is the most constant, rarely varying by more than 2 mmHg (0·2 kPa) from the mean; capillary pressure shows slightly greater variations and the arterial pressure considerably greater fluctuations. Thus in an individual examined at rest on several occasions, the systolic pressure may vary from 110 to 130 mmHg (14·6 to 17·3 kPa)

---

* By Pascall's third law (pressure increases with depth by $\Delta P = \rho g(\Delta h)$ where ΔP is pressure increase; ρ is density of the fluid, g is acceleration due to gravity, or 980 cm s^{-2}, and Δh is the depth in cm) we may convert mmHg pressure into fundamental SI units $P = \rho gh = 13 \cdot 6 \times 980 \times 0 \cdot 1 = 133$ Pa. Thus 1 mmHg pressure = 133 Pa, or, more conveniently, 0·13 kPa.

Although numerical values are to be expressed in SI units from 1976 onwards, it is most unlikely that blood pressure will ever be given in anything other than the mmHg, since it is measured with a mercury manometer. Indeed, there now exists an influential body called "The Society for the Protection of the Millimetre of Mercury"!

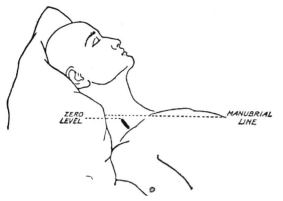

Fig. 34.11 Measurement of the venous pressure in man. The subject lies down with his head supported by pillows; the jugular vein is seen distended with blood up to a point which usually lies just below the level of the notch of the *manubrium sterni*. (Lewis.)

and the diastolic pressure from 60 to 80 mmHg (8 to 10.6 kPa); psychical factors are amongst the more important causes of these variations. Average values for the systolic blood pressure rise from 80 to 100 mmHg at the age of 5, to 115 mmHg at the age of puberty and to 140 mmHg (18.6 kPa) at the age of 60.

Figure 34.12 shows the blood pressure determined directly in various parts of the vascular circuit of the guinea pig, the rat and the frog, by the introduction of a micropipette into the appropriate vessel. It will be seen that there is no appreciable fall of pressure until the smaller arteries are approached, then the pressure falls rapidly until the capillaries are reached, when the

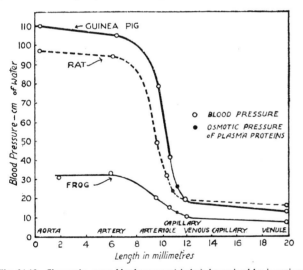

Fig. 34.12 Shows the *mean blood pressure* (circles) determined by inserting a micropipette into different parts of the mesenteric circulation of the guinea pig, rat and frog. In each curve the thin line represents the fall of pressure along the capillaries. The dots indicate determinations of the osmotic pressure of the plasma proteins in each species. (After Landis.)

pressure fall becomes more gradual. The fall of pressure as the blood traverses the veins is very small. The figure also shows that the pressure in the arterial limb of the capillary is higher, and in the venous limb lower, than the colloid osmotic pressure of the blood; fluid thus tends to pass out of the blood at one end of the capillary and to be absorbed at the other. In an active organ, as we shall see, there is considerable arterial dilatation, and increase in rate of blood-flow. The capillaries also dilate, and those previously closed, open up.

The effect of gravity

The above values of the blood pressure are all given for vessels lying at the same level as the heart. When the vessels lie below this level, then to the pressure which is imparted to the blood by the heart must be added the pressure due to gravity, that is the pressure exerted by a column of blood equal in height to the vertical distance of the vessels examined, from the heart. This relationship holds good for all the vessels of the perfectly flaccid limb; but in dependent limbs the venous pressure is reduced by the repeated movements which are usual during active life. Even the smallest movements empty the veins which, owing to the action of the valves, can only fill up from below. Thus in health the venous pressure in dependent limbs rarely rises very much above that of the atmosphere. This is important because if the venous pressure becomes greatly raised, then the capillary pressure is correspondingly raised and fluid tends to pass out of the capillaries and waterlog the tissue spaces. If the leg of a normal person is allowed to hang down for several hours without movement, it thus becomes oedematous. When the veins of the leg are distended by disease (varicose veins), the valves become incompetent, and after a day's work the feet become swollen because of the raised capillary pressure.

The effects of acceleration

Standing upright, the changes in arterial pressure due to gravity will approximate to a mean level of about 180 mmHg (24 kPa) in the feet as compared with 100 mmHg (13 kPa) at the heart level and about 80 mmHg (10.6 kPa) in the head. In aerobatic flight, however, accelerations will often give forces towards the feet many times gravity, say more than 5 g. Systolic pressures in the brain will fall to zero or below and the ischaemia of the brain will cause loss of consciousness, termed a "blackout".

Much experimental work on the tolerance of human subjects to g forces has been carried out using huge centrifuges. A healthy young pilot can withstand about 3 g for half a minute. Centrifugal force toward the feet is termed negative g and is counteracted by the use of the g-suit. This consists of a close-fitting rubber suit which covers the abdomen and lower limbs and which is inflated with compressed air automatically as g rises. This prevents pooling of blood in the lower limbs.

Other aerobatics produce positive g, i.e. the force is headwards, and blood pressure in the head is raised. This leads to the "red-out" in which vision becomes blurred and red due to the vascular congestion in the retina. Higher positive g forces lead to multiple haemorrhages into the brain substance due to the greatly increased transmural pressures.

Weightlessness (zero g)

Before space travel became possible many physiologists predicted that severe problems would be encountered by persons subject to zero g for any length of time. Circulatory effects, however, are negligible since the forces involved in maintaining the circulation are little different from those incurred when lying horizontal. There may be transient problems when returning to an environment of 1 g, since more circulatory work will be required to maintain the increased level of muscular output needed. The main difficulties associated with weightlessness have turned out to be due to sensory disorientation, and a persistent form of "motion-sickness" with nausea and vomiting has occasionally occurred.

35. Some General Circulatory Responses

In the course of life, people are likely to be subjected to various conditions, resulting from activity or accident, which will affect the circulatory systems as a whole. The circulation will respond to these conditions in various appropriate ways, which we will now describe.

The circulatory adaptations to changes in posture

When a recumbent subject stands up, blood pools in the veins of the lower part of the body, there is less in the great veins to fill the heart and cardiac output diminishes. Arterial blood pressure tends to fall. However the carotid sinus and depressor reflexes come into play in a fraction of a minute; there is reflex stimulation of the heart and blood vessels. In the upright state arterial pressure is unchanged and cardiac output is decreased, but this is offset by an increase in the total peripheral resistance. The heart beats a little more quickly.

The Valsalva manoeuvre

This is used to test the integrity of the baroreceptor vasomotor reflex. The subject closes his glottis and attempts to expire forcibly.* Arterial blood pressure rises transiently and then falls because of diminution in the venous return and decrease in cardiac output (Fig. 35.1). The fall in arterial pressure evokes the

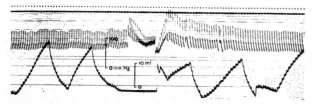

Fig. 35.1 (Upper curve) Arterial blood pressure in man. (Lower curve) Forearm blood flow. The beginning of the Valsalva manoeuvre is shown in the arterial blood pressure trace by an upward shift of about 40 mmHg, the end by a sudden decrease in pressure 11 s later. After the Valsalva, note the "over-shoot" of the arterial pressure and vasoconstriction in the forearm. (E. P. Sharpey-Schafer (1945), *J. Physiol.* 122, 353.)

baroreceptor reflex; heart rate increases as the record shows, and peripheral resistance increases and checks the fall in blood pressure.

After the Valsalva manoeuvre, the rush of blood through the heart in the face of a raised peripheral resistance causes an upward surge of the arterial blood pressure. Sudden stimulation of the baroreceptors causes the bradycardia now seen in the arterial tracing. The result of a Valsalva in a subject without a baroreceptor reflex is altogether different. There is no tachycar-

* For an excellent and comparative account of the physiology involved in coughing, defaecation and trumpet-playing, the reader should consult Sharpey-Schafer, *loc cit.*

dia during the Valsalva, nor any bradycardia afterwards. The arterial blood pressure fall, during the manoeuvre, proceeds unchecked by vasoconstriction and after the Valsalva there is no overshoot of the arterial pressure.

The mess trick and the fainting lark
It is possible, by combining the above procedure with a short period of hyperventilation beforehand, to produce a fall in blood pressure so marked that the brain blood supply falls to a low level and the subject loses consciousness. This effect is due to the concomitant washing-out of carbon dioxide altering the oxygen dissociation curve of blood to such an extent that the lowered pressure together with decreased oxygen liberation lead to severe cerebral hypoxia.

This party piece, the "fainting lark", is, however, not to be indulged in for pleasure, because pleasant though the sensations accompanying the return of consciousness may be, large numbers of cortical grey cells may be killed off by the hypoxia.

Circulatory adjustments in exercise

Higher centres of the brain can have a preparatory effect upon circulatory function, when the use of muscles is contemplated. Imminent exercise leads to subsidence of vagal inhibition of the heart, it beats more rapidly and the force of its contractions is augmented by sympathetic stimulation and by the action of adrenaline.

At the beginning of exercise, heart rate and cardiac output rise rapidly and in about three minutes will have reached levels related to the severity of the exercise. As is shown in Table 35.1, the increase in output is due partly to an increase in the frequency of the heart and partly to an increase in the amount expelled by each beat (stroke volume). X-ray pictures of the human heart show no change in the diastolic volume, and the increase in stroke volume results from the ventricles emptying more completely in systole. The diastolic volume of the ventricle of the resting untrained subject in the upright position is about 90 ml and the stroke volume 45 ml, leaving 45 ml behind—the residual volume. In very severe exercise, the ventricle empties almost completely and the stroke volume is nearly 90 ml. In trained subjects the heart hypertrophies and enlarges and in severe exercise the ventricle can discharge up to 200 ml per beat (Table 35.1).

The venous return, which in the steady state, of course must equal the cardiac output, is increased in proportion to the work done. Venous pressure, measured through a catheter in the great veins near the heart or in the right auricle, remains almost constant. In a given time, a much greater volume of blood can enter the heart because the duration of the diastolic pause is increased at the expense of the systolic pause and because the ventricle relaxes more rapidly and offers less resistance to the entering blood. Owing to the increase in the force of contraction, also, the ventricle empties more rapidly and completely from a given diastolic volume.

Although in exercise the venous pressure remains constant, there is probably an increase in the "effective filling pressure". This is the

Table 35.1 *The effect of exercise on the output of the heart (after Christiansen)*

Subject	Work performed *kp m min^{-1}	Oxygen consumption (1 min^{-1})	Pulse rate per min	Cardiac output (1 min^{-1})	Output per beat (ml)
Untrained female	0	0.24	77	4.6	60
	600	1.57	131	14.5	111
	720	1.79	145	17.4	120
	840	2.05	159	19.0	120
	960	2.45	168	23.8	142
Trained male	0	0.25	70	4.2	60
	720	1.93	118	16.5	140
	960	2.22	140	20.6	147
	1200	2.83	174	23.0	132
	1440	3.26	180	26.9	149
	1680	3.94	179	37.3	208

* 1 Kilopond (kp) ≈ 9.8N

difference between the pressures outside and inside the heart—that is, between the pressure outside the chest, and the intrapleural (intrathoracic) pressure. Since the intrapleural pressure is decreased owing to the increased respiratory movements, there is probably a corresponding increase in the effective filling pressure.

We must now consider the cause of the increase in venous return.

In the resting dog, stimulation of the heart causes a fall in venous pressure and only a small increase in output of the heart. It is therefore clear that in exercise the cause of the increase in venous return is to be sought in the peripheral vascular system. If cardiac output increases sixfold, the peripheral resistance must decrease to about one-sixth of its resting value if the mean arterial pressure is to rise only slightly. The decrease in resistance is mainly in the vessels of the skeletal muscles, and to a lesser extent in those of the skin. If cardiac output increases from 5 to 30 l min^{-1}, then the flow through the muscle vessels will increase from 1 to 22 or 23 l min^{-1}, and that through the skin vessels will increase from a fraction of a litre per minute to 2 or 3 l min^{-1}. The increase in flow through the skin helps to prevent the body temperature from rising. In spite of this enormous decrease in the resistance of the vessels in the muscles and skin of the limbs, there is little or no increase in the limb volume. This is due to the action of the "muscle pump" (Fig. 35.2). When we are at rest, a good deal of blood is contained in the veins of the limbs. In exercise, muscular contraction squeezes these veins and propels the blood towards the heart, the venous valves preventing reflux from the heart. Thus the increase in the amount of blood in the dilated arterioles is more than redressed by the decrease in the amount of blood in the limb veins.

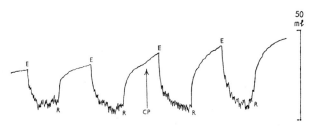

Fig. 35.2 Shrinkage of the calf of the leg due to the action of the "muscle pump".

Calf volume changes recorded with plethysmograph: downward movement of the record denotes shrinkage. E, rhythmic exercise of calf muscle for 10 s; R, rest for 10 s; CP, pneumatic cuff applied just above the knee and inflated to 90 mmHg (12 kPa); this was maintained to the end of recording. (Barcroft and Dornhorst.)

Vasoconstriction occurs in the viscera, and the volume of blood within their vessels decreases, with a corresponding increase in the amount of blood in the active circulation, for which accommodation must be found. Owing to the "muscle pump" and to this redistribution of the blood, the central parts of the venous system contain more blood than they do in the resting subject. This tends to raise the venous pressure, and thus to increase venous return to the heart. But the stimulated heart offers less resistance to the entering stream of blood, and this will tend to lower the venous pressure. As a result, there is a large increase in the venous return without any substantial rise in pressure in the neighbourhood of the right auricle.

The following imaginary experiment may help the reader to visualise the part played by the peripheral vascular system in increasing the venous return during exercise. Suppose that the peripheral vascular system of a resting animal is being perfused by a pump. Blood is pumped from the venous side through an oxygenator into the aorta. The output of the pump is regulated by the experimenter in such a way as to keep constant the pressure in the great veins (venous pressure) at the normal resting level. Now suppose that the animal begins severe exercise. Vasodilatation begins in the muscles and the arterial pressure falls. The "muscle pump" and the redistribution of blood from the viscera fill up the great veins and the venous pressure rises. The operator, watching the venous pressure, adjusts the pump so as to increase its output and bring back the venous pressure to the normal resting level. Several such adjustments will be necessary, increasing the output and work of the heart, until the redistribution of blood from the viscera is complete and steady conditions are established. Finally, the venous pressure will be constant at the normal resting level, the output and work done by the pump will be increased about sixfold, the mean arterial pressure will be the same as it was initially, or slightly raised, and the peripheral resistance will have fallen to one-sixth of its initial value.

In exercise, the heart, quickened and strengthened, automatically adjusts its output so as to maintain the venous pressure constant.

The effects of hypoxia on the circulation

The effects of oxygen lack (hypoxia) have been studied extensively owing to their importance in flying and space travel. When the oxygen pressure in the inspired air is reduced to one-half, the heart beats much faster (tachycardia) and the cardiac output and systolic pressure are increased. Reduction of the oxygen pressure to one-third of its normal value causes loss of consciousness; in most subjects, the "non-fainters", the circulatory changes just described are accentuated; in the others, the "fainters", vasovagal fainting occurs (see below).

The effect of haemorrhage

If blood is lost, and the volume in circulation is reduced, it is to be expected that the veins would be less well filled and that the output of the heart and the arterial pressure would fall. This does occur, particularly if the haemorrhage is severe, but the changes are smaller than might be expected owing to the existence of compensating mechanisms. Compensation is absent after section of the spinal cord in the neck and is thus effected through the central nervous system. If, in an anaesthetised dog, the blood flow through the gut, the limbs, the liver and the spleen are simultaneously measured, it is found that when, say, one-tenth of the blood volume is removed, the flow through the gut and the limbs is reduced, showing that the vessels constrict in these areas. The arterial pressure does not fall, but is maintained by an increase in peripheral resistance brought about by a reduced activity of the depressor and carotid sinus reflexes. On the other hand, the outflows in the veins from the liver and spleen are found to be temporarily increased and exceed the inflows through the arteries; the blood content of these organs is largely expelled into the great veins, partly compensates for that removed, and lessens the fall in cardiac output. In the rabbit and the cat another compensatory mechanism quickly comes into action; fluid is absorbed from the interstitial spaces through the capillary walls, diluting the blood and partially restoring its volume. In the dog and man dilution of the blood is slower, and in man is not complete until 24 to 72 h after removing 1 litre of blood. But if the quantity of blood lost is sufficient to reduce the cardiac output to 30 to 50 per cent below normal, the arterial pressure falls. After profuse haemorrhage in man, the fall may be profound enough to produce loss of consciousness through anaemia of the brain; there may be complete recovery, nevertheless.

When the circulation in the anaesthetised dog is so disturbed by bleeding that the mean arterial pressure has fallen to 30 or 40 mmHg and remained so for six hours, a return of all the blood removed does not lead to a lasting recovery of the circulation (irreversible shock). If, after severe haemorrhage, the irreversible state is to be prevented, prompt measures must be taken to restore the depleted blood volume. This can most effectively be done by transfusing fresh or stored blood from a healthy donor of the same or a compatible blood group. In an emergency, stored plasma or serum may also be used, but in any case the amount transfused must be adequate to restore arterial pressure to normal. Saline is useless for the purpose for, having no colloid osmotic pressure, it quickly passes out into the tissue spaces. It is obviously desirable also, to know the volume of blood in circulation. In most animals, the blood normally makes up from 5 to 8 per cent of the body weight, there being about 3 to 5 litres in an average man, and 500 to 1500 ml in an average dog of 6 to 20 kg weight.

Peripheral circulatory failure and shock

A condition resembling that seen after frank haemorrhage in which the blood pressure falls, and in which the reduction in cardiac output is to be ascribed not to cardiac weakness but to changes in the vessels or circulating blood volume, is described as peripheral circulatory failure or, more loosely, shock. The following are some of the more important examples.

1. *The vasovagal or fainting attack* is characterised by low blood pressure, slow pulse, pale cold sweating skin, and sometimes loss of consciousness. The slowing of the heart is effected through the vagus nerves and is abolished by atropine. The fall of blood pressure is due to vasodilatation in voluntary muscle effected through the nerves; the blood flow through the skin decreases; the cardiac output may fall. The intense pallor of the skin persists after the arterial pressure has returned to normal, and may be due to the copious secretion of the posterior pituitary which appears to occur during a fainting attack.

2. *Burns*. In severe burns there is a rapid and profuse loss of plasma into the burned and adjacent tissues. The blood volume falls and the haemoglobin content of the blood rises, and in severe cases the circulation may fail. Failure is prevented by transfusion of adequate amounts of plasma.

3. *Wound shock*. Circulatory failure frequently occurs after extensive wounds, and sometimes without clear evidence of severe blood loss, in which case it has been attributed to vasodilatation from release of a histamine-like substance. Experience in the war of 1939–45 confirmed experience in the war of 1914–18, in showing conclusively that wound shock is not a single entity; but the most important cause of peripheral circulatory failure after wounds is undoubtedly loss of blood, either externally, or into the tissues of the body. In fact the enormous saving of life after wounding in the second war was due to the provision of proper supplies of blood for early and adequate transfusion (amounts up to 7 litres have been given before and during operation) and the use of antibacterial agents.

36. The Control of the Vascular System

Introduction: Feedback systems

Most complicated machines need built-in control mechanisms to ensure that they function efficiently, or indeed function at all. In this chapter we shall be dealing with the variables in the cardiovascular system which are controlled and the way in which these controls operate. In Chapters 14 and 15 the nature and operation of basic feedback servomechanisms was discussed; the controls to be found in the circulatory system follow the general pattern and operate in the same ways. In general, as many as possible of the separate functions are kept within the optimal limits of operation by their feedbacks. In addition, it will often be necessary to exert over-riding control of these systems; the brain contains "nervous centres" which co-ordinate the activity of the circulatory system as a whole.

In many instances, the control mechanism operates within such fine limits that its operation has long been overlooked by physiologists. One example, the details of which are still hotly debated by physiologists, concerns the mechanism of control of respiration in muscular exercise. Do the large changes in pulmonary ventilation (and indeed heart rate) really stem from the minute changes in $P\text{CO}_2$ in circulatory blood? Or are other controls operative?

The variables to be controlled

The function of the cardiovascular system being to supply blood under pressure to various subcirculations in tissues, one would expect to find control systems working to keep the general arterial pressure level constant. There are, in fact, a number of these.

1. *Arterial pressure controls*. These are the carotid sinus and aortic reflexes.
2. *Heart rate controls*. The Bainbridge reflex (hotly disputed over the years!) controls heart rate.
3. *Myogenic reflexes*. Distension of arterial walls increases the muscle tone in the vessels.
4. *Hormonal controls*. A long-term control of arterial pressure is exerted by renin, a hormone secreted by the kidney.

The function of the circulation is also to maintain an adequate oxygen supply to tissues requiring it so we should predict the presence of homeostatic mechanisms controlling arterial $P\text{O}_2$. These also exist:

5. *Autoregulation of kidney blood flow*
6. *Autoregulation of coronary blood flow*
7. *Reactive hyperaemia in skin and muscle*
8. *Control of pulmonary blood flow*.

Carbon dioxide and tissue metabolite extraction also are controlled:

9. *Reactive hyperaemia in muscle*
10. *Regulation of cerebral blood flow*.

Other controls also operate upon various specific features of the cardiovascular system:

11. *Control of rate of blood flow*. Receptors in the veins control arteriolar tone. This could be looked on as a control of venous pressure.
12. *Control of cardiac output*. Starling's law (p. 255) is a mechanism, locally in heart muscle, which tends to stabilise cardiac output as venous filling of the heart alters. This appears to be a local control for equating the outputs of the left and right sides of the heart. One would not expect any servocontrol of cardiac output *per se*, since this is the "effector mechanism" working in the outputs of many of the other controls already listed.

The overriding control

Various forms of overriding control exist in both man-made servosystems and in biological controls. Usually these are integrating mechanisms which alter the "set-point" of the feedback in accordance with overall requirements, or which deal with abnormal conditions either within the system or in the external environment.

In the control of the circulation, the vasomotor centre in the medulla acts as the coordinating mechanism controlling the overall responses of the various systems. There is also hormonal control of certain of the control systems we shall be describing. For example, adrenaline is secreted by the adrenal medulla in severe exercise, or in emotional states, and is carried by the bloodstream to various parts of the body. One of its effects lies in the alterations it produces on the sensitivity of the carotid sinus and aortic baroreceptors. As we will find later in this chapter, this has general effects upon the level of arterial pressure.

ARTERIAL PRESSURE CONTROL: THE BARORECEPTORS AND CARDIOVASCULAR REFLEXES

As has already been pointed out, the correct functioning of the circulatory system depends critically upon having a high-pressure arterial system, the pressure in which is precisely controlled. The most important *cardiovascular reflexes* are responsible for this. Reflex control of blood pressure has been a topic for research for many years, going back to Marey, who promulgated a law that heart rate and blood pressure were reciprocally related. The conditions under which this statement is true are so limited that it is astounding that so much publicity has been given to it over the years. In 1806, Cyon and Ludwig found the *depressor nerve*, which when stimulated slowed the heart and lowered the blood pressure, whilst cutting it had the reverse effects.

226 HUMAN PHYSIOLOGY

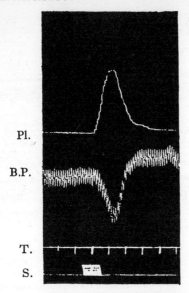

Fig. 36.1 The depressor reflex, producing vasodilatation and a fall in arterial pressure.

From above downwards: Volume (plethysmograph) of a loop of intestine, arterial blood pressure, time in 10 s intervals, signal showing period of stimulation of the central end of the vagus (containing the depressor fibres). (W. M. Bayliss.)

The slowing of the heart is largely but not entirely abolished by previous section of the vagi; it is thus due mainly to a reflex augmentation of vagal tone, and partly to a reflex inhibition of sympathetic tone. The fall of blood pressure is independent of slowing of the heart and is due to a vasodilatation that affects all the organs of the body except perhaps the brain (Fig. 36.1). Thus stimulation of the depressor nerve produces an increase in the volume of a limb or of a loop of intestine, and an increase in blood flow from the submaxillary gland. The physiological stimulus exciting the depressor reflex is a rise of pressure in the arch of the aorta.

The carotid sinus reflexes

It has long been known that in man, pressure over the bifurcation of the common carotid artery produces a sensation of faintness accompanied by slowing of the pulse and fall of blood pressure.* This effect has been shown to be due to stimulation of baroreceptors lying under the adventitia of the carotid sinus. If the sinus is compressed, or if the pressure of the blood inside is raised, or if the sensory nerve to it (a branch of the glossopharyngeal) is stimulated, the blood pressure falls and the heart slows—changes produced reflexly in a manner similar to those of the depressor reflex. Conversely, if the common carotid artery is compressed so as to produce a fall of pressure within the sinus, the heart accelerates and the blood pressure rises. The paths followed by both this and the depressor reflex are very similar. The afferent impulses entering the hind-brain through the glossopharyngeal and vagus nerves reach the cardioinhibitory, cardioaccelerator and vasomotor centres lying

---

* There is the story told by a lecturer to entertain his students, and which just could be true, that a certain tram driver always fainted at a particular sharp bend, but only on Sundays. It turned out that on Sundays he wore a stiff wing-collar, which as his tram negotiated the curve, pressed upon his somewhat too-sensitive carotid sinus....

close by in the medulla. The impulses sent out by these centres through the vagus and sympathetic nerves are thus modified in the way described.

If the carotid sinus and depressor nerves are cut, the blood pressure and pulse rate rise permanently. Blood pressure and pulse rate in man rise if the carotid sinus nerves are blocked by local anaesthesia (Fig. 36.2). In normal life, therefore, the con-

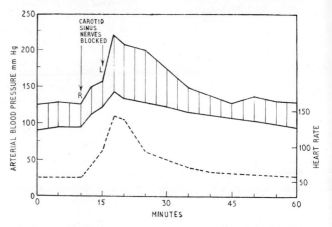

Fig. 36.2 Arterial blood pressure and heart rate in the human subject before, during and after blocking the carotid sinus nerves with local anaesthetic. After release of the vasomotor centre from the inhibitory impulses of these baroreceptors, the arterial pressure rose to 224/148 mmHg (systolic/diastolic), and the heart rate increased to 140 beats/min. (After Lampen, Kedzi and Kaufmann.)

stancy of the blood pressure and pulse rate is due largely to impulses ascending these nerves; any variation in blood pressure produces an inhibition or augmentation of these impulses, and so reflexly initiates changes which restore the blood pressure to its normal level.

When the arterial pressure falls, these reflexes produce an increase in sympathetic vasoconstrictor tone, mainly in the splanchnic area—that is to say, in the intestines, kidneys, liver and spleen—and to a much less extent in the skeletal muscles. The cerebral vessels and pulmonary vessels are not involved. These pressor effects are reinforced by the simultaneous release of adrenaline and noradrenaline. The two reflexes are thus of extreme importance in maintaining the distribution of blood to the tissues according to their needs; at the same time they prevent undue strain on the heart by keeping the blood pressure within safe limits.

EXPERIMENTAL OBSERVATIONS ON VASCULAR REFLEXES

We now briefly consider the experimental physiology which led to the discovery of these feedback mechanisms.

Cross-circulation experiments

The methods used by Heymans and his co-workers in investigating the functions of the carotid sinus are interesting as an example of physiological technique (Fig. 36.3). One or both carotid arteries of one dog B are perfused with blood from a second dog A (cross-circulation). The head of dog B is completely severed from its trunk, except for the spinal cord and vagus nerves. On raising the arterial pressure in dog A and thus in the carotid sinuses of dog B, the blood pressure in the trunk of dog B falls and the heart rate diminishes; the opposite change occurs when the blood pressure of dog A is lowered (Fig. 36.4). If now the suprarenal vein of dog B is anastomosed with the internal jugular vein of a third dog C (Fig. 36.3), then a fall of blood pressure in dog A

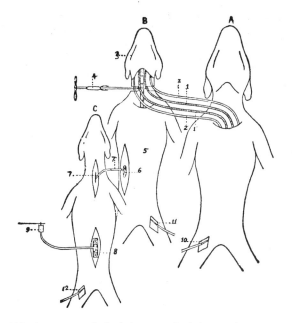

Fig. 36.3 Arrangement of animals in a cross-circulation experiment.

The head of dog B (3) is perfused from dog A by anastomosis of the carotid arteries and also of the jugular veins (1, 1', 2, 2'), the vertebral arteries and veins being tied, the muscles of the neck divided between ligatures, and the rest of the tissues compressed round the vertebral column by the *ecraseur* of Chassaignac (4). Blood from the suprarenal of B (6) is led into one of the jugular veins (7) of C, which has been adrenalectomised; the spleen of C (8) is enclosed in a plethysmograph, and contracts whenever adrenaline is secreted by the suprarenals of B. The arterial blood pressures of the three dogs are recorded from the femoral arteries 10, 11, 12. (From C. Heymans.)

produces, in addition to the effects mentioned, a contraction of the spleen of dog C (Fig. 36.4). Thus a fall of pressure in the carotid sinus of dog B leads in this dog to an increased secretion of adrenaline, as shown by the effect of blood from its suprarenal vein on the spleen of C. The effects are abolished by denervating the carotid sinuses of dog B.

The baroreceptors

As in all feedback servosystems the loop (or reflex arc) consists of *sensory receptors → afferent nerves → controlling centre → efferent nerves → effector system*. We must now consider these components in greater detail.

Baroreceptors* are found in: (1) the arterial system; (2) the pulmonary system; (3) the atriocaval junctions; (4) the ventricles (see Fig. 36.5).

Arterial baroreceptors, in the aortic arch, root of the right subclavian artery and at the junction of the thyroid artery and common carotid artery, signal the degree of extension of the artery wall and hence the pressure within the lumen.

In the *carotid sinus*, which is the region most extensively studied, one finds histologically that afferent endings of the glossopharyngeal nerve (the sinus nerve comes from it) are profusely distributed in the adventitia but not elsewhere.

Occlusion of both carotid arteries leads to a *rise* in systemic blood pressure—an effect abolished by section of the carotid sinus nerves.

Section of both carotid sinus nerves causes a rise in arterial pressure.

---

* *Baroreceptor*: an unpleasant word which, apart from being polyglot and not euphonious, denotes an ending responsive to pressure. It is not; it is merely a stretch-receptor which responds to any form of mechanical elongation of the arterial wall. However, we are now saddled with the word.

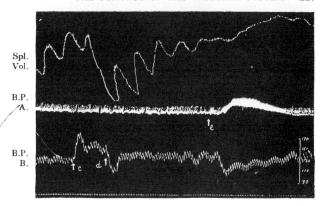

Fig. 36.4 The Regulation of the arterial blood pressure. (Cross-circulation experiment, see Fig. 36.3.)

From above downwards: Volume of the spleen of dog C; arterial pressure of dog A (perfusing the carotid sinus of dog B), arterial pressure of dog B.

At c the pressure in the carotid sinus of dog B was reduced by partially clamping the interconnecting artery, and at d the pressure was returned to the initial value. A fall in pressure in the carotid sinus leads to a reflex rise in pressure in the rest of the body, and a reflex secretion of adrenaline, as shown by the contraction of the spleen of dog C; a rise in pressure in the carotid sinus has the reverse effect. At e 0.1 mg of adrenaline was injected into the circulation of the perfusing dog, A, raising its arterial pressure and hence, also, the pressure in the carotid sinus of dog B. This resulted in a reflex fall in the arterial pressure of B, and an inhibition of the secretion of adrenaline, as shown by the dilatation of the spleen of dog C. (From C. Heymans.)

Perfusion of the carotid sinus with saline solutions at higher-than-arterial pressures confirms that the aforementioned effects are due primarily to pressure. It results in a fall of arterial pressure. (See Fig. 36.6.)

Perfusion of carotid sinus

Such experiments upon the effects of altering the pressure within the carotid sinus can be carried out by ligating all the

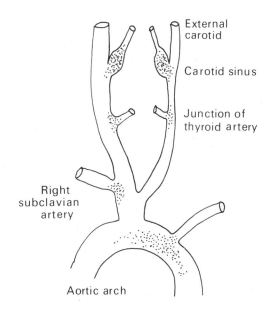

Fig. 36.5 Diagram showing the location of the baroreceptors in the arterial system. The endings, of the sinus (afferent) and aortic nerves, are profusely distributed in the adventitia of the vessels. They are purely stretch receptors, signalling length of fibres in the artery wall, since if the sinus is encased in plaster of Paris rises of pressure within it are not detected.

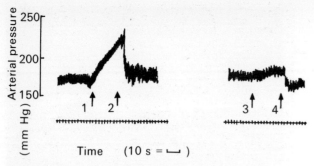

Fig. 36.6 Diagram showing experiment to illustrate the reflex effect on arterial pressure when the common carotid arteries are occluded. Dog, chloralose anaesthesia, aortic nerves cut. The records are of arterial blood pressure.

Between (1) and (2) both common carotid arteries were occluded. Rise in arterial pressure and increase in heart rate occurs. Both carotid sinus nerves divided, and between (3) and (4) experiment repeated. There is little or no effect. The fall in pressure at (4) is due to a "mechanical" fall in blood pressure. (Rough diagram after records published by Heymans and Bouckaert, 1931.)

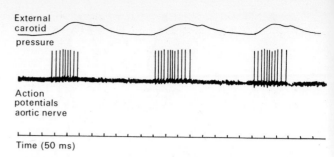

Fig. 36.7 The afferent nerve discharge from baroreceptors in the aortic arch. Records are of arterial pressure, action potentials in a single fibre of the aortic nerve and a time trace at 50 ms intervals. This is the electrical activity of a single receptor. There are many thousands in the aorta as a whole, and these have slightly varying thresholds and sensitivities of response. Graded responses to varying pressures are signalled by: (a) the number of fibres active (more when pressure is higher); and (b) the frequency of discharge of each ending (the higher the pressure the faster the discharge rate).

vessels branching off from the region of the bifurcation and then connecting the common carotid artery to a perfusion pump. It is found that the mean pressure within the sinus must reach about 60 mmHg (8 kPa) before any reflex inhibition of the activity of the cardiovascular centres in the medulla, leading to a fall in pressure in the arterial system, is apparent.

If the nerve action potentials in the sinus nerve are monitored, stimulation does in fact occur at pressures below 60 mmHg (8 kPa), but the total impulse density is insufficient to have any obvious reflex effects until a pressure of 60 mmHg is reached.

It should be noted that only one kind of receptor is found—it responds solely to a rise in intravascular pressure (or more correctly, also to rate of rise in pressure). There are no receptors signalling a fall in pressure.

The normal mean pressure of 140 mmHg (19 kPa) in the dog causes continual stimulation of the receptors, which of course wax and wane in *frequency* of discharge with each pulse beat. The reflex results of this give a "tonic" inhibition of sympathetic activity and a "tonic" discharge down the vagus—the so-called "vagal tone". Both these feedbacks tend to maintain the arterial pressure at a constant level.

Action potentials in the sinus nerve

Single nerve fibres, or filaments consisting of a small number of fibres, can be dissected out from the sinus nerve and put on suitable electrodes* for amplification and recording. A typical record from a single receptor is shown in Fig. 36.7.

As discussed in the context of "derivative control" (Chapter 14), these endings also tend to respond to rate of change in pressure rather than pressure *per se*. It is not until pressures of at least 200 mmHg (27 kPa) are reached that they respond equally to steady or to pulsatile pressure.

Cardiac and pulmonary baroreceptors

Atrial receptors

There are baroreceptors in the atria and these are of two types: (1) the type A endings which are stimulated by a rise of pressure

in the atrium (caused by atrial contraction, say); (2) the type B endings which give a burst of impulses in late systole. By contrast the systemic baroreceptors discharge in early or midsystole.

These receptors are more slowly adapting than are the carotid sinus and aortic ones. Thus their natural stimulus is distension of the atrium and it would appear that they are "volume" receptors. For example, they fire off at a great rate if venous return is augmented by a rapid injection of blood or saline into the inferior vena cava. (For the implications concerning control of body water, see p. 128.)

Ventricular receptors

Discovered electrophysiologically rather than anatomically, the ventricular receptors give an early systolic discharge of impulses when the ventricular walls are under tension (in the isometric phase, p. 252). Apparently they are mainly sensitive to the rate of rise of intraventricular pressure. Similar receptors can be found in the wall of the coronary artery.

Their functional significance is at present unknown, but their stimulation can cause reflex hypotension and slowing of the heart (the Bezold–Jarisch reflex).

The chemoreceptors

Endings of sensory nerves which respond to chemical influences, notably PCO_2, PO_2 and pH are also found. As these are primarily concerned with the control of respiration, we will defer their detailed consideration until the next section.

Any kind of anoxia stimulates the chemoreceptors. The reflex circulatory response to this, logically, is a rise in arterial pressure. Under normal conditions, though, chemoreceptor control plays very little part in the regulation of the circulation.

ARTERIAL PRESSURE CONTROL: THE EFFECTOR UNITS

We have so far considered the afferent side of the reflex arcs controlling the circulation and its function. We have also discussed the properties of the sensory endings. This sub-section now deals with the means by which the actual control is achieved, in other words the "output" or effector mechanisms.

As we have already seen, the flow through an organ (or

---

* The very best electrodes for this kind of work are said to be pig's bristles, boiled in saline and then soaked in saturated agar solution. Not all neurophysiologists would agree and would instead use silver–silver chloride wires.

through the body as a whole, i.e. the cardiac output) can be defined in terms of an equation similar to Ohm's law:

$$F = \frac{\Delta P}{R}$$

where F is flow rate, ΔP is driving pressure and R is the resistance to flow. For the whole body this equation is:

$$CO = \frac{BP}{TPR}$$

where CO is cardiac output, BP is arterial pressure and TPR is total peripheral resistance.

In general terms, flow in an organ can be changed in two ways, either by altering the driving pressure or by altering the peripheral resistance. For convenience, the multiple ways in which the efferent response of the vasomotor centre to the baroreceptor input brings about circulatory control, may be divided into these two categories:

1. *Cardiac effects*, i.e. changes in driving pressure.
2. *Peripheral vasomotor effects*, i.e. altered resistance.

Out consideration of the cardiac control mechanisms will be deferred until the chapter on the control of the heart.

Reflex vasomotor control

From the data already given for the blood pressure in different parts of the vascular circuit it will be seen that the main fall of pressure occurs in the small arteries and arterioles. The peripheral resistance is thus chiefly constituted by these vessels and its magnitude is dependent upon the strength of contraction of their smooth muscle coats. The state of contraction, or "tone" of the arterioles, may be modified for one of two purposes: to fulfil local metabolic requirements or to safeguard the circulation to the brain.

THE SITE OF PERIPHERAL RESISTANCE

At first sight, one might legitimately be surprised to learn that the peripheral resistance lies almost entirely in the arterioles and not in the capillaries. The clue to this state of affairs is to be found in Table 36.1 and in Fig. 36.8. It is that (a) the total cross-

Table 36.1 *Cross-sectional area and velocity at various parts of the vascular system. Representative values for a 70 kg man.*

Part of circuit	Total cross-section(cm^2)	Velocity(cms^{-1})
Aorta	4	20
Large arteries	6	15
Small arteries	8	10
Arterioles	8	10
Capillaries	3 000	0·2 mms^{-1}
Venules	2 000	0·3 mms^{-1}
Venae Cavae	8	10

sectional area of all the arterioles in the body is not very much greater than that of the aorta. This means that velocity of flow is still high: however, (b) since these arterioles are small vessels, the *surface* area of vessel walls which blood has to pass through is large. In Chapter 34 it was pointed out that this is an important factor in generating resistance to flow. In fact, resistance is

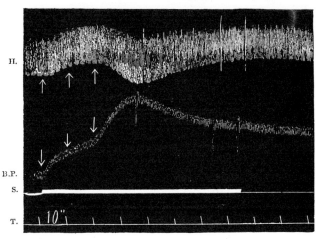

Fig. 36.8 The effect of stimulating the splanchnic nerve on the arterial blood pressure, and on the output and volume of the ventricles.

H, heart volume (a rise in the curve indicates an increase in volume); B.P., arterial blood pressure; S, signal showing duration of stimulation of the splanchnic nerve; T, time marker, showing 10 s intervals. Note that the first rise in arterial pressure is associated with an increase in the volume of the heart, owing to the greater power needed in order to expel the blood against the raised pressure, but that secondary rise, due to the *secretion of adrenaline*, is associated with a decrease in the volume. The sympathetic stimulation has produced an intense vasoconstriction of the splanchnic arterioles and this is the main cause of the rise in B.P. since it can be seen that CO is not much increased.

proportional to the inverse of the fourth power of the radius of the vessel:

$$R \propto \frac{1}{r^4}$$

where R is resistance to flow and r is radius of vessel.

Thus this combination of high velocity and small diameter vessels leads to arterioles being high-resistance vessels.

The capillaries, although very much smaller, offer little resistance to blood flow because they constitute an enormous cross-sectional area and the actual velocity of blood in them is very low.

The arterioles in many parts of the body may become constricted, or allowed to dilate, so adjusting the total peripheral resistance as to maintain the arterial pressure and thus the blood flow to the brain. A simple vasodilatation in any organ will lower the peripheral resistance and hence the arterial pressure. It is not surprising, therefore, that mechanisms exist which ensure that when vasodilatation occurs in one part of the body, this is compensated by vasoconstriction elsewhere. This control is initiated by the baroreceptors. The baroreceptors send nerve impulses to a group of nerve cells known as the "vasomotor centre" lying in the floor of the fourth ventricle close to the vagus nucleus. These in turn send impulses through the autonomic nerves to the blood vessels. (Fig. 36.8)

The vasomotor nerves

The nervous control of the blood vessels is chiefly effected through the sympathetic branches of the autonomic nervous system (Chapter 30). The action of the vasomotor nerves was discovered by Claude Bernard, who found in the rabbit that when the cervical sympathetic chain was divided on one side, the ear on that side became flushed and warm, remaining so for a considerable time. Conversely, stimulation of the cervical sympathetic produces pallor and coldness of the corresponding ear.

We now know that if the appropriate branches of the sympathetic nerves are stimulated under suitable conditions, then the blood vessels in all parts of the body (except perhaps the heart) constrict. The degree of constriction, however, varies in different organs. Thus stimulation of the appropriate sympathetic fibres produces intense narrowing of the vessels (vasoconstriction) of the skin and alimentary canal (Fig. 36.8),

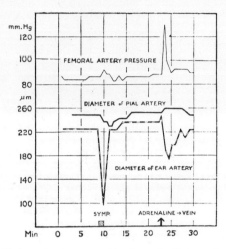

Fig. 36.9 The lower two curves record in microns the diameters of an artery of the pia mater of the brain, and of an ear artery, observed microscopically in a cat. The pial artery was seen through a glass window screwed into the skull. At the signal "symp" the cervical sympathetic trunk was stimulated and produced a pronounced constriction of the ear artery but only a slight narrowing of the pial artery. At the next signal 0·01 mg adrenaline injected into a vein produced a marked constriction of the ear artery and a small dilatation of the pial artery which is to be attributed to a passive effect of the coincident rise of blood pressure. (Forbes, Finlay and Mason.)

but only slight narrowing in those of the brain and lungs (Fig. 36.9). The vessels constricted by stimulation of the sympathetic are particularly the arteries and arterioles, to a less extent the capillaries and the veins. Section of the sympathetic nerves supplying an organ leads to an increase in the blood flow through it and it thus appears that normally there is a steady stream of constrictor impulses passing along these nerves to the blood vessels. Furthermore, since in the majority of organs the vessels cease to participate in the vasomotor reflexes after their sympathetic fibres have been cut, it seems that changes in vascular calibre of vasomotor origin are chiefly determined by an increase or decrease (inhibition) of sympathetic vasoconstrictor impulses.

Sympathectomy
Division of the sympathetic fibres supplying a limb, either by removal of the appropriate sympathetic ganglia or section of the preganglionic sympathetic fibres, is followed by vasodilatation, by loss of vascular responses to change of body temperature and loss of reflex sweating and pilomotor response. The vasodilatation chiefly affects the skin, muscle blood flow being little altered. With the progress of time, the vasodilatation subsides and the vessels are found to be abnormally sensitive to adrenaline, histamine and other vasoactive substances. Complete sympathectomy, by removal in separate stages of the chain of sympathetic ganglia on both sides, has been performed in the cat and dog. The completely sympathectomised cat is sluggish, and very susceptible to exposure to cold, to oxygen lack and to haemorrhage. The sympathectomised dog is normally active and its arterial pressure at rest is little below normal. It is evident that while in the normal animal the vasomotor nerves are extremely important in regulating the circulation, yet other probably chemical mechanisms can, in the dog, take over much of this function.

Reciprocal innervation and vasodilator nerves

It was thought at one time that the blood vessels generally were supplied by both sets of autonomic nerves—sympathetic (constrictor) and parasympathetic (dilator)—the two sets acting reciprocally. Thus an increase in flow would be brought about by a decrease in impulse frequency in sympathetic fibres and by a concomitant increase in impulse frequency in parasympathetic fibres. Later studies showed that direct parasympathetic vasodilator fibres probably occur only in the nervi erigentes supplying the erectile tissue of the genital organs. The vasodilatation that occurs in the salivary glands during parasympathetic nerve stimulation is not brought about by true parasympathetic vasodilator fibres, but by the action of vasodilator substances formed during stimulation of the secretory cells. Skeletal muscle vessels in the cat are supplied by sympathetic vasodilator fibres; but these do not act reciprocally with the vasoconstrictor fibres, but quite independently.

Venomotor control

Not only the sympathetic vasoconstriction of arterioles controls peripheral resistance. It is also of importance to consider the changes in "tone" of the walls of the venous system. Such changes prevent shifts of blood volume, rather than simply alter peripheral resistance. Particularly in the compensation for postural changes in blood pressure, venous mechanisms are at least as important as the carotid sinus reflexes.

As we have seen, most of the blood is in the veins, so that even small changes in venomotor tone will be as effective as relatively large changes in arterial or arteriolar calibre.

OTHER CARDIOVASCULAR REFLEX CONTROLS

As intimated earlier, a complex machine such as the circulation needs a multiplicity of control systems. We have described the main feedback control of arterial pressure and of the flow into different organs and tissues. There are, however, many other control systems, some known and presumably yet others to be discovered e.g. purinergic systems may be involved.

The Bainbridge reflex

Controversy in scientific circles is no less intense and no less acrimonious than in the political scene. In many ways too, it tends to be as irrational and unobjective. Bainbridge, in 1915, discovered a reflex which, coming from the *right* side of the heart (atrium), affected heart rate. He found that an intravenous infusion of blood or saline gave rise to a reflex tachycardia, which was abolished by bilaterally cutting the vagus.

Since that time, although appearing prominently in all the student textbooks, the Bainbridge reflex has proved to be disappointingly elusive to elicit and most workers came to believe it a figment of Bainbridge's imagination.

This controversy was recently resolved by the finding that whether or not tachycardia resulted from distension of the right atriocaval junction depended on the actual rate of the heart at the time. If initially the rate was below 130 min^{-1} (in the dog) tachycardia reflexly occurred; if above, bradycardia.

This answer has the merit of satisfying all concerned (and saving their scientific reputations), besides indicating a somewhat complex reflex system, probably directed towards homeostasis of heart rate.

Atrial and ventricular reflexes

The existence of receptors in the left atrium, pulmonary artery, pulmonary vein and in both ventricles has already been alluded to. Stimulation of these, both by stretch (i.e. pressure) and by various drugs (e.g. veratrine) gives a reflex decrease in heart rate, mediated by the vagus. Usually this results in a fall of arterial pressure as might be expected. Our present state of knowledge about these control mechanisms is insufficient to allow us to speculate about their function in the normal animal.

Reflexes from the lungs

In man, the *Valsalva manoeuvre* (expiration against the closed glottis) raises the intrapulmonary pressure and the blood flow to the limbs falls abruptly, indicating a mass, reflex vasoconstriction. At the same time a bradycardia develops.

One can demonstrate this on one's self remarkably easily by feeling the pulse and making brief expiratory efforts with the glottis closed. The pressure involved need not be very high.

DIVING MAMMALS

Irving has shown that seals can remain beneath the water for at least 12 min, much longer than the available oxygen would last. When immersed, these animals redistribute their cardiac output away from muscle, but to the brain, and the heart rate is at the same time decreased. In pearl divers much the same changes occur.

THE LOCAL CONTROL OF THE VASCULAR SYSTEM

There are a number of purely peripheral feedback controls which do not involve either nervous or hormonal pathways to and from the vasomotor centre.

Local control by vasodilator substances

The arterioles are relaxed, or dilated, locally in any organ or tissue when its activity increases and there is a greater demand for oxygen. The most important of these organs are the heart, the skeletal muscles and the digestive glands. Such an adjustment of the circulation is brought about chiefly by the production in the active tissues of vasodilator substances which act directly on the arterioles; the blood flow is increased and a greater supply of oxygen is made available. This will be discussed again later.

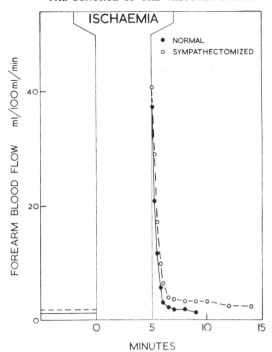

Fig. 36.10 Blood flow in the normal forearm and in the contralateral sympathectomised forearm before and after the arrest of the circulation in the arm for 5 min. (Diagram drawn from data obtained by Grant.)

Myogenic vascular tone

The arterioles of the skeletal muscles are endowed with very strong *myogenic tone*. It is sometimes called basal, inherent or intrinsic tone. While the skeletal muscle is at rest some intrinsic mechanism causes the vascular, plain muscle to contract spontaneously. If myogenic tone in the skeletal muscle vessels is relaxed, as in severe exercise, vascular resistance in the skeletal muscles may fall to 1/20 of its resting value with reduction of the total peripheral resistance to 1/5 of its resting value. In exercise, loss of myogenic vascular tone is accompanied by stimulation of the rate and force of the heart-beat so that there is an increase in the cardiac output which a little more than makes up for the fall in total peripheral resistance. Arterial blood pressure, in consequence, is maintained, indeed raised a little. Loss of myogenic tone in skeletal muscle vessels takes place soon after its arterial supply is cut off (Fig. 36.10) and therefore probably soon after death. Myogenic tone in resting voluntary muscle vessels is modified by the sympathetic nervous system but not to a very great extent.

37. Hormonal Regulation of the Circulation

We have so far dealt with nervous and local control systems involving, for the most part, nervous pathways. These all have a relatively rapid rate of action. Hormonal controls, while acting on a more widespread scale, also have longer time-constants of action.

Adrenaline and noradrenaline

These two substances are produced by activity of the sympathetic nervous system, are secreted by the adrenal gland, and are released at sympathetic nerve endings in blood vessels and the heart. In the resting human subject, minute amounts of both substances have been detected in the blood. During excitement and exercise, activity of the sympathetic nervous system increases and greater amounts of the sympathomimetic amines circulate in the blood. Their effects on the human circulation will now be briefly described.

When adrenaline is infused intravenously in man at about the rate corresponding to maximum secretion of the substance, the subject goes pale owing to constriction of the cutaneous vessels, and he soon becomes aware of his heart-beats (palpitation), Figure 37.1 shows the response of the general circulation. The heart often beats a little faster owing to stimulation of the pacemaker. There is a considerable rise in the systolic pressure, the mean pressure changes little, and there is a slight fall in the diastolic pressure. Cardiac output is increased. Since the output increases relatively much more than the mean blood pressure, it follows that the peripheral resistance must decrease. That is, in physiological doses in man, adrenaline causes an overall peripheral vasodilatation. This is because it dilates the

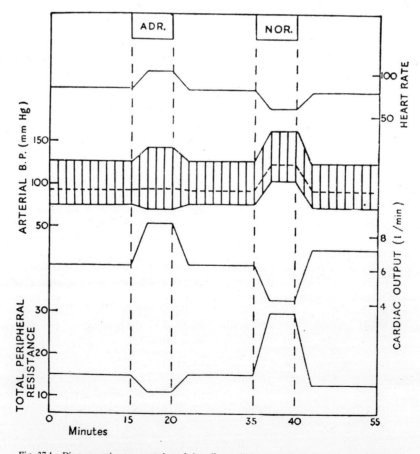

Fig. 37.1 Diagrammatic representation of the effects of intravenous infusions of adrenaline and noradrenaline on the heart rate, arterial blood pressure, cardiac output and total peripheral resistance, in man. The infusions were at the rate of 10 mg min^{-1}. (From Barcroft and Swan (1953), *Sympathetic Control of Human Blood Vessels*. Edward Arnold, London.)

splanchnic, skeletal muscular, and coronary vessels more than it constricts those of the skin, kidneys and other organs.

The action of noradrenaline is rather different. The subject pales, but feels no palpitation. Both systolic and diastolic pressures are raised but, since the cardiac output is decreased, noradrenaline must constrict the peripheral vessels strongly. It is interesting to note that the heart usually beats more slowly (bradycardia); and this is due to the large rise in arterial pressure and strong stimulation of the baroreceptors in the aortic arch and carotid sinuses. Reflex vagal inhibition swamps the rather weak direct excitatory effect of noradrenaline on the pacemaker.

In some medical and surgical emergencies, vasoconstrictor tone is markedly reduced and the arterial pressure may fall dangerously. Normal tone may be temporarily, and sometimes permanently, restored by intravenous infusion of noradrenaline. If necessary, infusion can be continued for several hours. When applied in large doses locally, adrenaline causes strong vasoconstriction. For this reason it is often added to local anaesthetic solutions to localise them near the point of injection and so prolong their effect.

Acetylcholine

Acetylcholine, the transmitter liberated at parasympathetic nerve endings is so efficiently destroyed by cholinesterase that the amount in the general circulation is negligible. When injected in large doses it causes transient arteriolar vasodilatation. Vasodilator substances such as *histamine* and *adenosine triphosphate* (ATP) occur in the tissues, but only in significant amounts in the general circulation. Large quantities of *post-posterior pituitary* extracts (i.e. large compared with any likely to be released from the gland) produce in man a slight rise of blood pressure and intense pallor of the skin, due to constriction of the capillaries. There is no evidence that the posterior pituitary takes part in maintaining capillary tone in normal conditions.

Renin

When the kidneys are deprived of blood, a hormone, *renin* (see p. 147), is secreted, which by various complex processes gives a prolonged rise in arterial pressure due to generalised vasoconstriction. This mechanism is described in more detail elsewhere.

THE VASOMOTOR CENTRE

The co-ordination of all the modes of controlling circulatory function is vested in the *vasomotor centre*. It is not primarily a homeostatic device, for the activity of the circulatory system has to meet the varying needs of the body. At one extreme might be the heavy demands for oxygen and heat removal occurring in severe exercise; at the other would be the lowered general level of activity during sleep. Figure 37.2 shows how the *resting* arterial pressure (systolic above; diastolic below) varies when recorded at five minute intervals throughout the day. It is interesting to note the wide limits over which pressure varies, quite apart from any changes brought about by exercise.

The maintenance of the general vasomotor tone is intimately dependent on the integrity of this centre. Section of the hindbrain below the level of these cells leads to generalised vasodilatation, and the blood pressure falls from, say, 120 to 80 mmHg (16 to 10.6 kPa). After several days, if the animal survives, the blood pressure may rise again almost to its previous level; destruction of the spinal cord reduces the blood pressure

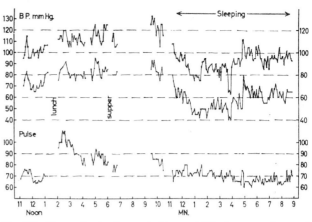

Fig. 37.2 Twenty-four hourly record of the arterial blood pressure of a normal subject. Note the fall in arterial pressure during sleep; it is accompanied by peripheral vasodilatation and by decrease in venous tone. (After Richardson, Honour, Fenton, Stott and Pickering.)

It is not stated what the subject was doing at about 10 p.m., but it is noteworthy that it resulted in the highest pressures of the day being recorded.

almost to zero. It thus appears that there are also vasomotor centres in the spinal cord, but in ordinary circumstances these are mainly controlled by that of the hindbrain (medulla); when this is destroyed, then the spinal centres gradually take over control. Electrical stimulation of the vasomotor centre, on the other hand, leads to generalised vasoconstriction and a rise of blood pressure. The normal activity of the centre in maintaining general vasomotor tone is, as we shall see, profoundly modified by the influence of the cerebral hemispheres, as well as by afferent impulses from the baroreceptors, and by chemical stimuli, the centre being stimulated by carbon dioxide and, indirectly, by inadequate oxygen supply.

Carbon dioxide and the vasomotor centre

The role of carbon dioxide in determining the activity of the cells of the vasomotor centres is well shown when the gas is excessively removed from the blood by overventilating the lungs. If the lungs of an anaesthetised cat are artificially overventilated with air, the blood pressure may fall in two minutes from 140 to 40 mmHg (18–5 kPa). If air containing 5 per cent carbon dioxide is substituted for the ordinary air previously used, the rate of ventilation remaining the same, the blood pressure returns to its original level.

After a cerebral haemorrhage the blood pressure may rise to 150 mmHg (20 kPa) and the pulse is slowed. The rise of blood pressure is usually attributed to stimulation of the vasomotor centre, arising from compression of the brain by the haemorrhage.

Anatomical location of vasomotor centre

The vasomotor centre is a bilateral group of neurones in the reticular substance of the lower third of the pons and upper two-thirds of the medulla as shown in Fig. 37.3.

Impulses are transmitted via the spinal cord to all vasoconstrictors in the body. Lateral parts of the centre are tonically active; the end result in a peripheral vasoconstrictor nerve might ordinarily be the passage of 1 to 2 impulses per second. It is called *sympathetic vasoconstrictor tone*. Proof is that total spinal anaesthesia which stops this activity causes a dramatic fall in arterial blood pressure.

The medial parts of the vasomotor centre are inhibitory in function, leading to vasodilatation since the normal, continuous vasoconstrictor tone is partly, or completely, released as required according to circumstances.

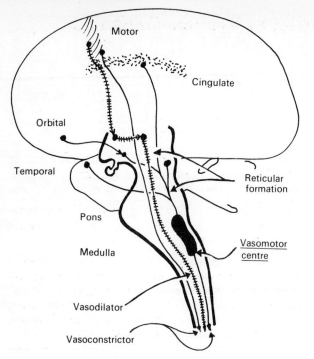

Fig. 37.3 Diagram showing areas in the brain which have a role in the nervous regulation of the circulation.

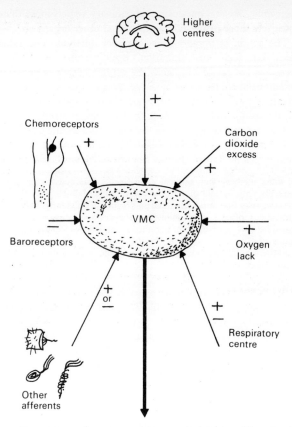

Figure 37.4 Factors affecting the activity of the vasomotor centre. In connection with the arrow shown from the chemoreceptors, it should be noted that chemoreceptors exert no effect of importance in normal animals breathing air in resting conditions. Daly and Scott (1963) found that the effect of carotid body excitation on heart rate was influenced by the consequent reflex respiratory stimulation. The tachycardia produced is due to sensory endings in the *lungs*, which, excited by the hyperpnoea, affected the vasomotor and cardiac centres.

In haemorrhage, however, the anoxia gives rise to vasomotor excitation via the chemoreceptors which helps to maintain the arterial pressure.

Control of the vasomotor centre by higher centres

Regions in the pons, mesencephalic and thalamic reticular formation and diencephalic areas can either excite or inhibit the activity of the vasomotor centre (*in toto* or in part).

The hypothalamus has powerful effects of both kinds. The cerebral cortex has both excitatory and inhibitory effects. There is also a direct pathway from the motor cortex to muscle, probably in the corticospinal tract, which bypasses the vasomotor centre and terminates on preganglionic sympathetic cells. This system is not tonic, but only causes generalised vasoconstriction when muscle activity is called forth by the motor cortex.

Summary

The vasomotor centre co-ordinates the various feedback servo-loops controlling blood flow through the body. Figure 37.4 gives a diagrammatic summary of these effects.

38. Characteristics of the Circulation in Various Organs

The metabolic and functional requirements in various parts of the body are different. In consequence the general features of the haemodynamics, the blood distribution and the circulatory controlling systems, differ in detail from one organ to another. For instance, the arterial pressures in the pulmonary circuit are considerably lower than in the systemic arteries because the peripheral resistance is lower and the required flow rate can be achieved with lower pressure gradients. In specialised organs, such as the kidney, other modifications to the normal vascular pattern are found. Before considering these adaptations to function in detail, it will be convenient to discuss briefly the methods available for measurement of blood flow rates through tissues.

Measurement of blood flow in the experimental animal

As in most physiological measurements, there are two important factors to be considered when assessing the accuracy of the procedure and the significance of any results obtained. First, the errors inherent in the physical measurement itself; second, the interference with normal function that is introduced by the process of making the measurement. In *animal experiments*, one of several methods may be used:

1. A cannula is tied into the vein draining the organ; the blood is collected for a given period of time and its volume measured.
2. An instrument for measuring the rate of the blood flow is interposed between the cut ends of the artery or vein supplying the organ; for example, a rotameter or an electromagnetic flowmeter. The rotameter is essentially a slightly funnel-shaped vertical glass tube containing a float. The height to which the float is lifted by the blood stream is an index of the rate of flow. To express it in absolute units (say, millilitres per minute), the instrument must be calibrated. The electromagnetic flowmeter works on the principle that an electromotive force is induced in any conductor (blood for example) which is moving in a magnetic field. The blood passes through a hole in a perspex block which is pierced on each side by the poles of a powerful magnet, and above and below by a pair of electrodes. The EMF developed between the two electrodes increases in direct proportion to the velocity of flow. After amplification, it is recorded by an ink-writing pen-recorder or by a cathode ray oscilloscope and camera. This instrument, also, must be calibrated, but it has the advantage of responding accurately to very rapid changes of flow.

The bubble flowmeter consists of a transparent tube, interposed in the bloodstream, into which a bubble of air can be injected and its passage, along a known length of the tube, timed.

The ultrasonic flow meter, used for measurement of pulsatile flow, depends on the physical principle that sound travels faster in a moving column of fluid, when its direction is with the current. A pulse of high frequency sound is timed (electronically) in a downstream and upstream direction and the difference gives a measure of the velocity of the stream.

Two other methods, while they do not actually measure the rate of blood flow, may be used to detect changes in flow, or calibre of the blood vessels:

3. The organ may be placed in an airtight box for recording changes in its volume (plethysmograph). These represent changes in the volume of blood in the organ, and may be ascribed to active changes in the calibre of its vessels so long as passive changes due to alterations of the general arterial and venous pressures can be excluded.
4. In a transparent tissue lying superficially, like the frog's web, or the conjunctiva, or in an organ that can be exposed, the vessels may be observed microscopically and their diameters measured.

Measurement of blood flow in man

In experiments on *human subjects*, clinical investigations, or in animal experiments when it is not possible to carry out extensive operative interference, an indirect method of flow measurement must be used. These indirect methods for determining blood flow depend on two general techniques. The first is applicable to an organ which either gives off or takes up a substance present in the blood flowing through it. If the concentrations of the substance in the blood going into the organ and coming out of it can both be measured, the volume of blood passing through the organ to carry a given amount of the substance can be easily calculated. The second method is based on the injection, into the blood flowing through a vessel, of a dye which mixes with the blood but does not leave the circulation. If a known amount of the dye is introduced in a given time, the blood passing the injection point during the same time will dilute the dye and this dilution can be measured easily by withdrawing a sample of blood at a point downstream in the vascular system.

THE FICK PRINCIPLE

The two methods for flow rate determination outlined above are applications of the so-called *Fick principle*. The word "principle" is hallowed by textbook tradition and is clearly used because the underlying idea is very simple, being in fact another way of stating the Law of Conservation of Mass.

If 1 g of a dye is injected into a blood vessel, say an artery, during the course of 1 min and at some point further along the vessel (mixing having taken place), a sample withdrawn from the vessel is found to have a concentration of the dye of 1 g l^{-1}, then clearly 1 litre of blood must have passed along the vessel during the 1 min. This can be expressed as an equation for the flow rate, F (in, say, litres per min).

$$F = \frac{Q_d}{V_d}$$

Where Q_d is the amount of dye injected per minute and V_d is the concentration of the dye found in the sample withdrawn from the vessel.

This argument can be extended to cover the state of affairs when dye is present in the arterial stream before the point of injection, simply by measuring not only the venous concentration of the dye, V_d, as before, but the arteriovenous difference in concentration, $(V_d - A_d)$. The Fick equation then becomes

$$F = \frac{Q_d}{(V_d - A_d)}$$

DIRECT FICK METHOD

To measure the blood flow through the lungs by this technique, a dye need not be used because ready-made substances are present in the blood in the form of oxygen or carbon dioxide. Blood can be easily analysed for these gases and the amount of them absorbed or given off in unit time can readily be measured. The volume of oxygen added to blood per minute as it passes through the lungs can be measured with a spirometer. The difference in concentration of oxygen can be determined by measurement of samples withdrawn from the arterial side (a systemic artery will do) and from the venous side (this must be from the right heart or pulmonary artery). The formula used is

$$F = \frac{Q_o}{A_o - V_o}$$

Q_o being the oxygen consumption per minute and $(A_o - V_o)$ the arteriovenous oxygen difference. Since the blood flow through the lungs is the same as it is through the heart, this technique is widely used in the measurement of cardiac output and we will discuss it later, in the chapter on the heart.

The practical details of the procedure are complicated by the fact that it is necessary to use a sample of *mixed venous blood* in order to arrive at a figure for the concentration of oxygen upstream of the lungs. This follows from the fact that blood samples withdrawn from the peripheral venous system vary considerably in their oxygen concentrations (depending on tissue utilisation of oxygen). Mixed venous blood samples are obtained by passing a *cardiac catheter* into the right heart or pulmonary artery from a peripheral vein, commonly the antecubital vein. This is not such a drastic procedure as it sounds, provided skilled operators perform the catheterisation although it must be stated that the operation, in common with all other surgery, does carry a mortality rate. The precise location of the tip of the catheter can be localised by means of viewing the patient on the X-ray screen or preferably by making use of the expected pressure changes to be found at different times in the cardiac cycle for various locations within the heart and its vessels.

The arterial sample comes from any convenient systemic artery, since blood composition does not change effectively until tissues are traversed. Oxygen consumption per minute necessitates volume measurement over quite lengthy periods, say 10 to 20 min, in order to obtain the necessary accuracy. It should be obvious from the foregoing account that flow measurements of this kind only represent the overall magnitude of blood flow (or cardiac output) during the course of a long time and must be interpreted accordingly. The *direct Fick method*, as the foregoing description is called, is the standard way of determining the cardiac output in man and is used mainly in clinical cardiological research. It is, however, subject to a number of errors, particularly during severe exercise, and repeated determinations or comparisons with other methods agree within about 30 per cent of the total flow.

INDIRECT FICK

An older and indirect method of estimating the oxygen and carbon dioxide contents of the mixed venous blood depended on the following reasoning. It is clear that if we stop breathing, the gas in the lungs will gradually tend to come into equilibrium with the venous blood, the oxygen content falling and the carbon dioxide content rising. Unfortunately, we cannot measure the composition of the venous blood as simply as this would suggest, because equilibrium takes too long to be established. If this method is to be used at all, the whole operation must be completed in a time less than that taken for any part of the blood to circulate once (about 23 s at rest); if a longer time is taken, then the composition of the venous blood will be altered, because such blood before traversing the tissues and returning to the lungs, will already have been equilibrated with an abnormal gas mixture in the lungs. The difficulty can be overcome in several different ways; the use of *intermittent rebreathing* being one of the best. In principle, the subject breathes in and out of a bag containing air, for about 10 s or so; the bag is then closed and the subject breathes from the open air for a while, so as to get the composition of the blood back to normal again. He then rebreathes from the bag for a further short period, and so on. During each rebreathing, the composition of the air in the bag approaches that of the mixed venous blood, and eventually reaches it. In practice, the bag is made to contain about 6 per cent carbon dioxide initially, so as to reduce the time required for equilibration.

THE FOREIGN GAS PRINCIPLE

The necessity for estimating the concentration of the reference substance X in the venous blood is avoided altogether if this substance is foreign to the body, and is removed completely as the blood passes through the tissues. No such substance has yet been discovered, however (except in the special case of the kidney, in Chapter 21). But if the whole sequence of procedure can be carried out in less than the circulation time—i.e. before any significant quantity of X has returned to the lungs in the venous blood—the same end is achieved. Although nitrous oxide and ethylene have been used as the foreign gas, acetylene appears to be the most suitable for this method, since it is harmless and easily estimated, diffuses readily through the lungs, and has a convenient and constant solubility in the blood. After emptying the lungs, the subject breathes quickly and deeply four times in and out of a rubber bag containing 2 litres of air and 0·5 litre of pure acetylene, a sample of the air being taken at the end of the last expiration. The lungs are thus filled with a gas mixture containing a known percentage of acetylene. Another sample of the gas is taken after two to six more breaths in and out of the bag, during which time some of the acetylene is carried away in the arterial blood. From the composition of the two samples, and from the total quantity of gases in the lungs and bag, the amount of acetylene absorbed and its average partial pressure in the alveolar air (and thus its concentration in the arterial blood) are determined.

THE HAMILTON DYE DILUTION METHOD

A known amount of the dye "cardiogreen" is rapidly injected into a vein. During its passage through the pulmonary circulation, it becomes evenly distributed throughout the bloodstream. Successive samples of arterial blood are collected, by means of a disc rotating at a known speed, and having attached to its circumference a large number of small tubes which in turn catch the blood flowing from an intra-arterial needle. Suppose, as an ideal simplification, that the dye first appears in the arterial blood at a time t_1 s after the intravenous injection, and disappears again at a time t_2 s. Then, if CO is the cardiac output in 1 s^{-1}, the total volume of the blood in which the dye becomes distributed is

$$CO(t_1 - t_2)$$

If the arterial concentration of the dye in the interval between t_1 and t_2 is A_d and $\dot{Q}_d$ g of dye were injected then

$$A_d = \frac{\dot{Q}_d}{\mathrm{CO}(t_1 - t_2)}$$

or

$$\mathrm{CO} = \frac{60 \dot{Q}_d}{A_d(t_1 - t_2)} \ \mathrm{l\ min}^{-1}$$

Actually, the calculation is more complicated than this, since the dye does not appear in the arterial blood and disappear again suddenly, nor is its concentration constant. The successive estimations of A_d must be plotted against time and the mean value of the arterial dye concentration and the best values of t_1 and t_2 determined from the curve. This method has the advantage that the subject is not required to co-operate with the experimenter in carrying out special respiratory procedures.

There are a number of variations on this theme. For instance, different dyes such as Evans blue (T1824) or iodocyanine green are used; or radioactive tracers such as albumen labelled with ^{131}I, or radioactive red cells can be used. The only requirement is that the indicator is not lost from the circulation in the time required to carry out the measurements (Fig. 38.1).

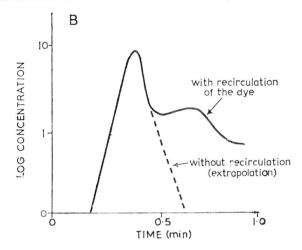

Fig. 38.2 Dye concentration curves.
When recirculation of dye occurs, it may do so early in the curve and the extrapolation is difficult. Experiments show that the logarithm of the downslope is linear; thus the curve can be completed, as in Fig. 38.1, and the area measured to give mean dye concentration as before.

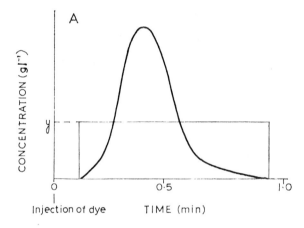

Fig. 38.1 Hamilton dye method.
Recording of the concentration of dye in the outflow of an organ. The amount of dye is given by the area under the curve and the time–concentration is the area as measured between two known times and shown as a rectangle. Here, mean concentration would be taken as y g l^{-1}. (The area under the rectangle = area under curve = $\Sigma V_d \Delta t$. Flow F is then

$$F = \frac{\dot{Q}_d}{\int_0^\infty (A_d - V_d) dt} = \frac{\dot{Q}_d}{\Sigma V_d \Delta t}$$

The technical problems in applying this technique concern recirculation and adequate mixing of the dye. In the case of measurement of pulmonary flow or cardiac output, the blood travelling via the coronary circulation will return to the arterial side within 15 s, so the measurements must be complete within this time or allowance must be made for recirculation in the calculation (see Fig. 38.2).

APPLICATIONS OF THE FICK METHODS OF
FLOW MEASUREMENT

Blood from any artery can be used for the estimation of A_d; for that of V_d the blood must come from the venous outflow of the organ under consideration. This principle is used for estimating the rates of blood flow through the human hepatic, renal, cerebral and coronary circulations. If the reference substance is dissolved in the plasma only, the value of F obtained is that of the rate of *plasma* flow; the total blood flow is obtained by dividing this by the relative volume of the plasma, which may be obtained by means of the haematocrit.

The human *hepatic blood flow* is estimated by using the dye bromsulphthalein as reference substance; this is excreted by the liver into the bile. The value of $\dot{Q}_d$ is given by the rate of intravenous infusion; A_d is obtained from an arterial blood sample; and V_d from a sample obtained by means of an X-ray-opaque non-wettable catheter introduced into an elbow vein and manipulated into the openings of one of the hepatic veins. The results, expressed in ml min^{-1}, are given in Table 38.1. Para-aminohippuric acid (PAH), or alternatively diodone, is used for estimating the human *renal blood flow*. The value of Q_d is obtained by estimating the amount excreted in a sample of urine collected during a known time. The value of V_d may be obtained from a sample of renal venous blood obtained by means of a catheter manipulated into one of the renal veins; but this is not usually done, since it is found that if the arterial concentration is not too high, the value of V_d is so small as to be negligible. PAH and diodone are removed from the plasma flowing through the kidney almost to completion. Table 38.1 shows the normal rate of renal blood flow. To estimate the *rate of the cerebral circulation*, the subject is made to breathe air containing a small proportion (about 15 per cent) of nitrous oxide. The nitrous oxide accumulates in the brain tissues, and after about 10 minutes a state of equilibrium is reached. The total quantity of nitrous oxide taken up by unit mass of brain is then given by the product of the solubility coefficient (discovered in separate, *in vitro*, experiments, or on experimental

Table 38.1 *Approximate distribution of a cardiac output of* 5·0 l min^{-1} *in a man at rest*

		Blood flow (ml min^{-1})	
	Weight kg	Total	Per 100 g tissue
Brain	1·5	750	50
Heart	0·3	150	50
Liver	1·5	1500	100
Kidneys (2)	0·3	1200	400
Skeletal muscles	25·0	750	3
Other organs	40·0	650	1·5

animals) and the partial pressure of nitrous oxide in the venous blood at the end of, say, the tenth minute. This quantity is equal to $10 \times (\dot{Q}/W)$, if the determination has lasted exactly 10 min, where W is the weight of the brain (not, of course, accurately known). The determinations of the arteriovenous concentration difference $(A_{N_2O} - V_{N_2O})$ is more complicated, since initially, when there is no nitrous oxide in the blood, it is zero, and finally, when equilibrium has been reached, with the tissues, it is again zero. It is necessary, therefore, to find the effective average value during the intervening period. This is best done by plotting the values of A_{N_2O} and V_{N_2O}, determined every few minutes, against time, as in Fig. 38.3. Any number of values of $(A_{N_2O} - V_{N_2O})$

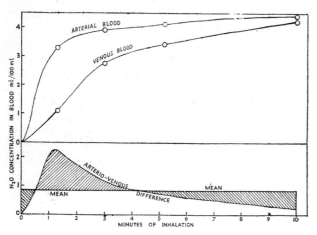

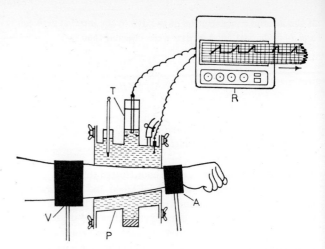

Fig. 38.4 Determination of the rate of blood flow in the human forearm by venous occlusion plethysmography.

The plethsmograph (P) is filled with water maintained at a temperature of 34–35°C. To record the blood flow, a pressure of 200 mmHg (27 kPa) is thrown into the lower cuff (A) to arrest the circulation in the hand. One min later a pressure of 60 mmHg (8 kPa) is thrown into the venous occlusion cuff (V). The arterial inflow is recorded on moving paper. After the experiment, the pen-recorder (R) is calibrated, the speed of the paper is ascertained, and the forearm volume found by water displacement. From the slope of the inflow tracing, the rate of the blood flow is calculated and expressed in millilitre per minute for each 100 ml of forearm.

The volume is recorded as a change in the electrical resistance between the water in the plethysmograph and a carbon rod partly immersed in the side-arm. The output of this transducer system (T) is amplified and recorded by a hot wire writing on heat sensitive paper.

Fig. 38.3 Typical curves showing the concentrations of nitrous oxide in the arterial and jugular venous bloods of a man during a 10 min period of inhalation of 15 per cent nitrous oxide. From curves such as these, the mean value of the arteriovenous concentration difference can be calculated, as is indicated in the lower part of the diagram. (After Kety and Schmidt.)

may be read from the smoothed curves drawn through these points and the arithmetic mean calculated; but in practice, it is sufficient to take the values at the end of each minute. We can now apply the equation of the Fick principle, but since we can only discover $\dot{Q}/W$, and not $\dot{Q}$, we can only measure F/W, i.e. the rate of blood flow through unit weight (usually 100 g) of brain tissue. The cerebral venous samples are obtained from a needle placed in the jugular bulb just below the exit of the internal jugular vein from the skull. The insertion of the needle is done under local anaesthesia, and is safe in expert hands. Table 38.1 shows the normal result. The nitrous oxide method has been used occasionally to estimate the human *coronary flow*. The venous blood samples are obtained from a catheter in the coronary sinus.

THE VENOUS OCCLUSION PLETHYSMOGRAPH

Part of an extremity is enclosed in a plethysmograph (Fig. 38.4) which is a rigid watertight case. Any change in the volume of blood in the part enclosed is transmitted to a sensitive volume recorder. The blood flow is measured by recording the rate of increase of volume during temporary occlusion of the venous drainage. This is done by throwing a pressure of about 60 mmHg (8 kPa) into a pneumatic cuff surrounding the limb just above the plethysmograph.

THE CIRCULATION IN THE CAPILLARIES

Joining the arterial and venous sides of the circulation are the capillaries, which are the thin-walled vessels in tissues where the interchange of oxygen, carbon dioxide, water, ions, metabolites and hormones takes place. The ultimate function of the remainder of the circulatory system is that of bringing these substances to and from the lungs, kidneys, liver, etc., to enable the exchange in the tissues to occur. A great deal of active research work is in progress at the present time, on capillary physiology, work no doubt stimulated in part by the fundamental nature of capillary function and also by the fact that recent advances in electron microscopy, microdissection and histochemistry make this an attractive field for investigation.

Organisation of the capillary network

The capillary network is the distributing system for supplying blood to the tissues. The flow through it is variable in direction, rate and amount and in general is adjusted rather closely to the needs of the tissues. A theoretical diagram of the anatomy of a capillary net is shown in Fig. 38.5 which has been compiled from numerous photomicrographs and cine films taken from different body regions.

Observations of capillary function have been made using mesentery, frog's web preparations and other very thin tissues mounted under the microscope. Blood tends to flow regularly only in the so-called *throughfare channels* between arterioles and venules. Otherwise the rate and direction of flow in the *true capillaries* is unpredictable. The *precapillary sphincters*, smooth-muscle thickenings at the mouths of most capillaries, contract and relax thus allowing less or more blood to pass through the capillary. The structure of the true capillaries has been the subject of hot dispute in the past—largely centred upon whether or not they are contractile. They are 8 to 10 μm in diameter and up to 750 μm long, are constructed of a single layer of endothelial cells held together by connective tissue and can be shown to contract down when prodded with a fine needle. However, no muscle fibres are present according to the most recent electron microscope findings.

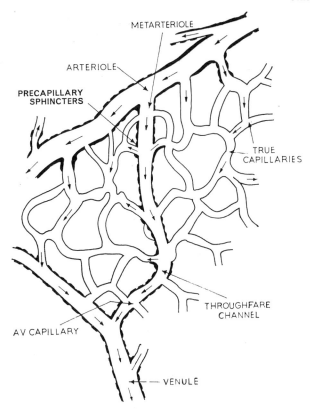

Fig. 38.5 Diagram of capillary network showing arteriole, metarterioles and capillaries. (After Chalmers and Zweifach, 1947.)

The surface area for exchange of substances between blood and tissues is enormous; it has been asserted that one muscle contains a capillary bed having the total area of a football field.

Venules form the collecting system; they are variable in diameter (say 10 to 100 μm) and the larger ones have an outside layer of smooth muscle. The contraction of this muscle plays a part in the control of outflow from the capillary bed.

Function of capillaries

The patency of capillary channels varies cyclically from time to time—a phasic form of activity denoted by the incongruous term *vasomotion*. The degree of opening of the vascular bed, of course, varies with activity in the tissue concerned; in resting muscle the number of active capillaries may be only 1 to 2 per cent of the total, a proportion greatly increased in exercise.

Capillary permeability, tissue fluid formation, lymph flow and oedema are dealt with in Chapter 18.

Capillary flow

Capillary flow can be studied experimentally. The total capillary surface area has been estimated at 70 cm^2 g^{-1} muscle (i.e. 500 m^2 in the muscles of a 70 kg man). Bone and connective tissue contain fewer capillaries than muscle. The resting flow rate is around 0·2 mm s^{-1}, but very variable.

We must enquire why resistance to flow is so low in capillaries yet their size is so small. According to Poiseuille's law, the pressure drop, P_D, across a vessel or vessels is given by

$$P_D = \frac{QL}{nr^4}$$

where Q is the flow, L is the length, n is the number of vessels in parallel and r is their radius.

Brief consideration will show that it is the *short length* of the capillaries, coupled with the low flow rates which are responsible for their low resistance (length is normally 500–750 μm).

Capillary pressures

Normal capillary pressure is not known because the act of measurement renders the vessel abnormal.

Direct cannulation gives an average mean pressure of 25 mmHg (3·3 kPa) in most tissues (arterial end, 30–40 mmHg; venous end, 10 mmHg).

Isogravimetric methods give slightly lower figures, i.e. 17 mmHg (2·3 kPa). Such methods are based on continuously weighing a section of mesentery or gut still with intact arterial supply. If the weight changes, fluid must be passing into or out of the tissues. Venous pressure is manipulated until there are no weight changes and the level required is taken as capillary pressure (since normally, in time, there is no net loss or gain of fluid from capillaries).

Isovolumetric methods make use of the same principle but measure tissue volume instead of weight.

THE VENOUS CIRCULATION

The vessels collect the blood which has passed through the tissue capillary network and then lead into the larger veins. Venous pressure cannot be given a fixed value, for it depends largely upon the particular vein concerned and also on whether the subject is standing or is in the horizontal position (Table 38.2). When standing, the pressure in the veins in the feet may rise to 100 mmHg (13·3 kPa), if the subject is at rest or if the valves are incompetent (see Fig. 38.6). This fact accounts for the

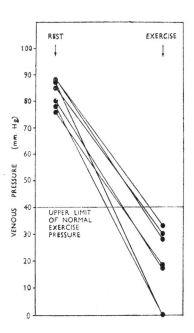

Fig. 38.6 Results showing the effect of exercise on the venous pressure in the foot. Each line shows the change in pressure for one subject.
During the exercise the subject marked time smartly, each foot being raised 9 inches, sixty times per minute. (From data of Walker and Longland.)

development of oedema of the feet and ankles in individuals who stand continuously, because the capillary pressure is thereby raised and re-absorption of tissue fluid is impeded.

Venous tone

The degree to which the smooth muscle in the wall of the venous system is contracted plays a part in determining the venous return to the heart. Other things being equal, the length of time that the blood spends in the venous system will be an inverse function of venous tone. Recent studies of venous tone in the human hand and forearm and in the hind limbs of a dog have shown that it increases in exercise due to increase in sympathetic venomotor nerve impulse frequency; it decreases during sleep. In haemorrhage also, it increases so much so that it may be impossible to get a needle into the lumen of a superficial vein.

Central venous pressure

This is the pressure near the openings of the venae cavae into the right auricle; it is the pressure which drives the blood into the heart. Stopping the heart causes a large rise in central venous pressure as blood accumulates in, and distends, the great veins. Other things being equal, central venous pressure is inversely related to the performance of the heart as a pump. Although cardiac output increases several fold in exercise, central venous pressure scarcely alters. The enormous increase in the venous return, of course, tends to raise central venous pressure; the fact that it does not rise is explained by the increase in the performance of the heart as a pump; increase in rate and the force of the heart beat enables the heart to transfer the blood more rapidly from the venous to the arterial side. As might be imagined, central venous pressure is influenced by alterations in the posture of the body and in venous tone. When blood is added to the circulation, most of it will be accommodated in the veins so that central venous pressure will increase. Heart failure is accompanied by a large increase in the pressure in the great veins.

Blood reservoirs

The spleen is peculiar in that blood is accommodated in it and whilst there it is also concentrated by loss of plasma. In the spleen pulp there is a bypass of the circulation; in the dog the spleen can accommodate one-fifth of the total volume of blood. The other reservoirs, the liver and portal system, the subpapillary plexus of the skin, and the great veins are, unlike the spleen, part of the general circulation, but are, like the spleen, capable of considerable variations in capacity; the liver and portal system (including the spleen) at rest contain about one-quarter to one-third of the total volume of blood in the cat and dog, the skin very much less. The capacities of these structures in man are not accurately known, but it is probable that the spleen is relatively less capacious than in the cat and dog.

THE PULMONARY CIRCULATION

The whole output of the right ventricle is delivered into the pulmonary artery at a pressure of 15 to 20 mmHg (2 to 3 kPa). The blood flow through the lungs is the same as the outflow through the aorta, i.e. at rest about $4 \, l \, min^{-1}$; in exercise it may rise to $30 \, l \, min^{-1}$ or more. The resistance offered by the pulmonary vessels appears to be low. Since there is no alternative route for the blood between the right and left sides of the heart, it is not surprising that the vasomotor supply is unimportant.

The chief factor modifying the pulmonary circulation is the pressure change accompanying respiration. The lungs are elastic structures which are kept open by the chest wall; if the chest is opened the lungs collapse. In consequence of this pull of the lungs, if a needle is thrust into the pleural cavity of man and connected to a manometer, this will normally register a pressure about 4 cm of water below that of the atmosphere (i.e. −400 Pa). During inspiration the lungs are further stretched and the intrapleural pressure falls by as much as 20 cm of water (2 kPa). These negative pressures affect all intrathoracic structures but particularly the pulmonary capillaries, the great veins and the chambers of the heart in diastole, whose walls are yielding; the effect on the thick-walled arteries is of less consequence.

The effects of respiration on the systemic arterial blood pressure are complex and variable. In man during quiet respiration of the thoracic type (i.e. mainly by the intercostal muscles) the blood pressure falls during inspiration and rises during expiration. This effect which is usual in man is probably due to mechanical changes in the pulmonary vessels; when the lungs expand the pulmonary vessels are pulled open and fill with blood, and diversion of this extra blood from its onward movement leads to a reduced flow into, and output from, the left ventricle. Occasionally in man when the breathing is abdominal in type (i.e. mainly by the diaphragm), the blood pressure rises during inspiration and falls during expiration.

During inspiration blood is aspirated into the great veins and heart from extrathoracic structures, and the rise of intra-abdominal pressure produced by descent of the diaphragm forces blood from the abdomen into the chest; in this way the filling and output of the heart would be increased during inspiration. X-ray photography, after intravenous injection of an opaque substance, shows that during inspiration the flow through the superior vena cava is increased, but in the dog, cat and rabbit, the inferior vena cava is constricted by contraction of the diaphragm; in these animals at least it is unlikely, therefore, that the flow through the inferior vena cava is increased during inspiration.

The pulse is accelerated during inspiration and slowed during expiration—a reflex effect which is abolished by section of the vagi.

Regulation of pulmonary circulation

The pulmonary arterial pressure does not vary a great deal in spite of the large changes in blood flow through the lungs, a fact indicating that the pulmonary vascular resistance decreases in inverse proportion to the blood flow. This change in the vascular bed consists of a passive dilatation of the arterioles and capillaries.

It should be noted that the regulation of the pulmonary vascular resistance is very largely passive in character and the sympathetic innervation plays only a small part.

Pulmonary vascular resistance is increased by breathing gas mixtures which stimulate respiration (e.g. low oxygen or high carbon dioxide contents)—an effect said to be separable from the increase in cardiac output and arterial pressure which also results. Local effects of this nature occur in the lung's circulation and one may look upon it as a compensatory mechanism which controls gas exchange in different parts of the lung. An underventilated alveolus, having in consequence a low P_{O_2} and a high P_{CO_2} will have in its local circulation a high peripheral resistance. Hence the local blood flow will be small and gas exchange across the pulmonary capillary membrane will proceed at a slow rate. It can be seen that this is a self-regulating homeostatic mechanism, to ensure even gas exchange throughout the lung. It is of considerable importance in pathological conditions affecting alveolar membranes, or giving rise to arteriovenous shunts in the lung.

The pulmonary venous pressure is about 5 mmHg (0·7 kPa). It depends almost entirely upon the left atrial pressure and in failure of the left side of the heart the venous pressure is raised, and in turn, with it, the capillary pressure. This leads to fluid exudation and pulmonary oedema.

Pulmonary capillary pressure is of extreme importance in normal lung function. In the systemic circulation, capillary pressure is around a mean of 30 mmHg (4 kPa) and, as we have already seen, the arterial end having a slightly higher value than this acts as a site of formation of tissue fluid. In the lung the formation of tissue fluid must be prevented. In fact, the low capillary pressure 10 to 7 mmHg (1·3 to 0·9 kPa) is well below the colloid osmotic pressure of 25 to 30 mmHg (3·3 to 4 kPa) in the plasma, so tissue fluid formation is minimal. Until the capillary pressure rises to several times its normal level, pulmonary oedema will not occur.

THE CEREBRAL CIRCULATION

While the cerebral vessels are to some extent influenced by vasodilator metabolites during cerebral activity, they are little affected by vasomotor impulses; persistent inquiry failed to reveal any vasomotor supply, until it was shown that stimulation of the cat's cervical sympathetic produced a slight narrowing of the pial arteries observed through a glass window screwed into the skull. The cerebral blood flow is thus determined in the main by the height of the arterial blood pressure, and it is rather surprising that this should be regulated exclusively by receptors lying outside the brain, in the carotid sinus and arch of the aorta. For if these receptors are excluded, alterations in blood flow to the brain lead to no reflex changes altering the height of the arterial pressure, unless the cerebral blood flow is so reduced that it produces asphyxia of the vasomotor centre. Table 38.2 gives some data about the cerebral circulation in man.

The importance and significance of the reflexes controlling blood pressure is now evident, for the brain is the master organ of the body and is extremely sensitive to reduction of its blood supply; if the blood flow to the brain ceases for five seconds consciousness is lost, and after twenty seconds epileptic twitching begins. By means of the carotid sinus and depressor reflexes the arterial blood pressure, and thus the cerebral blood flow, are maintained by appropriate regulation of the rate and force of the heart-beat and of the blood flow through organs other than the brain.

The normal cerebral flow is about 0.5 ml g^{-1} min^{-1} brain tissue; this respresents about 750 ml min^{-1}, or about one-fifth of the cardiac output. The cerebral oxygen usage is about 0.035 ml g^{-1} min^{-1} (or 45 ml min^{-1} *in toto*).

Control of cerebral blood flow

If the $P\text{CO}_2$ of carotid blood rises the brain undergoes vasodilatation (e.g. doubling the $P\text{CO}_2$ doubles the flow rate). A similar but less marked effect occurs when the $P\text{O}_2$ falls. This is another local autoregulation system as we have described in the blood vessels of the lung previously. The physiological value of a system such as this is fairly obvious; at high values of $P\text{CO}_2$, cortical neurones (and indeed all brain nerve-cells) become relatively inexcitable. A high local blood flow will wash out excess carbon dioxide (Table 38.3).

Table 38.3 *Cerebral blood flow (CBF) in normal human subjects breathing gas mixtures*

Gas mixture breathed	CBF (ml g^{-1} min^{-1})
Air	0·55
5% CO_2 in O_2	0·95
10% O_2 in N_2	0·75
Hyperventilation (at 25 l min^{-1})	0·35
100% O_2	0·45

Data from Kety and Schmidt (1948). Note:
1. Cerebral circulation is not greatly influenced by vasomotor nerves. Maximal stimulation of cervical sympathetic gives only 15% reduction in CBF.
2. $P\text{CO}_2$ in arterial blood is the most important control. High $P\text{CO}_2$ and low $P\text{O}_2$ dilate; low $P\text{CO}_2$ and high $P\text{O}_2$ constrict cerebral vessels.
3. Hyperventilation acts by lowering arterial $P\text{CO}_2$ also by alkalosis, Bohr effect and thus O_2 lack.
4. Hyperbaric oxygen (3–4 atm, say) constricts cerebral vessels.

The overall metabolic rate of the brain alters little; deep sleep, intense mental activity or the motor output occurring during exercise, do not give rise to measurable changes in metabolism or blood flow in any part of the brain. On the other hand, the flow in localised areas of brain alters very rapidly and concomitant with local neuronal activity. The experimental evidence for this is that the surface blood flow in vessels on the pia mater can be observed directly under the microscope and correlated, say, with seizure activity in the cells beneath. Also, measurements can be made electrometrically using gold, or gold-plated, wires inserted into the brain substance, in order to find the oxygen tension. At the same time, local neuronal firing can be recorded and it has been demonstrated that a period of intense local activity can raise the blood flow in the neighbourhood by as much as 50 per cent. Gold wires used in this type of experiment have been inserted into human patients for the determination of oxygen tensions in the brain in various forms of mental disease and it has been found that the oxygen tension tends to show a cyclic variation about a mean level with a period of about $\frac{1}{2}$–1 s.

It must be emphasised, however, that the mean cerebral blood flow is remarkably constant in spite of local changes. In the face of an alteration in the carotid systolic pressure of from 50 to 200 mmHg, cerebral blood flow does not alter. Pressures have to fall below about 50 mmHg before the brain becomes relatively ischaemic.

Table 38.2 *Venous pressure in man*

Vein	Pressure (cmH$_2$O)*	
	Mean	Range
Median basilic	97	50–148
Femoral	111	98–128
Abdominal	115	70–160
Dorsal metacarpal	130	70–170
Great saphenous	150	110–190
Dorsalis pedis	175	124–210

Values for an upright man; notice increase, due to hydrostatic gravitational effect, in lower limb. (From Burch, 1950, p. 109.)

* 1 cm H$_2$O ≃ 0·1 kPa.

The measurement of cerebral blood flow

The method used in experiments in humans is based on the Fick principle. It has been validated by comparison of results with simultaneously determined flow-rates measured in monkeys with bubble flowmeters.

An oxygen–nitrogen mixture containing 15 per cent N_2O is breathed by the subject for 10 min. Internal jugular and arterial blood samples are taken and the integral of the A–V nitrous oxide difference is found. Then:

$$CBF = 100 \frac{\Sigma N_2O \text{ taken up}}{(A-V)N_2O \text{ difference}} \text{ ml g}^{-1} \text{ min}^{-1}$$

The grey matter in the brain is the main oxygen consumer. It weighs less than 10 per cent of the whole body yet consumes about 20 per cent of the body's oxygen at rest.

Raised intracranial pressure

If the intracranial pressure rises acutely, arterial hypertension develops. This is a regulatory mechanism to maintain cerebral blood flow, if possible. The mechanism is believed to be that the rise in cerebrospinal fluid pressure (due to a tumour or haematoma, for example) gives rise to anoxia of the bulbar regions in the brain stem. This causes generalised excitation of the vasomotor centre with consequent vasoconstriction throughout the body and hence a rise in arterial pressure. If the cerebrospinal fluid pressure rises too much (above about 45 cm H_2O; 4·5 kPa) arterial pressure then falls due to depression of the bulbar centres by the oxygen lack.

Cerebral vascular accidents

Two types of cerebral vascular accident occur, particularly in patients having raised arterial pressure. About three-quarters of these are haemorrhages, which destroy the surrounding brain tissue; the remainder are arterial occlusions (cerebral thromboses). In either case paralysis, blindness and dementia may occur, the nature of the illness depending upon the site of the lesion.

THE CORONARY CIRCULATION

The heart muscle in mammals is supplied with blood from two coronary arteries arising from the aorta just beyond the semilunar (aortic) valves. The blood is returned to the right auricle by a number of openings of which by far the largest is the coronary sinus. The rate of the coronary blood flow in an experimental animal may be measured by inserting a cannula into the coronary sinus; the blood issuing represents, in the dog, three-fifths of the total flow through the whole coronary system.

The rate of flow in the coronary circulation has been measured in the human subject at rest and is about 150 ml min^{-1}. It is not possible to measure it during severe exercise; but to supply the heart with enough oxygen to produce the energy needed for a cardiac output of 30 l min^{-1}, the coronary flow would have to be nearly 1 l min^{-1}. Since the arterial blood pressure in severe exercise increases from 120/70 mmHg only to, say, 180/70 mmHg, the sixfold increase in coronary flow cannot be due to an increased perfusion pressure. It is probably due to the vasodilator action of metabolites produced by the heart muscle during a condition of lowered oxygen pressure. The effect of lowered oxygen partial pressure is illustrated in Fig. 38.7. When,

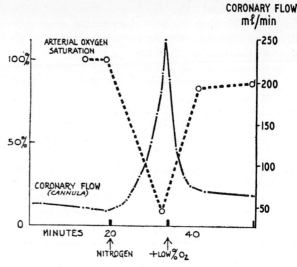

Fig. 38.7 The relation between coronary blood flow and oxygen saturation of the arterial blood.

Observations on a heart–lung preparation. Between the first and second arrows the lungs were ventilated with pure nitrogen, and from the second arrow onwards with nitrogen to which a little air had been added. The coronary flow varies inversely with the oxygen saturation of the blood. (Hilton and Eichholts.)

in the heart–lung preparation, the lungs were ventilated for a short time with nitrogen, the coronary blood flow increased about 10 times. The condition is not quite the same in exercise because the arterial blood is almost fully saturated with oxygen. Nevertheless the enormous increase in rate of oxygen usage by the heart must, at least temporarily, lower the oxygen saturation of the venous blood and the oxygen pressure of the tissue fluid bathing the plain muscle of the arterioles. This may relax them directly; or they may be relaxed by vasodilator metabolites diffusing in from the heart muscle.

The autonomic nervous control of the coronary vessels is probably weak and its action easily swamped by the effects of local deficiency of oxygen. Stimulation of the sympathetic causes vasodilatation, and stimulation of the vagus causes vasoconstriction; but these effects may be due to concomitant alterations in heart metabolism. In the fibrillating heart, which no longer beats, the opposite effects are obtained during nerve stimulation. Probably, therefore, the sympathetic has a constrictor action, and the vagus a dilator one.

For a short time at the beginning of systole, contraction of the heart muscle arrests the coronary circulation. Studies of the rate of inflow into one of the coronary arteries show that the flow is intermittent—most rapid in diastole, and stopped during the isometric contraction phase of the cardiac cycle, the coronary branches being compressed and occluded between the tightly contracting muscle strands. In this respect, cardiac muscle resembles skeletal muscle in which, too, the flow is arrested during strong contraction.

Certain hormones influence coronary blood flow. Adrenaline and noradrenaline (with weaker action) dilate the coronaries. Acetylcholine also dilates the coronary vessels. 5-Hydroxytryptamine is a powerful vasodilator of the coronary vessels.

A number of drugs are used to increase coronary flow in coronary heart disease.

Commonly employed is nitroglycerine, which acts both on the coronaries to dilate them, thus improving the blood supply to the heart muscle, and on the peripheral circulation as well, by lowering peripheral resistance and hence the workload of cardiac muscle.

Coronary occlusion and angina pectoris

The function of the heart, like that of every organ, is intimately dependent on its blood supply, and the arrangements that we have discussed are such that in health, increased work of the heart is accompanied by increased blood flow. If a coronary artery is suddenly blocked by a clot, then the patient may die at once from ventricular fibrillation or after some hours from congestive heart failure; in a large number of cases, particularly if the area deprived of its blood supply is small, the remaining healthy heart muscle is adequate to maintain the circulation at rest, the bloodless area is slowly converted into fibrous tissue and the patient recovers. Such small coronary occlusions are accompanied by intense substernal pain, probably due to the stimulation of sensory nerves in the heart itself by chemical substances released locally from the ischaemic muscle. If the coronary arteries are thickened and narrowed by disease, they are incapable of dilatation, and the circulation becomes inadequate to the demands of muscular work. In this condition substernal pain (angina pectoris) is produced on exercise, probably again by the release of metabolites from the inadequately oxygenated heart muscle. A somewhat similar pain known as intermittent claudication is experienced in the muscles of the legs on walking, when the arteries are narrowed or blocked by disease. The student may reproduce this pain by working the muscles of his forearm, after the circulation has been arrested by inflating a cuff on the upper arm to above systolic pressure. He may ascertain that the rate at which pain develops depends on the frequency and the force of the muscular contractions. On stopping work the pain remains present until the circulation is restored, when it quickly disappears. The pain thus seems to be due to stimulation of the nerve endings in the muscles by a substance released during muscular contraction, and normally removed by the circulating blood.

THE CIRCULATION IN MUSCLES

Around 40 per cent of the total weight of the body is muscle. In exercise, cardiac output goes up to 25 or 30 litres min^{-1} and of this probably 20 litres min^{-1} will go through the active muscle. The peripheral resistance is very high in muscle at rest.

The effect of exercise (local control)

The changes in muscle blood flow during exercise are of particular interest and are the outcome of two opposing factors—vasodilatation and mechanical compression of the vessels. In weak sustained contractions, the blood flow increases; but when the contraction is strong, as is that of the human gastrocnemius and soleus muscles, for example, when a person is standing tiptoe on one leg, the circulation in the muscle is almost arrested by the pressure of the taut tissue. It is not surprising that such contractions can only be kept up for three or four minutes. In rhythmic exercise, the muscle vessels dilate, but the pattern of flow depends upon the kind of movement. Blood flow is continuous if the contractions are weak, but intermittent if they are strong.

During exercise, muscle blood flow may rise tenfold or more (Fig. 38.8). The multilayered arterioles, the site of the main resistance, are widely dilated and the number of open capillaries increases many times, as do their diameters. Capillary surface area in the human gastrocnemius and soleus muscles may increase from the size of a handkerchief to that of a sheet.

As Fig. 38.8 shows, this vasodilatation occurs in sympathectomised subjects, and is due to a local factor. In exercise, it is probable that the greatly increased rate of oxygen usage leads to

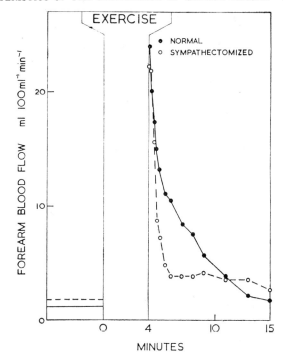

Fig. 38.8 Results showing that exercise causes vasodilatation in human muscles; the vasodilatation occurs in sympathectomised muscle and is due to the action of vasodilator metabolites. Venous occlusion plethysmography. The exercise was gripping a bar tightly in each hand for 40 s.

a reduction in the oxygen pressure in the neighbourhood of the arterioles. Like the coronary vessels, the muscle vessels will be relaxed, either directly by the lack of oxygen or by vasodilator metabolites from the muscle fibres, perhaps, for example, by adenosine triphosphate (ATP) or adenosine diphosphate (ADP). It is unlikely that alteration in the local acidity (pH), or in the local carbon dioxide or lactate concentration, is adequate to account for the vasodilatation. Thus the dominant feature of the control of the vessels supplying skeletal muscles, like that of the vessels supplying the heart, is a local control of the circulation in accordance with metabolic requirements.

Vasoconstrictor tone (reflexes in posture)

The vasomotor centre and sympathetic vasoconstrictor nerves maintain in resting skeletal muscle a small amount of constrictor tone. A change from the recumbent to the upright position brings reflexes into play to prevent fall in arterial blood pressure, and the vasomotor centre constricts the vessels in skeletal muscles. Reflex vasoconstriction also occurs during haemorrhage. In the common faint (vasovagal syndrome) the fall in arterial blood pressure is due to vasodilatation in the skeletal muscles, mediated by the sympathetic, and unaccompanied by compensating vasoconstriction elsewhere. In exercise, the effect of the vasomotor centre on skeletal muscle vessels is of little significance and easily swamped by the local vasodilator mechanism.

Observations on a policeman, aged 23, on whom bilateral lumbar sympathectomy was performed in order to prevent excessive sweating of the feet, also proved that relaxation of myogenic tone in the skeletal muscles is predominantly due to the action of metabolites released by the active muscle. On the day before the operation he twice ran 380 yd in 65 and 61 s respectively; on the 99th day after the operation, he again ran the same distance twice in 60 and 62½ s. His time was not affected by sympathetic denervation of his leg muscles.*

When the amount of adrenaline in the general circulation increases, the muscle vessels dilate; but they constrict when adrenaline is applied to them directly. That is to say, intravenous and intra-arterial infusions of adrenaline in man have opposite effects on the muscle circulation. The reason is not yet known.

Although there is decrease in pH, and an increase in PCO_2 and lactate in the venous effluent from active muscle (Fig. 38.9) it is vein draining the skin, nor in blood taken from a deep vein draining the forearm muscles. However, A–V oxygen difference is greatly diminished, signifying increase in muscle blood flow. This is mediated by sympathetic cholinergic vasodilator fibres and by an increase in the amount of circulating adrenaline.

Control of muscle blood flow

The blood flow through resting muscle varies from 0·01 to 0·1 ml g^{-1} min^{-1}. The hyperaemia of exercise together with the

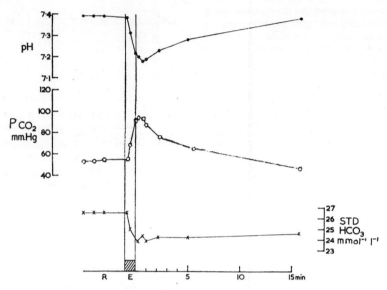

Fig. 38.9 pH, PCO_2 and standard bicarbonate in the venous blood draining the forearm muscles, before (R), during and after 1 min strong handgrip (E). (Barcroft, Greenwood and Rutt.)

unlikely that any or all of these bring about the loss of myogenic tone.

Occlusion of the blood flow to a muscle is followed by reactive hyperaemia (vasodilatation). The degree of hyperaemia depends directly on the duration and also whether it has been induced by arterial occlusion or venous occlusion. In the latter case the reactive hyperaemia is much less. This is thought to be due to the fact that myogenic relaxation occurs only when the arterial inflow is restricted and the tension of the smooth muscle fibres in the vessel walls is reduced thereby. In venous occlusion, on the other hand, the pressure in the vessels is higher than usual and the smooth muscle reacts by contracting—an effect which tends to counteract the dilator response to low PO_2 and accumulated metabolites.

Emotional factors and muscle blood flow

Sympathetic vasoconstriction also occurs in exercise in muscles that are not active. In fright, excitement, and emotional stress, muscle blood flow is increased in all skeletal muscles. Forearm blood flow is increased (Fig. 38.10); there is no change in the arteriovenous oxygen difference in blood taken from a superficial

increase in cardiac output may raise this figure to 0·5 ml g^{-1} min^{-1}.

Control of muscle blood flow, as already outlined in the previous two sections, is brought about by chemical, hormonal and nervous mechanisms.

Chemical factors are mainly the accumulation of metabolites and lowered PO_2, which dilates muscle vessels and increases flow. Hormones involved are adrenaline and noradrenaline (Table 38.4) while nervous control is mediated via the sympathetic innervation.

Figure 38.11 shows the effect of an infusion of adrenaline on the forearm blood flow. There is an initial transient fivefold increase followed by a return to about double the resting value for

Table 38.4 *The control of blood vessels in muscle*

	Vasoconstriction	Vasodilatation
Local factors		Metabolites (? vasodilator specific substance) ATP* ADP
Nervous	Sympathetic constrictor fibres	Sympathetic vasodilator fibres
Hormonal	Noradrenaline	Adrenaline Acetylcholine

Many influences control muscle blood vessels. Observe that the sympathetic nervous innervation can be either constrictor or dilator.

* This is one reason why ATP has been thought to be the purinergic transmitter.

---

* One wonders whether a less traumatic and possibly more effective treatment might not have been for the medical attendant to proffer advice on ways of washing the feet; however, had this been done instead, a delightful experiment would not have been recorded for posterity.

THE CIRCULATION IN OTHER ORGANS

The salivary glands

Stimulation of the parasympathetic nerve to the submandibular gland causes secretion and marked vasodilatation. Activity of the salivary secreting cells is accompanied by the release from them of a proteolytic enzyme, and in the presence of this enzyme, tissue fluid protein is hydrolysed, forming a polypeptide known as *bradykinin*. This substance is a potent vasodilator, and its action on the neighbouring vessels increases submandibular blood flow in accordance with the metabolic requirement. It is probable that local vasodilator mechanisms are brought into action during the secretion of all other digestive glands, but this is not known for certain.

The liver and portal system

The blood flow through the liver is very large; about half the blood flow through the inferior vena cava comes from this source. After leaving the intestinal capillaries the blood gathered into the portal vein traverses a second set of capillaries in the liver. Since the portal pressure in the dog is only about 8 cm of water (0·8 kPa), the resistance offered by the liver vessels must be very small. The liver is also supplied by the hepatic artery, which contributes about a quarter of the blood and 40 per cent of the oxygen supplied to the organ. The liver, the portal vein and the territory it drains ordinarily accommodate about one-third, or more, of the total blood volume. Experiment suggests that a large proportion of this is expelled in the early stages of haemorrhage. After the injection of adrenaline or stimulation of the sympathetic nerves the outflow from the liver exceeds its inflow, a large part of its blood being thus discharged into the great veins. It is likely, therefore, that the liver and portal system constitute a variable reservoir, whence blood is discharged to augment the inflow and output of the heart in conditions such as haemorrhage, emotion and exercise.

A piece of dog's colon, transferred to the outer abdominal wall with its nerve and blood supply intact, blanches at the beginning of exercise though, as exercise is continued, it slowly fills again with blood. This presumably illustrates what happens to the vessels of the whole gut in exercise.

The spleen

The branches of the splenic artery open into venous sinuses, which unite to form the splenic vein. Along the course of the artery and vein are perforations communicating with the spleen pulp, which contains red and white blood corpuscles in its network. In the dog the spleen has been brought to the exterior through an incision on the abdominal wall and left there for many months. Its size at rest indicates that it may hold one-fifth of the total blood volume. During haemorrhage, emotion, asphyxia and muscular exercise, the muscular capsule of the spleen contracts, and its blood content, which is exceptionally rich in red cells, is expelled into the general circulation. The reservoir function of the spleen appears to be less important in man.

The skin

Heat is lost from the skin by conduction, radiation and evaporation. The rate of heat loss is regulated by the temperature

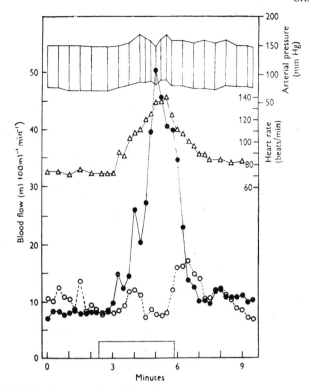

Fig. 38.10 Emotion and muscle blood flow.
The effect of severe emotional stress (at time of rectangle) is to produce a fivefold increase in forearm blood flow (●). There is little change in hand flow (○). The effects on arterial pressure and heart rate are also shown. (Blair, Glover, Greenfield and Roddie.)

the duration of the infusion. These changes are to a very large extent in the skeletal muscles. First stimulation of the β receptors relaxes the muscle vessels and this is very soon followed by stimulation of the α receptors and contraction, enough to reduce the flow to double its resting value. The blood vessels of the hand have no β receptors; adrenaline causes simple constriction.

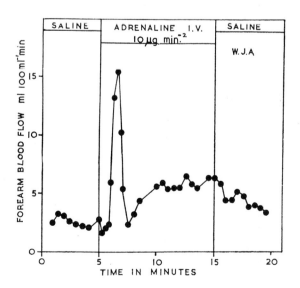

Fig. 38.11 The effect of an intravenous infusion of adrenaline on forearm blood flow. The changes are for the most part in the muscle vessels. The result of stimulation of the β receptors, marked vasodilatation, is soon reversed by vasoconstriction due to stimulation of the α receptors.

regulating centre acting through the sympathetic nerves to the cutaneous blood vessels and sweat glands. According to the manner of this regulation, the skin of the body can be divided into two areas. Thus the skin of the hands, feet, nose, lips and ears has to be considered separately from that of the forehead, trunk and limbs as far as the wrists and ankles.

When the room temperature is below about 20°C (68°F), the vessels in the hands, feet, lips and ears of an ordinarily clothed man are constricted—strongly so if the temperature is below about 16°C (about 60°F). The effect of a rise of environmental temperature on the blood flow through the skin of the hand is shown in Fig. 38.12. This flushing of the hand is chiefly due to a

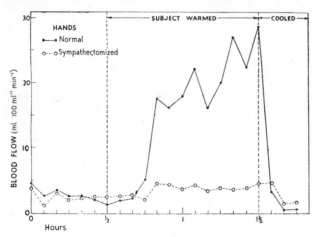

Fig. 38.12 Results showing that warming the body causes vasodilatation in the hand; this does not occur in the sympathectomised hand. It is mainly, if not entirely, due to the release from sympathetic constrictor tone. Venous occlusion plethysmography. The body was warmed by covering the subject with blankets and immersing the feet in water at 44°C, and cooled by removing the blankets and immersing the feet in water at 17°C. (Barcroft.)

release of the vasoconstrictor tone brought about by sympathetic nerves, because blocking the nerve supply by local anaesthesia causes a similar increase in flow. The chief factor in the regulation of the skin blood flow in accordance with changes in the environmental temperature is a central mechanism which responds to a rise in blood temperature by reduction of sympathetic tone, and *vice versa*. In this connection an important part is played by the arteriovenous communications, short wide connections between the arterioles and venules, which are very abundant in man in the nail beds, the skin covering the volar surfaces of the fingers and palm of the hand and in the corresponding sites of the foot. The changes in blood flow through the fingers which may occur in response to changes in body temperature are very large, ranging from 1 ml min^{-1} through each 100 ml of finger when the body is cool, to 100 ml min^{-1} when the body is hot.

The skin of the forehead, trunk and proximal parts of the limbs has very little, if any, sympathetic vasomotor innervation. Nevertheless, in a very warm subject marked vasodilatation occurs, though without much flushing because the vessels lie too deeply. This vasodilatation accompanies sweating. It is due to the vasodilator action of bradykinin, formed during sweat gland activity. It will be recalled that the vasodilatation which accompanies salivary secretion is also due to the formation of bradykinin.

The colour and temperature of the hand and face are closely related to the state of the skin circulation. The hot, pale hand, common in summer, is one in which blood flows rapidly through the deeply situated invisible arteriovenous anastomoses in the fingers and palms. The superficial capillaries whose contents give colour to the skin are narrow and poorly filled. The cold red hand seen in winter is one in which the arteriovenous anastomoses and arterioles are narrow and the capillaries dilated. The blood remains red, since little oxygen is removed by the cold tissues. The cold blue hand is one in which the arterial vessels are still further constricted, and the flow becomes so slow that an appreciable fraction of the oxygen content of the blood is removed even by the slowly metabolising tissues.

Although the vasomotor and temperature regulating centres predominate in the regulation of the skin circulation—directly through vasomotor nerves or indirectly through local vasodilator substances—there are some other responses which must be briefly described.

1. *Reactive hyperaemia.* This important response was first seen in the arm and leg. If the circulation to a warm limb is arrested for a few minutes and then released, a bright flush, reactive hyperaemia, at once suffuses the skin and then slowly fades. After circulatory arrest lasting 10 minutes the blood flow to the forearm may be increased to ten or twenty times the normal; both muscle and skin vessels share in this vasodilatation. The intensity and duration of reactive hyperaemia depend on the duration of circulatory arrest and on the temperature of the limb. The flush represents a dilatation of the minute vessels; it is independent of any central or local nervous mechanism and is due to the action of vasodilator substances formed locally and normally removed by the circulating blood. When, as is constantly happening, areas of skin and of subcutaneous tissue are rendered bloodless by supporting the weight of the body, they may be said to accumulate a blood-flow debt; reactive hyperaemia ensures the discharge of this debt as soon as blood is free to enter the tissues again.

2. *The white reaction.* This is chiefly of interest as the basis of an experiment performed by Lewis to show that human capillaries can contract actively. If the skin of the forearm or back is lightly stroked with the end of a ruler, the line of the stroke becomes marked by pallor. This is due to narrowing of the minute vessels (capillaries and venules), for these are the only vessels which come near enough to the surface of the skin for blood within them to be visible. Narrowing of the capillaries in this white reaction might be due to (a) constriction of the deeper arterioles and passive collapse of the more distal capillaries or (b) to active contraction of the capillary walls. Lewis showed that the white reaction can be induced in the forearm skin after the circulation in the arm has been arrested. In this case, the pallor cannot be due to constriction of the underlying arterioles, for this would, if anything, tend to increase capillary pressure; therefore the capillary walls themselves must contract.

3. *Triple response.* If the skin is injured by scratching, by lightly burning, by freezing or by pricking-in injurious substances such as hydrochloric acid or caustic soda, the point of injury is marked by reddening of the skin, which later gives place to wealing as fluid passes out of the capillaries and distends the tissue spaces of the skin. Around the local reddening is a diffuse red mottled flush, or "flare", which is due to the opening of the surrounding arterioles. The local *redness* due to widening of the minute vessels (capillaries and venules), the *weal* due to their increased permeability, and the *flare*, are the components of the

triple response of the vessels to injury. The whole response is independent of the central nervous system, being unchanged immediately after section of all nerves to the skin. After all the nerves have degenerated, however, the flare is absent; if the sympathetic supply alone has degenerated, the flare is unimpaired. The flare is an example of a local axon reflex through the sensory fibres. The fibres entering the posterior roots of the cord divide at their periphery into branches to the blood vessels and to the tissues. Injury to the skin stimulates the sensory branches, and the stimulus passes proximally to the point of bifurcation and back down the other branch to the arterioles (see Fig. 38.13). The whole of the triple response has been shown to be due to the release of a chemical substance from the injured skin, and this,

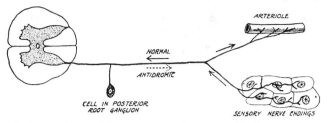

Fig. 38.13 Diagram of the nervous connections concerned in the axon reflex.

from its resemblance to histamine, has been termed "H substance". The triple response is, in the skin, the vascular basis of the phenomenon of inflammation.

39. The Heart: General Characteristics as a Pump

The heart is a simple type of pump, which operates on the principle of emptying and filling its chambers by alternate contraction and expansion of them. Valves limit the flow of blood in and out of these chambers to one direction, thus enabling the forcible muscular contraction to pump blood out. It should be noted that the valves are passive structures and open or close in response only to pressure differentials across their orifices. The two sides of the heart act independently from the mechanical point of view, but of course beat together.

Structure of the heart

The heart is divided longitudinally into the right and left hearts, each consisting of two communicating chambers, the auricle (or atrium) and ventricle. The capacity of each ventricle when fully relaxed is about 140 to 200 ml in man. The heart wall consists essentially of muscle (the myocardium), which has an inner covering (the endocardium), lined by endothelium, and an outer covering (the epicardium or visceral layer of the pericardium). Covering the heart is a fibrous sac, the pericardium, which at its attachment to the great vessels is reflected over the outer surface of the heart, thus leaving between its outer or parietal layer and its inner or visceral layer a potential space, the pericardial cavity. The pericardium is attached to the surrounding structures, and thus partially fixes the heart while it allows it such freedom of movement as is essential for its contraction.

The heart muscle consists of quadrilateral cells, which are joined longitudinally to form fibres and anastomose with neighbouring cells by short bridges. The properties which these muscle cells possess in common with other contractile tissues has been dealt with in Chapter 8 on cardiac muscle. Over the auricles the muscular wall is relatively thin, over the ventricles relatively thick; the wall of the left ventricle is four times as thick as that of the right. The thickness of the muscular wall of each chamber thus corresponds to the tension developed during its contraction. The muscle fibres of the right auricle are continuous with those of the left, those of the right ventricle are continuous with those of the left ventricle. The muscle fibres of the auricles, however, are separated from those of the ventricles by a fibro-tendinous ring, the auriculoventricular ring. The heart muscle is modified to form two important structures (Fig. 39.1). The first or *sinoauricular node* lies close to the junction of the superior vena cava with the right auricle and is about 2 cm long and 2 mm wide in man. It consists of a plexus of fine muscle fibres embedded in fibrous tissue. The second is the *auriculoventricular connection*, which forms the only functional junction between the muscular tissues of the auricles and the ventricles. This begins at the base of the interauricular septum close to the mouth of the coronary sinus as the auriculoventricular node, composed of slender interlacing muscle fibres. Continuous with the auriculoventricular node is the auriculoventricular bundle (of

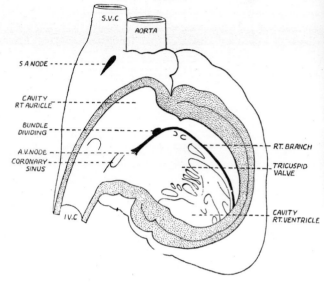

Fig. 39.1 A diagram of the human heart to show the sinoauricular node (S.A. node), the auriculoventricular node (A.V. node) and bundle of His. The walls of the inferior vena cava, right auricle and right ventricle have been partially removed to expose the septa. The cut surfaces are stippled.

His), which runs across the fibrous ring between auricles and ventricles and enters the interventricular septum, where it divides into right and left branches distributed to the appropriate ventricles. Each branch is continuous with a network of large, poorly striated cells, rich in glycogen, the Purkinje tissue, which forms a plexus under the endocardium of each ventricle.

The cavity of each auricle is separated from that of the corresponding ventricle by an *auriculoventricular valve*, a fibrous membrane covered with endocardium and arising from the auriculoventricular ring. On the right side the valve is divided into three flaps (*tricuspid*), on the left into two (*mitral*). To the ventricular aspects of the margins of these valves are attached tendinous chords (chordae tendineae), which terminate in nipple-like projections of the ventricular muscle (papillary muscles). These valves are so arranged that when blood flows from auricle to ventricle the valves lie flat against the ventricular wall. When the ventricular pressure rises above the auricular pressure the valves are floated out by eddies and seal the auriculoventricular openings; the chordae tendineae, aided by the contraction of the papillary muscles, prevent the valves from being thrust out into the auricular cavity. The openings of the right ventricle into the pulmonary artery and of the left ventricle into the aorta are each guarded by semilunar valves, consisting of three semicircular pockets whose cavities face away from the ventricles. The openings of the caval veins into the right, and of the pulmonary veins into the left auricle are unguarded; they are, however, sealed at the beginning of auricular systole by the contraction of the auricular muscle fibres surrounding them.

THE CARDIAC CYCLE

The cardiac cycle is one of those abstractions which delight physiologists. It lends itself to unending subdivision into various phases, each with its own eponymous identification ringing down the halls of fame. Fortunately, we need only consider a simplistic version.

The series of mechanical events, pressure changes, valve actions and electrical potentials which occur during one complete heart beat are collectively termed the *cardiac cycle*. It should be emphasised that the cardiac cycle is merely a convenient and logical descriptive system devised for the study of the heart's action.

The contraction of the heart can be seen in man by means of X-rays, and in experimental animals directly by opening the chest. The cardiac cycle begins with a simultaneous contraction of both auricles (*auricular systole*), which are seen to become paler and smaller in size. After a short pause both ventricles contract (*ventricular systole*), at first becoming paler and more rounded, then smaller in size. As the ventricles empty, the aorta and pulmonary artery fill. After contraction, each chamber relaxes (*diastole*) and then gradually fills, to empty again at its next beat. The hardening and change in shape which constitute the first phase of ventricular contraction are accompanied by a thrusting of the apical region of the ventricles against the chest wall. This thrust commonly moves the overlying intercostal space, and the movement, known as the cardiac impulse ("*apex beat*"), indicates the point at which the region of the apex of the heart lies. The position of the impulse has great importance clinically in indicating the size and position of the heart. Before a detailed description of the cardiac cycle and the pressure changes which occur during it is given, we must briefly deal with mechanical factors in heart muscle.

The contraction of heart muscle

The whole of the musculature of the auricles contracts at the same time, and is soon followed by contraction of the whole of the muscle of the ventricles. This property of simultaneous activation of all the muscle fibres comprising a chamber of the heart is of course necessary if the auricles and ventricles are to act as a pumping mechanism. Indeed, as we shall see later, when the synchronisation does not occur, as in fibrillation of the ventricle, the output of the heart falls to zero. Skeletal muscle, which is composed of large numbers of separate fibres (grouped into motor units) could be induced to contract synchronously and rhythmically as does the heart, by means of synchronous bursts of action potentials travelling in the nerves supplying it. This mechanism, however, does not account for the cardiac contractions, for cardiac muscle is a so-called *functional syncytium*, which means that the whole muscle behaves as if it were composed of a single cell. From the structural point of view, it is not a single cell but consists of many cells, electrically coupled through intercalated discs (see Chapter 8). It has the property of *rhythmicity*, i.e. it is able to generate its own rhythm of contraction without an input from the nervous system.

In common with other types of muscle, the contraction in cardiac muscle results from a membrane depolarisation due to the propagation of action potentials, both in the muscle fibres themselves and in muscle fibres specialised for the purpose. We will discuss the electrical events connected with the heart in more detail in the last part of this chapter; it is sufficient for the moment to state that the spread of activity over the wall of the auricle is two-dimensional, rather like waves on a pond produced by a stone dropped into it. In the thick-walled ventricle spread is three-dimensional.

The origin and spread of the heart-beat

A suprathreshold stimulus (an electrical pulse of some kind), applied at any point on the heart will give rise to electrical and mechanical activity which spreads throughout the muscle. However, in the normal body it can be shown that although all the heart is potentially capable of initiating the rhythm of contraction, only one region does in fact do so.

If a mammalian heart is excised, it will continue to beat for hours, provided that an adequate supply of warm oxygenated fluid of suitable composition is supplied to the muscle through the coronary vessels from the aorta. It is clear, then, that the origin of the heart-beat is independent of any connection with the rest of the body. Now the heart contains nerve ganglia, chiefly derived from the vagus, but even if these are dissected out in the cold-blooded heart, the beat continues; strips of auricular and ventricular mammalian muscle devoid of ganglia may contract rhythmically if placed in warm oxygenated Ringer–Locke solution. Further, in the chick embryo the heart begins to beat before it has received any nerves. Thus we may conclude that the beat originates in the heart muscle itself.

Although heart muscle is thus endowed with the property of contracting rhythmically, different parts of the heart behave differently in this respect. This is most easily shown in the classical experiments of Stannius on the heart of the frog (Fig. 39.2). If a ligature is tied tightly

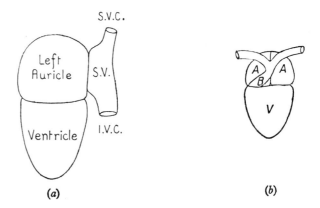

Fig. 39.2 Diagram of the frog's heart: (a) from the side; and (b) from the ventral aspect. The first Stannius ligature is tied tightly round the junction of the sinus venosus and right auricle (lying behind the left in the figure), the second between auricles and ventricle. S.V. = sinus venosus; S.V.C. = superior vena cava and I.V.C. = inferior vena cava; B = bulbus arteriosus.

around the junction of the sinus venosus with the auricles, the auricles and ventricle stop beating while the sinus continues at the same rate as before. After 5 to 30 minutes the detached part of the heart begins to beat, but at a slower rate than that of the sinus; the auricle contracts before the ventricle. If a ligature is now tied tightly between auricles and ventricle (second Stannius ligature), the auricles continue to beat as before, while the ventricle stops, after making a few rapid beats due to the stimulus of the ligature. The ventricle begins to beat again after about an hour, at a very slow rate. Thus, although each chamber of the heart is able of itself to contract rhythmically, the frequency of contraction varies from the sinus venosus at one end to the ventricle at the other; in the intact heart the rate of beating is that of the fastest chamber—the sinus venosus. This experiment thus suggests that in the frog the heart-beat begins in the sinus region and spreads over the chambers successively; the same is true of the mammalian heart.

Instead of recording mechanical movements of the heart, the spread of activity over the walls of the auricles and ventricles can be just as easily demonstrated by recording the cardiac action potentials. The technique of electrical recording, of course, has one big advantage, namely that the time-relations of the process can be measured with accuracy. The first experiments of this kind were carried out by Sir Thomas Lewis using a string galvanometer; nowadays amplifiers and cathode ray oscilloscopes are employed.

Recording from various parts of the heart's surface, the electrical changes associated with the excitation of these parts can be observed and accurately timed. In this way, it was shown by Lewis that the electrical change which accompanies contraction of the mammalian heart begins in the sinoauricular node, which lies in that part of the heart corresponding with the sinus venosus of the frog. From here the wave of electrical disturbance radiates in all directions over the auricular muscle, because heart muscle cells are a syncytium, and for a short distance up the great veins. When the wave arrives at the auriculoventricular node it passes thence along the auriculoventricular bundle through the Purkinje tissue and into the ventricular muscle. The rate of conduction through the Purkinje tissue is very rapid (500 cm s^{-1}) as compared with its rate through the ventricular muscle (50 cm s^{-1}). Thus in spite of their size all parts of the ventricles contract almost simultaneously. That the normal heart-beat actually originates in the sinoauricular node (the pacemaker) is confirmed by the fact that when this structure alone is warmed or cooled, the heart-beat quickens or slows respectively; when other parts of the heart are similarly warmed or cooled the frequency of the heart-beat is unaltered.

THE MECHANICAL PROPERTIES OF CARDIAC MUSCLE

(a) Development of tension

Within limits, the force of contraction of a heart muscle fibre is proportional to the initial (i.e. resting) length of the fibre. This relationship is a fundamental property of contractile systems. In the heart, it can be applied to the whole of the organ because the force of contraction determines cardiac output, while the initial length of the fibres is dependent upon the pressure in the chambers of the heart. Thus we can re-state the relationship in the form that cardiac output is directly proportional to the right auricular pressure (within limits). This is shown in Fig. 39.3.

(b) Duration of the contraction

In *cardiac muscle* the action potential lasts very much longer than it does in skeletal muscle, and the membrane does not recover until the contraction is almost completed. Correspondingly, the refractory period lasts almost as long as the whole period of contraction and relaxation. If two stimuli are applied in succession the second stimulus produces no effect at all unless it is applied so long after the first that the responses of the two are independent contractions occurring one after the other. *No summation* of contractions and *no tetanus* can be produced under normal conditions in heart muscle. If the second stimulus occurs during the relative refractory period, the contraction produced is usually smaller than a normal one, owing to imperfect recovery of the contractile process, as may be seen in Fig. 39.4.

Let us observe a spontaneously beating heart, and interrupt its rhythm by interpolating an electric shock applied to the ventricle. Figure 39.4 shows the ventricular contraction during such an experiment, in which

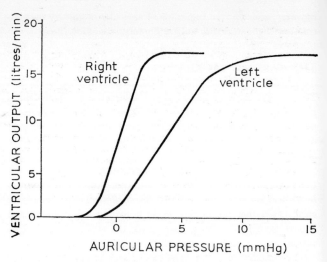

Fig. 39.3 Relation between cardiac output and auricular pressures in a normal heart.

a series of shocks was applied, each at a slightly later phase of the normal contraction. In the lower three records, the stimulus fell in the absolute refractory period and produced no effect. In the upper ones it fell later and produced a smaller or greater contraction according as it occurred sooner or later. This contraction took place, abnormally, before its time, and was accompanied by its own refractory period. It was followed by a pause known as the *compensatory pause*, before the next (normal) contraction, because one of the regular stimuli transmitted from the pacemaker found the ventricle in a refractory state and so produced no effect. Comparable effects are sometimes produced in the

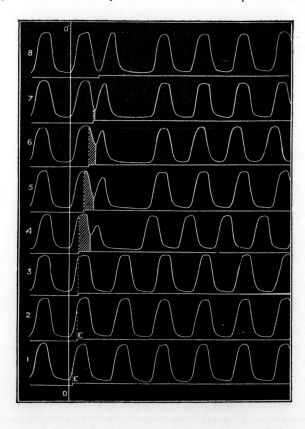

Fig. 39.4 Tracings of spontaneous contractions of frog's ventricle, to show refractory period and compensatory pause. An electrical stimulus is given at the step in the base-line (see text).

human heart by abnormal contractions (premature contraction, or extrasystole) originating at some irritable focus independently of the normal rhythm.

If the frequency of the stimuli initiated by the pacemaker (or applied artificially) is progressively increased, a time will come when the interval between successive stimuli is so short that it is less than the refractory period of, say, the ventricular muscle; the muscle will then suddenly respond only to alternate stimuli. If the frequency be further increased so that of three stimuli, the second and third fall within the refractory period accompanying the response to the first, the muscle will respond only to every third stimulus. Analogous events are observed in certain cases of "heart-block" in human disease as discussed later, in which the ventricle responds to only a fraction of an abnormally rapid sequence of stimuli coming from the auricle. If the frequency of stimulation be constant, and the duration of the refractory period sufficiently increased, a similar omission of the response to alternate stimuli may be observed.

If the rate of stimulation be increased gradually, the refractory period will itself become gradually shorter and so enable the heart to beat faster than it could have done had the rate of stimulation been suddenly increased. This may explain the fact that in certain conditions the human heart can attain a rate of 200 beats per minute; such a rate would be impossible without an actual shortening of the normal refractory period.

(c) The "all or none" properties of the heart muscle

Since the muscle is a functional syncytium, when excited *all* the muscle contracts. If the stimulus strength is below threshold, the fibres do not contract—there is no half-way state; gradation of contraction is not possible. At first sight this "all or none" law appears to be contravened by the statement in paragraph (a) about fibre length. Many factors will alter the force of contraction including fibre length, but, provided these factors are kept constant, the stimulation of heart muscle will produce either contractions of given strength or none at all. (See Chapter 8 for detailed discussion of heart muscle.)

PRESSURE CHANGES IN THE HEART DURING THE CARDIAC CYCLE

Before embarking on a detailed description of the pressure changes in the auricles and ventricles, it is of interest to see what methods are available for making the measurements of these pressures.

Recording techniques (see Fig. 39.5 for details)

The same methods are used in recording intracardiac pressure in experimental animals and in human subjects. They involve the introduction of catheters (sterile for human patients!) into a vein and thence into the various chambers of the heart. This, of course, confines observations to the right side of the heart in humans.

The requirements for an acceptable recording system are that it shall have a uniform response characteristic up to about 50 Hz. Hence U-tube manometers and simple tambours cannot be used. Sensitivity should be such that pressures between 2 to 200 mmHg (say 0 to 30 kPa) can be accurately reproduced as a tracing.

Various types of optical manometer have been used, consisting of a metal or glass diaphragm to which is cemented a small mirror. As the membrane is pressed inwards and outwards by pressure variations in the cannula attached to it, the mirror reflects a beam of light onto photographic recording paper.

Recently, pressure transducers have been constructed which convert the pressure changes into an electrical output which is then amplified and displayed either on the cathode ray tube or recorded permanently on heat-sensitive paper with a hot-wire pen. These transducers usually work on the strain gauge principle, variable capacitance measurement or on the variable reluctance of a coil into which is introduced a core.

An example of the latter type can now be constructed of dimensions so small that it fits into the tip of a catheter—doing away with one of

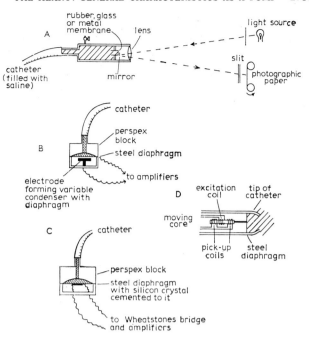

Fig. 39.5 Methods for measuring pressures.

A. Optical manometer.

B. Capacitance manometer. The pressure variations bend the steel diaphragm and change the capacity between it and the static electrode. A frequency-modulation phase discriminator (5) is used to detect and record the changes.

C. Strain-gauge transducer. Here the bending of the diaphragm is transmitted to a crystal of silicon cemented to it which forms one arm of a Wheatstone's bridge. The out-of-balance current is amplified and put onto a pen-recorder.

D. Variable reluctance transducer.

A little transformer is wound on a core which moves as the pressure changes bend the diaphragm. The centre coil is supplied with alternating current. As the core moves, the voltage in the two pick-up coils will vary with its actual displacement. After rectification and amplification the output is displayed on the oscilloscope or pen-recorder. This device is very small and fits into the tip of a cardiac catheter.

the problems inherent in the usual technique, namely the distortions introduced by recording at the end of a long and somewhat narrow tube.

The pressure changes accompanying the cardiac cycle

Although the heart consists of two pumps, the right and left hearts, these work simultaneously and in the same way, and it will be convenient to describe the pressure changes only of the left auricle and ventricle; those of the right auricle and ventricle are similar and simultaneous, though of less magnitude. Of the various events of the cardiac cycle whose time relations are shown in Fig. 39.6 the pressure changes in the heart and aorta and the change in ventricular volume will now be described. The pressure changes have been recorded by electrical or optical manometers connected by means of rigid tubes filled with fluid to cannulae thrust into chambers of the heart.

The intra-auricular pressure curve

This shows waves of rise of pressure (positive waves) corresponding to auricular systole (first positive wave) and the sudden closure of the auriculoventricular valves (second positive wave). The pressure then abruptly falls as the relatively emptied auricle relaxes. Blood flows into the auricles from the great veins, producing a gradual rise of pressure (third positive wave), which is interrupted when the auriculoventricular valves open and put the relaxed and relatively emptied ventricle into connection with

252 HUMAN PHYSIOLOGY

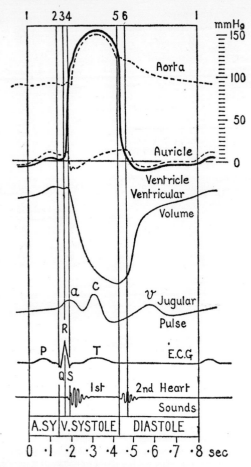

Fig. 39.6 The sequence of events in the cardiac cycle.
  The upper four curves have been taken from actual records obtained from the dog's heart; they represent the pressure changes in aorta (broken line), left ventricle (continuous line), and left auricle (broken line), and the curve of ventricular volume. The lower three curves, representing the jugular pulse, the electrocardiogram, and the heart sounds, have been reconstructed from data obtained on human subjects. The vertical lines represent the following events: 1 = auricular excitation (P wave of electrocardiogram); 2 = ventricular excitation (Q wave of electrocardiogram); 3 = auriculoventricular valves close; 4 = aortic valve opens; 5 = aortic valve closes; 6 = auriculoventricular valves open. A. SY represents the duration of auricular systole. (More usual times for the duration of auricular systole in man would be 0·10 s, of ventricular systole 0·24 s, and of diastole 0·46 s.).

the auricle. It will be noted that during diastole both intra-auricular and intraventricular pressures are below that of atmospheric pressure, which is represented as 0 mmHg in Fig. 39.6. This is not due to any sucking action of the heart, but to the transmission of the subatmospheric pressure in the thorax through the slack heart wall.

Ventricular and aortic pressure

While the ventricles are quiescent, blood is flowing into them from the auricles, and the intraventricular pressure is slightly lower than, and closely follows, the intra-auricular pressure. With the onset of ventricular contraction the intraventricular pressure rises abruptly until it exceeds the aortic pressure, the aortic valves now open, blood is forced into the aorta, and the two pressures mount together. As ventricular ejection begins to decline the ventricular and the aortic pressures begin to fall, at first slowly, then rapidly, as the ventricle passes into diastole. The aortic valves now close; the aortic pressure falls slowly as

blood flows out at the periphery and the ventricular pressure falls abruptly. The ventricular pressure now falls below auricular pressure, which has been rising owing to the venous inflow, the auriculoventricular valves open and blood flows into the ventricle, gradually raising its pressure until the next cardiac cycle begins.

The ventricular volume changes

The volume of the ventricles is slightly increased during auricular systole. The onset of ventricular contraction is associated with no diminution of volume, for the ventricle is now a closed cavity separated from the auricle by the auriculoventricular valves and from the aorta by the aortic valves. The first period of ventricular systole is thus a period in which the muscular contraction is isometric (associated with no change in length). With the opening of the aortic valves the ejection phase begins and the ventricular volume rapidly diminishes. The rate of ejection gradually lessens as the ventricle empties and ceases as the aortic valves close, and the ventricle passes into diastole. With the opening of the auriculoventricular valves blood enters the ventricles, whose volume increases rapidly.

THE ARTERIAL AND VENOUS PULSE

The arterial pulse

The sudden ejection of blood into the aorta that occurs with each beat of the ventricles produces a wave of increased pressure that is propagated along the arteries towards the periphery; this is known as the pulse wave, and it may be felt and recorded in any of the superficial arteries of the body as the pulse beat. The velocity at which this pressure wave is propagated may be determined by recording the times of its arrival at two different points such as the subclavian and radial arteries, and dividing the time differences by the distance between the two points at which measurements are taken. The pulse wave velocity varies in different subjects chiefly with the thickness and elasticity of the arterial wall, and since the arteries tend to become more rigid with advancing years (arteriosclerosis), the velocity increases from an average rate of $5·2$ m sec^{-1} at the age of 5 to an average of $8·6$ m sec^{-1} at the age of 80. It is important not to confuse the pulse wave, which is simply a wave of increased pressure, with the movement of the blood itself; the velocity of blood flow is nowhere greater than $0·5$ m sec^{-1} at rest and is considerably less in the smaller vessels.

THE FORM OF THE PULSE

The changes in pressure and their spacing in time that occur in a superficial artery such as the radial may be recorded using pressure transducers applied to the skin over the artery. With practice the main features exhibited by such a record can usually be ascertained with the finger. The main deflection in the pulse record, the primary wave (Fig. 39.7) is the result of the sudden distension of the aorta during ventricular systole. Following the primary wave are a number of secondary waves, which arise in several ways. The most constant and conspicuous secondary wave results from closure of the aortic (semi-lunar) valves. At the end of ventricular systole the pressure in the ventricle falls rapidly, and blood begins to flow back from the aorta, but is suddenly checked by closure of the aortic valves. This sudden check causes a rise of pressure in the aorta, and this wave travels down the arterial tree with the same velocity as the primary

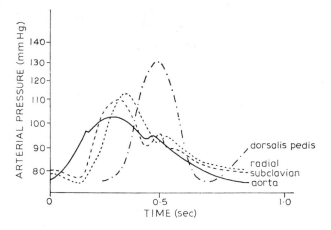

Fig. 39.7 Arterial pulses.
The form of the arterial pulse varies with the distance from the heart. Note the facts that: (a) there is a progressive delay the greater the distance; (b) pressure at the peak of the wave increases with distance; (c) The slope of the pressure front becomes steeper.

wave which it follows. This secondary wave is known as the dicrotic wave and is preceded by a notch, the *dicrotic notch*. From what has been said it will be realised that the upstrokes of the primary and of the dicrotic waves are separated by the same interval of time as the opening and closure of the aortic valves. In addition to those two waves there are a number of inconstant and small waves arising from the reflection of the primary wave by obstacles such as the bifurcations of the arteries.

Modifications of form of pulse wave
The form of the pulse wave is modified by such conditions as affect the discharge of the blood from the heart and its escape through the arteries. Thus, when the aortic valves are narrowed by disease (aortic stenosis) the distension of the aorta during systole is very slow and the pressure in the peripheral arteries rises slowly to its maximum; this slow-rising pulse is termed anacrotic. When, on the other hand, the aortic valves do not close properly (aortic regurgitation), the pressure in the aorta falls very quickly during diastole and is suddenly and greatly raised during systole; in this condition the upstroke of the primary wave is unusually sudden and its downstroke rapid; from the sudden thrust on the finger feeling the artery this pulse is described as "water hammer". In conditions of low pressure and rapid blood flow, as may occur in children and in fevers, the dicrotic wave may be so pronounced as to be easily felt at the wrist; the frequency of the heart may thus be mistaken for twice its true value.

The venous pulse

The polygraph records simultaneously on a moving paper strip-chart pulsations of the jugular vein and of the radial artery or heart's impulse. Chief interest in the records so obtained (Fig. 39.8) attaches to those of the venous pulse. In normal subjects this consists of three waves in each cycle, *a*, *c* and *v*. The *a* wave is due to auricular contraction and is caused by the arrest of venous inflow to the heart by constriction of the mouths of the great veins. The *c* wave indicates ventricular systole and is largely transmitted from the carotid artery. The *v* wave is of less importance and is largely due to slowing of the venous flow consequent on the filling of the heart in diastole. The object of the other tracings is to time ventricular systole and thus to identify the *c* wave of the jugular pulse; the *c* wave of the jugular pulse occurs about 0.1 s before the radial pulse, the difference in time corresponding to the difference in their distances from the heart. The most important feature of the jugular pulse is the *a*

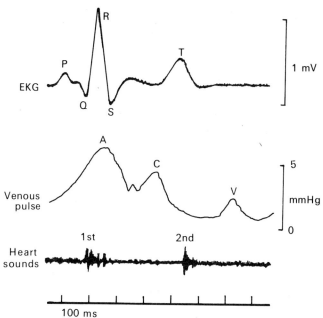

Fig. 39.8 The Venous pulse recorded from jugular vein.
Record shows, from above down: the EKG, venous pulse, heart sounds and a time trace.
Pressure in great veins is pulsatile and reflects events in right auricle and right ventricle, plus certain artefacts due to pressures transmitted from pulsing arteries nearby. Three main waves are described: *a* wave due to auricular contraction (occurs just after P wave of EKG); *c* wave due to transmitted carotid pulse; *v* wave due to opening of tricuspid valve and rise of pressure at commencement of right ventricular filling. EKG and heart sound traces are included to indicate timing of *a*, *c* and *v* waves.

wave. In the early stages of heart block where conduction through the bundle of His is impaired, the interval between auricular and ventricular contractions and that between the *a* and *c* waves is prolonged. In the later stages, the ventricle ceases to respond to each auricular systole, and the *a* waves are not always followed by *c* waves. Lastly, in auricular fibrillation, where co-ordinated contraction of the auricles has ceased, the *a* wave is absent.

The action of valves in the cardiac cycle

We have seen how the pressure varies during the cardiac cycle in the four chambers of the heart. For flow of blood to take place in one direction only, as a result of these pressure changes, valves must be present between the auricles and ventricles, and between the ventricles and the arterial tree.

The aortic and pulmonary valves prevent blood flowing back from the arterial tree during diastole. They each have three tough cusps making a triangular opening. The A–V valves, mitral and tricuspid, prevent reflux of blood in ventricular systole. The cusps of these valves are held back by the chordae tendinae which act like the cords of a parachute to prevent the valves from everting under the strain of full ventricular pressure. The details are shown in Fig. 39.9.

The valves open and shut entirely passively according to the pressure differences across them. When regurgitation of blood begins the flaps close; also various rather complex eddies are set up in the vicinity of the cusps which tend to close them slightly before any actual reflux of blood has taken place. These eddies are in part responsible for the heart sounds.

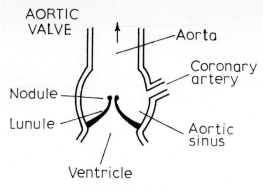
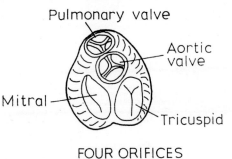

Fig. 39.9 The heart valves.

The normal heart sounds

The sounds associated with cardiac contraction may be heard by placing the ear on the chest wall over the heart; they are usually detected by placing over the heart a hollow metal or plastic cone connected through rubber tubing to ear-pieces (the stethoscope). Each contraction of the normal heart is associated with two sounds, the first prolonged and of low pitch, the second abrupt and of higher pitch. They are usually somewhat crudely imitated by the sounds "lubb", "dup". The first sound is produced in part by muscular contraction and in part by eddies set up by closure of the auriculoventricular valves, the second is produced by closure of the aortic valves. Occasionally a third heart sound is heard, and is said to be due to the floating up of the auriculoventricular valves as the ventricle fills with blood in the early part of a long diastole.

Murmurs

When the valves of the heart are diseased, the heart sounds are usually accompanied or followed by abnormal sounds termed murmurs. Murmurs occurring with the first sound or between it and the second are associated with ventricular systole and are termed systolic. Systolic murmurs are not infrequent in normal subjects when the action of the heart is augmented, as after exercise. Murmurs occurring after the second sound and between it and the next first sound are associated with ventricular diastole and are termed diastolic; they are purely pathological and are of first importance in detecting valvular disease of the heart. These murmurs are probably produced by the vibrations set up in the neighbourhood of diseased valves by the rapid stream of blood passing over them.

Valvular disease of the heart consists of narrowing of the orifice of the valve, *stenosis*, or an interference with proper watertight closure, *incompetence*. Both conditions lead to turbulent flow (since there is in each case a relatively narrow orifice for blood to pass and the critical velocity for laminar flow is exceeded). Sometimes a cusp is ruptured which then vibrates producing an adventitious sound which is musical in character.

Clinical examination of cardiac function includes the detection of abnormal heart sounds. The valve responsible can be determined by timing the murmur in relation to systole and diastole and the kind of sound heard differentiates between stenosis and incompetence.

40. The Control of Cardiac Output

By far the most important aspect of the physiology of the heart and its control is the manner in which its output is adjusted to suit the demand for blood by the various organs and tissues. This may be studied conveniently when the heart is isolated, many observations on the mammalian heart having been made with Starling's heart–lung preparation.

The heart–lung preparation. In studying the behaviour of the isolated heart it is of great advantage to leave the lungs in full functional connection with the heart, because the blood can be aerated and the necessary oxygen supplied to the heart by ventilating the lungs, and because the lungs remove vasoconstrictor substances which develop in shed blood and make perfusion of isolated mammalian preparations difficult.

The preparation is shown diagrammatically in Fig. 40.1, and is made briefly as follows. The venous reservoir having been filled with warm defibrinated dog's blood, the dog's chest is opened under artificial respiration; cannulae are tied into the brachiocephalic artery and superior vena cava and all the other systemic vessels (inferior vena cava, azygos vein, subclavian artery and aorta) are tied. The blood entering the heart from the venous reservoir must now pass from the aorta through the arterial cannula and artificial circulation. In the artificial circulation the two important features are (a) the air cushion (B, Fig. 40.1) consisting of an inverted bottle containing suitable compressed air which simulates the elastic reservoir provided by the aorta and larger arteries, and (b) the resistance *R* consisting of a thin rubber sleeve inside a glass tube containing air under the known pressure of a large reservoir with which it is connected; blood will only flow through the sleeve at a pressure higher than that of the air outside it, and thus the arterial pressure may be kept constant and independent of output. By these two devices the blood pressure in the aorta is prevented from falling too far during diastole and so the coronary circulation to the heart is well maintained.

In this preparation the nerves to the heart are severed and the heart beats at a constant rate, which may, however, be varied by altering the temperature of the blood (action on the sinoauricular node).

By using the heart–lung preparation, it is possible to vary independently the venous input to the heart, and the arterial pressure developed; the first by adjusting the height of the venous reservoir, or better by opening and closing a screw-clip on the tube between the reservoir and the vena cava, the second by adjusting the air pressure outside the rubber sleeve in the "arterial resistance". By varying the rate of venous inflow, the output of the heart can be varied smoothly over a very wide range; in this, it is totally unlike a rigid mechanical pump in which the output is fixed by the cross section of the cylinder, the throw of the crank and the number of strokes per minute. Variation of the arterial pressure, on the other hand, has no action on the output of the heart, within an upper limit set by the capability of the heart to do the necessary work, and a lower limit below which the heart is inadequately nourished through the coronary circulation. In this respect, it is quite similar to the rigid mechanical pump. The heart, in fact, acts very effectively in propelling the blood available in the venous system against any normal value of the arterial pressure; neither allowing the veins to become engorged by propelling too little, nor sucking them empty by propelling too much.

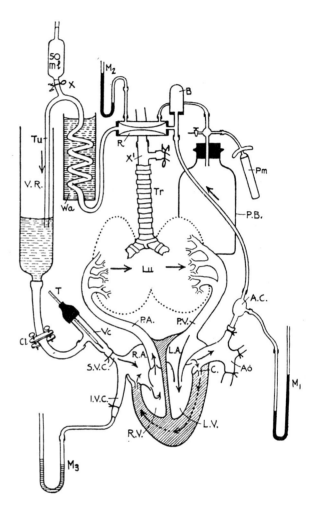

Fig. 40.1 The heart–lung preparation.

The direction of the flow of blood is shown by arrows. V.R., venous reservoir; Cl, clamp for adjusting the rate of inflow to the heart; Vc, venous cannula, with thermometer, T; S.V.C., superior vena cava; I.V.C, inferior vena cava, connected to the manometer M_3 for measuring the venous pressure; R.A., right auricle; R.V., right ventricle; P.A., pulmonary artery; Lu, lungs; Tr, trachea, with cannula X^1; P.V., pulmonary vein; L.A., left auricle; L.V., left ventricle; C, coronary artery; Ao, aorta, ligatured; A.C., arterial cannula in brachiocephalic artery, connected to the manometer M_1 for measuring the arterial pressure; B, elastic cushion; R, arterial resistance; pressure is applied to the outside of the sleeve by the pump Pm, is stabilised by the pressure bottle P.B., and measured by the manometer M_2; Wa, warming coil in hot water; X, clamp for admitting blood to the graduated vessel when it is desired to measure the output of the heart; Tu is then temporarily clamped.

Intrinsic heterometric autoregulation—Starling's law

Any variation in the output of the heart (at a constant frequency) must be accompanied by a parallel variation in the amplitude of the beat, i.e. in the change in length of the muscle fibres on contraction. Similarly, any variation in the arterial pressure must be accompanied by a parallel change in the force which the muscle must exert in order to expel the blood. The way in which these adjustments are made may be demonstrated by recording the volume changes of the ventricle by means of a *cardiometer*. In its simplest, and original, form this consists of a glass cup the shape of a wine glass, the stem of which is hollow and connected to a volume recorder; the open end of the cup is fitted with a rubber membrane in which a suitably sized hole has been burned. The ventricles are slipped through this hole, the edge of which grips lightly but securely the auriculoventricular groove.

Figure 40.2 shows the results of experiments on a heart–lung

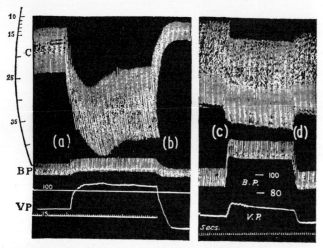

Fig. 40.2 Effect of changing (1) the venous inflow and (2) the arterial pressure on the volume of the heart.

C, cardiometer curve, the curved line at the side indicating the value of the cardiometer excursions, i.e. of alterations of ventricular volume, in millilitres; movement downwards indicates *increase* in volume; B.P., arterial pressure; V.P., venous pressure (water manometer) in the inferior vena cava. (Patterson, Piper and Starling.)

preparation: a movement downward in the cardiometer tracing, C, indicates an increase in the size of the heart; the volume at the end of diastole is thus given by the lower limit of each movement, and the volume at the end of systole by the upper limit; the total excursion gives the output of the heart per beat (stroke volume). When at (a) (left-hand record) the venous inflow is increased from 516 to 840 ml min^{-1}, the venous pressure rises from 95 to 145 mm H_2O (0.9 to 1.5 kPa), and the ventricles distend, at first putting out less blood than they receive. The diastolic volume of the ventricles thus gradually increases, and this is associated with a gradual rise in the output per beat, until finally the output equals the inflow and the heart ceases to dilate. The arterial pressure remains practically constant, rising from 124 to 130 mmHg only, since the arterial "resistance" is specially designed for this purpose. The heart remains at the new size until at (b) the venous inflow is suddenly reduced to 198 ml min^{-1}; the heart puts out more per beat than it receives, and the diastolic volume gradually falls. Finally, at a new and smaller diastolic volume the heart again puts out precisely what it receives, the venous pressure having fallen to 55 mm H_2O.

In the right-hand record of Fig. 40.2, the venous inflow is kept constant at 924 ml min^{-1}. When at (c) the arterial "resistance" is altered, so that the arterial pressure is raised from 98 to 128 mmHg, the ventricles at first fail to put out as much blood as they receive and so increase in size. The increase in diastolic volume is associated with a progressive recovery in the output per beat until the heart, at an increased diastolic volume, is again ejecting all that it receives. When the arterial pressure is again reduced, at (d), the ventricles at first expel more than they receive and so decrease in size.

Now the output per beat—which, in a steady state, must be equal to the inflow per beat—and the arterial pressure are the two factors which determine the work done by the heart on each beat; variations in either, as we have just seen, are accompanied by variations in the diastolic volume. It is found experimentally, moreover, that similar increases in the rate of work, whether produced by augmented output or by raised aortic pressure, are accompanied in a given heart by similar expansions of diastolic volume. The energy set free by skeletal muscle during contraction varies with the initial length of its fibres, and so it seems to be with the heart. When the diastolic volume of the heart is increased, the muscle fibres are stretched, the energy set free in systole is increased and the heart is able to perform more work. This relationship between the energy of contraction of the heart and its diastolic volume was formulated by Starling as "The Law of the Heart".

The effect of metabolites

The effect of accumulated metabolites during heterometric autoregulation has, under certain conditions, a further effect in increasing force of contraction. This may sometimes be large enough to allow the venous pressure to fall. In Fig. 40.2, the rise of the cardiometer tracing from its lowest point is due to homeometric autoregulation. The mechanism of this effect is thought to be that the increased work done by the heart muscle gives rise to leakage of potassium ions through the cell membranes and accumulating in the tissue fluid. Another theory implicates intracellular calcium ions.

Extrinsic regulation. The sympathetic nervous system

Stimulation of the sympathetic nerve supply to the heart, or the administration of adrenaline, increase the force of contraction of heart muscle *from a given length*. This will be seen from the results of a heart–lung preparation experiment shown in Fig. 40.3. Following the addition of adrenaline to the blood in the venous reservoir, the cardiometer tracing shows that the heart volume decreased. The heart performed the same amount of work from a smaller diastolic volume. It did so not only because, owing to the action of adrenaline, it beat faster, and the output per beat was therefore less, but also and chiefly because the amount of energy liberated from a given length was increased.

In Fig. 40.4 have been plotted diagrammatically the relationships between cardiac output and venous pressure for a number of doses of adrenaline. It will be seen from such a "family of curves' (after Sarnoff): (a) that increase in venous pressure increases cardiac output; and (b) that for any given venous pressure the cardiac output is greater during the action of adrenaline.

As previously stated, the inflow into the heart in the conventional heart–lung preparation is determined almost wholly by the resistance to the passage of blood past the screw-clip placed on the tubing between the venous reservoir and the heart (Fig. 40.4), this resistance being large compared with that opposed to the entry of blood into the heart itself. Accordingly, changes in frequency and strength of the beat can have little effect on the inflow and thus on the output of the heart. An altogether different situation arises, however, if the screw-clip is removed and the venous reservoir quickly lowered until the level of the

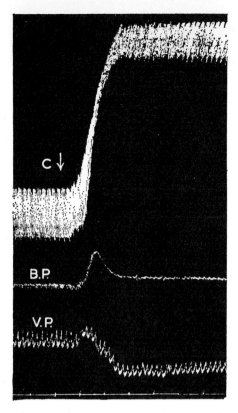

Fig. 40.3 Effect of adrenaline on the volume of the heart.

C, cardiometer curve; a rise indicates reduction in volume of the heart; V.P., pressure in the inferior vena cava.

At the arrow, 0·1 mg adrenaline was added to the blood in the venous reservoir.

The resistance to the venous inflow was not altered. The changes in arterial blood pressure reflect the large transient increase in output while the volume of the heart was falling; and a smaller permanent increase in output owing to the increased inflow produced by the rise in level of the blood in the venous reservoir and the fall in venous pressure. (Patterson.)

blood in the reservoir is the same as that in the venous manometer. In this circumstance, owing to the capacity of the reservoir, the level of blood and the venous pressure must remain nearly constant. The main resistance to the flow of blood into the heart is now that of the heart itself. In this preparation adrenaline causes a large increase in output, and this is not surprising as the heart fills much more readily (Fig. 40.4).

Changes in the performance of the heart mediated intrinsically by the Starling mechanism. When a standing subject lies down, blood is transferred from the swollen veins in the legs to the veins of the abdomen and chest. Central venous pressure, venous inflow and diastolic volume increase. Cardiac output increases (heterometric autoregulation). However, a rise in arterial blood pressure is prevented by the baroreceptors. These bring about reflex peripheral vasodilatation accompanied by some slowing of the heart and weakening of the force of contraction (extrinsic regulation of the force of ventricular contraction), causing a further rise in central venous pressure. In a few seconds all is in order, central venous pressure is raised, diastolic volume is increased, heart-rate is slowed, cardiac output is increased, arterial blood pressure is constant, total peripheral resistance has fallen.

Starling's law plays an important role in everyday life in maintaining the balance between the pumping of the right and left ventricles. This balance must be exact. Only a delicate adjustment of strength of contraction to the degree of filling can explain their maintained balance of output.

Sometimes a normal heart-beat is followed early in diastole by an extrasystole, there is then a compensatory pause or dropped beat and

THE CONTROL OF CARDIAC OUTPUT 257

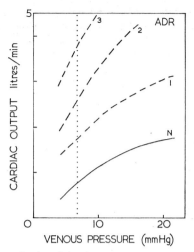

Fig. 40.4 Diagram showing relations between *venous* pressure and *cardiac* output in the heart–lung preparation.

N, before adrenaline; *1, 2, 3*, after the addition of successively greater amounts of adrenaline to the circulating blood, causing increase in rate and force of the heart beat. For a venous pressure of 8 cm of water (0·8 kPa) the cardiac outputs were 800, 1300, 2500 and 3500 ml min^{-1} respectively. Venous pressure in the healthy human heart is maintained at a much lower pressure.

normal rhythm is resumed. The extrasystole may be heard over the apex of the heart but not felt as a pulse at the wrist; the heart-beat after the compensatory pause is often so powerful as to be felt by the subject. The duration of diastole before the extrasystole is unusually short, the diastolic volume of the heart is small, the force of contraction is not enough to open the aortic valves. After the compensatory pause the diastolic volume is so large, and the beat so powerful, as to be felt as a "thump".

Changes in the performance of the heart mediated extrinsically in the animal by the sympathetic nervous system. Severe exercise is accompanied by an enormous decrease in vascular resistance in the skeletal muscles, by activity of the pumping action of the muscles on the blood in the veins, by venoconstriction and other factors which increase venous return fourfold or more. At the same time, or even before, the sympathetic nervous system quickens and strengthens the heart beat remarkably, so that the subject is aware of his beating heart (palpitation), the corresponding increase in the performance of the heart as a pump explains why there may be no increase in venous pressures or in end diastolic volume of the heart (X-ray). Heterometric autoregulation which many regard as synonymous with "Starling's law" is often not manifested in exercise.

Work done by the heart

The external work done by each ventricle during a single contraction is the sum of the potential and kinetic energy imparted to the expelled blood. In symbols

$$\text{Work done per minute} = (PV + \tfrac{1}{2}\rho V v^2)N$$

where P represents the mean aortic pressure, V the stroke volume or output per beat, v the velocity of blood in the aorta, ρ the density of the blood and N the number of beats per minute.

Assuming that the pressure developed in the right ventricle is one-sixth that of the left, the external work done by the whole heart per minute becomes:

$$V(\tfrac{7}{6}P + \rho v^2)N$$

The oxygen consumption of the heart beating at constant rate is found to be proportional to the external work done and thus to the diastolic volume of the heart. If the heart rate is varied, then it is found that the oxygen consumed per unit of work done is greater at a high rate of beat than at a low. Since the oxygen consumption is a measure of the total energy liberated, we may say that the efficiency of the heart, *external work done/energy set free*, is greater at low rates than at high. The efficiency of the isolated heart is usually about 0·2 (20 per cent).

The cardiac output in man

The output of the heart is remarkably constant for each individual under conditions of complete physical and mental rest, that is after the subject has been lying down and fasting for 10 or more hours (basal conditions). In different individuals the output is closely related to the surface area of the body; it varies from 4 to 5 l min^{-1} (2·5 l min^{-1} for each square metre of body surface), as estimated by the acetylene and dye injection methods, though rather higher values have been obtained by the method of cardiac catheterisation.

The cardiac output is reduced by a change from the recumbent to the standing position, which produces a fall in right auricular pressure. Small increases are produced by excitement (about 1 l min^{-1}), and by the ingestion of food and drink (up to 2 l min^{-1}); but it is in muscular exercise that the greatest increases are found. Even such slight exercise as flexing the thigh once every second doubles the cardiac output. With more severe exercise, such as cycling, running or swimming, the cardiac output increases proportionately to the rate of work, commonly reaching 20 l min^{-1}, and in trained individuals, values of over 30 l min^{-1} have been attained.

THE AUTONOMIC CONTROL OF CARDIAC FUNCTION

The innervation of the heart

The heart receives two sets of nerve fibres from the autonomic system, parasympathetic fibres from the vagus and sympathetic fibres. The vagal fibres terminate in ganglion cells in the heart, from which fibres pass to the sinoauricular and auriculoventricular nodes and the auricular muscle. Stimulation of the vagus slows the heart by action on the pacemaker, and also depresses the conduction from the auricle to the ventricle by action on the auriculoventricular node; it may also reduce the force of the auricular contractions, but in the mammal has little action on the ventricles themselves. The sympathetic fibres arise from cell stations in the middle and inferior cervical ganglia and terminate around the sinoauricular and auriculoventricular nodes and in the heart muscle. Stimulation of the sympathetic fibres quickens the heart by action on the pacemaker, facilitates conduction from auricle to ventricle, and augments the force of the auricular and ventricular contractions. It may thus be seen that the action of the vagus on the heart is almost the converse of that of the sympathetic (reciprocal innervation). Normally, impulses are passing to the heart along each set of nerves, for section of the vagus quickens the heart and section of the sympathetic slows it.

The control of the frequency of the heart-beat

The rhythm of the pacemaker is affected by impulses in the vagal and sympathetic nerves, by the concentration in the blood of adrenaline and thyroxine, and to some extent by the temperature of the blood. The nervous impulses come from the cardiac parts of the vasomotor centre. The activity of this centre depends upon local conditions and upon reflexes from the cardiopulmonary system and from receptors (*baroreceptors*) in the walls of the aorta and carotid artery which respond to a rise in arterial pressure. In the normal resting subject, the natural rhythm of the pacemaker is made slower by vagal impulses from the centre; if the vagal nerve endings are paralysed by atropine (belladonna), the heart-rate increases from about 70 to about 180 beats per minute. This is due largely to the fact that the baroreceptors are normally in a state of moderate excitation, one of their actions being to bring about a reflex slowing of the heart.

The heart-rate is *increased* in the following circumstances:

(a) In a warm environment, and by fevers.
(b) By haemorrhage, and when a recumbent subject stands upright. There is a reduction in venous return, owing to an actual reduction in blood volume, or to blood pooling in the lower part of the body; the cardiac output decreases and the arterial pressure tends to fall. The baroreceptors—and thus the cardiac centre—are less excited, there is a reduction in vagal inhibition and an increase in heart rate.
(c) By excitement, fright, exercise and lack of oxygen. This is due, at least initially, to an action of the higher centres of the brain on the vasomotor centres, reducing or abolishing the normal vagal inhibition of the heart. Later, in exercise, reflexes from the cardiopulmonary system, the increase in the amount of circulating adrenaline, and the increase in the temperature of the blood accentuate and prolong the increase in heart rate. Reflex action through the vasomotor centres contributes, also, to the effects produced by breathing air deficient in oxygen; but the site of the receptors has not yet been discovered. The arterial pressure rises both in exercise and during lack of oxygen; reflexes from the baroreceptors will thus tend to slow the heart, but their effect is swamped by the predominant accelerating action of the other factors.

The heart rate is *decreased* greatly during a fainting attack, and in some subjects when consciousness is lost as a result of severe lack of oxygen.

Increase in the intracranial pressure as, for example, by a cerebral tumour, often slows the heart; this is believed to be due to restriction of cerebral blood flow and to the effect of lack of oxygen on the vagal centre. Adrenaline injections in pregnant rabbits slow the fetal heart; this is because constriction of the uterine vessels causes oxygen lack in the fetus; the mechanism for accelerating the heart does not function, and so the effect of the oxygen lack on the pacemaker becomes manifest.

CARDIAC DYSFUNCTION

Cardiac failure

The function of the heart is to propel blood through the vessels at a rate fast enough to meet the metabolic requirements of the tissues; failure to do so may result either from a lesion of the heart itself or because the supply of blood to the heart from the great veins is inadequate (peripheral circulatory failure). Heart failure in man is frequently accompanied by a condition of stenosis (narrowing) or incompetence (leakiness) of one or more valves, the heart is then working at a mechanical disadvantage; yet it will readily be appreciated that the origin of heart failure is usually to be sought in disease of the heart muscle itself. The earliest symptom of heart failure is undue breathlessness on exertion. In advanced heart failure the patient is breathless at rest and more so in the recumbent than the sitting posture;

venous pressure in the cervical veins is increased, the liver is enlarged and extensive accumulations of fluid occur, in the tissues (*oedema*) of the legs and in the peritoneal cavity (*ascites*) and in pleural cavity (*pleural effusion*). In such an advanced case the cardiac output may be reduced to half the normal. This reduction is not the chief cause of the symptoms since a similar or greater reduction of cardiac output in peripheral circulatory failure is associated with an entirely different clinical picture. The chief factor producing the symptoms in cardiac failure is the rise in venous pressure which, on the left side of the heart, produces engorgement of pulmonary veins and capillaries, and by thus increasing the rigidity of the lungs leads to breathlessness and may lead to oedema of the lungs; on the right side of the heart the raised venous pressure distends the hepatic veins and thus enlarges the liver, and by raising capillary pressure favours the passage of fluid out of the capillaries into the tissues.

The most instructive example of cardiac failure in man is that seen in paroxysmal tachycardia in which the heart suddenly begins to beat at a rate of 180 to 200 beats per min. As we have seen, for a given output and arterial pressure, the rate of oxygen consumption by the heart rises as the frequency of the beat becomes greater. The flow of blood in the coronary vessels, moreover, occurs chiefly during diastole. If the duration of diastole is reduced excessively, the rate of supply of oxygen will fall, and may well become inadequate. At very high frequencies, then, even previously healthy hearts may after some hours display the phenomena of cardiac failure. With the onset of the paroxysm the patient gradually becomes breathless, the neck veins swell, the liver enlarges, and the cardiac impulse moves outwards for perhaps 5 cm (2 in). This movement of the impulse is due to the enlargement of the diastolic size (*cardiac dilatation*) consequent on the increased venous pressure. Even with such a cardiac dilatation the output of the heart is found to be reduced. With the end of the paroxysm, the neck veins collapse, the liver decreases in size and the heart's impulse returns to its normal position; the relationship between cardiac output, diastolic size and venous pressure once more becomes normal.

Usually both sides of the heart go into failure. However, in many patients, early failure occurs predominantly in the left side of the heart only. In rare cases right-sided failure occurs alone, particularly in congenital abnormalities of the heart.

A lowered cardiac output is by no means always present in cardiac failure. Failure quite often occurs in patients who have outputs well above normal levels. They are in failure because the auricular pressures are very high. This happens, usually as a result of peripheral circulatory maladjustment leading to a greatly increased venous return. Such conditions include (1) *arteriovenous shunts,* in which blood passes directly between the arteries and veins (bypassing the main peripheral resistance), (2) *beri-beri* in which the avitaminosis leads to generalised systemic vasodilatation, although usually myocardial weakness is superadded to the peripheral factors mentioned, so that the signs of congestion are even worse than they would otherwise be, and (3) systemic *hypertension*.

Heart block

This occurs clinically in cases in which the bundle of His is affected by some pathological process. The block may be either partial or complete. In partial heart block most, but not all, of the auricular contractions excite ventricular ones. Often every third or fourth auricular beat fails to do so; the radial pulse is then "regularly irregular". In complete heart block the auricles beat normally and the ventricles contract quite independently and much more slowly, thirty to forty times a minute. If complete heart block occurs suddenly, the ventricle may stop entirely for several seconds before taking up its independent rate of beat. During the period of complete ventricular quiescence, consciousness is lost, and an epileptiform seizure may occur, due to cessation of the circulation to the brain, and is termed the *Stokes-Adams syndrome*.

Extrasystoles

These may occur in normal subjects and usually arise outside the pacemaker (ectopically).

Auricular flutter and fibrillation

These two forms of disordered heart action in man are characterised by extremely rapid auricular beats as judged by the waves of the electrocardiogram. In auricular fibrillation the auricular waves occur irregularly at the rate of about 400 to 500 min^{-1}; in flutter, the waves are regular at 250 to 300 min^{-1}. The bundle of His is incapable of conducting impulses at such rates and a variable degree of heart block is always present, the ventricle usually beating at about 100 to 150, perhaps regularly in flutter, always irregularly in fibrillation.

Conditions analogous to flutter and fibrillation can be produced experimentally by stimulating the auricle with rapid rhythmic electric shocks, or by touching it with a brush dipped in aconitine. At first sight, the fibrillating auricle appears to have stopped beating. However, on close inspection the whole surface is seen to be writhing and shimmering. Numerous small waves replace the P-wave of the electrocardiogram. According to Lewis, in flutter and fibrillation the excitation wave circulates continuously through the auricular muscle around the mouths of the caval veins. But results obtained more recently by means of the high-speed camera and by the cathode ray oscillograph, have thrown some doubt on this classical explanation of flutter and fibrillation; the movements are not circus in character, but radiate from some fixed point in the auricle. This finding is supported by experiments on auricles made to fibrillate by applying aconitine. Cooling the spot to which the aconitine was applied abolished the fibrillation and restored the normal rhythm.

Ventricular fibrillation

This occurs in man after blockage of the coronary vessels and during electrocution. Owing to the cessation of ventricular contractions, it is rapidly fatal. It can be induced experimentally by direct electrical stimulation of the ventricles.

The immediate cause of death in *freshwater drowning* is usually ventricular fibrillation, which in this case is due to a rapid, large rise in the serum–potassium levels. Water is inhaled into the lung in the terminal struggling and, rapidly passing through the alveolar-capillary membrane, gives rise to dilution of the pulmonary blood. The consequent fall in osmotic pressure causes a massive haemolysis, with the liberation of potassium ions (among other things!) in sufficient amounts to increase the excitability of cardiac muscle to the level needed to initiate fibrillation. Note that saltwater drowning does not end in this sudden manner because the sea contains about 3·5 per cent of dissolved salts and in fact haemoconcentration occurs.

Defibrillation

Since fibrillating heart muscle will not act as a pump (no volume changes are occurring) it is clear that a method for restoring the normal beat would be a life-saving measure. A strong current passed briefly through two electrodes, one on each ventricle, will "defibrillate" the heart. It excites all the muscle fibres in both ventricles simultaneously and they subsequently become refractory. The heart is then quiescent for two or three seconds, whereupon it commences beating again, the pacemaker taking over the initiation of the beat (sometimes another irritable focus in the conducting system or in the ventricular muscle becomes the pacemaker). In the case of ventricular fibrillation occurring during open chest surgery, this method of restoring the beat is useful. Often, however, the cause of the original initiation of fibrillation is still present and after a variable interval, fibrillation sets in once more. Using electrodes placed on the heart, 50 V at 50 Hz usually gives a current flow of about 1 A and this is customarily employed in the operating theatre.

Electrical defibrillation can also be used through the chest wall but the number of instances where the outcome has been successful is small. This is because there is less than two minutes available in which to apply the current, otherwise lack of coronary blood renders the heart too feeble to respond when the normal rhythm is restored, and lack of blood supply to the brain for this length of time is fatal. With electrodes above and below the heart, on the anterior chest wall, about 500 V at 50 Hz is required. This gives a current flow of about 5 A (the resistance of the skin and rib cage is higher than cardiac muscle alone).

Cardiac massage

By alternately compressing and releasing the ventricles in the hand, the surgeon can maintain a cardiac output sufficient to keep a fibrillating patient alive. (This has been done for up to two hours.) Defibrillation may be successful at the end of this time. It is also possible to carry out cardiac massage in the intact subject by compressing the anterior chest wall with the hand, at regular intervals. The ventricles are thus squeezed between the sternum and the vertebral column. A detailed account of cardiac arrest and resuscitation will be found in Chapter 47.

41. Cardiac Electrophysiology

From the brief introduction to the electrophysiology of cardiac muscle given on page 72, it will have been clear that there are considerable differences between the properties of the action potential of cardiac muscle and of ordinary nerve fibres. The action potential of nerve is finished in 2 to 3 ms but in cardiac muscle it lasts for 300 ms or more (Fig. 41.1).

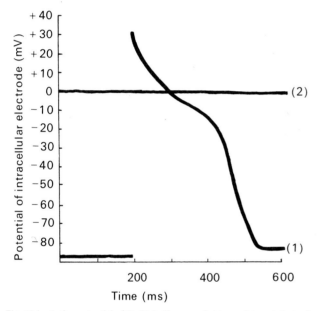

Fig. 41.1 Action potential of Purkinje fibre recorded by an internal electrode. Time intervals 100 ms. Ordinate: Inside potential relative to outside. Potentials above zero are positive inside; those below zero are negative inside. Resting potential = −90 mV. Peak of action potential = +31 mV. (Draper, M. and Weidmann, S. (1951) *J. Physiol. Lond.* **115**, 74.)

As in ordinary nerve fibres, there is an initial rapid potential reversal from the resting value of −90 mV to +30 mV. Following this phase, however, the membrane potential returns only to around −20 mV in the form of a quasiplateau. This lasts for up to 350 ms before a final repolarisation occurs. During the initial phase, the membrane resistance is greatly decreased; during the plateau it is, in fact, raised slightly above normal values.

The spike potential, as in nerve and muscle fibres, is due to the inward flux of sodium ions (positive feedback, see Chapter 14), resulting from the regenerative changes in sodium conductance. In the quasi plateau, calcium flux into the cell increases whilst potassium efflux almost balances it thus maintaining the membrane potential at a more or less constant level. The final repolarisation is due to cessation of calcium influx and continued potassium efflux.

In nerve, a net outward flow of potassium ions brings the membrane back to its resting potential level following the action potential. An increase in the amount of potassium outside the nerve should then reduce the net outward current and hence prolong the falling phase of the action potential. In nerve and muscle fibres, this does happen. However, one finds in cardiac muscle that the long-lasting part of the action potential is actually shortened by raising the external potassium concentration.

As an explanation, it has been found that the potassium permeability, P_K, is increased by an experimentally applied inward current, and decreased by an outward one. Raising the external potassium concentration would lower P_K and the potassium current would be increased even though the gradient itself was smaller.

Changes in potassium conductance, g_K, in heart muscle

From the information given above, it is clear that the membrane of the cardiac muscle cell behaves rather differently from nerve and skeletal muscle membrane. Detailed studies have been carried out on this topic and some of the results are summarised in Figs 41.2–41.6.

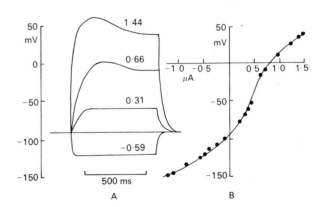

Fig. 41.2 Purkinje fibre in solution in which sodium was all replaced by choline. Two microelectrodes were put into a single cell, one passed current, the other recorded membrane potential in order to observe the relationship between potential and potassium conductance g_K.

Depolarisation of the fibre (current being passed outwards is positive, is shown as an upward deflection for 0·31, 0·66 and 1·44 μA, and produces depolarisation) *increases* the initial slope of E_m/I_m and thus *decreases* g_K. This is called anomalous rectification.

One interesting finding is that of *anomalous rectification*; this is a decrease in potassium conductance g_K with depolarisation.

Sodium conductance, g_{Na} changes

As we have already seen in peripheral nerve, the initial part of the action potential is dependent on sodium flux. The rate of rise of the potential is proportional to the value of the resting potential. The rate is slower in nerve fibres bathed in low sodium con-

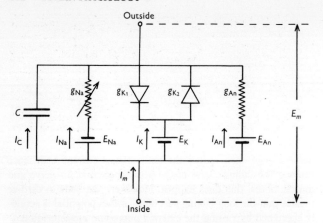

Fig. 41.3 Equivalent electrical circuit for Purkinje fibre membrane. (Noble, D. (1962) *J. Physiol., Lond.*, **160**, 317.)

On the basis of the results already quoted, Noble proposed the above model for the cardiac muscle fibre membrane. There are two potassium conductances, g_{K_1} and g_{K_2}, representing the passage of potassium ions through two separate channels. g_{K_1} is an instantaneous function of membrane potential and falls with depolarisation. g_{K_2} rises gradually as the membrane is depolarised.

In the diagram, g_{K_1} and g_{K_2} are represented as two parallel rectifiers, operating in reverse polarity.

Using empirical equations of the Hodgkin–Huxley type, the time course of the action potential of cardiac muscle could be predicted accurately using the above model. A sudden fall in E_m gives a sudden fall of g_K. This remains at its low level for over 100 ms and then slowly returns.

centration. This is also true in cardiac muscle fibre membrane, but apparently the g_{Na} does not return as rapidly to normal levels, remaining raised during the plateau. These changes have also been used to implement a computer simulation of action potential shapes for cardiac muscle. The agreement between these and the genuine articles is excellent (see Figs 41.4 and 41.5).

Action potentials from the pacemaker

If the auricle of a heart which is at rest is stimulated at any point, and the stimulus is above threshold, the whole sheet of cardiac muscle contracts, and change in the strength of the stimulus does not produce any change in the strength of the contraction. If electrodes are placed on the auricle, diphasic action potentials will be recorded, which are of the same size whatever the strength of the stimulus (above threshold) and wherever the electrodes are placed. This experiment may equally well be done on the ventricle. The whole ventricle will contract and action potentials, all of the same size, will be picked up from any point. The excitation is normally conducted to the ventricles from the auricles by means of the auriculoventricular bundle and the Purkinje tissue which spreads out over the interior surface of the ventricle. These consist of cardiac muscle cells, of rather a specialised kind, but not essentially different from those in the rest of the heart. Both the excitatory disturbance and the wave of contraction are thus conducted from one muscle cell to another in all directions over the whole heart. If any part of the heart contracts, it all contracts (except in abnormal conditions); there is no gradation of the contraction with gradation of the excitation, as different fibres are successively brought into action, as there is in skeletal muscle.

The response of the whole heart is thus "all-or-none", and this type of response was first noticed in heart muscle (by Bowditch in 1871). But it must be emphasised that many other factors,

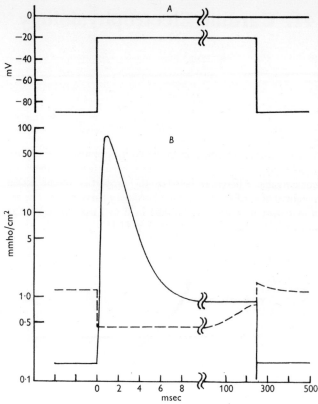

Fig. 41.4 Resting potential and ion conductance in a Purkinje fibre. A, Sudden depolarisation of resting potential. B. Computed conductances. Continuous line: sodium conductance. Interrupted line: potassium conductance. (Noble, D. (1962) *J. Physiol., Lond.*, **160**, 317.)

An abrupt change of E_m gives a prolonged fall of g_K which after 100 ms slowly returns to its normal level. When E_m is made to be normal again, g_K first is slightly raised and then falls to normal.

These are computed curves based on the Hodgkin–Huxley equations and appear to fit the observed pattern of events well. This fact of course need not necessarily mean that the model given in Fig. 41.3 is the only one to work or that it represents the true biological state of affairs.

such as stretching and drugs, can produce changes in the degree of contraction of cardiac muscle, and that variation in intensity of stimulus, with which the all-or-none type of response is alone concerned, may never occur except under experimental conditions.

It should also be noted that as a result of the cardiac action potential having a duration of up to 350 ms—a large part of each heart-beat—the muscle is necessarily refractory for this length of time. It therefore cannot be re-excited until the mechanical events of the previous contraction are over. A little thought will convince one that in a pump, alternating periods of contraction and relaxation are essential for its action, and that for this reason any condition resembling a tetanic contraction would be disastrous. The lengthy refractory period consequent upon the characteristics of the action potential, effectively ensures that summation of contraction can never occur.

The spontaneous rhythmic contraction of cardiac muscle is, as we have seen, an intrinsic property of the muscle and is not due to external stimulation. The property is shown in different degree by different portions of the heart, as is demonstrated by the experiments with the Stannius' ligatures. The excitable membrane of cardiac cells differs from that of skeletal muscle fibres in that it is not stable in the resting (polarised) condition,

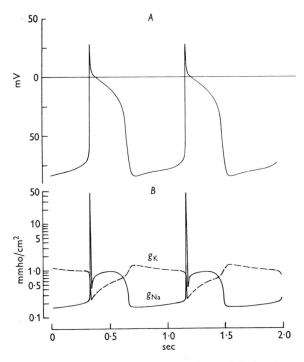

Fig. 41.5 Computed action potentials of Purkinje fibre and (below) conductance changes. Continuous line: sodium conductance. Dotted line: potassium conductance.

It will be observed that calcium conductance changes are not invoked in the explanation of the plateaux found in cardiac muscle fibre action potentials.
(Noble, D. (1962) *J. Physiol., Lond.* **160**, 317.)

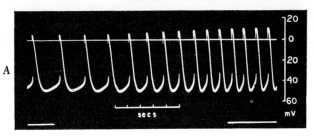

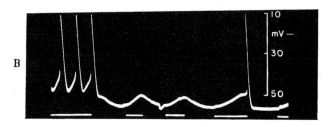

Fig. 41.6 Membrane potentials of frog's heart.

Intracellular recording from a "pacemaker" fibre in a spontaneously beating sinus venosus. (The upstrokes of the action potentials, being very rapid, have been lost in the reproduction.)

A. During the break in the lowest reference line, the vagosympathetic trunk was stimulated at $20\,s^{-1}$, the action of the vagus fibres being inhibited by atropine $1:10^6$. Note the increase in slope of the slow "pacemaker potential" from which the action potential arises, and the increased overshoot of the action potential.

B. During the four breaks in the reference line the vagus nerve was stimulated at $20\,s^{-1}$. (The tops of the action potentials have been cut off in the recording.) Note the hyperpolarisation, reduction in slope of the pacemaker potential and the rapid repolarisation, as shown particularly in the "escaped beat" at the beginning of the 4th stimulation period. (Hutter and Trautwein, 1956.)

but undergoes a relatively slow "spontaneous" depolarisation. When this depolarisation has reached a critical value, there is an "explosive" self-accentuating increase in permeability to sodium ions and potassium ions, successively, just as there is in a nerve fibre or skeletal muscle fibre, and a propagated disturbance is set up. The spotaneous depolarisation in the resting state occurs more rapidly in those cells which constitute the "*pacemaker*" and in consequence these cells drive the rest of the heart.

The SA node is spontaneously discharging; the AV node slightly less so and even lower excitability areas of the spontaneous kind are found in the bundle of His and in Purkinje fibres. The SA rhythm is normally dominant. The heart beats in an orderly sequence. If the SA node is damaged however, lower regions take over.

The question we must now ask is: How do the cells of the SA node act as pacemakers? This has been answered by intracellular recording from such cells and by observing sodium and potassium conductances during their spontaneous activity. Unlike other cardiac muscle cells, pacemaker cells have a resting potential (during diastole) which is not constant but gradually depolarises from, say, -80 mV to about -50 mV. At this point the threshold for spike generation is reached and a propagated action potential occurs. The cycle is repeated. This slowly changing depolarisation is called the *pacemaker potential*.

The rate of diastolic depolarisation is fastest in SA node cells and also the threshold is lowest there, thus the interval between successive spontaneous potentials from there will be the shortest. The SA node will therefore set the heart rate.

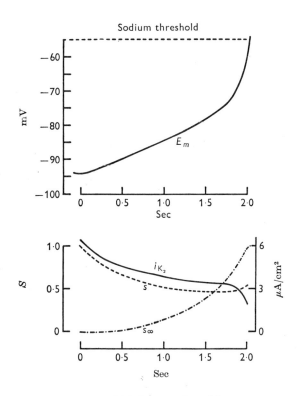

Fig. 41.7 Membrane potential during pacemaker activity.

(Bottom) i_{K_2} changes (calculated from voltage clamp results). (Noble, D. and Tsien, R. W. (1968) *J. Physiol., Lond.* **195**, 185.)

Recent work indicates how the pacemaker potentials are generated. At the end of the repolarisation of the membrane (i.e. at the start of diastole), E_m is extremely negative. There is an inward sodium current, in spite of g_{Na} being low. At first, it is about equal to the outward potassium current. The potassium conductance decays and potassium current continuously falls. Here we have the situation that the sodium inflow will gradually come to exceed the potassium outflow, so that the membrane will steadily become more depolarised (see Fig. 41.7).

Effects of autonomic stimulation on the pacemaker

Vagal stimulation slows the heart, mainly by the action of acetylcholine on the SA node. At rates of 10–20 s^{-1}, it abolishes the pacemaker potential, gives a hyperpolarisation and stops the heart.

Application of adrenaline, or stimulation of the sympathetic nerve fibres supplying the heart, increases the instability; the depolarisation occurs more rapidly and an action potential is initiated after a smaller delay, as is indicated in Fig. 41.6a; the frequency of the beat is thus increased. Application of acetylcholine or stimulation of the vagus nerve, on the other hand, increases the stability, accelerates the repolarisation of the membrane, shortens the action potential and increases the delay before the next beat is initiated, as indicated in Fig. 41.6b. The resting potential, also may be increased, so that a larger change in membrane potential is needed before the critical value for excitation is reached.

Administration of acetylcholine, or stimulation of the vagus nerve has been shown to increase the permeability of the excitable membrane to potassium ions. In consequence, the membrane potential is more firmly "locked" in a state of maximum polarisation and large depolarising currents are needed before the critical state of depolarisation is reached, at which the "explosive" increase in permeability to sodium ions occurs. Repolarisation of the membrane, after excitation, which results from the delayed increase in permeability to potassium ions, also occurs more rapidly, as may be seen in Fig. 41.8.

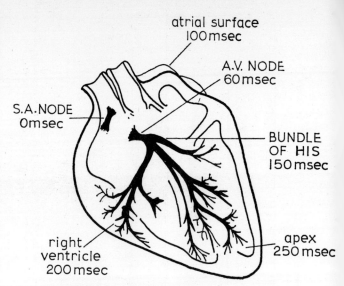

Fig. 41.9 Conduction rates and the pathways concerned. Figures, in milliseconds, are times after firing of the pacemaker that the depolarisation occurs at a point.

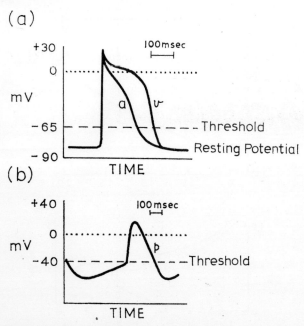

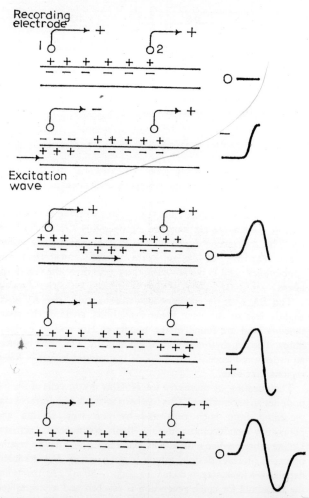

Fig. 41.8 Cardiac action potentials.
(a) Intracellular recording of membrane potentials in an atrial fibre, *a*, and ventricular fibre, *v*. The depolarisation is rapid but is succeeded by a plateau in the depolarised state, longer in the case of ventricular muscle.
(b) Similar recording from pacemaker fibre. Note the gradual spontaneous depolarisation occurring during diastole; when threshold is reached, a spontaneous action potential is generated.

Fig. 41.10 Recording from a potential source in a conducting medium. The wave of depolarisation travels from left to right and in doing so passes under each recording electrode in turn. This gives a '*diphasic*' record, quite unlike the actual variation with time, of the potential at the membrane's surface. The shape obtained depends among other things on the velocity of propagation of the impulse, its duration at any point on the membrane and the geometrical arrangement of the pick-up points with respect to the membrane.

THE ELECTROCARDIOGRAM (ECG)

Origin and spread of the cardiac impulse

Conduction of the impulse is by depolarisation spreading from one part of the heart to another in the same manner as in nerve or skeletal muscle. Reference to Fig. 41.9 shows that there is no specialised conduction system between the pacemaker (SA node) and the AV node; excitation spreads over the auricular muscle. In the human heart excitation spreads at about 1 m s^{-1}. The AV node contains fibres with an even slower conduction velocity, about 0.1 m s^{-1} (probably due to their small diameter, small fibres having a slower conduction velocity than large ones). The conduction in the Purkinje system is faster, about 5 m s^{-1}, a fact which tends to make all the ventricular muscle contract more or less simultaneously.

Generation of the ECG

The cardiac action potentials are large enough to be recorded from the body surface in humans and animals. In a volume conductor, which animal tissue is, the potential recorded at the surface of it does not resemble the generating waveform closely. Consideration of Fig. 41.10 will make this clear.

After this, it will not be surprising to learn that the electrocardiogram (ECG), as recorded between two points on the body surface, has a rather complex form; it may be considered as an esoteric example of a diphasic (or rather, multiphasic) action potential.

The actual shape is shown in Fig. 41.11. The P-wave, occurring first is due to auricular systole. After an isoelectric interval of about 200 ms, due to the conduction delay in the AV node and the bundle of His, the QRS complex follows. The QRS complex together with the T-wave are due to depolarisation in the muscle fibres of both ventricles. Figure 41.11 also shows how action potentials in the two ventricles tend to be of opposite polari-

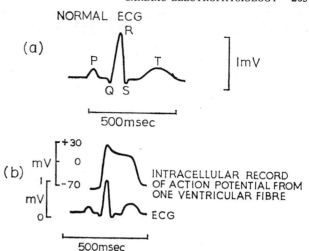

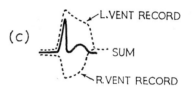

Fig. 41.11 (a) Normal ECG as recorded between R arm and L leg (lead II). (b) The same compared with the membrane changes in a single muscle cell on the same time scale. (c) Algebraic summation of action potentials due to R and L ventricles leading to a waveform similar to the normal ECG.

ty when recorded from the surface of the body since they travel in opposite directions in the muscle of each. Algebraic summation, taking into account the fact that the left ventricular action potential is larger than that of the right, gives a good approximation to the actual waveform of the QRST waves.

42. Clinical Electrocardiography

The foregoing account of cardiac electrical activity should suffice for a knowledge of the *physiology* of the circulatory system. The following notes are appended for general interest and as an indication of the ways in which electrophysiology can be applied to clinical diagnosis and treatment.

At the outset it should be made clear that the majority of phenomena in clinical electrocardiography are empirically determined.

The cardiac action potential may be considered as a somewhat complex dipole because it is moving and there are multiple sources and sinks of potential. Since it is in a conducting medium, electrodes placed in the volume conductor itself, or upon its boundaries, i.e. the skin, will pick up derived potentials as shown in Fig. 42.1.

Methods for recording the ECG

Peak voltages due to cardiac electrical activity will be around 1 mV between two widely spaced electrodes on the skin, e.g. between left and right wrists. Amplification of these potentials is therefore required. Early work was done using the "string galvanometer" in which the ECG current flowing along a silver-coated quartz fibre (=the "string") caused it to move in the magnetic field due to coils and pole-pieces according to the "right-hand rule". An optical system cast a shadow of this moving string on photographic paper or film. Modern electrocardiographs employ amplifiers (transistor) driving a heated stylus which traces a black line upon heat-sensitive paper (usually black paper covered with a thin layer of wax, coloured white; melting of the wax by heat leads to the black showing through).

Differential amplification is universally employed to counteract electrical interference arising from a.c. mains, etc. Three electrodes are applied to the subject, one being "earth", the recording of biological signals being between the other two. Since electrical interference ("hum") appears between the whole of the body (or biological preparation) and earth, this is not amplified. The required signal (the ECG) appears between the two other recording electrodes and is amplified.

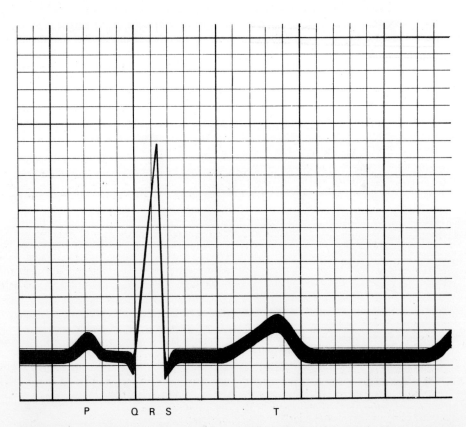

Fig. 42.1 The normal ECG. From skin derivations (e.g. lead I), amplitude, peak-to-peak should be approximately 1 mV. Durations of PR interval should be less than 0·2 s, QRS interval less than 0·1 s and the QT interval is usually around 0·4 s. Calibration: 1 small square = 0·1 mV; 40 ms.

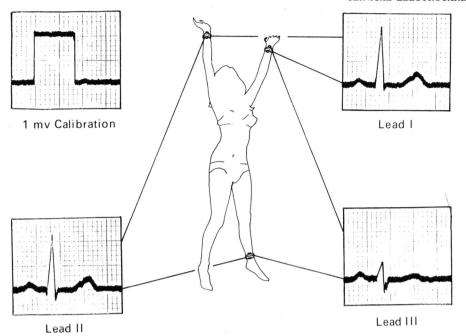

Fig. 42.2 Electrocardiograms from standard limb leads, I, II and III in normal subject.
Recording is routinely made upon heat-sensitive paper divided into a 1 mm and 5 mm grid. Calibration is arranged so that 1 mV at the amplifier input gives a deflection of 1 cm of the recording pen (top left tracing). Paper moves at $2 \cdot 5$ cm s^{-1}; 5 mm on the paper thus represents a time interval of $0 \cdot 2$ s.
Classical leads are: I—left arm to right arm; II—right arm to left leg; and III—left arm to left leg. Notice that the precise waveform of the ECG differs when recorded from the different leads.

For all normal purposes, the raw record is inspected and it is empirically known what appearances a healthy heart will give. Deviations, with skilled interpretation, lead to the diagnosis of cardiac disease. More complex analysis such as vector cardiography and the use of computer techniques, are used in research.

ECG leads
The heart may be looked upon as a complex, moving dipole in a more or less uniform volume conductor. The various components of the electrical waveform generated by the electrically active muscle will vary in amplitude, duration and polarity, according to where, in the volume conductor, the recording pickup leads are placed. This variation has diagnostic significance.

The classical limb leads are connected as follows:

Lead I	Right arm to left arm
Lead II	Right arm to left leg
Lead III	Left arm to left leg

Figure 42.2 shows typical normal records from these leads. Most of the early classical work upon the significance of the ECG, and the basic studies on the spread of the cardiac impulse was done using leads I, II and III.
Unipolar leads. Fashions change in the electrocardiographic world as rapidly as they do in the Paris salons. We now have unipolar recording in which an "indifferent" electrode is at zero potential (achieved by connecting all three above leads to one side of the input through a 5 KΩ resistor). An exploring, or "active" electrode can placed almost anywhere on the body surface and indeed within several of its orifices. Table 42.1 and Fig. 42.3 give more detail about these positions and the records normally to be obtained from them.

Table 42.1 *The unipolar leads*

Designation	Location of active electrode
V_L	Left arm
V_R	Right arm
V_C	Chest
V_F	Foot (i.e. left leg)
V_1 to V_6	Various positions on chest wall

XYZ leads. The latest electrode positions involve placing pairs of electrodes in three axes intersecting the heart, X or transverse, Y or vertical and Z or anteroposterior. Theoretically, it is possible to derive the waveform due to any possible electrode pair using these three and moves are afoot to standardise this system.

The normal ECG (see Figs 42.2 and 42.3)

The P-wave (or auricular complex), is a small, usually rounded, wave lasting for about $0 \cdot 1$ s and is the first event, electrically speaking, in each cardiac cycle. It corresponds with the impulse travelling from the SA node in the auricles. When the peak of the P-wave occurs, excitation has reached the AV node. Its size is a rough measure of the function of the auricular muscle.
Clinical conditions in which alterations from normal are found in the P-wave include:

(a) *Mitral stenosis.* The left auricle is usually hypertrophied and as a result the P-wave is larger than normal (sometimes it has a double peak).
(b) *Auricular fibrillation.* Since co-ordinated muscle activity in the auricle is lacking in this condition, the P-wave as such is absent. Usually, however, one can see low-amplitude, irregular

268 HUMAN PHYSIOLOGY

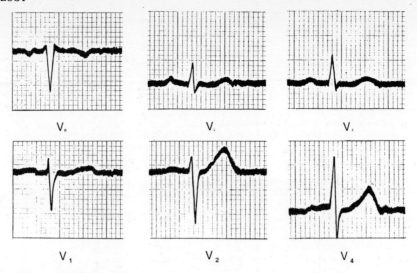

Fig. 42.3 Normal examples of records from unipolar leads V_R, V_L and V_F and unipolar precordial leads V_1, V_2 and V_4 to show differences in waveform of ECG as recorded from different sites. All recordings from the same subject.

waves in the record, related to the rapid excitation of the atria.

(c) *Auricular flutter.* Fast, but regular contractions (around 400 min^{-1}) occur in auricular muscle. These appear in the ECG as the f-waves, small but regular oscillations which take the place of normal P-waves.

(d) *Abnormal pacemaker.* If the cardiac impulse arises ectopically and does not follow the usual pathway of excitation (as on p. 264), the P-wave is altered. Its shape may be abnormal; it may be inverted, i.e. excitation travels in the reverse direction through the muscle fibres of the auricles. This may happen for isolated beats or continuously.

The P–R interval indicates the conduction time in the bundle of His. It is taken as the time interval between the start of the P-wave to the start of the R-wave (Fig. 42.1). In normal persons it varies between 0.1 to 0.15 s; the limit of normality is customarily taken to be 0.2 s. A longer interval than 0.2 s indicates a pathological delay in conduction through the bundle of His.

The ventricular complex (or QRST waves), as can be seen from Fig. 42.1, consists of a small downward deflection Q, a large upward R-wave, a small downward S deflection and, after an isoelectric interval, a T-wave. The length of QRS complex is usually about 0.08 s; longer than 0.1 s is abnormal and indicates

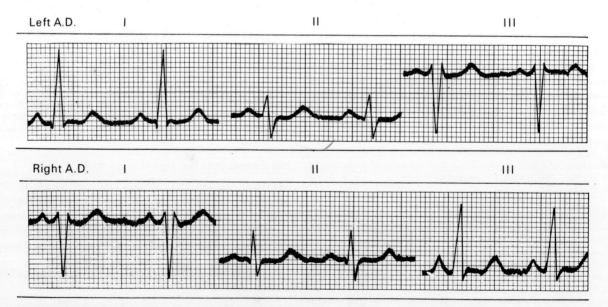

Fig. 42.4 Axis deviation (AD) (For normal records: see Fig. 42.2)

Records from two patients showing: (top recording). Left axis deviation; (bottom recording) right axis deviation. In each case ECG tracings from leads I, II and III are shown.

Left axis deviation is characterised by a tall R-wave in lead I and a large S-wave in lead III. This indicates a shift of the electrical axis of the heart to the left, either purely anatomically or because of left ventricular hypertrophy.

Right axis deviation is indicated by a large S-wave in lead I and a big R-wave in lead III.

Chronic hypertension or aortic incompetence gives rise to left axis deviation; mitral stenosis or emphysema leads to right axis deviation.

pathological slowing of conduction in the smaller branches of the bundle of His or in ventricular muscle fibres. The rising phase of the R-wave coincides with the start of ventricular systole; the conclusion of the T-wave occurs at the finish of ventricular systole.

Left ventricular hypertrophy, as occurs in hypertension, alters the ratio of the electrical output of the two sides of the heart. Thus, in lead I, for example, the size of the upward R-wave is increased and in lead III it is decreased and a large S-wave appears. In V_6, the R-wave is large. Changes in the opposite sense are found in right ventricular hypertrophy.

Axis deviation

As briefly explained on page 265 (Fig. 41.11), the QRS complex and the T-wave are generated as a result of opposing depolarisations in the right and left ventricles. The voltages due to each ventricle are similar but the time courses of electrical activity are not. At the start of the QRS complex, right and left ventricular depolarisations summate to form the R-wave since activity is travelling in the same direction in the interventricular septum and then in the apices of the ventricles. The downstroke of the R-wave is due to cancellation of the resultant depolarisations as they progress in opposite directions, back around the outer ventricular walls. Since the left ventricle is larger than the right, its muscle fibres are longer and therefore depolarisation lasts longer in it. This produces the T-wave, i.e. it is the unopposed electrical activity in the left ventricle.

Clearly, this electrical balance can be disturbed, either as described in the previous paragraph by muscle hypertrophy or by an actual anatomical displacement of the electrical axis of the heart. Abnormalities of this kind are termed axis deviation (Fig. 42.4).

The heart rate

The ECG is an accurate method for measurement of heart rate; even during severe exercise it has proved possible to strap tiny radio transmitters to subjects and, by telemetry, to obtain records of the response of the heart to muscle work. However, it is in the clinical diagnosis of disorders of cardiac rhythm that the ECG is especially useful.

Abnormal rhythms usually originate at an abnormal site for initiation of impulse formation.

The characteristics of cardiac arrhythmias depend on several factors:

(a) Any part of the myocardium and/or conducting system is able to initiate excitation and impulse spread.
(b) Heart is in the form of four separate muscle masses (auricles and ventricles), functionally joined together.
(c) Excitation spreads throughout the heart from an excited region.
(d) Spread through muscle or Purkinje fibres may follow an abnormal course.

Figure 42.5 shows the characteristic ECG records obtained from patients in whom an *ectopic focus* of excitation is originating the heart beat.

Paroxysmal tachycardia may be of auricular or ventricular origin, often reaching rates of over 150 min^{-1}. The ECG shows regularly and rapidly repeated complexes similar to those shown in Fig. 42.5.

Ventricular fibrillation consists of randomly travelling patterns of excitation in the ventricles; hence no co-ordinated muscle contraction occurs and the action of the heart as a pump ceases. Survival is not possible for more than 1 to 2 min. The

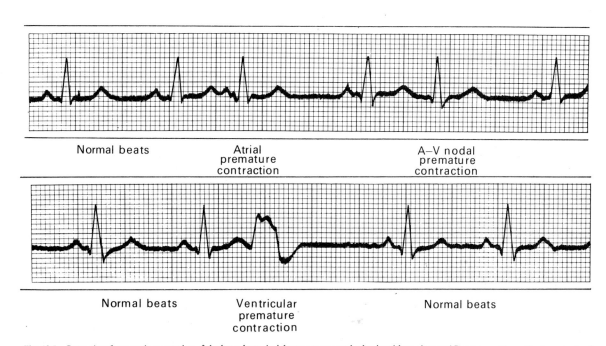

Fig. 42.5 Some sites for ectopic generation of the heart beat. Atrial premature systoles begin with an abnormal P-wave usually on the downstroke of the previous T-wave, a shortened P-R interval and have normal QRST complexes. A-V premature systoles have no P-wave, since although the atria contract, they do so simultaneously with the ventricles. Ventricular premature systoles have deformed and prolonged QRST complexes and no P-waves. Note, after each premature beat that there is a compensatory pause. (This pause allows greater diastolic filling and by Starling's law a stronger beat, which is felt by the patient often as an alarming thump in his chest, a so-called "palpitation").

Premature ventricular systoles may occur regularly as "coupled beats" or a "bigeminal rhythm."

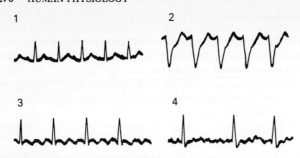

Fig. 42.6 Examples of abnormal rhythms.
1. Paroxysmal atrial tachycardia. Note P-waves are present.
2. Paroxysmal ventricular tachycardia.
3. Atrial flutter. Note f-waves at about 300 min^{-1}. The ventricles (due to their refractory period) respond to one in three atrial waves and the HR is around 100 min^{-1}.
4. Atrial fibrillation. Ventricular rhythm is "irregularly" irregular due to random excitation from the disorganised atrial muscle.

ECG shows irregular low-voltage waves. Figure 42.6 shows examples of disturbances of heart-rate as seen in the ECG.

Coronary occlusion

Although the life-expectancy at birth is now longer than it was in 1900, for a man aged 40 it is little better than it would have been then. In view of the major advances in all fields of medical care, this may seem surprising but the major cause of death in middle-age is coronary heart disease, the incidence of which is increasing. In the year 1966, for example, the male death rate from coronary heart disease was about 3 per 1000 of the population in England and Wales. It is higher than this now.

The pathological processes underlying this disease are well known, although the etiology is still a mystery. Coronary arteries are "end-arteries"; they have no overlap in distribution so that an *infarct* (=death of muscle fibres) follows occlusion of a coronary artery or one of its branches. Occlusion is usually the result of progressive narrowing by *atheroma*—a variable combination of changes in the intima of arteries consisting of focal accumulation of lipids, complex carbohydrates, blood, fibrous tissue and calcium deposits, and associated with medial changes.

Acute blocking of the coronary arteries produces characteristic changes in the electrocardiogram. Figure 42.7 shows what happens in the experimental animal, when the main artery supplying the muscle of the left ventricle is ligated with a piece of thread. Experimental lesions of the myocardium have also been produced by the local application of 0·2 mmol l^{-1} KCl solutions, which depolarises the muscle membrane.

The changes in the human ECG following a myocardial infarction are similar. During the first few hours the S–T segment is elevated.

After several days this S–T deviation decreases and the apparent T-wave becomes abnormal. If the S–T segment has been elevated, the T-wave sinks in this period and eventually is inverted.

The way in which the infarct generates these changes is not completely understood but it is clear that the *dying* muscle generates injury potentials and that the individual fibres are depolarised longer at the junctions of living and dead tissue, with

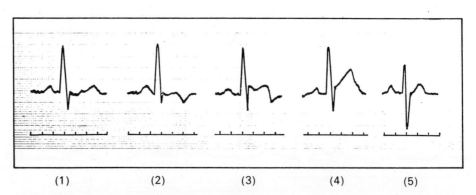

Fig. 42.7 ECG recorded direct from dog's heart during ligation of anterior descending branch of left coronary artery. One electrode silver chlorided disc on ventricular muscle; indifferent electrode in neck muscles. First record: normal control. Second record: 30 s following ligation; note inverted T-wave Third record: 2 min after ligation; the S–T segment is elevated. Fourth record: 5 min after ligation; the QRS complex is followed by a raised plateau-like potential occupying the whole S–T interval. Fifth record: after recovery.

These changes are similar to the ECG changes found in human patients after a coronary thrombosis.

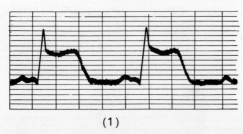 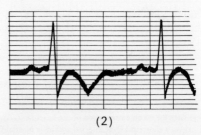

Fig. 42.8 ECG (1) immediately after and (2) three weeks after a coronary thrombosis affecting the anterior surface of the left ventricle. Records are from lead I. Note the elevated S–T segment, (in lead III this would be a depressed S–T segment in (1). In (2) the injured muscle has been replaced by fibrous tissue, and since the net electrical activity from the left ventricle is now less than that from the right the T-wave is inverted (see Fig. 41.11 for explanation).

each action potential. Hence depolarisation due to the left ventricle (when this is infarcted) will predominate, and will not entirely be cancelled out by the activity in the right. This leads to the raised S–T segment.

When all the damaged muscle has been replaced by electrically-inactive fibrous tissue (after a week or two) the net depolarisation occurring in the left ventricle will be less than normal and will terminate earlier. This will lead to the inverted T-wave. Figure 42.8 shows a typical recording from a patient having had a coronary thrombosis.

43. Mechanics of Breathing

M de B Daly

Continued vital activity of tissues demand the appropriate interchange of the gases, oxygen and carbon dioxide, between the tissues and the atmosphere. In its more usual sense, respiration means the operation of the special apparatus concerned with the absorption of oxygen by, and the removal of carbon dioxide from, the body as a whole; this is termed external respiration. Internal, or tissue respiration, on the other hand, is the local process of utilisation of oxygen and production of carbon dioxide by the tissue cells, and will be considered in a later chapter on Metabolism. In the present chapter we are mainly concerned with the process of external respiration.

STRUCTURE OF THE RESPIRATORY TRACT

Nose

Air enters and leaves the lungs by way of the buccal cavity, when the mouth is open, and the nasal passages. The latter are lined with a vascular mucous membrane characterised, except at the entrance to the nose, by ciliated columnar epithelium and scattered mucous glands. The incoming air is warmed and moistened, and is also freed of dust.

The nasal passages vary considerably from time to time with regard to the ease with which air can pass through them. The patency of the passages depends, in part, on the degree of engorgement of the submucous blood vessels. Increased resistance to air flow occurs when the blood vessels dilate, such as in response to exposure of the body to heat. Oedema of the mucosa due to infection also increases the obstruction to air flow. Drugs, such as adrenaline and ephedrine applied locally, have the opposite effect on air flow resistance.

The nasal mucosa also contains sensory nerve endings innervated by the trigeminal nerves and their stimulation leads to reflex effects discussed on page 297.

Tracheobronchial tree

Air entering the lungs then traverses the pharynx, the larynx and, lastly, the trachea and the main bronchi. The main bronchi subdivide into airways of diminishing diameter, the bronchioles and respiratory bronchioles (Fig. 43.1). The final branching network of the bronchial tree opens into the terminal air sacs with their alveolar saccules, or alveoli (Fig. 43.1). It is in the alveoli that the respiratory gases, oxygen and carbon dioxide, are able to make close contact with the blood in the capillaries.

The trachea and bronchi are kept permanently open by a series of cartilaginous rings embedded in their walls. These are partly made up of a dense fibroelastic membrane. The whole of the bronchial tree is richly supplied with elastic fibres, mainly disposed in a longitudinal direction. These fibres, along with the elastic tissues in the lungs, account for the recoil of the lungs and

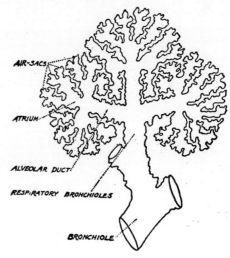

Fig. 43.1 Sectional diagram of a structural unit of the lungs. (After Miller.)

the bronchial tree which takes place during expiration (Fig. 43.2).

The respiratory passages are lined with ciliated and with mucus-secreting cells. The cilia produce a constant wave-like motion in the direction of the nasal and buccal cavities and are very efficient in expelling any foreign material that may come to rest on their surfaces.

The pharynx is a common pathway for air and food. In order to prevent the food going down the "wrong way", the aperture of the larynx is guarded by the epiglottis. During the process of swallowing, the arytenoid cartilages are closely approximated and pulled forwards towards the epiglottis, and at the same time breathing is inhibited (deglutition apnoea).

The lungs are so constructed that an almost instantaneous exchange of gases can take place between the air within them and the pulmonary blood passing through them. The pulmonary capillaries, although individually only about 8 μm in length and 8 μm in diameter, may altogether expose a surface area to the alveolar gases of approximately 70 m^2; the blood in passing through these capillaries is exposed for about 0.75 s (with the subject at rest), during which time gas exchange has to take place (Fig. 43.3). Interposed between the blood and the air in the alveoli are two delicate membranes each of one cell thickness, namely, the epithelium forming the alveolar wall and the endothelium of the capillary. The respiratory gases can readily diffuse across these membranes.

THE BRONCHIAL MUSCLES

The walls of the bronchi and bronchioles contain strips of plain muscle which run spirally round them forming a "geodetic"

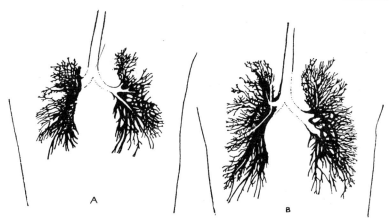

Fig. 43.2 X-ray photographs (retouched) of the bronchial tree of a young woman: (A) in full expiration; (B) in full inspiration. (Macklin.)

network.* Contraction of this muscle, therefore, narrows the bore and increases the resistance to the passage of air in and out of the lungs. Difficulty in breathing occurs when the contraction is excessive.

Blood supply

The whole of the bronchial tree, as far as the respiratory bronchioles, is supplied with blood, not by the pulmonary arteries, but by the bronchial arteries which are branches of the aorta. The blood is collected by bronchial veins and eventually drains into the right atrium; part of the blood is returned to the left atrium via the pulmonary veins.

Nerve supply

The *efferent* nerve supply is derived entirely from the autonomic nervous system. The vagus nerves are constrictor to the bronchioles whereas stimulation of the sympathetic nerves dilates them. The bronchioles also respond to the action of autonomic drugs; parasympathomimetic ones, such as acetylcholine or pilocarpine, constrict; and sympathomimetic drugs, adrenaline and isoprenaline, dilate them. There is also some evidence that the calibre of the bronchi and bronchioles increases during inspiration and diminishes again on expiration, thereby assisting the diaphragm and the intercostal muscles in renewing the air in the lungs. The mechanism for this is largely passive.

The *afferent* nerve supply to the laryngeal mucosa is the superior laryngeal nerve and to the bronchial mucosa is the vagus nerve. Stimulation of their nerve endings, by the presence of a foreign body or as a result of disease, reflexly causes coughing. From pulmonary stretch receptors in the bronchial tree, afferent fibres run up in the vagus nerves; the chief function of these receptors is to indicate to the respiratory centre the degree of expansion of the lungs. This will be referred to later.

Asthma
Under certain conditions the muscles in the bronchioles are stimulated by irritation in various parts of the body, especially the upper respiratory tract, and undergo spasmodic contraction. This results in the condition known as asthma, in which great difficulty is experienced in breathing, particularly in expiration, since this is normally a passive

---

* A "geodetic" line is the shortest distance between two points on a curved surface. A method of aircraft construction employs geodetic ribs since these give great strength to a fuselage, combined with light weight.

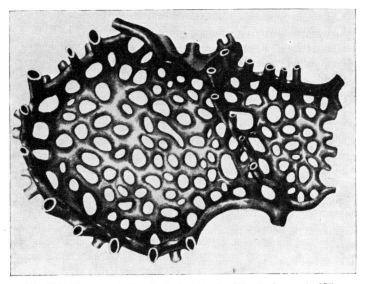

Fig. 43.3 The network of capillaries in the walls of the alveolar sacs ($\times$ 375). (From Miller, S. W. (1947) *The Lung* C. C. Thomas, Springfield.)

movement. Bronchial asthma is associated also with hyper-sensitivity to certain proteins, notably those in the pollens of some grasses. This is analogous to the sensitivity of an anaphylactic type responsible for urticaria and hay fever. Anaphylaxis in general appears to be associated with the production of a histamine-like substance and it may be significant in this connection that histamine has a powerful constricting action on the bronchioles in some animals. Relief is rapidly obtained on administration of adrenaline, isoprenaline and sometimes of antihistamines, when they bring about relaxation of the bronchial muscles.

RESPIRATORY MOVEMENTS

A constant renewal of air in the lungs is brought about by movements of the diaphragm and thorax, and this constitutes normal breathing. With inspiration the cavity of the thorax is enlarged, and the lungs enlarge as well to fill the increased space. As a result the capacity of the air passages of the lungs is increased, and air is drawn in through the trachea. Inspiration is immediately followed by expiration, which causes a diminution of the capacity of the thorax and expulsion of the air. During quiet breathing, expiration lasts 1·3 to 1·4 times as long as inspiration. Respiratory movements are to some extent under the control of the will, and one can, for instance, cease breathing for a time. It is impossible, however, to prolong this respiratory standstill for much more than about a minute, for the urge to breathe becomes excessive and against our will we are forced to breathe.

Inspiration

Inspiration is achieved, first by descent of the diaphragm which enlarges the thorax from above downwards. The diaphragm, which is innervated by the phrenic nerves, is the most important muscle of respiration although its exact contribution in normal quiet breathing varies widely in different subjects. During deep breathing it may be responsible for as much as 65 per cent of the total volume of air inspired. Secondly, the thorax is enlarged by elevation of the ribs, mainly the second to the tenth. Since each pair of corresponding ribs forms a ring directed obliquely from behind downwards and forwards, this upward movement of the anterior end causes elevation of the sternum and an increase in the anteroposterior diameter of the thorax. At the same time there is an outward or lateral movement of each rib which increases the transverse diameter of the thorax.

The action of the intercostal muscles is at present uncertain but it would appear that in man, at least, they are responsible for elevation of the ribs. When surface electrodes are placed over the intercostal muscles in the lower rib spaces, bursts of impulses occur only during inspiration in normal quiet breathing.

Expiration

The muscles of inspiration relax. The thoracic cage through its own elasticity diminishes in size and the walls of the chest are brought closer together assisted by the recoil of the bronchial tree and the lung tissues (see Fig. 43.2). In quiet breathing, expiration is probably entirely passive.

If a subject is observed in the supine position, it will be noticed that the abdominal wall rises and falls in each respiratory cycle. This is due to the movements of the diaphragm displacing the abdominal contents and is sometimes referred to as "abdominal respiration".

The *abdominal muscles* form a group of muscles having two important mechanical actions: first, raising the intra-abdominal pressure which results in the abdominal contents being pressed against the diaphragm and forcing it to ascend; and, secondly, drawing the lower ribs downwards and medially. These muscles show no electrical activity during quiet breathing. Activity appears, however, during the expiratory phase of respiration when breathing is increased, and in other circumstances involving voluntary expiratory manoeuvres, e.g. coughing, defaecation.

The *accessory muscles of respiration* comprise the scalene and sternomastoid muscles, the pectorals and serratus anterior, the rhomboids, trapezius and latissimus dorsi. For the ribs to be raised effectively during inspiration by the intercostal muscles, the thoracic inlet, i.e. the first rib, must be fixed. This is done by contraction of the scalene muscles which show electrical activity only during inspiration. Of the other accessory muscles of respiration, the sternomastoid muscles are probably the most important and show activity during voluntary inspiratory efforts and in patients with *dyspnoea*, that is, difficulty in breathing.

THE PLEURA AND THE PLEURAL CAVITIES

The lungs are enveloped in a closed membranous sac. The outer wall of the sac lines the chest wall and is called the *parietal* pleura. This is reflected at the root of the lungs, where it is continuous with the visceral layer which covers the lungs. The potential space between the layers is spoken of as the pleural cavity. It contains a small amount of fluid which acts as a lubricant. When the thorax alters in volume during the phases of respiration, the parietal and the visceral layers of the pleura normally maintain contact with each other. Even at the end of normal expiration the healthy adult lungs are in a stretched condition and this may be shown by the fact that if an opening is made in the pleural cavity, air rushes into the opening and the lungs collapse. The condition where air is present in the pleural cavity is called a pneumothorax. Since the lungs are always tending to collapse due to their elasticity, it is evident that they must exert a pull on the thoracic wall. The pressure in the pleural cavity is called the *intrapleural pressure*. It may be measured either by injecting a very small volume of air between the two layers of pleura and inserting a needle, connected to a suitable manometer, into this pocket of air or, more conveniently, by measuring the intra-oesophageal pressure, a tube being passed through the mouth for this purpose. The walls of the oesophagus are lax and so the pressure in its lumen is practically the same as the intrapleural pressure.

Owing to the pull of the lungs, the intrapleural pressure is below the pressure of the surrounding atmosphere and is called a "negative pressure" in consequence. In the expiratory position, it is 3–5 mmHg (0·4–0·7 kPa) below atmospheric pressure, that is, −3 to −5 mmHg; in normal inspiration it becomes −5 to −10 mmHg. If the lungs are more fully distended by a deep inspiration, the elastic forces are brought more into play, and the intrapleural pressure may amount to −30 mmHg (−4 kPa).

In forced expiration with the glottis closed (Valsalva's manoeuvre) and in other circumstances involving expiratory efforts such as straining, coughing and defaecation, enormous increases in intrapleural pressure occur. The maximum pressure that can be maintained voluntarily for 1–2 seconds is about 110 mmHg (15 kPa). Transient intrapleural pressures of 300 mmHg (40 kPa) may occur during severe coughing.

The lungs may be thought of as a pair of bellows with an elastic recoil provided by a spring (Fig. 43.4). To expand the bellows, we must pull on the handles with a force which is

Fig. 43.4 A simple model of the lungs represented by a pair of bellows and a spring. (From J. L. D'Silva, after R. V. Christie.)

sufficient: (1) to extend the spring; (2) to deform the material out of which the bellows are made; and (3) to overcome the resistance to air flow through the nozzle. The first part of the force depends on the amount to which the spring is stretched, i.e. on the volume of air in the bellows. The two other parts depend on the *rate* at which the bellows are being expanded, i.e. on the rate at which air is being drawn in. Similarly, when the lungs are expanded during inspiration, the respiratory muscles must overcome both an *elastic resistance*, analogous to that exerted by the spring, and a *non-elastic* or *"viscous" resistance*, exerted by the lung tissue and by the flow of air through the bronchial tree.

If the lungs were perfectly elastic, and the viscous resistance negligible, the change in intrapleural pressure (analogous to the force pulling on the handle of the bellows) would depend only on the change in the volume of the lungs; the pressure–volume relation would be the same whether the lungs were being expanded or allowed to collapse. Furthermore, during respiration, the greatest negative pressure would coincide with the maximum volume at the end of inspiration, and the least negative pressure with the minimum volume at the end of expiration; the cyclic changes in intrapleural pressure and in tidal air volume would be in phase with each other and the relationship between the two parameters would be linear (interrupted line in Fig. 43.6). Examination of the records in Fig. 43.5, however, shows that the two curves are not in phase; the changes in intrapleural pressure precede the changes in tidal air volume. This is because of the viscous resistance of the lungs. At the beginning of inspiration, the intrapleural pressure falls more rapidly than would be expected, since not only has it overcome the elastic resistance of the lungs, but also the viscous resistance, which increases as the rate of expansion of the lungs increases. In the same way, an additional force has to be exerted on the handle of the bellows to deform the material and draw air in through the nozzle.

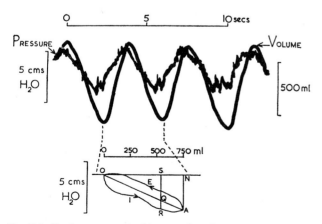

Fig. 43.5 Simultaneous records of intraoesophageal pressure (representing intrathoracic pressure) and tidal air volume in a normal subject at rest. (McIlroy, Marshall and Christie.)

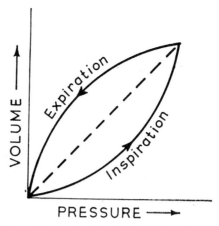

Fig. 43.6 Diagram showing the relationship between intrapleural pressure and tidal volume. The interrupted line shows the relationship were the lungs to comprise only an elastic resistance. The continuous line shows the relationship during inspiration and expiration with the viscous component added.

Conversely, when the rate of expansion becomes smaller, towards the end of inspiration, this additional fall of intrapleural pressure also becomes smaller until, at the end of inspiration (and again at the end of expiration) the lungs are momentarily at rest and the "viscous" forces vanish. During expiration, the rate of change of volume is reversed (the volume is decreasing, instead of increasing), the additional pressure necessary to overcome the viscous resistance is also reversed, and is subtracted from the pressure necessary to overcome the elastic resistance, instead of added to it; the pressure–volume relation is a closed loop, as shown in Fig. 43.6.

The slope of the time-related intrapleural pressure and tidal volume obtained under static conditions (interrupted line in Fig. 43.6) gives a measure of the *distensibility* or elastance (dP/dV) of the lungs and thorax. It is more usual to consider the *compliance* of the tissues (dV/dP) which is the reciprocal of the elastance, and is defined as the change in volume per unit change of pressure. The normal values are 0·1 l/cm H_2O (1 l kPa^{-1}) for the lungs and thorax, and 0·2 l/cm H_2O (2 l kPa^{-1}) for the lungs alone.

The existence of viscous, as well as elastic, resistance has two consequences of some importance. First, it contributes to the work which must be done and to the oxygen consumed by the respiratory muscles to ventilate the lungs; this oxygen consumption is added to that needed for more "useful" work and of itself necessitates an increase in the volume of air inspired. The work done in overcoming the viscous resistance is about 40 per cent of the total, which increases disproportionately as the respiratory minute volume increases. Normal values are 0·5 ml oxygen per litre of ventilation, or 0·5 kg m min^{-1} at rest; when a subject breathes maximally (about 180 l min^{-1}) the work of breathing amounts to 250 kg m min^{-1}. Secondly, the ratio of the viscous resistance to the elastic resistance determines the speed at which the lungs collapse. If, for some reason, the viscous resistance increases without a corresponding increase in the elastic resistance (the strength of the spring in Fig. 43.4), the lungs may not return to the initial expiratory state before the next inspiration begins. The volume of air in the lungs will become progressively greater (the spring will be increasingly stretched) until a new steady state is reached and the work needed to maintain the respiration will be greatly increased. This may occur during an attack of asthma.

FETAL RESPIRATION AND EXPANSION OF THE LUNGS AT BIRTH

The lungs of the fetus are airless and unexpanded; they contain a small amount of amniotic fluid. Only about 20 per cent of the right ventricular output passes through the pulmonary circulation, the greater part of the blood flow being short-circuited by an anastomotic channel between the pulmonary artery and aorta, the *ductus arteriosus*. It is now known that small but ineffective respiratory movements are made *in utero*. In animals these movements become accentuated when the mother is given a gas mixture containing little oxygen to breathe, and sometimes when the carbon dioxide content is increased. Increased respiratory movements also occur when the placental circulation is obstructed.

At birth, respiration is brought about by contraction of the diaphragm and intercostal muscles. At the same time, the pulmonary circulation is established although it is some hours before the ductus arteriosus closes completely. There are probably many factors coming into play in promoting the first breath. Of major importance in this connection are the impulses emanating from the skin and from proprioceptors in the muscles, tendons and joints after birth of the fetus into the atmospheric air. Not only is the fetus exposed to the cold air and to contact with surrounding objects, but its transfer from its aqueous environment subjects it to the strains put upon it by its own weight. Another factor which may be important in bringing about the first breath is the chemical change in the blood when the umbilical cord is tied or the placental circulation fails; the oxygen supply to the fetus is cut off and carbon dioxide produced by the body accumulates. Consequently a state of asphyxia results in which the arterial blood oxygen partial pressure (P_{O_2}) falls and carbon dioxide partial pressure (P_{CO_2}) rises. But how these changes affect respiration in the fetus is not fully understood.

The pressure necessary to inflate the fetal lungs for the first time is much greater than for subsequent breaths. This is due to the fact that in the fetal lung, two resistances have to be overcome: (1) that due to the cohering bronchiolar and alveolar surfaces which have to be separated by the residual volume of air, and (2) the resistance offered by the elastic tissue and smooth muscle of the lung parenchyma. The first of these is considerable and as a result pressures of up to -35 mmHg (4.6 kPa) must be applied to the outside of the fetal lungs to inflate them.

Surfactant

We have spoken of the elastic recoil of the lungs as being due to the presence of elastic tissue. Another important factor is the presence of a special surface film lining the alveoli, called surfactant, which contributes to the elastic force of the lungs. All the alveoli have a liquid–air interface, and quite irrespective of the presence of elastic fibres in the lungs, will tend to collapse just as a soap bubble does due to surface tension.

Surface tension

An atom or molecule in the centre of a solid or liquid is held on all sides by cohesive forces which may be of different kinds such as a chemical link between atoms or a metallic bond where positive ions of the metal are held in a sea of electrons. Near the surface of a medium, however, the force is not balanced because the surface particle is not completely surrounded by neighbours. As a result there is an attraction perpendicular to the surface tending to draw the particles inwards so that they will be at a higher potential energy than those in the bulk of the medium. The result of this imbalance of intermolecular forces is that the surface shrinks to its smallest possible area. The inward force upon atoms or molecules at a surface is mathematically equivalent to a tension in the surface, and the term *surface tension* has arisen to describe the phenomena associated with the high potential energy of such a region. Tension operates in the same plane as the surface, so that in a sphere the tension in the wall tends to make the sphere smaller. To prevent this happening there must be an equal and opposite force opposing it created by the pressure within the sphere. In a bubble, for instance, the tension in the wall contracting it is equal to that exerted by the pressure within the bubble. Then

$$P = \frac{2T}{r}$$

where r is the radius of the sphere (Law of Laplace). If the surface tension in a bubble remains the same, irrespective of its radius, the pressure required to inflate it diminishes as the radius increases and vice versa. This explains why, when two different sized bubbles are put in communication with each other, the small one empties into the large one.

If the lung is considered to be made up of millions of bubbles of varying size representing the alveoli, it is inherently an unstable system, for the air in the smaller alveoli would empty into the larger ones with the result that all the alveoli would either be collapsed or hyperinflated. The presence of surfactant lining the alveoli prevents this from happening. The surface lining, however, must do more than simply alter the surface tension by the same amount in all alveoli. Alveoli vary in size throughout the lungs, some have a radius 3–4 times that of others, so that if all air spaces are to be kept inflated, the pressure would also have to vary 3–4 times. In alveoli which freely communicate with each other, this clearly cannot happen. The explanation lies in the fact that the surface tension of the alveolar lining decreases as the area of the surface film diminishes, and this property ensures that all alveoli, irrespective of their size, are stable.

The alveolar surface lining in contact with the alveolar gas is an insoluble lipoprotein layer about 5 nm thick. Between this layer and the alveolar epithelium is a layer of saline—dispersible lipoprotein 30–200 nm thick. This is the "living complex" or "alveolar surfactant" which acts as a reserve from which the insoluble "lining film" is formed. There is still some doubt about the identity of the alveolar cells responsible for forming surfactant. It appears in human fetal lungs between the 21st and 24th week of intrauterine life.

Respiratory distress syndrome of the newborn

This condition occurs in a minority of newborn babies, more commonly prematures. They rapidly develop difficulty in breathing with an indrawing of the thoracic cage with each inspiration, and impairment of the oxygenation of the blood. The mortality rate is about 50 per cent. Examination of the lungs of such cases post-mortem shows that they are collapsed and, significantly, there is an absence of the lung lining complex. It would appear, therefore, that the lungs, instead of remaining inflated after the first breath, collapse due to the increased alveolar surface tension consequent upon the deficiency of surfactant.

Sounds associated with the movements of the lungs

There are two distinct sounds associated with the movement of air into and out of the normal lungs, both of which may be heard by placing the ear on the chest or, better, by using a stethoscope. The first of these is a fine rustling noise, occurring during inspiration and the beginning of expiration, known as *vesicular breathing*; the second, heard only when the stethoscope is placed over one of the larger air passages, is louder and rougher, like a whispered "hah", and is known as *bronchial breathing*.

44. Lung Volumes, Ventilation and Diffusion

M de B Daly

The primary subdivisions of the lung volume

The diagrammatic representation of a tracing of respiratory movements shown in Fig. 44.1 was taken from a subject who, after a few normal quiet respirations, took the deepest inspiration that he could and then expired to the limits of his ability. The upward stroke represents inspiratory and the downward stroke expiratory movements. The lung volumes may be divided into two main divisions, the "volumes" which can be breathed by the subjects, these being dynamic volumes, and the "capacities", each of which includes two or more of the primary volumes.

Values for the volumes are obtained by breathing in and out of a spirometer such as that shown in Fig. 44.2. This is a volume recorder on the lines of a gasometer; a cylindrical bell, closed at the upper end, is immersed in a tank of water, and counterbalanced. Air blown through a pipe passing through the water-seal raises the bell and the distance the bell travels is recorded by a writing-point on a moving paper. The spirometer is calibrated by injecting into it known volumes of air.

The volume that is inspired and expired during quiet breathing is referred to as the *tidal volume* and varies from 350 to 600 ml in different subjects. The tidal volume multiplied by the frequency of the respirations per minute gives the total volume of air breathed per minute; this is known as the *respiratory minute volume*. At rest this is usually $4-8 \, l \, min^{-1}$. The volume of the deepest inhalation that can be taken at the end of a normal inspiration is called the *inspiratory reserve volume* and is about 2500 ml. The deepest exhalation possible at the end of a normal expiration is called the *expiratory reserve volume* and is about 1500 ml.

Vital capacity

When a subject, after the deepest inspiration, expires the largest volume that he can into a spirometer, the volume of air that he expires is termed his *vital capacity*. Reference to the diagram in Fig. 44.1 shows that this volume is the sum of his tidal, inspiratory and expiratory reserve volumes and amounts to 4–5 litres. This amount is not an expression of the total volume of air

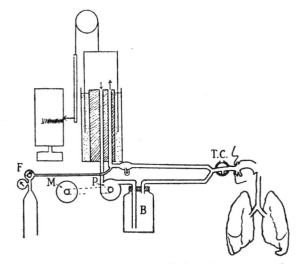

Fig. 44.2 Spirometer arranged in a closed circuit respiratory system to determine lung volumes. T. C., tap; B, bottle with absorbent for CO_2 (soda–lime or caustic potash solution); P, air-circulating pump driven by motor, M; volume of the system maintained constant by running in O_2 from cylinder F at the same rate as usage. (Redrawn from Herrald and McMichael.)

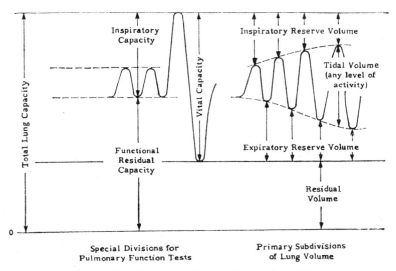

Fig. 44.1 Diagram showing lung volumes and capacities.

in his lungs, since when he has completed his forced expiration there still remains about 1,500 ml of the *residual volume*. To determine the total capacity of the lungs this residual volume must be estimated and added to the vital capacity. Even in the collapsed lung a small volume of the air is trapped; it is this which gives the collapsed lungs their buoyancy in water and is used in medicolegal investigations to ascertain whether the lungs have ever expanded and so determine whether a child breathed after birth.

The vital capacity is affected by the volume of blood in the pulmonary circulation. It is smaller when the subject is supine than it is when he is standing, the larger volume of blood in the pulmonary vessels presumably encroaching on the air capacity of the lungs. In certain forms of heart failure, also, particularly in left ventricular failure where the pulmonary vessels are congested, there is a reduction in vital capacity. These conditions are often accompanied by respiratory distress or difficulty in breathing (dyspnoea), but the mechanism by which it is brought about is still uncertain.

The *inspiratory capacity* is the maximum volume of gas that can be inspired from the resting expiratory level and is the sum of the inspiratory reserve volume and the tidal volume (Fig. 44.1).

The volume of gas remaining in the lungs at the resting expiratory level is known as the *functional residual capacity*. Reference to Fig. 43.6 will show that this is the sum of the residual volume of the lungs and the expiratory reserve volume.

Methods of recording respiration

Ideally, the record of the respiratory movements should allow us to discover (a) the frequency of respiration, (b) the depth of respiration and (c) the degree of expansion of the lungs at any moment, even if respiratory movements have stopped. There are only two ways in which this can be done. The first, which is shown in Fig. 44.2, is to connect the mouth of the subject (or the tracheal cannula of an anaesthetised animal) with a spirometer to remove continuously the carbon dioxide from the expired air, and to add oxygen to make up for that used in the metabolism—to make use, in fact, of exactly the same apparatus as is used for determining the metabolic rate, as will be described later.

The simplest forms of apparatus for recording the respiration in man do not give accurate measurements of the total ventilation, but are useful, nevertheless, for many purposes. The *stethograph*, as now commonly used, consists of a piece of large diameter rubber tubing, stoppered at both ends and connected with a tambour by a side tube at the centre; this is tied round the chest or abdomen, the movements of which distort it, and drive air into or out of the tambour. Many other devices for recording the movements of the chest or abdomen have also been described.

Pulmonary function tests

These tests "should now be performed on every patient with known or suspected cardiopulmonary disease, just as haemoglobin determination, blood pressure measurement and urine analysis are performed routinely..." (Comroe). They provide an understanding of the functional changes in the lungs and assist in the detection of impairment of lung function but do not of course tell us where the lesion is or what it is if one is present. The results of such tests must be considered in relation to others, e.g. the history and clinical examination of the patient, X-ray evidence, etc., and then they enable a more complete assessment of the clinical condition of the patient. In effect these tests must be considered analogous to those carried out to determine, for instance, renal function.

To carry out pulmonary function tests, the only equipment needed is a spirometer and recorder.

1. *Lung volumes and capacities*. These are determined as discussed above. In certain patients it is also useful to obtain values for arterial blood PO_2, PCO_2 and pH as an indication of the function of the lungs.

2. *Forced vital capacity (FVC)*. This is the gas volume expired after a maximal inspiration, with expiration being as rapid and complete as possible. Normal values are equivalent to those of the vital capacity, 4·5–5·5 l.

3. *Maximal voluntary ventilation (MVV)*. The patient breathes as deeply and as quickly as possible for a period of 15 s and the expired volume is measured and converted to litres per minute. Normal values vary from 125–170 l min^{-1}, but what is more important and significant in any one patient is a gradual change in the value one way or the other. The maximum voluntary ventilation is a test of function of motor neurones, respiratory muscles, lung volume and tissue, and airway resistance.

4. *Forced expiratory volume (FEV)*. In this test the patient inspires fully and then breathes out forcibly and maximally. The volume is recorded on a rapidly moving kymograph. The volumes expired at the end of 1, 2 and 3 seconds are determined from the record and expressed as a percentage of the total volume expired. The forced expiratory volume is designated $FEV_{1.0}$, $FEV_{2.0}$ to indicate the value determined after 1 and 2 seconds respectively. Normal values are: $FEV_{1.0}$ 83 per cent, $FEV_{2.0}$ 94 per cent, $FEV_{3.0}$ 97 per cent. Examples from a normal subject, a patient with "restrictive" movement of the thorax, and a patient with partial airway obstruction are shown in Fig. 44.3.

5. *Forced expiratory flow (FEF)*. This is the average rate of air flow for a specified volume segment of a record of a forced expiratory volume. Usually the volume measured is between 200 ml and 1200 ml and the flow rate is expressed in litres per minute. The normal $FEF_{200-1200}$ is 400 l min^{-1}. The normal forced inspiratory flow ($FIF_{200-1200}$) is also 400 l min^{-1}.

COMPOSITION OF THE RESPIRED AIR

Expired air

The tissues of the body use oxygen for the oxidation of various materials and in consequence produce carbon dioxide. A man weighing 70 kg consumes about 250 ml oxygen per min and produces about 200 ml carbon dioxide per min. The blood reaching the lungs contains more carbon dioxide and less oxygen than arterial blood, and in passing through the lungs therefore gives off carbon dioxide and takes up oxygen through an interchange with the air in the alveoli. This air is continually renewed by breathing, and hence expired air contains less oxygen and more carbon dioxide than that which is inspired. The composition of expired air, however, is not constant.

A sample of expired air is obtained by making the subject breathe through valves so arranged that he breathes in from the atmosphere and out into an airtight bag, known as a *Douglas bag*, from the name of its first user. Samples of air for analysis are drawn off from the bag and the volume of air collected in the bag in a given time is measured by pressing out its contents through a gas meter.

The expired air of a normal resting subject at sea-level contains 3·0–4·5 per cent of carbon dioxide and 16·0–17·5 per cent of oxygen. A comparison of the composition of inspired (atmospheric) and expired air is shown in Table 44.1.

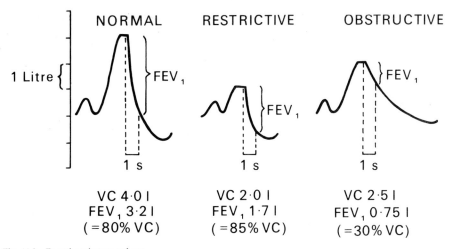

Fig. 44.3 Forced expiratory volume.
Three records taken from spirometer tracings (inspiration upwards: expiration downwards). After a deep inspiration, the subjects breathed out as rapidly and forcibly as they could. The total volume expired is the *vital capacity*, and the volume expired in the first second is the *forced expiratory volume* in one second (FEV_1). The "restrictive" record was obtained from a patient with kyphoscoliosis; the "obstructive" record from a patient with partial obstruction of the airways. Although the vital capacity is more severely reduced in the patient with kyphoscoliosis than in the patient with emphysema, the FEV_1 is normal. (Campbell, Dickinson and Slater.)

Table 44.1 *Composition of inspired and expired air*

	Inspired air (%)	Expired air (%)
Oxygen	20·95	16·4
Carbon dioxide	0·04	4·1
Nitrogen (including argon)	79·01	79·5

As mentioned above, the volume of oxygen used up per minute is larger than the volume of carbon dioxide added to the expired air. The total volume of nitrogen leaving the lungs is, however, the same as that taken in. In the above analysis, the nitrogen percentage is seen to be higher in the expired air. This is because the volume at NTP of air expired is less than that inspired, owing to the disappearance of a certain amount of oxygen without the production of a corresponding amount of carbon dioxide, so that the relative amount of nitrogen is slightly increased.

Alveolar air

This is defined as the gas in the alveoli which participates in gas exchange. Of the 500 ml of air drawn into the lungs during an average breath, only about 350 ml reach the alveoli. The other 150 ml are contained in the conducting airway from the nose down to the respiratory bronchioles. This volume is known as the *respiratory dead space* and does not participate in gas exchange in the lungs. Hence alveolar air must contain more carbon dioxide and less oxygen than mixed expired air which consists of alveolar plus unchanged dead space air.

At the end of an expiration the dead space has been swept out by, and remains filled with, alveolar air. Based on this fact, a sample of alveolar air may be obtained for analysis in the following way. A piece of rubber or plastic tubing, 25 mm diameter and 1·5 m long, is fitted with a mouthpiece, near to which is connected a gas-sampling tube which is provided with a tap at each end. Before an experiment, the sampling tube is evacuated. The subject of the experiment applies a nose-clip and after breathing normally through the mouth a few times, puts his mouth to the tube at the end of a normal inspiration, expires quickly and deeply and closes the mouthpiece with his tongue. The tap of the sampling tube is then turned, and the air near the mouthpiece, which is that last expelled from the lungs, rushes into it. The tap of the tube is then turned off, and the gas sample removed for analysis. A similar sample is then taken, in which the subject expires deeply at the end of a normal expiration. This sample will contain slightly more CO_2 and less O_2 than that obtained at the end of inspiration. The mean of the two samples is taken as the average composition of the subject's alveolar air.

An alternative method is to sample only the last portion of gas leaving the lungs during a normal expiration. This is the so-called end-tidal sample and values for gas tensions obtained by the two methods agree closely. The end-tidal method, however, can only be satisfactorily used in conjunction with gas analysers which continuously measure and record the gas tensions at the level of the mouth. These analysers depend on the paramagnetic properties of oxygen and the infrared absorption of carbon dioxide.

The significance of the composition of alveolar air will be better understood by reference to Fig. 44.4. This is a diagrammatic representation of what happens to the partial pressure or tension of the carbon dioxide and oxygen in the alveolar air and in the blood as it circulates through the lungs. (The partial pressure or tension of a gas is denoted by the symbol P followed by the suffix, thus: P_{O_2}, P_{CO_2}, P_{N_2}). Let us suppose that a

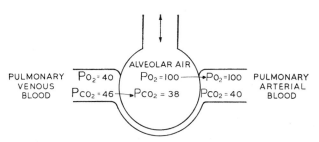

Fig. 44.4 Diagrammatic representation of alveolar air (see text).

sample of the venous blood entering the lungs by the pulmonary artery contains carbon dioxide at a pressure of 46 mmHg (6 kPa). As this sample circulates through the pulmonary capillaries, some of the carbon dioxide will diffuse out into the air in the alveoli and will continue to diffuse until there is no further pressure difference between blood and alveoli. If the pressure of carbon dioxide in the alveolar air is 40 mmHg (5·3 kPa), the pressure of carbon dioxide in the blood in the distal end of the pulmonary capillary will also reach this value. This means that the carbon dioxide pressure in the alveolar air is almost the same as that of the arterial blood leaving the lungs.

The average composition in volumes per cent of alveolar air for adult men at rest, at 760 mmHg (101·1 kPa) barometric pressure, is shown in Table 44.2, together with the partial pressures exerted by the constituent gases. It is better to express the data regarding respiratory gases in terms of partial pressures, as in the last two columns of this table, than in terms of percentage composition, as this is more meaningful since diffusion is dependent among other factors on the difference in the partial pressure of a gas on the two sides of the alveolar membrane and not on the difference in the content of the gases. From Dalton's law we know that the total pressure exerted by a mixture of gases is equal to the sum of the separate pressures which each gas would exert if it alone occupied the whole volume. Alveolar air is in contact with wet tissues and is saturated with aqueous vapour; its partial pressure, depending only on the temperature (47 mmHg; 6·2 kPa, at 37°C) must be included as a part of the total (i.e. barometric) pressure. The following example will explain the calculation.

Table 44.2 *Composition of alveolar air*

	Volumes (per cent) (dry)	Partial pressures (mmHg) (wet)	(kPa) (wet)
Carbon dioxide	5·6	40	5·32
Oxygen	13·8	99	13·17
Nitrogen	80·6	574	76·34
Water vapour	0·0	47	6·25
Total	100·0	760	101·08

Suppose that a sample of alveolar air had the composition (dry): $CO_2 = 5·6$ per cent; $O_2 = 13·7$ per cent; $N_2 = 80·7$ per cent by volume, and that the barometric pressure was 755 mmHg at the time. Then, since the aqueous vapour pressure = 47 mmHg, the remaining gases together contribute (75·5 − 47) = 708 mmHg, and of this

$$P_{CO_2} = \frac{5·6}{100} \times 708 = 40 \text{ mmHg}$$

$$P_{O_2} = \frac{13·7}{100} \times 708 = 97 \text{ mmHg}$$

$$P_{N_2} = \frac{80·7}{100} \times 708 = 571 \text{ mmHg}$$

These are the pressures exerted by each gas on any boundary surface.

Dead space

As has already been mentioned, this refers to those parts of the respiratory tract—the nose, pharynx, trachea, bronchi and bronchioles—which act as a conduit for the passage of gases to the alveoli and in which there is no gas exchange. This is known as the *anatomical dead space*. The *physiological dead space* includes the anatomical dead space and two additional volumes: (1) the volume of inspired gas which ventilates alveoli which receive no pulmonary capillary blood flow, and (2) the volume of inspired gas which ventilates alveoli in excess of that volume required to arterialise the blood in pulmonary capillaries. In the normal subject, these two additional volumes are negligible and so the anatomical and physiological dead spaces are equal.

The volume of dead space cannot be measured directly, but can be calculated using Bohr's equation. This states simply that the gas expired from the lungs is a mixture of gas from the dead space and from the alveoli; if two of these (the expired air and alveolar air) are known, the third (the dead space air) can be calculated. From Tables 44.1 and 44.2, we can take the following representative figures: (1) CO_2 in alveolar air, 5·6 per cent; (2) CO_2 in expired air, 4·1 per cent. We will take the volume of tidal air as 600 ml (all measured dry and at room temperature). Then the total quantity of carbon dioxide expired in each breath is $(600 \times 4·1)/100 = 24·6$ ml. But this quantity of carbon dioxide is contained in $24·6 \times (100/5·6) = 440$ ml of alveolar air; so that the 600 ml of expired air has the same composition as 440 ml of alveolar air mixed with 160 ml of inspired air, containing no carbon dioxide. The dead space consequently would have a volume of 160 ml. It must be remembered that this is the volume of the gas in the dead space measured dry and at a lower temperature than that at which it was when in the body. But since we wish to measure the volume of a cavity within the body, we must correct this gas volume to body temperature (37°C) and saturate with water vapour (47 mmHg).

The volume of the dead space varies, not only in different individuals, but even in the same individual according to posture and other factors. The most reliable measurements indicate that normally the dead space volume varies between 100 and 250 ml.

In disease, the physiological dead space may greatly exceed the anatomical dead space when ventilation in parts of the lungs is greater than that required to arterialise the blood.

Alveolar ventilation

We have seen how the product of the tidal volume and frequency of breathing gives a value for the respiratory minute volume, or the turnover of air as measured at the mouth. It must be realised, however, that part of the inspired air at each breath never reaches the alveoli, but remains in the dead space and is in consequence expired again without undergoing any change. The respiratory minute volume, therefore, gives little indication of the turnover of air in the *alveoli*, which is of vital importance when considering how effectively oxygen is supplied to the alveoli and carbon dioxide is got rid of during breathing. The turnover of air in the alveoli is known as the *alveolar ventilation* and is the product of the tidal volume minus the dead space times frequency of breathing. To illustrate how changes in breathing may affect the respiratory minute volume and alveolar ventilation differently, we may consider the following example. A subject has a tidal volume of 500 ml, respiratory frequency 16 min^{-1} and dead space 150 ml. His respiratory minute volume = $500 \times 16 = 8000$ ml min^{-1}. On the other hand, his alveolar ventilation = $(500 − 150) \times 16 = 350 \times 16 = 5600$ ml min^{-1}. Now, if the subject doubled his tidal volume (to 1000 ml) and halved his frequency (to 8 min^{-1}), his respiratory minute volume would remain unchanged. His alveolar ventilation, however, would increase to 6800 ml min^{-1}. This emphasises the importance of the measurement of the volume of inspired air entering the alveoli rather than that entering the upper respiratory tract as a measure of the turnover of air in those parts of the lungs concerned with gas exchange.

THE MECHANISM OF GASEOUS EXCHANGE BETWEEN THE LUNGS AND THE BLOOD

In the lungs, carbon dioxide leaves the bloodstream and enters the air in the lung alveoli, and oxygen leaves the alveoli and enters the blood. Since in doing so the gases have to pass through two membranes, the walls of the alveoli and the walls of the capillaries, the question arises as to whether the interchange of gases between the alveolar air and the blood in the pulmonary capillaries can occur by the purely physical process of diffusion. To answer this question we must study the P_{O_2} and P_{CO_2} in the alveolar air, in blood coming to the lungs and in the blood leaving the lungs. If the process is one of diffusion, the partial pressures and hence the flow of oxygen must be, in descending order: alveolar air, arterial blood, venous blood; and those of carbon dioxide in the reverse order.

The results of determinations made on man under normal conditions are summarised in Fig. 44.4 and show that the P_{O_2} in the arterial blood is very close to that in the alveolar air. The P_{O_2} in the alveoli is about 100 mmHg (13.3 kPa), and the venous blood, therefore, with a P_{O_2} of 40 mmHg (5.3 kPa) flowing through the alveoli, will rapidly take up oxygen from them and approach the point of saturation.

The P_{CO_2} of venous blood is about 46 mmHg (6.1 kPa) and that of alveolar air is about 40 mmHg. This pressure gradient will tend to cause a flow of carbon dioxide from the blood into the alveoli.

It is now generally agreed that the respiratory gas exchange is accomplished by a process of simple diffusion, the direction and extent of which depends almost entirely upon the difference in tension or pressure on the two sides of the alveolar membrane, that is, the molecules of oxygen pass from a region of high partial pressure to one of lower partial pressure.

Diffusion

In the lungs of man, about 3 litres of alveolar air, corresponding to the functional residual capacity (see Fig. 44.1), surround about 100 ml blood contained in the capillary network of the pulmonary vascular bed. The pulmonary capillaries provide an effective area of about 40 m^2, and a freely permeable membrane less than 1.5 μm thick, separating the blood from the alveolar air. This area however, is not the total alveolar area nor the total pulmonary capillary area, but represents that of functioning alveoli which are in contact with pulmonary capillaries in which blood is flowing. Nor is the area constant in any individual. It increases, for instance, in exercise.

The various structures comprising the alveolar membrane are shown in Fig. 44.5, from which it may be seen that an oxygen molecule entering a red blood corpuscle from the alveolus must pass through the layer of surfactant, alveolar epithelial lining, capillary endothelium, plasma and the red cell membrane. Diffusion of gases across the alveolar membrane will be impaired if there is any thickening due to disease or if there is an excess of fluid in the alveoli (pulmonary oedema).

The diffusion of gases across the alveolar membrane is very rapid, and at normal tensions of oxygen and carbon dioxide equilibrium between alveolar gas and pulmonary capillary blood is complete within 0.3 s. This compares with 0.75 s that the blood remains in the pulmonary capillaries in the resting subject.

The volume of gas which is capable of passing across the alveolar membrane is known as the *diffusing capacity of the lungs*. It is defined as the quantity of gas transferred each minute for each millimetre of Hg difference in partial pressure of the gas in the alveolar air and in the pulmonary capillary blood. The diffusing capacity is (1) proportional to the total area available for diffusion, (2) inversely proportional to the average thickness of the alveolar membrane, (3) proportional to the ease with which the gas diffuses through the type of tissue comprising the alveolar membrane, and (4) proportional to the solubility of the gas in the alveolar membrane. The solubility of carbon dioxide in watery solutions (and tissues) is very much greater than that of oxygen, and this explains why almost the same volume of carbon dioxide as oxygen diffuses across the pulmonary membrane in spite of the fact that the gradient for carbon dioxide is only 6 mmHg (46–40), as against 60 mmHg (100–40) for oxygen.

The normal value for diffusing capacity of young adults is 20–30 ml oxygen per min for each mmHg P_{O_2} difference. It more than doubles during exercise on a treadmill due to an increase in surface area for diffusion caused by opening up of additional pulmonary capillaries and to dilatation of others already patent. The maximum diffusing capacity decreases with age in the later decades of life, probably due to a reduction in the number of pulmonary capillaries.

Distribution of blood and gas in the lungs

We have so far discusssed gas exchange in the lungs as if all the alveoli had an equal ventilation and an equal pulmonary capillary blood flow. This, however, is not strictly true because neither is really uniform so that the supply of air and blood is never perfectly matched in any region of the lungs, even in the lungs of normal healthy man. In this connection the important factor is the *relation* between alveolar ventilation to capillary blood flow in different regions of the lungs, for as we shall see this has a bearing on the degree of oxygenation of the arterial blood.

The distribution of inspired gas in upright man is not equal in all parts of the lungs. Expressed in terms of ventilation per unit volume of the lung it is greatest at the bases of the lungs and diminishes up the lung so that it is smallest at the apices. This has been demonstrated by administering a single breath of radioactive carbon dioxide (Fig. 44.6). The rise in counting rate, measured by scintillation counters over different parts of the chest, is determined by the ventilation of the lung in the counting field, and its volume. During the subsequent 15 second breath-holding period, radioactive gas can only be removed by the pulmonary capillary blood flow. The slope of the counting rate tracing (clearance rate) is a measure of regional blood flow. This technique has shown that not only does the ventilation decrease up the lung, but the blood flow as well. A comparison of the two is shown in Fig. 44.7, from which it may be seen that the reduction in blood flow to the lungs is greater than that of the ventilation. This is also evident from a consideration of the relationship between ventilation and pulmonary capillary blood flow in different parts of the lungs, which may be expressed in terms of a ratio, called the ventilation–perfusion ratio. The symbol $\dot{V}_A$ is used to denote the mean alveolar ventilation per min

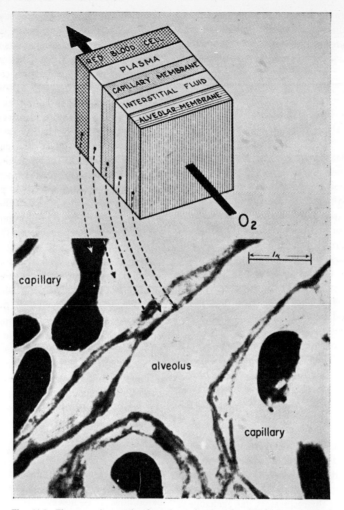

Fig. 44.5 Electron micrograph of rat lung showing the tissues through which oxygen must pass from the gas phase in the alveolus until it combines with haemoglobin within the red cell blood (× 20 000). No attempt has been made in the diagram to portray the relative thickness of the different structures. (From Comroe *et al.* 1967, after Low, 1958.)

and $\dot{Q}$ the mean pulmonary capillary blood flow. The ratio is then $\dot{V}_A/\dot{Q}$. The normal alveolar ventilation in resting man may be taken as 4 l min^{-1} and the pulmonary blood flow 5 l min^{-1}.

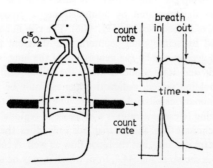

Fig. 44.6 Method of measuring regional ventilation and blood flow in the lung with radioactive carbon dioxide. Each pair of scintillation counters over the chest is connected to a recorder. The subject takes a single breath of labelled gas and holds his breath for about 15 s. The counting rate at the end of inspiration is proportional to the ventilation of the lung in the counting field and its volume. The slope of the tracing during breath-holding (clearance rate) measures the regional blood flow. (West, 1963.)

Then, by dividing one by the other, a value for the ventilation–perfusion ratio of 0.8 is obtained.

Returning now to Fig. 44.7, when the ratio is calculated for different parts of the lungs, different values are obtained because of the local variations of ventilation and blood flow. The ventilation–perfusion ratio is low (0.6) at the base of the lung and high (>3.0) at the apex. This means that the alveoli at the base of the lung are slightly overperfused in relation to their ventilation, whereas alveoli at the apex are underperfused in relation to their ventilation.

What is the significance of this ratio?

When considered for the lungs as a whole (ventilation 4 l min^{-1}; blood flow 5 l min^{-1}; $\dot{V}_A/\dot{Q}$ 0.8) the calculated value may give a false impression as to what is happening in individual parts of the lungs and therefore to pulmonary function. An extreme example of this is as follows. If all the ventilation were to go to the left lung with no blood flow, and all the blood flow were to go to the right lung with no alveolar ventilation, the patient would quickly die of asphyxia! But calculation shows that the ventilation–perfusion ratio is normal ($\dot{V}_A/\dot{Q} = 0.8$). What is important, therefore, is not so much the absolute value

LUNG VOLUMES, VENTILATION AND DIFFUSION 283

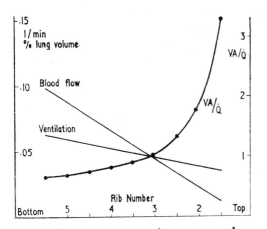

Fig. 44.7 The distribution of ventilation ($\dot{V}A$), blood flow ($\dot{Q}$), and ventilation–perfusion ($\dot{V}A/\dot{Q}$) ratio in different parts of the upright human lung. Note that because the blood falls more rapidly than ventilation with distance up the lung, the ventilation–perfusion ratio rises, at first slowly, then more rapidly. (West, 1965.)

The value of the alveolar, and hence the pulmonary end-capillary, PO_2 is about 100 mmHg (13·3 kPa), and this level is determined by a balance between the rate at which fresh oxygen (at a PO_2 of about 150 mmHg (20 kPa) in room air) enters the alveoli and the rate it disappears from the alveoli via the blood. Thus, if the alveolar ventilation decreases while the oxygen consumption remains constant, the alveolar (and end-capillary blood) PO_2 will fall. In the same way, if the rate at which oxygen is removed from the alveoli is increased by augmenting the pulmonary capillary blood flow, while the alveolar ventilation is maintained constant, the alveolar PO_2 will be reduced, and vice versa. The alveolar (and end-capillary blood) PO_2 depends therefore on the *ratio* of the alveolar ventilation to blood flow, or the ventilation–perfusion ratio. In fact this ratio controls not only the alveolar PO_2, but in a similar way the PCO_2 as well.

The presence of some underventilated alveoli in relation to their blood flow, even in normal lungs, has important implications in so far as the degree of oxygenation of arterial blood is concerned, because blood leaving these areas will not be fully saturated with oxygen. Relative hyperventilation of other areas cannot raise the PO_2 of mixed arterial blood enough to compensate for this. The reason is that the dissociation curve of oxyhaemoglobin is almost horizontal between values of PO_2 of 95 and 125 mmHg (12·6 and 16·6 kPa), and so blood leaving relatively hyperventilated areas at a PO_2 of, say, 125 mmHg

for the ventilation–perfusion ratio for the whole of the lungs, but whether the normal ratio exists in all parts of the lungs. When it does not, then pulmonary function may be impaired and lead to arterial hypoxaemia. How this occurs will now be discussed.

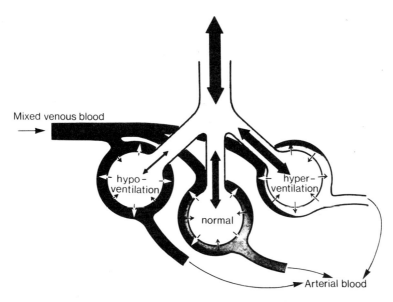

Fig. 44.8 Diagrammatic representation of possible variations in the ventilation–perfusion ratio ($\dot{V}A/\dot{Q}$) and respiratory quotient (R) in different parts of the lungs, and their effects on the PO_2, PCO_2 and PN_2 in alveoli and blood (see table below) (After Comroe, 1962.)

	Pulmonary artery Mixed venous blood	Alveoli			Mixed alveolar gas	Mixed pulmonary vein (arterial) blood
		Hypo-ventilated	Normal	Hyper-ventilated		
PO_2	40	92	100	121	104	97
PCO_2	46	42	39	34	39	40
PN_2	575	579	574	558	570	575
$\dot{V}A/\dot{Q}$		0·68	0·8	1·7		
R		0·68	0·8	1·3	0·80	

Values for gas tensions in mmHg.

contains hardly any more oxygen than blood leaving alveoli with a normal ventilation–perfusion ratio. The result is that the partly unsaturated blood from relatively underventilated alveoli lowers slightly the PO_2 and saturation of the mixed blood from all parts of the lungs. This largely accounts for the fact that the normal arterial haemoglobin saturation is only 97 per cent and not 100 per cent. Gross unevenness of ventilation in relation to blood flow is the commonest cause of arterial hypoxaemia seen in patients in hospital wards.

With regard to carbon dioxide, the dissociation curve, unlike that for oxygen, has no horizontal portion in the physiological range (Fig. 13.1). The CO_2 content of blood leaving relatively underventilated alveoli will be compensated by the lower CO_2 content of blood leaving relatively overventilated alveoli. The mixture of blood from all parts of the lungs, that is arterial blood, may therefore have a normal carbon dioxide content and PCO_2.

In *summary*, therefore, normal lungs contain alveoli which are normally ventilated in relation to their blood flow, alveoli which are relatively overventilated and alveoli which are relatively hypoventilated. The effects of these differences in the ventilation blood flow in different parts of the lungs on alveolar and blood gas tensions are shown diagrammatically in Fig. 44.8.

Alveolar-arterial PO_2 and PCO_2 gradient

The unevenness of alveolar ventilation in relation to blood flow means that although complete equilibrium is reached between blood and gas in each alveolus, the arterial blood PO_2 is slightly lower than alvcolar PO_2. This is known as the alveolar-arterial blood PO_2 gradient and in man is about 4 mmHg (0.5 kPa). The gradient for carbon dioxide is considerably less, about 0.5 mmHg (0.07 kPa) (Fig. 44.7).

45. The Regulation of Breathing

Every movement of inspiration involves coordinated contraction of a number of muscles as a result of a discharge of impulses along their motor nerves. The extent to which these muscles contract determines the volume of air taken into the lungs. During expiration these impulses diminish or cease altogether and the inspiratory muscles relax. The elastic recoil of the lungs and chest wall is therefore largely responsible for expulsion of the air from the lungs, but active expiration may take place under certain circumstances and then a number of expiratory muscles contract. The respiratory rhythm and the extent of contraction of the two groups of muscles are regulated by the central nervous system, principally the pons and medulla, from which the impulses to the muscles of respiration are derived, in such a way as to keep the composition of the gases in the alveoli and arterial blood practically constant. To a large extent this is due to afferent impulses from chemically sensitive receptors reaching the specialised areas of the brain controlling respiration, so that any deviation in the P_{O_2} or P_{CO_2} is immediately compensated by alterations in breathing.

The overall control of breathing can be looked upon as a servocontrol system, the general principles of which have been discussed in Chapter 14. A model depicting the organisation of breathing in such terms is shown in Fig. 45.1. The central respiratory mechanism may be represented as comprising two functional parts, a computer and a servomechanism. The output from the computer and the input of the servomechanism provide the "set" or required ventilation (ventilatory demand), and the output of the servomechanism provides sufficient motor nerve activity to attain this ventilation (motor demand). The diagram shown in Fig. 45.1 should be kept in mind in the discussions of the various aspects of the control of breathing that follow.

In this chapter we discuss first the rhythmic control of breathing by the central nervous system, and then the ways in which respiration is modified by nervous and chemical factors.

Discharge in efferent nerves

The regulation of breathing is brought about by alterations of the activity of the groups of motoneurones in the spinal cord which control the respiratory muscles. These motoneurones are in turn controlled by groups of nerve cells in the brain which are collectively known as the respiratory centres.

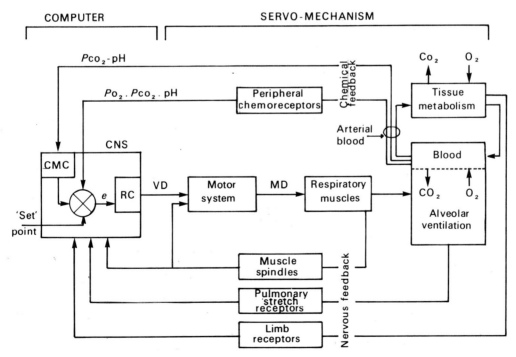

Fig. 45.1 Representation of the organisation of breathing as a control system. The performance of each breath depends on proprioceptive information ("nervous feedback") from receptors in the respiratory muscles and airways. Long-term stability of the blood gases is determined by information from central medullary chemoreceptors (CMC) and peripheral chemoreceptors ("chemical" feedback).

CMC, central medullary chemoreceptors: e, error signal; VD, ventilatory demand; MD, motor demand; RC, respiratory centres.

The main muscles of respiration are the diaphragm, innervated by the phrenic nerves which originate from the spinal roots C_2, C_3 and C_4, and the intercostal muscles which receive their innervation via the intercostal nerves. Each act of inspiration involves a discharge along these and a number of other nerves, e.g. the facial to the muscles moving the *alae nasi* and the vagus to the muscles of the larynx. Thus many segmental levels are involved in the innervation of the muscles of respiration and the respiratory act is integrated not only in the brain stem but also in the spinal cord.

Studies of the activity of motor units in the diaphragm and intercostal muscles have shown that there is activity throughout the whole phase of respiration, but that it is greater in inspiration than in expiration. During expiration the rate of firing is slow and is responsible for the thorax being maintained in a state of partial inspiration. The periodic increases and decreases in volume of the thorax are therefore superimposed on an underlying postural tone. The expiratory muscles also exhibit tonic activity. The act of inspiration begins, therefore, on a background of tonic activity in both the inspiratory and expiratory muscles, but whereas the activity of those fibres supplying inspiratory muscles which are in tonic contraction increase their rate of firing, that of expiratory neurones is reciprocally inhibited. Furthermore, as inspiration proceeds, new motor units come into action, or are "recruited", as shown in Fig. 45.2, so that the inspiratory act gains force as it proceeds. Then, when inspiration reaches a peak, it is abruptly terminated by various factors which control the depth of respiration. As expiration occurs, the units which maintain inspiratory tone revert to their former steady rate of firing, whereas the tonic discharge in expiratory units returns.

The control of respiration is integrated at two levels in the central nervous system; firstly in the pons and medulla by the respiratory centres, and secondly at a spinal level. The respiratory centres are considered below, but in the spinal cord it is the respiratory motoneurones that have the integrative function. These are large cells in the anterior horn of the grey matter, and in the cervical region they innervate the diaphragm; in the thoracic region they innervate internal and external intercostal muscles. Cells in both regions innervate accessory muscles of respiration.

Studies with microelectrodes have demonstrated that their membrane potential is about -60 mV. Each cell has a number of afferent endings on its surface and some of these afferents when activated cause depolarisation of the membrane with the result that the cell fires off and the impulses cause contraction of some respiratory muscle fibres. On the other hand, other afferent fibres when activated cause hyperpolarisation of the cell membrane and then the cell is inhibited and stops firing. These fluctuations in potential of the cell membrane can be observed during normal respiration and are called "central respiratory drive potentials" (Fig. 45.3). Since the potentials continue after cutting the dorsal

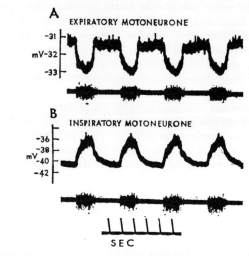

Fig. 45.3 The 'central respiratory drive potentials'. Upper traces in A and B, intracellular d.c. recordings from thoracic respiratory motoneurones; lower traces, diaphragm electromyogram (EMG). B recorded approximately 15 min after A. The records have been aligned above each other according to the diaphragm EMG. (Sears, 1964.)

roots of adjacent segments of the spinal cord, they must be largely central in origin. These respiratory motoneurones are therefore the site of convergence of a large number of excitatory and inhibitory pathways of central and reflex mechanisms which affect the size and timing of the central respiratory drive potentials. Of importance in the control of breathing, therefore, are

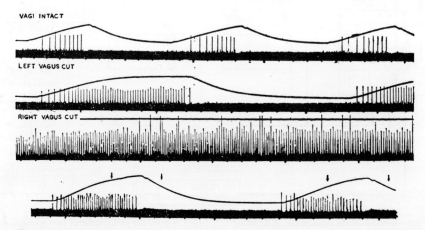

Fig. 45.2 The effect of cutting the vagus nerves on respiratory activity in a cat after section of the brainstem immediately below the pneumotaxic centre. The lowest record shows the production of apneusis.

Each record shows, from above downwards: respiratory movements (upstroke = inspiration); nerve impulses in two phrenic motoneurones; time in 0·2 s. (Pitts, 1942.)

not only the impulses reaching the respiratory motoneurones from the respiratory centres, but those from the periphery, such as evoked by stretch receptors, which give rise to *segmental reflexes*.

RESPIRATORY CENTRES

Section of the brainstem above the level of the pons in the dog and cat does not affect breathing, but respiratory activity ceases altogether when all but the lower third of the medulla is removed. This indicates that the centres controlling respiration lie in the pons and upper part of the medulla.

Several methods have been used to locate these centres. The first is one already mentioned, namely, to transect the brain stem at different levels and observe the effects on respiration. Another is to stimulate electrically various parts of the brain through a fine electrode placed in position by a mechanically operated device. The actual position of the tip of the electrode is checked histologically after the experiment and correlated with the respiratory response observed by stimulation of that area. A third method is to analyse the electrical activity of neurones in specific parts of the brainstem.

Our present knowledge of the location of the respiratory centres is based largely on investigations of this sort. These so-called "centres" are nerve cells arranged not in discrete groups as the term would imply, but diffusely in certain areas of the reticular formation of the pons and medulla. It is believed that there are four such centres, and these have been called: the *inspiratory* and *expiratory* centres in the medulla, and the *apneustic* and *pneumotaxic* centres in the pons.

Medullary centres

The *inspiratory centre* is in the ventral reticular formation immediately over the cephalic four-fifths of the inferior olive at the level of the entrance of the vagus nerve (Fig. 45.4). Electrical stimulation of this region causes deep inspiration involving both the thorax and diaphragm.

The *expiratory centre* consists of neurones which intermingle with those of the inspiratory centre but extend slightly higher up in the medulla and lie more dorsally and laterally in the reticular formation (Fig. 45.4). Electrical stimulation of the expiratory centre causes cessation of respiration in the expiratory position and contraction of the expiratory muscles.

Pontine centres

Apneustic centre. If the vagus nerves are cut to exclude impulses from pulmonary stretch receptors initiating the Hering-Breuer reflex and the brainstem sectioned in the lower pontine region, rhythmic breathing ceases and is replaced by prolonged inspiratory spasms. This type of breathing is called apneusis (Figs. 45.2 and 45.5). On the other hand, if further transections of the brainstem are made more caudally so as to involve the upper part of the medulla, rhythmic respiration returns. Transection below a level 2 mm caudal to the border between the pons and medulla results in abolition of all respiratory movements, as already stated.

The origin of the apneustic type of breathing has been a matter of controversy for many years, but it now appears to be due to uninhibited activity of an apneustic centre situated in the middle and caudal regions of the pons. This centre dominates

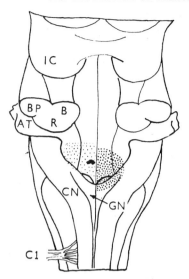

Fig. 45.4 Dorsal view of the brainstem (cat) with cerebellum removed, showing the projection of the medullary respiratory centres on the floor of the fourth ventricle. To avoid overlapping, the *expiratory* centre is shown only on the left, and the *inspiratory* centre only on the right. IC, inferior colliculus; BP, brachium pontis (middle cerebellar peduncle); B, brachium conjunctivum (superior peduncle); R, restiform body (inferior peduncle), AT, acoustic tubercle; CN, cuneate nucleus; GN, gracile nucleus; Cl, first cervical root. (Pitts, Magoun and Ranson, 1939.)

the inspiratory centre in the medulla, but can be inhibited experimentally by electrical stimulation of the central (rostral) end of the cut vagus nerve.

A further point which emerges from these observations is that when the medullary centres are cut off from all afferent impulses, rhythmic respiration continues. This is interpreted as indicating that these centres are inherently rhythmical. But it must be mentioned that the rhythmic respiration seen under these conditions (transection of the brain at a level of the upper part of the medulla combined with division of the vagus nerves) is not normal breathing; it is of a gasping type in which each inspiration is maximal and involves all inspiratory muscles.

The *pneumotaxic centre* lies in the upper part of the pons and has been localised accurately by electrical stimulation to the *locus caeruleus* and the neighbouring dorsolateral reticular formation of the isthmus of the rostral pons. The neurones fire predominantly during inspiration but are not inherently rhythmical. The activity is dependent on impulses from the inspiratory centre.

Functional organisation of the respiratory centres. The neurones of the medullary respiratory centres are closely related synaptically and excitation of a small part of one may lead, through the synaptic interconnections, to activity of the whole of that centre. It is assumed that the normal respiratory rhythm originates within this medullary neuronal network through a self-re-exciting mechanism. The apneustic centre, however, overrides the medullary centres and under certain circumstances discharges tonically to maintain an inspiratory spasm without interruption and is clearly so powerful as to be able to obliterate the normal rhythmicity of the medullary centres.

In normal breathing the apneustic centre is rhythmically inhibited by two mechanisms, firstly, by impulses from pulmonary stretch receptors which are excited during inspiration, and secondly, by impulses from the pneumotaxic centre.

The present concept of the organisation of the respiratory

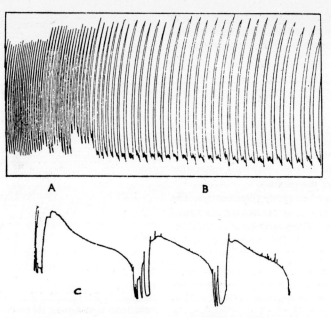

Fig. 45.5 Respiratory movements of a cat, showing the different types of respiration.
A. Normal respiratory movements (inspiration upwards).
B. The same after section of the vagi.
C. The apneustic type, with prolonged inspiratory tonus.

centres is shown diagrammatically in Fig. 45.6, and the sequence of events is briefly as follows: (1) The activity of the inspiratory centre, dominated by the apneustic centre, brings about inspiration. (2) Impulses from the inspiratory centre excite the pneumotaxic centre, and at the same time the frequency of impulses from pulmonary stretch receptors increases as the lungs inflate. (3) At the height of inspiration impulses from the pneumotaxic centre and from the pulmonary stretch receptors cause inhibition of the inspiratory centre through an action largely on the apneustic centre. (4) Expiration is thereby brought about passively. Then as the inhibitory effects of the activity of the pneumotaxic centre and pulmonary stretch receptors wane, the cycle of events starts all over again.

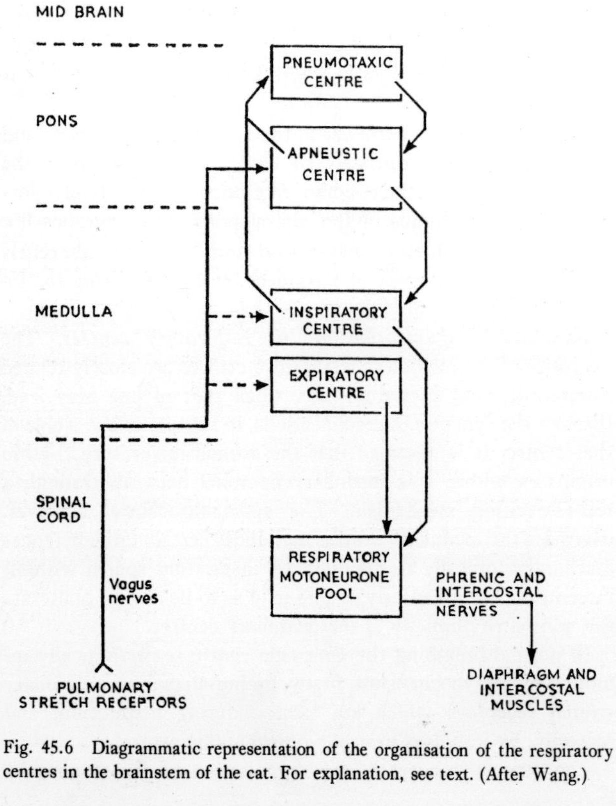

PROPRIOCEPTIVE CONTROL OF RESPIRATORY MUSCLES

We have so far discussed the control of respiratory movements as if they depended entirely on the activity of the respiratory centres sending impulses via the descending tracts to the spinal motoneurones of the principal respiratory muscles. These muscles are, of course, skeletal muscles which in the limbs are known to be affected not only by the direct pyramidal tracts, but also by the stretch reflex. It is now evident that this reflex can also be elicited from the intercostal muscles and that it plays an important part in the control of respiratory movements.

The stretch reflex is elicited in a limb muscle by a sudden pull on the muscle, and this causes an immediate contraction of the same muscle. The response is reflex in nature, being abolished by cutting the corresponding dorsal or ventral root. The sensing element in muscle is the muscle spindle, and its stretch receptor, the annulospiral ending, fires impulses which pass by nerves to the anterior horn cells in the spinal cord to form a monosynaptic reflex. The impulses then pass down the motor nerves to the same muscle, and the resulting contraction and shortening of the muscle causes the muscle spindle to shorten as well, thereby releasing the stretch on its receptor. The afferent impulses then cease. Histologically these muscle spindles have been shown to be present in the intercostal muscles and also, though in fewer

Fig. 45.6 Diagrammatic representation of the organisation of the respiratory centres in the brainstem of the cat. For explanation, see text. (After Wang.)

numbers, in the diaphragm. Their physiological characteristics are similar to those in limb muscles.

The muscle spindle is not a simple stretch receptor, however, and its physiology is more complicated than is indicated in the paragraph above. It will suffice to give a brief résumé here of its dominant features.

The spindles in limb muscles are arranged in parallel with the skeletal (extrafusal) muscle fibres and have a low mechanical threshold to stretch. Their primary afferent (1A) fibres are connected at a spinal level so as to excite the α-motoneurones of the same muscle via the monosynaptic pathway. This is a "feedback" servoloop. Within the connective tissue capsule of the muscle spindle lie spindle (intrafusal) muscle fibres innervated by the small motor or gamma fibres which leave the spinal cord by way of the ventral roots. The motor outflow from the spinal cord consists therefore of two effectively distinct motor systems: (1) the α-motor fibres innervating extrafusal skeletal muscle fibres and constituting Sherrington's "final common path" and (2) the γ-efferent or fusimotor fibres innervating exclusively the intrafusal muscle fibres of the muscle spindles.

The muscle spindle, lying parallel to the skeletal muscle fibres, acts as a detector of changes in length of the extrafusal muscle fibres of the parent muscle, and through reflex pathways brings about contraction (or relaxation of an already contracted muscle) to oppose any further lengthening (or shortening) of the muscle. The sensitivity to any change in length of the muscle spindle may be altered by means of the activity of the fusimotor fibres changing the length of the intrafusal muscle fibres. For instance, if increased activity is evoked in the intrafusal fibres through their γ-efferent innervation, the sensory element will generate more impulses and reflexly increase the rate of discharge in the α-motor fibres causing contraction of the extrafusal muscle fibres. As the muscle shortens, the stretch on the muscle spindle diminishes and its afferent discharge diminishes. In fact, the sensory element is all the time signalling a difference in length, or a misalignment, between the muscle spindle and the main skeletal muscle fibres.

During spontaneous breathing it has been found that there is an increased discharge of impulses not only in α-motor fibres to the intercostal muscles, but also in the fusimotor fibres. On this evidence, it could be postulated that the increased α-motoneurone activity responsible for inspiration is the result of an increased discharge from muscle spindles consequent upon the augmented activity in fusimotor fibres. Alternatively, the α-motoneurones may be "driven" by impulses from the respiratory centres during inspiration in spite of a decreasing excitatory input from the muscle spindles, the increased discharge in the fusimotor fibres being insufficient to prevent a reduction in the discharge from the muscle spindle as a result of shortening of the extrafusal fibres. Studies of the impulse activity in afferent fibres from intercostal muscle spindles show clearly that during contraction of the external intercostal muscles in the phase of inspiration, the muscle spindle increases its rate of firing. This means that intrafusal contractions through fusimotor fibre activity dominate over extrafusal contractions. Selective block of the fusimotor fibres results in a pronounced reduction in the discharge from muscle spindles during inspiration. There is still little information about the central mechanisms whereby the increased activity in the fusimotor fibres is brought about during inspiration.

It has been suggested that another role of the muscle spindles in the control of respiratory movements is to adjust automatically the force of contraction of muscle in such a way that any "demand" for a certain tidal volume is achieved in spite of variations in load. For instance, if the load on an intercostal muscle is increased by imposing an inspiratory resistance, the muscle will not shorten so much as against a smaller load, and the passive shortening (unloading) of the muscle spindles will be less, with the result that their discharge will be greater. This is similar in effect to imposing a stretch on the muscle spindles and the extra depolarisation of the α-motoneurones so evoked would summate with the concurrent "central respiratory drive potentials" to increase the discharge frequency of the active motoneurones and to recruit others into activity. This is spoken of as a "load compensating reflex".

Further information about the role of the muscle spindles in the control of breathing has been obtained from observations on human patients in whom certain cervical or thoracic sensory roots were cut on one side to relieve pain caused by cancer. Division of the posterior roots of the cervical segment from which the phrenic nerve arises or of the 4th–7th thoracic nerves led to a temporary paralysis of the diaphragm or temporary decreased activity of the intercostal muscles respectively. These observations provide further evidence that the sensory input from muscles to the spinal cord is an important factor exciting the anterior horn cells either directly or by facilitating impulses descending from higher parts of the central nervous system.

Another mechanism by which the force of muscular contraction can be modified reflexly is through the Hering-Breuer reflex operating by way of alterations in intercostal muscle spindle activity. The sequence of events is as follows. If the trachea of an anaesthetised rabbit is temporarily occluded at the height of inspiration there is prolongation of the subsequent expiratory pause and powerful contraction of the expiratory muscles. The impulse discharge increases not only in α-motor fibres to expiratory muscles, as would be expected, but also in expiratory fusimotor fibres as well. All these effects do not occur if the vagus nerves have been cut. More specifically, however, it is found that the increased activity in α-motor fibres is reduced or abolished after cutting the dorsal roots indicating that it is the afferent discharge from muscle spindles which is exciting or facilitating monosynaptically the motoneurones.

An important point which emerges from all this is that so far as the control of muscle spindle activity is concerned, the "load compensating reflex" and the vagal reflex behave as complementary systems.

THE CHEMICAL AND REFLEX CONTROL OF BREATHING

One factor in the control of breathing is the sensitivity of chemoreceptors in the medulla to changes in the carbon dioxide pressure. This is known as the *chemical* control. Normal breathing in the resting subject continues its even tenor so long as the carbon dioxide pressure in the alveolar air, and so in the arterial blood leaving the lungs, remains constant at about 40 mmHg (5·3 kPa). If the $P\text{CO}_2$ falls below this level, breathing is inhibited, until carbon dioxide accumulation once more restores it to the normal value. On the other hand, a rise in the alveolar $P\text{CO}_2$ acts as a respiratory stimulant and the pulmonary ventilation is increased in an effort to maintain the alveolar $P\text{CO}_2$ at the normal value, or nearly so. The part played by oxygen in the chemical control of breathing is, at sea-level, a relatively minor one.

In addition, the activity of the respiratory centres is affected by reflex mechanisms. The afferent pathways for these are:

1. The carotid sinus nerves and the aortic (depressor) nerves carrying impulses from peripheral arterial chemoreceptors,

the carotid body and aortic bodies, situated in the bifurcation of the common carotid artery, and in the arch of the aorta, respectively.
2. The vagus nerves carrying impulses from stretch receptors in the lungs.
3. Afferent fibres from proprioceptors in the limbs.
4. Afferent fibres from receptors of various kinds in the skin and the mucous membranes of the respiratory tract.

CHEMICAL CONTROL

Central chemoreceptors

The increase in pulmonary ventilation brought about by inhalation of a carbon dioxide-enriched atmosphere is due to stimulation of cells within the central nervous system. For a long time it was thought that these cells were the same medullary neurones that constitute the medullary respiratory centres. There is now good evidence, however, that they are anatomically quite separate from the centres coordinating respiration and since they serve a specific function, have been called *chemoreceptors*, being sensitive to carbon dioxide and to hydrogen ions. To distinguish them from the peripheral arterial chemoreceptors in the carotid bodies and aortic bodies, they are usually referred to as the central or medullary chemoreceptors.

The chemosensitive cells or their sensory nerve endings are located on the lateral surface of the upper part of the medulla, near the exit of the 9th and 10th cranial nerves (Fig. 45.7). Their localisation is based on experiments in which solutions, to which have been added carbon dioxide, hydrogen ions or drugs such as acetylcholine or nicotine, have been applied locally to various parts of the surface of the medulla. Hyperventilation occurs when application is made to the lateral surfaces. The chemosensitive elements have not yet been studied histologically but they send afferent impulses to the medullary respiratory centres. In contrast to the peripheral arterial chemoreceptors, they do not appear to be stimulated by decreased P_{O_2} (hypoxia).

The central chemoreceptors, lying superficially on the brain surface, can be influenced more by changes in the composition of the cerebrospinal fluid, especially [H^+], than by that of the blood. When carbon dioxide is inhaled, the arterial blood P_{CO_2} increases; and CO_2, being freely diffusible, passes rapidly across the blood–brain and blood–cerebrospinal fluid barriers with the result that within a few seconds the cerebrospinal fluid P_{CO_2} rises. But whereas the increase in arterial blood [H^+] is small, that in the cerebrospinal fluid is far greater. This is because cerebrospinal fluid lacks a protein buffer, such as haemoglobin in blood, to accept H^+ ions. When a rise in arterial blood P_{CO_2} occurs, therefore, it takes a little time for changes in the P_{CO_2} and [H^+] of the cerebrospinal fluid bathing the central chemoreceptors to take place, and this accounts for the delay in the respiratory response reaching its maximum.

A diagrammatic representation of the central H^+ receptor is shown in Fig. 45.8. It will be noted the diffusion of HCO_3^- and Cl^- ions between blood and cerebrospinal fluid is also depicted. This is an active transport mechanism which comes into operation when prolonged changes in arterial blood and cerebrospinal fluid P_{CO_2} takes place. This is considered in more detail on page 303, but an example may be cited briefly. In chronic hypoxia the hyperventilation due to the low oxygen pressure reduces the

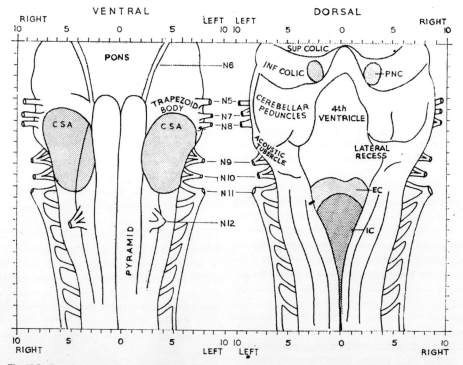

Fig. 45.7 Dorsal and ventral views of the medulla and pons of the cat. Stippled areas (CSA) represent the regions of respiratory chemosensitivity to high P_{CO_2} and H^+, nicotine or acetycholine. Areas PNC, pneumotaxic centre in region of locus ceruleus; areas IC and EC, inspiratory and expiratory centres in reticular formation; region AP, area postrema located superficially in fourth ventricle. *Abscissa* and *ordinate*, stereotaxic co-ordinates (mm). (Mitchell, 1966.)

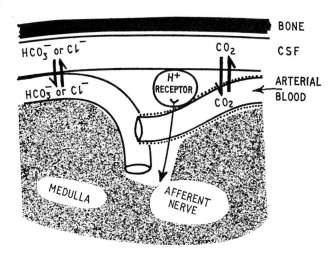

Fig. 45.8 Diagram of a central chemoreceptor or H⁺ receptor near the surface of the medulla. It is influenced by the $P\text{CO}_2$ and $[H^+]$ of the cerebrospinal fluid (CSF) and by the arterial $P\text{CO}_2$. There is a ready diffusion of CO_2 molecules across the walls of the capillaries, but not of other molecules due to the blood–CSF barrier (represented by the interrupted lines). An active exchange of HCO_3^- and Cl^- occurs across some capillaries.

arterial blood and hence the cerebrospinal fluid $P\text{CO}_2$. As the cerebrospinal fluid is unbuffered the H^+ ion concentration is reduced and this depresses the central chemoreceptors to carbon dioxide, a state of affairs which would be detrimental to the organism in the face of a poor oxygen supply. However, bicarbonate ions diffuse out of the cerebrospinal fluid to restore the $HCO_3^-/P\text{CO}_2$ ratio and hence the H^+ ion concentration. The threshold of the central chemoreceptors is lowered and hyperventilation continues in spite of the lowered arterial blood $P\text{CO}_2$.

Role of central chemoreceptors in normal breathing
When the arterial blood $P\text{CO}_2$ is reduced, for instance by voluntary overbreathing, respiration ceases for a short period of time. It is believed therefore that a certain level of carbon dioxide in the blood is necessary for normal rhythmic respiration. Cessation of breathing may also be produced in the anaesthetised animal by applying procaine, a local anaesthetic, to the central chemoreceptors and this observation suggests that the normal activity of the respiratory centres is dependent on a certain level of continuous discharge of nerve impulses from the chemoreceptors.

The way in which the medullary chemoreceptors respond to carbon dioxide can be demonstrated by a number of simple experiments.

1. *Asphyxia.* The proper aeration of the blood may be interfered with by allowing a subject to rebreathe repeatedly in and out of a small bag. In this condition not only does the alveolar and arterial blood $P\text{CO}_2$ increase, but the $P\text{O}_2$ decreases. If the subject's respiratory movements are recorded, they will be seen to increase gradually in depth and frequency. This is known as *hyperpnoea*.

2. *Hypercapnia.* In asphyxia, each factor, the increased $P\text{CO}_2$ and decreased $P\text{O}_2$, have certain effects on breathing and it is important, therefore, to distinguish *oxygen deficiency* (*hypoxia*), and excess of carbon dioxide or *hypercapnia*. These two states can be separated in a simple experiment. If a subject is made to rebreathe from a bag containing room air, respiration becomes noticeably increased when the concentration of carbon dioxide in the bag reaches about 3 per cent and that of the oxygen falls to about 17 per cent. If the experiment is repeated, but this time with a soda lime tower situated between the subject's mouth and the bag to absorb the carbon dioxide and prevent it accumulating in the bag, there is no appreciable increase in respiration until the oxygen concentration in the bag falls to about 14 per cent. The carbon dioxide concentration in the bag remains, of course, zero. Finally, if the subject is made to breathe 100 per cent oxygen and the carbon dioxide is again allowed to accumulate, the hyperpnoea becomes intolerable when the carbon dioxide concentration reaches 8–9 per cent, despite the fact that the oxygen concentration is abnormally high. These experiments demonstrate two things: first, that the hyperpnoea of rebreathing is due largely to accumulation of carbon dioxide, and second, that a reduction in the concentration of oxygen in the inspired air stimulates breathing.

3. Another experiment demonstrating the way in which the medullary chemoreceptors respond to carbon dioxide and maintain a constant alveolar $P\text{CO}_2$ is to add small quantities of carbon dioxide to the inspired air. The typical effect is shown in Fig. 45.9. The smallest effective concentration of carbon dioxide in

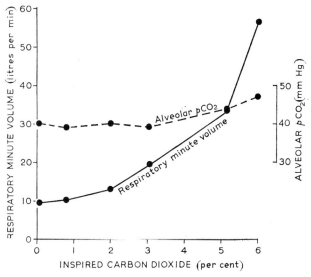

Fig. 45.9 The effect of increasing the concentration of carbon dioxide in the inspired air on the respiratory minute volume (continuous line) and alveolar $P\text{CO}_2$ (interrupted line) of a man.

the inspired air is about 1 per cent (7·6 mmHg; 1·0 kPa) and this causes a measurable increase in respiratory minute volume, although the subject is unaware of it. As the carbon dioxide concentration of the *inspired air* is increased the respiratory response also increases until at about 3 per cent his respiratory minute volume is double the normal value when he is breathing room air. It will be noted in Fig. 45.9, however, that breathing this concentration of carbon dioxide failed to cause a measurable increase in the alveolar $P\text{CO}_2$. As the inspired carbon dioxide concentration is raised still further, larger increases in respiratory minute volume occur and a rise in alveolar $P\text{CO}_2$ is also evident. But the point to be emphasised here is that the alveolar $P\text{CO}_2$ when breathing 6 per cent carbon dioxide is very much less than the value would have been had no increase in pulmonary ventilation occurred. In this connection, J. S.

Haldane found that changing his own inspired gas from room to 3·8 per cent carbon dioxide increased his pulmonary ventilation by 258 per cent, but the alveolar carbon dioxide concentration rose only from 5·62 to 5·97 per cent, equivalent to an increase in alveolar $P\text{CO}_2$ from 40 to 42·5 mmHg (5·3 to 5·7 kPa). Without such an increase in pulmonary ventilation, his alveolar CO_2 concentration would have risen to about 9 per cent which is equivalent to an alveolar $P\text{CO}_2$ of about 64 mmHg (8·5 kPa).

This experiment demonstrates that the central medullary chemoreceptors are very sensitive to carbon dioxide, an increase of only 1·6 mmHg (0·2 kPa) in alveolar $P\text{CO}_2$ being sufficient to double the respiratory minute volume. It should be emphasised, however, that the normal carbon dioxide content of the atmospheric air (0·04 per cent) is too small to have any measurable effect on respiration and is in no way responsible for the maintenance of normal respiration.

The importance of the *partial pressure* of carbon dioxide rather than its percentage can be seen by examination of Fig. 45.10. Here can be seen the effect of alterations in the barometric

Sensitivity to carbon dioxide

Sensitivity is determined by measuring the respiratory minute volume and the arterial blood or alveolar $P\text{CO}_2$ simultaneously under steady state conditions and is calculated as the change in minute volume per mmHg change in arterial blood or alveolar $P\text{CO}_2$. At levels of alveolar $P\text{CO}_2$ above about 42 mmHg (5·6 kPa), the relationship is a linear one and the average value for healthy men is 2·5 $1 \text{ min}^{-1} \text{ mmHg}^{-1}$ change in alveolar $P\text{CO}_2$. The sensitivity, however, is not a constant one; it is depressed by drugs such as morphine and barbiturates, and increased by a rise in body temperature and also by a fall in arterial blood $P\text{O}_2$. Figure 45.11 shows the effect of increasing hypoxia (deficiency

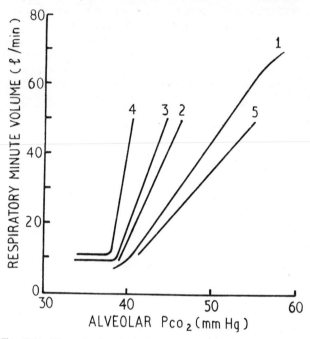

Fig. 45.11 The main factors altering the slope of the curve relating the respiratory minute volume to alveolar $P\text{CO}_2$. *1*, Control; *2*, noradrenaline; *3*, hyperthermia; *4*, hypoxia; *5*, hyperoxia (oxygen breathing). (After Perkins, 1963.)

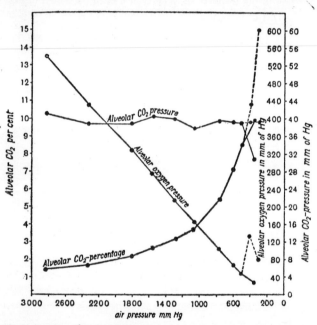

Fig. 45.10 The effect of alterations in the barometric pressure on the pressure of carbon dioxide, the percentage of carbon dioxide, and the pressure of oxygen in the alveolar air of a man. The dotted lines show the results obtained when oxygen was added to the air breathed. Note the constancy of the carbon dioxide pressure until the barometric pressure was reduced to values less than 500 mmHg (corresponding to a height of about 10 000 ft), and the respiration was stimulated by oxygen lack. (Boycott and Haldane.)

pressure, both above and below the normal value, on the alveolar partial pressure and percentage of carbon dioxide, and on the alveolar pressure of oxygen. It will be observed that while the alveolar $P\text{O}_2$ falls steadily as the atmospheric pressure falls, as might be expected, the percentage of carbon dioxide rises steadily and in such a way as to maintain the alveolar $P\text{CO}_2$ constant. This relation breaks down when the atmospheric pressure falls below 500 mmHg or 66·5 kPa (corresponding to an altitude of about 10 000 ft and an oxygen pressure of 105 mmHg or 14·0 kPa), owing to the stimulating action of oxygen deficiency. This inspired $P\text{O}_2$ would be equivalent to breathing about 14 per cent oxygen at sea level.

in oxygen) on the relationship between respiratory minute volume and alveolar $P\text{CO}_2$ in man. It will be noted that there is a gradual increase in sensitivity to carbon dioxide as the alveolar $P\text{O}_2$ falls.

The effect on breathing of a reduction in alveolar $P\text{CO}_2$

A fall in the alveolar $P\text{CO}_2$ can be brought about by forced breathing. Due to increased alveolar ventilation there is also a rise in alveolar $P\text{O}_2$. After the period of forced breathing, there follows a period of no breathing (*apnoea*), after which it starts again, shallow at first, then gradually returns to its original level. More rarely it is found that the period of apnoea is followed by a period during which a few breaths are taken, and then a second apnoeic period. This "periodic breathing" continues with shortening periods of apnoea until normal breathing is resumed. Amongst the interesting questions which we have to answer are:

Why are respiratory movements inhibited after forced breathing?

Why is there sometimes a waxing and waning rhythm or periodicity?

The period of respiratory inhibition or apnoea which follows the period of forced breathing is due to the inhibitory effect of the low alveolar $P\text{CO}_2$ for the following reasons.

1. During the forced breathing, the carbon dioxide is washed out of the lungs and its pressure is lowered.
2. If such a subject repeats the forced breathing with a gas mixture containing carbon dioxide (approximately 4·5 per cent), apnoea will not develop since the alveolar $P\text{CO}_2$ remains at approximately the resting value during the period of forced breathing.
3. It can be shown that the rise in alveolar $P\text{O}_2$ is not to blame, since gas mixtures containing a high percentage of oxygen have no inhibitory effect on breathing.

The explanation of the "periodic breathing" is as follows. During the first period of apnoea, the blood $P\text{O}_2$ falls to a level where it acts as a respiratory stimulus, despite the fact that the blood $P\text{CO}_2$ is still below normal, and would thus exert an inhibitory influence. It is to satisfy this oxygen deficiency that the subject begins to breathe again. A few breaths suffice to satisfy the oxygen requirements, so that once again the low alveolar and blood $P\text{CO}_2$ could exert its inhibitory influence, and thus a second apnoeic period follows. This periodicity continues until the alveolar $P\text{O}_2$ and $P\text{CO}_2$ return to normal.

Reflexes from peripheral chemoreceptors

Peripheral arterial chemoreceptors are situated in two locations: in the carotid body at the bifurcation of the common carotid artery and in the aortic bodies in the region of the arch of the aorta. They are supplied by the carotid sinus nerve, a branch of the glossopharyngeal, and by the aortic nerve, a branch of the vagus, respectively. These receptors are not to be confused anatomically with the pressoreceptors (baroreceptors) in the carotid sinus and arch of the aorta.

The carotid and aortic bodies consist of epithelial cells which are surrounded by a rich network of sinusoidal blood vessels (Fig. 45.12). For its size, the carotid body has a greater supply of blood than any other organ of the body, equivalent to 20 ml min^{-1} g^{-1} tissue, or about four times that of the thyroid gland and 40 times that of the brain.

The chemoreceptors, as their name implies, are sensitive to changes in the chemical composition of the arterial blood. They are stimulated by:

1. *Hypoxic (arterial) hypoxia.* This is a reduction in the $P\text{O}_2$ of the arterial blood.
2. *Stagnant hypoxia.* If the arterial blood pressure falls the chemoreceptors are stimulated even although the $P\text{O}_2$ and the $P\text{CO}_2$ of the arterial blood does not change. This is due to the fact that the fall in blood pressure reduces the blood flow through the chemoreceptors and causes local *stagnant hypoxia* whereby there is greater extraction of oxygen from the blood resulting in a reduction in tissue $P\text{O}_2$. Such a mechanism comes into operation during severe haemorrhage. Other factors causing a reduction in carotid body blood flow, such as stimulation of the sympathetic vasoconstrictor supply to the organ, or noradrenaline, also stimulates chemoreceptors.
3. *Histotoxic hypoxia.* Chemicals, such as cyanide, which poison the respiratory enzyme systems, also stimulate chemoreceptors.
4. *Increased arterial* $P\text{CO}_2$. In the presence of a normal arterial

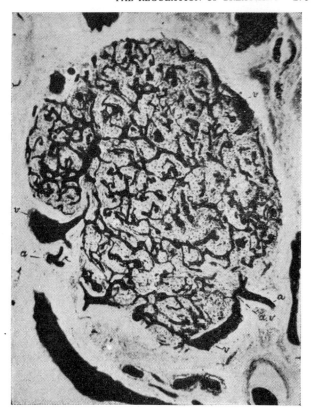

Fig. 45.12 Carotid body of the adult cat injected with gelatine-carmine showing enormous blood supply, arteries (*a*), veins (*v*) and arterio-venous anastomoses (a.v.). (De Castro, 1940.)

$P\text{O}_2$, carbon dioxide is a relatively weak stimulant of the chemoreceptors.

5. *Decreased arterial pH.* Again, a weak stimulant of chemoreceptors in the presence of a normal arterial blood $P\text{O}_2$.

The activity of the chemoreceptors may be studied by recording the electrical activity in fibres of the carotid sinus or aortic nerve. Such studies indicate that, in anaesthetised animals breathing room air, the chemoreceptors are only slightly active (Fig. 45.13a).

In man, changing the ventilating gas from room air to 100 per cent oxygen causes a temporary reduction in respiratory minute volume of about 10 per cent. This response is interpreted as being the result of withdrawal of the resting chemoreceptor "drive" to respiration, and its evanescent nature is due to the fact that the reduction in breathing causes, in turn, an increase in blood $P\text{CO}_2$ and a decrease in pH, which cause the minute volume to return to its original value. Artificial hyperventilation of an animal on 100 per cent oxygen will cause the resting carotid body discharge to cease and this is evident from Fig. 45.14. This shows that at an arterial $P\text{O}_2$ of, say, 100 mmHg (13·3 kPa) and $P\text{CO}_2$ of 40 mmHg (5·3 kPa), there is a discharge equivalent to about 7 per cent of maximum. Increasing the arterial $P\text{O}_2$ to greater than 200 mmHg (26·6 kPa) and lowering the $P\text{CO}_2$ to 20 mmHg (2·7 kPa) results in cessation of the discharge.

Lowering the arterial $P\text{O}_2$ by ventilation with a gas mixture of low oxygen content increases the afferent discharge of impulses

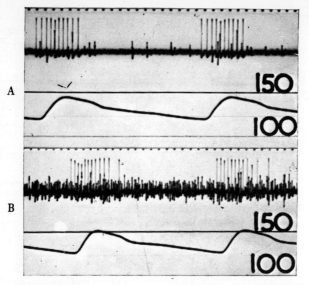

Fig. 45.13 Afferent impulses in a few fibres of the carotid sinus nerve. The large action potentials are from a single baroreceptor fibre firing synchronously with the anacrotic wave in the blood pressure record; the small potentials are those from chemoreceptor fibres. Cat spontaneously breathing air in A, and 10 per cent O_2 in N_2 in B. Note the increase in chemoreceptor activity during hypoxia. Blood pressure calibration in mmHg. Time trace, 50 Hz. (Heymans and Neil, 1958.)

from the carotid body (Fig. 45.13b). At any given arterial P_{CO_2} the discharge increases as the P_{O_2} falls.

The responses of the carotid bodies to carbon dioxide are shown in Fig. 45.14, from which it may be seen that increasing the arterial P_{CO_2} at any given P_{O_2} augments the chemoreceptor discharge. However, within the "physiological range" of arterial P_{CO_2}, when the arterial P_{O_2} is greater than 100 mmHg (13·3 kPa), the increased discharge is relatively small and this is in keeping with the finding that the *respiratory* response to carbon dioxide is not materially altered by denervation of the chemoreceptors. At lower levels of arterial P_{O_2}, the same increase in carbon dioxide may have an enhanced effect.

When the chemoreceptors are perfused with blood of low P_{O_2} and high P_{CO_2}, the frequency of discharge increases so that combined effects of low oxygen and high carbon dioxide tensions are equal to, or greater than, the algebraic sum of their separate effects. Thus when the arterial P_{O_2} is lowered to a certain value, the chemoreceptor response will be greater if the P_{CO_2} is elevated at the same time compared to the response at constant P_{CO_2} (Fig. 45.14). If the P_{CO_2} falls, on the other hand, the response to lowered P_{O_2} will be less. There is, therefore, an *interaction* between oxygen and carbon dioxide on the chemoreceptors. This fact may explain, at least in part, a similar interaction that oxygen and carbon dioxide have on breathing in unanaesthetised man. With reference to Fig. 45.11, a given increment in alveolar P_{CO_2} produces a much greater increase in respiratory minute volume during ventilation hypoxia (curve 4) than during room air breathing (curve 1).

Reflexes from the carotid bodies

These may be studied by isolating the chemoreceptors from the circulation and perfusing them with arterial blood from a donor animal. Changing the gaseous composition of the blood in the donor, and hence in the recipient animal's carotid bodies, evokes changes in respiratory minute volume in the recipient. Lowering the blood P_{O_2} and, to a lesser extent, increasing the blood P_{CO_2} or H^+ ion concentration causes an increase in the rate and depth of breathing. Denervation of the chemoreceptors by cutting the carotid sinus nerves abolishes the response, indicating that it is reflex in nature.

Alternatively, if in an intact animal the arterial blood P_{O_2} is lowered by administering a gas mixture with reduced oxygen content, an increase in respiratory minute volume results. This hyperpnoea is due entirely to a *reflex* stimulation via the chemoreceptors, for inhalation of the same gas mixture after cutting the afferent nerves supplying these receptors causes only

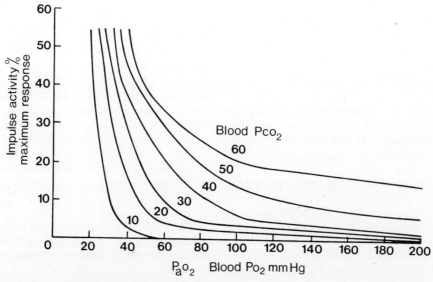

Fig. 45.14 The relationship between the impulse discharge in nerve fibres from the carotid body and arterial blood P_{O_2} at different levels of arterial P_{CO_2} (Hornbein, 1968.)

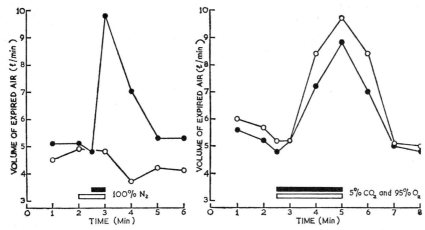

Fig. 45.15 The effect of denervation of the carotid sinuses on the respiratory response of an unanaesthetised dog to inhalation of nitrogen and of carbon dioxide. Both experiments were made on the same animal: ●, normal, ○, denervated. (Gemmill and Reeves.)

depression of respiration (Fig. 45.15). This latter effect is due to a central action. The effects of "denervation" of the carotid and aortic bodies on the ventilatory response to hypoxia are similar in man. It will be noted in Fig. 45.16 that after blocking transmission in the 9th and 10th cranial nerves with a local anaesthetic in a normal conscious human subject the hyperpnoea due to inhalation of 8 per cent O_2 in N_2 was completely abolished.

On the other hand, inhalation of gas mixtures containing small quantities of carbon dioxide, say 5 per cent, also causes hyperpnoea, but this response is unaffected by division of the carotid sinus nerves, as indicated in Fig. 45.15. The reflex mechanism, therefore, appears to play only a small part in the control of respiration by changes in arterial blood $P{CO_2}$, the main action being a direct one of carbon dioxide on the central medullary chemoreceptors.

The mechanism of the effect of carbon dioxide on respiration requires a little more explanation in the light of studies of action potentials in chemoreceptor fibres, which have shown that a small increase in impulse discharge occurs as the arterial blood $P{CO_2}$ is raised. How is it, therefore, that interruption of these impulses does not modify the *respiratory response* to hypercapnia? The position may be summarised by saying: (1) the chemoreceptor discharge produced by raising the arterial blood $P{CO_2}$ is relatively small compared to that caused by lowering the arterial blood $P{O_2}$ by inhalation of 7–10 per cent oxygen in nitrogen; (2) the main effect of carbon dioxide is on the central chemoreceptors, and this overshadows the reflex component; and (3) increasing the arterial blood $P{CO_2}$ of itself tends to diminish the effects of reflexes by an action on central synapses. Thus the small discharge of impulses from chemoreceptors evoked by inhalation of carbon dioxide become blocked, and therefore ineffective, somewhere along the reflex nervous pathway.

Role of the chemoreceptors in respiratory depression

Certain drugs, such as morphine and the barbiturates, depress the respiratory centres and the respiration becomes very much slower. There is an increase in depth, but this does not fully compensate for the decrease in frequency, and the respiratory minute volume is diminished. In consequence, there is a rise in the alveolar $P{CO_2}$ and in the arterial blood $P{CO_2}$, and a fall in the alveolar and blood $P{O_2}$. The sensitivity of the central chemoreceptors to carbon dioxide is reduced and it can be shown experimentally that if a subject is given morphine, his respiratory response to an increase in the carbon dioxide concentration in the inspired air becomes smaller than it was before.

In large doses, the sensitivity of respiration to carbon dioxide may be so depressed that breathing is maintained only by the chemoreceptor reflex mechanism stimulated by the reduction in arterial blood $P{O_2}$. In these circumstances, administration of oxygen may have the unexpected, and undesirable, effect of stopping respiration altogether, due to withdrawal of this reflex "drive" to the respiratory centres.

The part played by afferent impulses from the lungs in the control of breathing

Nerve endings are present in the bronchi and bronchioles which are sensitive to distension of the lungs, but are insensitive to changes in the partial pressures of oxygen and carbon dioxide. They afferent fibres run in the vagus nerves.

In 1868, Hering and Breuer showed that interruption of

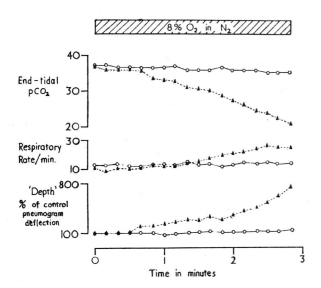

Fig. 45.16 The effects of "denervation" of the carotid and aortic bodies by bilateral block of the IX and X cranial nerves on the response to hypoxia in a healthy conscious human subject. Closed triangles (▲), control observations; open circles (○), during "denervation". (Guz et al. 1966.)

breathing by blocking the respiratory passages during inspiration and expiration had marked effects upon the pattern of breathing, but that when the vagus nerves were cut these effects were absent. A few years later, in 1889, Head again demonstrated this reflex (now known as the Hering-Breuer reflex), by observing the effect of inflation of the lungs on the movements of the diaphragm in the rabbit. The rabbit is unique in that its diaphragm is so arranged that it is possible to separate that part which is attached to the ensiform cartilage from the remainder, without damaging the blood supply or nervous connections. The active movements of this strip follow exactly those of the remainder of the diaphragm, and can easily be recorded without interference from movements of the chest or diaphragm as a whole, whether active or passive. The essence of Head's results is shown in the accompanying Fig. 45.17.

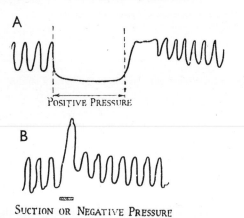

Fig. 45.17 Recordings of contractions of a slip from the diaphragm of a rabbit. (Upward trend of lever represents contraction of the slip.)
  A. Between the arrows a positive pressure was applied to the lungs. The contractions of the slip, and inspiratory movements, were inhibited.
  B. A "momentary" diminution in the volume of the lungs (suction) was produced during normal respiration, as indicated. An increased contraction of the slip, i.e. an increased inspiratory effort, resulted. (After Head.)

In the first tracing (A) a positive inflating pressure has been applied to the trachea, and it can be seen that while the pressure is applied the activity of the diaphragm is inhibited. In the lower figure (B) suction applied for a very short period causes an immediate contraction of the strip of the diaphragm. These effects were abolished when the vagus nerves were cut.

It can be shown that action potentials are set up in the vagus nerve by inflating the lungs, and that the rate of discharge of the impulses is roughly proportional to the degree of inflation. The frequency of impulses in a single vagal fibre from a cat during normal breathing waxes and wanes in rhythm with the inspiratory and expiratory phases of breathing. The impulses from the lungs fall to a minimum at the completion of expiration.

The function of these afferent impulses in the vagus nerve in normal breathing is to signal the depth of inspiration to the respiratory centres and allow expiration to take place, after an adequate tidal volume has ventilated the lungs. Double vagotomy in an otherwise intact animal removes this signal and thus the inspiratory phase is prolonged, and breathing becomes slower and deeper (see Figs 45.2 and 45.5); a normal respiratory minute volume, however, is maintained. Thus when the medullary inspiratory centre transmits its impulses to the inspiratory muscles, it is subjected, via the apneustic centre, to an afferent discharge from the lungs which progressively increases in intensity until it inhibits the inspiratory activity and initiates the expiratory phase, allowing the chest wall to relax in a passive manner. Consequently a cycle of (a) inspiratory activity and (b) progressive inspiratory inhibition takes place during normal quiet breathing. It has already been pointed out that when the brainstem is transected between the pneumotaxic and apneustic centres combined with division of both vagus nerves, rhythmic respiration ceases and apneustic type of breathing occurs. If such a brainstem section is made when the vagus nerves are intact, rhythmic breathing will continue, due to the part the vagus nerve plays in the cycle of events just described.

In experiments carried out on man at operation, it has been found that there are some differences in the Hering-Breuer reflex compared with animals. Inflation of human lungs inhibits inspiration and probably promotes expiration; this effect is abolished by blocking transmission in the vagus nerves. But there is absence of the inflation reflex activity within the normal tidal volume range so it appears that the Hering-Breuer reflex is weaker in man than in animals. Pulmonary stretch receptor activity is present in man at the functional residual capacity and increases with inspiration, as in animals.

TO SUMMARISE:

In *quiet breathing*, the respiratory centres initiate inspiratory activity. Afferent impulses in the vagus nerve signal the degree of expansion of the lungs. This afferent vagal discharge (which increases as inspiration progresses) inhibits the inspiratory activity and thus the respiratory muscles relax and expiration takes place. The Hering-Breuer reflex in man is weak.

Cortical control

Breathing is under voluntary control, but this control is limited to the extent that if voluntary changes in respiration are brought about which alter the chemical composition of the blood, then the latter exerts its effects on breathing which overcome those produced voluntarily. For instance, one can hold one's breath voluntarily but only for a maximum period of about 1 minute. Then one is forced to breathe due to rise in arterial blood $P\text{CO}_2$ and fall in arterial blood $P\text{O}_2$.

Reflexes from muscle and joint receptors

In experimental animals and in human subjects, passive movements of a limb result in an increase in pulmonary ventilation even when the circulation is cut off by a pressure cuff to prevent metabolic products from the muscles entering the circulation and hence reaching the respiratory centres. In man, it was found that passive movement of one leg at the knee 100 times per minute with the circulation occluded increased the respiratory minute volume by 40 per cent. Stimulation of such receptors may play an important part in the increased respiration occurring in muscular exercise.

Protective reflexes

Stimulation of cutaneous afferent nerves produces an increase in both rate and depth of respiration. Presumably, the afferent fibres which mediate the sensation of pain and temperature are mainly responsible for this increased breathing. Most of us have observed how a cold shower "takes our breath away".

Reflexes arising from stimulation of receptors in the respiratory tract are concerned with the protection of the tract itself. During *swallowing*, respiration is reflexly inhibited by impulses running in the glossopharyngeal nerve from the postpharyngeal wall.

Inhibition of breathing also occurs as a result of stimulation of mucous membranes of the nasal passages by irritant gases such as ether, chloroform and halothane, or by water. The typical effect is shown in Fig. 45.18. Slowing of the heart and peripheral

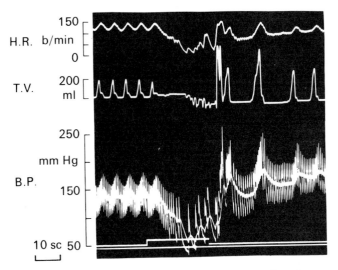

Fig. 45.18 The effects of stimulation of the nasal mucous membrane by water localised to the nose of an anaesthetised dog. HR, heart rate (beats/min); TV, tidal volume (inspiration upwards); BP, arterial blood pressure (phasic and mean). Note the apnoea, profound bradycardia and fall in blood pressure, and the post-stimulus hypertension.

vasoconstriction also occur. The sensory nerve endings of the trigeminal nerve are involved. In other instances, stimulation of these endings causes *sneezing*, a modified respiratory act. *Coughing* follows similar irritation of the mucous membranes of the pharynx, larynx, trachea and bronchi. The afferent pathway is in the glossopharyngeal nerve from the pharynx, and in the vagus nerve from the larynx, trachea and bronchi. The explosive quality of the act of coughing is the result of the initial closure of the glottis, which only opens after the beginning of the expiratory phase.

The influence of changes in blood pressure

A rise or fall in arterial blood pressure causes a diminution and an increase in respiratory minute volume respectively. By employing suitable experimental techniques, these changes in respiration have been shown to be due largely to reflexes from baroreceptors in the carotid sinus and arch of the aorta. Such reflexes probably play a minor role in the nervous control of breathing in the normal animal; they must not be confused, however, with the very important part played by the baroreceptors in the reflex regulation of the blood pressure.

CATECHOLAMINES

Adrenaline or noradrenaline infused intravenously in conscious man or anaesthetised animals in doses giving blood concentrations within the normal physiological range cause an increase in respiratory minute volume. Such doses have little or no effect on arterial blood pressure. The hyperventilation is due predominantly to a constrictor action of these amines on the carotid body vessels, diminishing carotid body blood flow and thereby stimulating the chemoreceptors by stagnant hypoxia.

Large doses (outside the normal physiological range) of adrenaline and noradrenaline cause apnoea in anaesthetised animals. This is largely reflex in nature due to stimulation, by the profound rise in arterial pressure, of the carotid sinus and aortic arch baroreceptors.

Other factors modifying breathing

There are many circumstances in which the normal pattern of breathing is modified and yet are quite unrelated to the need for an adequate alveolar ventilation. Some examples, viz. swallowing, sneezing and coughing, have already been discussed.

Speaking, singing or whistling all involve changes in breathing which interrupt the normal rhythmic pattern. The act of singing or speaking can only be carried out while air is being expelled from the lungs, so that at intervals it is necessary to take in air during which time there is no phonation.

Another procedure modifying breathing is Valsalva's manoeuvre. Experimentally this consists of voluntarily closing the glottis, or the mouth and nose together, and then attempting to expel the air from the lungs by contraction of the abdominal and thoracic expiratory muscles. The intrathoracic and intraabdominal pressures rise, sometimes to values exceeding 100 mmHg (13.3 kPa). In effect a similar manoeuvre is carried out during defaecation. These high intrathoracic pressures do not place any added stress on the airways or alveoli. The reason for this is that during the Valsalva manoeuvre the pressures in the airways and alveoli and the intrapleural pressure are all equal, so that the difference in pressure across the walls of these structures (the transmural pressure) remains the same. There are, however, profound effects on the circulatory system. The high

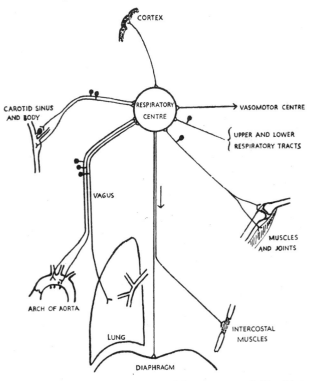

Fig. 45.19 Diagrammatic representation of the nervous control of breathing.

intrathoracic pressure reduces the venous return to the right side of the heart which in turn causes a fall in cardiac output and arterial blood pressure. This reduction in blood pressure is partly compensated by reflex peripheral vasoconstriction. On cessation of the Valsalva manoeuvre, there is a sudden rise in cardiac output against the increased peripheral vascular resistance so that the blood pressure temporarily "overshoots" above the control until it is restored by the baroreceptor mechanisms.

The various factors concerned in the nervous control of breathing are summarised diagrammatically in Fig. 45.19.

46. Examples of Respiratory Adjustments

M de B Daly

Hyperventilation

In hyperventilation, one form of which involves a subject voluntarily overbreathing, large volumes of air are moved in and out of the lungs with the result that carbon dioxide is washed out thereby reducing the alveolar CO_2 concentration and $P{CO_2}$. The fall in the alveolar $P{CO_2}$ increases the diffusion of carbon dioxide from blood to the lungs and so the arterial blood CO_2 content and $P{CO_2}$ fall as well.

The relationship between alveolar ventilation and alveolar $P{CO_2}$ (which is the same as the blood $P{CO_2}$) is shown in Fig. 46.1. It will be noted that at a normal alveolar $P{CO_2}$ of 40 mmHg (5·3 kPa) the alveolar ventilation is 4·5 l min^{-1}. By doubling ventilation, the alveolar $P{CO_2}$ falls to half this value (20 mmHg). It will also be seen that changes in alveolar $P{O_2}$ occur as well. Doubling alveolar ventilation increases the turnover in the alveoli of oxygen and consequently the $P{O_2}$ rises to a value closer to that of oxygen in atmospheric air (about 150 mmHg; 20 kPa). Referring again to Fig. 46.1, the alveolar $P{O_2}$ increases from 105 to 125 mmHg (14 to 16·6 kPa). The blood $P{O_2}$ will, of course, rise to a similar level, but in spite of this the arterial blood will carry very little extra oxygen. The oxygen saturation remains about the same (Fig. 46.1) and this is because in this range of $P{O_2}$, the oxyhaemoglobin dissociation curve is almost flat.

As a result of the lowering of the arterial blood $P{CO_2}$, the blood becomes more alkaline with a rise in pH (Fig. 46.1).

The above changes in alveolar and blood $P{O_2}$ and $P{CO_2}$ will occur during hyperventilation when unaccompanied by any change in the body's metabolism. We must distinguish this from hyperventilation which takes place during muscular exercise, and here respiratory adjustments are such that the increase in alveolar ventilation balances the metabolic need for oxygen. The

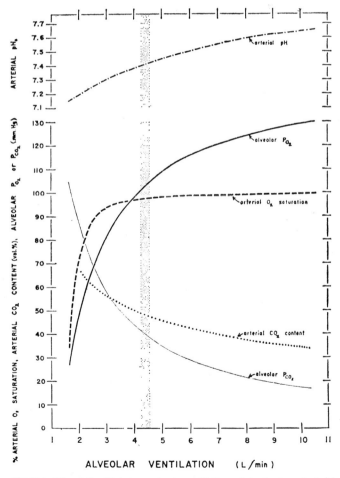

Fig. 46.1 The relationship between alveolar ventilation and alveolar gas and arterial blood O_2, CO_2 and pH. (Comroe *et al.*, 1962.)

increased volumes of oxygen and carbon dioxide diffusion across the alveolar walls results therefore in there being little or no change in the alveolar and blood PO_2 and PCO_2.

Hypoventilation

Figure 46.1 also shows what happens to the alveolar and blood PO_2 and PCO_2 during hypoventilation. This can be induced voluntarily for a limited period of time or may result from depression of the respiratory centres, e.g. by drugs. There is an increase in alveolar and blood PCO_2 and a reduction in PO_2.

Panting

An increase in body temperature causes rapid shallow breathing (*panting*) in many animals (dog, cat, sheep, rabbit and cattle) though not in man. The frequency of breathing rises to 200–300 breaths per min, but the tidal volume decreases so that the alveolar ventilation, and hence the alveolar and blood-gas tensions, remain almost constant. At these high respiratory frequencies, ventilation of the dead space becomes a large proportion of the total ventilation.

Breath holding

If a subject holds his breath after a normal expiration, he can maintain an apnoea for up to about 1 minute. He then reaches his "breaking point" at which the urge to breathe becomes dominant. During the apnoeic period the alveolar and blood PO_2 fall and the PCO_2 rises. The breath-hold time can be increased by an initial period of voluntary hyperventilation to "wash out" carbon dioxide, by breathing 100 per cent oxygen, or by a combination of the two. At the breaking point typical values for alveolar PO_2 and PCO_2 are 65 and 48 mmHg (8·7 and 6·4 kPa) respectively.

The question arises as to the mechanism of the "desire to breathe" sensation which occasionally can give rise to distressing symptoms. If a subject at his breaking point is asked to breathe, not room air, but a gas mixture which maintains his alveolar PO_2 and PCO_2 at the breaking-point level, the "desire to breathe" sensation is relieved in about two breaths. This indicates that the changes in blood-gas tensions, while probably contributing to the symptoms, are not entirely responsible for them. As a subject nears his breaking point during breath holding, there are increasing contractions of the diaphragm though not to the extent as to cause breathing until the breaking point is actually reached. These "frustrated" contractions of the diaphragm, as they are called, are engendered reflexly by afferent impulses running in the vagus nerves and are believed to be the origin of the "desire to breathe" sensation.

By contrast the sensation of "inability to get enough air" which occurs during the rebreathing manoeuvre, and is associated with accumulation of carbon dioxide, is completely different in quality. The major muscles of respiration are probably not involved in producing this sensation; it depends on the development of hyperpnoea and on the integrity of the vagus nerves, and receptors in the lungs are therefore probably involved.

THE HYPERPNOEA OF MUSCULAR EXERCISE

The respiratory minute volume increases with the demand for oxygen; in fact, there is a linear relationship between oxygen

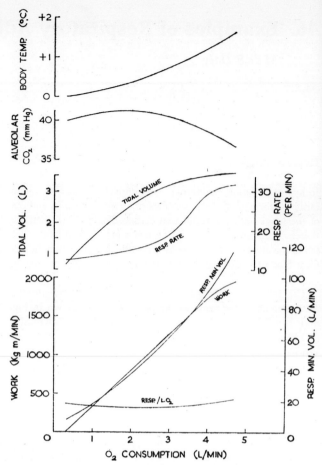

Fig. 46.2 The respiratory responses to exercise on a bicycle ergometer. All the data was obtained from the same experiment on six physically fit young men. (Data from Christensen, 1932.)

consumption and the volume of air breathed under steady-state conditions when the work is moderate in amount (Fig. 46.2). For any given amount of work, there are two phases in the response of respiration. There is an initial rapid increase in breathing which occurs before any metabolites could reach the respiratory centres and is probably due to stimuli of nervous origin (Fig. 46.3). This is followed by a slow rise in respiration to its final

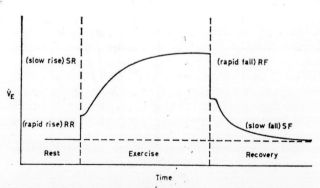

Fig. 46.3 Components of the ventilatory response to exercise (after Dejours, 1963). $\dot{V}_E$ is the respiratory minute volume, and RR and RF represent the neural components of the drive to respiration at the beginning and end of exercise. SF is due to the humoral component continuing to act during the period of recovery whilst SR is the result of interaction between the neural and humoral components. (Cotes, 1966.)

value which depends on the severity of the exercise. Maximum values for the respiratory minute volume are of the order of 80 to 120 l min^{-1}.

This second phase is difficult to explain. Ever since Haldane stressed the importance of the chemical control of respiration by carbon dioxide, much attention has been given to its possible role in the production of the hyperpnoea of muscular exercise. The increase in ventilation has been attributed to the rise in $P\text{CO}_2$ and reduction in pH of the arterial blood due to the increased oxidation and to the accumulation of lactic acid. Such a view is untenable, however, because the hyperpnoea usually occurs without any demonstrable alteration in the arterial blood $P\text{O}_2$ or $P\text{CO}_2$. Furthermore, the maximum ventilatory response to increasing the inspired carbon dioxide concentration in the resting subject is considerably smaller than that produced by severe exercise. Nevertheless, there is strong evidence that this second phase is due to humoral factors.

The possible mechanisms responsible for the hyperpnoea of muscular exercise may be classified under two headings:
 Neural component
 Humoral component.

1. Neural component

The immediate rapid increase in ventilation and the rapid fall on cessation of exercise (Fig. 46.3) are almost certainly of neural origin as they occur too quickly to be accounted for by a blood-borne agent.

Mechanoreceptors in muscles and joints. Passive movements of a limb in man may increase pulmonary ventilation by 40 per cent due to stimulation of receptors in the limb. This could therefore be a contributory mechanism increasing respiration in exercise.

Central irradiation. Impulses arising from higher centres to the respiratory centres has been suggested as a cause for the hyperpnoea. In subjects exercising after hypnosis and suggestion that it was going to be pleasant and effortless, it was found that they ventilated less compared with the control value before hypnosis. The alveolar $P\text{CO}_2$ was about 8 mmHg (1 kPa) higher, indicating that the lower ventilation was not due to a reduction in the level of metabolism.

2. Humoral component

Carbon dioxide. It might appear, because the alveolar $P\text{CO}_2$ does not alter during exercise, that carbon dioxide is not concerned in the hyperpnoea. But recent evidence suggests that the respiratory response to inhalation of small quantities of carbon dioxide is greater at rest immediately after a period of exercise than at rest before exercise. The threshold of the central chemoreceptors for carbon dioxide may in some way be reduced during exercise with the result that at the normal level of arterial blood $P\text{CO}_2$, the respiratory minute volume is elevated.

Low oxygen. Blood returning from exercising limbs is much more venous than normal so that the mixed venous blood in the right atrium and pulmonary artery has a lower $P\text{O}_2$ than normal and a higher $P\text{CO}_2$. The possibility that there might be chemoreceptors in the veins or pulmonary circulation responding to a low $P\text{O}_2$ or high $P\text{CO}_2$ has been considered but there is little evidence for such chemoreceptors.

Considering now the $P\text{O}_2$ in *arterial* blood: there is general agreement that this does not alter appreciably, at least during mild and moderate exercise, and certainly the changes, as small as they are, are not sufficient to stimulate the arterial chemoreceptors in the carotid and aortic bodies.

Taking the evidence as a whole, there is still very little to indicate that in exercise a lowering of the arterial $P\text{O}_2$, which is confined to venous blood, is a cause of the hyperpnoea. Yet it must be borne in mind in this connection that the respiratory minute volume is very closely related to oxygen consumption and moreover, when, during exercise, oxygen is substituted for atmospheric air, there is a reduction in breathing. The possibility that low oxygen is in some way a stimulus to breathing in muscular exercise cannot yet be dismissed too lightly.

Hydrogen ion concentration. There is now good evidence that there is a small increase in hydrogen ion concentration of arterial blood during exercise. Since this is not associated with any appreciable change in arterial $P\text{CO}_2$, it must represent a change in the amount of fixed acid in the blood (metabolic acidosis). It has been calculated that the observed changes in arterial pH may account for up to 60 per cent of the exercise hyperventilation, the site of action being probably the central and peripheral chemoreceptors.

Body temperature. When the temperature of the body at rest is raised artificially, pulmonary ventilation is increased. The body temperature also increases during exercise. Experimentally it has been shown that a steady raised body temperature increases the sensitivity of respiration to carbon dioxide and lowers the threshold (Fig 45.10). The hyperpnoea of muscular exercise may therefore be due, at least in part, to an effect of the rise in body temperature causing an increase in the sensitivity to carbon dioxide.

Other humoral agents. Evidence has been obtained for the release of a humoral agent from exercising muscle, which on gaining access to the carotid circulation causes stimulation of breathing. The nature of this agent is not known.

Catecholamines (adrenaline and noradrenaline) released from the suprarenal medulla may also be involved as they are known to stimulate breathing in physiological doses through an action on the carotid bodies (see Fig. 45.0).

Pulmonary ventilation is adjusted closely to the metabolic needs of the body so as to maintain a perfect, or near perfect, homeostasis. In all probability, this is brought about not by any one single factor, but by a combination of several contributory causes.

HYPOXIA

Whenever for any reason the cells of the body do not have, or are unable to utilise, sufficient oxygen to carry on normal function, they are said to be suffering from an oxygen deficiency or hypoxia. Hypoxia may be classified into four types: hypoxic, anaemic, stagnant and histotoxic.

1. *Hypoxic hypoxia* occurs when there is defective oxygenation of the blood in the lungs causing a reduction in the arterial oxygen pressure. This may result from two causes: (a) reduction in the inspired air $P\text{O}_2$ such as occurs from the addition of an inert gas such as nitrogen or to a fall in the total atmospheric pressure associated with ascents to high altitude; (b) reduction in respiratory minute volume through depression of the respiratory centres by various drugs including anaesthetics, or paralysis of the respiratory muscles.

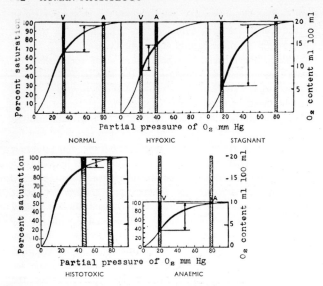

Fig. 46.4 Diagram illustrating the oxygen dissociation curves in various types of hypoxia. The vertical columns, representing the arterial (A) and venous (V) blood, indicate the amount of oxygenated haemoglobin (shaded portion) and reduced haemoglobin (black portion). The perpendicular arrows denote the volume of oxygen per unit volume of blood delivered to the tissues. (See also text.)

Under normal conditions, as shown in Fig. 46.4, the blood leaves the lungs about 97 per cent saturated with oxygen and reaches the tissues containing about 19.5 volumes per cent (i.e. 195 ml l^{-1} blood) of oxygen (at a pressure of 80 mmHg; 10.6 kPa). Here about 5 volumes per cent (50 ml l^{-1}) are abstracted during rest, so that mixed venous blood contains about 14 volumes per cent (140 ml l^{-1}) of oxygen (around 70 per cent saturated and at a pressure of 35 mmHg; 4.7 kPa). The gradient between arterial and capillary P_{O_2} under normal conditions is therefore about 45 mmHg (80 minus 35). During hypoxia when the arterial oxygen saturation is, say, 75 per cent (P_{O_2} 40 mmHg; 5.3 kPa), there is still about 15 volumes per cent of oxygen available to the tissues. Since the latter require 5 volumes per cent, the remaining 10 volumes in venous blood is held at a relatively low P_{O_2} of about 25 mmHg (3.3 kPa). This results in an arterial–capillary P_{O_2} gradient of only 15 mmHg (2.0 kPa) and a lowering of the oxygen pressure at which the cells have to metabolise.

2. *Anaemic hypoxia.* This type is caused by haemorrhage or anaemia. As will be seen in Fig. 46.4, the arterial P_{O_2} is normal and the haemoglobin is 97 per cent saturated. But because the amount of haemoglobin per unit volume of blood is considerably reduced, the oxygen content of the blood is diminished by a similar proportion. A large part of the oxygen supply to the tissues must therefore be delivered at a lower pressure than normal unless the rate of the bloodflow is increased.

3. *Stagnant Hypoxia* occurs in conditions in which the circulation rate through the tissues is slowed. Although the oxygen content, haemoglobin saturation and oxygen pressure of the arterial blood are normal (Fig. 46.4), the venous blood is considerably more reduced than usual resulting in a low tissue oxygen pressure. This is simply due to the blood spending longer time in the capillaries and, in consequence, a large volume of oxygen per unit volume of blood is extracted.

4. *Histotoxic hypoxia* occurs when the respiratory enzyme systems in the tissues are poisoned, e.g. by cyanide. The cells are, therefore, unable to utilise the oxygen carried to them and, as a result, the venous blood has a high oxygen content (Fig. 46.4).

Acute hypoxia

We have seen that when the inspired oxygen concentration is lowered, hyperventilation results, though little change in breathing occurs until the inspired concentration falls to about 14 per cent (equivalent to 100 mmHg; 13.3 kPa) and this corresponds to an alveolar P_{O_2} of about 60 mmHg (8 kPa). When the alveolar P_{O_2} falls below 60 mmHg, the respiratory minute volume increases and this is due entirely to a reflex drive from the carotid and aortic chemoreceptors. However, the chemoreceptors are much more sensitive to a decrease in arterial blood P_{O_2} than is indicated by the change in respiratory activity disclosed by measurement of the respiratory minute volume. Whereas the respiratory response to inhalation of gas mixtures of lower oxygen content does not begin until the arterial blood P_{O_2} falls to 60 mmHg (8 kPa; arterial oxygen saturation of 90 per cent), the electrical activity in chemoreceptor fibres increases with arterial oxygen tensions below 100 mmHg. This is due to the fact that the respiratory response evoked by the chemoreceptor drive is antagonised by three opposing factors: (1) a direct depressant effect of hypoxia on respiration by a central mechanism; (2) depression of the central chemoreceptors due to a lowering of the arterial blood P_{CO_2} by washing out of carbon dioxide through the reflex hyperpnoea; and (3) alkalosis resulting from an increased amount of reduced haemoglobin. Partially reduced haemoglobin is able to mop up more hydrogen ions than is oxyhaemoglobin and the effect of the slight alkalotic change in the blood antagonises the chemoreceptor drive. Thus, the respiratory response to hypoxia would be much greater if these three antagonistic mechanisms could be prevented. We can illustrate the effect of maintaining the alveolar P_{CO_2} constant during hypoxia by taking the following example. A human subject breathing room air has a resting ventilation of 9 l min^{-1} and an alveolar P_{CO_2} of 36 mmHg (see lower part of curve A in Fig. 46.5). He is then made hypoxic by lowering his alveolar P_{O_2} to about 47 mmHg and, as a result, his respiratory minute volume increased to 13 l min^{-1} and his alveolar P_{CO_2} fell to about 27 mmHg (see lower part of curve B). His alveolar P_{CO_2} was then raised by adding carbon dioxide to his inspired air, the alveolar P_{O_2} being maintained at the same level as before. It was found that his breathing was unaffected until the alveolar P_{CO_2} was artificially raised to 30–33 mmHg, then with further increases in alveolar P_{CO_2} the respiratory minute volume increased linearly (curve B). When the alveolar P_{CO_2} had reached its control level of 36 mmHg, the respiratory minute volume was 36 l min^{-1}. This experiment demonstrates that the ventilatory response to hypoxia is much greater if the concomitant reduction in alveolar P_{CO_2} is prevented by administering carbon dioxide.

Figure 46.5 shows one other interesting phenomenon. Curves A, B and C represent the relationship between respiratory minute volume and the alveolar P_{CO_2} at three different levels of alveolar P_{O_2}, 169 and 110, 47 and 37 mmHg respectively. It will be noted that the slopes of curves B and C, in acute hypoxia, are greater than that of curve A, in normal conditions of air breathing. This means that the sensitivity of the respiratory

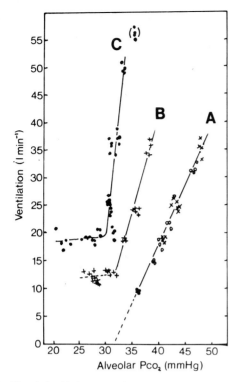

Fig. 46.5 The relationship between pulmonary ventilation and alveolar $P\text{CO}_2$ (37°C; prevailing barometric pressure saturated). Alveolar $P\text{O}_2$: (Curve A) ×, 168.7 ± 2.1 mmHg; ○, 110.3 ± 1.9 mmHg. (Curve B) +, 47.2 ± 1.5 mmHg. (Curve C) ●, 36.9 ± 1.3 mmHg. (Nielson and Smith, 1952.)

mechanism to carbon dioxide is greater during acute hypoxia than during air breathing. In other words, the effects of oxygen and carbon dioxide excess are not simply additive, but there is a positive interaction between them, and this is of some physiological importance in asphyxia when hypoxia and hypercapnia occur together.

Chronic hypoxia

At high altitudes, the percentage of oxygen in the atmosphere is the same as at sea-level, but since the barometric pressure is reduced, the partial pressure of oxygen is diminished to a comparable degree. The effects on man of oxygen deficiency depend not only on the altitude to which he ascends, but also on the rate of ascent.

THE MOUNTAIN CLIMBER

The first signs and symptoms of oxygen lack occur when the healthy mountaineer ascends slowly to about 12 000 ft (3700 m) above sea-level, corresponding to a pressure of 480 mmHg (63.8 kPa). They are headache, nausea, vomiting and a feeling of lassitude and are regarded as characteristic of "mountain sickness". Mental features such as a feeling of well-being, exhilaration, talkativeness and sometimes emotional outbursts of laughing or crying, and development of fixed ideas may be evident. These effects wear off as the climber becomes "acclimatised" after a few days. The altitude to which man may climb varies with the individual, but without the additional use of oxygen is in the region of 28 000 ft (8600 m) (close to the summit of Mount Everest). The limit is determined by the pressure of oxygen in his alveoli.

It is found that with increasing altitude, the pulmonary ventilation increases and there is a reduction in both the alveolar $P\text{O}_2$ and $P\text{CO}_2$ (Fig. 46.6). The augmented pulmonary ventilation is the result of stimulation of the peripheral arterial chemoreceptors by the lowered arterial blood $P\text{O}_2$, and not of the diminished barometric pressure *per se*. The hyperpnoea in turn causes the observed fall in alveolar $P\text{CO}_2$, the increased elimination of carbon dioxide leading at first to an increase in alkalinity of the blood and to the excretion of an alkaline urine.

Acclimatisation to hypoxia

After a period of time at high altitude, processes of adaptation come into operation which improve the delivery of oxygen to the tissues and enable the mountaineer to withstand the effects of hypoxia better than would otherwise be the case. The term acclimatisation is given to these processes of adaptation. This is well illustrated by the fact that climbers can attain heights of about 29 000 ft (8900 m); in a relatively fast ascent, such as that made in 1875 by Tissandier and his two colleagues in a balloon to the same height, two of the party succumbed.

During acclimatisation several changes occur:

1. *There is an increase in the number of red blood cells and haemoglobin* due to a stimulant action of hypoxia on the blood-forming organs. After several weeks' duration at high altitude, values up to 8 million red cells mm^{-3} and 21 g haemoglobin per decilitre of blood have been recorded. The increased amount of haemoglobin raises the oxygen capacity of the blood so that at any given $P\text{O}_2$ the arterial blood will contain more oxygen. The main advantage of this to the organism is that more oxygen is given up to the tissues for a given fall in oxygen saturation and $P\text{O}_2$. This means that the tissue $P\text{O}_2$ will be higher than would otherwise be the case.

2. *The increase in respiratory minute volume* is maintained in spite of the persistent low $P\text{CO}_2$. The pH of the blood, however, is partially or wholly restored by the renal excretion of bicarbonate. The question arises as to how the increased ventilation is maintained because administration of oxygen to an acclimatised subject at altitude causes only a slight reduction in breathing and increase in alveolar $P\text{CO}_2$, the latter still remaining well below the sea level value of about 40 mmHg (5.3 kPa). This is in contrast to the effect of giving oxygen in *acute* hypoxia, when the respiratory minute volume is restored to normal. It is apparent therefore that when the hypoxic stimulus is removed in the acclimatised subject by giving 100 per cent oxygen, respiration remains "driven" by an alveolar, and hence arterial blood, $P\text{CO}_2$ which is considerably lower than normal under sea-level conditions. It is also of interest, in this connection, that if an acclimatised subject is brought down quickly to sea-level, his respiratory minute volume is higher than his control value observed before exposure to chronic hypoxia.

These findings suggest that the chemical threshold of the central chemoreceptors becomes altered during acclimatisation. It is believed that at altitude the hypoxia at first excites the peripheral arterial chemoreceptors which, by stimulating breathing reflexly, reduces the arterial blood $P\text{CO}_2$. Carbon dioxide being readily diffusible, the cerebrospinal fluid $P\text{CO}_2$ is quickly reduced to a similar extent. But the cerebrospinal fluid contains no buffers and so its pH rises. However, the active transport of bicarbonate out of the cerebrospinal fluid restores a normal $HCO_3^-/P\text{CO}_2$ ratio. From the point of view of the central chemoreceptors (or H$^+$ receptors) this reduction in cerebrospinal fluid HCO_3^- permits ventilation to increase and the $P\text{CO}_2$ to fall

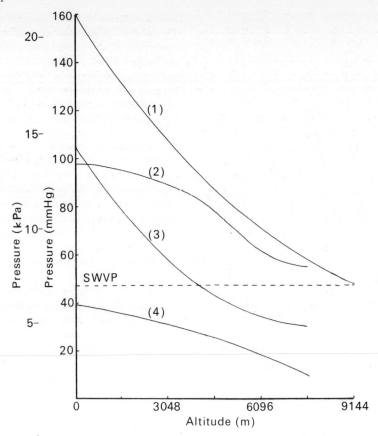

Fig. 46.6 Curves showing (1) the atmospheric Po_2, (2) the arterial oxygen saturation, (3) the alveolar Po_2, and (4) the alveolar Pco_2; at all altitudes between sea-level and 30 000 ft (9200 m). (After Grow and Armstrong.)

during the sojourn at altitude more than during administration of gases of comparable inspired oxygen tension acutely at sea-level. The threshold of the central chemoreceptors to carbon dioxide is lowered therefore during the process of acclimatisation; the sensitivity to carbon dioxide, however, is unchanged.

These adjustments in the pH of the cerebrospinal fluid occur more quickly than those in the pH of arterial blood through renal compensation.

3. *The increases in heart rate and cardiac output* associated with acute hypoxia are well maintained during acclimatisation. The increased bloodflow in tissues is another important adaptive mechanism helping to compensate for the reduction in tissue Po_2 consequent upon the fall in arterial Po_2. Redistribution of circulating blood also takes place, so that the more vital tissues will receive a priority of supply.

The above mechanisms are involved in helping to maintain the internal environment of the tissue cells as close as possible to that at sea-level. In this connection it is of interest that at high altitude there is a marked reduction in the gradient between the Po_2 of the ambient air and the Po_2 of the arterial blood. At sea-level, the gradient is 159−100 = 59 mmHg (7·9 kPa): at 20 000 ft (6100 m), for instance, it is 74−40 = 34 mmHg (4·6 kPa). There is a corresponding reduction in the gradient between the ambient air Po_2 and the mean capillary blood Po_2.

Hypoxia in Flying

It has already been pointed out that the climber who ascends slowly and has time in which to become acclimatised has been able to reach heights of 28 000 ft (8600 m). When an aircraft pilot climbs rapidly, there is no time for these processes to take place and it is, therefore, impossible for him to reach such altitudes without the use of oxygen. Should he expose himself to the atmosphere at 25 000 ft (7600 m) for as long as 10 minutes he is likely to die.

The effects of oxygen deficiency first show themselves at about 5000 ft (1500 m) by increased breathing. Above 12 000 ft (3700 m), mental and physical functions are impaired and over-confidence develops. At 18 000 ft (5500 m), circulatory changes occur, producing an increase in pulse-rate and blood pressure. The senses of touch, pain, vision and hearing are impaired. At these altitudes the flyer may observe how much brighter the day appears and how much louder the engines sound when he breathes from his oxygen mask. Between the heights of 18 000 and 30 000 ft (5500 and 9200 m), unconsciousness may occur—which is sudden in onset and without warning—followed by paralysis and death.

The effects of oxygen deficiency are affected by bodily activity. During quiet walking oxygen consumption may be three times that of the resting individual, and the pilot should, therefore, reduce his movements to a minimum.

The changes that occur in the pressures of the lung gases in the high altitude flyer can be seen in Fig. 46.6. The alveolar air pressures shown were taken at heights up to 25 000 ft (7600 m), under conditions where oxygen was not inhaled; 25 000 ft (7600 m) is the highest altitude to which a subject can fly without the use of oxygen. The top curve represents the P_{O_2} in the atmosphere; the next curve represents the oxygen saturation of the blood in the lungs; and the third curve represents the alveolar P_{O_2}. This last curve does not follow a course parallel to that of the atmospheric P_{O_2} because of the hyperventilation caused by the low oxygen pressure; for the same reason the alveolar P_{O_2} (bottom curve) falls, instead of remaining constant. The horizontal broken line represents the water vapour pressure in the lungs which remains constant at 47 mmHg (6·3 kPa).

If the effects of hypoxia are to be avoided, it is essential that pilots breathe 100 per cent oxygen above about 10 000 ft (3000 m). This will maintain an adequate alveolar and arterial P_{O_2} up to altitudes of about 40 000 ft (12 200 m). At this height the barometric pressure is 140 mmHg (18·6 kPa), the aqueous vapour pressure is 47 mmHg, and the alveolar P_{CO_2}, say, 30 mmHg, i.e. lower than normal due to hyperventilation. The alveolar P_{O_2}, therefore, must be: $(140 - 47 - 30) = 63$ mmHg (8·4 kPa). This is an oxygen pressure just sufficient to maintain the pilot orientated and in control of his aircraft, but at altitudes above 40 000 ft (12 000 m) it is necessary to raise the pressure by the use either of a sealed flying-suit filled with oxygen under slight pressure, or of a sealed cockpit also under pressure. Using sealed cockpits, oxygen insufficiency ceases to be a limiting factor in flying at high altitudes and in spacecraft.

The rarefied atmosphere at high altitudes produces certain other effects upon the body apart from the respiratory effects due to the low oxygen pressure. Speech has a nasal quality, which is partly due to the inability of the rarefied atmosphere to vibrate the vocal cords. Foreign bodies can only be expelled with difficulty, and a cough fails to dislodge particles from the respiratory mucous membranes, so that the cause of the coughing remains and pilots under these circumstances are subjected to continuous irritation. The low barometric pressure causes distention of abdominal organs, which contain gas; this may cause pain and discomfort. Severe frontal sinus pain may develop for the same reason.

CYANOSIS

Cyanosis may be defined as the bluish colouration imparted to the skin and mucous membranes by the presence of reduced haemoglobin in the blood of the superficial blood vessels. It depends on the absolute amount of reduced haemoglobin in the capillary blood and not on the relative proportions of reduced haemoglobin and oxyhaemoglobin. It has been found that 50 g l^{-1} of reduced haemoglobin in capillary blood, or an oxygen unsaturation of 67 ml l^{-1} is about the threshold level at which cyanosis appears. An anaemic subject, therefore, who has a total of less than 50 g of haemoglobin per litre—i.e. an oxygen capacity of less than 67 ml l^{-1}—does not ordinarily become cyanotic. Thus the presence of cyanosis means that the tissues are hypoxic, although the absence of cyanosis does not necessarily indicate that there is no hypoxia.

The *causes* of cyanosis are numerous and are most important from the standpoint of diagnosis and treatment of the patient.

1. *Conditions in which the alveolar P_{O_2} is reduced.* We have already discussed the problem concerning the mountaineer and the aviator at high altitude where the lowered inspired P_{O_2} causes hypoxia which may be severe enough to produce cyanosis through a diminished alveolar P_{O_2}. Another cause is a reduced respiratory minute volume due either to failure of the respiratory mechanism or to obstruction of the airway. The respiratory centre is depressed, for instance, by overdose of sedative drugs; peripheral respiratory failure occurs as a result of degeneration of the lower motor neurones to the muscles of respiration, as in poliomyelitis. Obstruction may be caused by foreign bodies lodging in the trachea or bronchi. In the case of complete obstruction of a bronchus to one lobe, absorption of air behind the obstruction gradually takes place leading to *collapse* of the lobe. The blood passing through the lobe will not therefore be oxygenated. *Asthma* is another form of partial obstruction of the airways so that the movement of air in and out of the lungs is hindered by contraction of the bronchial muscle.

2. *Conditions in which there is impaired diffusion across the alveolar membrane.* This occurs in pathological states in which the alveolar membrane is abnormally thick due to inflammatory processes. The presence of oedema fluid in the alveoli also hinders diffusion and, in consequence, increases the gradient of the oxygen pressure between alveolar air and arterial blood. A similar state of affairs occurs in *emphysema* in which there is destruction of many of the septa between alveoli causing a reduction in the area available for gas exchange.

3. *Conditions in which there is abnormal reduction of haemoglobin in the systemic capillaries.* This may be due either to an increase in oxygen utilisation by the tissue or to a decrease in blood flow through it. The latter is the more usual cause of cyanosis in this group and gives rise to stagnant hypoxia. It may occur *locally* as a result of intense vasoconstriction through cooling a part of the body, or it may be *generalised* as in advanced heart failure and other conditions in which the peripheral circulation is abnormally sluggish.

4. *Conditions in which there is an abnormal mixture of arterial and venous blood.* In many forms of congenital heart disease, there is cyanosis due in part to direct mixture of venous with arterial blood through abnormal communications between the two sides of the heart. In other conditions which have already been referred to, such as obstruction to the air passages involving a part or whole of one lung, arterial blood will be a mixture of oxygenated and venous blood if the pulmonary blood flow through these areas is maintained. In point of fact, the pulmonary circulation gradually closes down in the collapsed part of the lung thereby reducing the volume of venous blood mixing with arterial blood and hence diminishing the cyanosis.

5. *Conditions in which there is alteration in the haemoglobin.* A number of drugs and poisons are capable of converting the iron in the haemoglobin from the di- to the trivalent form causing the formation of *methaemoglobin*. This is dark in colour and so gives rise to cyanosis. In contrast with the conditions giving rise to cyanosis enumerated in (1), (2) and (4) above, the arterial P_{O_2} in methaemoglobinaemia is normal and, in consequence, there will be no stimulation of respiration reflexly through the chemoreceptors. The oxygen-carrying power of the blood is, however, reduced in proportion to the amount of methaemoglobin present. *Carbon monoxide poisoning* may also be mentioned here, since there is a similar loss of oxygen-carrying power of the blood. There is, however, no cyanosis, since carboxyhaemoglobin is red in colour. Carbon monoxide

combines with haemoglobin with an affinity about 250 times that of oxygen: 0·1 per cent in the air produces a concentration of 150 to 200 ml per litre in the blood. Very small concentrations in the air are therefore sufficient to produce severe symptoms.

THE THERAPEUTIC VALUE OF OXYGEN

In treating certain forms of hypoxia, the administration of high concentrations of oxygen is a measure of the utmost value. By increasing the alveolar PO_2 the haemoglobin saturation of the blood leaving the pulmonary capillaries will be increased in conditions such as those in which the alveolar PO_2 is reduced and in which there is impaired diffusion across the alveolar membrane. There is relief of the cyanosis and a lessening of the dyspnoea associated with these conditions. Oxygen is also of extreme value in carbon monoxide poisoning for the high oxygen pressure displaces carbon monoxide from the blood.

In other forms of hypoxia, oxygen therapy is of less value. In anaemic hypoxia, for instance, the blood leaves the lungs with its haemoglobin fully saturated with oxygen, so the only benefit would be derived from the increased amount of oxygen carried in physical solution in the plasma. For a similar reason, oxygen is usually of less use in stagnant hypoxia.

Hyperbaric oxygen therapy

Attempts are now being made to treat certain conditions in which the tissues are hypoxic, by increasing the quantity of oxygen in the blood higher than can be achieved by administering 100 per cent oxygen. This is done by increasing the ambient pressure to 2 atm (200 kPa). The patient is put in a cylindrical chamber, sometimes large enough to accommodate an operating team as well, and is given oxygen to breathe. The quantity of oxygen dissolved in the plasma increases linearly with the arterial PO_2 by 0·003 ml oxygen mmHg^{-1} PO_2 100 ml^{-1} blood (0·23 ml kPa^{-1} l^{-1}); breathing 100 per cent oxygen at 1 atm pressure (alveolar PO_2 of 673 mmHg; 89·5 kPa) it increases to 20 ml l^{-1} blood, while breathing oxygen at 2 atm it is about 42 ml l^{-1} blood at an arterial PO_2 of about 1400 mmHg (186 kPa). The arterial oxygen content, being the volume of oxygen chemically combined with haemoglobin plus that dissolved in the plasma, is 195 + 42 = 237 ml l^{-1} blood breathing 2 atm of oxygen compared to 195 + 3 = 198 ml l^{-1} blood in the patient breathing room air at one atmosphere pressure. This method is proving beneficial in the treatment of coronary thrombosis, various forms of peripheral vascular disease and coal-gas poisoning.

Respiration at high atmospheric pressures

When a man exposes himself to air at high pressures, he may develop certain symptoms either during exposure to the increased pressure or when he has returned to normal atmospheric pressure.

Increased air pressure is met in deep-sea diving, in caissons and in submarine escape apparatus (the pressure within the submarine is at 1 atm). Every 33 ft of seawater means an additional pressure of 1 atm (100 kPa), so that as a diver exposes himself to increased pressure the volume of respiratory gases *dissolved* in his blood plasma and tissues increases in proportion to the raised partial pressures of these gases in the alveolar air (Henry's law). Thus at 2 atm pressure, twice as much gas will be dissolved in his blood and tissues as at 1 atm. The gases we have to consider are oxygen and nitrogen; the alveolar PCO_2 remains almost constant, and so carbon dioxide is not a gas which presents serious problems in this connection.

Nitrogen

The dangers of nitrogen under increased pressure arise from two facts: (1) nitrogen diffuses relatively slowly through living membranes and (2) it is about five times as soluble in fat as in water. Consequently, not only does it take a considerable time for the body to absorb the extra nitrogen at any given high atmospheric pressure, but the elimination of the extra nitrogen when decompression occurs is also prolonged. If a diver ascends too quickly to a lower pressure, the nitrogen may come out of solution and form bubbles of gas in his blood and tissues, in particular in the central nervous system on account of its high fat content. These bubbles are apt to lodge in capillaries and obstruct the flow of blood, giving rise to localised symptoms due to asphyxia and to distension of the tissues. These symptoms are pains in the muscles and joints, loss of cutaneous sensation and paralysis through involvement of the central nervous system ("caisson disease", "bends", "diver's palsy"). The severity of the symptoms depends on the pressure attained, the length of time spent by the subject at that pressure and on the rate of ascent. In severe cases, death may result from multiple air emboli in the heart and brain.

In order to prevent sudden evolution of gas, divers must ascend slowly, so that the tissues have time to get rid of their excess nitrogen via the lungs, without the formation of bubbles. Treatment of a case of caisson disease involves recompression in a pressure chamber to dissolve the gas bubbles, followed by slow decompression to atmospheric pressure.

Another method of preventing the formation of nitrogen bubbles is to replace atmospheric nitrogen by helium, since helium is an inert gas and less soluble in fat than in nitrogen. In practice, the diver is given a helium–oxygen gas mixture to breathe. This has another advantage, namely, it lessens "nitrogen narcosis", a condition associated with the onset of euphoria, hilarity, impaired mental activity and an increased difficulty in concentrating on and performing an allotted task. These effects, though they are very slight, are observed when air is breathed at 3 atm pressure; they begin to handicap the subject at 4 atm, and may rapidly make a man helpless at 10 atm (one mega Pascal, 1 MPa).

A similar condition of bends may occur in occupants of high-flying aircraft. Machines which are intended to fly at high altitude are fitted with pressurised cabins in which the pressure is maintained at an equivalent altitude of about 8000 ft (2440 m) (barometric pressure 560 mmHg; 74·5 kPa). This enables an adequate oxygen pressure to be maintained in the lungs of the crew and passengers. Should an accident occur, resulting in the pressure in the cabin being suddenly reduced to the ambient pressure, nitrogen bubbles may form in the blood, and cause caisson disease.

Oxygen

At sea-level, gas mixtures containing up to 60 per cent of oxygen may be inhaled without danger. Pure oxygen (100 per cent) breathed for more than 24 hours may produce mental dullness and evidence of pulmonary congestion. Newborn infants are particularly susceptible to oxygen poisoning: there is proliferation of the retinal vessels into the vitreous humour with excess formation of fibrous tissue (retrolental fibroplasia), which may lead to permanent blindness. When oxygen is to be used in infants for resuscitation purposes, the concentration should be limited to 40 per cent in the inspired air.

Inhalation of pure oxygen at 4 atm pressure produces signs and symptoms of oxygen poisoning which include convulsions, unconsciousness and a fall of blood pressure. These were thought to be due to the increased amount of oxygen dissolved in the blood, so that when the blood passes through the tissues, less oxygen is lost from combination with haemoglobin. As we have seen in Chapter 13, haemoglobin becomes more acid when it combines with oxygen and is, therefore, less ready to surrender base for carbon dioxide combination. This results in a reduction in the amount of carbon dioxide removed from the tissues so that the effects of oxygen excess may be due to accumulation of carbon dioxide in the tissues. If this is the correct explanation, the carbon dioxide pressure should rise in the venous blood. More recent work indicates that, at these high oxygen pressures, the carbon dioxide pressure in cerebral venous blood, and hence in brain tissue, does rise, but that the rise is small and can in fact be excluded as an important contributing cause of oxygen poisoning in man. A direct toxic effect of oxygen on tissue enzyme systems, in particular those containing sulphydryl groups, is the most likely explanation.

Oxygen poisoning can be a danger in diving unless the correct gas mixture is breathed compatible with the depth of dive, for as the ambient pressure increases, so does the oxygen pressure in a given mixture of gases. The depth to which a diver may go when breathing a gas will depend on its oxygen concentration as follows:

100% O_2	Dive limited to 26 ft (8 m)
60% O_2	Dive limited to 69 ft (21 m)
40% O_2	Dive limited to 120 ft (37 m)
21% O_2	Dive limited to 260 ft (80 m)

It may be seen from this that in dives greater than 260 ft (80 m) the subject must progressively decrease his inspired oxygen concentration below 21%. Thus at a depth of 1000 ft (305 m), where the ambient pressure is about 31 atm, or 23 500 mmHg (3·125 MPa), a gas mixture containing as little as 0·7% oxygen would give the diver an inspired P_{O_2} of 165 mmHg.

47. Artificial Respiration and Cardiac Resuscitation

M. de B. Daly

Artificial respiration

There are many circumstances in which respiratory movements cease temporarily. Death will follow from asphyxia unless fresh oxygen can be supplied, and the excess carbon dioxide washed out, by some means of artificial ventilation of the lungs. The cause of the respiratory failure can often be removed in this way, and complete recovery results. Common instances of such circumstances are drowning, electric shock, asphyxia from smoke, etc., in fires, carbon monoxide poisoning, certain diseases of the central nervous system which result in paralysis of the respiratory muscles, and the failure to breathe in the newborn.

In patients undergoing certain surgical operations it is desirable to produce complete muscular "relaxation" which is done by administering a drug which blocks transmission at the neuromuscular junction. Since all skeletal muscles, including the diaphragm and intercostal muscles, are paralysed by such drugs, respiration must be maintained artificially.

In the emergencies mentioned above, it is usual for respiration to fail before the heart stops beating. As asphyxia gradually develops, however, not only does it adversely affect the heart, but it also depresses the medullary vasomotor centre. It is absolutely essential, therefore, that oxygen should be supplied to the tissues as quickly as possible to prevent deterioration of the heart and circulation, and to give the respiratory centres every opportunity of recovering their normal rhythmic activity. Before starting artificial respiration in such emergencies, however, it is necessary first to ensure that the patient's airway is not obstructed; no time should be wasted before loosening clothing and removing foreign bodies and water from the mouth and upper respiratory tract.

EMERGENCY METHODS

1. *Expired air, or 'mouth-to-mouth'', method* (Fig. 47.1). This is now recommended by the St John Ambulance Association of the Order of St John and the British Red Cross Society as the method of choice in an emergency.

In this method, first practised by Elisha (II Kings, Chap. 4, verse 34), the operator blows air into the patient's lungs intermittently. The operator kneels to one side of the head of the patient, who is lying in the supine position. With one hand resting on the patient's forehead and occluding his nose, the patient's head is fully extended, the object of this manoeuvre being to prevent the tongue falling back and occluding the airway. The other hand pulls the chin downwards so as to open the patient's mouth. The operator then takes a deep breath and applies his wide open mouth to the patient's and blows air into the patient's lung. During expiration, the operator withdraws his head so as to enable him to take another breath and to allow expulsion of air from the patient's lungs by means of the elastic recoil of his lungs and chest wall. The cycle is then repeated about 15 times per minute. The main advantage of this method

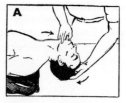
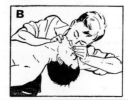
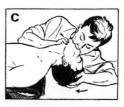

Fig. 47.1 The mouth-to-mouth or expired air method of artificial respiration. A, The patient's head is fully extended to prevent the tongue falling back and obstructing the airway. B, The operator pulls the chin down to open the mouth with his right hand, pinches the patient's nose with the other, and applies his own mouth to the patient's at the same time forcing air into the lungs. The operator then removes his mouth to allow the patient to passively exhale. C, An alternative way of occluding the patient's nostrils by means of the operator's cheek. (From *First Aid Manual*, St John Ambulance Association of the Order of St John.)

is that one knows immediately if there is any obstruction in the patient's airway; furthermore, one can tell from movements of the patient's chest wall approximately how much air is entering the lungs.

2. *Arm-lift back-pressure (ALBP) method of Holger-Nielsen* (Fig. 47.2). This method has certain advantages (see below) and

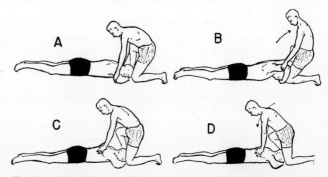

Fig. 47.2 Holger-Nielsen method of manual artificial respiration. A, Placing hands for arm lift. B, Arm lift. C, Placing hands for back pressure. D, Back pressure.

is as effective in ventilating the lungs as the mouth-to-mouth method. It may have to be used in cases involving facial injuries. The patient lies in the prone position with his arms above his head and elbows flexed so that one hand rests on the other; his head, turned to one side, lies on the uppermost hand. The operator kneels on his left knee at the patient's head, with his right foot near the patient's left elbow (A). Grasping the patient's arms just above the elbows, the operator rocks backwards, raising the patient's arms until a resistance is felt (B). The arms are then dropped, and the operator, placing his hands just below the scapulae (C), rocks forwards, keeping his arms straight, until his arms are in a vertical position, at the same time maintaining a steady pressure on the patient's chest (D). The movements of lifting and compression occupy about $2\frac{1}{2}$ seconds each and should be repeated about 15 times a minute.

3. *Silvester's method.* When for any reason the mouth-to-mouth method cannot be used, Silvester's method may be carried out as an alternative. It has the advantage that it allows external cardiac compression to be carried out conveniently and at the same time if required.

The patient is laid on his back and the operator kneels astride his head, grasps his wrists and crosses them over the lower part of the chest. At the same time the operator rocks his body forward and presses down on the patient's chest. The pressure is then released and with a sweeping movement the patient's arms are drawn backwards, upwards and outwards as far as possible. This sequence is repeated about 15 times per minute.

These methods of artificial respiration have been established as the best of many which from time to time have been advocated in emergencies. The two main criteria on which success of a particular method is based are, firstly, it must be capable of producing an adequate pulmonary ventilation as indicated by direct measurement with a spirometer and by estimations of the arterial blood PO_2, PCO_2 and pH. Secondly, the method must be capable of being carried out by non-medical personnel for prolonged periods of time, so that it should be simple and not require excessive physical effort on the part of the operator. In this connection, it is of interest that the oldest established method has been proved experimentally to be the best.

When applying artificial respiration, the administration of oxygen may be beneficial. The addition of small quantities of carbon dioxide, however, is not recommended; patients requiring resuscitation almost certainly have a high arterial blood PCO_2, and to administer to them more carbon dioxide may depress the respiratory centres, rather than stimulate them, due to its narcotic action in high doses.

An important consideration in the choice of method of artificial respiration is its effect on the patient's circulatory system. Any method adopting the principle of intermittent positive pressure breathing, whether applied by means of a pump, or by the mouth-to-mouth emergency method, causes the mean intrapleural pressure to rise *above* atmospheric pressure and, in consequence, abolishes the "thoracic pump" aiding venous return to the heart. As a result, the cardiac output falls considerably, particularly if the patient is in a state of shock. In the mouth-to-mouth method, it is essential therefore not to overinflate the lungs. On the other hand, any method adopting the principle of inflating the lungs by expansion of the thorax from without, thereby making the intrathoracic pressure more "negative" as, for instance, in raising the arms in the Holger-Nielsen method, improves venous return to the heart, and hence increases cardiac output.

Cardiac arrest and resuscitation

The term *cardiac arrest* is used to mean failure of the heart's action to maintain a circulation of blood.

Sudden cessation of the heart beat is a dramatic and often catastrophic emergency. The urgency with which attempts should be made to restart the heart beating again is directly related to the susceptibility of the brain to hypoxia. The brain cannot withstand cutting off its blood supply for any length of time. It has a high oxygen consumption, 33 ml kg^{-1} min^{-1} cerebral tissue, and a high rate of blood flow equal to 540 ml kg^{-1} min^{-1} or about 15 per cent of the resting cardiac output. Thus cessation of the cerebral circulation in man leads to loss of consciousness in a few seconds; after 3 minutes, cerebral damage occurs which may result in personality changes and physical disability, the severity of which increases with the duration of hypoxia. It will be evident therefore that it is of paramount importance to institute resuscitation therapy immediately.

One form of cardiac arrest is spontaneously reversible. An example of this is excessive vagal stimulation due to pressure over the carotid sinus regions in sensitive individuals. Such cases do not require any immediate treatment, because the heart starts beating spontaneously.

A second form of cardiac arrest is not spontaneously reversible and can occur as a result of many causes, viz. coronary thrombosis, pulmonary embolism, electrocution, drowning, during administration of anaesthetic agents, and in anaphylactic shock. In all these cases, the heart stops beating in one of two ways, cardiac asystole or ventricular fibrillation. In asystole, the heart is motionless, soft, relaxed and blue in colour. The coronary veins are engorged with dark blood and are very prominent. In ventricular fibrillation, on the other hand, there is ventricular movement, not as a coordinated beat, however, but as a fine or course irregular uncoordinated writhing of the whole of the muscle, which can be felt as a "bag of worms" when the fibrillation is vigorous. The heart muscle is pale and cyanotic giving it a lavender hue. In both cardiac asystole and ventricular fibrillation the blood pressure rapidly falls to about 20 mmHg (2.6 kPa) or less, and there is complete cessation of blood flow. Outwardly the only way of distinguishing between these two forms of cardiac arrest is by studying the electrocardiogram.

Diagnosis. Cardiac arrest may occur quite suddenly and unexpectedly. Thus this condition must be suspected when a person suddenly loses consciousness and collapses. The absence of the pulse in a large artery, such as a carotid artery, is the essential feature in making the diagnosis. A convulsion may be the first phenomenon noticed, and the pupils are often dilated initially. The heart sounds are absent and respiration may have ceased or may be of a gasping type. The skin will have a greyish-white appearance due to cessation of the cutaneous circulation. The blood pressure will not be recordable. It must be stressed, however, that in the event of the carotid pulse being absent in a collapsed patient, resuscitation should be started immediately.

Principles of treatment

When cardiac arrest has been diagnosed it must be ascertained whether failure of respiration has occurred as well. Then the treatment of such a patient must be carried out with three principles in mind: (1) the circulation of the blood must be assisted artificially by compressing the heart rhythmically; (2) oxygenation of the blood must be carried out by any immediately available method; and then (3) attempts should be made to start the heart beating spontaneously.

CLOSED-CHEST CARDIAC COMPRESSION

The technique of closed-chest cardiac compression is now the recommended method in emergencies. It has the advantage that it can be carried out without the use of any equipment and by people with little experience. In principle, the heart is rhythmically compressed between the sternum and the vertebral column thereby squeezing the blood into the pulmonary artery and aorta.

In practice, the first thing the operator should do is to apply three sharp blows with the closed fist over the sternum. This *may* restart the heart when cardiac asystole is present, especially in the absence of gross heart disease or hypoxia. The operator then kneels to one side of the patient, who is placed in the supine position on a rigid surface, preferably with the legs elevated to aid venous return. The heel of one hand is placed over the lower end of the sternum and the other hand is pressed vertically downwards on top of it, and at a rate of about once a second (Figs 47.3 and 47.4). The pressure should be sufficient to move

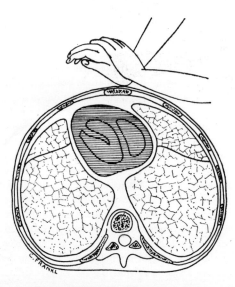

Fig. 47.3 Cross-section of the thorax showing how pressure on the sternum compresses the heart against the vertebral column. (Milstein (1963) *Cardiac Arrest and Resuscitation*, Lloyd-Luke, London.)

the sternum 3–4 cm towards the vertebral column, and the operator may find that the best position is one in which he can use his own body weight in applying pressure. Between each compression pressure is removed to allow the chest to expand and the heart to refill. It is important not to perform cardiac compression too fast or there will be insufficient time for adequate filling.

In children it is usually sufficient to apply this form of resuscitation by gentle pressure with the palm of one hand only, and in infants pressure with the thumb alone is adequate.

When failure of respiration occurs at the same time as cardiac arrest, an emergency method of artificial respiration must be carried out as well as external cardiac compression because the latter alone produces a quite inadequate pulmonary ventilation. Thus a single-handed operator should start by inflating the patient's lungs with two or three breaths by the mouth-to-mouth method and then commence cardiac compression with interruptions every minute to ventilate the lungs again.

In summary this method of closed-chest cardiac compression has the advantages that it can be applied anywhere, without the use of any equipment and by a trained first-aid worker; it is not necessary to open the thorax, so that if spontaneous respiratory movements are re-established the patient can breathe normally.

The method is not without its disadvantages, however. Firstly, various complications may arise: (1) Fractures of ribs or costal cartilages may occur, especially in patients suffering from diseases of the skeleton and in those with a very rigid chest wall. This damage is relatively unimportant if a death is prevented. Fractures are usually due to the method being applied incorrectly. (2) There can be bruising of the cardiac muscle and even rupture of the heart. (3) Laceration of the liver, by exerting too much pressure over the xiphisternum, can result in a fatal haemorrhage.

Secondly, the method will not correct ventricular fibrillation if present. It is then necessary to use a "defibrillator" which consists of passing an electric current through the heart with two electrodes, one placed on the upper end of the sternum, the other just below the left nipple. In principle, electric defibrillation renders all the fibrillating muscle fibres refractory at the same time. There is then a short period of asystole before a pacemaker initiates a coordinated contraction from which a normal cardiac rhythm ensues.

The cardiac output produced by external cardiac compression is considerably below the normal resting level even under optimal conditions, and the blood pressure may only rise to 60–70 mmHg systolic. It is inevitable that the heart receives some bruising by the very nature of the procedure, so that the sooner a spontaneous heart beat can be restored the better. Nevertheless cases are recorded where external cardiac compression has been performed for up to 2 hours and the patients have left hospital physically well.

TRANSTHORACIC OR DIRECT CARDIAC COMPRESSION

An alternative method of artificially assisting the circulation is, immediately following cardiac arrest, to open the thorax and apply compression directly to the heart. But whereas closed-chest cardiac compression can be carried out by anyone trained in the use of the technique, direct cardiac compression should only be attempted by medically qualified personnel. It also requires efficient and careful control of the airway and ventilation of the lungs, and it cannot be practised outside the environment of a hospital.

Direct cardiac compression should be employed in cases in which the closed-chest method is being ineffective, or if ventricular fibrillation is suspected and no defibrillator is available.

A long incision is made with a scalpel in the fourth intercostal space on the left side and the ribs forced apart to gain access to the heart. To carry out cardiac compression the heart is grasped from behind with the right hand and compressed against the deep surface of the sternum. The left hand is placed over the sternum so that counterpressure can be applied. Rhythmic compression about once a second is carried out, and this usually produces an adequate cerebral circulation. The heart may start

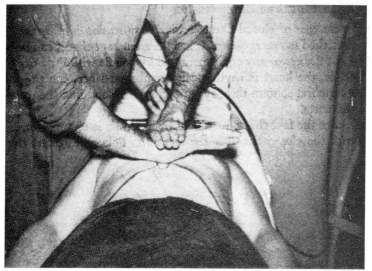

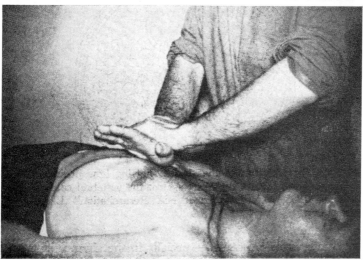

Fig. 47.4 Two views showing the method of applying closed-chest cardiac massage. The clavicles, sternum, xiphisternum and costal margins are outlined. (Milstein (1963) *Cardiac Arrest and Resuscitation*. Lloyd-Luke, London.)

beating after 20–30 such compressions, but if not, the pericardium must be opened and the ventricles compressed between the palmar surfaces of the two hands. It is necessary to pause between each compression to allow diastolic filling of the heart.

Restoration of the heart beat

When it is evident that closed-chest compression is being successfully applied, there being a resonable pulse felt in the carotid region, attention should be given to restoring the heartbeat. This is best done after removing the patient to hospital because treatment depends on whether the heart is in a state of asystole or is fibrillating, for which an electrocardiogram is required to distinguish between the two. When the heart is asystolic, a spontaneous beat may be induced by injection of calcium chloride (5 ml of a 10 per cent solution) or adrenaline (0·2–1·0 ml of a 1 in 1000 solution) into the heart intravenously administered during closed-chest cardiac massage. Adrenaline, however, has the disadvantage that it may cause ventricular fibrillation. Since cardiac arrest causes a rapidly increasing metabolic acidosis, attempts to restart the heart are often more successful if preceded by an intravenous infusion of 100 mmol l^{-1} of sodium bicarbonate.

Electric defibrillation. In cases of ventricular fibrillation, the best method of restoring the heart beat is to "defibrillate" it by means of an electric shock. Defibrillation may be carried out externally as described above. An alternative but much less commonly used method is to open the chest through a rib space and to apply the electric shock by means of two large electrodes placed on either side of the heart (for details, see p. 260).

48. Digestion in the Mouth and Stomach

R. A. Gregory

Most of the food we eat is unsuitable for direct use by the cells of the body, either because it is solid or colloidal and therefore cannot pass through the wall of the intestine into the blood, or because, although diffusible, it is in some form which the cells cannot at once assimilate.

The effect of digestion is to resolve the different foodstuffs into simple components which will pass easily into the cells of the intestinal mucosa and from there via the circulation into the cells which are to make use of them. Polysaccharides must be broken down into monosaccharides, fats into fatty acids and glycerol, and proteins into their constituent amino-acids. A few substances, such as the fat-soluble vitamins which have no value as a source of energy but are of vital importance to the body, may have to be rendered soluble in water before they can be absorbed.

This necessary and radical transformation of the food is effected by the enzymes contained in the digestive juices which are poured into the gut by the various glands situated in or near it, whenever food is eaten. It is noteworthy that all these enzymes are *hydrolytic* in their action, and the minimum of energy is wasted during the process of digestion. Absorption of the products of digestion, which proceeds coincidently with their liberation throughout the small intestine, is also a highly efficient process; the material which is finally collected in the colon for excretion has little food value and consists largely of cellulose, bacteria and debris from the intestinal mucosa.

Besides the products of digestion, the water, salts and organic constituents of the juices themselves must be absorbed. The total volume of digestive juices secreted daily is not accurately known, but it is estimated to be some 4 to 9 litres in man, i.e. of the same order as the volume of circulating blood. There is thus a very large daily "turnover" of water and salts between the blood and intestinal lumen, and if reabsorption of the fluids is prevented (e.g. loss through vomiting or diarrhoea) *dehydration* of the body tissues quickly ensues, fluid being withdrawn from these into the blood to maintain its volume.

The secretory work of the digestive glands

Vasodilatation occurs in all the digestive glands when they are active (Fig. 48.1), so that during the digestion of a meal there is a great increase in the blood flow through the portal circulation. A ready supply of water and salts is thus assured for the production of secretions; but the enzymes and other organic constituents, e.g. mucin, are probably prepared from "raw materials" in the blood by the gland-cells themselves. Many of the cells in the digestive glands contain granules or droplets which are apparently antecedents of organic constituents of the juice. These accumulate during inactivity and are discharged during secretion, particularly if this is prolonged; these cellular changes can be correlated to some extent with the amounts of enzyme, mucin or other organic material found in the juice (Fig. 48.2).

The act of secretion is not merely a washing-out of preformed constituents from the cell by fluid filtered off from the blood; the submandibular gland can produce saliva against a pressure much higher than that in the arteries (Ludwig) and during secretion its usage of oxygen and sugar is increased (Barcroft).

Apart from the synthesis of organic materials by gland-cells *osmotic* work may be done during secretion; for instance, the parietal cells of the gastric glands concentrate hydrogen ions about three million times in preparing the acid of the gastric juice from blood.

Innervation of the digestive glands and alimentary tract

The existence of "secretory" nerves was discovered by Carl Ludwig (1851) who stimulated the lingual nerve, a branch of which (the *chorda tympani*) supplies the submandibular gland, and found that it caused the secretion of saliva. It has since become abundantly clear that all the digestive glands receive a dual innervation from the autonomic nervous system (Chapters 30 and 31) namely, vasodilator and "secretory" fibres from the parasympathetic division, and vasoconstrictor and (possibly) inhibitory fibres from the sympathetic division. The former are distributed to the abdomen in the vagus and pelvic visceral nerves; the latter for the most part run from the autonomic ganglia to the viscera along the walls of the large arteries.

There is a similar dual nerve supply to the smooth muscle of the alimentary tract, the parasympathetic fibres increasing the motor activity, and the sympathetic fibres depressing it, so that in general the state of activity of an organ may be said to represent the resultant of the influence of the two systems. However, the extent to which the motor and secretory functions of the digestive tract are normally controlled by these nerves remains problematical; for instance complete denervation

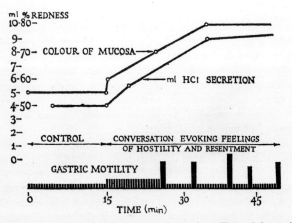

Fig. 48.1 Contractions of the human stomach (gastric motility) and changes in acid secretion and vascularity (colour) of the mucosa accompanying feelings of hostility and resentment aroused by conversation. The subject was a man ('Tom') who had a large gastric fistula. The intensity of the red colour of the mucosa was expressed in arbitrary units. The strength of the contractions of the stomach is indicated by the height of the black bars. (Wolf and Wolff, *Human Gastric Function.*)

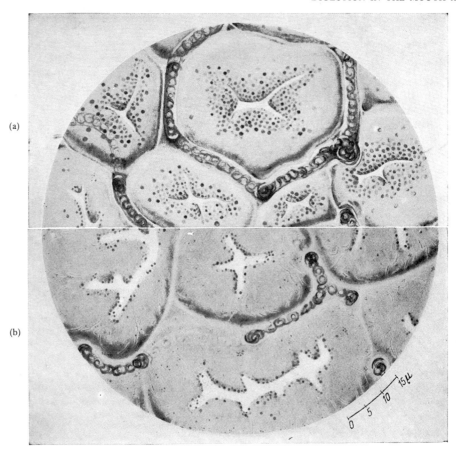

Fig. 48.2 Portions of the unstained living pancreas of a white mouse (a) after 24 hours fast; the cells contain plenty of zymogen granules, and (b) after 3 hours stimulation of secretion. Most of the intracellular material has been discharged. (Hirsch.)

of an intestinal loop produces a striking increase in tone, motility and spontaneous secretion, but a rapid recovery occurs and in a few days the behaviour of this denervated loop is almost indistinguishable from normal.

In addition to the motor or efferent autonomic fibres mentioned, *afferent* fibres carry sensory impulses from all parts of the tract to the central nervous system; the reflex arcs formed by these with the autonomic nerves play an important part in the activities of the gut. Sensations from the normally functioning digestive tract are almost entirely absent, apart from fullness of the stomach after a meal and of the rectum before defaecation, which are readily appreciated. Excessive distension or strong contractions of the intestines, particularly in the presence of inflammation or poor blood-supply, gives rise to *pain* which is griping or colicky in character, is poorly localised and may be "referred" to areas supplied by somatic nerves entering the spinal cord in the same segment as the visceral afferents.

The ultimate "centre" in the brain for visceral afferents and efferents appears to be the *hypothalamus* which thus exerts a general influence over the motor, secretory and vascular reactions of the entire alimentary tract; damage to or experimental interference with this region of the brain produces changes in secretion and motility of the stomach and intestines.

The hypothalamus has connections with the thalamus and cerebral cortex; and stimulation or destruction of certain areas of the cerebral cortex causes changes in motility and secretion of the alimentary tract. Furthermore, the thalamus is well known to be concerned with the perception of the painful or pleasurable quality of sensations. Means thus exist by which disagreeable or pleasant emotions may influence the working of the digestive tract. A good example of this is afforded by the experiment depicted in Fig. 48.1, which is taken from the study by Wolf and Wolff of the daily variations in gastric function of a laboratory technician ("Tom"). This man had had from childhood a large gastric fistula, permitting inspection of the interior, withdrawal of contents, etc. It is common everyday experience that pain, fear, anger, resentment or worry are potent causes of "indigestion", and similar upsets of the gastrointestinal tract.

Saliva

The prompt response of the salivary glands to the sight, smell, or even the anticipation of appetising food, is familiar to everyone as "watering of the mouth". This is a "conditioned" reflex, i.e. one which has become established by training and experience and in which the cerebral centres play an important part. The stimuli received by the special sense organs are conveyed to the cerebral cortex; from there they are relayed to cells in the medulla which form the "salivary nuclei" (they lie in the reticular formation in the floor of the fourth ventricle) and from these cells fibres run to the various glands. Salivation is also brought about when food is actually eaten, by direct stimulation of sensory end-organs in the mucosa of the mouth, tongue and pharynx, from which impulses are transmitted to the salivatory nuclei. This reflex is "unconditioned"; it is present from birth and does not involve the higher cerebral centres. Thus, reflex salivation is readily produced in a decerebrate cat by introducing acid, alcohol, etc. into the mouth.

Saliva has a pH of about 6·8, is fairly well buffered and contains a lubricant *mucin* and (in man) the enzyme *ptyalin*, which breaks down starch into a mixture of dextrins and maltose. The main functions of saliva are to moisten and lubricate the food, thus preparing it for swallowing, and to dissolve its soluble constituents, so that the flavour is appreciated and the secretion of saliva itself and of other digestive glands thereby stimulated.

Swallowing

Each mouthful of food is chewed and mixed with saliva until it forms a pulpy mass or "bolus" suitable for swallowing, and is collected from time to time on the surface of the tongue for this purpose. Swallowing begins with a quick contraction of the tongue muscles which propels the bolus past the faucial pillars into the pharynx; from then onwards, its progress is beyond voluntary control and is accomplished by a rapid and complicated series of movements which constitute the "swallowing reflex" and are coordinated by a "centre" in the medulla.

As the bolus enters the pharynx, the soft palate is approximated to the posterior pharyngeal wall by contraction of its muscles, so as to prevent entry of food into the nasal passages. At the same time the larynx is brought upwards and forwards under the shelter of the base of the tongue, raising and opening the upper end of the relaxed oesophagus, which the bolus now enters. The epiglottis may turn backwards to guard the entrance of the larynx; and the risk of food entering the air-passages is lessened by a reflex approximation of the vocal cords and momentary inhibition of respiration, which also form part of the reflex.

The swallowing reflex is touched off by contact of the food with areas of the fauces, pharynx and tonsils which are very sensitive to tactile stimulation and from which impulses travel to the medullary centre.

Having travelled through the upper third of the oesophagus in a fraction of a second, the bolus is now carried the rest of the way much more slowly by an advancing ring-like contraction of the smooth muscle of the oesophagus. If the bolus is soft and well-lubricated, it reaches the cardiac sphincter at the entrance to the stomach in a few seconds; but if dry, it may take a minute or so and *secondary waves* (which give rise to a painful sensation in the chest) may arise in the oesophagus and force it along.

Liquids, owing to the impetus given them by the act of swallowing and the effect of gravity, outstrip the oesophageal wave and arrive at the cardiac sphincter in a second or two, where they wait for the arrival of the oesophageal wave. When this approaches the cardiac sphincter the latter relaxes before it and the food enters the stomach.

Gastric digestion

X-ray examination of the human stomach after eating meals made radio-opaque by the addition of barium sulphate, shows that there are wide variations among apparently normal persons in the position, shape and motility of the stomach. An ordinary meal begins to leave the stomach less than thirty minutes after it is eaten, and although the rate of gastric emptying varies with the size of the meal and its consistency and composition, gastric emptying is usually completed in four to five hours.

While the food remains in the stomach it becomes mixed with the gastric juice, the secretion of which from the millions of tubular glands buried in the mucosa starts within a few minutes of eating.

Composition of gastric juice

The juice is really a mixture in variable proportions of the individual secretions of the various types of cell present in the glands. Heidenhain (1878) first recognised the *chief cells*, which contain pepsinogen, and the *parietal cells*, which secrete HCl. Both are absent from the pyloric region of the stomach; the glands there contain *mucous cells* which produce mucus, and between the pyloric region and the rest of the stomach there is a transitional zone where pyloric mucoid cells are mingled with chief and parietal cells. Besides the pyloric glands, mucus is secreted by cells in the necks of the glands elsewhere and also by the cells of the surface columnar epithelium.

The parietal cells are believed to secrete a fluid which is isotonic with blood, contains most of the water of the gastric juice and is a practically pure solution of hydrochloric acid, the strength of which as secreted (140 millimoles per litre in man) is constant whatever the rate of its formation. However, the acidity of the gastric juice is generally much lower than this maximal value, owing to neutralisation by the bicarbonate of the mucous secretion and buffering by the proteins, peptones and polypeptides of the gastric contents.

There is good evidence that the hydrogen ions of the gastric juice are derived by splitting water molecules, the hydroxyl ions remaining being neutralised by carbonic acid, with the formation of bicarbonate ions. The chloride ions of the gastric juice pass in from the blood so as to preserve electrical neutrality. The carbonic acid is formed by the hydration of carbon dioxide, catalysed by the enzyme *carbonic anhydrase*, a large amount of which is found in the parietal cells.

The chief cells probably contribute a scanty non-acid secretion; it contains pepsinogen, which is activated by acid, forming the proteolytic enzyme pepsin. The mucous cells produce a jelly-like fluid which contains much mucus and is faintly alkaline owing to the presence of bicarbonate.

Although stimulation of the vagus causes the secretion of a juice containing acid, enzyme and mucus indicating that the cells concerned all have a secretory innervation from the vagus, they can respond to some extent independently of one another to other forms of stimulation (mechanical or chemical) so that the final composition of the gastric juice may show wide variations. Thus the drug *histamine* or the hormone *gastrin* (see later) are powerful stimulants of the parietal cells, providing a juice of high acidity and containing little pepsin or mucus; while mechanical or chemical irritation of the mucous membrane causes a profuse flow of mucus, with comparatively little acid or pepsin. No selective stimulus for the chief cells is yet known.

Pepsin in acid solution breaks down proteins into peptones and proteoses, which are fairly large fractions of the original molecule; some amino acids are liberated, but the further breakdown of proteins and the above derivatives is accomplished later by the enzymes of the pancreatic and intestinal juices.

Stimulation of gastric secretion

Pavlov (1902) and his pupils were the first to show clearly that the secretion of gastric juice which starts within a few minutes of eating a meal occurs whether the food actually enters the stomach or not, and is due to a combination of "conditioned" and "unconditioned" reflexes similar to those causing the flow of saliva under the same conditions. A dog was provided by a previous surgical operation with a gastric fistula for the collection of gastric juice and an oesophageal fistula so that the food which was swallowed did not enter the stomach but fell out of the opening in the neck (Fig. 48.3). A few minutes after the animal was thus "sham-fed" there began a flow of gastric juice, which could be stopped by cutting the gastric branches of the

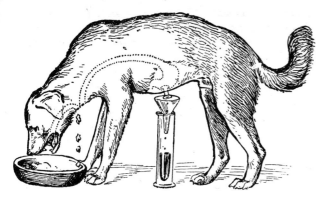

Fig. 48.3 A dog with oesophageal and gastric fistulae.

The food consumed is seen dropping out of the open end of the anterior portion of the oesophagus; the animal is fed through the opening in the posterior portion. The gastric fistula consists simply of a tube flanged at each end, stitched into the wall of the stomach at one end, and into the abdominal wall at the other. (Höber.)

vagus, or paralysing them by the injection of the drug atropine. Sham-feeding was not always necessary to elicit secretion; in intelligent animals, the mere sight, smell or sounds associated with the arrival of food were sufficient.

These findings have been confirmed and extended by experiments on human subjects who have become accustomed by training to swallow and retain without discomfort a stomach-tube for withdrawal of the gastric juice; and occasional opportunities have also arisen of making similar and more extensive experiments on patients who, usually on account of an oesophagal stricture, have been provided by means of an operation with a gastric fistula for feeding. The classical example is that of Alexis St Martin, an Indian "runner" at a trading station in Michigan, USA, who was left, as the result of a gunshot wound, with a large gastric fistula. The observations and experiments made upon him (1825–33) by his physician, William Beaumont, have become famous. More elaborate studies of a similar kind have since been made by Carlson (1916) and by Wolf and Wolff (1943).

The reflex response to a "sham" meal gradually ceases in about an hour; but if the swallowed food is allowed to enter the stomach in the usual way, to be digested by this juice (and later by the pancreatic and intestinal enzymes), gastric secretion is augmented and prolonged for three hours or more; in fact, long after the meal has been forgotten.

This continued secretion resulting from digestion of the food was first shown by Pavlov by means of the famous "Pavlov pouch" (Fig. 48.4). The secretion from this "miniature stomach" always runs closely parallel with that of the remainder, so that it becomes possible to follow the course of secretion in the main stomach during digestion, without interfering with it in any way. An earlier and simpler type of pouch, the "Heidenhain pouch" was invented by Pavlov's teacher, Heidenhain (1879). It lacks a vagal nerve supply and so can only respond to *humoral* agents, i.e. substances such as hormones carried to it in the circulation. The response of a Heidenhain pouch when the dog is fed a meal commences later than that of the Pavlov pouch and is much smaller, indicating the importance of vagal excitation in the response of the gastric glands to a meal. The circulating hormones which stimulate the Heidenhain pouch (and, of course, the Pavlov pouch, or intact stomach) after a meal come from cells situated in the pyloric glands of the stomach itself and also from cells situated in the mucosa of the upper intestine (chiefly duodenum).

It has long been known that the introduction of food or its digestion products directly into the small intestine also stimulates gastric secretion. The existence of this "intestinal phase" of gastric stimulation has been proved in dogs by making the entire stomach into a pouch at an aseptic operation, joining the oesophagus directly to the duodenum. After recovery from the operation, when the animal eats a meal, this passes straight into the small intestine and is there digested; a considerable secretion of gastric juice from the pouch occurs.

Gastrin

Scattered among the mucus-secreting exocrine cells in the mid-portion of the pyloric glands are wedge-shaped endocrine cells which contain the hormone gastrin (Edkins, 1905). During gastric digestion the hormone is secreted into the bloodstream and, returning to the fundic glands, stimulates the parietal cells to secrete acid. The release of this hormone can be demonstrated by providing a dog with two gastric pouches, one of the pyloric region and the other of the acid-secreting (fundic) region, and dividing all nervous connexions between them. When the pyrloric pouch is stimulated by distending it with saline or meat extract, the fundic pouch secretes. Such local stimulation of the pyloric mucosa occurs when the swallowed food enters the stomach and is digested by the "vagal" juice; there occurs a

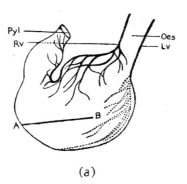
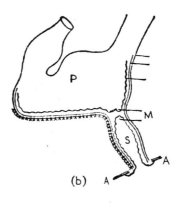

Fig. 48.4 The Pavlov pouch.

The left diagram (a) shows the line of the incision, A–B, into the gastric wall. Oes, oesophagus; R_V., L_V., right and left vagus nerves; Pyl., pylorus.

The right diagram (b) shows the operation completed (partly sectional). S, main portion of stomach; P, pouch; M, partition of mucous membrane between the two cavities; A–A, abdominal wall.

prompt rise in the level of gastrin in the circulation. A second important factor in the release and action of the hormone is the reflex vagal excitation which commences with eating, and is sustained by further reflexes originating in the stomach and also from the upper small intestine. This vagal excitation is conveyed not only to the fundic glands but also to the pyloric glands, including the "gastrin cells". Its effect on the latter is to increase their responsiveness to local stimulation by the digesting food, so that a greater release of the hormone occurs; in the fundic glands the concomitant vagal excitation to the parietal cells increases their responsiveness to the circulating gastrin, so that a much greater secretion of acid occurs.

As gastric secretion proceeds, the contents of the stomach, including the pyloric region, become increasingly acid, and this inhibits progressively the release of gastrin, so that the level of the hormone in the circulation falls to a lower level. This "negative feedback" mechanism is an important means of controlling the acidity of the gastric contents. Gastrin is also found in the duodenal mucosa; it is probably liberated during digestion there, contributing to the "intestinal phase" of stimulation.

Gastrin has been isolated from the pyloric mucosa of man and several other mammalian species. The major amount of the hormone present is in the form of a peptide ("little gastrin") having 17 amino acid residues. There is also present in small amount a large peptide ("big gastrin") having 34 residues, the C-terminal 17 of which are identical with "little gastrin". Both forms of the hormone are released into the circulation on stimulation of the gastrin cells, but unexpectedly it is "big gastrin" which usually predominates in the circulating blood. The explanation for this may lie partly in the fact that "big gastrin" is removed from the circulation much more slowly than is "little gastrin". Both forms of the hormone are extremely potent stimulants of gastric acid secretion; the active portion of the gastrin molecule is the C-terminal tetrapeptide Trp-Met-Asp-Phe-NH$_2$, the remainder having no known activity of its own.

Histamine, a potent stimulant of parietal cell secretion, is present in all parts of the gastric and intestinal mucosa, particularly in the vicinity of the parietal cells; it is also found in the gastric juice. The function of this substance is uncertain; it may play some role in the stimulation of the parietal cells.

Gastric inhibition
Besides the stimulation of gastric secretion from the small intestine, the presence there during gastric emptying of food from the stomach brings into operation nervous and hormonal mechanisms tending to inhibit gastric secretion and emptying. Several constituents of the duodenal contents, such as protein digestion-products, acid, fats and sugars, can be shown to cause gastric inhibition when introduced into the duodenum. The nature of the hormonal component of this inhibition is not entirely clear. Secretin, the duodenal hormone which stimulates pancreatic secretion (p.319) is also an inhibitor both of the release of gastrin from the "gastrin cells" and of its action on the parietal cells; and there has been isolated from the duodenal mucosa a peptide similar in structure to secretin ("gastric inhibitory peptide") which inhibits gastric secretion. Thus the secretory response to a meal observed in a "Pavlov pouch" represents the integrated effects of several mechanisms, both nervous and hormonal, of excitatory and inhibitory character, originating in the brain, the stomach itself, and in the duodenum.

MOVEMENTS OF THE STOMACH

Some hours after the meal the normal human stomach is empty apart from a small and variable quantity of gastric juice, saliva, mucus, etc. and its walls are in a state of tonic contraction. When the swallowed food enters it, a *receptive relaxation* occurs as the result of a nervous reflex and the food slides down into the most dependent portion. Soon, as indicated in Fig. 48.6, ring-like contractions appear in the body of the stomach and slowly move towards the pyloric sphincter, becoming deeper as they pass into this region where the muscle is stronger (W. B. Cannon, 1898). As digestion and emptying proceed, the strength and frequency of the contractions increase to a maximum which varies with the size and nature of the meal, and then gradually decline. Each wave occupies the stomach for about half a minute, and as many as four may be seen at the same time during the height of digestion. These contractions serve to mix the food with the gastric juice and, particularly in the pyloric antrum, provide the propulsive force for the passage of gastric contents at intervals into the duodenum.

Although the stomach relaxes when food enters, the tone is gradually regained, so that by the time most of the food has left

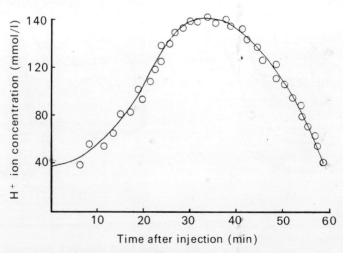

Fig. 48.5 H$^+$ ion concentration in samples of gastric juice following the intravenous injection of 5 μg gastrin (at time 0) in a human subject.

DIGESTION IN THE MOUTH AND STOMACH 317

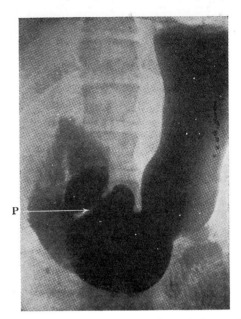

Fig. 48.6 Radiograph of a human stomach after a "barium meal" showing peristaltic waves. P, pyloric sphincter. (F. Haenisch in A. E. Barclay, *The Digestive Tract*.)

the stomach, a high tone is again present, with small regular fluctuations, termed a "tonus rhythm". This continues for a few hours after the stomach has emptied and then, if the next meal is not forthcoming, gives place to contractions similar in type to those normally seen in the filled stomach, but much more powerful. These occur in groups, lasing for about half an hour, and at intervals of two to three hours. Their incidence coincides with a sensation of hunger; as they become stronger, definite pain—"pangs of hunger"—is felt with each contraction (Carlson, 1919).

Emptying of the stomach

For many years it was believed that gastric emptying was chiefly controlled by the pyloric sphincter, which was supposed to remain closed for most of the time in the face of the gastric contractions, opening briefly at intervals to allow exit of some gastric contents. However, direct observations of the behaviour of the stomach and pyloric sphincter in human subjects by means of X-rays and the gastroscope, and experiments on trained conscious animals in which the regions concerned have been made accessible by the surgical preparations of fistulae, show that the sphincter has no such independent role, but behaves like the pyloric antrum of which it is anatomically a part. In fact, the three regions, pyloric antrum, sphincter and duodenal cap act as a single coordinated physiological unit. As a gastric wave passes over each in turn, the antral contraction expels food through the still relaxed sphincter into the duodenal cap; but this is brought to an end by the closely following contraction of the sphincter, and before the antrum and sphincter have relaxed, the contraction of the duodenal cap occurs, expelling the food down the duodenum. The effect is that of a "gastric pump" (Quigley, 1943), regurgitation from duodenum to stomach being prevented by the slightly persistent contraction of antrum and sphincter; not every gastric wave results in this complete cycle of contractions so that only a proportion of the waves which arrive at the pyloric antrum cause the exit of gastric contents.

Gastric emptying thus depends fundamentally upon the propulsive activity of the gastric muscle and the coordination of the three regions mainly concerned; both these factors are controlled to a large extent from the duodenum. There are many substances besides fat, such as acid, hypertonic solutions and protein digestion-products, which retard gastric emptying when they are introduced into the duodenum by causing reflex inhibition of the gastric musculature including the pyloric sphincter (Fig. 48.7). In fact, by means of this mechanism, a con-

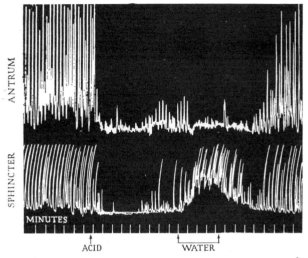

Fig. 48.7 Record showing the effect on the pyloric antrum and sphincter in a conscious dog of: (Top) injecting acid into the duodenum (20 ml N/10 HCl); and (Bottom) distending the duodenum with water (30 cm pressure). Acid relaxes both the antrum and sphincter; distension of the duodenum beyond the physiological range of pressure caused contraction and spasm (rise of baseline).

Gastric peristalsis was recorded by a balloon in the antrum, and contractions of the sphincter by a second balloon in the sphincter. (Thomas, Crider and Mogan.)

stant restraining influence is normally exercised from the duodenum on gastric tone and motility; if it is prevented from operating during gastric emptying, gastric motility is greatly increased and the stomach empties abnormally rapidly (Fig. 48.8).

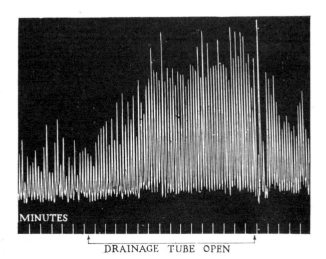

Fig. 48.8 Record showing the effect of duodenal drainage on gastric peristalsis in a conscious dog provided with a high duodenal fistula (exclusion of enterogastric reflex). (Thomas, Crider and Mogan.)

Towards the end of gastric emptying, particularly of a fatty meal, the contraction cycles are weak, the pressures in antrum, sphincter and duodenum are nearly equal, and the sphincter is open most of the time; such conditions are favourable for the regurgitation of intestinal juices and bile into the stomach, and evidence of this is afforded by the presence of these in samples of the gastric contents withdrawn by a stomach-tube.

Vomiting

This is a reflex act involving the muscles of the diaphragm and abdominal wall and those of the stomach and oesophagus. It is coordinated by a "centre" in the medulla, which may be stimulated by irritation of any part of the digestive tract, by impulses from the semicircular canals (sea-sickness) or by disturbance of the centre itself (e.g. by cerebral tumours or the action of drugs such as apomorphine). A more or less prolonged sensation of *nausea* usually precedes retching and vomiting; it is marked by pallor, sweating, salivation and partial or complete inhibition of the gastric musculature; anti-peristalsis in the small intestine has been observed radiographically in human subjects. Nausea may culminate in *retching*, which consists of a series of inspiratory-like efforts accompanied by closure of the glottis, the stomach becoming compressed between the diaphragm and the contracted abdominal muscles; the gastric contents are finally ejected through the relaxed cardiac sphincter and oesophagus. The larynx is drawn up as in swallowing and elevation of the soft palate also occurs; this largely prevents egress of the vomitus by the nose.

49. Intestinal Digestion

R. A. Gregory

As the stomach contents pass at intervals into the duodenum, they meet and mix with secretions from the pancreas, liver and intestinal glands, which complete the digestion of proteins, fats and carbohydrates, as the food passes down the intestine. The products are absorbed simultaneously into the portal and lymphatic circulations.

Secretion of pancreatic juice

The collection of pancreatic juice from a conscious dog by means of a cannula tied into the pancreatic duct, was first carried out by Regnier de Graaf (1664) and the method was revived nearly 200 years later by Claude Bernard, who gave the first description of the properties of the juice. Animals provided by a previous surgical operation with such pancreatic fistulae remain in excellent health indefinitely, provided the juice is returned to the intestine daily, and not lost to the animal. For some purposes, however, collection of the juice for a few hours after cannulation of the duct in an anaesthetised animal is more suitable.

The acinar cells, which secrete the pancreatic juice, are apparently all of the same type; they produce a secretion containing a number of enzymes and having a pH of 8·0–8·4 with a bicarbonate content which is approximately 1–5 times that in the blood, increasing with the rate of secretion. The amount of chloride present is approximately inversely proportional to the rate of secretion, so that the sum of the concentrations of the two ions HCO_3^- and Cl^- remains about the same (Fig. 49.3). The pancreas contains the enzyme *carbonic anhydrase* which is presumably concerned in the formation of the bicarbonate in the juice.

As in the case of the saliva and gastric juice, a reflex mechanism exists for the provision of pancreatic juice when a meal is eaten, and again it is the vagus which carries the secretory fibres to the gland (Fig. 49.1); the secretion is scanty, but rich in enzymes. A much greater flow of juice occurs, however, as the gastric contents are passed into the duodenum,

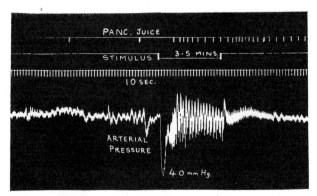

Fig. 49.1 Pancreatic secretion produced in an anaesthetised dog by stimulation of the vagus nerves in the neck.

Note the effect on the heart-rate (cardio-inhibitory fibres in the vagus), the long latent period before secretion commences (40 s) and the scanty response. (Gregory, unpublished.)

and cause the liberation from the intestinal mucosa into the circulation of the hormone *secretin* which excites a copious and watery secretion from the pancreas. The agent chiefly responsible for the release of secretin is *the acid* in the gastric contents (Fig. 49.2). Since the bicarbonate in the pancreatic juice

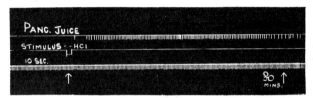

Fig. 49.2 Pancreatic secretion produced in an anaesthetised dog by the injection of acid (50 ml of N/10 HCl) into the duodenum.

Note the copious and persistent secretion, compared with that given by vagus stimulation (Fig. 49.1) in similar circumstances. (Gregory, unpublished.)

neutralises the acid in the duodenal contents, there takes place what is in effect a process of titration by which the pH of the duodenal contents is regulated, the "indicator" being circulating secretin. A second hormone, *cholecystokinin-pancreozymin*, is also liberated along with secretin by the presence of fats, bile and protein digestion-products; this, like vagal excitation, causes the secretion of pancreatic juice containing bicarbonate and enzymes, so that intestinal digestion is accomplished in an approximately neutral medium (pH 6·5).

Neurohormonal interactions

Vagal excitation, in addition to its direct effect upon the pancreatic acinar cells, also increases their responsiveness to the duodenal hormones secretin and cholecystokinin-pancreozymin; it also increases the liberation of these hormones which is brought about by local stimulation of the duodenal mucosa. These effects are reminiscent of the actions of the vagus on the release and action of gastrin. Furthermore, it has been shown that the hormones secretin and cholecystokinin-pancreozymin potentiate each other's actions upon pancreatic secretion, so that the effects on water, bicarbonate, and enzyme secretion observed when both hormones are acting together (as is normally the case) are greater than when the same amounts of each hormone are administered separately.

Figure 49.3 shows the effect of an injection of secretin on the flow of pancreatic juice in a human subject; the juice was withdrawn by means of a stomach tube which was swallowed and allowed to pass into the duodenum.

Secretin
This was the first hormone to be discovered. The fact that the entry of the acid gastric contents into the duodenum excited pancreatic secretion was well known to Pavlov and his contemporaries, but was ascribed to a reflex. However, in 1902, Bayliss and Starling showed that

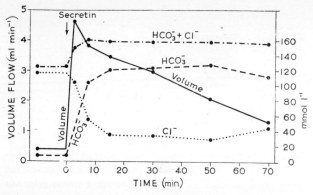

Fig. 49.3 Human pancreatic secretion evoked by intravenous administration of secretin at time 0. The volume-rate and bicarbonate concentration are increased; the chloride concentration is decreased. (Lagerlöf, 1942.)

dilute acid still excited pancreatic secretion when placed in a *denervated* loop of small intestine, so that the effect must be mediated by way of the circulation. Intravenous injection of acid was without result; but the injection of an acid extract of the intestinal mucosa caused a copious secretion of pancreatic juice; and the active principle, secretin, has been isolated (Jorpes and Mutt, 1962) and identified as a peptide containing 27 aminoacid residues.

Cholecystokinin-pancreozymin
Ivy and Oldberg (1928) discovered that emptying of the gall-bladder which occurs when food, particularly fat, enters the duodenum was due to the release of a hormone which they named "cholecystokinin": they distinguished it from secretin but were unable to identify it chemically. In 1943, Harper and Raper discovered that the stimulation of pancreatic enzyme secretion caused by food in the duodenum was due to the release of a hormone distinct from secretin, which they named "pancreozymin". The Swedish chemists Jorpes and Mutt (1966) finally isolated from duodenal mucosa a single peptide containing 33 amino acid residues which possessed the properties of both cholecystokinin and pancreozymin; it is a potent stimulant both of gall-bladder contraction and pancreatic enzyme secretion, and is usually referred to as cholecystokinin-pancreozymin (CCK-PZ). The C-terminal portion of CCK-PZ is identical with the active tetrapeptide of gastrin (p. 316) and these two hormones therefore have an almost identical range of physiological effects. Thus gastrin, besides its action on the parietal cell, is also quite a strong stimulant of pancreatic enzyme secretion and a very weak stimulant of gall-bladder contraction. Conversely, CCK-PZ is a stimulant of gastric acid secretion.

Actions of pancreatic juice

The pure juice is almost without action on most varieties of protein, the powerful proteolytic enzyme it contains being present in an inactive form *trypsinogen*. This is rapidly converted into the enzyme *trypsin* when the juice mixes with the intestinal juice which contains an enzyme-like activator *enterokinase*. Thus formed, trypsin acts upon all proteins and their digestion products, converting them finally into amino-acids and polypeptides. Pancreatic *diastase* breaks down starch into maltose, while the *lipase* also present hydrolyses the fats into fatty acids and glycerol.

The bile

The entry of the gastric contents into the duodenum provides the stimulus for the appearance there of the bile, whose importance for digestion lies chiefly in the fact that the bile salts and lecithin it contains are valuable aids in the emulsification, digestion and absorption of the fats of a meal. It is also necessary for the efficient absorption of iron and of the fat-soluble vitamins.

Hepatic bile, a neutral golden-yellow slightly syrupy fluid, is secreted by the hepatic cells, and there is little evidence that its production is normally under nervous control. Between meals the tone of the sphincter-like muscle around the duodenal end of the common bile duct is relatively high, and the bile flows into the relaxed gall-bladder where it is rapidly concentrated by the activity of the mucosa, becoming more viscid, very dark and slightly acid.

When a meal is eaten, a little bile is sometimes reflexly expelled from the gall-bladder into the duodenum; but the main emptying occurs when the hormone *Cholecystokinin-pancreozymin* (above) is liberated, causing slow contractions and emptying of the gall-bladder, with relaxation of the common bile-duct sphincter. Fat is particularly effective in liberating the hormone and hence causing emptying of the gall-bladder (Fig. 49.4).

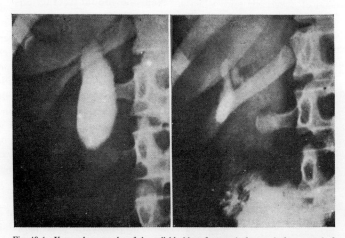

Fig. 49.4 X-ray photographs of the gall-bladder of a man before and after a meal of fat.

Tetraiodophenolphthalein was injected intravenously 14 hours before the first photograph was taken (Graham-Cole test). The second photograph was taken 20 min after the meal of fat. The discharge of the contents of the gall-bladder in response to the presence of fat in the duodenum has filled the cystic and common ducts with dye, and they can be seen, in the second photograph, forming a loop above the gall-bladder; in some cases the hepatic duct becomes filled also. (Ivy.)

After the gall-bladder has emptied, hepatic bile may flow directly into the duodenum for a time until digestion there is over; gradually the tone of the sphincter increases and the bile is once more diverted into the gall-bladder until the next meal.

Cholecystography. This clinical test of gall-bladder function depends on the fact that tetrabromphenolphthalein and similar compounds are opaque to X-rays and are excreted in the bile after oral or intravenous administration. They are concentrated in the gall-bladder and so enable it to be visualised by X-rays. If a meal rich in fat is then fed, the emptying of the gall-bladder may be recorded by serial radiographs (Fig. 49.4).

The increase in the rate of flow of bile from the liver which occurs during the digestion of a meal is to a small extent due to the presence in the portal blood of the hormones gastrin, secretin and cholecystokinin-pancreozymin, which increase the hepatic secretion of water and electrolytes, particularly bicarbonate. However, the chief stimulus to bile flow appears to be the bile salts themselves. These are absorbed from the intestinal contents in the ileum, and returning to the liver are resecreted (the "enterohepatic circulation"). In each "circuit" about 10 per cent of the bile salts is lost into the faeces, and this is made good by the synthesis in the liver of new bile salts from cholesterol, so that the total "bile salt pool" (about 2–4 g in man) remains constant. If extra bile salts are administered to an animal or human subject, the "pool" returns to its normal size in a few days, apparently because the synthesis of new bile salts by the liver is inhibited and the 10 per cent loss into the faeces is not restored. In disease of the ileum, where it become necessary to remove this part of the small intestine, the bile salts are largely lost to the body and the production of bile salts by the liver falls to a low level which presumably represents the response of that organ to a maximal stimulus—the fall to zero of the quantity of bile salt in the enterohepatic circuit. The nature of the "homeostatic mechanism" by which the hepatic cells "sense" the quantity of bile salt in circuit and adjust their activities accordingly, is not understood.

The bile salts are derivatives of the steroid cholic acid, and are thus allied structurally to cholesterol and the sex and adrenal-cortical hormones (see Chapter 32). A number of different bile acids exist, but only a few are present in the bile of a particular species. A small amount only is present as the acid itself; the rest is in the form of a compound of the bile acid with the base taurine or the amino acid glycine. Taurocholic and glycocholic acids are present in human and ox bile: the dog, sheep and goat have only the former, the hog only the latter.

The bile pigments. The haemoglobin of worn-out red blood corpuscles is broken down by the cells of the reticuloendothelial system, notably those of the liver (Kupffer cells), spleen and bone marrow, through the stages of haemochromogens which still contain the iron and globin of the original haemoglobin molecule, to the *bile pigment biliverdin*, and its reduction product *bilirubin*, which is iron- and protein-free. The latter is set free into the blood, contributing to the yellow colour of normal plasma, and taken up from it by the liver, to be excreted in the bile. In some circumstances, e.g. in starving dogs, biliverdin is excreted by the liver in place of bilirubin.

The intestinal juices

The digestive juices contributed by the glands present in the wall of the small intestine come from *Brunner's glands* in the first inch or so of the duodenum, and *Lieberkuhn's glands,* which are found throughout the small and large intestines. Both secretions are alkaline and contain a good deal of mucus; the stimulus for their appearance seems to be local mechanical and chemical excitation of the mucosa by digesting food.

The juice as ordinarily obtained—e.g. by distension with a balloon of an isolated loop of intestine (Thiry-Vella loop, Fig. 49.5)—contains small amounts of a variety of enzymes; but similar enzyme activity is also demonstrable in extracts of the intestinal mucous membrane, and the invariable presence, in such samples of juice, of cast-off mucosal cells, leucocytes, etc. has given rise to the suspicion that most of its varied digestive properties may be due to *intracellular* enzymes liberated from the debris. If this material is rapidly removed from cat's intestinal juice by centrifuging it immediately after collection, the only enzymes found in appreciable amounts are lipase, amylase and enterokinase. (Florey.)

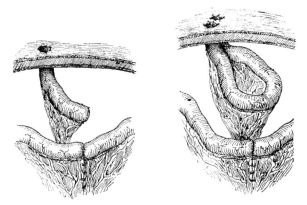

Fig. 49.5 The Thiry and Thiry-Vella intestinal loops. (From Markowitz, *Textbook of Experimental Surgery.*)

THE MOVEMENTS OF THE INTESTINES

The mixing of the food with the intestinal secretions, and its passage through the alimentary canal, is accomplished by the intestinal movements; these are nicely coordinated with the progress of digestion and absorption so that both are virtually completed by the time the colon is reached.

A good way to gain some idea of the normal pattern of the intestinal movements (and incidentally, to study them experimentally) is to open, under warm saline, the abdomen of a decerebrate or lightly anaesthetised animal at the height of digestion. The inhibition caused by cold and drying is thus avoided; and the movements of the coils of intestine, as they float outside the abdominal cavity, may be recorded by attaching them to levers writing on a smoked drum (enterograph), or by taking moving pictures which are analysed later; or balloons may be inserted into the intestine and connected to volume or pressure recorders.

Many other methods have been used for study of the intestinal movements; the more fruitful are probably those which utilise as a subject a conscious trained animal previously operated upon to render accessible the required region of the intestine (e.g. the Thiry-Vella loop).

Bayliss and Starling (1899) discovered the basic mechanism in the gut wall by which food is moved through the digestive tract. They showed in an anaesthetised dog that localised distension or stimulation of the intestine excited contraction above, and inhibition below. This double response, named by them "peristalsis" moved down the bowel, carrying with it the contents (Fig. 49.6). The mechanism depends upon a local reflex in the nerve-plexuses present in the intestinal wall, and for this reason has also been termed the "myenteric reflex" (Cannon). Many variations on this fundamental pattern of propulsive

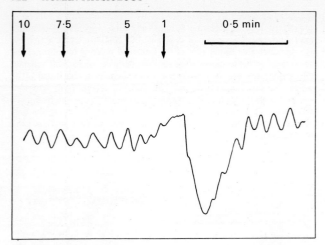

Fig. 49.6 Peristalsis. The passage of a bolus along the small intestine. Contractions of the longitudinal muscle coat, upward deflexion indicating relaxation.

A bolus of soap and cotton wool was inserted into the small intestine 10 cm above the point from which records were taken. At the arrows, the bolus was 7.5, 5, and 1 cm above this point. When the bolus arrived about 5 cm away, the rhythmic contractions ceased and the muscle tone was inhibited. The bolus was forced past by a strong contraction behind it. Dog, lightly anaesthetised.

movement are seen in different circumstances; for instance "pendular movements" consists of a rhythmical lengthening and shortening of a segment of intestine which is caused by gentle waves of contraction which travel down the bowel for a short distance at about 2–5 cm per second and occur about 10–12 times a minute. Another type of movement often seen is "segmentation"; a portion of intestine becomes occupied more or less simultaneously by several localised contractions. After a few seconds these are replaced by new contractions in the intervening regions, so that the intestine is divided into a fresh set of segments (Fig. 49.7). By this means the food is mixed with the

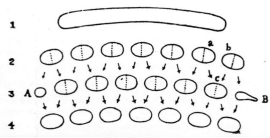

Fig. 49.7 Segmentation in small intestine.
 1. Loop of intestine before segmentation begins.
 2. The loop is cut into little ovoid pieces by contraction of the circular muscle.
 3. A moment later, each segment is divided into two parts, as shown by the dotted lines, and neighbouring parts, as *a* and *b* in line 2, run rapidly together and merge to form new segments, as *c* in line 3. The end pieces A and B are left small and move to and fro.
 4. The process is repeated, with a return to the condition shown in line 2. (Cannon.)

digestive juices and brought into intimate contact with the mucous membrane. A third type of movement which may be seen from time to time is the "rush wave", in which a rapidly moving contraction wave transfers the contents to a lower section of the bowel; such waves may occur frequently and vigorously in irritation or inflammation of the intestinal mucosa (enteritis).

The movements just described, and variations of them, form the normal complicated pattern of intestinal activity. But we do not yet understand very well how they are coordinated; for instance, what determines the appearance of a given type of movement in some region of the gut, its intensity and range of influence, why it gives place to some other movement after a time, and finally how the movements as a whole are kept in step with the progress of digestion, so that the food moves along neither too quickly nor too slowly.

The intestine as a whole shows a descending gradient of activity throughout its length. After a meal, the duodenum and jejunum show great and varied activity, but as the ileum is approached, the bowel becomes more and more quiescent, the terminal ileum making only occasional movements as it gradually fills with the residue of digestion.

Filling and emptying of the colon

As the stomach empties, the ileum is stimulated reflexly (*gastroileal reflex* of Hurst), to pass on its semifluid contents into the caecum by sustained "stripping" contractions which occur every few minutes and persist while food remains in the stomach. All movement of the caecum is inhibited and it relaxes to receive the ileal contents. Gradually the caecum, ascending and transverse colons are filled, without obvious peristalsis or other movement, the general appearance being one of "impressive immobility" (Hardy). There is a wide range of variation among normal persons in the time taken for different regions of the colon to become filled; but as soon as caecum and ascending colon are well filled, the saccular folds known as haustra appear (Fig. 49.8a), and by their slow filling and emptying knead the contents and aid the absorption of salts and water.

This slow and irregular process of filling is interrupted two or three times a day by a "mass movement" (Fig. 49.8b). Starting usually about the middle of the colon, the haustra disappear and the colon becomes shortened and flattened by a rapidly advancing powerful contraction, and its contents are moved on bodily into the descending and pelvic colons in a few seconds usually without any subjective sensations whatever.

In most people the rectum is almost empty until just before the urge to defaecate or "call to stool" comes (commonly after breakfast) which is caused by a mass movement distending the rectum with faeces. The attainment in this way of a certain degree of distension initiates afferent impulses which are sent to a "defaecation centre" in the sacral region of the spinal cord. In newborn animals and infants, or after complete transection of the spinal cord, efferent impulses then return from this centre, and produce contractions of the terminal colon, relaxation of the sphincters, and involuntary or "automatic" defaecation. In normal adults, filling of the rectum is appreciated in consciousness; if circumstances are suitable the sacral centre is "permitted" and even "encouraged" by the cerebral cortex to operate as described above. Expulsion of the faeces is assisted by "straining", i.e. raising the intra-abdominal pressure by expiring against a closed glottis and contracting the abdominal muscles; emptying of the anal canal is completed by contracting the levator ani muscles, which also restores the averted mucous membrane. If on the other hand defaecation would be inconvenient, activity of the sacral centre is inhibited by the cerebral cortex. The tone of the anal sphincters increases and that of the colon decreases; the faeces in the rectum move back into the colon and the desire to defaecate disappears.

INTESTINAL DIGESTION 323

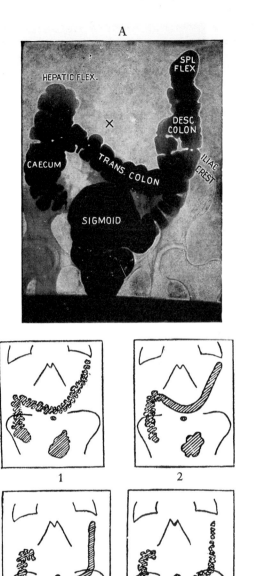

Fig. 49.8 A. Radiograph of the human colon after a barium enema, showing haustra.
B. A diagram (Holzknecht) of a "mass movement". (From A. E. Barclay, *The Digestive Tract*.)

Colonic secretion

The colonic mucosa contains very large numbers of mucous cells; and Florey has shown that a secretion of mucus, accompanied by vasodilatation and contractions of the muscle, is produced by stimulation of the parasympathetic nerve supply, the pelvic visceral nerves. Stimulation of the sympathetic supply causes vasoconstriction and inhibition of movements without secretion.

ABSORPTION OF THE DIGESTION PRODUCTS

The villi

The columnar epithelium of the small intestine is specially adapted for absorption of the products of digestion by the presence of the villi. These are finger-like projections of the surface, about 0·5 mm long, containing a strand of muscle from the muscularis mucosae, blood-vessels, nerves and a central lymph vessel termed a lacteal; the surface available for absorption is thus greatly increased. Between the villi open the mouths of the intestinal glands.

During fasting, the villi are shrunken and motionless; but during digestion they swell up, due to the increased blood and lymph flow through them, and contract rhythmically and independently of each other (Fig. 49.9). These movements are probably of value in maintaining a good circulation through each villus and ensuring that this is constantly brought into contact with fresh portions of the intestinal contents. The water-soluble products of digestion, such as the amino acids and sugars, are absorbed into the portal venous bloodstream and so pass through the liver before gaining the general circulation.

The pressure in the portal vein is about 20 mmHg (2·7 kPa), which is higher than the hydrostatic pressure of the intestinal contents. The intestinal wall, however, is impermeable to colloids, so that unless there is an appreciable colloid osmotic pressure within the intestine, water may well be absorbed as a result of the colloid osmotic pressure of the plasma proteins. The end result of the digestive processes is the breakdown of all colloidal material into crystalloidal; finally, therefore, the intestinal contents exert no colloid osmotic pressure. But in the intermediate stages there may well be a substantial colloid concentration, and water may be drawn in from the blood. Similarly, any substances which can diffuse through the intestinal wall will do so if their concentrations within the intestine are greater than those in the blood; if there is no such concentration gradient in either direction, they will be carried through with the water. There is reason to believe that some substances of relatively small molecular weight normally leave the intestine in this way. On the other hand, other substances are transferred from the intestine to the blood proportionately more rapidly when in low concentration than in high by a "facilitated" process; or they may be absorbed

Fig. 49.9 Portion of a cinematograph film following the movements of the villi in the living intestine of the dog.
The interval between each frame is approximately one second. Note that the villus indicated by the arrow in the right-hand frame becomes progressively shorter until it can only just be seen in the third and fourth frames from the right; it then becomes longer again, and is only a little shorter in the last frame than it is in the first. (Kokas and Ludany.)

even when the intestinal concentration is less than the blood concentration, i.e. "up" the concentration gradient. Some secretory process, or "active transport", must be involved and metabolic energy, oxidative or glycolytic, is needed. As examples, we may mention urea, xylose and erythritol, which appear to be absorbed by diffusion only, while glucose and most amino-acids are absorbed, at least partly, by active transport. Sodium and chloride ions are absorbed rapidly, active transport (the "sodium pump") being involved, but calcium, magnesium and sulphate ions are absorbed very slowly.

Such feats on the part of the intestinal epithelium are reminiscent of those performed by the kidney tubule cells in producing urine from the glomerular filtrate and the intracellular mechanisms involved are no doubt similar.

Absorption of fat

During the absorption of a fatty meal, the lymphatics draining the intestine can be seen to be filled with a creamy fluid, *chyle*, which consists of lymph loaded with globules of neutral fat. This very early observation (Asellius, 1622) gave rise to the natural assumption that the emulsified fat in the intestines was absorbed unchanged, much as oil soaks through paper; but Claude Bernard's discovery (1846) of the powerful lipase present in the pancreatic juice originated the view which is generally accepted today, that the greater part of the fat is hydrolysed in the small intestine before absorption.

Most of the dietary fat is in the form of triglycerides. In the duodenum, an oil-in-water emulsion is formed, and in this the bile salts play an essential role; pancreatic lipase acts at the oil–water interface and produces fatty acids and monoglycerides. These substances, together with the bile salts, form stable water-soluble molecular aggregates, or "micelles", which can take up other water-insoluble constituents of the intestinal contents, such as the fat-soluble vitamins. The micelles enter the absorptive cells of the intestinal mucosa, leaving behind the bile salts, which are absorbed in the terminal ileum (p. 321). In the intestinal cells, triglyceride is resynthesised and the fat droplets so formed become enclosed in a layer of protein and lecithin forming "chylomicrons". These pass into the central lacteal of the villi (Fig. 49.10) and are ultimately discharged into the systemic venous blood via the thoracic duct. Most of the fat in the blood after a meal is in the form of these minute droplets, 0.5–1.0 μm

Fig. 49.10 A photomicrograph of the small intestine of a rat, illuminated with ultraviolet light, during the absorption of fat containing vitamin A. This is fluorescent, and is visible in the epithelium, central lacteal of the villus and submucosal lymphatics. (Popper and Greenberg.)

in diameter, which may be counted to follow the progress of fat absorption.

Fatty acids of shorter chain length which are more readily soluble in water are absorbed into the portal venous blood like the products of protein and carbohydrate digestion. Thus, although almost all of the stearic (C_{18}) and palmitic (C_{16}) acids of fat are found in the intestinal lymph after feeding, only about one-half of lauric (C_{12}) and one-fifth of decanoic (C_{10}) acids take this route, the remainder entering the portal blood.

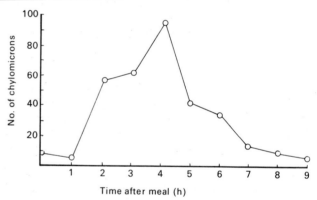

Fig. 49.11 Curve showing number of chylomicrons in a standard field (under the microscope) after a fatty meal. In this subject, the peak was reached after 4 h.

50. General Metabolism

D. H. Smyth

The tissues forming the animal body are composed of chemical substances all of which are derived from its environment, and during the animal's life these substances are returned to that environment many times over. In spite of this dynamic equilibrium the living animal maintains its individuality and indeed life is the continuity of this individuality. At death the continuity is broken, and the elements which composed the animal body become once more part of the environment. This maintenance of the individual depends on an elaborate series of chemical processes and the combination of all these reactions is called metabolism. These chemical reactions can be regarded mainly as of two kinds. The first is concerned with the building up of the complex substances of the body tissues out of simpler substances. These processes are responsible for the maintenance of the tissues by replacement of the loss due to wear and tear and in addition to this, in the young animal, they are responsible for growth and development. These synthesising reactions (sometimes grouped together under the term *anabolism*) do not supply the energy which the body needs for carrying out its functions; indeed, they themselves require a certain amount of energy from some other source. The source of this energy, and also of the energy which the body needs for its other activities, is provided by a second set of chemical reactions which are grouped together under the term *catabolism*. In these reactions more complex chemical substances are broken down into simpler ones and this disintegration which is mostly of an oxidative nature, is accompanied by liberation of energy. The complex substances, capable of yielding energy on oxidation, are taken into the body as food and it has already been seen how the breaking-down process begins in the intestine. Digestion, however, is not an oxidative process, and is accompanied by liberation of only a very small fraction of the energy of the food substances. The main liberation of energy takes place after the food has been absorbed into the blood stream and it is those changes subsequent to absorption that the term metabolism is generally applied.

The general aim of metabolic studies is to determine how the chemical energy of the food substances is utilised in contraction of muscles, secretion of glands, transmission of impulses along nerves, growth of tissues and the other activities characteristic of the living animal. The present position of the problem is that a very great deal is known about the chemical reactions which occur and the amount of energy made available, but less is known about how this energy is used by the tissues for their purposes. Consequently most of this chapter is devoted to an account of the chemical processes occurring in metabolism, the methods used for their study, the end products produced and the energy relations involved in these processes.

GENERAL METHODS OF METABOLIC STUDIES

The actual oxidative processes take place in the separate cells which compose the body, but the provision of the metabolic fuel, and the utilisation of the energy liberated, is only possible when the cells are organised into tissues, organs, whole animals and indeed societies of animals. On the other hand, many of the chemical reactions cannot easily be studied in the whole animal, but only in isolated organs or tissues. Hence it is necessary to study metabolic problems at a series of different levels of biological organisation, i.e. to study the metabolism of individual cells, of isolated tissues, or organs, and of whole animals. From the disconnected pieces of information obtained from these different sources we try to form a composite picture of the whole process of metabolism. Each type of metabolic experiment has its own special use. In the case of the whole animal we can administer substances by mouth, intravenously, intraperitoneally or subcutaneously, collect the waste products in the urine, estimate chemically the changes in the blood, measure the gaseous metabolism and thus study the energy relationships under physiological conditions. In particular, studies on the whole animal are essential for investigation of the effect of hormones on metabolism and the total nutritional requirement of the individual. Under this heading must also be placed the mass of clinical observation which has contributed greatly to our knowledge of physiological processes in the field of metabolism. Experiments with various organs make possible a rather fuller study of the chemical changes undergone by the food substances and also bring the oxidative processes into relationship with the special activity of each particular organ. The use of isolated tissues has made possible a very detailed study of the enzyme systems of the body and of the chemical changes brought about by these, even although the tissues in the experiments are not working under physiological conditions. The study of the individual cells has shown that many of the enzymes can be located in definite cells and in some cases in definite positions in the cell. Until recently the kilocalorie (kcal) was always used to express energy changes in living tissues. The introduction of SI units requires the joule (J), kilojoule (kJ), or megajoule (MJ) to be used.

The respiratory quotient

Whatever method we use to investigate metabolism we consider the amount of oxygen used, the amount of carbon dioxide formed, the nature and amount of the foodstuff oxidised and the energy liberated, and all these must bear a strict relationship to each other. Since, of these quantities, the amounts of carbon dioxide and of oxygen are often the simplest to measure, and also because they give easily obtainable quantitative information about metabolism, it is of great importance to understand their full significance.

If the food substance being oxidised is glucose, the reaction (or rather the sum total of a large number of intermediate reactions) is as follows:

$$C_6H_{12}O_6 + 6O_2 = 6CO_2 + 6H_2O$$

Since the volumes of gases are proportional to the numbers of molecules present, the volume of carbon dioxide formed will be exactly equal to the volume of oxygen used. The ratio of carbon dioxide produced to oxygen used, the respiratory quotient, in this case is equal to one. In a similar way, if equations for the oxidation of fats are written down the respiratory quotient can be calculated for each case; thus for triolein

$$C_{57}H_{104}O_6 + 80O_2 \rightarrow 57CO_2 + 52H_2O$$

The respiratory quotient in this case is 57/80, or 0·71. In the case of proteins, it is not possible to work out the respiratory quotient in this simple way, but from a knowledge of the percentage composition of the protein, and the products of oxidation in the body, it is possible to calculate that the respiratory quotient is about 0·8.

The respiratory quotient for all forms of carbohydrate is 1·0. The exact value of the respiratory quotient will vary for different fats and proteins, but by taking a mean value for those normally present in the food it can be determined that the respiratory quotient for a diet consisting of fat would be 0·71 and for protein 0·81. A knowledge of the respiratory quotient therefore gives us information about the type of food which is actually being oxidised, and since the energy liberated by a certain amount of oxygen depends on the nature of the substance oxidised, it is evident that, from the amount of oxygen used and from the respiratory quotient, much information can be obtained about metabolic activities. The significance of the respiratory quotient will be considered more fully later in connection with the metabolism of the whole animal.

Metabolism of the cells

The study of the metabolic machinery of the cell in relation to its structure is called histochemistry or cytochemistry, and the methods used are largely a combination of histological and biochemical techniques. By disintegration of cells and subsequent centrifuging it has been possible to divide the cell contents into different fractions, and to determine the content of enzymes and other substances in these fractions. As a result of this work, information is gradually being built up on the metabolic activities of the various subcellular particles and the relation of these to each other. For example, it is known that the enzymes concerned with the oxidative processes in the cells are located in the mitochondria.

Metabolism of isolated tissues

In the higher animals respiration is often divided into external respiration and internal respiration. External respiration includes all the processes which result in oxygen being brought to the tissues and carbon dioxide being removed. In contrast to this, internal respiration includes the processes which take place inside the cells of the body resulting in the oxidation of the food substances and the formation of the waste products. The study of internal respiration is largely carried out by investigating the chemical changes which take place in isolated tissues. The remarkable thing is that most living tissues, disintegrated to a greater or lesser degree by mincing or slicing and deprived of their blood supply, can still, under suitable conditions, take up oxygen, carry out chemical reactions and produce carbon dioxide. The requisites for respiration in isolated tissues are a supply of available oxygen, the maintenance of normal body temperature, and the provision of a suitable fluid medium in which the tissue is suspended. The media usually employed are balanced salt solutions (see Chapter 16) and the other essential conditions are achieved by the use of respirometers, which are mostly developments of the type used by Warburg.

The tissue, either minced or sliced, is put into a glass vessel which can be attached to a manometer. The manometer is fixed on a stand so that it, together with the attached cup, can be shaken continuously with the cup immersed in a water bath at constant temperature. A centre compartment in the cup contains a small piece of filter paper soaked in caustic soda, and this absorbs the carbon dioxide produced. After introducing the tissue suspended in a suitable saline medium, the whole apparatus is filled with oxygen, and the shaking apparatus is then set in motion. Readings are taken at regular intervals and the consumption of oxygen is indicated by the change of pressure in the manometer. The manometer cups are usually provided with one or two side arms into which reagents can be put which are to be added during the course of the experiment. Analysis of the contents of the cups can also be made at the end of the experiment and the products of metabolism estimated. There are many modifications of the manometers and cups and it is possible by various techniques to measure the respiratory quotient or the changes taking place anaerobically.

It is important to appreciate both the possibilities and the limitations of experiments with minced and sliced tissues. There is no doubt that the conditions are highly unphysiological, the tissue is deprived of its blood supply and its normal relations with other tissues. The chemical reactions which can be demonstrated by these techniques do not necessarily take place in the tissues under more physiological conditions and furthermore, we derive no information about how the tissues utilise the energy liberated by the reactions which occur. At the same time the method is invaluable for studying the various enzyme systems present in the cells, and for finding out what substances actually take part in the chemical reactions in the tissues. It is not too much to say that the enormous advances in many fields of biochemistry have been due in great part to experiments of this kind.

From the study of minced and sliced tissues it has been found that the food substances oxidised do not combine directly with oxygen to form the final products of oxidation, but pass through a number of intermediary stages. One of the methods of studying these intermediary stages is to interfere with the normal process of metabolism by blocking the series of reactions at some point so that intermediary substances, which would normally be further oxidised, accumulate in sufficient amount to allow their detection and estimation. A number of such blocking agents or inhibitors is known which will specifically put out of action certain enzymes in the tissues, while allowing others to function normally, and by the use of these many of the intermediate products of metabolism have been identified. A few of the outstanding results of this kind of work are referred to in later sections, but for a more complete account of the chemical changes undergone by the food substances, the reader is referred to textbooks of biochemistry.

Metabolism of isolated organs

It is often useful to study the metabolism of some particular organ in relation to its functional activity, e.g. to study kidney metabolism in relation to the formation of urine or cardiac metabolism in relation to the pumping activity of the heart. For such purposes it is necessary to use whole organs instead of isolated tissues, but it is also necessary to separate in some way the metabolism of the organ being investigated from the metabolism of the animal as a whole.

Perfusion experiments. One method of attacking such a problem is to supply the organ with nutrient material and oxygen by an artificial perfusing system instead of by its own circulation. Heparinised blood is usually employed for the purpose and this is saturated with oxygen, either by passing it through the lungs or through an oxygenator, and is then pumped through the vessels of the organ to be studied. By such a system organs can be kept "alive" for a number of hours after complete isolation from the rest of the body. The metabolism of heart, lungs, liver, kidney, brain and limbs have all been studied in this way. The gaseous metabolism is measured by estimating the volume of oxygen taken up by the blood and the volume of carbon dioxide given off, or alternatively it may be measured by estimating the content of oxygen and carbon dioxide in the blood supplying and leaving the organ and also the rate of blood flow. In some cases, e.g. heart and voluntary muscle, the physical work done can be measured and compared with the rate of metabolism; in other cases, e.g. the kidney, the osmotic work can be studied. In some organs, however, such as the brain, we know very little about the quantitative relation between functional activity and metabolism.

Metabolism of organs in situ. Another method of studying the metabolism of individual organs is to leave the organ *in situ*, and measure the oxygen and carbon dioxide content of the arterial and venous blood together with the rate of blood flow. For this purpose it is very useful to have an apparatus which will register continuously the amount of oxygen in the arterial and venous blood.

This can be done by means of two *oximeters* in which the degree of oxygenation of the blood is measured by means of photoelectric cells, in terms of the optimal transmittance at certain appropriate wavelengths. A continuous graphic record may thus be obtained of the oxygen content of the arterial and of the venous blood.

The metabolism of the intact animal

When the metabolism of the whole animal is considered it is usually not in terms of the intermediary products of metabolism but rather of the sum total of the metabolic reactions, i.e. the amount of food and of oxygen used and the amount of carbon dioxide and waste products formed. It is also possible to measure the total heat production of the animal and the amount of physical work carried out, so that balance sheets can be prepared of the total intake and output of energy in all its various forms; thermal, chemical, mechanical, etc. Very careful measurements of these energy relations have shown that the animal body behaves exactly like all other chemical or mechanical systems as regards the law of the conservation of energy, in that the total energy produced is equal to the total energy supplied. The vital activities of the body are essentially a transference of energy from one form to another. The particularly "vital" part of the process is that some of these energy transformations can only take place in living tissues and so far have not been imitated in non-biological systems.

The energy value of the foods

In order to prepare our complete balance sheet we require to know the energy values of the foods taken. For this purpose we assume that oxygen is freely available and therefore consider the amount of energy capable of being liberated by oxidation of the food. This can be determined outside the body by means of the bomb calorimeter.

This consists of a strong steel chamber which can be sealed by a tightly fitting lid. Into the chamber a measured quantity of the food substance is introduced and the whole apparatus filled with oxygen at high pressure. The bomb calorimeter is now placed in a known volume of water at a certain temperature. Combustion of the contained substance is initiated electrically and the amount of heat produced is estimated from measurement of the rise in temperature of the surrounding water.

By means of the bomb calorimeter it is found that 1 g of carbohydrate produces 4·1 kilocalories (17 kJ), 1 g of fat 9·2 kilocalories (39 kJ) and 1 g of protein 5·3 kilocalories (22 kJ). These are the energy values when combustion is complete. In the body complete combustion of carbohydrate and fat takes place, but in the case of protein the end product, urea, is still capable of further oxidation, though not inside the body, and hence the energy value of the urea must be subtracted from that of the protein. Making these corrections the values for the three types of food are: carbohydrate and protein each 4·1 kcal g^{-1} and fat 9·2 kcal g^{-1}. Knowing these values and also the amount of each food substance in the diet, we can calculate the total energy provided by the food. This knowledge of the energy content of food is essential in making up diets for people under various conditions.

Energy production from oxygen intake

The food taken into the body gives the total energy intake but it does not tell us the rate of metabolism at any one time, since the food is not all used immediately but may in part be stored. The rate of metabolism can be derived from the oxygen intake, for we know the oxygen intake only keeps pace with the immediate metabolic needs. If we wish to know the metabolic rate approximately, we can calculate it from the oxygen consumption by assuming that each litre of oxygen used in the body yields 4·8 kilocalories (20 kJ). Thus in the human subject where there is an oxygen consumption of 300 ml min^{-1}, the energy production would be 87 kilocalories per hour (0·37 MJ h^{-1}). This degree of accuracy is in fact sufficient for many purposes.

If we require to know the metabolic rate more accurately, we must consider the fact that the energy liberated by the consumption of 1 litre of oxygen in the body is not a constant figure of 4·8 kilocalories, but varies with the kind of food being oxidised. The food taken is a mixture of protein, fat and carbohydrate, so that in addition to determining the oxygen consumption we have to determine what proportions of these three kinds of foods are present in the diet. This can be done in the following way.

The total nitrogen in the urine is estimated over a known time. By multiplying this value by 6·25 one can calculate the total amount of protein metabolised during that period. (The average amount of nitrogen in dietary proteins is 16 per cent.) We can further calculate the amount of oxygen needed to oxidise this protein in the body and the amount of carbon dioxide produced. These amounts of oxygen and carbon dioxide are now subtracted from the total amounts of oxygen and carbon dioxide involved in metabolism, and the resultant figures gives the oxygen used and the carbon dioxide produced in non-protein metabolism. Since there are now only two substances to be dealt with, fat and carbohydrate, and since we know the respiratory quotient corresponding to each, we can calculate the proportions of each necessary to give the respiratory quotient of the non-protein metabolism. In practice one gets the result from tables already worked out for each possible respiratory quotient (Table 50.1). Such tables also give the energy production per litre of

Table 50.1 *The relation between the respiratory quotient, the relative amounts of fat and carbohydrate oxidised, and the energy production of the non-protein metabolism*

Non-protein respiratory quotient		1·00	0·95	0·90	0·85	0·80	0·75	0·718
Percentage total O_2 consumed by carbohydrate		100	82	65	47	29	11	0
Grams foodstuffs per litre O_2	Carbohydrate	1·23	1·01	0·80	0·58	0·36	0·14	0
	Fat	0	0·09	0·18	0·27	0·36	0·45	0·50
	Total	1·23	1·10	0·98	0·83	0·72	0·59	0·50
Kilocalories per litre O_2		5·05	4·99	4·94	4·88	4·83	4·77	4·74

oxygen for each respiratory quotient so that the total energy production can readily be calculated once the total oxygen consumption and the non-protein respiratory quotient are known.

Since the respiratory quotient for protein is intermediate between that for fat and for carbohydrate it is often considered sufficiently accurate to neglect altogether protein metabolism, and treating the whole metabolism of the animal as non-protein metabolism, to make the calculations accordingly.

It is also possible to make the necessary calculations about the gaseous exchanges and the type of food being oxidised from the following equations.

Carbohydrate (g) = $4·12\ CO_{2m} - 2·91\ O_{2m} - 2·54\ U_N$

Fat (g)   = $1·69\ O_{2m} - 1·69\ CO_{2m} - 1·94\ U_N$

Protein (g)   = $6·25\ U_N$

Energy (cal)   = $3·78\ O_{2m} + 1·16\ CO_{2m} - 2·98\ U_N$

where CO_{2m} is the number of litres of carbon dioxide produced, O_{2m} is the litres of oxygen used, and U_N is the urinary nitrogen expressed in grams.

In making calculations from the respiratory quotient it is assumed that the amount of carbon dioxide given out in the expired air is the same as that being produced by the tissues. While this is true if the carbon dioxide output is considered over long periods, it is not necessarily so over short periods. During strenuous exercise lactic acid accumulates in the blood and this results in liberation of some of the carbon dioxide stored in the blood as bicarbonate, so that the amount of carbon dioxide expired by the lungs is greater than the amount which is being formed by the tissues. In such conditions, the respiratory quotient can in fact exceed 1·0. Similarly, in the period of recovery after exercise when lactic acid is being removed from the blood, the amount of carbon dioxide expired may be less than the amount formed by the tissues, as some is being retained by the blood to form bicarbonate. Another factor which can contribute to changes in the respiratory quotient is interconversion in the body of the various food substances.

Methods of determining metabolic rate in the intact animal

Two different principles are used for this purpose—direct calorimetry and indirect calorimetry. In the former the object is to measure the heat production directly, while in the latter the heat production is calculated from the gaseous exchange.

DIRECT CALORIMETRY

This involves very elaborate and expensive apparatus and is therefore little used in routine measurement. For some purposes, however, it gives information not obtainable in any other way, e.g. if we wish to draw up a complete balance sheet of the energy exchanges in the body. The subject is placed inside a large calorimeter, which in the case of the human subject is a small thermally-insulated room known as the Atwater–Benedict respiration calorimeter (Fig. 50.1). The calorimeter is in fact

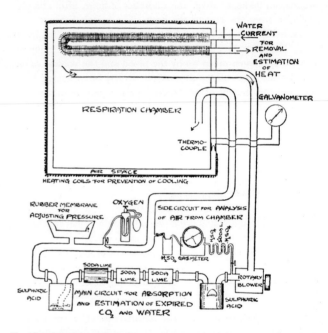

Fig. 50.1 Diagram to illustrate the principle of the Atwater-Benedict respiration calorimeter.

The upper part represents the calorimeter, in which the subject is placed, and his heat production measured in terms of the rise in temperature and rate of flow of the cooling water (in practice the cooling pipes are carried right round the chamber).

The lower part represents the Benedict respiration apparatus, in which the carbon dioxide produced by the subject is absorbed by the soda-lime, and the oxygen used is replaced from the cylinder through a meter (not shown) at such a rate as to keep the rubber membrane of the pressure equaliser in a constant position. The gas is dried in the sulphuric acid bottles before entering the carbon dioxide absorbers, since soda-lime absorbs water as well as carbon dioxide. Any water that may be evolved from the soda-lime in its reaction with carbon dioxide is absorbed in the second sulphuric acid bottle.

When the respiration apparatus is used for indirect calorimetry only, the tubes leading to the calorimeter are connected to a mouthpiece, which is held between the subject's teeth, his nose being closed by a clip. (Parson's *Fundamentals of Biochemistry*.)

constructed so that the measurement of heat production is combined with the indirect method of measuring metabolism, i.e. by the gaseous exchanges; and these accessories are shown in the diagram. An energy balance drawn up from procedures of this kind is shown in Table 50.2.

Table 50.2 *Energy balance sheet for human subject*

I. *Indirect calorimetry*

Food ingested[1]		Class of food	Food oxidised[2]	
Weight (g)	Energy (MJ)		Weight (g)	Energy (MJ)
79.2	1.88	Protein	64.8	1.54
59.6	2.39	Fat	117.8	4.72
201.0	3.54	Carbohydrate	226.3	3.98
339.8	7.81	Total	408.9	10.24
+69.1	+2.43	← From body stores (i.e. excess of food oxidised over food ingested)		
408.9	10.24	*Total*	408.9	10.24

II. *Direct calorimetry*

	MJ
Heat produced, as measured	9.80
Potential energy of urine	0.38
Total heat produced	10.18

III. *Balance sheet*

	MJ
Heat production as calculated from results of indirect calorimetry	10.24
Heat production as observed by direct calorimetry	10.18
Difference	0.06
	(i.e. 0.6%)

[1] Corrected for losses in digestion and absorption.
[2] Calculated from respiratory exchange.

INDIRECT CALORIMETRY

The principle of calculating the metabolic rate from the oxygen consumption has been discussed above. The problem at the moment is the technique of measuring the gaseous exchanges. Two different principles are used for this purpose and these are distinguished as closed methods and open methods.

CLOSED METHODS

These in general require a less mobile equipment and hence are limited in their use. In the simpler types of apparatus for the human subject the expired air, collected from a mouthpiece supplied with valves, passes over soda-lime which absorbs the carbon dioxide. The soda-lime is usually contained in a recording spirometer, and from this the subject rebreathes his own expired air freed from carbon dioxide. During this process the volume of the air in the circuit diminishes at a rate equal to the consumption of oxygen and by recording the volume change by means of the spirometer the rate of oxygen consumption can be measured. It will be noted that in such a method no account is taken of the respiratory quotient and the metabolism is calculated by assuming an average for the respiratory quotient.

By means of more elaborate circuits it is possible to measure the carbon dioxide production as well. In such cases the air is circulated by a pump and passes through containers with sulphuric acid to absorb moisture, and with soda-lime to absorb carbon dioxide. These can be weighed periodically and hence the production of water and carbon dioxide measured. In these systems the oxygen consumption is measured by adding oxygen at a known rate so as to replace the amount used. This is most conveniently done by having in the circuit, at some point, a sensitive rubber membrane, which will indicate alterations in the total pressure and hence in the volume of the system. Such circuits are often used for small animals. For application to man, Benedict's apparatus is employed. For more complete metabolic estimations in man, this system may be combined with the Atwater–Benedict respiration calorimeter.

By means of such apparatus a complete balance sheet of the energy exchanges in man can be made, and since much can be learned from the study of this, the results of one such experiment are given in Table 50.2.

OPEN METHODS

In these the subject breathes in from the atmosphere, but by means of a mouthpiece containing valves the expired air is collected in a Douglas bag. The volume of air expired in a given time is measured and a sample is taken for determination of the percentage of oxygen, nitrogen and CO_2 present. The oxygen utilisation is obtained by calculating the amount of oxygen in the expired air in a given time and subtracting this from the amount of oxygen in the inspired air in the same time. Similarly, the CO_2 produced is the amount of CO_2 in the expired air from which has been subtracted the amount of CO_2 in the inspired air (an almost negligible quantity). In making these calculations, particularly for oxygen, it must be remembered that the volume of inspired air is not the same as the volume of expired air, owing to the fact that the volume of oxygen taken in is greater than the volume of CO_2 given out, unless the RQ is exactly 1. It is the volume of expired air which is measured, but since the amount, although not the percentage of N_2 in the inspired and expired air, is the same, the volume of inspired air can be calculated as follows:

Vol. of inspired air =
$$\text{vol. of expired air} \times \frac{\text{percentage } N_2 \text{ in expired air}}{\text{percentage } N_2 \text{ in inspired air}}$$

The open methods of measuring gaseous exchanges have the merit of great mobility, and this enables metabolic rate to be measured in a great variety of conditions. In recent years much improved forms of the open method have been devised, in which the whole apparatus is more portable and very little resistance is applied to the flow of expired air, and one of the best of these is Wolff's integrating motor pneumotachograph (the IMP).

The basal metabolic rate

The rate of metabolism depends on the amount of physical work which the body does and, therefore, can be reduced if the subject remains completely at rest. But even when all unnecessary movements are stopped a certain amount of energy is used in maintaining the body temperature and in providing for the needs of circulation, respiration and other vegetative processes, and this cannot be further reduced without damage to the tissues. This amount of metabolism is considered the basal level, on which extra metabolic activities are superimposed when the body undertakes more work. It is called the basal metabolic rate, or more usually the BMR.

Although the term BMR is very firmly fixed, there is a case for replacing this by the term "resting metabolism". This is what is actually measured and makes no assumptions about what activities are basal. One of the functions of this resting metabolism in warm blooded animals is to keep the body temperature above that of its surroundings and, since the loss of heat from the body takes place from the surface exposed to the atmosphere, it is not

Table 50.3 *The basal metabolic rates of various animals*

Animal	Weight in kg	Megajoules produced per day per kg of weight	per m² of surface
Horse	441	0.05	3.98
Pig	128	0.08	4.53
Man	64.3	0.13	4.38
Dog	15.2	0.22	4.36
Mouse	0.018	0.89	4.69

surprising to find that the BMR is more closely related to the body surface than to the body weight. This is well illustrated in Table 50.3, which gives the BMR for animals of different sizes expressed per unit weight and per unit body surface.

The area of the body surface is not an easy quantity to measure directly. In man it is usually obtained from the height and weight by means of the following formula of Du Bois:

$$S = 0.007184 \times W^{0.425} \times H^{0.725}$$

where S is the body surface in square metres, W the weight in kilograms, and H the height in centimetres. It can be obtained more easily from nomograms based on this formula (Fig. 50.2).

In a well-nourished man, the BMR has been found to be about 7.14 MJ per day. This is equivalent to about 0.17 MJ per square metre body surface per hour, or to about 4 kJ h^{-1} kg^{-1} body weight. It varies with sex and with age, being higher in males and young people. It is usual to express the BMR of an individual as a percentage increase or decrease above or below these standard values. In normal individuals, the values lie within about 15 per cent of the standard values.

Since the metabolic activities of the body can be increased by taking food or by performance of work, it is very important in measuring the BMR that the subject should be at complete mental and physical rest and should be in a condition of fasting for about twelve hours. Extremes of temperature, previous diet, previous exercise, emotion or menstruation may all have some effect on the result.

The thyroid gland and the basal metabolic rate. Of the pathological causes of alteration in the BMR, the commonest is abnormal activity of the thyroid gland discussed in Chapter 32. In diseases where there is thyroid deficiency, as in cretinism and myxoedema, there is a low BMR and this can be increased by administration of thyroid extract. In exophthalmic goitre there is hyperthyroidism and here the BMR is considerably increased. Surgical removal of part of the thyroid gland leads to a fall in the BMR along with alleviation of the other symptoms. The active principle in thyroid extract which is responsible for the effect on metabolism is thyroxine. The stimulating effect of this and related compounds on metabolism has been used for their biological assay. Since pure thyroxine is administered in quantities of the order of milligrams, which may increase the metabolic rate by up to 1 MJ per day, its action is clearly independent of any specific dynamic action it may have as an amino acid. This could not amount to more than a small fraction of a calorie.

Specific dynamic action

Under basal conditions a certain amount of metabolism is going on at the expense of the body tissues and stores, resulting in a

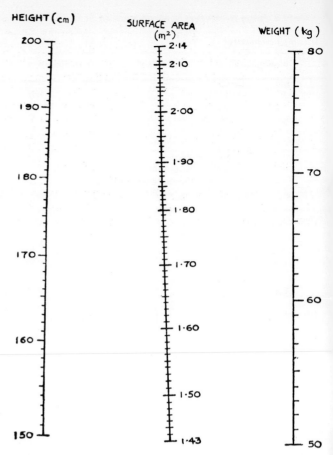

Fig. 50.2 Alignment chart for calculation of the area of the human body surface from the height and weight. (W. A. M. Smart.)

certain rate of energy production. If food is given to a fasting animal it is used to replace the body tissues as metabolic fuel, but in addition it causes an increase in the metabolism. For example, if sufficient protein is given to cover the calorie needs of basal metabolism, it is found that the metabolic rate rises by about 100 kJ and the protein intake must therefore be increased, if it is to cover metabolic requirements. This effect is called the *specific dynamic action*. It is shown by fat and carbohydrate also, but to a lesser extent. The exact cause of the specific dynamic action is uncertain, but is thought to be due to chemical changes carried out by the body on the food substances preparatory to oxidation. The specific dynamic action of protein has been most studied on account of its greater magnitude. If certain amino acids are injected into the bloodstream, they produce a specific dynamic action equal to that of the corresponding amount of protein, so that the effect cannot be due to digestion. The specific dynamic action of protein is prevented by previous removal of the liver, and this would suggest that the effect is probably concerned with deamination of the amino acids or formation of urea rather than with stimulation of cellular metabolism generally.

The use of isotopes in metabolic studies

A new phase in the study of metabolism began with the introduction of isotopes as a means of following the chemical changes which take place in the body. Many elements can exist in more than one form and these forms are known as isotopes. The isotopes vary in atomic weight, but are chemically indistinguishable from each other, and if metabolisable compounds are prepared containing the isotopes, many experiments have shown that the animal body does not treat these in any way differently from the corresponding compounds containing the ordinary form of the element. For example, amino acids containing nitrogen of atomic weight 15 (^{15}N) follow the same course of chemical changes as do those containing ordinary nitrogen of atomic weight 14. Two kinds of isotope are available for metabolic studies. There are, first, the stable isotopes which are distinguished by the mass spectrograph and, second, the radioactive isotopes, recognised by the radiations which they emit. Both kinds of isotope are extensively used, but, in general, the radioactive ones are preferred, when possible, on account of the simpler techniques for their estimation. The use of isotopes enables certain substances, or even parts of the molecule of the substance, to be "labelled", so that they can again be recognised after passing through various chemical transformations. Sometimes two isotopes are introduced into one molecule, e.g. the carbon chain of an amino acid may be labelled with the radioactive ^{14}C, while the amino group may be labelled with the stable ^{15}N or one of the hydrogen atoms replaced by deuterium. One of the great advantages of isotope studies is that chemical reactions can be investigated in whole animals under natural physiological conditions. These reactions could previously be investigated only on isolated tissues or organs, or under abnormal conditions where substances not normally metabolised by the body were used, or where inhibitors had to be used to stop metabolic processes at some particular stage.

51. The Metabolic History of the Food Substances

D. H. Smyth

In the preceding chapter the general methods of metabolic studies have been discussed and we can now turn to the special problems of the various substances in the food. The account given here is only a very bare outline of the chemical changes involved in metabolism; and for details, text-books of biochemistry must be consulted.

The energy-liberating substances are proteins, fats and carbohydrates. These are broken down in the alimentary tract, and from them are formed the substances which can be regarded as primarily available for metabolism, i.e. amino acids, fatty acids and glucose. All these undergo a series of chemical transformations which result ultimately in the formation of carbon dioxide and water. (In the case of protein some other end products are also formed, e.g. urea.) These chemical transformations can be divided into two stages, and one of these stages—the citric acid cycle—is common to all the different types of food substances. The pre-citric acid cycle stage is quite different for different food substances; it is therefore convenient to consider the pre-citric acid cycle stage for fats, proteins and carbohydrates, and then to consider subsequently the citric acid cycle which is common to them all (Fig. 51.3). The important connecting link between the two stages is a substance called coenzyme A, which therefore occupies a very special place in metabolism.

Coenzyme A. This is a complex substance containing adenosine and pantothenic acid (a substance related to one of the B vitamins) joined by two molecules of phosphoric acid. It exists in two forms, coenzyme A and acetyl coenzyme A, and its function can be regarded as a carrier of acetyl groups (CH_3CO—), which it is able to accept from one compound and transfer to another. Coenzyme A contains an —SH group, and since this is the part which combines with the acetyl radical, the free and combined forms of coenzyme A are often abbreviated to H—S—CoA, and CH_3COS . CoA. In general, the final reaction in the pre-citric acid cycle stage of metabolism is the formation of acetyl-coenzyme A. This transfers the acetyl group to the citric acid cycle and coenzyme A is re-formed, to take part in the further formation of acetyl coenzyme A.

Protein metabolism

The protein taken in the food is absorbed from the intestine after breakdown to small peptides and amino acids. Some of these amino acids are subsequently resynthesised to protein to help to replace the worn tissues of the body, or to produce hormones or enzymes. The remaining amino acids are used to supply energy. The replacement of the worn tissues can be carried out only by protein supplied in the food, whereas the energy supplying function of the protein can be replaced by fat or carbohydrate. Since protein is the most expensive part of the diet, and under many conditions the part most likely to be scarce, it is important to determine precisely how far it can be replaced by the other food substances, and what is the minimum protein intake. This can be discovered by studying the nitrogen balance of the body.

Nitrogen equilibrium

When an animal is living on an adequate diet and maintaining its weight at a constant level, it takes in a certain amount of nitrogen in the protein of the food, and loses the same amount of nitrogen in the excreta. Such an animal is said to be in nitrogen equilibrium. The nitrogen in the excreta comes partly from the nitrogen of the food, and partly from the breakdown of protein in the animal's own tissues, this being replaced again from the food. If now the animal be given a diet with ample fat and carbohydrate, but completely lacking in protein, it continues to excrete some nitrogen from its own tissues, and since this is not replaced it loses weight and ultimately dies. If after a short period of nitrogen starvation protein be added to the food in known amounts and at the same time the nitrogen loss estimated, it can readily be determined how much nitrogen must be given in the food to bring the animal back to nitrogen equilibrium. The result of such an experiment is given in Table 51.1.

Table 51.1 *Establishment of nitrogen equilibrium in a dog after starvation*

Food	Nitrogen in food (g)	Nitrogen in excreta (g)	Difference (g)
Starvation	0	4·00	−4·00
100 g meat	4·10	5·56	−1·46
140 g meat	5·74	6·50	−0·76
165 g meat	6·77	7·22	−0·45
185 g meat	7·59	7·80	−0·21
200 g meat	8·20	8·73	−0·53
230 g meat	10·24	10·58	−0·34
360 g meat	11·99	12·05	−0·06
410 g meat	15·58	14·31	+1·27
360 g meat	13·68	13·62	+0·06
Starvation 3rd day	0	4·03	−4·03

From such an experiment two important conclusions can be drawn. First, the minimum protein requirement for the particular animal can be seen. In the example given, an intake of about 360 g of meat is required to make the intake of protein balance the loss. But another important result appears. The amount of nitrogen which a starving animal excretes is not the amount which is needed to maintain it in nitrogen equilibrium, for it can be seen that by the time it has reached nitrogen equilibrium it is taking in and excreting about three times the amount excreted in the starvation state. The principal reason for this is the following. The protein given in the food does not supply the amino acids in the proportion required by the body, so that food containing a considerable excess of some may have to be given to supply a sufficient quantity of others. The amount of protein needed to maintain nitrogen equilibrium depends, therefore, on the type of proteins supplied; and the value of proteins is estimated from the amounts they contain of certain of the amino acids, which the body cannot manufacture for itself, and which are, therefore, regarded as essential in the diet. If an animal be fed on a protein deficient in any one of these, it cannot be kept in nitrogen equilibrium no matter how much of the protein is supplied. An essential amino acid has been defined as one which cannot be synthesised by the organism out of materials ordinarily available at a speed commensurable with the normal requirements. The essential amino acids are determined for the rat are: lysine, valine, tryptophan, methionine, histidine, phenylalanine, leucine, isoleucine, threonine and arginine. In the human subject arginine and histidine are not essential in the adult; while in some species glycine is essential at a certain stage of growth. From the practical aspect, the important point is that proteins can be divided into those of high biological value (first-class proteins) and low biological value (second-class proteins), the first group containing the essential amino acids in proportions approaching those required by the body. The term "first-class protein" usually means protein of animal origin (proteins of milk, cheese, meat, eggs, etc.).

THE BREAKDOWN OF PROTEIN

The amino acids not used for synthesis of body tissues are oxidised with liberation of energy. The first stage in the process is deamination, i.e. the removal of the nitrogen-containing group from the rest of the molecule. If amino acids are injected into the bloodstream they are rapidly removed and can be partly recovered from various tissues. Of these the liver has been found to take up the largest quantity. It can further be shown that the amino acids taken up by the liver gradually disappear, suggesting that they undergo some transformation. At the same time as the amino acids disappear from the blood there is a rise in the concentration of blood urea. It is possible to keep an animal alive for some time after removal of the liver and, if in such a preparation amino acids be injected, they are only removed slowly from the blood and there is no increased formation of urea. These experiments together with much other evidence show that the liver plays a very important part in the formation of urea from amino acids. Urea is one of the end products of protein metabolism. It does not undergo further change and is excreted in the urine. A small fraction of the amino groups is not excreted in this form, but as ammonia which is formed largely in the kidney itself, the amount being related to the acidity of the urine.

Formation of urea

If ammonium salts containing ^{15}N be fed to animals the isotope appears in the urea, so that formation of ammonia is probably a preliminary process in the formation of urea. The method of formation of urea from ammonia is still a matter of discussion, but it seems very likely that it depends on a cyclical series of reactions involving ornithine, citrulline and arginine. Ornithine combining with carbon dioxide and ammonia forms first citrulline and then arginine, which is hydrolysed by the enzyme arginase to form urea and ornithine.

Fate of the non-nitrogenous residues

The non-nitrogenous part of the amino acid left after deamination undergoes oxidation, the ultimate products being carbon dioxide and water. In carnivorous animals much of the energy needed by the body is obtained from this source, but in man the amount obtained is relatively small, depending on the excess protein in the diet over the protein minimum. The immediate product of deamination is a keto acid but the type of keto acid will vary according to the amino acid from which it is derived. All of them are ultimately oxidised by the citric acid cycle, but before they reach that stage, the metabolism proceeds by two different routes. Some of the keto acids, e.g. pyruvic acid, are known to be intermediaries in carbohydrate breakdown and could thus provide a route by which the further breakdown of protein might follow the line of carbohydrate metabolism, or by which proteins could cause formation of carbohydrates. It can be shown that this does actually happen under certain conditions. In a diabetic animal, or in an animal poisoned with *phlorrhizin*, there is a great loss of sugar from the body. (Phlorrhizin is a drug which causes elimination of sugar through the kidneys without any increase in the blood sugar concentration.) In such an animal, the carbohydrate stores are rapidly depleted. If now carbohydrates be withheld from the diet but protein be given it is found that the excretion of glucose continues, and this must have been derived from protein.

The amino acids which can give rise to glucose during metabolism are glycine, alanine, valine, serine, cystine, arginine, ornithine, threonine, proline, hydroxyproline, aspartic acid and glutamic acid. Some amino acids, e.g. leucine, phenylalanine and tyrosine, do not give rise to glucose, but to acetoacetic acid. This is a product of fat metabolism, which under certain conditions accumulates in large quantities in the blood, and leads to a condition called ketosis. (See sections on fat and carbohydrate metabolism.) For this reason, leucine, phenylalanine and tyrosine are said to be ketogenic, while those amino acids which give rise to glucose are said to be antiketogenic. A few amino acids, lysine, histidine and tryptophan, form neither glucose nor acetoacetic acid.

Creatinine

This is another end-product of protein metabolism. It is formed from creatine (methyl guanidine acetic acid) of which it is the anhydride. Creatine is present in the tissues in the form of creatine phosphate and, in this form, it plays a part in the chemical processes responsible for muscular contraction. Creatinine is regarded as a waste product of muscle metabolism and is always a normal constituent of the urine. Creatine is not normally present in the urine but in some cases, for reasons not understood, it may be a urinary constituent. It often appears in the urine of women, but its relation to the menstrual cycle is uncertain.

Nucleoproteins

These are proteins found especially in the nuclei of cells and they form a small part of the dietary protein. They consist of a protein conjugated with nucleic acid, this latter substance being a combination of a purine or pyrimidine base with phosphoric acid and a pentose. The purines present are adenine and guanine, the pyrimidines, thymine and cytosine. The nucleoproteins are broken down during digestion and among the products are the purine and pyrimidine bases. The latter after absorption are completely oxidised. The purine bases undergo deamination and partial oxidation to uric acid. In man, this is an end-product of nucleoprotein metabolism and is excreted in the urine, but in most mammals it undergoes a further stage of oxidation to allantoin.

Exogenous and endogenous metabolism

The main end products of protein metabolism differ considerably in different states of nutrition. The urea output of the body is fairly closely dependent on the amount of protein in the diet. On the other hand the amount of creatinine excreted is almost constant, even during large fluctuations of dietary protein. Uric acid occupies an intermediate position as regards the relation between the amount excreted and the protein intake. The different behaviour of these substances suggests that protein metabolism can be divided into the metabolism of the protein fuel supplied by the diet and the metabolism of that supplied by the tissues. Since the amount of creatinine does not vary with the diet it is thought to be an index of tissue metabolism, or "endogenous" metabolism, while urea is thought to be an index of "exogenous" metabolism. Experiments with isotopes have shown, however, that protein metabolism cannot be divided in this way. Animals were fed with leucine and glycine containing deuterium attached to the carbon chain and ^{15}N in the amino group of the molecule. When the excreta were collected and examined, it was found that only a small amount, about one-third, of the isotope had been eliminated from the body. An examination of the different tissues showed that the proteins of the blood, and the proteins of most of the organs, contained isotopic leucine and glycine. Since deuterium as well as ^{15}N was present in the tissue proteins, it proved that not only the amino group but the whole molecule of the amino acid given in the food had been incorporated in the tissue proteins. The animals had not gained weight during the process, so that it was not a question of retention of amino acids to build up extra body tissues. The only possible conclusion was that the amino acids given had replaced some of the leucine and glycine previously present in the body proteins. This showed that there is a constant synthesis and breakdown of body protein with replacement of the amino acid molecules by new ones derived from the dietary protein. It is thus not possible to separate the exogenous and the endogenous metabolism, as there is a dynamic equilibrium between the amino acids in the body tissues and those in the bloodstream. Further investigation showed that there was not only replacement of amino acids in the tissue proteins, but also replacement of the nitrogen in the amino acids. When isotopic leucine or glycine was fed, other amino acids isolated from the tissues contained the ^{15}N. This indicated that the ^{15}N supplied in the leucine and glycine had been used for the synthesis of other amino acids to supply new units for the tissue protein. Even when abundant quantities of some particular amino acid are supplied in the diet, synthesis of this amino acid still occurs.

Transamination. The transference of the amino group is called transamination. It involves a keto acid as well as an amino acid, and in many cases glutamate and oxaloacetate are involved. The enzyme responsible is often called GOT (glutamate-oxaloacetate transaminase) and its concentration in the blood is usually increased when muscle tissue is being broken down. As this happens when the muscle blood supply is reduced, e.g. in coronary thrombosis, the estimation of the serum GOT has clinical importance.

Metabolic acidosis and alkalosis

Most kinds of protein contain sulphur and phosphorus atoms in organic (un-ionised) combination, and some also contain chlorine atoms. When these molecules are completely oxidised, sulphate and phosphate ions (and perhaps chloride ions) are released into the body fluids, the negative charges being derived from hydrogen atoms which are oxidised to hydrogen ions. These hydrogen ions combine with bicarbonate ions and form carbon dioxide which is eliminated in the lungs. The net result, therefore, is the replacement of bicarbonate ions in the body fluids by sulphate and phosphate ions, so that according to the Henderson–Hasselbalch equation (Appendix II), if the alveolar carbon dioxide pressure were constant, the acidity of the blood would rise (the pH would fall). Actually, since the respiratory centre is activated by an increase in acidity of the blood (Chapter 45), there would be an increase in the respiratory minute volume, the alveolar carbon dioxide pressure would fall and the increase in acidity would be diminished. The essential point, however, is the reduction in the bicarbonate concentration of the body fluids (the "alkali reserve") and it is this that is called *acidosis*.

For experimental purposes, this type of acidosis may be produced to almost any desired extent, by ingesting ammonium sulphate or ammonium chloride. The ammonia is converted into urea in the liver and hydrogen ions, together with sulphate ions or chloride ions, are released into the body fluids.

Many kinds of vegetable food, on the other hand, notably fruits, contain substantial quantities of salts of organic acids. The acids undergo oxidative metabolism, with the formation of bicarbonate ions, so that the alkali reserve is increased and an alkalosis results. On a mixed diet, of course, the reduction of bicarbonate consequent on eating protein will be compensated by an increase on eating fruit and vegetables; but in man, the former usually preponderates, and there is a net gain of hydrogen ions, which are excreted in the urine.

LIPID METABOLISM

The term "lipid" is applied to a number of different classes of substance which occur in the animal body and in the diet. These are: (1) the simple triglycerides of fatty acids (the fats proper), (2) the sterols and their esters with fatty acids and (3) the more complex phospholipids and cerebrosides. Of these, the fats proper form the greatest bulk of the tissue and dietary lipids and most of this section will be devoted to them.

Fats

The fatty acids present in the body fat are restricted to those with even numbers of carbon atoms in the molecule. Of these, all members of the series from acetic acid (2 carbon atoms) to stearic acid (18 carbon atoms) have been found, but by far the most important quantitatively are the 16 and 18 carbon atom fatty acids, palmitic ($C_{15}H_{31}COOH$) and stearic ($C_{17}H_{35}COOH$) together with the unsaturated oleic acid ($C_{17}H_{33}COOH$). These three in the form of their glycerol esters make up most of the body fats.

Part of the fat exists in the form of non-esterified fatty acids (NEFA) and although forming only a small fraction of the total fat this is the metabolically active form of the lipid.

Unsaturated fatty acids. More highly unsaturated fatty acids, linoleic, linolenic and arachidonic, occur in the body and play an important, though unknown, part in metabolism. They cannot be synthesised in the body and, if not supplied in the diet of rats, symptoms of deficiency appear in the form of skin disturbances.

The physical properties of the body fat depend on the varying proportions of the constituent fatty acids and are more or less characteristic for each species. This constancy of body fat is, however, dependent on dietary habit and if an animal be starved so as to reduce its fat stores and then fed with an unusual type of fat, it is quite easy to alter the nature of its body fat as regards physical and chemical properties, e.g. melting point, iodine number, saponification value, etc.

The storage of fat

There are certain parts of the body where the fat content can be greatly altered and such parts like the omentum and the subcutaneous tissues can act as fat depots. When an animal is putting on weight, fat is laid down in these stores and, when it is living on its reserves, the fat in these parts diminishes before other parts of the body are called on to contribute their share of metabolic fuel. The source of this depot fat we have just seen is the dietary fat, of which a part is used directly for oxidation and part is laid down as storage in the fat depots. If fat containing deuterium is fed to animals, it can be shown that the deposits do not consist of a static deposit of storage fat, but that the fat laid down is constantly being used and replaced.

The adipose tissue is of two kinds—brown and white, the white being much larger in amount. The brown adipose tissue has a much richer blood supply, and when the animal is exposed to cold and requires to mobilise fuel rapidly this comes from the brown adipose tissue.

It has also been well established that the non-fat part of the diet can contribute to the fat stores. This was demonstrated by the classical experiments of Lawes and Gilvert in 1852. Young pigs were fed on a diet of barley containing very little fat and it was found that the amount of body fat present when the animals were killed was greater than could have been obtained from the fat supplied or even the fat and protein together, thus proving that carbohydrate can be converted into fat. Whether or not fat can be derived in the body from protein is uncertain.

Interconversion of fat and carbohydrate

The formation of fatty acids from carbohydrates takes place by addition of 2-carbon units and the substance involved in this addition is acetyl coenzyme A, which is produced from pyruvic acid. The fatty acid thus formed combines with glycerol to form neutral fat. The process of fat production from carbohydrate requires the participation of certain vitamins of the B group; and aneurin, riboflavin and pantothenic acid are all probably involved.

One important aspect of the interconversion of fat and carbohydrate in the body is the effect on the gaseous exchanges. While the respiratory quotient usually depends only on the type of food being oxidised, it will also be affected by the interconversion of fat and carbohydrate. Inspection of the formula of glucose and a typical fat shows that the latter contains fewer atoms of oxygen per carbon atom than the former and hence, when fat is being formed from carbohydrate, oxygen will be freed for use in general metabolism and so less will be taken into the body from the lungs. Since the carbon dioxide production and excretion remains unchanged the respiratory quotient will be abnormally high. An example of this process is seen in the behaviour of those animals which hibernate. At the end of summer when they are building up large reserves of fat from carbohydrate to last over the winter, the respiratory quotient is abnormally high, and may reach a value of $1\cdot5$.

The oxidation of fat

The fats utilised by the body normally undergo complete oxidation with formation of carbon dioxide and water. Evidence as to how oxidation proceeds was obtained by Knoop's method of feeding to animals fatty acids containing a benzene ring attached to the carbon chain.

If benzoic acid is fed to animals it is found to be conjugated with glycine to form hippuric acid, and this is excreted in the urine.

$$C_6H_5 . COOH + NH_2 . CH_2 . COOH = C_6H_5 . CONH . CH_2 . COOH$$

If phenylacetic acid is fed, the excretory product is phenaceturic acid, which again represents a conjugation with glycine.

$$C_6H_5 . CH_2 . COOH + NH_2 . CH_2 . COOH = C_6H_5 . CH_2 . CONH CH_2 . COOH$$

If phenyl derivatives of higher fatty acids are fed, it is found that the product of excretion is always either hippuric acid or phenaceturic acid, depending on whether there is an even or odd number of carbon atoms in the side-chain. Thus hippuric acid was produced from $C_6H_5 . COOH$, $C_6H_5 . CH_2 . CH_2 . COOH$ and $C_6H_5 . CH_2 . CH_2 . CH_2 . CH_2 . COOH$, while phenaceturic acid was given by $C_6H_5 . CH_2 . COOH$, $C_6H_5 . CH_2 . CH_2 . CH_2 . COOH$ and $C_6H_5 . CH_2 . CH_2 . CH_2 . CH_2 . CH_2COOH$. These findings suggested that during the course of oxidation of the fatty acid chain the carbon atoms are split off in pairs, or in other words, oxidation takes place at the β-carbon atom. Hence this suggested method of fat oxidation was called β-oxidation.

The details of the reactions of β-oxidation cannot be discussed here, but coenzyme A plays a very important role. In general the process can be summarised by saying that one molecule of fatty acid combines with a molecule of coenzyme A thus:

$$R . CH_2CH_2CH_2COOH + HS . CoA \rightarrow R . CH_2CH_2CH_2COS . CoA$$

where R represents the rest of the molecule. After various intermediate stages this is oxidised to form a keto acid $R . CH_2CO . CH_2COS . CoA$.

This reacts with another molecule of coenzyme A thus:

$$R . CH_2CO . CH_2COS . CoA + HS . CoA \rightarrow R . CH_2COS . CoA + CH_3COS . CoA$$

The result is that two carbon atoms are split off, with formation of acetyl coenzyme A. This process proceeds until only a four-carbon compound is left, which is $CH_3CO . CH_2CoS . CoA$, i.e. acetoacetyl coenzyme A. The end-products are thus a number of molecules of acetyl coenzyme A, together with one molecule of acetoacetyl coenzyme A, and these products are normally oxidised by the citric acid cycle.

Ketosis

In certain conditions the oxidation of fat is incomplete and certain products appear in the urine, β-hydroxybutyric acid, acetoacetic acid and acetone. These substances are called "ketone bodies", and the condition in which they appear is called ketosis. It occurs typically in diabetes mellitus, but it also occurs in less serious disturbances such as fasting, and severe vomiting. Of these ketone bodies, it is known that acetone is formed from acetoacetic acid, and that the primary substances are the four-carbon atom substances, β-hydroxybutyric acid and acetoacetic acid. Probably acetoacetic acid is formed first, but the two substances are known to be interconvertible in the liver. The appearance of ketone bodies in fat metabolism is always related to a lowered oxidation of carbohydrate by the tissues, and this may be brought about either by lack of carbohydrate as in fasting, or by inability of the tissues to oxidise carbohydrate as in diabetes. When the formation of ketone bodies is very great it may be sufficient to lower the alkali reserve of the blood and produce a condition of acidaemia (movement of pH of blood towards the acid side).

THEORIES OF KETONE BODY FORMATION

The acetyl coenzyme A is metabolised by various different routes, the two most important of which are by the citric acid cycle and by formation of acetoacetic acid. The acetoacetic acid formed is metabolised by the tissues so that the concentration of ketone bodies is maintained at a very low level. If there is a reduction in the amount of acetyl coenzyme A entering the citric acid cycle then more acetoacetic acid is formed than can easily be used up by metabolism and ketosis results. Since oxaloacetate is necessary for incorporation of acetyl coenzyme A in the citric acid cycle the availability of oxaloacetate is an important factor in ketosis. Oxaloacetate is a product of carbohydrate metabolism and hence a certain minimum rate of carbohydrate metabolism is essential to prevent ketosis. Ketosis is thus seen to result from excessive utilisation of fat relative to utilisation of carbohydrate.

Control of fat metabolism

The breakdown and synthesis of fats are under hormonal control. The blood concentration of NEFA is decreased by insulin due to diminution in the release of fatty acids from adipose tissue. The concentration of NEFA is increased by adrenaline, which mobilises fat from the fat depots. Growth hormone, adrenocorticotrophic hormone, thyroid-stimulating hormone, and glucagon also stimulate fat mobilisation.

CARBOHYDRATE METABOLISM

The greater part of the carbohydrate in the body consists of the glycogen stored in the tissues, principally in the liver and in the muscles, and of the glucose in the blood and tissue fluids. The total amount of carbohydrate present in a well-nourished human body, of which about one-half is present in the liver as glycogen, is equivalent to less than one day's supply at the normal rate of consumption. It must be remembered, however, that carbohydrate taken into the body can be stored as fat, and also that carbohydrate can be formed in the body from non-carbohydrate sources.

The carbohydrate absorbed from the small intestine from a normal diet consists largely of three monosaccharides—glucose, fructose and galactose. They are all convertible to glycogen in the liver, and the glycogen so formed breaks down, under the influence of the enzyme systems present in liver, to give glucose which may therefore be considered as the form in which carbohydrate is available to the tissues.

The equilibrium between glycogen and glucose

Many years ago Claude Bernard showed that the glycogen of the liver, but not that of the muscles, could break down in the body to give glucose and that, except after a recent meal containing carbohydrate, the blood leaving the liver contained more sugar than that entering the organ. Other organs of the body, and in particular the muscular tissues, appeared to be continuously absorbing glucose from the blood flowing through them. These important observations established the fact that although the concentration of the sugar in the blood falls only slightly during post-absorptive conditions or even during a long fast, this was not because the tissues ceased to absorb glucose from the blood under these conditions. The relative constancy of the blood glucose concentration was to be ascribed to the fact that the amount of glucose withdrawn from the blood by the tissues in general was counterbalanced by the amount of glucose liberated into the circulation by the liver under these conditions. The importance of the power of the liver to secrete glucose is dramatically emphasised by the fact that if the liver is removed from a dog, the blood sugar concentration falls rapidly and death ensues within a few hours unless glucose is administered in large amounts. Undoubtedly some part of the glucose secreted by the liver comes from the glycogen present in it, but the stored liver glycogen is quite insufficient to supply the body's needs for very long and, as will be discussed below, glucose can be formed in the liver from non-carbohydrate sources, a process called gluconeogenesis.

After the ingestion of carbohydrate food the blood glucose concentration rises from its normal post-absorptive value of 0.08 to 0.1 per cent to a value in the neighbourhood of 0.15 per cent. As the blood glucose concentration rises, the liver secretes less glucose until, when the concentration reaches about 0.12 per cent, the liberation of glucose by the liver ceases altogether. At blood glucose concentrations above this, the liver begins to absorb glucose from the bloodstream and the liver glycogen content begins to rise. It is clear, therefore, that the glycogen in the liver represents, in part at least, a storehouse for carbohydrate coming from the food. The glucose which is thus stored during the temporary period of plenty is liberated by the liver during the lean period of post-absorptive conditions. The ability of the liver to adjust its output of glucose to the requirements of the body is sometimes described as its homeostatic function.

Although the skeletal muscles also absorb more glucose when the blood glucose concentration rises after carbohydrate food, the glycogen which accumulates in these tissues is not reconverted to blood glucose during post-absorptive conditions. The glycogen of muscle may be oxidised or it may be converted to lactic acid, a process called glycolysis, but it never forms glucose in the body; this contrast with events in the liver is to be ascribed to differences between the enzyme systems of liver and muscle tissues.

The process of the formation of lactic acid from glycogen in skeletal muscle is one which does not involve the addition of oxygen nor the elimination of hydrogen from the system. It is accordingly not an oxidation process and can take place under the largely anaerobic conditions which exist in skeletal muscles during exercise. If the exercise is light, any lactic acid formed may be in part reconverted to glycogen in the muscles themselves, but if the exercise is severe most of the lactic acid may escape into the blood stream and be then converted to glycogen in the liver. Therefore, in the presence of the liver, muscle glycogen can give rise to blood glucose indirectly. These relationships will be clear from the diagram.

It should be noted that just as the formation of lactic acid from glycogen in the muscles liberates energy, the reformation of

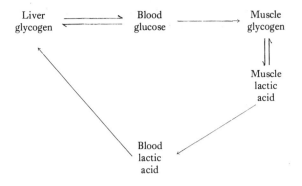

glycogen from lactic acid requires the addition of energy to the system. In the muscles any energy thus required can be provided, under aerobic conditions, by oxidation of part of the lactic acid formed or equivalent glucose or glycogen.

Gluconeogenesis

The amount of glycogen stored in the liver is insufficient to maintain the normal level of blood sugar for more than a short time, and glucose is in fact formed by the liver from various non-carbohydrate sources, a process called gluconeogenesis. These sources are alcohols such as glycerol, many amino acids, and a number of organic acids, e.g. fumaric, malic, succinic, etc., formed in metabolism. Of these, the amino acids are the most important, and hence the dietary protein can be considered an important source of blood glucose. The process of gluconeogenesis is inhibited by insulin, and stimulated by the hormones of the anterior pituitary gland (see below).

During starvation or deprivation of carbohydrate, the utilisation of carbohydrate by the muscles and other tissues is diminished and some part of their energy requirements is met by oxidation of fats. Nervous tissue, however, is peculiarly dependent on a constant supply of carbohydrate or similar substances and hepatic gluconeogenesis is of particular importance in providing the nervous system with one of its requisite metabolites. It should be mentioned that the brain (and the heart also) can oxidise lactic acid as well as glucose and thus utilise the lactic acid formed anaerobically in skeletal muscle during vigorous exercise.

The regulation of the blood sugar concentration

The accurate control of blood sugar concentration is of the utmost importance in the economy of the human body. If the concentration rises above about 0.18 per cent (10 mmol l^{-1})—the renal threshold—the reabsorption of sugar from the renal tubules is incomplete and glucose is lost to the body in the urine. If, on the other hand, the concentration falls below about 0.04 per cent (2.2 mmol l^{-1}), the central nervous system becomes disturbed. In man there is at first extreme hunger and fatigue, with general sweating; later, delirium and profound coma are produced. In rabbits and certain other mammals, severe convulsions occur in addition just before the coma and death from respiratory failure; these are known as *hypoglycaemic convulsions*. If the condition is not relieved by the administration of glucose, death may ensue. Even if death does not occur as the result of a temporary lowering of the blood sugar concentration, lesions may appear in the brain which cause permanent functional damage.

The mechanism which ensures that the blood sugar concentration normally varies only within the relatively narrow range of 0.08 per cent to 0.16 per cent (4.5–8.9 mmol l^{-1}), is largely hormonal in character, adrenaline, insulin and the growth hormone of the anterior pituitary being concerned.

(a) *Adrenaline*
When the blood sugar concentration falls below about 0.07 per cent (3.9 mmol l^{-1}), the excitation of nervous centres in the lower part of the brain causes sympathetic stimulation and a release of adrenaline from the adrenal glands. The adrenaline so released stimulates a breakdown of liver glycogen to glucose, the liberation of which tends to arrest the fall of blood sugar concentration. The general sympathetic stimulation may assist this process. Under the influence of adrenaline, muscle glycogen breaks down to lactic acid which may be carried to the liver for reconversion to glycogen and glucose. These several processes tend to prevent a fall in the blood sugar concentration by rapidly mobilising the glycogen stores of the body.

During muscular exercise there is a general stimulation of all the sympathicoadrenal system, so that the rate of breakdown of glycogen in the liver is accelerated on just those occasions when glucose is being used most rapidly in the muscles. As far as we know, however, adrenaline does not directly affect gluconeogenesis in the liver, but can only bring about the breakdown of glycogen which has been stored in that organ.

It was shown many years ago by Claude Bernard that puncture of the floor of the fourth ventricle in the rabbit results in hyperglycaemia and glycosuria; this is known as *diabetic puncture*. The hyperglycaemia is due to the excitation of fibres that run into the sympathetic system, with a consequent outpouring of adrenaline.

(b) *Hormones of the adrenal cortex*
The adrenal cortex plays a part in carbohydrate metabolism, as is shown by the fact that if it is removed or diseased the blood sugar concentration falls, and there is muscular weakness, in addition to other abnormalities discussed in Chapter 32.

Cortisol is the most important hormone of the adrenal cortex influencing carbohydrate metabolism. By reducing protein synthesis it makes more amino acids available for glycogen formation, and hence is part of the defence against hypoglycaemia.

(c) *Hormones of the pancreas*
The two hormones taking part in the regulation of blood sugar are insulin, produced by the beta cells of the islets of Langerhans, and glucagon, produced by the alpha cells. When the blood sugar concentration rises as a result of the ingestion of carbohydrate the liver responds by first diminishing and then abolishing the entry of sugar into the bloodstream and then proceeds to absorb part of the excess in the bloodstream. The ability of the liver to suspend the production of sugar and inhibit gluconeogenesis, and to promote glycogen storage at the expense of the excess glucose in the blood, is dependent on the presence of insulin.

When the blood sugar concentration rises after a carbohydrate meal the excess glucose is stored as glycogen in the muscles as well as in the liver, while in the muscles the rate at which sugar is oxidised may increase under these conditions. All these processes for the storage and utilisation of glucose depend on the availability of insulin, and there is good evidence that the secretion of insulin by the pancreas is stimulated by a rise in the concentration of sugar in the blood; the controlling action appears to be the direct effect of the excess sugar concentration on the cells of the pancreas, although it is possible that a nervous reflex is also involved. Apart from the formation of glycogen, a further process for the storage of energy from the excess ingested carbohydrate is the conversion of glucose to fat (glycerides) which are then stored in the fat depots. In general, therefore, insulin may be said to promote those metabolic processes which cause glucose to leave the bloodstream (conversion to glycogen or fat, or promotion of carbohydrate oxidation) and to inhibit gluconeogenesis. All these processes tend to lead to a fall of the blood sugar concentration and if a large dose of insulin is given to a normal person in a post-absorptive state, fatal hypoglycaemia may develop. While many theories have been put forward it is not yet possible to explain the manifold actions of insulin in molecular terms.

The other pancreatic hormone is glucagon, which in general terms opposes the action of insulin, and can be regarded as tending to prevent hypoglycaemia. Its role is complicated by the existence of a somewhat similar substance, enteroglucagon, produced by the gut. Both stimulate secretion of insulin, and possibly enteroglucagon in this way prepares the tissues for the hyperglycaemia follows ingestion of glucose.

(d) *Anterior pituitary hormones*

An animal from which the anterior pituitary gland has been removed becomes abnormally sensitive to the hypoglycaemic action of insulin. Conversely, a normal animal to which anterior pituitary extract is administered becomes highly insensitive to the action of a small dose of insulin in lowering the blood sugar concentration. These observations show that in some respects anterior-pituitary secretion and insulin act antagonistically.

The hormone chiefly involved is the growth hormone, which depresses carbohydrate oxidation in the muscles and increases accumulation of glycogen.

Diabetes in man

If a normal man takes 50 g of glucose by mouth and the blood sugar concentration is determined at intervals afterwards, the type of curve shown in Fig. 51.1 is obtained. The rise of the blood sugar concentration stimulates the secretion of insulin and the mechanism which we have considered in the previous section comes into play, with the result that much of the glucose is stored as glycogen; the blood sugar concentration thus falls again without having reached so high a value that glucose is excreted by the kidneys, i.e. the renal threshold is not exceeded and glycosuria is absent. Such a test is called a *glucose tolerance* test and is of much value clinically in the diagnosis of *diabetes mellitus*. Occasionally glycosuria may follow the ingestion of a large amount (150–200 g) of glucose by a normal person, but this is not considered to be necessarily indicative of diabetes mellitus; it is described as *alimentary glycosuria*.

When a glucose tolerance test is performed on a diabetic person, the blood sugar concentration may be very high initially and may rise excessively after glucose administration, failing to fall again to the original value for many hours. The mechanism for the storage of glycogen in the liver and muscles is defective, while glucose oxidation is also subnormal; much of the carbohydrate taken in with the food is therefore lost in the urine as glucose. The depression of carbohydrate utilisation causes the production of excess ketone bodies in the liver with the development of ketosis and ketonuria. The appearance of ketone bodies in the urine is an indication of the gravity of the disease, a slight

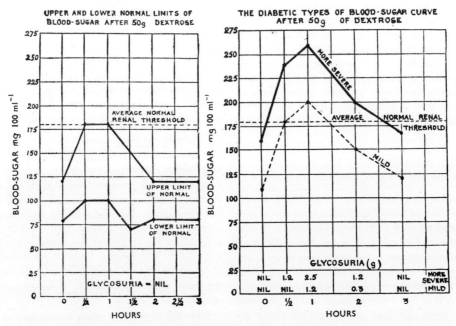

Fig. 51.1 The effect of the ingestion of 50 g glucose on the concentration of sugar in the blood.
 (Left) *Normal subjects*. The concentration is not increased excessively and returns to the initial value within 2 h. No sugar is excreted in the urine.
 (Right) *Diabetic subjects*. The concentration rises to a considerably higher value and takes 3 h to return to the initial value. Considerable quantities of sugar are excreted in the urine.
 The figures under 'glycosuria' indicate the total quantity of sugar excreted in the sample of urine collected at that time. (Harrison's *Chemical Methods in Clinical Medicine*.)

glycosuria being of no great importance. If the ketone bodies continue to accumulate, the patient ultimately dies in a coma, if he has not already succumbed to infection, to which he becomes very liable. Daily subcutaneous injections of insulin, however, can relieve most of the symptoms and the patient can lead a normal life; omission of the insulin is rapidly followed by a reversion to the diabetic condition. Excess insulin, as we have seen, leads to hypoglycaemia and serious consequences, so that doses must be carefully controlled; immediate subcutaneous injection of glucose, or a large quantity given by mouth, quickly alleviates the symptoms should an overdose be given accidentally.

In recent years considerable advances have been made in the production of slowly-absorbed insulin preparations. These have the great advantage that fewer and larger injections can be given, and these more nearly simulate the physiological secretion of insulin by the pancreas.

THE CHEMISTRY OF CARBOHYDRATE METABOLISM

The details of the chemical reactions involved in breakdown of glucose and glycogen must be sought in textbooks of biochemistry and only some general outlines will be discussed here. Like the other food substances the metabolism of glucose can be divided into two stages, the pre-citric acid cycle stage and the citric acid cycle, the connecting link being acetyl coenzyme A.

Glycolysis

The pre-citric acid cycle stage of carbohydrate metabolism is a series of reactions often included under the term glycolysis, although this term strictly applies to the production of lactic acid from glucose and glycogen. The relation between the oxidative pathway for glucose metabolism and the glycolytic pathway is seen in Fig. 51.2. The reactions can start either from glucose taken in from the bloodstream or from glycogen stored in the tissue. Both are converted to glucose-6-phosphate and this undergoes a series of changes which result in formation of pyruvate. In aerobic conditions the pyruvate forms acetyl coenzyme A, which goes into the citric acid cycle, while in anaerobic conditions pyruvate is reduced to lactate. This formation of lactate provides an important source of energy in muscle, when work is taking place at a faster rate than can be covered by current oxygen intake.

THE CITRIC ACID CYCLE

The citric acid cycle, or Krebs' cycle, is the final common pathway for oxidation of proteins, fats and carbohydrates. The preliminary reactions result in formation of acetyl coenzyme A, and this is the substance which enters the citric acid cycle. The details of the reactions of the cycle must be sought in textbooks of biochemistry, and only a very bare outline is given here, which is shown schematically in Fig. 51.3. Acetyl coenzyme A (two-carbon unit) reacts with oxaloacetate (four-carbon unit) to form citrate (six-carbon unit). Citrate undergoes a series of reactions with formation and breakdown of a large number of substances and ultimate formation of oxaloacetate again. During the course of the cycle carbon dioxide appears twice as an end product, so that the six-carbon skeleton of citric acid is reduced first to a five-carbon skeleton and then to the four-carbon

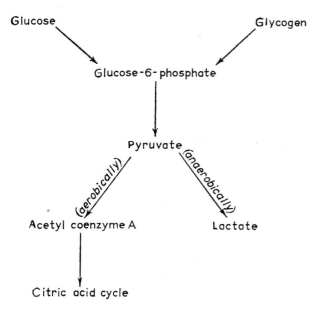

Fig. 51.2 Carbohydrate metabolism in tissues. The intermediate stages between glycogen and glucose-6-phosphate, and between glucose-6-phosphate and pyruvate are omitted.

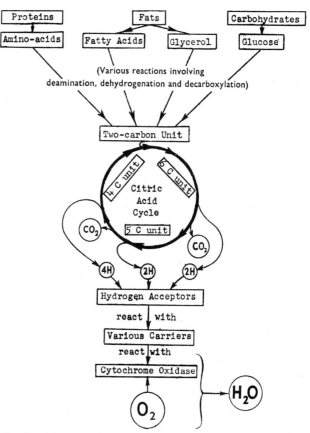

Fig. 51.3 Schematic representation of the probable course of breakdown of the food substances.

The citric acid cycle is shown diagrammatically in a much abbreviated form. The various six-, five- and four-carbon compounds actually taking part form at least nine stages in the cyclical process.

skeleton of oxaloacetate. The re-entry of another molecule of acetyl coenzyme A starts the cycle off again. The carbon dioxide produced by the cycle is excreted by the lungs. The other end product is hydrogen and, as seen in Fig. 51.3, each complete operation of the cycle results in the formation of four pairs of hydrogen atoms, which are ultimately oxidised to water by the oxygen taken into the body by respiration. This final oxidation of the hydrogen is, however, in itself a complex process involving a number of different stages. The hydrogen is not able to react with molecular oxygen but is "accepted" by substances called coenzyme 1 and coenzyme 2, which are pyridine nucleotides. These substances can exist in either oxidised or reduced forms and the oxidised form can accept hydrogen and hence become reduced. The reduced forms of the coenzymes can in turn pass on their hydrogen to substances called cytochromes, which can in their turn exist in oxidised or reduced forms. The reduction of the cytochromes re-oxidises the coenzymes which are thus available to accept more hydrogen from the substances of the cycle. Finally, the cytochromes are re-oxidised by passing on their hydrogen to cytochrome oxidase and this substance is able to pass on hydrogen to molecular oxygen. The final products of the cycle are thus water and carbon dioxide.

METABOLISM AND ENERGY LIBERATION

The oxidation of food substances results in the production of water and carbon dioxide and the release of energy, which is partly utilised by the various tissues in the processes of growth, secretion, muscular contraction, etc., while the rest appears as heat. The way in which the energy becomes available is one of the fundamental problems of metabolism and it is only in recent years that a tentative answer has been formulated. It now seems fairly certain that the "useful" part of the energy is very largely derived from one chemical reaction, the dephosphorylation of adenosine triphosphate (usually referred to as ATP). The removal of one phosphate radical from ATP results in the formation of adenosine diphosphate (ADP). The ADP formed can, in suitable conditions, combine again with a phosphate radical to form ATP. The phosphorylation process requires energy and the dephosphorylation process makes energy available. The amount of energy associated with this type of reaction is relatively large and, hence, the term "active (or energy-rich) phosphate" has been used in connection with it. Since the dephosphorylation of ATP is the primary reaction which makes energy available to the tissues, the function of the other reactions of the food substances can be regarded as supplying the energy necessary for ATP synthesis; a considerable fraction of the energy liberated by oxidative reactions, even though they yield energy, can take part in ATP synthesis and, furthermore, if the process is to be carried out efficiently the amount of energy liberated by a particular oxidative reaction should not be too large in relation to the amount of energy needed for ATP synthesis. This makes more understandable the apparently complicated pathway of metabolism of the food substances. The large number of reactions which occur ensures that the energy is liberated in small amounts, which are suitable for an efficient synthesis of ATP, and some of the reactions are directly "coupled" with ATP synthesis.

From the biological point of view the "efficiency" of metabolism can be measured in terms of the ATP produced; in fact, much more ATP is generated at the citric acid stage than in the preliminary stages. Some of the ATP generated is required to maintain the metabolic reactions and, if this is subtracted from the ATP generated, the amount available for biological work is obtained, which can be called the net yield of ATP. In the case of glucose, for each molecule metabolised under aerobic conditions, the net yield in the glycolytic reactions is six ATP compared with thirty ATP in association with the citric acid cycle. Most of the ATP is generated in the oxidation of hydrogen by the pyridine nucleotide–cytochrome system. In anaerobic conditions where the pyruvate is reduced to lactate, the glycolytic reactions yield only two moles of ATP.

One of the current problems of biochemistry is the question of the control of the rate at which chemical reactions occur in the body. In some cases hormones exert an effect while in others it is products of metabolism. It is now known that a very important role is played by cyclic adenosine monophosphate (cAMP), but details of this are outside the scope of an elementary survey of metabolism.

52. Nutrition

D. H. Smyth

The natural stimuli that lead to the choice of appropriate foods are hunger, thirst and appetite, guided by the sensations of sight, taste and smell. Under the conditions of twentieth-century civilisation where food is, in general, not produced in the place where it is consumed, and where for one reason or another adequate supplies of food are likely to be periodically restricted, it is necessary to replace to some extent the operation of natural desires by carefully calculated allocation of the different food constituents. Nutrition is the science on which such attempts are based and embraces a very wide field of study. In deciding a nutritional policy for a population the following factors would have to be considered: (1) the amounts of the different food constituents required by the human body in different working and climatic conditions; (2) the amounts of these food constituents in the different food materials used; (3) the availability of food in terms of labour and space required for production, bulk and value of food in relation to transport, cost of food in relation to wages level, ability to import and export, etc.; (4) the storage and preparation of food so that the highest nutritive value is maintained; and (5) general propaganda and education in regard to the preparation and use of foods. Of these aspects of nutrition, only the first two belong to the province of physiology, and accordingly only these will be considered here.

The six essential classes of substances required in the diet are proteins, fats, carbohydrates, vitamins, mineral salts and water. The first problem is to know the amounts of each of these required by the body under different conditions, and the second to know the amounts of each of these in the food materials or, more accurately, to know the amounts of these absorbed from the intestine. Substances originally present in the food but destroyed by cooking or not absorbed from the intestine will have no food value. In thinking of the food requirements we can divide these into foods supplying energy for bodily activity, and foods necessary for some other purpose. In the first case we are concerned with making up a certain amount of energy which is required for metabolism and, in this respect, the different energy-providing foods are within certain limits interchangeable. In the second case we have to think of specific functions of food constituents which cannot be replaced by giving other types of food. To the energy-providing foods belong proteins, fats, and carbohydrates; and the only other dietary constituent likely to be a source of energy is alcohol. The mineral salts, the vitamins, and water belong to the group of food substances with specific functions, but to these we must also add the proteins and fats, as these, in addition to supplying energy, have other roles, which cannot be taken over by any other food constituent.

THE ENERGY REQUIREMENTS

When the body is at rest it requires an energy production of about 0.17 MJ m^2 h^{-1}, which for an average adult amounts to about 7 MJ in 24 hours. When any physical activity is undertaken the energy consumption increases and the greater the physical work carried out the greater will be the need for energy. While mental work can cause a feeling of great fatigue and even hunger, it does not cause an increase in energy consumption apart from any physical movements accompanying it. The requirements are, therefore, dictated by physical effort. There are two main ways in which the needs can be assessed: (1) calorimetry, or measurement of the metabolic rate of the body in the appropriate conditions, and (2) the dietary survey, or measurement of the amount of food consumed.

The techniques for calorimetry have already been discussed in Chapter 50, and, in general, indirect calorimetry, i.e. measurement of the oxygen consumption, is employed. Modern techniques have enabled oxygen consumption to be be determined by subjects at work or play of many kinds and, as a result, fairly accurate figures are now available for energy requirements in these conditions.

Dietary survey

The dietary survey is an important approach to nutritional problems because it expresses energy needs in terms of actual food used, and also gives information about wastage in preparation, storage and distribution. The survey can be carried out at different levels. There is the detailed survey on one individual, and this is the most accurate as regards actual intake, provided the subject is intelligent and co-operative. Then there is the family survey, in which a trained observer attempts to assess the food intake of a normal family living under their usual conditions. In addition there is the survey at national level, where the total food production, imports, exports and consumption are assessed. These dietary surveys will give information of a rather different kind from that given by calorimetry, but this kind of information is equally important in assessing nutritional problems as a whole.

As a result of techniques of these kinds the requirements of individuals can be drawn up and, in doing so, it is usual to divide the day into three periods of eight hours each, i.e. a period of rest in bed, a period of work, and a period of non-occupational activities, and to assess the energy needs for each of these. This can be expressed in relation to the "reference man" or "reference woman", and these terms apply to an average individual in standard conditions. The needs of these individuals are shown in Tables 52.1 and 52.2 (from Food and Agriculture Organisation of the United Nations, 1957); corrections are then made for deviations from the standard age, weight and environmental temperature. In correcting for age in adults the percentages of the allowances for the reference man and woman are taken: 97 per cent for age 30–40; 94 per cent for age 40–50; 86 per cent for 50–60; 79 per cent for 60–70; and 69 per cent for ages over 70.

Table 52.1 *The energy expenditure of a "reference man"*

(Weight, 65 kg. Age, 25 years. Mean annual environmental temperature, 10°C)

	MJ day^{-1}
A. 8 h. Working activities: mostly standing (overall rate 10·5 kJ min^{-1})	5·0
B. 8 h. Non-occupational activities:	
1 h washing, dressing, etc. at 12 kJ min^{-1}	0·7
1½ h walking at about 6 km h^{-1} at 22 kJ min^{-1}	2·0
4 h sitting activities at 6 kJ min^{-1}	1·5
1½ h active recreations and/or domestic work at 22 kJ min^{-1}	2·0
C. 8 h. Rest in bed at basal metabolic rate	2·0
Total	13·2

Table 52.2 *The energy expenditure of a "reference woman"*

(Weight, 55 kg. Age, 25 years. Mean annual environmental temperature, 10°C)

	MJ day^{-1}
A. 8 h. Working activities in the home or in industry: mostly standing (overall rate, 7·5 kJ min^{-1})	3·7
B. 8 h. Non-occupational activities:	
1 h washing, dressing, etc. at 10·5 kJ min^{-1}	0·6
1 h walking at about 5 km h^{-1} at 15 kJ min^{-1}	0·9
5 h sitting activities at 6 kJ min^{-1}	1·8
1 h active recreation and/or heavier domestic work at 15 kJ min^{-1}	0·9
C. 8 h. Rest in bed at basal metabolic rate	1·8
Total	9·7

For adjusting for different weights the following formulae can be used:

$$\text{For man} \quad E = 3.42 + 0.154\,W$$
$$\text{For woman} \quad E = 2.44 + 0.131\,W$$

where E is the number of MJ and W the weight in kg. To correct for temperature the energy requirement is increased by 8 per cent for each 10° fall in environmental temperature from the standard, 10°C, and decreased by 5 per cent for each 10° rise. This is based on the fact that man is better able to protect himself from the effects of cold than of heat. Occupation naturally is one of the most important factors in providing for the individual and in some heavy occupations the daily energy requirement may amount to more than 20 MJ. For an excellent discussion of the energy expenditure of the human subject in many conditions, the reader is referred to *Energy, Work and Leisure* (Durnin & Passmore).

The requirements of children have to be considered separately and it must be remembered in children particularly that the range of activities vary enormously in individuals. Table 52.3 gives figures for the energy requirements based on recommendations from various sources.

Energy content of food

The determination of the energy content of food by the bomb calorimeter has been discussed on page 327, and the figures

Table 52.4 *Recommended dietary allowance (restricted allowances in brackets)*

(*Food and Nutrition Board, National Research Council*)

	MJ	Protein (g)	Calcium (g)	Iron (mg)
Man (70 kg):				
Sedentary	10·5			
Moderately active	12·6	70	0·80	12
Very active	18·9		(0·56)	(8·5)
Woman (56 kg):				
Sedentary	8·8			
Moderately active	10·5	60	0·80	12
Very active	12·6		(0·56)	(8·5)
Pregnancy (latter half)	10·5	85	1·5	15
Lactation	12·6	100	2·0	15
Children up to 12 years:				
Under 1 year[4]	0·4 MJ kg^{-1}	4 g kg^{-1}	1·0	6
1–3 years[5]	5·0	40	1·0	7
4–6 years	6·7	50	1·0	8
7–9 years	8·4	60	1·0	10
10–12 years	10·5	70	1·2	12
Children over 12 years:				
Girls, 13–15 years	11·8	80	1·3	15
16–20 years	10·0	75	1·0	15
Boys, 13–15 years	13·4	85	1·4	15
16–20 years	16·0	100	1·4	15

[1] Tentative goal towards which to aim in planning practical dietaries; can be met by a good diet of natural foods. Such a diet will also provide other minerals and vitamins, the requirements for which are less well known. The restricted allowances are probably adequate for adults other than nursing or expectant mothers.

[2] Requirements may be less if provided as vitamin A; greater if provided chiefly as the pro-vitamin, carotene.

[3] One mg thiamin equals 333 iu; 1 mg ascorbic acid equals 20 iu.

[4] Needs of infants increase from month to month. The amounts given are for approximately 6–8 months. The amounts of protein and calcium needed are less if derived from milk.

[5] Allowances are based on needs for the middle year in each group (as 2, 5, 8, etc.), and for moderate activity.

NUTRITION 343

Table 52.3 *Energy requirements of children*

Recommendation of Davidson and Passmore (1966)

	Age (in years)	MJ day^{-1}
Children	1–3	5.5
Children	4–6	7.1
Children	7–9	8.8
Children	10–12	10.5
Boys	13–15	13.0
Girls	13–15	10.9
Males	16–19	15.1
Females	16–19	10.1

usually taken are carbohydrate and protein 17.2 kJ g^{-1}, and fat 38.6 kJ g^{-1}. Most foods eaten are, in fact, mixtures of these, and when energy need has to be translated in terms of actual food, reference is made to food tables which give the required values for the actual food material eaten. It is usual to find that by far the greatest part of the energy supply in the human diet is provided by carbohydrate, as this is usually the cheapest and most plentiful source. When food is scarce, the aim is to supply sufficient protein and fat dictated by other needs (see following sections), to calculate the megajoules provided by these and to make up the rest with carbohydrate. A good average diet, e.g. as recommended by the British Medical Association Report in 1933, might make up the calories as follows:

100 g protein	1.7 MJ
100 g fat	3.9 MJ
500 g carbohydrate	8.4 MJ

There is no doubt that many people live reasonably healthy lives making up a greater proportion of the energy with carbohydrate, and the general rule is that the cheaper the diet, the more carbohydrate it contains.

SPECIFIC REQUIREMENTS

In the following paragraphs are discussed the specific requirements for different food substances. Figures are given in Table 52.4 of the amounts of the various constituents which are regarded as desirable in the diet. In considering these figures it must be borne in mind that they are somewhat arbitrary and often more in the nature of a generous guess rather than an accurate knowledge of the amounts which are required. Since the effects of deprivation of certain food constituents, particularly the vitamins, may not become apparent for a long time, it is obvious that the difficulties of assessing accurately the minimum requirements of the human subject for any one food factor are very great. It is easier to be sure that a certain intake is adequate than to know what is the threshold requirement for health, and probably for this reason figures tend to be somewhat too high.

Vitamin A[2] (iu)	Thiamin[3] (B^1) (mg)	Riboflavin (mg)	Nicotinic acid (mg)	Ascorbic acid[3] (mg)	Vitamin D (iu)
5000 (3500)	1.5 (1.1)	2.2 (1.5)	15 (10.5)	75 (52)	6
	1.8 (1.3)	2.7 (1.9)	23 (13)		
	2.3 (1.6)	3.3 (2.3)	23 (16)		
5000 (3500)	1.2 (0.8)	1.8 (1.3)	12 (8)	70 (49)	6
	1.5 (1.1)	2.2 (1.5)	15 (10)		
	1.8 (1.3)	2.7 (1.9)	18 (13)		
6000	1.8	2.5	18	100	400 to 800
8000	2.3	3.0	23	150	400 to 800
1500	0.4	0.6	4	30	400 to 800
2000	0.6	0.9	6	35	6
2500	0.8	1.2	8	50	
3500	1.0	1.5	10	60	
4500	1.2	1.8	12	75	
5000	1.4	2.0	14	80	6
5000	1.2	1.8	12	80	6
5000	1.6	2.4	16	90	6
6000	2.0	3.0	20	100	

[6] Vitamin D is undoubtedly necessary for older children and adults. When not available from sunshine, it should be provided probably up to the minimum amounts recommended for infants.

Further recommendations. The requirement for iodine is small; probably about 0.002 to 0.004 mg per day for each kg of body weight. This amounts to about 0.15 to 0.30 mg daily for the adult, which is easily met by the regular use of iodised salt; the use of this salt is especially important in adolescence and pregnancy.

The requirement for copper for adults is in the neighbourhood of 1.0 to 2.0 mg per day. Infants and children require about 0.05 mg per kg body weight. The requirement for copper is approximately one-tenth of that for iron.

The requirement for vitamin K is usually satisfied by any good diet. Special consideration needs to be given to newborn infants. Physicians commonly give vitamin K either to the mother before delivery or to the infant immediately after birth.

They must be interpreted as something to be aimed at, rather than as a carefully determined minimum need.

Protein in the diet

In addition to supplying energy, protein is necessary for the building up of new tissues and replacing of used ones. It has been seen in Chapter 51 (p. 332) that a minimum amount of protein must be supplied and also that this must contain certain essential amino acids. If even one of these is not present, the body proteins will be broken down to supply it and hence it will not be possible to keep the animal in nitrogen equilibrium. From the dietary point of view it is important to assess the nutritional value of different kinds of proteins and this has been done in various ways. One way is to give a certain protein in the diet and make calculations about the fraction of the nitrogen retained by the body. This fraction is called the biological value of the protein and varies widely with different dietary proteins, e.g. egg albumen has a value 94 per cent, milk protein 85 per cent, gluten in wheat as low as 40 per cent. Another method of expressing quantitatively the value of a protein is the chemical score. The amino acid content of a protein is determined and expressed as a percentage of that of an ideal protein, i.e. one which contains all the essential amino acids in the correct proportions. The lowest value found for any essential amino acid is taken as the chemical score. There is a reasonable correlation between the biological value and the chemical score measured in these ways. A simpler approach to the problem is to classify proteins as first class (animal protein) and second class (vegetable protein) as generally, although there are exceptions, vegetable protein is of less biological value.

Recommendations have been made in various forms about dietary protein. The League of Nations (1936) suggested a minimum of 1 g per kg body wt per day, with part of this comprising animal protein. The British Medical Association (1950) recommended that 11 per cent of the energy of the food should come from protein (14 per cent in the case of pregnancy, lactation, children and adolescents). The recommendations of the US National Research Council are included in Table 52.4.

Fat in the diet

The fat of the diet is used like the carbohydrate and part of the protein to supply energy by oxidation, but in this respect it has certain advantages over the other types of food. One gram of fat on oxidation will yield more than twice as much energy as the same amount of carbohydrate or protein and, furthermore, the fat is taken in a concentrated form in the food, whereas in the case of protein and carbohydrate foods, a large part of the bulk of the food is composed of water. Fat has another important role in metabolism in that it is the only form of food which can be stored in large amounts as such. This stored fat is valuable in the protection of the body against cold since it is stored partly below the skin where it forms an insulating layer. The food fat has an irreplaceable function in acting as a solvent for the fat-soluble vitamins and also for providing the body with the indispensable unsaturated fatty acids. It has been seen (on p. 317) that fat inhibits the movements of the stomach and, on account of this property, fat taken in the diet prevents the onset of hunger for a longer time. It is rather striking that, in spite of these important and definite roles of fat in the body economy, few figures are available to suggest what are the minimum fat requirements. The usual aim is to supply 100 g daily.

In recent years in the more highly developed countries the possibility has been widely considered of the possible harmful effects of excessive fat in the diet, particularly if it is highly saturated. The evidence for these effects is still conflicting.

Carbohydrate

As this usually forms the bulk of the human diet and is the cheapest form of energy-providing food, it is never likely to form too small a fraction of the total requirement. The idea that it has no specific function is, however, erroneous. The functioning of the citric acid cycle depends on a constant supply of oxaloacetate (Chapter 51, p. 339); this is derived from dietary carbohydrate and, to a lesser extent, from some amino acids. These can be grouped together as the antiketogenic substances and, in order that ketosis should be avoided, the antiketogenic substances being metabolised should amount to at least half the ketogenic substances, which are comprised mainly of fat.

Mineral salts

The important inorganic substances which are essential in the diet are calcium, sodium, potassium, magnesium, iron, phosphorous, iodine and chlorine. In addition, smaller amounts of many other elements are required; copper, bromine, cobalt, zinc, etc. These do not liberate any energy on oxidation, but they are responsible for the maintenance of the normal function of many parts of the body. The concentration of these substances necessary in the body fluids is usually small, but, on the other hand, there is a constant loss of small amounts in the urine and other body secretions, and this loss must be replaced. The minerals present in highest concentration in the body fluids are sodium and chloride. In the case of most of the mineral requirements, the amounts likely to be in the diet will be more than adequate for the body's needs and there will be excretion of the excess in the urine. The substances which demand special attention are sodium chloride, iodine, calcium and iron. Sodium chloride is only likely to become deficient when there is a great loss of sweat from the body, as in conditions of working in very hot atmospheres, or in certain abnormal conditions, when there is a repeated loss of gastric juice or of intestinal secretions by persistent vomiting or diarrhoea. Iodine deficiency only occurs regionally and may cause abnormal thyroid metabolism; this is considered in Chapter 32. The two minerals to be considered in ordinary nutritional problems are calcium and iron.

Calcium

The recommended calcium intake can be seen from Table 52.4. Calcium is most abundant in milk and cheese, and is contained only in very small quantities in bread and meat. It is, however, a common practice to fortify bread with increased amounts of calcium. There is a considerable loss of calcium in the intestine as part of the calcium of the food is not absorbed. For this reason, the calcium required in the diet is much greater than that actually needed by the tissues. Calcium has a great diversity of functions in the body, and reference to other chapters will show its relation to heart-beat, clotting of blood, clotting of milk, permeability of membranes, neuromuscular excitability and bone formation. Pregnancy and lactation make specially heavy demands, and the dietary calcium should be specially considered in these conditions.

Iron

The chief function of iron is in connection with the haemoglobin and the cytochrome in the tissues. In women there is a periodic loss of iron with menstruation, but in man the loss of iron from the body is extremely small. It will be recalled that disintegration of the red cells is followed by excretion of the iron-free bile pigment, while the iron is used again for haemoglobin production. The daily amount recommended, 10 to 15 mg, certainly does not represent the amount lost from the body. It seems that most of the iron taken in the food is not absorbed, but a certain surplus is necessary in order that a small fraction should be available. Deficiency of iron is associated with anaemia.

THE VITAMINS

It is well known that animals cannot be maintained in good health on diets which will supply the necessary calories together with protein, fat and mineral requirements, if certain accessory food factors, the vitamins, are absent from the diet. There are, at present, a great many different substances which have been recognised as vitamins. The usual conception of a vitamin is an organic substance which is necessary for health, including growth in young animals, and which does not act by supplying energy. Most of the vitamins have a known composition and chemical formula and many of them can be synthesised. Several of the vitamins, especially of the B group, are known to take part in the oxidative processes of the body, either as coenzymes or carriers (aneurin, nicotinic acid, riboflavin).

The vitamins can be classified according to their solubility in fats or in water. The following shows the different members of each of these groups:

Fat-soluble	*Water-soluble*
Vitamin A (retinol)	Vitamin B group
Vitamin D_2 (calciferol)	Thiamine (aneurin)
Vitamin D_3	Riboflavin
Vitamin E (α-tocopherol)	Pantothenic acid
Vitamin K	Niacin
	Pyridoxine
	B_{12} (cobalamin)
	Biotin
	Choline
	Folic acid
	Vitamin C (ascorbic acid)
	Vitamin P (citrin)

The human dietary requirement of the more important vitamins is given in Table 52.4.

Vitamin A

Deficiency of vitamin A in the diet leads to cessation of growth, loss of weight and decreased resistance to infection. There is keratinisation of the epithelium in the eye, respiratory tract and genitourinary tract. In the human subject there is xerophthalmia and night-blindness. The vitamin is related to β-carotene, and represents half the molecule of this with addition of an alcoholic group. β-Carotene can be regarded as a forerunner of the vitamin and can replace it in the diet, if large quantities are given. Neither β-carotene nor vitamin A can be synthesised in the animal body, but they are ingested with green plants and are found in the fatty tissues, especially the liver. There is a chemical relation between vitamin A and rhodopsin (visual purple) which probably explains the connection between vitamin A and night-blindness.

Sources: butter, egg yolk, carrots, spinach, and particularly fish-liver oils.

Thiamine (aneurin)

This is the antineuritic part of the vitamin B complex, and its absence leads to polyneuritis in the pigeon and rat, and to beri-beri in man. Lesser degrees of deficiency cause fatigue, loss of appetite, dyspnoea on exertion, and neuritis. In the pigeon and rat, bradycardia (slowing of the heart) is characteristic. Experimentally and clinically, symptoms of deficiency occur on a diet composed mainly of polished rice, as the vitamin is present in the outer part of the grain. The pyrophosphoric acid ester of thiamine is co-carboxylase, which forms part of the enzyme system for metabolising pyruvate, and in deficiency of the vitamin the pyruvate concentration of the blood is raised. Thiamine is also essential for the conversion of carbohydrate into fat.

Sources: yeast, wheat germ, pulses, meat.

Riboflavin

Deficiency of riboflavin in the diet causes disturbances in the mouth and tongue, in the cornea and in the skin. In combination with a protein and phosphoric acid it forms flavoproteins, enzymes concerned with tissue respiration.

Sources: yeast, liver, meat, eggs.

Niacin

This is the pellagra-preventing factor of the vitamin B group. Pellagra is a disease characterised by diarrhoea and skin disturbances in the human subject. In dogs, deficiency of the vitamin causes black-tongue. Either nicotinic acid or its amide will prevent these disturbances, and the term niacin is used to include both these substances. Chemically, nicotinic acid amide forms part of the molecule of the phosphopyridine nucleotides, known as coenzymes I and II, or as DPN and TPN, substances which are known to take part in the oxidative processes in the body.

Sources: yeast, meat, fish, wheat flour, liver.

Pyridoxine

Deficiency of this substance has been found to produce a disturbance in young rats, which resembles pellagra, but which cannot be cured with nicotinic acid: it can be obtained from rice bran. Pyridoxine forms part of the enzyme systems necessary for protein metabolism.

Pantothenic acid

When rats are fed on diets deficient in the B complex with added aneurin, nicotinic acid, riboflavin and pyridoxine they develop a greying of the skin, which can be prevented by addition of pantothenic acid to the diet. Deficiency of pantothenic acid also leads to a pellagra-like dermatitis in chickens. The function of pantothenic acid in human nutrition is not known, but it is thought to be essential. Pantothenic acid enters into the constitution of Coenzyme A which plays a major role in metabolism (Chapter 51).

Vitamin B$_{12}$ (cobalamin)

This is the anti-pernicious anaemia factor, present in liver, and is identical with the "extrinsic factor". The molecule contains about 4 per cent of cobalt. It plays an essential role in the production of red cells.

Folic acid

Derivatives of folic acid are involved in the transfer of one-carbon units in biological synthesis. It is specially important in the production of red cells.

Other vitamins of the B group

Other substances which have been found to be effective in replacing deficiencies in experimental diets are choline, folic acid, biotin, p-aminobenzoic acid and inositol. Choline is one of the constituents of the phospholipid, lecithin. If it is not present in the diet, changes occur in the liver which are chiefly characterised by excessive fat deposition. The addition of choline to the diet prevents this, and choline is thus said to exert a "lipotropic action". Little is known about the human requirements of the other substances.

Vitamin C (ascorbic acid, antiscorbutic vitamin)

It has been known for several hundred years that scurvy could be prevented by including fresh fruits in the diet, and it was found in 1932 that the antiscorbutic substance in fresh fruits was ascorbic acid. In the animal body ascorbic acid is found in the adrenal cortex. On oxidation it readily forms dehydroascorbic acid, and can be easily reformed from this by reduction, but the biological significance of this is unknown. Vitamin C deficiency is associated with disturbance in the formation of the enamel in the teeth, and with the process of calcification of bone. Claims have been made for the value of ascorbic acid in treating or preventing a range of diseases from the common cold to schizophrenia, but these still await substantiation.

Sources: fresh fruits and vegetables, especially the citrus fruits, oranges and lemons.

Vitamin D (antirachitic vitamin)

Absence of vitamin D in the diet gives rise to the characteristic appearance of rickets, a disease of children associated with softening of the bones and hence giving rise to abnormal shapes of the parts of the skeleton which have to bear the weight of the body. Rickets is associated with an abnormal metabolism of calcium and phosphorus, and it can be prevented by giving vitamin D. The antirachitic property is possessed by a number of substances, but of these the most important are called vitamin D$_2$ and vitamin D$_3$. Vitamin D$_3$ (cholecalciferol) is the naturally occurring vitamin, while vitamin D$_2$ (ergocalciferol) is produced from ultra-violet irradiation of ergosterol. Vitamin D$_3$ can also be produced by ultraviolet irradiation, the precursor in this case being 7-dehydrocholesterol. This substance is present in the skin and the beneficial effect of sunlight in preventing rickets is due to formation of the vitamin. Vitamin D is the vitamin which is most likely to be inadequate in the diet and hence the widespread habit of giving to children cod liver oil or other vitamin D source.

Sources: butter, cream, eggs, but especially fish-liver oils.

Vitamin E (α-tocopherol)

Deficiency of vitamin E in rats gives rise to failure of the reproductive organs. In male rats there is deterioration of the testis and in the female, death of the fetus. In other animals deficiency may be accompanied by muscular disturbances. Little is known definitely about the requirements of the human subject, and the value of treatment with vitamin E for prevention of abortion is still a matter of some dispute. Massive doses of vitamin E have been used in the treatment of coronary artery disease.

Sources: cereals, especially oats and wheat, liver and eggs.

Vitamin K (anti-haemorrhagic vitamin)

Deficiency of vitamin K in the diet of chickens produces a haemorrhagic disturbance associated with a lowering of the amount of prothrombin in the blood (Chapter 72). In the human subject a vitamin K deficiency can be produced when there is an absence of bile in the intestine, e.g. in case of a biliary fistula. The vitamin does not have any effect on such haemorrhagic diseases as haemophilia or purpura and, as far as is known, is only related to formation of prothrombin.

Sources: it occurs more abundantly in plants than in animals. Cabbage, spinach, cauliflower are good sources, while milk is very poor.

ALCOHOL

Alcohol is capable of oxidation by the animal body, but it is usually taken not so much for the purpose of providing energy as for the effect it produces on the higher centres of the central nervous system. The relaxation of rigid self-control and discrimination which it produces in suitable doses is found by many people to increase the enjoyment of congenial company, and its widespread use lends interest to a consideration of its metabolism by the tissues of the body.

Alcohol differs from other energy-providing foods in that it can be absorbed from the stomach, although the rate of gastric absorption is much less than that from the intestine. After absorption most of it is metabolised, the remainder being excreted either in the urine or in the expired air. When taken in small doses so that the concentration in the blood does not rise above a certain level, most of the alcohol is oxidised without any accompanying pharmacological action.

The concentrations of alcohol in the blood, expired air and urine has considerable practical interest in relation to tests for drunkenness. In Britain it is now an offence to drive a vehicle while the proportion of alcohol in the blood is in excess of 80 mg dl^{-1}. An instrument called the *breathalyser* operates on the principle that the concentration of alcohol in expired air containing alcohol comes into contact with potassium dichromate, the dichromate is reduced by the alcohol and changes colour. The concentration in the urine is also proportional to the amount in the blood, and urine tests are also used for determining the amount of alcohol in the blood.

The oxidation of 1 g of alcohol in the body yields 29 kJ of energy. Alcohol metabolism proceeds at a practically constant rate for any one individual, varying usually from 6 to 10 g of absolute alcohol per hour. The food value of alcohol is, however, limited in spite of the fact that it is quickly absorbed and requires no digestion, partly because there is no storage mechanism and partly because of the inconvenient effects of alcohol on the central nervous system.

If alcohol is taken in larger doses it acts on various parts of the body, chiefly on the central nervous system. The effect is mainly dependent on the concentration in the blood, although it is also influenced by the rate at which the concentration is attained. Hence the effects of a large dose of alcohol will be reduced if it is taken with food and particularly with fatty foods which delay the emptying time of the stomach. Thus cocktails before a meal are more potent than liqueurs containing a similar quantity of alcohol after the meal, and the practice of starting a meal with hors d'oeuvres with a high content of fat and oils is conducive to the retention of a discriminating palate throughout the course of that meal. The relation between blood concentration and pharmacological action is of considerable medicolegal interest, as the symptoms of alcoholism can be roughly related to the concentration of alcohol in the blood at the particular time (Fig. 52.1).

the other. Although usually related, they are more or less distinct sensations. The hunger feeling is referred to the epigastrium and may be accompanied by contractions of the stomach. Attempts have been made to relate hunger to some measurable change in the blood, e.g. to the blood sugar concentration. That there is no simple relationship is shown by the observations that injection of insulin produces a fall in blood sugar which is accompanied by a sensation of hunger, while on the other hand hunger is a common symptom of diabetes where the blood sugar is raised. Appetite is still more difficult to relate to any physiological or biochemical basis. It is more susceptible than hunger to other influences such as emotion, habit or artificial stimulation by attractively prepared food. Under conditions of modern life other factors besides hunger and appetite take part in the selection of food (e.g. in children training and example, and in adults advertising and propaganda). Fashions, fads and the cultivation of a discriminating palate also play a part in the choice of diet.

It is now recognised that the hypothalamus (see Chapter 32) plays an important part in the regulation of food intake. Experimental lesions in the hypothalamus in rats may produce a voracious appetite, and such animals rapidly increase in weight; other lesions may abolish appetite. Electrical stimulation of the

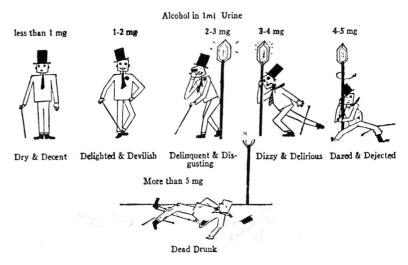

Fig. 52.1 The relation of the degree of intoxication to the concentration of alcohol in the urine.
The concentration of alcohol in the urine is approximately the same as that in the blood and tissues, except during absorption from the alimentary canal, when the concentrations are changing rapidly. (From Emil Boger Emerson's *Alcohol and Man*, by permission of Macmillan.)

Alcohol has a definite diuretic action. If 50 g ethyl alcohol are taken in 250 ml of water, it is followed by an output of urine of 600 to 1000 ml in two to three hours. The mechanism of the diuresis may be depression of the hypothalamic centre with consequent decrease in the secretion of the antidiuretic hormone of the pituitary gland. Alcohol is neither concentrated nor diluted by the kidney, and the concentration in the urine may therefore be used as a rough measure of the concentration in the blood plasma.

HUNGER AND APPETITE

The sensations of hunger and appetite are ill-defined feelings of "emptiness" in the one case and pleasant anticipation of food in

hypothalamus (in unanaesthetised goats) has led to excessive feeding. The most probable basis for hypothalamic control of appetite is that the hypothalamus responds in some way to the total amount of fat in the body, with increase in fat content depressing food intake. How this mechanism functions is quite unknown.

How far the sense of taste is a reliable nutritional guide is a problem of considerable interest. A good deal of experimental work in this field has been done on animals suffering from deprivation of some body constituent, and given the opportunity to select a diet from a large choice of substances with a view to testing their ability to make good the deficiency. Removal of the adrenal gland causes a fatal loss of sodium from the body, and it was found that adrenalectomised rats chose sodium salts out of a number of substances available and by this process outlived a control group of adrenalectomised rats to which sodium was not given. Sodium-deficient sheep, also, choose to drink salt solutions rather than water. In another set of experiments, rats from which the parathyroid glands had been removed increased their intake of calcium lactate, and thereby prevented the fall in blood calcium which, accompanied by tetany, usually supervenes in parathyroidectomised animals.

One of the important principles of animal physiology is the maintenance of the "internal environment" of the body cells and many reflex processes contribute to this end. It has been suggested that the ability to select diets suitable to physiological need is an example of the behaviour regulators acting towards the common goal of stabilising the composition of the body fluids, so essential for the continued activity of the cells (physiological "homeostasis").

Section 4:

Information: Its Reception and Processing

53. General Properties of Receptors

Introduction

Animals cannot behave purposefully in their environment, nor can they control their own bodily functions in the absence of information about the outside world or about the state of their bodies. Some of this information controls reflex responses; for example posture is largely controlled by reflexes originating in sensory receptors within muscles, tendons and joints. Part of the information, generated by sensory receptors, may reach consciousness.

Sensation and receptor mechanisms

The subjective response of sensation is our only experience of the activity, in our own bodies, of receptor mechanisms, and it is of importance to understand the relation between a sensation and the receptor activity which gives rise to it. We can take an example from the visual system. Light, coming from various parts of the outside world, excites retinal receptors. These initiate impulses which travel via various cell bodies, synapses and nerve fibres, eventually to reach the visual cortex. At all the fibre-to-cell junctions the message becomes modified before it reaches the cortex, where it is generally presumed that the entirely unexplained processes in consciousness and sensation take place.

From this example, we can say certain things about sensation. First, it is and can only be a subjective response. Second, a sensation cannot include more information about an object or event than has been transmitted by the physical events outside the body and by the activity of the receptors and the nervous system. Third, sensation may be a product, not only of the information from one particular kind of receptor, but also of that from other kinds of receptor of the same general type but differing in detail. Fourth, the intensity and quality of a sensation depend on the nature and intensity of other sensations, on the general activity of the central nervous system at the time, and on its antecedent activity associated with previous experience. One can diminish a sensation of pain by applying a counter irritant; one can fail to hear quite loud sounds if one is deep in thought; and one can fail to observe, by touch or sight, a familiar object in its accustomed place. The selection and synthesis by the central nervous system of the information transmitted to it from a variety of receptors forms the basis of the wide range and subtlety of sensation of which the human mind is capable. Sensations depend as much on the central nervous system as on the receptors, and great care must be taken to scrutinise the interpretation when sensation is used as an index of receptor activity.

The most striking thing about our sensations is their clarity and detail, but it must be remembered that there are real limitations; for example, "sounds" with a frequency greater than 20 kHz are inaudible, and the effect of doubling the intensity of a sound wave is not to increase its loudness by a factor of two, but by a variable factor, which is quite different for high and low notes. It is the purpose of this chapter to consider how the receptors are able to transmit so much information to the central nervous system and to describe also some of their limitations.

RECEPTOR UNITS

The first step in considering the problem of the transmission of information to the central nervous system is to consider the behaviour of the individual unit, the word unit being used to describe a single afferent (sensory) axon and the one or many receptors with which it is directly connected and the whole lying "*in situ*". It seems probable that all types of unit, including those in the complex organs of the so-called "special senses", behave in fundamentally the same way; however, most experiments on single units have been done in places other than the organs of special sense.

In this chapter, we shall mainly consider the series of events which result in the generation of sensory impulses. We will analyse the mechanisms involved in production of sensation by *adequate stimulation* of receptors.

The specificity of receptors

By the term adequate stimulation, we mean "the form of stimulation to which a receptor has the lowest threshold". For instance, sound waves pass to the inner ear and stimulate the cells of the organ of Corti; the sensation of sound results. Light reaches the retina and the sensation of light results. It is undoubtedly true, though, that many receptors will respond to other kinds of stimulation in addition to their "adequate stimulus" (sound, light, in the examples quoted). Pressure on the eyeball, for instance, stimulates the retinal receptors, but the boxer who sees "stars" can hardly be said to have had his eye adequately stimulated, because in this case the stimulus must be very strong before it has any result.

Single afferent nerve fibres can be dissected out from nerve trunks and separated from other fibres which have different connections. When this is done, it is usually found that activity appears in response to excitation of the receptors from which the fibre is derived; and usually, it appears much more readily when one particular type of energy is used, e.g. mechanical, thermal, chemical or light, than when other types are used. Thus the receptors of the retina normally respond only to light, and those associated with the senses of taste and smell respond only to chemical stimulation. In general, therefore, each kind of receptor unit is specially sensitive to one form of energy; receptor units are "specific" in that each unit exhibits a particular pattern of sensitivity. One class of unit which does not show big differences in its sensitivity to different forms of energy but shows a distinct sensitivity pattern of great importance, is that associated with the sensation of pain. These units may have different patterns of sensitivity but in general they respond to high intensities of stimulation by all types of energy—that is levels which are nearly or actually damaging. When receptor units, which are specially sensitive to one form of energy, are excited by a high intensity of some other form, the sensation aroused is that expected of the normal stimulus, demonstrating that the sensation is dependent on the connections of the unit and not on the nature of the stimulus that excited it. A good example of a sensation being aroused by an abnormal stimulus is to be found in the flash of light that is seen if the eyeball is struck or if an electric current is passed through it.

The old concept that a given receptor responded only to a certain stimulus was embodied in the "law" of Johannes Müller (Law of Specific Nerve Energies). As the situation is understood today, though, receptor specificity is much less rigid than the old ideas would permit. A muscle spindle responds not only to muscle stretching but also to temperature changes. When this phenomenon is also found to occur in most skin mechanoreceptors, the question arises as to whether part of the sensations of heat and cold depend on receptors other than those specifically categorised as temperature receptors.

The effects of stimulus intensity

The response that can be recorded from a single afferent fibre when the area it innervates is stimulated, consists of one or many impulses. These impulses, like all nerve impulses, are all-or-nothing; it is clear, therefore, that information can only be carried either by the time relations of impulses in any one fibre, or by the interrelations of the activity of a number of different units. The simplest type of response is found in the signalling of steady states by certain units; these include temperature sensitive units, some mechanically sensitive units, some units in the retina and the chemically sensitive receptor units associated with smell. Such a unit, when in a steady state, sets up a train of impulses which lasts as long as the stimulus; the frequency of the impulses depending on the strength of the stimulus (Fig. 53.1). These frequencies are repeatable so that, for example, whenever a limb is in a given posture, a given receptor in the joint capsule will discharge at a particular frequency. This is important, since the nervous system can only act consistently if the signals it receives are consistently related to the stimulus.

Natural stimulation produces a train of impulses. So we can see that the input to the central nervous system is one involving *frequency modulation* (not amplitude modulation as the action potential is an all-or-nothing event) for providing a gradation in the response to a stimulus of varying strength.

This naturally prompts the question: "How does the central nervous system recognise stimuli as being of different intensities?" As can be seen from Fig. 53.1, a weak stimulus induces a low-frequency discharge, a strong one a high frequency. However, a further mechanism for signalling intensity exists in most sense organs, in that many sensory fibres are usually connected to them. Thus, strong stimulation increases not only the frequency of firing in single afferents but in addition, increases the number of afferents which are active. This phenomenon is called *recruitment*, quite an apt term considering the general preoccupation with militaristic thinking at the time when these basic physiological discoveries were made.

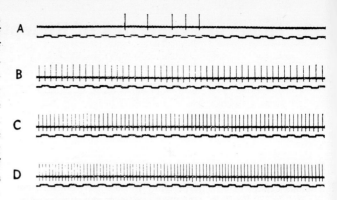

Fig. 53.1 The relation of frequency of discharge to stimulus strength. Records of impulses from a single fibre of the incisor nerve of an anaesthetised rabbit while different weights were hung from the tooth. A, 5 g (threshold). B, 10 g. C, 20 g. D, 50 g. All records 1 min after the weight was applied. Note occasional irregular impulses in A.

Below each record are time marks in 1/10th second.

All the action potentials are seen to be of the same size, irrespective of their frequency and thus of the strength of stimulation, illustrating the all-or-nothing law. (From a record kindly prepared by Mr A. R. Ness.)

Adaptation

The response is not quite as simple as that just described if the stimulus is applied abruptly and then left constant; for example, if a muscle stretch receptor (muscle spindle) is suddenly stretched to a new length, the impulse discharge starts at a high frequency, which declines at first rapidly and then more slowly until the steady frequency associated with the particular degree of stretch is reached. This decline of frequency after the onset of a steady stimulus is called "adaptation". The frequency at the beginning of such a discharge depends not only on the final size of the stimulus, in this instance the amount of stretch, but also the rate at which the final level is reached, in this example, the velocity of the stretch. Discharges from receptor units which behave in this way are shown in Fig. 53.2.

A large number of receptors do not respond to steady stimuli at all. They respond to a suddenly imposed steady stimulus by a discharge at a frequency which is initially high but which falls rapidly to zero (Fig. 53.2); and often give another burst of impulses when the stimulus ends.

As an extreme case, the large capsulated end organs, known as Pacinian corpuscles, often respond to a steady pressure with a single impulse at the moment the pressure is applied and another when it is released. Such receptors which respond only to a change of state are called "rapidly adapting", or "phasic", receptors; they are to be found both among those responding to mechanical stimuli and those responding to light stimuli.

The annulospiral endings in the muscle spindle to be described later, are examples of phasic receptors. Figure 53.3 shows how such an ending responds to sinusoidal stretching. Note that the discharge occurs when the slope of the sine-wave is steepest, i.e. the "*velocity*" of stretching is at a maximum. The significance of this type of response has been discussed on page 106.

A little thought about one's own sensations will immediately make clear the importance of signalling "*change*" of state; one responds vigorously to sudden changes of sound, while being oblivious to a steady noise; and if one wants to feel an object carefully one keeps one's hand moving and passes it over the

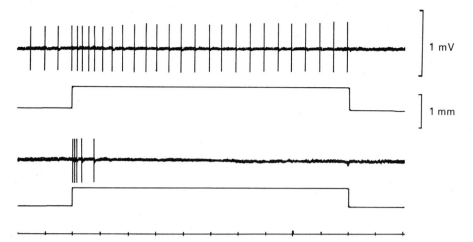

Fig. 53.2 Adaptation in two receptors.

Recordings are from two single fibre preparations, the top being from a slowly adapting stretch receptor, the bottom from a rapidly adapting stretch receptor. The lower trace of each pair shows the extent and duration of the stretch applied (approximately the same in both cases). Time trace is at 100 ms intervals; calibration bars are 1 mV and 1 mm stretch.

The slowly adapting receptor discharges for the duration of the stimulus. Note that the frequency of the sensory discharge is greater just after the onset of stretch than it is at the end of the period of stimulation. This receptor also had a resting discharge.

The rapidly adapting receptor responds to the stretch with a short burst, which is present only at the beginning of the stimulation.

small irregularities of the surface, thus allowing the rapidly adapting receptors to respond.

Mechanisms of impulse initiation

Physiologists have long been interested in the way in which a sensory impulse is initiated by a receptor, although it is only comparatively recently that the problem has been studied at the cellular level. Successful investigation of the transducer mechanism in receptors depends on being able to make electrical recordings very close to the receptor site, and naturally, from within the receptor cell or fibre itself if this is possible. It is not in mammals, because the cells of origin of sensory terminals in the dorsal root ganglion are too far from the receptor to be useful. In the crayfish, however, stretch receptor cells are found actually sitting on muscle fibres and they are large enough to permit the easy introduction of microelectrodes into them. Figure 53.4 shows the structure of such endings and the type of recording one can obtain from them if they are stretched.

Properties of the generator potential

The generator potential is a local potential and has the following characteristic features:

1. It is not propagated, remaining localised to the immediate vicinity of the region of the neurone which originates it. Electrotonic spread, with a typical electrotonic decrement, occurs to adjacent regions of the nerve fibre.
2. The response is graded, i.e. the degree of depolarisation depends on the strength of the stimulus.
3. There is no refractory period.
4. Local anaesthetics (e.g. procaine), which will block nerve conduction, have no effect on the generator potential. Also tetrodotoxin, which abolishes the nerve action potential, has little or no effect on the generator potential.
5. When the magnitude of the generator potential reaches a threshold value, the sensory nerve fibre itself is stimulated thereby and a propagated action potential is generated.
6. There is a proportionality between the amplitude of the generator potential and the frequency of the afferent discharge (Fig. 53.5).

The way in which the generator potential is produced is not at present understood. It would appear that the terminal membrane is particularly sensitive to deformation; physical changes of some kind must be evoked in the membrane which in turn give rise to a change in its sodium conductance.

The evidence for the depolarisation being due to an inward flux of sodium ions was obtained in the isolated pacinian corpuscle (Fig. 53.6) by J. A. B. Gray, who found that the generator potential can be depressed by lowering the sodium concentration of the bathing fluid. However, other ions are also implicated as has since been discovered in the crayfish stretch receptor.

However, whatever the mechanisms involved in setting up these potentials, they must be highly sensitive; a rod cell in the retina can be activated by a single light-quantum, i.e. the smallest possible quantity of light; movements of the basilar membrane of molecular or of even atomic dimensions are adequate to set up impulses in responses to sound; the threshold movement on the outside of a Pacinian corpuscle (not on the sensitive element) is probably about $0.2\ \mu\mathrm{m}$, i.e. below the limit of resolution by the best light microscope.

In view of this high sensitivity, it is not surprising to find that the energy received by a receptor, in the form of light or mechanical work, for example, is too small to account for the electrical energy which is released by the receptor, and which can be recorded as the receptor potential. The external event must therefore, release energy from an internal store. It appears that receptors store energy in the form of concentration gradients of ions across membranes. The stimulus causes a

Fig. 53.3 Derivative response in a muscle spindle.

Response to sinusoidal stretching of a single annulo-spiral ending in gastrocnemius of anaesthetised cat. (Top trace) Action potentials recorded from single dorsal root fibre, ventral roots intact. (Bottom trace) Mechanical displacement of tendon, stretch upwards. Frequency approx 10 Hz, amplitude of spikes 1 mV, stretch 1 per cent of resting length of muscle.

Note that response occurs when velocity of stretch is at a maximum. From Lippold, O. C. J. (1973) *Origin of the Alpha Rhythm*. Churchill-Livingstone: Edinburgh.

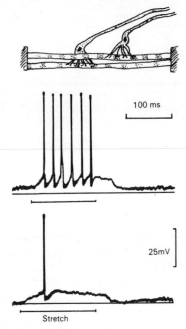

Fig. 53.4 The response of crayfish stretch receptors.

Receptors are found paired in the abdominal segments, each having a small muscle on which the stretch receptor neurone sits with its dendrites attached. One is rapidly and the other slowly adapting. By virtue of the large cell size and its proximity to the active nerve terminals, a microelectrode inserted into the soma will record local electrical events.

Typical recordings during stretch are shown below the structural diagram.

Stretch is indicated by the bar; top record is from a stronger stretch than the bottom one. Both the generator potential, a slow depolarisation of about 10–15 mV, and the action potential resulting from it can be seen. The generator potential takes place in the terminals; the action potential is initiated in the axon and travels along it; the soma is also invaded.

change in the permeability of this membrane and allows the ions to move down their electrochemical gradients.

Generator potentials and adaptation

Now that we know the stages in the transducer process, can we decide at which step adaptation occurs? The evidence, in mechanoreceptors at least, points to adaptation being a property of the mechanical and not the electrical systems. The steps in the transducer process are as follows:

1. Mechanical stress deforms the whole receptor organ
2. The deformation is transmitted to the nerve terminal
3. The nerve membrane is deformed
4. The membrane is depolarised (generator potential)
5. The generator potential gives rise to the action potential

Figure 53.7 shows an experiment upon the tenuissimus muscle of the cat to investigate this question. A single muscle spindle has been stimulated by stretch and by passing a small polarising current through it. Only in the first case does adaptation occur. The process must therefore be in stages (1) and (2) above.

Further evidence is afforded by the fact that the generator potential itself declines with time, even though the degree of stretch is unaltered after the initial change in length. This effect can be seen in the lowest of the three generator potentials illustrated in Fig. 53.5 (this record was the result of a larger and more abrupt stretch than the two above it).

The production of trains of impulses

We do not at present fully understand why a slowly adapting receptor should discharge repeatedly for the duration of a constant stimulus. The generator potential is thought to initiate the action potential at the first node of Ranvier. This action potential, besides travelling centripetally up the axon, also travels in the opposite direction, invades the terminals and depresses the generator potential below the threshold required to fire off another impulse. As the terminal membrane repolarises again, the generator potential takes over; the interval required before threshold is reached depends on the degree of depolarisation reached by the generator potential. Figure 53.8 illustrates this explanation.

Transmission of information

We have already briefly mentioned how the intensity of a stimulus is signalled. However, the time course of the discharge cannot convey any information about the site of the stimulus or its nature; such information reaches the central nervous system through the organisation of particular fibres, each of which has its own particular properties. There are thus two general ways in which information is signalled: in time, and by distinguishing between different units; or one can say in time and space. These two general ways are correlated with the two types of summation found in the central nervous system, temporal summation between successive impulses and spatial summation between impulses in neighbouring fibres.

Summary

It is now possible to summarise the variables available for transmitting information:

FACTORS DEPENDING ON TIME
(a) The intervals between impulses
(b) The duration of the discharge

FACTORS DEPENDING ON THE SPECIFIC PROPERTIES OF THE RECEPTOR UNITS
(a) The specificity of the unit to a particular type or types of stimulus and its rate of adaptation, which decides whether it is sensitive to a steady state or a change of state.
(b) The size of the receptive field and its position in the body.
(c) The sensitivity of each unit and, arising from this, the number of units active for a given stimulus.

A pattern of nerve impulses is therefore set up by each stimulus, however complex, and this pattern involves both the timing of each impulse in each fibre and also the spatial pattern formed by those units which are active and those which are not. For every stimulus, which can be distinguished from other stimuli, this pattern must be unique in that it must differ from

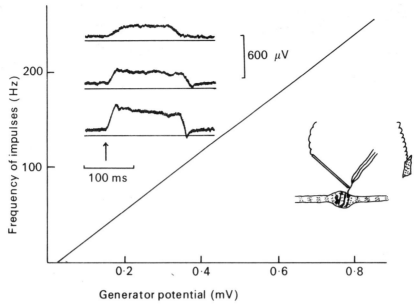

Fig. 53.5 Relation between generator potential and discharge frequency.
Top left inset shows recordings, made extracellularly, of the generator potential in a muscle spindle from the tenuissimus of the cat. Stretch applied at arrow; action potentials abolished by previous application of 0·01 per cent procaine hydrochloride. Recordings obtained, as shown in bottom right inset, between active electrode near to terminal and an indifferent electrode at a distance.
Graph shows relation between sensory ending depolarisation (generator potential) and firing frequency in the sensory axon from a normal preparation. (Data from Lippold, Nicholls and Redfearn, 1960.)

the patterns set up by each of the other stimuli. The physical properties of the tissues in which the receptors lie perform a major role in distributing the energy of the stimulus amongst the different receptors and also in determining the time course of any disturbance occurring in the immediate vicinity of each receptor. Examples of this will appear in the following chapters. The mechanics of the ear determine the distribution of activation amongst the receptor units of the cochlea; it will be seen that, as the intensity of the sound increases, the mechanical disturbance spreads further and so more units are activated. As the frequency of the sound wave changes, the position of the maximum disturbance moves. In the vertebrate eye the light is focused by an optical system onto an array of receptors; each unit probably responds with an increasing impulse frequency to increasing intensity of illumination at that point; in some species, such as man, receptors having different sensitivities to different wavelengths are intermingled and the combined output of groups of such different units enables the wavelength, as well as the intensity and position, of the light to be determined.

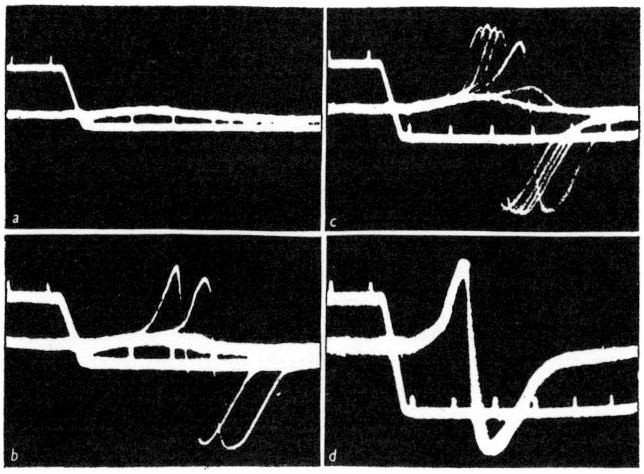

CHEMICAL RECEPTORS

There are different kinds of receptor which are sensitive to the presence of various chemical substances in the fluid around them. Many of these, such as the *chemoreceptors* in the brain and carotid body, concerned in the regulation of breathing, and the *osmoreceptors* concerned in the control of the concentration of the urine, are of great importance in the life of man as well as that of all animals. Certain specific chemical substances, however, when present in or near the appropriate kinds of receptor may arouse the sensations of "taste" and "smell". In this respect, the two sensations are superficially related; but when examined in more detail, it is seen that there are very considerable physiological differences between them. There is thus a fundamental distinction between the working of internal chemoreceptors and external chemoreceptors, namely that the latter transmit information leading to a sensation such as taste or smell, whereas the former do not.

Fig. 53.6 Receptor potentials and impulses in Pacinian corpuscle.
Each picture is of 20 superimposed traces; the top beam (at the left) signals the amplitude and time course of the mechanical stimulus that was applied to the corpuscle and also shows 1 ms time interval. The stimulus strength was increase between each record. The receptor potential can be seen to increase in size with the stimulus and the impulses increase in number. In (b) and (c) 2 and 5, respectively, of the stimuli gave rise to impulses, which "took off" at different values of the receptor potential, owing to slight variations of the threshold. In (d) every stimulus results in an impulse. (Gray and Sato (1953) *J. Physiol., Lond.*)

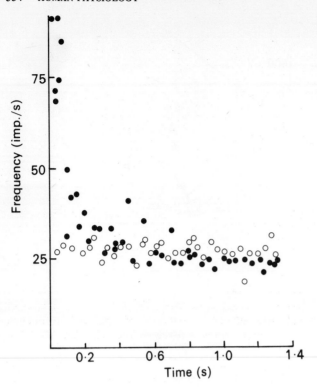

Fig. 53.7(a) Relationship between frequency of firing and time after stretch. Single muscle spindle from tenuissimus of cat.

Stretch (●) response shows adaptation; passage of polarising current (○) gives rise to maintained firing without decrement. This shows adaptation in the muscle spindle to be the result of mechanical events and to be not due to non-linearities in the electrical events. (Lippold, Nicholls and Redfearn (1960) *J. Physiol., Lond.* **153,** 214.)

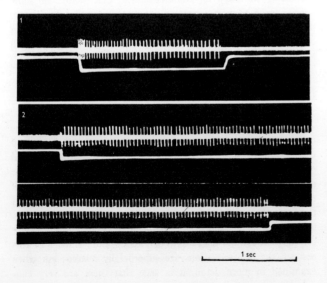

Fig. 53.7(b) The effect of stretch (1) and d.c. (2) on the same muscle spindle. In (1) lower beam records displacement; there is a high-frequency burst initially, followed by a steady discharge. In (2) depolarisation (lower beam) gives rise only to a maintained discharge. The bottom records are the continuation of the middle ones.

THE EXTERNAL CHEMORECEPTORS

The *gustatory* and *olfactory* receptors enable us to sample the chemical nature of the environment, at least insofar as the latter is either volatile, water soluble or in some other way comes into contact with the tongue and the nose.

Taste

The mucous membrane of the epiglottis and soft palate, and of the tip, sides and root of the tongue, contain special receptor organs known as "taste-buds". In these are the chemoreceptors associated with "taste". All "tastes" can be divided into four (or perhaps six) groups; the sour, the salt, the bitter and the sweet, to which are sometimes added the metallic and the alkaline. It is significant that although many substances give rise to mixed sensations of taste, it is nearly always possible to distinguish the components, and it is impossible to create an entirely new taste by combining any or all of the "pure" tastes. "Smell", as we shall see, differs very markedly in this respect. It must be remembered that in most cases the actual flavour of any substance present in the mouth depends upon the excitation of olfactory receptors almost as much as on the excitation of taste-buds.

Sourness is a sensation aroused by all solutions containing hydrogen ions in sufficient concentration. For the minerals acids, such as HCl, the threshold concentration is at a pH of about 4. The organic acids, such as acetic, and also carbonic acid, appear more sour than would be expected from their hydrogen ion concentration, probably owing to the greater ease with which they penetrate through cell membranes.

Saltness is a sensation aroused by the salts in the strong acids, particularly the monobasic acids. The least concentration of NaCl which can be tasted is about $0\cdot 02$ mmol l^{-1} ($0\cdot 12$ per cent).

Bitterness is a sensation aroused by many substances with a very wide range of chemical composition, but above all, of the alkaloids such as strychnine and quinine. Salts of magnesium, calcium and ammonium have a bitter taste, and so have ether and most glucosides. The threshold concentration for strychnine is about $0\cdot 000\,06$ per cent.

Sweetness is a sensation aroused by the sugars, and also by a number of other completely unrelated compounds, e.g. beryllium salts, lead acetate, chloroform, many amino-acids and saccharin. The least concentration of cane sugar which can be tasted is about $0\cdot 5$ per cent, whereas that of saccharin is only about $0\cdot 001$ per cent.

These four sensations are not equally easy to arouse in all parts of the tongue, some being most easily aroused in some parts, and others in other parts. There are even substances such as magnesium sulphate and dulcamarin (the glucoside from bitter-sweet) that give rise to different sensations when applied in different places. The receptors through which come the information on which these sensations are based must vary in their sensitivities to the different classes of substance; but records from single receptor units in various species of experimental animals indicate that there is not a distinct class of receptor for each of the four sensations. Thus in the cat there are receptors responding predominantly to acids; others responding to acids and to salts; still others responding to acids and to substances like quinine; and some responding to sucrose. We do not know that the receptors in man have the same sensitivities as those in the cat, but such a pattern of sensitivities contains all the

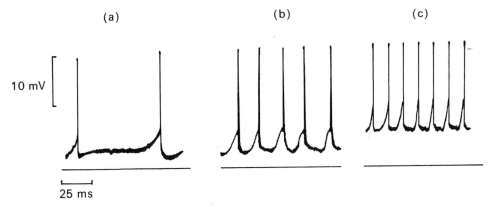

Fig. 53.8 Presumed mechanism of rhythmic firing.
Intracellular recording shows generator potential in a slowly adapting ending, with impulses superimposed. As the depolarisation increases (a) to (c), the time interval between individual strikes becomes shorter and the amount by which the antidromic invasion of the terminals by the spike tends to discharge the generator potential progressively lessens.

variability required to transmit the necessary information; there is a unique combination that responds to each class of substance.

TASTE RECEPTORS (TASTE-BUDS)

Taste receptors are ovoid structures found on the tongue and also scattered diffusely in the mucous lining of the buccal cavity (Fig. 53.9). Each taste-bud has a number of elongated (10 μm)

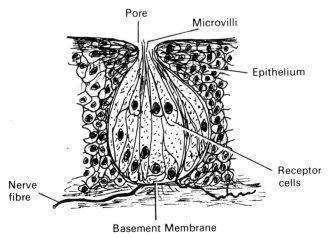

Fig. 53.9 Structure of a taste-bud.

receptor cells, whose apical tips project through the surface epithelium at the *gustatory pore*. Chemicals dissolved in the saliva come into contact with the receptor cell membrane via this pore.

Electron micrographs show that the tips of the receptor cells are in the form of *microvilli*, 0·1 to 0·2 μm diameter and 2 to 3 μm long, which poke out through the pore. Myelinated nerve fibres innervate the receptors via the base of the taste bud, and a plexus in this region.

An interesting physiological problem arises in respect of these receptor cells. Radioisotopic thymidine, which becomes incorporated into dividing chromosomal material, reaches a peak concentration in these cells at about 100 hours following injection. This must mean that they have a very short life and are constantly being replaced. How are the nerve terminal connections between receptor cells and nerve filaments maintained?

Electrophysiology

Generator potentials have been found in the receptor cell (resting E_m of 50 to 90 mV). Introduction of chemicals into the pore gives positive potential shift in the cell, proportional to concentration.

Action potentials can be recorded from the nerve supplying the taste bud. These show the phenomenon of adaptation.

The modality of taste is important in the life of the animal as it controls nutrition and hence the internal environment. Rats deprived of a dietary constituent will select food containing it, presumably because they are able to taste it. Adrenalectomised animals will go for salt-containing foods, preferentially. Experimental cutting of the afferent nerves from the taste buds abolishes this behaviour.

The sensations of taste

The sensation of taste is modified by a number of factors, both peripheral and within the central nervous system. For example, the maximum sensitivity of taste receptors lies within the range of temperatures between 35° to 40°C. There is also considerable individual variability with regard to sensitivity to various substances although a complete selective loss of taste for a particular modality of taste is unknown. There is, however, a condition which is inherited as a Mendelian recessive, in which persons cannot perceive phenylthiourea which normally has a very strong bitter flavour (normal threshold is 2×10^{-5} M, but those who cannot taste it normally may require 0·01 M solutions before they are aware of it). Approximately half of those of Caucasian origin cannot taste phenylthiourea, although their thresholds to ordinary bitter materials does not differ from normal. This fact has been used by anthropologists and others engaged in ethnic and genetic studies to follow the migration and mixing of populations over the world.

Adaptation occurs in taste sensation and in order to gain maximum satisfaction from pleasant foods these must be moved around in the mouth. Long-term adaptation also occurs and curry addicts require stronger and stronger mixtures as time goes by.

Smell

The organs for smell are situated in the upper parts of the nasal cavity. Odorous substances in the inspired air dissolve in the mucus covering the sensitive cells, diffuse into the hairs which protrude from the cells into the mucous layer, and so excite the receptors.

In contrast with the sense of taste, it is quite impossible to classify the various types of smell into definite components; each substance has its own distinctive smell. There are certain general resemblances, however, and it has been suggested that odorous substances can be grouped into the spicy, the flowery, the fruity, the resinous or balsamic, the burnt and the foul. Unlike taste, again, the combination of two or more smells may produce a completely new smell, which cannot be analysed into its components. One smell, again, can mask or neutralise another (the action of perfumes in this connection is well known), and this can take place even if the two odorous substances are applied to different nostrils.

One of the peculiarities of the sense of smell is its rapid "fatigue"; air which initially has a powerful smell may seem quite odourless within a few minutes. Recovery is equally rapid. This "fatigue" only applies to the particular substance exciting it, and another substance, even though it has a very similar smell, may be perceived normally. Different substances "fatigue" the sensory apparatus at different rates, but for any given substance, the rate of "fatigue" increases with the intensity of the smell. Some people are completely deficient in the sense of smell and many are incapable of smelling certain substances which have a strong odour to others (hydrocyanic acid is a typical instance). This deficiency may be congenital or acquired.

Action potentials in the olfactory bulb and the olfactory area of the brain of experimental animals have been recorded by Adrian. These potentials indicate the activity of cells on which impulses from thousands of primary units have converged; this activity, therefore, represents a stage in the analysis of the information. Nevertheless, it seems clear that different kinds of odorous substance may affect preferentially different groups of receptor. The number of such different groups appears to be very large, and each group may respond to a number of different substances; there is no indication of the existence of any analysis in terms of a small number of "standard" components. It was found, however, that on the whole, water-soluble substances, such as acetone or amyl acetate, excite preferentially receptors in the anterior part of the olfactory epithelium (in the rabbit), while oily (lipoid-soluble) substances, such as cedar wood oil or benzene, excite receptors in the posterior part. The discharge set up by oily substances also had a longer latency, a less abrupt onset and longer duration than those set up by water-soluble substances. The discrimination of smells may thus depend both on the spatial distribution of the activated receptors and on the temporal characteristics of the response. Fatigue of olfactory sensation is a central phenomenon; the receptors respond indefinitely.

54. Cutaneous Receptors

One of the functions of the skin is to mediate the reception of information about the environment. Throughout the thickness of the skin and over the whole body are receptor organs of various kinds.

The sensory modalities

The skin receptors can respond to a number of different types of stimulation; in turn these responses lead to the perception of different sensations. These have been termed *sensory modalities*; examples are touch, pain, temperature, hair-movement, etc.

In general terms, it has been found that the histological structure of the various types of nerve ending found in the dermis and epidermis, is related to function. Some receptors are simply bare nerve terminals. More complex types have the nerve terminals encased in corpuscular structures. Diagrams are shown in Fig. 54.1.

Figure 54.2 is a diagram of the histological appearance of the skin in the human fingertip. The distribution of the various endings is fairly typical of skin in general, but of course the density of endings per square millimetre will be much less in most other skin areas.

Touch

The sensation of touch is generated when light pressure is applied to the skin. Areas sensitive to light pressure, stroking with a brush, etc. are widespread throughout the skin; in certain areas sensitivity to touch is greater than in others. The tips of the fingers are the most sensitive regions.

Receptors subserving touch are slowly adapting; they usually (but not always) are attached to myelinated afferent nerve fibres.

Receptive fields

Single sensory fibres attached to receptors subserve an area of skin called the "receptive field". For touch, these are generally small, and in the form of discrete areas which respond. The histological structures associated with touch have in some cases been identified as Meissner's corpuscles (or sometimes as Merkel's discs). Often, it has been found that only myelinated fibres with free nerve endings are present in a touch-sensitive receptive field; it must be deduced therefore that these are responsive to touch.

The sensory discharge from touch receptors can be increased

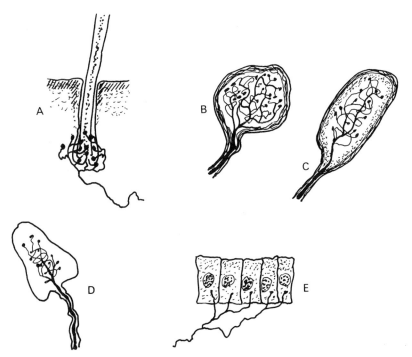

Fig. 54.1 Diagrammatic representation of endings in skin of typical human.
The histology of skin endings is variable and it is likely that transitional structures between those shown do occur.
A: Terminals and fine nerve branches at hair follicle. B: Krause end-bulb. C: Golgi-Mazzoni corpuscle (a pressure receptor). D: Meissner's tactile corpuscle. E: Free nerve endings in the cornea (pain receptors).

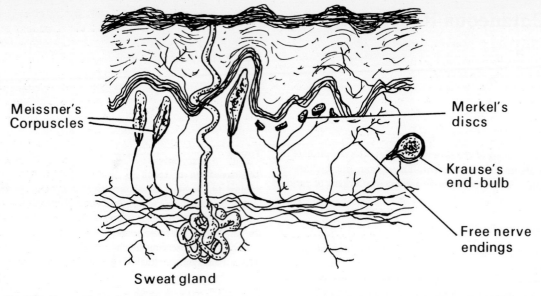

Fig. 54.2 Nerve endings in human skin.
The subpapillary plexus runs horizontally in the dermis sending branches of Meissners corpuscles, Krause's end-bulbs and Merkel's discs. Many free nerve endings are present, both in the dermis and epidermis.

by cooling the skin; sometimes it is decreased or abolished by warming, and in this respect the receptors resemble muscle spindles (pp. 361 and 362).

Hair receptors

Hairs, besides having thermal insulating properties, are most important as sensory receptors.

How sensitive they are can be judged from an experiment in which one of the authors was recording from the giant fibre in the ventral nerve cord of the cockroach. The results were complicated by the fact that small currents of air set up by the experimenters holding a conversation three metres away from the preparation, were stimulating the hair receptors on the anal circae.

The base of the hair follicles (Fig. 54.1a) are surrounded by a network of small nerve fibres. These are rapidly adapting and respond to the bending of a hair with a short burst of impulses, as shown in Fig. 54.3.

The receptive fields of the hair receptors are different from those of touch receptors; they are much larger and often overlap.

Temperature receptors

Temperature-sensitive receptors are found in all parts of the skin and in some mucous membranes. Certain single units investigated in detail have been found to set up a steady discharge whose frequency depends solely on the temperature of the receptor; for a given receptor there is a temperature at which the frequency of the discharge is a maximum, and the frequency declines at temperatures on both sides of this maximum. There are two main groups of receptor, those which have a maximum discharge around a temperature of 30°C, and those which have a maximum around 40°C. A particular combination of frequency of discharge in each group of units is thus uniquely related to a particular value of temperature. Information from a single group of receptors having a temperature-frequency characteristic of the type mentioned would always be ambiguous, since it gives the same frequency of discharge at two separate temperatures. When a man is in a "comfortably warm" state, the temperature of his skin ordinarily lies between 25° and 35°C, that of the temperature receptors being a few degrees higher; in these con-

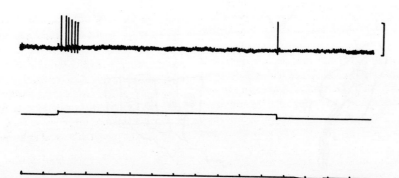

Fig. 54.3 Adaptation in a hair receptor.
Bending a hair (with the time course shown in the middle trace) gives a rapidly adapting response in its afferent fibre. The hair is kept bent during the course of about 95 ms; when released, an "off-discharge" results.
Time cal. = 10 ms. Voltage bar = 500 μV.

ditions, a rise of temperature causes one group of receptors to discharge more rapidly, and the other group to discharge less rapidly, while a fall in temperature has the converse effect. Both groups of receptors respond with a burst of impulses, over and above the steady response, if the temperature is suddenly changed; the "30°" receptors respond in this way to a sudden fall but not normally to a sudden rise of temperature, and the "40°" receptors respond to a sudden rise of temperature. These higher frequencies resulting from change of temperature, decline rapidly and within seconds, or at most one or two minutes, have reached their steady level as shown in Fig. 54.4.

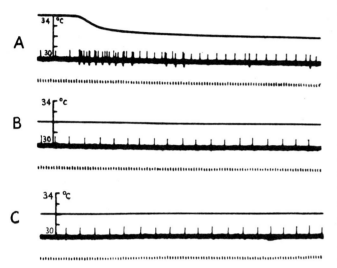

Fig. 54.4 The impulse discharge, in two nerve fibres from temperature receptors in a cat's tongue, showing adaptation. The upper line signals the temperature, which falls at the beginning from 34° to 32°C. At the bottom are time marks in 1/50th second. The discharge in one fibre (diphasic potentials) adapts to zero, while the other (monophasic potentials) adapts to a steady frequency. The records were taken at the following times: A, zero; B, 1 min; C, 15 min. (Hensel and Zotterman, *Acta physiol. scand.* 1951.)

Certain other points of interest arise if we consider the sensations aroused from human skin as a whole instead of considering the responses of single units from experimental animals. If an area of skin is tested for these sensations using metal rods with small tips (say, diameter 1·5 mm), which can arouse the sensation called "hot" or "cold", it is found that there are certain areas where it is especially easy to arouse either the sensation of "warmth" or that of "cold". In these areas the sensation is easily obtained in the centre, but less easily in the immediate surroundings.

Provided that the testing rod is left long enough in contact with the skin, temperature differences can be felt on most parts of the skin owing to the conduction of heat to or from the sensitive spots. The areas associated with the sensation of "warmth" do not coincide with those associated with the sensation of "cold"; neither are they equal in number. Certain "mucous membranes" are relatively insensitive to stimulation by raising the temperature, so that it is possible, for instance, to drink liquids which are hotter than the skin can bear, and the same also applies to medicinal douches.*

---

* It should, of course, be appreciated that very hot tea is usually drunk with a "slurping" action which provides a blast of cooling air to pass over the tea before it is swallowed. Also, the mouth contains plenty of saliva which dilutes and cools the hot tea.

If one hand is placed in hot water and the other in cold, and if they are allowed to remain until the sensations of cold and warmth have subsided and then both hands are placed in tepid water, the hand that was in the hot water will now feel cold and the one that was in the cold feel hot. This effect is probably not due simply to an initial adaptation of the receptors themselves. Like some other peculiarities of the sensation of temperature, it may be due partly to the cutaneous vasomotor changes that accompany changes in the temperature of the surroundings. The temperature of the deeper layers of the skin is rarely the same as that of the superficial layers, and the magnitude, and even sign, of the difference depends largely upon the rate of blood flow in the various parts of the cutaneous circulation. The temperature sensitive receptors in the two hands may thus be at different temperatures, even though both hands are in water of the same temperature. But it is possible, also that there is some central adaptation; the sensation at any moment will not be determined solely by the signals sent in by the receptors at that moment.

At one time it was considered that the Ruffini corpuscles were sensitive to warmth and the Krause end bulbs to cold. This has since proved not to be the case and in many places sensitive to temperature change, only bare nerve endings can be found. The problems in finding a receptor exclusively responding to temperature shifts, are aggravated by the fact that many other types of receptor are incidentally affected by temperature. The only way to show that an ending is purely temperature-sensitive, is by testing it for other modalities (such as tactile, stretch, etc.) and showing that it is unresponsive. Even this approach is imperfect, since for various reasons, a tactile receptor, may, at the time of testing, not respond to stimulation.

Deep pressure

The sensation induced in deeper-lying structures by stronger pressure than is required to activate touch receptors, is probably mediated by numbers of different nerve terminals. Deep sensation is produced by the stimulation of pressure receptors in the hypodermis. These have a relatively high threshold and so only respond when the pressure applied is considerable. For strong forces, other endings in tendons, muscles, bones and joints will be stimulated so that the sensations aroused will be a mixture of a number of modalities.

Vibration

It is unlikely that separate endings, solely responsive to vibration will be found. Vibration stimulates many different receptors to discharge. Some touch receptors, for example, may be driven by a tuning-fork applied to the skin at up to 400 impulses per second. Pacinian corpuscles are highly sensitive to vibration. When a vibrator (of the type used in beauty salons to massage away unwanted adiposity) is applied to a muscle, a contraction of up to half the maximal voluntary strength can result. This is due to activation of the muscle spindles by the vibration, giving rise to a reflex response.

PAIN

A detailed account of pain and pain mechanisms is given in Chapter 80. "Pain" is a sensation associated with high intensity stimulation of any type, though any particular pain is likely to have characteristic sensation related to the particular stimulus arousing it. As stimulus intensity increases, the frequency of firing in certain individual tonic units increases; in any range of intensity only certain units will be in their working range, some will be subthreshold and others may have already reached their maximum rate of firing. High intensities of stimulus are therefore signalled by the activation of those units having the highest working ranges, and hence it is the activation of this group of units which is associated with the sensation of pain. Unpleasant sensations may be aroused in other ways but these are not usually described as pain; for example, vibrations of certain frequencies can be unpleasant even though each individual stimulus in the sequence is of very low intensity and is signalled as such by the receptor units.

Two types of pain, *fast* and *slow*, are usually described. This distinction was first made from results of subjective experiments, but analysis of the responses in single primary receptor fibres has shown that receptor units with a high threshold for all types of energy fall into two groups with different velocities of conduction of the nerve impulses. The faster group have a conduction velocity comparable to that of the slower low threshold mechanically sensitive units concerned in the sensations of "touch" and "pressure"; while the slower group are among the slowest of all sensory fibres.

A special word must be said about visceral pain. Most of the information from visceral receptors never reaches consciousness, but on occasion a sensation of pain may be aroused; when it is, the pain may be referred by the nervous system to the skin belonging to the same segment as the visceral nerve responsible.

This is known as "*referred pain*". For example, gall-bladder pains are often localised in the right shoulder, kidney pains in the groin. One finds that the receptor nerves to joints, to the muscles working those joints and to the covering skin are all supplied from the same spinal cord segment. In joint injuries it is usual to find impaired movement due to fixation by the muscles, and also pain in the overlying skin. Visceral sensations are usually painful and are of the same type whatever their origin. Two very common causes of visceral pain are (1) prolonged contraction of plain muscle such as occurs in a ureter when partly obstructed by a stone, and (2) the stretching of organs such as the mesentery.

The receptors responsible for pain sensation

The actual nervous structures responsible for the sensation of pain have yet to be clearly identified. Probably most are free nerve endings. We know that some forms of stimulus can give rise to pain without other sensations being involved.

This has been shown by combining animal and human experiments. In the human subject, it can easily be determined if a particular form and strength of stimulation is painful. When the identical stimulation is now given to an animal, sensory discharges can be recorded in skin nerve filaments. The receptors involved have a high threshold and do not respond to weak stimuli; they connect to the central nervous system via small myelinated delta fibres or small unmyelinated C fibres.

There are only free nerve terminals present in the cornea. The only type of sensation derived from the cornea is pain. Temperature, touch and pressure cannot be felt.

55. Muscle Receptors

Introduction

Skeletal muscle, when it contracts, is responsible for all the voluntary movements made by the body. The majority of these movements are very closely controlled in terms of their strength and duration, in order to produce the highly skilled responses necessary for survival. As we have seen in Chapter 15, complex nervous pathways arranged in the form of servocontrol systems are provided in all skeletal muscles. In this chapter we will discuss the nature and properties of the sensing elements (or receptors) found in muscle.

Receptors sensitive, as these are, to the position of the body and its parts are termed *proprioceptors*.

MAMMALIAN PROPRIOCEPTORS

In mammals, the proprioceptors are the *muscle spindle*, the *Golgi tendon organ* and various types of *joint receptor*. Most muscles have spindles and tendon organs, although in a few species no spindles are present in extraocular muscles.

Histology of the muscle spindle

The muscle spindle is a fusiform, encapsulated structure, consisting of several small modified striated muscle fibres and attached nerve endings. The fibres are called *intrafusal* fibres to distinguish them from the main muscle's *extrafusal* fibres. Each spindle consists of one to twenty (usually six) intrafusal fibres each 1 to 5 mm long and attached at their ends to tendon or endomysium. They are thus in parallel with the main muscle—a most important point to note when considering their mode of functioning.

The striation of the intrafusal fibres disappears in their middle part, or equator. In certain fibres this region is swollen and contains nuclei and is called the nuclear bag. Fibres like this are called *nuclear bag fibres*. Other fibres are thinner, and often shorter in length, and have nuclei disposed in a chain along the length of the fibre. These have no central swelling and are termed *nuclear chain fibres*.

NUCLEAR BAG FIBRES

Nuclear bag fibres are innervated by large sensory nerves which terminate in a spiral fashion around the equatorial regions, the so-called *annulospiral endings*. The nerve fibres are 15 to 20 μm in diameter, conduct at 80–120 ms^{-1} and are group 1A fibres.

Smaller nerve fibres also innervate the nuclear bag fibre, but at a distance from the bag, nearer to the *polar regions*. The fibres concerned here are group II (Table 55.1).

NUCLEAR CHAIN FIBRES

The nuclear chain fibres have their main innervation via thinner myelinated group II fibres both for their equatorial and polar endings, although often the equatorial endings are of group I afferents.

Fusimotor nerve supply

The intrafusal muscle fibres naturally have a motor nerve supply (Fig. 55.1). These are called *intrafusal*, or *gamma*, fibres and are around 5 μm in diameter.

The first evidence that there were two distinct kinds of innervation in muscle came from the work of Eccles and Sherrington in 1930 (for details see Fig. 55.2). They were making cross sections of various peripheral nerves and found that in *motor* nerves to cat's muscle, there was a bimodal distribution in size of the fibres. One group was of 3 to 8 μm in diameter, the other 9 to 17 μm; there was no overlap.

It was later shown that if one stimulated a ventral root, two separate motor action potentials were produced in the muscle nerve. If the stimuli were weak, an action potential with a conduction velocity of 76 ms^{-1} resulted (i.e. fibres with diameters of 12–13 μm). Strong shocks brought in additional fibres, conducting at 27 ms^{-1} (i.e. diameters of 4–5 μm).

Effect of stimulation of a motor nerve

By using suitable techniques, it is possible to stimulate the large (alpha) and small (gamma) motor nerves to a muscle separately.

Table 55.1 *Details of the nomenclature of nerve fibres as discussed in this chapter*

Group	Diameter and velocity	Origin	Reflex
IA (A, α)	12–20 μm: 70–120 ms^{-1}	Low threshold stretch	Myotatic reflex; tendon jerk
IB (A, α)	12–20 μm: 70–120 ms^{-1}	High threshold stretch	Tendon reflex—lengthening reaction
II (A, β)	5–12 μm: 30–70 ms^{-1}	Skin touch and pressure	Flexor withdrawal reflex
III (A, δ)	2–5 μm: 12–30 ms^{-1}	Muscle, skin pain receptors	Flexor withdrawal reflex
IV (C fibres)	0.5–1 μm: 0.5–2 ms^{-1}	Pain receptors	Flexor withdrawal reflex

Note that terminology is a bit confused. Group IA is also termed Aα, group IV are alternatively C fibres, etc. (i.e. 5–12 μm diameter; conducting at 30–70 ms^{-1}) and they end in *flower spray endings*, so called because of their characteristic appearance. In both types of sensory innervation, the fibres lose their myelin near the spindle, the terminal regions being bare with expanded parts along their length.

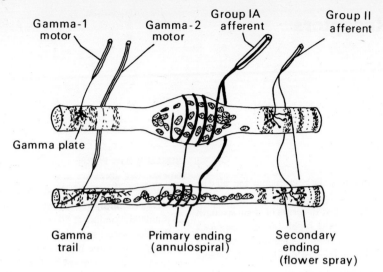

Fig. 55.1 Simplified diagram of central region of a muscle spindle.

The two types of intrafusal fibre are shown: above is a nuclear bag fibre, below a nuclear chain fibre. Their motor innervation, gamma-1 and gamma-2 fibres ending in gamma plates and gamma trails respectively, is shown on the left.

The afferent group IA fibres coming from primary endings (annulo-spiral) are shown in the middle, and on the right are group II fibres from secondary (flower spray) endings.

When large fibres alone are stimulated, the main muscle (extrafusal fibres) contracts. At the same time, if the sensory discharge from the spindle is monitored, there is a pause in the train of impulses. This is a logical outcome, bearing in mind the already emphasised fact that spindles are disposed in parallel with the main muscle. When the main muscle contracts, tension on the spindle is unloaded.

Stimulation of the small fibres alone does not lead to the generation of external tension by the muscle. However, the discharge from spindles is increased. This is due to the polar regions undergoing contraction, stretching the equatorial regions and hence activating the group IA terminals.

A third group of fibres has recently been discovered. These are the *beta fibres* which are thought to innervate both extrafusal and intrafusal muscle fibres, a branch from one nerve fibre going to each. Recent evidence seems to show that there are relatively few such fibres in most muscles.

The mode of action of muscle spindles

Electrophysiological studies indicate that the responses of the primary and secondary afferents from the spindles are different.

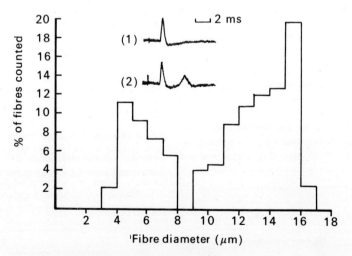

Fig. 55.2 Bimodal distribution of fibre size in a motor nerve.

Distribution histogram of fibre diameters in deafferented nerve to biceps femoris (cat). Notice that there are two distinct populations of efferent fibre (there are no sensory fibres left). (The original work was done by Eccles, J. C. and Sherrington, C. S. (1930), *Proc. R. Soc. B* **106,** 326.)

Insert (1) shows the effect of a weak shock given to a muscle nerve (cut central end of soleus nerve) at its sacral ventral root. In (2), the stimulus strength was increased to bring in a smaller, late potential due to gamma fibre activation. Conduction velocities 76 ms^{-1} and 27 ms^{-1} respectively. (Results originally reported by Kuffler, S. W., Hunt, C. C. and Quilliam, J. P. (1951), *J. Neurophysiol.* **14,** 29.)

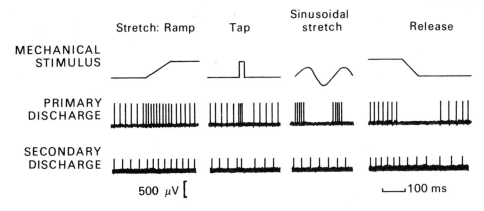

Fig. 55.3 Comparison of the responses of primary and secondary endings to various stimuli.

The responses are from a single primary afferent (middle line) and a single secondary afferent (bottom line) from a muscle moderately stretched and having ventral roots cut to abolish fusimotor activity.

Responses of both types of ending when recorded under static conditions are in general similar (i.e. during flat parts of stimulus record). When looked at *during* the process of stretching, it can be seen that the primary endings respond to rate of stretching. This is the *dynamic* response of the spindle; little or no dynamic component can be observed in the responses of the secondary type of ending.

When the muscle is stretched, the primary endings give a response both to the actual *length* of the muscle and to the *rate* at which it is being stretched, i.e. the velocity of extension. The secondary endings, on the other hand, signal the instantaneous length of the muscle. Figure 55.3 shows how primary and secondary endings respond to various stimuli and should make these points clear.

When recorded under static conditions of stretch, the responses of primary and secondary endings are broadly speaking the same. The two types have a similar relationship between discharge frequency and the extension of the muscle. If, however, the respective responses are examined actually during the processes of stretching, it is clear that the primary endings have a burst of high frequency impulse generation at the instants when velocity of shortening is at a maximum. This is the so-called *dynamic* component of the response to stretch. It is generally thought that the dynamic response of the primary endings is due to the spiral terminals being situated in the relatively non-viscous equatorial region of the spindle. The poor dynamic component found in secondary discharges is conversely due to their being sited well out on the polar regions of the intrafusal fibre, where the viscous properties are the same as the whole of the remainder of the fibre.

The function of the motor supply to the spindles seems to be twofold.

1. Both for nuclear bag and nuclear chain fibres a mechanism must be supplied which adjusts the length of the spindle as a whole during any time when the extrafusal muscle is contracting. It should be clear that any kind of stretch receptor will be useless unless it is in slight tension at any given muscle length. If there is slack to be taken up, spindles will be unable to signal length or velocity. Evidence for this action of the fusimotor nerve supply is afforded by the work of Vallbö who recorded from human motor nerves whilst muscles were voluntarily contracted. He found that the initial event in any type of movement was a burst of action potentials travelling to the muscle concerned in the alpha motor fibres. Following this was a train of action potentials in the fusimotor nerves, which acted to shorten the spindles *pari passu* with the extrafusal fibres (so that they could still be responsive).

2. The sensitivity of the afferent endings is determined by the degree of fusimotor activation. The γ_1 fibres which end in "gamma plates" on the polar regions of nuclear bag fibres set the dynamic sensitivity. The γ_2 fibres, which terminate in "gamma trails" on the nuclear chain fibres, control the static sensitivity. In both cases, provided extrafusal conditions are constant, fusimotor discharge causes an increased afferent discharge from the spindle, just as one would expect on the basis that the two ends, if contracting, will pull on the equatorial sensory regions and excite the receptors.

Supposing however, the muscle be made to contract while fusimotor stimulation is given, the situation is different. If γ_2 fibres are stimulated, there is no dynamic response; if γ_1 fibres are stimulated the dynamic response is enhanced. Thus it would appear that central control of the proportion of dynamic or static response arising from a given muscle stretch can occur.

This dual efferent control system can also be demonstrated electrophysiologically. Stimulation of the γ_1 fibres (often termed *dynamic* fusimotor nerves) induces a local, graded and non-propagated electrical response in intrafusal muscle fibres. Excitation of γ_2 fibres (*static* fusimotor nerves) induces propagated action potentials in intrafusal fibres. At present there is controversy as to whether propagated and non-propagated depolarisations occur in the same or different intrafusal muscle fibres.

56. The Auditory System

Sounds are pressure waves travelling through air (or through fluids and solids). The auditory system is a very effective form of distance reception developed in animals for the detection, analysis and interpretation of sounds.

In man, the sense of hearing and a particular use of the ability to analyse air vibrations has been developed—the ability to discriminate the subtleties of speech—so providing a method of communication between individuals. The first steps in the reception and analysis of the information contained in these vibrations (or sound waves) are carried out in the ear.

The role of the ear in the reception of sound is to set up patterns of nerve impulses in such a way that for every distinguishable sound there is a pattern which is different from that set up by any other sound. These patterns are transmitted to the central nervous system. This is referred to in Chapter 53. The first step is for the energy of the stimulus, in this case the vibrations in the air, to be transmitted to and distributed amongst the receptors. These receptors then produce electrical changes which are directly related to the mechanical changes in their immediate vicinity; these electrical changes set up nerve impulses in the fibres, which run in the auditory nerve. A major factor in determining the pattern of nerve impulses in the fibres of the auditory nerve is, therefore, the way in which the mechanical changes are distributed amongst the receptors.

PHYSICAL PROPERTIES OF SOUND

If a tuning fork is struck, the "prongs" of the fork are set into vibration, that is, they move backwards and forwards with great rapidity at one moment coming nearer to one another, at the next farther away. If, now, a small mirror is attached to one such prong, so that a beam of light reflected from it moves with the prong, it will be possible, by arranging that this beam falls on to a moving strip of photographic film, to obtain a record of the movements of the prong over a given time. In other words, the record will constitute a graph showing the position of the prong (ordinates) against time (abscissae). Such a graph is shown in Fig. 56.1. The completed record is said to show the *wave-form* of the tuning fork vibrations, and the distance between the resting position of the fork and its position of maximum deflection gives a measure of the *amplitude* of swing of the fork; where this distance is large, the vibrations are said to be of large amplitude, and where it is small, of small amplitude. It is easily demonstrable that a fork vibrating at greater amplitude will produce the same number of waves in a given time, but the amplitude of each wave will be greater. When the wave-form of a good tuning fork is examined in this way, it may be shown that the curve thus reproduced is almost exactly the same as that obtained by plotting the sine of an angle against the angle itself. Such a wave is thus known as a *sine-wave* and the vibrations of the fork which give rise to it as *sinusoidal*.

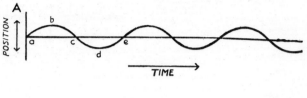
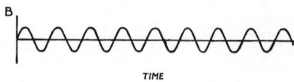

Fig. 56.1 Waveform of vibrations of tuning forks.
A. *a b c d e* constitutes one complete cycle. B. Frequency of fork three times that of A. C. Composite waveform produced by sounding forks A and B simultaneously.

A similar record may be obtained without a mirror on the fork. When the prongs of the fork vibrate, air pressure changes are set up in their vicinity, and these in turn set up pressure changes nearby; thus the sound travels through the air in the form of waves, the direction of wave motion being the same as that in which the sound is travelling, unlike waves on the surface of water, whose motion is at right angles to their direction of travel; these propagated pressure variations are perceived by man as sound. These pressure changes can then be converted, by means of a microphone, into electrical changes having an identical time course; these electrical changes are easily amplified and displayed on a cathode ray oscilloscope. Such a record will be found to be exactly as Fig. 56.1a, except that in this case the ordinates represent *air pressure changes*. The number of complete cycles, *a b c d e*, in one second of time is known as the *frequency* of the fork or note and is usually measured in *hertz* (Hz). The wire set into vibration by striking the note middle C on a piano has a frequency of 256 Hz, that is, it performs 256 complete double vibrations in one second.* *Pitch* is a quality of sensation related to the frequency of the wave, and the higher the frequency the higher is the pitch.

If, now, a fork of three times the frequency of the first fork be set into vibration, the resulting oscilloscope record will be as shown in Fig. 56.1b. Should forks of both of these frequencies be

---

* A frequency of 256 Hz for the note middle C is traditional in physics. Pianos and other musical instruments are normally tuned to the international standard pitch (A = 440 Hz), corresponding, when adjusted for equal temperament, to a frequency of 261·6 Hz for middle C.

set going at the same time, the resultant wave will be that shown at c in Fig. 56.1. Thus, it is evident that the wave-form of the air vibrations set up by two forks sounding simultaneously is quite different from that obtained from either fork alone. Most of the sounds commonly met with in daily life have an infinitely more complicated wave-form than this, the *quality* of a sound depending on its wave-form. The wave-form of the vowel "ah" is shown in Fig. 56.2. According to Fourier's theorem, however,

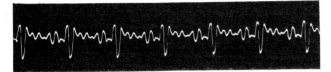

Fig. 56.2 Waveform of an English vowel of the type "ah". (D. B. Fry from *Science and Speech*.)

just as the relatively complicated wave-form of C is the result of adding together A and B, so the wave-form of any periodic vibration, however complicated, may be resolved into a series of simple waves, consisting of a *fundamental* with the lowest frequency found, and a number of *overtones* or *harmonics*, whose frequencies are multiples of that of the fundamental. Each of these single component frequencies will be of the same wave-form as Fig. 56.1, i.e. sinusoidal, and in the case of highly complicated sounds, the number of these components may be very great.

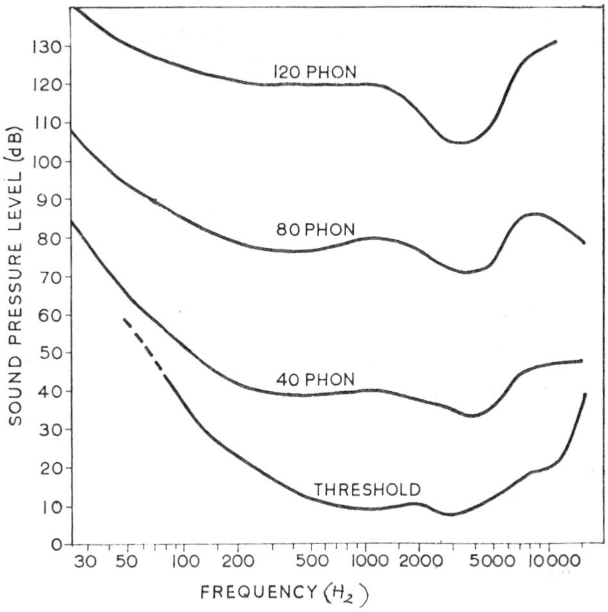

Intensity of sound and air pressure

A just perceptible sensation may be produced by sound whose intensity is somewhat less than 10^{-12} W cm^{-2}; the variation in air pressure is then about 10^{-9} atm, and the air molecules vibrate with an amplitude less than one-tenth of their own dimensions.* In quite ordinary circumstances, however, we are exposed to sounds with intensities between a thousand times and a thousand million times greater than this. It is more convenient, therefore, to avoid the use of very large figures by adopting a logarithmic scale (analogous to the pH scale for hydrogen ion concentration). The unit is called the *bel*†, and the difference in bels between the intensity levels of the two sounds is the common logarithm of the ratio of the two actual intensities; a tenfold increase in the intensity of a sound is an increase of 1 bel in the intensity level, a 100-fold increase is an increase of 2 bels, and so on. For practical purposes, however, such a unit is inconveniently large, and the *decibel* is more often used. This is simply one-tenth of a bel, and represents an increase in intensity of roughly 25 per cent. This is, as it happens, about the least change in intensity that can ordinarily be detected.

Since the power and hence the intensity of a sound is proportional to the square of the variations in air pressure, the difference in intensity level between two sounds, in decibels, is given by twenty times the common logarithm of the ratio between the two sound pressures. Note that the decibel is a ratio. It must have a base standard quoted. In clinical practice the latter is taken as threshold intensity (at each frequency) of an "average" person, as shown in Fig. 56.3. "Hearing loss" is then

---

* Sound is radiated power. *Intensity* is power (watts) passing across unit area (1 cm²) at right angles to direction of propagation. In practice, variations in air pressure are measured with a suitable microphone.

† Yet another eponymous unit; the bel honours the memory of Alexander Graham Bell, inventor of the telephone.

Fig. 56.3 The variation of intensity with frequency (a) for just detectable (threshold) tones and (b) for tones of equal loudness (at three different levels).

The intensity is measured in terms of the sound pressure level, in decibels above a pressure of 20 μPa. (Robinson and Dadson, from Kaye and Laby, *Tables of Physical and Chemical Constants* 11th edition, 1957.)

the intensity increase, in decibels, required to enable a patient just to hear the sound.

Physical properties and sensation

Frequency, wave-form and intensity are physical properties of the propagated waves of pressure change, whether they can be "heard" or not; pitch, quality and loudness are of psychological significance, and describe characteristics of the sensation evoked by a sound wave. Pitch and quality are, for all practical purposes, determined uniquely by frequency and wave-form respectively. The pitch associated with a given frequency may, however, depend somewhat on the intensity, and the quality associated with a given wave-form may depend on both the intensity and the frequency of the fundamental component. The loudness of a sound of given intensity, on the other hand, may differ very considerably according to its component frequencies; and the discrepancy is of sufficient importance to require special discussion in the next section.

AUDITORY ACUITY

Threshold

When the intensity of a sound is continuously decreased, it reaches a value where no sensation of sound is produced. The smallest intensity of sound required to produce a sensation is said to be on the *threshold of audibility*. As shown in Fig. 56.3, the physical intensity (sound pressure level) of a threshold sound varies very considerably with its frequency. The ear is not uniformly sensitive throughout the range of frequencies; it requires a smaller intensity of sound at 1000 to 4000 Hz to produce a just perceptible sensation than it does at frequencies higher or lower than these. At 100 Hz and at 10 000 Hz the threshold intensity is nearly 1000 times greater than it is at 3000 Hz.

The curve of threshold intensities given in Fig. 56.3 represents the average values obtained from a group of young people, aged 18–25, using one ear only and applying the sound through an earphone. This is the method ordinarily used for testing deficiencies in hearing. But we usually hear sounds in the open air, using both ears; in such "free field" conditions, the threshold of audibility is some 5 to 10 decibels (depending on the frequency) below the values plotted in Fig. 56.3.

If the intensity is adequate, a pure tone may be heard and recognised as such, if its frequency lies between some 20 and some 20 000 Hz. However, many kinds of animal, such as dogs, can hear notes with frequencies well above 20 000 Hz, which are inaudible to man. Beyond these limits of frequency, a "sound", if sufficiently intense, may be felt by parts of the body other than the ear, but it will not, strictly speaking, be "heard". Within these limits of frequency, a very intense sound is not only heard, but also felt by the ear, often a somewhat painful sensation; it is then said to be on the *threshold of feeling*.

Loudness

The loudness of a pure tone may be measured in terms of the physical intensity of the tone, expressed in decibels above the physical intensity of the same tone when it can just be heard—i.e. at the threshold of audibility—as given, for example, by the lowermost curve in Fig. 56.3. Tones of different frequency, but of the same physical intensity, therefore, will not necessarily have similar intensity levels (each above its own threshold), and will not have the same loudness. Loudness cannot, therefore, be measured in terms of physical intensity, for complex sounds.

We get around this difficulty by adjusting the intensity of a pure tone of 1 kHz to be the same as that of the sound of noise being investigated (as judged by a "normal" person or panel of normal persons). This is then the unit of subjective loudness and is called the "*phon*". As indicated in Fig. 56.3, subjective loudness differs considerably from physical intensity.

In practice, sound intensity meters are used (which measure sound power directly). Their frequency characteristic is adjusted to match that of the human ear. Table 56.1 gives approximate values for loudness of various noises.

Table 56.1 *Loudness in phons (approx. values)*

Faintest audible	4
Whisper	20
Quiet street	40
Conversation	60
Busy street	80
Diesel lorry, 10 m away	110
Painful sound	130

Discrimination thresholds

Normal persons can detect very small differences between the frequencies of two tones. The minimum perceptible difference (*absolute difference threshold*) is, for most subjects, around 3 Hz. This value is independent of frequency up to about 500 Hz; it rises over 1 kHz (Fig. 56.4).

Auditory acuity for intensity difference is poorer than for frequency. Five to ten per cent increases in loudness can just be detected by most people. Intensity discrimination is less good for low notes or low intensities.

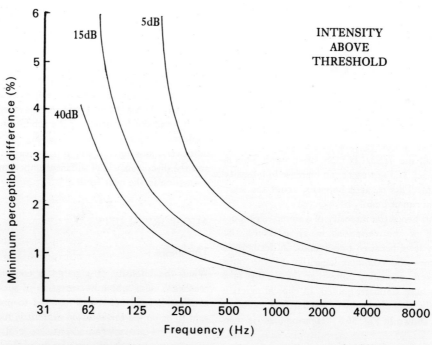

Fig. 56.4 The minimum differences in frequency which can be detected at various frequencies and at various intensities above the threshold (expressed as a percentage of the frequency of the stimulus). At intensities greater than 40 dB, the changes in sensitivity are small. The values given are for listening with one ear; if both ears are used, the minimum perceptible differences are smaller. (From data by Shower and Biddulph.)

Localisation of sound

The direction from which a sound originates can be recognised; two mechanisms are responsible.

1. Sound has a finite velocity, so that unless a source of sound is equidistant from the two ears, a given vibration will not arrive at both ears simultaneously; a sound from a source at an angle of 45 degrees from the sagittal plane, for example, will reach one ear approximately 0·4 ms before it reaches the other. If the two earpieces of a set of headphones are wired separately, so that clicks may be presented to the two ears with a small time difference between them, it is found that the direction from which the sound appears to come varies with this interval, provided that it is not greater than about 1 ms.
2. For high frequency tones, above 3000 Hz there will be a difference of intensity at the two ears due to the "shadow" effect of the hear; the sound is localised to the side where it is loudest. Both methods are probably used together for localising the most common sounds, which contain components of many different frequencies.

Many animals have another way of localising sounds. They rotate their pinnae until the apparent intensity is greatest, when the pinnae will be facing the source of sound.

The ability to localise the direction of sound is poorly developed in man, particularly at high and low frequencies. It is rarely possible to be more accurate than within 10 degrees in the horizontal plane and much more than this in the vertical plane.

Auditory fatigue

If the ear be subjected to prolonged loud sounds, it suffers a transient loss of sensitivity, i.e. it becomes fatigued. This means that the intensity of a note, if it is just to be perceived, must be made greater after such stimulation than before. A sufficiently intense stimulus, applied to a single ear only, causes a loss of sensitivity in both ears, that is to say, the fatigue is binaural. The loss in the nominally unstimulated ear is not produced by the residue of sound reaching it (due to bone conduction, slight leaks, etc.), as is shown by the following experiment. Suppose a sound of intensity only 70 decibels (dB) above the threshold be applied to the right ear, then no loss of acuity will be found in either right or left ear. If however, the intensity of the fatiguing sound is raised, although still applied only to a right ear, the left ear, as well as the right will suffer a loss of sensitivity. Yet the residue of sound reaching the left ear is at least 60 (dB) below that in the right ear, and thus can exert no fatiguing effect on it. It follows, therefore, that stimulation of one ear, if sufficiently intense, lowers the sensitivity of both ears; this effect is probably central. The loss in the stimulated ear, however, is somewhat greater than that in the unstimulated one. The greater fatigue of the stimulated ear is presumably due to a loss of sensitivity in more peripheral mechanisms. Frequencies above 1 kHz produce much more fatigue than do the lower ones.

Deafness

The tests for deafness normally employed are the spoken voice, whisper, the tick of a watch, and tuning forks of various frequencies. Such tests are innaccurate, but in experienced hands can give a fair indication of the degree of loss of hearing. A more accurate instrument, with which the sensitivity over almost the whole auditory spectrum may be tested, frequency by frequency, is the *audiometer*. In this, pure tones generated by a high quality oscillator are presented to the ear by an earphone; intensity is controlled by the operator and the point at which the patient can no longer hear the sound is estimated.

The threshold of audibility at 10 kHz and above increases rapidly with age. Usually there is also a small loss down to 1 kHz.

CAUSES OF DEAFNESS

Deafness may result from several types of disorder.

(a) *External ear obstruction.* The external auditory meatus is obstructed by a wax plug, by dirt or by inflammation and swelling of the meatal wall (*otitis externa*).

(b) *Middle ear disease.* The ossicles are prevented from functioning properly, a condition which frequently follows a nasal catarrh, and starts as an inflammation of the middle ear (*otitis media*), the infection entering via the Eustachian tube. Later, a pathological condition of the bone round the inner ear may develop (*otosclerosis*). This mainly results in the formation of new bone in the neighbourhood of the oval window which causes fixation of the stapes footplate to the bony capsule, with subsequent loss of hearing. In the later stages, the organ of Corti often shows degeneration.

(c) *Inner ear disease.* This usually involves loss of function of the organ of Corti or of the cochlear nerve, and is thus sometimes referred to as "nerve deafness". It may be due to:

(i) *Injuries to the inner ear*, such as boilermakers' deafness (*chronic labyrinthine concussion*);

(ii) *Diseases of the inner ear and auditory nerve*, due to local haemorrhages or inflammation or to general disease (e.g. syphilis, malaria, rheumatism). The commonest cause of inner ear disease, however, is middle ear disease.

External and middle ear deafness may often be distinguished from inner ear (nerve) deafness by making use of the phenomenon of "bone conduction". If the auditory meatus be carefully plugged, a tuning fork will be heard without difficulty if its stem is placed on the bones of the head (e.g. the mastoid bone). The vibrations from the fork set up similar vibrations in the cochlea, transmission taking place through the bones of the skull. It is clear, therefore, that if air-conducted sound is not heard, whilst bone-conducted vibrations are readily audible, loss of middle ear function must be suspected. This diagnosis may often be strengthened by inspection of the tympanic membrane, which, in most cases of middle ear disease, presents an abnormal appearance.

DEAF AIDS

Deaf aids are made in many forms. They all depend on raising the intensity of the received sound to above the incident level. Electronically, this is achieved by means of a microphone, amplifier and an earpiece. Compensation for selective hearing loss in specified frequency ranges can be incorporated in such a device, merely by altering the amplifier gain at the frequencies concerned. It is usually quite difficult technically to avoid some unwanted distortion; for this and other reasons, hearing aids may often be unsatisfactory. In general, the better quality instruments tend to be the most expensive.

FUNCTIONAL ANATOMY OF THE EAR

There are three stages in the transmission of the mechanical energy from the air to the receptors. The first is concerned with the transmission of the energy in the form of sound waves in air. This part is anatomically the external ear and ends with a diaphragm, the tympanic membrane Fig. 56.5. In the second stage the energy is transmitted by a lever system and this occurs anatomically in the middle ear. In the last stage the energy is transmitted predominantly through water and this takes place in the cochlea; this is anatomically part of the inner ear, the other parts of which are not concerned with hearing, but with orientation and acceleration, and are dealt with in the following chapter.

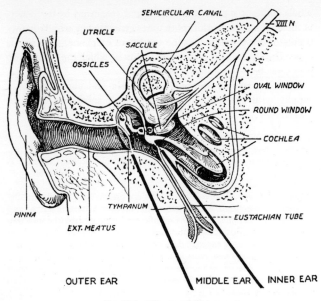

Fig. 56.5 Diagram of the ear.

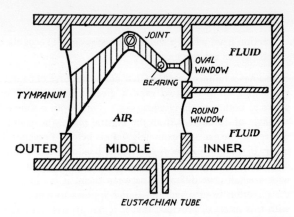

Fig. 56.6 Model illustrating the method of transmission of vibrations from the outer to the inner ear. (Modified from Beatty.)

The external ear

Sound conduction is in air through the external ear. In many mammals the *pinnae* collect sound waves, and can be moved, so as to be in the optimum position. This not only increases the collecting power of the ear, but may also serve as a means of localising the source of the sound. In man the pinnae serve little purpose and sounds can be heard almost as well without them. The external meatus is, in man, 25 mm long and conducts the waves to a diaphragm, the *tympanic membrane*, which can move in and out as the pressure just outside it goes up and down; remember that a sound wave consists of a succession of increases and decreases in air pressure. The external meatus is protected by hairs and wax, and in spite of both, small children succeed in jamming innumerable and highly unlikely objects therein. It is cunningly made smaller in diameter than the little finger, so that this cannot be used for the extraction of wax.

The middle ear

The tympanic membrane is conical in shape, with its concave surface facing outwards and downwards. Attached to the apex of the cone on its inner side is one of the small bone ossicles of the middle ear. There are three of these called the *malleus, incus* and *stapes*, but they function as a single unit and will be treated as such. Figure 56.6 shows how this lever system links the inner side of the tympanic membrane to the liquid-filled inner ear at the *oval window*; a bony "piston" lies in the window and is "sealed" in by means of a membrane. The position of the join and the bearing on the system are such that displacements at the oval window are about one-third of the displacements at the tympanic membrane. The force of the thrust at the oval window is consequently three times that at the tympanic membrane. The force at the oval window is spread over an area of $3 \cdot 2$ mm^2 whereas that at the tympanic membrane is spread over 65 mm^2. The pressure changes on the liquid just inside the oval window are therefore considerably greater than the air pressure changes outside the tympanic membrane. The increase in pressure and decrease in displacement is an example of "impedance matching", of which the matching of a loudspeaker to an amplifier is a familiar example. The purpose of this matching is to obtain the maximum transfer of power from one system to another; in this example a maximum transfer is required from air which is light and compliant to liquid which is heavy and stiff. In man the "transformer ratio" is 18 to 1; in the cat it is 60 to 1.

The lever system has two small muscles associated with it. One called the tensor tympani keeps the tympanic membrane taut; if its tendon is cut, the cochlea microphonics (see opposite page) fall by about one fifth. The stapedius has a similar action on the membrane of the oval window. High sound intensities cause reflex contractions of these muscles—a protective reflex, since displacements of the membranes would be reduced for a given sound pressure.

The Eustachian tube

External atmospheric pressure can change, apart from sound waves, e.g. in flying, diving, etc. Pressure in the middle ear is equalised, and bursting of the eardrum is prevented by a communication, the *Eustachian tube* which connects the middle ear to the nasopharynx. Swallowing or yawning opens the "valve" at the mouth end of the tube and should be indulged in during take-off and landing by air travellers. Small babies who normally will screech with pain at these times can, with profit, be given a dummy or their bottle.

The inner ear (cochlea)

This represents the crucial stage which determines the pattern of activation of the receptors and hence the way in which the information is transmitted in the form of nerve impulses. Although in mammals the cochlea is arranged in a spiral, it can, for functional purposes be regarded as a straight tube divided down its length by a membranous partition, the basilar membrane (see right side of Fig. 56.6). The receptors are arranged along the length of this partition. At the end of the tube there is a small open connection between the two compartments as indicated in Fig. 56.6. This opening allows slow changes of pressure to equalise on both sides of the partition, but has no effect on the rapid changes of pressure resulting from sound waves. Displacements due to sound pressure changes pass through the partition itself and end at the round window (Fig. 56.6). The liquid in the cochlea has mass and hence, by its inertia, opposes the displacements, which the pressure changes tend to produce. The higher the frequency of the pressure change the greater will be the opposition to displacement exerted by the mass of the

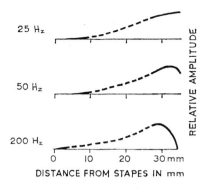
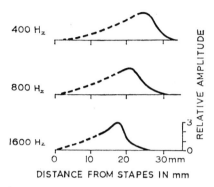

Fig. 56.7 The displacement of the basilar membrane in response to vibrations of different frequency.

Vibrations of constant amplitude but with different frequencies were impressed on the fluid within the cochlea, and the amplitude of the oscillatory displacement of the basilar membrane was measured at different distances from the stapes (basal end of the cochlea). At low frequencies, the amplitude is greatest at the apical end of the membrane; as the frequency is increased, the position of maximum amplitude shifts towards the basal end.

In order to observe the more basal parts of the membrane, the uppermost turns of the cochlear spiral had to be removed. This may be expected to affect the displacement of the parts observed at those frequencies at which the parts removed would themselves have undergone displacement; these observations are represented by the broken lines. Control observations showed, however, that in the conditions of measurement the error was quite small. (von Bekesy, J. acoust. Soc. Am. 1949.)

liquid (i.e. the greater its reactance). One might therefore expect that high frequency waves would spread along the cochlea for shorter distances than low frequency ones. In general this is what happens, but since the distribution of displacement along the partition is of fundamental importance and its mechanical basis is complex, some space must now be given to describing certain experiments.

Displacements of the basilar membrane

Von Bekesy has made direct observations on the cochlear partitions of fresh human cadavers after grinding away the bone over the apex of the cochlea. He looked at the displacements of the partition and his results are summarised in Fig. 56.7. The lines in each of these diagrams represents the maximum displacement occurring at different distances from the oval window (stapes). The actual movements of the partition are in the form of waves which travel from the oval window along the partition. The plotted lines can be thought of as joining the peaks of the wave on one side.

At 25 Hz, the wave travels along the whole length of the partition and increases in amplitude. As the frequency of incident sound rises, the peak displacement of the membrane moves backwards towards the oval window. Note that above frequencies of 200 Hz, the end part of the basilar membrane does not move at all. At 3 kHz only a very short segment of the partition is in motion.

THE AUDITORY RECEPTOR SYSTEM

The next stage in the transducer process which converts sound into neuronal information is the conversion of basilar membrane displacements into nerve impulses.

Organ of Corti

On the basilar membrane stand the two rods of Corti (see Figs. 56.8 and 56.9 for the histological details).

Internal to the inner rod is a single row of hair cells; external to the outer rod are three or four rows of hair cells. This adds up to about 3 500 inner and 20 000 outer hair cells along the length of the cochlea. These hair cells are supplied by afferent fibres from the cochlear division of the acoustic (eighth) nerve. The cell bodies of these fibres are in the acoustic spiral ganglion (equivalent to the dorsal root ganglion).

The hair cells, 8–12 μm in diameter, have 4 μm long hairs projecting from them and these are embedded in the tectorial membrane, a stiff gelatinous structure. The tunnel of Corti lies between the two rods and is filled with perilymph (not endolymph!).

The endolymphatic potential

Von Bekesy thrust a microelectrode into the cochlea. He found the scala media to be around 100 mV positive with respect to the scala vestibuli. The inside of Reissner's membrane was 30 mV negative to perilymph (scala vestibuli). The voltage in the scala tympani (as would be expected since the two are in continuity), was the same as in the scala vestibuli. Hence Reissner's membrane separates two fluids of differing compositions and potential.

Endolymph is a high-potassium fluid (like intracellular fluid; K^+ = 137 mmol l^{-1}; Na^+ = 15 mmol l^{-1}; Cl^- = 108 mmol l^{-1}). Perilymph resembles extracellular fluid (K^+ = 5 mmol l^{-1}; Na^+ = 154 mmol l^{-1}; Cl^- = 121 mmol l^{-1}). We cannot therefore explain the big potential across Reissner's membrane in terms of ionic concentration differences, for they are the wrong way round.

The endolymphatic potential is oxygen dependent, abolished by sodium cyanide, and is changed markedly by displacement of the basilar membrane. It is thought that the stria vascularis is the source of the potential and that the forces on the hair cells changing with basilar movement cause the observed fluctuations. The membrane potential across hair cells is thus augmented by the 100 mV and this may represent a mechanism for increasing receptor sensitivity.

Cochlear microphonic potentials (the Weaver–Bray phenomenon)

Weaver and Bray, in 1930, found that electrodes placed on the eighth nerve would record, with reasonable clarity, sound reaching the ear. A popular student demonstration involves utilising this phenomenon as part of a telephone system. These

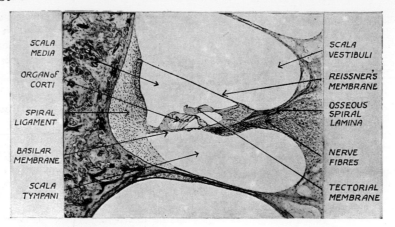

Fig. 56.8 Section through cochlea of a dog.
The organ of Corti is shown (vaguely; a drawing appears in Fig. 56.9). The scala media contains endolymph, a high-potassium fluid; the scala vestibuli and scala tympani contain perilymph, a high-sodium fluid. Across Reissner's membrane exists a potential of about 100 mV which, however, can not be explained on the basis of the ionic concentration differences, for the endolymph is *positive* with respect to the perilymph. Between the scala media and scala tympani appears the cochlear microphonic potential, a voltage which fluctuates with the incoming sound waveform as if the ear were a microphone. (Magnification ×75. Photograph kindly supplied by Professor C. S. Hallpike.)

potentials were at first thought to be nerve action potentials, but it was subsequently shown that they were microphonic potentials originating in the cochlea itself and spreading electrotonically to the nerve (and also to most other surrounding tissues).

The microphonic potentials are resistant to ischaemia, anaesthetics etc. and do not have any latent period of refractoriness. They thus behave as if they were purely mechanically generated, much like piezo-electricity. Bekesy has concluded that the source is in the hairs of the hair cells, which during sound activation move with respect to the tectorial membrane. Tasaki, using microelectrodes, has found that the basal turn of the cochlea responds to all frequencies of sound; only the low frequencies will excite the apical part. Local degeneration of the organ of Corti results if an animal is exposed to prolonged loud noises; the cochlear microphonic potential due to the frequency

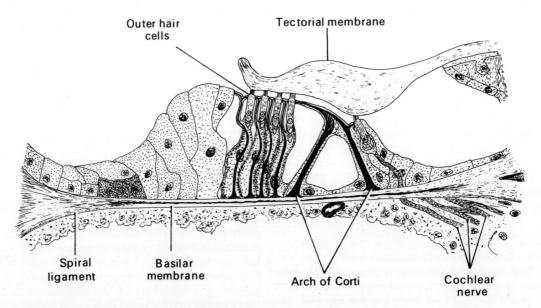

Fig. 56.9 Diagram of radial transection of the organ of Corti.
The organ of Corti is a ridge of specialised epithelial cells, extending throughout the length of the cochlea. It is composed of hair cells (receptors) and various supporting cells. Although the supporting cells are separated by large intercellular spaces, their upper surfaces are contiguous and, together with the hair-cells, form a continuous free surface (the reticular membrane).

Within the arch of Corti is the inner tunnel, extending the length of the cochlea, bounded below by the basilar membrane and above by the inner and outer pillar cells. The inner tunnel is filled with perilymph and the nerve fibres of the cochlear nerve traverse it (not shown).

The hairs of the inner and outer hair cells are embedded in the tectorial membrane, which is a secretion of protein having much the same composition and structure as epidermal keratin. The cochlear microphonic potential is developed between the tectorial membrane and the basilar membrane, and is thought to be due to relative movement of the hairs and tectorial membrane.

band involved is also abolished. In experiments such as these, the histological appearance enables a map of tonal localisation to be made. Various animals have been bred which have congenital absence (or underdevelopment) of the organ of Corti, e.g. albino cats and waltzing mice. These have no cochlear microphonic potentials.

Increasing negativity of the scala media during the microphonic potential is thought to cause an outward current in the hair cell membrane and this in turn generates impulses in the cochlear nerve.

Significance of cochlear microphonic potential

When carefully examined, the cochlear microphonic potential is seen to be complex. It consists of several waves, as shown in Fig. 56.10. The

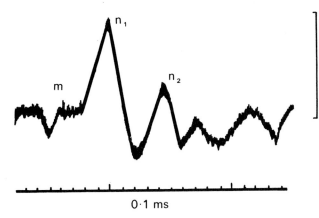

Fig. 56.10 Response of auditory nerve.

A brief click given to an anaesthetised cat gives rise to the cochlear microphonic potential m, followed by nerve fibre responses n_1 and n_2. The component m is unaffected by anaesthetics and death, and is a "non-biological" transducer action. The neural components, n_1 and n_2 are sensitive to all the variables which affect nerve impulses.

first wave, m is said to be due to the potential developed by the displacement of hair cells relative to the basilar membrane. The n_1 and n_2 waves are the result of activity in the first-order neurones. The implications of finding these potentials are obscure and subject to controversy at the moment. It used to be thought that they were generator potentials, but it will be clear to the reader by now that generator potentials depend on intact and living cell membranes in sensory terminals for their production.

Action potentials in the eighth nerve

How can sounds with frequencies up to 15 or 20 kHz be appreciated by the brain, yet nerve fibres are only able to conduct impulses at frequencies up to about 1 kHz? The answer is fairly simple. Single fibres will not follow frequencies of over 1 kHz, but the auditory nerve is composed of many thousand fibres. A high frequency tone will excite many fibres, each of which will probably respond to peaks of the sound wave at submultiples of its frequency (say every third or fourth peak). The total number of active fibres, however, will include some active at each peak—not necessarily the same fibres each time. Thus a minimum of 16 fibres would be required to convey the sensation of pitch corresponding to a 16 kHz source. There is thus a temporal pattern of nerve activity, as well as a spatial pattern, which is related to the frequency of the sound wave.

If the displacement of the cochlear partition and the associated microphonic potentials are direct precursors of the excitation of the receptors, one would expect the pattern of varying amplitude with distance from the oval window to be reflected in the behaviour of the primary receptor units. It is very difficult to record from such units, but it has been done by Tasaki. When he recorded from a primary unit, selected at random, he obtained a result such as that shown in Fig. 56.11. With

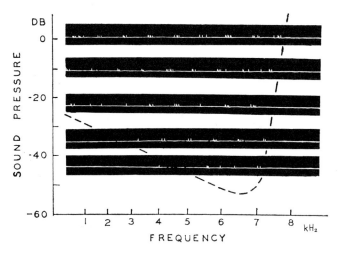

Fig. 56.11 Action potentials from a single fibre in the auditory nerve of a guinea pig.

Each line, from left to right, gives the responses to tones of short duration (tone pips) in an ascending series of frequencies, as given in the scale at the bottom, at a constant value of the intensity. The intensity is greatest in the top line, and becomes progressively smaller in the lower lines. The broken line shows the relation between the threshold intensity and the frequency, for this particular receptor unit. The sensitivity is greatest between 6000 and 7000 Hz, falls off gradually as the frequency is reduced and rapidly as the frequency is increased. (Tasaki, I. (1954). *J. Neurophysiol.* **17**, 97.)

a stimulus of very low intensity, he obtained a response only at one frequency, presumably that which caused the largest displacement at that particular point at which the receptors were located. With larger stimuli, he obtained responses from greater and greater ranges of frequency, the extension of the range being mainly towards the lower frequencies. This is what would be expected from the mechanical evidence, since every lower frequency would displace the basilar membrane from the oval window up to some point beyond the receptor from which the recording was made; higher frequencies, which displace only a short length of the membrane would, on the other hand, leave the receptor in a quiescent part of the membrane. (Compare in Fig. 56.7, for example, the displacement at different frequencies of a point on the membrane 25 mm from the stapes.) The shape of the curve in Fig. 56.11 is, in fact, entirely in accordance with the direct observations on the displacement of different parts of the membrane at different frequencies. Thus we see that the greater the intensity of a sound, the greater the length of the membrane in which receptors will be excited; but that the distribution of these excited receptors along the length of the membrane will be different according to the frequency of the sound.

The auditory system can detect differences in frequency of an incoming sound of less than 0.5 per cent over much of the middle range (Fig. 56.4), and simultaneously it can detect differences in intensity of about 10 per cent. These observations imply that there must be a considerable number of distinct impulse patterns in the auditory nerve. Clearly no single receptor unit can transmit all this information; in fact a single unit can only supply a small amount. It has just been pointed out that the timing of impulses, particularly at certain frequencies, is tied to the phases of the microphonic potential; when this is at its most striking, with one impulse to each cycle at frequencies from say 300 to 1500 Hz there can be little or no change in the discharge as the intensity of the sound is changed. Furthermore, a unit will tend to respond in a similar way to low intensity at its optimum frequency and to a higher intensity at some other frequency. The same problem is found in the skin, where a single receptor unit will respond in the same way to a small force in the centre of its receptive field as to a large force further away. The required amount of information to make all the necessary distinctions is provided in both situations by the use of large numbers of units each placed in a slightly different position.

In general, it can be deduced from the experiments described above, that as the intensity increases the number of units active increases. With a change of frequency certain characteristics of the whole pattern, probably in particular the maximum density of active units, moves in relation to the whole population of units. It has been possible to demonstrate this in rather more detail in a situation with skin. It is the task of the central nervous system to analyse these factors and those of timing referred to before. Impulse patterns from receptor systems are the input to the central nervous system. It is a major function of a nervous system to analyse step by step these inputs and to extract from them information required either for storage or to determine immediate action.

Theories of hearing

No theory at present is entirely satisfactory and fits all the facts—a common thing in physiology. The problem here will probably not be solved until we can record from single cochlear units—a most difficult task technically.

The idea that the cochlea played an essential part in frequency discrimination by causing different receptors to fire at different frequencies of vibration is much older than the evidence given here. The first of these *place theories*, as they are called, was put forward by Helmholtz; this was his *resonance theory*. He postulated that the basilar membrane behaved as a set of strings, each of which was set to resonate at a different frequency. The theory as he proposed it is not consistent with modern findings, but the important point in his idea is that the displacements of various parts of the basilar membrane are different at different frequencies. There is an entirely different type of evidence which has long suggested a place theory. High intensity sounds can, in experimental animals and through industrial injury in man (as in "boilermakers' disease", for example) cause damage to a localised part of the basilar membrane. That such damage can be seen post mortem is merely confirmation of what is known of the mechanical behaviour of the cochlea. It does not tell us whether or not the brain finds the analysis so made essential. There is, however, some evidence that such injuries are accompanied by an inability to appreciate sounds that have a frequency which would cause a maximum displacement in the injured region of the basilar membrane.

Efferent control of the auditory input

In more and more systems we are finding that peripheral control of sensory input is exerted by an output from the brain. Efferent activity in the auditory nerve has an effect on the afferent output from the cochlea.

Galambos in 1956 showed that the impulse activity in the afferent fibres of the eighth nerve, in response to a given sound, is markedly lessened by electrical stimulation of the efferent olivocochlear bundle (which travels in the acoustic nerve). This stimulation has no effect on the magnitude of the cochlear microphonic potentials recorded at the same time as the afferent action potentials. These might be completely abolished if stimulation is strong enough. The sensitivity of the auditory system can presumably be altered by afferents in much the same way as muscle spindle discharge is controlled by the fusimotor nerves. The mechanisms by which this is brought about are entirely unknown.

57. The Labyrinth

The term labyrinth, or vestibular apparatus, describes the parts of the inner ear and associated structures which have the function of maintaining posture and equilibrium.

Functional anatomy

Each labyrinth lies in the petrous bone in close anatomical relation with the inner ear, with which, however, it has no physiological connections (Fig. 57.1). It consists of three specialised structures each containing mechanically sensitive receptors. (1) The *utricle* encloses small calcareous bodies (the otoliths) lying against the projecting filaments of the receptors (hair cells). The receptors are activated by the weight of the otoliths and so provide information as to the orientation of the head with respect to gravity; they are also activated by linear accelerations. (2) The *saccule*, which also contains otoliths, is perhaps also concerned with signalling the position of the head in space. In the frog, the saccule is concerned only with the detection of low frequency vibration. In the cat, there are apparently two separate systems which signal tilts from side to side, and fore and aft, respectively, but they have not yet been identified anatomically. (3) The *semicircular canals* lie in three planes at right angles to each other and are filled with endolymph. Each canal contains a *cupula*, which consists of hair cells whose bases are embedded in a gelatinous matrix in which they are in contact with a number of nerve endings. When the head is rotated, the endolymph in the semicircular canals, owing to its inertia, lags behind; the cupola is dragged through the endolymph and thus displaced and distorted (see Figs. 57.2; 57.3).

Figures 57.2 and 57.3 show the details of the receptors found in the ampullae (*cristae*) and in the utricle and saccule (*maculae*).

PHYSIOLOGY OF THE SEMICIRCULAR CANALS

Electrical studies have shown that the nerve endings in the cupola discharge impulses when the head is at rest. When the cupola is displaced in one direction the frequency of discharge is increased, and when it is displaced in the opposite direction the frequency is decreased (Fig. 57.4). As a result, rotations in both directions may be signalled. In steady rotations, the discharge of impulses behaves as though the cupola were displaced during the period of acceleration, and then drifted slowly back to its resting position during the period of steady rotation. Suddenly stopping the rotation of the head will displace the cupola in the opposite direction, and although movements of the endolymph die down in about three seconds, the cupula seems to take about 25 to 30 seconds to return to its resting position. During this time the subject will experience a sensation of rotation in the opposite direction.

Reflexes based on information provided by the receptors in the labyrinths are important, not only in the maintenance of posture, but also in controlling the position of the eyes, as will be discussed in the next chapter.

Planes of orientation

The superior and posterior canals are orientated at angles of 45 degrees to the transverse and medial planes. Thus either

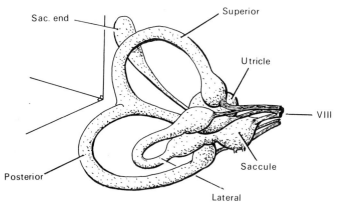

Fig. 57.1 Structural relations of human labyrinth.

The horizontal (external, lateral), superior (anterior) and posterior (inferior) semicircular canals are within bone. They are continuous with the utricle and filled with endolymph. At the junction with the utricle each canal has an enlargement, the ampulla, which contains the receptor organ called the crista ampullaris, shown in Fig. 57.2. Afferent branches of the 8th nerve leave the receptors.

The utricle and saccule are anterior to the semicircular canals and communicate via the ductus endolymphaticus. Receptor organs of the utricle and saccule are called maculae, and are shown in Fig. 57.3.

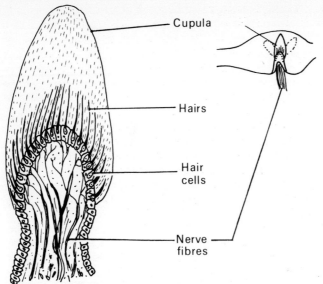

Fig. 57.2 Schematic diagram of crista ampullaris.

The ampulla contains the receptor organ, crista ampullaris, a ridge of columnar cells having hair filaments which are embedded in a gelatinous material called the cupula. The cupula extends from the crista to the opposing wall of the ampulla; it is a moveable mass which is mechanically distorted when there is any displacement of endolymph within the canals.

Inset shows mode of stimulation of semicircular canals. The cupula, on its transverse ridge the crista, completely blocks the ampulla of the membranous canal. Movements of endolymph cause the gelatinous cupula to swing, as shown dotted. Hair cells are then activated and impulses are generated in the 8th nerve.

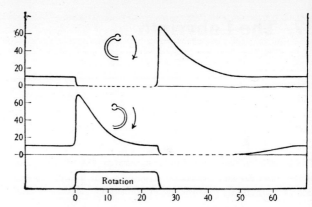

Fig. 57.4 Frequency of action potentials from the horizontal semicircular canal of a cat when acceleration and deceleration are separated by an interval of steady rotation.

There is a steady discharge during rest which is suppressed by rotations in one direction, and augmented by rotation in the opposite direction. Cessation of the rotation leads to an after discharge, or a silent period according to the direction of the rotation. (Adrian.) x-axis = time in seconds; y-axis = impulses per second.

transverse or medial rotation of the head gives rise to compound excitation of the canals. A transverse, downward rotation to the right will displace endolymph upwards and to the left; this compresses the left superior and posterior ampullae and releases the right superior and posterior ampullae. These effects give a sensation of downward, right-hand rotation. Stopping the movement (by inertia of the endolymph) gives rise to opposite sensations. Similar effects are produced by accelerations and decelerations in other planes.

Effects of stimulation of the labyrinths

Stimulation of the semicircular canals gives rise to feelings of vertigo, nausea and various autonomic responses. These are in addition to the changes in extraocular muscle tone (nystagmus) and effects upon postural (antigravity) musculature. Stimulation can be achieved experimentally, using Barany's rotating chair (a company director's or even a typist's chair will suffice).

Caloric stimulation

While bodily rotation is the usual method employed to stimulate the semicircular canals, caloric stimulation has the advantage that the vestibular system on each side can be tested separately. Clinical diagnosis in vestibular disorders makes use of the method. Warm (46°C) or cold (15°C) water is poured into the external auditory meatus. If the head is held at 60 degrees backwards, the horizontal canals are vertical, and since the water causes a greater temperature change in the canal near the meatus convection currents in the endolymph are set up. With cold water, the induced flow in the endolymph is away from the ampulla; with warm, towards it. The crista is stimulated, vertigo occurs and horizontal nystagmus. If the head be upright, as explained above, both sets of canals on the one side are stimulated. This results in rotatory nystagmus which has the rapid component away from the "cold" ear and towards the "warm" ear.

Electrical stimulation

An oscillator, capable of delivering between 100 μA to 2 mA into 10 kΩ can be connected to an electrode (saline pad) on one mastoid process and an indifferent. With frequencies around 0·5 to 2 Hz, it is possible to stimulate the vestibular apparatus. As the current is increased, subjective effects make their appearance. At first the environment appears to sway gently from side to side, the amplitude of this movement increasing as the current is increased. The swaying is at the same rate as the oscillator frequency. Nystagmus at the same frequency can be detected. As the current is further increased, the subjective

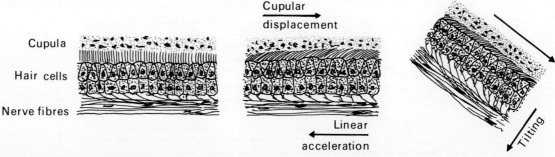

Fig. 57.3 The maculae (otolith-organs).

The receptor organs of the utricle and saccule are similar to those found in the three semicircular canals, and are called maculae. Embedded in the cupula of this receptor, however, are small particles of calcium carbonate, termed otoliths. The macula of the utricle is horizontal when the head is upright; the saccule is then vertical. The nerve terminals connect to the central nervous system via the eighth nerve.

Shown above are the effects of linear acceleration and tilting, in producing forces in shear which bend the hairs to excite the receptors.

effects are accompanied by muscle contractions which may eventually become violent enough to throw the subject on the floor.

Sensations of vertigo, nausea and vomiting ensue if the procedure is maintained for any length of time. Presumably the posterior semicircular canals are being stimulated in this experiment.

Vertigo

This is a sensation of "spinning", which if severe will cause loss of balance and falling to the floor. The direction of the spinning sensation is dependent on which semicircular canal is being activated, the vertigo being in the direction opposite to the displacement of endolymph (and in the post-rotatory state, opposite to the original rotation).

Autonomic disturbances accompany the vertigo and include nausea, vomiting, pallor, sweating, cardiac and vasomotor changes giving hypotension.

Nystagmus

This is a repetitive, jerky eye-rotation, usually having a rapid component in one direction and a slow return in the opposite one. The rapid movement is due to an efferent component of the vestibular reflex; its direction is used in the description of the condition. Horizontal rotation gives horizontal nystagmus, medial vertical and transverse rotatory nystagmus.

Falling reaction

The antigravity (postural) muscles exhibit increased tone on the side towards which endolymph flows. Thus the subject falls in the direction opposite to the direction of rotation. This is a compensatory reflex.

PHYSIOLOGY OF THE UTRICLE AND SACCULE

The maculae are stretch receptors and the stimulus to which they respond is the pull of gravity. The otoliths pull on the hairs and hair cells, these are deformed and set up impulse-trains in the vestibular nerve.

As one would expect, laterally tilting the head affects the sac-

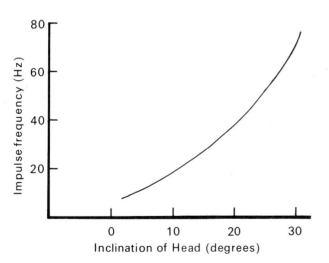

Fig. 57.5 Response of otolith organs.

Relation between head tilt and impulses in the vestibular nerve. Head of decerebrate cat is tilted laterally by amounts shown on *x*-axis. Impulse frequency is recorded from right vestibular nerve while head is being tilted to the right. As the tilt increases, impulse frequency rises. When the head is tilted to the left, the discharge in the right nerve ceases.

(Experiment first performed by Adrian E. D. (1943), *J. Physiol.* **101**, 393.)

cules. If the head is rested on the left shoulder (say), the left cupula hangs downwards, pulls on its macula which is then stimulated (maximally, if the head is held 90 degrees from its upright position; see Fig. 57.5).

It can be seen in Fig. 57.5 that these endings respond purely to the degree of tilt; they are not dynamic endings. In the labyrinth, then, we have the dynamic and static receptors as separate mechanisms.

Ventral and dorsal flexing of the head on the neck excites the utricles. Maximal excitation of the receptors occurs with the head held either fully forward or fully back. The working of the system can be investigated by recording from branches of the vestibular nerve. It is found that tilts of as little as two degrees will stimulate the receptors. The receptors show little or no adaptation.

Extirpation of the labyrinths

Bilateral removal of the labyrinths in the experimental cat or monkey produces severe disability at first, but after several days the visual system takes over and the animal behaves normally. Only if blindfold will the animal fail to right itself on falling, etc.

Unilateral removal leads to various disturbances, mostly the result of unopposed action of the normal labyrinth. There is deviation of the eyes (one inwards and outwards, the other downwards and inwards), nystagmus, lateral flexion of the head and various inequalities in postural tone.

MOTION SICKNESS

Motion sickness is an age-old malady; the Greeks were familiar with it, Julius Caesar, Lord Nelson, Charles Darwin and now Russian astronauts have suffered from its devastating effects. The word "nausea" comes from the Greek *naus*, a ship.

Irwin, in 1881, was the first one to use the term motion sickness, an appropriate description as it can occur in cars, trains, aircraft, on the sea, in space and even upon camels—as Lawrence of Arabia was only too well aware. The 1939–45 war, during which many men were transported by sea and air, was the stimulus for research upon the condition.

INCIDENCE

Most individuals can be made sick, if the movements applied to them are severe enough. In large ships, 25 to 30 per cent of passengers become sick within two to three days of an ocean passage. Possibly 5 per cent of persons may at times suffer from car-sickness; for modern jet air-transport no figures are available but the number would be small.

CLINICAL PICTURE

The signs consist of pallor, nausea, cold sweating and vomiting. The pallor and sweating are autonomic phenomena; it is said that the nausea and vomiting are not. Secondary effects are observed on pulse rate, blood pressure, gastrointestinal motility, salivation and often there is a fall in body temperature.

ETIOLOGY

No theory so far advanced in explanation of the condition satisfactorily explains it. It is clear that the vestibular mechanism is involved, for in patients without it motion sickness never occurs. It is not known whether the semicircular canals or the otoliths or both are necessary to produce motion sickness. The eyes also play a part in generating nausea and vomiting, as can be shown by people who have been sick in the wide-screen cinema showing films taken from high-speed cars or aerobatic aircraft. Experimentally, subjects can be made motion-sick by putting them in a "tilting room", even though they themselves are immobile.

Body position greatly influences the degree of motion sickness suffered in given circumstances. The incidence is much lower in the supine (but not the prone) position and also if the head is fixed, as by a high-backed car seat. Movement of the head during variable motion (e.g. on ship, even in a lift) will produce severe vertigo.

Theories to explain motion sickness

Most theories fall into two categories; sickness is due to:

1. Discord or conflict in sensory inputs
2. Overstimulation of the vestibular input

The conflict theory holds that the visual (or other) inputs during motion, conflict with the vestibular inputs. Evidence for this is supposed to be that observing the horizon at sea tends to reduce the incidence of motion sickness. Conversely, the presence of no external visual clues about the character of the motion (as in passengers confined below decks), makes things worse. However, it is difficult to accept this view because some exceedingly confusing stimuli, e.g. reversing spectacles, static tilted rooms, etc. do not cause sickness. Equally, some stimuli that would appear to cause relatively little sensory conflict, e.g. ships, swings, vertical accelerators, do in fact produce a high incidence of symptoms. Conflicting sensory input in persons without both labyrinths causes no sickness.

The overstimulation theory. Motion sickness is held to be due to "irradiation" from the overexcited equilibratory centres to the vomiting centre. Certainly the incidence of motion sickness is directly related to the intensity of the stimulus (in an experimental situation). However, there seem to be optimum rates and frequencies of motion required to give sickness; above these, comparatively vigorous stimulation fails to produce the syndrome. Mild stimuli, such as the rotating room (at 3 rev/min.) are very effective, while the incidence of sickness is lowest at high frequencies and accelerations.

Conclusions

No theory is satisfactory at the moment. Both the theories mentioned suppose motion to be a noxious stimulus but even if this is accepted, motion sickness remains puzzling from a teleological or evolutionary standpoint. Many animals also suffer from motion sickness (including fish in a tank transported by sea!). Does the phenomenon have any survival value? Possibly this could have been true in a common sea-dwelling ancestor, but at the moment we still have to regard the condition as an anomaly.

Treatment

The efficacy of the supine posture and the beneficial effect of the avoidance of head movements have been mentioned. Various drugs are effective, but the reasons for this are unknown. Antihistamines, belladonna, phenothiazines and amphetamines are all effective to varying degrees.

Belladonna was tried because it was thought it would decrease stomach motility. It is now known that it does not operate in this way. Dramamine was accidentally discovered through being given for "hives"* to a woman who had suffered all her life from car sickness.

---

* "Hives"—a skin eruption, or "inflammation of the bowel".

58. Vision

Before we consider in detail the physiology of the visual system, it will be necessary to review some of the physical properties of light and of optical systems.

PROPERTIES OF LIGHT

Light is the visible part of the spectrum of electromagnetic waves; its wavelength is 400 to 800 nm, and blue light has the shortest and red light the longest wavelengths. Electromagnetic radiation has the properties of both matter (i.e. a stream of corpuscles) and waves, being in the form of *photons*, or "packets", of energy travelling with a speed of 186 000 miles per second.

The wave properties of light enable it to be refracted, reflected and focused.

Refraction: image formation

Light rays striking an object are either absorbed by it, reflected by it or transmitted through it, and often all three processes take place. Transmitted light rays are bent or *refracted* according to the angle at which they are incident to the surface and to the refractive index of the material composing the object.*

Lenses owe their action of focusing light and producing images to this property of refraction (see Fig. 58.1). Light rays which reach the centre of the convex lens are perpendicular to its surface and are bent very little; rays at the periphery strike the air–glass interface at an angle and are bent towards the optical axis as shown in Fig. 58.1. For a spherical lens, rays meet at a point on the other side of the lens to form an image.

ABERRATION

Rays traversing the periphery of a lens do not come to a focus at the same point at those going through its centre. This defect is called *spherical aberration*. Also, because refractive index depends upon the wavelength of the light, different coloured light is brought to a focus at slightly different distances from the lens. This is called *chromatic aberration*. The optical system of the eye suffers from both these defects to a small extent. The glass optical systems used in cameras, microscopes and telescopes can be largely corrected for both types of aberration by the use of multielement lenses made of glasses with slightly differing refractive indices and by using lens shapes which marginally depart from the spherical.

Power of a lens

It is important to be able to define and measure the strength of a lens, i.e. its *power*, and also the position of the image it forms in various circumstances.

Parallel light rays are brought to a focus at the *focal point* of the lens and the distance of this point from the centre of the lens

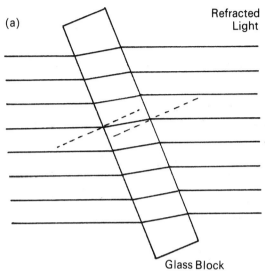

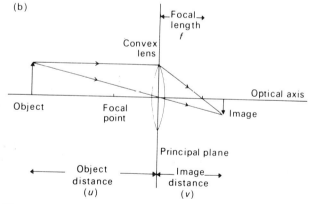

Fig. 58.1 (a) Bending of light rays as they pass through a block of glass (having a higher refractive index than air).
(b) Formation of image by spherical convex lens. Construction of image when lens properties and object distance are known.

is called its *focal length*. The strength, or power, of any lens is really the degree to which it can bend light rays. The more powerful a lens, the greater the angle through which light rays are bent by it. Thus the focal length is shorter the more powerful is the lens. In ophthalmic optics, the power of any lens is expressed in *dioptres*, the reciprocal of the focal length in metres.

$$D = \frac{1}{f}$$

where D is power in dioptres and f is focal length in metres.

---

* Strictly speaking, this is the ratio of the refractive indices of the object itself and the surrounding medium. Refractive index is the ratio of the sines of the angles of incidence and refraction at the interface between two media in which the velocity of light differs.

Images due to lenses

It is quite easy to determine the position of an image formed by a lens; all that is needed is the distance between the object and the lens together with the focal length of the latter. The construction can be geometric, as shown in Fig. 58.1, or one may use the lens formula

$$\frac{1}{v} + \frac{1}{u} = \frac{1}{f}$$

where v is the image distance and u is the object distance.

For example, the image distance for an object placed 75 cm from a lens of power 20 D would be found as follows:

$$\frac{1}{0 \cdot 75} + \frac{1}{u} = 20$$

Multiply both sides by 0·75

$$\frac{0 \cdot 75}{0 \cdot 75} + \frac{0 \cdot 75}{u} = 20 \times 0 \cdot 75$$

$$u = 0 \cdot 054 \text{ m}$$

For an object at infinity, the image is produced at the focal point, as can be seen in the following example with a 10 D lens:

$$\frac{1}{v} + \frac{1}{u} = \frac{1}{f},$$

$$\therefore \frac{1}{v} + \frac{1}{\infty} = 10$$

$$\therefore v = 0 \cdot 1 \text{ m}$$

In the eye, for practical purposes, any object more than 6 m away can be considered to be at infinity and the light rays coming from it are parallel.

SIZE OF IMAGE

The magnification, m, of any lens system is given by the ratio of the object distance and image distance

$$m = \frac{v}{u}$$

THE OPTICAL SYSTEM IN THE EYE

A receptor system which can measure the intensity, position and wavelength of light is one which can give an animal a very large amount of detailed information about its environment. Our own experience of human vision, the conscious appreciation of information received through the eyes, can tell us immediately of the great amount of information we can derive in this way.

As in any receptor system, the role of an eye is to set up patterns of nerve impulses in such a way that for every distinguishable pattern of light reaching the eye there is a pattern of nerve impulses different from that set up by any other light stimulus. The first stage is for the external energy—in this instance light—to be transmitted to and distributed among the receptors; in eyes, such as the human eye, this is done by the optical system which focuses the light on the array of receptors, the retina. Figure 58.2 is a general diagram. The receptors almost certainly produce electrical changes related to the light intensity falling on them; in retinal receptors a stage in converting the light energy into electrical energy is the absorption of the light

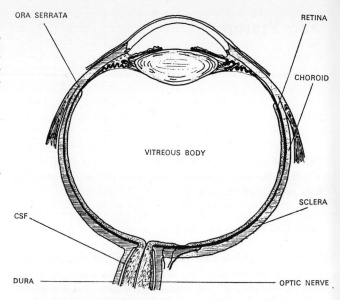

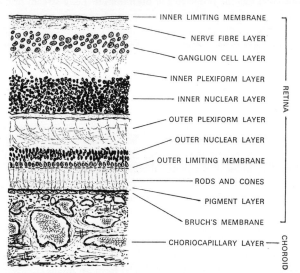

Fig. 58.2 (a) Horizontal section of right eye.
(b) Cross section through retina and choroid near posterior pole of globe. (From Lippold (1973), *Origin of the Alpha Rhythm*. Churchill-Livingstone, Edinburgh.)

energy by a *photochemical* substance. The electrical changes in the receptors then set up the nerve impulses.

The first step towards setting up patterns of nerve impulses which are related in a meaningful way to any particular light stimulus, is to distribute the light in a suitable manner among the receptors. In vertebrate eyes, as well as in many molluscs (e.g. octopus), this is done by forming a focused image on the retina. The focusing may depend largely on the corneal surface, as in man, or on a lens, as in fishes; there are even a few examples of the image being formed by a spherical mirror (e.g. scallop). These eyes with focused images (excluding the few focused by mirrors), are sometimes called "camera eyes" to distinguish them from the "compound eyes" of, for example, the insects. The analogy to a camera is clear, but there is a difference. In a good camera the image must have a uniform high quality all over the film. In the human eye, as has been shown in the section on vision, only the fovea has the capacity to transmit further all the information available in a well-focused image; the fovea is moved over any area it is desired to look at in detail, as for example in scanning down a line of print. If there is a focused image there must be a means of adjusting the focus for objects at different distances from the eye. In mammals, as will be described in the second subsection, this is done by altering the shape of the lens. But this is not the only method found in animals; in some the eye is moved backwards and forwards as in a camera.

In man the image is formed as a result of a refraction at the corneal surface and in the lens.

It is important to remember that the strongly curved outer surface of the *cornea is the chief refracting surface of the eye*. The lens is very powerful if examined in air, but since it is suspended in fluids whose refractive indices are only a little less than its own, it loses most of its power when in the eye. The lens is concerned with the fine focusing of the image on the retina. Application of the laws of optics to the refracting surfaces and the refractive indices of the eye media shows that the retinal image must be inverted as in a camera. Such an inverted image can be seen in the excised eye of an albino rabbit by looking at it from behind, through the non-pigmented sclera.

If the retinal image is reversed (i.e. made upright) experimentally, by means of reversing spectacles, the visual sensation is reversed. After some days, however, the sensation may "de-reverse" and appear upright again, despite the inversion of the retinal image from normal. This reversal is helped by clues from other senses.

Before reaching the retina the rays of light have to pass through the cornea, the intraocular fluid, the lens and the vitreous body. The absorption of visible rays by these eye media is small except in old age, when there is an appreciable absorption of the shorter spectral waves by the lens. Infrared and ultraviolet radiations, both of which are capable of damaging the retina, are absorbed by the eye media before reaching it. Both these types of radiation are harmful to other structures in the eye. The infrared rays, for instance, can cause cataract of the crystalline lens (e.g. among glass-blowers), while the ultraviolet rays cause intense inflammation of the outside surface of the eye (film-star's eye; snow blindness). The retina is protected against intense visible radiation by changes in the pupil diameter, the iris (Fig. 58.2) contracting reflexly under these conditions and dilating again more slowly in the dark. Even when the pupil is fully contracted, the absorption of visible light by the retina at the site of a very bright image may lead to a rise of temperature sufficient to cause a burn resulting in local blindness. This condition is not uncommon in those who unwisely observe an eclipse of the sun without adequate protection by dark glasses.

The image on the retina is formed by what, in geometrical optics, is called a "thick lens". The size and position of the image can thus be calculated from knowledge of the cardinal points of the optical system. The most useful of these is the *nodal point*, an imaginary point at the optical centre: it lies at the geometrical centre of a thin lens, but in the eye it is in the crystalline lens near the posterior surface, 15 mm in front of the retina (Fig. 58.3). A pencil of light directed towards the nodal point behaves as though it had passed through undeflected, although this does not mean that the pencil actually follows this path. Its use is in calculating the size of the retinal image, when the size and distance of the external object are known. The calculation is made by the use of the principle of similar triangles. It can be applied, for instance, to the calculation of the distance between the blindspot (optic disc) and the region of most distinct vision (fovea). A circle and, 60 mm to its left, a cross are made on a piece of paper (Fig. 58.4) and, looking at the cross with the right eye only, the

Fig. 58.4 Figure for illustrating the presence of the blind spot. Look at the cross with the right eye only and hold the book at about 22 cm from the eye.

paper is moved away from the eye. At about 220 mm distance the circle will be invisible because its image will fall on the blindspot, at which point the sensitive elements in the retina are absent. By "similar triangles" the distance between the fovea and the centre of the optic disc works out to be 4 mm, a result which agrees with histological observation.

The *anterior focal point* is 13 mm in front of the cornea, and a pencil of rays which has passed through it is refracted parallel to the optic axis when it meets the *principal plane*, which lies 2 mm behind the anterior surface of the cornea (Fig. 58.5). Using

Fig. 58.5 The changes in the Purkinje images during accommodation.
A three-point source of light forms mirror images at the anterior surface of the cornea and at the anterior and posterior lens surfaces. During accommodation (right) only the anterior lens image changes. Notice the contraction of the pupil. (Fincham.)

these quantities in conjunction with the nodal point, the position of the image within the eye can be determined.

The eye behaves like other optical instruments in that it is subject to some of their defects. Among these can be mentioned

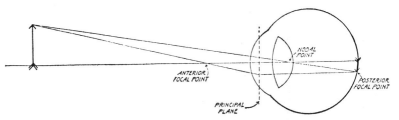

Fig. 58.3 The formation of the retinal image in the schematic eye.

spherical aberration. The power of a spherical lens is greater at the periphery than at the centre, that is to say, the image formed by its periphery is nearer to the lens than the image formed by its centre. In the eye the outline of the refracting surfaces is slightly hyperbolic and in this way spherical aberration is diminished. *Chromatic aberration* of a lens system is due to the fact that the short (blue) waves of the spectrum are bent at a refracting surface through a greater angle than are the long (red) waves, and so come to an earlier focus. Actually the eye, on looking at a point source of light, is focused on the middle wavelengths, giving the light a halo of red and blue. These halos can be seen by looking at a small light through a piece of cobalt glass which transmits only blue and red rays.

The focusing of the image in mammals

A normal mammalian eye, directed on a distant object, will focus the image on the retina; its *far point* of distinct vision is said to be at infinity, and this is the position of rest for the eye. When the eye is directed on to a near object the image is focused by altering the shape of the lens; the nearest point at which objects can be focused sharply on the retina is known as the *near point*. The process of altering the focus from infinity towards a nearer object is sometimes known as *accommodation*; when the eye is adjusted to focus the near point, it is said to be at *maximum accommodation*.

The changes that occur in the lens of the human eye can be observed by means of the Purkinje images (Sanson's images). These are images of a source of light, held slightly to one side of the subject's eye, formed by reflection at the principal refracting surfaces. The anterior surface of the cornea acts like a convex mirror and forms a bright uninverted image. The anterior surface of the lens functions similarly, but the image is faint and often very difficult to see, and normally is larger than that formed by the cornea. The posterior surface of the lens acts as a concave mirror; the image is inverted and much smaller than the others. When the eye focuses alternately for far and near objects a marked change occurs in the image from the anterior lens surface; there is a very slight change in the image from the posterior lens surface, and no change at all in the image from the cornea, showing that this structure is unchanged during focusing (Fig. 58.5). In order to focus near objects the curvature of the anterior lens surface becomes smaller. When the eye is focused on infinity, the radius of curvature of the anterior surface of the lens is about 10 to 12 mm; when focused on the near point, the radius of curvature is reduced to about 6 mm. The radius of curvature of the posterior surface changes only slightly, from about 6 mm to about 5.5 mm.

ELASTICITY OF LENS
An excised lens has surfaces as highly curved as in the eye when accommodated for near vision. The reason for this is that the lens is an elastic body which, for distant vision, is stretched and flattened as a result of the method by which it is supported; and for near vision is released and allowed to thicken. The lens substance is held under pressure by a capsule, which is held around its edge by an elastic ligament. This pulls the capsule out in one plane, so tending to flatten the lens. If the tension provided by the ligaments is reduced, the lens becomes more curved. The capsule does not have a uniform thickness and hence the changes of curvature are greater in some parts than in others. The tension can be taken off the capsule by the contraction of a circular muscle running around the lens near the line of attachment of the ligament.

The increase in lens curvature during near vision is accompanied by a reduction in size of the pupil (Fig. 58.5). The reduction in the optical aperture improves the depth of focus. The eyes, in man and other mammals with binocular vision, converge slightly when viewing a near object. The change in lens curvature, the change in pupil diameter and the slight convergence and divergence of the eyes are maintained at any moment in correct adjustment by means of appropriate neural control systems. The pupil diameter is, of course, adjusted in relation to the light intensity as well as the distance of the object on which vision is focused. The muscles responsible for altering lens shape and pupil diameter are supplied by parasympathetic fibres (in the IIIrd nerve) and the neuromuscular transmitter is in each case acetylcholine. This means that blocking agents such as atropine and homatropine cause relaxation of the muscles and hence the eye is focused at infinity and the pupil is dilated. These substances are effective if dropped in the eye and hence are used in clinical practice. Eserine and pilocarpine have the opposite effects and can be used to counteract the effects of atropine.

Optical defects in the eye

The range over which the human eye can focus in young people (under 25) is equivalent to 10 dioptres (D). With increasing age the lens hardens and the range over which the eye can focus diminishes; by the age of 50, this range has been reduced, on the average, from 10 to 1.8 D. In a normal young person the near point is around 10 cm, while at 50 it is around 55 cm.

Myopia

In some people the image of an object at infinity is not focused on the retina with the eye relaxed. If this image is in front of the retina it means that the focal length of the whole lens system needs to be increased or the power in dioptres decreased, if infinity is to be focused. Such people can focus near objects and are *near-sighted* (sometimes called *myopic*).

Myopia is usually due to an abnormally long eyeball so that the distance between the lens and the retina is increased, or rarely it can be due to a greater than normal curvature of the cornea or the lens. The use of a concave spherical lens of appropriate power will correct the condition (see Fig. 58.6).

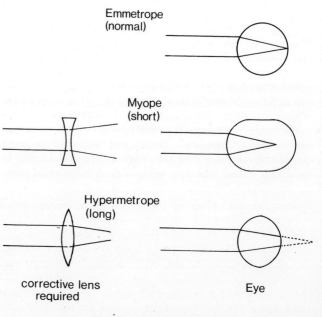

Fig. 58.6 Optical defects in the eye.

Hypermetropia

Hypermetropia (or *far-sightedness*) is due to an abnormally short eyeball, so that the image is formed behind the retina in the resting eye. Some degree of correction is possible by physiological mechanisms. A hypermetropic subject can partly accommodate his lens, thus bringing an object into focus. This leads to fatigue of eye muscles and also results in disparity between the muscle effort required to accommodate and to give convergence on the object.

The patient finds that close work, such as reading, etc., becomes a "strain" after any length of time because the near point is further from the eye than normal and the continual effort required to focus the print gives rise to unpleasant ocular sensations. The use of convex spectacles remedies the defect.

Astigmatism

Abnormal curvature of the cornea is known as astigmatism. The anterior surface is ellipsoid (i.e. it is a partially cylindrical lens) and so light in one plane is focused at a different distance from that in another plane. A cylindrical lens of appropriate dimensions is used to correct the defect.

THE VISUAL PROCESS IN THE HUMAN

Threshold

Vision is the conscious sensation aroused by light and must not be confused with any other nervous activity associated with light. The threshold of vision varies according to conditions, in particular on the position on the retina, on the wavelength of the light, on the state of dark adaptation of the eye, on the area of the test illumination, and on the duration of the test. Each of these will be considered in turn in the subsections which follow.

Absolute threshold

By the absolute threshold is meant the smallest amount of light which, under ideal conditions, can arouse a sensation. If a short flash of light is used to illuminate the peripheral parts of the eye it is possible to measure the least quantity of light energy, incident on the cornea, which can just be seen. The light has to pass through the parts of the eye shown in Fig. 58.2; it passes through the transparent cornea, through the aqueous solution (*aqueous humour*) behind the cornea, then through the lens and the transparent jelly-like material (*the vitreous body*) which fills the main part of the eyeball and finally reaches the retina, which lines the inner surface of the back of the eye. Measurement of the optical properties of these media ahows that if we use the green light to which the eye is most sensitive, about one-half of the light incident on the cornea will reach the retina; about one-fifth of that incident on the retina will be absorbed by the rods and its energy made available for their excitation. According to the quantum theory of radiation, light energy can be emitted or absorbed only in packets of a certain size, known as *quanta* (or *photons*), the size of each quantum increasing as the wavelength becomes smaller. Since we know the wavelength of the light used, we can convert the threshold quantity of light energy into the threshold number of light quanta. Now, in threshold conditions, by definition, the light will be seen in about 50 per cent of the trials. But, in these conditions, the number of quanta absorbed per unit area of the retina (about 250 per square centimetre in 0·1 s) is very much smaller than the number of rods per unit area (about 14 million per square centimetre): thus according to the laws of probability it will be exceedingly rare for any single rod to absorb more than one quantum—certainly in very much less than 50 per cent of the trials. Thus it must be possible for a rod to become activated if it absorbs a single quantum of light energy—the smallest amount which can exist.

There is no sensation of vision, however, unless several quanta—probably between 5 and 10—are absorbed within the retinal action time by a group of rods which are all connected together, i.e. are included in one receptor unit. This figure is based on rather indirect evidence and is still uncertain.

The retina

The central retina contains the fovea (Fig. 58.2), a small depression on to which the centre of interest of the visual field is normally focused. This part of the retina has the advantage of having less tissue in front of it than the remainder of the retina. Curiously, vertebrate eyes, unlike the other group of "camera eyes" found in molluscs, have evolved with several layers of nerve cells and nerve fibres lying between the source of light and the receptors themselves, which are at the back of the retina. In the fovea the nerve cells and fibres tend to be displaced to the side, so giving more direct access of light to the receptors. In the human fovea these receptors are all *cones*, a name for one of the two main types of receptor; the others are *rods*. As one moves towards the periphery of the retina, the proportion of rods to cones increases. It is the peripheral part of the retina which can give rise to a sensation at the lowest levels of illumination; a common experience is to see a faint light in the periphery of one's visual field and then to fail to be able to see it when one looks directly at it. There appears to be an association between low thresholds for sensation and the presence of rods; nocturnal animals have a preponderance of rods in their eyes. On the other hand, the human fovea which is purely cones has not got a low threshold for sensation. Cones appear to be associated with colour vision and rods incapable of providing the necessary information for such sensations; this is suggested by the reverse of the argument just given associating rods with a low threshold for sensation. Taken together, these and certain other facts imply considerable differences between vision at low intensities (*night vision*, or *scotopic vision*) and high intensity vision (*day vision*, or *photopic vision*). At low intensities we cannot perceive colour, detail is poor and the eye is most sensitive to light of a wavelength about 500 nm (would give sensation of green); vision is best in the periphery of the visual field. At high intensities we can perceive colour and detail and the eye is most sensitive to light at 560 nm (appears yellow); the fovea is the most important part of the retina for this type of vision.

Spectral sensitivity

The eye can only detect a narrow band of electromagnetic waves and within the band there are big differences in sensitivity; sensitivity is the reciprocal of the intensity required to give a specified response. These differences are shown in Fig. 58.7. There are two curves given in this figure, that on the left for vision in dim light (rod vision) and that on the right for daylight vision (cone vision). It will be seen that in dim light the maximum sensitivity is at 505 nm and that the sensitivity has dropped to about 25 per cent at 560 nm which is the wavelength at which sensitivity is maximum in daylight. The longer wavelengths, normally associated with sensations of orange and red, are virtually undetected in dim light. The shape of the curves in Fig. 58.7 are almost certainly directly dependent on the properties of

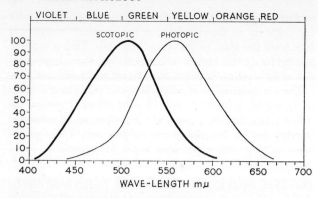

Fig. 58.7 Scotopic and photopic luminosity curves.
Abscissae: Wavelength in nm. Ordinates: Relative sensitivity of the eye, the sensitivity at the most effective wavelength (505 nm for scotopic conditions, and 560 nm for photopic conditions) being made, arbitrarily, equal to 100. (Crawford and Thomson.)

the photochemical substances found in the rods (scotopic) and cones (photopic).

Dark adaptation

It is common experience that on going from light to dark, dim lights cannot be seen, but that after a time they can be seen with ease. This process is known as dark adaptation. When a subject has been adapted to a bright light and this light is then turned off, his threshold can be measured by exposing him to test lights. When this is done curves, such as those in Fig. 58.8, are ob-

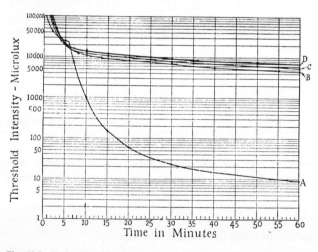

Fig. 58.8 Dark-adaptation curves of one normal (A) and three congenitally night-blind (B, C and D) subjects.
1 lux = 1 metre-candle = 0·1 foot-candle (very nearly). (After Dieter.)

tained. The different curves given in Fig. 58.8 are for different individuals; they were all obtained away from the fovea. The top part of the curve (0–5 min) is probably due to the adaptation of the cone vision. The second phase of the curve (5–60 min), is probably associated with rod vision. Only in curve A from the normal subject is the second phase present; in the night-blind retina, rods are absent. At long wavelengths there is no dim-light vision and hence almost certainly no excitation of rods; the adaptation curve to test lights of these wavelengths is of the pure cone type as found in the fovea for all wavelengths.

Figure 58.8 illustrates the time course of adaptation on changing from light to complete darkness. It will be noted that changes are slow: full adaptation takes more than 45 minutes. Clearly one would expect that for any constant level of general illumination there will be a steady level of adaptation. This may be observed by measuring what is known as the *difference threshold*, the smallest increase in intensity which can just be detected. This is considered further below in the section on visual discrimination.

It now appears certain that the large changes in sensitivity which occur during adaptation (Fig. 58.8) are largely due to changes in the nervous pathways. The bleaching of the photochemical substances was always thought to be the main reason for the decrease in sensitivity as light intensity increases and this may be a factor.

The area and duration of the stimulus

The intensity of light required to stimulate decreases as area increases over small areas (Fig. 58.9, left). Under certain conditions the intensity (energy per unit time per unit area) is inversely proportional to area; i.e. for a constant time energy is constant regardless of area. This is shown by the straight lines drawn through the points on the left of the figure. Under certain conditions this phenomenon is certainly a function of the nervous connections of the receptors.

A similar finding regarding the duration of a light is also shown in Fig. 58.9 (right). Again on the left of the figure the straight line shows where the intensity (energy per unit time per unit area) is inversely proportional to the duration of the stimulus; i.e. for a constant area, energy is constant regardless of time.

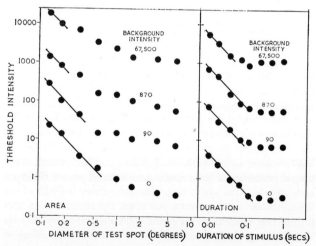

Fig. 58.9 Spatial and temporal summation in the human eye.
The threshold intensity, or the increment threshold, is plotted against the size of the test spot (on the left), the duration of exposure being constant at 0·93 s; and against the duration of the test stimulus (on the right), the size of the test spot being constant, with an area subtending 27·6 deg² at the eye (diameter subtending 5·9 deg).
The sloping lines indicate the relation to be expected if the threshold were inversely proportional to the area (Ricco's law); or inversely proportional to the duration (Bunsen–Roscoe law). The brightest background had an illumination of about 1 lux, and was thus only moderately bright.
The intensities, both threshold and background, are expressed in units of 1000 quanta (507 nm) in each (degree)² arriving at the eye in each second. (From Barlow, redrawn.)

Visual discrimination

Visual discrimination is concerned with the ability to distinguish one detail of a pattern of light and shade from another. The ability to make such a distinction will depend on their separation and the relative intensities reflected from the two parts.

VISUAL ACUITY

Visual acuity is defined as the angular resolving power of vision; that is to say the least angular separation between two contours that can be distinguished from each other visually. The best test of this is to use an array of parallel lines or rectangles alternately light and dark of equal width; there should be maximum contrast between them. The array can be viewed from different distances; or one can have different arrays in each of which the lines are of different width, all of the arrays being viewed from the same distance. With either method the angular separation of the light and dark lines subtended at the eye can be altered. If the eye looks directly at such a test object the image falls on the fovea. When tested in this way it is found that the smallest separation of lines which can be detected gives an image on the retina in which the lines are separated by a distance which is the same, or little more than, the diameter of a cone, about 3 μm. Assuming that the discrimination is truly simultaneous, the spacing of the cones would be expected to set one limit to visual acuity. (If the eye were to scan, in the way that is familiar from television, this would not necessarily be true.) Since it is vision which is being tested, the information must be transmitted to the brain and it is likely that there are as many nerve fibres in the optic nerve carrying information from the foveal cones as there are cones. With this kind of test the acuity is little dependent on level of illumination over a wide range of relatively high intensities. Some test objects commonly used for measuring acuity are at the same time tests of a threshold and if these are used, the level of illumination has a greater effect on the apparent acuity. "Landolt's C" and letter cards are of this type. The acuity achieved by the eye appears, therefore, to be very close to the limit set by the spacing of the array of receptors. It is interesting that the acuity achieved is also close to the limits imposed by the optics of the system.

In contrast to this high acuity achieved by the fovea, the acuity of the extrafoveal retina is much worse than either the optics or the fineness of the pattern of rods and cones would lead one to expect. The explanation must be that the information is lost between the receptors and the optic nerve. There are one hundred million receptors and only one million optic nerve fibres, so it is the latter which will limit the amount of information which can reach the brain.

Discrimination of one intensity from another is measured as a *difference* (or *incremental*) *threshold*; this is the smallest difference in intensity, from that of the background illumination, which can be detected. The relation between the difference threshold and the level of illumination is not simple except under special conditions. If the test field is large, the duration long and the background intensity greater than a thousand times the absolute threshold, then usually, but not always, the difference threshold, ΔI, is proportional to the background intensity, I. The ratio $\Delta I/I$ may under these conditions have a value of about 0·03. This figure is quoted, despite the qualifications, in order to give some idea of the order of intensity discrimination obtained in vision.

COLOUR VISION

The sensation aroused by a light needs three terms to define it; the brightness, the hue and the saturation. In this section we are concerned with the latter two. Hue is best illustrated by the sensation associated with a pure spectral colour, i.e. with a light of a single wavelength; this ranges from red through yellow and green to blue and violet. Hues of this kind are highly saturated; on the other hand, white, grey and black are three brightness levels of the completely unsaturated situation, which consequently has no hue at all. Between the saturated hues of the spectral colours and the unsaturated neutral tone there are a series of shades of varying hues and varying saturations; for example, there is a series of shades running from green through greyish green to greenish grey to grey. Human vision can distinguish 120 different hues and about 1000 different shades.

There is one very important finding about colour vision, which dominates ideas in this field. If two lights, each of a single wavelength, one in the red and the other in the green, are shone on to a white screen and their intensities varied, it is possible to produce on the screen colours having any hue lying between these two in the spectral sequence. If a third, blue, light is added to the mixture it is possible to match, by an adjustment of intensities, almost any shade of light projected on to another screen; there are a few shades which cannot be matched. If one moves one of the three standard lights and mixes it with the "unknown" shade and then matches this mixture with the mixture of the other two standards, it is possible to match the few shades which could not be mixed before; this procedure simply introduces the possibility of a minus intensity. It is possible to use any three standards, but a red, a green and a blue are commonly used as with these the number of negatives is minimal and the accuracy of the matching is high. It must be emphasised that these experiments are concerned with the mixing of lights of known wavelength; the mixing of pigments involves many different questions. The importance of such experiments is that the three characters of the sensation, its brightness, its hue and its saturation can be completely controlled by adjusting three knobs controlling the intensity of three

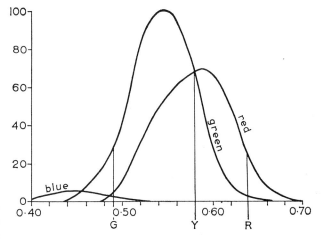

Fig. 58.10 Konig and Dieterici's suggested response curves for three channels in the mechanism for colour vision (corrected by Judd).

Abscissae: Wavelength in μm. Ordinates: Relative sensitivities of three hypothetical receptor systems at different wavelengths, the maximum sensitivity of the "green" system being made equal to 100. (From Walters, *Proc. Roy. Soc.*)

lights. If a further control is added nothing further is achieved, since the new variable can always be exactly defined in terms of the other three. If the number of variables is reduced to two, it is not possible to match all shades. Thus there are three variables or degrees of freedom in the visual system connected with colour vision and there are not more nor less. This means that there must be three independent channels in the mechanism taken as a whole from the receptors to the highest levels of the nervous system; it does not necessarily imply three types of receptor.

Attempts have been made to estimate the sensitivities of the three channels (Fig. 58.10). Such attempts can only indicate possibilities; they are based on matching experiments such as those described and others, including matching after adaptation to lights of particular wavelengths. Vertical lines drawn on Fig. 58.10 indicate the relative sensitivities of the three mechanisms to certain particular wavelengths. These curves are sensitivity curves (relative reciprocal of intensity to produce a standard response under different conditions, here wavelength). Response curves (relative size of response to a stimulus of constant intensity under different conditions) do not necessarily have the same shape. The curves presented do, however, indicate, in a general way, that each wavelength will arouse a unique combination of responses from the three channels.

The ability to match all shades with three standards does break down if very small fields (less than 30 min of arc) of illumination are used on the fovea. Under these conditions colour vision is less good and two standards is sufficient.

Colour blindness

Defects in colour vision do occur. There are those people who are totally colour blind, but they are rare. Many more are able to match all the shades they can see by mixing two, instead of three standards. They are dichromatic, instead of the normal trichromatic. In these people there appear to be only two channels available. Most cases are not as severe as this and have anomalous colour vision, possibly one or more of the channels having abnormal characteristics.

The detection of deficiencies in colour vision is not a task which can be undertaken light-heartedly, since there are all grades of severity, and when a defect is present in a mild form it may only be possible to detect it when the conditions are made difficult. A subject may, for instance, show no signs until the area which he is required to recognise is very small, feebly illuminated, or the colour mixed with a lot of white light, as happens in a fog. In the Board of Trade test the candidate is required to name the colour of a small illuminated area of variable size. This is the basis of the Eridge-Green colour perception lantern. Another method is to present the subject with a skein of wool, and ask him to pick out the one which matches it from a pile of variously coloured skeins. In the third method, which is the most convenient for rapid use, a series of cards (Ishihara charts) are printed with figures in coloured spots on a background of similarly shaped spots, but in a colour which is liable to look the same as that of the figure to a colour defective. The background spots may also be arranged to include a figure which is not obvious to a normal person, but which may be seen clearly by a colour defective.

THE RETINAL RECEPTOR SYSTEM

Once the light has been distributed to the receptors, the next stage in transmitting information about the environment to the brain is to convert the light energy into electrical energy. This is the task of the retinal receptors. It is probable that in the vertebrate retina, as in other receptor situations, the receptors first produce a potential change whose size is related to the intensity of light at the receptor; this graded potential change is probably responsible for setting up nerve impulses. The pattern of these impulses, both in time in any one fibre and on the spatial distribution of active and inactive fibres, must be uniquely related to a particular pattern of light intensity and wavelength reflected from the environment. Direct records from receptors of the vertebrate eye have not been obtained with any certainty and hence we must rely on less direct methods. The energy of the light is absorbed by a photochemical substance and these are discussed in the first subsection. In the second section reference is made to light receptors in invertebrate eyes. In the final section the responses of the higher order cells in the vertebrate retina will be considered.

Photochemical substances

The purpose of a photochemical substance is to provide enough energy for the next process in the chain to occur and to provide it when and only when it receives energy from the arrival of a quantum of light. There are a number of different substances known, but the one best known and studied in most detail is *rhodopsin* (visual purple) which can be extracted from the rods of a wide range of vertebrates. In studying this and other photochemical substances much use has been made of the *absorption spectrum* of the substance. Lights of different wavelengths will have different effects on the molecules of the substance. Those wavelengths which have the greatest effect on the molecule will be those that are giving up most energy in the substance, and hence are those for which least light passes right through the sample. If a solution of the substance is put in a light beam from a source of monochromatic light, the absorption at each wavelength may be measured.

Rhodopsin

Rhodopsin is almost certainly the substance responsible for light absorption in the rods of man. Not only can it be extracted from the rods and is the only known photochemical substance to be extracted, but its absorption spectrum is very nearly identical in position and shape to the dim-light sensitivity curve (Fig. 58.7). The agreement is closer if the absorption spectrum of the substance is measured *in situ*, that is with the molecules highly concentrated and laid down in an organised way.

Rhodopsin consists of retinene with a protein called opsin. Retinene is formed by oxidation of vitamin A and is the aldehyde of this substance. Retinene has a number of isomers of which two are important in this context. The retinene attached to opsin in rhodopsin is known as 11-*cis* retinene. The absorption of light energy by rhodopsin is thought to change this to another isomer, the all-*trans* retinene. All-*trans* retinene does not remain attached to opsin and the molecule breaks up. All-*trans* retinene, as would be expected from this, does not form rhodopsin in the presence of opsin. Rhodopsin is formed, however, by opsin in the presence of 11-*cis* retinene. The latter may be derived *in vivo* from vitamin A. An enzyme has been found to catalyse the change from the 11-*cis* to the all-*trans* form, but this does not appear to be important in the intact human eye.

Other pigments

A most interesting finding has come from the development of a technique for measuring the absorption spectra of single cones in the isolated retinae of primates. Three types of cone each giving a different absorption spectrum were found and the peaks of the curves were at 445 nm (blue), 535 nm (green) and 570 nm (red) (cf. Fig. 58.10). If further investigation shows these to be homogeneous groups and does not reveal any further categories, it would seem that the number of channels required for colour vision is limited to three at the level of the photochemical substances.

Most of the photochemical substances found have come from rods, since the quantities in cones are generally much less and hence their pigments are not easily detected by extraction methods. In mammals the substances which have been found have all had maxima to their absorption spectra close to 500 nm (cf. Fig. 58.7). Most freshwater fish have at least one photochemical substance with absorption maximum between 510 nm and 550 nm. An interesting group are the deep-sea fish, many of whom have substances with maxima between 475 nm and 490 nm; it is these wavelengths which are able to penetrate to great depths.

Photoreceptor activity

Photoreceptor activity has been investigated in a single part of the compound eye of an arthropod called *Limulus*. In this the nerve cell is surrounded by a number of supporting cells, which appear to contain the photosensitive mechanism. It has been possible to record the electrical activity of the nerve cell in the region of the receptor and it has been found that this cell is depolarised when light falls on the receptor. These slow potential changes are related in a graded manner to the intensity of the illumination falling on the receptor, in a manner similar to that described for receptors in general in Chapter 53. It has been shown in these photoreceptors that the potential change results from a change in the permeability of the cell membrane and in this respect also they are like other receptors. Impulses are also seen to be discharged by these nerve cells and the frequency of the impulses is related to size of the slow potential change and hence to the intensity of the light. Electric currents related to illumination have also been found in the receptors of octopus eyes; in this case there are no supporting cells and this is clearly a response of the receptor itself.

NEURONAL MECHANISMS IN THE RETINA

It has already been pointed out that our knowledge of impulse activity from the retina derives from units which are not the primary receptor units, but from the nerve cells in the retina which are of higher order. These responses therefore are comparable to those of cells which occur in the central nervous system in the case of other receptor systems in mammals. Most of the results come from the ganglion cells and their axons, which form the fibres in the optic nerve. The responses of these cells result from the interaction of considerable numbers of receptors.

The overall activity of the whole optic nerve can be observed by means of electrodes placed on it, in suitable kinds of experimental animal. As one would expect, the relation between the intensity of the stimulus (in this case light) and the number of impulses passing up the optic nerve in a given time is similar to that for other receptor systems. Within limits, the frequency of the optic nerve impulses appears to depend on the total amount of light stimulating the eye. In other words, in order to obtain a given response the product of the intensity of the light, the duration of the illumination and the area of retina illuminated, must be kept constant. This, as we have seen, applies also to the subjective sensations. The frequency of the impulses, moreover, rises with time after the start of the illumination, reaches a peak and then declines; the greater the intensity, the higher the peak, the sooner it is reached and the more rapidly it decays.

If special techniques are used to eliminate the effects of all eye movements the sensations of vision also decline. It is clear that a part of this adaptation must occur in the retina.

The subjective sensation of brightness (as observed in man) thus seems to follow in a general way the frequency of the impulses in the whole optic nerve (as observed in experimental animals). But the more detailed content of the sensation—the detection of spatial and temporal patterns, and of colour—must be conveyed by signals in the separate nerve fibres from different receptor units.

By the use of microdissection methods and of microelectrodes, it is possible to record the action potentials in single optic nerve fibres, the ganglion cells from which they arise and (probably) the bipolar cells. These experiments have shown that in both the frog and the cat, there are two main types of response. The impulses may increase in frequency when the light is switched on, and decrease in frequency, or cease altogether, for a short time after the light is switched off (the "on" response, Fig. 58.11a): or the impulses may decrease in frequency, or cease, when the light is switched on, and increase in frequency after it is switched off (the "off" response, Fig. 58.11b). This "off" response can be

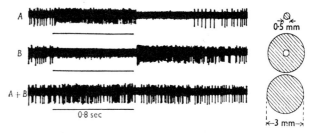

Fig. 58.11 Action potentials from a cell in the inner nuclear layer of a cat's retina.

A. On illuminating a spot 0·5 mm diameter in the centre of the receptive field of the cell; "on" response. B. On illuminating a ring 3 mm diameter surrounding the 0·5 mm spot illuminating in A; "off" response. A + B. On illuminating both the central spot and the peripheral ring; both responses are present but each is partially inhibited by the other.

The patterns of light falling on the retina are shown on the right of the figure. (Brown and Wiesel (1959), *J. Physiol.*)

suppressed immediately by turning the light on again. The impulses usually appear at a high frequency for a fraction of a second after the light is switched on, or off, and then settle down to a steady lower frequency, or may cease. Some nerve fibres give a short burst of impulses both at "on" and at "off" (the "on–off" response). This type of response is characteristic of the rapidly adapting type of receptor; the response is determined by the change in the illumination, whether this is an increase or a decrease as, for example, would accompany movements of light and shade across the retina. Such movements, particularly in the peripheral regions, are particularly effective in attracting attention.

Observations on single fibres have demonstrated conclusively a fact that may be inferred from other evidence, and has already been mentioned. The receptive field connected to a single nerve fibre in the optic nerve may be very large, and action potentials may be set up in a given nerve fibre by illuminating rods or cones spread over a wide area; the sensitivity decreases markedly towards the periphery of the field, so that it has no sharp boundary. Conversely, a very small spot of light, stimulating only a very few receptor cells, will produce impulses in several different fibres, so that the receptive fields of different units must overlap considerably.

If the eye is fully dark-adapted, there is summation between the effects of illuminating any one part of the receptive field with that of illuminating any other part; the threshold light intensity falls as the area illuminated rises in the central part of the whole receptive field by some units, but not by all. If the eye is not fully dark-adapted, the discharge set up by illuminating one spot in the receptive field may be inhibited when another spot is illuminated simultaneously. The discharge obtained from a unit in the centre of the receptive field, when that part is illuminated, is regularly inhibited when the peripheral part is illuminated as well, and vice versa. This is illustrated in Fig. 58.11. In the cat, one and the same unit may give, say, an "on" discharge when the central part of its receptive field is illuminated, and an "off" discharge when the peripheral part is illuminated (as in Fig. 58.11), or vice versa. This does not affect the mutual inhibition.

Spectral sensitivities

The least amount of light energy which will produce a just detectable response in an optic nerve fibre varies with the wavelength of the light. If the eye examined is dark-adapted, the spectral sensitivity of all the receptors has a maximum at about 500 nm, the curve has much the same shape as the human scotopic visibility curve (Fig. 58.7), and is just what would be expected if rhodopsin were the photochemical substance in action. If the eye is light-adapted and if the retina contains a sufficient proportion of cones, the spectral sensitivity of some of the units will have a maximum sensitivity at about 560 nm and the curve resembling the human photopic visibility curve. Granit, who has been responsible for much of this work on single optic nerve fibres, called these curves the scotopic and photopic *dominator curves*. In addition, light-adapted eyes are found to contain units whose spectral sensitivity curves have more complex shapes, some with humps and others with sensitivity maxima at two different wavelengths; some curves, also, even though they have only one peak, are much narrower than the dominator curves. These results are explained by Granit as being due to the existence of *modulator curves*, of which there are three kinds: those with maxima in the red part of the spectrum (about 600 nm); those with maxima in the green (about 530 nm) and those with maxima in the blue (about 450 nm). The precise position of the maximum sensitivity varies somewhat from unit to unit, but always lies in one or other of these three regions. These complex responses from the ganglion cells may result from the interactions of responses from receptors containing the different photochemical substances referred to above, one for rods and three for cones.

The control of eye movements

The eyes must be accurately directed if detail is to be appreciated in the visual image. The control of the eye position is chiefly initiated from signals from the retina itself. However, compensation must be made for movements of the head or of the whole body; the labyrinthine organs play a role in bringing about the necessary compensation of the position of the eyes.

The adjustment of the direction in which the eye is looking is extremely precise but is never perfect, even though there are no complicating movements either of the head or of the object looked at. Unless special attention is given to maintaining an exact fixation, the image on the retina is continually shifting slightly, by a few minutes of arc at intervals of a few tenths of a second. This has, in fact, some advantages. Acuity of vision depends on the detection of small differences in the illumination falling on neighbouring cones, but the electrophysiological evidence suggests that many of the receptor units behave as if they were rapidly adapting. Slight movements of the image from one to another would thus facilitate its detection. Colour vision, also, necessitates the illumination of relatively large numbers of receptor units, either simultaneously or in rapid sequence. The anomalous types of colour vision associated with very small areas of illumination can only be detected when fixation is very careful and deliberate.

Depth perception

Stereoscopic vision depends on the fusion of images from the two eyes. The sense of depth obtained depends on the slight difference of viewpoint, when the two eyes are both fixated on the same point. Precise positioning of the eyes is necessary if there is to be fusion of the image and double vision avoided. Accurate control mechanisms are needed for all these purposes. The detectors are the eyes themselves, the control mechanisms in the central nervous system and the effectors are the eye muscles.

The pupil

When a bright light is encountered, the pupil constricts. This is called the *light reflex* and involves retinal receptors, pretectal region and ciliary ganglion which gives the contraction of the sphincter pupillae muscles. Illumination of only one eye constricts both pupils, the *direct* light reflex and the *consensual* light reflex. The *Argyll-Robertson pupil* is found in tabes dorsalis and is a condition in which the light reflex is lost although the pupils will still constrict when the eyes undergo accommodation.

The function of the light reflex is to protect the retina from too intense illumination. In normal persons, the pupil can constrict down to a diameter of about 1·5 mm from a maximum of about 8 mm, a change in area (and hence of incident light) of about thirty times. It should be remembered, however, that the retina itself is able to adapt to changes in light intensity over a range of about ten billion times.

VISUAL PERCEPTION
(Central visual pathways)

Sight plays the predominant role in man's perceptual world, a fact presumably arising (in the evolutionary sense) out of his tree-dwelling ancestry. It has been suggested that around 90 per cent of the information we receive about our environment is carried along the optic nerves.

Neuronal processing in the retina

Table 58.1 shows that the input of information from around nine million photoreceptors is whittled down to pass along a nerve tract, the optic nerve consisting of only a quarter of a million fibres.

Table 58.1 *The composition of the rat's retina*

Region	Total neurones
Rods	9 180 000
Cones	120 000
Bipolar neurones	3 530 000
Optic nerve fibres	260 000

From Lashley, K. S. (1950), *Symp. Soc. exp. Biol.* No. 4. New York. C.U.P.

It is known that a great deal of information processing occurs in the retina itself and that the messages transmitted along the optic nerve are considerably condensed. Part of this processing has already been described (p. 385) and the concept of visual fields has been explained. The area of retinal surface able to excite a particular ganglion cell is defined as the visual field of the latter.* Since retinal receptors are only a few micrometres in diameter but visual fields are up to 1 mm across, a great deal of convergence must take place between the receptors and the ganglion cells. Figure 58.12 is a diagram of the connections of retinal neurones and indicates how such convergence could occur.

---

* Around such a ganglion-cell visual field, is a ring of receptors which does not, on its own, affect the firing of the ganglion cell but does do so when both the visual field and the ring around it are stimulated. This phenomenon means that the real boundaries of fields cannot be mapped exactly. They would become smaller in bright illumination and larger during dark adaptation. (See Kuffler, S. W. (1953), *J. Neurophysiol.* **16**, 37.)

Visual receptive fields in the frog

Visual receptive fields were first investigated in the frog by H. K. Hartline, who found three distinct types, quite apart from their size or location. He termed them "*on-fields*", "*off-fields*" and "*on–off-fields*".

THE ON-FIELD

Figure 58.11 gives the characteristics of the three different kinds of visual field. The on-field shown in (a) is the area of retina supplying a ganglion cell which, when illuminated, causes the cell to discharge. The rate of discharge depends directly upon the intensity of illumination and on the area of the field which is illuminated. The cell stops firing when the illumination ceases.

THE OFF-FIELD

This is just the opposite of an on-field. The ganglion cell normally shows a continuous discharge which is inhibited by light falling on the retinal visual field. Firing rate falls gradually with time in the dark but the ganglion cell is only completely silent when its receptive field is illuminated (Fig. 58.11b).

THE ON–OFF-FIELD

Here the properties of the two previous types of field are combined. A light shone on the retina activates the ganglion cell both when it is switched on and when turned off. The response is short-lasting in both cases; in conditions either of darkness, or of steady illumination, the discharge decays and the neurone is silent.

The analysis of these fields can be taken further. Interaction occurs inside one field. For an off-field, one gets a larger response when the periphery and the centre of the field are darkened, i.e. the response taken over the field as a whole is additive. The converse is true for an on–off-field. The response to central illumination of the visual field is considerably reduced if simultaneously the periphery is illuminated. This is also true for the opposite situation when the lights are switched off. One obtains the maximum response from an on–off-field when the

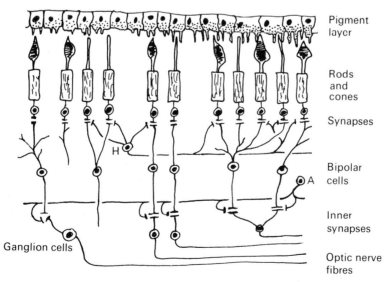

Fig. 58.12 Section through mammalian retina, showing type of connections between neurones. The convergence described in the text can be observed since four optic nerve fibres are supplying eleven photoreceptors. In the whole retina the figure would be nearer to one fibre for 36 receptors. Note that light is entering the retina from below in this diagram.
H = horizontal cell; A = amacrine cell.

difference in illumination between the centre and the periphery of the field is greatest. The meaning of this phenomenon appears to be that *boundaries*, where there is contrast in intensity of illumination, are particularly good at stimulating the retinal ganglion cells.

Mammalian visual fields

On the whole, the details of retinal organisation vary in the species, and all the vertebrates so far investigated seem to have different kinds of visual field. The rabbit, for example, has eight types of visual field. A common type of field, and one found frequently in all mammalian retinae, is one having an *on* centre and an *off* periphery, being roughly circular in shape (Fig. 58.13).

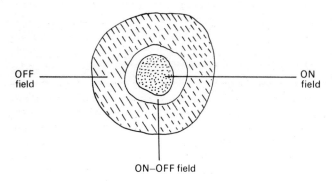

Fig. 58.13 A typical mammalian retinal visual field. If the centre and the periphery of the field are simultaneously illuminated, the responses of the photoreceptors are cancelled out at the level of the ganglion cells and no discharge proceeds up the optic nerve.

The periphery, illuminated alone, gives an off-discharge; the centre an on-discharge. See text.

The important functional property that is shown by these visual fields is the antagonism between the centre and the periphery. If the centre and the periphery are illuminated at the same time, the discharges from the receptors are equal and opposite, so at the ganglion cell level, no net response occurs, although a purely peripheral illumination will give an off-response while a central one will give an on-response. Between the two regions is usually an annulus where on–off-responses can be obtained by illuminating it.

Anatomical basis for concentric visual fields
Dowling and Boycott in 1966 analysed sections of the retina under the electron microscope and put forward a wiring diagram which would explain the known features of concentric visual fields in the mammal (Fig. 58.14).

In the mammalian retina, the ganglion cells are "spontaneously" active, even without any light falling on the retina. For an on-centre, off-periphery unit, excitation of the ganglion cell is produced via the direct bipolar–ganglion junctions whenever the central rods are illuminated. Inhibition of the ganglion cell is produced via inhibitory amacrine–ganglion junctions if the periphery of the visual field is stimulated by light.

Stimulation of both the centre and the periphery sums algebraically and only a weak (or no) response occurs. From the far-periphery of the field, there are effects mediated by amacrine–amacrine junctions.

Relatively simple modifications of this scheme can explain other receptive field modifications. Off-centre, on-periphery ganglion cells would be connected with the excitatory and inhibitory synapses reversed.

Other, more complicated, types of visual field are found in mammalian retinae, particularly that of the rabbit which appears to undertake much visual analysis in the eye which in cats or primates takes place in the visual cortex. Receptive fields are found in the rabbit that detect moving borders, even distinguishing the direction of movement.

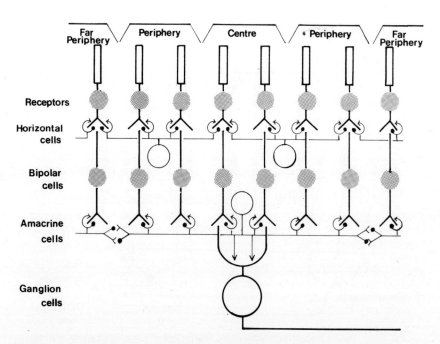

Fig. 58.14 A "wiring diagram" for a ganglion cell receptive field based on the anatomy and physiology of the retina. ⅄, excitatory synapse; ⊥, inhibitory synapse. See text for details. The synaptic contacts between horizontal processes and receptor terminals and amacrine processes and bipolar terminals are represented as reciprocal contacts; the others as one-way junctions. (Boycott, 1966).

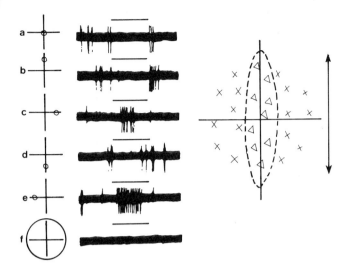

Fig. 58.15 Responses of a cell in the cat's striate cortex to a 1° spot of light. Receptive field located in the eye contralateral to the hemisphere from which the unit was recorded, close to and below the area centralis, just nasal to the horizontal meridian. No response evoked from the ipsilateral eye. The complete map of the receptive field is shown to the right. X, areas giving excitation; △, areas giving inhibitory effects. Scale, 4°. Axes of this diagram are reproduced on left of each record. a, 1° (0·25 mm) spot shone on the centre of the field; b–e, 1° spot shone on four points equidistant from centre; f, 5° spot covering the entire field. Background illumination 0·17 log mc Stimulus intensity 1·65 log mc Duration of each stimulus 1 s. Positive deflexions upward.

(Hubel, D. H. and Wiesel, T. N. (1959), Receptive fields of single neurones in the cat's striate cortex. *J. Physiol.* **148**, 574–591.)

side, and is a vertical strip in which off(inhibitory)-responses occur, surrounded by an area giving on(excitatory)-responses. As in the case of the retinal ganglion cell, these inhibitory and excitatory inputs summate at the cortical neurone. The illumination of a large part of the excitatory region gives a correspondingly large response; illuminating a small part gives a small response. When both excitatory and inhibitory regions are stimulated by light the resultant effect is very small, or there is none, because the two cancel out. This is shown in the last line (f) of Fig. 58.15, where a large light spot is used to stimulate both the excitatory surround and the inhibitory centre. There is no effect upon the neurone which had previously responded to the smaller light spot as shown by the records above. By the same token, diffuse lighting of the retina is without effect on these cortical neurones.

Response to illuminated bars and edges

The concentric visual field is by far the most common type found in the ganglion cells of the retina in cats and monkeys. But this system is not present in the cortex. A cortical neurone, as explained above, "sees" the external environment "through" a slit which has a particular orientation (vertical in the example shown in Fig. 58.16). Sometimes the slits are inhibitory with an excitatory surround; others are the opposite. Whenever a feature of the external visual world has an area of increased intensity of illumination whose image is bar shaped and is in the correct orientation, the cortical neurone will give a response. Figure 58.16 shows how this is tested experimentally. A lighted slit is exhibited on a screen placed in front of the anaesthetised cat

Central visual processing

Let us first see how the retina is connected with the visual cortex. A simplified diagram of the relevant anatomy is given in Fig. 60.1. The retinal photoreceptors are connected to the optic nerve via synapses with bipolar cells and amacrine cells and ultimately to the ganglion cells whose axons are the optic nerve. The optic nerve fibres travel to the *lateral geniculate nuclei* on the two sides via the *optic chiasma*—a partial cross-over point where fibres originating in the temporal half of the retina go straight through to the lateral geniculate on the same side; those coming from the nasal retina cross over and are distributed to the lateral geniculate of the opposite side. Thus the field of view is transmitted to the brain in such a way that either half of it is exclusively represented by cells on the opposite hemisphere. One might almost expect to be aware of a discontinuity in one's perception of the visual field where the two fields overlap (or just don't meet). But of course this does not happen!

Visual fields in cortical neurones: simple fields

As we have already seen, it is possible to record the electrical activity from single ganglion cells. It is also possible to do the same in lateral geniculate and in visual cortical neurones. Interesting work has been carried out by Hubel and Wiesel, using the latter technique to analyse the way in which the cortical neurones process information. Figure 58.15 is from an experiment in which a 1 degree spot of light was shone into a cat's eye. The electrical records (middle column) give the responses of a single cell in the contralateral striate cortex when the light was switched on at the locations shown in the left-hand column.

A map of the cell's receptive field is shown on the right-hand

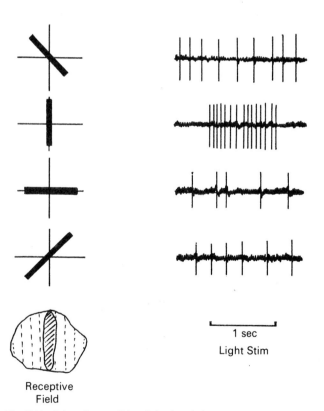

Fig. 58.16 Orientation specificity of visual cortical neurone.
A bar of light is shown to an anaesthetised cat for 1 s periods as indicated. Only when the bar is vertically orientated with respect to the cat's retina does the cell give a discharge. The receptive field for this neurone is shown with the central slit being excitatory (cross-hatched) and the surround inhibitory (dotted).

Fig. 58.17 Response of same neurone, as illustrated in Fig. 58.16, to movement. Top line: Bar is moved horizontally giving a response during the time it crosses the visual field. Bottom line: Vertical movement is without effect.

whose eyes have been immobilised. As the slit is rotated, the neuronal discharge is recorded and it is observed that the maximum response is to a vertical slit (in this case). As well as responding to a bar, such a cortical neurone will respond to an "edge", i.e. a linear discontinuity of intensity.

Response to movement

As well as analysing orientation of bars and edges, cortical neurones are sensitive to movement, again in a preferred direction (Fig. 58.17). The reason why a bar when moved along the

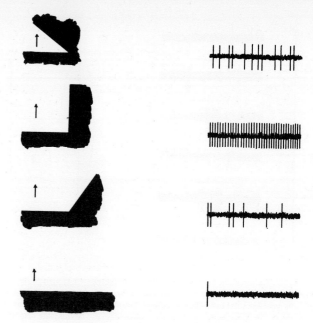

Fig. 58.19 Visual cortical neurone having a hypercomplex field. This cell would only respond maximally when a 90° corner was moved across its field in a certain direction. (Original experiments were done by Hubel and Wiesel, 1965.)

Complex visual fields

The visual receptive fields that we have discussed here can be mapped using small spots of light. Hubel and Wiesel also discovered more complicated visual fields in cortical neurones which could only be found using various stationary and moving shapes such as slits, edges, bars, and moreover only when these were orientated in the correct plane. They are termed *complex* fields.

It might be thought that there is little difference between simple and complex fields from the foregoing (apart from the fact that complex fields do not respond to light spots). They differ in the major property that the bar or edge gives its response from any part of the field. (A simple field is not like this; the stimulus has an optimal orientation and must be in the middle of the field.)

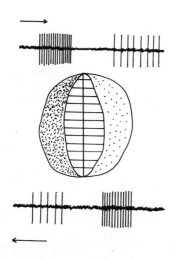

Fig. 58.18 Directionally sensitive field. The central slit is surrounded by two peripheral fields which are excitatory but of differing sensitivities. The degree of stippling indicates the strength of the response to a given light intensity. The top record shows the response of the visual cortical neurone to a left/right movement; below is the response to a right/left movement.

length of the central slit does not activate the cortical neurone, is that the two ends of the bar of light extend over into the inhibitory flanks and cancel out any excitation from the slit. To explain how a neurone analyses "direction", Hubel and Wiesel have supposed that the receptive field has a central inhibitory strip and on either side of it are two excitatory areas. In addition it was postulated that these two areas had a different sensitivity. Stimulation of one would give a much bigger response than the other would do. A little thought and a consideration of Fig. 58.18 should convince the reader that a mechanism of this nature would distinguish left to right from right to left movement of the illuminated bar.

Hypercomplex fields

There are yet further steps in the hierarchy of cortical neuronal analysis. There are cells called *hypercomplex* neurones which will only respond to highly specific "trigger" stimuli such as an edge, bar or line which is stopped at one or at both ends. Some of these cells respond only to corners, as shown in Fig. 58.19 recordings from a hypercomplex cell which gives an optimal response to a lighted, right-angled corner, anywhere in the visual field.

Hubel and Wiesel found even more specific cells than these—the so-called *higher-order hypercomplex cells*—which only gave a response to a feature in the visual field such as a "tongue" of light anywhere in it and arranged in any configuration or orientation (although if the tongue were too fat, the response tended to fall off).

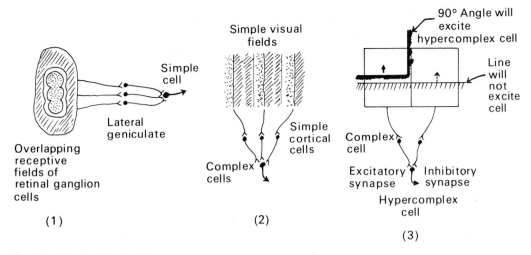

Fig. 58.20 Hypothetical "wiring" diagrams for explaining the analytic mechanisms found in visual processing. (Hubel, D. H. and Wiesel, T. N. (1962), receptive fields, binocular interaction and functional architecture in the cat's visual cortex. *J. Physiol.* **160**, 106–154.)

(1) Many geniculate cells (only three shown) have overlapping concentric fields with on-centres in a line. All of these feed one cortical simple cell via excitatory synapses. The cortical cell will thus have a field in the form of a slit which is excitatory with an inhibitory surround.

(2) A large number of such simple cells are connected to one complex cell. The simple fields are strips with excitatory centres and inhibitory edges, i.e. alternating regions of excitation and inhibition. Provided a light stimulus is within the field of the simple cells and if it is an edge in the correct orientation (vertical) at least one of the simple cells will give a response. Thus a complex cell will be activated.

(3) Only a corner at 90° will activate a hypercomplex cell. Two complex cells are shown connected to the hypercomplex cell via an excitatory and an inhibitory synapse. A line activates both equally and at the hypercomplex cell level, the effects cancel out.

Summary

We find neurones in the cortex which respond to spots of light in specific parts of the visual field. As we ascend the neuronal hierarchy, the trigger stimulus required to activate the neurone concerned becomes successively more specific. Specifically orientated edges, bars etc. are needed to activate complex cells. At the same time as the stimulus itself is required to be more specific, its actual location on the retina becomes less important. We end with cells responding to complicated shapes, placed anywhere (within limits) on the retina. Indeed, some workers claim to have recorded, in the monkey, from one visual cortical neurone which would only respond to a shape resembling a hand. Four fingers would not activate it; five would! Neurophysiologists can now begin to tell us how the mechanisms are operating in cortical neurones, which lead ultimately to visual perception.

Columns in the visual cortex

The visual cortex has been examined histologically following microelectrode recording from neurones. It has been found that the cells of the striate cortex are arranged in columns—any one column containing cells having a similar function. Such a column might be around 0.5 mm in diameter, and usually arranged at right angles to the surface of the brain. A single column contains the simple, complex and the hypercomplex cells which are involved in the processing of information originating in one part of the retina. The columns themselves are also arranged in an orderly fashion. Rows of columns are found which have cells in them with receptive fields whose orientation changes regularly from one column to the next.

CELL CONNECTIONS IN THE VISUAL CORTEX

We do not yet know how the neurones in the cortex are connected up to give rise to the phenomena outlined above, but some intelligent guesses can be made. In general terms, all the known facts about visual analysis can be explained in terms of the known properties of cortical neurones.

One model is as follows:

1. The slit-shaped field of the simple cell is due to several geniculate neurones having their fields overlapping, in a row, and activating a single simple cell.
2. Complex visual fields arise from several simple cells being connected to one complex cell. The simple field is a strip divided into inhibitory and excitatory regions. Thus an edge (or bar) will activate the complex cell (a) if it is orientated correctly, i.e. along the long axis of any of the simple fields, and (b) wherever it is provided it covers at least one of the simple fields.
3. Hypercomplex fields can be explained if one supposes that the outputs of two complex cells are passed on to one hypercomplex cell, one being excitatory, the other inhibitory. The highly specific nature of the responses of higher-order hypercomplex cells can be explained by suitable arrangements of excitatory and inhibitory inputs from complex cells. Figure 58.20 should make the above considerations clear. It must be noted however, that rather little evidence as yet has been found to support this model.

59. The Central Nervous System: I

P. A. Merton

ANATOMY

Because the anatomy of the central nervous system is commonly less familiar, even to medical readers, than that of the rest of the body, all the structures subsequently mentioned in these chapters (except for a very few which are specifically noted) are briefly described in this section or are depicted in Figs. 60.1, 60.2, 60.3 and 60.7. A more thorough knowledge can be acquired from the books listed in the Bibliography, but the non-medical reader who may be daunted by the large volume of detail in such works should remember that, although it is axiomatic that physiology must always rest on a firm basis of anatomy, so little of the physiology of the brain is understood at present that a relatively small amount of anatomical knowledge will be found adequate to support it. Outside the skull there are probably no organs or tissues in the human body of which the function or, in the case of rudiments, the significance is not understood, at least in a general way, but in the brain the student must not allow himself to be nonplussed to find numerous large masses of nerve cells for which factual information about function is negligible. He may be warned, however, that speculation on these matters, masquerading as fact, in anatomical and physiological texts, is an old if little recognised branch of science fiction.

The central nervous system consists of the spinal cord, the part within the spinal column, and the brain, the part within the skull. The spinal cord retains the segmental structure of the primitive nerve cord, giving rise in each segment to a pair of spinal nerves, each formed from separate dorsal (sensory) and ventral (motor) roots (Fig. 59.1). Inside the cord is a core of nerve cells (grey matter) surrounded by the tracts of nerve fibres (white matter) that carry nerve impulses up and down the cord.

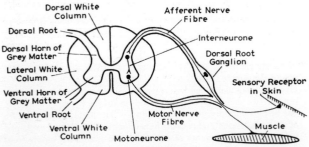

Fig. 59.1 Diagrammatic cross-section of the spinal cord showing the H-shaped grey matter, the columns of white matter surrounding it, and the components of a simple reflex arc.

The spinal cord is the seat of various reflexes, protective, muscular, visceral, sexual, etc and of various autonomic nervous mechanisms concerned in standing, walking, etc. It also transmits sensory information to the brain, and executes the orders the brain sends it. In the higher vertebrates the spinal cord becomes increasingly subordinate to the brain, and the tracts of white matter running to and from the brain occupy a large part of it. The position of the principal spinal tracts is indicated in Figs. 60.2, 60.3 and 60.7.

The vertebrate brain develops from three hollow swellings at the cranial end of the primitive neural tube, the forebrain, midbrain and hindbrain vesicles. The olfactory and optic nerves take origin from the front and back respectively of the forebrain; all the other cranial nerves belong to the mid- and hindbrain (Fig. 59.2). The main external features of the mammalian brain result from elaboration of this basic structure in two regions. (1) From the front of the forebrain two further hollow swellings develop, the cerebral hemispheres (Fig. 59.2). The original cavity of the forebrain vesicle becomes known as the IIIrd ventricle and its extensions into the hemispheres are the two lateral ventricles. (2) In the roof of the front part of the hindbrain develops a large unpaired structure, the cerebellum. Tracts of nerve fibres running transversely to the cerebellum form a conspicuous bulge, the pons, on the ventral surface of the hindbrain. Behind the pons, where it tapers into the spinal cord, the hindbrain is known as the medulla oblongata, or medulla for short. The cavity of the hindbrain is named the IVth ventricle; it is connected to the IIIrd ventricle by the aqueduct, the narrowed cavity of the midbrain vesicle. The part of the brain around the IIIrd ventricle (the diencephalon), the midbrain and the hindbrain, excluding the cerebellum, are referred to jointly as the brainstem.

The cerebral hemispheres and the cerebellum are the parts of the brain that increase most in size as the evolutionary scale is ascended. In man they are so large that they hide most of the rest of the brain (Fig. 59.3). The cerebral hemispheres and the cerebellum are the only places in the central nervous system where grey matter lies on the surface forming a "cortex". Elsewhere the grey matter is in the middle with the connecting tracts of white matter around it. Underneath the cerebral cortex, thick bundles of white matter connect the parts of the cortex to each other and to other parts of the brain. A massive tract, the corpus callosum, runs between the two cerebral hemispheres above the roof of the IIIrd ventricle (Fig. 59.3). The main tracts connecting the cerebral cortex to the lower parts of the brain and to the spinal cord pass between the basal ganglia and the thalamus (see below), where they are known as the internal capsule, and emerge on the ventral surface of the midbrain to form the cerebral peduncles.

The basal ganglia, or corpus striatum, is a mass of grey matter than develops in the floor of the lateral ventricle. It is the highest centre for motor functions, apart from the cerebral cortex. The thalamus is a large group of nuclei in the wall of the IIIrd ventricle. It is the highest centre for sensory activity apart from the cerebral cortex. Below the thalamus the floor of the IIIrd ventricle descends to meet Rathke's pouch growing up from the roof of the buccal cavity, to form with it the pituitary gland. The grey matter in the walls of the IIIrd ventricle immediately above the pituitary constitutes the hypothalamus.

The brainstem contains a central core of grey matter, continuous with that in the spinal cord, from which arise the cranial nerves (Fig. 59.2), but the segmental origin and the division of each nerve into motor and sensory roots seen in the spinal cord are not in evidence in the cranial nerves. Among the cranial nerve nuclei and the other specifically named structures in the brainstem there is also a large region where small groups of cells lie in a network of nerve fibres running in all directions, hence known as the reticular formation (Fig. 60.7). This part of the brain is fashionable among neurophysiologists, who do not always confine the term to the true reticular formation of the anatomists. Many functions are attributed to the reticular formation. The respiratory and vasomotor centres are also claimed as part of the reticular formation.

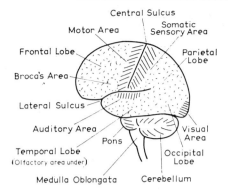

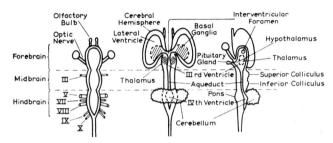

Fig. 59.2 Idealized mammalian brain in two stages of development. Left: the three primitive brain vesicles in horizontal section and the principal cranial nerves. Centre and right: a later stage, in horizontal and in midline section (cranial nerves, except optic nerves, omitted).

Cranial nerves: III (oculomotor) motor to most of the eye muscles; V and VII (trigeminal and facial) sensory and motor to face and mouth; VIII (auditory) consists of cochlear and vestibular portions from the organs of hearing and balance respectively; IX and X (glossopharyngeal and vagus) motor and sensory to the alimentary and respiratory tracts and to the heart.

CATEGORIES OF NERVOUS ACTION

The central nervous system is the organ which directs the behaviour of animals, all those responses and activities by which an animal reacts to its environment and attempts to master it and flourish. In man and many other animals, nearly all behavioural activities are carried out by the contractions of skeletal muscle. (This is not always so; the skunk and the electric fish have other ways of getting the better of their enemies.) In addition to its outwardly directed activities the central nervous system has the job of running most of the internal systems of the body; it controls, for instance, the endocrine, digestive, respiratory, blood circulatory and excretory systems. Most of these internal duties are discussed in the chapters dealing with those systems; in this chapter we are mainly concerned with the central nervous system as it determines the external behaviour of the animal through its muscles.

An animal's reactions to changes in its environment may differ greatly in complexity. Shining a light in the eye causes the pupil to constrict; this is one of the simplest of nervous reactions, and historically one of the first that was clearly understood to involve a nervous pathway from the sensitive surface to the central nervous system and back to the effector muscle. At the other end of the scale the production of a Ph.D. thesis may equally clearly be a response to a change of environment. Those reactions that seem to be automatic and immediate, such as blinks, knee jerks or sneezes, are termed "reflexes". The other main category of muscular actions is those that, because they involve conscious thought and decision, are called "voluntary movements".

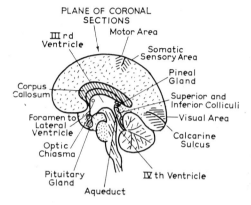

Fig. 59.3 Lateral view and median sagital section of the human brain, with "projection areas" marked.

Above the pituitary gland and optic chiasma the lateral wall of the IIIrd ventricle is formed by the hypothalamus, above that by the thalamus. The arrow indicates the plane of the coronal (frontal) sections in Figs. 60.2 and 60.7.

Typically reflex is something the subject can't help doing, like coughing during a concert, or choking if someone touches the back of his throat with a feather, whereas voluntary actions are actions he could have refrained from if he had wished; hence the satirical definition of a voluntary movement as a movement a patient makes when he is told to by a neurologist.

The line between reflex and voluntary is very indistinct. There are all levels of behaviour between the most voluntary and unpredictable, such as writing a Ph.D. thesis, and the most automatic, such as the pupillary reaction to light. Reading a book is obviously a voluntary activity, looking up when someone enters the room is practically automatic, blinking when a threatening movement is made towards the face, although it may be possible to prevent it by a strong effort of will, is so automatic that it is generally considered a reflex, blinking when an object touches the cornea is an indubitable reflex.

No one knows whether the nervous mechanisms of the most elaborate voluntary actions and of the simplest reflexes differ in principle or only in complexity and detail. No rational classification of the activities of the central nervous system is therefore possible in the present state of knowledge, so whether we call some particular manifestation of nervous action voluntary or reflex, or by some other term, is purely a matter of descriptive convenience, and usages are bound to differ in so complex a subject.

An important category of semi-automatic reactions that are not usually called reflex is those involved in the expression of emotion. Emotional movements of the face are certainly different from voluntary movements in their nervous mechanism because, as we shall see, disease of the brain may paralyse voluntary movements of the face such as screwing up the eyes, baring the teeth, etc., and yet the patient may smile normally if amused.

So far we have spoken as if all behaviour were a reaction to some environmental stimulus direct or remote but, of course, in ordinary life a good many of a man's or an animal's voluntary acts are likely to be in response to internal stimuli; thirst, hunger and sexual and other instincts. Instances of simpler types of activity that appear to arise spontaneously, without any obvious external stimulus, are seen with blinking and with eye movements. Blinking may sometimes be reflex, as when something touches the cornea, or it may be a voluntary act. Blinking, however, goes on irregularly all the time while we are awake and for most of these occasions there is no apparent external stimulus and certainly no voluntary factor; its function is presumably to prevent the cornea from becoming dirty or drying up. Such blinking certainly appears to be spontaneous and, indeed, cannot for long be suppressed by the will. This urgency is even more in evidence with the incessant shifting of the direction of gaze that goes on all the time we have our eyes open. It is much more difficult to avoid moving the eyes than it is to avoid blinking. The need to move the fixation point at frequent intervals is probably connected with the fading of contrast which is observed if the retinal image is not allowed to move. At all events it cannot be because the world is full of interesting objects that provide a succession of irresistible reflex stimuli, for spontaneous eye movements if anything increase when looking at a blank surface; indeed, in the dark it appears to be impossible to keep the eyes still.

Nothing is known of the nervous mechanisms responsible for such apparently spontaneous actions, and they will not be discussed further. They are mentioned partly to emphasise that the nervous system may take a much more active part in shaping events than is implied in the simple idea of reflex action.

It will also be evident from the foregoing that there are nervous actions of which it is no use asking, "Is it a reflex?", expecting an answer yes or no. Thus, as we have seen, blinking may be reflex; more usually it is apparently spontaneous, but it may be voluntary or even emotional in origin. It is best thought of as just a specific type of muscular activity, that may arise in a variety of ways.

These remarks serve to introduce the ubiquitous and important subject of posture. Posture is maintained by the same muscles that are used to carry out the ordinary outwardly-directed behavioural activities (reflex, voluntary, etc.) of the animal, but otherwise it is more akin to the internal homeostatic mechanisms such as breathing. Readers will be familiar with Claude Bernard's teaching that stable surroundings for the cells of the body as regards temperature, pH, ionic composition, etc. are prerequisite for a life of unrestricted activity, but a land-living mammal is just as disabled if it cannot stand up as if, owing to faulty breathing, the pH of its blood is wrong. Correct posture is not so much a behavioural activity itself as an initial state which is necessary before an animal can act effectively. As with breathing, etc., it is best not to ask whether posture is reflex or spontaneous but to regard it as a state of the body which is controlled by automatic mechanisms sensitive to the direction of gravity, the pressure of the ground on the feet, etc., in the same way that breathing is controlled by the concentration of carbon dioxide in the blood, etc. No one thinks of breathing as a reflex response to a smack on the buttocks at birth, nor is standing merely a response to the sound of the breakfast gong. Both activities so long outlast the stimulus, that how they are maintained has become of much more interest than how they were initiated. (Standing is not a reflex response to gravity, for gravity acts very constantly and since animals sometimes lie down and sometimes stand up it cannot be gravity that determines which they do. It is true that when for other, usually inscrutable, reasons an animal stands up the direction of gravity determines how it does it, but that is not the same thing.)

Locomotion (walking, running, etc.) is another complex action whose mechanism deserves to be studied in its own right, quite apart from the stimulus that sets it going. It is closely related to posture and subsequently the two will be dealt with together.

This completes the introductory discussion of the types of nervous activity that have to be taken account of in these chapters. Nothing has been said of such things as social behaviour or psychology for the very good reason that we are so far from coming to grips with the neurological mechanisms involved that they are not yet part of the subject matter of physiology.

Since Sherrington's famous, but for many people almost unreadable, book *The Integrative Action of the Nervous System* (1906, reprinted 1947) it has been conventional to emphasise integration or coordination as functions of the central nervous system, so much so that some authors apparently feel that to describe a nervous centre as exercising an integrating or coordinating influence over some function or other absolves them from attempting any description of what it does. Sherrington himself, who was always prone to use words in highly specialised senses of his own, clearly used the word "coordination" with a very extended meaning, as the following sentences, with which he opens an article on the brain, show. "The nervous system has as its function the coordinating of the activities of the organs one with another. It puts the organs into such mutual relation that the animal reacts as a whole with speed, accuracy and self-advantage, in response to the environmental agencies which stimulate it." This quotation can be regarded as defining "coordination". For Sherrington, it is apparently that activity which is the function of the nervous system. Other passages show that he used "integration" in much the same sense. To avoid misunderstanding, the words "integration" and "coordination" are not used in these chapters.

REFLEX ACTION

The word "reflex" implied originally that nervous messages travelling up sensory nerves were "reflected" in the central nervous system and passed out again into motor nerves. Ideas of this kind were current in the seventeenth century; an instance of reflection given by Descartes is the involuntary blink a man gives if a threatening gesture is made towards his face. Some of the first experiments were made by Stephen Hales, the perpetual curate of Teddington, Middlesex, in about 1730. He used a spinal frog, i.e. a frog that has been decapitated so that the spinal cord is the only part of the central nervous system remaining. When such a preparation is suspended the legs hang down limply, but if one foot is pinched that leg is drawn up as if to remove the foot from the injurious stimulus. This withdrawal reaction (the flexion reflex) is lost if the spinal cord is destroyed, a result which strongly suggests that the nervous pathway from the skin of the foot to the flexor muscles runs via the spinal cord.

Hales' experiment does not definitely exclude other more far-fetched possibilities, such as that the site of the reflexion is really in the lumbar nerve plexus but that the nerve plexus is under the influence of the spinal cord and ceases to reflect if the cord is destroyed. (Something not dissimilar is now known to occur in the axon reflex from skin on to cutaneous blood vessels.) But such alternatives were made superfluous by the celebrated experiments of Magendie and Bell on the functions of the spinal nerve roots, at the beginning of the nineteenth century. They showed that, if a ventral root was cut through, excitation of the peripheral stump caused widespread muscular contraction, but excitation of the central stump connected to the spinal cord was without obvious effect; in particular the animal exhibited no signs of sensation. The peripheral stump of a divided dorsal root gave no muscular movements (and, of course, no signs of sensation), but the central stump caused obvious pain and elicited muscular movements (which disappeared if the ventral roots were divided too). These experiments showed that the ventral roots were exclusively motor, with no sensory function, and that the dorsal roots were exclusively sensory but could excite movements by "reflexion" in the spinal cord.

Before Magendie and Bell it was not clear that motor and sensory nerves were distinct. The real significance of their experiments was to show that the nervous system uses separate channels for input and output—the most important single principle in the organisation of the nervous system. Nowadays the idea of motor and sensory fibres is so familiar that it is easy to forget that Magendie and Bell discovered their separate existence and did not merely demonstrate an anatomical fact about their mode of origin from the spinal cord. As regards reflex action, the fact that motor and sensory fibres maintained a complete separation of function until they reached the spinal cord showed that the connection between them, the site of reflexion, must be in the cord.

The anatomical parts of the simplest idealised reflex pathway (Fig. 59.1) are: a sensory receptor connected to a sensory (afferent) nerve fibre running to the central nervous system, a motor (efferent) nerve fibre running from the central nervous system to a muscle, and a synaptic connection in the central nervous system between the sensory fibre and the motoneurone. This synaptic connection may either be direct or, more usually, via one or more additional nerve cells called interneurones. In all real reflex pathways, of course, many receptors and many sensory and motor fibres are involved.

The central connections between input and output are seldom simple enough to give the impression that sensory impulses are merely "reflected" into motor channels. An instance that might suggest reflexion is the constriction of the pupil when light is shone into the eye. With the majority of reflexes, however, the pattern of discharge of motor impulses differs, often very widely indeed, from the pattern in the sensory nerves that excite them. A relatively simple reflex is the corneal reflex: a touch on the cornea, e.g. with a wisp of cotton wool, causes a blink. The sensory impulses for this reflex travel in the ophthalmic division of the trigeminal (Vth cranial) nerve to the medulla (see Fig. 59.2) and motor impulses leave from nearby in the facial (VIIth cranial) nerve which supplies the orbicularis oculi—the muscle responsible for blinking. There is clearly a formal sense in which impulses may be said to find their way from the cornea via the facial nerve nucleus back to the orbicularis, but such a mode of description certainly tends to conceal the significant fact that a blink is an entity, almost an all-or-none event, a stereotyped pattern of motor discharge which differs little whatever type of stimulus is used to evoke it. Hence it is more illuminating to say that touching the cornea triggers off a blink, rather than that it reflexly excites the orbicularis. As Hughlings Jackson (1835–1911), the greatest speculative mind in the history of neurology, put it; even in the simplest reflexes the nervous system "thinks" in terms of movements, not muscles—the corneal reflex results in a blink, not just in some contraction of the orbicularis. Nothing is known of the details of the motor mechanism that organises a blink.

Several examples of more complicated reflexes are found at the head end of the digestive and respiratory tracts: swallowing, vomiting, sneezing, coughing, etc. In swallowing the problem is to get a mouthful of food or drink rapidly into the oesophagus without any entering the nose or the larynx. When voluntary activity is in abeyance, e.g. in stuporose human beings or in animals from whom the cerebral hemispheres have been removed (decerebrate preparation), reflex swallowing can be induced by placing small pieces of food on the back of the tongue. Vomiting, sneezing, etc. are also self-contained sequences of muscular acts which are triggered by sensory stimuli.

A different type of complexity is seen with micturition in the spinal dog. In a dog in which the spinal cord has been functionally separated from the brain by transection in, say, the thoracic region, the bladder, after some days, empties itself reflexly when it is full. Emptying of the bladder is itself another example of a definite act triggered by sensory impulses rather than a reflexion of these impulses to the effector organ. But what is of particular interest is that the spinal dog, when it has finished passing its water, may wag its tail, just like any ordinary dog. Thus this elaborate gesture involving many muscles, none of them anything to do with the bladder, is clearly a part of the reflex act of micturition.

The examples so far discussed make it evident that on the motor side the reflex centres of the central nervous system deal in terms of complete muscular acts. It is natural to ask if there is any corresponding generalisation to be made about the sensory side of reflex action. Take the case of the reflex blink in response to a threat. If we speak of this reflex as a matter of impulses finding a pathway from retina to orbicularis muscle we shall be on safe ground, but this formulation disregards the fact that only a spatiotemporal pattern of impulses in many thousands of optic nerve fibres which the brain recognises as a threatening object rapidly approaching the face will evoke a blink. This is known to be a task that requires the highest centres for sensory interpretation in the cerebral cortex, for the blink in response to a threat from one side may be abolished by a small lesion of the opposite parietal lobe (see Fig. 59.3) which causes no detectable disturbance of vision (p. 408). Evidently in this reflex by far the most difficult part is the process of recognising the threat; this is presumably why the delay after a threat is made before the blink occurs (variously known as the latent period or the reaction time) is of the order of 250 milliseconds, whereas the blink to a touch on the cornea takes only some 40 milliseconds. For the purposes of this reflex, therefore, impulses in the optic nerve are of no importance as such, but only become important when their message has been decoded into whatever pattern of nervous processes it is that constitutes the nervous representation of a threat; the threat is the significant thing, not the impulses. To adapt Hughlings Jackson, the central nervous system "thinks" in terms of the causative physical events not in terms of sensory receptors.

The blink reflex is an extreme example of this principle because the process of recognition is so complex and the motor mechanism relatively so simple, but at lower levels in the central nervous system sensory recognition is still important, chiefly in reflexes initiated from the skin. Elsewhere it is not so in evidence: e.g. when impulses from distention receptors in the bladder arrive at the spinal cord, presumably they can only mean one thing; no problem of recognition arises. But the skin is played upon by many diverse influences, and the appropriate reflex responses can only be made if the central nervous system can distinguish the various stimuli.

In a spinal dog or cat two opposite reflexes can be elicited from the pads of a foot. Pressure by a flat object on the pads causes a brief powerful extension of hip, knee and ankle, the extensor thrust reflex (tentatively identified by Sherrington as an element in galloping), whereas pinching a pad (or any other form of painful stimulation) gives the withdrawal or flexion reflex of the limb, already mentioned in the frog. The stimulus for an extensor thrust presumably excites touch and pressure receptors. A pinch which elicits flexion excites pain fibres as well as touch and pressure. Experiment shows, however, that any noxious stimulus which is likely to cause pain will elicit the flexion reflex, so the essential thing about a pinch must be the pain, and this must be easy enough to recognise because it is carried by special pain fibres. Protective reflexes caused by noxious stimuli always seem to get priority in the nervous system (for obvious practical reasons), whence it appears that the recognition of pain carries with it an automatic refusal to recognise other stimuli (in this case touch and pressure) or in some other way to overrule them.

The recognition of the stimulus that calls for an extensor thrust is not so straightforward a matter, however. It is not merely a matter of impulses in touch and pressure fibres in the absence of pain. Sherrington never succeeded in eliciting an extensor thrust reaction by electrical stimulation of the sensory nerves supplying the pads although there is no reason to doubt that he was exciting touch and pressure fibres. (Pain fibres, being of small diameter, require a larger electrical stimulus so it is not difficult to avoid exciting them.) It seems likely that recognition of pressure on the pads depends on a distinctive pattern of discharge from touch and pressure receptors in different positions on the pads, and also perhaps on a distinctive distribution in time of the impulses from each receptor, and that electrical stimulation fails to imitate this spatio-temporal pattern of discharge.

At all events this difficulty in eliciting reflexes by electrical excitation of nerve trunks is a widespread and interesting phenomenon that has attracted attention since the middle of the nineteenth century. With large stimuli reflexes due to excitation of pain fibres are obtainable but, generally speaking, no other reflexes can be elicited by electrical stimulation of skin nerves (muscle nerves are different, as is discussed on p. 401). The explanation offered is that for most skin stimuli recognition depends on characteristic patterns of discharge in several nerve fibres which cannot be reproduced by electrical stimuli to nerve trunks, whereas pain stimuli are recognised because they travel in pain fibres and the pattern of discharge is unimportant; hence shocks large enough to excite pain fibres successfully excite pain reflexes.

Subjective observations by human subjects lend plausibility to this interpretation. Repetitive electrical stimulation of cutaneous nerve trunks is easily performed with an induction coil (Fig. 4.3) through pad electrodes applied to the skin over the nerve. With shocks that are not so large as to cause pain all that is felt is a vague tingling and sensation of tightness over the area of skin distribution of the nerve, not unlike "pins and needles". The feelings aroused never correspond to any sensation that could result from natural stimulation of the skin surface. This failure to evoke recognisable sensations cannot be due to any inadequacy in the total number of impulses but must be because of their unnatural arrangement. Apparently the reflex centres of the spinal cord likewise cannot make sense of the synchronous volleys of impulses that result from electrical stimulation of skin nerves and they therefore give no reflex responses. With stronger shocks the situation alters; in the human subject pain sensations are aroused which are recognised as similar to those caused by ordinary painful stimuli; likewise in animals ordinary pain reflexes are elicited.

The general conclusions from the foregoing are that the reflex centres only recognise patterns of afferent impulses which correspond to naturally occurring physical stimuli, and that these stimuli, when recognised, only elicit definite movements, each of which is an appropriate response to its stimulus. Nothing definite is known of the details of the connections between nerve cells that are responsible for the properties summarised in this paragraph.

Spinal reflexes

The central nervous system is so immensely complex that to analyse how it works is bound to be extremely difficult. On the intact animal it is not possible to proceed by the classical scientific method of varying only one factor at a time while all others are held constant, because an animal's response to a particular change in its environment often depends on things that happened to it long before the experiment started. Even its reflex responses tend to be variable and unpredictable, and may furthermore be disturbed by apparently spontaneous activity. These difficulties can be minimised by confining attention to the reflex responses of the spinal cord after it has been severed from the brain to exclude voluntary and spontaneous interference. Often it is more convenient to remove the forebrain only, after section through the midbrain at the level of the colliculi, under temporary anaesthesia. Such *decerebrate* preparations breathe spontaneously (unlike spinal animals which have lost the medulla oblongata) and maintain a good blood pressure.

For a time after spinal transection it is impossible to elicit any, or almost any, reflexes from the isolated part of the spinal cord below the section. This is called *spinal shock*. After shock passes off, the reflexes tend to become brisker and more easily provoked than they were originally. The higher the animal is in the evolutionary scale the more profound the shock, the longer it lasts, and the smaller the number of reflexes that eventually reappear, although ultimately some of these reflexes may be greatly enhanced. The decapitate frog recovers from spinal shock in a minute or two. In the cat or dog some reflexes reappear in a matter of minutes; most of those that are going to appear at all do so in a few days, but the ease with which they can be elicited may go on increasing for months after. In the spinal monkey and in cases of accidental division of the spinal cord in man (spinal man) the return of responsiveness does not begin for several days and progresses very slowly.

The nature of spinal shock is still unknown; it is clear that shock is a bad name for something that may last for years and that grades imperceptibly into enhancement. It is not the mechanical damage that causes shock for, in an animal that has recovered from shock, a second section immediately behind the first causes no return of shock. Sherrington showed in the monkey that at a time when the lower limbs were entirely unresponsive to massive stimuli (such as repetitive electrical stimulation of large nerve trunks or extensive burning of the sole of the foot) which would normally cause violent limb reflexes, movements could still be obtained with ease by exciting electrically the tracts of the spinal cord exposed at the cut surface. This suggests that the spinal neurones are not so much inexcitable as deprived of excitation normally reaching them from higher levels in the nervous system. (This would explain why there is no trace of shock in any part of the nervous system in front of the section.) The more advanced the animal the more its behaviour is dictated by its brain and the larger the deficit of excitation in the spinal cord after it is cut off from cerebral influences. During recovery certain reflex pathways, in some unknown way, increase their power of exciting and overcome the deficit. The larger the deficit the longer it takes to fill. At any rate in quadrupeds, much of the excitation that is missing comes from the midbrain and hindbrain; for decerebrate preparations, lacking the forebrain only, are usually held not to show signs of shock. They are certainly much more active than spinal preparations. That is another of their advantages.

The reflexes that recover after spinal section seem only to be those that in the ordinary life of the animal are little or not at all under voluntary control. In man almost every action is voluntary or requires voluntary permission, and a spinal man is left, below the level of the lesion, with little more than flexion reflexes (p. 398), tendon jerks (p. 401) and a number of visceral (sympathetic and parasympathetic) responses, notably automatic emptying of the bladder and rectum, sexual reflexes (erection and seminal emission) and vasomotor responses to pain and to a full bladder Many of these become unnaturally brisk and easily provoked. In a proportion of cases a stretch reflex (p. 420) appears and after some years brief periods of spinal standing may be possible. In a spinal dog or cat, apart from similar visceral responses there are a large number of protective reflexes (of which the best known are the flexion reflex and the scratch reflex (p. 399), some rather halfhearted attempts at standing and walking, several fragmentary reflexes of less easily recognisable purpose, such as the extensor thrust and a stretch reflex which appear to be elements of postural or locomotor mechanisms or of the nervous mechanisms that control muscular contraction, and little else.

It is a price that the experimenter has to pay for getting all the variables of the experiment under his control that the isolated spinal cord of the ordinary laboratory mammals does not have a very large or interesting repertoire of reflexes, but to offset this the simplicity of spinal reflexes has been turned to great advantage for purposes of analysis, and most of what is known of the detailed behaviour of nervous pathways has come from the study of a few spinal reflexes. Sherrington, who pioneered in this field in the last years of the nineteenth century, was impelled by the belief that reflexes were the units of nervous action and that more complicated types of behaviour would prove to be compounded of simple reflexes. Although what he and his pupils discovered about individual reflexes and the way reflexes interact (for instance the way in which the protective flexion reflex overrides, or inhibits, other reflexes) is of very great interest and importance, the attempt to show that complex acts like walking are made up of a succession of simpler reflexes was not equally successful and may (as is further discussed on pp. 399 and 405)

prove to be mistaken. Hence the reader must not be disappointed to find that a knowledge of what goes on in the simpler spinal reflexes, of the nature of synaptic action and so forth, does not throw much light on the mechanism of more complex reflex actions, let alone on the mode of functioning of the higher parts of the brain.

The reflexes which have been investigated in greatest detail and from which most generalisations about reflexes are conventionally drawn are the flexion reflex, the scratch reflex and the stretch reflex. All these reflexes can be obtained in spinal animals, but it should be born in mind that, when similar or related reactions are studied in intact animals or human subjects or in decerebrate animals, it is not in general possible to be certain that the reflex pathway remains exclusively spinal.

The flexion reflex

The flexion or withdrawal reflex of the limb to a painful stimulus has already been referred to more than once; some of its characteristics as seen in the cat will now be described more fully. The reflex exhibits a threshold; to elicit it any stimulus must reach a certain intensity. The interval between the time of application of the stimulus and the start of the reflex contraction (the latency of the reflex) is of the order of 30 milliseconds, much longer and more variable than the latency of contraction when a stimulus is applied to a motor nerve. A stimulus is more effective the larger the area it covers and the longer it lasts (spatial and temporal summation). As the stimulus is increased in intensity the flexion movement becomes more vigorous and more extensive; thus a weak stimulus may cause movement of toes and ankle only, while with a very strong stimulus the knee and hip flex as well (irradiation). With strong stimuli the limb may remain flexed for a second or more after the stimulus ends (after-discharge). The words in brackets are the special terms introduced by Sherrington.

In the most clear-cut demonstrations of the above properties Sherrington and his colleagues used electrical stimulation of peripheral nerves to elicit the reflex and recorded the contractions of an individual muscle (tibialis anterior) by detaching its tendon from its insertion and connecting it instead to an isometric lever (Fig. 6.1). The reflex latency could be measured easily with this arrangement. After making allowance for conduction time to and from the spinal cord the delay in the cord itself was found to be a minimum of about 5 ms (usually it is very much longer). Spatial summation was shown by applying shocks to two nerves, both shocks just too weak by themselves to elicit a reflex (sublimal), but successful if applied together. When both shocks were large enough to produce small reflex contractions, applying them together gave a tension greater than the sum of the tension produced by stimulating the two nerves separately (spatial summation again). When, however, both shocks were increased to give the largest possible reflex contractions, applying both together often gave a tension little greater than with either separately (occlusion) because the muscle was already giving a nearly maximal contraction with the single stimuli. Temporal summation was shown by applying two sublimal shocks in rapid succession to the same nerve; when the interval between them was less than some 20 milliseconds a reflex contraction resulted.

The fully developed flexion reflex is accompanied by extension of the opposite leg, the knee and ankle straighten out and the limb is thrust backwards at the hip. This *crossed extensor* reflex has a much longer latency (40–100 milliseconds) than the flexion reflex proper, and once it starts it takes a second or two to build up to its greatest strength (recruitment), in contrast to the rapid onset of flexion. Crossed extension still occurs when the dorsal roots of that limb have been cut. The crossed extensor reflex probably represents the first part of a step away from the noxious stimulus.

Temporal summation, as described above for the flexion reflex, implies that the first shock causes a change in excitation somewhere on the pathway of the reflex which persists for several milliseconds to summate with the effect of the second shock. Sherrington called this persisting change "central excitatory state". It seems to have been taken for granted that the alterations of central excitatory state during the flexor reflex took place in the motoneurones of the flexor muscles—in the "motoneurone pool", as it was termed. Electrical recording with fine electrodes inside the spinal cord has now shown, however, that the afferent nerve fibres that elicit the flexion reflex (and in fact all skin afferents), end on interneurones in the dorsal horn of grey matter (Fig. 59.1) and do not themselves run through to the motoneurones in the ventral horn.

A reflex contraction of tibialis anterior can be produced by exciting any one of a number of peripheral nerves in that leg. Sherrington pictured impulses from these various sources "converging" on to tibialis motoneurones, which formed a "final common path". (As we have just seen, the site of convergence is more likely to be the interneurones of the dorsal horn.) The concept of the final common path found wide application in his writings, but nowadays, although true in a certain formal sense, it is a less illuminating doctrine than it previously appeared. For one thing, by implying that in reflex action afferent impulses find their way through the spinal cord eventually to emerge via the motor roots, it speaks the language of simple reflexion and therefore tends to distract attention from the view preferred here (and endorsed incidentally by Sherrington himself) that the spinal cord "thinks" in terms of organised movements; other objections will appear later.

Withdrawal reflexes in man

Withdrawal reflexes in man can be elicited from the leg of a healthy subject by painful stimuli applied to the skin. Kugelberg and his co-workers have shown that the precise kind of movement obtained varies with the site of stimulation. As a convenient and reproducible stimulus they used a burst of high voltage electric shocks delivered through a pair of electrodes pressed against the skin. With the stimulus applied to the underside of the big toe the response obtained was flexion at all joints: upward movement of the toes, flexion of the ankle, knee and hip and forward flexion of the trunk, as illustrated in Fig. 59.4. When the stimulus was moved further back on to the sole of the foot the toes moved downwards instead of upwards, the other joints flexing as before. The same response can be obtained by firmly stroking the sole. With the electrical stimulus under the heel the toes again moved downward but now the ankle extended too, the other joints again flexing. With the stimulus on the buttock, there was extension of the trunk and hip, some flexion of the knee, extension of the ankle, and downward movement of the toes, as shown in Fig. 59.4.

The effect of these movements in each case is to withdraw the stimulated point from the stimulus; but in addition they contain components which are clearly not part of the withdrawal, downward movement of the toes and extension of various joints.

In a standing subject, the effect of these components of the reflex would be to press the toes downwards and assist to maintain standing. This only does not occur when the stimulus is applied

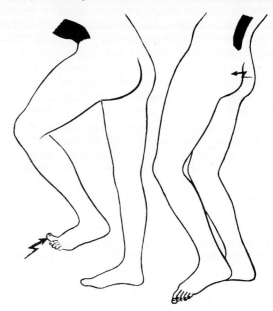

Fig. 59.4 Withdrawal responses to painful stimuli applied to the underside of the big toe and to the buttock. The filled-in areas mark the trunk muscles that contract. (Kugelberg, Eklund and Grimby (1960) *Brain*, **83**, 394.)

to a toe, when it would be inconsistent with effective withdrawal. Hence it is thought that the reflex responses of the human leg to painful stimuli have the dual function of defence with maintenance of posture. These responses, unlike withdrawal reflexes in spinal animals, may not be wholly spinal; reflex pathways to and from the brain cannot be ruled out.

When the corticospinal motor tracts (p. 412) from the brain (without which a man cannot stand) are damaged, the components of the withdrawal response which appear to assist standing disappear. Electrical stimuli to the sole of the foot in such patients gave exactly the same response (i.e. upward movement of the toes and flexion at all other joints) as stimulation of the underside of the toe in healthy subjects. Firm stroking of the sole of the foot in such patients also causes an upward movement of the toes. The same response is found in infants before they begin to stand up, and before the corticospinal tracts receive their myelin sheaths. The interpretation usually put on these facts is that long pathways from the brain modify the behaviour of the spinal reflex arcs. An alternative interpretation, hinted at above, is that they establish new reflex arcs to the brain and back, which dominate or suppress the primitive spinal responses. We shall return to this possibility in connection with the stretch reflex.

In spinal man an upgoing big toe accompanied by flexion of the leg is one of the first reflexes to appear during recovery from spinal shock. Over a period of months the threshold of the reflex falls and flexion becomes more vigorous, until eventually violent flexion may result from trifling stimuli anywhere on the limb.

The response of the big toe to a firm stroke on the sole of the foot with a blunt point (the plantar response) is of great importance in clinical diagnosis, as an upgoing big toe is often the first unequivocal evidence of disease of the motor pathways (pp. 412, 414). The sign was first described by Babinski in 1896. An upgoing big toe is part of the general flexion reflex of the limb, but, although morphologically a movement of flexion, in anatomical nomenclature it is extension. Hence an upgoing big toe is referred to by neurologists as an extensor plantar response. A healthy plantar response is said to be flexor.

The scratch reflex

Some days or weeks after spinal section in the cervical region, a dog will respond to irritation of the skin over its shoulder by scratching at the place with its hind foot. The movements made (Fig. 59.5) closely resemble the familiar scratching of a normal dog, consisting of a series of rhythmical strokes of the whole leg at a frequency of about five per second. A good form of stimulus is a series of light touches with a pointed object. A single touch never succeeds; several dozen (at a rate of a few per second) may be necessary if they are all applied at one spot, but the total number diminishes if two or more spots are touched simultaneously; the most effective method is to touch a row of spots consecutively, thus imitating the pattern of stimulation normally caused by a moving flea. Once scratching has begun it continues for some seconds after the stimulus is withdrawn. (Readers are invited to apply to the above properties the Sherringtonian terminology given with the flexion reflex.)

The scratch reflex cannot be elicited by electrical excitation of cutaneous nerve trunks or of dorsal roots; apparently such volleys are not recognised as objects to be scratched at; but small shocks applied through a fine pin (electric flea) pushed just into

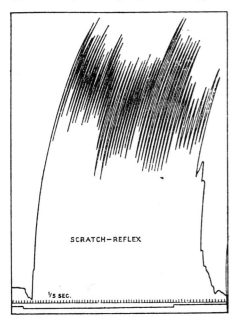

Fig. 59.5 The scratch reflex in the spinal dog. The amount of the hip flexion recorded on a smoked drum by means of a string connected to the writing lever. The reflex was elicited by electrical stimulation of a point on the skin on the animal's back. The period of stimulation is shown by the descent of the signal line beneath the time scale. (Sherrington. *The Integrative Action of the Nervous System* New York 1906; Cambridge 1947.)

the epidermis are successful. Not very surprisingly perhaps the frequency of scratching movements proves to be independent of the frequency of the electric flea bites that cause them. This is one of the usually quoted formal differences between (repetitive) movements elicited reflexly and by direct excitation of a motor nerve. Another is fatigability: if the scratch reflex is elicited several times from the same point of skin it becomes weaker and more difficult to obtain. It may be useful to the animal to disregard a source of irritation it does not succeed in removing, so possibly the phenomenon is more akin to adaptation than to what is ordinarily thought of as fatigue. At all events the "fatigue" is not of the motor part of the reflex mechanism, for if the stimulus is moved to another point on the skin the reflex immediately returns in full strength. (It is not in general true that reflexes are readily fatigable. The cough reflex may persist with undiminished sensitivity after it has wholly exhausted the subject. The stretch reflex (below) is also believed to be less fatigable than the muscles it employs.) If, during a scratch reflex, a painful stimulus is given to the leg, scratching immediately ceases and is replaced by a flexion reflex. This is an example of a protective reflex taking precedence by actively inhibiting a less urgent reflex.

The mechanism which generates the rhythm of scratching is not known. The frequency does not depend on the strength of the stimulus, or on the amplitude of the beat. Sherrington showed that if the amplitude is reduced by a carefully graded painful stimulus to a leg nerve, not quite strong enough to inhibit scratching completely, the frequency is unaltered. He also found that cutting the dorsal roots of the segments supplying the limb reduced the accuracy with which the scratching was directed to the point irritated, but did not alter the frequency. These experiments suggest that the rhythm is determined by a pacemaker in the spinal cord rather than by a sequence of flexor and extensor reflexes set up alternately by the moving limb itself; we shall see later (p. 405) that there is similar evidence for the slower rhythm of stepping.

The stretch reflex and the tendon jerk

So far we have dealt only with reflexes of obvious functional significance, elicited from the skin, the nose and throat, the cornea, the bladder, etc. In all cases the stimulus for the reflex would also have caused a conscious sensation in a normal human subject—pain, touch, tickle in the throat, etc. We now come to a very different reflex found in muscles. Muscles, like skin, are supplied with numerous sensory nerve fibres connected to sensory receptors, but unlike skin the impulses set up by these receptors do not (according to the view adopted here) give rise to conscious sensations. The knowledge we possess of the position of our limbs in space, position sense, comes primarily from sense-endings in and around the joints, not from muscle.

The evidence for this statement is, first, that the eyes, which of course have no joint receptors, are without position sense, as was clearly understood by Helmholtz a century ago. Yet human eye muscles contain numerous sense-endings, including muscle spindles, which must necessarily be stimulated by these manipulations. Secondly, in the limbs position sense is lost, or greatly impaired, when the tissues surrounding a joint are rendered anaesthetic, even though the muscles acting at the joint, whose length alters when the joint is passively moved, are outside the anaesthetic region.

Strictly, receptors and nerve fibres that do not cause sensations should not be called sensory, but this usage is accepted along with "sensory root" for dorsal root, etc., and causes no trouble if care is taken to make the intended meaning clear when ambiguity might arise. Other familiar examples of sense-endings that do not affect consciousness are the blood-pressure receptors in the carotid sinus and the vagal receptors signalling inflation of the lungs.

Sense endings in muscles and those associated with joints were classed together as "proprioceptors" by Sherrington. The term is not used here as it tends to obscure the distinction that ought to be drawn between their functions.

The receptors in muscles elicit reflexes, but, again unlike the reflexes from the skin, they are not of unequivocal function. The most important and best understood muscular reflex is the stretch reflex: when a muscle is stretched by pulling on it a reflex contraction is set up which opposes the pull. The stretch reflex can be demonstrated on intact animals and human beings; it only reappears in the spinal dog after some weeks and in spinal man after a year or two. In a decerebrate preparation, however, the stretch reflex is present in the extensor muscles at once; indeed it is exaggerated, leading to a state known as *decerebrate rigidity*, which is mentioned again on pages 404, 415

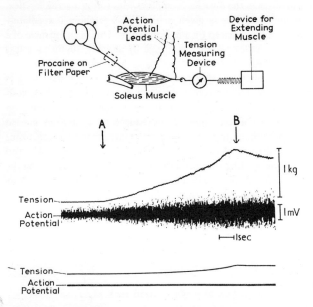

Fig. 59.6 The stretch reflex in the soleus muscle of a decerebrate cat. The diagram above shows the experimental arrangements. The bones of the leg are clamped to a stand and the severed tendon of the soleus muscle connected to a device for recording muscle tension. To stretch the muscle the whole tension recording device is moved by means of a motor driven screw. The electrical response of the muscle is led off by electrodes on the muscle. (The application of procaine refers to Fig. 59.9.)

The records below show the rise in tension that occurs when the muscle is extended 13 mm at a rate of 1·7 mm per second. Extension starts at A and stops at B. At the same time a great increase in the electrical activity in the muscle occurs, showing that the increase in tension is associated with an active reflex contraction of the muscle. The lower records, taken during a similar extension after the stretch reflex had been abolished, show the absence of electrical response and the much smaller rise in tension caused by the passive elastic properties of the muscle. (P. B. C. Matthews, unpublished records.)

The stretch reflex was named, and its properties in the decerebrate cat described, by Liddell and Sherrington in 1924. It is chiefly present in the extensors, the muscles that straighten the limbs. When such a muscle, e.g. the soleus, one of the ankle extensors, is cut away from its insertion and connected via its tendon to a device for measuring tension it is found that any attempt to elongate the muscle results in a rapid rise in the tension it develops, and at the same time electrodes on or in the muscle record numerous action potentials (Fig. 59.6). This active contraction is reflex, because it disappears if either the motor nerve to soleus or the appropriate dorsal or ventral roots are cut through. Stretching the muscle after cutting its nerve then only causes a much smaller rise in tension due to the simple passive elastic properties which muscle, like any other piece of soft tissue, possesses. It is clearly established by other experiments that it is the change in length of the muscle that excites the stretch reflex and not the change in tension. Except when reflex excitability is artificially raised the stretch reflex is private to the muscle stretched, no other muscles contract; and it persists in that muscle when the whole skin of the limb and every other muscle in the limb is denervated by systematic section of the nerves.

In human patients the neurologist believes that he detects the absence, presence or exaggeration of stretch reflexes by assessing the "sense of passive resistance" he obtains when he moves the joints of the limb about while the patient relaxes and initiates no voluntary movements of his own. To express the results the word "tone" is used. The normal degree of passive resistance is called normal tone; diminished sense of passive resistance is diminished tone, the muscles are hypotonic or flaccid; increased sense of passive resistance is increased tone with hypertonic or spastic muscles. It cannot be regarded as certain that all normal or pathologically altered tone is due to stretch reflexes. We return to the stretch reflex in human muscle at the end of the next chapter. This chapter mainly sticks to the stretch reflex in animals.

Opinions about the function of the stretch reflex are undergoing revision. It was once supposed that the stretch reflex was a characteristic of postural antigravity muscles, but it is now clear that the stretch reflex is active during reflex and voluntary as well as during postural contractions, and occurs in flexor as well as in extensor muscles. The view adopted here is that the stretch reflex is part of the general nervous machinery controlling any muscular contraction. The role of the stretch reflex in posture is discussed on pp. 405 and 420

The tendon jerk, of which the best known example is the knee jerk, was discovered and put into use in medicine independently by Erb and by Westphal in 1875. It is now believed to be a transient manifestation of an element in the stretch reflex which, as we have seen, was only described in its full-blown form a half a century later. To elicit a knee jerk the subject relaxes with his knee bent and the tendon below the knee cap is struck a sharp blow with a rubber-covered hammer. The extensor muscles of the knee respond with a brief twitch-like contraction, the jerk. The jerk illustrated in Fig. 59.7 is an ankle jerk. The jerk is reflex, for it disappears if either the sensory or motor fibres to the muscles are interfered with. The latency is very short, about 20 milliseconds for the human knee jerk, little more than is needed for conduction time to the spinal cord and back. Experiments on animals show, in fact, that the central delay is less than a millisecond, time for only one synaptic junction to be passed. The afferent impulses which excite the reflex arise in the fleshy part of the muscle, not in the tendon; striking the tendon is effective only because it causes a sudden rapid slight elongation of the muscle. The initial rapidity of stretch is the essential feature; slower stretch of larger amplitude never elicits a jerk. The initial rate of stretch needed to elicit a tendon jerk is probably greater than is normally imposed on the muscles by movements of the joints, or than is used by physiologists to elicit the stretch reflex.

Several other muscles with conveniently accessible tendons exhibit tendon jerks, notably the ankle extensors (the Achilles tendon) and the biceps and triceps muscles of the upper arm. The tendon jerks are of great clinical importance because alterations in them are one of the most sensitive objective indexes available of disease of the nervous system. They disappear if any part of their reflex arc is put out of action; and they are increased if the long motor tracts (p. 412) connecting the brain with the spinal cord are damaged, probably for the same (unknown) reasons that reflexes are augmented after complete spinal section.

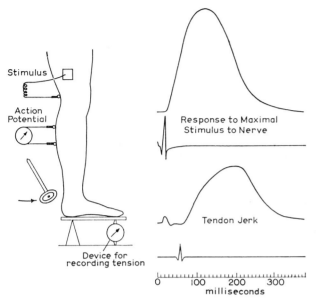

Fig. 59.7 The human ankle jerk. The reflex, which is obtainable in almost all healthy persons, is elicited by striking the tendon of the ankle extensors (the soleus and gastrocnemius muscles) a sharp blow with a rubber-covered hammer; it consists of a twitch contraction of these muscles. To record the contraction, the root rests on a board hinged under the heel; contraction of the ankle extensors causes an increase in the downward pressure exerted by the front of the foot on the board, which is recorded on a cathode ray oscilloscope. The action potential of the extensors is led off by surface electrodes on the calf and a record of it appears below the tension record.

The oscilloscope time base is triggered by contact of the hammer with the skin over the tendon. The blow itself causes a brief rise in tension which appears as a small hump before the reflex twitch begins. The latency of the reflex in this subject, measured from the contact of the hammer with the skin to the start of the action potential, is 47 milliseconds. Other evidence shows that of this some 35 milliseconds is nerve conduction time from the muscle to the spinal cord and back again, together with a brief central delay.

For comparison a maximal motor twitch in the extensors, elicited by an electrical stimulus to the motor nerves in the popliteal fossa, is shown above. The time base in this record is triggered by the stimulus. The latency is much shorter and the action potential and twitch tension are larger. But the duration of the action potential and the twitch are much the same as in the reflex jerk, showing that the reflex discharge itself must be a highly synchronous volley of nerve impulses. (P. A. Merton, unpublished records.)

More is known about the nervous mechanism of the stretch reflex than about any other reflex, but even here nothing like a complete account can yet be given, and several fundamental properties still lack explanation. The receptors that excite the reflex are the muscle spindles. Muscle contains two special types of receptor: the Golgi tendon organs lie in the tendons and in tendinous bands and aponeuroses within the muscle, and respond to a rise of tension whether this is due to contraction of the muscle fibres or to externally applied stretch. The muscle spindles are bundles of modified muscle fibres (intrafusal fibres) with sensory endings wrapped around a short length of the bundle (Fig. 59.8). Impulses are set up when the sensory portion elongates. The muscle spindles lie among the ordinary (extrafusal) muscle fibres and share their attachments; they therefore change length only when the muscle changes length. Hence they signal changes of muscle length. The muscle may shorten in length either because it contracts, in which case the tension rises, or because the load is reduced, in which case the tension falls. Thus tension and length can alter independently in an actively contractile structure (which they cannot do in a

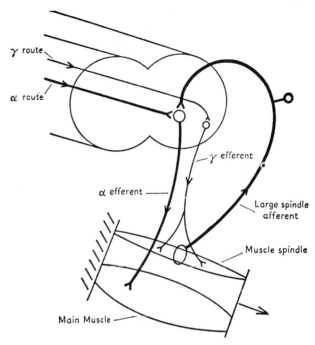

Fig. 59.8 Diagram of the mechanism of the stretch reflex. When the muscle is pulled on, the sensory portion of the muscle spindle is stretched. As a result impulses are sent up the spindle afferent fibre, which reflexly excite the motoneurone of the muscle. Impulses are thus sent down the large motor fibre (a efferent) to the muscle and it contracts.

The sensory portion of the muscle spindle (which is itself noncontractile), is also stretched when the contractile ends of the muscle spindle contract. This they do when motor impulses reach them via their special motor nerves, the γ efferents. Contraction of the ends of the muscle spindle, due to impulses arriving along the γ route, reduces the amount of externally applied stretch that has to be applied to elicit the stretch reflex; or it may even initiate contraction of the main muscle, via the stretch reflex pathway, without any applied stretch. (Hammond, Merton and Sutton (1956) *British Medical Bulletin*, **12**, 214.)

passive structure such as a spring), so separate sense organs are needed to measure length and tension.

Impulses from spindle sensory endings travel in afferent fibres conducting with a wide range of velocities. The fast ones are among the fastest in the body; they belong to the a division of the A group (Fig. 4.15), which also includes the motor fibres. The spindle afferents run right through the dorsal horn of grey matter to make synaptic contact directly with the dendrites and cell bodies of the motoneurones of the muscle from which they come, thus forming what is called the monosynaptic reflex arc. In a tendon jerk the sudden extension produced by the tap sets up a synchronous volley in the fastest monosynaptic afferents which, when it arrives, excites a substantial fraction of the motoneurones belonging to the muscle stretched, and a twitch results.

A similar reflex occurs if the monosynaptic afferents are excited by an electrical stimulus instead of by a tendon tap. Such a *monosynaptic reflex* can seldom be obtained merely by applying electric shocks to a motor nerve, because a shock large enough to excite the monosynaptic afferents also excites the motor fibres which are in the lowest threshold, fastest conducting group too. As usual, impulses are propagated in both directions from the point of stimulation, so that as well as a volley to the muscle another passes backwards up the motor fibres to the spinal cord (an antidromic volley) at the same time as the volley goes up the monosynaptic afferents. The antidromic motor volley has the effect of inhibiting the motoneurones (antidromic block), so no reflex gets through. To avoid this in animal experiments it is conventional to cut the ventral roots so that the antidromic volley does not reach the cord, and to record the action potential of the reflex volley leaving the cord by placing electrodes on the central stump of the ventral root.

In the human subject, a carefully adjusted electrical stimulus to the popliteal nerve behind the knee (see Fig. 59.7) can be made to excite monosynaptic afferents from the ankle extensors without exciting motor fibres. The resulting monosynaptic reflex is known as an "H reflex" after Paul Hoffmann, who discovered it in 1918. H reflexes are rarely to be seen in other muscles.

In the decerebrate cat, the stretch reflex proper is excited by both fast conducting ("primary") and slow conducting ("secondary") spindle afferents. The participation of the secondary spindle endings has only been established in the last few years. The monosynaptic arc, which is certainly the pathway for the tendon jerk, is open to both fast and slow afferents in the stretch reflex; but it is thought likely that slower spinal pathways involving interneurones are also in use. In primates a long-latency stretch reflex via the cerebral cortex is a distinct possibility, but one which we need not consider at this stage. Thus on at least two counts the stretch reflex is a more complicated affair than the tendon jerk. The terminology should take account of this; hence it is best not to refer to tendon jerks as stretch reflexes, although strictly they are a variety of reflex response to stretch.

We now come to the key fact which lifts the stretch reflex onto a higher plane of interest than the other reflexes so far considered. It is that the muscle spindles which excite the stretch reflex are themselves contractile, the only contractile sense organs in the body. This has been known since 1894, but its full significance is not yet understood. The intrafusal muscle fibres, except in the zone where the sense-endings lie, are essentially fine striated muscle fibres. They receive a motor innervation from motoneurones lying near the ordinary extrafusal motoneurones in the ventral horn, and when motor impulses reach them they contract. The sensory portion is non-contractile or less contractile, so that when the remainder of the spindle (the two "poles") contracts the sensory part is extended. This causes an increased discharge of sensory impulses, just as if the extension of the sensory ending had been due to extension of the whole muscle. (Contraction of the spindles does not cause the muscle as a whole to develop any appreciable tension, for the total cross section of intrafusal fibre is minute.) The motor nerve fibres that run to the intrafusal muscle are of smaller diameter than those going to extrafusal fibres. They belong to the γ division of the A conduction velocity group, and for this reason they are generally called "γ motor fibres"; those to the main contractile (extrafusal) muscle fibres are "α motor fibres".

Extensions of the sensory endings caused by contraction of the spindles and by stretch of the muscle are additive. It follows that, if the spindles are in a state of steady contraction owing to a stream of impulses reaching them by the γ fibres, less stretch ought to be necessary to elicit the stretch reflex than when the spindles are relaxed. Experiment has shown that this is so. It has been found by direct recording from single α and γ motor fibres, and in other ways, that even when a muscle is quite relaxed (no α discharge) there is a continuing discharge of impulses in γ fibres. The discharge is much more intense in a rigid decerebrate cat than in a spinal cat; the former has readily obtainable stretch reflexes, the latter little or none. The intermediate stages can be demonstrated elegantly by making use of the property of the drug procaine of blocking conduction in small nerve fibres before it blocks large ones (Chapter 4). When a drop of procaine solution of the right strength is put on a bared motor nerve it is possible slowly to block the γ fibres without affecting the α fibres. While the γ fibres to an extensor muscle of a decerebrate cat are being blocked in this way the extension required to elicit a stretch reflex increases progressively as block deepens (Fig. 59.9). The spindles are relaxing and a greater stretch is needed to produce the same rate of spindle discharge.

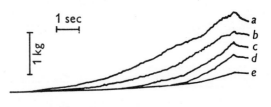

Fig. 59.9 The effect of paralysing the muscle spindles on the stretch reflex of a decerebrate cat. The figure shows superimposed tracing of the tension developed in the soleus muscle during a series of identical extensions, each of 15 mm at a rate of 1·5 mm/sec. The duration of each extension is indicated by the signal line beneath the records. The experimental arrangements were as in Fig. 59.6. The spindles were gradually paralysed by blocking the γ efferents with procaine. Procaine solution (0·2 per cent) was applied to the motor nerve to soleus on a piece of filter paper (see Fig. 59.6).

Record a was obtained before applying procaine; records b, c and d were obtained when the nerve had been exposed to procaine for 4·5, 6·5 and 10 minutes respectively; record e shows the tension developed in the passive muscle when the stretch reflex had been completely abolished by another method.

After record d the tetanic tension developed by the muscle to electrical stimulation of the nerve central to the procaine block was no less than it was before applying procaine. Hence the diminution in the stretch reflex was not due to blockage of α motor fibres. (P. B. C. Matthews (1959) *Journal of Physiology*, **147**, 547.)

The degree of contraction of the spindles is said to control the "bias" of the stretch reflex. It is suspected that much of the unresponsiveness in spinal shock may be due to lack of bias, "spindle paralysis", due to a cutting off of excitation from the higher parts of the nervous system to the γ motoneurones; conversely, decerebrate rigidity is thought to be associated with "spindle cramp". Whether the increased stretch reflexes and tendon jerks found in muscles rendered spastic by disease of the long motor tracts in man are also due to increased spindle bias is another matter to which we shall return (p. 403).

So far nothing has been said about the part played by the Golgi tendon organs. Various pieces of evidence show that impulses from the tendon organs can inhibit the motoneurones of the muscle from which they come, but it is not known to what extent this occurs in the normal stretch reflex. The nearly straight-line relation between extension of the muscle and the tension it exerts, seen in Figs. 59.6 and 59.9, has suggested that the function of the stretch reflex is not merely to pull back as hard as may be when the muscle is stretched, but is to set this particular relation between length and tension. How this is achieved is not known, but clearly a mechanism that causes the muscle to develop a particular tension at each length might require information about tension from the tendon organs as well as information about length from the spindles. Such a function, however, is at the moment speculative; the accepted role for the tendon organs is in the non-linear properties of the stretch reflex. If the knee of a decerebrate cat or a spastic man is forcibly bent, at first the exaggerated stretch reflex resists powerfully, but at a certain point resistance collapses. This is called the "clasp-knife" response. It is attributed to impulses from the Golgi tendon organs. There is some evidence that the inhibition acts on the γ motoneurones causing a sudden relaxation of the muscle spindles. The significance of the response is not clear; it is sometimes said to be protective against excessive tension, but this (as Sherrington himself saw) it clearly is not, for it may occur at tensions which must often be exceeded in ordinary life and are obviously not dangerous. Clasp knifing has not been noticed, but has not been specifically searched for, in healthy human subjects when powerful stretch reflexes are evoked by stretching muscles that are already engaged in voluntary contractions (p. 421).

THE SERVO THEORY OF MUSCLE CONTRACTION

It has long been clear that the stretch reflex, by feedback action, confers valuable self-regulating properties on the muscle. Thus in a steady postural contraction the muscle would automatically adjust to an increase in load by contracting harder as it was put under strain. There would also be automatic compensation for fatigue, for if the muscle flagged and started to give under the load, this again would call up extra contraction to arrest the process. To an engineer the ordinary stretch reflex is a servomechanism tending to hold the muscle at a certain demanded length by using negative feedback from change-of-length detectors in the muscle. It was proposed in 1952 that the advantages of this system could be retained during movement if the muscle shortening was driven via the stretch reflex arc by causing the spindles to contract. Impulses coming down the γ-route (Fig. 59.8) would make the spindle poles contract and thereby stretch the sense endings and activate the stretch reflex. The muscle would contract and, if not prevented by the prevailing load, would go on shortening until it had offset the contraction of the spindle poles and slowed down the spindle discharge again. Such a system is a "follow-up" length servomechanism, the stretch reflex automatically turning on more contraction to cause the main muscle to follow changes in the length of the spindle poles. Contraction of the spindle is equivalent to changing the demanded length. Thus we arrive at the idea of a servomechanism based on the stretch reflex which allows the motor centres to order a certain change of length, which furthermore will be carried out with the same automatic adjustment of tension to load that is observed in the ordinary stretch reflex. In many actions what is required is a change of limb position, i.e. changes of muscle length. A muscle, on the other hand, is essentially a device for producing tension; feedback from the muscle spindles on to the motoneurones appears to modify the characteristics of the system so that when excited via the γ-route it tends instead to produce changes of length, by servo action.

There is certainly a great deal that is not yet discovered about the functions of muscle sense organs, and how they are employed to make the muscles do what is required of them. We are not likely to be led astray in thinking that one of their functions is to confer new properties on the muscle by feedback action (in the same way that feedback confers new properties on an amplifier), even if we are not yet sure quite what those new properties are. At all events this statement ought to make clear why the stretch reflex is coming to be thought of as part of the muscles' own nervous machinery; and why, when this machinery depends on excitation of the γ motoneurones, in addition to the α motoneurones, it is missing the point (if not exactly incorrect) to think of all reflex and voluntary pathways as converging on to the "final common path" of the α motoneurones.

As to evidence for the servo theory, it has often been observed that an increased rate of discharge in gamma fibres precedes the acceleration of α-discharge during reflex and other movements in experimental animals. Furthermore, recording from single spindle afferents during reflex contractions in ordinary decerebrate animals has shown that the spindle discharge accelerates too during shortening. If the spindle were not contracting in advance of the main muscle, they would slow down, for the effect of the shortening would be to unload the sensory region (as can be understood from Fig. 59.8). Another cogent finding is that, with the experimental technique of Figs. 59.6 and 59.9, eliciting a crossed extensor reflex (p. 397) in soleus has the effect of shifting the curve of tension against time in a stretch reflex to the left, instead of to the right as occurs with procaine block. Hence the bias on the stretch reflex increases during the crossed extension. In human subjects, the remarkable technique developed by Hagbarth and Vallbo for recording from single nerve fibres in the median nerve has shown that a vigorous acceleration of muscle spindles occurs during voluntary contractions, although not in advance of the α discharge to the main muscle at the start of a contraction, as the theory would require.

All these observations confirm that spindles contract during movement, but they are inconclusive for they do not allow us to gauge how important this is in driving the muscles. Recent experiments on human subjects, to which we return in the next chapter, reveal that the servo theory as described above needs elaboration and as regards the initiation of voluntary movement was wrong, as Vallbo's results show.

Direct and servo activation. The two kinds of decerebrate rigidity

It is anyway clear that not all muscular contractions are based on the stretch reflex. It is well known that many spinal actions can be obtained after cutting all the dorsal roots to the moving limb, which necessarily abolishes all stretch reflexes in it. Examples already given are the crossed extensor reflex (p. 397), the scratch reflex (p. 399) and spinal walking (p. 397). Voluntary movements can also be carried out by a deafferented limb. In these instances all the excitation must be reaching the α motoneurones direct. Granit and his collaborators have shown that even with the dorsal roots intact the nervous system has the choice of employing this direct α route instead of turning on the stretch reflex by exciting the γ motoneurones. Which route is employed appears to depend on the cerebellum. The evidence for this derives from experiments on decerebrate rigidity, which must now be discussed.

Decerebrate rigidity develops in the cat, dog, etc. when the midbrain is sectioned behind the superior colliculus. As described by Sherrington in 1898, such rigidity disappears in a limb if the dorsal roots to that limb are cut because, as we now know, it depends on exaggerated stretch reflexes. Much later it was observed that, if the anterior lobe of the cerebellum was removed too, an outwardly similar rigidity developed, but it was not abolished by section of the dorsal roots. It, therefore, could not be due to exaggerated stretch reflexes. The mechanisms in the brainstem and cerebellum responsible for these phenomena are not understood. (Magoun and Moruzzi and their colleagues have investigated the excitatory and inhibitory effects on spinal reflexes which can be obtained by stimulation of two regions in the reticular formation. An explanation of decerebrate rigidity is sought in terms of a disturbance in the balance of such antagonistic reticular influences.) At all events it is clear that there are two distinct kinds of decerebrate rigidity, and it turns out there are two kinds of reflex movements also. Single fibre recording has shown, as already described, that during reflex movements vigorous contractions of the muscle spindles occur in the ordinary decerebrate animal; after removing the anterior lobe of the cerebellum practically indistinguishable reflexes can be elicited but the spindles remain passive. Thus the reflex centres may activate the muscles in either of two ways. Probably varying combinations of the two routes are normally employed. Recent experiments on human subjects have given some information on the question (p. 420).

It is a matter of conjecture what the advantages of this double system are to an animal. The direct route may be better for starting movements because of the extra delays involved in conduction time with the γ-route. The desirable properties that the stretch reflex confers on a muscle have been emphasised already. It has been suggested that the dysmetria and failure of joint fixation in human cerebellar ataxy (see p. 417) may be due to a loss of the type of contraction driven through the γ motoneurones, to spindle paralysis in fact. One of the ways in which the cerebellum supervises muscular activity may be by making sure that the best route, the α-route or γ-route, is used to activate the muscle.

Reciprocal innervation

A common feature of movement, discovered and analysed at length by Sherrington around the turn of the century, is that when a limb moves, the muscles in it that would oppose the movement are caused to relax. This is known as *reciprocal innervation*; it is seen very clearly if a flexion reflex is elicited in the leg of a decerebrate cat. The rigid extensor muscles of the knee immediately relax; this occurs even if the flexors of the knee are detached from their insertion so that the knee is not moved, or even if the motor nerves to the flexors are cut so that they do not contract at all. The mechanism of reciprocal innervation appears to be in the spinal grey matter, but it is assisted as usual by reflexes from the periphery. Thus intracellular recording from motoneurones has shown that impulses from muscle spindles in flexor muscles inhibit extensor motoneurones. Movements elicited by electrical stimulation of the motor cortex also exhibit reciprocal innervation.

Reciprocal innervation was of great importance in the development of neurophysiology because it revealed to Sherrington the large part played by inhibitory processes in nervous action. Not all types of muscular action, however, employ reciprocal innervation. It is very clear that flexors and extensors can readily be made to contract together by voluntary effort, and the same thing occurs automatically to fix the joints in standing (p. 405). In these instances the limbs are stationary; reciprocal innervation comes into play when they move.

POSTURE AND LOCOMOTION

As already emphasised it is unrewarding to think of posture and locomotion as reflex responses; how they are carried out is the interesting question—largely unanswered, it is hardly necessary to add. The basic mechanisms for both standing and walking lie in the spinal cord, for spinal dogs and cats if allowed enough time to recover from shock will stand, and the hind limbs, supported clear of the ground, will perform indubitable walking movements in response to such stimuli as a pinch on the buttocks. Spinal standing and walking are imperfect performances; in order to stand and walk in a normal manner a cat or dog requires not only the spinal cord but the brainstem up to the midbrain. In an ordinary rigid decerebrate preparation the midbrain is sectioned between the superior and inferior colliculi. If the section is in front of the superior colliculus, so as to leave behind the red nuclei (Fig. 60.7), decerebrate rigidity does not result and the preparation (known as a midbrain animal) stands with a normal distribution of contraction between the extensors and flexors of the limbs. Why this should be is not known.

The midbrain is also necessary because it is involved in various reflexes that keep the animal balanced on its legs and restore it if it is thrown off balance. A spinal animal falls over very readily. These righting reflexes and other related reflex adjustments of posture are named after Magnus, who described them in detail in the 1920s and demonstrated the various sensory channels they employ. In the midbrain animal the most important are the vestibular balance organs in the labyrinth of the inner ear; they reflexly cause the head to be kept upright. The body reflexly follows the head, but for this to occur the angle between head and body must be known; this is signalled by joint receptors in the neck vertebrae. With both labyrinths destroyed a midbrain animal can still right itself using information from pressure receptors in its flanks. In most intact animals the eyes are of at least as much importance as the balance organs for righting; the rabbit is an exception—its eyes are of no help. The foregoing applies to quadrupeds; in the monkey and in man the basal ganglia are said to be necessary for effective righting and postural reactions to occur (p. 416).

Man has a more difficult balancing task than four-footed animals, particularly if he stands with his feet together; he uses his eyes and differential pressure sensations from the soles of his feet. If sensation in the feet is impaired (as it is for instance in alcoholic neuropathy) the patient falls over if he stands with his feet together and then shuts his eyes (Romberg's sign). The labyrinths do not provide sufficiently sensitive information to enable him to balance in this situation. In fact, urban man is little disabled by loss of labyrinthine sense; he can walk and run normally, recover his balance if he is bumped into, and even ride a bicycle. His chief disabilities, as regards equilibrium, are difficulty in walking over uneven or soft ground, particularly in the dark, and a dangerous loss of orientation in water. Such patients are also apt to complain that they cannot see distinctly unless they stand still. This is because of the loss of the important vestibulo-ocular reflexes whose function is to steady the eyes by compensatory movements when the head turns, so that fixation of external objects can be automatically held.

Apart from the general brainstem reflexes concerned with balance there are also individual mechanisms in each limb for throwing the muscles into an appropriate state of contraction when an animal is placed on its feet. Thus when a midbrain animal is held up in the air its limbs hang limply, but as soon as a paw makes contact with the ground the whole limb stiffens into a standing posture; both flexor and extensor muscles contract to fix the joints. This is called the positive supporting reaction. The stimuli immediately responsible are pressure on the pads of the foot and stretch of the muscles to the digits and feet. The positive supporting reaction, however, is by no means just a stretch reflex, for many of the muscles that contract are not stretched at all.

The importance of the stretch reflex in standing has probably been overemphasised in the past. The attitude of a decerebrate cat was interpreted as an exaggeration of normal standing, and the fact that the extensor rigidity is due to greatly augmented stretch reflexes suggested that normal standing could be accounted for in terms of stretch reflexes of a more appropriate vigour. This argument has lost its force now it is known that there is another form of decerebrate rigidity, in which the attitude of the animal is the same, but which does not depend on the stretch reflex (p. 399). Furthermore, decerebrate rigidity differs significantly from normal standing in that rigidity involves only extensors whereas both extensors and flexors contract in standing. The truth may prove to be that internal motor mechanisms, so far unknown, cause the appropriate muscles to contract and, as in many other actions (see p. 400), the contracted muscles exhibit stretch reflexes. The stretch reflex is obviously well adapted to oppose the forces that gravity applies to the muscles used in standing. On this view stretch reflexes remain of importance in standing but standing involves much more than the existence of stretch reflexes in extensor muscles.

For locomotion there is better evidence that the mechanism that produces the contractions of the flexors and extensors of a limb (in this case in sequence) does not depend on reflexes from the limb. A chronic spinal cat with all dorsal roots severed behind the spinal section, except those belonging to the tail, will still make walking movements of the legs when the tail is pinched. This strongly suggests that the organisation of walking is internal, and that walking is not of the nature of a chain reflex, in which movement of a limb forward sets up a reflex to carry it back and so on. Reflexes, however, are no doubt of great importance in modifying and making precise the movements of walking, although definite evidence as to how exactly they do it is lacking.

60. The Central Nervous System: II

P. A. Merton

VOLUNTARY ACTION

A voluntary action like picking up a pencil involves first seeing and recognising the pencil and then making the appropriate movement to take hold of it. This description, it will be seen, is of the same form as the description of reflex action that was arrived at in the previous chapter. Quite possibly it glosses over some essential difference between voluntary and reflex action, but it takes a good deal of the mystery out of both to emphasise their formal similarity. Of course, voluntary actions are often immensely more complex than reflex actions, and they are attended by consciousness. Physiology has nothing to say about the nature of consciousness except that it is something that seems to accompany the most elaborate reactions of an animal to its environment. It may mean that some fundamentally different kind of process is occurring or it may not. Whichever is the case it certainly makes voluntary actions, in spite of their complexity, easier to think about than reflexes because we have, or believe we have, an immediate subjective awareness of the steps involved in a voluntary action (recognition, the decision to make a particular movement, etc.) that can only be reached by indirect argument in the case of a reflex.

In voluntary action the recognition of what is going on in our surroundings and the decision what to do about it both take into account an enormous store of memories of what the messages from our sense organs signified in the past and what the results of previous muscular efforts were. When a large number of factors have to be weighed up simultaneously in this way the brain seems to choose to do the job in a thin sheet of nerve cells the area of which, rather than the thickness, increases as the complexity of the problem increases. Such sheets of grey matter develop on the surface of the brain and are hence called *cortex*, from the Latin word meaning "outer shell". Conscious voluntary activity attends events taking place, mainly at any rate, in the cerebral cortex; as the old writers put it, it is the organ of mind. In ascending the evolutionary scale the area of the cerebral cortex increases in step with increasing complexity of behaviour, ending with the enormous cerebral cortex of man.

The cerebral cortex has tackled its manifold tasks by subdividing and spreading them out over its surface, the more complex the task the larger the area devoted to it. In man there are large areas devoted to vision, hearing, muscular movement, sensations from the skin surface, etc. In picking up a pencil, there is evidence that seeing the pencil and recognising it are associated with activity in and around the visual area of the cortex, and that the movements of the arm to pick it up are the responsibility of the area concerned with movement (the motor area). Thus it would appear that the steps into which we subjectively divide the action of picking up the pencil are in fact related to nervous processes occurring in distinct regions of the cerebral cortex. That this is so has only become clear within the last hundred years or so, but already the facts of functional localisation in the cortex are so familiar that it is easy to forget that they are not merely an anatomical matter. Just as the idea of separate motor and sensory nerves is now taken so much for granted that we are apt to forget that their discovery was not at the time just a matter of the anatomy of the spinal roots, so too we tend to think that modern neurology has merely answered the anatomical questions: Where are functions A, B and C localised in the cortex? But a 120 years ago it was not known what the functions of the cortex were, not whether they were generalised or localised, so the questions could not have been asked.

Thus the prime importance of the observations which showed a localisation of cortical function was to reveal what kinds of function the cortex performs. The story, indeed, is by no means complete yet; there are large areas of which we only know in the most general way what they do, and some functions (e.g. vision) seem to be much more definitely localised than others (e.g. memory, which some people think is not really localised at all, see p. 424), but the main outlines seem clear enough. Roughly speaking, the back part of the cerebral cortex is sensory in function and the front part motor, the dividing line being the central sulcus. In the back half of the cortex are areas devoted to vision, hearing, smell and taste, and to sensations from the skin and the body generally (somatic sensations). These areas (except smell and taste) are shown in Fig. 59.3. They are called the primary or receiving areas because messages from the respective sense organs are actually brought by nerve fibres to these areas. The cortex around and between the receiving areas seems to be concerned with interpreting what comes in, with building up a single picture of the external world from the information supplied by all the senses and with assessing the significance of what is going on. One highly specialised aspect of this general function with which special areas are associated is the comprehension of speech and the written word.

The front half of the cortex is concerned with the execution of motor acts and in a general way with planning. The detailed orders to the muscles to perform particular movements are sent out by the cortex just in front of the central sulcus. Again speech gets special treatment; its motor organisation occupies a patch of cortex (Broca's area) in front of the main motor area, in the left hemisphere.

The motor area and the four sensory areas (somatic, visual, auditory and olfactory) are the only parts of the cortex which have large and obvious tracts of nerve fibres connecting them to other parts of the nervous system; for this reason they are sometimes referred to collectively as projection areas; the parts in between, which are principally connected to other parts of the cortex, are called association areas; formerly they were called silent areas because electrical stimulation or surgical removal of these areas, particularly in animals, is often without obvious effect. In lower mammals the projection areas occupy a large part of the cortex but in man the proportion is quite small. A man's eyes, ears and hands are not much better than those of a chimpanzee, nor are the absolute sizes of the cortical areas to which they project greatly different; he is superior chiefly because of the far greater use he makes of the information from his sense organs and the far greater skill with which he uses his hands; this he owes to his enormously expanded association areas.

In man, and indeed in nearly all vertebrates, the most important source of information about the environment is vision. What we see occupies so much of our attention that with the great majority of mankind it can be assumed that if their eyes are shut they will not be attending to anything; conversely, although we can, by an effort of will, disregard many sounds, smells and bodily sensations, it is very difficult indeed not to pay attention to what is passing before our eyes. The dominance of vision makes it natural to begin the more detailed discussion of the cerebral cortex with vision, more particularly since the anatomical peculiarities of the visual pathway have effects which are felt throughout the whole of the nervous system.

The visual system

In man it is established with great certainty that damage to the occipital lobes causes blindness, and that lesions (i.e. areas of damage) elsewhere in the cerebral cortex never have this result. Hence we speak of a visual sensory area in the occipital lobes. When a lesion is small and circumscribed, blindness is limited to a particular part of the visual field. This result could not have been foreseen, for supposing, speaking metaphorically, the visual cortex had functioned like a lens to focus nerve impulses from the retina on to consciousness, then putting out of action one part of it would have caused a general dimming of the whole field and it would not have mattered which part was involved, only the total area would count. There may be functions of the brain, such as memory, in which the same disability might be produced by lesions in different areas of the cortex in this way and in these cases it is sometimes written that they are functions of the cortex "as a whole". We shall return to the question later but vision at any rate is very clearly not like that; blindness only results from cortical lesions if they are in the occipital lobe and blindness in a certain part of the field of vision can only be produced by a lesion in a certain area within the visual cortex.

Electrical stimulation of a point on the visual cortex in man causes the subject to have the illusion that he sees a patch of light at some position in the visual field. Brindley has investigated the properties of these "phosphenes" with an implanted array of stimulating electrodes on the visual cortex of a blind person, put there as a visual prosthesis. The mapping of the visual field onto the cortex, so observed, is similar to that deduced earlier from the results of injury.

Optic nerve fibres from the retina do not run direct to the visual cortex but go to the lateral geniculate bodies, two nuclei lying on the under side of each thalamus at the base of the cerebral hemispheres; from the neurones in them large tracts of nerve fibres arise, called the optic radiations, which curve backwards to the occipital cortex (Fig. 60.1). Now over most of the cerebral hemispheres the cortical grey matter is remarkably uniform in naked eye appearance, consisting of a featureless sheet some 2 mm thick, but the area to which the visual radiation runs is distinguished by a conspicuous white band (of nerve fibres) in the middle of the grey matter, called the stria of Gennari; hence the visual area is also known as the striate cortex.

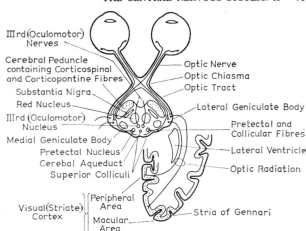

Fig. 60.1 The human visual pathways. Horizontal sections, roughly to scale, are shown of the right occipital lobe and of the upper part of the midbrain adjoining the thalamus. Some non-visual structures in the midbrain are also labelled.

The pretectal and collicular fibres are separate optic nerve fibres and not branches of geniculate fibres as the diagram has it.

The function of the stria is unknown. In man, part of the striate area is on the tip of the occipital pole but the greater part is out of sight in the deep groove where the medial surfaces of the two hemispheres face each other.

By no means all optic nerve fibres go to the lateral geniculate bodies and thence to the cortex. In lower animals like fish, frogs and birds the main visual pathway runs to the optic lobes on the dorsal side of the midbrain, and even in man 20–30 per cent of all fibres from the retina go past the geniculate bodies and end in the same region of the midbrain, known in higher animals as the superior colliculi (superior corpora quadrigemina), or in the grey matter deep to the colliculi (pretectal nuclei). Animal experiments indicate that the pretectal fibres are responsible for the contraction of the pupil when a light is shone in the eye, but there is no evidence about the function of the very numerous collicular fibres in man. It is a fair surmise that they are involved in automatic control of eye movements but only a surmise. The visual projection to the cerebellum is via the superior colliculi.

Even in primitive vertebrates visual impulses from external objects to the right-hand side of an animal are fed to the left half of the brain, and vice versa. The reason for this crossing over, or *decussation*, is not properly understood although it is often thought to be a consequence of the reversal of the optical image caused by the camera-like structure of the vertebrate eye with its single lens. To remember the scheme on which the different parts of the retinae are projected on to the human visual cortex it is convenient to think first of an imaginary man with a single central eye. The natural method seems to be for the upper part of the retina to be connected to the upper part of the visual area and the lower part to the lower part; the right-hand side to the right-hand half of the visual area, i.e. to the right occipital pole; and the left-hand side to the left occipital pole. The two real eyes both see almost the same field and each has the same connections as the hypothetical single eye, which means that fibres from the nasal halves of each retina have to cross in the optic chiasma. As a matter of fact, things are not quite so simple because although it is true that fibres from the right-hand halves of both retinae go to the right-hand occipital cortex, with upper retinal fibres keeping to the top and lower retinal fibres to the bottom, a reversal has occurred such that the central part of the field (the macula) is not represented *medially* as it ought to be, if the above scheme were carried through, but *laterally* and the peripheral retina is represented medially, i.e. on the opposed surface of the hemispheres (Fig. 60.1). Thus if we imagine an object moving from left to right in front of a man who gazes steadily straight ahead, the area of cortical activity representing the object starts on the right striate cortex deep in the groove where it faces its fellow; it then moves posteriorly and laterally on to the tip of the occipital pole; when the object is dead ahead the active area jumps over to the opposite pole and moves medially and anteriorly down on to the medial surface again.

Part of the evidence for the above statements is that division of one optic tract, or removal of the whole of one occipital lobe in man, results in complete blindness of the opposite half of the visual field of both eyes (a hemianopia). The vertical division of the visual field between left and right cerebral hemispheres revealed in this way seems to be quite sharp and splits even the macula, the part of the retina we use when looking straight at an object, straight down the middle. It is perhaps rather surprising that, when a small object moves from one side to the other of the central line and its area of cortical representation jumps from one hemisphere to the other, there is no trace of discontinuity in our sensation. Of course the two visual areas are connected together by enormous numbers of nerve fibres running in the corpus callosum (Fig. 59.3), and these connections may be important in welding them into a functional whole (but see p. 424). But division of the whole corpus callosum has been performed several times in man with remarkably little effect at all, and in particular with no conspicuous visual symptoms.

We have seen that the retinae are, as it were, mapped on to the visual cortex, but nothing has been said as to scale. Here we come on the most notable instance of the general rule that the area of cortex devoted to a part of the body depends on its importance and not on its size. The total area of visual cortex is not far different from the area of the two retinae but something like half the total area of visual cortex is probably devoted to the macula, the central part of the retina about 1 mm in diameter where visual acuity is highest. Thus the macula occupies a very much larger area on the cortex than it does on the retina.

When a lesion of the striate cortex causes a scotoma (a localised area of blindness) vision is lost as a whole in that area; it never happens, for instance, that colour vision is lost selectively or that threshold to perception of light remains unaltered, but objects cannot be perceived; all these functions are affected together, although in partially damaged areas not necessarily to the same degree. Scotomata, whether of cortical origin or due to damage to the retina or other parts of the visual pathway, can often only be found by careful examination, for patients are frequently unaware of them. Indeed with blindness of cortical origin they may deny that they have any trouble in seeing when they are grossly disabled. It is a general characteristic of cortical lesions, which goes under the name of anosognosia, that patients may lack insight into their condition and may not complain of their disability. This may seem less odd if we reflect that no one would expect a man who had had the whole of his cerebral cortex removed to realise how stupid he was. To remember what it was like to be intelligent he would have to be intelligent still. Similarly the visual cortex is not only the part of the cortex that does the seeing; it seems also to be the part that knows what seeing is.

The parts of the occipital lobe around the striate area and the posterior part of the parietal lobe adjoining it appear to be concerned with interpreting what we see with the striate cortex. The symptoms of disease in this region (collectively known as visual agnosia) are very various and fascinating, but there is only space to mention a few. To start with one that has been mentioned already, damage to the posterior part of the parietal lobe may abolish the reflex blink to a threatening gesture. Such a patient may have no detectable interference with vision and certainly sees a fist approaching, but his brain fails to interpret it as a threat, at any rate in time to do anything about it. Other lesions may upset judgement of distances. The patient may state that he clearly sees a chair placed in front of him, and yet when invited to step forward he walks straight into it and appears surprised to find it there. In these cases the disability may not be confined to a failure to judge distance, but may extend to a general inability to grasp the spatial relations of objects to each other and to himself. With his eyes open he may be unable to find his way about a house he has lived in for years, and paradoxically may manage better blindfold, feeling his way. If five pennies are put on the table before him he cannot count them because, as it were, he never knows where he has got to. Another manifestation of disease in this region is the failure to recognise things which nevertheless are seen clearly enough. The patient, when shown various common objects such as a box of matches or a pair of scissors, etc., may not be able to put names to them or describe what they are used for, although he can do so at once when allowed to handle them. The frequently associated inability to read will be discussed later under the subject of speech.

The ultimate problem of how the information contained in the optical image focused on the retina is conveyed to the brain and handled there has begun to yield in recent years to the technique of recording from single neurones in the visual pathway. Retinal ganglion cells in the cat respond to, or are inhibited by, light falling on the receptors in a circular patch surrounding the cell, i.e. they have circular or concentric receptive fields (Chapter 58). The next neurones on the pathway, the cells of the lateral geniculate body (whose axons end in the visual cortex) are likewise influenced from concentric fields in the retina. In the visual cortex, Hubel and Wiesel have discovered that the great majority of cells do not respond at all to general illumination of the retina, and only poorly to small circular patches of light, which are efficient stimuli for retinal ganglion cells and for the geniculate cells. For cortical cells the preferred stimuli are linear, bright lines of light on a dark background, or dark bars on a bright ground, or straight edges (boundaries between light and dark areas). Each cell responds best to a line stimulus with a particular orientation, and is more or less insensitive to a stimulus perpendicular to this optimal orientation. Cortical cells fall into three classes: "simple" cells, which respond to line stimuli (suitably orientated) presented at one particular position in the visual field only. Simple cells respond to moving line stimuli only when the line passes across this position. "Complex" cells are not so positionally sensitive; they respond to suitably orientated line stimuli anywhere within a certain area of the visual field. And they give a sustained discharge to such stimuli moving across this area. "Hypercomplex" cells differ from complex cells in that the line or boundary that excites them at the preferred orientation must not extend right across the receptive field, but must have an end (or a bend) or, in some cases, must have both ends (or two bends) within the receptive field.

The great majority of these cortical visual cells can be driven by similar stimuli from either eye separately. Barlow and his colleagues found that, when both eyes were in use, such "binocular units" responded best when the object that excited them was at a particular distance from the cat. This critical distance varied from cell to cell over a wide range. Here would seem to be a neural mechanism for distance estimation.

Hubel and Wiesel also found that the visual cortex is divided into functional columns, each about half a millimetre across, extending through the whole thickness of grey matter. All the responsive cells in a column have the same optimal orientation for line stimuli, but they differ in whether they prefer bright lines, dark bars, or edges. In each column some of the cells are simple and some are complex. The centres of their receptive fields are scattered over a region of retina comparable in size to the receptive field of a single complex cell. Neighbouring columns respond to stimuli in the same region of retina, but the preferred orientation varies randomly from column to column.

In the human subject, responses of the visual cortex to flash illumination of the whole visual field can be recorded with surface electrodes on the scalp. Similar "evoked responses" can also be detected to movement of quite small continuously illuminated patterned objects (e.g. a grating of light and dark bars). To see these small responses clearly, repetitive stimulation is employed and successive responses are added up in a computer. This multiplies the size of the evoked response while at the same time averaging out the spontaneous irregular activity of the brain. Such averaging techniques, originally developed by Dawson in 1951, are nowadays widely used for the study of small evoked potential changes in all parts of the nervous system.

The auditory and olfactory pathways

The auditory area of the cortex is in the temporal lobe, in man in the part opposite the bottom end of the central sulcus (Fig. 59.3). Impulses from the ears reach the auditory cortex via the VIIIth nerve nucleus in the medulla and the medial geniculate body (Fig. 60.1). In addition to this geniculate pathway there are a large number of fibres that run to nuclei in the medulla and midbrain, notably to the inferior colliculus (Fig. 60.2). The automatic turning of the head and eyes towards a source of sound is suspected to be one function of these pathways (compare the visual fibres to the superior colliculus). The geniculate pathway is mainly crossed but a substantial fraction is not; each geniculate body and hence each auditory cortex is in connection with both ears. The result is that, in contrast to vision, damage to one auditory area or to the auditory pathway on one side does not cause marked deafness. As lesions large enough to affect both pathways are likely to be fatal, deafness is a rare symptom of disease of the brain. Because of this much less is known about hearing than about vision, even the position of the auditory area is not clearly delimited in man.

Electrical recording from the exposed cerebral cortex of the anaesthetised dog has shown that impulses excited by notes of different frequencies go to different points on the auditory cortex. The frequencies are spread out along a strip, high notes in front and low notes behind. Each octave occupies about 2 mm of the strip. In man a jump of an octave in pitch gives the same subjective impression in whatever part of the scale it occurs. The frequency mapping in man may be quite different from that in the dog, but it is certainly very suggestive that in the dog a jump of an octave always moves the point of arrival of impulses the same distance along the cortex.

Signals from the balance organs of the inner ear play little part in conscious life, unless powerful stimulation makes us giddy or seasick (see Chapter 57). The cortical area dealing with vestibular sense appears to be in the temporal lobe near the auditory area.

The parts of the cortex concerned with smell (and taste) are hidden away underneath the cerebral hemispheres. The olfactory nerves go to the olfactory bulbs under the frontal lobes and these are connected to the olfactory area of the cortex in the uncinate region under the temporal lobe. Smell is a relatively unimportant sense in man, and the olfactory parts of the brain are poorly developed. It still has survival value by occasionally giving an invaluable danger signal of a fire or a gas-leak, but loss of smell is a small disability compared with the loss of any of the other senses, although enjoyment of food is much diminished for it depends largely on the sense of smell. Uncinate epilepsy is mentioned later (p. 423).

Somatic sensation

Afferent nerve fibres from sense-endings in the skin and the interior of the body are gathered together into peripheral nerve trunks and run to the spinal cord via the dorsal roots. The cell bodies of these nerve fibres lie in ganglia on the dorsal roots (Fig. 59.1). These particular cell bodies are merely trophic centres; nothing of nervous interest happens in a dorsal root ganglion, impulses just pass through as if it were an ordinary piece of nerve. On entering the cord afferent fibres divide into ascending and descending branches; from these branches arise collateral branches that run into the grey matter for reflex purposes.

Ascending branches that carry messages destined for the brain, to arouse conscious sensation, do one of two things (Figs. 60.2 and 60.3), either they immediately enter the dorsal column (on the same side they enter) and run straight up to the dorsal column nuclei in the medulla, or, after ascending for a segment or two, they terminate by making synaptic contact with nerve cells in the grey matter of the dorsal horn. The axons of these cells run all the way to the thalamus, the main sensory nucleus of the brain,

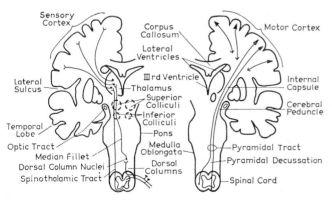

Fig. 60.2 Diagrammatic coronal (frontal) sections of the human brain to show the sensory and motor pathways. Plane of section indicated on Fig. 59.3. Roughly to scale except that the diameter of the brainstem and spinal cord is exaggerated. Some structures, shown here but not labelled, are labelled on Fig. 60.7. For spinal tracts see also Fig. 60.3. The outlines of the colliculi on the dorsal surface of the midbrain shown dotted. The median fillet is also known as the medial lemniscus.

which lies near the midline at the base of the cerebral hemispheres; to get there they cross the midline and run up the opposite side of the spinal cord in the ventrolateral column of white matter. The tract is called the spinothalamic tract, but it is not a clearly demarcated area of white matter like the tracts in the dorsal columns. The fibres composing it are mixed in among a much larger number of descending fibres and other ascending fibres.

The axons that arise from the cells of the dorsal column nuclei in the medulla also cross the midline and run to the opposite thalamus. In the medulla they form a conspicuous slab-like band of white matter called the medial lemniscus. Thus all nerve impulses that give rise to conscious sensation find their way to the opposite thalamus; either they run up the dorsal column on the same side as they enter the cord and cross over immediately above the dorsal column nuclei, or they cross within a few segments of entering the cord and run up the opposite spinothalamic tract. These facts are important because the two pathways do not carry the same types of sensation. The spinothalamic pathway conveys sensations of pain, temperature and crudely localised mechanical contact. The dorsal columns are responsible for all the more refined types of touch sensation, such as we possess pre-eminently in our finger tips, and for sensations of position and movement of the limbs.

The result of this segregation of sensory pathways is that lesions of the spinal cord other than complete division seldom cause a complete loss of all forms of sensation in any part of the body, such as occurs when a peripheral nerve is divided. As a rule spinal disease causes a *dissociated sensory loss* in which some forms of sensation are lost and some retained. The clearest instance is when injury or disease sections either the right or the left half of the spinal cord, giving rise to what is known as Brown–Séquard's syndrome. Below the level of hemisection the patient suffers a loss of dorsal column sensation on the side of the lesion and a loss of spinothalamic sensation on the opposite side.

The dorsal column loss shows itself by an inability to localise contacts: if one of his fingers is touched the patient will feel it but be unable to say which finger it is; again the ability to distinguish two points from one when the tips of a pair of compasses are pressed on the skin is greatly impaired, and the points may have to be opened out several inches before they are recognised as two. Position sense is lost and the patient cannot say what position his joints are in or whether they are being flexed or extended by an examiner. The loss of discriminative tactile sense and the loss of position sense together prevent him from recognising the size, shape and texture of objects which are put in his affected hand; this important symptom is called astereognosis and it makes the affected hand useless for most purposes.

The spinothalamic loss on the opposite side of the body makes the patient insensitive to pain, however aroused, and incapable of distinguishing hot from cold objects. No straightforward tactile loss is detectable, for dorsal column sensation remains on that side, but the loss of the spinothalamic type of tactile sensibility can sometimes be detected by the absence of sensations of itching and tickling. Itch and tickle are curious quasi-painful sensations that are believed to be carried in the spinothalamic tracts. The principal symptoms, therefore, are astereognosis and loss of position sense on the side of the section and loss of pain and temperature sensation on the opposite side—a dissociated loss. Simple contact, e.g. a brush with a piece of cottonwool, is

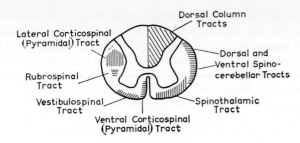

Fig. 60.3 The approximate positions of the principal tracts in the human spinal cord, ascending tracts on the right, descending tracts on the left.

felt everywhere because it can travel by either the direct or the crossed pathway.

Clinically, the existence of two distinct kinds of sensory loss called "dorsal column" and "spinothalamic" is not doubted, but recently it has been questioned whether lesions strictly confined to the dorsal columns give the full-blown picture of a "dorsal column" sensory loss. One possibility is that other tracts lateral to the dorsal horn of grey matter are involved too. Another awkward fact is that in patients in whom the spinothalamic tract on one side has been sectioned by anterolateral cordotomy to relieve pain, pain sensibility may return on the opposite side below the cut after an interval of a year or more. In some cases a second cordotomy on the same side as the first has again relieved the pain, but this may not be the whole story.

Both the ascending sensory pathways eventually end in the ventral part of the thalamus; from these a large tract of fibres (forming the posterior part of the internal capsule) runs to the strip of cerebral cortex behind the central sulcus, the (somatic) sensory area (Fig. 59.3). Destruction of the sensory area in man causes a loss of sensibility over the opposite half of the body; but, whereas loss of one occipital lobe gives rise to permanent complete blindness in the crossed half-field (p. 407), even very extensive damage in the sensory area does not cause permanent complete anaesthesia on the opposite side of the body. Except, perhaps, for a day or two after an acute lesion such patients are found still to appreciate pain, differences of temperature, and ill-localized contact on the opposite side of the body. The interpretation of this phenomenon is controversial but the simplest view, originally put forward by Head and Holmes, is that consciousness of these sensations depends in some way on activity in the thalamus. All the more highly developed forms of sensation require the integrity of the sensory area of the cortex. Roughly speaking, sensations associated with the thalamus are of the type carried in the spinothalamic tracts, and sensations permanently lost after cortical damage are those carried in the dorsal columns. The former tend to be associated with feelings of pleasure or of unpleasantness, warmth, cold and pain (see Chapters 30, 31 for further evidence that the central parts of the brain are concerned in emotion); but cortical sensations, such as distinguishing two compass points from one, are without emotional colour.

Lesions of the thalamus itself, besides causing a loss of sensation, may lead (after an interval of some weeks) to spontaneous and intractable pain, referred to the opposite side of the body; or it may happen that, without spontaneous pain, a stimulus such as a light pin-prick may cause disproportionately severe pain. Such symptoms, which constitute part of the *thalamic syndrome* almost never occur with lesions of the cortex, which is another reason for thinking that the thalamus is particularly concerned in appreciation of pain.

Lesions localised to a part of the sensory area cause loss of sensation in a particular part of the body in the same way that small lesions of the occipital cortex cause a localised patch of blindness in the visual field. In man the extent of the sensory area and the way the body is represented on it have been learnt mainly from the results of local damage, and by stimulation of the exposed cortex at operation. The brain substance itself and its immediate coverings are insentient, so that operations on the brain can be performed with only a local anaesthetic to insensitise the skin. It is then a simple matter to stimulate a point on the brain electrically by passing a low voltage 50 cycles alternating current between a round-tipped wire electrode resting on the cortex and an indifferent plate electrode on the skin elsewhere. When the electrode is on the post-central cortex the patient has a sensation which he feels in some part of the opposite side of his body. The electrical stimulus presumably does not imitate the normal pattern of activity caused by sensory impulses arriving at the cortex, for the sensation is indefinite and unfamiliar, not unlike the sensation from electrical stimulation of a cutaneous nerve (Chp. 4), but there is no reason to doubt that it indicates which part of the body normally sends sensory impulses to the point being excited. (With strong stimulation the referred sensations may become painful, an observation which suggests that the cortex is not wholly unconcerned in the appreciation of pain.) In this way it has been found that the body is mapped upside down on the sensory area; messages from the foot and leg go to the top of post-central cortex, the middle region is concerned with the hand and arm, and the bottom with the face. As with vision the scale of mapping is very uneven; important parts, like the fingers, have a large area devoted to them, the skin of the back a very small area.

In animals the sensory area can be mapped by recording electrically from the places which receive impulses when particular parts of the body are touched. The animal is fairly deeply anaesthetised to cut down the noisy background of spontaneous activity in the cortex. Then a fine wire inserted into the cortex will detect sensory impulses arriving. Interestingly enough, although they have crossed at least two synaptic junctions (in the dorsal column nuclei and the thalamus) the trains of impulses from the sense-endings are little modified and what arrives at the cortex is very much like what can be recorded from sensory fibres in peripheral nerve. The difference is that with parts of the body, such as the back, in which the cortex is not much interested, several or many sense-endings may share a single projection fibre, so that

Fig. 60.5 The somatic sensory area in the pig. The whole area is devoted to the opposite half of the snout. (Adrian (1943), *Brain* vol. 66, p. 89.)

touches at widely separated points, supplied by different spinal nerves, all cause activity in the same cortical fibre. Fig. 60.4 shows Adrian's map of the sensory area in the monkey, which is not unlike that of a man.

In other animals the rule that only important matters get a hearing in the cortex is sometimes even more obvious than in the monkey and man. A pig uses its snout not only for finding its food but also for digging it up; it is a pig's principal executive organ; the legs are little more than props for the body. Adrian found that the snout, so far as he could tell, was the only part of the body that sent sensory impulses to the cortex. The whole somatic sensory area was snout (Fig. 60.5). And the absolute area of cortex devoted to the pig's snout was larger relative to its size than the area of human cortex concerned with the hand. Other mammals also devote much of the sensory area to the nose and lips, particularly when they bear vibrissae (the long hairs with which cats and rats feel their way in the dark), but if the legs are used for fighting or digging they also are represented.

As regards the crossing over of sensory pathways by which sensory impulses from one side of the body are carried to the thalamus and cortex of the opposite side, the accepted explanation is that this occurs in order that they should be dealt with on the same side of the brain that deals with vision for their side of the body. The mammalian brain is dominated by vision, and because the visual representation of external objects on the cortex is crossed (the left striate area dealing with the right half of the visual field and so on) everything that has to be correlated with vision crosses too; that in-

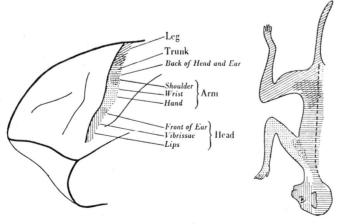

Fig. 60.4 The somatic sensory area in a rhesus monkey, showing the large area devoted to the hand and face. The trunk region includes the back of the neck and head. (Adrian (1941), *Journal of Physiology*, Vol. 100, p. 159.)

cludes not only somatic sensation but hearing and the motor pathways, too. It lends support to the correctness of this view that when, exceptionally, somatic sensations are to be correlated with olfaction, whose pathway does not decussate, the somatic pathway remains uncrossed too. Thus Adrian found in ruminants (the sheep and goat) in whom smell is probably the main guide to what is to be eaten next, that the sensory area for the lips was uncrossed. The remainder of the body had a crossed representation as usual. In the cat and in other animals in which vision is used in feeding, the cortical representation of the lips was crossed.

In man the symptoms of damage to the somatic sensory area, say the region concerned with the hand, are in the first place loss of all delicate, discriminative skin sensations by which objects are recognised and fine manipulations directed. The patient also tends to be inattentive to the affected hand, so that if quantitative testing of sensation is attempted, performance is irregular. If touched simultaneously on both hands he may fail to notice the touch on the affected hand, although he does so all right if it is touched alone. It seems that the affected hand tends to slip out of his mind; indeed with large lesions it may go out of his mind altogether, giving rise to the bizarre symptom that the patient when shown the affected hand denies that it belongs to him. He has apparently lost the part of the brain that deals with his conception of his hand. This is analogous to the patient with damage to the occipital lobes who is not aware that he cannot see. Needless to say, a severe cortical sensory loss renders the hand useless, and although it is not paralysed, to an onlooker it may appear to be, for the patient attempts nothing with it.

Voluntary action and the motor cortex

The orders that the cerebral cortex issues to the muscles of the body leave from a strip of cortex in front of the central sulcus called the motor area (Fig. 59.3). If the motor area is damaged or removed the patient develops a voluntary paralysis on the opposite side of the body. He knows what he wants to do but when he tries to move the affected parts nothing happens. Lesions elsewhere in the cortex never have this result; this is the first reason for regarding the precentral cortex as an area specially devoted to voluntary movement.

Movements of different parts of the body are dealt with by different parts of the motor area; the arrangement matches that in the sensory area on the opposite side of the central sulcus, leg and foot at the top, arm and hand in the middle and face at the bottom. Figure 60.6 shows the motor area of a chimpanzee. The human motor map is known to be very much the same but a detailed map of a whole healthy human motor area is not available. As in the sensory cortex there are large areas for important parts like the hand and only small areas for parts like the buttocks which have little to do directly with voluntary activities. Since on the whole important parts are small and large parts unimportant, it turns out that roughly speaking the area of cortex devoted to a part of the body is inversely proportional to its bulk. It is presumably because sensation, particularly touch, is of such great importance in guiding movement that the cortical area which controls the movement of a part lies next door to the area to which it sends its sensory messages. In fact the motor and sensory areas to some extent run into each other, for when stimulating points on the exposed human cortex electrically it is found that sensations (referred to the opposite side of the body) are sometimes felt when the electrode is on the pre-central

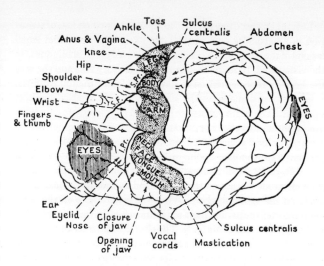

Fig. 60.6 The motor area of a chimpanzee mapped by electrical stimulation. There is much overlapping of the regions and their subdivisions which the diagram does not attempt to indicate.

Conjugate eye movements can be obtained by stimulating the visual area in the occipital lobes and also from the frontal lobe in front of the motor area. These movements (not mentioned in the text) differ in their characteristics from movements obtained from the ordinary precentral motor area. (Sherrington, *The Integrative Action of the Nervous System.* New York, 1906; Cambridge, 1947.)

cortex, and movement sometimes results when the electrode is on the post-central cortex.

Movements of the opposite side of the body elicited by electrical stimulation are a hallmark of the motor area. The ease with which movements can be obtained by electrical stimulation of the motor area, compared with the impossibility of obtaining movements from most other places in the cortex, is the second reason for linking the pre-central area with the control of voluntary movement. The movements occur in the same part of the body which would become paralysed if the region of cortex around the stimulating electrode were excised. Thus the way in which the parts of the body are mapped on to the motor cortex can be discovered by electrical stimulation.

The motor area is connected to the spinal cord by numerous fibres running in the corticospinal or pyramidal tract (Fig. 60.2). The pyramidal tracts descend in the internal capsule between thalamus and basal ganglia; they occupy part of the cerebral peduncles on the ventral surface of the midbrain, pass through the pons and emerge to form two swellings, the pyramids (not shown in any of the figures), on the ventral surface of the medulla. Most of the fibres cross the midline in the pyramidal decussation to enter the opposite lateral column of the spinal cord (Fig. 60.3). A fraction (of unknown function) remain uncrossed and pass down the cord on the same side, mainly in the ventral column. This uncrossed ventral pathway does not extend beyond the upper thoracic segments of the cord in man. Other fibres initially taking a similar course branch away to the motor nuclei of the cranial nerves (corticobulbar fibres).

The largest and fastest conducting fibres arise from exceptionally large neurones called Betz cells, of a size not found elsewhere in the cortex, but these are in a minority; by far the greater number of fibres from the motor area arise from ordinary small cortical cells. The pyramidal tracts (defined as the tracts that form the medullary pyramids) also contain very many corticospinal fibres (of obscure function) from the so-called premotor areas in front of the motor areas and from the sensory areas and adjacent parts of the parietal lobes behind them. In all there are more than half a million fibres in each pyramidal tract in man, but only some 30 000 Betz cells in each motor area.

The corticospinal and corticobulbar tracts represent only a small proportion of the efferent fibres from the motor cortex. There are also very large numbers of fibres carrying messages to the pons, for the cerebellum (p. 416), and other numerous *extrapyramidal* fibres to the motor nuclei at the base of the brain (p. 415 and Fig. 60.7).

The motor area was historically the first part of the cerebral cortex that was clearly shown to have a specific function. As so often in the central nervous system, the first evidence came from the study of human patients. The central nervous system in its bony case is so difficult for the experimenter to get at, and the results of operating on it are so difficult to evaluate in animals, who cannot be given instructions to attempt various movements or be asked to describe what sensations they have, that in many instances the first clues to the function of a part of the brain have come from the destructive experiments that disease performs in man. In fact to this day the parts of the brain about which we are most ignorant, such as the basal ganglia of the cerebral hemispheres or the olives in the medulla (see Fig. 60.7), are mainly those which do not suffer localised clear-cut destruction in disease.

The first definite indication of a localisation of motor function in the cortex came with Broca's publication in 1861 of two cases of loss of speech associated with lesions in the lower pre-central region of the left hemisphere. More precise information came from another direction. With the cerebral cortex it is not only destructive lesions that provide evidence, for the cortex is also prone to those paroxysmal bursts of overactivity which are the cause of epileptic fits (convulsions). This sort of instability seems to be peculiar to the cortical grey matter. The unsolved question of what happens in the cortex during an epileptic fit is a matter of physiology, not pathology, for anyone will have a fit if his brain is stimulated sufficiently violently. A convulsion triggered off by passing 50 Hz alternating current through the head is a standard and effective method of treatment in various psychiatric disorders, while in normal infants convulsions not uncommonly usher in an acute infective illness. In sufferers from epilepsy similar convulsions occur spontaneously. In a typical fit the whole cortex is involved more or less simultaneously, the patient loses consciousness and then has a generalised convulsion. Much less often convulsive movements begin in one part of the body only and spread to involve the remainder of that side of the body and then the other side.

Epilepsy is usually not associated with any changes in the naked-eye or microscopical appearance of the brain, but sometimes it is precipitated by obvious damage, a cerebral tumour or the scars of a brain injury. In the early 1860s Hughlings Jackson observed that convulsions starting unilaterally (still known as Jacksonian fits) were associated with damage to the pre-central region of the opposite hemisphere. When the fit began in the foot the lesion was at the top of the pre-central region; with lesions in the middle the hand was involved first, and so on. The commonest sites for the fits to start were the feet, the fingers and hand, and the corner of the mouth. From these facts Jackson inferred that the pre-central gyrus is the cortical centre for movements, and that within it the body is represented upside down with the largest areas concerned with movements of the important "leading parts". The manner of spread of the convulsion, the "march" as he called it, reflected the spread of epileptic activity over the motor cortex, involving one area after another.

In 1870, Fritsch and Hitzig showed in the dog that electrical stimulation of the pre-central cortex would cause movements of the opposite limbs. Previous to this it had been generally held that the cerebral cortex was electrically inexcitable. Before the end of the century electrical stimulation had been used to map out the motor area in many species of animals and in man.

The paralysis caused by a lesion of the motor cortex is of a highly distinctive character and distribution. The actions lost or most severely weakened are the most voluntary and least automatic, the most accurately graded and individual movements in which the limb is usually employed, the sort of movements out of which skilled actions such as writing or driving a car are made up. Movements of this type employ predominantly the distal parts of the limbs, particularly the hands, and it is here that the paralysis makes itself most felt; the proximal parts are less affected, while bilateral movements of the trunk are preserved. As already mentioned (p. 394) expressive and emotional movements of the face tend to be much less affected than movements made to order.

Within the hand itself the same principle applies; the most highly specialised movements are the most paralysed. The ability to move single fingers independently of the others is the first thing to go, so that the patient may lose manual dexterity completely while retaining a powerful grip. The same muscles move each finger whether they are flexed one at a time or all together, but the one movement is lost and the other is not. From this and many similar instances it is clear that it is not the muscles themselves that are paralysed but only certain movements carried out by them. The motor cortex, even more obviously than the spinal cord, "thinks" in terms of movements, not muscles.

Electrical stimulation of the human motor cortex exposed at operation under local anaesthesia causes movements of just the type that are lost when the same area is removed. They predominantly involve the distal parts of the limbs; fingers and toes, wrist and ankle, and include a proportion of movements of the thumb and individual fingers of just the sort that a patient with a cortical palsy cannot make. Similar movements can be obtained after the grey matter has been removed, by stimulating the cut ends of the fibres in the white matter underneath; hence stimulation of the cortex reveals mainly the properties of the efferent pathways and if, as seems probable, the cortex itself has the function of organising complete actions from combinations and sequencies of these fragmentary movements, this activity is not put into motion by the type of electrical stimulation employed.

Some writers have argued that movements are represented in the motor cortex in the same sense that movements could be said to be represented in the skin; appropriate patterns of excitation can call forth elaborate muscular actions, but simple electrical stimulation of the skin or the cortex only discovers the most elementary. The analogy is a good one, for corticospinal fibres end mainly, and cutaneous fibres end exclusively, on interneurones and not on the motoneurones themselves. Both appear to activate the spinal mechanisms that synthesise functional movements out of contractions of individual muscles. The ease with which movements of the digits are obtained by cortical stimulation is partly a property of the spinal cord, for these are also the movements most readily obtained in spinal men and spinal monkeys by mechanical stimulation of the skin (e.g. the extensor plantar response, p. 398) or by electrical excitation of the sensory nerve fibres. Thus the great importance of the distal parts of the limbs in voluntary life is reflected in the spinal motor mechanisms, and it is the large area of cortex connected to these mechanisms that cortical stimulation reveals.

In man voluntary paralysis is commonly the result of rupture of fibres leaving the motor cortex by haemorrhage from a burst blood vessel near the internal capsule. The patient has an "apoplexy", or "stroke", i.e. suddenly has his senses and his power of motion taken from him, and is subsequently found to be paralysed down one side of his body (hemiplegia). The plantar response (p. 398) on the paralysed side becomes extensor (upgoing big toe) from the onset. As usual after any sudden damage to the brain, a good deal of recovery of function takes place in the next few weeks, possibly due to recovery of structures that were put out of action but not actually killed. Initially the paralysed limbs are quite flaccid but, in a typical case, if voluntary movements do not recover they gradually become stiff (spastic) owing to exaggeration of the stretch reflexes (p. 399) and the tendon jerks (p. 400) are increased. There is evidence that the spasticity is not due to destruction of corticospinal fibres; thus it is said that following division of the corticospinal tract in the medullary pyramid in a monkey, the paralysed limbs remain flaccid. With a lesion in the internal capsule many fibres other than corticospinal fibres are divided. The important ones for spasticity may be those from the motor and premotor areas to motor nuclei of the extrapyramidal system. Why division of such fibres should cause spasticity is unknown. Spastic paralysis with increased tendon jerks also results from damage to the spinal cord short of complete division (paralysis in both legs due to a spinal lesion is called paraplegia). How this type of spasticity is related to that due to disease in the brain is not known. In view of these uncertainties, it is safest not to speak of spasticity and increased jerks, when they occur with lesions of the brain or spinal cord, as indicative of damage to the pyramidal tract, but to use some non-committal term such as "long motor tracts".

Another expression used is "upper motor neurone", dating from the days when the voluntary motor pathway was considered to consist of two types of neurones only, the upper motor neurone (the Betz cell in the motor cortex with its axon) connected to the lower motor neurone (the motor cell in the ventral horn of the spinal grey matter with its motor axon running to a muscle). Spastic paralysis with increased tendon jerks and an extensor plantar response are spoken of as signs of an upper motor neurone lesion. Lesions of the lower motor neurone (e.g. in poliomyelitis) cause a flaccid paralysis with loss of tendon jerks and eventual wasting of the denervated muscles. There is no wasting to speak of in an upper motor neurone lesion.

As pointed out by Hughlings Jackson, destruction of nervous tissue can only be the direct cause of loss of function, of purely negative symptoms such as paralysis. Positive symptoms such as spasticity and the change in character of the plantar response must be due to the action of surviving parts of the nervous system that are "released" from influences that previously controlled them. Release phenomena appear to fall into two categories, those that come on immediately after a lesion (e.g. the extensor plantar response after a stroke, or decerebrate rigidity after section of the midbrain in an animal) and those that develop after an interval of weeks or more (e.g. spasticity, and the thalamic syndrome, p. 410). The hypersensitivity of denervated muscle to acetylcholine (Chp. 5) may also be considered as a release phenomenon of the second class. The relationship of clinical spasticity to the somewhat similar state of the muscles in decerebrate rigidity in animals is not understood.

Speech

Speech is a form of motor activity peculiar to man. Inability to speak which is not caused by muscular paralysis or other local disease of the larynx, tongue, etc., is called aphasia. Patients with aphasia are often unable to understand what is said to them, or to read and write or to do arithmetic. These related disabilities are conventionally dealt with under the heading of aphasia. In right-handed persons aphasia is caused by lesions of the left cerebral hemisphere. Damage to the cortex in front of the motor area for the face, at the bottom of the pre-central gyrus (Broca's area, Fig. 59.3), causes a loss of speech with no loss of comprehension of other people's speech or of writing, a so-called motor aphasia. Damage to the parietal and temporal lobes in the region between the receiving areas for vision and hearing gives rise to a receptive aphasia; the patient cannot understand speech and may not be able to read. In this type of aphasia he may talk jargon and fail to recognise that what he is saying is meaningless. Instances of demonstrably pure motor or pure receptive aphasia are seldom met with in practice. Although difficult to assess in an aphasic it is thought that intelligence generally suffers. Thus aphasia may be only one symptom of a general inability to use symbols in thinking.

One of the many interesting things about aphasia is that the use of words to express emotion is often preserved. The patient who cannot get out the simplest sentence to express a thought, may yet be able to swear or to express approbation and disapprobation with a simple yes and no. We have already noted that movements of the face to express emotion may likewise be preserved after damage to the cortical motor pathways.

The frontal lobes

The frontal lobes are the parts of the cerebral hemispheres extending forwards of the central sulcus. They therefore include the motor and pre-motor areas already discussed. Extensive damage or removal of one or both frontal lobes ahead of the motor and premotor areas does not cause any immediately obvious disability. Such a patient displays no defect of sensation, no paralysis of movement or striking loss of skill and no marked impairment of memory or intellect. A bizarre sign is the "grasp reflex", in which an otherwise apparently normal individual may, like a baby, clench his first on a finger drawn across his palm and be wholly unable to let go. This may be the only objective sign of a contralateral frontal tumour. Loss of bladder control, with urgency or incontinence, is another little-recognised feature of frontal lobe lesions.

More subtle and more remarkable are changes in the patient's personality, which are much more conspicuous if both frontal lobes are involved. He is no longer able to plan ahead or to pursue any but the most immediate goals. Even if cautious and farsighted before the operation he is now happy-go-lucky. His insensibility to the consequences of his acts may be embarrassingly apparent in his social behaviour. He makes jokes in bad taste and quarrels readily with a blithe unawareness of the awkwardness he causes. These facts suggest that the frontal lobes are concerned with the long-term planning of behavioural activity, with appreciating the consequences of various courses of action and with the emotional concomitants of this sort of activity, such as anxiety.

The operation of pre-frontal leucotomy, cutting the white matter which contains the fibre tracts connecting the frontal lobes with the rest of the brain, may be of benefit to certain types of mentally sick patient. The general tendency of leucotomy is to make the patient carefree (see also p. 425).

THE EXTRAPYRAMIDAL SYSTEM

It is easy to gain the impression that there is little more to voluntary movement than the discharge of motor impulses down the pyramidal tract to the spinal motor mechanisms. The very large contribution that subsidiary mechanisms make to ensuring successful execution of the desired act tends to be lost sight of because, like the perfect servant, we are not conscious of it. The parts of the brain involved (i.e. all the subcortical motor nuclei and tracts, excluding the pyramidal tract) are referred to collectively as the extrapyramidal system. The cerebellum, which comes within this definition is, however, usually treated apart.

The structures comprised (Fig. 60.7) are the basal ganglia (the

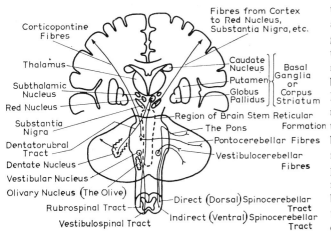

Fig. 60.7 The extrapyramidal system and the cerebellum. The nuclei are shown on a diagrammatic section of the human brain in the plane of the brain stem as in Fig. 60.2 with, here, the outline of the cerebellum and one dentate nucleus superimposed. Some structures not labelled here are labelled in Fig. 60.2; for spinal tracts see also Fig. 60.3.

The extrapyramidal nuclei are connected to each other and to the thalamus by numerous bundles of nerve fibres (not shown). The main tracts running to them are the dentatorubral from the cerebellum (which contains many fibres to the thalamus too) and fibres from the cerebral cortex to the red nucleus and substantia nigra. There are also direct connections between the cerebral cortex and the corpus striatum. Likewise the efferent connections of the extrapyramidal nuclei to the spinal cord arise mainly from the red nucleus and reticular formation. The reticulospinal tract (not shown) runs near the rubrospinal tract.

The olives are large nuclei of unknown function (only the left olive is shown), which receive large tracts of fibres from the neighbourhood of the red nuclei and the thalami (the thalamo-olivary or central tegmental tracts, not shown). They have efferent connections to the opposite cerebellar cortex and to the spinal cord by the olivospinal tracts (not shown), which run near the vestibulospinal tracts.

The cerebellum lies dorsal to the hindbrain connected to it by three peduncles (not distinguished in the diagram) on each side; superior, containing mainly the large dentatorubral tract and the indirect spinocerebellar tract; middle, the largest in man, consisting of pontocerebellar fibres from the pons; and inferior, containing principally the direct spinocerebellar tract, vestibulocerebellar fibres, and (not shown) fibres from the dorsal column nuclei and the olives to the cerebellar cortex, and efferent fibres (not shown) from the central nuclei other than the dentate to the medulla.

corpus striatum) and the brainstem nuclei closely connected with the basal ganglia, notably the subthalamic nucleus, the substantia nigra and the red nucleus, to which should probably be added some of the cells in the reticular formation of the brainstem (see also p. 393). Some nuclei in the front part of the thalamus too, although formerly thought to be sensory in function, have an equal claim to be regarded as part of the extrapyramidal motor system. They have extensive connections in both directions with the basal ganglia and with the motor area of the cerebral cortex; they send many fibres to the olives and receive many from the dentate nucleus of the cerebellum.

In lower vertebrates in which the cerebral cortex is little developed the basal ganglia are the highest motor centres and the rubrospinal tract from the red nucleus is the principal tract to the spinal cord. As the cortex and pyramidal tracts increase in size the rubrospinal tract becomes less important and in man it is small. At the same time the amount an animal can do, after the pyramidal tracts are sectioned in the medulla, decreases. A cat moves about and performs many acts normally, but movements of individual limbs, pawing objects, etc., are lost. The monkey loses all delicate limb movements, but after some weeks it can stand and move about unwillingly. Man, no doubt, could manage even less.

These activities are little more than might from analogy with the cat and dog, be attributed to action of the midbrain alone and scarcely seem to require the large development of the basal ganglia in primates. The evidence of disease in man suggests that the basal ganglia, having resigned the leading role in the motor system, have been taken into partnership by the cortex; but exactly what they contribute is very obscure. Brindley and Lewis showed that, after severing the pyramidal tracts in the medulla in a monkey, the motor map obtained by electrical stimulation of the cortex was the same as that found immediately beforehand; but that larger stimuli were required to obtain the movements. Each point stimulated gave movements in the same part of the body as before, but the threshold was higher. Hence we must envisage that the motor cortex has access to the musculature by two quite separate routes, pyramidal and extrapyramidal (of which the pyramidal route greatly increases in importance as we ascend the animal scale). Understanding of the extrapyramidal route is backward, largely because disease of the extrapyramidal system tends unfortunately to be diffuse and to affect unevenly many of the nuclei, partially destroying them. (It is clear at any rate that unilateral disease of the basal ganglia results in symptoms on the opposite side of the body.)

The commonest affliction of the extrapyramidal system is paralysis agitans (Parkinson's disease: see also p. 487, Chapter 78) in which the essential lesion is probably a loss of dopamine-containing cells in the substantia nigra, whose axons run to the corpus striatum. In this condition the limbs become stiffened at all joints by contraction of both flexors and extensors and develop tremor at a frequency of three or four oscillations per second. The muscular contractions involve the stretch reflex but tendon jerks are not increased and the clasp-knife response is not obtained. This type of stiffness is called *rigidity* to distinguish it from the *spasticity* ensuing on damage to the long motor tracts (p. 414). What a physiologist calls decerebrate rigidity would be decerebrate spasticity to a clinical neurologist.

In Parkinson's disease the patient is disabled because he cannot get his voluntary movements going (akinesia); all movements become cramped and slow and he walks with tiny shuffling steps. In the early stages the first things noticed, even before rigidity and tremor develop, are slowness in manual tasks, such as getting dressed, a dead-pan expressionless face, and a failure to swing the arms when walking unless attention is given to them. There is also an impairment of various postural and righting mechanisms similar to those described by Magnus in animals (p. 404). The patient has difficulty in turning from the supine to the prone position and in rising from the ground or from a chair. While standing he is easily thrown off balance by a push or by throwing his head back to look at something above him. It appears that many automatic elements in movement are lost and are slowly replaced by rigidity and tremor. Rigidity and tremor, in Hughlings Jackson's terminology, are release symptoms and they must be due to overaction of surviving nervous tissue. What is overacting in Parkinson's disease may perhaps be the surviving parts of the extrapyramidal nuclei, for destruction of the globus pallidus, or of part of the thalamus, by stereotaxic surgery, has been found not infrequently to relieve the rigidity. The most recent and successful therapy is based on the discovery that the dopamine content of the basal ganglia is diminished. Administration of L-dopa, which is converted to dopamine in the brain, may cause a dramatic improvement in the symptoms, including the akinesia.

There are facts which suggest that a patient with Parkinson's disease is not so much incapable of moving as prevented from doing so. It is of great interest that some gravely disabled cases may, if suddenly and powerfully motivated, move swiftly and effectively. Thus a patient who has great difficulty in feeding himself or walking may deftly catch a ball thrown to him, or jump out of the way of a car about to run him down. Gordon Holmes saw "a man who could scarcely walk in his waking state wander about easily in a period of somnambulism". Such dramatic temporary recovery would be unthinkable in patients with damage to the pyramidal pathways or the cerebellum.

In other patients disease of the corpus striatum results in involuntary movements of the limbs, which may take the form of elaborate writhing movements (athetosis) or of irregular jerkings (chorea), which may be sufficient violence to damage the limbs. Which precise structures have to be destroyed to produce the rigidity–tremor syndrome and which ones the involuntary movements of athetosis or chorea is not known. In 1974 it was discovered that in an hereditary form of chorea, Huntington's chorea, the γ-aminobutyric acid (GABA) content of the basal ganglia is diminished. GABA is almost certainly an inhibitory transmitter in the central nervous system (p. 48). Its action is blocked by picrotoxin, and it is therefore of interest that injections of minute doses of picrotoxin directly into the caudate nucleus of normal rats results in choreiform jerks of the opposite limbs.

THE CEREBELLUM

The cerebellum appears early in evolution as a piece of private brain belonging to the vestibular-lateral line system of sense organs, which are believed to be important in the control of swimming. In primitive animals the cerebellum consists of two separate outgrowths, one on each side of the medulla, each connected to the vestibular nerve on that side. In higher animals they enlarge and fuse in the midline, but the connections to the cerebellum from sense organs remain predominantly uncrossed, so that each half of the cerebellum deals with the half of the body on the same side.

In higher animals the cerebellum has become the civil service of the motor system; having started with the job of seeing that swimming was properly carried out having regard to all the circumstances communicated to it by the vestibular-lateral line system, it later took over the executive side of voluntary movements as well and uses a wider range of sense data to see that they are carried out properly too.

The cerebellum reaches its greatest size in the higher mammals with their elaborate voluntary motor activities. In them it is not the vestibular parts of the cerebellum that have developed but mainly phylogenetically newer portions, the enlargement of which parallels the great development of the cerebral hemispheres. In man the vestibular parts are completely overshadowed.

In mammals the cerebellum receives afferent impulses from the vestibular organs, from the limbs and trunk (via the direct and indirect spinocerebellar tracts (Fig. 60.7) and to a lesser extent via the dorsal columns), and from the eyes and ears. The visual and auditory connections are via the superior and inferior colliculi respectively. Information about what the motor cortex is doing is sent by numerous fibres that run near the corticospinal tract as far as the pontine nuclei, whence fresh fibres forming the pons and the large middle cerebellar peduncle are distributed to the opposite cerebellar cortex. There is also a large and presumably important pathway from each olive to the opposite cerebellar cortex, but its function, like that of the olives themselves, lies purely in the realms of speculation. The cerebellar cortex consists of a uniform thin sheet of grey matter the area of which is enormously increased by deep and elaborate folding. The efferent fibres do not arise directly from the cortex but from central masses of grey matter, of which the largest are the dentate nuclei, one on each side. The principal efferent tracts run in the superior peduncles to the red nuclei and the thalamus; others go via the inferior peduncles to motor centres in the reticular formation in the medulla.

In the cat and monkey afferent impulses from the hind limbs go to the anterior part of the cerebellum (lobulus centralis and front part of culmen), from the fore limbs to the part behind this (culmen and sometimes part of lobulus simplex) and from the face to the part behind this again (lobulus simplex). Figure 60.8 shows these areas in the monkey. The pathways to the cerebellum are uncrossed, the left fore-limb sends impulses to the left half of the culmen, and so on. Discharges to the limb areas are most easily produced by movement of joints and muscles and by pressure on the pads of the feet, but light touch is sometimes effective. The arrival of impulses is detected in the same way as in the sensory area of the cerebral cortex (p. 411). As with the cerebral cortex the pattern of impulses that arrives is very much like what it is in a peripheral nerve fibre. Auditory and visual stimuli cause activity behind the face area.

The cortico–ponto–cerebellar pathway as a whole projects to a larger area than the afferent pathways just described, but within this area the hind-limb, fore-limb and face areas of the motor cortex are connected to the cerebellar areas that receive afferent impulses from the same parts of the body (Fig. 60.8). It is worth noting that the mapping of the body on to the cerebellar surface continues across the morphologically important division between anterior and middle lobes, which occurs between culmen and lobulus simplex. Space forbids mention of other,

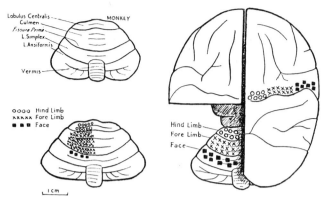

Fig. 60.8 Regions of the monkey's cerebellum connected with the limbs and face, and with the motor cortex.

Top left: The anatomical divisions of the dorsal surface of the monkey's cerebellum. Bottom left: The regions in which impulses can be detected by electrical recording when various parts of the body are moved or pressed upon. Right: The regions in which discharges originating in the limb and face areas of the motor cortex can be detected arriving in the cerebellum via the cortico-ponto-cerebellar fibres. (Adrian (1943), *Brain* Vol. 66, p. 289.)

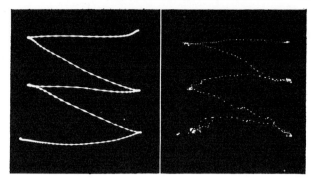

Fig. 60.9 Records to illustrate cerebellar ataxy. A flashing light was attached to the tip of a forefinger and its movements photographed in a dark room by leaving the camera shutter open. The light flashed 25 times per second; thus the record shows not only the track of the forefinger but also, by the spacing of the flashes, its velocity.

The task was to move the forefinger accurately between two columns of illuminated points (which do not show up in the photograph), some 75 cm apart. The patient had right-sided cerebellar ataxy. A is the record from the unaffected forefinger, B that from the right forefinger. Record B shows intention tremor as each target point is approached, and irregularities of rate and direction in between. (Gordon Holmes (1939), *Brain*, Vol. 62, p. 1.)

perhaps equally important, projections to other parts of the cerebellar cortex.

The cerebellum is wholly concerned with the execution of muscular acts; it has no say in what is to be done, policy-making is for the cerebral cortex. The results of damage to the cerebellum are disturbances of posture and disorganisation (ataxy) of movement. If the lesion is on one side of the cerebellum the symptoms are on the same side of the body. In man nothing that goes on in the cerebellum reaches consciousness and damage to the cerebellum does not affect memory, intelligence, character, or any other aspect of our mental life. Although it is richly supplied with afferent information, damage to the cerebellum causes no sensory loss. Neither is the patient paralysed; he is not unable to initiate movements, but the movements he begins do not come off properly. At rest little or nothing is observed to be wrong. The characteristic disturbances appear when the patient is asked to perform some action, e.g. to touch an object in front of him with his index finger. He starts off without difficulty, but his hand soon deviates from the correct line and may change direction and velocity several times in an irregular jerky manner. His finger may miss the target altogether, often by overshooting it, and wobbles irregularly as it approaches or even after it has made contact, a symptom called intention tremor (Fig. 60.9). The patient knows that his arm is moving incorrectly and may complain spontaneously that he is unable to control it; he tries to move it in one direction, he says, and it goes off in another. As already implied this is not due to loss of sensation in the arm, in particular it is not due to loss of position sense; the patient knows where his arm is well enough, and he does not have to look at it to find out. Gordon Holmes observed that a patient with one-sided cerebellar ataxy who could not hold his hand steady in front of him could, nevertheless, with his eyes shut, place the index finger of his good hand on the wavering index finger of the ataxic hand. Cerebellar ataxy is not aggravated by shutting the eyes, unlike the superficially similar sensory ataxy caused by lesions of the dorsal columns and elsewhere.

Examination shows that the muscles in the ataxic limb are contracting and relaxing at the wrong moments and exerting the wrong force when they do contract (dysmetria). Much of the trouble is due to failure of the muscles to steady the shoulder girdle and elbow; the arm swings like a barn door in the wind and the intention tremor is partly due to the patient's voluntary attempts to compensate for this lack of the normal postural fixation of the proximal joints. At rest the arm offers no resistance to passive movement; the tendon jerks, however, are normal. These observations show that the cerebellum has the task of making the automatic modifications of posture that we take for granted when we make voluntary movements, and for arranging that the numerous muscles involved in a movement contract at the right time and with the right force. One possible way in which the cerebellum supervises muscular contractions is discussed on pages 418–420. The cerebellum also ensures that successive components of a complex act run smoothly into each other; in cerebellar ataxy there is particular difficulty with alternating movements, delays occurring at each turning point; similarly, in speaking, words are broken up and each syllable pronounced separately (scanning speech). How deeply the cerebellum is involved in the actual learning of motor skills (e.g. playing the piano or throwing a cricket ball), a task usually, but on little evidence, assigned tacitly to the cerebral cortex, is a question not yet answered (see also pp. 423–425).

THE FUNCTIONAL ARCHITECTURE OF THE CEREBELLUM[*]

Structure

Most parts of the brain contain cells and fibre tracts whose wiring diagrams are but poorly understood. The complexity is baffling. Recently, however, the cerebellum has been investigated in much detail and detailed neuronal circuits have been worked out.

The cerebellum in some respects is like the cerebral hemispheres; it has a covering of grey matter, greatly convoluted

---

[*] This section is by O. C. J. Lippold.

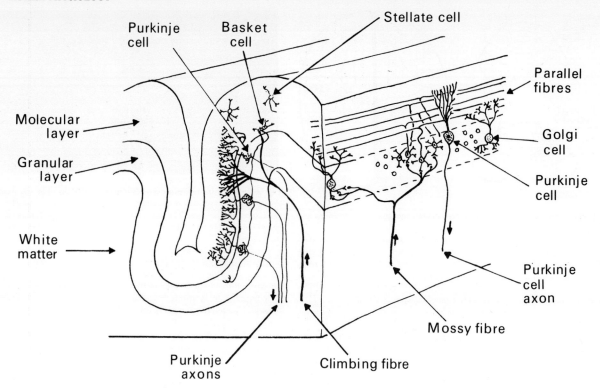

Fig. 60.10. Cells in the cerebellar cortex. Incoming information is transmitted in *mossy* fibres and *climbing* fibres; it is mainly proprioceptive in nature. The mossy fibres are branched and widely distributed; climbing fibres end locally. One type of cell, the Purkinje cell, provides the output of the cerebellum, which is entirely inhibitory.

and infolded, then white matter with great fibre tracts within it and, centrally, yet more nuclei of grey matter. As we have seen, its function is to coordinate muscular activity and because of this it has inputs from proprioceptors all over the body, in addition to inputs from most of the special senses such as the visual system and hearing. Inputs also come from the motor cortex.

This input is all fed into the cerebellum via two major fibre types, known as *mossy fibres* and *climbing fibres*. The mossy fibres, as might be imagined from the name, are widely branched within the cerebellar cortex, whereas the climbing fibres have strictly localised terminals (Fig. 60.10).

The output from the cerebellum is transmitted via the Purkinje cell axons only. Purkinje (or P-cells) are large neurones, having globular cell bodies up to 60 μm diameter, and although they have only one axon each, the dendrites are profuse and have very many branches. These branches are all in a single plane, and running at right angles to this plane are numerous small axons, called parallel fibres since they run together, looking like telephone wires. The parallel fibres make very many synaptic contacts with the P-cell dendrites, possibly a quarter of a million on each P-cell.

The cells of origin of the parallel fibres are fairly small, lying between the P-cells and the cerebellar white matter, and are called granule cells. Each granule cell gives off several short dendrites which have synaptic contacts with ascending mossy fibres and its single axon then runs up towards the cerebellar surface. When it reaches the surface, the axon makes a T-junction and then becomes the parallel fibre already referred to. This region of the cerebellar cortex is termed the molecular layer.

The molecular layer also contains many stellate cells, each about a quarter of the size of a P-cell, and having numerous dendrites spreading through the molecular layer. One kind of stellate cell such as this is called the basket cell, whose axon runs at right angles to both parallel fibres and P-cells and eventually ends in synaptic terminals upon the axon hillocks of 20 to 30 P-cells.

The final type of neurone in the cerebellar cortex is the Golgi cell. It is a large multipolar cell, and is distributed among the granule cells in large numbers. Its axons and dendrites are widely branched throughout the molecular and granular layers.

Mode of operation

Can we say how this complicated system works? Eccles[*] has, to a considerable extent, solved the puzzle by detailed work with neurophysiological recording techniques. The first startling fact that can be shown is that the output of the cerebellum, via the P-cell axons, is entirely inhibitory. This usually comes as a great surprise to most people, who expect the activity of a large part of the brain such as the cerebellum to be a complex mixture of excitation and inhibition. The P-cell output runs to neurones in the cerebellar nuclei and brainstem which pass it on to the spinal cord, basal ganglia, etc. These neurones in the cerebellar nuclei and brainstem nuclei are normally tonically active at frequencies up to 100 Hz; the cerebellar P-cell output inhibits this tonic discharge. The result is that muscular activity, which would be produced by this tonic discharge, is inhibited and, taking into

---

[*] Eccles, J. C., Ito, M. and Szentágothai, J. (1967), *The Cerebellum as a Neuronal Machine*, Springer Verlag, New York.

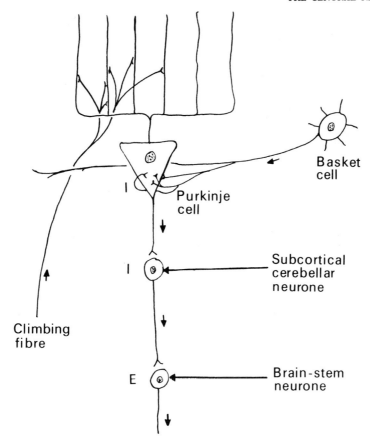

Fig. 60.11. Cell connections in the cerebellum. The Purkinje cell has one axon (its output) but a profusely branching dendritic tree, all the branches being in one plane. At right angles are running (like telephone wires) many fine axons coming from the granule cells. The granule cells are activated by synapses with ascending mossy fibres (the input). Climbing fibres (the other input) terminate directly on the Purkinje cell dendritic tree. Basket cells have synapses on the axon hillock regions of several Purkinje cells each. Inhibitory synapses are marked I; excitatory, E.

consideration the input to the cerebellum, it could be thought of as a giant negative feedback system (Fig. 60.11).

We now have to enquire where the input to this feedback system originates. As already described, the input to the cerebellum is carried in the mossy fibres and the climbing fibres. These have synaptic connections with granule cells and Golgi cells. When several granule cells are activated, it follows that parallel fibres transmit the nerve impulses. The parallel fibres have the dendritic branches of P-cells hanging on them (and making, *en passant*, synaptic contacts with them). Parallel fibre activity excites P-cells.

The molecular layer contains basket cells, whose axons are perpendicular to the parallel fibres and also make synaptic contact with the P-cells. These synapses are inhibitory. This means that if a band of parallel fibres is activated by an input to the cerebellum, the P-cells in the band are excited, but in addition, all the P-cells outside it are inhibited (Fig. 60.12).

There is sharpening of the edge of this band of excited P-cells by the action of the Golgi cells. The Golgi cell dendrites are widely distributed among the parallel fibres and so parallel fibre activity also excites Golgi cells in the neighbourhood. The multitudinous branches of these Golgi cells are widely dispersed in the granular layer and they make inhibitory synaptic contact with the granule cells. Therefore Golgi cell feedback will cut out the less strong mossy fibre activity on both sides of the main band of parallel fibre excitation. The focusing effect of the Golgi cells is very much like the phenomenon of lateral inhibition to be found in the *Limulus* eye (Fig. 60.13).

The input to the cerebellar cortex is in the form of a continuous barrage of impulses from the proprioceptors all over the body coming in through the mossy fibres. The P-cells are thus given complicated and always-changing patterns of excitation. These patterns spread to and fro over the surface of the cerebellar cortex, interact with each other and, to some extent, are a topographical representation of what is going on in the musculature of the body.

As well as this everchanging picture of P-cell activation, there is the input in the climbing fibres. The climbing fibres differ from mossy fibres in the pattern of their distribution within the cerebellar cortex. They only go to one P-cell each. Their terminals are excitatory. No matter how strong is the ambient P-cell inhibition, an impulse in a climbing fibre will cause the P-cell to fire at least once. The less strong the P-cell inhibition, though, the larger the number of impulses it will discharge in response to climbing fibre activation.

These patterns of P-cell activation are distributed to the

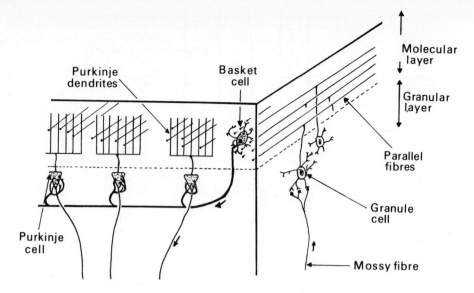

Fig. 60.12. Inhibitory action of basket cells. An input from mossy fibres activates granule cells which send impulses along their parallel fibres. The latter, via their synapses on basket cells, excite basket cell synapses which are in contact with the Purkinje cell axon hillock regions, thus inhibiting Purkinje cell firing.

cerebellar nuclei, brainstem motor nuclei and to the spinal cord as waxing and waning bursts of inhibition. This inhibition serves to coordinate and impose fine control upon limb movement in such a way that well-balanced and integrated voluntary acts can take place as already described earlier in this chapter. Eccles (loc. cit.) likens the process to sculpturing in stone:

"Spatio-temporal form is achieved from moment to moment by the impression of a patterned inhibition upon the 'shapeless' background discharges of the subcerebellar neurones, just as an infinitely more enduring form is achieved in sculpture by a highly selective chiselling away from the initial amorphous block of stone."

SERVO ACTION AND THE STRETCH REFLEX IN HUMAN SUBJECTS

Although historically, the servo theory of muscle contraction arose from experiments done on human subjects in 1948, it was, as we have seen in the previous chapter, expressed in terms of stretch reflex mechanisms derived from animal experiments and the early attempts at validation were performed on animals. Direct human evidence was much to be desired and in 1971 servo responses were demonstrated in human subjects. Their pursuit has led to a considerable extension and modification of the earlier ideas. The muscle used was the flexor pollicis longus; it lies in the forearm and is the only muscle that flexes the top joint of the thumb. The technique was to get the subject to bend the thumb repeatedly against a constant load, standardising the rate of flexion by giving him a tracking task, and unexpectedly to interfere with some of the movement. According to the servo theory he ought to react automatically to an increased load by turning on more contraction, and vice versa for an unexpected decrease in load. Such indeed proved to be the case. A characteristic set of results is illustrated in Fig. 60.14A. The upper records are integrated action potentials from the muscle. The lower records give the angle of flexion of the terminal phalanx. There were four possible situations, which were selected at random: either the movement might continue to the end with a constant load (a control, labelled C), or it might be abruptly halted by a suitable increase in load (H), or driven back by imposing a sudden force, eliciting a stretch reflex (S), or allowed to accelerate by suddenly reducing the load, eliciting a release response (R). To smooth out irregularities 16 records in each situation were collected and added up in a computer to get the average. The figure shows that about 50 msec after a perturbation is encountered the muscle's activity alters in the expected way, i.e. by increasing when the movement is halted (H) and more so when it is reversed (S), and by decreasing when the load is reduced (R). The latency of 50 msec is too short for these responses to be voluntary attempts at compensation, hence the conclusion that they are manifestations of a true automatic servo mechanism. They all have the same latency which is that of the stretch reflex (S), as it ought to be if the servo is based on the stretch reflex.

When this experiment is tried with different values for the constant initial load against which the thumb flexes, a new fact emerges, namely that the size of the servo responses is scaled to the magnitude of initial load. With ten times the initial load reversing the movement produces a stretch response ten times larger, and similarly for halt and release. Carried to the limit in the opposite direction, suddenly extending a relaxed thumb gives no stretch reflex at all, or only a small one. Thus what we may call the "gain" of the servo is roughly proportional to the ambient load. Engineers apparently do not design servo systems with this property, but in physiological terms it makes sense; the vigour of the corrective reactions is adjusted to suit the delicacy or otherwise of the task in hand. Nothing of the kind was envisaged in the original theory or perceived in the animal experiments. In decerebrate rigidity the gain of the stretch reflex is set at a non-zero value, or, if it changes, the preparation is regarded as unstable.

If the gain of the stretch reflex is zero in relaxed muscle, contractions cannot be initiated, as originally proposed, merely by causing the spindles to contract. Nothing will happen unless the gain is turned up at the same time. Vallbo's recordings from single human spindles, as already mentioned (p. 403), also appear to rule out initiation by spindle contraction.

Another new finding, not foreshadowed in the animal experiments, is that servo responses are attenuated or abolished by anaesthetizing the thumb itself (without interfering with the mus-

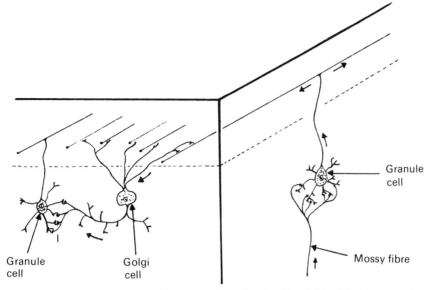

Fig. 60.13. Golgi cells exert lateral inhibition, i.e. they have a focusing effect. Golgi cell dendrites are widely distributed upon parallel fibres. Parallel fibre activity excites Golgi cells, whose axons are inhibitory to granule cells. This inhibitory feedback will abolish any weaker mossy fibre activity at the edges of an excited region. Cerebellar cortex is activated by widespread input from the proprioceptors of the body via the mossy fibres. Thus Purkinje cells are affected by always changing, complicated patterns of excitation, reflecting the action of the muscles of the body.

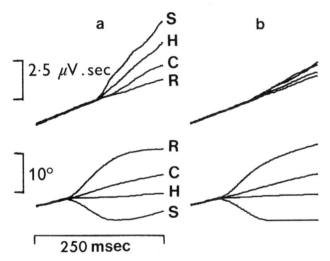

Fig. 60.14 Servo responses from the long flexor muscle of the human thumb and the effect thereon of anaesthesia of the thumb.

The proximal phalanx of the thumb is held in a clamp. The distal phalanx, by pressure of the pad on a crank arm, moves, or is moved by, a low inertia electric motor the shaft of which is coaxial with the axis of rotation of the top joint of the thumb. To begin with, in all movements, a constant current in the motor causes it to offer a constant opposition to movement. The load can be altered suddenly by changing the current in the motor, introducing perturbations. A potentiometer on the motor shaft measures the angular displacement of thumb. The subject is given a tracking task to standarize the flexion movements he makes.

The electrical activity of the flexor pollicis longus is recorded by surface electrodes over the lower part of the belly of the muscle in the forearm. The action potentials are full-wave rectified and integrated. The effect of this is that a constant activation of the muscle gives an integrated record of constant slope. If the activity changes the slope changes in proportion.

Flexion movements are 20° in extent and last 1·0 s, if the subject tracks accurately. Recording only takes place for 250 msec in the middle of the movement. The various perturbations, which are detailed in the text, are all applied 50 msec after the start of a recording sweep. The electrical records are above, with the corresponding displacement records below. Each trace is the average of 16 trials, as described in the text.

(a), records taken with a normal thumb; (b), records taken after anaesthetising the thumb by a pneumatic tourniquet round the wrist for 1·5 h (Marsden, Merton and Morton, *Nature*, vol. 238, p. 140, 1972.)

cle, which lies in the forearm). This is shown in Fig. 60.10B. It is not known whether it is the skin or the joint receptors which have to be put out of action to get this effect, but the skin seems the more likely candidate. Particularly with a freshly anaesthetized thumb the subject is conscious that he needs to make a larger effort of will to move his thumb than he normally does. He misses his servo. At these times the contraction must be largely driven through the α route; but during movement in normal circumstances the α component does not appear to be large, for good release responses are seen at all levels of initial load. If the muscle can be silenced by a sudden release, the simplest interpretation is that the contraction was dependent on spindle excitation via the γ route. From Vallbo's results, however, the α route is likely to be important at the start of a movement.

The effect of anaesthesia is equivalent to a reduction of servo gain, but it is probably not the case that the gain is normally adjusted to the load by pressure sense from the pad of the thumb, for there appears also to be an increase of internal gain when the muscle is fatigued, in circumstances where the pressure on the thumb is unaltered. The servo treats the decreased muscle contractility in fatigue as equivalent to an increase in load; the result, it turns out, is a better compensation for fatigue than would be expected on the original theory.

An unexpected feature of servo action in the thumb is the long latency, 40–50 msec, as compared with 20–25 msec for the time to and from the spinal cord (as evidenced by tendon jerk latencies). One attractive explanation of this extra delay is that the stretch reflex arc for the thumb runs to the cerebral cortex and back. Phillips, from his work on the rapid corticospinal pathway ending monosynaptically on the motoneurones of the baboon's hand and the knowledge that impulses from muscle spindles have a pathway (later found also to be very rapid) to the cortex, suggested that in the case of the primate's hand the spinal stretch reflex arc had "been overlaid in the course of evolution by a transcortical circuit". This theory remains speculative, although, in man, it has so far survived a number of experimental tests which might have ruled it out.

In proximal arm and shoulder muscles in man, biceps and infraspinatus, there appear to be two components in the stretch reflex, one of spinal latency and one compatible with a cortical pathway. In infraspinatus neither component is affected by anaesthesia of the arm up to the axilla. Susceptibility to peripheral anaesthesia may thus be a special feature of the stretch reflex in hand muscles. If the stretch reflex from these muscles is transcortical, the cortex would be a natural place for tactile messages from the digits to facilitate it, if that is what happens. It must always be remembered that although the human experiments in many ways confirm the predictions of the original servo theory based on spindle contraction, that is no guarantee that the theory is correct, for other possible mechanisms cannot be ruled out.

The new knowledge of the stretch reflex discussed in this and the preceding chapter helps with the interpretation of clinical symptoms, so far on a purely speculative basis. In the days when the tendon jerk and the stretch reflex were supposed to have the same reflex pathway, special pleading was required when they did not alter hand-in-hand. They generally did in spastic paralysis, e.g. in hemiplegia, but sometimes in atypical cases they did not. It was also puzzling that (as described earlier in the relevant section of this chapter) the tendon jerks were commonly of normal briskness in both Parkinson's disease and with cerebellar lesions, although in the former the muscles were rigid and in the latter hypotonic. Now that several differences between the mechanism of the stretch reflex and the tendon jerk have emerged, it is not difficult to see how they may vary independently. Furthermore, the stretch reflex itself apparently has more than one component and this may be relevant to the different types of increased tone, spasticity and rigidity, encountered. Finally in hypertonic states we have to consider not only the possibility of spindle cramp, mentioned on p. 404, but also of distortions of the new-found gain control.

61. The Central Nervous System: III

P. A. Merton

Learning and memory

A man's present actions may be influenced by what happened around him, by what he thought and felt, and by what he learnt to do many years ago. The question of how and where the brain stores memory traces of these things is of great general interest, but so far only hints of the answer are available. As regards how, it has been difficult to find electrical phenomena, action potentials, changes in synaptic excitability, etc. which last longer than a few seconds. It has therefore been suggested that memories depend on nerve impulses continually circulating over closed pathways, the nature of the memory being determined by the pattern of the pathway. This theory runs into the difficulty that memories cheerfully survive events such as profound anaesthesia or epileptic fits which would be expected to stop or disorganise the electrical activity of the brain. For these reasons it seems probable that long-lasting structural changes, will prove to be involved, using the term in the widest sense to include such things as changes in the amount of an enzyme in synaptic terminals. The possible nature of these changes is a matter of speculation only, at the present. In the last few years, careful experiments in animals appear to have demonstrated changes lasting for hours or even days in the efficiency of transmission of nerve impulses along certain pathways in the brain. But it is uncertain what structures are involved in these changes and whether they have anything to do with laying down memories.

A fact that any theory has to account for is that recent memories are more easily lost than old ones. It is a commonplace that old people may recall endless details of their childhood after they can no longer remember the events of middle age, or even of a week ago. Similarly after a head injury with concussion the patient may remember nothing of the events leading up to the accident; he has what is called a *retrograde amnesia*; commonly it goes back a few minutes or an hour or two, but in severe cases it may extend to several weeks or even years before the accident. Retrograde amnesia is one instance of a general principle, ennunciated by Hughlings Jackson, that the most recently acquired and the most highly organised activities are the first to be lost when the brain is damaged. A hemiplegic loses the ability to write while he can still pick up his bread and butter. The extensor plantar response (p. 398) he exhibits represents a return to the infantile form of the reflex. The emigré who develops a degree of aphasia reverts to his mother tongue, which he may not have spoken for years.

In addition to their retrograde amnesia, patients recovering from head injury pass through a phase when, although behaviour is normal and consciousness not clouded in any way, they are not laying down new memories, and when fully recovered cannot recall what happened during this period of *anterograde amnesia*. A familiar instance is seen when a football player is mildly concussed. After a momentary loss of consciousness, followed by a short period of confusion, he rises and plays on, apparently normal. After the game it is found that he has a brief retrograde amnesia and, in addition, has no recollection of what he did for several minutes after he came to again. With more severe injuries there is a rough correlation between the durations of the retrograde and anterograde amnesias.

These facts suggest that all memory traces have a tendency to fade, but this is normally counteracted by a mechanism which consolidates memory traces. This mechanism is temporarily put out of action by head injury. No new memories are consolidated, and the least consolidated (i.e. the most recent) old memories fade irretrievably before the consolidating mechanism restarts. The fact that retrograde amnesia may be of any duration up to several years, shows that memories remain subject to consolidation indefinitely.

There is evidence that the mechanisms that organise the formation and recall of memories are in the under part of the temporal lobes near the cortical area for smell and taste. (This is not to say, of course, that the actual memory traces are laid down in this region.) It has been known for a century that in epilepsy beginning in the under part of the temporal lobe (uncinate fits) the patient may experience feelings that what is going on around him is intensely familiar; it has all happened before (*déjà vu*). Or he may vividly relive incidents in his past life. Some twenty years ago it was found that if the anteromedial parts of both temporal lobes are removed (as was done in an attempt to cure epilepsy) new memories cannot be formed. The patient cannot remember what happened even a few seconds ago and lives literally for the moment. Intellect, sensation, and other cortical activities appear unaffected. A point of great interest is that in such patients it is found that new motor skills can be learned and retained. There is also evidence from a single patient that a motor skill, learning to typewrite, acquired during a period of dense retrograde amnesia due to a subsequent head injury, was not lost afterwards. It seems that the learning of motor skills employs a different nervous mechanism from the learning of ordinary memories. Those not afraid of speculation (and there are few people who are when it comes to the brain) will recall the hint dropped a few pages ago about motor skills and the cerebellum.

Loss of memory is an early symptom of generalised shrinkage of the cortical grey matter when it occurs in old age or in various forms of dementia. The demented patient forgets past events, names, faces, words, the time, where he is, what he was going to do next, and so on. He also loses his code of social behaviour, his power of reasoning and his ability to cope with new situations. One might say he has forgotten how to behave, how to think and how to adjust. Many of these disabilities are presumably due to loss of the learned activities of some of the special areas of the cerebral cortex already described. A patient with damage to the frontal lobes can be said to have forgotten how to behave. The aphasic cannot recall names or words. After damage to the parietal lobe a patient may forget where he is. Regional cortical lesions of this kind, however, seem never to cause a loss of memory for past events. It is therefore sometimes suggested that memory for past events is a function of the cortex as a whole.

Support for this view came from Lashley's experiments with rats in the 1920s on the learning and retention of simple habits, such as threading a maze to obtain food. He excised various parts of the cerebral cortex and came to the conclusion that it did not matter which part he removed; the impairment in learning appeared to depend only on the total area of cortex removed. The only sign of localisation was that tasks involving visual pattern recognition required the visual areas. Lashley's results could be explained if the learning process involved every part of the cortex.

Information on this question and on other fundamental issues in learning and memory has come from the split-brain preparation pioneered by Sperry and Myers. In their leading experiment the main fibre connections between the two cerebral hemispheres of a monkey were severed by midline section of the corpus callosum. The optic chiasma was also divided in the midline. For the right eye this left only optic nerve fibres running to the right hemisphere, and likewise for the left eye. One eye, say the left, was covered with a patch while the animal was trained to make some visual discrimination with the right eye, e.g. between a square and a circle (see under "conditioned reflexes" below). When the patch was then shifted to the right eye, and the same task was presented to the left eye, there was no evidence of transfer of learning, i.e. it took as long to train the animal to make the discrimination with the left eye as it did originally with the right eye. Hence the memory traces of the task learnt with the right eye were confined to the right hemisphere. When the experiment was repeated without section of the corpus callosum, the learning of the task transferred readily from one eye to the other.

Cutting the corpus callosum at various intervals after learning the task showed that the memory traces were at first confined to the hemisphere into which the visual information came, but after some days another set of memory traces had been laid down in the other hemisphere; for, after cutting the corpus callosum at that interval, the task was performed equally well with either eye. Such doubling-up, however, is not found with all tasks in the monkey and is little in evidence in man, in whom the hemisphere associated with speech is "dominant" to the other; e.g. a patient whose corpus callosum has been divided can only read print which he sees in the half of his visual fields belonging to his dominant hemisphere. (It is remarkable that, in man, division of the whole corpus callosum, the largest fibre tract in the brain, gives rise to no disabilities that obtrude into everyday life. Indeed, none were detected at all until a few years ago.)

Conditioned reflexes

Reflexes such as the knee jerk, or constriction of the pupil when light is shone in the eye, are innate, but many other automatic actions are learnt. If a dog is given food and at the same time a bell is rung and this is repeated several times at intervals of, say five minutes, it is found that after some time the dog will salivate when the bell is rung, without any food being offered. The dog's nervous system has learnt to associate the bell with food and a new reflex is established. The famous Russian physiologist, Pavlov, called the process of acquiring such reflexes "conditioning" and the reflexes "conditioned reflexes". Innate reflexes were called "unconditioned". Most reflexes that depend on visual and auditory recognition by the cerebral cortex are probably conditioned, not innate. Salivation when food is placed in the mouth is an unconditioned reflex. In addition, all normal dogs, if hungry, will salivate when merely shown meat. Pavlov showed that this is a conditioned reflex, for if dogs are reared without ever eating meat they do not salivate when shown it for the first time, although if allowed to eat it they salivate at once. In man such reflexes as the blink to a threatening movement, and an enormous number of other automatic responses to events signalled by the eyes and the ears, are doubtless conditioned reflexes established in childhood. Only the very simplest conditioned reflexes remain or can be established after removal of the cerebral cortex.

In the first half of this century Pavlov and his school intensively studied conditioned reflexes, working chiefly with reflexes causing salivation in the dog. One parotid duct was transplanted to the skin of the cheek so that the saliva could be collected, and its amount measured. Thus the experiments were made roughly quantitative. Conditioned reflexes are easily prevented by any disturbance. Thus if someone walks into the experimental room, or if a buzzer is sounded, the conditioned secretion of saliva to ringing of a bell is likely to fail on that occasion. For this, and other reasons, the dogs were put in sound-proof rooms with the experimenter invisible outside, and all manipulations, etc. were done by remote control. The prevention of conditioned reflexes by miscellaneous interruptions is called "external inhibition". The other sort, "internal inhibition", occurs, for example, if (having established a conditioned reflex) the bell is rung at intervals but food never follows. After some repetitions the flow of saliva dwindles and ceases. It would not be very surprising if the association were lost under these circumstances, but there is good evidence that the reflex is not so much lost as actively suppressed. For instance, it recovers very rapidly if food is presented with the bell once or twice. This is called "reinforcement". It also recovers spontaneously if the dog is left for some hours. But most striking of all, it reappears at once if a stimulus, such as an electric buzzer, of the kind that usually causes external inhibition, is given at the same time as the bell. It seems that the first thing external inhibition acts on is the last thing the animal learnt, namely the association of the bell with no food which overrode the original reflex. Hence the conclusion that the original reflex was inhibited, not lost.

Animals can be trained to make sensory discriminations in their conditioned reflexes. For example, if a musical note of a certain pitch is established as a conditioned stimulus, then initially many other notes will also excite the reflex. But if the first note alone is reinforced (i.e. followed by food) while the others never are, sooner or later only the original note is effective. Pavlov thought of this as an internal inhibition to the other notes. When an indifferent disturbing stimulus is given, such as the buzzer, discrimination tends to be lost; a wider range of notes again causes the reflex (internal inhibition again counteracted by external inhibition).

If notes closer and closer together are used the animal eventually has difficulty in distinguishing them and when faced with the task of doing so may display symptoms very like those of human anxiety neurosis. In his later years Pavlov devoted much work to these experimental neuroses. It was the observation by Fulton and his associates in America that a chimpanzee with such a neurosis lost her anxiety after removal of the frontal lobes that led to the development of the operation of prefrontal leucotomy by the Portuguese surgeon, Moniz (p. 415).

In addition to Pavlovian conditioning of reflexes there is another type of experimental learning situation, originally called "trial and error" learning, but now known as "operant conditioning" or "instrumental learning". When a hungry cat is placed in a cage with food outside, it will struggle to get out. If there is a lever which opens the door, sooner or later the cat will accidentally press the lever and get out. If this procedure is repeated over and over again, the time taken before the lever is pressed diminishes, until finally the cat operates the lever directly it is placed in the cage. This basic experiment lends itself to endless elaboration, e.g. the cat can next be trained to distinguish between a lever with a square end and one with a circular end.

Both conditioned reflexes and operant conditioning have been very widely used by experimental psychologists for testing the performance of sense organs in animals. If a dog can be trained to salivate to one note and not to another half a semitone below it, it obviously has pitch discrimination at least as good as that. It is important to note that the reverse does not follow. In this type of experiment the animal has not only to distinguish the two stimuli but also to remember the first stimulus for several minutes. Dogs happen to have an excellent memory for pitch, but many experiments claiming to show that various animals do not have colour vision may merely mean that they have no colour memory.

Pavlov's work was important because it was the first experimental investigation of higher nervous function in animals. He was careful not to use subjective words like perceive, desire, etc. in describing the results of the experiments. A great deal was learnt about the rules of habit formation in the cerebral cortex, but only recently has a start been made at investigating the neuronal mechanisms involved.

The electrical activity of the cerebral cortex

Nerve impulses can be detected arriving in the projection areas of the cortex and leaving the motor area, but what goes on in between is mysterious. When recording electrodes are placed on the surface of the brain potential oscillations are recorded all the time, even when no sensory impulses are arriving, unless the animal is deeply anaesthetised. In a normal conscious man such brain waves can be detected through the skull by electrodes resting on the scalp; the record is called the electroencephalogram (EEG). Over most of the head it is usual to find regular more or less sinusoidal waves, some 50 microvolts (5×10^{-5} V) in size with a frequency of about 10 Hz. These waves, known as the *alpha rhythm*, were first described by Hans Berger in 1929. The alpha rhythm is present only when the eyes are shut; with the eyes open it is replaced by lower voltage, irregular activity (Fig. 61.1). It appears that the alpha rhythm is characteristic of an awake but inattentive brain. Thus even with the eyes closed the alpha rhythm may disappear if the subject's attention is engaged, if he tries to do mental arithmetic, or if he is unexpectedly touched. Normally it is difficult to be inattentive with the eyes open, for vision is the dominent sense, but if the scene is made uninteresting, for example by wearing strong spectacles which blur everything, or by looking at a blank screen, the alpha rhythm returns. If inattention goes so far that the subject falls asleep, the 10 Hz rhythm dis-

appears and is replaced by irregular higher voltage waves of much lower frequency, 1–3 Hz. At intervals during any period of prolonged sleep these slow waves vanish and the EEG exhibits low voltage fast activity, similar to what is seen in the waking attentive state. These phases, during which the subject remains deeply asleep (hence the name, paradoxical sleep), are discussed below.

The alpha rhythm must be due to the synchronised beating of very large numbers of cortical neurones, for if few neurones were involved or if large numbers were beating asynchronously at different frequencies, the resultant potentials would not be large enough to detect through the skull. The detailed mechanism of the alpha rhythm is not understood, but it seems to be an example of a general tendency to synchronised rhythmical oscillation that is shown by many large masses of nerve cells when they are not doing anything in particular. Remarkably similar waves can be recorded from the optic ganglion of a water beetle, and, like the alpha rhythm, they disappear when a light is shone into the eye (Fig. 61.1).

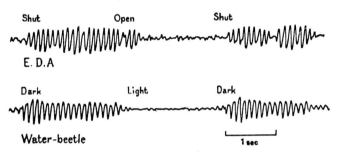

Fig. 61.1 The electroencephalogram of a human subject (E.D.A.), recorded with electrodes on the scalp, showing the prominent α rhythm at about 10 cycles/sec with the eyes shut, and its disappearance when the eyes are open.

The lower record shows the very similar rhythm obtained by leading from the optic ganglion of a water beetle in the dark, which again is blocked by allowing light to enter the eye. (Adrian and Matthews, *Brain*, vol. 57, p. 355, 1934.)

Nothing very definite can be made out in the cortex when attention is aroused by the arrival of a sensory message. Leading directly from the cortex in animals has shown that the synchronised activity of large regions is broken up and replaced by an indecipherable shifting pattern of waves of varying amplitude and frequency.

The human electroencephalogram is important in medicine. During an epileptic fit very large waves occur all over the brain. Many epileptics continue to show some abnormal waves in between attacks, and their discovery is often useful in diagnosis. Distinctive abnormalities also occur around cerebral tumours and other lesions, and assist in their location.

The cerebellar cortex also shows spontaneous rhythmical waves, but of a much higher frequency, some 250 Hz. They cannot be picked up by electrodes on the scalp in man.

Consciousness and sleep

Throughout these chapters it is taken for granted that conscious happenings in the mind—sensations, emotions, voluntary efforts, etc.—are associated with definite physical events in the brain. This is a reasonable working assumption until it is shown to be wrong, and the resultant mixture of physical and psychological terms it encourages (as when we speak of a tract of nerve fibres conveying sensations of pain) saves space and is not likely to be misleading unless it is taken out of context. The nature of the interaction between mind and brain is so far a matter for philosophers, but physiologists have something to say about what parts of the brain are involved in conscious life.

A distinction must be drawn between the state of consciousness, whether we are conscious or unconscious, and the content of consciousness—what we are conscious of. The content of consciousness appears to depend mainly on the normal activities of the cerebral cortex. With the emphasis changing from moment to moment as our attention shifts, we are ordinarily aware of what we see and hear and feel and of what we are thinking and doing, and so on. As described in previous sections it looks as if, when the region of cortex dealing with one of these activities is lost, that side of consciousness drops out; it ceases to be a possible object of attention and the conscious self is that much diminished. Were all the cortex removed we should presumably be conscious only of the crude skin sensations attributed to activity in the thalamus; and with the thalamus gone too, of nothing. Thus the content of consciousness seems to depend mainly on how much of the cerebral cortex is there, or rather on how much of it is in a normally responsive state. It does not respond normally to sensory impulses arriving if an anaesthetic has been given or if an epileptic fit is in progress, and at these times the subject is unconscious.

In ordinary life our state of wakefulness or sleep seems to be determined by events at the level of the brainstem. It has been known for a long time that tumours growing in this region may cause pathological sleepiness and that mechanical interference around the midbrain at operation under local anaesthetic is liable to cause a rapid lapse into unconsciousness. On the other hand, large pieces of cortex can be removed at operation under local anaesthetic without any apparent alteration in mental state. Probably it is for the same reason that a punch on the jaw which transmits the shock to the base of the brain is a much more effective way of knocking a man out than hitting him on the top of the head.

The precise location and mode of action of these lower centres is at the moment the subject of much research. Magoun, Moruzzi and their followers urge the claims of the reticular formation of the brain stem to be the region responsible for keeping the cortex awake. The principal evidence is firstly that lesions destroying the grey matter around the aqueduct of the midbrain in animals leads to somnolence. This is not due to interference with the sensory tracts running to the thalamus, for it occurs when these are not damaged; conversely, lesions interrupting the sensory tracts but not the reticular formation do not cause sleepiness. Secondly, electrical stimulation of this region in anaesthetised animals converts the electrical activity of the cerebral cortex from the pattern characteristic of sleep to that of wakefulness (see p. 425). Similar "arousal reactions" cannot be obtained from elsewhere in the midbrain.

There is evidence, then, for the view that normal wakefulness depends on the action of neurones in or near the reticular formation. The next question is what determines whether the reticular activating system is itself active or quiescent. In the nineteen-twenties Hess showed elegantly that stimulation of the lower part of the thalamus (and of several other regions at the base of the brain) could cause sleep. He used freely moving conscious cats with electrodes chronically implanted into the thalamus. Appropriate stimulation caused a cat to look sleepy, circle round, lie down and go to sleep, exactly like a normal cat going to sleep. There was no question that the cat was merely knocked unconscious by the current. These experiments suggested that falling asleep may be an active process. There is evidence in man, too, that falling asleep is more than something that happens when all sensory inputs are reduced to a minimum, although this assists. In the mysterious disease of narcolepsy the patient is seized with irresistible drowsiness during the day, perhaps several times a day, and rapidly falls asleep. After a minute he can be aroused and is once more normal. Jouvet has now found in the cat a group of neurones in the midline of the brainstem, the raphe nucleus (not shown in the figures), which appear to be able to suppress the reticular activating system. If they are destroyed, the animal becomes strikingly wakeful, sleeping only 10 per cent of the time, instead of the 70 per cent or so which is normal for a cat. These neurones contain a high concentration of 5-hydroxytryptamine or serotonin. If it proves that the raphe nucleus is what controls the reticular activating system in the normal sleep cycle, we are still left with the problem of its own cyclical activity. The relation of the raphe system to the structures higher up the brainstem stimulated by Hess is not known.

As was hinted in the previous section, sleep itself can no longer be regarded as a single state. Ordinary sleep, in which the subject makes no limb movements and exhibits slow waves in his EEG, is interrupted several times a night by periods of paradoxical sleep, lasting for a few minutes, during which there are jerky movements of the limbs, sudden eye movements under closed lids, and low voltage fast waves in the EEG characteristic of wakefulness. But, judged by the intensity of stimuli necessary to wake him, the subject is more deeply asleep during paradoxical sleep than during ordinary sleep. During paradoxical sleep in animals, strong bursts of activity can be detected in many parts of the brain, including the optic radiations. There is evidence that this activity originates in the pontine region of the brainstem. It can scarcely be irrelevant that the periods of paradoxical sleep are the times when human subjects dream, as discovered by waking them up when their EEG indicates that they are in this phase. For these reasons ordinary sleep is also known as slow-wave sleep, or light sleep; paradoxical sleep is also known as fast-wave sleep, rapid-eye-movement sleep, deep sleep and dreaming sleep.

During light sleep in the cat, the head is held up by a "tonic" contraction of the neck muscles (the activity in which can be monitored by recording their action potentials). On passing into paradoxical sleep, the head falls forward, electrical activity in the neck muscles ceases, and spontaneous eye movements, begin. Jouvet has made the remarkable discovery that a cat, decerebrated by removal of the forebrain and midbrain after section of the brainstem at the level of the upper border of the pons and kept alive by skilled treatment, oscillates regularly for weeks between paradoxical sleep, with flaccid neck muscles and roving eyes, and a state of apparent wakefulness, during which the neck muscles are tonically active. No paradoxical sleep is seen, however, if the section is behind the pons. Hence there appears to be a centre for paradoxical sleep, with inherent cyclical activity, in this region. This has been identified as the locus caeruleus, a bilateral nucleus in the pons (not shown in the figures) whose neurones contain noradrenaline.

SUMMARY OF CHAPTERS 59 TO 61

The two main categories of nervous action are reflex and voluntary. Subsidiary categories are spontaneous and quasi spontaneous actions (e.g. breathing, blinking), the expression of emotions, postural and locomotor activities.

Reflexes are of all grades of complexity, some (e.g. conditioned reflexes) involving the cerebral cortex; but they have usually been studied in the spinal cord, because of the relative ease with which the conditions of the experiment can be controlled.

If the spinal cord is sectioned, to cut off the hind portion from the unpredictable influence of the higher nervous centres, a state of spinal shock ensues, much more severe in the higher mammals than in the frog. There is no spinal shock, however, in the much used decerebrate preparation, in which after a section at the level of the midbrain, the forebrain is removed.

Sherrington and his pupils mainly drew their conclusions from experiments on a small selection of reflexes; the flexion or withdrawal reflex (a manifestation of which is the Babinski or extensor plantar response in man), the scratch reflex of the spinal dog, and the stretch reflex (to which the clinically important tendon jerks are closely related). The most deeply analysed of these is the stretch reflex, evoked by impulses from the muscle spindles. The complexity of the stretch reflex is due to the fact that the muscle spindles themselves are under efferent control and contract during reflex action (and probably in voluntary and postural action too). The stretch reflex may have both spinal and cortical components.

Righting reflexes and other postural reactions were described in detail by Magnus, but their internal mechanism is not known; and the same goes for the mechanisms responsible for locomotion.

Conscious voluntary action is the province of the cerebral cortex, different functions being localised in different parts of the cerebral hemispheres. The visual area, which receives impulses from the retina (via the lateral geniculate nucleus), is in the occipital lobe. Damage to the visual area causes blindness. The cortical area for hearing is in the temporal lobe; the area for taste and smell is underneath this lobe.

The important region that receives sensory messages from the body surface, etc., is in the postcentral gyrus. After removal of this cortical area in man, all delicate sensation on the opposite side of the body is lost; but crude sensations are retained. Impulses to the brain to arouse conscious sensation ascend the spinal cord in two pathways: fine tactile sense and position sense run uncrossed in the dorsal columns; sensations of pain, temperature and crude contact are conveyed in the crossed spinothalamic tract.

The motor area is in the precentral gyrus. Damage to this region causes voluntary paralysis on the opposite side of the body, while electrical stimulation gives movements of the same parts. The motor area is connected to the spinal cord by the large corticospinal or pyramidal tract. There are also large extrapyramidal pathways. Damage to these motor pathways (an "upper motor neurone lesion") causes spastic paralysis with exaggerated tendon jerks, an extensor plantar response, and no muscular wasting.

In front of the motor area on the left side (in right-handed people) is the cortical motor area for speech. The receptive side of speech is dealt with in the parietal and temporal lobes. Damage to any of these regions results in loss of speech, aphasia.

The function of the extrapyramidal motor nuclei (the basal ganglia, etc.) is not well understood. In Parkinson's disease, in which they are damaged, there is clumsiness and slowness in voluntary movement, with muscular rigidity and tremor.

The cerebellum in man is mainly concerned with the subconscious side of voluntary movement. Damage leads to a general disorganisation of voluntary movement, with intention tremor and muscular hypotonia, known as cerebellar ataxy, the symptoms being on the same side of the body as the lesion.

All muscular movements (reflex, voluntary, etc.) probably have various motor mechanisms in common, e.g. reciprocal innervation, and feedback mechanisms based on the stretch reflex.

Learning and memory are cortical functions, which are associated with the temporal lobes. Memory theories have to take into account the clinical facts of retrograde and post-traumatic amnesia, and the results of experiments on animals and men after section of the corpus callosum.

Conditioned reflexes (and operant conditioning) are again mainly a cortical phenomenon, widely used for the study of sensation and learning in animals.

The electrical activity of the cortex can be recorded in man by electrodes on the scalp; the resulting record is the electroencephalogram (EEG).

Although conscious life as we know it depends on activity in the cortex, the state of wakefulness or sleep is determined by centres at the base of the brain and in the brainstem. There appear to be two phases of sleep, ordinary light sleep being interspersed with periods of deep or paradoxical sleep.

Section 5:

Growth and Reproductive Physiology

Section 5. Introduction

One of the major attributes of all living matter is its ability to reproduce. In the higher forms the simple process of reproduction has become inextricably mixed with the complications of sex, to form a mechanism for the intermingling of genes in new individuals.

In mammals, the developing young pass the early part of their lives within the actual body cavity of the mother, protected in this way from the physical and chemical vicissitudes of the external environment and provided, from her bloodstream, with a supply of food and oxygen. The developing embryo, formed from the union of the ovum and sperm lies in the uterus, a thick-walled, muscular organ situated in the lower abdomen. The embryo has a blood circulatory system which comes into intimate relationship with the circulation of the mother, the maternal and embryonic blood vessels producing a structure called the placenta. It is through this organ that food, oxygen and other substances diffuse from the mother to the fetus and through which carbon dioxide and waste products pass in the opposite direction. During pregnancy, the fetus and the placenta associated with it continue to grow and during this growth process the uterus itself enlarges to accommodate its expanding contents. Pregnancy, which in the human lasts for about forty weeks, is terminated by a series of contractions of the muscular wall of the uterus, a process which ultimately leads to the expulsion of the fetus and the placenta. For a time after birth, the young are fed wholly or partly by means of milk secreted by the mother's mammary glands.

The nervous system coordinates body function, a fact known before the time of Hippocrates, but the details of other coordinating and regulating mechanisms have been elucidated since. Among the first of these mechanisms known, was the influence exerted by the primary reproductive organs, or gonads, on various aspects of sexual function. Subsequent investigations indicated that a relatively complex system of circulating chemical substances, hormones, acted as controlling agents in the activity of the reproductive system. The pituitary gland was found to have the primary controlling function in the series of physiological processes which occur in reproduction. A number of these hormones are now known to exist, some having been isolated chemically, yet in many cases detailed explanations for the observed phenomena are still lacking or imperfectly understood.

During early embryonic life, the reproductive system develops identically in the male and female, although the genetic makeup of the embryo has already determined its sex. The genital ridge, which later is to give rise to either the male or female primary reproductive organs, is differentiated by the fifth week of intrauterine life. The germinal epithelium, which forms the germ-cells, is the top layer of this ridge and between the eighth and tenth week differentiates into either ovarian tissue of the female or testicular tissue of the male.

The influence of hormones which originate in the gonads, even at this early developmental stage, can be shown by removing the gonads. If this is done, in either male or female embryos, a female type of internal and external genital system develops. Thus the conclusion may be drawn that the testis secretes a hormone, the presence of which is necessary for differentiation into the male pattern.

After a period of growth, in man 11 to 15 years, puberty occurs. At this time, in the female the ovaries begin to produce and release ova, and in the male the testes form spermatozoa. Physical and psychological changes take place resulting in the development of the secondary sexual characteristics. Girls become an interesting and attractive shape (to boys) as a result of enlargement of the breasts, deposition of subcutaneous fat and changes in the bony structure of the pelvis. Pubic and axillary hair appear in both sexes although the distribution is slightly different. In boys, the beard grows, the voice breaks, and typically male skeletal and muscular growth takes place. These changes occur in response to the presence in the body of increased amounts of the sex hormones—androgens in the male oestrogens in the female—secreted by the testes and ovaries respectively.

Growth

Growth is in many respects a similar physiological process to reproduction; at least it is complementary. Many of the basic control systems are common to both. When growth is played out in adulthood, new individuals must be produced to maintain the species.

62. Human Growth

The study of the process of growth is of considerable importance in physiology and medicine, for it gives us further insight into individual differences and into the abnormalities which occur in disease. It is not the uncontrolled growth of a carcinoma that is hard to explain, so much as the incredible control of growth patterns that results in such a beautifully perfect object as an eye.

Growth curves

Figure 62.1 is a growth curve. It shows (a) the height of a boy, plotted every six months from birth to 18 years old, and (b) the increments in the height between each observation. Growth may be thought of as a form of motion; hence curve (a) represents distance moved while curve (b) is one of velocity. The velocity curve is a better measure, at a given time, of the "state" of the individual since it represents the growth process actually taking place, and is not dependent to a great extent on preceding growth. The biochemical reactions involved in growth are likely to be paralleled with the velocity curve rather than be related to the size of the individual at a given time. Acceleration may be an even better measure, e.g. of the spurt in growth at puberty.

The record illustrated in Fig. 62.1 is the oldest study of growth of a child, and it was done from 1759 to 1777 by Count Philibert de Montbeillard on his son. One can observe that the velocity of growth in height decreases from birth, apart from a short spurt during adolescence. Since the time of de Montbeillard, it has been shown that the peak growth velocity is achieved during intrauterine growth, at about four months following conception, and that the rate decays smoothly thereafter. Growth curves between 36 to 40 weeks of intrauterine life show a slowing, which is thought to be due to the restriction of the uterus. Twins show this slowing at an earlier age when their combined weight is equal to that of a single 36-week fetus. This slowing mechanism enables a genetically large fetus in a small mother, to be successfully brought to term and delivered.

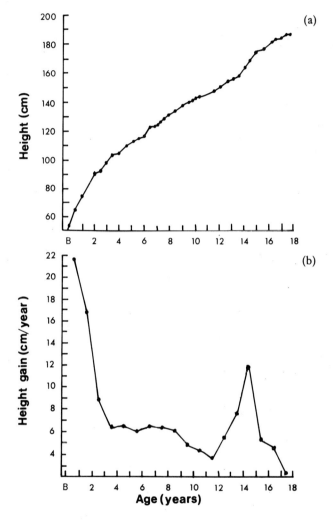

Fig. 62.1 Growth in height of de Montbeillard's son from birth to 18 years, 1759–77.

Mathematical expressions for growth

Growth is an extremely regular process (despite certain statements to the contrary that it proceeds jerkily) and many attempts have been made to fit mathematical curves to the data. Truth in such an approach is elusive, for either the equation has so many irrelevant parameters that it is meaningless and cannot be interpreted biologically, or simple solutions only apply to the data from which they were originally constructed and do not describe the general situation. The difficulty would appear to be that the measurements usually made upon growth are highly complex, biologically speaking. The height of a child, for example, is the sum of the lengths of the shin, thigh, trunk, neck and the head height—all of which have separate and different growth curves.

If one measures simply the increase in length with time of the tibia, for instance, it is still far from clear what biological assumptions must be made as a basis for a growth curve.

Assuming that cells are continuously dividing at a given rate leads to the increment being proportional to the measurement. Equally one might assume that a given number of cells already present are adding material to themselves. This leads to increments being constant with time. Such simple ideas are unlikely to be true, and a complex mixture of factors such as the proportion of the two varying with time independently and their two rate-constants decreasing with time, might be nearer the mark.

Such curve fitting, however, does assume practical importance in research about the effects of illness on growth or upon genetic factors in growth.

Longitudinal and cross-sections

So far we have considered the growth of a *single* individual. Other measures are possible. For instance, in a cross-sectional study, each individual in a group is measured once only. Different individuals are thus used to obtain the growth measurements at different ages. In a *longitudinal* study, each individual is measured at each age as it grows up.

Cross-sectional studies are easy to carry out, can be done with large samples, and give information about the standards for height and weight reached by normal children at each age. Any individual differences in growth, e.g. the spurt at adolescence, are obscured.

Longitudinal studies are laborious but are essential in revealing the details of genetic or hormonal influences on growth, or interesting information such as the relation between growth and educational achievement, psychological development, etc.

Growth of different tissues

The dimensions of muscles, organs such as the liver and kidney and the bones, on the whole, follow the growth curve for height. Other tissues behave differently, as shown in Fig. 62.2. The

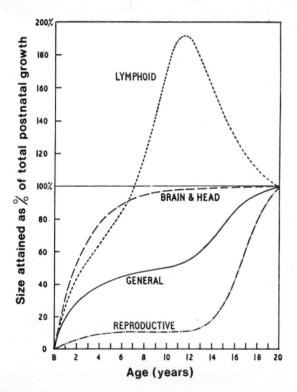

Fig. 62.2 Growth curves of different parts and tissues of the body, showing the four chief types. All the curves are of size attained, and plotted as percentage of total gain from birth to 20 years, so that size at age 20 is 100 on the vertical scale.

Lymphoid type: Thymus, lymph nodes, intestinal lymph masses. Brain and head type: Brain and its parts, dura, spinal cord, optic apparatus, cranial dimensions. General type: Body as a whole, external dimensions (except head) respiratory and digestive organs, kidneys, aortic and pulmonary trunks, musculature, blood volume. Reproductive type: Testis, ovary, epididymis, prostate, seminal vesicles, Fallopian tubes.

reproductive organs, for example, follow a different curve from that of general somatic and skeletal growth. Before puberty the increase in size is slow; at adolescence it is rapid. The brain, skull, ears and eyes develop earlier than other parts of the body. Lymphoid tissue (tonsils, adenoids, appendix, spleen, lymph glands, etc.) has yet another characteristic curve. It reaches a peak in size just before puberty and actually decreases in size thereafter, to reach the adult figure at about 20 years of age. Body fat has an even more complex growth curve, and differs in males and females.

Organisation of growth

At the moment, we know very little about the way the intricate growth patterns which occur during development are organised. Growth in a single dimension is regular; different tissues and parts of the body grow at different rates. It is these differences which give rise to the characteristics of the morphology of the different species (man has longer legs relative to the arms than do apes, although at birth this distinction is not apparent).

Growth of skeletal muscle

In adult life, and indeed after the fourth month of intrauterine life, no new muscle fibres are laid down. Growth of muscle thus depends on: (a) increases in length and diameter of existing fibres; (b) increases in the amount of non-contractile structures and cells within the muscle belly, i.e. connective tissue and fibrous tissue.

The increase in diameter of a muscle fibre involves the synthesis of sarcoplasm and myofibrils. Although the bulk of whole muscles vary greatly between species (cf. whale and mouse) there appears to be an upper limit for the diameter of single fibres, so that the species differences are mainly due to the number of fibres contained in a given muscle.

Work hypertrophy of muscle follows its prolonged usage; this is an adaptation to environmental factors and little is known about the ultimate mechanism which generates the effect. Isometric contractions (as in weight-lifting) are, in general, productive of greater enlargement of fibres than are isotonic contractions. Indeed, a long-distance runner may actually suffer a decrease in leg muscle girth on commencing training, presumably as connective tissue, etc. decreases in amount within his musculature and subcutaneous fat disappears.

Androgenic hormones are important in the development of greater muscular size and strength in the male.

Disuse atrophy occurs when muscles are not used, have divided tendons or are denervated. Little is known about the factors involved. The atrophied fibres are replaced by fibrous tissue. Often fibrous tissue, after it is laid down, undergoes a progressive shortening in length and there is further disability due to the muscle "contracture".

Growth of skin

The growth of skin, *pari passu* with that of the whole body, poses some interesting biological questions, particularly with regard to the manner in which it is so carefully matched to the expansion of the underlying tissues. Experiments appear to indicate that tension, in the plane of the skin, is a stimulus to growth of new tissue within the old. Cells in the basal-cell layer seem to be stimulated to divide at an increased rate. Fibrous and connective tissue is laid down later.

Sometimes the expansion is too rapid, as is often the case in pregnancy, and the outer layers of skin then become thinned, shiny and tend to split. The *striae gravidarum* arise in this way.

Growth of fat

Fat growth and metabolism assume great importance in the civilised world of today. Half of the world's population is too fat; the others are suffering from malnutrition (although it should be recognised that these two problems may often coexist).

Fat is deposited in connective tissue as adipose cells (described on p. 11). These cells first appear at the 34th week of intrauterine life and their number and size increase until about one year post-natally. There is some evidence that in adult life no new fat cells are formed (like muscle and nerve cells) but that the deposition of fat reserves in the body is brought about by filling out pre-existing adipose cells with the excess fat (from the diet). If this is true, it probably means that if a child is not allowed to put on too much fat and become overweight before the age of 1 year, when adult he will not suffer from the periodic necessity to "slim" in order to be healthy and sexually acceptable. At all events, there are considerable differences between the amount of subcutaneous excess fat that individuals accumulate and it is clear that these are not all cases resulting from metabolic or purely dietary factors.

At adolescence, the characteristic bodily contours in male and female are due in large measure to sex-dependent, distinctive, distribution of subcutaneous fat. Presumably, sex hormone secretion determines where the fat shall be deposited, if not how much.

A fortune awaits the researcher who finds a simple method of subcutaneous fat control.

Growth of the nervous system

A number of interesting physiological questions arise in connection with growth of the nervous system. We are here not concerned with the developmental aspects, or with the embryology of the nervous system, so it must be stated that in post-natal life relatively little growth takes place in the peripheral or central nervous system as a result of cell division. However, with growth neurones become larger as the somatic regions enlarge; an anterior horn cell may be almost a metre in length eventually.

As far as cerebral cells are concerned, the usual view is that all these neurones are laid down well before birth, and that the continued growth and the development of function in the brain, concern only the connectivity of those cells. Consideration of the facts that there are about 10^{11} cells in the cortex alone, and that each of these makes many thousands of contacts with other neurones, would lead one to conclude that the precise pattern of nerve fibres within the grey matter cannot be laid down genetically as there would be insufficient carrying capacity for this information in the genetic code. Thus the growth taking place in the grey matter must be to some extent random in its organisation or, more likely, must depend upon the functional activity going on within the nervous system.

Tissue cultures of neurones, and explants of nerve cells, examined with time-lapse photography, reveal weird and beautiful slow-motion movements to be in progress the whole time. The growing ends of nerve processes may be observed making contacts with other nerve cells and it is a matter for further research to determine the identity of chemical, electrical, or other factors at work in controlling and directing these growth movements.

A current, widely held theory to account for learning and memory invokes such growth processes as being responsible for the formation of new functional pathways which subserve the plastic changes in neuronal activity responsible.

It is also commonly believed that with advancing years the neurones in the grey matter inexorably and rapidly die off. Estimates vary between 1000 and 100 000 cells per day. Lest the reader be alarmed by the latter figure, he should note that it merely represents a millionth of his total stock of cells!

However, these estimates are based on the light microscopic examination of sections of brain stained by the Golgi method which picks out only a small proportion of neurones present. What factors determine whether a cell is stained or not are quite unknown. Thus if we find progressively diminishing counts of neurones in our sections as older animals are examined, this may only be the result of a smaller proportion of cells being actually stained in the older specimens.

As a limb grows the nerve trunks within it must elongate to accommodate this increase in length between the spinal column and the distal end organ (muscle or sensory receptor). It is usually said that traction on the nerve fibres is the ultimate stimulus for this but in some respects it would be unlikely, for the fibres in nerve trunks are much coiled to enable them to be elastic when the limb is moved. As anyone who has dissected out single nerve fibres knows, these are not under tension in the normal body.

The adolescent growth spurt

As we have seen, growth rates reach their peak in early embryonic life, but the adolescent growth spurt is an obvious phenomenon and hence more dramatic. Associated with it is the onset of sexual maturity. Figure 62.3 shows velocity curves of

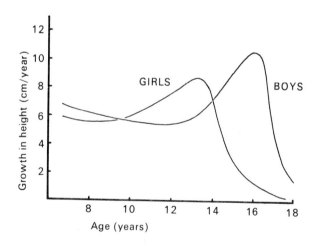

Fig. 62.3 The growth spurt at puberty. Growth rates (cm/yr) for girls and boys.

(Figures 62.1, 62.2 and 62.3 are all from Tanner (1962), *Growth at Adolescence*, 2nd edn. Blackwell, Oxford.)

growth during adolescence for boys and girls. The characteristics are:

(a) The spurt has a sudden onset.
(b) Rate of growth accelerates from 6 to 10 cm per year (fetus grows at 50 cm per 9 months).
(c) Girls start the spurt earlier.
(d) The spurt is larger in boys.

Men have, eventually, a larger stature than women since they grow for longer because of factors (c) and (d).

These events, occurring at the time of puberty, are presumed to result from the activity of the hypothalamus in releasing pituitary hormones (growth hormone and gonadotrophic hormones which control sexual development).

The control of growth

We must distinguish between *growth*, the process of increase in size of the body (or tissue, organ etc.) and *differentiation*, the process of specialisation of cells into different organs or tissues with differing functions.

Growth itself is controlled by several factors, although it must be stated at the outset that remarkably little is yet known about them.

1. *Genetic factors*. The size and form of an individual depend largely upon his or her genetic makeup. With adequate nutritional and environmental conditions, the full growth potential dictated by the inherited characteristics will be realised.

2. *Dietary factors*. Deficient diet (see Chapter 52) will result in varying types and degrees of growth impairment.

3. *Disease*. Most illness causes a depression in growth rate. A good diagnostic measure of chronic disease in growing children can be the observation of their "auxodromes" (or height–weight velocity patterns) since these reveal departures from a smooth path, often before other clinical symptoms supervene. Endocrine or metabolic diseases are examples.

After the child has recovered, the growth lost is made up and the velocity curve is brought back to where it would have been.

4. *Endocrine control*. As mentioned above, endocrine disorders may profoundly alter growth patterns. The cellular basis of growth is one of incorporation of protein (and hence nitrogen) into body structure. Pituitary growth hormone, thyroid hormone and androgens all influence protein metabolism (see Chapter 50).

63. The Physiology of the Female Reproductive System

The female organs of reproduction consist of the *ovaries* and their accessory organs: the *uterus* in which the fertilised ovum is retained during pregnancy, the *Fallopian tubes* (or oviducts) which convey the shed ova to the uterus and the *vagina* and *genitalia*. The *mammary glands* develop partially at puberty, increase their growth during pregnancy and become functional during the first part of the extrauterine life of the offspring.

THE OVARIES

The two ovaries, each the size and shape of a walnut in the human female, puzzled the early anatomists, who could not believe that mammals had eggs. Regner de Graaf, in 1672, described follicles (the Graafian follicles) in the ovaries of sheep, rabbits, cows, cats and women; de Graaf in fact, mistakenly thought that these follicles were the eggs and it was not until 1827 that von Baer discovered the smaller ova within the ovarian follicles.

At birth, the ovary consists of a stroma of spindle-shaped cells covered by a layer of cuboidal epithelial cells, the *germinal*

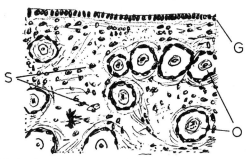

Fig. 63.1 Section of cortex of an infantile ovary with primordial ova. G, germinal epithelium; O, ovum with nucleus; S, stroma of ovary.

epithelium. Embedded in the stroma are many *primordial follicles*, which arise as downgrowths of germinal epithelial cells occurring during fetal life. Each of these primordial follicles is about 200–300 μm in diameter, has a large central single cell 45 μm in diameter with a pale, vesicular nucleus, mitochondria and Golgi apparatus. An external cell membrane is replaced by 10 to 20 follicular cells arranged round its periphery (Figs. 63.1, 63.2, 63.3, 63.4).

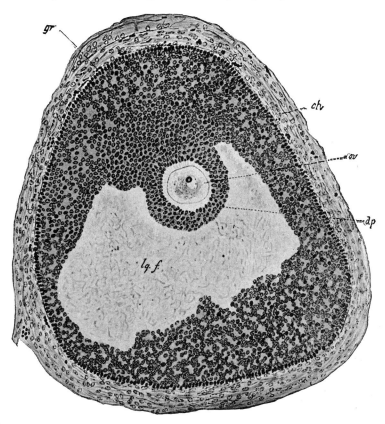

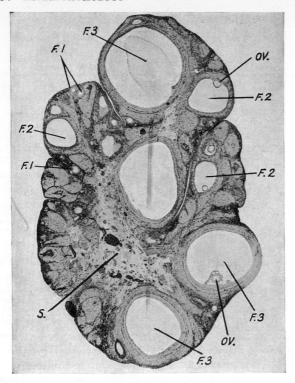

Fig. 63.3 Section through the ovary of a ferret just before ovulation. F.1, very immature follicles; F.2, developing follicles; F.3, mature follicles, OV, ovum; S, stroma tissue. (From a photograph supplied by Dr Parkes.)

In the two ovaries of a young woman aged 22 years, Haggström was able to count as many as 420 000 primordial follicles. Normally, during the reproductive life of a woman, about 400 of these follicles ripen and ultimately discharge their ova. The remainder, in varying stages of their development wither, to become the so-called "atretic follicles"—simply small masses of fibrous tissue laid down as a physiological response to dead cells. It is not known why the ovum dies in this way or what factors are involved in differentiating between those which die and those which survive.

Maturation of follicles

Maturation of the follicles is a process beginning at puberty. A mass of cells formed by mitotic division of follicular cells now surrounds the ovum, to form a cavity filled with a fluid called the *liquor folliculi*. The epithelium surrounding the ovum is separated into two parts: the *membrana granulosa*, which is a thin layer of syncytium composed of finely granular polyhedral cells surrounding the whole follicle, and a solid mass of follicular cells in which the ovum is embedded—the *discus proligerus*—projecting into the liquor folliculi. Outside the granulosa cells the stroma of the ovary forms a capsule of two layers. The outer layer, or *theca externa*, is of tough fibrous tissue, while the *theca interna* is a thin layer of secretory cells having a rich blood supply. The cells of the *theca interna* are very important, as we shall see later.

When completely ripened, the ovum itself is well over 100 μm in diameter, and is contained within a distinct membrane, the *zona pellucida*, through which pass small canals, while the whole follicle may reach 5 or 6 mm across.

Ovulation

The mature follicle enlarges and moves near to the external surface of the ovary, bulging it outwards and as the intrafollicular pressure rises, the surface layers get thinner until eventually the blood vessels are compressed. A small region at the apex ruptures and liquor folliculi, cell debris, together with the ovum are discharged into the peritoneal cavity. This sequence of events is termed ovulation. In some animals this process of ovulation takes place at definite seasons (bitch); in others ovulation only occurs after coitus (rabbit); in the human female ovulation usually occurs alternately from each ovary, every month or so. Unless ovulation does occur in this way, the follicle—even if apparently fully mature—degenerates and is turned into a fibrous atretic follicle.

Corpus luteum formation

Once the ovum has been discharged, the remaining portions of the ruptured follicle turn into a structure known as the *corpus luteum*, which in addition to the *theca interna* has an important controlling function.

The mammal, unlike the frog or the fish, is still responsible for the developing egg which must be given nourishment and protection within the mother's uterus. The uterine wall must be adapted to house the ovum; also the mammary gland has to be stimulated to give milk after the fetus has been born. One of the main mechanisms involved in this complex chain of responses to the developing fetus is brought about by the secretion of a hormone called *progesterone*. At first, the main source of this hormone is to be found in the corpus luteum, formed in the cavity left in the ovary after ovulation by rapid proliferation of the remaining membrana granulosa cells. These quickly increase, both in size and number, so that a dense mass completely fills the original follicle. At first these cells are grey in colour; if fertilisation has occurred they continue to grow, reaching a maximum size about 10 to 12 weeks later (Fig. 63.4).

The developing cells soon have a yellow colouration (orange in humans) due to the presence in them of a *lipochrome* (lutein). There also develops a rich blood supply. In the absence of fertilisation, the corpus luteum undergoes degenerative changes about three weeks after it was first formed. Animals which produce a single offspring at a time naturally have a single corpus luteum since only one ovum has been shed into the peritoneum. Others, with multiple litters, have a corpus luteum for each ovum.

The function of the corpora lutea remained a mystery; even in the early years of this century it was still possible for a thesis to be written which gave as many as twenty-five different, and all quite incorrect, hypotheses regarding their function. However, they have the usual characteristics of endocrine tissues. They are large cells having a granular cytoplasm, a profuse blood supply and they communicate with the rest of the body via blood vessels only. We will consider the physiology of progesterone later.

Extirpation of the ovaries

In adults, following removal of both ovaries there is atrophy of the whole genital apparatus (uterus, vagina, external genitalia). Menstruation ceases. Vasomotor abnormalities, similar to those occurring during the menopause, are common, i.e. flushing of the skin of the neck, face and chest (hot flushes), sweating and feelings of suffocation.

The breasts may either increase or decrease in size, depending upon

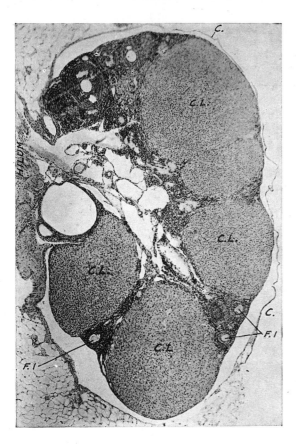

Fig. 63.4 Section through the ovary of a mouse during di-oestrus. F.1, immature follicles; C.L., corpora lutea; C., capsule of ovary. (From a photograph supplied by Dr Parkes.)

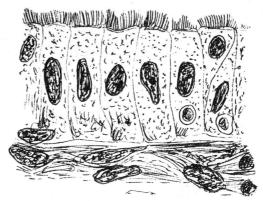

Fig. 63.5 Epithelium of Fallopian tube from human female aged 25, showing ciliated epithelial cells and submucosa. Specimen obtained at midinterval, (×600.)

whether fat deposition in the region more than offsets the actual loss in glandular tissue which takes place. Obesity of the body in general usually develops and often causes secondary psychological problems.

Strange to say, sexual desire may be unaffected or even heightened following ovariectomy. This can be taken to indicate that although sexual behaviour is modified by hormonal action, it is basically determined by nervous (instinctive/reflex/emotional) factors. Heightened sexual interest in women whose ovaries have been removed can often be traced to the fact that in intercourse they no longer fear pregnancy.

Considerable mental upset often results from removal of the ovaries, but this must not be hastily ascribed to lack of hormonal influences on the brain, for the purely psychological shock of the mutilation itself is a potent determinant of psychiatric disturbances.

Before puberty, removal of both ovaries may have remarkably little effect. Not much is known about the clinical picture because it is a surgical procedure that is relatively rarely resorted to. Puberty does not occur and menstruation does not begin. Secondary sexual characteristics do not develop.

THE FALLOPIAN TUBES

The rupture of the follicle leads to the ovum being discharged into the abdominal cavity. This procedure is in fact less hazardous for the ovum than one might imagine. In the first place, the open internal end of the oviduct is applied to the surface of the ovary at the instant of ovulation and, in addition, the abdominal cavity is not a vast open space but is closely packed with intestines and other viscera. A flow of the serous fluid found within the abdominal cavity towards the funnel-shaped extremity of the Fallopian tube has been demonstrated using Indian ink particles. Nevertheless, there are unexplained mysteries concerning the mode of transport of ova (they are non-motile, of course) since pregnancy has frequently occurred following the removal of one ovary and ligation (tying with surgical thread) of the opposite tube.

The oviducts are called Fallopian tubes after Gabriele Fallopio (1523–62) chiefly because he was not the first to describe them although he put forward a most attractive theory to explain their function, namely that they acted as ventilators to disperse noxious fumes from the uterus. The structure of the Fallopian tubes is complex and the interior is folded into numerous, high branched ridges of mucous membrane so that in cross-section the appearance is like a honeycomb. The whole length of the tube, about 12 cm, is lined with cilia. In these spaces, the spermatozoa and ovum meet and fertilisation occurs (Figs. 63.5 and 63.6).

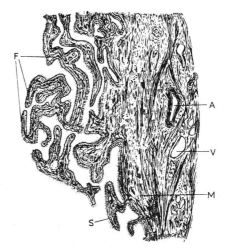

Fig. 63.6 Part of cross section of Fallopian tube from a 25 year old woman. S, mucosa; M, muscular layer, showing bundles of smooth muscle cut both longitudinally and lengthwise; A, artery; V, vein. Note the extremely folded processes of mucous membrane covered with ciliated epithelium and containing blood vessels (F). (×40.)

Transport of the ovum

How are the ova transported from the ruptured follicle to the uterus? There is some difficulty in attributing the passage of the ovum to any single agency. It is known that the lashing of cilia gives rise to a general flow from within outwards; the tube, being muscular, undergoes squirming and peristaltic-type movement. However, some animals have oviducts without cilia.

It has been found that, in all mammals, the ova take 3 to 4 days to make the passage of the oviduct. That this might have interesting implications is evident when one considers that the oviduct of the sow is at least forty times as long as that of the mouse! The explanation is now thought to be due to a delay in the ovum's passage through the middle portion of the tube, where the honeycomb-like structure is at its thickest. This is the probable reason why it is that if a large number of cross-sections of tubes are cut, it is only in this middle region that ova are ever found.

We can therefore assume that, in addition to its function of transport, the Fallopian tube acts as a temporary nidus for the ovum. Presumably a short-term retention of this nature would increase the chances of fertilisation taking place and also would give the uterus time to respond to the hormonal secretion of the newly formed *corpus luteum*. It can indeed be shown that progesterone has certain effects on the lining of the uterus, that in the absence of these changes implantation of the ovum cannot occur and, in addition, if the ovum arrives in the uterus two or three days too early, it is not viable.

THE UTERUS

The uterus is part of the excretory passages; here the ovum develops into the fetus until the time of delivery. In the human, it is a single cavity and, together with the vagina, is the development of the embryonic Müllerian ducts, fused in the midline. In other mammals the fusion occurs only at the vagina and the uterus then has two "horns". Developmental abnormalities can lead to this situation occurring in the human. The human uterus is a pear-shaped organ having a small cavity and a thick muscular wall. The body of the uterus consists of a fundus (the larger, rounded upper end), an isthmus and the cervix, the cylindrical lower part with its canal. The wall consists of a great mass of smooth muscle, the *myometrium*, and an inner lining, the *endometrium*.

A great deal is known about the myometrium, in particular its response to drugs, because in virgin small rodents it provides an extremely easily set-up test organ. When suitably perfused, it will continue to contract and relax for long periods and the action of vast numbers of relevant or irrelevant drugs can be investigated or, if necessary, quantitatively assayed. The adult non-pregnant uterus is continually undergoing cycles of contraction and relaxation. During the pre-ovulatory phase, these become more rapid and are at maximum frequency at about the time of ovulation. Following ovulation they then become slower again and more irregular. Changes in the endometrium will be discussed in the section on menstruation.

The embryo passes into the uterus on the fourth or fifth day following ovulation. It is at the 4-cell stage of development by this time.

Implantation

The developing embryo becomes attached to the uterine wall, and is said to be implanted, by the seventh day after ovulation. The lining endometrium of the uterus is prepared for this by hormonal action—particularly by progesterone—which produces the enlargement and secretion of its tubular glands without which the ovum would be unable to survive or become attached. An unfertilised ovum does not become implanted.

64. The Oestrous and Menstrual Cycles

THE OESTROUS CYCLE

The sexual, and indeed most, behaviour of females, shows cyclic changes varying in detail in different species. One complete cycle of such changes is known as an oestrous cycle. These recurrent periods of sexual excitement (or "heat", as they are popularly known) were first named *oestrus* by Walter Heape in 1901. The word originates in Virgil's description of the insect "called the gadfly, a brute with a shrill buzz that drives whole herds crazy...."

In many species, there is only a single oestrous period in a sexual season. The quiet period between each season is termed *anoestrus*. In other species the oestrous cycles occur more or less continuously; the short interval between two cycles is termed *dioestrus*. The precise arrangement of these sexual activities is timed to ensure that in the wild state, offspring are produced at times when the environment is least inclement. It appears to be subject to fairly rapid adaptation (as is found in the domestic animals) to different environmental conditions. Humans do not show any clear alternation between sexual rest and activity (oestrus and anoestrus); neither are there any marked behavioural changes with each sexual cycle.

Many women show the syndrome of "pre-menstrual tension". This consists in varying degrees of irritability, lassitude, depression and occasionally frankly psychotic changes, lasting for two or three days just before the commencement of a menstrual period. Statistical surveys show that women are more accident-prone and untidy at these times and even athletic performance or examination results suffer. It is held to be the result of water retention in extracellular spaces, including the brain, because there is a rapid weight gain which is lost with a diuresis at the time of menstruation, and is due to low oestrogen levels in conjunction with the sudden fall in progesterone which occurs at this time in the cycle. Nowadays, treatment is often efficacious and consists in administering oral diuretic compounds or graded replacement therapy with progesterone derivatives.

The functional significance of the behavioural changes during the oestrous cycle is that these are patterns that will ensure successful mating. Therefore, just following ovulation, sexual excitement occurs, the female will receive the male, coitus and fertilisation will occur. If this course of events is not followed through, the whole cycle is repeated. This is the basic plan. Of course there are many variations on this theme in the animal kingdom (e.g. the guinea-pig female cannot mate, even if it wanted to, other than during oestrus, since a skin grows over the vaginal orifice and only breaks down at oestrus). Human females are probably more easily aroused sexually at around the time of ovulation.

The psychological manifestations of oestrus are brought about by the actions of oestrogens (from the follicles). Quite normal heat can be brought on by giving oestrogen to ovariectomised female animals. In other words, provided that oestrogens are administered, the phenomena of heat are independent of the ovaries—a fact which assumes importance in the therapeutic use of oestrogen, especially in veterinary practice.

In the female guinea-pig, priming with small amounts of progesterone is necessary before the female exhibits the mating response. In some domestic animals so-called "quiet heat" occurs; all the physiological events take place (ovulation, histological changes, etc.) but the mating behaviour is absent. This phenomenon is responsible for the near-extinction of a certain Swedish breed of cattle, for the bulls have been unable to decide when the cows were on heat.

THE MENSTRUAL CYCLE

A particular variation of the oestrous cycle occurs in primates. Mating will occur in primates at any stage of the sexual cycle; the common phenomena of heat are suppressed and behavioural changes with the cycle are minimal. In the middle of the cycle there is a period of breakdown of the uterine wall and bleeding occurs. The relationship between menstruation and oestrus naturally interested early physicians greatly and also proved confusing to them, because mammals have one prominent feature—*oestrus*—while primates have another—*menstruation*. Are these processes comparable?

The menstrual cycle is the interval between the onset of two periods of uterine breakdown. The modal length (i.e. the most frequently observed duration) is approximately 28 days. The average length is longer by two or three days. In order to see the relation between the oestrous cycle and menstruation we need to know whether the menstrual flow is comparable with oestrus. In other words, does it accompany ovulation?

The time of ovulation in the menstrual cycle

The generally accepted view is that ovulation takes place about the middle of the menstrual cycle. At first, this conclusion emerged from observations of the cyclic changes in the histology of the endometrium of the human uterus but now is based on a great deal of evidence obtained both in women and apes.

Direct evidence was obtained from the inspection of ovaries at *operations* undertaken at various times in the cycle. It is possible to see with the naked eye whether ovulation has occurred recently. *Single matings* at known times, resulting in pregnancy, enable the time of ovulation to be dated within a day or two and finally live ova have been recovered during abdominal operations in women (Allen, 1930) and also in monkeys.

Indirect evidence includes many observations which indicate that the first half of the menstrual cycle is associated with physiological parameters differing from those occurring in the second half.* These suggest that some abrupt change is taking place in the middle of the cycle. Examples are:

---

* Spontaneous activity in rats can be increased by injecting oestrogen. During the menstrual cycle, use of a pedometer indicates that women show two peaks in the amount of spontaneous walking about that they do. One peak (for obscure reasons) is during menstruation, the other is at the time of ovulation.

1. *The endometrium* in both monkeys and humans is subject to cyclic changes. The spontaneous contractile activity of the human uterus varies with the two phases (see p. 439). Hormone levels in the blood and urine of women show characteristic differences during the first and second halves of the cycle.

2. *The vaginal pH* is usually lower during the first half of the cycle than it is during the second.

3. *The basal body temperature* is lower in the first part of the cycle. The transition occurs abruptly and there is a sudden rise of about 0·5°C at the presumed time of ovulation.

4. Some women have a short-lived attack of pelvic pain in the "midinterval", often accompanied by slight "intermenstrual" bleeding. The evidence that this is due to the process of ovulation comes from a small series of cases whose ovaries have actually been examined (at operation) at the height of the pain and show clearly that ovulation has just occurred.

In view of the evidence that ovulation takes place about the middle of the menstrual cycle, it is clear that the menstrual cycle and the oestrous cycle are not strictly comparable. We may look upon the menstrual cycle as a modified form of the general type of sexual cycle, in which, in addition to the changes leading to ovulation, a phase of uterine bleeding and breakdown occurs.

Phases of the menstrual cycle

Although the primary event in the female sexual cycle is ovulation, there is no easy method in women of determining when this has occurred, so it is customary to time the menstrual cycle from the first day of the menstrual flow. Ovulation occurs, on this basis, halfway through the cycle, at about the fourteenth day (Fig. 64.1).

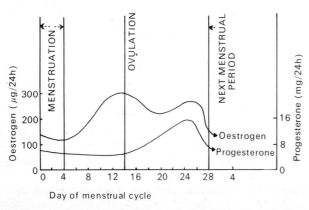

Fig. 64.1 Course of events in the human menstrual cycle. (G. W. Corner.) Graphs show levels in plasma of oestrogen and progesterone.

Control of menstrual periodicity

It is known that menstruation can occur without preceding ovulation. Hence the length of the two phases of the menstrual cycle, and the duration of the menstrual cycle as a whole, is not necessarily related to ovulation or to the processes involved in maturation of the follicle. Such "anovulatory" cycles do not have a different length from those with ovulation so that it is reasonable to assume that ovulation and the formation of a *corpus luteum* are not the primary factors underlying the cyclical rhythm of the uterus.

It is, of course, difficult to determine the precise mode of operation of the factors responsible for the menstrual rhythm, because of the impossibility of predicting the exact duration of a particular cycle. However, emotional disturbances can often alter the menstrual periodicity; in some individuals the onset of the menstrual period is delayed and in others it is accelerated following psychological stresses of various kinds. Presumably, since the length of the post-ovulatory phase is relatively constant such psychological stress is acting to alter the time of ovulation. There is experimental evidence, obtained in monkeys, that this effect is probably mediated by nervous pathways, because severing the preganglionic fibres to the ovaries does give rise to temporary disturbances of the ovarian and also the uterine cyclical changes.

Other mechanisms also influence uterine periodicity for, in the monkey and in women, it is found that uterine bleeding occurs regularly in the absence of the ovaries provided that a constant threshold dose of oestrogen is given. Experiments suggest that the pituitary gland is not concerned in this although the adrenal cortex definitely plays an important part.

The way in which this effect is brought about is not known but possibilities include:

(a) The adrenal cortex cyclically secretes small amounts of oestrogenic hormone* (this would summate with the maintained constant dose given).

(b) Similarly, the cyclical secretion by the adrenal of androgens or progesterone would give a regular neutralisation of the injected oestrogen.

(c) Cyclical activity of the adrenal cortex might give rise to changes in the distribution of intracellular and extracellular water in the endometrium, thus altering its sensitivity to the pre-existing, artificially administered oestrogen.

We will return to this problem of the interaction of hormones and the regulation of cyclical uterine activity in the section on the endocrine control of the sexual cycle.

Composition of menstrual fluid

Menstruation has been recorded in all human races and anthropoid apes. The periodic flow of mucus and blood from the uterus begins at about 14 years of age, the time of puberty, and continues throughout the female reproductive life until the age of cessation, the *menopause*.

The number of days for which the flow lasts varies, but is usually from 3 to 7 days. The amount of blood and debris lost varies from a total of 50 ml in virgins† to 200 ml in matrons, and in composition it is different from circulating blood as it contains serum, uterine secretions, mucosal cells and occasional tissue fragments. The flow from the external os of the uterus is not continuous but occurs at 3 to 5 min intervals with the uterine contractions. Normal menstrual fluid does not coagulate as does blood; it is partly haemolysed. The presence of clots in the fluid indicates excessive bleeding, or *menorrhagia*.

---

* An aberration of the normal sexual cycle may occur in cattle, thought to be due to adrenal dysfunction. This is *nymphomania* and is characterised by sterility and a continuous desire to mate—if necessary with all the other cows. The condition has also been observed in bitches.

† The volume of blood lost varies in any one woman, as well as between women, and like the time of onset of the bleeding is also subject to psychological factors such as stress, emotion etc.

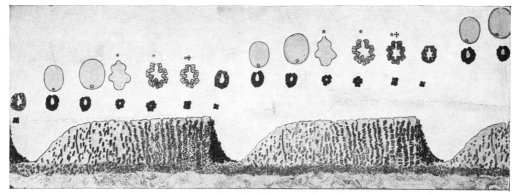

Fig. 64.2 Diagram showing, from left to right, the changes in the human endometrium (below) and ovary (top row, this month's follicle or corpora lutea; lower row, last month's) during two complete menstrual cycles. (After Schröder.)

Changes in the endometrium during the menstrual cycle

Since, as we have seen, the endometrial lining of the uterus undergoes, at the menstrual period, a structural breakdown with bleeding it follows that considerable changes must be apparent in the histological structure of the mucosa during the course of the cycle (Fig. 64.2).

In the course of a normal cycle, the thickness of the mucosa increases from about 0·5 mm just after the menstrual period to 5·0 mm at the beginning of the phase of breakdown. The histological changes in the endometrium are well known because it is a common clinical procedure (D&C—dilation and curettage) to remove small portions of the mucosa and these may be examined by the usual histological techniques.

The changes involved can be categorised according to the time in the cycle.

1. *Phase of repair* (*or proliferation*). This lasts from the first to the tenth day (day 1 being the first day of the menstrual period). Immediately after the endometrium has been shed, re-epithelialisation begins, even before the bleeding has stopped. The tissues shed comprise the whole thickness of epithelium, submucosa and part of the stratum vasculare, so that after menstruation there is left only a thin layer of endometrial stroma and the basal regions of the uterine glands. The only epithelial cells are those lining the ducts of the glands and it is these cells, at the torn-off mouths, which spread over the endometrial surface to form a new lining, which is complete within the first four or five days of the cycle. The endometrium increases in thickness as a result of proliferation of stromal cells and growth of endometrial glands and blood vessels. By the time that ovulation occurs, the endometrium is about 3 mm in depth.

2. *Secretory phase.* Following ovulation the thickness of the endometrium increases further, due mainly to increases in size and tortuosity of the glands, the development of secretion in them and to the increased size of the stromal cells already laid down. The blood vessels also become more tortuous and the blood flow through the tissue is augmented. In the stromal cells, glycogen and lipids are deposited.

These changes prepare the uterine lining to receive the developing embryo, the blood supply, the stored nutrient materials and the general secretions providing ideal conditions for its survival. At first the embryo simply absorbs the uterine secretions; later, after implantation, the trophoblast cells digest and absorb the endometrium itself to provide nutrition until the placental transfer mechanism is functional.

3. *Phase of breakdown* (*menstruation*). During the last day of the menstrual cycle there is a rapid involution in the endometrium due to spasm of the small spiral arteries supplying the mucosa. Vascular stasis and vasoconstriction lead to necrosis and breakdown of the blood vessels. Not all blood vessels are affected simultaneously by the constriction and there are transitory periods of local vasodilatation during which bleeding takes place.

The mechanism responsible for the breakdown process has been investigated by an elegant method involving the transplantation of a fragment of endometrium into the anterior chamber of the eye (of a female monkey) where direct microscopical observation is then possible. The commonly held theory is that, primarily, dehydration of the mucosa leads to shrinkage and hence the spiral arteries are compressed and necrosis follows. However, it is unlikely that this is the whole explanation, for menstruation occurs in certain South American monkeys in whom there are not any spiral endometrial vessels.

It has been suggested that the uterus produces a "menstrual toxin" which is responsible for the vascular damage leading to menstruation but this remains to be proved.

During menstruation, large numbers of leucocytes are released from the endometrial surface, as one of the many mechanisms responsible for rendering the uterus resistant to infection, in spite of the large amount of tissue destruction which occurs at this time.

Experimental evidence on menstruation

(a) Removal of both ovaries (in apes, in women) leads to bleeding after several days, irrespective of the time of the menstrual cycle during which removal was undertaken. We may deduce that bleeding is in some way related to the removal of an ovarian factor, probably hormonal.

(b) If a series of oestrogen injections is given after ovariectomy, no bleeding follows. Bleeding is dependent on oestrogen deprivation, for it follows cessation of the oestrogen therapy.

(c) A monkey having had both its ovaries removed is then given oestrogen. The proliferative phase of endometrial development occurs.

At this stage oestrogen administration is stopped and progesterone is substituted. No menstrual flow takes place; instead, after a while, the progestational phase of the endometrium is found. If the giving of progesterone is stopped, the endometrium, deprived of both oestrogen and progesterone now undergoes menstruation.

The experiment described in (c) above must closely resemble the state of affairs pertaining in the normal female sexual cycle. The early development of the endometrium (proliferative phase) is controlled by oestrogen, the subsequent progestational phase by progesterone. Withdrawal of progesterone, as the corpus luteum fades, gives rise to the onset of the menstrual flow.

We are now in a position to consider the hormones concerned and their actions in greater detail.

65. The Hormonal Control of Sexual Cycles

I. THE OVARIAN HORMONES

The oestrous cycle and, of course, the menstrual cycle are controlled and timed by the action of hormones. As we have seen in the chapter on endocrines, relatively complex feedback mechanisms control the output levels of various hormones. The pituitary gland constitutes the central controlling system and the interrelationships between the anterior lobe and the ovaries bring about the cyclical changes in sexual function that we have already discussed.

The anterior pituitary gland secrets two gonadotrophic hormones:

1. *Follicle stimulating hormone* (FSH). This hormone stimulates unripe follicles to mature and thus indirectly controls the amount of oestrogen produced by the ovary.
2. *Luteinising hormone* (LH). This hormone controls the formation of the corpus luteum and hence the production of progesterone by the body.

Two types of female sex hormones are secreted by the ovaries—*oestrogens* and *progesterone*. Oestrogenic hormones stimulate growth and cell division in certain tissues and organs and are responsible for the development of the primary sexual organs and the secondary sexual characteristics in females. Progesterone is responsible for the secretory changes which occur in the endometrium to enable it to receive the fertilised ovum and also for the growth of the secretory cells in the mammary glands prior to lactation.

A. OESTROGENS

The oestrogens are secreted, in appreciable amounts, only by the ovaries in normal, non-pregnant, females. In pregnancy and certain pathological states, however, oestrogenic substances may be secreted in large amounts by other tissues. In the human, six oestrogenic compounds are known to be present in circulating blood but, of these, only three are considered to have functional significance: *oestradiol-17β* and *oestrone* are found in relatively large amounts in the venous outflow from the ovaries and *oestriol* is their oxidation metabolic end-product. All these substances are steroids and are derived from cholesterol, the primary compound in their synthesis being 17α-hydroxy-progesterone.

Formulae of oestrogens

The structure of oestrogens is based on oestratriene (not found in nature). It is a steroid and is based on the cyclopentenophenanthrene ring. (A biochemistry text should be consulted for details of the numbering; meanings of suffixes -*ene*, -*ane*, -*ol*, and -*one*.)

Ring A, having the three double bonds is aromatic; the 3-hydroxyl group is phenolic. Oestradiol-17β is the oestrogen actually secreted by the ovary:

Oestratriene

Oestradiol-17β

Less active metabolic products appear in the urine:

Oestrone

Oestriol

and others such as 16-hydroxyoestrone and its various conjugation products with sulphates, glucuronic acid, etc.

The ovarian oestrogens combine in the plasma with plasma protein, about 75 per cent being in combination and the remainder in free form.

The liver inactivates oestrogens, as can easily be demonstrated by transplanting the ovaries into the spleen so that their venous outflow all has to pass through the liver. In these circumstances, no oestrogenic activity can be detected in the experimental animal, although the ovaries are active and functional.

Oestrogens are secreted by the *theca interna* cells of the developing follicle, and since their development is controlled by the pituitary gonadotrophins it is clear that the level of oestrogen in the circulating blood is ultimately determined by the action of the anterior pituitary gland. The corpus luteum also secretes some oestrogen. Oestrogens can also be extracted from the placenta, the adrenal cortex, the testes* and are found in the urine of both males and females. Large amounts are found in the urine of pregnant women and, for some curious and unexplained reason, in the urine of stallions.

Oestrogens in the body

If oestrogens are injected experimentally, it is found that approximately 10 per cent subsequently appears in the urine. One can therefore determine the amount secreted, naturally, by the ovaries, i.e. it will be about ten times the amount appearing in the urine. An adult human female will produce about 1 mg each month; the ovaries must therefore secrete about 10 mg. The products mentioned are excreted in the urine throughout the menstrual cycle, peak concentration being reached at about the time of ovulation—presumably because the follicle is largest at this time. Appreciable amounts are secreted during the second half of the cycle, indeed a second peak is usually found not long before menstruation which must indicate that the corpus luteum also secretes some oestrogen.

Oestrogens are secreted by the kidneys in increasing amounts during pregnancy; during the last two months the daily excretion may be as much as 45 mg. The source is the placenta; the amounts of circulating hormone are proportional to the size of the placenta; when the placenta is removed or expelled with the fetus during birth, levels rapidly fall to normal again.

---

* The significance of the oestrogen content of the adrenal glands and the testes is not known. They are released, normally, only in small amounts and may simply represent byproducts in steroid metabolism and synthesis.

Function of oestrogens

As already mentioned, oestrogens are mainly concerned with the regulation of cell division and growth, particularly of the sexual organs. Before puberty, only small amounts are secreted, but after, large amounts are produced under the influence of pituitary gonadotrophins. Their effect is to increase the size of the Fallopian tubes, uterus, vagina and external genitalia until they reach adult dimensions. The breasts and skeletal growth are also affected.

The nature of these widespread bodily effects of oestrogens can be studied by the administration of these substances to ovariectomised animals.

EFFECTS OF OESTROGENS ON THE UTERUS

The administration of oestrogen gives rise to an increase in size of the uterus as a result of cellular proliferation. Removal of the ovaries, as might be expected, has the opposite effect of decreasing the size of the uterus. The mucosa is also affected by oestrogens. Its thickness and vascularity are increased and its cellular water and electrolyte content rises.

In the absence of oestrogen, uterine muscle is quiescent; when oestrogens are given the myometrium undergoes rhythmical contractions and its sensitivity to oxytocin is enhanced. These effects are, in fact, due to the action of oestrone on the resting membrane potential of uterine muscle which is increased (with resultant changes in threshold and excitability).* The effects of oestrogens on the Fallopian tubes are similar.

EFFECTS ON THE VAGINA

Oestrogen administration, in the ovariectomised animal, results in cell division and cornification in the vaginal mucosa. There is, in addition, deposition of glycogen and increased vascularity. Detailed changes in histological structure of the vaginal mucosa can be detected by taking vaginal smears for microscopical observation. In animals, the time in the oestrous cycle can be determined. In the human these changes are less marked but are still useful in determining when ovulation has occurred.

METABOLIC EFFECTS

The main action of oestrogens is on the state of hydration of tissues and on the sodium and chloride balance. Retention of sodium and chloride ions occurs in the kidney, urine volume is less and hence water-loading tends to occur. There is also an increase in general anabolic reactions, including protein retention (positive nitrogen balance), calcium and phosphate retention and increased deposition of fat in subcutaneous tissues.

EFFECTS ON BONE

Oestrogens give rise to an increase in osteoblast activity and at puberty the female skeleton is affected in a characteristic way. Early epiphyseal union occurs as a result of oestrogenic activity with the result that female skeletal growth, in general, stops at an age two or three years earlier than in the male. A female eunuch is, on this account, several inches taller than her normal sister because of the absence of oestrogen production.

OTHER EFFECTS

The water content of skin and subcutaneous tissues is increased by oestrogens; it decreases sebaceous gland secretion. Vaginal secretion is stimulated; this is important (a) in coitus and (b) in the prevention of vaginal infection.

Oestrogen lowers blood cholesterol levels. This is thought to be the explanation for the lower incidence of atheromatous degeneration in the arterial system of females found up to the menopause.

Behavioural effects: minute quantities of oestrogen implanted in the anterior part of the hypothalamus cause typical oestrous and mating behaviour in experimental ovariectomised animals. Presumably there are similar actions in humans, but since the whole gamut of sexual behaviour is so strongly overlaid by psychological attitudes conditioned by upbringing, etc. in childhood, it is less obvious than in animals. For example, sexual libido is often still marked in females after the menopause.

Mechanism of action at the cellular level

Very little is known about the way in which hormones act on biochemical reactions. Many hormones are known to raise oxygen consumption; for example, the human placenta treated with oestrogen will convert acetate and pyruvate to CO_2.

In the rat uterus, the following events take place when a single injection of oestrogen is given. Water is imbibed for the first 2 h. Glycine and formate are incorporated into nucleic acid. For the next 18 h there is an accumulation of RNA, resulting in protein synthesis. Up to 72 h later, DNA synthesis occurs.

Pyridine nucleotide transhydrogenase is an intracellular enzyme which catalyses the reaction:

$$NAD + NADPH_2 \rightarrow NADH_2 + NADP$$

Oestrogens increase the activity of the enzyme. Particularly, the oestrogen-dependent form of the enzyme is found in those cells which are target organs for oestrogens.

Oestrogens are also known to have effects upon the polymerisation of acid mucopolysaccharides which will change the structure and properties of the ground substance in connective tissue.

Synthetic oestrogens

A number of potent oestrogenic substances have been synthesised which do not necessarily resemble naturally occurring oestrogens structurally.

Stilboestrol is an example of a group of compounds known as *dihydroxystilbenes*.

General formula   Stilboestrol

Other similar compounds are hexoestrol, benzoestrol and dienoestrol.

Ethinyloestradiol is a modified form of the normally occurring steroids. These artificial steroids are active when given by mouth, in contradistinction to the natural hormones which are not, since they are rapidly inactivated by the liver.

Ethinyloestradiol

Clinical use includes replacement therapy, suppression of lactation, treatment of certain male hormone-dominated carcinomata and in the contraceptive pill. Hormones of this kind are also in use in animal husbandry for various purposes (such as the production of large amounts of tender but tasteless meat).

B. PROGESTERONE

Progesterone is secreted by the corpus luteum and by the placenta in pregnant women.

---

* Action potentials only occur in normal uterine muscle or in muscle from an ovariectomised female if it is treated with oestrogen. The rise in membrane potential produced by oestrogen only takes place if calcium ions are present. The general effects of oestrogen are to increase excitability in endometrial cells and to give rise to increased spontaneous activity.

The substance is an intermediary product in the synthetic pathway for other steroids such as cortisol or testosterone and it may be that significant amounts escape during steroid metabolism from the adrenal cortex and testis to account for the fact that *pregnanediol* (the metabolite) appears in male urine. Progesterone also occurs in small amounts in the first two weeks of the menstrual cycle and its source is likely to be in a similar leakage.

In the normal menstrual cycle, the circulating level of progesterone (as is evident from the quantities of pregnanediol appearing in the urine) increases abruptly just after ovulation and persists at a raised level until just before menstruation. Secretion rates are around 5 mg per day in the follicular phase going up to a peak of about 25 mg per day in the midluteal phase.

Function of progesterone

Progesterone is mainly concerned with the preparation of the uterine wall for implantation of the fertilised ovum and preparation of the mammary glands for lactation. It is secreted, as we have already seen, by the *theca lutein* cells and these are controlled by the anterior pituitary; here again is an example of the overall influence of the pituitary gland on sex hormone levels.

THE EFFECT OF PROGESTERONE ON THE UTERUS

Here the effects are "secretory". By this we mean that the uterine glands are stimulated, they become tortuous, the lumen filled with secretion and water accumulates in the tissues. This so-called "progestational response" of the endometrium occurs after it has previously been primed with oestrogen. In the absence of this initial priming the changes brought about by progesterone are much less marked.

An additional effect on the uterus is that the contractions of the myometrium are inhibited by the presence of progesterone. This is a physiological mechanism to prevent premature expulsion of the developing embryo, and is brought about by hyperpolarisation of the uterine smooth muscle cell membrane which leads to the expected decreased excitability. A similar effect on motility is to be found in the Fallopian tubes.

EFFECTS ON MAMMARY GLANDS

The proliferation of alveolar secretory cells is promoted by the action of progesterone but only if oestrogen is present at the same time. Milk secretion does not actually occur, however, until the additional action of prolactin further stimulates the breast. Swelling of the breasts occurs as a result of increased subcutaneous tissue fluid accumulation. The main stimulus for lactation to begin is the drop in progesterone level at the time of birth.

EFFECTS ON WATER BALANCE AND ELECTROLYTES

The effects of progesterone are very similar to those of aldosterone (see Chapter 32)—sodium, chloride and water reabsorption from the distal tubules are increased. Progesterone can replace deoxycorticosterone in the therapy of adrenal cortical insufficiency. However, in the normal individual there is competitive inhibition between the action of progesterone and aldosterone, presumably because both hormones are combining with the same receptor sites in the kidney tubules. Hence in the normal, progesterone has the paradoxical effect of actually reducing sodium and chloride retention because its effects are very much less potent, although it still blocks the action of normally secreted aldosterone.

Progesterone also has a slight stimulating effect on protein breakdown; it has a slight elevating effect on basal body temperature, the mechanism of which is unknown and, as one would expect from its steroid structure, it has effects which mimic androgens and adrenal corticoids in nature.

Progesterone as an antagonist to oestrogens

In certain respects progesterone may be regarded as an antagonist to oestrogens. For example, in some species it may decrease the excitability of the myometrium (spontaneous electrical activity is reduced, membrane potential is raised, sensitivity to oxytocin is decreased).

Progesterone inhibits ovulation in human females, in all probability by decreasing the neuronal activity in the particular hypothalamic neurones which control the secretion of gonadotrophic hormones (see Chapter 32). In pregnancy ovulation is inhibited by progesterone—first originating in the corpus luteum of pregnancy, and after the third month from the developing placenta.

SUMMARY

We may summarise the actions of the primary sex hormones, oestrogen and progesterone, as follows.

In response to pituitary secretion of gonadotrophins, the follicles and corpora lutea in the ovaries are respectively stimulated to secrete oestrogens and progesterone. The amounts vary cyclically during the menstrual cycle as shown in Fig. 65.1.

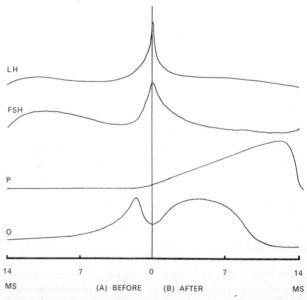

Fig. 65.1 Blood levels of oestradiol (O), progesterone (P), luteinising hormone (LH), and follicle stimulating hormone (FSH) during a normal menstrual cycle. (A) is from the 14th day before ovulation at time O and (B) is from ovulation to the 14th day after.

FSH levels are at their lowest during the luteal phase, O and P being then high. FSH rises during menstruation (MS), and the early follicular phase. It declines in the second half of the follicular phase, while follicle is rapidly growing and blood oestrogen levels are rising. A mid-cycle pre-ovulatory peak of FSH coincides with a peak in LH levels. FSH then falls again.

LH levels are low in the luteal phase and rise during menstruation and the early follicular phase.

Actual plasma concentrations are not given for clarity. (O varies from 80 to 350 pg ml^{-1}; P from 0 to 10 ng ml^{-1}; FSH from 5 to 25 miu ml^{-1}; and LH from 10 to 100 miu ml^{-1})

Primary effects of oestrogens are to cause cellular proliferation and growth of tissues of the sexual organs and of other reproductive tissues.

Primary effects of progesterone are concerned in promoting secretory changes in the endometrium, preparing the uterus for implantation of the fertilised ovum.

Formula of progesterone

Progesterone is a diketone, having strong absorption at 240 nm. It has the following formula:

Progesterone

II. HYPOTHALAMUS AND ANTERIOR PITUITARY

The anterior pituitary gland controls the cyclic nature of the secretion of hormones by, and the physiological activity of, the ovaries. Through its effects on ovarian hormone secretion, the pituitary therefore controls the functioning and the structure of the rest of the reproductive system (including the female secondary sexual characteristics).

The central nervous system controls the activity of the pituitary gland via the hypothalamus. In turn the control exerted by the hypothalamus is in large part dependent on blood-circulating levels of the primary hormones (oestrogen, progesterone): the higher the level, the lower the output of pituitary hormone. We have here, then, yet a further example of a feedback control mechanism. Whether this mechanism alone is enough to generate the oestral periodicity is unknown.

HYPOPHYSECTOMY
Feedback control systems are customarily investigated by cutting the loop involved and/or injecting artificial signals into the loop such as ramp, square or sine-functions. This servoloop can be opened by *hypophysectomy* (removal of the pituitary).

In immature animals, the ovaries (and hence all the reproductive apparatus) remains in the infantile stage.

In adult females, cyclic ovarian activity ceases. A transplant of anterior pituitary gland in an immature animal induces rapid growth of the ovaries, precocious ovulation and corpus luteum formation. In adult hypophysectomised animals, cyclic activity returns.

The pituitary gonadotrophins

Two hormones acting on the ovaries are secreted:

1. Follicle stimulating hormone (FSH)
2. Luteinising hormone (LH)

These are glycoproteins. In humans they have a molecular weight of 6700 and 25 000 respectively.

Control of pituitary secretion of gonadotrophins

As we have already discussed in Chapter 32 on endocrinology, pituitary secretion is controlled by the central nervous system. In the rabbit, for instance, the act of coitus releases an ovum at a timed brief interval later, as a result of nervous activity causing the pituitary gland to secrete gonadotrophin. In humans, purely emotional factors can disturb the menstrual rhythm and can reduce fertility.

Hypothalamic factors
The hypothalamic neurones liberate FSH-releasing factor and LH-releasing factor, at appropriate times. These factors reach the anterior pituitary gland via the hypophysial portal vessels.

It is of interest to note that the removal of the pituitary gland and its subsequent transplantation into the experimental animal elsewhere leads to gonadal atrophy, i.e. the output of gonadotrophins from the gland is inadequate. Normal gonadal function, however, continues if the removed pituitary is reimplanted in the sella turcica since it becomes revascularised there and is once more subject to hypothalamic control.

Figure 65.2 shows the effects of injecting FSH and LH into immature animals.

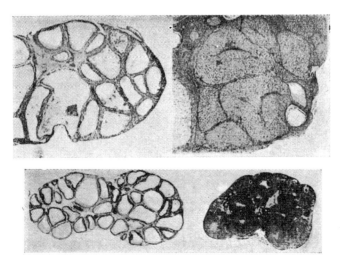

Fig. 65.2 To illustrate the effect of the follicle-stimulating and luteinising hormones on immature ovaries.
 (1) Follicle-stimulating effect on rat, from FSH fraction of 1 g sheep pituitary. No luteinisation.
 (2) Ovary of rat after unfractionated (FSG and LH) pituitary of sheep, 0·1 g. Full of corpora lutea.
 (3) Rabbit after FSH fraction of 2·5 g sheep pituitary.
 (4) Rabbit after 2·5 g whole sheep pituitary.
 (After Hisaw *et al.*)

CYCLIC ACTIVITY OF HORMONES

In summary, the events in the oestrous cycle consist of development and maturation of follicles in the ovaries, ovulation (with certain behavioural changes) and the growth of corpora lutea to control progestational changes in the endometrium. In the absence of the implantation of a fertilised ovum the cycle is repeated; in primates breakdown and shedding of the endometrium also occurs.

These repeated cyclic changes in reproductive activity in the female are brought about by changes in circulating levels of the two primary sex hormones, oestrone and progesterone. These levels are set by complex interactions between the ovaries, the anterior lobe of the pituitary and its secretion of gonadotrophins, and the nervous system (see Chapter 32 for details of hypothalamic control of hormones).

The exact way in which the pituitary regulates the ovarian hormones is unknown, but work in rodents seems to show that there are extensive feedback mechanisms by which the circulating blood levels of oestrogens and progesterone affect the pituitary output of FSH and LH respectively.

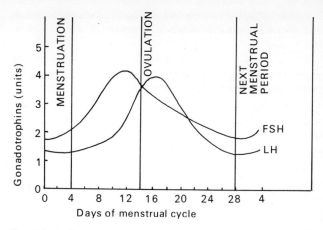

Fig. 65.3 Cyclic changes in anterior pituitary secretion of gonadotrophins (FSH and LH), during one menstrual cycle. The concentrations of hormone in the circulating blood are given as arbitrary units because there is at present little firm evidence for the precise values.

This figure should be compared with Fig. 65.1 which shows the results of gonadotrophin levels on ovarian hormone secretion.

The experimental evidence for this follows the lines we have already discussed in the chapter on endocrinology. This is a simple concept, and consists of administering either pituitary hormone (or implanting pituitary glands) in immature experimental animals. Ovaries examined later show that follicles have been stimulated, ovulation has occurred and luteinisation taken place. Some of the extracts produce mainly growth of follicles; others give rise to ovulation and corpora lutea in addition. It is, moreover, found that a small dose of LH augments the ovarian response to FSH, so one can say that it is essentially a synergistic action and that balanced amounts of each hormone are necessary to produce the observed effects. FSH and LH are mucoproteins, secreted by distinct cell types in the peripheral and central parts of the pars distalis of the pituitary respectively.

Pituitary ovarian feedback mechanisms

Pituitary FSH combined with a small amount of LH gives rise to ovarian production of oestrogen. This raises the blood level of oestrogen, which in turn has a feedback action, which results in increased output of LH and decreased output of FSH. Ovulation and luteinisation then occur. As the corpus luteum regresses, both progesterone and oestrogen secretion fall off. The pituitary, in consequence, again begins to secrete more FSH, and thus the cycle is repeated. These cyclic variations in hormone level have been shown experimentally in circulating blood and in the urine. Figure 65.3 shows the presumed blood levels of FSH and LH during one menstrual cycle. (For a summary of hormone interrelationships in both male and female see Fig. 68.2.)

Positive feedback

In Chapter 14, the role of feedback mechanisms in biological control was discussed and we have seen that positive feedback results in an "explosive" event that can only happen once, certainly unless there is some kind of resetting mechanism operative. The mechanism of sodium permeability changes during the generation of the nerve action potential was cited as one example.

Another example is to be found in the mechanism of ovulation control by the pituitary. It is known that on the day before ovulation a sudden and large output of LH is released by the pituitary (not shown in Fig. 65.3). The explanation usually put forward is that when the circulating level of oestrogen is high, the response of the hypothalamus to progesterone is reversed. Normally it is a negative feedback (the more progesterone present, the lower the output of LH); at this time it becomes positive. Thus an explosive secretion of large amounts of LH takes place. When ovulation has occurred, oestrogen levels fall away, the sensitivity of the hypothalamic neurones reverses to normal again, and LH secretion shuts down.

Area responsible for cyclic release of gonadotrophins

Although it appears that LH is released from the pituitary cyclically, inducing periodic ovulation in the female, the secretion of gonadotrophin in the male seems to proceed at a constant level. This is evident because grafting ovaries into a castrated male does not cause ovulation but merely results in ovaries containing large follicles; grafted vaginal epithelium, in the same way, shows persistent cornification. In contrast, the same experiments in spayed females do show cyclic activity.

Why is there this difference? A clue was provided, in fact, as long ago as 1930, when Pfeiffer found that if one did the above experiment using male rats castrated at birth, ovarian grafts would indeed show cyclic activity; and simply transplanting a testis into a female at birth would destroy its rhythmicity later.

These findings would suggest that an intrinsic cyclic mechanism does exist in the female, but that male hormone converts the cyclic pattern into a steady one. It is interesting to observe that more recent work demonstrates that a single minute dose of testosterone will have this effect in the female for the remainder of its life, but only if it is injected during the first 10 days of life. It looks as if there is a critical period in which androgens can interfere, permanently, with this cyclic mechanism.

It might be supposed that this experimental effect, and by inference the seat of the cyclic mechanism, is located in the ovary or the pituitary. This is found not to be the case however, for grafting either a pituitary or an ovary from one of the injected, androgenised, acyclic females mentioned above, is followed by a return to normal cyclic function of the graft in question. Hence, by exclusion, we arrive at the conclusion that the hypothalamus must have been affected and that it is the seat of the cyclic discharge pattern. Presumably the differences between the sexes, as far as the hypothalamus is concerned, are developed at around the first 10 days of extrauterine life.

Induction of ovulation artificially

In the human, it is possible to inhibit ovulation by giving progesterone but it is much more difficult to induce ovulation artificially by the injection of gonadotrophins. One imagines that this is because a finely graded amount of FSH and LH is needed for growth of the follicle and that the proportions of the two hormones must be appropriately altered at the correct stage of growth in order to induce ovulation. It is interesting to note that follicles may be stimulated hormonally and fail to rupture. Cysts in the ovary can be produced in this way and a possible cause of the clinical condition of follicular cysts in the ovary, is an endocrine imbalance.

One cause of infertility in women can be traced to a low pituitary gonadotrophin secretion, which may lead to anovulatory cycles, irregular menstruation or to *amenorrhoea* (lack of menstruation). In such patients, it is sometimes possible to induce ovulation by the administration of *human chorionic gonadotrophin* (HCG). FSH is usually given for a period first.

The skill of the gynaecologist is tested not by his ability to produce ovulation, so much as in his limiting ovulation to a reasonable number of ova at one time. In the early days of this kind of treatment, a couple barren for years were often presented with quintuplets or sextuplets as a result of overenthusiastic hormone therapy.

Recently a synthetic drug, *clomiphene*, has been found which will induce ovulation. It apparently has this action via the hypothalamus, which is stimulated to produce LH-releasing factor.

Bromocriptine (see Chapter 67) will induce ovulation in 10 to 15% of patients with secondary amenorrhoea; the mechanism is obscure.

66. Fertility Control

Thomas Malthus, in his famous book *An Essay on Population*, published in 1799, pointed out the need for a balance between the numbers of a population and the amount of food required. Cautious estimates made by expert demographers indicate that the present world population of $2\frac{1}{2}$ billion, will become 4 billion by AD 2000 and 6 billion by AD 2050. The agriculturalists at the 1954 Rome Conference on Population claimed that the food production of the world might be doubled, but not increased more. It is clear, therefore, that the control of population is perhaps the most important social problem at the present time.

Contraceptive methods*

Contraception is employed (a) to postpone the first pregnancy, (b) to space out children and (c) to prevent conception occurring at all.† Methods in common use depend on one, or a combination, of four different principles:

1. Occlusive and impermeable barriers to the passage of sperm into the uterus (rubber diaphragms worn by the woman; rubber sheaths by the man).
2. Chemical spermaticides which immobilise or kill the spermatozoa in the upper part of the vagina or cervical canal.
3. Mechanical irritant devices placed within the uterine cavity which render implantation and embedding of the fertilised ovum impossible.
4. The oral contraceptive pill.

The relative efficacy of these methods is shown in Table 66.1.

Table 66.1 *The efficacy of contraceptive methods*

Method	Number of pregnancies per 100 women-years of use
Vaginal douche	18–36
Jelly or cream	6–49
Withdrawal	3–38
Safe period	14–35
Condom	6–28
Diaphragm	3–34
Intrauterine device	2–8
Oral tablets	0–3

---

* AMA Committee on Human Reproduction, *J. Amer. Med. Ass.* **194**, 462 (1965).

† It must be understood that the fear of pregnancy and its consequent disastrous effects on the life and behaviour of women was a very common and potent cause of marital disharmony. Among the major contributions of modern medicine are the discoveries of contraceptive techniques of proven efficacy, that lighten this psychological burden which had to be borne by previous generations of women. Moreover, the advent of really efficient contraceptive techniques have resulted in the separation of the sexual act from the process of procreation—an entirely new and liberating phenomenon bound to have far-reaching effects on human behaviour and interrelationships.

Oral contraception (general principles)

Since this is illustrative of applied physiological principles, oral methods of fertility control will be considered in more detail. The administration of progesterone was known for many years to inhibit ovulation in experimental animals. When this effect was studied in humans it was found that ovulation could be prevented in a group of normal women by giving them 300 mg of progesterone daily between day 5 and day 24 of the menstrual cycle. However, this represented a large amount of the hormone to administer and there were unpleasant side effects, apart from the fact that it was not completely effective. In 1955 clinical studies were undertaken with orally active progestational substances called 19-nor steroids. Field trials were carried out in Puerto Rico which showed the contraceptive effectiveness of these compounds (nor-ethisterone and nor-ethynodrel) particularly when combined with a small amount of oestrogen. (Twenty-one tablets were given, one each day as a course, at the end of which time progesterone withdrawal bleeding occurred.)

The "sequential pill" consists of giving oestrogen alone for 15 days, then oestrogen plus progesterone for 5 days. This is thought to inhibit ovulation by stopping secretion of both FSH and LH.

The method of choice is to give a continuous low dose of *gestagen* (artificial, orally active progesterone-like compounds) during the whole menstrual cycle. This has been found to cause complete infertility without inhibiting ovulation.

When administration of oral contraceptives is stopped there is usually a "rebound" increase in fertility, possibly due to raised gonadotrophin secretion (left over from the period of drug administration).

Physiological action

The precise way in which oral contraceptives work is not known at present. Primarily, they are considered to act by the inhibition of ovulation as a result of depression of the pituitary output of gonadotrophin. Oral progestogens prevent the rise (and peak at the fourteenth day) in concentration of LH in the blood. FSH is unaffected. Oestrogens, in small doses, have the opposite effect on LH secretion, i.e. it is increased, but in larger amounts oestrogenic compounds suppress ovulation by lowering secretion of both LH and FSH.

Other effects on the reproductive system are produced by oral contraceptives, which play an important part in their action. The viscosity of cervical mucus increases at the midcycle; the endometrium is thin and hypoplastic, both factors which militate against fertilisation and successful implantation of the ovum if it has been fertilised.

There are, in a small minority of women, unpleasant side effects experienced with oral contraceptives, such as nausea, vomiting, tenderness in the breasts, headache and various mild psychiatric disturbances. These are, in the main, the same kind of unpleasant symptoms as occur in the early months of pregnancy. Long-term adverse effects are said to include the possibility of an increase in frequency of cervical cancer in women regularly taking the pill and also an increase in the incidence of thrombosis of peripheral blood vessels due to hypercoagulability of the blood. In fact, studies extending over ten years show a slightly decreased incidence of cervical smears giving cytological evidence of carcinomatous change.

However, the risk of thromboembolic episodes is significantly increased by oral contraceptive use of oestrogens (now abandoned) in high concentrations. Nevertheless, to put this matter in correct perspective it must be pointed out that the risks of pregnancy are much greater than those due to long-term taking of the pill.

THE MENOPAUSE

Menstruation continues for 30 to 40 years and then ceases, usually gradually. Cessation is termed the *menopause* and the period immediately preceding this, when ovarian sexual function is declining, is termed the *climacteric*. At this time, follicles in the ovary are depleted, the menstrual periods may become profuse (menorrhagia), painful (dysmenorrhea), and irregular; they ultimately cease completely. Blood oestrogen levels fall, urinary excretion therefore falls and in virtue of the interrupted feedback to the pituitary due to lack of ovarian function, gonadotrophin secretion (and excretion in the urine) is augmented. This gonadotrophin is mainly FSH, and the role of feedback in its production is shown by the fact that when oestrone is injected the amount of it in the plasma is reduced.

One must conclude, therefore, that the primary cause of the menopause is exhaustion of the ovaries. The pituitary continues to function and produces more FSH than normally. Unpleasant symptoms often accompany the menopause, presumably as the result of hormonal changes, and include vasomotor effects (hot flushes), sweating and other autonomic disturbances. Psychiatric disturbance is relatively common during the climacteric, also as a result of hormonal imbalance. These symptoms are often treated by small doses of oestrogens. This is called hormone replacement therapy or HRT, and the main problem is to find an oestrogenic substance which can be taken by mouth for many years, and which is as close as possible structurally to the naturally occurring hormone. The need for HRT in menopausal women is determined by observing the state of the vaginal mucosa (which is hormone-dependent) using the "cornification index".

67. Pregnancy, Parturition and Lactation

Fertilisation

Due to the delayed passage of the ovum in the Fallopian tube, the spermatozoa meet the ovum here and fertilisation takes place, normally in the ampulla. The ovum is surrounded by many follicle cells embedded in an intercellular substance which is largely *hyaluronic acid*. Spermatozoa secrete an enzyme called *hyaluronidase*, which breaks down this ground substance and enables them to penetrate the outer layers of cells.

When the fertilising spermatozoon enters the vitellus, the ovum undergoes cell division with the production of the *female pronucleus*. The *male pronucleus* is the head of the spermatozoon, the two unite and this is followed immediately by the first mitotic division of the embryonic nucleus to give the first two daughter cells. As soon as one spermatozoon penetrates the zona pellucida, a chemical change occurs over the surface of the zona preventing the entry of any other spermatozoa. This block may fail, but if several spermatozoa penetrate the ovum at once, the normal events leading to fusion of the male and female pronuclei do not occur, and fertilisation fails (Fig. 67.1).

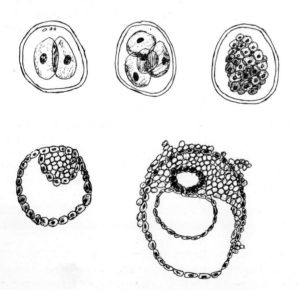

Fig. 67.1 Development of fertilised ovum. Upper row: Segmentation, 2, 4 and many cell stages. Lower left: Blastocyst. Lower right: Developing embryo.

Implantation

The fertilised ovum is transported to the cavity of the uterus where it remains free, on the uterine wall for several days. This time, of course, coincides with the height of the secretory, progestational phase of endometrial development and the uterine glandular secretion provides nutriment and a bland environment for the developing embryo. Not all the cells in the embryo divide at the same rate; those on the outer surface are smaller in size and form the trophoblast.

Embedding in the endometrium occurs in the early blastocyst stage. It does so in virtue of the trophoblast secreting proteolytic enzymes which digest the endometrial tissue with which it comes into contact.* It sinks into the uterine wall like a hot shot into a piece of butter, but as it sinks it is all the time getting larger and hence does not fall out again.

Chorionic gonadotrophin

The cells of the trophoblast also secrete "chorionic" gonadotrophin, which passes into the maternal circulation and prevents degeneration of the corpus luteum. Because of continued hormonal stimulation of the corpus luteum, progesterone is produced in gradually increasing amounts and the endometrium develops into a *decidua*. Also, menstruation is inhibited for the same reason, a state of affairs termed the "physiological amenorrhoea of pregnancy". Chorionic gonadotrophin passes easily through the kidneys and appears in the urine of the mother.

Pregnancy tests

Since human chorionic gonadotrophin appears in the urine within 10 to 12 days after ovulation, some form of bioassay should enable this fact to be put to applied use as a pregnancy test. The amount of urinary gonadotrophin reaches a peak at about the seventh week of pregnancy and is very largely composed of LH, so that a possible biological test for it would be whether or not ovaries in experimental animals show corpus luteum formation when the urine is injected into them.

The *Ascheim-Zondek test* depended on the fact that chorionic gonadotrophin induces hyperaemia (increased vascularity) in immature mouse ovaries. The *Friedman test* involved the production of ovulation by the intravenous injection of pregnant women's urine into the rabbit, and the *Hogben test* utilised the Xenopus toad, which ovulates when immersed in urine containing gonadotrophin. A thriving industry used to centre on testing urine samples for gonadotrophin using these toads.

The *Ascheim-Zondek* took 5 days to give a result and the *Hogben* 24 h.

Pregnancy diagnostic services are now available confidentially and by return of post, using an immunological test requiring only an hour or so to perform. Antibodies to HCG can be made by injecting rabbits with it and taking the serum. A haemagglutination test can then be used to test for HCG in a sample of urine from a possibly pregnant woman. Either sheep red-cells or minute latex "ballotini" are coated with normal HCG. These will then agglutinate if mixed up with the anti-HCG serum, as prepared above from the rabbit.

---

* *Chorion epithelioma.* The trophoblast of the developing embryo burrows into the endometrium until a balance is struck between tissue destruction and new growth of the trophoblastic villi. In some circumstances this equilibrium fails and the result is an overgrowth of the chorionic cells. In this way tumours arise ranging from the comparatively benign in the uterus to the overwhelmingly malignant.

These tumours give rise to large amounts of chorionic gonadotrophin and in consequence the ovaries are often filled with lutein cysts and the pregnancy tests may be positive at 1:1000 dilution. Death ensues from haemorrhages or from secondary deposits of the chorionic tumour in the lungs and elsewhere, often within only a few weeks of the onset.

If we have now previously mixed the anti-HCG serum with urine from a pregnant female containing much HCG, all the antibodies will be consumed. They will then not cause agglutination of the sensitised sheep's red cells. Thus if we find a urine sample that causes agglutination, it comes from a non-pregnant woman.

The decidual reaction

While the trophoblast is making living space for the embryo and absorbing food materials, reactive changes are taking place in the endometrium. Initially this occurs in the vicinity of the ovum, but later the whole uterine lining takes part. "The decidua of pregnancy" is the result:

1. There is proliferation and hyperplasia of the endometrial stromal cells, to form large polygonal cells, closely packed and relatively lacking in intercellular fibrillary material. Thus the endometrium is thicker and softer.
2. The uterine glands alter. Superficially they are small but in the deeper layers they become tortuous and dilated.
3. The decidua thus is divided into two layers, a superficial compact layer and a deep, spongy layer.

These changes adapt the endometrium to combine with the fetal tissues of the placenta for the purpose of transfer of oxygen, carbon dioxide and food materials between maternal and fetal circulatory systems.

The placenta

The embryonic mesoderm sends processes containing fetal blood vessel loops between the inner and outer layers of the trophoblast, to form the primary or anchoring villi. Later, by budding and branching from these, *secondary villi* are formed whose ends project into the intervillous blood spaces. These secondary villi are very numerous and comprise most of the villous septum covering the placental area. Hence, over a large area, maternal and fetal blood is separated by very thin connective tissue and a thin layer of trophoblast. The *discrete placenta* is formed by the third month of pregnancy.

THE PLACENTA AS AN ENDOCRINE ORGAN
The placenta has two major functions:

(a) It enables exchange to occur between the maternal and fetal blood although these do not mix.
(b) It has endocrine functions directed towards maintaining a suitable environment for the developing fetus.

The placenta secretes oestrogens, progesterone and gonadotrophins and analysis shows that the fetal parts, in particular the trophoblast cells, have the highest concentration of them, so it is reasonable to suppose that these cells are secretory. The high hormone levels produced by the placenta may sometimes affect the embryo itself. For instance, it is not unusual for human newborn males to have hypertrophied mammary glands. This may also be one cause of male pseudohermaphrodism (see p. 463).

Extrauterine changes in pregnancy

From the fifth or sixth week onwards for a period of weeks or months, many pregnant women often experience frequent episodes of nausea and vomiting, particularly on rising each morning. This *morning sickness* varies from a minor inconvenience to a severe disturbance (*hyperemesis gravidarum*) which may necessitate the termination of pregnancy. The cause is unknown but is probably the result of a hypersensitivity to foreign protein—the chorionic gonadotrophin—which is partly of paternal origin. Other unpleasant symptoms may occur such as pain in, and enlargement of, the breasts, frequency of micturition and digestive upset. Frequently, there are periods of craving for unusual foods and, although it is said by some to represent an attempt by nature to secure supplies of some essential mineral or vitamin, there is no evidence for this view. This condition becomes definitely pathological when materials such as household coke are consumed with relish. Changes occur in the activity of the central nervous system, presumably as a result of hormonal factors operating. Emotional outbursts, irritability and depressive episodes are common, but it must be remembered that pregnancy is essentially a physiological process and in consequence many pregnant women alter for the better in their attitudes.

Body weight increases due to water retention, as well as the added bulk of the fetus, membranes, amniotic fluid and the uterus. Fat is deposited. The distension of the anterior abdominal wall is accommodated by splitting of the skin which shows up as the so-called *striae gravidarum*. Symptoms and signs of increased intra-abdominal pressure may include oedema of the legs, dyspnoea, palpitations and the dilatation of superficial veins of the legs (due to mechanical obstruction of the femoral veins). Obvious changes also occur in the mammary glands which hypertrophy and which exude a clear (or in multiparae* a milky) fluid which can be expressed in the latter half of pregnancy. Pigment is deposited in the areola round the nipple, which turns from pink to brown during the first pregnancy, and also in other regions of the skin. Occasionally, unsightly patches of pigmentation occur on the face (chloasma uterinum) but fortunately fade after the child is born.

Nitrogen is retained by the mother in excess of the needs of the fetus; the same applies to calcium. Often the red cell count falls and there is a leucocytosis. Other blood changes are a lipaemia, cholesterolaemia and a reduction in the alkali reserve. Cardiac output rises progressively in the second half of pregnancy—presumably a physiological mechanism required to maintain adequate oxygen supply to the fetus.

Changes in the uterus in pregnancy

The uterus has a number of functions during the course of pregnancy. At first it acts as a suitable site for the fertilised ovum to pass its early developmental stages; later its wall is adapted to implantation and to the reception of the placenta. During pregnancy it enlarges to keep pace with the growing fetus and eventually it plays a large part in the expulsion of the fetus during parturition.

Thus the virgin uterus in the human is about 5 ml in capacity; at the end of the period of gestation it is 5 or 6 litres. The increase in weight likewise is from 50 g to 1 kg. All structures hypertrophy, particularly the myometrium. Each muscle fibre increases in length five to ten times and its diameter about five times. There is also a hyperplasia (increased number of fibres). These changes are due to: (1) The action of oestrogens. There is a slight hypertrophy during oestrus; similar to that occurring in female castrates. (2) Distension of the uterus also causes hypertrophy, even if produced by an inflated balloon within the lumen.

Uterine contractions
In the first half of pregnancy there are continuous and regular contractions of the myometrium once or twice per minute.

---

* A "multiparous" woman is one who has borne children previously.

These gradually increase in amplitude during pregnancy until, during the last two weeks, the contractions resemble those of parturition but on a smaller scale. The uterus is sensitive to posterior pituitary secretion (ADH in humans; oxytocin in guinea-pig) and the response to a given amount of hormone increases throughout pregnancy.

There are three factors causing this increased excitability:

(a) There is a steady rise in circulating oestrogen levels throughout pregnancy. These stimulate the myometrium.
(b) During the latter months of pregnancy progesterone levels decrease. Progesterone inhibits the stimulating effect of oxytocin.
(c) Distension stimulates muscle directly.

HORMONAL CONTROL OF PREGNANCY

In the pregnant female the anterior lobe of the pituitary gland is hypertrophied. The thyroid and adrenal cortex also increase in size. The urine contains large quantities of oestrogens, pregnanediol (the excretion product of progesterone) and chorionic gonadotrophins. The corpus luteum does not regress in the presence of an implanted fertilised ovum, possibly due to raised oestrogen production by it and its gonadotrophins. The precise way in which this persistence is brought about is unknown but it is clear that this phenomenon is the initial trigger in the sequence of hormonal changes which control pregnancy.

The activity of the corpus luteum continues up to the third or fourth month of pregnancy, after which time the glandular cells undergo gradual involution. Progesterone is produced and gives rise to the following effects:

(a) It enables pregnancy to continue because menstrual cycles and ovulation are inhibited.
(b) It inhibits uterine motility.*
(c) It maintains the progestational stage of endometrial development and has a similar action on the formation of the placenta.
(d) Alveolar development of the mammary gland is stimulated.

By the third month the placenta takes over the function of producing progesterone and also produces large amounts of chorionic gonadotrophins, oestrogens and ACTH. It is the main source of oestrogens in pregnancy.

The hormonal interactions in pregnancy are complex and not yet fully understood. One of the main reasons for this lies in the fact that there is considerable confusion in the experimental evidence, a state of affairs not helped by the undoubted fact that the various hormones do have different actions in different species.

The pituitary gland

A great deal of information on the role played by the various endocrine glands in pregnancy can be obtained by their removal and observation of the consequent effects.

---

* In cases of habitual abortion it used to be the practice to administer large doses of progesterone on the basis that this would remedy a natural deficiency of it, deemed to be responsible for the premature termination of pregnancy. The current view is that habitual abortion is often due to chronic bacterial infection of the reproductive tract. The treatment, sometimes successful, is to give a comprehensive course of a wide-spectrum antibiotic.

In some animals the whole pituitary gland is necessary for gestation to continue and it cannot be removed at any stage of pregnancy (e.g. rabbit, dog). In others it can be removed without obvious effect on the pregnancy (cat, mouse, rat).

The posterior lobe is not necessary for normal parturition.

The ovary

Presence of at least one ovary is necessary for the normal continuance of gestation in all species during the first week because presumably the process of embedding will not occur in the absence of its progesterone secretion. After that, in some species (rat and mouse) removal terminates pregnancy; in others (women, mares) the ovaries can be dispensed with without interfering with the pregnancy.

The placenta

In mice, rats and monkeys when the placenta is fully established, the fetus can be removed without disturbing it and the placenta will remain and be delivered at the time for normal parturition. Meanwhile, the extrauterine signs of pregnancy persist although they disappear if the placenta is removed.

This suggests that pregnancy is a complex maternal syndrome, having a definite duration controlled by the placenta.

Summary

1. Changes in the uterus are controlled by oestrogens and progesterone.
2. Activity of the corpus luteum is controlled by chorionic gonadotrophin and hence oestrogens (and uterine contents later in pregnancy).
3. Activity of the ovary is controlled by the anterior pituitary. In pregnancy ovulation is inhibited.
4. Activity of the anterior pituitary affects uterine contents by an unknown mechanism.
5. The pituitary gonadotrophin secretion increases until just before parturition.

PARTURITION

We have seen that the uterus is quiescent and not undergoing violent contractions during the first half of pregnancy. From the thirty-sixth week onwards the contractions increase in frequency, strength and duration, having a marked effect. They compress the fetus, particularly downward, and as a result it assumes the correct anatomical position for expulsion. The internal os of the uterus begins to dilate and the pregnant woman, at sometime between the thirty-sixth and fortieth week, is amazed to discover that her stomach has apparently become smaller again. This is due to the infant's head "engaging" in the pelvis, i.e. it sinks into the pelvic canal (Fig. 67.2).

Stages of labour

With the onset of "labour" (midwives' terminology for parturition) uterine contractions now occur strongly and are accompanied by painful sensations presumably due to pressure within the myometrium rising above arterial pressure and so giving rise to ischaemic pain. These "pains" (midwives' term for contractions) occur at first every $\frac{1}{2}$–1 h, then every 3–5 min lasting

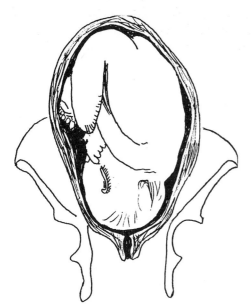

Fig. 67.2 First position of the vertex; left occipito-anterior (LOA). Vertex presentations occur in 96 per cent of all cases. There are four types, according to the position of the back of the foetus or of its occiput. If the vertex is in the uterine fundus, the presentation is abnormal and is said to be a breech presentation.

about ½–1 min each, and result in the final length of uterine muscle being a little shorter with each one.

Labour is divided into three stages.

First stage
The cervix is dilated by these continual contractions and usually—but not necessarily—the membranes rupture, allowing escape of amniotic fluid (midwives' "breaking of the waters").

Second stage
Full dilatation of the cervix marks the beginning of the 2nd stage. This is characterised by longer, stronger contractions which force the infant down through the pelvic canal. Abdominal muscles and diaphragm also contract with the uterus and the infant is finally expelled from its hitherto comfortable nidus to an inclement world (see Fig. 67.3).

Third stage
After a rest of 5 to 10 min, the uterus again contracts vigorously and the placenta, with the ruptured membranes, is expelled; the uterus then contracts down and venous sinuses are thus closed up.

Parturition in the human primigravida (mother having her first baby) lasts on average 15 to 20 h. The next labour lasts a mere 7 to 10 h and may indeed, to the consternation of all concerned, be very much more rapid than this. After several children (multipara) two or three good contractions only may do the trick!

The puerperium

After delivery (a period called the puerperium) the uterus involutes and lactation begins after about 24 to 48 h. Many authorities hold that menstruation does not begin until lactation has finished or has been terminated. This is undoubtedly not the case, and in many women the periods begin two or three months after birth even though lactation is proceeding normally.

CONTROL OF PARTURITION

Factors determining the end of pregnancy and the onset of parturition are unknown. However there is good evidence that the end of gestation is accompanied by large changes in the hormone balance. The major changes are (1) decrease in luteal hormone secretion; (2) decrease in circulating oestrogens.

Progesterone

There is direct evidence that parturition is preceded by a fall in progesterone production. The pregnanediol excreted in the urine decreases before parturition. In cases of habitual abortion there is often a lower pregnanediol excretion than normal. It is possible to prolong artificially the duration of pregnancy in rabbits if the luteal phase is also prolonged by giving progesterone or by giving gonadotrophin. In fact, the rabbit near term can be given gonadotrophins, a new corpus luteum is stimulated in the ovary and parturition does not begin until this corpus luteum has regressed.

Oestrogens

Circulating oestrogen levels reach a peak just before birth of the fetus. There is a sudden increase just before parturition (Fig. 67.4).

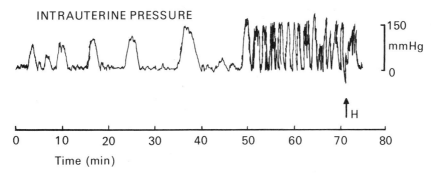

Fig. 67.3 Record of the intrauterine pressure of a woman in labour. The contractions are relatively infrequent up to the 50 min mark, when the first stage ends and the second begins; they immediately become more frequent and more violent. The head of the child was born after 75 min, at the moment indicated by the sudden fall in the pressure below the previous minima, and the body, just after the last rise in pressure. (Bourne and Burn.)

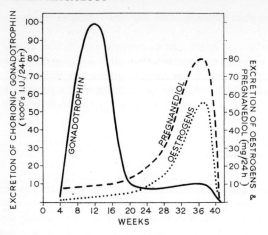

Fig. 67.4 Urinary excretion of chorionic gonadotrophin, pregnanediol and oestrogens during pregnancy.

It has been suggested that the uterus is sensitised to the action of oxytocin by the presence of oestrogen and that oxytocic activity is inhibited on uterine muscle by the presence of progesterone. This might represent a possible mechanism to account for the onset of labour in view of the hormone changes described above but the difficulty in accepting such an explanation is that by no means all species have these changes nor do the actions of oxytocin occur in all species.

The latest view is that the fetus itself times the onset of parturition. The mechanism postulated still involves an increased maternal circulating blood-level of oxytocin, but it has been shown that this originates in the fetal pituitary gland (see opposite page).

Prostaglandins have also been implicated.

Distension of the uterus is an obvious stimulus which might, upon reaching a certain level, start labour. However, this is not so because in multiple human pregnancy the uterus and contents are very large; in animals, after removal of the fetus, the placenta is of negligible bulk, but in both instances parturition occurs normally at the right time. Several other mechanisms to account for the onset of parturition have been postulated and it is by no means impossible that a combination of numerous stimuli is in fact responsible. In biological terms, such a system would have a greater safety margin. Parturition can take place after dorsal section of the spinal cord and even after complete destruction of the lumbosacral region.

LACTATION AND THE MAMMARY GLANDS

Development of mammary glands in pregnancy

The mammary glands enlarge in pregnancy due to the combined effects of oestrogens and progesterone. In the majority of mammals, oestrogens increase the development of the duct system; progesterone influences the proliferation of alveoli. Since these hormones are ineffective in producing this mammary development in hypophysectomised animals, it is now generally agreed that growth hormone and adrenocortical hormones are essential as well. Milk secretion is initiated by the sudden drop in progesterone levels at birth. Although the pituitary is not necessary for the continued secretion of milk in certain species, in the human necrosis of the pituitary (which may follow severe post-partum haemorrhage: Sheehan's syndrome) abolishes lactation.

Lactation

In women, *colostrum* is secreted by the breasts at the end of pregnancy and is a clear, straw-coloured fluid, possibly largely an ultrafiltrate of plasma from the as yet inactive alveoli. It is popularly held to have a laxative function but upon consideration of the fecal proclivities of the newborn infant, any mechanism such as this would seem to be gilding the lily. True milk does not appear until some three days after parturition.

Hormonal control of lactation

There are two separate processes involved in lactation. The first is the secretion of milk and the second is its expulsion from the mammary gland. The secretion of milk begins with the initiation of secretion, or *lactogenesis*, which normally occurs two or three days after the child has been born.

Lactogenesis is hormonally controlled; psychological factors may affect it. Low levels of oestrogen in the circulation stimulate the pituitary to produce prolactin, the hormone which causes the alveoli to secrete milk. Throughout pregnancy and whilst the mammary glands are developing in size, the effect of low oestrogen levels is inhibited by the circulating progesterone. At parturition the placental source of progesterone disappears and its circulating level abruptly falls. Thus the oestrogen is then able to act upon the pituitary to cause it to produce prolactin. On the contrary, it is a well known fact that high oestrogen levels inhibit lactation. Stilboestrol, an artificial oestrogenic substance, used to be given to mothers to prevent lactation by virtue of this action, but this is done no longer because (a) it is suspected of having carcinogenic properties and (b) because lactation ceases naturally if the baby is not put to the breast.

Galactopoiesis, the actual process of milk secretion is also controlled hormonally, probably by the thyroid hormone and the growth hormone. Prolactin is important in lactogenesis, less so in galactopoiesis. Thyroid hormone is sometimes used in lactating women to improve the amount of milk secreted.

Bromocriptine

Bromocriptine (2-bromo-alpha-ergocryptine; CB154) is a prolactin inhibitor first developed for use in reproductive physiological experiments. It is a dopaminergic receptor stimulator with only weak vasoconstrictor effects. Bromocriptine slows dopamine turnover in the tuberoinfundibular nuclei and inhibits prolactin secretion by the pituitary. Prolactin is unique in being controlled by an inhibiting factor, PIF which has turned out to be dopamine.

Bromocriptine has been used in various clinical conditions such as to suppress lactation and in the hypogonadism due to hyperprolactinaemia. Why oversecretion of prolactin should cause infertility is not known but prolactin induced block of gonadotrophins at the ovaries must be in part the cause. A rare condition due to the same mechanism occurs in the male and again bromocriptine treatment restores fertility. The drug has also been used to treat acromegaly and Parkinson's disease, which is due to a nigro-striatal dopamine deficiency. Further clinical applications may evolve now that it has been realised that dopamine is an important transmitter peripherally as well as in the CNS.

Secretion and expulsion of milk

The process of suckling (or emptying the breast in any other way) initiates nervous reflexes which lead to: (a) further milk secretion; (b) expulsion of milk already in the gland.

Expulsion is due to the reflex secretion of oxytocin from the posterior pituitary gland, which stimulates contraction of the myoepithelial cells in the duct system. Children are normally weaned at six to nine months of age; but lactation and suckling can be continued for several years and, indeed, this is often the case in certain Eastern cultures or among famine-stricken populations.

Clinically it is found that failure of lactation may occur and is often due initially to failure of milk ejection. In such cases, an injection of oxytocin at the beginning of breast-feeding may greatly increase the milk yield, at the time and thereafter. Many of these cases have a psychological origin. A cow, for instance, will "hide her milk" if she is frightened or is disturbed by the presence of a different milkmaid. The calf's presence, on the other hand, gives rise to an increase in milk flow. The presence of the child, or hearing its cry, will stimulate a nursing mother's milk flow, and many complex conditioned reflexes are often built up in the process.

Milk

Milk is a white opaque fluid containing lactose and caseinogen in solution and various salts and fats in suspension.

Milk ferments, if allowed to stand, by the action of the *Lactobacillus*, lactose being hydrolysed to lactic acid. The resultant increase in acidity denatures the caseinogen and it "curdles" or eventually coagulates as in cheese. Cream is formed on the top of milk because the fat droplets rise, a process which can be hastened with a centrifuge (cream separator). Violent agitation of cream (churning) gives a phase reversal. Cream is a suspension of fat globules in a watery medium, while butter is a suspension of watery droplets in fat.

The chemical composition of milk is not the same in all mammals. Cow's milk, for instance, contains less lactose and more caseinogen and salts than does human milk. This difference is largely the reason why it is often difficult to raise babies on cow's milk and why its composition must be altered by dilution and altering curd formation before it can be used in artificial feeding. The caseinogen is the difficult factor. It is not due merely to the difference in amount in cow's milk but also to the difference in its nature. On addition of acid, cow's caseinogen gives a solid curd; human milk is flocculent.

Other constituents of milk

Milk contains numerous other constituents of variable importance. It is a source of vitamins for the child, particularly A, D and riboflavin. Cow's milk is deficient in vitamin C, although human milk contains an adequate amount. It follows that orange juice, etc. must be provided for bottle-fed babies. If the mother's dietary intake of vitamins is deficient, this will be also true in the milk.

Antibodies and globulins of a protective nature are present in milk, also in colostrum.

There are two factors which may cause iron deficiency, and consequently anaemia in infancy. These are inadequate dietary iron and inadequate stores at birth. About half the iron stores are laid down in the last month of fetal life, so that premature birth, for any reason, leads to an iron deficiency state in the child. A growing child needs 0.5 to 1.0 mg Fe per day and this is not supplied by either human or cow's milk. A normal child weighing 4 kg has enough iron to last at least six months. A 2.5 kg baby only has enough iron to grow to 4 kg and will then need iron supplements, as it will not be on solid food by this time.

Milk is an excellent culture medium for micro-organisms, either provided by the cow (or indeed human) secreting it or introduced during its handling. Such pathogens as *Brucella abortus* and *tubercle bacilli* are examples of the former; *typhoid* and various *streptococci* of the latter. The advantages of breast feeding are obvious.

THE INITIATION OF PARTURITION

Research[*] has clearly shown that the fetus is actively involved in the processes which control the length of gestation. If the pituitary of a fetal lamb be removed, the pregnancy is prolonged indefinitely. The injection of adrenocorticotrophin (ACTH) or glucocorticoids into fetal lambs shortens pregnancy. It has recently been found that there is a rise in fetal plasma corticosteroids in the ten days before birth and, moreover, that this rise is not related to any changes in the maternal levels of these substances. The mechanism leading to this increased secretion is unknown; possibly an increase in circulating levels of ACTH in the fetus is responsible.

Anterior pituitary and adrenal cortex in the fetus

The infusion of ACTH mentioned above achieves its result when given ($10 \mu g\ h^{-1}$) over a period of 4–5 days. We do not yet know anything about normal fetal ACTH levels.

The fetal adrenal undergoes hypertrophy during the last ten days of gestation; this is prevented by hypophysectomy so presumably ACTH and perhaps other pituitary hormones are needed for it. There is no clear-cut evidence on this mechanism in the human fetus, but the general view is that pregnancy is prolonged in those cases where the adrenals are small due to anencephaly. In premature infants (before 36 weeks) the adrenals are usually larger than normal. Cortisol levels in cord blood are ten times higher in normal deliveries than they are in elective caesarian sections or in deliveries induced by oxytocin.

Oestrogens

The levels of maternal urinary oestrogens, as measured for several weeks before parturition are related to uterine contractility, and to the length of gestation. The mechanism of this appears to be that 17β-oestradiol in the mother causes the release of prostaglandin F into the uterine vein which in turn sensitises the myometrium to the action of circulating oxytocin. There is a considerable increase in oestrogenic substances in the 24 h before normal or premature parturition, caused by ACTH infusion into the fetus (as described above).

Oxytocin

Oxytocin has been used for a long time past to initiate labour. There is, however, no evidence that circulating maternal levels of oxytocin change at the time of parturition or that the hormone plays any part in the initiation of normal labour. In the fetus, however, it is found that oxytocin levels are much higher (in the umbilical artery) when normal vaginal delivery has occurred than when elective caesarian section has been done. The release of oxytocin appears to be associated with the process of labour, being at its highest level during the expulsive phase. (Origin in the fetus is deduced because there is a higher level in the umbilical artery than the vein; fetal levels are much higher than maternal ones.)

Progesterone

It has always been thought that progesterone plays a large part in the maintenance of pregnancy, but the theory is based upon work involving species in which the corpus luteum is the major source of progesterone. In these species, removal of the corpus luteum at any stage in pregnancy results in abortion. Large doses of progesterone block the release of prostaglandin F; also progesterone withdrawal causes myometrial activity leading to parturition.

However, in the human it is difficult to accept the theory that progesterone is of importance since one cannot show a fall in plasma levels of it before labour commences.

---

[*] Liggins, G. C. (1969) in *Foetal Autonomy*, Ciba Foundation Symposium (Eds. G. E. W. Wolstenholme and H. O'Connor). Churchill, London.

Prostaglandins

In the sheep prostaglandin $F_2\alpha$ has been identified in the placenta and myometrium. Changes in its blood level correlate with the strength of uterine contractions occurring at the time of assay, maximal concentrations being found at the end of the second stage and during expulsion of the fetus.

It is found that the changes in prostaglandin F in venous blood leaving the uterus are closely correlated with 17β-oestradiol levels. This fact has led to the formulation of the hypothesis that fetal oestrogen (entering the maternal circulation via the placenta) could act as the trigger for synthesis and release of prostaglandins.

Uterine contractility

Recent work appears to show that uterine contractions are determined by the intracellular levels of cyclic AMP and GMP. Motility is inhibited by increased cyclic AMP and enhanced by increased cyclic GMP. Beta adrenergic fibres stimulate cyclic AMP production; oxytocin and prostaglandin $F_2\alpha$ raise cyclic GMP levels. They possibly act through the myometrial cell membrane, high AMP levels hyperpolarising and high GMP levels depolarising it.

Indomethacin, which inhibits prostaglandin synthesis, abolishes myometrial activity. Oxytocin, moreover, has no effect if prostaglandin synthesis is first inhibited. Indomethacin prolongs pregnancy in rhesus monkeys.*

If strips of uterine muscle from a pregnant human are put in an organ bath and then stretched, prostaglandin E_2 is released into the bathing fluid. This could be the basis for a positive feedback mechanism operative at the termination of pregnancy.

---

* It is of interest to note that women on large and long-term doses of aspirin, which inhibits prostaglandin synthesis, tend to have pregnancies lasting longer than do matched controls.

68. The Male Reproductive System

The male reproductive system produces spermatozoa for fertilisation of the ovum. Spermatozoa are produced in the *testes*; the rest of the male genital system consists of the necessary ducts for storage and transport of spermatozoa to the exterior, and glands for producing secretions in which the sperm are suspended. The testes are organs of internal secretion and elaborate male sex hormones under the primary control of the pituitary gland.

Spermatogenesis

Very large numbers of spermatozoa are produced by the testes—several millions daily—although for fertilisation to occur very much smaller numbers are needed (e.g. 50–100 reaching the surface of the ovum). Development occurs in the *seminiferous tubules*, which form most of the volume of the testis. The seminiferous tubules are formed by a number of layers of cells with a thin connective tissue membrane outside. The cells in the outermost layers are called *spermatogonia*. These divide and give rise to *first-order spermatocytes*, and since the progress of development is inwards (towards the lumen of the tubule) these cells are then more superficial. A further division produces *second-order spermatocytes*, which on a third division gives the *spermatids*. These, without division, mature into spermatozoa (Fig. 68.1).

Between the developing spermatogonia are masses of columnar cells, called *Sertoli cells*, extending as far as the lumen, on which the spermatozoa hang until they are ripe for release and ejaculation.

Spermatogenesis begins at puberty and, just as might be expected, the control of this process is in large measure brought about by the anterior pituitary, which secretes the gametogenic hormone. This has been shown to be the same hormone as the female FSH.

Endocrine function of the testes

The *interstitial cells* in the testes are mesodermal in origin, being typically epithelial in nature with mitochondria, granules, rods and prosecretion granules in their cytoplasm. They have a well-developed, pale-staining nucleus, and occur mainly around the blood vessels. In mammalian species the development of interstitial cells is controlled by the *interstitial cell stimulating hormone*, or ICSH, which is the same as LH in the female.

In certain species, including man, the testes are permanently active; spermatogenesis is continuous. In other species there are

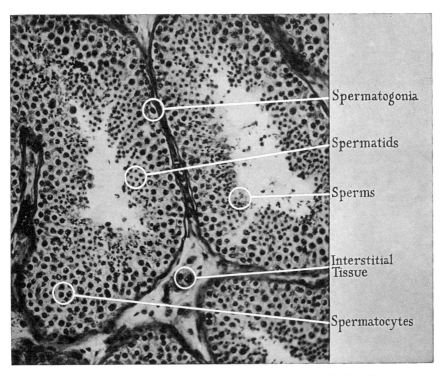

Fig. 68.1 Microscopic section of human testis, showing seminiferous and interstitial tissue. (From a preparation by Mr K. C. Richardson.)

sexual seasons, and activity in the testes alternates with periods of rest. In man, spermatogenesis gradually decreases from the age of about 50 years onwards.

The hormone testosterone (an androgen) is secreted by the interstitial cells. This fact can be shown by tying the vas deferens, whereupon the seminiferous tubules degenerate, only the Sertoli cells remaining. The interstitial cells, on the other hand, undergo hypertrophy and hyperplasia. Hormone levels in blood are unaltered, or even enhanced, and the secondary sexual characteristics, which are controlled by testosterone, are unaffected. Small doses of X-irradiation have a similar effect; larger doses cause atrophy of the interstitial cells as well and the various phenomena of castration ensue.

Action of androgens

Androgens stimulate spermatogenesis. If the pituitary is removed in an experimental animal, the testes atrophy; this effect can be prevented by the administration of androgens. A compound is said to be androgenic if it has a "masculinising" or "virilising" effect. The main androgens are:

1. *Androsterone*, which can be synthesised and was first extracted from urine.
2. *Testosterone*, obtained by testicular extraction. It is about 10 times as active as androsterone.
3. *Dehydroisandrosterone*, less active and also found in urine.
4. Various other androgenic steroids which occur in the adrenal cortex.

All these substances are rapidly inactivated by the liver and hence are not suitable for oral administration. The synthetic derivative, methyltestosterone, however, is active when given by mouth.

Testicular hormones have specific sexual actions and widespread general effects.

The primary sexual action is on the development of male genital organs and the male secondary sexual characteristics.

The sexual action of androgens can be investigated by (a) administration of androgens in the fetus (gives rise to male organs or intersexuality), (b) in immature animals (when precocious puberty is induced) and (c) in male castrates or females (when masculinisation is produced).

Androgens, if administered before puberty, at first increase growth but as the ossification of epiphyseal cartilages occurs prematurely, normal height is not exceeded. The penis, prostate and seminal vesicles develop more rapidly. All bodily hair grows to the adult male distribution. There are also changes in the skin, the larynx enlarges, with deepening of the voice, muscles develop and all the psychological changes characterising maleness appear. Figure 68.2 gives a summary of the hormonal factors which are concerned in the male and female reproductive systems.

Testicular hormones

These are steroids based on the C_{19} steroid ring, having oxygen containing groups at the C-3 and C-17 positions. The structures of the three active compounds, testosterone, androsterone and dehydroepiandrosterone are as follows:

Testosterone   Androsterone   Dehydroepiandrosterone

Testosterone bears a close resemblance to progesterone being an α,β-unsaturated ketone. Testosterone is the most potent of the known androgens.

Removal of testes (castration)

If the testes are removed *before puberty*, the pubertal changes do not occur. In man, the penis remains small, seminal vesicles, prostate and Cowper's glands remain undeveloped and the secondary sexual characteristics fail to appear.

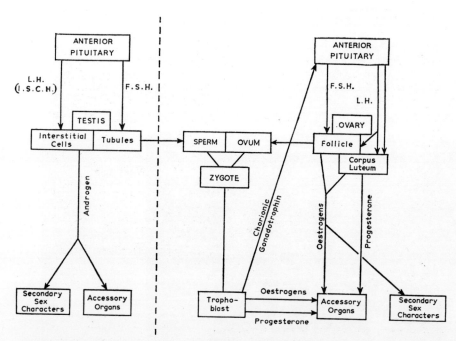

Fig. 68.2 Diagrammatic summary of the interrelationships of the principal hormones concerned with reproduction, male and female. LH, leuteinising hormone; FSH, follicle stimulating hormone; ICSH, interstitial cell stimulating hormone.

Castration *after puberty* has less effect. The seminal vesicles regress but the other changes of puberty such as penile enlargement and alterations in the larynx persist. Erection and copulation may continue to be undertaken for years in some post-pubertal castrates, although *libido* (the sex urge) is usually diminished to a greater or lesser extent. Frequently, castrates of this kind tend to be "effeminate" in nature, but intellectual development is not disturbed by castration and it is well known that many castrates are of outstanding intellectual achievement.

The removal of one testicle and most of the other is not followed by the usual signs of castration. Only 2 to 5 per cent of intact testicular tissue is sufficient to maintain normal hormonal function, a fact due to the remaining interstitial cells undergoing hyperplasia owing to pituitary action. Just as we have seen in the female, in the male there are feedback mechanisms which ensure adequate secretion of gonadotrophins. Thus in the partial castrate an excess of ICSH is produced and the cells left are overstimulated.

Testicular insufficiency

Total castration gives rise to the condition of *eunuchism*; relative testicular insufficiency results in *eunuchoidism*. These conditions, apart from testicular removal or injury, result from various factors:

(a) *Primary disturbances* in the testes; e.g. castration, congenital absence of testicles, and tumours or infections (e.g. mumps). In all these conditions, both endocrine and spermatogenic functions are abolished or diminished. ICSH secretion is increased; testosterone is decreased.

(b) *Seminal insufficiency.* In these conditions (e.g. Klinefelter's syndrome) seminiferous activity is reduced or absent but the interstitial cells are normal. Cryptorchism or undescended testicle is accompanied by lack of spermatogenesis because the germinal epithelium degenerates if it is kept at body temperature. In the scrotum, the temperature is from 1 to 8°C lower than in the abdomen. Heating the scrotum beyond its normal temperature quickly injures the germinal cells and prolonged pyrexia may be followed by temporary sterility. For a few months recovery is possible, but ultimately the damage is irreversible. Thus, unless cryptorchism is treated before the onset of puberty, sterility ensues and cannot be corrected by surgery later.

(c) *Secondary testicular insufficiency* occurs as a result of hypopituitarism; in all cases both interstitial cells and seminiferous tubules are atrophied. Causes are, for example, complete pituitary destruction by tumours, selective insufficiency of gonadotrophin secretion, and inhibition by excess circulating oestrogens.

Treatment of eunuchoidism and eunuchism is by giving methyl-testosterone by mouth, in regular maintenance doses.

Testicular hyperfunction

Androgens or gonadotrophins when injected produce testicular hyperfunction, although it is interesting to note that contrary to expectation, injections of testosterone soon abolish spermatogenesis. Prolonged treatment in men gave first a reduction, then total abolition, of sperm with marked histological abnormality in the seminiferous tubules. It was surmised that this was due to the testosterone having its action in stopping the anterior pituitary secretion of LH, a fact confirmed when it was found that urinary excretion of gonadotrophins had ceased. Recovery took 18 months!

There is also a control mechanism for adjusting sperm production to the demand. In the absence of sexual activity (coitus or masturbation), production gradually falls off to a low level. Continuous sexual activity increases the rate of production accordingly. It is likely that this control is mediated via the pituitary ICSH secretion.

Various studies (e.g. Kinsey Report) show that the great majority of male humans have regular and frequent "sexual outlets" (*sic*), a term covering normal heterosexual intercourse, homosexual activity and masturbation. Masturbation (self-stimulation) occurs particularly at puberty and for several years thereafter, and appears to be a necessary process (in maintaining spermatogenesis) in the absence of the heterosexual activity less easily available for the adolescent and young adult in modern society. The only morbid results of this practice are likely to be of psychological origin engendered by superstitious fears and by unfortunate propaganda. Sexual behaviour to be regarded as normal in this respect, depends greatly upon age group, social class, general health, etc., but between 14–20 years of age it is claimed that 99

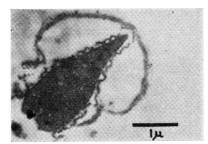
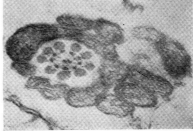
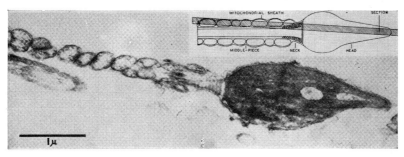

Fig. 68.3 Human spermatozoa.
 Top left: Longitudinal section through head of human spermatozoon. Acrosome and enveloping membranes have separated from surface of sperm head (× 20 000). Top right: Transverse section through middle piece of human spermatozoon which may be immature (× 80 000). Bottom: Longitudinal section through head and middle piece of human spermatozoon (× 27 000). Small inset diagram shows approximate plane of section. (Rothschild (1958) *The Human Spermatozoon. Br. med. J.* i, 301.)

per cent of individuals masturbate. Average rates quoted are five to six "outlets" (of any kind) per week, with the peak frequency at 18 years of age; a low, but presumably normal rate would be 1 per month, and a high rate 5 per day.

Spermatozoa

Mature spermatozoa of different species show enormous morphological variation. The human spermatozoon (Fig. 68.3) consists of a head 4·6 μm long and 2·6 μm wide and of a caudal appendage 35 to 45 μm long. The head contains the nucleus at its base. The caudal appendage includes the midpiece, tail and terminal filament. The midpiece is about the same length as the head and is traversed by the axial filament, covered by a sheath. Around this is coiled the helical filament of five or six turns contained within a cylindrical sheath. The tail is 30 to 40 μm long.

The life span of sperm, during which they are capable of fertilising the ovum is only 24 hours, although they can be frozen ($-169°C$) and stored and still be fertile on rewarming. Sperm concentrations of less than 5×10^6 per ml of ejaculate in the human generally result in sterility. One billion (10^9) sperm are produced in the male for every single ovum produced by the female!

Environmental conditions influencing spermatozoa

The most important is hydrogen ion concentration, the optimum pH for preservation being a little over 7·0. Other ions may exert an important but less fully investigated effect. As the pH rises up to 8·5, sperm motility increases but this exhausts the small available supply of energy. It is often said that epididymal spermatozoa are quiescent but this is not true; when on operation on men with blockage of the tail of the epididymis (caused by gonorrhoea), the distended head of the epididymis is incised and a drop of fluid from it is examined under the microscope, highly active spermatozoa are often found. Prostatic secretion (pH 7–8) increases, vaginal secretion (acid) diminishes and cervical secretion (alkaline) increases motility.

Body temperature is optimal for the motility, a lower temperature for the preservation of spermatozoa. The significance of the natural secretions which they meet is not known for certain, though there is little doubt that these do have some effect on the fertilising capacity of the sperms. It has clearly been demonstrated that spermatozoa must pass a certain minimal time within the Fallopian tubes, during which they undergo an effect called *capacitation*, before they become capable of fertilising an ovum.

69. Coitus (Copulation)

This is the act of union whereby the male deposits spermatozoa in the genital tract of the female. Coitus is attended by excitement which culminates in the *orgasm*—a paroxysm of sensation largely contributed by sensory elements in the glans penis in the male and accompanied by the ejaculation of semen. The degree of sexual excitement experienced by females varies considerably in different subhuman species but it is doubtful if, in most, orgasm is experienced at all. Among women, great variability of orgastic experience is encountered; in some it is intense, in others totally absent.

Male erection

The nervous basis for coitus is a spinal reflex and the act can occur after section of the spinal cord in the dorsal region, when there is complete absence of sensation. The two essential parts of the act are erection, which enables the penis to be inserted into the vagina, and ejaculation. *Erection* is the result of distension with blood of the venous sinuses of the corpus spongeosum and of the corpora cavernosa, whose resistant fibrous capsules then render the penis hard and rigid. This is brought about by dilatation of the helicine arteries of the penis, as a result of which inflow of blood into the corpora cavernosa increases, while through compression of veins the outflow of blood is hindered. Associated with this is relaxation of the smooth muscles in the trabeculae of the fibrous tissue. Stimulation of the pelvic nerves (second, third and fourth sacral segments) initiates erection and their section abolishes it. Stimulation of sympathetic fibres from the lumbar region is said to constrict the vessels of the penis and make it flaccid. The afferent side of the reflex arc conveys sensory impulses from the penis* but superimposed upon the basic reflex arc are the effects of impulses from the higher nervous centres, by means of which many other stimuli, such as sight, smell, sound, as well as the results of purely cortical activity, such as thought, memory and so on, can cause erection. At the same time, many stimuli acting through the association areas of the brain can exert an inhibitory effect on the erection reflex, either preventing its occurrence or abolishing it once it has begun. This inhibitory mechanism is held responsible for most causes of impotence. It also provides the means whereby some control of a voluntary nature can be exercised so that, for example, whereas in some circumstances various stimuli may evoke erection, in others where erection would be undesirable it can be deliberately prevented. Involuntary inhibition of an erection in progress may result from disturbing influences of all sorts occurring at an inopportune moment.

A further important factor affecting erection is the presence of male sex hormones (testosterone). Erections occur in boys long before puberty, in *eunuchs* (men whose testes have been removed) and in *eunuchoids* (men whose testes have never developed properly); hence, testosterone is not essential for erection. On the other hand, many eunuchoids complain of infrequent and imperfect erections which are rendered normal by appropriate treatment with male hormone. Moreover, if given in excessive doses, testosterone may induce a state of more or less continuous erection (called *priapism*). On the other hand, the administration of testosterone to men who are impotent but who secrete normal amounts of male hormone is almost invariably without any effect whatsoever. It would seem, therefore, that testosterone facilitates the normal erection reflex, reducing the threshold of the stimuli necessary to excite it; it is powerless, however, to overcome the effects of inhibition exerted by the higher centres and it is quite clear that these inhibitory stimuli are prepotent since, coming in circumstances which have already excited erection, they can abolish it.

Psychological factors

The performance of the sexual act may be greatly enhanced by psychological factors. The mental consideration of, or remembrance of, sexual matters (or perusal of pornographic literature or photographs) facilitate erection and attainment of an orgasm. Dreams of an erotic nature often result in the involuntary erection and ejaculation of semen (nocturnal emissions); these are common in teenage boys and perhaps may reflect the fact that sperm production at that age exceeds the rate at which it is released by normal sexual activity.

Ejaculation

When the stimuli which excite erection are sufficiently intense and sufficiently prolonged, they set in train a remarkable series of nervous and muscular effects, culminating in orgasm and ejaculation. Pulse and respiratory rates increase and blood pressure rises; there is a general development of muscular tension throughout the body and rhythmic movements, especially of the pelvic region, occur and increase in speed. At the climax, or orgasm, tensions are released, ejaculation occurs and the body then rapidly returns to its normal state. *Ejaculation*, like erection, is brought about primarily by a spinal reflex and once the reflex has been set in motion it is beyond the reach of voluntary inhibitory control. In this it is unlike the subjective accompaniments, since these can be enhanced by influences acting on the higher centres of the brain. Preceding the actual ejaculation, the stimuli, which will eventually produce it, cause reflex increased secretion of the accessory sex glands so that, in some men, clear fluid (mainly derived from the glands of Littré) may escape from the urethral meatus. The discharge of impulses from

---

* Impulses also from nearby "erogenous" zones may also add to the totality of afferent input. For example, stimulation of anus, scrotum and perineum add to sexual sensation. Considerable sexual stimulation results from internal structures, e.g. a full bladder or irritated areas of the reproductive tract. An enlarged prostate may result in almost continual sexual stimulation and desire, a notorious feature to be observed in many lecherous old men. Aphrodisiac drugs, such as cantharidin, increase sexual desire by causing irritation of the bladder and urethra.

the spinal centre eventually causes rhythmic contractions of the vasa deferentia, seminal vesicles and prostate, thereby expelling the contained spermatozoa and accessory secretions. The seminal fluid so formed is ejected from the urethral opening in a series of spurts, varying in number from two or three to perhaps a dozen. The first spurt is usually devoid of spermatozoa, being composed chiefly of secretion from the urethral glands and Cowper's glands. An intermediate fraction of the ejaculate is rich in spermatozoa, the remainder consists largely of seminal vesicle and prostatic gland secretion. The smooth muscles involved in ejaculation are supplied by the presacral nerve (sympathetic) and its section will therefore lead to sterility.

The orgastic sensation is variable in its intensity and duration. It may be local, confined to the perineal region and the penis, and simply be a sensory barrage of pleasant nature accompanying each ejaculatory spurt. At its best, it can be a tremendous sensory experience apparently generated by practically every muscle in the body and occupying the whole of consciousness.

The female sexual act

In the female, similar events lead to the orgasm. Erectile tissue in the vulva and around the vagina becomes engorged as a result of sexual stimulation and secretion of the various glands provide the necessary lubrication for coitus.

As in the male, both local stimulation of the sexual regions and psychological stimulation are very important in generating the orgasm. It is hard to evaluate the factors concerned in "sexual-drive" in the human female; the sex hormones and adrenal cortical hormones are certainly important. Nevertheless, many female children learn that sexual matters have rigid "taboos" of far-reaching nature attached to them, and that normal sexual activity is immoral and definitely not to be enjoyed. It is therefore not surprising that after such "training", much of the woman's normal sex-drive is inhibited. Frigidity of this nature is notoriously hard to treat psychiatrically.

Female erection

Mechanical and irritative stimulation of the perineal region, sexual organs and urinary tract originate sexual sensory impulses, much as they do in the male. *The clitoris* is extremely sensitive in this respect.

Around the vaginal introitus in the labia minora and passing forward into the clitoris, is erectile tissue identical with that of the penis. It is supplied by the nervi erigentes (parasympathetic from the sacral plexus). The arteries to this erectile tissue are dilated during sexual stimulation, and as a result the introitus swells and clasps the penis. This in turn, during copulatory movements, stimulates the shaft of the penis. Bartholin's glands, supplied also by the parasympathetic, then secrete large amounts of mucus for purposes of lubrication. Without the combined output of the male glands of Littré and the Bartholin's glands the friction between the glans penis, and the vagina and cervix, becomes painful and the achievement of an orgasm somewhat prolonged, difficult or even impossible.

Female orgasm

Local and cerebral cortical sexual stimulation gradually build up to a climax of sensory experience causing the female orgasm. It is accompanied by rhythmical contractions of the perineal muscles which, as in the male, give rise to a whole spectrum of indescribable sensory barrages.

The human female who regularly experiences full orgasms is known to be more fertile than the female who does not, or who has been artificially inseminated. These facts are held to indicate a direct physiological importance of the orgasm. A common view is that during the orgasm, uterine and Fallopian tube motility is increased, a factor which would tend to promote rapid transport of sperm to the vicinity of the waiting ovum. However, the evidence may also be interpreted as showing merely that hormonal imbalance, for example, both reduces fertility at the same time as it decreases the likelihood of a woman achieving an orgasm.

Sexuality and human behaviour

It will be apparent that the behavioural, emotional and general psychological aspects of sexual matters in the human species far exceed, in scope and magnitude, the requirements of a purely reproductive process. After all, reproduction as a biological act proceeds quite satisfactorily in all other species without the elaborate and far-reaching sexuality that accompanies the process in humans. In particular, the human spends an inordinate proportion of his waking hours in the direct or indirect pursuit of sexual gratification.

The view of Desmond Morris is that the secondary manifestations of sex in the human have evolved as a mechanism for cementing the family bond, in other words for ensuring that once a male and female form a pair, this pairing is maintained for a long period of time. The necessity for a stable family unit is imposed by the comparatively slow rate at which the human offspring matures. The complex nature of human brain function, and the enormous load of information that the developing brain must absorb, imply a necessarily lengthy process of learning and education. The parents must, in order to play an active role in these processes, remain securely bonded together and it is the gratification and rewards inherent in this heightened sexuality which have a great deal to do with family stability.

70. Sex Determination and Differentiation

The factors determining the sex of an individual are primarily genetic. The nuclei of the ovum and spermatozoon each contain only half the number of chromosomes (the *haploid* number) present in the remaining cells of the body but by their union they form a cell whose nucleus has the normal (*diploid*) sum. Chromosomes in all cells except the gametes are therefore paired and so are the genes which they carry. For every maternal gene which affects, for instance, eye colour, there is in the complementary chromosome a corresponding paternal gene which influences it in the same or a different way. In the latter event, the final eye colour will be decided by the "dominant" gene of the pair and the "recessive" gene will be powerless until the next generation. Apart from this kind of genetic inequality, the half set of chromosomes in an ovum is similar to that in other ova and to that in half the spermatozoa of members of the same species. In the other half of the spermatozoa, one chromosome is modified. This is called the Y chromosome and is smaller than the alternative chromosome, which is called X, X and Y being the "sex chromosomes" (Fig. 70.1). Half the unions between ova and spermatozoa result, therefore (theoretically), in cells having nuclei with two exactly paired sets of chromosomes. These cells divide and differentiate to form individuals bearing ova, i.e. females. The other half of the union yields cells in whose nuclei one member of one pair of chromosomes differs slightly from its fellow. These cells divide and differentiate to form individuals producing spermatozoa, i.e. males. Females are said to be *monogametic* and males *digametic* (in moths and birds the same principle holds good but the female is the digametic member).

Genetic sex differentiation

A survey of the animal kingdom shows that *secondary sexual characters* (variations in parts of the body other than in the gonads and accessory organs which characterise animals as male or female) are governed by two agencies, respectively genetic and chemical. The first is found in its purest form in the insect world.

We have seen (Fig. 70.1) that the combination of two similar sex chromosomes (XX) yields a female and of dissimilar sex chromosomes (XY) a male. The accepted explanation is that the X chromosome carries a factor making for femaleness which, when doubled, is sufficiently strong to balance the tendency of the combined remaining chromosomes (*autosomes*) to produce maleness. When single, as in the XY combination, it is not strong enough. In insects, the form of all sex variable parts of the body emerges according to the state of this balance. For

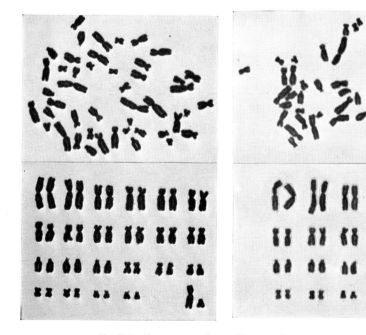

Fig. 70.1 Chromosomes of normal human cells.
Left: Male. Right: Female. The 44 autosomes of each cell have been selected and arranged in pairs in descending order of size, leaving the two sex chromosomes, XY in the male cell and XX in the female cell, as shown at the bottom right-hand corners. (From photographs kindly provided by Professor L. S. Penrose.)

example, the shape of a wing is not determined solely by the coincidence of a single gene in a paternal chromosome with the corresponding gene in the corresponding maternal chromosome, but by the combined effect of several genes scattered through each complementary half set of chromosomes. The gene, or genes, which can influence wing shape and which reside in the second sex chromosome have the "casting vote" on sex form.

Sometimes in a genetically female zygote (XX) a fault occurs in the first mitosis of the fertilised egg and the X chromosomes of one of the daughter cells are altogether lost. Since each daughter cell gives rise to one half of the body, all the cells in, say, the left half are, sexually XX and in the right half OO. The result of this accident is an insect in which the left half of the body in all its sex-variable parts is female and the right half male. Other degrees of genic imbalance which give a range of intersexual forms can be produced experimentally in insects.

Chemical sex differentiation

The findings of embryologists, many in lower vertebrates and many depending on experiments difficult to perform and to interpret, suggest the following generalisation.

In vertebrate embryos, the primitive gonad is bipotential, the cortical part being capable of developing into an ovary and the medullary part into a testis. The influence of the genetic factor is exerted at a very early stage, causing the suppression of the medullary part with development of the cortex in the case of females and of the reverse situation in the case of males. It will be appreciated that a failure of the normal balance between "femaleness" sponsored by the X chromosomes, and the "maleness" by the autosomes, can, in certain circumstances, lead to a disturbance of the above-described process so that either both cortex and medulla develop to form an ova-testis (as found in some true hermaphrodites) or a testis may develop on one side and an ovary on the other, or various other possible combinations (as have also been described in human true hermaphrodites) may occur. Moreover, it is possible for an ovary to develop in a genetic male or a testis in a genetic female in the same kind of way and such gonadal anomalies are believed to exist in certain humans. Such gonads are defective to a varying extent.

Further experimental studies, in which the gonads in the "indifferent" stage have been removed from embryos or have been destroyed *in situ* by X-irradiation, have demonstrated that, whereas the removal of the gonads in a genetic female does not disturb the normal development of female accessory sex organs, if those of a genetic male are removed the individual develops as though it were a female. From this it has been concluded that the embryonic testis secretes a substance which is necessary for the development of the male accessory organs (i.e. Wolffian duct derivatives, penis and scrotum) and for the suppression of the female counterparts, while in its absence the female structures (i.e. Müllerian duct derivatives and vulva) develop and the male structures are suppressed. There is experimental evidence that treating embryos at a sufficiently early stage with sex hormones may modify the development of the primitive gonad. One of

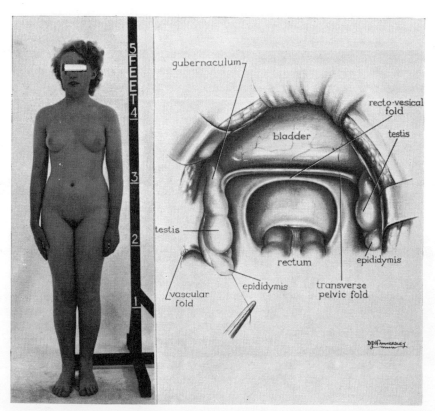

Fig. 70.2   Male pseudohermaphroditism.

A patient, 17 years old, with "testicular feminisation". Well-developed breasts and external feminine characteristics (in fact she won a beauty competition) but no pubic or axillary hair and infantile vulva. At laparotomy, no internal female genitalia were found but undescended testicles were present. Sex chromatin negative. (Case of Dr C. N. Armstrong.)

nature's best-known experiments of this kind is found in cattle when the circulations of twins of genetically opposite sex communicate with each other during the early indifferent stage of development. The male twin develops normally (possibly because its ovarian rudiment disappears before it can be stimulated) but the "female" twin, a so-called "free-martin" is extensively modified. Sterile testes and male genital ducts are formed, the ovaries and female ducts being suppressed. The external genitalia are indeterminate, usually of a rudimentary female type, though the clitoris may be enlarged.

Although differentiation of mammalian sex characters is thus vested mainly in the sex hormones, the genetic foundation on which these work is often subject to modification. This is well seen in the plumage of some birds but is also apparent in man. Many traits, such as cephalic index, presence or absence of palmaris muscle, and height, about which we say "most men are taller than most women", and so forth, are thought to be fundamentally genetic in origin.

Intersexuality occurs in man in three forms: true hermaphroditism, chromosomal intersexuality and male and female pseudohermaphroditism. The first is rare and implies the presence of gonadal tissue of both sexes in the same individual. It has been mentioned above. Chromosomal intersexes are individuals with sex-chromosome aberrations. One group, with XO constitution (45 chromosomes only), consists of hypogonadal "females" with dysgenetic gonads (chromatin-negative Turner's syndrome). Another, with XXY constitution (47 chromosomes) consist of hypogonadal males with abnormal testes (chromatin-positive Klinefelter's syndrome). Reduplication of sex chromosomes may also give rise to XXX, XXXYY and similar anomalies, most of the individuals so affected being mentally defective as well as showing sexual abnormalities. A further group consist of individuals believed to be "mosaics", some nuclei having one sex chromosome constitution and others another. In the pseudohermaphrodites, the sex of the gonad is indicated by the adjective "male" or "female", while the external genitalia and the secondary sex characters partake to a greater or less extent of the nature of those associated with the opposite sex.

In *male pseudohermaphrodites*, testes are present but are usually undescended, and although they show apparently normal interstitial cells, spermatogenesis does not occur. The external genitalia are maldeveloped to a greater or less extent. Thus the urethral meatus opens under the glans penis or at the base of the penis or several centimetres behind in the perineum (hypospadias). Sometimes the penis is so poorly developed as to resemble merely an enlarged clitoris. The internal genital organs appropriate to the female are sometimes found in a rudimentary form. In some individuals, flat male breasts are found, but in others well-developed breasts of female type are found. Indeed, in one group of these patients (sometimes referred to as the "testicular feminisation syndrome") the individuals appear to be normally developed females except that pubic and axillary hair are usually absent and the vagina is short and ends blindly, no uterus being present (Fig. 70.2). In some intersexual individuals it may be impossible to determine, on clinical grounds, whether they are male or female pseudohermaphrodites except on the basis of hormone studies, biopsy of the gonads or nuclear sex studies.

The commonest cause of *female pseudohermaphroditism* (Fig. 70.3) is a genetically determined disturbance of adrenal function arising in embryonic life and leading to the production of excessive amounts of an-

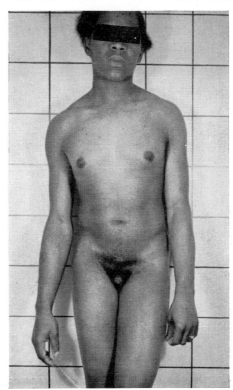

Fig. 70.3 Female pseudohermaphroditism.

A patient aged 24 who complained of genital abnormalities and primary amenorrhoea. Bodily configuration male, with fairly well-developed muscles, no breasts and a narrow pelvis. Phallus measured 3 cm and urethra opened at its base. No vaginal orifice. Very high excretion of adrenal steroid metabolites, reduced to normal by prednisone (which suppresses pituitary ACTH secretion). Laparotomy revealed ovaries, uterus and tubes and a cervix opening into a vagina, access to which was gained by incising the perineum, so constructing a vaginal orifice. The phallus was amputated. With continued prednisone treatment, menstruation began within a month, has continued normally and the patient has conceived and delivered a normal infant. (Case of Dr G. I. M. Swyer.)

drogenic steroids by the adrenal cortex. These, though not interfering with the development of the ovary or of the internal female genital apparatus, cause a great enlargement of the clitoris, so that it resembles a hypospadiac penis. They may also produce varying degrees of vaginal maldevelopment, and cause the early appearance of hirsutism and of masculinisation. In a small proportion of these individuals, the adrenal defect leads to excessive sodium loss, so that, unless treated, the infant dies within the first few weeks of life. Very rarely, a non-adrenal form of female pseudohermaphroditism is encountered, in which the only defect appears to be enlargement of the clitoris. This has been observed to result from treatment of the mother with male hormone during pregnancy but it is also known to occur occasionally after treatment with progesterone-like substances or even when there has been no treatment at all. In these cases, it is believed that the embryo must have been abnormally sensitive to the slight androgenic action of the maternal or administered progesterone-like bodies.

Section 6:

Compensatory Adjustments and Reactions to Injury

Section 6. Compensatory Adjustments and Reactions to Injury

INTRODUCTION

In order to keep alive we must be in close contact with the outside world—with the air, water, food, and all kinds of solid objects which it contains. It is a matter of common knowledge that in doing so we may suffer injury. Our skins and the mucous membranes of our noses, mouths and throats may be damaged by cuts and grazes, or by excessively hot or cold materials; but for the most part, these injuries are of little consequence. Even if an injury is sufficiently deep to open a blood vessel, little real harm is done provided that the shed blood clots properly; if it does not, the loss of blood is inconvenient, to say the least, and may become so great as to be fatal. Clotting, or *coagulation*, of the blood, then, is an important component of the reactions to injury which we shall consider.

The surrounding world, also, contains large numbers of micro-organisms, protozoa, bacteria, and viruses. These may enter the body through the skin and the mucous membranes, particularly if damaged; some of them become parasites, multiply in the body fluids and cause disease. The various processes by which such an invasion is opposed, form the other group of reactions to injury which we shall consider. These reactions are brought into play, also, by the presence in the body fluids of different kinds of "foreign" material, many of which are not micro-organisms, nor any kind of living organism, and not necessarily injurious. As a result of studying these reactions, we have learnt that each individual is not only unique in its physical structure and appearance, but also in its chemical composition; cells of one individual, if transferred to another, are treated as "foreign", potentially injurious, and destroyed. One cannot, in general, replace organs or tissues lost by one individual by those taken from another individual, although there are a few exceptions. Fortunately, the red blood cells are among these exceptions, and are less "unique" than are those of other kinds of cell; those of any individual may be placed into one or other of a small number of groups. The blood of one individual may be "compatible" with that of another, so that blood lost by one may be replaced by that of the other.

In addition to these reactions to injury and infections, we shall also discuss, in this section, certain compensatory adjustments. Where the mechanisms concerned throw some light on normal physiology, the pathophysiology of some disease processes will also be considered.

71. Defence Reactions (Inflammation, Immunity, Healing)

The study of disease processes is relevant in the consideration of the normal physiological working of the body. The basic reactions to various types of injury, infection or deficiency are themselves physiological; moreover the changes brought about by diseases often help to reveal the true nature of physiological mechanisms.

We may, for the purposes of this book, consider *disease* to be a departure from the normal in the functions and/or structure of some part of the body. These departures may be due to *congenital* (inborn) causes, to *trauma* (physical injury), to *deficiency* (nutritional, metabolic or due to the malfunction of endocrines), to *infection* with microorganisms and to *poisoning*. In most of these instances, damage occurs to the cells of the body; their function is impaired, they may die or the affected cells may recover.

INFLAMMATION

Cells, comprising living tissue, when acted upon by noxious influences in their environment, react in a specific manner. As we go higher in the phylogenetic scale we see that this reaction becomes more complex and that complicated mechanisms have been developed in order to combat the injury.

This reaction to injury is called *inflammation* and is an integral part of the majority of diseases. It may be rapid in its evolution, i.e. *acute*, or it may be of slower or continuous development, i.e. *chronic*. The ingress of micro-organisms into tissues, particularly when these are *pathogenic* (i.e. cause disease) must be countered and the organisms destroyed or removed; in addition the injury caused by them must be repaired. Thus the natural history of the inflammatory process involves two distinct, but related parts, the *defence reaction* and *repair*.

The changes occurring in inflammation can easily be studied under the microscope and varied as this reaction is, the main features of the process are constant and it is generally believed that the way in which infection is overcome is fundamentally similar in very many diseases.

The word inflammation is derived from *inflammare* (L., to burn), descriptive in that the signs are redness, heat, swelling, pain and loss of function.

Macroscopic appearances

A hard stroke made on the skin elicits the "triple response" first noted by Sir Thomas Lewis in 1927. This consists of:

1. The red reaction
2. The flare
3. The wheal

The red reaction is due to dilated precapillary sphincters, not mediated by nerves, and the effect of histamine or polypeptides released from the damaged region.

The flare is due to arteriolar dilation as well. Skin temperature is raised because the blood flow locally is much increased. Application of local anaesthetic prevents the flare occurring; thus it is mediated by nervous activity (the axon reflex).

The wheal is a raised area of skin resulting from increased capillary permeability due to the damage, leading to oedema (swelling).

The foregoing events are a mild form of inflammation (sunburn is an example). The local increase in skin temperature, the redness and the swelling are all more marked in more severe inflammation, and pain may be generated. The swelling may proceed to blister formation. Loss of function may result from reflex inhibition of movement, pain, or the limitations due to the swelling.

Microscopic events

Immediately following the injury, a transient contraction of arterioles occurs, then they and the capillaries and venules dilate. Blood flow is at first increased thereby, but later falls, due to loss of fluid into the tissue spaces as the permeability increases. Blood viscosity rises in the regions of stasis.

Leucocytes begin to stick to endothelium in the damaged vessels, later to penetrate in large numbers into the tissue spaces. Platelets also behave in the same way, forming clumps or thrombi.

Localisation

Successful resistance to infection entails that it must be localised—if the distribution of the infecting agent is widespread, the risk to life is increased. It is usual for the infection to be "walled off" by a fibrin barrier formed in lymphatics and in interstitial spaces (much like a blood clot).

This barrier can be demonstrated experimentally by observation of its formation under the microscope or by injecting trypan blue (a) into the region, when it remains there or (b) into the surrounding tissues, when it fails to penetrate into the area.

Antibodies also accumulate in the infected area. The response of the tissues depends markedly upon the nature of the infecting organism. Some bacteria characteristically cause spreading infections and it can be shown that these do not elicit the rapid production of fibrin in the tissues. For example *streptococci* release an enzyme, *fibrinolysin*, which is responsible for the delay in formation of the fibrin barrier.

Methods for studying inflammation
Blood flow.
Skin or tissue temperature can be measured to provide an indication of relative changes in blood flow. Direct microscopic observation is possible;

transparent chambers can be used or the frog's mesentery can be examined with transmitted light.

Capillary permeability

The intravascular injection of dyes or minute particles allows an estimate of passage of molecules etc. into tissue fluid. Albumen can be labelled with radioactive iodine and used in the same manner.

Leucocyte migration

White cells move in the tissue spaces, usually towards inflamed areas. They can be studied experimentally within the rabbit's ear transparent chamber. Chemotaxis can be demonstrated in vitro by passing cultured bacteria, or a toxic extract, through a small pipette into a suspension of leucocytes. Time lapse photomicrographs can be used.

The leucocyte response in inflammation

In the inflamed region, leucocytes migrate by diapedesis from the capillaries and congregate in the extracellular spaces. *Chemotaxis* (directed movement) and *phagocytosis* (ingestion of particulate matter, commonly bacteria) are basic properties of white cells, first discovered by Metchnikoff in 1901. Most leucocytes first "stick" to the endothelium although red cells and lymphocytes do not do so. As the inflammation progresses, the endothelium becomes gelatinous and the cells then penetrate it to the extracellular spaces. In acute infections the main cell type involved is the polymorphonuclear. In chronic disease, on the other hand the main cellular component of the reaction is lymphocytic or monocytic. In some infections e.g. tuberculosis and leprosy or slow viral infections of the CNS, multinucleate giant cells that appear probably consist of coalesced monocytes. Eosinophils are particularly common in allergic manifestations such as in the nasal mucosa of hay fever sufferers or in urticarial wheals of the skin. Mast cells which are tissue basophil cells near small blood vessels, release heparin and histamine.

Capillary permeability

In injury and infection capillary permeability is markedly increased and protein gets into the tissue spaces. This causes oedema locally because of the altered osmotic relationships across the vessel walls.

The increased permeability is caused by:

1. Specific toxins or chemical and physical agents directly acting on the capillary wall.
2. Specific chemical mediators liberated by injury such as histamine, 5-hydroxytryptamine, and a polypeptide, bradykinin. These highly active substances are normally present in the tissues in an inactive form. Massive release of them can lead to death from anaphylactic shock. Histamine is released by mast cells; bradykinin by trypsin from an inactive precursor.

Lymphatics

Lymphatic drainage is increased in an inflamed region, whenever permeability increases. Fibrin is often deposited in the tissue spaces and helps to localise bacterial invasion.

Pain

The chemical mediators mentioned above also give rise to the stimulation of pain endings; this has a protective function.

REACTIONS TO "FOREIGN" MATERIALS IN THE BODY FLUIDS

When "foreign" or abnormal substances or living organisms enter the body, they are removed, and any deleterious actions produced are opposed, by a number of processes which are often referred to collectively as "defence reactions". Particulate matter, whether living or dead, is removed by absorption (*phagocytosis*) in certain kinds of *leucocyte*, which circulate in the blood, and in certain tissue cells belonging to the *reticuloendothelial system*. Certain substances, usually but not necessarily associated with, or derived from, foreign organisms, react in various ways with substances already present in the body fluids of the animal or person invaded (or specially formed in response to the invasion), the reactions being of an unusual nature and called "immune reactions"; they were first studied in connection with the fact that after a person has recovered from certain kinds of disease, he may subsequently be unharmed by—or immune to—further attacks by the particular microorganism which causes the disease.

WHITE CELLS OR LEUCOCYTES

These are usually classified into three groups: the granular cells (sometimes called *polymorphs*, but now often called *granulocytes*), the *lymphocytes* and the *monocytes*.

The *granulocytes* are called "polymorphs" (polymorphonuclear leucocytes) because in stained smears of blood their nuclei are seen to be divided into two, three or four lobes by deep indentations. They are subdivided, according to the staining reactions of their granules, into:

Neutrophils, with very fine cytoplasmic granules, showing no striking affinity for acidic or basic stains (the name "polymorph" is often restricted to this group of granulocytes);

Eosinophils, with very coarse granules, stained bright red by eosin, the nucleus being nearly always bi-lobed; and

Basophils, with coarse granules having an affinity for basic dyes and hence stained blue by the stains commonly used for blood.

The red bone marrow is the normal site of production of granulocytes and in it can be recognised immature granular cells in two developmental stages. The younger are known as myeloblasts and the older as myelocytes. Myeloblasts have no granules, but myelocytes have granules in which the tendency to become eosinophilic, basophilic or neutrophilic can be easily seen. The nuclei of the myelocytes, however, are only slightly indented. These immature forms of granular blood cells may appear in the circulating blood in blood diseases, called *leukaemias*.

The *lymphocytes* of normal blood are the smallest of the white blood cells, the small forms having about the same diameter as a red cell. They are characterised by an almost round, densely staining nucleus surrounded by a narrow zone of cytoplasm free from granules. The lymphocytes are produced in the lymphoid tissue from larger cells with pale nuclei.

The *monocytes* (*transitional cells* or *large mononuclears*) have large oval or bean-shaped nuclei; both nuclei and cytoplasm are pale staining.

Normal human blood contains 5000 to 10 000 leucocytes per mm^3. When the number is lower than normal, a state of *leucopenia* is said to exist—as, for example, in typhoid and influenza. Most general infections, on the other hand, are accompanied by a striking increase in the number of circulating leucocytes, a condition known as *leucocytosis*.

The white cell count

The number of leucocytes per cubic millimetre of blood (the *white cell count*) is estimated in the *haemocytometer* in much the same way as is that of the red blood cells. The blood is diluted 20-fold, in a special pipette, with a fluid which contains acetic acid to dissolve the red cells, and suitable stains for the white cells. A drop of the suspension so formed is placed on the special microscope slide, forming a layer exactly $1/10$ mm deep. When the cells have settled, the number lying within a square $1/16$ mm^2 in area is counted. These squares are formed by lines ruled at the corners of the area containing the much smaller squares used for the red cell count. Several counts are made, and the average count is multiplied by 160 (the cells counted came from $1/160$ mm^3 of suspension) and by 20 (the dilution factor) to give the number of cells per mm^3 of the original blood.

The estimation of the proportion of each type of leucocyte in a particular sample of blood can be done on any properly stained blood smear, and is known as the *differential white count*, approximate values for normal human blood being:

Granulocytes	neutrophils,	70 per cent	
	eosinophils,	1 per cent	of total
	basophils,	0·5 per cent	white cell
Lymphocytes		24 per cent	count
Monocytes		4 per cent	

The leucocytes are motile, and can creep over surfaces and make their way out of the blood into the tissue fluids, through interstices in the blood vessels. The lymphocytes, in particular, circulate from the blood stream into the tissue spaces, whence they are carried to the lymph nodes with the lymph, and then back again into the blood stream by way of the thoracic duct.

The reticuloendothelial system

The loose, or reticular, connective tissue which surrounds the blood vessels, for example, and provides support and attachments for the abdominal organs, contains, besides the fibroblasts which are responsible for the reticular structure, other kinds of cell called mast cells, plasma cells and macrophages. The same, or very similar, kinds of cell are found in the endothelial lining of the sinusoids of the liver, spleen and bone marrow, and in the lymph glands. These cells, in their various locations, constitute the *reticuloendothelial system*, and they are closely related to the leucocytes of the blood. The whole subject is very confused, but it is thought that the basophil leucocytes may well be identical with the mast cells of the reticuloendothelial system; and lymphocytes seem, in certain conditions, to develop into macrophages, with monocytes perhaps as intermediate stages.

Phagocytosis

The most important of the cells which ingest bacteria and other foreign particles, living or dead, are the macrophages of the reticuloendothelial system. When dyes, such as trypan blue or lithium carmine, or particulate matter such as Indian ink, are injected intravenously, they are taken up for the most part by these macrophages, particularly by those in the liver known as Kupffer cells. When there has been a local invasion of bacteria, as in an infected wound of the skin, there is a great accumulation of neutrophil granulocytes (polymorphs), which leave the bloodstream in the infected region, ingest the bacteria, and release proteolytic enzymes—perhaps as a result of their own disintegration; these enzymes break down cells which have been killed by the infection. The liquid known as pus is made up of these broken down cells, together with the remains of the leucocytes, and possibly some erythrocytes; it is sealed off from the body fluids in general by the proliferation of fibroblasts, which form the scar tissue.

IMMUNE REACTIONS

Foreign substances, when introduced into the body, act as *antigens*, and lead to the development of *antibodies*, which react with them in various ways. If the antigen is a protein in solution—such as serum protein of a different kind of animal—the antibody which is formed precipitates the antigen, a reaction which can be easily observed in a test-tube; the antibody is thus termed a *precipitin*. Antigens, however, may also be present on, or released by, living cells such as bacteria or red blood cells. The corresponding antibody will then: (a) make these cells stick together in clumps, or agglutinate, when it is called an agglutinin; or (b) lead to their destruction or lysis, when it is called a *lysin* (or more specifically, for example, a *bacteriolysin* or a *haemolysin*); or (c) neutralise the poisonous (toxic) effects of the substances released by bacteria, when it is called an *antitoxin*. Antibodies of all kinds seem to be produced chiefly by lymphocytes; but at the site of an infection there is a great accumulation of eosinophil granulocytes, and these also may play some part in the antigen–antibody reactions.

The antigen–antibody reactions are highly specific, most kinds of antibody reacting only with one particular kind of antigen; serum proteins from different species, which cannot be distinguished by chemical methods, can be distinguished by the way they react with an antibody prepared against one of them.

It is the presence of the antibodies formed in response to an invasion of bacteria, or other microorganisms, which is responsible for the subsequent failure of further infection to produce the disease—an immunity which may last for many years. Many of these antibodies can be prepared artificially, or the animal, or man, can be induced to develop them itself without having to undergo the full rigours of an actual attack of the disease; their study has consequently become an important part of medical science, known as *Immunology*.

Antibody formation

Antigenic substances trigger the division of the lymphocytes which give rise to appropriate antibodies. It is this ability of the body to produce an unlimited selection of these molecules which is responsible for the resistance to disease. The main antibody molecules found in the circulation are the gamma globulins, of which there are more than 10 000 different kinds, although they all have similar structural properties and a molecular weight of 150 000. They are made up from a long chain of amino acids and it is the sequence in which the various acids are arranged that leads to the infinite variety of antibodies which are found.

Each gamma globulin molecule consists of four subunit polypeptide chains (Fig. 71.1). Two are the same and have a molecular weight of 25 000; these are the light or L-chains. The remaining two are of double the molecular weight and are the heavy or H-chains. The L- and H-chains are folded in a way enabling them to interact to give half-molecules which in turn combine forming the complete gamma globulin molecule. Part of the chain structure keeps the whole molecule together, whilst the remainder is available for folding into different shapes which

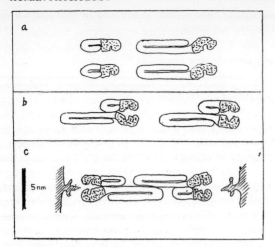

Fig. 71.1 Gamma globulin antibody molecules consist of four polypeptides, 2 L-chains and 2 H-chains (a). These form half-molecules (b) which combine to form the whole structure of the antibody (c).

The white parts are identical and are the linking portion of the molecule; the stippled parts are the folded regions, having variable amino acid sequences, which form zones interacting with and inactivating the antigen.

dovetail into, and interact with, a specific antigen to inactivate it.

Where do antibodies originate?

The *Clonal Selection* theory of the immune response was originated by Macfarlane Burnet. It is based on the idea that when certain cells having the property of antibody synthesis meet antigen, they multiply forming specific colonies (*clones*) of cells manufacturing the specific antibody.

The *Instructional* theory supposes that each of the antibody-forming cells contains the antigen, which then behaves like a template from which antibody molecules are synthesised to fit, much like the way a lock and key do.

Direct experimental proof has now been found by Nossal, that the clonal selection process occurs *in vivo* when lymphocytes respond to antigens. It was first shown that antigens are never found within the antibody-forming cells. Later, using a new method for the detection of marker chromosomes in isolated, single, antibody-forming cells he established that non-antibody-forming lymphocytes are able to stimulate the initiation of *germinal centres*, acting as sources of long-lasting memory cells having a specific reactivity for an antigen, when this is injected.

Antigen-sensitive cells

The antigen itself first acts on a special target cell of lymphoid origin termed the "antigen-sensitive" cell. Only about 1 in 20 000 to 1 in 50 000 of the population of lymphoid cells exposed to antigen respond in this way. If two distinct antigens are applied, the total number of responding cells is exactly doubled, a fact which appears to indicate that each antigen affects a separate group of antigen-sensitive cells.

Following antigenic stimulation, each of the antigen-sensitive cells divides sequentially about eight or nine times (mitotically) producing a clone of progeny. During the first day or two these daughter cells remain close together and form a recognisable antibody-forming focus.

By using a highly immunogenic protein, flagellin (prepared from *Salmonella typhi*) labelled with radioiodine, the distribution and fate of injected antigen can be followed in detail. Two types of antigen capture occur in the lymph nodes as revealed by electron micrographic autoradiographs. In the medullary part of the node, macrophages ingest the antigen into a vacuole and it then becomes enclosed in a membrane to form a lyzosomal structure. In the cortical region, the antigen is trapped in lymphoid follicles, and following this, lymphoblast cells make their appearance in the form of typical germinal centres. Such a germinal centre created by the trapping of an antigen is still capable of dealing with other unrelated antigenic materials in the same way, so that it appears that one centre can manufacture a number of different strains of memory cell for different antigens, simultaneously.

Lymphocytic transformation

Micromanipulation methods and the examination of marker chromosomes, so that labelled lymphocytes in an injected dose can be traced, have shown that donor lymphocytes eventually turn into large plasma blast cells which produce the antibody molecules.

SUMMARY OF THE BASIC FEATURES OF THE IMMUNE RESPONSE

1. The antigen–antibody reaction is specific, and is due to chemical groups on the antigenic molecule, especially polar groups.
2. The number of antibody molecules produced very greatly exceeds the number of antigenic molecules introduced into the body.
3. A second dose of antigen evokes a larger and faster antibody response than did the first. This can be considered as a type of memory and is termed the *anamnestic response*.
4. When antigens are introduced at an early stage in the individual's development, the immune response is absent or small. This is called immunological tolerance.
5. The immune response, in normal conditions, is to foreign molecules. The body can distinguish between "non-self" and "self". *Autoimmune diseases* occur when this recognition of self-markers breaks down and an immune response develops to an individual's own cells.

Immunosuppression can be artificially induced with drugs such as azathioprine (Imuran) or X-irradiation, and consists in a damping down of the normal development of immunity.

Complement

Suppose a specific haemolysin has been produced in one animal by repeated injections of the red cells of another animal; if serum containing the specific haemolysin is heated to 56°C its haemolytic power is lost. The addition of fresh serum from almost any normal animal, however, restores the haemolytic power. Since the added normal serum would not, by itself, have haemolysed red cells, it is evident that two factors are concerned in this type of haemolysis: (1) a specific antibody which is stable to heat (thermostable), and (2) a non-specific factor which is destroyed by heat (thermolabile), known as the "complement"; this is present in any normal serum. Red cells which have been treated with heated serum (containing the antibody which would have haemolysed them if the complement had not been destroyed) are said to be sensitised, since, if they are introduced into a solution containing the complement haemolysis follows immediately.

Not all types of reaction between antibodies and antigens require complement for their completion, but nearly all antigen–antibody compounds have the property of absorbing complement, and thus making it inactive or *fixed*. Complement fixation can be used as a test for the presence of either antibody or antigen. The most famous of these complement-fixation tests is the *Wassermann reaction* (or the WR), which is a test for the presence of the syphilitic antibody in the blood of a patient suspected of suffering from syphilis. A standard amount of antigen and a standard amount of complement are added to the heated serum of the patient. If the syphilitic antibody is present, it reacts with the antigen; complement is fixed, and sensitised red cells added subsequently are not haemolysed. Curiously enough, the standard antigen used in the Wassermann test has nothing to do with syphilis, but is prepared from an alcoholic extract of heart muscle. Presumably this empirical antigen, unexpectedly discovered in the course of controls on the Wassermann reaction, has the same chemical configuration as some antigen in the causal organism of syphilis, for exhaustive tests have only confirmed the usefulness of the reaction in the diagnosis of syphilis.

Anaphylactic shock

If a single injection of an antigen is followed after an interval of about 10 to 14 days by an injection of a second dose of the same antigen, the consequence is a profound, and often fatal, collapse, due to a very low blood pressure resulting from dilatation and increased permeability of the capillaries; in some animals, asphyxia is induced by intense constriction of the bronchi. The first dose of the antigen clearly rendered the animal hypersensitive, instead of immune, the hypersensitivity being associated with a low content of antibodies in the blood. A widely supported view of the mechanism of anaphylactic shock supposes that the free circulating antibodies due to the first injection are sufficient to neutralise only a part of the second dose of antigen; the remainder of the antigen reacts with antibodies which are attached to tissue cells. This reaction damages the tissue cells in some way, leading to the production of histamine, which is known to produce, when injected intravenously, a train of events very like anaphylactic shock. If the animal in the hypersensitive state is given a series of injections of the antigen, each too small to produce shock, it will, in time, become desensitised, presumably because an adequate supply of circulating antibodies is developed.

On rare occasions a condition resembling anaphylactic shock results from an intravenous injection of an antitoxin. This is usually due to the patient being hypersensitive to the horse serum from which most antitoxins are prepared.

Allergy

There is a mild type of anaphylactic response, with much less violent manifestations of hypersensitivity, known as an "allergic reaction". This may follow the consumption of mussels, lobsters, strawberries or several common foodstuffs by certain persons who are said to be "sensitive"; or it may follow contact of the skin or mucous membranes of the respiratory passages with the pollen of certain grasses (in hay fever), the hairs of certain animals, or even the close presence of these animals. The nature of the reactions produced depend on the nature of the antigen (or "allergen"), and may vary considerably from one person to another. Characteristic reactions are nettle-rash or "urticaria" —i.e. the eruption of wheals on the skin; congestion and excessive irritability of the mucous membranes of the nose and pharynx; and constriction of the bronchioles, producing asthma.

In general, the anaphylactic response is restricted, in these cases, to the local cells which have come into contact with the specific allergen. The fact that the reaction is not a general one accounts for the response being less violent.

HEALING

We have discussed the initial defence reactions of tissues to trauma or invasion by microorganisms. These responses also initiate the processes of repair. Macrophages invade the fibrin clot to remove cellular breakdown products, and the fibrin itself. New blood vessels grow into the damaged tissue in the form of sprouting capillaries, which later produce arterioles, capillaries and venules. Vasomotor nerve innervation is brought in with them. Collagen fibres condense in the muco-polysaccharide ground substance as a result of the activity of fibroblasts. The fibrous tissue in scars such as this usually contracts because collagen fibres tend to shorten with time after their formation.

A number of specific and non-specific factors influence the healing process. At the site of repair, remaining infection or a poor blood supply will prevent or delay it. General factors affecting healing include hormonal and nutritional status. Ascorbic acid deficiency (scurvy) prevents the proper development of collagen and so hinders repair; in scurvy, scar tissue is of little mechanical strength. Protein lack has a similar effect. Large doses of corticosteroids impair healing because of their interference with protein synthesis, although it should be noted that small amounts of topically applied steroids promote the healing process by virtue of their suppression of the inflammatory response. They are much used in dermatology for this action.

72. The Coagulation of Blood

When a blood vessel is opened and the blood flows over the surrounding tissues, it normally sets to a jelly in the course of five to ten minutes. It is then said to have *clotted*, or *coagulated*. If the clot is collected and washed free from red cells, it is found to consist of an interlacing network of fine white fibres of protein material to which the name *fibrin* is given. If a blood clot is allowed to stand, it slowly shrinks (retraction or syneraesis) and a pale yellow liquid called *serum* is squeezed out. Serum closely resembles plasma, but lacks a protein constituent which can be precipitated from plasma by half-saturation with sodium chloride. This protein is called *fibrinogen*, because it changes into fibrin during the process of clotting. If precipitated fibrinogen is redissolved in 2 per cent sodium chloride, and a little fresh serum added, a clot is formed in much the same way as it is in plasma. Fresh serum therefore contains a substance which induces fibrinogen to clot; this substance is known as *thrombin*.

Thrombin is an enzyme, and may be isolated in a moderately pure state by extracting fresh serum with alcohol, or by various other more elaborate procedures. Unclotted blood treated in the same way yields no thrombin, so that thrombin does not exist in circulating plasma, but is formed some time after the blood is shed. Plasma, therefore, must contain a substance which can change into thrombin under the proper conditions. This substance is known as prothrombin. It is ordinarily formed continuously in the liver, an adequate supply of a special vitamin (vitamin K) being necessary. If the supply is inadequate (owing to an inadequate diet, for example) there is partial failure of blood clotting.

Vitamin K deficiency

There is a deficiency of vitamin K, also, when, owing to disease, the bile duct is obstructed; it is then not properly absorbed from the alimentary canal. The formation of prothrombin is impaired, as might be expected, when the liver is damaged as a result of disease or poisons such as chloroform. In all these types of inadequate liver function, however, the failure of blood clotting does not result only, or even chiefly, from an inadequate supply of prothrombin, but also from an inadequate supply of factors V and VII.

A failure of blood clotting, generally similar to that produced by inadequate liver function, is produced by administration of a substance known as *dicoumarol*. This is a derivative of the substance coumarin, which is responsible for the odour of new-mown hay; dicoumarol may be formed from it if sweet clover is improperly cured before being stacked. Animals eating this hay then develop a haemorrhagic disease.

Mechanism of clotting

Since blood does not normally clot in the circulation, we infer that the conditions necessary for the conversion of prothrombin into thrombin, and the initiation of the process of clotting, are: (1) the contamination with tissue juices which takes place as blood flows from a wound, and (2) the contact with foreign surfaces such as the skin and the vessels in which the blood is collected. In these conditions, it appears that a substance called *thromboplastin* is released. This is most probably a second enzyme, and is present in extracts of most tissues, since the addition of such extracts to blood increases the rate of clotting. But it is derived particularly from very small cellular elements normally present in the blood, called *platelets* or *thrombocytes*. These are round or oval disc-shaped bodies, 2 to 3 μm in diameter, with granular cytoplasm, but no nucleus. They are believed to be derived from large multinuclear cells of the bone marrow called megakaryocytes. Cytoplasmic processes of these cells become pinched off and pass into the blood as platelets. Normal blood contains from 250 000 to 450 000 platelets per mm^3. Within a few seconds after blood is shed, its platelets agglutinate, that is, clump together, and then, more slowly, disintegrate. Consequently, in smears of normal blood the platelets are to be seen only in clumps, often consisting of as many as thirty or forty platelets. The clumping of platelets at the site of bleeding in small vessels tends to plug the wound in the vessel and thus helps to restrain haemorrhage; their subsequent disintegration releases thromboplastin, and so hastens blood clotting. If the early stages of clotting are observed by means of an ultramicroscope, needles of fibrin are seen to form in the immediate neighbourhood of clumps of disintegrating platelets, and to grow out from these centres until the whole of the blood is enmeshed in the network.

Agglutination and disintegration of platelets proceeds rapidly when blood is in contact with some kinds of "foreign" surface—such as glass and other substances which are wetted by blood—and only slowly if blood is shed through a cannula into a receptacle, made of a suitable plastic such as polythene, or coated with silicone or paraffin wax; these surfaces are chemically inert and are not wetted by the blood. If the platelets disintegrate rapidly, clotting is rapid; if they remain intact for some time, clotting is delayed, but can be brought on rapidly if disintegrated platelets are added.

A further condition for the conversion of prothrombin to thrombin, and for the release of thromboplastin from the platelets, is the presence of calcium ions; if these are removed from freshly shed blood, by precipitation as calcium oxalate, by combination in an un-ionised state as calcium citrate, or by "chelation" with ethylene diamine tetracetate (EDTA) clotting does not occur.

Thromboplastin

It is probable that prothombin is not converted into thrombin directly by the substance called "thromboplastin" (also known as *thrombokinase*); and the thromboplastin of tissue extracts is probably not identical with that of the platelets (at one time called *cytozyme*). The name "prothrombinase" has been given to the substance derived from tissue extracts which is responsible for the formation of thrombin; it has not been established that this same substance appears in the platelets in the absence of tissue extracts. This, however, does not affect the essential nature of the sequence of reactions which brings about clotting.

The process of blood clotting, then, may be regarded as resulting primarily from the interaction of three substances normally present in the blood—fibrinogen, prothrombin and calcium ions—with a fourth substance—thromboplastin—released from the platelets as a result of their disintegration. This release, moreover, is accelerated by the presence of thrombin, which thus indirectly accelerates its own formation from prothrombin; the reaction sequence is autocatalytic, and once started, proceeds at a progressively increasing rate. This accounts for the fact that even in the presence of all the factors necessary for clotting, there is a delay of several minutes before any change occurs, and then there is a rapid appearance of the clot. The whole reaction sequence is summarised, schematically, in Table 72.1.

BLOOD CLOTTING FACTORS

Thirteen "factors" have been found which take part in the clotting process (roman numerals in international notation*). These are as follows:

Factor I.	This is fibrinogen, a soluble protein of mol. wt. 330 000. It forms fibrin when acted upon by thrombin.
Factor II.	Prothrombin, the inactive precursor of thrombin. Formation depends on vitamin K, in liver. Thromboplastin converts it to thrombin.
Factor III.	Thromboplastin. Converts prothrombin to thrombin (also called variously prothrombin activator, thrombokinase, tissue factor, tissue extract, in order to confuse medical students). Factors V, VII, X, Ca^{2+} and phospholipid essential.
Factor IV.	Calcium.
Factor V.	Labile factor (see factor III above).
Factor VI.	Nil. (Given originally to a factor which was later found not to have a separate existence.)
Factor VII.	Proconvertin. Required for formation of thromboplastin by tissue extracts.
Factor VIII.	Antihaemophilic factor (AHF). Haemophilia is due to lack of factor VIII. It is needed for formation of factor III from constituents in blood. (Also called antihaemophilic globulin, thromboplastinogen, plasma thromboplastic factor, platelet cofactor I.)
Factor IX.	Christmas factor. Needed for formation of factor III from blood. Absence gives a disease similar to haemophilia (Christmas disease).
Factor X.	Stuart Power factor.
Factor XI.	Plasma thromboplastin antecedent. Required for formation of factor III.
Factor XII.	Hageman factor. This factor is activated by glass and plastic surfaces. It is part of the system for forming factor III.
Factor XIII.	Fibrin stabilising factor. It is a plasma protein which polymerises fibrin (so that it forms insoluble threads).

---

* The precise details can be found in the Report of the International Committee on Nomenclature of Blood Clotting Factors (1959). *J. Amer. Med. Ass.* **170,** 325.

PHYSIOLOGY OF CLOTTING

Table 72.1 summarises the events now thought to take place in the clotting process.

Inhibitors of clotting

Circulating blood does not clot, a fact which may reasonably be attributed to the normal stability of platelets. But there is good reason to believe that "positive" inhibition of clotting also occurs, being due to the presence of naturally occurring anticoagulant substances, the most important of which is *heparin*. This acts both in preventing the formation of thrombin from prothrombin, and in antagonising the action of thrombin on fibrinogen. Heparin was originally isolated from the liver (hence the name), but was later shown to be contained particularly in the mast cells of the reticulo-endothelial system. These cells are present in considerable numbers in the connective tissue which surrounds the small blood vessels, and it is possible that they release heparin into the blood stream.

Prevention of clotting

Shed blood can be preserved in the fluid state in the following ways:

1. By defibrination. The blood, while clotting, is stirred with some object, e.g. a bundle of feathers—to which fibrin will adhere. The fibrin is thus removed as it forms and the red cells are left suspended in serum.
2. By precipitating the calcium by the addition of sodium or potassium oxalate (0.1 to 0.3 per cent).
3. By removing calcium ions by the addition of sodium citrate (0.2 to 0.4 per cent), fluoride or EDTA. Since calcium ions are necessary not only for the change of prothrombin to thrombin, but for the disintegration of platelets, oxalate and citrate solutions can be used to preserve platelets intact.
4. By cooling the blood to 0°C, which retards clotting almost indefinitely and preserves the platelets.
5. By preserving blood from contact with surfaces which it wets.
6. By the addition of heparin.
7. By the addition of certain azo dyes, e.g. chlorazol fast pink, which are thought to act partly as antithromboplastins.
8. By the addition of hirudin, a material obtained from leech heads which acts as an antithrombin.
9. By the addition of suitable concentrations of almost any neutral salt, e.g. one-seventh saturation with $MgSO_4$.

Blood clotting is of great value when localised near bleeding points due to injury or surgery. If, however, clots become detached from such points, or arise spontaneously, and travel as emboli in the blood circulation, they become a great danger. Often they lodge in the lungs (pulmonary embolism). Moreover, clots may form in diseased blood vessels and by blocking them, may impair vital functions, as in coronary thrombosis or cerebral thrombosis. In embolism or thrombosis, dicoumarol or more commonly synthetic anticoagulants are sometimes administered. When blood is passed through external apparatus and back into the body, as in many kinds of physiological experiment on animals, or when an "artificial heart" is used to maintain the patient's circulation during operations on the heart, clotting is likely to be initiated by the materials used in the apparatus. In these circumstances, heparin is injected to prevent the undesirable clotting.

Blood withdrawn from a donor, for subsequent transfusion to a patient who needs it, is ordinarily prevented from clotting by the use of citrate, heparin being relatively expensive. Citrate, however, cannot be used to prevent clotting in the whole of the blood in the circulation, since the complete removal of calcium

Table 72.1 *Details of clotting mechanism (activation of both intrinsic and extrinsic systems)*

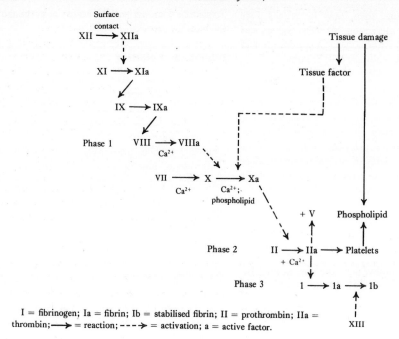

I = fibrinogen; Ia = fibrin; Ib = stabilised fibrin; II = prothrombin; IIa = thrombin; ⟶ = reaction; ---⟶ = activation; a = active factor.

ions would have disastrous effects on a great many of the bodily functions. It may be desirable, on the other hand, to promote clotting in order to check haemorrhage when blood vessels have been opened during surgery or as a result of accident; it is usually sufficient to provide a large area of "foreign" surface by applying a swab of cotton, but in severe cases the swab may be soaked in a solution of thrombin. Certain surgical procedures involve the use of a gelatin sponge for this purpose, having the advantage that it is subsequently re-sorbed and can be left *in situ*.

Haemophilia

The problem of checking haemorrhage may become really serious in certain persons, known as "bleeders", who suffer from an inborn disease in which blood clotting is so slow that haemorrhage from a relatively trivial wound may be profuse enough to be dangerous. *Haemophilia* is an hereditary defect which is manifested almost exclusively in males, but inherited through the mother. A person suffering from this disease—or more strictly, group of diseases—is found to lack none of the primary components of the clotting reactions (as summarised in Table 72.1). Detailed study of these reactions, however, and progressive purification of the known components, both in normal blood and in the bloods of haemophilic subjects, has revealed that the abnormality is a deficiency of factor VIII. The closely related Christmas disease* is due to lack of factor IX.

Bleeding time in these patients is not longer than normal because a pin-prick in the skin is sealed by capillary contraction. The considerable increase in coagulation time though, may lead to fatal bleeding following minor surgery or dental extraction. Factors VIII and IX can be prepared by freeze-drying normal fresh plasma and given as a short-term protective measure to haemophiliacs.

---

* Christmas disease—yet another example of the love of eponymous terminology manifested by the medical profession, but unusual in this case because the name of the first-observed patient suffering from the disease is immortalised.

Blood clotting may be defective, also, owing to a reduction in the number of platelets in the blood (*thrombocytopenia*), or, very rarely, owing to a lack of fibrinogen. These deficiencies may be congenital, or may result from infection.

Measurements of clotting time

The clotting time of normal blood may vary from 4 to 60 min, depending on the method used to measure it. To get consistent results, the following conditions must be controlled:

1. *Temperature.* Clotting time increases as the temperature decreases.
2. *The manner of obtaining the blood.* Blood drawn from a vein clots more slowly than blood from a skin puncture, which allows more contamination by tissue juices.
3. *Agitation of the blood.* Agitation hastens clotting.
4. *Cleanliness of apparatus.* The cleaner the apparatus, the slower the clotting.

A simple method is to collect a few drops of blood from a puncture in the lobe of the ear on a clean watch glass, which is then covered by another watch glass to limit evaporation. The fluidity of the drop is tested from time to time by gently tipping the watch glass. The time from the shedding of the blood till the first signs of clotting appear under these conditions is about 4 to 8 min at 20°C, for normal human blood.

Detection of abnormalities in the clotting reactions starts by the use of such simple methods. But if clotting is found to be abnormally slow, more elaborate procedures must be used in order to establish the nature of the defect.

The Dale-Laidlaw apparatus consists of a short glass capillary whose ends are partly sealed to retain a small lead shot. Blood, introduced by capillarity, fills the tube which is then rocked to and fro at intervals. The shot ceases to move when clotting occurs and provides an accurate end-point. Normal clotting times with this method range from 5 to 15 minutes.

Bleeding time

Blood loss is also controlled by a mechanism involving the contraction of vessel walls. A ragged tear of a blood vessel (as opposed to a clean, sharp incision) stimulates its smooth muscle directly and the opened end is completely or partially closed off. Capillary bleeding is controlled in this way very rapidly and as a rule the haemorrhage is arrested well before clotting has had time to be effective. Even in haemophiliacs, in the absence of clotting, haemorrhage from *small* vessels is rarely dangerous.

The bleeding time is measured by making a small (standard) incision into the skin of a finger, continuously blotting the resultant drop of blood with a filter paper, until the bleeding stops. The normal bleeding time is 2 to 5 min.

Prothrombin time

It is useful to know the amount of prothrombin in the blood, and there is a well-defined relationship between prothrombin *concentration* and the prothrombin *time*. The latter is found experimentally by immediately treating a sample of the patient's blood with oxalate to prevent conversion of any prothrombin into thrombin. An excess of calcium ions and tissue extract is mixed with the oxalated blood and the time from mixing to coagulation is measured. This is the prothrombin time, the precise value depending on the conditions of the test, being usually about 10 to 15 s.

Thrombosis and embolism

A clot, formed abnormally in any blood vessel is termed a *thrombus*. If such a clot is friable, or occurs in a site where the blood-flow past it is able to break off parts of it, these *emboli* are carried in the circulation until they lodge in a narrow part of the system, usually blocking the vessel as a result. For instance, emboli formed as a result of a clot in the great veins or the right side of the heart, pass into the lung causing, "*pulmonary embolism*". These thrombi are usually started in two ways. The intimal lining of the heart or the vascular system becomes "rough" as a result of disease (atherosclerosis, infection or trauma) thus triggering the deposition of platelets at the site, followed by the formation of a clot as already described. When blood is stagnant (*stasis*) the minute quantities of thrombin always being formed reach a local concentration high enough to start the clotting process.

An example of the first mechanism is *coronary thrombosis* where platelet and fibrin deposits gradually narrow the lumen of the coronary artery or one of its branches. Eventually, the vessel suddenly collapses and the ventricular muscle is deprived of its blood supply—often with fatal results.

The second mechanism occurs as a result of blockage of blood flow in any part of the body. It is seen commonly in bedridden patients (often after abdominal surgery or parturition) when the legs are immobilised or kept in the flexed position for long periods of time. Large clots can then occur during the course of an hour or two with the considerable risk that they will be carried to the heart and then to the pulmonary arteries, blocking them. If both arteries are blocked, death is immediate unless "*pulmonary embolectomy*" (clot removal) can be carried out in the operating theatre. Smaller clots may occlude only part of a lobe of the lung and a clinical picture of pain, difficulty in breathing and coughing blood (*haemoptysis*) ensues.

A possible mechanism connecting vascular stasis and intravascular clotting (put forward by P. C. Malone) is that the low oxygen tension in endothelial cells (in a region of stasis) damages them so that platelets, which are phagocytic, attack them and form the basis of a thrombus. Polymorphs are thought to do the same. The phagocytes themselves, having arrived via arteries from the lung, are oxygenated and hence fully functional.

Anticoagulants

The injection of 1 mg kg^{-1} (patient's body weight) of heparin prolongs the clotting time to about ten times the normal value, having an action lasting for four or five hours. It is destroyed by an enzyme called *heparinase*. Overdosage of heparin, as revealed by massive, uncontrollable haemorrhages, can be neutralised by giving *antiheparins* (e.g. protamine or hexadimethrine bromide) —substances which combine with heparin, inactivating it.

Heparin is used in clinical practice for the prevention of clotting when the patient is being maintained on kidney machines or heart lung machines.

Synthetic anticoagulants, such as dicoumarol, having a longer action than heparin, are used in cases of arterial or venous thrombosis where the prolonged depression of the ability of the blood to clot is required. They act by competing with vitamin K, in the prothrombin formation system*.

Fibrinolysis

Clots formed in tissues must eventually be removed in the course of the repair process. *Fibrinolysis*, the dissolution of the clot, occurs as the result of the action of *fibrinolysin* (or plasmin). Plasma contains *plasminogen*, which is its inactive precursor. The conversion is due to activators, extrinsic or intrinsic.

Normal blood remains clotted for several weeks. Stress, exercise, or adrenaline injection cause a subject's blood clot to undergo rapid fibrinolysis (due to activation of intrinsic factor).

There are also inhibitors of fibrinolysis in plasma. Steroids, phenformin (an antidiabetic drug) enhance fibrinolysis; EACA and trasylol are inhibitors.

---

* Iatrogenic Addison's disease may be caused by anti-coagulant therapy. Haemorrhage occurs in the adrenal cortex which is destroyed.

73. The Blood Groups

Blood transfusion

It is well known that effects of severe loss of blood are usually best countered by transfusion of blood from another individual. Early attempts at such transfusion often had disastrous results, owing to the fact that the injected red cells clumped together (*agglutinate*) in large masses which block certain of the capillaries in the body (Fig. 73.1); the cells then haemolyse, and the liberated haemoglobin is in part converted to bilirubin, with consequent jaundice, and in part excreted by the kidneys; the secretion of urine is impaired, or may even stop. When such effects follow the transfusion, the blood of the donor is said to be *incompatible* with that of the recipient. This incompatibility was explained when it was discovered that human serum may contain antibodies which act on the red cells of certain other individuals, making them stick together (agglutinins) or break up (lysins); these are termed "naturally occurring" antibodies, since they have not been formed in response to the presence of known antigens. To be susceptible to agglutinins, the red cells must contain agglutinogens (i.e. antigens) with which the agglutinins react. The experimental facts were found to be explicable by the hypothesis that two kinds of agglutinogen, A and B, are to be found in human red cells. In some bloods the red cells contain agglutinogen A, in others they contain B, in others both A and B together, and in still others they contain neither. Thus blood can be classified into four groups according to whether or not their cells contain the agglutinogens, A, B, AB or O. Similarly, it is postulated that there are in human sera two agglutinins, α and β, which react respectively with agglutinogens A and B. Obviously, in any normal blood, the corresponding agglutinins and agglutinogens which would react with each other cannot be present at the same time. Consequently, only in O blood are α and β agglutinins to be found together. In A blood only β agglutinin is present, in B blood only α, while in AB blood neither α nor β is present.

Determination of blood groups

The blood of a particular man can be easily assigned to its proper group if samples of serum from blood of groups A and B are available as is indicated in Fig. 73.1, which shows the effect of serum from each group on cells of each of the other groups. In blood transfusion it is always desirable to use a donor of the same group as the recipient, but in emergency it is considered allowable to use any donor whose cells are not agglutinated by the serum of the recipient. The donor's cells are exposed to the full effect of the recipient's serum, whereas the donor's serum is diluted by the greater volume of the recipient's blood, and hence is not likely to harm the recipient's cells. As can be seen in Fig. 73.1, cells of group O are not agglutinated by any type of serum; people with blood of group O are thus called *universal donors*. Similarly, serum from group AB will not agglutinate cells of any

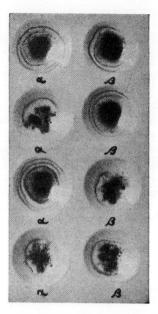

Fig. 73.1 The agglutination of red blood corpuscles, and the four blood groups in man.

Four large drops of serum of group B (containing agglutinin α) and four large drops of serum of group A (containing agglutinin β) were placed on the slide. To each of the top two drops was added a small drop of blood of group O; to the second two, a drop of blood of group A; to the third two, a drop of blood of group B; and to the bottom two, a drop of blood of group AB. Agglutination only occurred when the agglutinogens on the corpuscles met the corresponding agglutinins in the serum. (Lattes's *Individuality of the Blood*.)

group; people with blood of group AB are thus called *universal recipients*.

More detailed and extensive studies of blood groups have shown that the matter is much more elaborate and complicated than was at first thought. Agglutinogen A consists of two parts or varieties A_1 and A_2; and there may be agglutinogens on human red cells which may be classified into at least ten groups or systems, in addition to the ABO system, and to which identifying letters and names have been given. These are of little consequence in transfusion, since normal human sera rarely contain the corresponding agglutinins; they are responsible, however, for the fact that on rare occasions two bloods are found to be unexpectedly incompatible.

Inheritance of blood groups

Blood groups are inherited according to Mendelian laws, the presence of each of the agglutinogens on the red cells being decided by the presence of a certain gene in the chromosomes of the individual concerned.

Different races of people tend to have different blood groups. In the people of western and north-western Europe, for example, groups A and O are the most common, and the proportion of the population who are of group A becomes smaller as we proceed east across Asia; in Central Asia and India, group B is the most common, while the Indians of Central and South America may be almost entirely of group O. Study of the blood groups, therefore, particularly when those of the systems other than ABO are included, is of great value in anthropology, indicating possible inter-relations between different races of mankind.

In southern England (in the natives) the blood groups have the following distribution:

Group O, 46%; Group A, 40%; Group B, 10%; Group AB, 3·5%.

The three genes responsible for inheritance of the classical blood groups are called A, B and O. The actual group of any individual (phenotype) is due to two genes, one received from the mother and one from the father. Genes A and B are co-dominant; O is recessive. Thus to anti-A (or anti-B) serum both persons with the genetic makeup of AA or AO (or BB and BO) will react as group A (or B). The relation between phenotype and genotype is as follows:

Phenotype (blood group)	Genotype
A	AA or AO
B	BB or BO
AB	AB
O	OO

This can be utilised in instances where paternity is disputed. Clearly an O-group parent can give only O genes; a group A only A and O; or a group B only B and O, etc. The MN groups and the various rhesus factors will help further. These tests cannot prove who is the father but they may well show that a given person cannot be the father.

Calculation of gene frequency

A table can be constructed showing the relation between the frequency of occurrence of the A, B, O genes in the population as a whole and the frequency of the blood groups (Table 73.1).

If the lengths along the sides O, A and B are constructed to represent the gene frequency (O = 68%; A = 26%; B = 6%) then the areas are the blood group frequency. It should be remembered that persons of group A will comprise both genetic AAs, AOs and OAs.

Gene frequency can be calculated from the known blood group frequency as follows:

Let p, q and r be the frequency of occurrence of the genes A, B and O respectively in a population.

Then
$$p + q + r = 1$$
Frequency of gene O as $\bar{O} = r^2$ (from Table 73.1)
$$\bar{A} = p^2 + 2pr$$
$$\bar{B} = q^2 + 2qr$$
$$\overline{AB} = 2pq$$

Then
$$r^2 + p^2 + 2pr + q^2 + 2qr^2 + 2pq = 1$$

The above equation will enable the frequency in the population of any gene (p, q, r) to be found.

$$\bar{O} = r^2$$
$$\therefore \quad r = \sqrt{\bar{O}} \quad (1)$$
$$\bar{A} + \bar{O} = p^2 + 2pr + r^2$$
$$= (p + r)^2$$
$$\therefore \quad p + r = \sqrt{(\bar{A} + \bar{O})}$$
$$\therefore \quad p = \sqrt{(\bar{A} + \bar{O})} - \sqrt{\bar{O}} \quad (2)$$
and $\quad q = \sqrt{(\bar{B} + \bar{O})} - \sqrt{\bar{O}} \quad (3)$

Indications for blood transfusion

Blood transfusions are given for the two major effects: (1) the volume of circulating blood is increased; (2) a supply of oxygen-carrying red cells is added to the circulation. Also, substances essential for coagulation, serum proteins, leucocytes and platelets and antibodies are provided.

Conditions in which transfusions are given may be classified as follows:

1. Haemorrhage (traumatic, local disease, surgery)
2. Traumatic shock
3. Preparatory for operations
4. Haemorrhagic and blood diseases
5. Certain poisons, infections and other diseases

If more than about one-third of the blood is acutely lost, death will usually result unless the blood is replaced by transfusion. If the blood loss is slower than this, possibly two thirds of the blood can be lost without difficulty.

During haemorrhage, initially the compensatory mechanisms maintain more or less normal arterial pressure. Eventually, failure of the vasoconstrictor system occurs, the blood pressure falls and the pulse becomes feeble and rapid. The patient will not often survive a depression of the systolic blood pressure below 80 mmHg (approx. 10 kPa) or of the diastolic below 40 mmHg (or 5 kPa) for longer than one hour. If the systolic pressure falls much below 100 mmHg (13 kPa) and the pulse rises above 120 min^{-1}, this is an indication for immediate transfusion.

The rhesus (Rh) blood group system

The Rh blood group system differs from the ABO because normally there are no circulating antibodies to the Rh antigens. Red cells from rhesus monkeys were injected into rabbits; after a brief interval the rabbits responded by forming antibodies to the injected cells. It was found that when this rabbit serum is mixed

Table 73.1

		Genes from mother		
		O	A	B
		68%	26%	6%
Genes from father	O 68%	68%		
	A 26%			
	B 6%			

with human red cells, agglutination occurred in about 85 per cent of the samples. Such persons are Rh-positive. Their serum contains no antibody however; the remaining 15% not agglutinating are Rh-negative and also contain no antibody. However, in certain circumstances (q.v.) Rh-negative persons can develop antibodies.

The main Rh antigens are C, c, D, d, E and e. Inheritance is by simple Mendelian laws. All these antigens are in theory capable of giving rise to antibody formation in a subject lacking the particular antigen; however, it is only the D antigen which normally does this.

Rhesus factors

The group of agglutinogens known as the *Rh factors* were first discovered on the red cells of rhesus monkeys (hence the abbreviation), but they also occur on human cells in most individuals (85 per cent of Europeans). The corresponding agglutinins do not normally occur, even in the 15 per cent of Rh-negative persons. *Injection of Rh-positive blood into a Rh-negative person will, however, lead to the production of the agglutinins.* A later transfusion, therefore, of Rh-positive blood will have serious consequences. As with the other agglutinogens, the presence or absence of the Rh factors is inherited. The fetus within a Rh-negative mother may, by inheritance from the father, be Rh-positive. The agglutinins are then formed in the mother's blood, pass into that of the fetus, and lead to destruction of the red cells, and usually miscarriage, or death of the infant shortly after birth. The remedy is to remove the agglutinins by complete replacement of the infant's blood with that from a normal person (exchange transfusion).

Direct cross-matching

It is thus advisable before transfusion to test directly the donor's cells against the recipient's serum, and the recipient's cells against the donor's serum, a procedure known as the *cross agglutination* test. This is a precaution not only against errors in grouping, but also tests for the presence of the Rh factors and of the other agglutinins which may occasionally be present.

ROULEAUX FORMATION AND SEDIMENTATION VELOCITY

Red blood corpuscles in plasma nearly always show a tendency to come together with their broad surfaces in apposition, thus forming aggregates which look like rolls of coins and hence have been named *rouleaux*. In practice it is never difficult to distinguish these orderly rouleaux of ten or twenty cells from the disorderly clumps of thousands of cells found in the agglutination reactions. The tendency to form rouleaux varies among different individuals and different species, and largely determines the suspension stability of the blood, i.e. the time required for the red cells to sediment down completely in blood (made incoagulable) which is allowed to stand. This is due to the fact that the rate at which a system of suspended particles settles, increases with the size of the particles, other factors remaining constant. The formation of rouleaux may be regarded as a mild, and reversible, kind of agglutination; it occurs more readily when the concentration of fibrinogen or globulin in the plasma is increased, and it may be related to the presence of antibodies and to the occurrence of immunity reactions. An increased sedimentation velocity has been observed to accompany most inflammatory diseases, and also to accompany pregnancy; its measurement, therefore, is of diagnostic interest.

Sedimentation sometimes occurs so rapidly that a clear layer of plasma is left before coagulation begins; the clot is thus partly free from corpuscles and forms what is known as the *buffy coat* on the surface of the corpuscular mass. This fact has been known since the days of the Greeks, and very largely formed the basis of the practice of blood-letting as a cure for all diseases. It was thought that the buffy coat was formed by the foul matter in the blood that was responsible for the disease; the more blood one could remove, therefore, the quicker would the patient recover. Accidents sometimes happened, however, for instance through the physician mistaking the normal effect of pregnancy for evidence of a pathological condition.

74. The Pathophysiology of Shock

Shock: definition

The circulation, as we have seen in Chapter 36, is beset by numerous servocontrols, mechanisms which protect its normal functioning in the face of adverse environmental and other factors, to a surprising degree. However, a time comes, sooner or later, when the applied stress is too great for any compensation to correct it, and the self-same protective mechanisms then lead the circulatory system into self-destruction. To begin with, this vicious spiral known as "shock" can, by vigorous measures, be reversed; in later stages the changes are irreversible and end in death.

Shock is defined as the "abnormality arising from inadequate propulsion of blood into the aorta" (and hence the inadequate supply of capillary blood to the tissues). The condition of shock is characterised by deterioration of cellular function and a steadily increasing failure of the circulation leading to an irreversible sequence of events.

THE TYPES OF SHOCK

The term shock should be restricted to a syndrome affecting the circulation that involves prostration, hypotension, collapse of the veins and all the other signs of circulatory collapse. This then excludes entities such as spinal shock, anaphylactic shock, shell shock, etc.

Primary shock

An episode of acute hypotension is termed "primary shock". Of reflex origin, it is due to trauma, pain and injuries such as blows on the head. Fainting (syncope) may occur as a result of neurological, psychological or circulatory abnormalities; it is often due to sudden failure of the reflex control of the circulation so that vasodilatation of the splanchnic vessels leads to abdominal pooling of blood and therefore a greatly reduced venous return with consequent brain anoxaemia. Primary shock is usually transient and reversible.

Secondary shock

Here, the normal function of cells in tissues has deteriorated, and if no external assistance is given it, the circulation ends up in an irreversible state leading to death. Evidence for the presence of this irreversible condition is a progressive fall in the arterial pressure and collapse of the veins. Primary shock existing for any length of time proceeds to secondary shock.

CAUSATION OF SHOCK

The important causes of shock are listed below; most involve the loss of significant amounts of blood or plasma.

1. Haemorrhage
2. Exudation of plasma from burns, chemical or other damage
3. Vomiting, diarrhoea or other loss of secretions
4. Disturbed water and electrolyte balance
5. Capillary or venous engorgement (reduced blood volume)

SYMPTOMS OF SHOCK

Clinical examination of the shocked patient reveals pale, cold skin due to peripheral vasoconstriction. Sweating is found because sympathetic stimulation occurs. Often cyanosis is seen as a result of peripheral circulatory stasis. The appearance is drawn, ill-looking and the face is sunken—all the result of decreased extracellular fluid which flows into the circulation to attempt to counteract the fluid loss there.

Blood pressure is low, pulse pressure decreased, and pulse rate is rapid (clinicians categorise the pulse in shock as rapid, "thin" and "thready"—euphonious wording hallowed by tradition, if not by clarity).

Respiration is rapid and shallow and body temperature falls. Reflexes and activity of the nervous system in general are depressed.

STAGES OF SHOCK

1. Early events (compensated shock)

The loss in fluid volume means that there is a disparity between circulating blood volume and the volume of the circulatory bed. Venous return is reduced, cardiac output decreases and thus arterial pressure is low, so that tissues receive a reduced oxygen supply. As would be expected from a consideration of the mode of action of compensatory mechanisms, these act to restore the blood pressure via baroreceptors, catecholamine secretion and resultant increases in arteriolar tone. Similar reflex activity increases heart rate, and at the same time respiration rate.

Attempts are made to replace the lost fluid volume through a reduction in capillary pressure altering peripheral fluid balance, the general tendency being towards increased passage of extracellular fluid into the capillaries. The changes described may be sufficient to compensate for the fluid loss and arterial pressure remains relatively constant.

2. Incomplete compensation

The second stage of shock is characterised by a progressive fall in arterial pressure. Compensatory mechanisms are no longer able to deal with the advancing disturbance. Vasoconstriction can proceed no further and the total peripheral resistance begins to decrease again. This is the initiation of a cycle of positive feedback from which recovery is unlikely unless intervention of some kind can interrupt it. The decreased peripheral resistance further lowers pressure; in turn this lessens tissue oxygen supply and still further decreases the ability of the arterioles to maintain the peripheral resistance (oligaemic shock).

The underlying causation of this vicious spiral is the generalised tissue hypoxia which leads to interference with the function of tissue enzyme systems. There is accumulation of lactates and pyruvates; phosphate synthesis fails. Potassium then leaks out of the cells as their metabolic processes, including the sodium pump, run down.

Effects on the kidney
Most of the secondary effects of shock upon the function of the various organs are due to the stagnant anoxia interfering with them. In the kidney, glomerular filtration fails because of reduced blood flow and as a result of anoxic damage to the cells of the glomerular tuft causing them to become more permeable. Tubular lesions follow; the syndrome is one of renal failure, anuria and hence uraemia. In addition, metabolic acidosis follows because the kidney can no longer deal with excess hydrogen ions.

The *crush syndrome* (injuries in which the limbs are crushed) gives rise to even more severe renal damage. This is because mechanically damaged muscle liberates myoglobin, various proteins and products of tissue autolysis into the bloodstream. These are then carried to the kidney where the renal tubules are blocked with it—a condition called lower nephron nephrosis.

Effects on the heart and brain
Damage to the myocardium results from the reduced oxygen supply—a further positive feedback which lowers cardiac output, and coronary supply still more.

Brain blood supply tends to be maintained at the expense of that to the remainder of the body.

3. Irreversible shock

It has been general practice to classify shock into these two main categories, namely *reversible* (or compensated) shock and *irreversible* shock. It should be noted that the term irreversible does not mean that the shock cannot be successfully treated. It is a term used because if one replaces the patient's blood volume, there is not a corresponding return to normal physiological functioning of the circulation. In reversible shock a transfusion to replace the depleted blood volume results in the rapid return of normality. Traumatic shock in an older patient (or in a young one who has been shocked for some time) results in a clinical condition much more difficult to treat. Many of such patients will die, even though all the available therapeutic measures have been tried.

The basic difference between reversible and irreversible shock lies in the reaction of the microcirculation (arterioles, capillaries and venules). Shock results in tissue hypoxia. If prolonged and severe, this hypoxia paralyses the vascular smooth muscle of arterioles and metarterioles before the venules, a state of affairs resulting in overfilling the capillaries with blood. Plasma leaks out into the interstitial spaces.* A significant fall in useful blood volume thus occurs ("blood out of currency", as it was termed by Cannon). Indeed, unless the process is halted by treatment, the peripheral vessels become filled with a thick "sludge" of red cells, which may lead to actual thrombosis of such vessels. This stage is often called *normovolaemic* shock, to indicate the fact that the patient is still in shock even though the blood volume has been restored to normal by transfusion.

HAEMORRHAGE

The pathophysiological effects of haemorrhage depend not only on the amount of blood lost, but also greatly upon the rate at which it is lost. Compensation can be enormous, but takes time to reach its maximum effect; a sudden loss of blood is therefore a more serious threat than a much larger haemorrhage taking place over a longer time.

Immediate effects of severe haemorrhage

Experimental study of the effects of withdrawing 15 to 20 per cent of the blood volume (750 to 1200 ml) in volunteers has been made. During the venesection the face turned pale, hands became cold and sweaty. Systolic pressure fell by around 10 mmHg (about 1 kPa), diastolic was unaltered but heart-rate rose by 15 to 30 beats per minute. Venous pressure fell.

Then, between one and four minutes after the start of the venesection, the subjects showed signs of circulatory collapse —in other words the compensatory mechanisms were proving inadequate. Blood pressure fell considerably, heart rate fell to 36–50 beats per minute and the skin became grey in colour. Subjects were very weak, nauseated, sweating and had blurred vision. Figure 74.1 illustrates these changes.

The sequence of events is as follows: loss of blood decreases blood volume; venous pressure (hence right auricular pressure) falls. Since venous return is diminished, cardiac output falls. Thus one would expect blood pressure to fall. Its initial fall is small however. This means that peripheral resistance rises (vasoconstriction of skin and splanchnic vessels, secretion of adrenaline). Heart rate rises (baroceptor reflex from fall in arterial pressure).

Restoration of blood

Compensation for blood loss occurs in two stages, *immediate* and *long-term*. First, the water then the plasma protein and last the haemoglobin and red cells are replaced. Figure 74.2 gives the essential details of this process.

THE MECHANISM OF SHOCK

Various theories to account for the irreversible changes already mentioned have been put forward, based on animal experiments. The major feature is fluid loss from the circulatory bed (due to haemorrhage, plasma loss in burns, or prolonged vomiting and diarrhoea). This triggers the changes already discussed.

Nervous factors

The effect of afferent impulses (including those in pain fibres) from the sites of burns, trauma etc. have been studied. The vasoconstriction which results from the operation of such reflexes may later cause the irreversibility. Most clinicians now do not emphasise the importance of these mechanisms because prolonged, intense experimental stimulation of sensory afferents

---

* Large infusions of blood (or substitutes) are of only temporary benefit. This is because the anoxia of vessel walls renders them more permeable than usual and the added fluid (and often cells) is rapidly lost into the tissue spaces.

THE PATHOPHYSIOLOGY OF SHOCK 479

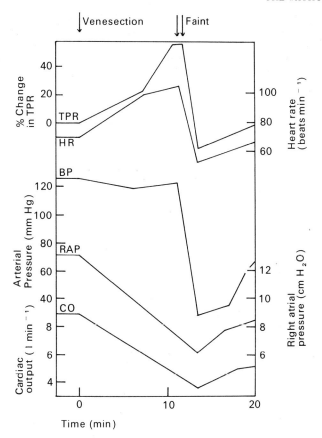

Fig. 74.1 Circulatory changes in severe experimental haemorrhage in man.
About 1·2 litres were withdrawn from an arm vein during the period from 0 to 10 min. Fainting occurred at the end of the blood loss, and is accompanied by a sudden fall in pressure, slowing of the heart and a large decrease in total peripheral resistance.

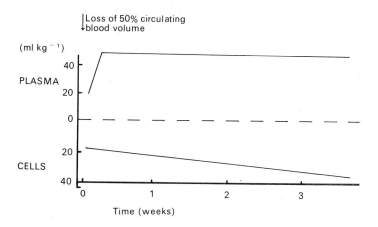

Fig. 74.2 Recovery of plasma and red cell volume following haemorrhage in man.
At the end of a venesection of about 50 per cent of circulating blood volume, blood haemoglobin level was normal, plasma and cell volumes were each reduced to about 20 ml kg^{-1}. During recovery, plasma volume was restored within 48 h with a concomitant fall in haemoglobin from the normal 15 g/100 ml to about 8 g/100 ml which recovered slowly, pari passu with the red cell volume as shown by the lower line during the subsequent three weeks.

(e.g. sciatic, vagus) does not give rise to shock. It also appears that total peripheral resistance is fairly well maintained during the early phases of shock (oligaemia) and the initial part of the irreversible phase (normovolaemia). Therefore, the progressive fall in pressure after transfusion must be the result of other factors.

Chemical factors

At the site of an injury, or in the damaged circulatory bed following blood loss, humoral substances are released into the circulation and influence its function. Some of these are pressor in action; they presumably take part in the compensation process. Catecholamine secretion would be an example. Renin is also secreted during shock.

Not all chemicals appearing in the circulation during shock have beneficial actions. For example, tissue damage (including anoxic damage occurring in blood loss) releases potassium ions which will depress the myocardium, histamine which greatly increases capillary permeability and other vaso-depressive substances such as adenosine compounds and bradykinin. Proteolytic enzymes and bacterial exotoxins also add to the malfunction of the microcirculation. There is not only a local effect due to these substances, they are also carried by the blood to remote structures where similar destructive effects result. This can be regarded as yet another example of a deleterious positive feed-back.

Treatment of shock

Treatment of shock is directed towards restoration of normal cardiac output. The obvious immediate measure is to give a transfusion of blood or a suitable substitute, according to the kind of fluid loss. The amount transfused must be controlled by measuring central venous pressure; in practice the transfusion is stopped when the latter reaches about 10 cm H_2O (say 1 kPa). Overtransfusion may lead to pulmonary oedema whilst undertransfusion will delay recovery and will increase the possibility of irreversible changes supervening.

There is as yet no definitive measure for treating irreversible shock. Some clinicians use sympathomimetics such as noradrenaline, to stimulate alpha receptors and cause vasoconstriction. This is seen as helping the circulation by raising arterial pressure. It may however, also initiate ischaemic damage in the vasoconstricted regions and thus begin the positive feed-back disasters mentioned in the previous paragraph.

Other clinicians will use vasodilators such as phenoxybenzamine which dilate the peripheral resistance vessels and hence improve tissue perfusion. Also, the lowered peripheral resistance might reduce the load on the heart and actually raise cardiac output. Central venous pressure should again be monitored since the vasodilator drugs affect all types of vessel including the venules and veins.

Steroids in large doses are sometimes used for their peripheral vasodilatory action; beta-stimulating drugs such as isoprenaline are also employed in cases where a depressed cardiac output is evident although central venous pressure is normal.

It used to be a cardinal feature of the treatment of shock that the patient be kept warm. This was in spite of the demonstration as long ago as the Great War, 1914–1918, that casualties who had remained in the open and in the cold, tended to have a better chance of survival than those who were warmed with hot-water bottles and blankets. It is now known that it is most important not to warm up shocked persons; this merely increases tissue oxygen demand and accentuates the positive feedback damage already discussed.

75. High Blood Pressure

The servoregulatory systems which control the pressure within the arterial system, keeping it constant around 100–120 mmHg (13–16 kPa), i.e. resting normal systolic pressure, have been described in Chapter 36. This mechanism is subject to long-term error in the form of persistently-raised arterial blood pressure, a disease of middle-aged and elderly persons, known as *essential hypertension*. Since it has been known for many years that anoxaemia of the kidney is compensated by hormonal mechanisms which elevate the arterial pressure, it used to be thought that the basis for essential hypertension also lay in the kidneys. Recent work however, indicates that the renal involvement is secondary and that abnormally strong responses to environmental stimuli may initiate the disorder.

At present, the explanation for the increased arterial pressure in hypertension is unknown, largely because of the numerous interacting mechanisms involved in the development of the disease.

Pathophysiological effects of hypertension

An arterial pressure of 80–100 mmHg is adequate for normal tissue requirements in the resting, healthy human. A hypertensive may have pressures of 200–300 mmHg (up to 40 kPa). The consequences of this depend on the sustained level to which arterial pressure is raised.

(a) *Heart*. The increased load imposed by a raised arterial pressure of itself will not lead to cardiac failure. Heart muscle will undergo compensatory hypertrophy until it is up to 25 per cent larger than before.

(b) *The arteries*. High systemic pressures damage the walls of small arteries; they become *sclerotic* and of narrowed lumen. This leads to impaired tissue blood-flow. Effects are apparent in the heart leading to failure and in the brain leading to brain haemorrhage, etc. If these arterial lesions affect the kidney, a vicious circle is set up; renin is secreted because of the anoxaemia and the arterial pressure is further elevated as a result (malignant hypertension; often fatal within a few months of onset).

(c) *Hormonal conditions*. Cushing's disease, due to excessive production of cortisone and the other glucocorticoids by an adrenal tumour, usually leads to hypertension. The same happens if excessive amounts of cortical hormones are given therapeutically.

(d) *High salt intake*. A high dietary intake of salt is thought to play a role in the development of hypertension, particularly in persons with pre-existing kidney disease.

Experimental hypertension

When the renal artery is constricted with a clamp (but is not entirely obliterated), a progressive elevation of blood pressure occurs. The concomitant administration of salt and/or corticosteroids (which cause salt retention) makes it more severe. Salt in large quantities given with desoxycorticosterone will, in fact, cause hypertension in normal animals. The mechanism of this effect of renal impairment, has not yet been fully worked out. It may, at first, be due to increased cardiac output, an increase in plasma volume and venoconstriction. Later, a component of importance is the increased arterial resistance due to changed salt and water content of the walls of small arteries and arterioles. Also, an increase in sympathetic vasoconstrictor tone occurs.

Malignant hypertension can also be produced in the experimental animal; raised levels of plasma renin are found, and the kidneys show arteriolar necrosis. In the later stages of experimental renal hypertension, the functional disturbance becomes contradictory. The cardiac output falls to normal; there is no change in heart rate or blood volume. The only factor which maintains the elevated arterial pressure is the increase in peripheral resistance. At this stage, denervation of the kidneys, or denervation of the heart, or cutting the spinal cord, does not affect the hypertension; this has led to the theory that some undefined humoral agent is responsible. The long-term release of very small amounts of renin from the kidney, either through anoxaemia of the kidney or as a result of the direct stimulation of the sympathetic nerve supply to the juxtaglomerular apparatus (this would give aldosterone secretion and hence sodium and water retention), may be the mechanism.[*]

The role of the nervous system

What are the effects, in the experimental animal, of denervating the carotid and aortic baroceptor zones? Of course, one would expect that as any rise in pressure would no longer set-off impulses, reflexly resulting in lowering it again, a sustained elevation would occur. The experiment does indeed produce chronic high blood pressure. However, after the hypertension has persisted for some time, an adaptation of the responses occurs and the reflex is set to operate at higher pressure levels.

A different method for arriving at a similar functional result involves making a rigid cast (with plastic material) around the baroceptor regions. This of course will mimic hardening of the arteries, a given transmural pressure leading to a much reduced baroceptor discharge. Thus arterial pressures reach much higher values before reflex action can control them. It is possible that a neurogenic mechanism such as this, having as a basis reduced receptor sensitivity, might underlie the human disease in some cases.

Whatever the cause of essential hypertension eventually turns out to be, it appears that a vicious circle develops in which the

---

[*] Attempts to demonstrate this in patients with essential hypertension have proved abortive. There is usually no electrolyte imbalance and renal blood flow and function are usually normal. Most patients have no apparent abnormality in the renin-angiotensin system.

raised arterial pressure gives rise to renal damage that in turn raises the blood pressure still further. Such positive feed-back ultimately will kill the patient.

Genetic and environmental factors have been implicated as causative. The blood pressure in monozygotic twins is very similar. Activity, diet and stress appear to be important aetiological variables but at the present time little substantial information is available upon which to base an all-embracing theory.

One plausible hypothesis suggests that the putative hypertensive initially has exaggerated autonomic responses to emotional stress, including sympathetic-mediated renal vasoconstriction. This causes sodium retention, increased extracellular fluid volume, and increased cardiac output (to tissues other than the kidney). As a compensatory mechanism, there tends to be autoregulatory vasoconstriction which increases the total peripheral resistance, and therefore mean arterial pressure.

If such a state of affairs is maintained, secondary changes in vessel walls take place, in the kidney to give renal hypertension, and in baroceptor sites to reduce receptor sensitivity, with consequent initiation of the positive feed-back previously emphasised.

Clinical picture

In early cases, or in mild degrees of hypertension the condition is symptomless. When it is more severe, dizziness and headaches occur. The latter are typically occipital and are worse on awakening.

Signs include left ventricular hypertrophy, and dyspnoea. Visual defects are commonly found (*hypertensive retinopathy*—visible in the ophthalmoscope). Renal failure is a late complication in the disease. Attacks of focal cerebral ischaemia occur, giving epileptic fits and fainting attacks. Angina of effort commonly occurs.

Treatment

The physiological principles of treating hypertension, are those involving a reduction in arterial blood pressure. Most antihypertensive drugs reduce sympathetic tone in the peripheral resistance vessels.

Guanethidine and bethanidine reduce the amount of noradrenaline in sympathetic nerve terminals. They also prevent release of transmitter. Methyl dopa impairs noradrenaline synthesis. Thiazide diuretics are also given. Their mode of action is unclear but probably they also have a direct relaxing effect on the vascular wall. Mild hypertension, particularly if accompanied by angina, is often treated with beta-blocking sympatholytic drugs (e.g. propranolol) which lower cardiac output.

76. Hepatic Failure

Jaundice

Defective function of the liver often leads to jaundice—the yellow discoloration of sclerae and skin due to accumulated bile pigment. Since the liver has many functions apart from dealing with the breakdown products of haemoglobin, there will be other signs of hepatic failure, quite apart from jaundice.

Jaundice appears if bilirubin excretion fails to keep pace with its formation. Therefore there are two basic reasons for the appearance of jaundice—a too rapid formation of bilirubin (as in haemolytic anaemia), or an obstruction to biliary removal (as in hepatic disease or in obstruction of the bile duct). The yellow pigmentation is not obvious until the plasma concentration of bilirubin exceeds 0.02 g l^{-1}.

Production of bilirubin

Bilirubin is the end result of haemoglobin breakdown from dying red cells in the reticuloendothelial cells of the liver. The first stage consists of splitting the haemoglobin molecule into haem and globin, the iron being removed to bone marrow as transferrin. Haem minus iron is protoporphyrin, which splits enzymatically to bilirubin transported in plasma conjugated with the albumen there.

Removal of bilirubin

In the liver, bilirubin is conjugated with glucuronic acid; obviously when conjugated in plasma with albumen it cannot pass into the kidney tubules; after processing in the liver it can do so. In the liver it is excreted into the bile canaliculi. The enzyme glucuronyl transferase catalyses the conjugation; at birth there is a deficiency of this enzyme which is the main reason for the so-called "physiological jaundice" of the first days of life.

Liver excretion

In obstructive jaundice, plasma bilirubin is raised. Urine is dark; faeces pale. The excretion of all biliary constituents is halted. Loss of bile salts gives steatorrhoea due to fat malabsorption and vitamin deficiencies (especially vitamin K). Alkaline phosphatase is normally excreted in bile; its plasma level will be raised.

Synthesis in the liver

Plasma protein, prothrombin, fibrinogen, angiotensinogen are manufactured by the liver. The conversions involving glucose and glycogen, synthesis of glucose from non-carbohydrates, fat breakdown are all carried out in the liver. Detoxification, urea production and steroid conjugation also occur there.

Hepatic failure usually comes about as a consequence of chronic damage, and can affect all these systems.

Cirrhosis of the liver

This is the commonest cause of hepatic failure. It consists of disruption of the normal structure of the liver, and its replacement by fibrous tissue. A number of different pathological processes may lead to this, such as alcoholism (possibly the accompanying nutritional defect is responsible), malnutrition, various poisons, and autoimmune disease.

FUNCTIONAL SIGNS OF HEPATIC FAILURE

Jaundice is a late and minor sign. Hypoglycaemia may occur; also the glucose tolerance curve may be high and have a long duration due to reduced glucose uptake by the liver. Plasma cholesterol rises. Plasma proteins are low, apart from gamma-globulin (synthesised elsewhere) so the patient often has oedema (low plasma oncotic pressure). Clotting factors are not formed (factors II and VII) so abnormalities of blood coagulation occur.

Detoxication is usually impaired. Thus hormonal overactivity appears to occur since circulating hormones are not removed as usual. Aldosterone, ADH are examples, and give water retention, with oedema and ascites. Drug metabolism is impaired and certain compounds, e.g. morphine, must not be given to patients with hepatic failure. Fever often accompanies hepatic failure and is thought to be due to the accumulation of bacterial and other pyrogens, perhaps originating in the intestinal flora, which would otherwise be removed.

Cardiac output is usually increased in chronic hepatic failure; this is due to the excess aldosterone and ADH mentioned above. Also, anaemia often co-exists.

Liver function tests

Bromsulphthalein (BSP) excretion is measured by injecting 5 mg kg^{-1} at zero time. A sample is withdrawn from the other arm after 45 min. In the absence of hepatic failure not more than 10 per cent of the dye should now be present.

Total serum bilirubin should be less than 2 to 7 $\mu\text{mol l}^{-1}$. Serum protein (total) should be 60–80 g l^{-1}. Prothrombin time should be 10–14 s in normals. Serum cholesterol should be less than 3.5–6.0 mmol l^{-1}.

Neuropsychiatric abnormalities

Some kinds of hepatic failure give rise to neurological and psychiatric abnormality, possibly due to toxic effects of substances normally removed by the liver (e.g. ammonia, especially if the subject is on a high protein diet).

The abnormalities commonly encountered are coma, personality changes, increased reflexes and tremor. These effects progress to paraplegia and dementia.

77. Transplantation

A piece of skin, or even a whole organ, may be removed from an individual and transferred, or *transplanted*, to some other place in the same individual, where it may survive and retain its function. But if it is transferred to another individual, even of the same species of animal, it will not survive, but, after a week or two, will be destroyed and removed just as if it were a "foreign" substance, and with the same kind of inflammatory reactions. This is due to the formation of antibodies which react with the antigens of the transplanted tissue. If the animal on which the skin, for example, is grafted has been "immunised" ("sensitised" would describe the phenomena better in this instance) by a previous graft from the same animal, or the injection of a suspension of living cells from this animal, the graft will be destroyed very rapidly, the necessary antibodies being already present. But if the donor and recipient of the graft are identical twins, i.e. have developed from the same fertilised egg-cell, and have the same genes in their chromosomes, a graft from one will "take" in the other. Studies of this kind lead to the conclusion that no two individuals (apart from identical twins) have cells and body fluids of precisely the same chemical composition, any more than they have precisely the same facial appearance or fingerprints.

Genotypes and grafting

Some people live to a great age, appearing simply to "wear out". Most people, however, in common with motor cars, washing machines etc., break down due to the failure of a certain part. This failure may have serious consequences, and even may result in death, if a vital organ is affected. Occasionally the loss of an organ can be made up for by the taking over of its functions by another (e.g. when one kidney is damaged) or by replacement therapy (e.g. injections of insulin in a diabetic).

Modern surgical techniques now offer the possibility of transplanting tissues and even organs from one person to another or, occasionally from other species to human beings. Apart from live transplants it is now also possible to insert artificial organs either temporarily or permanently into patients. The main problem to be surmounted in obtaining successful grafts is in overcoming the immune response which destroys the graft when it is introduced into the recipient. Most of the antigenic proteins found in red cells which lead to transfusion reactions are present also in the remaining cells of the body; in addition there are many others. Thus foreign cells when grafted, set up immune reactions, e.g. the α and β agglutinins will kill off any cells containing type A or type B antigen, while Rh antigen in the graft itself will stimulate Rh antibody formation in the recipient, this antibody later attacking the graft. There are very many similar antigens, and the recipient can be looked upon as responding in just the same way to transplanted foreign cells as to pathogenic bacterial infection.

Types of transplant

1. *Autotransplants* are pieces of tissue or organs which are grafted from one part of the body to another. Considerable use is made of autotransplants in orthopaedic and plastic surgery (cartilage, bone, tendon and skin). Since the genotype is the same, reactions do not occur and provided that blood supply to the graft is adequate, all the grafted cells survive.

2. *Homotransplants* are taken from another person (or animal of same species—in experiments). They survive for several days or weeks, then to be destroyed by the recipient's immune response. Even so, some homotransplants are useful, e.g. to act as a "scaffolding" for new bone formation as in bone transplants, or to overcome a temporary crisis, as in blood transfusions.

3. *Heterotransplants* are grafts from one species to another and usually are destroyed even more rapidly than homotransplants.

Corneal transplantation

Certain tissues which do not have cells (or the cells are not vital for the purposes of the graft) can often be transplanted from one individual to another successfully. Examples are cornea, tendon, fascia and bone. The first experiment in human corneal grafting was carried out in 1789 by a Frenchman, Pellier de Quengsy, who used a glass disc set in a silver ring sewn into the cornea.

However, it was not until 1905 that a successful corneal tissue graft was achieved.

At the present time about 80 per cent of all grafts are successful (and of course the failures can be grafted again). The success of corneal grafts, in comparison with that of other organs, is due largely to the fact that the cornea is avascular, and hence the antigens of the cornea never reach the recipient in sufficient amounts to evoke an immune reaction which would cause the graft to be rejected, or to go opaque. Also the antigenicity is probably low.

An interesting new development is the "odonto-kerato-prosthesis" which consists of a plastic cylinder cemented within a very thin, ring-shaped, cross section of the patient's own tooth root. The tooth, since it is an autograft, is accepted by the cornea and the plastic cylinder acts as a pinhole lens.

Immunosuppression

As we have seen, only homografts from an identical twin will survive, because only then are the antigenic proteins of both donor and recipient alike (since they are determined by identical genes from the single fertilised ovum). This of course limits the possibilities of organ transplanting on any useful scale, although the first successful skin-graft between monozygotic twins was carried out as long ago as 1920.

The stimulus of the experimental work on immunological tolerance carried out mainly by Sir Peter Medawar, has led to a great variety of attempts at the clinical transplantation of tissues and organs with increasing success recently. This success is wholly due to the new techniques of suppressing the immune reaction in the recipient, which have developed as a direct result of Medawar's work.

Mechanisms of action of immunosuppressives

Immunosuppressive agents reduce all components of the homograft immune reaction (onset is delayed; intensity is reduced and lymphatic proliferation is less). It is difficult to be certain about the role played in suppression by these mechanisms, because we are still ignorant of the manner in which the immune response itself is built up.

(a) *Inhibition of mitosis*
Reduced mitotic frequency has been shown to occur after administration of most immunosuppressive drugs. This is responsible for the reduced lymphoid activity and antibody output observed.

(b) *Reduced antigen capture*
Only cortisone has this effect. It impairs the whole defence reaction of the host including phagocytosis and hence uptake of antigen from the donor.

(c) *Inhibition of DNA synthesis*
Various drugs affect the different stages in nucleic acid synthesis, e.g. 6-mercaptopurine inhibits the first step in purine synthesis—the conversion of phosphoribosylpyrophosphate to phosphoribosylamine. X-irradiation, the alkylating agents and antibiotics such as actinomycin C act on the end stages of nucleic acid synthesis.

(d) *Alteration of nucleic acid bases*
Antimetabolites (such as 6-thioguanine) mimic the natural metabolites and hence replace them in the ultimately produced nucleic acids. Thus in immunologically competent cells in which RNA has been altered in this way, antibody of impaired combining power, or of anomalous structure is produced.

(e) *Destruction of cells*
Many immunosuppressive agents give wholesale cellular destruction, although presumably the important factor here is the recipient's lymphocytic killing.

Azathioprine (AZT, Imuran) is believed to act by inhibiting purine synthesis at three stages in the pathway. (Incorporation of glutamine; conversion of inosinic acid to adenoyl succinic acid and in the subsequent conversion of this to adenosine monophosphate.)

The commonly used examples in clinical practice are azathioprine, actinomycin C, and prednisone. The disadvantage of using these drugs is that the patient's resistance to any infection is reduced or absent and a delicate balance must be achieved between dosage levels required to suppress the immune reaction sufficiently to prevent rejection of the graft and yet retain a modicum of protection from infectious disease.

Kidney grafting

In 1954 the first successful human kidney transplantation was performed and the rapid restoration of this patient's health was a potent stimulus to further work. The homograft was from an identical twin. The current era in kidney grafting began in 1961 with the use of azathioprine as an immunosuppressive agent in a grafted patient who survived for 36 days, but who died from toxic effects because very little was known at that time about the optimum use of this drug. At the present time there are several thousand patients who have had successful renal grafts. Numbers of centres have been set up in various parts of the world and the average one-year survival rate is about 70 per cent; for two years or more it is about 60 per cent.

This successful outcome is mainly the result of the realization that the patient with kidney failure can only be treated surgically if in good general health. Thus artificial kidney units have been established in conjunction with the transplantation units and *intermittent haemodialysis* is usually carried out for long enough beforehand to ensure that the patient is fit to receive the graft. Haemodialysis (developed by Kolff 30 years ago) consists in connecting the patient via a pump to a dialysing chamber in which blood circulates on one side of a cellophane membrane and the substances normally removed from the plasma by the kidney pass into the dialysate on the other side of the cellophane. The amounts of the various urinary constituents removed can be controlled by the composition of the dialysing fluid. For instance the dialysing fluid will be made up to contain the same amount of Na^+ and K^+ as does plasma (142 mmol l^{-1}; 5 mmol l^{-1} respectively) and no net exchange will occur. On the other hand the dialysate will not contain any urea or creatinine, whereas normal plasma does. Therefore the urea and the creatinine are extracted by the dialyser.

Intermittent dialysis is performed on patients by permanently inserting two plastic silicone cannulae into the arm or leg. One is to give access to an artery, the other to a vein and when the patient is not being dialysed the two are connected together to form an arteriovenous shunt having a continuous rapid flow within it. This prevents clotting, but such shunts are the main problem encountered in looking after the patients. The site must be kept scrupulously clean, it cannot be bathed and it must be protected from injury. Patients have to be connected to the machine for about 10–12 h, two times a week and usually in order to interfere as little as possible with normal life, dialyses are performed at night. The patient comes in the late afternoon and remains on the machine until the next morning. (He has an evening meal, sleeps and has breakfast before going to work the next day!) Home dialysis is being increasingly commonly performed, using the "Kiil" or "Multipoint" dialysers.

In *peritoneal dialysis* several litres of dialysing fluid are introduced via a catheter into the peritoneal cavity and removed a few hours later. Here, the peritoneal membranes act as the dialyser. This method is much less effective in terms of plasma clearance than haemodialysis. It also is liable to infection and the peritoneum eventually fibroses so that exchange does not occur.

The surgical technique of renal transplantation is well established. The renal vessels are anastomosed, extraperitoneally to the iliac vessels and the ureter is joined to the bladder. After the transplant, dialysis may still be required for a time. *Tissue typing* (a process similar to blood grouping) is carried out to select compatible donors. A major problem lies in the provision of a large enough number of donors, since the majority of grafts are now taken from cadavers.

A National Organ Matching and Distribution Service has been set up in Bristol, England, which keeps data on computer files of the renal failure patients waiting for a donor kidney. When one becomes available, the computer picks out the best-matched recipient. Around 500 transplants a year are carried out in the United Kingdom; the waiting list for kidneys runs at between 750 to 1000 patients.

78. Physiological Tremor

All contraction of voluntary muscle is overlaid by tremor, in the form of minute oscillations. In a recording of muscle activity tremor can be seen as a trace of fine rhythmic movements superimposed upon the record of the contraction itself. The amplitude of these fluctuations is between 1 to 2 per cent of the total movement produced by the contraction and the predominant frequency, in man, is around 10 Hz.

Various explanations have been put forward to account for physiological tremor. There were theories that the generator of tremor might be the heartbeat, or the alpha waves of the brain, or the synchronised activity of groups of motoneurones in the spinal cord, or some kind of filtering action occurring in muscle itself.

Oscillation in the stretch reflex arc

It is now generally accepted that the origin of the 8 to 12 Hz component of physiological tremor is in the feedback loop controlling voluntary muscle. This loop tends to go into oscillation as a byproduct of its controlling action.

The human body, as a necessary condition of existence in a continually changing and for the most part hostile environment, must reach and maintain internal stability. Various nervous reflexes are known to underly this self-corrective process, and have been described throughout the book in the context of the physiological systems under discussion.

The stretch reflex is one example of a feedback servomechanism which has been described in detail in various places in this book already, and we have seen how it acts to maintain the length of a muscle constant, in the face of different external loads.

Errors in control systems

Servomechanisms are often prone to overshoot the correct, required, error-free output and they give rise to trains of oscillations. Control theory, of course, predicts oscillation of this nature, in any system if it is underdamped; a common feature of most biological control systems is that they are underdamped. As far as voluntary muscle is concerned, there would appear to be a compromise here. With overdamping, the control system can achieve accuracy, but at the expense of speed of response. On the other hand, an underdamped system gives greater speed of response but at the risk of inaccuracy due to overshooting and hunting (oscillation, or tremor) in the response. It seems that most people can tolerate about 2 per cent of their muscle activity being in the form of tremor—the price to be paid for speed of response.

Experimental evidence

What evidence can be put forward to support the servoloop theory? The first clue leading to this explanation came from the work of Halliday and Redfearn nearly twenty years ago when they used Fourier analysis to produce frequency spectra of tremor in normal subjects and tabetic patients. In the former, a well-defined peak was found between 8 to 12 Hz; in the latter this peak was absent. Since *tabes dorsalis* (tertiary syphilis) is a disease process in which the dorsal roots are functionally severed, their result indicated that an intact servoloop was necessary for physiological tremor to occur.

Once it was realised that tremor had much the same nature as an oscillation in a control system, it became accessible to the kind of detailed investigation in mathematical terms that engineers employ in the study of servo mechanisms. There are three main ways of attacking the problem:

1. Opening, or interrupting the feedback loop at a suitable point.
2. Modifying the properties of the loop.
3. Introducing inputs into the loop, such as step-functions, ramps, sine waves, etc.

Opening the servoloop
If the reflex arc is cut on its afferent side, muscular activity can still occur but in the absence of feedback control. Figure 78.1

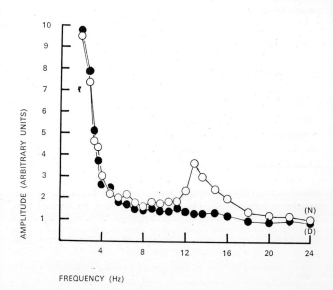

Fig. 78.1 Frequency spectra derived from tibialis anterior tension record in anaesthetized cat. Open circles (N) show record before deafferentation. Closed circles (D) are taken after deafferentation. The peak at 15 Hz is absent in the second curve, but the irregular activity remains.

shows the results of cutting the appropriate dorsal roots in an anaesthetised cat. Muscle movement, of a quasivoluntary nature was induced by electrical stimulation of the motor cortex and the limb displacement subjected to frequency analysis. This showed, in the control case, a well-defined tremor peak (around 12–15 Hz

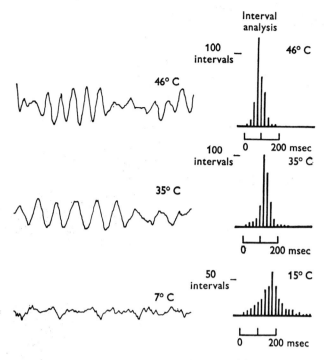

Fig. 78.2 The effect of arm temperature on frequency of finger tremor. Arm was immersed in water bath at stated temperatures. Tremor of forefinger was recorded using beam of light and phototransistor. From Lippold (1973). *Origin of the Alpha Rhythm.* Churchill–Livingstone, London.

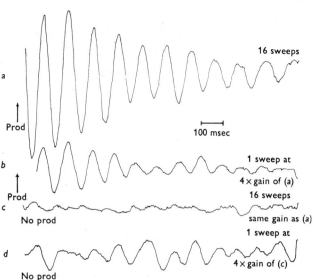

Fig. 78.3 In normal subjects the finger can be made to oscillate by itself, 5 to 20 times, by prodding it once. A brief (30 msec) stretch given 80 msec before the arrows gives a phase-locked damped train of waves of the same frequency as normal tremor. Recording method (*b*) Fig. 5.16. Trace (*a*) Average of 16 sweeps; 256 counts per address. (*b*) One sweep; 64 counts per address (i.e. at 4 × the gain of (*a*)). (*c*) Control. No prodding. Average of 16 sweeps; 256 counts per address. (*d*) Control. No prodding. 1 sweep. 64 counts per address.

in the cat) whereas after the deafferentation, this component in the analysis was abolished.

Modifying loop properties

If tremor is due to oscillation in the servoloop, altering the delay in the loop should alter tremor frequency. The delay may be altered by warming or cooling a limb, since the properties of muscle, especially its twitch-time (the major part of the delay in the loop) are changed thereby. Figure 78.2 shows that cooling slows and warming speeds up physiological tremor.

Introducing inputs to servoloop

A mechanical prodder (a moving coil driven by an oscillator or step-function generator applied to the limb) can be used to alter abruptly the position of a limb, in this case the forefinger.

It turns out that a prod of the freely extended finger, lasting about 30 ms and of around 1 mm amplitude will generate a train of five or more tremor waves (oscillations). These waves have the same frequency as tremor in a given subject, respond in the same way to factors such as cooling and warming and are accompanied by bursts of action potentials in the extensor muscle just like those found in normal physiological tremor (Fig. 78.3).

This evidence would indicate that the 8–12 Hz component of tremor is due to self-sustaining oscillations in the feedback loop.

In the interpretation of these results, we must be clear what we mean by the term physiological tremor. Tremor in voluntary movement can be generated by a number of mechanisms such as the beating of the pulse, pathological tremors (such as intention tremor of cerebellar disease, the tremor of Parkinsonism, etc.). All these other causes of tremor have a different frequency spectrum from physiological tremor and a different mechanism of production. In identifying and examining physiological tremor we are specifically concerned with the ripple, at 8 to 12 Hz, that is superimposed on the voluntary contraction of a specific muscle and arises only from this activity.

A summary of the events concerned is given in Fig. 78.4.

ABNORMAL TREMOR

1. Parkinson's disease (see also Chapter 60)

The syndrome of 5–7 Hz tremor at rest, bradykinesia* and extrapyramidal rigidity which occurs characteristically in middle-aged and elderly persons is called Parkinson's syndrome, or paralysis agitans. Many cases developed in persons who had encephalitis during the 1917–20 pandemic of influenza but they are now less often seen. Also some drugs produce a Parkinson-like syndrome. Most cases are idiopathic.

The essential pathological changes are loss of nerve cell bodies from the pigmented nuclei of the brain stem, e.g. the substantia nigra and locus ceruleus. The neurones which are left in these regions often have large spherical, laminated bodies in their cytoplasm (Lewy bodies) and neurofibrillary tangles are seen, more commonly in the post-encephalitic type.

Substantia nigra neurones are dopaminergic and their axons end in the basal ganglia, especially the putamen and globus pallidus where they are inhibitory. There is, in Parkinson's disease, a considerable reduction of dopamine here.

As treatment, the precursor of dopamine, L-3,4-dihydroxy-phenylalanine (L-Dopa) is usually given to patients with Parkinsonism, since it penetrates the blood–brain barrier and raises dopamine levels in the brain, although dopamine itself does not easily pass the barrier.

Most patients have benefit from L-Dopa therapy and

---

* The patient is slow to initiate any movement. It is difficult to start walking, for example, and once started it is often difficult to stop again.

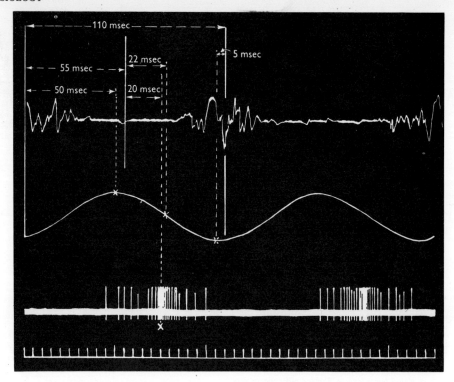

Fig. 78.4 The relation, in human calf muscle, between electrical activity (top), mechanical record (middle), and spindle activity (whose time relations are calculated from a reflex time of 35 msec). The action potential bursts are spaced at 110 msec intervals. The muscle is fully shortened at 50 msec later which equals a phase lag of 170°. At the calculated time of greatest spindle discharge it is seen that the velocity of lengthening of the muscle is at a maximum. Under these conditions, spindles therefore are sensitive to velocity. The displacement record shows an increase in muscle length as a downward deflection.

A similar diagram, during cooling, leads to the same conclusions, the interval between the bursts then being 140 msec. (From Lippold, Redfearn and Vučo, 1957.)

although the tremor is not improved markedly, the rigidity, which constitutes the most distressing facet of the complaint, usually is.

Various drugs may induce Parkinsonism, e.g. reserpine (which depletes central stores of 5-HT and dopamine). Chlorpromazine (Largactil) and haloperidol also give rise to Parkinsonism, possibly because they block catecholamine receptors in the brain stem. Acetylcholine is probably the excitatory transmitter in the corpus striatum, so that release of inhibition (as described above) causes Parkinsonian symptoms as also would anticholinergic drugs. This proves to be true and most anticholinergics aggravate Parkinson's disease. Conversely, anticholinesterase drugs and atropine have long been known to improve the symptoms.

For many years it has been observed that if a Parkinsonian has a stroke, tremor and rigidity are improved on the affected side. The surgical undercutting of the cortex and later procedures such as generating lesions in the globus pallidus and thalamus derived from this observation. Such surgery presumably interrupts some extrapyramidal pathways and removes an abnormal feedback path operating in the disease, for usually clinical benefit to the tremor and rigidity results, although regrettably the improvement is often transient and the symptoms will return in several months.

In spite of the large volume of work which has been carried out upon Parkinson's disease, the mechanism whereby the 5 Hz tremor is generated is still as obscure as it ever was. As long ago as 1929, Walshe showed that voluntary movement could be made normal by the injection of 1 per cent procaine solution into the affected muscles. The beneficial effect lasted several hours. It is now thought that this finding indicates that there is an exaggeration of fusimotor tone in the disease and when the gamma motor fibres are blocked by the local anaesthetic, this tone is cut off, the tremor and rigidity consequently disappearing. It has also been found that an arterial cuff, inflated upon a Parkinsonian limb, decreases the abnormal tremor and rigidity, sometimes for days following a period of ischaemia of not more than half an hour. Presumably a similar mechanism is responsible.

2. Huntington's chorea

This disease (which is inherited as a Mendelian dominant) has a characteristic clinical picture which involves the appearance, in the fourth or fifth decade of life, of choreic movements (irregular, rapid spontaneous limb movements). This progresses to athetoid movements (writhing, grimacing and characteristic sideways head movements). The origin of the chorea and athetosis is cortical, for they disappear if the pyramidal tracts are damaged. It has been suggested that, in this condition, an inhibitory pathway from Brodmann's area 4s to the thalamus via the caudate nucleus is blocked. This might also lead to lack of cerebellar inhibition of cortically initiated movements. Division of the dorsal roots makes the disease worse.

Dementia usually develops, associated with mood disturbances and impaired judgement. In some families the dementia appears first; in others the intellect is unimpaired but the chorea and athetosis are dominant clinical signs. Examination of the brain reveals moderate to severe parietal and frontal cortical atrophy, neurones being lost and much gliosis being present. Brain stem nuclei and the caudate nucleus are also much affected.

79. Changes in the Fetal Circulation at Birth

The fetal circulation

The placenta is the fetal lung, as far as function is concerned. At birth, respiratory function is taken over by the lungs, a changeover which means considerable alteration in the paths taken by the circulating blood.

After oxygenation in the placenta (to about 80 per cent saturated) blood travels in the umbilical vein to the liver. Part of it goes to the left side of the liver; part goes directly, via the *ductus venosus* to the inferior vena cava. The rest of the liver blood comes from the portal vein, so that when final admixture in the inferior vena cava is complete it is about 50 to 60 per cent saturated. Most of the blood passes into the left auricle via the *foramen ovale* (joins IVC and LA). The rest goes to the right ventricle, the output of which is into the aorta through the *ductus arteriosus* (joins PT and aorta). Thus the two ventricles are working in parallel. About a quarter of the output of the right ventricle goes through the fetal lungs. Figure 79.1(a) summarises the layout of the fetal circulation.

Changes at birth

At delivery of the infant, the umbilical cord is tied off by the midwife, or chewed through by the mother; in either case this results in a brief rise in arterial pressure due to the rise in total peripheral resistance. Now, the lack of oxygenated blood from the placenta leads to asphyxia of the infant which responds by gasping, coughing, heaving, then screeching which finally ends in more or less normal respiration being established.

The struggling causes a large fall in intrathoracic pressure, 75 mmHg (10 kPa) have been measured and as the lungs expand to fill this region of lowered pressure, air enters them and in addition the pulmonary vascular resistance is enormously reduced. Pulmonary blood flow will suddenly increase at this stage by five to ten times its previous amount. The direction of flow in the ductus arteriosus becomes reversed, as one would expect from the new pressure relationships between the left and right sides of the heart. After about 48 h the ductus arteriosus closes and is only found as a small tough fibrous cord in the child. The foramen ovale closes more or less at birth, because it is constructed as a valve and is open in fetal life simply because pressure is higher in the inferior vena cava. When the increase in pulmonary blood flow occurs, this pressure gradient is reversed, the valve shuts and later is fused with the atrial wall by fibrous tissue. Figure 79.1(b) illustrates the changed circulation.

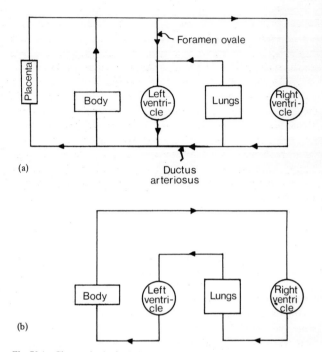

Fig. 79.1 Changes in the fetal circulation at birth. Top diagram shows the path that is taken by blood in the fetus. Bottom diagram shows adult system. At birth, the placenta is cut off and the foramen ovale and ductus arteriosus close up.

FETAL BLOOD

In the early fetus, all its blood contains haemoglobin of the so-called fetal type. Adult blood gradually appears during intrauterine life, accounting for about 10 per cent of the total at the 25th week and 20 per cent at birth. At age six months, an infant's blood is almost entirely of the adult type.

Fetal haemoglobin (distinguished spectrographically and electrophoretically) has an unusual oxygen dissociation curve. Its curve is considerably shifted to the left, as compared with normal (of course maternal) blood. Because of this, fetal blood can take up larger volumes of oxygen at low partial pressures of oxygen than does adult blood (at 20 mm Po_2, fetal haemoglobin is 70 per cent saturated; adult blood would be only 20 per cent saturated). This state of affairs enables the fetus to extract oxygen from the mother's blood in the placenta much more easily than would otherwise be the case.

80. Pain

Pain is a common experience yet physiologists, psychologists and clinicians who are involved with the problems of pain find it very difficult to give a satisfactory definition. Pain and pleasure are the main driving forces of animal behaviour and in spite of the great deal of attention psychologists have paid to the phenomena of motivation, we still know comparatively little about pain. It used to be thought that pleasure attended the stimulation of "beniceptors" and pain that of "nociceptors", both being types of sensory receptor. We now know more about the neurophysiology and psychology of pain and the recent "gate theory" seems to promise a new approach to treatment of patients with intractable pain even though as a theory it is almost certainly wrong.

The psychology of pain

There are people born without the ability to feel pain, a rare condition but one which illustrates the physiological importance of pain. These people suffer extensive burns, tissue injuries of all kinds and often bite into their own tongues whilst eating. They usually have severe pathological changes in the spine, bones and joints and usually die at an early age from extensive infections.

At the opposite end of the scale are persons who feel severe pain in the absence of any adequate painful stimulus. This condition may follow injury or amputation, where the pain is felt in the phantom limb, or it may be purely spontaneous and the patient has severe generalised pain sensations of no obvious origin.

At the present time we have no good explanation for the mechanisms subserving pain although, on the face of it, a simple method for investigating the problem would seem to be the application of noxious stimuli to a peripheral structure and the following of the pain impulses through the nervous pathways in the spinal cord and up to the brain.

However, the mechanisms involved are more complicated than this because it is generally not possible to stimulate pain receptors on their own. When the experimenter tries to activate pain receptors he must use a strong stimulus which will also activate receptors mediating other sensations such as touch, pressure, warmth or cold. In addition the noxious stimulus will activate many other mechanisms such as changes in sweating and skin blood flow. We still do not know whether the message signalling pain is one involving just the impulses in a specific pain pathway or whether the brain makes use of the whole pattern of afferent activation which is associated with noxious stimulation.

The fact that pain has a clear biological value might lead us to suppose that the pain level produced by tissue damage would be in proportion to the amount of that damage. On the whole this is not true and the severity of pain felt in any given circumstances appears to depend strongly on other factors such as previous experience, and the emotional and cultural significance of all the events involved at the time. Pain is a very variable phenomenon and differs widely between persons and in differing cultures. Eric Hiscock,* visiting Fiji in *Wanderer III* wrote:

"In August, the Hindus were to perform their annual fire-walking ceremony.... For the past two days a great fire of timber has been kept burning in that place, and the heat from the ash, which now and again glowed evilly, almost white, was terrific. When a cigarette carton was thrown on to it by an onlooker it did not burn in the ordinary way; a small jet of flame shot up as though from celluloid and the carton vanished instantly.

Some of the men were bare from the waist up, except for a yellow wreath around the neck, and faces, chests and arms had dabs of yellow on them; others wore shirts. The majority were dreadfully disfigured. Through the softest and tenderest parts of their flesh, in throat and breast and back, were thrust skewers and steel knitting needles, and some—most sickening sight—had such things driven right through their faces, entering one cheek and emerging from the other. One, an elderly man, had a toasting fork through his face, the three prongs protruding from his right cheek and the handle from his left; through his open hanging mouth I could see the steel shaft passing from one cheek to the other....

Suddenly one of the fire-walkers stepped off the grass on to the embers and walked right through the pit; he moved fast. Others followed singly or in pairs. Some walked at a normal pace, others plodded slowly, scuffling in the ashes from which wicked little tongues of pale flame licked, while yet others actually danced their way across. The only woman remained a long time on the ashes, swaying from side to side."

Psychophysics of pain

The variability in perception of pain we have already mentioned has, in the past, been thought to derive from different "pain thresholds" in different persons. The evidence indicates that all people have a very similar threshold to sensation; an electric shock, that will just give a detectable sensation, is of the same voltage for all people. This finding is true in the laboratory but may not be so in real life; for example in the heat of battle a soldier may receive a severe injury without being aware of it at the time it happened. Pain tolerance is very variable, even under experimental conditions in the laboratory and appears to depend on ethnic factors as well as many others.

Pavlov showed that the normal response of an experimental dog to an electric shock (violent attempts to escape) could be abolished by giving food to the animal immediately after each shock. The dog then wagged its tail, salivated and showed no outward signs of distress.

In laboratory experiments on humans, it can be shown that distraction† lessens the severity of pain reported during elec-

---

* Eric C. Hiscock (1956): *Around the World in Wanderer III*. London, Oxford University Press.

† The complexity of pain perception is illustrated by the finding that strong auditory stimulation by white noise will, in some patients, sufficiently distract them to enable the dentist to perform tooth extraction without the need to give any anaesthetic. But this technique does not work in the majority of patients.

trical stimulation, or on the other hand the insertion of the word "pain" into the instructions given to the subject would often considerably lower the level of electrical stimulation that was required to evoke pain as a sensation.

PHYSIOLOGY OF PAIN

At one time it used to be thought that overstimulation of any type of sensory endings gave rise to pain. Then the pendulum swung the other way and it was thought that specific pain endings were entirely responsible for initiating the sensations of pain. It now appears that the truth lies somewhere between these two extremes.

Receptors commonly respond to more than one modality of stimulation, for example muscle spindles respond to stretching of the muscle and to cooling of it. Usually the response to one modality is at a low threshold, whilst to other modalities the receptor is much less sensitive. It seems as if the message about the sensory input is coded as a pattern of impulses in a multiplicity of pathways. Pain is probably mediated by various pathways rather than by a simple system of receptors activating a single route for impulses travelling to the brain.

Pain receptors

Lynn (1975)* puts pain receptors into four classes:

1. High threshold mechanoreceptors with axons falling within the A-delta group.
2. High threshold mechanoreceptors with unmyelinated axons.
3. Polymodal nociceptors with unmyelinated axons, the most numerous group in limb nerves.
4. Units, sensitive to both heat and mechanical stimulation, having thinly myelinated axons.

In skin, only these specialised types of nociceptor have the necessary properties to transmit information about painful stimuli. None of them appears to fire spontaneously in the absence of tissue damage.

Synaptic connections of nociceptor afferents

Although the primary afferent inflow to the spinal cord is fairly well understood, the way in which its synaptic connections are organised is not.

Two types of cell have been found in the spinal cord of experimental animals (usually cats) which are activated by nociceptive stimulation of the skin. These are (1) "specific" units which only respond to noxious stimuli and (2) units which respond to noxious as well as other stimuli ("wide dynamic" range or "common carrier" units).

The specific units were first characterised in lamina 1 of Rexed and responded only when A-delta fibres in the dorsal root were stimulated; they are probably connected therefore only to the high threshold mechanoreceptors (of group 1 above). Specific cells have now also been found elsewhere in the spinal cord.

The units with a wide response range are mostly found in lamina 5 of the dorsal horn and seem to be activated by both A and C fibres of cutaneous nerves. These cells are greatly affected by other parameters such as the state of the experimental animal, descending impulses, or anaesthetic depth; they may even be inhibited by afferent impulses in the largest A fibres (i.e. the equivalent of light mechanical stimulation). The gate theory of pain (Melzack and Wall, 1965) assigns a key role for these lamina 5 cells.

Projections to higher centres

The precise pathways subserving pain are not fully known. The fibres run in the spinoreticular and spinothalamic pathways which seem to have a fairly widespread origin in cells throughout the spinal cord. In monkeys a fairly large proportion of the fibres come from laminae 1 and 5. There is no evidence that the primary afferent collaterals running up in the dorsal columns carry information about pain. This contradicts the finding that cats with bilateral division of the dorsal columns do not react to painful stimuli, but the answer may lie in the fact that there are some post-synaptic fibres also running in the dorsal columns and they may originate in the lamina 5 cells mentioned above.

Brain mechanisms

It has been widely supposed that the fibres subserving pain sensation end in a "pain-centre" in the brain. Yet this concept cannot explain much of the contradictory findings about the sensations of pain unless one calls the brain in its entirety a pain centre. The thalamus, hypothalamus, brain stem, limbic system and cerebral cortex are all involved in the perception of pain. Illustrative of the difficulties are the facts that a tumour of the thalamus may give rise to severe, continuous intractable pain—the thalamic syndrome. Lesions of the frontal cortex, on the other hand, abolish the unpleasant aspects of severe pain and the patient rarely complains about the pain although if asked he will usually admit that the operation has not abolished it. He is still aware of the sensory component of pain but it has no unpleasant effect associated with it.

THEORIES FOR PAIN

The specific theory

The usual theory put forward to explain pain is the so-called "Specific Theory" of von Frey. He supposed that specific pain receptors in the tissues send impulses along pain fibres to a pain centre in the brain. The receptors were identified as free nerve endings connected to A and C fibres which travelled in the anterolateral spinothalamic tract to the thalamus (where pain sensation was reputed to be felt).

The reasons for not accepting this theory are as follows:

1. The C fibres and their free nerve endings are stimulated by many kinds of innocuous stimuli as well as by those which might be presumed to give pain.
2. There are very few sensory fibres which are activated solely by noxious stimuli.
3. Various operations have been carried out to relieve intractable pain, the objective being to divide the "pain pathways" as envisaged above. The most useful of these is cordotomy in which the anterolateral part of the spinal cord is severed. This produces a loss of pain sensitivity on the side opposite the lesion, but often such relief is only short-lived (months). The pathways concerned with pain are evidently variable.

The pattern theory

This theory states that patterns of ascending impulses are set up by any intense stimulation of all kinds of receptors. In terms of pain, nerve endings and their fibres are non-specific.

---

* Lynn, B. (1975) Physiology of the activation of pain. *Prog. Neurol. Surg.* **7**.

As already stated, however, there is very good evidence that specialised endings and their corresponding fibres exist for signalling the various sensory modalities, including pain. The cornea, for example, has only free nerve endings present when investigated histologically; the only sensation derived from it is that of pain, although there is now dispute about this finding.

The gate control theory of pain

Melzack and Wall (1965) have put forward a new theory. There are three input systems to the spinal cord:

1. Substantia gelatinosa (SG) cells in the spinal cord dorsal horn.
2. Dorsal column fibres.
3. The first central transmission (T) cells. These are also in the dorsal horn.

The SG cells are a "gate-control" system. They control the afferent inflow to the brain, and prior to the impulses reaching the T cells.

The afferent patterns, when they get through the gate and pass up to the brain, initiate further descending impulses which modify the activity of the gate.

The T cells are responsible for the transmission of patterns of impulses to the brain which give the sensation of pain.

A burst of impulses in large afferent fibres produces a burst of firing in T cells which is immediately inhibited. Simultaneously SG cells are activated.

A similar burst in small afferent fibres gives rise to a prolonged response of T cells whilst SG cells are inhibited. Because SG cells have inhibitory connections with afferent nerve terminals the effects of both large and small fibre inputs to the T cells is cut off by negative feedback when the former are active. Small fibres will stimulate T cells and hence SG cells are inhibited. T cells will remain active. Thus the gate is opened by small fibre inputs; is closed by inputs involving large fibres.

If descending activity in the spinal cord is blocked by cooling it, the nociceptive input to the brain can be shown to increase greatly. This experiment shows that the brain itself initiates nervous activity controlling the gate.

When the output of T cells becomes large enough, pain is felt, or at least this is the presumption—there are no direct tests. The degree of afferent activation required to produce a sensation of pain depends on a number of factors:

(a) The steady basal discharge level in the afferent system.
(b) The proportion of large and small fibres activated by the noxious stimulus.
(c) The degree of activation of the descending pathways from the brain.

This gating process probably occurs at several levels in the pathway from periphery to the brain. The substantia gelatinosa (laminae 2 and 3) is the most likely site of the spinal gating mechanism. The properties of cells in lamina 5 suggest that they are the T cells.

Implications of the gate control theory

When introduced in 1965 the gate control theory of pain was hailed as a breakthrough in this field of physiology. However, a number of discrepant observations soon became obvious and it is not now generally believed that the small and large fibre inputs behave in this manner. Having said this, it is nevertheless clear that a considerable peripheral modification of pain sensation can occur at the spinal level. It is likely that the major part of this originates in downcoming activity in the spinal cord which controls the input from spinal to cortical and thalamic levels, just like any other sensory modality. Pain mechanisms are not exceptional.

Testing the theory in man

Nathan and Rudge (1974)* tested the theory in experimental subjects. If the large fibres of peripheral nerves are stimulated through the skin, the pain induced by C-fibre activation should cease or be reduced, according to the theory. They induced C-fibre pain in their subjects by pressure, repeated pinprick, cold and heat in the upper limb. The limb had previously been kept ischaemic with an arterial cuff for 25 min to ensure that only C fibres were indeed stimulated by the procedures. When this pain was at its height, the relevant peripheral large fibres were stimulated through the skin. The subjects noticed no changes in the quantity nor the quality of the pain.

Other experiments have involved the pain threshold to a radiant heat stimulus before and during an electrical stimulus. The electrical stimulation caused no change in threshold, either if all the peripheral fibres were conducting or if the large fibres were blocked. These results have been taken by most physiologists to indicate that the gate theory is wrong.

Pathological conditions

In certain neuropathies there is a selective removal of large myelinated fibres in the peripheral nerves; others are characterised by the opposite—removal of small fibres. The gate control theory predicts in the former a state in which all stimuli would cause pain; in the latter a state in which pain could not be produced. Clinical evidence does not show this and one must conclude that the theory is untenable.

However, in spite of the gate control theory now being widely regarded as erroneous, the ideas involved have stimulated much research which otherwise would not have been carried out. Moreover, it still provides a useful framework within which to consider the phenomena of pain as seen clinically.

An excellent review of this topic is by Nathan (1976).†

ACUPUNCTURE ANALGESIA

Acupuncture is the Chinese technique of inserting long fine needles into special regions of the body for curative purposes. Recently it has been used to produce analgesia strong enough to enable surgical operations to be performed, such as abdominal and chest surgery. Opinion is divided at present concerning its mode of operation. Many consider it to be merely a form of hypnosis (or suggestion) and of little real value. It might be explained in terms of the central biasing mechanism. The stimulation of particular nerve endings could initiate an input to the biasing system which would then close the appropriate gates.

Acupuncture is carried out by twiddling the needles continuously or by passing an electric current between them, both processes which should give rise to a fairly strong sensory, and probably painful, stimulus. At the moment we must keep an open mind on this subject.

---

* Nathan, P. W. and Rudge, P. (1974) Testing the gate-control theory of pain in man. *J. Neurol., Neurosurg. Psychiat.* **37**, 1366–1372.

† Nathan, P. W. (1976) The gate control theory of pain. A critical review. *Brain* **99**, 123–158.

CHEMICAL MEDIATION OF PAIN

Many substances are known to cause pain. It is also possible that one of the steps in the transducer process at receptors involves a chemical transmitter. In long-lasting pain, such as that produced by burning or physical damage, it is likely that a chemical substance is liberated by the damaged tissues and mediates the initiation of nerve activity.

The production of pain using chemical substances has been studied by forming little blisters on the skin and using the exposed base for testing. Nerve fibres in the epidermis will be destroyed but deeper ones are intact and exposed. The following produce pain: $10\,\mu g\,ml^{-1}$ of acetylcholine, $1\,g\,l^{-1}$ KCl, HCl and lactic acid at pH less than 3, hypotonic and hypertonic NaCl. Histamine in low concentrations ($1\,\mu g\,ml^{-1}$) gives itching; in higher concentrations ($100\,\mu g\,ml^{-1}$) it gives pain.

In blood there are two substances which will give pain: (a) 5-hydroxytryptamine (5-HT), which occurs in platelets; (b) plasma kinins (bradykinin), which are polypeptides formed by proteolytic enzymes such as trypsin, plasmin or by contact between plasma and foreign surfaces.

The pain of tissue damage may be due to the release of histamine from mast cells or 5-HT from disintegrating platelets.

Hyperalgesia

Hyperalgesia is the term denoting an increased sensitivity to noxious stimulation; a typical example is the phenomenon in sunburn where light touch on the affected skin generates pain. It is primary when the damaged area itself coincides with the hyperalgesic area; it is secondary when at a distance.

Lewis (1936) considered hyperalgesia to be the consequence of liberation of P substance in the damaged area, to account for primary effects. Secondary hyperalgesia he thought due to diffusion of the P substance into surrounding areas.

DEEP PAIN

When pain endings in structures below skin level are stimulated, the effect can be described as an aching pain. Headache, toothache or stomach ache are examples. Thermal stimulation of the skin, if strong enough (warming to 50°C or cooling below 10°C), evokes this aching pain. A feature which distinguishes it from pricking (superficial) pain is its tendency to spread. If a finger is placed in ice water, the painful sensation eventually spreads to the other fingers and some way up the arm.

Headache

The usual type of headache is not due to any events in the brain substance, which is quite insensitive to pain. It is due to tension put upon the meninges and perhaps the blood vessels in them. Variations in intracranial pressure alter the tension on the meninges and give rise to pain. Headache accompanies lumbar puncture if the procedure alters CSF pressure and the pain, due to either raised or lowered intracranial pressure can be relieved by restoring the pressure to normal. The headaches of space-occupying lesions within the skull (astrocytoma, meningioma, etc.) are due to raised intracranial pressure. It should also be noted that many extracranial sources will give rise to pain felt in the head such as sinusitis and spasm of various muscles in the head and neck region.

Ischaemic muscle pain

Pain develops in a limb after its blood supply is completely occluded and its muscles are contracted. A hand-gripping contraction, made once per second, gives pain after about 30 seconds which becomes unbearable after about a minute. It is a deep pain and is diffusely felt throughout the arm.

The pain is directly related to activity in the muscle, for if the grips are made at 2 s intervals it takes twice as long to come on. If the cuff is maintained above arterial pressure the pain remains of the same severity even though all muscle action is stopped. Immediately the cuff is released, the pain disappears. Lewis suggested that muscular activity releases a pain-producing (P) substance which diffuses into the tissue spaces and is normally removed by the bloodstream. It accumulates during ischaemic exercise and when a threshold concentration is exceeded, gives rise to ischaemic pain.

Patients with narrowed blood vessels in the leg (Buerger's disease) often have pain in the legs during walking, usually relieved by standing still. The blood supply to the leg muscles is only just adequate for their resting needs, and with exercise P substance accumulates and ischaemic pain results.

The pain of coronary insufficiency results in similar fashion, giving rise to the typical history of cardiac pain arising with exertion and passing off with rest. In an acute infarct of coronary muscle, pain arises in the same way from the ischaemic muscle but in addition 5-HT will be released from platelets and potassium from the damaged muscle. Plasma kinins are also probably released. The pain wears off only when the damaged muscle is replaced by fibrous tissue and the nerve endings are dead.

Referred pain

Referred pain from viscera and deep structures is felt in addition to, or in the absence of, pain from the region concerned. It is felt in an area remote from that which is being stimulated but usually supplied by the same, or an adjacent, segment of the spinal cord.

Patients with heart disease frequently develop referred pain in the shoulder and root of the neck. They usually show a pattern of "trigger areas" in the chest and shoulder. Pressure in these commonly gives an intense pain which may last for hours. Patients without heart disease have similar trigger areas, which also give discomfort when pressed (but not severe or long-lasting pain). These patterns of trigger areas are consistent from person to person and enable accurate diagnosis of disease to be carried out on the basis of their distribution. The injection of local anaesthetic into the trigger areas abolishes the referred pain and quite often the pain from the diseased organ as well. Sometimes the pain disappears permanently. A treatment for the disabling pain of angina pectoris which spreads down the left arm causing the patient to keep it immobile, is to infiltrate the appropriate trigger areas with novocaine. The relief from the pain enables the patient to use his arm normally for quite long periods of time.

It is assumed that these trigger zones produce a continuous input to the CNS and this summates with the input from the diseased organ to give pain in both areas.

One mechanism is that afferent pain activity spreads to nearby areas. T cells have a restricted receptive field. They are, in addition, marginally affected by inputs from a larger area. Normally, this diffuse input is inhibited by the gate control, but may trigger T-cell firing if the input is at a high enough level or if too few large fibres are active to close the gate. Anaesthesia acts by blocking the spontaneous impulses. Support is given to this idea by the finding that small visceral afferents project to lamina 5 cells. The spread of referred pain to distant regions implies that the gate can be opened by nervous activity coming from anatomical areas remote from the normal nerve supply to the gate. Pressure on the back of the head may, in some patients, trigger off pains in their phantom legs.

LOCALISATION OF PAIN

To be able to locate the source of pain is biologically important if this leads to positive survival action. It is also important in the diagnosis of disease in clinical practice.

Thus pain is accurately localised when it arises in superficial structures such as the skin, less accurately from bones and joints and pain from viscera usually cannot be localised at all. Skin is richly innervated, particularly on the palmar surfaces of the fingers and thumb; less richly on the back. Accuracy of localisation varies accordingly. The main form of localisation is in terms of the particular spinal segment involved; finer degrees of accuracy are superimposed upon this. The subjective analyses of the localisation of pain are the function of the cerebral cortex.

RELIEF OF PAIN

Drugs have been used in the relief of pain for many centuries, but their mode of action remains obscure. Anaesthetics work by abolishing nerve conduction. Local anaesthesia does so at the level of peripheral nerves; cocaine, procaine and lignocaine block conduction through neurones by raising their threshold without depolarising them (resting potential is unaltered; amplitude of action potentials are reduced or they are abolished; conduction rate is slowed).

It has been suggested that these drugs act by diminishing the ionic permeability of the permeability barrier round the neurone and in the membranes of the Schwann cells. They interact with the polar groups of the lipoprotein molecules. 0·005 per cent cocaine solutions abolish action potential generation at the nodes of Ranvier.

Spinal anaesthesia works the same way, the local anaesthetic agent being introduced into the spinal CSF. General anaesthetics mostly work by blocking central neuronal activity by gradual depolarisation.

Analgesics are pain-relieving drugs such as morphine, codeine and thebaine which are derivatives of phenanthrene, or drugs like papaverine, narcotine, etc. which are derivatives of isoquinoline.

The action of these substances upon the CNS is curiously irregular, since it depresses some centres and stimulates others. A dose of 10 mg of morphine in man produces sleepiness and removes all painful and unpleasant sensations. Such doses do not depress higher brain centres, although the power of continuous concentration seems to be impaired. The touch threshold is unaltered yet pain thresholds are greatly raised by morphine. It has a selective action on medullary centres in that respiratory centres and cough are depressed whilst vagal centres, especially the vomiting centre, are stimulated.

Morphine receptors have been found in the brainstem and the natural transmitter is thought to be a polypeptide called enkephalin, which may be a fragment of a larger peptide.

Lesions of the central tegmental tracts in the cat produce hyperalgesia; electrical stimulation of the tract and adjacent structures produce analgesia. This may be profound and during such stimulation major surgery can be performed without any other kind of anaesthesia. This analgesia may last five minutes longer than the period of stimulation. The neurones are part of the system that is selectively activated by morphine as detailed above. Morphine may act by exciting fibres that have an inhibitory influence over the somatic input via the gates. It is also known that electrical stimulation of this area results in a widespread release in the cortex of gamma-aminobutyric acid which is believed to be the inhibitory transmitter in the brain. This system of neurones is probably the central biasing system already referred to.

Endorphins are naturally-occurring morphine-like compounds which have similar central effects on pain, presumably through transmitter action on certain mid-brain neuronal systems.

Epilogue

The functions of the various organs in the body have been described in the preceding chapters with rather special emphasis on the properties of the organs or physiological systems concerned. In living animals and man, however, an organ or tissue rarely increases or decreases its activity without affecting the activities of many other organs or tissues. These are all controlled by the nervous and endocrine systems so as to promote, as far as possible, the well-being and stability of the animal as a whole and the preservation of the species.

To a large extent this control is directed towards maintaining at steady values the volume, chemical composition (content of oxygen, carbon dioxide, glucose, salts, pH, etc.) and temperature of the body fluids (the "internal environment"), as already described in several chapters; any or all of these are likely to be disturbed by change of activity in one part of the body and restored by compensatory changes in other parts. It was this kind of regulatory process to which the term "physiological homeostasis" was originally given. The concept, however, may be extended to many aspects of our relations with the external world; for example, maintenance of the characteristic posture and orientation with respect to gravity. These, of course, are not necessarily constant. We are capable of movement and of adjusting our behaviour according to changes in the environment, and in general this involves coordinated actions directed towards maintaining continued existence in the "normal" state. If we are threatened by some object in the external world, our muscles are activated for self-preservation by taking evasive or hostile action as instinct, experience and instruction dictate; the consequent disturbance of the internal environment brings into play all the necessary restorative processes. These will include not only those involving, for example, the cardiovascular and respiratory systems, but also the acquisition, intake, digestion and absorption of food and water to replace the stores used up.

The first step in the maintenance of homeostasis is the detection by appropriate receptors of small changes in the various states and conditions of the external and internal environments. The messages, or "signals", which they originate pass to various parts of the nervous and endocrine systems where they are modified and coordinated and sent out again (reflected) to various effectors, e.g. muscles and secreting glands, which can change the position and movement of the body and the physical and chemical states of the internal environment. The changes that result affect the receptors which originated the signals, that is, there is feedback from the effectors to the receptors as well as reflex control of the effectors by the receptors. It is an important feature of such a system that the feedback should be "negative", so arranged that any change which disturbs the steady state induces a response which acts so as to reduce the disturbance and restore the initial or "standard" conditions. The actions of the effectors then cease, or become stabilised, only when the receptors cease to originate signals, or originate some standard pattern of signal; they are continuously monitored and adjusted so as to conform to a coordinated pattern. Movement of a leg, for example, involves much more than the excitation and contraction of certain leg muscles; the nervous system "thinks" in terms of coordinated movements so that there is an appropriate and varying excitation of some muscles and inhibition of others, monitored by signals from many kinds of receptor. Likewise, secretion of the trophic hormones of the anterior pituitary gland is reduced or suppressed by the hormones which they cause to be secreted by the target organs; this negative feedback, from target organ to anterior pituitary gland, keeps the amount of these hormones in the body relatively constant. Homeostasis is maintained, therefore, by the action of systems which are essentially the same as those known to engineers as "servosystems".

Imagine a man hunting or reaping for his food supplies; or, a more likely occupation for our readers, imagine him in some active game or sport. His eyes are directed so that the images of the object aimed at lie on the foveae—a good example of "automatic following" by means of a servosystem. His arm muscles are activated so as to direct, say, the arrow, reaping-hook or implement of sport until the pattern of signals from his eyes reaches the desired standard condition such that he will hit the prey, cut the corn or drive the ball. His leg muscles may be activated so as to produce the movements of running. There is thus a continuous adjustment of contractions of opposing muscles, while balance and posture are maintained by the excitation and inhibition of most of the other muscles in the body; all being controlled by combining information derived from the eyes, the receptor organs for balance (semicircular canals, etc.) and receptors in the skin, joints and muscles themselves. Unless this control is smooth and accurate, brought about by well coordinated activity of many parts of the nervous system, energy will be wasted in checking needless movements and in failure to achieve the goal desired.

The active muscles use oxygen and glucose and produce carbon dioxide and heat more rapidly than they did when at rest. The flow of blood through them is increased and, through the action of the vasomotor control system, that through the temporarily unimportant abdominal organs is decreased in compensation. The temperature regulating centres ensure that the flow through the skin is also increased so that the extra heat generated is dissipated. By the action of the "muscle pump" and the secretion of adrenaline, the output of the heart and flow of blood through the lungs are increased; more oxygen is removed from the lungs and more carbon dioxide delivered to them. The increased amount of carbon dioxide in the body, the decreased amount of oxygen, and other associated departures from the "standard" state, excite the respiratory centres; more air is breathed in and out of the lungs, extra oxygen is supplied and more carbon dioxide carried away. The fall in blood sugar concentration leads to a release of glucose from the glycogen stores of the liver. Water is lost from the body in the dissipation of heat, particularly if the exercise is severe enough to cause sweating; excitation of the osmo-receptors results in secretion of antidiuretic hormone by the posterior pituitary gland and the kidneys excrete as little water as possible until the loss is made good by drinking. All these homeostatic processes, moreover, are adjusted continuously, mainly through the autonomic nervous system and secretion of hormones, as conditions change throughout the period of exercise and subsequent recovery. Nearly every organ of the body, therefore, is affected by any form of severe exercise; and, indeed, the exploration of the consequences of exercise on various organs has been one of the most fertile fields of experimental inquiry in physiology.

Appendix I: Units

UNITS OF QUANTITY

The unit of quantity is the mole (mol), and is the number of atoms in 12g of carbon (6×10^{23}). The atomic weight of any element contains this number of atoms, and the molecular weight of any substance contains this number of molecules. It is important to specify the particles concerned (atoms, ions, molecules); for example one mole of NaCl molecules will give rise to two moles of ions when in solution. The millimole (mmol) is 6×10^{20} particles, the micromole (μmol) is 6×10^{17} particles.

The unit of volume is the cubic decimetre (dm^3), equivalent to one litre (l). Concentrations are expressed as moles per litre (mol l^{-1}), millimoles per litre (mmol l^{-1}) or micromoles per litre (μmol l^{-1}) and are referred to as molar (M), millimolar (mM) or micromolar (μM) in strength respectively. The molar system can only be used if the molecular (or ionic) weight of the substance concerned is accurately known. Proteins etc. have molecular weights not yet fully worked out and solutions of these are specified as grams per unit volume, usually grams per litre (g l^{-1}) or sometimes grams per decilitre (g dl^{-1}).

PRESSURE

The SI unit of pressure is the pascal (Pa) which is one newton per square metre. In medicine, the kilopascal (kPa) is used for gas tensions and as the unit of pressure of body fluids etc. One kilopascal is 7·5 mm of mercury. Since blood pressure is normally measured using a column of mercury it appears at the moment (1978) that blood pressure will be expressed in mmHg for a considerable time to come. Gas tensions however will be expressed in kPa. Barometric pressure (BP) is normally around 750 mmHg which is approximately 100 kPa. This has the advantage over mmHg as a unit because partial pressures of gases in kPa at sea level are equal to their percentage composition in a gas mixture.

HEAT AND ENERGY

The unit of heat and energy is the joule, which is 1 newton-metre (Nm) or 1 watt-second. One calorie is 4·2 kilojoules (kJ).

CONVERSION TABLES:

PRESSURE

mmHg	kPa
1	0·13
10	1·33
20	2·67
30	4·00
40	5·33
50	6·67
60	8·00
70	9·33
80	10·67
90	12·00
100	13·33
110	14·67
120	16·00
130	17·33
140	18·67
150	20·00

ENERGY

Cal	kJ
1	4·19 joules
100	0·42 kilojoules
200	0·84
300	1·26
400	1·67
500	2·09
600	2·51
700	2·93
800	3·35
900	3·77
1000	4·19
2000	8·37
3000	12·56

MOLECULAR WEIGHTS

Hydrogen	2·0
Nitrogen	28·0
Oxygen	32·0
Water	18·0
Carbon dioxide	44·0
Sodium chloride	58·5
Sodium bicarbonate	84·0

CONCENTRATIONS

1 millimole per litre	milligrams per litre
Na^+	23.0
NaCl	58.5
K^+	39.1
$NaHCO_3$	84.0
KCl	74.6
$KHCO_3$	100.1
Cl^-	35.5
Ca^{2+}	40.1
Mg^{2+}	24.3
HCO_3^-	61.0

NORMAL VALUES IN BODY FLUIDS

Blood

Alkali reserve	50 to 70 ml CO_2/dl plasma
Calcium (serum)	
non-diffusible	4 to 5 mg/dl
diffusible	5 to 6.5 mg/dl
	(= 2.5 to 3.25 mmol/litre)
Bilirubin (serum)	0.2 to 0.9 mg/dl
Mean corpuscular haemoglobin	= $\dfrac{\text{Hb in g l}^{-1}\text{ blood}}{\text{RBCs in millions/mm}^3}$
	= 29 ± 2 pg
Cholesterol	
esterified (plasma)	130 mg/dl
esterified (cells)	10 mg/dl
Corpuscular volume (PCV; haematocrit)	
males	40 to 55%
females	35 to 45%
Mean corpuscular volume (MCV)	= $\dfrac{\text{corp. vol ml/l blood}}{\text{RBCs in millions/mm}^3}$
	= $87\,\mu m^3$
Mean corpuscular haemoglobin concentration (MCHC)	= $\dfrac{\text{Hb in g/litre blood}}{\text{PCV}}$
	= $34 \pm 2\%$
Plasma electrolytes	
calcium	2.4–2.9 mmol/litre
potassium	3.9–5.0 mmol/litre
chloride	100–105 mmol/litre
sodium	135–150 mmol/litre
Haemoglobin	
at birth	19.5 ± 5.0 g/dl blood
males	16.0 ± 2.0 g/dl
females	14.0 ± 2.0 g/dl
pH	7.3 to 7.5
Nitrogen (non-protein in plasma)	30 to 45 mg/dl
Phosphorus (as P) inorganic	2.5 to 3.5 mg/dl
Proteins in plasma	
albumin	3.3 to 5.7 g/dl
globulin	1.5 to 3.0 g/dl
fibrinogen	0.2 to 0.5 g/dl
total	5.5 to 7.6 g/dl
Red blood cells	
males	5.4 ± 0.8 millions/mm^3
females	4.8 ± 0.6 millions/mm^3
Plasma folate	5 to 20 ng/dl
Plasma iron	50 to 175 μg/dl
PBI (protein-bound iodine)	4 to 8 μg/dl
White blood cells/mm^3	5,000 to 10,000
neutrophils	50 to 60%
eosinophils	1 to 3%
basophils	0 to 1%
lymphocytes	25 to 35%
monocytes	3 to 10%
Platelets	150,000 to 500,000
Glucose (fasting)	60 to 120 mg/dl
Urea	25 to 50 mg/dl
Blood Volume (per kg body wt)	
males	75 ml/kg
females	65 ml/kg

Cerebrospinal Fluid

Lymphocytes	0 to 10 mm^{-3}
Calcium	2.2 to 2.7 mmol/litre
Chloride	120 to 130 mmol/litre
Potassium	2.4 to 3.2 mmol/litre
Sodium	140 to 150 mmol/litre
Glucose	50 to 100 mg/dl
Protein	10 to 50 mg/dl

Urine

Electrolytes	
calcium	2 to 20 mmol/24h
chloride	50 to 250 mmol/24h
potassium	50 to 100 mmol/24h
sodium	40 to 350 mmol/24h
17–ketosteroids	
males	10 to 30 mg/24h
females	5 to 15 mg/24h
Reaction (pH)	4.5 to 7.5
Volume	5 to 30 dl/24h
Urea	10 to 30 g/24h
Urea clearance	50 to 100 ml blood/min

Faeces

Total fat	10 to 20% w/v
Free fatty acid	7.5 to 15% w/v
Neutral fat	2.5 to 5% w/v

Gastric secretion

Fasting volume	10 to 60 ml/h
pH	0.9 to 1.5
basal concentration	25 mmol/litre
max histamine stim.	50 mmol/hr

Appendix II: Formulae

HENDERSON-HASSELBALCH EQUATION.

The dissociation or equilibrium constant $K_{H_2CO_3}$ of H_2CO_3 is given by

$$K_{H_2CO_3} = \frac{[H^+][HCO_3^-]}{[H_2CO_3]} \quad (1)$$

In practice the carbonic acid (H_2CO_3) is in equilibrium with the dissolved CO_2 of the body fluids. Thus the equilibrium constant is changed to take into account the following equation:

$$CO_2 + H_2O \rightleftharpoons H_2CO_3$$

and the new equilibrium constant K' is given by

$$K' = \frac{[H^+][HCO_3^-]}{[CO_2 \text{ in solution}]} \text{ or } \frac{[H^+][HCO_3^-]}{0 \cdot 03 P_{CO_2}}$$

This equation is transformed into the Henderson-Hasselbalch equation by taking the logarithm of both sides.

$$\log K' = \log H^+ + \log \frac{[HCO_3^-]}{0 \cdot 03 P_{CO_2}} \quad (2)$$

$$\text{or } pH = pK' + \log \frac{[HCO_3^-]}{0 \cdot 03 P_{CO_2}} \quad (3)$$

In normal plasma, $pK' = 6 \cdot 1$; $[HCO_3^-] = 24$ mmol l^{-1} and [CO_2 in solution] = $0 \cdot 03 P_{CO_2} = (0 \cdot 03 \times 40) = 1 \cdot 2$ mmol l^{-1}

Hence

$$\frac{[HCO_3^-]}{0 \cdot 03 P_{CO_2}} = \frac{24}{1 \cdot 2} = \frac{20}{1}$$

Substituting in (3),

$$pH = 6 \cdot 1 + 1 \cdot 3$$
$$= 7 \cdot 4.$$

NERNST EQUATION.

For potassium ions in equilibrium across a semi-permeable membrane, the potassium equilibrium potential E_K is given by

$$E_K = \frac{RT}{ZF} \log_e \frac{[K_o^+]}{[K_i^+]}$$

where F is the Faraday constant (coulombs per mole of charge)

Z is the valency of potassium
R is the universal gas constant
T is absolute temperature
K^+o, K^+i concentrations of potassium ions outside and inside the cell respectively.

Substituting and converting to logarithms to the base 10

$$E_K = 61 \log_{10} \frac{[K_o^+]}{[K_i^+]} \text{ mV at } 37°C.$$

If $E_K = 0$, the equilibrium condition for ions is reduced to that for neutral substances and the concentrations inside and outside the cell are equal. The Nernst equation can be written for any species of ion present.

FICK PRINCIPLE

O_2 used up is found with Douglas bag or spirometer;
arterial O_2 by arterial puncture;
venous O_2 by catheterization of right ventricle.
Pulmonary blood flow is equal to cardiac output.

Pulmonary blood flow

$$= \frac{\text{ml } O_2 \text{ used/min}}{\text{a--v } O_2 \text{ difference in ml/100 ml}} \times 100 \text{ ml/min}$$

RESPIRATORY QUOTIENT (RQ)

$$RQ = \frac{CO_2 \text{ expired/min}}{O_2 \text{ used/min}}$$

DIFFUSING CAPACITY OF LUNGS FOR OXYGEN (DO_2)

$$DO_2 = \frac{A_L \times dO_2}{t_i}$$

where A_L is total area of diffusion surface
t_i is distance diffused
dO_2 is oxygen diffusion coefficient.

Bibliography

General Topics

Hodgkin, A. L., Huxley, A. F., Feldberg, W., Rushton, W. A. H., Gregory, R. A., and McCance, R. A. (1977) *The Pursuit of Nature*. Cambridge: Cambridge University Press.

Robson, C. (1973) *Experiment, Design and Statistics in Psychology*. Harmondsworth: Penguin Education.

Medawar, P. B. (1967) *The Art of the Soluble*. London: Methuen.

Section 1: Specialised Cells and Tissues

Lockwood, A. P. M. (1974) *The Membranes of Animal Cells*. London: Edward Arnold.

Wilkie, D. R. (1977) *Muscle*. London: Edward Arnold.

Katz, B. (1966) *Nerve, Muscle and Synapse*. New York: McGraw-Hill.

Wintrobe, M. M. (1967) *Clinical Haematology*. 6th edn. London: Henry Kimpton.

Section 2: The Cellular Environment and its Control

Davson, H. (1964) *General Physiology*. 3rd edn. London: Churchill.

Maxwell, M. H. and Kleeman C. R. (1962) *Clinical Disorders of Fluid and Electrolyte Metabolism*. New York: McGraw-Hill.

Harvey, R. J. (1974) *The Kidneys and the Internal Environment*. London: Chapman and Hall.

Catt, K. J. (1971) *An ABC of Endocrinology*. Boston: Little, Brown.

Section 3: Systems for Transport in the Body

Gregory, R. A. (1962) *Secretory Mechanisms of the Gastro-Intestinal Tract*. London: Arnold.

Hurst, J. W. and Logue, R. B. (1970) *The Heart, Arteries and Veins*. 2nd edn. New York: McGraw-Hill.

Comroe, J. H. (1965) *Physiology of Respiration*. Chicago: Year Book Medical Publishers.

Lippold, O. C. J. (1968) *Respiration. A Programmed Course*. San Francisco: Freeman.

Section 4: Information: Its Reception and Processing

Nathan, P. (1973) *The Nervous System*. Harmondsworth: Pelican.

Matthews, P. B. C. (1972) *Mammalian Muscle Receptors and their Central Actions*. London: Edward Arnold.

Lance, J. W. (1970) *A Physiological Approach to Clinical Neurology*. London: Butterworth.

Eccles, J. C. (1968) *The Physiology of Nerve Cells*. Cambridge: Cambridge University Press.

Section 5: Growth and Reproductive Physiology

Fraser Roberts, J. A. (1967) *An Introduction to Medical Genetics*. 4th edn. London: Oxford University Press.

Williams, R. H. ed. (1968) *Textbook of Endocrinology*. London: W. B. Saunders.

Wolstenholme, G. E. W. and Knight, J. (1972) *Lactogenic hormones*. CIBA Foundation Symposium. Edinburgh: Churchill Livingstone.

Baird, D. T. and Strong, J. A. (1972) *Control of Gonadal Steroid Secretion*. Edinburgh: *Pfizer Medical Monographs*. Edinburgh University Press.

Section 6: Compensatory Adjustments and Reactions to Injury

Roitt, I. M. (1978) *Essential Immunology*. 2nd edn. Oxford: Blackwell Scientific Publications.

Hardisty, R. M. and Ingram, G. I. C. (1971) *Bleeding Disorders: Investigation and management*. Oxford: Blackwell Scientific Publications.

Pickering, G. W. (1968) *High Blood Pressure*. 2nd edn. Edinburgh: Churchill Livingstone.

Lippold, O. C. J. (1973) Descriptive and functional characteristics of Physiological Tremor. In *The Origin of the Alpha Rhythm*. Chapter 5. Edinburgh: Churchill Livingstone.

Index
by Barbara Cogdell

Figures in **bold type** refer to pages on which illustrations or tables occur.

A band, **56, 57,** 61
Abdominal muscles, 274
Aberration, 377, 380
Acclimatisation to hypoxia, 303
Accommodation in nerve, 22, 34
Accommodation of the eye, 380
Acetoacetic acid, 333, 335–336
Acetone bodies, 206
Acetyl-coenzme A, 332, 335–336, 339–340
Acetylcholine, 39–43, 45, 48, 52, 179, 183–184
   and Parkinson's disease, 488
   detection of, **180**–181
   effect on circulation, 233
Acetylene, use as foreign gas, 236
Acid, effect in duodenum, **319**–320
   production in the body, 157
Acid-base abnormalities, detection of, 154
   balance, 146, 147
   pairs, **152,** 153
   regulation, 152–155
Acidaemia, 336
Acidosis, 96, 159–160
   metabolic, 301
Acids, 152
   relative contribution to buffering, **155**
Aconitine, 259
Acoustic nerve, **368,** 369, **370,** 371–372
Acromegaly, 204
Actin, **57,** 58, 59, 61
   in smooth muscle, 76, **77,** 80
Actinomycin C, 485
Action potential, **24**–27, **32**–34, 37
   in muscle, 39–40, 58, 59
Acupuncture analgesia, 493
Adaptation, 111, 350–**351,** 352, **354, 358, 359**
Addison's disease, 13, 197
Adenohypophysis, see pituitary gland, anterior lobe
Adenosine diphosphate (ADP), 54, 64–**66,** 340
Adenosine triphosphatase (ATPase), 87
Adenosine triphosphate (ATP), 1, 31–32
   effect on circulation, 233
   in muscle contraction, 54, 58, 64–**66,** 80
   releasing energy to tissues, 340
Adipose tissue, 10, **11,** 206, 335, 431
Adolescent growth, 429, 430, **431**–432
Adrenal cortex, 146, 147, 192–197
   and resistance to stress, 193–195
   in the fetus, 453
   influence on uterine periodicity, 438
   role in carbohydrate metabolism, 337
Adrenal function, abnormalities, 197
Adrenal gland, structure of, 191, **192**
Adrenal medulla, 191–192
   abnormalities, 192
   and sympathetic system, 173, 174, 175, 183, 191
   secretion of, 181
Adrenalectomy, 192–193

Adrenaline, 39, 173, 174, 175, **182**–183
   and cardiac output, 256–**257**
   and cardiovascular reflexes, 226
   concentration, 191–192
   detection of, 179, 181
   effect in circulation, **232**–233
   effect on heart rate, 258, 311
   effect on muscle blood vessels, 244–**245**
   effect on renal blood vessels, 146
   secretion by adrenal medulla, 191
Adrenergic blocking agents, 182–183, 184
   fibres, 181, **182,** 183
   transmission, 181
Adrenocorticotrophic hormone (ACTH), 196, 203, 450, 453
Afferent nerve fibres, 172, 349–350, 395, 402
Afterdischarge of reflexes, 397
After-hyperpolarisation, 47
Agglutination of red blood cells, **474,** 476
Agglutinins, 467, 474
Agglutinogens, 474, 476
Akinesia, 415
Albumen, 185, 186, 187
   in connective tissue, 10–11
   serum, 119
Albustix, 148
Alcohol concentration in blood, 346–**347**
   intake, 346–347
   oxidation, 346–347
Aldosterone, 146–147, 170, 192, 193, **194,** 195
   secretion, 159
Alkalosis, 159–160
All-or-none law, 25–27, 32, 37
   in muscle fibres, 45
Allergy, 469
Alloxan, **206**
Alpha blockers, 184
Alpha motoneurones, 361, 363, 402, 403, 422
Alpha motor pathway, 111–**112**
Alpha receptors, **177, 182**–183
Alpha rhythm, 425
Altitude, effect on respiration, 303–305
Alveolar air, **279**–281
   -arterial Po_2 and Pco_2 gradient, 284
   carbon dioxide partial pressure, **291**–293, 294, **299**–303
   membrane, 281, **282**
      impaired diffusion causing cyanosis, 305, 306
   oxygen partial pressure, **292**–293, **299**–300
      and cyanosis, 305
   ventilation, 280–284, **299**
Alveoli, 272, 276, 281
Amenorrhoea, 445
Amino acids, 320, 330, 331, 332–334, 337
   essential, 333, 344
γ-aminobutyric acid (GABA), 48, 51, 52, 416, 495
Ammonia as a buffer, 154

   system in the kidney, 157–**158,** 333
Amnesia, 423
Amphipathic protein, **87**
Amplification resulting from positive feedback, 110
Ampulla of semicircular canals, **373,** 374
Amygdala, 177
Amylase, 321
Anabolism, 325
Anaemia, 83, 90, 97, 117, 345
Anaesthetics, 495
Analgesia, 495
Anamnestic response, 468
Anaphylactic shock, 466, 469
Androgens, **194,** 197, 430, 432, 444, 456, 457
Androsterone, 456
Aneurin, see thiamine
Angina pectoris, 243, 482, 494
Angiotensin, 147, 193
Angiotensinogen, 483
Anions, action on tissues, 115
   in body fluids, **7,** 120
Ankle jerk, **401**
Annulospiral endings, see muscle spindles annulospiral endings
Anode break excitation, 23, 33
Anorexia nervosa, 178
Anosognosia, 408
Anovulatory cycles, 438
Anoxia, 96
   stimulation of chemoreceptors, 228
Antagonist muscles, 68
Antagonistic innervation, dual, 172, 312
   see also reciprocal innervation of blood vessels
Anterior pituitary gland, see pituitary gland, anterior
Antibodies, 467–469, 474
   produced by lymph nodes, 124
   to transplants, 484–485
Anticoagulants, 473
Antidiuretic hormone (ADH), 120–121, 127, 128–129, 142, 143, 145, 204–**205**
Antidromic impulses, 44, 48, **49,** 402
Antigen-sensitive cells, 468
Antigens, 467–469, 474
   in transplants, 484–485
Antihaemophilic factor (AHF), 471, 472
Anti-haemorrhagic vitamin, see vitamin K
Anti-inflammatory effect, 195
Antiketogenic substances, 344
Antipyretics, 170
Antirachitic vitamin, see vitamin D
Antiscorbutic vitamin, see vitamin C
Antitoxin, 467, 469
Anuria, 135
Aorta, 212, 213, **214**
Aortic bodies, 293, **295**
   nerve, 293
   pressure, 252

Aortic Bodies (cont.)
   reflex, 225, 226
   regurgitation, 253
Apex beat, 249
Aphasia, 414, 423–424
Apneusis, **286**, 287, 288
Apneustic centre, 287–**288**, 296
Apnoea, 292–293, 300
Apocrine glands, 14, 18, 169
Appetite, 178, 347–348
APUD cells, **209**
Apudomas, 209
Arachnoid, 125
Arcuate arteries, 131
Areolar tissue, 10
Arousal reactions, 426
Arterial blood pressure, 218–221, 229, **233**, 481–482
   and cardiac output, 255–**256**
   control of, 110, 225–**229**
   effect of glomerular filtration, 135–**136**
   effect of haemorrhage, 224
Arterial-capillary oxygen pressure gradient, 302
Arterial carbon dioxide pressure, 291, 293–295, **299**–300
   oxygen pressure, 293–**294**, 299–302
   pH, 293, 294, **299**, 301
   pulse, 252, **253**
Arteries, 210–213, **214, 215**
   and age, 215
   effect of hypertension, 481
Arterioles, 213, **214**, 229, **239**
Arteriosclerosis, 252
Arteriovenous communications in skin, 246
   shunts, 259
Artificial respiration, 308–309
Ascites, 188, 259
Ascorbic acid, 11, 192, **343**, 345, 346
Asphyxia, 291, 303, 308
Astereognosis, 410
Asthma, 273–274, 275, 305
Astigmatism, 381
Ataxy, **417**
Atherosclerosis and sympathetic hyperactivity, 175
Athetosis, 416, 488
Atmospheric pressure, effect on alveolar gas pressures, **292**
Atretic follicles, 434
Atrial receptors, 228, 231
   reflexes, 231
Atrium, see auricle
Atropine, 176, 183
   effect on heart rate, 258
   effect on pupil, 380
Attachment plaques in smooth muscle, 76
Atwater-Benidict respiration calorimeter, **328**, 329
Audibility, threshold of, 365, 367
Audiometer, 367
Auditory acuity, 365–367
   cortex, 409
   fatigue, 367
   input, efferent control of, 372
   pathways, 409
   system, 364–372
Auricle, 248–250, **264**
   stretching to produce diuresis, 128
Auricular flutter, 259, 268, **270**
   pressure, **250**, 251
Auriculoventricular node, **248**, 250, 262, 263, **264**, 265
Auscultation, **219**
Autoimmune diseases, 468
Autonomic control of cardiac function, 258
   influence on bronchioles, 273
   innervation of digestive glands and alimentary tract, 312–313

nerve fibre diameters, 37
nerves, 172
nervous sytem, 172–178
   chemical transmission in, 179–184
   stimulation, effects on pacemaker, 264
Autoregulation in the kidney, 136–**137**
Autosomes, 461
Autotransplants, 484
Auxodromes, 432
Azathioprine, 485
Azide, 31

Bainbridge reflex, 225, 230–231
Barany's rotating chair, 374
Barbiturates, effect on presynaptic inhibition, 49
   effect on respiration, 295
Baroreceptor vasomotor reflex, 222
Baroreceptors, 110, 226, **227**–**228**, 229
   control of breathing, 297
   effect on heart rate, 258
Basal ganglia, 392, 404, 415–416, 487
Basal metabolic rate (BMR), 329–**330**
Basement membrane, **2**, 3
Bases, 152
Basilar membrane, 351, 368–**369**, 370
Basket cells in cerebellum, **418**, **419**, **420**
Basophils, 466, 467
Bends, 306
Beniceptors, 491
Beri-beri, 259, 345
Beta blockers, 184
Beta receptors, **177**, **182**–183
Betz cells, 412–413, 414
Bicarbonate, 100–101, 102, 103
   buffer system, 153–154, 155
   control, 156
   in pancreatic juice, 319
   reabsorption by kidney tubules, 157, **158**
Bile, 84, 185, 186, 320–321
   pigments, 185, 186, 321
   salts, 185, 186, 187, 321, 324
Bilirubin, 84, 186, 321. 474, 483
Biliverdin, 186, 321
Binocular cells in visual cortex, 409
Biological value of a protein, 344
Biotin, 345
Bipolar cells, see retinal bipolar cells.
Bladder, 149, **150**–151
Blastocyst, 448
Bleeding time, 473
Blind spot, **379**
Blindness, caused by cerebral lesions, 407, 408
Blinking, 394, 395
blockade in phagocytosis, 188
Blood, abnormal mixture of arterial and venous, 305
   -brain barrier, 126
   buffer substances in, 155
   clotting, see coagulation of blood
   factors, 187, 470, 471, **472**
   composition of, 119, 499
   effect of freezing, 117
   flow distribution, 95
      hydrodynamics of, 216–221
      in skin following injury, 465
      rate, 211, 225
      through muscles, affected by temperature, 166
      through the skin, 161, 165–166, 175
      to glands, 19
      measurement, 235–**238**
   -gas analysis, 91–92
   glucose level and growth hormone, 203
   groups, 474–476
   pH, 499
   plasma, see plasma
   pressure, **218**–221

influence on breathing, 297
   see also arterial blood pressure
reservoirs, 240
restoration following haemorrhage, 478, **479**
serum, see serum
sugar concentration, 337–338
transfusion, 224, 471, 474, 475
   following shock, 478, 480
vessels, 213–214
   critical closing pressure, **218**
   pressure-flow relations in, 217–**218**
   rheology of, 214–215
   viscosity, 216–**217**
volume, 118, 224, 499
Blushing, 175
Body core, heat produced by, 162
   fluids, **118**
   composition of, 119–120, 499
   control system for volume of, **128**
   distribution of, 118–121
   formation of, 122–126
   properties of, 114–117
   total volume of, 119
   volumes, methods of estimating, 118–119
   surface, **330**
   weight, 330
Bohr effect, 92
Bohr's equation for calculating dead space volume, 280
Bone, **14**–16
   effect of oestrogens on, 441
   effect of parathyroid on, 208
   marrow, 14, 84
Botulinum toxin, 43
Bowman's capsule, 132
   pressure in, 135
Bradykinesia, 487
Bradykinin, 19, 245, 246, 466
   as a mediator of pain, 494
Brain, blood supply of, 211
   growth, **430**
   human, **393**
   slices, 115–116
   stages of development, **393**
Brain stem, 392, 426
   nuclei, 415
      damage in Parkinson's disease and Huntington's chorea, 487, 489
   postural reflexes, 404–405
Breath holding, 300
Breathalyser, 346
Breathing, bronchial, 276
   chemical and reflex control of, 289–290
   effects of forced, 292–293
   mechanics, 272–276
   nervous control of, 297
   regulation of, **285**–**297**
   vesicular, 276
Breech presentation of fetus, 451
Broca's area, 406, 414
Bromocriptine, 445, 452
Bromsulphthalein (BSP), 186, 237, 483
Bronchi, 272
Bronchial circulation, 213
   muscles, 272–274
   tree, **273**
Bronchioles, **272**–273
Bronchodilatation, 174
Brown-Séquard's syndrome, 410
Brunner's glands, 321
Brush-border, 5
Buccal cavity, 272
Buffer action, 152–153
   interactions, 157
   ratio, 154

substances in the blood, 99, 102, 155
systems, 153–154, **155**
Buffers, 152
Buffy coat, 476
Bungarotoxin, 41, 43
Burns, effect on circulation, 224

Caecum, 322, **323**
Caisson disease, 306
Calcitonin, 15, 198, **208**
Calcium chloride injection to restore heart beat, 311
Calcium intake, **342**, 344–345
  ion channels in muscle membrane, 41
  ions, action on tissues, 115
    concentration in blood, 207
    concentration in extracellular fluid, 41
    effect on sodium and potassium permeabilities, 34
    entry during an action potential, 34
    role in blood clotting, 470–471
    role in muscle contraction, 59, 72, 74
  metabolism and bone, 15, 208
Calorigenic effect, 199
Calorimetry, **328**, **329**, 341
Capacitation of spermatozoa, 458
Capacity of nerve membrane, 29
Capillaries, 212, 213, **214**, 229, **239**
Capillary circulation, 238–239
  interchanges, **122–123**
  permeability, 124
    and inflammation, 466
  pressure, **122**, 123, **220**, **221**, 239
Carbamino haemoglobin, 101–102
Carbohydrate, 319, 327, 332
  in diet, 341, 343, 344
  metabolism, 333, 336, **339**
    in muscle, 65
  oxidation, **328**
  respiratory quotient, 326, 328
Carbon dioxide and the vasomotor centre, 233
  as an acid, 99
  content in blood, 99
  dissociation curves, **100**, 101, 102
  effect on the affinity of haemoglobin for oxygen, 92
  effects of concentration in inspired air, **291**–292
  extraction, 225
  formed in metabolism, 325–326, 328, 329
  liberation from blood into air, **102**
  partial pressure (P_{CO_2}), 99, 153–154, 156, 279–**280**, 281, **283**, 284
    in control of breathing, 289–292
  sensitivity, 292
  solubility coefficient, 153–154
  transport, 99–103
  volume in alveolar air, **280**
Carbon monoxide poisoning, 93–94, 305–306
Carbonic acid, 99, 100, 153–154
Carbonic anhydrase, 101, 102, 153
  in pancreas, 319
  in parietal cells, 314
Carboxyhaemoglobin, 93–94, 305
Cardiac action potentials, 71–**72**, 250, 261–262, **264**, 265
  arrest, 309–310
  asystole, 309
  catheter, 236, 251
  compression, **310–311**
  cycle, 249, 251–**252**
  dilatation, 259
  electrophysiology, 261–265
  failure, 258–259
  glycosides, 31
  massage, 260
Cardiac muscle, contraction of, **72**, 249, 250

contraction, interval-duration relation, 74
  interval-strength relation, 73–74
development of tension, 250
electrical stimulation, 71
energetics of, 75
fibres, 55, **71**
  conduction velocity of, 72
mechanical properties, 74–75, 250
potassium conductance, 261, **262**, 263
refractory period, **250**
rested state contractions, 73
sodium conductance, 261–**262**
spontaneous activity, 72, 262–263
structure of, 55, **71**
Cardiac output, 211, 225, 229, 255–260
  effect on oxygen supply, 95
  and exercise, **223**
  distribution of, **210**, 237
  in artificial respiration, 309
  increase during hypoxia, 304
  measurement, 236–237
Cardiac resuscitation, 309–311
  sphincter at entrance to stomach, 314
Cardioaccelerator centre, 226
Cardiogreen, 236
Cardioinhibitory centre, 226
Cardiometer, 256
Cardiovascular failure, 97
  reflexes, 211, 225–226
  regulation and hypothalamus, 178
  system, control of, 225–231, 232–234
Carotene, 12
Carotid body, **293–295**
  reflexes, 294–295
Carotid sinus baroreceptors, 211
  nerves, **226**, 227, 228, 289, 293, **294**
  perfusion of, 227, 228
  reflexes, 222, 224, 225, 226
Cartilage, 14, 16
Castration, 456–457
Catabolism, 325
Cataract, 379
Catecholamines, 181
  and respiration, 297, 301
Cathodal stimulation, 23, 27
Cations in body fluids, **7**, 120
Caudate nucleus and Huntington's chorea, 488–489
Caveolae, 76
Cell membrane, 1, 3, 5–**6**, 7–9
  metabolism, 326
  nuclei, 2
  structure, 1
Cells, 1–6
  adhesion between, 3–4
  changes in volume of, 9
  secretory, 4–5
Cementum, 16, **17**
Central inhibition, 48
  nervous system, anatomy of, 392–393
  venous pressure, 240
Centrosomes, 3
Cerebellar cortex, 416, **418–420**
  nuclei, 417
  structure, 417–418
Cerebellum, 392, 403–404, 416–420, 427
  lesions of, 417, 422
  mapping of the body onto, 416, **417**
  mode of operation, 418–420
Cerebral blood flow, **237**, 241–242
  circulation, cessation of, 309
  cortex, 406, 427
    see also individual regions of cortex
    association areas, 406
    electrical activity of, 425
    functional localisation in, 406

primary or receiving areas, 406
  projection areas, 406
hemispheres, 392, **393**
oxygen usage, 241
vascular accidents, 242
Cerebrospinal fluid, 119
  composition of, 125–126, 499
  formation of, 125
  partial pressure of carbon dioxide in, 290–291
  pressure, 242
  volume, 126
Cervix, 436
Chemical score of a protein, 344
  mediation of pain, 494
  transmission, evidence for, 179
Chemoreceptors, 228, 353
  central or medullary, 289–292, 295
  effect of acclimatisation, 303–304
  peripheral, 289, 293–295, 302
Chemotaxis of leucocytes, 466
Chief cells in stomach, 314
Children, dietary allowances, **342**, **343**
Chloasma uterinum, 449
Chloride equilibrium potential, 30
  ions, 29–34
  permeability, 33
    during IPSP, 48
  pump, 30
  shift, 100–101, 102
Chlorpromazine, 488
Cholecystography, 321
Cholecystokinin-pancreozymin, 319–321
Cholesterol, 321
  in cell membrane, 6, 8
Choline, 179, 345
Cholineacetylase, 183
Cholinergic fibres, 45, 181, 183
Cholinesterase, 39, 179, 180
Chorea, 416, 488
Chorion epithelioma, 448
Chorionic gonadotrophin, 445, 448, 450, **452**
Choroid plexuses, 125
Christmas factor, 471, 472
Chromatin, 2
Chromosomes, 2, **461**
Chyle, 324
Chylomicrons, 11, **324**
Cilia, 3, **5**
  in fallopian tubes, **435**
Circulation and the kidney, 146–147
  design of, 212–221
  general function of, 210–211
Circulatory collapse, 477–478
  failure and shock, peripheral, 224
  responses, 222–224
Cirrhosis of the liver, 188, 483
Citrate, 471
Citric acid cycle, 332, 339–340
Claudication, intermittent, 243
Clearances, see kidney, clearances
Climacteric, 447
Climbing fibres of cerebellum, **418**, **419**
Clitoris, 460
Clomiphene, 445
Clonal selection theory of immune response, 468
Closed methods of measurement of metabolic rate, 329
Clotting time, 472
Coagulation of blood, 464, 470–473
  prevention, 471
Cobalamin, see vitamin B_{12}
Cocaine, 37, 495
Cochlea, 367–369
Cochlear microphonic potentials, 369–371
Codeine, 495

Coenzyme A, 332, 335
Coitus, 459–460
   effect on sperm production, 457
Cold exposure and thyroid secretion, 200
Collagen, 10–11, 12, 120
   in arterial walls, 215
   in bone, 14, 15
   in teeth, 16
Colliculus, inferior, 409
   superior, 407
Colloid, 198
Colloids and intestine, 323
Colon, filling and emptying, 322, **323**
Colonic secretion, 323
Colostrum, 452, 453
Colour blindness, 384
   index, 90
   vision, **383–384**, 386
Compensatory pause, **250**
Competitive inhibition, 180
Complement, 468–469
Complex cells in the visual cortex, **389**, **390**, 408
Compliance of lung tissue, 275
Conditional reflexes, 424–425
Conditioning, operant, 425
Conduction velocity in nerve fibres, **23–24**, 26, 27–28, **36**, 37
Condylactis toxin, 33–34
Cones in retina, 381, 384–385, **387**
Connective tissue, **10–11**
Consciousness, 425–426
Contraceptive methods, 446–447
Contractile component of skeletal muscle, **61, 62**
Contracture of smooth muscle, 80
Control systems, mathematical explanation, 107–108
   operation of, 109–113
   theory of, 104–108
Control theory and physiological tremor, 486
Copulation, see coitus
Corium, 12
Cornea, 360
   inflammatory processes in, **195**
Corneal reflex, 395
   surface, 378–379
   transplantation, 384
Coronary circulation, **237–238**, **242–243**
   heart disease, 175, 243, **270–271**, 471. 473
   sinus, 242
Corpus callosum, 392, 408, 424
Corpus luteum, 434, **435**, 436, 450
   secretion of progesterone, 441
Corpus striatum, see basal ganglia
Corti, organ of, 369, **370**–371
Cortical control of breathing, 296
Corticospinal pathway, rapid, involved in stretch reflex, 422
Corticosterone, 193, 195–196
Corticotrophin releasing factor (CRF), 196, 203
Cortisol, 193, **194,** 195–196
   influence on carbohydrate metabolism, 337
Cortisone, 485
Coughing, 273, 274, 297
Countercurrent theory in the kidney, 141–**142**, 144–145
Cranial nerves, **392–393**
Creatine phosphotransferase reaction, 64–**65**, **66**
Creatinine, 333–334
   clearance, 137, 138
Crenation, 84
Cretinism, **198**, 201, 330
Cross agglutination test, 476
Cross bridges, 62
Cross-circulation experiments, 226–**227**
Cross-sectional studies of growth, 429–430

Crossed extensor reflex, 397–398, 403
Crush syndrome, 478
Cryptorchism, 457
Cupula, 373, **374**
Curare, 39, 41, 43, 46, 184
Cutaneous nerves, fibre composition, **35**, 36–37
   receptors, **357–360**
   thermoreceptors, 163, 164, 165, 166
Cyanide, 31
Cyanosis, 305–306
Cyclic 3', 5'-adenosine monophosphate, 142–143, 190, 191, 204, 454
Cytochromes in muscle, 66
Cytozyme, 470

Dale-Laidlaw apparatus, 472
Dale's principle, 45, 48
Damping, in a control system, **109**
Dark adaptation, **382**
Deaf aids, 367
Deafferentation, 403–404
Deafness, 367
Deamination, 333
Decerebrate preparations, 396, 400, 402, 427
   rigidity, 400, 402, 403–404, 405
Decibels, 365
Decidua of pregnancy, 448, 449
Decussation, 407
Deep pain, 494
Defaecation, 322
Defence reactions, 465–469
Defibrillation, 260, 310, 311
Defibrination, 471
Deglutition apnoea, 272
Dehydration, 312
Dehydroepiandrosterone, 456
Dementia, 423–424
Dendrites, action potentials in, 46–47
Denervation of muscle, 41–43
Dental caries, 17–18
Dentate nuclei, 416
Dentine, 16, **17**
Deoxycorticosterone, 146
Deoxyribonucleic acid (DNA), 2, 18
Dephosphorylation, 340
Depolarisation, 32–33, 38
Depressor nerves, 289
   reflex, 222, 224, 225–**226**
Depth perception, 386
Desmosome, **4**
Detoxication, 185, 188
Detrusor muscle, 150–151
Diabetes insipidus, 120, 129, 204
   mellitus, 129, 131, 159, 206–207, 335, **338–339**
Diabetic puncture, 337
Dialysis, 485
Diaphragm, 274, 286, 287, **288**, 289, 296
   and "desire to breathe" sensation, 300
Diastase, 320
Diastole, 249
Diastolic pressure, 218–**219**, **220**–221
   volume, 222
Dicoumarol, 470, 471, 473
Dicrotic notch, 253
Dietary allowances, recommended, **342**, **343**
   survey, 341
Differentation, the specialisation of cells, 432
Diffusing capacity of the lungs, 281, 500
Diffusion in the kidney, 136
Digestion in the intestine, 319–324
   in the mouth and stomach, 312–318
   products, absorption of, 323–324
Digestive glands, 312
   juices, 312
Dihydroxystilbenes, 441

Dilation and curettage (D & C), 439
Dinitrophenol (DNP), **31**
Diodone, 237
   clearance, 138
Dioptres, 377
2, 3-diphosphoglycerate (DPG), 96, 97
Diploid number, 461
Disease, 465
   affect on growth, 432
Discus proligerus, 434
Dissociation constant, 153
Disuse atrophy, 430
Diuresis, 127, 128, 129, 135, 142, 143
Diuretic action of alcohol, 347
Diuretics, 146
Diving, 231, 306–307
Dominant gene, 461
Dominant hemisphere, 424
Donnan membrane equilibrium, 8–9, 101
L-Dopa (3, 4=dihydroxyphenylalanine), 4, 416, 487
Dopamine, 52, 181, **182**, 452, 487–488
   -containing cells in the substantia nigra, 415
   content of basal ganglia, 416
Dorsal column nuclei, 409–410
Dorsal root ganglion, 409
   potentials, 50–**51**, 53
Douglas bag, 278, 329
Dreaming, 426
Drinking, 129
   centres, 129
Drowning, 259
Drunkenness, 346, **347**
Ductus arteriosus, 276, 490
Duodenum, 316–318, 319–322
Dura mater, 125
Dye injection use in blood flow measurement, 235–**237**
Dysmenorrhea, 447
Dysmetria, 417
Dyspnoea, 274, 278, 306

Ear, anatomy of, 367–369
   role of, 364
   threshold of feeling, 366
Eccrine glands, 13, 169
Ectopic focus of heart beat, **269**
Efferent nerve fibres, 172, 395
Ejaculation, 459–460
Elastic fibres in the skin, 12
   tissue and blood vessels, 214–215
Elastin, 10, 120
Electrical synapses, 52
   stimulation of nerve, 22
Electrocardiogram (ECG), **265**, **266**, 267–269
   recording methods, 266–**267**, **268**
Electrocardiography, clinical, 266–271
Electrocution, 259
Electroencephalogram (EEG), **425**, 426, 427
Electrolyte balance and adrenal cortex, 193
Electrolytes in body fluids, see ionic composition of body fluids
   influence of progesterone on, 442
Electrophoresis, 40, 43, 48
Electrotonic spread of current, **27**, 35
Elephantiasis, 125
Embden-Meyerhof pathway, 65–66
Embolism, 471, 473
Embryo, 436
Emotion, 410
   expression of, 393–394, 413, 414
Emotional effects on muscle blood flow, 244, **245**
Emphysema, 305
Enamel of teeth, 16, **17**, 18
End-plate, 39–41
   potential, **40–41**

506   INDEX

miniature, **41**, 52
Endocardium, 248
Endocrine glands, 18, 185, 189–190
   secretions, 189
   system, amplification by, 189, 196
     and population density, 209
     general properties of, 189–190
Endolymph, 119
Endolymphatic potential, 369
Endometrium, 436, 438, **439**, 446
Endoplasmic reticulum, 1–2, **3**, 18
Endorphins, 495
Energy balance sheet for human subject, **329**
   content of food, 342–343
   of heart contraction, 256
   production from oxygen intake, 327
   requirements, 341–**343**
   units of, 498
   value of foods, 327
Enkephalin, 495
Enophthalmos, 175
Enteritis, 322
Enteroglucagon, 338
Enterohepatic circulation, 186
Enterokinase, 320, 321
Enzymes, 312
   secreted into duodenum, 319
Eosinophils, 195, 466, 467
Epidermis, 12–13
Epiglottis, 272, 297, 314
Epileptic fits, 413, 425
Error-rate control in feedback mechanisms, 106–107
Error signals in control systems, 106, 109, 486
Erythroblasts, 84
Erythrocytes, *see* red blood cells
Erythropoeisis in the kidney, 148
Erythropoietin, 96
Eserine, 39, 180–181
   effect on pupil, 380
Ethinyloestradiol, 441
Eunuchs and eunuchoids, 457, 459
Eustachian tube, **368**
Evaporation from skin, 168–169, 171
Excitation-contraction coupling, 59, 60, 72, 74
Excitatory post-synaptic potential (EPSP), 44–45, **46**
Exercise and muscle blood flow, **243**
   circulatory adjustments in, 222–**223**
   effect of training, 69
   effect on blood, 102–103
   effect on oxygen dissociation curve, 93
   maximum power output during, 68–**69**
   respiratory adjustments during, 299–301
Exocrine glands, 19, 185
Exocytosis, 41
Exophthalmic goitre, 330
Exophthalmos, 174
Exostoses, 188
Expiration, 274, 275, 285–286, 288
Expiratory centre, **287, 288, 290**
   reserve capacity, **277**
Expired air, 278–**279**
Extensor thrust reflex, 396, 397
Extracellular fluid, (ECF), 104, 118, 119, 120–121, 122
   volume, 9, 119
Extrafusal muscle fibres, 56
   see also skeletal muscle fibres
Extrapyramidal rigidity, 487–488
   system, 413, **415**–416, 427
Extrasystole, **73**, 257, 259
Eye, anatomy of, **378**
   intensity of light required to stimulate, **382**
   media, 379
   movements, 386, 407
   optical defects in, **380**–381

   optical system in, 378–381
   range of focus in human, 380
   spectral sensitivity of, 381–382, 386

Faeces, 322, 499
   water output, **127**
Fainting, 224, 226, 243, 477
   lark, 222
Falling reaction, 375
Fallopian tubes, 433, **435**–436
Fanconi syndrome, 159
Fat absorption from intestine, **324**
   and carbohydrate interconversion, 335
   balance, 11
   causing emptying of gall-bladder, **320**–321
   growth, 431
   metabolism control, 336
     effect of kidney on, 147–148
   oxidation, 326, **328**, 335, 344
   respiratory quotient, 326, 328
   storage, 335, 344
Fatigue, xvi
   of reflexes, 399
   muscle, 44, 422
Fats, 319, 320, 324, 327, 332
   in diet, 341, 343, 344
Fatty acids, 320, 324, 332, 334–335
Feedback, derivative, **106**–107, 109, 111
   in hormonal systems, 189
   mechanisms, 104–107, 109
   negative, 106, 109, 110, 111
   positive, 106, 110
Feedforward, positive, 111
Female reproductive system, 433–436
   sexual act, 460
Fertilisation, 448
Fertility control, 446–447
Fetal blood, 490
   blood, dissociation curve, 93
   circulation, changes at birth, **490**
   respiration, 276
Fetus, influence on length of gestation, 453
Fever, 170
Fibrillation of heart, 259, 267, 269, **270**, 309
   effect on coronary vessels, 242
   of muscle fibres, 43
Fibrin, 470, 471
   barrier in lymphatics, 465, 466, 469
Fibrinogen, 119, 185, 187, 470–471, 483
Fibrinolysis, 188, 473
Fibroblasts, 10, 11
Fick principle, 235–238, 500
Fight, flight and fright response, 174
Filtration pressure, effective, 122
   in glomerulus, 135, 136
Flagellae, 5
Flexion reflex, 394, 396, 397–399
Flow diagram of servocontrol system, **106, 107**
Flower spray endings, **111**
Fluid compartments of the body, 118
Fluorodinitrobenzene (FDNB), 66, 67
Fluoride addition to drinking water, 18
Flushing, 175
Flying and hypoxia, 304–305
Folic acid, 83, 346
Follicle stimulating hormone (FSH), 203, 204, 440, 443–445
   releasing factor, 204, 443
   variation during menstrual cycle, **442**
Follow-up servo, 111–112
Food substances, metabolic history of, 332–340
   specific requirements of, 343–345
Foramen ovale, 490
Forced expiratory flow (FEF), 278
   expiratory volume (FEV), 278, **279**

   vital capacity (FVC), 278
Foreign gas principle, 236
Fourier analysis in study of tremor, **486**
Fovea, 378–379, 381, 383
Free nerve endings as pain receptors, 492
   in the skin, **357**, **358**, 360
Frequency modulation of nerve impulses, 27, **350**
Frontal lobes of cerebral cortex, 414–415
Fructose, 336
Functional residual capacity, **277**, 278
Fusimotor neurones, *see* gamma motoneurones

Gain, in a control system, 109
Galactopoiesis, 452
Galactose, 336
Gall-bladder, 185, 186, **320**–321
Gametes, 461
Gametogenic hormone, 455
Gamma globulins, 467–**468**
Gamma motoneurones, 361–363, 402, 403, 422
   and Parkinson's disease, 488
Gamma motor pathway, 111–**112**
Ganglia autonomic, 172–**173**, 183–184
Ganglion cells, *see* retinal ganglion cells
Gas analysis, 91
Gas exchange in lungs, 272, 278–**279**, 281–284
Gaseous exchanges, measurement of, 328–329
Gastric digestion, 314
   emptying, 314, 316–318
   fistulae, 314–**315**
   function, influence of emotions, **312**, 313
   inhibition, 316
   juice, 314–315, 499
   secretion, stimulation of, 314–315
Gastrin, 314, 315–**316**, 320–321
Gastroileal reflex, 322
Gastrointestinal function controlled by hypothalamus, 178
Gate control theory of pain, 493
Generator potentials, 351–**352**, 353, 354
   initiating action potentials, 352, **353**
Gene frequency, occurrence of A, B and O genes in population, 475
Genes, 461
Genitalia, 433
Genotypes and grafting, 484
Germinal epithelium, **433**
Gestagen, 446
Gland secretion, nervous control of, 19
Glands, 18–20
   see also individual glands
Globin, 89, 483
Globulin in connective tissue, 10–11
   serum, 119
Globus pallidus, 416
Glomerular capillaries, 131, 132, 134
   filtration, 134–136, 139
   rate, 137–138, 139, 147
   fluid, method for obtaining sample of, **134**
   membrane, hydrostatic pressure across, 135
Glomerulotubular balance, 146
Glomerulus, 131, **132, 133**, 134–138, **194**
   "pores" in, 134–**135**
Glossopharyngeal nerve, 226, 227
Glucagon, 206, 207
   and regulation of blood sugar, 337–338
Gluconeogenesis, 336–337, 338
Glucose, 332, 333, 336–338, 339
   concentration in the blood, 120
   effects of ingestion, **338**
   oxidation, 326
   produced by liver, 185, 187
   synthesis by the kidneys, 148
   tolerance, 338
Glucuronic acid, 483

Glutamic acid, 52
Glutamine, source of ammonia, 158
Glycerol, 320, 335, 337
Glycine, 48, 49
Glycogen, 336–338, 339
    breakdown, 66
    store in liver, 187, 188, 206
Glycolysis, 65, 66, 69, 336, 339
Glycolytic enzymes, abnormalities, 97
Glyconeogenesis, 187
    reduced capacity of, 193
Glycosuria, 206, **338**–**339**
    alimentary, 131
Goblet cells, 19
Goitre, **200**
Golgi apparatus, **1**, **2**, **3**
Golgi bodies, 2
Golgi cells in cerebellum, **418**, 419, **421**
Golgi-Mazzoni corpuscle, **357**
Golgi tendon organs, 361, 401, **402**–**403**
Gomori staining substance, 204
Gonadal hormones, effect on bone, 15
Gonadotrophic hormones, 203, 432, 440–445, 446, 447
Gonads, 428
Gooseflesh, 13, 174
Graafian follicles, **431**, **433**
Granulation tissue, 195
Granule cells in cerebellum, **418**, 419, **420**, **421**
Granulocytes, *see* polymorphonuclear leucocytes
Grasp reflex, 414
Gravity, effect on blood pressure, 221
Grey matter, 406, 407
Growth, 428–432
    curves, **429**
Growth hormone (GH), 15–16, 203, 432
    and regulation of blood sugar, 338
    -releasing factor (GHRF), 203
Growth of different tissues, **430**
Gustatory receptors, 354
    pores, 355

H reflex, 402
H zone, **56**, **57**
Haem, 89, 483
Haematin, 89
Haematocrit, 83, 499
Haemochromogens, 89
Haemocyanin, 89
Haemodialysis, 485
Haemodynamics, 217–221
Haemoglobin, 85, 89–90
    abnormal reduction in systemic capillaries, 305
    abnormalities, 97
    affinity for oxygen, 95, 96, 97–98
    and cyanosis, 305
    as a buffer, 99, 155
    breakdown, 186, 483
    combination of carbon monoxide with, 93–94
    combination of oxygen with, 90–92, 93
    concentration in blood, 89–90, 95, 96, 499
    increase during acclimatisation, 303
    reduced, 99–100, 102, 305
Haemolysin, 467, 468
Haemolysis, **85**–**86**, 117
Haemolytic jaundice, 86
Haemophilia, 471, 472
Haemorrhage, 83, 128, 129, 233, 475
    cause of shock, 477, 478, **479**
    effect on circulation, 224
Hageman factor, 471
Haircells of cochlea, 369, **370**
    in labyrinth, 373, **374**, 375
Hair receptors, **357**, **358**
Hairs, **13**

Haldane apparatus for gas analysis, 91, 92
    effect, 100
    haemoglobinometer, 90
Haloperidol, 488
Hamilton dye dilution method, **236**–**237**
Haploid number, 461
Haustra, 322, **323**
Haversian systems, **14**, 15
Hay fever, 274
Head injury, 423
    position, 373, **375**, 404
Headache, 494
Healing, 469
Hearing, discrimination thresholds, 366
    loss, 365
    testing for deficiences, 366
    theories of, 372
Heart, 210, 212, 248–254
    abnormal rhythms, **269**–**270**
    axis deviation of, **268**, 269
    beat, origin and spread of, 249–250
        restoration of, 311
    block, 259
    effect of hypertension, 481
    failure, producing oedema, 124
    innervation, 258
    -lung preparation, **255**–**256**
    murmurs, 254
    muscle, *see* cardiac muscle
    rate, 258
        control, 225
        increase during hypoxia, 304
        measured by electrocardiogram, 269
    pressure changes, **251**–**252**
    sounds, 254
    work done by the, 257
    valves, 248, 252, 253, **254**
Heat balance in the body, 161
    distribution, 162
    loss, **161**, 162, **165**, 168, 245–246
    production, **161**, 162, 168
        measurement, **328**, 329
    stroke, 164
Hemianopia, 408
Hemiplegia, 414, 422, 423
Henderson-Hasselbalch equation, 153, 500
Henle's loop, 132, 141–142, 143, 146
Henry's law of solution, 153
Heparin, 466, 471, 473
Hepatectomy, 187
Hepatic artery, 186
    failure, 483
    vein, 186
Hepatocytes, 185, 188
Hering-Breuer reflex, 289, **296**
Hermaphroditism, 462–463
Heterometric autoregulation, 255–256, 257
Heterotransplants, 484
High altitudes, adjustments to, **97**
High blood pressure, 481–482
His, bundle of, 72, 248, 259, 263, **264**, 268–269
Histamine, 11, 124, 233, 274, 465, 466, 469
    as mediator of pain, 494
    effect on gastric juice, 314, 316
Holger-Nielsen method of artificial respiration, **308**–**309**
Holocrine glands, 13, 18
Homeostasis, xvi, 104, 496–497
    and the kidney, 130
Homotransplants, 484, 485
Hooke's law and blood vessels, **214**–**215**
Hormonal conditions and hypertension, 481
    control of circulation, 225, 232–234, 242
    control of gland secretion, 19–20
    control of pregnancy, 450

    control of sexual cycles, 440–445
    factors concerned with reproduction, **456**
Hormone replacement therapy, 447
Hormones, 185
    and heat production, 168
    and the environment, 208—209
    control of secretion of, 110
    destruction, 188, 189
    specificity, 189
Humoral effect of hyperpnoea of exercise, 301
Hunger, 178, 347–348
    pangs, 317
Huntington's chorea, 488–489
Hyaluronic acid, 120
Hyaluronidase, 448
Hydrocephalus, 126
Hydrochloric acid secreted in stomach, 314, **316**
Hydrogen ion concentration, action on tissues, 115
    concentration in body fluids, 152
    excretion by kidney, **157**, 158
    receptors in medulla, 290, **291**
Hydrolytic enzymes, 2, 5, 312
5-hydroxytryptamine (5HT), 52, 426, 466
    as a mediator of pain, 494
Hyperaemia, reactive, 244, 246
Hyperalgesia, 494, 495
Hyperbaric oxygen therapy, 306
Hypercapnia, 291, 295
Hypercomplex cells in visual cortex, **390**–**391**, 408
Hyperemesis gravidarum, 449
Hypermetropia, 381
Hyperoxia, **292**
Hyperpnoea, 291, 294–295
    of muscular exercise, 300–301
Hypertension, 75, 259, 481–482
    essential, 481
Hypertensive retinopathy, 482
Hyperthyroidism, 199
Hypertrophy of muscle fibres, 69
Hyperventilation, 156–157, 299–300
Hypoglycaemia, 337, 338, 339
Hypoglycaemic convulsions, 337
Hypophysectomy, 203, 443
Hypotension, acute, 477
Hypothalamic control of anterior pituitary, 200, 202, 203–204, 443–445
    control of autonomic system, 176–178
    control of posterior pituitary, 203, 204–205
    control of temperature, 162, 163–164, 168
    influence on alimentary tract, 313
    thermoreceptors, 163–164, 166
Hypothalamus, 128, 129, 196, 392
    and food intake, 347
    and vasomotor centre, 234
Hypothermia, artificial, 170
Hypoventilation, 156, 300
Hypoxaemia, arterial, 283–284
Hypoxia, 97, 290–291, **292**, 295, 301–305, 306
    acute, 302
    anaemic, **302**, 306
    chronic, 303
    effect on circulation, 223–224
    histotoxic, 293, **302**
    hypoxic, 293, **301**–**302**
    stagnant, 293, 297, **302**, 306

I band, **56**, 57
Ileum, 321–322
Image, focusing of, **379**, 380–381
    formation, 377–378
Immune reactions, 466, 467–468
    response and transplants, 484
Immunological tolerance, 468
Immunology, 467
Immunosuppression, 468, 484–485

Implantation, 436, 448
Impotence, 459
Incompetence of heart valves, 254, 258, **268**
Indomethacin, 454
Infarct, 270
Infertility, 445
Inflammation, 195, 247, 465–466
Inhibition of conditioned reflexes, 424
Inhibitory post-synaptic potential (IPSP), **47**
Injury, 464
  potential, **24**
Inotropic state of the heart, 74, 75
Inspiration, 274, 275, 285–286, 288
Inspiratory capacity, **277**, 278
  centre, **287–288**, **290**, 296
  reserve volume, **277**
Instructional theory of immune response, 468
Insulin, 11, 206–207, 337–339
Intention tremor, 417
Intercalated discs, 71, 249
Intercostal muscles, 274, 286, **288**–289
  nerves, 286, **288**
Intermediate junction, **4**
Intermenstrual bleeding, 436, 438
Internal capsule, 392, 410, 412
Internal environment, 104, 496
Interneurones, 48, 50–51, 395
Intersexuality, 463
Interstitial cell stimulating hormone (ICSH), 455
Interstitial fluid, 114–115, 118, 119, 122–123
  proteins, 119, 123, 125
Intestinal juices, 321
  movements, 321–**322**
Intestines, neurohormonal interactions in, 319
Intracellular fluid, (ICF), 118, 119, 120
  fluid volume, 119
  proteins, 120
Intracranial pressure, 121, 242, 494
  raised, 126
Intrafusal muscle fibres, 361–363
  see also muscle spindles
Intraocular fluid, 119
Intrapleural pressure, 223, 274–**275**
Intrauterine growth, 429
Inulin clearance, 137–138
Iodide, 199, 200
  clearance, 199
  intake, 199, 344
Iodoacetate (IAA), 61, 66
Ionic basis of nerve action, 28–34
Ionic composition of body fluids, **7**, 28–**29**, 37, 120
Ions, action on tissues, **115**
Iris, 379
Iron deficiency in infants, 453
  for synthesis of haemoglobin, 84
  intake, **342**, 345
Irradiation of reflex responses, 397
Ischaemia, 211, 221
  role in nerve blocking, 37
Ischaemic muscle pain, 494
Ishihara charts, 384
Islets of Langerhans, 206, 207
Isoprenaline, 179, 182
Isoquinoline, 495
Isotonic solutions, 85, 116
Isotopes, use in metabolic studies, 330–331

Jaundice, 86, 186, 187, 474, 483
Jaw muscles, 16
Joint injuries, 360
  receptors, 361, 399
Jugular vein, point of collapse, 220
Juxtaglomerular apparatus, **133**, 146, 147, **194**

Keratin in teeth, 16, 18

Keratinisation, 12
Keto acids, 333, 334
Ketone bodies, 335–336, 338–339
Ketosis, 159, 333, 335–336, 338
Kidney, action of aldosterone, 193
  and antidiuretic hormone, 128–129
  and high blood pressure, 481–482
  arteriovenous oxygen difference in, 131
  blood flow, see renal blood flow
  clearances, 137–138, 139–**140**
  collecting ducts, 142
  control of reaction, 157–159
  distal convoluted tubule, **132**, 142, 143–**144**, 147, **157**
  effect of shock on, 478
  failure, 121
    producing oedema, 124
  function, control of, 146–148
  grafting, 485
  hypertrophy, 147
  innervation of, 131–132, 146
  in newborn and old age, 147
  investigation of function in man, 148
  medulla, osmotic gradient in, 141–142, **144**
  non-excretory functions of 147–148
  physiology of tubules, 139–145
  proximal convoluted tubule, **132**, 134, 141, 143, **144**, 146
  regulation of blood pressure, 147
  secretion and reabsorption of water and ions, 143–**144**
  stones, 149, 208
  structure, **131**–133
Knee-jerk, **111**
Knoop's method of feeding animals fatty acids, 335
Krause end bulbs, **357**, 359
Krebs' cycle, see citric acid cycle
Krebs' solution, 114–115
Kupffer cells, 185, 188, 321, 467

Labile factor, 470, 471
Labour, stages of, 450–451
Labyrinth, **373**–376, 404
Lactation, 452–453
  dietary allowance during, **342**, **343**, 344
Lactic acid, 65, 69, 75, 103, 206, 328, 336–**337**, 339
Lactogenesis, 452
Lacunae in bone, **14**
Laplace's law, 214–215
Larynx, 272
Lateral geniculate nuclei, 387, 407, 408
Law of specific nerve energies, 350
Learning, 423–424, 427
Lecithin, 320
Lens, elasticity of, 380
  in the eye, 378–380
Lenses, physical properties of, 377–379
Leucocyte migration and inflammation, 466
Leucocytes, see white blood cells
Leucocytosis, 466
Leucopenia, 466
Leucotomy, 415
Leukaemias, 466
Lewy bodies, 487
Liberkuhn's glands, 321
Light, properties of, **377**–378
  reflex, 386
Lignocaine, 495
Limbic system, 176–177
Lipase, 320, 321, 324
Lipid metabolism, 334–336
Lipids in membranes, 86, 87–88
Lipid soluble compounds, 8
Lipolysis, 11
Lipotropic action, 346

Liquor folliculi, **433**, 434
Liver, 185–188
  and formation of urea, 333
  and synthesis of bile, 321
  as blood reservoir, 240, 245
  as glycogen store, 336
  circulation of, 185–186, **237**, 245
  effect on specific dynamic action, 330
  function, abnormalities, 188
  materials stored in, 188
  secretion of glucose, 336
  store of thyroxine, 200
Load compensating reflex, 289
Local circuit theory of transmission, **27**
Locke's solution, **114**
Locomotion, 394, 404–405
Locus caeruleus, 287, **290**, 487
Longitudinal studies of growth, 429–430
Lumbar puncture, 125, 126
Lung effect of inflation on respiration, 296
  expansion at birth, 276
  movement, sounds associated with, 276
  ventilation and diffusion, 281–284
  volumes and capacities, **277**–278
Lungs, 95, **272**, 274–276
  afferent impulses from, 295–296
  distribution of blood and gas in, 281–284
  simple model of, 274–**275**
  water loss from, **127**
Lutein, 434
Luteinising hormone (LH), 203, 440, **443**–445, 455
  -releasing factor (LHRF), 203, 443
  variation during menstrual cycle, **442**
Lymph, factors affecting the flow of, 123–124
  formation, 123–124
  nodes, **124**, 468
  protein content, 123
Lymphagogues, 124
Lymphatic pump, 123
  system, 123–124
Lymphatics and inflammation, 465, 466
  draining the intestine, 324
Lymphocytes, 83, 123, 466, 467, 468
Lymphocytic transformation, 468
Lymphoid tissue growth, **430**
Lysins, 467, 474
Lysosomes, 2, 188

Macrophages, 10, 469
Macula, central part of retina, 408
Macula densa, **133**, 146, 147, **194**
Maculae, in inner ear, **374**, 375
Magnesium ions to block calcium channels, 41
Male erection, 459
  reproductive system, 455–458
Malnutrition, 124
Malocclusion, 18
Malpighian layer in skin, 12
Malpighian pyramids in kidney, 131
Maltose, 320
Mammary glands, 433, 452
  influence of oxytocin on, 205
  influence of progesterone on, 442
Mannitol, 141, 147
Mass spectrometry, **90**, 91
Mast cells, **10**, 11, 466, 467, 471
Masturbation, 457–458
Maximal voluntary ventilation (MVV), 278
Mean cell haemoglobin (MCH), 82, 499
  haemoglobin concentration (MCHC), 82, 499
  volume (MCV), 82, 499
Mechanoreceptors, 350, 352, 492
Medial lemniscus, 410
Medulla oblongata, 392
Medullary respiratory centres, **287**

HUMAN PHYSIOLOGY 509

Megakaryocytes, 470
Meissner's corpuscles, **357**, **358**
Melanin, 4, 12–13
Melanocyte stimulating hormone (MSH), 13, 204
Membrana granulosa, **433**, **434**
Membrane permeability, 6
  transport, 7–9
Membranes, cell, 86–**88**
Memory, 423–424, 427
Menarche, 447
Meningitis, 126
Menopause, 434, 438, 447
Menorrhagia, 438, 447
Menstrual cycle, 437–439, **442**, **444**, 447
  fluid, 438
Merkel's discs, 357, **358**
Merocrine glands, 18
Metabolic acidosis and alkalosis, 159, 334
  clearance rate, 188
  rate, methods of determining, 328–329
    of the brain, 241
  studies, 325–331
Metabolism and energy liberation, 340
  endogenous and exogenous, 334
  general, 325–331
  involvement in heat production, 162, 168
  resting, 329
Metabolites, in heart, 242, 256
Methaemoglobin, 89, 305
Methyltestosterone, 456, 457
Micropuncture of kidney tubules, 141, 142
Microvillae, 5
Micturition, 149–151, 176, 395
Milk, 452–453
Mineral salts in diet, 341, 344
Mitochondria, **1**, **2**, **3**
  in cardiac muscle, 71
Mitral valve, 248, 253, **254**
Molecular weights, 498
Monocytes, 466, 467
Monoglycerides, 324
Monosynaptic reflex, 401–402
Morning sickness, 449
Morphine, 495
  effect on respiration, 295
Mossy fibres in cerebellum, **418**, 419, **420**, **421**
Motion sickness, 375–376
Motoneurone pool, 398
Motoneurones, intracellular recording from, 44, **46**, **47**
Motor cortex, 406, 412–414, 427
  electrical stimulation of, 412, 413, 415
  epileptic activity in, 413
  lesions in, 413
  projection map, **412**, 413
Motor nerves, 395
  fibre composition, 37
  recordings from human, 112
Motor pathways, **409**
Motor unit, 58, 68
Mountain sickness, 303
Mouth-to-mouth artificial respiration, **308**
Mucin, 19, 312, 313
Mucous cells, 19, 314
Multiparae, 449
Muscarine, 39, 183
Muscle, 54–81
  see also cardiac muscle, skeletal muscle and smooth muscle
  biochemistry of contraction, 64–66
  circulation in, **237**, 243–245
  classification of, 54–**55**
  contraction, 54
    energy sources for, 54
    free energy released during, 67–68

isometric, 60, **62**–63, 67
isotonic, 60, 63–**64**, **65**, 67
cross innervation of, 44
differences between skeletal and cardiac, **70**
experiments on isolated, 54
force-velocity relation, 62, 63, **75**, 80
length-tension relation, 62, **63**, **64**, 74–75
measurement of tension and length, **60**
pump, **223**
receptors, 361–363
servocontrol of, 109, 111–113
spindle bias, 402, 403
spindles, **111**–113, 352, 361–363, 401, 402, 403
  afferent nerves from, 361, **363**, 403
  annulospiral endings, **111**, **112**, 350, 361, **362**
  effect of paralysing on stretch reflex, **402**
  in respiratory muscles, 288–289
  isolated preparation of, 116
  nuclear bag fibres, 361, **362**, 363
  nuclear chain fibres, 361, **362**, 363
  recordings from human, 420
  response to vibration, 359
tetanus, 61, 66, 68
twitch, **60**–61, 66
twitch-tetanus ratio, 62
Muscles of respiration, accessory, 274
Muscular exercise, 68–69
Myelinated nerves, 34–37
Myeloblasts and myelocytes, 466
Myenteric reflex, 321
Myocardial contractility, 74, 75
Myocardium, 248
Myoepithelial cells, 205
Myofibrils, **56**, **57**
Myogenic contractions of muscle, 72, 78
Myogenic vascular tone, 225, 231
Myoglobin, 66, 93
Myograph, 23
Myometrium, 436
Myopia, **380**
Myosin, **57**, 58, 61, 76, **77**, 80
Myxoedema, 201, 330

Narcolepsy, 426
Nasal passages, stimulation of mucous membrane of, 297
Nausea, 318
Neck muscles, during sleep, 426
Nephrons, **132**, 146
  blood supply of, 132
Nephrectomy, 147
Nernst equation, 8, 29, 500
Nerve block, **28**, 37
  deafness, 367
  fibre diameters, 36–37, **362**
    relation to conduction velocity, 28, 34–37, **361**
  fibres, **21**, **361**
  specificity of, 183
  methods of stimulating, 22
  -muscle preparation, **22**, **23**
  site of excitation, 23
  trunk, use in study of individual nerve fibres, 21
Nervous action, categories of, 393–394
  system growth, 431
Neubauer chamber, 82
Neural effect on hyperpnoea of exercise, 301
Neurohypophysis, *see* pituitary gland, posterior lobe
Neuromuscular block, 44
  junction, **42**
  transmission, 39–44, 52
Neurone theory, 44
Neurosecretion, 203
Neuroses, experimental, 424–425
Neutrophils, 466, 467

Nexuses, 71, 76, **77**, **79**
Niacin, 345
Nicotinamide-adenine dinucleotide (NAD), 65
Nicotine, 39, 183–184
  effect on autonomic ganglia, 172
Nicotinic acid, **343**, 345
Nitrogen balance of the body, 332–333
  in the urine, 130–131, 327–328
  partial pressure (P_{N_2}), **280**, **283**
  starvation, 332
  under high pressure, dangers of, 306
  volume in alveolar air, **280**
Nitroglycerine, 242–243
Nitrous oxide use in blood flow measurements, 237–**238**, 242
Nociceptors, 491, 492
Nodes of Ranvier, 35–36
Non-esterified fatty acids (NEFA), 334, 336
Non-myelinated fibres, 35–37, 51, 360
Non-nitrogenous residues of amino acid breakdown, 333
Non-polar compounds, 8
Non-protein respiratory quotient, 328
Noradrenaline, 52, 173, 174, 175, 179, 181
  concentration in blood and adrenal medulla, 191–192
  effect on circulation, 226, **232**–233
  effect on renal blood vessels, 146
  secretion by adrenal medulla, 191
  storage of, 181
  sympathetic transmitter, 181
  synthesis, **182**
Normoblasts, 84
Normovolaemic shock, 478, 480
Nose, 272
Noxious stimuli, 396, 491
Nucleolus, 2, **3**
Nucleoproteins, 333
Nutrition, 341–348
Nymphomania, 438
Nystagmus, 374, 375

Obesity, 11
Occipital lobes, 407
Occlusion of teeth, 16
Occlusion in reflexes, 397
Odontoblasts, 16, 17
Odonto-kerato-prosthesis, 484
Oedema, 123, 124, 125, 221, 240
  during inflammation, 11, 465, 466
  in cardiac failure, 259
Oesophagus, 274–275, 314
Oestradiol-17β, 440, **442**
Oestratriene, 440
Oestriol, 440
Oestrogens, 110, 437–441, 444, 446, 450–453
  producing hypertrophy of the uterus, 449
  progesterone acting as antagonist to, 442
Oestrone, 440, 441, 444
Oestrous cycle, 437, 444
Olfactory bulb and area of the brain, 356
  pathways, 409
  receptors, 354, 356
Oligaemic shock, 478, 480
Olives, 416
Open methods of measurement of metabolic rate, 329
Opsin, 384
Optic chiasma, 389, 407, 424
  disc, 379
  nerve, 385–386, 387, 389
  radiations, 407
Oral contraception, 446–447
Organ baths, 116
  metabolism, 327

Orgasm, 459, 460
Osmoreceptors, 127, 128–129, 205, 353
Osmotic pressure, 7, 9, 85, 116
  and solute concentration, 84–85
Ossicles of middle ear, **368**
Osteoblasts, 14–16
Osteoclasts, 14–15
Osteocytes, 14–15
Otitis externa, 367
Otitis media, 367
Otoliths, 373, **374, 375**
Otosclerosis, 367
Ouabain, 31–32
Oval window of middle ear, **368**
Ovalocytes, 84
Ovarian follicles, **433**–435
  hormones, 440–443
  pituitary feedback mechanisms, 444
Ovaries, **433–435, 439**
  in pregnancy, 450
  removal of, 434–435, 439
Oviducts, see fallopian tubes
Ovulation, 434, 438, 444
  artificial induction of, 445
  time in menstrual cycle, 435, **436**, 437, **438**
Ovum, 433–434, 435–436, 461
  fertilised, **448**
Oxidative phosphorylation in mitochondria, 1
  processes, 65, 68–69, 325–326
Oximeters, 90, 327
Oxygen administration, 160
  capacity of blood, 89
  concentration, use in blood flow measurement, 236
  consumption, 325–326, 327–329
    and volume of air breathed, **300**
    of heart, 257
  content of blood, 90
  debt produced by exercise, 69, 103
  dissociation curves (equilibrium curves), 90–93, **96**, 103
    in hypoxia, **302**
  meter, 91
  partial pressure (P_{O_2}), 90, 279–**280**, 281, **283**–284
  control of, 225
  poisoning, 306–307
  release, factors promoting, 96
  requirements of tissues, **210**
  supply, **95**
  therapeutic value of, 306, 309
  transport, 95–98
    disorders of, 96–98
  uptake, disorders of, 97
  volume in alveolar air, **280**
Oxygenation, 89
Oxyhaemoglobin, 89, 99–100, 102
Oxytocin, 204, **205**, 452–453

P substance, 45, 52, 494
Pacemaker, 250, 251, 258, 263–264
  abnormality, 268
  cells in smooth muscle, 78–79
  cells of the heart, 72, **73**
  potentials, **263**
Pacinian corpuscles, 350, 351, **353**, 359
Packed cell volume (PCV) 82, 83, 118, 499
Pain, 350, 359–360, 410, 491–495
  -centre, 492
  during inflammation, 466
  receptors, 357, 360, 492
Palpitation, 257
Pancreas, 206–207
  and regulation of blood sugar, 337–338
Pancreatectomy, 206
Pancreatic juice, 319–320

Panting, 300
Pantothenic acid, 345–346
Papillary muscles, 70
Papilloedema, 126
Para-aminohippuric acid, 237
Paradoxical sleep, 425, 426
Parallel elastic component in muscle, **61**, 62, 74
Paraplegia, 414
Parasympathetic nerve supply to muscles of lens and pupil, 380
Parasympathetic nervous system, 172, 173, **174,** 176–**177**, 179, 183
Parathormone, 15, **208**
Parathyroid glands, 207–208
Parenchymal cells, 185
Parietal cells in stomach, 314, 315–316
Parkinson's disease, 415–416, 422, 487–488
Paroxysmal nocturnal haemoglobinuria, 86
Partial pressure of a gas, 279–280
Parturition, 450–452, 453–454
  role of oxytocin, 205, 452
Pattern theory of pain, 492–493
Pavlov pouch, **315**, 316
Pellagra, 345
Penis, 459
Pepsin and pepsinogen, 314
Perfusion experiments, 115–116, 327
Pericardial fluid, 119
Pericardium, 248
Perilymph, 119
Periosteum, 14, 16
Peripheral resistance, 211, 216, 223, 224, 229
Peristalsis, 79, 321, **322**
Peritoneal fluid, 119
Permissive effects, 189
Perspiration, insensible, 168
pH, 152, 153–155
Phagocyte index, 188
Phagocytes in lymph nodes, 124
Phagocytosis, 4, 466, 467
Phantom limb, 491
Pharynx, 272, 314
Phasic receptors, 350, **351**
Phenanthrene, 495
Phenotype (blood group), 475
Phenylthiourea, 355
Pheromones, 14, 208–209
Phloridzin, 138
Phon, 366
Phosphenes, 407
Phosphoric acid buffer system, 154, 155
Photons, 377, 381
Photopic vision, 381, **382**, 386
Photoreceptor activity, 385
Phrenic nerves, 286, **288**
Physiological tremor, 109, 486–489
Pia mater, 125
Picrotoxin, 49–50, 51, 416
Pilomotor reflex, 168
Pinnae, 367, **368**
Pinocytosis, 4, **5**
Pins and needles, 34, 37
Pitch, 364, 365
Pituicytes, 203
Pituitary dwarf, 204
Pituitary gland, 196, **202**–205, 392
  and lactation, 452
  anterior lobe, 202, 203–204
    and gonadotrophin secretion, 440–441, 443–445
    and regulation of blood sugar, 338
    blood supply of, 202
    control of testes, 455, 457
  effect on thyroid, 199, 200
  in pregnancy, 450

  in the fetus, 453
  pars intermedia, 202
  posterior lobe, 202, 203, 204–205
pK', 152, 153–154
Placenta, 428, 449, 450, 490
  secreting oestrogen, 440
  secreting progesterone, 441
Plantar response, 398–399, 414
Plasma, 82, 118, 119, 122
  calcium, 207–**208**
  cells, 467
  clearance by kidney, 137, 139, **140**
  composition, **114, 130**, 499
  electrolytes, 159, 499
  flow, 237
  glucose, 206–207
  kinins, 494
  oncotic pressure, 122
  osmotic pressure, 122–123
  proteins, 119–120, 122, 123, 483, 499
    as buffers, 155
    in kidney, 147
    "separated" and "true", 101
  specific gravity of, 82
  volume, 118
Plasminogen, 473
Platelet count, 83
Platelets, 470–471, 473
Plethysmograph, 235, **238**
Pleura and pleural cavities, 274
Pleural effusion, 259
  fluid, 119
Pneumotaxic centre, 287–**288, 290,** 296
Pneumothorax, 274
Poiseuille's law, **215**, 216
Polycythaemia, 83, 97
Polymorphonuclear leucocytes (polymorphs), 83, 466, 467
Pons, 392, 412, 426
Pontine respiratory centres, 287
Portal vein of liver, 186, 245, 323
Positive feedback in shock, 477–478, 480
Posture, 394, 398, 404–405
  effect on circulation, 222, 243
Potassium equilibrium potential, 29, 30, 31, 32
  ion permeability during action potential, 32–34, 37–38
  ions, 29–34
    action on tissues, 115
    in liver, 187
Precapillary sphincters, 238, **239**
Precipitin, 467
Prednisone, 485
Pregnancy, 428, 448–450
  and thyroid secretion, 200
  dietary allowance during, 342, 343, 344
  skin growth in, 430
  tests, 448–449
Pregnanediol, 442, 450, 451, **452**
Pre-menstrual tension, 437
Pressure, units of, 498
Presynaptic inhibition, 49–51, 53
Pretectal nuclei, 407
Priapism, 459
Primigravida, 451
Primordial follicles, 433–434
Procaine, 351, 402, 495
Proconvertin, 470, 471
Progestational response, 442
Progesterone, 438–444, 446, 450–453
  variation during menstrual cycle, **442**
Projection areas of cerebral cortex, 406
Prolactin, 203, 204, 452
Pronuclei, male and female, 448
Proprioceptors, 361, 399

Prostaglandins, 148, 452, 453–454
Protein, 319, 320, 327, 330, 332
  breakdown, 333
  buffers, 154
  in body fluids, 119–120
  in diet, 341, **342**, 343, 344
  in membranes, 86–88
  metabolism, 327, 332–334
  proton acceptor and donor groupings, 154
  respiratory quotient of, 326
Proteinuria, 134–135
Prothrombin, 185, 187, 470–471, 483
  time, 473
Prothrombinase, 470
Protoporphyrin, 483
Pseudohermaphroditism, **462, 463**
Ptosis, 175
Ptyalin, 313
Puberty, 447
Puerperium, 451
Pulmonary blood flow, 95
  blood pressures, 240–241
  capillaries, 272, **273**, 281
  capillary blood flow, 281–284
  circulation, 212, 213, 240–241
  embolism, 471, 473
  function tests, 278
  gas exchange, 95
  oedema, 241, 281
  stretch receptors, 273, 287–**288**
Pulp cavity, 17
Pulse, 252–253
  anacrotic, 253
  pressure, 218, **220**
Pupil, 379, 380, 386, 407
Purkinje cells in cerebellum, 418–**420**
Purkinje fibres in heart, 72, 248, 250, 262, 263, 265
  fibres, action potentials in, **261**
Purkinje images, **379**, 380
Pus, 467
Pyelography, 148
Pyloric glands of stomach, 314, 315–316
  sphincter, 316–318
Pyramidal tract, 412
Pyridoxine, 345
Pyrogens, 170, 197

Radiation effect on bone, 16
Rami communicantes, 173
Raphe nucleus, 426
Reabsorption by kidney tubules, 139, 141
Reaction, control of, 156–160
  of body fluids, 152
Rebreathing, intermittent, 236
Receptive fields, 352, 357
Receptor sites at end plate, 41
  units, 349–353
Receptors, 349–363
  adequate stimulation, 349
  chemical, 353–356
  mechanisms of impulse initiation, 351
  specificity of, 349, 492
  transducer process, 351, 352
Recessive gene, 461
Reciprocal innervation of blood vessels, 230
  of limb muscles, 404
Recruitment of afferents, 350
  in reflexes, 397
Rectum, 322
Recurrent collaterals, 48–49
Red blood cell count, 83
  diameter, 83, 84
  ghost, 85, 87
  indices (MCH, MCHC, MCV, PCV), 82–83, 499
  membrane, 86–**88**
  shape, 84
  volume, 118
Red blood cells, 82–88
  concentration of, 499
  formation of, 83–84
  fragility, 86, 117
  glucose content, 85
  isotonic solutions with respect to, 116
  lifetime, 83
  osmolarity, 85
  specific gravity of, 82
  volume changes, **116**
Red nucleus, 415, 416
Reference man or woman, 341–**342**
Referred pain, 360, 494–495
Reflex action, 394–403, 426–427
  see also under individual reflexes
Reflex arc, 111
  monosynaptic, 111
Reflexes, definition of, 393–394
  affecting breathing, 296–301
  through thermoregulation centres, 166
Refractory period, 34
Regeneration of nerve fibres, 183
Reinforcement, 424
Releasing factors, 196, 202, 203–204
Release phenomena, 414, 416
Renal blood flow, 131, 136, 138, **237**
  calyces, **131**, 149
  colic, 149
  papilla, **131**
  parenchyma, 132
  pelvis, **131**, 149
Renin, 133, 147, 193, 225
  effect on circulation, 233
  secretion during hypertension, 481
Renshaw inhibition, 48–**49**, 53
Reproductive organs, growth of, **430**
Reserpine, 184, 487
Residual volume of the lungs, **277**, 278
  of the ventricle, 222
Resistance of an axon, longitudinal, 28, 36
  of lungs, 275
  to blood flow, 216–217
Respiration at high atmospheric pressures, 306–307
  effect on circulation, 240
  external and internal or tissue, 272, 326
  methods of recording, 278
Respiratory acidosis and alkalosis, 159–160
  adjustments, 299–307
  centres, 285–288, 289, 295, 296
    role in buffering the blood, 155
  control of reaction, 156–157
  dead space, 279, 280
  depression, 295
  distress syndrome of the newborn, 276
  drive potentials, central, **286**, 289
  failure, 308
  minute volume, 277, 280, **291–292**, 294, 296, 297, 299–**303**
    and cyanosis, 305
    increase during acclimatisation, 303
  movements, 274, 288
  muscles, motoneurones of, 285–287, **288**
    proprioceptive control of, 288–289
  passages, heat loss from, 168, 169
  quotient (RQ), **283**, 325–326, **328**, 335, 500
  rhythm, central control of, 287
  tract, 272
Respired air composition, 278–281
  difference in volumes of inspired and expired air, 329
Respirometers, 326
Resting expiratory level, 278
Resting potential, 24, **25**, 28–31, 37

Retching, 318
Reticular fibres, 10
Reticular formation, 287, **290**, 393, 426
Reticulin fibres, 3
Reticulocyte count, 83
Reticulocytes, 84
Reticuloendothelial system, **187**, 188, 466, 467
Retina, 378, **381**, **387**
  convergence in, 387
  neuronal mechanisms in, 385–386, 387–**388**
Retinal bipolar cells, 385, **387**, 389
  ganglion cells, 385–386, **387**, 389, 408
    visual fields of, **385**, 387–**388**
  receptor system, 384–385
  receptors, adequate stimulus, 349
  photochemical substances in, 378, 382, 384–385
Retinene, 384
Retinol, see vitamin A
Reversing spectacles, 379
Rhesus (Rh) blood group system, 475–476
  factors, 476
Rheumatoid arthritis, 195
Rhodopsin, 384, 386
Rhythmic firing in receptors, **355**
Riboflavin, **343**, 345
Ribonucleic acid (RNA), 2, 18
Ribosomes, 2, **3**
Rickets, 16, 346
Rigidity, 415–416
Rigor in muscle, 61
Rigors in fever, 170
Ringer's solution, **114**–115
Rod cells in retina, 351, 381, 384, **387**
Romberg's sign, 404
Rouleaux formation of red blood cells, 82, 84, 476
Rubrospinal tract, 415
Ruffini corpuscles, 359

Saccule, **368**, 373, 375
Saline, physiological, 114–115
Saliva, 312, 313–314
Salivary glands, circulation in, 245
  reflex, 175, 176
Salt intake and hypertension, 481
  loss by sweating, 170
Saltatory conduction, 34–36
Sarcolemma, 56
Sarcomeres, 56, **57**
Sarcoplasmic reticulum in cardiac muscle, 71
  in skeletal muscle, 58, 59
  in smooth muscle, 76
Satiety centre, 178
Scar tissue, 467, 469
Sclerosis, 481
Scotomata, 408
Scratch reflex, 397, **399**, 403
Scurvy, 346, 469
Sebaceous glands, **13**
Sebum, 13, 18
Secondary sexual characteristics, 447, 456, 461
Secretin, 316, 319–321
Secretion from digestive glands, 312
Secretory cells, 18–20
"Secretory" nerves, 312
Sedimentation rate of blood, 82, 476
Semicircular canals, **368**, 373–374
Seminiferous tubules, **455**, 456
Sensation, 349
  deep, 359
  from digestive tract, 313
Sensory cortex, 410, 427
  effects of stimulation, 411
  lesions in, 411–412
  projection maps, **411**
Sensory discrimination, 424

loss, dissociated, 410
modalities, in the skin, 357
nerves, 395
pathways, **409**
Series elastic component in muscle, **61**
Serotonin, 426
Sertoli cells, 455, 456
Serum, 82, 470, 474
Servoloop theory of muscle contraction, 403–404, 420–422, 486–487, **488**
Servomechanisms, 110–113, 496
Servo responses, effects of anaesthesia, 420–422
in humans, 420–422
Sex chromosomes, **461**
determination and differentiation, 461–463
function controlled by parasympathetic, 176
hormones, 428
secreted by adrenal cortex, 197
Sexual reflexes, 397
Sexuality and human behaviour, 460
Scotopic vision, 381, **382**, 386
Sham-feeding, 314–**315**
Sham rage, 176
Shivering, 161, 162, 163, 168, 170
Shock, pathophysiology of, 477–480
reversible and irreversible, 478
Sickle cell anaemia, 84
Silvester's method of artificial respiration, 309
Simple cells in the visual cortex, **389**–390, 408
Sinoauricular node, 72, **248**, 250, 263, **264**, 265
Sinusoids in adrenal cortex, 192
in liver, 185, 186
in pituitary, 202
Skeletal muscle, 54–55, 56–69
blood supply to, 211
dimensions, 62, 63–64
effect of nicotine, 183
effect of temperature, 62, 64, 67, 68
efficiency of, 68
energetics, 67–68
fibres, **56**
growth of, 430
load, 63
mechanical properties, 61–64
strength-duration curve, 58, **59**
strength-response curve, **59**
structure of, 54–55, 56–58
types of, 43–44, 61, 66
Skin, 11–14
blood flow, 211, 245–247
colour of, 12–13
growth, 430
heat loss from, 161, 162
receptors, 357–360
temperature, 161, 165, 358–359
during inflammation, 465–466
water loss from, **127**
Sleep, 425–426, 427
Sliding filament hypothesis, 58
Small intestine, 319–322, 323–324
influence on gastric secretion, 315, 316
Smell, 355–356
Smooth muscle, **55**, 76–81
action potential, **78**–79
cells, 55, 76, **77**, **78**–79
chemical and electrical stimulation, 80
innervation, 76, 80
multiunit, 55, 76, 81
myofilaments, 76
operational range of lengths, 80
slow waves, 78–79
single unit, 55, 76, 78–81
speed of contraction, 80
spontaneous activity (Rhythmic contractions), 78–79

tension production, 80
tetanus, 80, 81
tone, 78
Sneezing, 297
Sodium chloride intake, 344
Sodium deficiency or excess in the body, 121
equilibrium potential, 30, 32
ion conductance in nerve, as a positive feedback system, 110
feedback in the kidney, 147
regulation, 193
permeability during an action potential, **31**, **32**–34, 38
ions, action on tissues, 115
in nervous tissue, 29–34
pump, 8–9, **31**–32
Soma, action potentials in, 46
Somatic nerves, 172
sensation, 409–419
sensory cortex, **411**
system, relation to autonomic system, 176
Somatotropin, see growth hormone
Sound, localisation of, 367
physical properties of, **364–365**
sensation of, 365
variation of intensity with frequency, **365**, 366
waves, coding of frequency and intensity, 371–372
Fourier analysis of, 365
Sounds, loudness of various, **366**
Snow blindness, 379
Spasticity, 400, 402, 414
Specific dynamic action, 330
Specific theory of pain, 492
Spectrin, 86
Speech, 414
Sperm production, control of, 457
Spermatogenesis, 455–456
Spermatozoa, 448, 455, **457**, 458, 461
Spherocytes, 84, 86
Sphygmomanometer, 219
Spinal animals, 394, 396–397, 398
cord, 392, 394–395, 396–397, 427
excitatory and inhibitory pathways in, **45**
principal tracts of, **410**
role in control of respiration, 286
nerve roots, 392, 395
nerves, 392
reflex arc, **392**
reflexes, 396–403, 426–427
shock, 396–397, 427
walking, 397, 403, 404
Spinocerebellar tracts, 416
Spinothalamic tract, 410, 427
Spirometer, **277**, 278
Spironolactones, 146
Spleen, 240, 245
Split-brain preparation, 424
Squid nerve, 24, **25**, 28, 29, 30
Staircase phenomenon, 73
Stannious ligatures, **249**, 262
Starch, 320
Starling's law of the heart, 64, 75, 225, 255–256, 257
Stasis, 473
Stenosis, 253, 254, 258, 267, **268**
Stercobilin and stercobilinogen, 84
Stereoscopic vision, 386
Steroids, synthetic, 195, 196
Stethograph, 278
Stethoscope, 254
Stilboestrol, 441
Stimulus intensity, the effects of, **350**
Stomach, 314–316
movements, 316–318

Stones in renal tract, 149, 208
Strength-duration curve, for the stimulus required to excite a nerve, 22–23
Streptococci, 465
Stress, responses to, 174–175, 191, 193
Stretch reflex, 397, 399–403, 405, 420–422, 427
arc, oscillation in, 486
gain of, 420–422
role in respiratory movements, 288–289
Striae gravidarum, 430, 449
Stroke volume, 222, 256
Stuart Power factor, 471
Strychnine, 47, 48, 50
Subliminal stimulation, 397
Submandibular gland, 312
Substantia gelatinosa, 493
Substantia nigra, 415, 487
Subthalamic nucleus, 415
Summation in reflex responses, 397
of muscle twitches, **60**, 61
of post-synaptic potentials, 46–47
spatial and temporal in receptor units, 352
Sunburn, 195
Supraoptic nucleus, 128, 204
Surface tension, 276
Surfactant, 276
Swallowing, 297, 314
reflex, 395
Sweat, composition of, 169–170
glands, 13–14, 168–169, **358**
Sweating, **127**, 129, 161, 163, 168–170, 175
Sympathectomy, 174, 230, **231**
Sympathetic block, effect on temperature, 166
control of blood vessels, 229–**230**
effects on pacemaker, 258, 264
nervous system, 172, 173, **174**–175, 176, **177**, 179, 183
stimulation of the heart, 73, 256–257, 258
transmitters, 181
vasoconstrictor tone, 233
Sympathin, 179
Synapse, 44–45, 47, 52
Synapses, axoaxonic, 50–51
Synaptic transmission, 44–53
vesicles, 41, **42**, 45
Syncope, 477
Syncytium, 4
functional in the heart, 71, 74, 249, 250, 251
functional in smooth muscle, **79**
Syneraesis, 470
Synergists, 48, 68
Synovial fluid, 119
Systemic circulation, 212
Systole, 247
Systolic pressure, 218–**219**, **220**–221

Tabes dorsalis (tertiary syphilis), 486
Tachycardia, 223–224
paroxysmal, 259, 269, **270**
Target organs, blood flow through, 189
Taste, 347–348, 354–355
buds, 354–**355**
Teeth, 16–18
Temperature, 161–164
control by vascular mechanisms, 165–167
effect on tremor, **487**
effective, 171
of body during menstrual cycle, 438
of body, effect of climate on, 170–171
of body, effect on respiration, 301
receptors, 163, 350, 358–**359**
Temporal lobes of cerebral cortex, 423
Tendon jerks, 397, 400–402, 422
Tension transducer, 60
Testes, **455**–457

Testicular hormones, 456
  hyperfunction, 457
  insufficiency, 457
Testosterone, 444, 456, 457, 459
Tetanus, see muscle tetanus
Tetanus toxin, 47
Tetany, 34, 207
Tetraethylammonium (TEA), 34
Tetrodotoxin (TTX), 33–34, 41, **43**
Thalamic syndrome, 410, 492
Thalamus, 392, 409–410, 426
Thebaine, 495
Theca externa, 434
  interna, 434, 440
  lutein cells, 442
Thermoreceptors, 163–164
Thermoregulatory process, systems involved in, **162**
  system, performance characteristics of, 168–170
Thiamine, **343**, 345
Thirst, 127, 129, 170
Thiry-Vella loop, **321**
Thoracic duct, 123
Threshold, nerve, 22, 32, 34, 37
  of sensation, 491
Thrombin, 470–472
Thrombocytes, see platelets
Thrombocytopenia, 472
Thromboplastin (thrombokinase), 470–471
Thrombosis, 471, 473
Thyrocalcitonin, see calcitonin
Thyroglobulin, 198, 199
Thyroid gland, **198**–201, 208
  and basal metabolic rate, 330
  removal of, 198
  secretion, abnormalities, 200–201
Thyroid stimulating hormone (TSH), 200, 203, 432
Thyroid stimulator, long-acting (LATS), 201
Thyrotoxicosis, 98
Thyrotropin-releasing factor (TRF), 200, 203
Thyroxine (T_4), 199, 200
  and metabolism, 330
  effect on heart rate, 258
Tidal air volume, **275**, **277**, 296
Tight junction, **4**
Tissue abnormalities, 98 ,
  cultures of neurones, 431
  metabolism, 326
  pressure, 124
  typing, 485
Titratable acidity, 157
Titration curves, **153**, 155
α-tocopherol, see vitamin E
Tone, degree of passive resistance, 400
Toothache, 17
Touch, 357
Trabeculae, 15, 16
Trachea, 272
Tracheobronchial tree, 272
Transamination, 334
Transfer function of a control system, 108
Transferrin, 483
Transmitter, criteria for being a, 181
Transmitters, 39–41
  in central nervous system, 52
  in spinal cord, 45, 48–49
  quantal release of, 41, 45, 52
Transplantation, 484–485
Transport maximum of secretion by kidney tubules, 139, **140**–141, 148
Transport proteins, 188, 196, 200
Transverse tubular system (T-system), **57**, 58, 59, 71
Tremor, see physiological tremor
Tricarboxylic acid cycle, in mitochondria, 1, 65
Tricuspid valve, **248**, 253, **254**

Trigger zone of motoneurone, 46–47
Triglycerides, 324
Trigone, 149
Triiodothyronine (T_3), 199, 200
Triple response, 246–247, 465
Trophoblast, 448, 449
Tropomyosin, **57**, 58, 59
Troponin, **57**, 58, 59
Trypsin and trypsinogen, 320
Turbulence, 216, 217, **218**
Tympanic membrane, 367–368
Tyrode's solution, 114
Tyrosine, 4

Ultimobranchial body, 207, 208
Ultrafiltrate of blood plasma, 134, 135
Universal donors and recipients, 474
Unsaturated fatty acids, 334, 344
Urea, 327, 332, 333–334
  clearance, 137
  in blood as an indication of kidney damage, 148
  in body fluids, 120
  in urine, 130
  production by liver, 185, 187
Ureter, 79, 149
Urethra, 150, 151
Urinary excretory channels, 149
Urine, 130–131, 134–138
  abnormal constituents in, 131
  acidification of, 146, 157
  alkalinisation of, 146
  composition of, **130**, 499
  concentration, 141, 142–143
  flow, **139**, 142
  osmolal concentration of, 142
  water output, **127**
  retention with overflow, 151
  secretion, nervous influence, 146
Urinometer, 148
Urobilinogen, 84
Urochrome, 130
Urticaria, 274, 469
Uterine contractions, 449–**451**, 454
  muscle, effect of hormones, 79
Uterovesical junction, **149**
Uterus, 433, 436
  changes in pregnancy, 449–450
  effect of oxytocin, 205
  influence of oestrogens on, 441
  influence of progesterone on, 442
Utricle, **368**, **373**, 375

Vagal influence on alimentary tract, 312, 315–316
  influence on heart, 258
  influence on pacemaker, 264
  influence on pancreas, **319**
  stimulation producing cardiac arrest, 309
  tone, 228
Vagina, 433, 436
  influence of oestrogens on, 441
Vaginal pH, 438
Vagus nerve, 176, 226, 295–296
Vagus nerves, effect of section on respiration, **286**, 287, **288**, 289, 296
Valsalva manoeuvre, **222**, 231, 274, 297–298
Van Slyke method of blood-gas analysis, 92
Varicose veins, 188, 221
Vasa recta, 131
Vascular bed, volumes of various parts, **214**
Vasoconstriction, 229, 230
  following sympathetic stimulation, 175
  in skin, **166**
  peripheral in shock, 477
Vasoconstrictor tone, 243–244

Vasodilatation, 211, 229, 230
  in muscle blood vessels, 243
  in skin, 166, 169
  produced by parasympathetic system, 175, 176
Vasodilator metabolites in muscle, 242, 243, 244
  nerves, 230
  substances, 231, 233
Vasomotion, 239
Vasomotor centre, 110, 225, 229, 233–**234**
  control by sympathetic system, 175
  control of skin blood vessels, 165–167, 175
  control, reflex, 229–230
  nerves, 229–230
Vasopressin, see antidiuretic hormone
Veins, **212**, 213, **214**
  valves in, 213
Venae cavae, 212, **214**
Venae comites, 165
Venomotor control, 230
Venous blood mixed, 236
  circulation, 239–240
  inflow and cardiac output, 255–**256**
  pressure, **220**, **221**, 222–223, 239–240, **241**
  pulse, **253**
  return, 223
  tone, 240
Ventilation-perfusion ratio, 281–**283**
Ventilation rate, 156
Ventricle, 125, 248–250, **264**, 392
Ventricular pressure, 252
  receptors, 228, 231
  reflexes, 231
  volume changes, 252
Venules, 213, **214**, 239
Veratridine, 128
Vertex presentation of fetus, 451
Vertigo, 374–376
Vestibular apparatus, 373–375
  apparatus, removal of, 375–376
  system, caloric stimulation of, 374
  sense area of the cortex, 409
Vestibulo-ocular reflexes, 404
Vibration, 359
Villi in intestine, **323**
Visceral pain, 360
  reflexes, 397
Viscometer, 217
Viscosity, 216
Vision, threshold of, in the human, 381–382
Visual acuity, 383, 386
  agnosia, 408
  cortex, 389–**391**, 408–409, 427
    columns in, 391, 409
    evoked responses in human, 409
    mapping of retinae onto, 407–408
  discrimination, 383
  fields in cortical neurones, **389**, 389–391, 408–409
  perception, 386–391
  processes in the human, 381–384
  processing, central, 389–**391**
  receptive fields, **385**, 386, 387–388
  system, 378–391, 407–409
Vital capacity, **277**–278
Vitamin A, 188, **343**, 345
  role in bone development, 16
Vitamin B group, 345–346
Vitamin B_{12}, 83, 345, 346
Vitamin C, 345, 346
  role in bone development, 16
Vitamin D, 16, **343**, 345, 346
Vitamin E, 345, 346
Vitamin K, 187, 345, 346, 470, 471
Vitamins in diet, 341, 345–346
  in milk, 453
Voltage clamp technique, 32–34

Volume receptors, 128, 129
Voluntary action, 406, 412, 427
  movements, 393
  muscles, heat production by, 162
  paralysis, 412, 413, 414
Vomiting, 318

Wallerian degeneration, 44
Warfarin, 187
Wassermann reaction (WR), 469
Water balance, **127, 128**
  balance, disturbance of, 129
  balance, influence of progesterone on, 442
  content of the body, 119, 127
  deficiency, 120
  distribution in the body, 127–129
  excess, 120–121
  intake, **127**, 341
  intoxication, 121, 129, 170
  output, **127**
  vapour partial pressure, **280**
  volume in alveolar air, **280**
Weaver-Bray phenomenon,
  see cochlear microphonic potentials
Weightlessness, 221
White blood cells, 2, 11, 466–467, 499

White cell count, 83, 467
White matter, 392
White reaction, 246
Whole animal metabolism, 327
Withdrawal reflex, *see* flexion reflex
Wolff's integrating motor pneumotachograph (IMP), 329
Work hypertrophy of muscle, 430
Wound shock, 224

Z line, **56, 57**, 61
Zona pellucida, 434
Zymogen granule, **2**